STANDARD ATOMIC WEIGHTS OF THE ELEMENTS 2001
Based on relative atomic mass of $^{12}C = 12$, where ^{12}C is a neutral atom in its nuclear and electronic ground state.[†]

Name	Symbol	Atomic Number	Atomic Weight	Name	Symbol	Atomic Number	Atomic Weight
Actinium*	Ac	89	(227)	Molybdenum	Mo	42	95.94(1)
Aluminum	Al	13	26.981538(2)	Neodymium	Nd	60	144.24(3)
Americium*	Am	95	(243)	Neon	Ne	10	20.1797(6)
Antimony	Sb	51	121.760(1)	Neptunium*	Np	93	(237)
Argon	Ar	18	39.948(1)	Nickel	Ni	28	58.6934(2)
Arsenic	As	33	74.92160(2)	Niobium	Nb	41	92.90638(2)
Astatine*	At	85	(210)	Nitrogen	N	7	14.0067(2)
Barium	Ba	56	137.327(7)	Nobelium*	No	102	(259)
Berkelium*	Bk	97	(247)	Osmium	Os	76	190.23(3)
Beryllium	Be	4	9.012182(3)	Oxygen	O	8	15.9994(3)
Bismuth	Bi	83	208.98038(2)	Palladium	Pd	46	106.42(1)
Bohrium	Bh	107	(264)	Phosphorus	P	15	30.973761(2)
Boron	B	5	10.811(7)	Platinum	Pt	78	195.078(2)
Bromine	Br	35	79.904(1)	Plutonium*	Pu	94	(244)
Cadmium	Cd	48	112.411(8)	Polonium*	Po	84	(209)
Cesium	Cs	55	132.90545(2)	Potassium	K	19	39.0983(1)
Calcium	Ca	20	40.078(4)	Praseodymium	Pr	59	140.90765(2)
Californium*	Cf	98	(251)	Promethium*	Pm	61	(145)
Carbon	C	6	12.0107(8)	Protactinium*	Pa	91	231.03588(2)
Cerium	Ce	58	140.116(1)	Radium*	Ra	88	(226)
Chlorine	Cl	17	35.453(2)	Radon*	Rn	86	(222)
Chromium	Cr	24	51.9961(6)	Rhenium	Re	75	186.207(1)
Cobalt	Co	27	58.933200(9)	Rhodium	Rh	45	102.90550(2)
Copper	Cu	29	63.546(3)	Roentgenium	Rg	111	(272)
Curium*	Cm	96	(247)	Rubidium	Rb	37	85.4678(3)
Darmstadtium	Ds	110	(271)	Ruthenium	Ru	44	101.07(2)
Dubnium	Db	105	(262)	Rutherfordium	Rf	104	(261)
Dysprosium	Dy	66	162.500(1)	Samarium	Sm	62	150.36(3)
Einsteinium*	Es	99	(252)	Scandium	Sc	21	44.955910(8)
Erbium	Er	68	167.259(3)	Seaborgium	Sg	106	(266)
Europium	Eu	63	151.964(1)	Selenium	Se	34	78.96(3)
Fermium*	Fm	100	(257)	Silicon	Si	14	28.0855(3)
Fluorine	F	9	18.9984032(5)	Silver	Ag	47	107.8682(2)
Francium*	Fr	87	(223)	Sodium	Na	11	22.989770(2)
Gadolinium	Gd	64	157.25(3)	Strontium	Sr	38	87.62(1)
Gallium	Ga	31	69.723(1)	Sulfur	S	16	32.065(5)
Germanium	Ge	32	72.64(1)	Tantalum	Ta	73	180.9479(1)
Gold	Au	79	196.96655(2)	Technetium*	Tc	43	(98)
Hafnium	Hf	72	178.49(2)	Tellurium	Te	52	127.60(3)
Hassium	Hs	108	(277)	Terbium	Tb	65	158.92534(2)
Helium	He	2	4.002602(2)	Thallium	Tl	81	204.3833(2)
Holmium	Ho	67	164.93032(2)	Thorium*	Th	90	232.0381(1)
Hydrogen	H	1	1.00794(7)	Thulium	Tm	69	168.93421(2)
Indium	In	49	114.818(3)	Tin	Sn	50	118.710(7)
Iodine	I	53	126.90447(3)	Titanium	Ti	22	47.867(1)
Iridium	Ir	77	192.217(3)	Tungsten	W	74	183.84(1)
Iron	Fe	26	55.845(2)	Ununbium	Uub	112	(285)
Krypton	Kr	36	83.798(2)	Ununhexium	Uuh	116	(292)
Lanthanum	La	57	138.9055(2)	Ununpentium	Uup	115	(228)
Lawrencium*	Lr	103	(262)	Ununquadium	Uuq	114	(289)
Lead	Pb	82	207.2(1)	Ununtrium	Uut	113	(284)
Lithium	Li	3	6.941(2)	Uranium*	U	92	238.02891(3)
Lutetium	Lu	71	174.967(1)	Vanadium	V	23	50.9415(1)
Magnesium	Mg	12	24.3050(6)	Xenon	Xe	54	131.293(6)
Manganese	Mn	25	54.938049(9)	Ytterbium	Yb	70	173.04(3)
Meitnerium	Mt	109	(268)	Yttrium	Y	39	88.90585(2)
Mendelevium*	Md	101	(258)	Zinc	Zn	30	65.409(4)
Mercury	Hg	80	200.59(2)	Zirconium	Zr	40	91.224(2)

[†]The atomic weights of many elements can vary depending on the origin and treatment of the sample. This is particularly true for Li; commercially available lithium-containing materials have Li atomic weights in the range of 6.939 and 6.996. The uncertainties in atomic weight values are given in parentheses following the last significant figure to which they are attributed.

*Elements with no stable nuclide; the value given in parentheses is the atomic mass number of the isotope of longest known half-life. However, three such elements (Th, Pa, and U) have a characteristic terrestial isotopic composition, and the atomic weight is tabulated for these. **http://www.chem.qmw.ac.uk/iupac/AtWt/**

GOB
Chemistry ◆ Now™

Completely integrated with this text!

What do you need to learn now?
Take charge of your learning with GOB ChemistryNow™!

This powerful online learning companion helps you manage and make the most of your study time. This collection of dynamic technology resources gauges your unique study needs to provide you with a *Personalized Learning Plan* that will enhance your problem-solving skills and understanding of core concepts. Designed to optimize your time investment, **GOB ChemistryNow**™ helps you succeed by focusing your study time on the concepts you need to master.

Look for references throughout the text that lead you to **GOB ChemistryNow**. They direct you to the corresponding media-enhanced activities within the program. Precise page-by-page integration enables you to go beyond reading about chemistry—you'll actually experience it in action!

Easy to use
The **GOB ChemistryNow** system includes two powerful assessment components:

WHAT DO I KNOW?
This diagnostic *Exam-Prep Quiz*, based on the core concepts in each chapter, gives you an initial assessment of your understanding of core concepts.

WHAT DO I NEED TO LEARN?
A *Personalized Learning Plan* outlines key elements for review. With a click of the mouse, **GOB ChemistryNow's** unique activities allow you to:

- Create a *Personalized Learning Plan* or review for an exam using the *Exam-Prep Quiz* web quizzes

- Explore chemical concepts through tutorials, simulations, and animations

- View *Active Figures* and interact with text illustrations. These *Active Figures* will help you master key concepts from the book. Each figure is paired with corresponding questions to help you focus on chemistry at work to ensure that you understand the concepts played out in the animations.

By providing you with a better understanding of exactly what you need to focus on, **GOB ChemistryNow** saves you time and brings you a step closer to success!

Make the most of your study time—
log on to **GOB ChemistryNow** today!

Get started today at **http://www.thomsonedu.com**

If an access card came with this book, you can start using this resource right away by following the directions on the card.

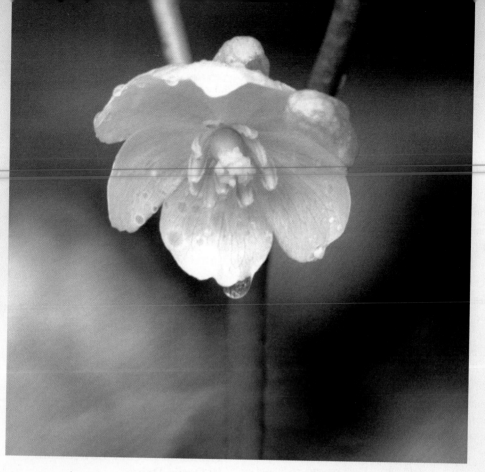

INTRODUCTION TO
General, Organic, and Biochemistry

EIGHTH EDITION

FREDERICK A. BETTELHEIM

WILLIAM H. BROWN Beloit College

MARY K. CAMPBELL Mount Holyoke College

SHAWN O. FARRELL Olympic Training Center

THOMSON

™

BROOKS/COLE

Australia • Brazil • Canada • Mexico • Singapore
Spain • United Kingdom • United States

THOMSON
BROOKS/COLE

Introduction to General, Organic, and Biochemistry, Eighth Edition
Frederick A. Bettelheim, William H. Brown, Mary K. Campbell, Shawn O. Farrell

CHEMISTRY ACQUISITIONS EDITOR: Lisa Lockwood
DEVELOPMENT EDITOR: Sandra Kiselica
ASSISTANT EDITOR: Ellen Bitter
EDITORIAL ASSISTANT: Lauren Oliveira
TECHNOLOGY PROJECT MANAGER: Ericka Yeoman
MARKETING MANAGER: Amee Mosley
MARKETING ASSISTANT: Michele Collela
MARKETING COMMUNICATIONS MANAGER: Bryan Vann
PROJECT MANAGER, EDITORIAL PRODUCTION: Teri Hyde
CREATIVE DIRECTOR: Rob Hugel
ART DIRECTOR: Lee Friedman
PRINT BUYER: Barbara Britton
PERMISSIONS EDITOR: Kiely Sisk

PRODUCTION SERVICE: Thompson Steele, Inc.
TEXT/COVER DESIGNER: Diane Beasley
PHOTO RESEARCHER: Jane Sanders
ILLUSTRATOR: J/B Woolsey and Thompson Steele, Inc.
COVER IMAGE: Thomas G. Barnes @ USDA-NRCS PLANTS
   Database / Barnes, T.G. & S.W. Francis. 2004.
   Wildflowers and Ferns of Kentucky. University Press
   of Kentucky
COVER PRINTER: Phoenix Color Corp.
COMPOSITOR: Thompson Steele, Inc.
PRINTER: Quebecor World-Dubuque

Photographs on pp. v–xii © Charles D. Winters

Library of Congress Control Number: 2005935780

ISBN-13: 978-0-495-01197-2
ISBN-10: 0-495-01197-5

Thomson Higher Education
10 Davis Drive
Belmont, CA 94002-3098
USA

We dedicate this edition to our friend,

colleague, and masterful teacher,

FREDERICK BETTELHEIM,

who left this world doing something he loved.

He was ever curious about his surroundings

and ever mindful of the

power of chemistry.

Contents in Brief

Contents

Preface

Perceiving order in the nature of the world is an ontological—not just pedagogical—need. It is our primary aim to convey the relationship among facts and thereby present a totality of the scientific edifice built over the centuries. In this process we marvel at the unity of laws that govern everything in the ever-exploding dimensions: from photons to protons, from hydrogen to water, from carbon to DNA, from genome to intelligence, from our planet to the galaxy and to the known universe. Unity in all diversity.

In writing the preface of the eighth edition of our textbook, one cannot help but ponder the shift in paradigm that occurred during the last 30 years. From the slogan of the 1970s, "Better living through chemistry," to today's epitaph, "Life by chemistry," one is able to sample the change in the focus. Chemistry helps not just to provide the amenities of good life but lies at the very core of our concept and preoccupation with life itself. This shift in emphasis demands that our textbook, which is designed primarily for the education of future practitioners of health sciences, should attempt to provide both the basics and the scope of the horizon within which chemistry touches our life.

The increasing use of our textbook made this new edition possible, and we wish to thank our colleagues who adopted the previous editions for their courses. Testimony from colleagues and students indicates that we managed to convey our enthusiasm for the subject to students, who find this book to be a great help in studying difficult concepts.

In the new edition, we strive further to present an easily readable and understandable text. At the same time, we emphasize the inclusion of new relevant concepts and examples in this fast-growing discipline, especially in the biochemistry chapters. We maintain an integrated view of chemistry. From the very beginning in the general chemistry part, we include organic compounds and biochemical substances to illustrate the critical principles. The progress is characterized by ascension from the simple to the complex. We urge our colleagues to advance to the chapters of biochemistry as fast as possible, because there lies most of the material that is relevant to the future professions of our students.

Dealing with such a giant field in one course—and possibly the only course in which our students get an exposure to chemistry—makes the selection of the material an overarching enterprise. Even though we tried to keep the book to a manageable size, we recognize that we included more topics than can be covered in a two-semester course. Our aim was to provide enough material from which each instructor can select the topics he or she deems important. We organized the sections so that each of them can stand independently. As a consequence, leaving out sections or even chapters will not create fundamental cracks in the total edifice.

We have increased the number of topics covered and provided a wealth of new problems, many of them challenging and thought-provoking.

Audience

As the previous editions, we intend this book for nonchemistry majors, mainly those entering health sciences and related fields, such as nursing, medical technology, physical therapy, and nutrition. It can also be used by students in environmental studies. In its entirety, it can be used for a one-year (two-semester or three-quarter) course in chemistry; alternatively, parts of the book can be used in a one-term chemistry course.

We assume that the students using this book have little or no background in chemistry. Therefore, we introduce the basic concepts slowly at the beginning and increase the tempo and the level of sophistication as we go on. In addition, we progress from the basic tenets of general chemistry to organic and finally to biochemistry. We consider this progression to represent an ascent in terms of both practical importance and sophistication. Throughout we integrate the three parts by keeping a unified view of chemistry. We do not consider general chemistry sections to be the exclusive domain of inorganic compounds, so we frequently use organic and biological substances to illustrate general principles.

While teaching the chemistry of the human body is our ultimate goal, we try to show that each subsection of chemistry is important in its own right, besides being required for future understanding.

Chemical Connections (Medical and Other Applications of Chemical Principles)

A list of all the applications used in this text can be found on pages xxiii–xxv.

The Chemical Connections features contain applications of the principles discussed in the text. Comments from users of the earlier editions indicate that these boxes have been especially well received and provide a much-requested relevance to the text. For example, in Chapter 1, students see how cold compresses relate to waterbeds and to lake temperatures (Chemical Connections 1C). Up-to-date topics appear throughout the eighth edition, including coverage of anti-inflammatory drugs such as Vioxx and Celebrex (Chemical Connections 21H). Another example is the discussion of obesity and uncoupling in the electron transport chain (Chemical Connections 27A). In Chapter 30, which deals with nutrition, students get a new look at the Food Guide Pyramid (Chemical Connections 30A). The ever-present issues related to dieting are described in Chemical Connections 30B. In Chapter 31, students learn important implications of the use of antibiotics (Chemical Connections 31D) and encounter a detailed explanation of the important and often controversial topic of stem cell research (Chemical Connections 31E).

The presence of the Chemical Connections features provides for a considerable degree of flexibility. If an instructor wants to assign only the main text, the Chemical Connections do not interrupt continuity, and the essential material will be covered. However, because they enhance core material, most instructors will probably wish to assign at least some of them. In our experience, students are eager to read the relevant Chemical Connections, even without assignments, and they do so with discrimination. Given the large number of features, an instructor can select those that best fit the particular needs of his or her course. We provide problems at the end of each chapter for all of the Chemical Connections.

Integrated Media and OWL Homework Management

GOB
Chemistry · Now™

The use of presentation tools, the Internet, and homework management tools has expanded significantly over the past several years. Because students are most concerned with mastering the concepts on which they will ultimately be tested, we have made changes in our integrated media program. Our new web-based, assessment-centered learning tool, **GOB ChemistryNow,** was developed in concert with the text. Throughout each chapter, icons with captions alert students to media resources that enhance problem-solving skills and improve conceptual understanding. In GOB ChemistryNow, students are provided with a Personalized Learning Plan—based on a diagnostic Pre-Test—that targets their study needs and helps them to visualize, organize, practice, and master the material in the text.

In addition to GOB ChemistryNow, we have worked closely with our colleagues at the University of Massachusetts in the development of a new **OWL** (Online Web-based Learning system) GOB course. Class-tested by thousands of students, OWL is a customizable and flexible web-based homework system and assessment tool. This fully integrated testing, tutorial, and course-management system features thousands of practice and homework problems. With both numerical and chemical parameterization built right in, OWL provides students with instant analysis of and feedback on homework problems, modeling questions, molecular-structure–building exercises, and animations created specifically for this text. This powerful system maximizes the students' learning experience and, at the same time, facilitates instruction.

Metabolism: Color Code

The biological functions of chemical compounds are explained in each of the biochemistry chapters and in many of the organic chapters. Emphasis is placed on chemistry rather than physiology. Because we have received much positive feedback regarding the way in which we have organized the topic of metabolism (Chapters 27, 28, and 29), we have maintained this organization in the eighth edition.

First we introduce the common metabolic pathway through which all food will be utilized (the citric acid cycle and oxidative phosphorylation). Only then do we discuss the specific pathways leading to the common pathway. We find this approach to be a useful pedagogical device, and it enables us to sum the caloric values of each type of food because its utilization through the common pathway has already been learned. Finally, we separate the catabolic pathways from the anabolic pathways by treating them in different chapters, emphasizing the different ways the body breaks down and builds up different molecules.

The topic of metabolism is a difficult one for most students, and we have tried to explain it as clearly as possible. As in the previous edition, we enhance the clarity of presentation by the use of a color code for the most important biological compounds discussed in Chapters 27, 28, and 29. Each type of compound is screened in a specific color, which remains the same throughout the three chapters. These colors are as follows:

ATP and other nucleoside triphosphates

ADP and other nucleoside diphosphates

The oxidized coenzymes NAD^+ and FAD

The reduced coenzymes NADH and $FADH_2$

Acetyl coenzyme A

In figures showing metabolic pathways, we display the numbers of the various steps in yellow.

In addition to this main use of a color code, other figures in various parts of the book are color coded so that the same color is used for the same entity throughout. For example, in all Chapter 23 figures that show enzyme–substrate interactions, enzymes are always shown in blue and substrates in orange.

Features

[NEW] **Key Questions**   We employ a new Key Questions framework for this edition to emphasize key chemical concepts. This focused approach guides students through each chapter by using section-head questions.

Chemical Connections   More than 150 essays describe applications of chemical concepts presented in the text.

Summary of Key Reactions   An annotated list summarizes new reactions introduced in each organic chapter (Chapters 10–19) along with an example of the reaction and the section in which it is introduced.

[NEW] **Summary of Key Questions**   End-of-chapter summaries have been completely revised to reflect the Key Questions framework. At the end of each chapter, the Key Questions are restated and the summary paragraphs that follow highlight the concepts associated with those questions.

[NEW] **Looking Ahead Problems**   At the end of most chapters, we have added one or more "challenge problems" designed to show the application of principles in the chapter to material in following chapters.

[NEW] **Tying It Together and Challenge Problems**   At the end of most chapters, we have included problems that build on past material as well as problems that will test students' knowledge of the material.

[NEW] **How To Boxes**   These features emphasize the skills students need to master the material. They include topics such as "How to Do Unit Conversions by the Factor Label Method" (Chapter 1) and "How to Balance a Nuclear Equation" (Chapter 3). The importance of these skills is emphasized in the new GOB ChemistryNow website.

Molecular Models   Ball-and-stick models, space-filling models, and electron density maps are used throughout the text as aids to visualizing molecular properties and interactions.

Margin Definitions   Many terms are defined in the margin to help students learn terminology. By skimming the chapter for these definitions, students can achieve a quick summary of its contents.

Margin Notes   Additional bits of information, such as historical notes and reminders, complement nearby text.

Answers to In-Text and Odd-Numbered End-of-Chapter Problems   Answers to these selected problems are provided at the end of the book. Detailed worked-out solutions to these same problems are provided in the Student Solutions Manual.

Glossary   The glossary at the back of the book provides a definition of each new term along with the number of the section in which the term is introduced.

GOB
Chemistry ⚛ Now ™

[NEW] **GOB ChemistryNow (http://www.thomsonedu.com)**   This new book-specific website correlates directly with the content of this book and contains additional problems, tutorials, and animations for students to practice and review for quizzes and tests. The same material can also be found on the CD that accompanies this book.

[NEW] **GOB OWL**   This web-based homework system gives students extra practice in problem solving and assessment.

Style

Feedback from colleagues and students alike indicates that a major asset of this book is its style, which addresses students directly in simple and clear phrasing. We continue to make special efforts to provide clear and concise writing. Our hope is that this readability facilitates the understanding and absorption of difficult concepts.

Problems

Many of the problems in this edition are new. In particular, the number of more challenging, thought-provoking questions has been increased. The end-of-chapter problems are grouped and given subheads in order of topic coverage. In the last group, "Additional Problems," problems are not arranged in any specific order. New sets of problems include "Tying It Together," "Challenge Problems," and "Looking Ahead."

The answers to all in-text problems and to the odd-numbered end-of-chapter problems are given at the end of the book. Answers to all in-text and end-of-chapter problems are supplied in the Instructor's Manual.

Support Package

For Students

- **GOB ChemistryNow.** Developed in concert with the text, GOB ChemistryNow is a web-based, assessment-centered learning tool. It was designed to assess students' knowledge of the material on which they will be ultimately tested. In each chapter, icons with captions alert students to tutorials, animations, and coached problems, which enhance problem-solving skills and improve conceptual understanding. In GOB ChemistryNow, students are provided with a Personalized Learning Plan—based on a diagnostic Pre-Test—that targets their study needs. A PIN code is required to access GOB ChemistryNow and may be packaged with a new copy of the text or purchased separately. Visit **http://www.thomsonedu.com** to register for access to GOB ChemistryNow. It is also available on the CD that accompanies this book.

- **Student Study Guide,** by William Scovell of Bowling Green State University. This ancillary includes reviews of chapter objectives, important terms and comparisons, focused reviews of concepts, and self-tests.

- **Student Solutions Manual,** by Mark Erickson (Hartwick College), Shawn Farrell, and Courtney Farrell. This ancillary contains complete worked-out solutions to all in-text and odd-numbered end-of-chapter problems.

- **OWL (Online Web-based Learning System) for the GOB Course.** Developed at the University of Massachusetts, Amherst, class-tested by thousands of students, and used by over 200 institutions and 50 thousand students, OWL is a customizable and flexible cross-platform web-based homework system and assessment tool for Introductory, General, Allied Health, Liberal Arts, and Organic chemistry courses. The OWL Online Web-based Learning system provides students with instant analysis and feedback on homework problems, modeling questions, and animations to accompany select Brooks/Cole chemistry textbooks. With deep parameterization of chemical structures, OWL offers thousands of questions. This powerful system maximizes the students' learning experience and, at the same time, reduces faculty workload and helps facilitate instruction. A fee-based code is required for access to the OWL database selected. OWL is only available for use within North America. Visit **http://owl.thomsonlearning.com/demo**.

- **Laboratory Experiments for General, Organic, and Biochemistry, sixth edition**, by Frederick A. Bettelheim and Joseph M. Landesberg. Forty-eight experiments illustrate important concepts and principles in general, organic, and biochemistry: 11 organic chemistry experiments, 17 biochemistry experiments, and 20 general chemistry experiments. Many experiments have been revised, and a new addition is an experiment on the properties of enzymes. All experiments have new Pre- and Post-Lab Questions. The large number of experiments allows sufficient flexibility for the instructor to select those most pertinent to his or her course.

For Instructors

Supporting instructor materials are available to qualified adopters. Please consult your local Thomson Brooks/Cole representative for details.

Visit **http://www.thomsonedu.com** to

- Locate your local representative
- Download electronic files of text art and ancillaries
- Request a desk copy

- **Instructor's Manual**, by Mark Erickson (Hartwick College), Shawn Farrell, and Courtney Farrell, contains worked-out solutions to all in-text and end-of-chapter problems.

- **Multimedia Manager for Organic Chemistry: A Microsoft® PowerPoint® Link Tool** is a digital library and presentation tool that is available on one convenient multiplatform CD-ROM. With its easy-to-use interface, you can take advantage of Brooks/Cole's presentations, which consist of text art and tables from the text. Available in a variety of e-formats that are easily exported into presentation software or used on web-based course support materials, it allows you to customize your own presentation by importing your personal lecture slides or other material you choose. The result is an interactive and fluid lecture that truly engages your students. A set of more than 1000 PowerPoint lecture slides by William Brown is also included.

- **ilrn Testing: Electronic Testing System**, by Stephen Z. Goldberg (Adelphi University), contains approximately 1000 multiple-choice problems and questions representing every chapter of the text. Available online and on a dual-platform CD.

- **Printed Test Bank**, by Stephen Z. Goldberg (Adelphi University), contains approximately 50 multiple-choice questions for each of the 31 chapters in this text.

- **JoinIn™ on TurningPoint® CD-ROM.** Thomson Brooks/Cole is pleased to offer book-specific JoinIn content for Response Systems, allowing you to transform your classroom and assess your students' progress with instant in-class quizzes and polls. You can pose book-specific questions and display students' answers seamlessly within the Microsoft PowerPoint slides of your own lecture, and in conjunction with the "clicker" hardware of your choice. Enhance how your students interact with you, your lecture, and each other. Consult your Brooks/Cole representative for further details.

- **Overhead Transparencies.** A set of 150 overhead transparencies in full color is available for the eighth edition. They include figures and tables taken directly from the text.

Table 1.3 Some Conversion Factors Between the English and Metric Systems		
Length	**Mass**	**Volume**
1 in. = 2.54 cm	1 oz = 28.35 g	1 qt = 0.946 L
1 m = 39.37 in	1 lb = 453.6 g	1 gal = 3.785 L
1 mile = 1.609 km	1 kg = 2.205 lb	1 L = 33.81 fl oz
	1 g = 15.43 grains	1 fl oz = 29.57 mL
		1 L = 1.057 qt

For people who have grown up using English units, it is helpful to have some idea of the size of metric units. Table 1.3 shows some conversion factors.

Some of these conversions are difficult enough that you will probably not remember them and must, therefore, look them up when you need them. Some are easier. For example, a meter is about the same as a yard. A kilogram is a little over two pounds. There are almost four liters in a gallon. These conversions may be important to you someday. For example, if you rent a car in Europe, the price of gas listed on the sign at the gas station will be in Euros per liter. When you realize that you are spending a dollar per liter and you know that there are almost four liters to a gallon, you will realize why so many people take the bus or a train instead.

B. Volume

Volume is space. The volume of a liquid, solid, or gas is the space occupied by that substance. The base unit of volume in the metric system is the **liter (L).** This unit is a little larger than a quart (Table 1.3). The only other common metric unit for volume is the milliliter (mL), which is equal to 10^{-3} L.

$$1 \text{ mL} = 0.001 \text{ L}$$

$$1000 \text{ mL} = 1 \text{ L}$$

One milliliter is exactly equal to one cubic centimeter (cc or cm^3):

$$1 \text{ mL} = 1 \text{ cc}$$

Thus there are 1000 cc in 1 L.

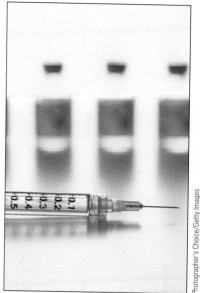

Hypodermic syringe. Note that the volumes are indicated in mL.

C. Mass

Mass is the quantity of matter in an object. The base unit of mass in the metric system is the **gram (g).** As always in the metric system, larger and smaller units are indicated by prefixes. The ones in common use are

$$1 \text{ kilogram (kg)} = 1000 \text{ g}$$

$$1 \text{ milligram (mg)} = 0.001 \text{ g}$$

The gram is a small unit; there are 453.6 g in one pound (Table 1.3).

We use a device called a balance to measure mass. Figure 1.3 shows two types of laboratory balances.

There is a fundamental difference between mass and weight. Mass is independent of location. The mass of a stone, for example, is the same whether we measure it at sea level, on top of a mountain, or in the depths of a mine. In contrast, weight is not independent of location. **Weight** is the

Figure 1.3 Two laboratory balances.

Courtesy of Brinkmann Instruments, Co.

force a mass experiences under the pull of gravity. This point was dramatically demonstrated when the astronauts walked on the surface of the Moon. The Moon, being a smaller body than the Earth, exerts a weaker gravitational pull. Consequently, even though the astronauts wore space suits and equipment that would be heavy on Earth, they felt lighter on the Moon and could execute great leaps and bounces during their walks.

Although mass and weight are different concepts, they are related to each other by the force of gravity. We frequently use the words interchangeably because we weigh objects by comparing their masses to standard refer-

CHEMICAL CONNECTIONS 1A

Drug Dosage and Body Mass

In many cases, drug dosages are prescribed on the basis of body mass. For example, the recommended dosage of a drug may be 3 mg of drug for each kilogram of body weight. In this case, a 50 kg (110 lb) woman would receive 150 mg and an 82 kg (180 lb) man would get 246 mg. This adjustment is especially important for children, because a dose suitable for an adult will generally be too much for a child, who has much less body mass. For this reason, manufacturers package and sell smaller doses of certain drugs, such as aspirin, for children.

Drug dosage may also vary with age. Occasionally, when an elderly patient has an impaired kidney or liver function, the clearance of a drug from the body is delayed, and the drug may stay in the body longer than is normal. This persistence can cause dizziness, vertigo, and migraine-like headaches, resulting in falls and broken bones. Such delayed clearance must be monitored, and the drug dosage adjusted accordingly.

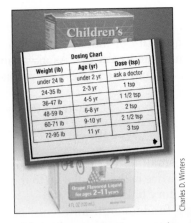

Charles D. Winters

This package of Advil has a chart showing the proper doses for children of a given weight.

- **Instructor's Manual to Accompany Laboratory Experiments.** This manual will help instructors in grading the answers to questions and in assessing the range of experimental results obtained by students. The Instructor's Manual also contains important notes for professors to tell students and details on how to handle the disposal of waste chemicals.

- **WebCT/Blackboard GOB ChemistryNow** can be integrated with WebCT and Blackboard so that professors who are using one of those platforms can access all of the Now assessments and content, without an extra login. Please contact your Thomson Brooks/Cole representative for more information.

Organization

General Chemistry (Chapters 1–9)

Chapter 1, Matter, Energy, and Measurement, serves as a general introduction to the text and introduces the pedagogical elements that are new to this edition.

In **Chapter 2, Atoms,** we introduce four of the five methods we use throughout the text to represent molecules: We show water as a molecular formula, a structural formula, a ball-and-stick model, and a space-filling model. We introduce electron density maps, a fifth form of representation, in Chapter 4. To draw the correlation between electron configuration and the Periodic Table, we use a miniature version of the Periodic Table to show the locations of groups of elements as they are discussed. The discussion of electron configuration and the Periodic Table in Section 2.7 focuses on the underlying organization by *s, p, d,* and *f* blocks.

We have moved **Nuclear Chemistry** forward to **Chapter 3** so that it now follows the discussion of electron configuration in Chapter 2.

The material in **Chapter 4, Chemical Bonds,** begins with a discussion of ionic compounds, followed by a discussion of molecular compounds. A new addition to **Chapter 5, Chemical Reactions,** is *How to Balance a Chemical Equation*. This How To box illustrates a step-by-step method for balancing an equation. We discuss the energetics of chemical reactions here and throughout the text only in terms of how exothermic or endothermic they are.

In **Chapter 6, Gases, Liquids, and Solids,** we present intermolecular forces of attraction in order of increasing energy—namely, London dispersion forces, dipole–dipole interactions, and hydrogen bonding.

Chapter 7, Solutions and Colloids, opens with a listing of the most common types of solutions, followed by a discussion of the factors that affect solubility, the most common units for concentration, and colligative properties.

Chapter 8, Reaction Rates and Chemical Equilibrium, provides the basis for understanding chemical reactivity in its most fundamental terms.

Chapter 9, Acids and Bases, now introduces the use of curved arrows to show the flow of electrons in organic reactions. Specifically, we use them here to show the flow of electrons in proton-transfer reactions. We have dropped the topics of normality and equivalents from the discussion of the stoichiometry of acid–base reactions. The major theme in this chapter is the applications of acid–base buffers and the Henderson-Hasselbalch equation.

Organic Chemistry (Chapters 10–19)

Chapter 10, Organic Chemistry, introduces the characteristics of organic compounds and the most important organic functional groups.

In **Chapter 11, Alkanes,** we introduce the concept of a line-angle formula and continue using these forumulas throughout the organic chemistry chapters (Chapters 11–19). These structures are easier to draw than the usual condensed structural formulas and are easier to visualize. Section 11.9B describes the halogenation of alkanes. The following section focuses on chlorofluorocarbons (the Freons) and their replacements, as well as haloalkane solvents. A Chemical Connections box describes the environmental impact of Freons.

A Chemical Connections box in **Chapter 12, Alkenes and Alkynes,** examines ethylene, a plant growth regulator. In Chapter 12, we introduce the concept of a reaction mechanism through the hydrohalogenation and acid-catalyzed hydration of alkenes. In addition, we present a mechanism for the catalytic hydrogenation of alkenes and later, in Chapter 18, show how the reversibility of catalytic hydrogenation leads to the formation of "*trans* fats." This introduction to reaction mechanisms is meant to demonstrate to students that chemists are interested not only in *what* happens in a chemical reaction, but also in *how* it happens. Section 12.5, on terpenes, has been shortened considerably.

Chapter 13, Benzene and its Derivatives, now follows immediately after the treatment of alkenes and alkynes. Our discussion of phenols includes phenols and antioxidants. In this subsection, we treat autoxidation of unsaturated fatty acid hydrocarbon chains as a radical-chain mechanism leading to hydroperoxidation, and we discuss how substituted phenols such as BHT and BHA act as antioxidants.

Chapter 14, Alcohols, Ethers, and Thiols, discusses the structures, names, and properties of alcohols first, and then gives a similar treatment to ethers, and finally to thiols.

In **Chapter 15, Chirality: The Handedness of Molecules,** the concept of a stereocenter and enantiomerism is introduced slowly and carefully using 2-butanol as a prototype. We then treat molecules with two or more stereocenters, emphasizing how to recognize a stereocenter by the fact of four different groups on a tetrahedral carbon stereocenter, and how to predict the number of stereoisomers possible for a particular molecule. We also explain the *R,S* convention for assigning absolute configuration to a tetrahedral stereocenter. The intent of this discussion is to acquaint students with the fact that a designation of (R)- or (S)- in the name of a compound, means that the compound has handedness, that its handedness is known, and that the compound referred to is a single enantiomer.

In **Chapter 16, Amines,** a new section entitled "Epinephrine: A Prototype for the Development of New Bronchodilators" traces the development of new asthma medications from epinephrine as a lead drug to albuterol (Proventil).

To **Chapter 17, Aldehydes and Ketones,** we have added a discussion of $NaBH_4$ as a carbonyl-reducing agent with emphasis on it as a hydride transfer reagent. We then make the parallel to NADH as a carbonyl-reducing agent and hydride transfer agent.

We have divided the chemistry of carboxylic acids and their derivatives into two chapters. **Chapter 18, Carboxylic Acids,** focuses on the chemistry and physical properties of carboxylic acids themselves. We have added a brief discussion of *trans* fatty acids and of omega-3 fatty acids and the significance of their presence in our diet.

Chapter 19, Carboxylic Anhydrides, Esters, and Amides describes the chemistry of these three important functional groups with emphasis on their acid-catalyzed and base-promoted hydrolysis, and reactions with amines and alcohols.

Biochemistry (Chapters 20–31)

Chapter 20, Carbohydrates, is the first of the biochemistry chapters. It begins with the structures and nomenclature of monosaccharides, their oxidation and reduction, and the formation of glycosides. The chapter concludes with a discussion of the structures of disaccharides, polysaccharides, and acidic polysaccharides.

Chapter 21, Lipids, covers the most important features of lipid biochemistry, including membrane structure and the structures and functions of steroids. We have updated the coverage of anti-inflammatory drugs in the section on thromboxanes and leukotrienes.

Chapter 22, Proteins, covers the many facets of protein structure and function. It gives an overview of how proteins are organized, beginning with the nature of individual amino acids, and describes how this organization leads to their many functions. These basics are needed to lead students into the sections on enzymes and metabolism. The topics of prions, proteomics, and allosterism have been expanded from the seventh edition.

Chapter 23, Enzymes, covers the important topic of enzyme catalysis and regulation. The focus is on how the structure of an enzyme leads to the vast increases in reaction rates observed with enzyme-catalyzed reactions. Specific medical applications of enzyme inhibition are included as well as an introduction to the fascinating topic of transition-state analogs and their use as potent inhibitors.

In **Chapter 24, Chemical Communications: Neurotransmitters and Hormones,** we see the biochemistry of hormones and neurotransmitters. The health-related implications of how these substances act in the body are a main focus of this chapter.

In **Chapter 25, Nucleotides, Nucleic Acids, and Heredity,** students are introduced to DNA and the processes surrounding its replication and repair. Students will have heard about the Human Genome Project, but the information gathered cannot be understood without a solid grasp of how nucleotides are linked together and the flow of genetic information that occurs due to the unique properties of these molecules. The sections on the types of RNA have been greatly expanded, as our knowledge increases daily about these important nucleic acids. The uniqueness of an individual's DNA is described in a Chemical Connections box that introduces DNA fingerprinting and explains how forensic science relies on DNA for positive identification.

Chapter 26, Gene Expression and Protein Synthesis, continues where Chapter 25 leaves off, showing how the information contained in the DNA blueprint of the cell is used to produce RNA and eventually protein. The focus here is on how organisms get the most "bang for the buck" by controlling the expression of genes through transcription and translation. The chapter ends with the timely and important topic of gene therapy—our attempt to cure genetic diseases by giving an individual a gene he or she is missing.

Chapter 27, Bioenergetics: How the Body Converts Food to Energy, is an introduction to metabolism that focuses strongly on the central pathways—namely, the citric acid cycle, electron transport, and oxidative phosphorylation.

In **Chapter 28, Specific Catabolic Pathways: Carbohydrate, Lipid, and Protein Metabolism,** we address the details of carbohydrate, lipid, and protein breakdown, concentrating on the energy yield.

Chapter 29, Biosynthetic Pathways, starts with a general consideration of anabolism and proceeds to carbohydrate biosynthesis in both plants and animals. Lipid biosynthesis is linked to production of membranes, and the chapter concludes with an account of amino acid biosynthesis.

In **Chapter 30, Nutrition,** we take a biochemical approach to understanding nutrition concepts. Along the way, we look at a revised version of the Food Guide Pyramid, and debunk some of the myths about carbohydrates and fats. Chemical Connections boxes expand on two topics that are often important to students—dieting and enhancement of sports performance through proper nutrition.

Chapter 31, Immunochemistry, covers the basics of the human immune system and explains how we protect ourselves from foreign invading organisms. The section on the innate immune system has been expanded from previous editions, but considerable time is also spent on the acquired immunity system that is more well known by the public. No chapter on immunology would be complete without a description of the human immunodeficiency virus, and this topic has been expanded compared to the seventh edition. The chapter ends with a foray into the controversial topic of stem cell research—our hopes for its potential and concerns for the potential downsides.

Chapter 32, Blood Chemistry and Body Fluids, can be found on the website that accompanies this book at **http://now.brookscole.com/gob8**.

Acknowledgments

The publication of a book such as this one requires the efforts of many more people than merely the authors. We would like to thank the following professors who offered many valuable suggestions for this new edition.

We are especially grateful to David Shinn and to Mark Erickson (Hartwick College), who read page proofs with an eye for accuracy and checked the wording and accuracy of problem answers.

We give special thanks to Sandi Kiselica, our Senior Development Editor, who has been a rock of support through the entire revision process. We appreciate her ability to set challenging but manageable schedules for us as well as her constant encouragement as we worked to meet those deadlines. She has also been a valuable resource person. We appreciate the help of our other colleagues at Brooks/Cole: Lisa Lockwood, Acquisitions Editor; Teri Hyde, Production Manager; Ellen Bitter, Associate Editor; and Nicole Barone of Thompson Steele.

We also greatly appreciate the time and expertise of our reviewers who have read our manuscript and given us helpful comments:

MADELINE ADAMCZESKI, San Jose City College
DAVID S. BALLANTINE, Northern Illinois University
BARRY BOATWRIGHT, Gadsden State Community College
DANIEL FREEMAN, University of South Carolina
LISA HOMESLEY, Cabrillo College
DELORES B. LAMB, Greenville Technical College
CRAIG MCCLURE, University of Alabama, Birmingham
DOUGLAS RAYNIE, South Dakota State University
FRED J. SAFAROWIC, Passaic County Community College
STEVEN M. SOCOL, McHenry County College
GARON C. SMITH, University of Montana

Health-Related Topics

Key:

ChemConn = Chemical Connections Box number
Sect. = Section number
Prob. = Problem number

CHAPTER **1**

Matter, Energy, and Measurement

KEY QUESTIONS

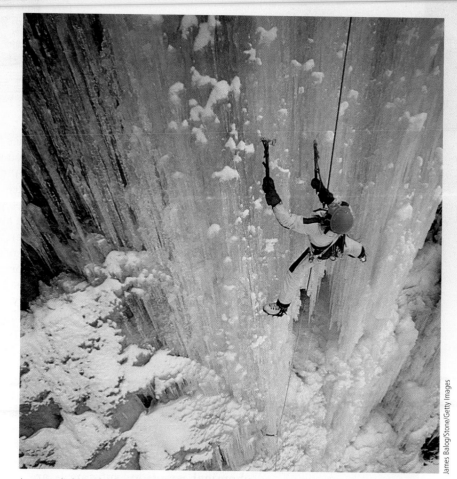

A woman climbing a frozen waterfall in British Columbia.

James Balog/Stone/Getty Images

GOB
Chemistry⚛Now™

Look for this logo in the text and go to GOB ChemistryNow at **http://now.brookscole.com/gob8** or on the CD to view tutorials and simulations, develop problem-solving skills, and test your conceptual understanding with unique interactive resources.

1.1 | Why Do We Call Chemistry the Study of Matter?

The world around us is made of chemicals. Our food, our clothing, the buildings in which we live are all made of chemicals. Our bodies are made of chemicals, too. To understand the human body, its diseases, and its cures, we must know all we can about those chemicals. There was a time—only a few hundred years ago—when physicians were powerless to treat many diseases. Cancer, tuberculosis, smallpox, typhus, plague, and many other sicknesses struck people seemingly at random. Doctors, who had no idea what caused any of these diseases, could do little or nothing about them. Doctors treated them with magic as well as by such measures as bleeding, laxatives, hot plasters, and pills made from powdered stag horn, saffron, or gold. None of these treatments was effective, and the doctors, because they came into direct contact with highly contagious diseases, died at a much higher rate than the general public.

A woman being bled by a leech on her left forearm; a bottle of leeches is on the table. From a 1639 woodcut.

Courtesy of the National Library of Medicine

Medicine has made great strides since those times. We live much longer, and many once-feared diseases have been essentially eliminated or are curable. Smallpox has been eradicated, and polio, typhus, bubonic plague, diphtheria, and other diseases that once killed millions no longer pose a serious problem, at least not in the developed countries.

How has this medical progress come about? The answer is that diseases could not be cured until they were understood, and this understanding has emerged through greater knowledge of how the body functions. It is progress in our understanding of the principles of biology, chemistry, and physics that has led to these advances in medicine. Because so much of modern medicine depends on chemistry, it is essential that students who intend to enter the health professions have some understanding of basic chemistry. This book was written to help you achieve that goal. Even if you choose a different profession, you will find that the chemistry you learn in this course will greatly enrich your life.

The universe consists of matter, energy, and empty space. **Matter** is anything that has mass and takes up space. **Chemistry** is the science that deals with matter: the structure and properties of matter and the transformations from one form of matter to another. We will discuss energy in Section 1.8.

It has long been known that matter can change, or be made to change, from one form to another. In a **chemical change,** more commonly called a **chemical reaction,** substances are used up (disappear) and others are formed to take their places. An example is the burning of the mixture of hydrocarbons usually called "bottled gas." In this mixture of hydrocarbons, the main component is propane. When this chemical change takes place, propane and oxygen from the air are converted to carbon dioxide and water. Figure 1.1 shows another chemical change.

Matter also undergoes other kinds of changes, called **physical changes.** These changes differ from chemical reactions in that the identities of the substances do not change. Most physical changes involve changes of state—for example, the melting of solids and the boiling of liquids. Water remains water whether it is in the liquid state or in the form of ice or steam. The conversion from one state to another is a physical—not a chemical—change. Another important type of physical change involves making or separating mixtures. Dissolving sugar in water is a physical change.

When we talk about the **chemical properties** of a substance, we mean the chemical reactions that it undergoes. **Physical properties** are all properties that do not involve chemical reactions. For example, density, color, melting point, and physical state (liquid, solid, gas) are all physical properties.

GOB
Chemistry⚛Now™
Click *Chemistry Interactive* to see an example of a **chemical change in action**

(a)   (b)   (c)

Charles D. Winters

GOB
Chemistry‑₊‑Now™

Active Figure 1.1   A chemical reaction. (*a*) Bromine, an orange-brown liquid, and aluminum metal. (*b*) These two substances react so vigorously that the aluminum becomes molten and glows white hot at the bottom of the beaker. The yellow vapor consists of vaporized bromine and some of the product of the reaction, white aluminum bromide. (*c*) Once the reaction is complete, the beaker is coated with aluminum bromide and the products of its reaction with atmospheric moisture. (*Note:* This reaction is dangerous! Under no circumstances should it be done except under properly supervised conditions.) **See a simulation based on this figure, and take a short quiz on the concepts at http://now.brookscole.com/gob8 or on the CD.**

1.2 | What Is the Scientific Method?

Scientists learn by using a tool called the **scientific method.** The heart of the scientific method is the testing of theories. It was not always so, however. Before about 1600, philosophers often believed statements just because they sounded right. For example, the great philosopher Aristotle (384–322 BCE) believed that if you took the gold out of a mine it would grow back. He believed this idea because it fitted in with a more general picture that he had about the workings of nature. In ancient times, most thinkers behaved in this way. If a statement sounded right, they believed it without testing it.

About 1600 CE, the scientific method came into use. Let us look at an example to see how the scientific method operates. The Greek physician Galen (200–130 BCE) recognized that the blood on the left side of the heart somehow gets to the right side. This is a fact. A **fact** is a statement based on direct experience. It is a consistent and reproducible observation. Having observed this fact, Galen then proposed a hypothesis to explain it. A **hypothesis** is a statement that is proposed, without actual proof, to explain the facts and their relationship. Because Galen could not actually see how the blood got from the left side to the right side of the heart, he came up with the hypothesis that tiny holes must be present in the muscular wall that separates the two halves.

Up to this point, a modern scientist and an ancient philosopher would behave the same way. Each would offer a hypothesis to explain the facts. From this point on, however, their methods would differ. To Galen, his explanation sounded right and that was enough to make him believe it, even though he couldn't see any holes. His hypothesis was, in fact, believed by virtually all physicians for more than 1000 years. When we use the scientific method, however, we do not believe a hypothesis just because it sounds right. We test it, using the most rigorous testing we can imagine.

Galen.

Hypothesis A statement that is proposed, without actual proof, to explain a set of facts and their relationship

Courtesy of the National Library of Medicine

William Harvey (1578–1657) tested Galen's hypothesis by dissecting human and animal hearts and blood vessels. He discovered that one-way valves separate the upper chambers of the heart from the lower chambers. He also discovered that the heart is a pump that, by contracting and expanding, pushes the blood out. Harvey's teacher, Fabricius (1537–1619), had previously observed that one-way valves exist in the veins, so that blood in the veins can travel only toward the heart and not the other way.

Harvey put these facts together to come up with a new hypothesis: Blood is pumped by the heart and circulates throughout the body. This was a better hypothesis than Galen's because it fitted the facts more closely. Even so, it was still a hypothesis and, according to the scientific method, had to be tested further. One important test took place in 1661, four years after Harvey died. Harvey had predicted that because there had to be a way for the blood to get from the arteries to the veins, tiny blood vessels must connect them. In 1661 the Italian anatomist Malpighi (1628–1694), using the newly invented microscope, found these tiny vessels, which are now called capillaries.

Theory The formulation of an apparent relationship among certain observed phenomena, which has been verified. A theory explains many interrelated facts and can be used to make predictions about natural phenomena. Examples are Newton's theory of gravitation and the kinetic molecular theory of gases, which we will encounter in Section 6.6. This type of theory is also subject to testing and will be discarded or modified if it is contradicted by new facts.

Malpighi's discovery supported the blood circulation hypothesis by fulfilling Harvey's prediction. When a hypothesis passes the tests, we have more confidence in it and call it a theory. A **theory** is the formulation of an apparent relationship among certain observed phenomena, which has been verified to some extent. In this sense, a theory is the same as a hypothesis except that we have a stronger belief in it because more evidence supports it. No matter how much confidence we have in a theory, however, if we discover new facts that conflict with it or if it does not pass newly devised tests, the theory must be altered or rejected. In the history of science, many firmly established theories have eventually been thrown out because they could not pass new tests.

The scientific method is thus very simple. We don't accept a hypothesis or a theory just because it sounds right. We devise tests, and only if the hypothesis or theory passes the tests do we accept it. The enormous progress made since 1600 in chemistry, biology, and the other sciences is a testimony to the value of the scientific method.

You may get the impression from the preceding discussion that science progresses in one direction: facts first, hypothesis second, theory last. Real life is not so simple, however. Hypotheses and theories call the attention of scientists to discover new facts. An example of this scenario is the discovery of the element germanium. In 1871, Mendeleev's Periodic Table—a graphic description of elements organized by properties—predicted the existence of a new element whose properties would be similar to those of silicon. Mendeleev called this element eka-silicon. In 1886, it was discovered in Germany (hence the name), and its properties were truly similar to those predicted by theory.

On the other hand, many scientific discoveries result from **serendipity,** or chance observation. An example of serendipity occurred in 1926, when James Sumner of Cornell University left an enzyme preparation of jack bean urease in a refrigerator over the weekend. Upon his return, he found that his solution contained crystals that turned out to be a protein. This chance discovery led to the hypothesis that all enzymes are proteins. Of course, serendipity is not enough to move science forward. Scientists must have the creativity and insight to recognize the significance of their observations. Sumner fought for more than 15 years for his hypothesis to gain acceptance because people believed that only small molecules can form crystals. Eventually his view won out, and he was awarded a Nobel Prize in chemistry in 1946.

1.3 | How Do Scientists Report Numbers?

Scientists often have to deal with numbers that are very large or very small. For example, an ordinary copper penny (dating from before 1982, when pennies in the United States were still made of copper) contains approximately

29,500,000,000,000,000,000,000 atoms of copper

and a single copper atom weighs

0.00000000000000000000000023 pound

which is equal to

0.000000000000000000000104 gram

Many years ago, an easy way to handle such large and small numbers was devised. This method, which is called **exponential notation,** is based on powers of 10. In exponential notation, the number of copper atoms in a penny is written

$$2.95 \times 10^{22}$$

and the weight of a single copper atom is written

$$2.3 \times 10^{-25} \text{ pound}$$

which is equal to

$$1.04 \times 10^{-22} \text{ grams}$$

The origin of this shorthand form can be seen in the following examples:

$$100 = 1 \times 10 \times 10 = 1 \times 10^2$$

$$1000 = 1 \times 10 \times 10 \times 10 = 1 \times 10^3$$

What we have just said in the form of an equation is "100 is a one with two zeroes after the one, and 1000 is a one with three zeroes after the one." We can also write

$$1/100 = 1/10 \times 1/10 = 1 \times 10^{-2}$$

$$1/1000 = 1/10 \times 1/10 \times 1/10 = 1 \times 10^{-3}$$

where negative exponents denote numbers less than 1. The exponent in a very large or very small number lets us keep track of the number of zeros. That number can become unwieldy with very large or very small quantities, and it is easy to lose track of a zero. Exponential notation helps us deal with this possible source of mathematical error.

When it comes to measurements, not all the numbers you can generate in your calculator or computer are of equal importance. Only the number of digits that are known with certainty are significant. Suppose that you measured the weight of an object as 3.4 g on a balance that you can read to the nearest 0.1 g. You can report the weight as 3.4 g but not as 3.40 or 3.400 g because you do not know the added zeros with certainty. This becomes even more important when you do calculations using a calculator. For example, you might measure a cube with a ruler and find that each side is 2.9 cm. If you are asked to calculate the volume, you multiply $2.9 \times 2.9 \times 2.9$. The calculator will then give you an answer that is 23.389 cm^3. However, your initial measurements were only good to a tenth of a centimeter, so your final answer cannot be good to a thousandth of a centimeter. As a scientist, it is important to report data that have the correct number of **significant figures.** A detailed account of using significant figures is presented in Appendix II. A discussion of accuracy, precision, and significant figures can be found in laboratory manuals [see Bettelheim and Landesberg, *Laboratory Experiments,* sixth edition (Experiment 2)].

EXAMPLE 1.1

Multiply:

(a) $(4.73 \times 10^5)(1.37 \times 10^2)$ (b) $(2.7 \times 10^{-4})(5.9 \times 10^8)$

Divide:

(c) $\dfrac{7.08 \times 10^{-8}}{300}$ (d) $\dfrac{5.8 \times 10^{-6}}{6.6 \times 10^{-8}}$ (e) $\dfrac{7.05 \times 10^{-3}}{4.51 \times 10^5}$

Solution

The way to do calculations of this sort is to use a button on scientific calculators that automatically uses exponential notation. The button is usually labeled "E."

(a) Enter 4.73E5, press the multiplication key, enter 1.37E2, and press the "=" key. The answer is 6.48×10^7. The calculator will display this number as 6.48E7. This answer makes sense. We add exponents when we multiply, and the sum of these two exponents is correct $(5 + 2 = 7)$. We also multiply the numbers, 4.73×1.37. This is approximately $4 \times 1.5 = 6$, so 6.48 is also reasonable.

(b) Here we have to deal with a negative exponent, so we use the "+/-" key. Enter 2.7E+/-4, press the multiplication key, enter 5.9E8, and press the "=" key. The calculator will display the answer as 1.593E5. To have the correct number of significant figures, we should report our answer as 1.6E5. This answer makes sense because 2.7 is a little less than 3, and 5.9 is a little less than 6, so we predict a number slightly less than 18; also the algebraic sum of the exponents $(-4 + 8)$ is equal to 4. This gives 16×10^4. In scientific notation, we normally prefer to report numbers between 1 and 10, so we rewrite our answer as 1.6×10^5. We made the first number 10 times smaller, so we increased the exponent by 1 to reflect that change.

(c) Enter 7.08E+/-8, press the division key, enter 300, and press the "=" key. The answer is 2.36×10^{-10}. The calculator will display this number as 2.36E − 10. We subtract exponents when we divide, and we can also write 300 as 3.00×10^2.

(d) Enter 5.8E+/-6, press the division key, enter 6.6E+/-8, and press the "=" key. The calculator will display the answer as 87.878787878788. We report this answer as 88 to get the right number of significant figures. This answer makes sense. When we divide 5.8 by 6.6, we get a number slightly less than 1. When we subtract the exponents algebraically $(-6 - [-8])$, we get 2. This means that the answer is slightly less than 1×10^2, or slightly less than 100.

(e) Enter 7.05E+/-3, press the division key, enter 4.51E5, and press the "=" key. The calculator displays the answer as 1.5632E-8, which, to the correct number of significant figures, is 1.56×10^{-8}. The algebraic subtraction of exponents is $-3 - 5 = -8$.

Problem 1.1

Multiply:

(a) $(6.49 \times 10^7)(7.22 \times 10^{-3})$

(b) $(3.4 \times 10^{-5})(8.2 \times 10^{-11})$

Divide:

(a) $\dfrac{6.02 \times 10^{23}}{3.10 \times 10^5}$

(b) $\dfrac{3.14}{2.30 \times 10^{-5}}$

1.4 | How Do We Make Measurements?

In our daily lives we are constantly making measurements. We measure ingredients for recipes, driving distances, gallons of gasoline, weights of fruits and vegetables, and the timing of TV programs. Doctors and nurses measure pulse rates, blood pressures, temperatures, and drug dosages. Chemistry, like other sciences, is based on measurements.

A measurement consists of two parts: a number and a unit. A number without a unit is usually meaningless. If you were told that a person's weight is 57, the information would be of very little use. Is it 57 pounds, which would indicate that the person is very likely a child or a midget, or 57 kilograms, which is the weight of an average woman or a small man? Or is it perhaps some other unit? Because so many units exist, a number by itself is not enough; the unit must also be stated.

In the United States, most measurements are made with the English system of units: pounds, miles, gallons, and so on. In most other parts of the world, however, few people could tell you what a pound or an inch is. Most countries use the **metric system,** a system that originated in France about 1800 and that has since spread throughout the world. Even in the United States, metric measurements are slowly being introduced (Figure 1.2). For

The label on this bottle of water shows the metric size (one liter) and the eqivalent in quarts.

Metric system A system of units of measurement in which the divisions to subunits are made by a power of 10

Figure 1.2 Road sign in Massachusetts showing metric equivalents of mileage.

Table 1.1	Base Units in the Metric System
Length	meter (m)
Volume	liter (L)
Mass	gram (g)
Time	second (s)
Temperature	°Celsius (°C)
Energy	calorie (cal)
Amount of substance	mole (mol)

GOB
Chemistry⚛Now™
Click *Mastering the Essentials* to practice using the **Metric System**

example, many soft drinks and most alcoholic beverages now come in metric sizes. Scientists in the United States have been using metric units all along.

Around 1960, international scientific organizations adopted another system, called the **International System of Units** (abbreviated **SI**). The SI is based on the metric system and uses some of the metric units. The main difference is that the SI is more restrictive: It discourages the use of certain metric units and favors others. Although the SI has advantages over the older metric system, it also has significant disadvantages. For this reason U.S. chemists have been very slow to adopt it. At this time, approximately 40 years after its introduction, not many U.S. chemists use the entire SI, although some of its preferred units are gaining ground.

In this book we will use the metric system (Table 1.1). Occasionally we will mention the preferred SI unit.

A. Length

The key to the metric system (and the SI) is that there is one base unit for each kind of measurement and that other units are related to the base unit only by powers of 10. As an example, let us look at measurements of length. In the English system we have the inch, the foot, the yard, and the mile (not to mention such older units as the league, furlong, ell, and rod). If you want to convert one unit to another unit, you must memorize or look up these conversion factors:

$$5280 \text{ feet} = 1 \text{ mile}$$

$$1760 \text{ yards} = 1 \text{ mile}$$

$$3 \text{ feet} = 1 \text{ yard}$$

$$12 \text{ inches} = 1 \text{ foot}$$

All this is unnecessary in the metric system (and the SI). In both systems the base unit of length is the **meter (m).** To convert to larger or smaller units we do not use arbitrary numbers like 12, 3, and 1760, but only 10, 100, 1/100, 1/10, or other powers of 10. This means that *to convert from one metric or SI unit to another, we only have to move the decimal point.* Furthermore, the other units are named by putting prefixes in front of "meter," and *these prefixes are the same throughout the metric system and the SI.* Table 1.2 lists the most important of these prefixes. If we put some of these prefixes in front of "meter," we have

$$1 \text{ kilometer (km)} = 1000 \text{ meters (m)}$$

$$1 \text{ centimeter (cm)} = 0.01 \text{ meter}$$

$$1 \text{ nanometer (nm)} = 10^{-9} \text{ meter}$$

Table 1.2	The Most Common Metric Prefixes	
Prefix	Symbol	Value
giga	G	$10^9 = 1,000,000,000$ (one billion)
mega	M	$10^6 = 1,000,000$ (one million)
kilo	k	$10^3 = 1000$ (one thousand)
deci	d	$10^{-1} = 0.1$ (one-tenth)
centi	c	$10^{-2} = 0.01$ (one-hundredth)
milli	m	$10^{-3} = 0.001$ (one-thousandth)
micro	μ	$10^{-6} = 0.000001$ (one-millionth)
nano	n	$10^{-9} = 0.000000001$ (one-billionth)

ence masses (weights) on a balance, and the gravitational pull is the same on the unknown object and on the standard masses. Because the force of gravity is essentially constant, mass is always directly proportional to weight.

D. Time

Time is the one quantity for which the units are the same in all systems: English, metric, and SI. The base unit is the **second (s):**

$$60 \text{ s} = 1 \text{ min}$$
$$60 \text{ min} = 1 \text{ h}$$

E. Temperature

Most people in the United States are familiar with the Fahrenheit scale of temperature. The metric system uses the centigrade, or **Celsius,** scale. In this scale, the boiling point of water is set at 100°C and the freezing point at 0°C. We can convert from one scale to the other by using the following formulas:

$$°F = \frac{9}{5} °C + 32$$

$$°C = \frac{5}{9} (°F - 32)$$

The 32 in these equations is a defined number and is, therefore, treated as if it had an infinite number of zeros following the decimal point. (See Appendix II.)

EXAMPLE 1.2

Normal body temperature is 98.6°F. Convert this temperature to Celsius.

Solution

$$°C = \frac{5}{9} (98.6 - 32) = \frac{5}{9} (66.6) = 37.0°C$$

Problem 1.2

Convert:
(a) 64.0°C to Fahrenheit   (b) 47°F to Celsius

Figure 1.4 shows the relationship between the Fahrenheit and Celsius scales.

A third temperature scale is the **Kelvin (K)** scale, also called the absolute scale. The size of a Kelvin degree is the same as that of a Celsius degree; the only difference is the zero point. The temperature −273°C is taken as the zero point on the Kelvin scale. This makes conversions between Kelvin and Celsius very easy. To go from Celsius to Kelvin, just *add* 273; to go from Kelvin to Celsius, *subtract* 273:

$$K = °C + 273$$
$$°C = K - 273$$

Figure 1.4 also shows the relationship between the Kelvin and Celsius scales. Note that we don't use the degree symbol in the Kelvin scale: 100°C equals 373 K, not 373°K.

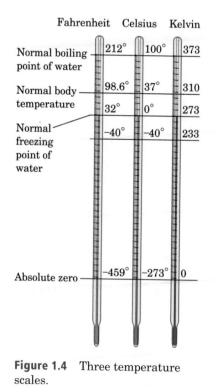

Figure 1.4   Three temperature scales.

Why was $-273°C$ chosen as the zero point on the Kelvin scale? The reason is that $-273°C$, *or 0 K, is the lowest possible temperature*. Because of this, 0 K is called **absolute zero.** Temperature reflects how fast molecules move. The more slowly they move, the colder it gets. At absolute zero, molecules stop moving altogether. Therefore, the temperature cannot get any lower. For some purposes it is convenient to have a scale that begins at the lowest possible temperature; the Kelvin scale fulfills this need. The Kelvin is the SI unit.

It is very important to have a "gut feeling" about the relative sizes of the units in the metric system. Often, while doing calculations, the only thing that might offer a clue that you have made an error is your understanding of the sizes of the units. For example, if you are calculating the amount of a chemical that is dissolved in water and you come up with an answer of 254 kg/mL, does your answer make sense? If you have no intuitive feeling about the size of a kilogram or a milliliter, you will not know. If you realize that a milliliter is about the volume of a thimble and that a standard bag of sugar might weigh 2 kg, then you will realize that there is no way to pack 254 kg into a thimble of water, and you will know that you made a mistake.

1.5 | What Is a Handy Way to Convert from One Unit to Another?

Factor-label method A procedure in which the equations are set up so that all the unwanted units cancel and only the desired units remain

Conversion factor A ratio of two different units

We frequently need to convert a measurement from one unit to another. The best and most foolproof way to do this is the **factor-label method.** In this method we follow the rule that *when multiplying numbers we also multiply units, and when dividing numbers we also divide units.*

For conversions between one unit and another, it is always possible to set up two fractions, called **conversion factors.** Suppose we wish to convert the weight of an object from 381 grams to pounds. We are converting the units, but we are not changing the object itself. We want a ratio that reflects the change in units. In Table 1.3, we see that there are 453.6 grams in 1 pound. That is, the amount of matter in 453.6 grams is the same as the amount in 1 pound. In that sense, it is a one-to-one ratio, even though the units are not numerically the same. The conversion factors between grams and pounds therefore are

$$\frac{1 \text{ lb}}{453.6 \text{ g}} \quad \text{and} \quad \frac{453.6 \text{ g}}{1 \text{ lb}}$$

To convert 381 grams to pounds, we must multiply by the proper conversion factor—but which one? Let us try both and see what happens.

First let us multiply by 1 lb/453.6 g:

$$381 \text{ g} \times \frac{1 \text{ lb}}{453.6 \text{ g}} = 0.840 \text{ lb}$$

Following the procedure of multiplying and dividing units when we multiply and divide numbers, we find that dividing grams by grams cancels out the grams. We are left with pounds, which is the answer we want. Thus 1 lb/453.6 g is the correct conversion factor because it converts grams to pounds.

Suppose we had done it the other way, multiplying by 453.6 g/1 lb:

$$381 \text{ g} \times \frac{453.6 \text{ g}}{1 \text{ lb}} = 173,000 \frac{\text{g}^2}{\text{lb}}$$

When we multiply grams by grams, we get g^2 (grams squared). Dividing by pounds gives g^2/lb. This is not the unit we want, so we used the incorrect conversion factor.

HOW TO ...

Do Unit Conversions by the Factor-Label Method

One of the most useful ways of approaching conversions is to ask three questions:

- What information am I given? This is the starting point.
- What do I want to know? This is the answer that you want to find.
- What is the connection between the first two? This is the conversion factor. Of course, more than one conversion factor may be needed for some problems.

Let's look at how to apply these principles to a conversion from pounds to kilograms. Suppose we want to know the weight in kilograms of a woman who weighs 125 lb. We see in Table 1.3 that there are 2.205 lb in 1 kg. Note that we are starting out with pounds and we want an answer in kilograms.

$$125 \text{ lb} \times \frac{1 \text{ kg}}{2.205 \text{ lb}} = 56.7 \text{ kg}$$

- The weight in pounds is the starting point. We were given that information.
- We wanted to know the weight in kilograms. That was the desired answer, and we found the number of kilograms.
- The connection between the two is the conversion factor in which the unit of the desired answer is in the numerator of the fraction, rather than the denominator. It is not simply a mechanical procedure to set up the equation so that units cancel; it is a first step to understanding the underlying reasoning behind the factor-label method. If you set up the equation to give the desired unit as the answer, you have made the connection properly.

If you apply this kind of reasoning, you can always pick the right conversion factor. Given the choice between

$$\frac{2.205 \text{ lb}}{1 \text{ kg}} \quad \text{and} \quad \frac{1 \text{ kg}}{2.205 \text{ lb}}$$

you know that the second conversion factor will give an answer in kilograms, so you use it. When you check the answer, you see that it is reasonable. You expect a number that is about one half of 125, which is 62.5. The actual answer, 56.7, is close to that value. The number of pounds and the number of kilograms are not the same, but they represent the same weight. That fact makes the use of conversion factors logically valid; the factor-label method uses the connection to obtain a numerical answer.

GOB
Chemistry ⚛ Now™
Click *Mastering the Essentials* to practice using the **Factor-Label Method**

The advantage of the factor-label method is that it lets us know when we have made an incorrect calculation. *If the units of the answer are not the ones we are looking for, the calculation must be wrong.* Incidentally, this principle works not only in unit conversions but in all problems where we make calculations using measured numbers.

The factor-label method gives the correct mathematical solution for a problem. However, it is a mechanical technique and does not require you to think through the problem. Thus it may not provide a deeper understanding. For this reason and also to check your answer (because it is easy to make mistakes in arithmetic—for example, by punching the wrong numbers into a calculator), you should always ask yourself if the answer you have obtained is reasonable. For example, the question might ask the mass of a single oxygen atom. If your answer comes out 8.5×10^6 g, it is not reasonable. A single atom cannot weigh more than you do! In such a case, you have obviously made a mistake and should take another look to see where you went wrong. Of course, everyone makes mistakes at times, but if you check you can at least determine whether your answer is reasonable. If it is not, you will immediately know that you have made a mistake and can then correct it.

Checking whether an answer is reasonable gives you a deeper understanding of the problem because it forces you to think through the relationship between the question and the answer. The concepts and the mathematical relationships in these problems go hand in hand. Mastery of the mathematical skills makes the concepts clearer, and insight into the concepts suggests ways to approach the mathematics. We will now give a few examples of unit conversions and then test the answers to see whether they are reasonable. To save space, we will practice this technique mostly in this chapter, but you should use a similar approach in all later chapters.

In unit conversion problems, you should always check two things. First, the numeric factor by which you multiply tells you whether the answer will be larger or smaller than the number being converted. Second, the factor tells you how much greater or smaller than the number you start with your answer should be. For example, if 100 kg is converted to pounds and there are 2.205 lb in 1 kg, then an answer of about 200 is reasonable—but an answer of 0.2 or 2000 is not.

EXAMPLE 1.3

The distance between Rome and Milan (the largest cities in Italy) is 358 miles. How many kilometers separate the two?

Solution
We want to convert miles to kilometers. Table 1.3 shows that 1 mi = 1.609 km. From this we get two conversion factors:

$$\frac{1 \text{ mi}}{1.609 \text{ km}} \quad \text{and} \quad \frac{1.609 \text{ km}}{1 \text{ mi}}$$

Which should we use? We use the one that gives the answer in kilometers:

$$358 \text{ mi} \times \text{conversion factor} = ? \text{ km}$$

This means that the miles must cancel, so the conversion factor 1.609 km/ 1 mi is appropriate.

$$358 \text{ mi} \times \frac{1.609 \text{ km}}{1 \text{ mi}} = 576 \text{ km}$$

Is this answer reasonable? We want to convert a given distance in miles to the same distance in kilometers. The conversion factor in Table 1.3 tells us that in a given distance the number of kilometers is larger than the number of miles. How much larger? The actual number is 1.609, which is approximately 1.5 times larger. Thus we expect that the answer in kilometers will be about 1.5 times greater than the number given in miles. The number given in miles is 358, which, *for the purpose of checking whether our answer is reasonable,* we can round off to, say, 400. Multiplying this number by 1.5 gives an approximate answer of 600 km. Our actual answer, 576 km, was of the same order of magnitude as the estimated answer, so we can say that it is reasonable. If the estimated answer had been 6 km, or 60 km, or 6000 km, we would suspect that we had made a mistake in calculating the actual answer.

Problem 1.3

How many kilograms are in 241 lb? Check your answer to see if it is reasonable.

EXAMPLE 1.4

The label on a container of olive oil says 1.844 gal. How many milliliters does the container hold?

Solution

Table 1.3 shows no factor for converting gallons to milliliters, but it does show that 1 gal = 3.785 L. Because we know that 1000 mL = 1 L, we can solve this problem by multiplying by two conversion factors, making certain that all units cancel except milliliters:

$$1.844 \text{ gal} \times \frac{3.785 \text{ L}}{1 \text{ gal}} \times \frac{1000 \text{ mL}}{1 \text{ L}} = 6980 \text{ mL}$$

Is this answer reasonable? The conversion factor in Table 1.3 tells us that there are more liters in a given volume than gallons. How much more? Approximately four times more. We also know that any volume in milliliters is 1000 times larger than the same volume in liters. Thus we expect that the volume expressed in milliliters will be 4 × 1000, or 4000, times more than the volume given in gallons. The estimated volume in milliliters will be approximately 1.8 × 4000, or 7000 mL. But we also expect that the actual answer should be somewhat less than the estimated figure because we overestimated the conversion factor (4 rather than 3.785). Thus the answer, 6980 mL, is quite reasonable. Note that the answer is given to four significant figures.

Problem 1.4

Calculate the number of kilometers in 8.55 miles. Check your answer to see whether it is reasonable.

EXAMPLE 1.5

The maximum speed limit on many roads in the United States is 65 mi/h. How many meters per second (m/s) is this speed?

Solution

Here we have essentially a double conversion problem: We must convert miles to meters and hours to seconds. We use as many conversion factors as necessary, always making sure that we use them in such a way that the proper units cancel:

$$65 \frac{\text{mi}}{\text{h}} \times \frac{1.609 \text{ km}}{1 \text{ mi}} \times \frac{1000 \text{ m}}{1 \text{ km}} \times \frac{1 \text{ h}}{60 \text{ min}} \times \frac{1 \text{ min}}{60 \text{ s}} = 29 \frac{\text{m}}{\text{s}}$$

Estimating the answer is a good thing to do when working any mathematical problem, not just unit conversions.

Is this answer reasonable? To estimate the 65 mi/h speed in meters per second, we must first establish the relationship between miles and meters. As in Example 1.3, we know that there are more kilometers than miles in a given distance. How much more? As there are approximately 1.5 km in 1 mi, there must be approximately 1500 times more meters. We also know that in 1 hour there are $60 \times 60 = 3600$ seconds. The ratio of meters to seconds will be approximately 1500/3600, which is about one half. Therefore, we estimate that the speed in meters per second will be about one half of that in miles per hour or 32 m/s. Once again, the actual answer, 29 m/s, is not far from the estimate of 32 m/s, so the answer is reasonable.

As shown in these examples, when canceling units we do not cancel the numbers. The numbers are multiplied and divided in the ordinary way.

Problem 1.5

Convert the speed of sound, 332 m/s to mi/h. Check your answer to see whether it is reasonable.

GOB
Chemistry Now™

Click *Coached Problems* to practice doing unit conversions by the **Factor-Label Method**

1.6 | What Are the States of Matter?

Matter can exist in three states: gas, liquid, and solid. **Gases** have no definite shape or volume. They expand to fill whatever container they are put into. On the other hand, they are highly compressible and can be forced into small containers. **Liquids** also have no definite shape, but they do have a definite volume that remains the same when they are poured from one container to another. Liquids are only slightly compressible. **Solids** have definite shapes and definite volumes. They are essentially incompressible.

Whether a substance is a gas, a liquid, or a solid depends on its temperature and pressure. On a cold winter day, a puddle of liquid water turns to ice; it becomes a solid. If we heat water in an open pot at sea level, the liquid boils at 100°C; it becomes a gas—we call it steam. If we heated the same pot of water on top of Mount Everest, it would boil at about 70°C due to the reduced atmospheric pressure. Most substances can exist in the three states: They are gases at high temperature, liquids at a lower temperature, and solids when their temperature becomes low enough. Figure 1.5 shows a single substance in the three different states.

(a) (b) (c)

Charles D. Winters

Figure 1.5 The three states of matter for bromine: (*a*) bromine as a solid, (*b*) bromine as a liquid, and (*c*) bromine as a gas.

The chemical identity of a substance does not change when it is converted from one state to another. Water is still water whether it is in the form of ice, steam, or liquid water. We discuss the three states of matter, and the changes between one state and another, at greater length in Chapter 6.

1.7 | What Are Density and Specific Gravity?

A. Density

One of the many pollution problems that the world faces is the spillage of petroleum into the oceans from oil tankers or from offshore drilling. When oil spills into the ocean, it floats on top of the water. The oil doesn't sink because it is not soluble in water and because water has a higher density than oil. When two liquids are mixed (assuming that one does not dissolve in the other), the one of lower density floats on top (Figure 1.6).

The **density** of any substance is defined as its *mass per unit volume*. Not only do all liquids have a density, but so do all solids and gases. Density is calculated by dividing the mass of a substance by its volume:

$$d = \frac{m}{V} \quad d = \text{density}, \quad m = \text{mass}, \quad V = \text{volume}$$

Water

Carbon
tetrachloride

Diethyl ether

Water

Figure 1.6   Two separatory funnels containing water and another liquid. The density of carbon tetrachloride is 2.961 g/mL, that of water is 1.00 g/mL, and that of diethyl ether is 0.713 g/mL. In each case the liquid with the lower density is on top.

EXAMPLE 1.6

If 73.2 mL of a liquid has a mass of 61.5 g, what is its density in g/mL?

Solution

$$d = \frac{m}{V} = \frac{61.5 \text{ g}}{73.2 \text{ mL}} = 0.840 \frac{\text{g}}{\text{mL}}$$

Problem 1.6

The density of titanium is 4.54 g/mL. What is the mass, in grams, of 17.3 mL of titanium? Check your answer to see whether it is reasonable.

EXAMPLE 1.7

The density of iron is 7.86 g/cm³. What is the volume in milliliters of an irregularly shaped piece of iron that has a mass of 524 g?

Solution
Here we are given the mass and the density. In this type of problem, it is useful to derive a conversion factor from the density. Since 1 cm³ is exactly 1 mL, we know that the density is 7.86 g/mL. This means that 1 mL of iron has a mass of 7.86 g. From this we can get two conversion factors:

$$\frac{1 \text{ mL}}{7.86 \text{ g}} \quad \text{and} \quad \frac{7.86 \text{ g}}{1 \text{ mL}}$$

The spillage of more than 10 million gallons of petroleum in Prince William Sound, Alaska, in March 1989 caused a great deal of environmental damage.

As usual, we multiply the mass by whichever conversion factor results in the cancellation of all but the correct unit:

$$524 \text{ g} \times \frac{1 \text{ mL}}{7.86 \text{ g}} = 66.7 \text{ mL}$$

Is this answer reasonable? The density of 7.86 g/mL tells us that the volume in milliliters of any piece of iron is always less than its mass in grams. How much less? Approximately eight times less. Thus we expect the volume to be approximately 500/8 = 63 mL. As the actual answer is 66.7 mL, it is reasonable.

Problem 1.7

An unknown substance has a mass of 56.8 g and occupies a volume of 23.4 mL. What is its density in g/mL? Check your answer to see whether it is reasonable.

The density of any liquid or solid is a physical property that is constant, which means that it always has the same value at a given temperature. We use physical properties to help identify a substance. For example, the density of chloroform (a liquid formerly used as an inhalation anesthetic) is 1.483 g/mL at 20°C. If we want to find out if an unknown liquid is chloroform, one thing we might do is measure its density at 20°C. If the density is, say, 1.355 g/mL, we know the liquid isn't chloroform. If the density is 1.483 g/mL, we cannot be sure the liquid is chloroform, because other liquids might also have this density, but we can then measure other physical properties (the boiling point, for example). If all the physical properties we measure match those of chloroform, we can be reasonably sure the liquid is chloroform.

We have said that the density of a pure liquid or solid is a constant at a given temperature. Density does change when the temperature changes. Almost always, density decreases with increasing temperature. This is true because mass does not change when a substance is heated, but volume almost always increases because atoms and molecules tend to get farther apart as the temperature increases. Since $d = m/V$, if m stays the same and V gets larger, d must get smaller.

The most common liquid, water, provides a partial exception to this rule. As the temperature increases from 4°C to 100°C, the density of water does decrease, but from 0°C to 4°C, the density increases. That is, water has its maximum density at 4°C. This anomaly and its consequences are due to the unique structure of water and will be discussed in Chemical Connections 6E.

B. Specific Gravity

Because density is equal to mass divided by volume, it always has units, most commonly g/mL or g/cc or (g/L for gases). **Specific gravity** is numerically the same as density, but it has no units (it is dimensionless). The reason is that specific gravity is defined as a comparison of the density of a substance with the density of water, which is taken as a standard. For example, the density of copper at 20°C is 8.92 g/mL. The density of water at the same temperature is 1.00 g/mL. Therefore, copper is 8.92 times as dense as water, and its specific gravity at 20°C is 8.92. Because water is taken as the standard and because the density of water is 1.00 g/mL at 20°C, the specific gravity of any substance is always numerically equal to its density, provided that the density is measured in g/mL or g/cc.

Specific gravity is often measured by a hydrometer. This simple device consists of a weighted glass bulb that is inserted into a liquid and allowed to float. The stem of the hydrometer has markings, and the specific gravity is read where the meniscus (the curved surface of the liquid) hits the marking. The specific gravity of the acid in your car battery and that of a urine sample in a clinical laboratory are measured by hydrometers. A hydrometer measuring a urine sample is also called urinometer (Figure 1.7). Normal urine can vary in specific gravity from about 1.010 to 1.030. Patients with diabetes mellitus have an abnormally high specific gravity of their urine samples, while those with some forms of kidney disease have an abnormally low specific gravity.

Figure 1.7　Urinometer.

EXAMPLE 1.8

The density of ethanol at 20°C is 0.789 g/mL. What is its specific gravity?

Solution

$$\text{Specific gravity} = \frac{0.789 \text{ g/mL}}{1.00 \text{ g/mL}} = 0.789$$

Problem 1.8

The specific gravity of a urine sample is 1.016. What is its density, in g/mL?

1.8 | How Do We Describe the Various Forms of Energy?

Energy is defined as the capacity to do work. It can be described as being either kinetic energy or potential energy.

Kinetic energy (KE) is the energy of motion. Any object that is moving possesses kinetic energy. We can calculate how much energy by the formula $KE = 1/2mv^2$, where m is the mass of the object and v is its velocity. This means that kinetic energy increases (1) when an object moves faster and (2) when a heavier object is moving. When a truck and a bicycle are moving at the same velocity, the truck has more kinetic energy.

Potential energy is stored energy. The potential energy possessed by an object arises from its capacity to move or to cause motion. For example, body weight in the up position on a seesaw contains potential energy—it is capable of doing work. If given a slight push, it will move down. The potential energy of the body in the up position is converted to the kinetic energy of the body in the down position moving to the up position. Work is done against gravity in the process. Figure 1.8 shows another way in which potential energy is converted to kinetic energy.

An important principle in nature is that things have a tendency to seek their lowest possible potential energy. We all know that water always flows downhill and not uphill.

Several forms of energy exist. The most important are (1) mechanical energy, light, heat, and electrical energy, which are examples of kinetic energy possessed by all moving objects, whether elephants or molecules or electrons; and (2) chemical energy and nuclear energy, which are examples of potential energy or stored energy. In chemistry the most important form

Potential energy is stored in this drawn bow and becomes kinetic energy in the arrow when released.

Figure 1.8 The water held back by the dam possesses potential energy, which is converted to kinetic energy when the water is released.

Potential energy ready to do work

Kinetic energy doing work

An example of energy conversion. Light energy from the sun is converted to electrical energy by solar cells. The electricity runs a refrigerator on the back of the camel, keeping the vaccines cool so that they can be delivered to remote locations.

of potential energy is chemical energy—the energy stored within chemical substances and given off when they take part in a chemical reaction. For example, a log possesses chemical energy. When the log is ignited in a fireplace, the chemical energy (potential) of the wood is turned into energy in the form of heat and light. Specifically, the potential energy has been transformed into thermal energy (heat makes molecules move faster) and the radiant energy of light.

The various forms of energy can be converted from one to another. In fact, we make such conversions all the time. A power plant operates either on the chemical energy derived from burning fuel or on nuclear energy. This energy is converted to heat, which is converted to the electricity that is sent over transmission wires into houses and factories. Here we convert the electricity to light, heat (in an electrical heater, for example), or mechanical energy (in the motors of refrigerators, vacuum cleaners, and other devices).

Although one form of energy can be converted to another, the *total amount* of energy in any system does not change. *Energy can be neither created nor destroyed.* This statement is called the **law of conservation of energy.***

1.9 | How Do We Describe Heat and the Ways in Which It Is Transferred?

A. Heat and Temperature

One form of energy that is particularly important in chemistry is **heat.** This is the form of energy that most frequently accompanies chemical reactions. Heat is not the same as temperature, however. Heat is a form of energy, but temperature is not.

The difference between heat and temperature can be seen in the following example. If we have two beakers, one containing 100 mL of water and the other containing 1 L of water at the same temperature, the heat content of the water in the larger beaker is ten times that of the water in the smaller beaker, even though the temperature is the same in both. If you were to dip your hand accidentally into a liter of boiling water, you would be much more severely burned than if only one drop fell on your hand. Even though the water is at the same temperature in both cases, the liter of boiling water has much more heat.

---

*This statement is not completely true. As discussed in Sections 3.8 and 3.9, it is possible to convert matter to energy, and vice versa. Therefore, a more correct statement would be *matter-energy can be neither created nor destroyed.* However, the law of conservation of energy is valid for most purposes and is highly useful.

CHEMICAL CONNECTIONS 1B

Hypothermia and Hyperthermia

The human body cannot tolerate temperatures that are too low. A person outside in very cold weather (say, $-20°F$ [$-29°C$]) who is not protected by heavy clothing will eventually freeze to death because the body loses heat. Normal body temperature is $37°C$. When the outside temperature is lower than that, heat flows out of the body. When the air temperature is moderate ($10°C$ to $25°C$), this poses no problem and is, in fact, necessary because the body produces more heat than it needs and must lose some. At extremely low temperatures, however, too much heat is lost and

body temperature drops, a condition called **hypothermia**. A drop in body temperature of 1 or $2°C$ causes shivering, which is the body's attempt to increase its temperature by the heat generated through muscular action. An even greater drop results in unconsciousness and eventually death.

The opposite condition is **hyperthermia**. It can be caused either by high outside temperatures or by the body itself when an individual develops a high fever. A sustained body temperature as high as $41.7°C$ ($107°F$) is usually fatal.

As we saw in Section 1.4, temperature is measured in degrees. Heat can be measured in various units, the most common of which is the **calorie,** which is defined as the amount of heat necessary to raise the temperature of 1 g of liquid water by $1°C$. This is a small unit, and chemists more often use the kilocalorie (kcal):

$$1 \text{ kcal} = 1000 \text{ cal}$$

Nutritionists use the word "Calorie" (with a capital "C") to mean the same thing as "kilocalorie"; that is, 1 Cal = 1000 cal = 1 kcal. The calorie is not part of the SI. The official SI unit for heat is the **joule (J),** which is about one-fourth as big as the calorie:

$$1 \text{ cal} = 4.184 \text{ J}$$

B. Specific Heat

As we noted, it takes 1 cal to raise the temperature of 1 g of liquid water by $1°C$. **Specific heat (SH)** is the amount of heat necessary to raise the temperature of 1 g of any substance by $1°C$. Each substance has its own specific heat, which is a physical property of that substance, like density or melting point. Table 1.4 lists specific heats for a few common substances. For example, the specific heat of iron is 0.11 cal/g $\cdot$ $°C$. Therefore, if we had 1 g of iron at $20°C$, it would require only 0.11 cal to increase the temperature to $21°C$. Under the same conditions, aluminum would require twice as much heat. Thus cooking in an aluminum pan of the same weight as an iron pan would require more heat than cooking in the iron pan. Note from Table 1.4 that ice and steam do not have the same specific heat as liquid water.

Table 1.4   Specific Heats for Some Common Substances

Substance	Specific Heat (cal/g $\cdot$ $°C$)	Substance	Specific Heat (cal/g $\cdot$ $°C$)
Water	1.00	Wood (typical)	0.42
Ice	0.48	Glass (typical)	0.22
Steam	0.48	Rock (typical)	0.20
Iron	0.11	Ethanol	0.59
Aluminum	0.22	Methanol	0.61
Copper	0.092	Ether	0.56
Lead	0.038	Carbon tetrachloride	0.21

CHEMICAL CONNECTIONS 1C

Cold Compresses, Waterbeds, and Lakes

The high specific heat of water is useful in cold compresses and makes them last a long time. For example, consider two patients with cold compresses: one compress made by soaking a towel in water and the other made by soaking a towel in ethanol. Both are at 0°C. Each gram of water in the water compress requires 25 cal to make the temperature of the compress rise to 25°C (after which it must be changed). Because the specific heat of ethanol is 0.59 cal/g · °C (see Table 1.4), each gram of ethanol requires only 15 cal to reach 25°C. If the two patients give off heat at the same rate, the ethanol compress is less effective because it will reach 25°C a good deal sooner than the water compress and will need to be changed sooner.

The high specific heat of water also means that it takes a great deal of heat to increase its temperature. That is why it takes a long time to get a pot of water to boil. Anyone who has a waterbed (300 gallons) knows that it takes days for the heater to bring the bed up to the desired temperature. It is particularly annoying when an overnight guest tries to adjust the temperature of your waterbed because the guest will probably have left before the change is noticed, but then you will have to set it back to your favorite temperature. This same effect in reverse explains why the outside temperature can be below zero (°C) for weeks before a lake will freeze. Large bodies of water do not change temperature very quickly.

It is easy to make calculations involving specific heats. The equation is

$$\text{Amount of heat} = \text{specific heat} \times \text{mass} \times \text{change in temperature}$$

$$\text{Amount of heat} = \text{SH} \times m \times \Delta T$$

where ΔT is the change in temperature.
We can also write this equation as

$$\text{Amount of heat} = \text{SH} \times m \times (T_2 - T_1)$$

where T_2 is the final temperature and T_1 is the initial temperature in °C.

EXAMPLE 1.9

How many calories are required to heat 352 g of water from 23°C to 95°C?

Solution

$$\text{Amount of heat} = \text{SH} \times m \times \Delta T$$

$$\text{Amount of heat} = \text{SH} \times m \times (T_2 - T_1)$$

$$= \frac{1.00 \text{ cal}}{g \cdot °C} \times 352 \text{ g} \times (95 - 23)°C$$

$$= 2.5 \times 10^4 \text{ cal}$$

Is this answer reasonable? Each gram of water requires one calorie to raise its temperature by one degree. We have approximately 350 g of water. To raise its temperature by one degree would therefore require approximately 350 calories. But we are raising the temperature not by one degree but by approximately 70 degrees (from 23 to 95). Thus the total number of calories will be approximately $70 \times 350 = 24{,}500$ cal, which is close to the calculated answer. (Even though we were asked for the answer in calories, we should note that it will be more convenient to convert to 25 kcal. We are going to see that conversion from time to time.)

Problem 1.9

How many calories are required to heat 731 g of water from 8°C to 74°C? Check your answer to see whether it is reasonable.

EXAMPLE 1.10

If we add 450 cal of heat to 37 g of ethanol at 20°C, what is the final temperature?

Solution

The specific heat of ethanol is 0.59 cal/g · °C (see Table 1.4).

$$\text{Amount of heat} = \text{SH} \times m \times \Delta T$$

$$\text{Amount of heat} = \text{SH} \times m \times (T_2 - T_1)$$

$$450 \text{ cal} = 0.59 \text{ cal/g} \cdot {}^\circ\text{C} \times 37 \text{ g} \times (T_2 - T_1)$$

We can show the units in fraction form by rewriting this equation.

$$450 \text{ cal} = 0.59 \frac{\text{cal}}{\text{g} \cdot {}^\circ\text{C}} \times 37 \text{ g} \times (T_2 - T_1)$$

$$(T_2 - T_1) = \frac{\text{amount of heat}}{\text{SH} \times m}$$

$$(T_2 - T_1) = \frac{450 \text{ cal}}{\left[\dfrac{0.59 \text{ cal} \times 37 \text{ g}}{\text{g} \cdot {}^\circ\text{C}}\right]} = \frac{21}{1/{}^\circ\text{C}} = 21{}^\circ\text{C}$$

(Note that we have the reciprocal of temperature in the denominator, which gives us temperature in the numerator. The answer has units of degrees Celsius.) Since the starting temperature is 20°C, the final temperature is 41°C.

Is this answer reasonable? The specific heat of ethanol is 0.59 cal/g · °C. This value is close to 0.5, meaning that about half a calorie will raise the temperature of 1 g by 1°C. However, 37 g of ethanol needs approximately 40 times as many calories for a rise, and $40 \times \frac{1}{2} = 20$ calories. We are adding 450 calories, which is about 20 times as much. Thus we expect the temperature to rise by about 20°C, from 20°C to 40°C. The actual answer, 41°C, is quite reasonable.

Problem 1.10

A 100 g piece of iron at 25°C is heated by adding 230 cal. What will be the final temperature? Check your answer to see whether it is reasonable.

EXAMPLE 1.11

We heat 50.0 g of an unknown substance by adding 205 cal, and its temperature rises by 7.0°C. What is its specific heat? Using Table 1.4, identify the substance.

Solution

$$SH = \frac{\text{Amount of heat}}{m \times (\Delta T)}$$

$$SH = \frac{\text{Amount of heat}}{m \times (T_2 - T_1)}$$

$$SH = \frac{205 \text{ cal}}{50.0 \text{ g} \times 7.0°C} = 0.59 \text{ cal/g} \cdot °C$$

The substance in Table 1.4 having a specific heat of 0.59 cal/g · °C is ethanol.

Is this answer reasonable? If we had water instead of an unknown substance with SH = 1 cal/g · °C, raising the temperature of 50.0 g by 7.0°C would require 50 × 7.0 = 350 cal. But we added only approximately 200 cal. Therefore, the SH of the unknown substance must be less than 1.0. How much less? Approximately 200/350 = 0.6. The actual answer, 0.59 cal/g · °C, is quite reasonable.

Problem 1.11

It required 88.2 cal to heat 13.4 g of an unknown substance from 23°C to 176°C. What is the specific heat of the unknown substance? Check your answer to see whether it is reasonable.

SUMMARY OF KEY QUESTIONS

SECTION 1.1 Why Do We Call Chemistry the Study of Matter?

- **Chemistry** is the science that deals with the structure of matter and the changes it can undergo. In a **chemical change** or **chemical reaction,** substances are used up and others are formed.
- Chemistry is also the study of energy changes during chemical reactions. In **physical changes** substances do not change their identity.

SECTION 1.2 What Is the Scientific Method?

- The **scientific method** is a tool used in science and medicine. The heart of the scientific method is the testing of **hypotheses** and **theories** by collecting facts.

SECTION 1.3 How Do Scientists Report Numbers?

- Because we frequently use very large or very small numbers, we use powers of 10 to express these numbers more conveniently, a method called **exponential notation.**
- With exponential notation, we no longer have to keep track of so many zeros, and we have the added con-

venience of being able to see which digits convey information (**significant figures**) and which merely indicate the position of the decimal point.

SECTION 1.4 How Do We Make Measurements?

- In chemistry we use the **metric system** for measurements.
- The base units are the meter for length, the liter for volume, the gram for mass, the second for time, and the calorie for heat. Other units are indicated by prefixes that represent powers of 10. Temperature is measured in degrees Celsius or in kelvins.

SECTION 1.5 What Is a Handy Way to Convert from One Unit to Another?

- Conversions from one unit to another are best done by the **factor-label method,** in which units are multiplied and divided.

SECTION 1.6 What Are the States of Matter?

- There are three states of matter: **solid, liquid, and gas.**

SECTION 1.7 What Are Density and Specific Gravity?

- **Density** is mass per unit volume. **Specific gravity** is density relative to water and thus has no units. Density usually decreases with increasing temperature.

SECTION 1.8 How Do We Describe the Various Forms of Energy?

- **Kinetic energy** is energy of motion; **potential energy** is stored energy. Energy can be neither created nor destroyed, but it can be converted from one form to another.

SECTION 1.9 How Do We Describe Heat and the Ways in Which It Is Transferred?

- **Heat** is a form of energy and is measured in calories. A calorie is the amount of heat necessary to raise the temperature of 1 g of liquid water by 1°C.
- Every substance has a **specific heat,** which is a physical constant. The specific heat is the number of calories required to raise the temperature of 1 g of a substance by 1°C.

PROBLEMS

Chemistry Now™

Assess your understanding of this chapter's topics with additional quizzing and conceptual-based problems at **http://now.brookscole.com/gob8** or on the CD.

A blue problem number indicates an applied problem.

■ denotes problems that are available on the GOB ChemistryNow website or CD and are assignable in OWL.

SECTION 1.1 Why Do We Call Chemistry the Study of Matter?

1.12 The life expectancy of a citizen in the United States is 76 years. Eighty years ago it was 56 years. In your opinion, what was the major contributor to this spectacular increase in life expectancy? Explain your answer.

1.13 Define the following terms:
(a) Matter   (b) Chemistry

SECTION 1.2 What Is the Scientific Method?

1.14 ■ In Table 1.4 you find four metals (iron, aluminum, copper, and lead) and three organic compounds (ethanol, methanol, and ether). What kind of hypothesis would you suggest about the specific heats of these chemicals?

1.15 ■ In a newspaper, you read that Dr. X claimed that he has found a new remedy to cure diabetes. The remedy is an extract of carrots. How would you classify this claim: (a) fact, (b) theory, (c) hypothesis, or (d) hoax? Explain your choice of answer.

1.16 ■ Classify each of the following as a chemical or physical change:
(a) Burning gasoline
(b) Making ice cubes
(c) Boiling oil
(d) Melting lead
(e) Rusting iron
(f) Making ammonia from nitrogen and hydrogen
(g) Digesting food

SECTION 1.3 How Do Scientists Report Numbers?

Exponential Notation

1.17 Write in exponential notation:
(a) 0.351   (b) 602.1   (c) 0.000128   (d) 628122

1.18 Write out in full:
(a) 4.03×10^5 (b) 3.2×10^3
(c) 7.13×10^{-5} (d) 5.55×10^{-10}

1.19 Multiply:
(a) $(2.16 \times 10^5)(3.08 \times 10^{12})$
(b) $(1.6 \times 10^{-8})(7.2 \times 10^8)$
(c) $(5.87 \times 10^{10})(6.62 \times 10^{-27})$
(d) $(5.2 \times 10^{-9})(6.8 \times 10^{-15})$

1.20 Divide:
(a) $\dfrac{6.02 \times 10^{23}}{2.87 \times 10^{10}}$ (b) $\dfrac{3.14}{2.93 \times 10^{-4}}$

(c) $\dfrac{5.86 \times 10^{-9}}{2.00 \times 10^3}$ (d) $\dfrac{7.8 \times 10^{-12}}{9.3 \times 10^{-14}}$

(e) $\dfrac{6.83 \times 10^{-12}}{5.02 \times 10^{14}}$

1.21 Add:
(a) $(7.9 \times 10^4) + (5.2 \times 10^4)$
(b) $(8.73 \times 10^4) + (6.7 \times 10^3)$
(c) $(3.63 \times 10^{-4}) + (4.776 \times 10^{-3})$

1.22 Subtract:
(a) $(8.50 \times 10^3) - (7.61 \times 10^2)$
(b) $(9.120 \times 10^{-2}) - 3.12 \times 10^{-3})$
(c) $(1.3045 \times 10^2) - (2.3 \times 10^{-1})$

1.23 ■ Solve:
$$\frac{(3.14 \times 10^3) \times (7.80 \times 10^5)}{(5.50 \times 10^2)}$$

1.24 ■ Solve:
$$\frac{(9.52 \times 10^4) \times (2.77 \times 10^{-5})}{(1.39 \times 10^7) \times (5.83 \times 10^2)}$$

Significant Figures

1.25 How many significant figures are in the following:
(a) 0.012 (b) 0.10203
(c) 36.042 (d) 8401.0
(e) 32100 (f) 0.0402
(g) 0.000012

1.26 How many significant figures are in the following:
(a) 5.71×10^{13} (b) 4.4×10^5
(c) 3×10^{-6} (d) 4.000×10^{-11}
(e) 5.5550×10^{-3}

1.27 Round off to two significant figures:
(a) 91.621 (b) 7.329
(c) 0.677 (d) 0.003249
(e) 5.88

1.28 ■ Multiply these numbers, using the correct number of significant figures in your answer:
(a) 3630.15×6.8
(b) 512×0.0081
(c) $5.79 \times 1.85825 \times 1.4381$

1.29 Divide these numbers, using the correct number of significant figures in your answer:
(a) $\dfrac{3.185}{2.08}$ (b) $\dfrac{6.5}{3.0012}$ (c) $\dfrac{0.0035}{7.348}$

1.30 Add these groups of measured numbers using the correct number of significant figures in your answer:
(a) 37.4083
  5.404
  10916.3
  3.94
  0.0006

(b) 84
  8.215
  0.01
  151.7

(c) 51.51
  100.27
  16.878
  3.6817

SECTION 1.4 How Do We Make Measurements?

1.31 In the SI system, the second is the base unit of time. We talk about atomic events that occur in picoseconds (10^{-12} s) or even in femtoseconds (10^{-15} s). But we don't talk about megaseconds or kiloseconds; the old standards of minutes, hours, and days prevail. How many minutes and hours are 20 kiloseconds?

1.32 ■ How many grams are in the following:
(a) 1 kg (b) 1 mg

1.33 Estimate without actually calculating which one is the shorter distance:
(a) 20 mm or 0.3 m
(b) 1 inch or 30 mm
(c) 2000 m or 1 mile

1.34 For each of these, tell which answer is closest:
(a) A baseball bat has a length of 100 mm or 100 cm or 100 m
(b) A glass of milk holds 23 cc or 230 mL or 23 L
(c) A man weighs 75 mg or 75 g or 75 kg
(d) A tablespoon contains 15 mL or 150 mL or 1.5 L
(e) A paper clip weighs 50 mg or 50 g or 50 kg
(f) Your hand has a width of 100 mm or 100 cm or 100 m
(g) An audiocassette weighs 40 mg or 40 g or 40 kg

1.35 You are taken for a helicopter ride in Hawaii from Kona (sea level) to the top of the volcano Mauna Kea. Which property of your body would change during the helicopter ride:
(a) height (b) weight (c) volume (d) mass

1.36 Convert to Celsius and to Kelvin:
(a) 320°F (b) 212°F (c) 0°F (d) −250°F

1.37 Convert to Fahrenheit and to Kelvin:
(a) 25°C (b) 40°C (c) 250°C (d) −273°C

SECTION 1.5 What Is a Handy Way to Convert from One Unit to Another?

1.38 Make the following conversions (conversion factors are given in Table 1.3):
(a) 42.6 kg to lb (b) 1.62 lb to g
(c) 34 in. to cm (d) 37.2 km to mi
(e) 2.73 gal to L (f) 62 g to oz
(g) 33.61 qt to L (h) 43.7 L to gal
(i) 1.1 mi to km (j) 34.9 mL to fl oz

1.39 Make the following metric conversions:
(a) 96.4 mL to L (b) 275 mm to cm
(c) 45.7 kg to g (d) 475 cm to m
(e) 21.64 cc to mL (f) 3.29 L to cc
(g) 0.044 L to mL (h) 711 g to kg
(i) 63.7 mL to cc (j) 0.073 kg to mg
(k) 83.4 m to mm (l) 361 mg to g

1.40 You drive in Canada where the distances are marked in kilometers. The sign says you are 80 km from Ottawa. You are traveling at a speed of 75 mi/h. Would you reach Ottawa within one hour, after one hour, or later than that?

1.41 The speed limit in some European cities is 80 km/h. How many miles per hour is this?

1.42 Your car gets 25.00 miles on a gallon of gas. What would be your car's fuel efficiency in km/L?

SECTION 1.6 What Are the States of Matter?

1.43 Which states of matter have a definite volume?

1.44 Will most substances be solids, liquids, or gases at low temperatures?

1.45 ■ Does the chemical nature of a substance change when it melts from a solid to a liquid?

SECTION 1.7 What Are Density and Specific Gravity?

1.46 The volume of a rock weighing 1.075 kg is 334.5 mL. What is the density of the rock in g/mL? Express it to three significant figures.

1.47 The density of manganese is 7.21 g/mL, that of calcium chloride is 2.15 g/mL, and that of sodium acetate is 1.528 g/mL. You place these three solids in a liquid, in which they are not soluble. The liquid has a density of 2.15 g/mL. Which will sink to the bottom, which will stay on the top, and which will stay in the middle of the liquid?

1.48 The density of titanium is 4.54 g/mL. What is the volume, in milliliters, of 163 g of titanium?

1.49 A 335.0 cc sample of urine has a mass of 342.6 g. What is the density, in g/mL, to three decimal places?

1.50 The density of methanol at 20°C is 0.791 g/mL. What is the mass, in grams, of a 280 mL sample?

1.51 The density of dichloromethane, a liquid insoluble in water, is 1.33 g/cc. If dichloromethane and water are placed in a separatory funnel, which will be the upper layer?

1.52 A sample of 10.00 g of oxygen has a volume of 6702 mL. The same weight of carbon dioxide occupies 5058 mL.
(a) What is the density of each gas in g/L?
(b) Carbon dioxide is used as a fire extinguisher to cut off the fire's supply of oxygen. Do the densities of these two gases explain the fire-extinguishing ability of carbon dioxide?

1.53 Crystals of a material are suspended in the middle of a cup of water at 2°C. This means that the densities of the crystal and of the water are the same. How might you enable the crystals to rise to the surface of the water so that you can harvest them?

SECTION 1.8 How Do We Describe the Various Forms of Energy?

1.54 On many country roads you see telephones powered by a solar panel. What principle is at work in these devices?

1.55 While you drive your car, your battery is charged. How would you describe this process in terms of kinetic and potential energy?

SECTION 1.9 How Do We Describe Heat and the Ways in Which It Is Transferred?

1.56 ■ How many calories are required to heat the following (specific heats are given in Table 1.4)?
(a) 52.7 g of aluminum from 100°C to 285°C
(b) 93.6 g of methanol from −35°C to 55°C
(c) 3.4 kg of lead from −33°C to 730°C
(d) 71.4 g of ice from −77°C to −5°C

1.57 If 168 g of an unknown liquid requires 2750 cal of heat to raise its temperature from 26°C to 74°C, what is the specific heat of the liquid?

1.58 The specific heat of steam is 0.48 cal/g·°C. How many kilocalories are needed to raise the temperature of 10.5 kg steam from 120°C to 150°C?

Chemical Connections

1.59 (Chemical Connections 1A) If the recommended dose of a drug is 445 mg for a 180 lb man, what would be a suitable dose for a 135 lb man?

1.60 (Chemical Connections 1A) The average lethal dose of heroin is 1.52 mg/kg of body weight. Estimate how many grams of heroin would be lethal for a 200 lb man.

1.61 (Chemical Connections 1B) How does the body react to hypothermia?

1.62 (Chemical Connections 1B) Low temperatures often cause people to shiver. What is the function of this involuntary body action?

1.63 (Chemical Connections 1C) Which would make a more efficient cold compress, ethanol or methanol? (Refer to Table 1.4.)

Additional Problems

1.64 The meter is a measure of length. Tell what each of the following units measures:
(a) cm^3 (b) mL (c) kg (d) cal
(e) g/cc (f) joule (g) °C (h) cm/s

1.65 A brain weighing 1.0 lb occupies a volume of 620 mL. What is the specific gravity of the brain?

1.66 ■ If the density of air is 1.25×10^{-3} g/cc, what is the mass in kilograms of the air in a room that is 5.3 m long, 4.2 m wide, and 2.0 m high?

1.67 Classify these as kinetic or potential energy:
(a) Water held by a dam
(b) A speeding train
(c) A book on its edge before falling
(d) A falling book
(e) Electric current in a light bulb

1.68 The kinetic energy possessed by an object with a mass of 1 g moving with a velocity of 1 cm/s is called 1 erg. What is the kinetic energy, in ergs, of an athlete with a mass of 127 lb running at a velocity of 14.7 mi/h?

1.69 A European car advertises an efficiency of 22 km/L, while an American car claims an economy of 30 mi/gal. Which car is more efficient?

1.70 In Potsdam, New York, you can buy gas for US$2.02/gal. In Montreal, Canada, you pay US$0.60/L. (Currency conversions are outside the scope of this text, so you are not asked to do them here.) Which is the better buy? Is your calculation reasonable?

1.71 Shivering is the body's response to increase the body temperature. What kind of energy is generated by shivering?

1.72 When the astronauts walked on the Moon, they could make giant leaps in spite of their heavy gear.

(a) Why were their weights on the Moon so small?

(b) Were their masses different on the Moon than on the Earth?

1.73 Which of the following is the largest mass and which is the smallest?

(a) 41 g

(b) 3×10^3 mg

(c) 8.2×10^6 μg

(d) 4.1310×10^{-8} kg

1.74 Which quantity is bigger in each of the following pairs:

(a) 1 gigaton : 10 megaton

(b) 10 micrometer : 1 millimeter

(c) 10 centigram : 200 milligram

1.75 In Japan, high-speed "bullet trains" move with an average speed of 220 km/h. If Dallas and Los Angeles were connected by such a train, how long would it take to travel nonstop between these cities (a distance of 1490 miles)?

1.76 The specific heats of some elements at 25°C are as follows: aluminum = 0.215 cal/g · °C; carbon (graphite) = 0.170 cal/g · °C; iron = 0.107 cal/g · °C; mercury = 0.0331 cal/g · °C.

(a) Which element would require the smallest amount of heat to raise the temperature of 100 g of the element by 10°C?

(b) If the same amount of heat needed to raise the temperature of 1 g of aluminum by 25°C were applied to 1 g of mercury, by how many degrees would its temperature be raised?

(c) If a certain amount of heat is used to raise the temperature of 1.6 g of iron by 10°C, the temperature of 1 g of which element would also be raised by 10°C, using the same amount of heat?

1.77 Water that contains deuterium rather than ordinary hydrogen (see Chapter 3) is called heavy water. The specific heat of heavy water at 25°C is 4.217 J/g · °C. Which requires more energy to raise the temperature of 10.0 g by 10°C, water or heavy water?

1.78 One quart of milk costs 80 cents and one liter costs 86 cents. Which is the better buy?

1.79 Consider butter, density 0.860 g/mL, and sand, density 2.28 g/mL.

(a) If 1.00 mL of butter is thoroughly mixed with 1.00 mL of sand, what is the density of the mixture?

(b) What would be the density of the mixture if 1.00 g of the same butter were mixed with 1.00 g of the same sand?

1.80 Which speed is the fastest?

(a) 70 mi/h

(b) 140 km/h

(c) 4.5 km/s

(d) 48 mi/min

1.81 In calculating the specific heat of a substance, the following data are used: mass = 92.15 g; heat = 3.200 kcal; rise in temperature = 45°C. How many significant figures should you report in calculating the specific heat?

1.82 A solar cell generates 500 kilojoules of energy per hour. To keep a refrigerator at 4°C, one needs 250 kcal/h. Can the solar cell supply sufficient energy per hour to maintain the temperature of the refrigerator?

1.83 The specific heat of urea is 1.339 J/g · °C . If one adds 60.0 J of heat to 10.0 g of urea at 20°C, what would be the final temperature?

Special Categories

Three special categories of problems—Tying It Together, Looking Ahead, and Challenge Problems—will appear from time to time at the ends of chapters. Not every chapter will have these problems, but they will appear to make specific points.

Tying It Together

1.84 Heats of reaction are frequently measured by monitoring the change in temperature of a water bath in which the reaction mixture is immersed. A water bath used for this purpose contains 2.000 L of water. In the course of the reaction, the temperature of the water rose 4.85°C. How many calories were liberated by the reaction? (You will need to use what you know about unit conversions and apply that information to what you know about energy and heat.)

1.85 You have samples of urea (a solid at room temperature) and pure ethyl alcohol (a liquid at room temperature). Which technique or techniques would you use to measure the amount of each substance?

Looking Ahead

1.86 You have a sample of material used in folk medicine. Suggest the approach you would use to determine whether this material contains an effective substance for treating disease. If you do find a new and effective substance, can you think of a way to determine the amount present in your sample? (Pharmaceutical companies have used this approach to produce many common medications.)

1.87 Many substances that are involved in chemical reactions in the human body (and in all organisms) contain carbon, hydrogen, oxygen, and nitrogen arranged in specific patterns. Would you expect new medications to have features in common with these substances, or would you expect them to be drastically different? What are the reasons for your answer?

Challenge Problems

1.88 If 2 kg of a given reactant is consumed in the reaction described in Problem 1.84, how many calories are liberated for each kilogram?

1.89 You have a water sample that contains a contaminant you want to remove. You know that the contaminant is much more soluble in diethyl ether than it is in water. You have a separatory funnel available. Propose a way to remove the contaminant.

Atoms

Image of atoms by SEM (scanning electron microscope).

GOB
Chemistry ⚛ Now™
Look for this logo in the chapter and go to GOB ChemistryNow at **http://now.brookscole.com/gob8** or on the CD for tutorials simulations, and problems.

2.1 | What Is Matter Made Of?

This question was discussed for thousands of years, long before humans had any reasonable way of getting an answer. In ancient Greece, two schools of thought tried to answer this question. One group, led by a scholar named Democritus (about 460–370 BCE), believed that all matter is made of very small particles—much too small to see. Democritus called these particles atoms (Greek *atomos,* meaning "not to cut"). Some of his followers developed the idea that there were different kinds of atoms, with different properties, and that the properties of the atoms caused ordinary matter to have the properties we all know.

Not all ancient thinkers, however, accepted this idea. A second group, led by Zeno of Elea (born about 450 BCE), did not believe in atoms at all. They insisted that matter is infinitely divisible. If you took any object, such as a piece of wood or a crystal of table salt, you could cut it or otherwise divide it into two parts, divide each of these parts into two more parts, and continue the process forever. According to Zeno and his followers, you would never reach a particle of matter that could no longer be divided.

Today we know that Democritus was right and Zeno was wrong. Atoms are the basic units of matter. Of course, there is a great difference in the way we now look at this question. Today our ideas are based on evidence. Democritus had no evidence to prove that matter cannot be divided an infinite number of times, just as Zeno had no evidence to support his claim that matter can be divided infinitely. Both claims were based not on evidence but on visionary belief: one in unity, the other in diversity. In Section 2.3 we will discuss the evidence for the existence of atoms, but first we need to look at the diverse forms of matter.

2.2 | How Do We Classify Matter?

Matter can be divided into two classes: pure substances and mixtures. Each class is then subdivided as shown in Figure 2.1.

A. Elements

An **element** is a substance (for example, carbon, hydrogen, and iron) that consists of identical atoms. At this time, 116 elements are known. Of these, 88 occur in nature; chemists and physicists have made the others. A list of the known elements appears on the inside back cover of this book, along with their symbols. Their symbols consist of one or two letters. Many symbols correspond directly to the name in English (for example, C for carbon, H for hydrogen, and Li for lithium), but a few are derived from the Latin or German names. Others are named for people who played significant roles in the development of science—in particular, atomic science (see Problem 2.11). Still other elements are named for geographic locations (see Problems 2.12).

B. Compounds

A **compound** is a pure substance made up of two or more elements in a fixed ratio by mass. For example, water is a compound made up of hydrogen and oxygen, and table salt is a compound made up of sodium and chlorine. There are an estimated 20 million known compounds, only a few of which we will meet in this book.

Figure 2.1   Classification of matter. Matter is divided into pure substances and mixtures. A pure substance may be either an element or a compound. A mixture may be either homogeneous or heterogeneous.

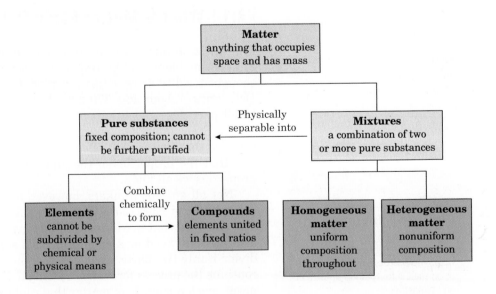

CHEMICAL CONNECTIONS 2A

Elements Necessary for Human Life

To the best of our knowledge, 20 of the 116 known elements are necessary for human life. The six most important of these—carbon, hydrogen, nitrogen, oxygen, phosphorus, and sulfur—are the subjects of organic chemistry and biochemistry (Chapters 10–31). Carbon, hydrogen, nitrogen, and oxygen are the "big four" in the human body. Seven other elements are also quite important, and our bodies use at least nine additional ones (trace elements) in very small quantities. The table lists these 20 major elements and their functions in the human body. Many of these elements are more fully discussed later in the book. For the average daily requirements for these elements, their sources in foods, and symptoms of their deficiencies, see Chapter 30.

Table 2A Elements and Their Functions in the Human Body

Element	Function	Element	Function
The Big Four		**The Trace Elements**	
Carbon (C) Hydrogen (H) Nitrogen (N) Oxygen (O)	The subject of Chapters 10–19 (organic chemistry) and 20–31 (biochemistry)	Chromium (Cr)	Increases effectiveness of insulin
		Cobalt (Co)	Part of vitamin B_{12}
		Copper (Cu)	Strengthens bones; assists in enzyme activity
The Next Seven		Fluorine (F)	Reduces the incidence of dental cavities
Calcium (Ca)	Strengthens bones and teeth; aids in blood clotting	Iodine (I)	An essential part of thyroid hormones
Phosphorus (P)	Present in phosphates of bone, in nucleic acids (DNA and RNA); and involved in energy transfer	Iron (Fe)	An essential part of some proteins, such as hemoglobin, myoglobin, cytochromes, and FeS proteins
Potassium (K)	Helps regulate electrical balance of body fluids; essential for nerve conduction	Manganese (Mn)	Present in bone-forming enzymes; aids in fat and carbohydrate metabolism
Sulfur (S)	An essential component of proteins		
Chlorine (Cl)	Necessary for normal growth and development	Molybdenum (Mo)	Helps regulate electrical balance in body fluids
Sodium (Na)	Helps regulate electrical balance in body fluids	Zinc (Zn)	Necessary for the action of certain enzymes
Magnesium (Mg)	Helps nerve and muscle action; present in bones		

A compound is characterized by its formula. The formula gives us the ratios of the compound's constituent elements and identifies each element by its atomic symbol. For example, in table salt the ratio of sodium atoms to chlorine atoms is 1:1. Given that Na is the symbol for sodium and Cl is the symbol for chlorine, the formula of table salt is NaCl. In water, the combining ratio is two hydrogen atoms to one oxygen atom. The symbol for hydrogen is H, that for oxygen is O, and the formula of water is H_2O. The subscripts following the atomic symbols indicate the ratio of the combining elements. The number 1 in these ratios is omitted from the subscript. It is understood that NaCl means a ratio of 1:1 and that H_2O represents a ratio of 2:1. You will find out more about the nature of combining elements in a compound and their names and formulas in Chapter 4.

Figure 2.2 shows four representations for a water molecule. We will have more to say about molecular models as we move through this book.

Figure 2.2 Four representations of a water molecule.

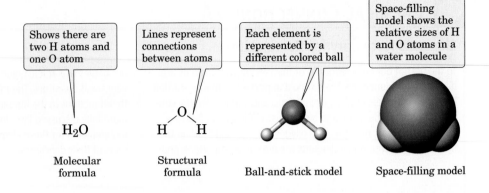

Shows there are two H atoms and one O atom	Lines represent connections between atoms	Each element is represented by a different colored ball	Space-filling model shows the relative sizes of H and O atoms in a water molecule

H_2O

Molecular formula

H——O——H

Structural formula

Ball-and-stick model

Space-filling model

EXAMPLE 2.1

(a) In the compound magnesium fluoride, magnesium (atomic symbol Mg) and fluorine (atomic symbol F) combine in a ratio of $1:2$. Write the formula of magnesium fluoride.
(b) The formula of perchloric acid is $HClO_4$. What are the combining ratios of the elements in perchloric acid?

Solution

(a) The formula is MgF_2. We do not write a subscript of 1 after Mg.
(b) Both H and Cl have no subscripts, which means that hydrogen and chlorine have a combining ratio of $1:1$. The subscript on oxygen is 4. Therefore, the combining ratios in $HClO_4$ are $1:1:4$.

Problem 2.1

Write the formulas of compounds in which the combining ratios are as follows:
(a) Sodium : chlorine : oxygen, $1:1:3$
(b) Aluminum (atomic symbol Al) : fluorine (atomic symbol F), $1:3$

C. Mixtures

A **mixture** is a combination of two or more pure substances. Most of the matter we encounter in our daily lives (including our own bodies) consists of mixtures rather than pure substances. For example, blood, butter, gasoline, soap, the metal in a wedding ring, the air we breathe, and the earth we walk on are all mixtures of pure substances. An important difference between a compound and a mixture is that the ratios by mass of the elements in a compound are fixed, whereas in a mixture the pure substances can be present in any mass ratio.

For some mixtures—blood, for example (Figure 2.3)—the texture of the mixture is even throughout. If you examine a mixture under magnification, however, you can see that it is composed of different substances.

Other mixtures are homogeneous throughout, and no amount of magnification will reveal the presence of different substances. The air we breathe, for example, is a mixture of gases, primarily nitrogen (78%) and oxygen (21%).

An important characteristic of a mixture is that it consists of two or more substances, each having different physical properties. If we know the physical properties of the individual substances, we can use appropriate physical means to separate the mixture into its component parts. Figure 2.4 shows one example of how a mixture can be separated.

Figure 2.3 Mixtures. (*a*) A cup of noodle soup is a heterogeneous mixture. (*b*) A sample of blood may look homogeneous, but examination with an optical microscope shows that it is, in fact, a heterogeneous mixture of liquid and suspended particles (blood cells). (*c*) A homogeneous solution of salt, NaCl, in water. The models show that the salt solution contains Na^+ and Cl^- ions as separate particles in water, with each ion being surrounded by a shell of four or more water molecules. The particles in this solution cannot be seen with an optical microscope.

(*a*)   (*b*)   (*c*)

Figure 2.4 Separating a mixture of iron and sulfur. (*a*) The iron–sulfur mixture is stirred with a magnet, which attracts the iron filings. (*b*) Much of the iron is removed after the first stirring. (*c*) Stirring continues until no more iron filings can be removed.

2.3 | What Are the Postulates of Dalton's Atomic Theory?

Around 1805 an English chemist, John Dalton (1766–1844), put forth a scientific atomic theory. The major difference between Dalton's theory and that of Democritus (Section 2.1) is that Dalton based his theory on evidence rather than on a belief. First, let us state Dalton's theory, and then see what kind of evidence supported it.

1. All matter is made up of very tiny, indivisible particles, which Dalton called **atoms.**

2. All atoms of the same element have the same chemical properties. Conversely, atoms of different elements have different chemical properties.

3. Compounds are formed by the chemical combination of two or more different kinds of atoms.

4. A **molecule** is a tightly bound combination of two or more atoms that acts as a single unit.

Atom The smallest particle of an element that retains the chemical properties of the element. The interaction among atoms accounts for the properties of matter.

A. Evidence for Dalton's Atomic Theory

The Law of Conservation of Mass

The great French chemist Antoine Laurent Lavoisier (1743–1794) discovered the **law of conservation of mass,** which states that matter can be neither created nor destroyed. Lavoisier proved this law by conducting many experiments in which he showed that the total mass of matter at the end of the experiment was exactly the same as that at the beginning. Dalton's theory explained this fact in the following way: If all matter consists of indestructible atoms, then any chemical reaction simply changes the attachments between atoms but does not destroy the atoms themselves.

In the following illustration, a carbon monoxide molecule reacts with a lead oxide molecule to give a carbon dioxide molecule and a lead atom. All of the original atoms are still present at the end; they have merely changed partners. Thus the total mass after this chemical change remains the same as it was before the change took place.

Carbon monoxide   Lead oxide   Carbon dioxide   Lead

The Law of Constant Composition

GOB
Chemistry ·✦· Now™
Click *Chemistry Interactive* to see an example of **Chemical Change on the Nanoscale**

Another French chemist, Joseph Proust (1754–1826), demonstrated the **law of constant composition,** which states that any compound is always made up of elements in the same proportion by mass. For example, if you decompose water, you will always get 8.0 g of oxygen for each 1.0 g of hydrogen. The mass ratio of oxygen to hydrogen in pure water is always 8.0 to 1.0, whether the water comes from the Atlantic Ocean or the Missouri River or is collected as rain, squeezed out of a watermelon, or distilled from urine. This fact was also evidence for Dalton's theory. If a water molecule consists of one atom of oxygen and two atoms of hydrogen, and if an oxygen atom has a mass 16 times that of a hydrogen atom, then the mass ratio of these two elements in water must always be 8.0 to 1.0. The two elements can never be found in water in any other mass ratio.

B. Monatomic, Diatomic, and Polyatomic Elements

Some elements—for example, helium and neon—consist of single atoms that are not connected to each other—that is, they are **monatomic elements.** In contrast, oxygen, in its most common form, contains two atoms in each molecule, connected to each other by a chemical bond. We write the formula for an oxygen molecule as O_2, with the subscript showing the number of atoms in the molecule. Six other elements also occur as diatomic molecules (that is, they contain two atoms of the same element per molecule): hydrogen (H_2), nitrogen (N_2), fluorine (F_2), chlorine (Cl_2), bromine (Br_2), and iodine (I_2). It is important to understand that, under normal conditions, free atoms of O, H, N, F, Cl, Br, and I do not exist. Rather, these seven elements occur only as **diatomic elements** (Figure 2.5).

Figure 2.5 Some diatomic, triatomic, and polyatomic elements. Hydrogen, nitrogen, oxygen, and chlorine are diatomic elements. Ozone, O_3, is a triatomic element. One form of sulfur, S_8, is a polyatomic element.

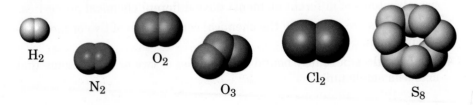

CHEMICAL CONNECTIONS 2B

Abundance of Elements in the Human Body and in the Earth's Crust

The table shows the abundances of the elements present in the human body. As you can see, oxygen is the most abundant element by mass, followed by carbon, hydrogen, and nitrogen. If we go by number of atoms, however, hydrogen is even more abundant in the human body than oxygen.

The table also shows the abundances of elements in the Earth's crust. Although 88 elements are found in the Earth's crust (we know very little about the interior of the Earth because we have not been able to penetrate into it very far), they are not present in

anything close to equal amounts. In the Earth's crust as well as the human body, the most abundant element by mass is oxygen. But there the similarity ends. Silicon, aluminum, and iron, which are the second, third, and fourth most abundant elements in the Earth's crust, respectively, are not major elements in the body. While carbon is the second most abundant element by mass in the human body, it is present to the extent of only 0.08% in the Earth's crust.

Table 2B The Relative Abundance of Elements in the Human Body and in the Earth's Crust, Including the Atmosphere and Oceans

| Element | Percentage in Human Body | | Percentage in Earth's Crust by Mass |
	By Number of Atoms	By Mass	
H	63.0	10.0	0.9
O	25.4	64.8	49.3
C	9.4	18.0	0.08
N	1.4	3.1	0.03
Ca	0.31	1.8	3.4
P	0.22	1.4	0.12
K	0.06	0.4	2.4
S	0.05	0.3	0.06
Cl	0.03	0.2	0.2
Na	0.03	0.1	2.7
Mg	0.01	0.04	1.9
Si	—	—	25.8
Al	—	—	7.6
Fe	—	—	4.7
Others	0.01	—	—

Some elements have even more atoms in each molecule. Ozone, O_3, has three oxygen atoms in each molecule. In one form of phosphorus, P_4, each molecule has four atoms. One form of sulfur, S_8, has eight atoms per molecule. Some elements have molecules that are much larger. For example, diamond has millions of carbon atoms all bonded together in a gigantic cluster. Diamond and S_8 are referred to as **polyatomic elements.**

2.4 | What Are Atoms Made Of?

A. Three Subatomic Particles

Today we know that matter is more complex than Dalton believed. A wealth of experimental evidence obtained over the last 100 years or so has convinced us that atoms are not indivisible but rather consist of even smaller particles called subatomic particles. Three subatomic particles make up all

There are many other subatomic particles but we will not deal with them in this book.

Table 2.1 Properties and Location within Atoms of Protons, Neutrons, and Electrons

Subatomic Particle	Charge	Mass (g)	Mass (amu)	Mass (amu); Rounded to One Significant Figure	Location in an Atom
Proton	+1	1.6726×10^{-24}	1.0073	1	In the nucleus
Electron	−1	9.1094×10^{-28}	5.4859×10^{-4}	0.0005	Outside the nucleus
Neutron	0	1.6749×10^{-24}	1.0087	1	In the nucleus

Proton A subatomic particle with a charge of +1 and a mass of approximately 1 amu; it is found in a nucleus.

Atomic mass unit (amu) A unit of the scale of relative masses of atoms: 1 amu = 1.6605×10^{-24} g. By definition, 1 amu is 1/12 the mass of a carbon atom containing 6 protons and 6 neutrons.

Electron A subatomic particle with a charge of −1 and a mass of approximately 0.0005 amu. It is found in the space surrounding a nucleus.

atoms: protons, electrons, and neutrons. Table 2.1 shows the charge, mass, and location of these particles in an atom.

A **proton** has a positive charge. By convention we say that the magnitude of the charge is +1. Thus one proton has a charge of +1, two protons have a charge of +2, and so forth. The mass of a proton is 1.6726×10^{-24} g, but this number is so small that it is more convenient to use another unit, called the **atomic mass unit (amu),** to describe its mass.

$$1 \text{ amu} = 1.6605 \times 10^{-24} \text{ g}$$

Thus a proton has a mass of 1.0073 amu. For most purposes in this book, it is sufficient to round this number to one significant figure and, therefore, we say that the mass of a proton is 1 amu.

An **electron** has a charge of −1, equal in magnitude to the charge on a proton, but opposite in sign. The mass of an electron is approximately 5.4859×10^{-4} amu or 1/1837 that of the proton. It takes approximately 1837 electrons to equal the mass of one proton.

Like charges repel, and unlike charges attract. Two protons repel each other just as two electrons also repel each other. A proton and an electron, however, attract each other.

| Two protons repel each other | Two electrons repel each other | A proton and an electron attract each other |

Neutron A subatomic particle with a mass of approximately 1 amu, and a charge of zero; it is found in the nucleus

A **neutron** has no charge. Therefore, neutrons neither attract nor repel each other or any other particle. The mass of a neutron is slightly greater than that of a proton: 1.6749×10^{-24} g or 1.0087 amu. Again, for our purposes, we round this number to 1 amu.

These three particles make up atoms, but where are they found? Protons and neutrons are found in a tight cluster in the center of an atom (Figure 2.6). This part of an atom is called the **nucleus.** We will discuss the nucleus in greater detail in Chapter 3. Electrons are found as a diffuse cloud outside the nucleus.

B. Mass Number

Each atom has a fixed number of protons, electrons, and neutrons. One way to describe an atom is by its **mass number,** which is the sum of the number of protons and neutrons in its nucleus. Note that an atom also contains electrons but because the mass of an electron is so small compared to that of protons and neutrons (Table 2.1), electrons are not counted in determining mass number.

> Mass number = the number of protons and neutrons in the nucleus of an atom.

For example, an atom with 5 protons, 5 electrons, and 6 neutrons has a mass number of 11.

EXAMPLE 2.2

What is the mass number of an atom containing:
(a) 58 protons, 58 electrons, and 78 neutrons?
(b) 17 protons, 17 electrons, and 20 neutrons?

Solution

The mass number of an atom is the sum of the number of its protons and neutrons in its nucleus.
(a) The mass number is $58 + 78 = 136$.
(b) The mass number is $17 + 20 = 37$.

Problem 2.2

What is the mass number of an atom containing:
(a) 15 protons, 15 electrons, and 16 neutrons?
(b) 86 protons, 86 electrons, and 136 neutrons?

Figure 2.6 A typical atom. Protons and neutrons are found in the nucleus, and electrons are found as a cloud outside the nucleus. Notice how small the nucleus is compared to the size of an atom.

C. Atomic Number

The **atomic number** of an element is the number of protons in its nucleus.

> Atomic number = number of protons in the nucleus of an atom

At the present time, 116 elements are known. These elements have atomic numbers from 1–116. The smallest atomic number belongs to the element hydrogen, which has only one proton, and the largest (so far) to the as-yet-unnamed heaviest known element, which contains 116 protons.

If you know the atomic number and the mass number of an element, you can properly identify it. For example, the element with 6 protons, 6 electrons, and 6 neutrons has an atomic number of 6 and a mass number of 12. The element with atomic number 6 is carbon, C. Because its mass number is 12, we call this atomic nucleus carbon-12. Alternatively, we can write the symbol for this atomic nucleus as $^{12}_{6}C$. In this symbol, the mass number of the element is always written in the upper-left corner (as a superscript) of the symbol of the element, and the atomic number in the lower-left corner (as a subscript).

Atomic numbers for all the known elements are given in the atomic weight table on the inside back cover. They are also given in the Periodic Table on the inside front cover.

$$\text{Mass number (number of protons + neutrons)} \nearrow \quad ^{12}_{6}C$$
$$\text{Atomic number (number of protons)} \nearrow$$

If you know the name of the element, you can look up its atomic number and atomic weight from the atomic weight table on the inside back cover. Conversely, if you know the atomic number of the element, you can look up its symbol from the Periodic Table on the inside front cover. The atomic number of each element is shown in red.

EXAMPLE 2.3

Name the elements given in Example 2.2 and write the symbols for their atomic nuclei.

Solution

(a) This element has 58 protons. We see from the Periodic Table on the inside front cover of the text that the element of atomic number 58 is cerium, and its symbol is Ce. An atom of this element has 58 protons and 78 neutrons and, therefore, its mass number is 136. We call it cerium-136. Its symbol is $^{136}_{58}$Ce.

(b) This atom has 17 protons, making it a chlorine (Cl) atom. Because its mass number is 37, we call it chlorine-37. Its symbol is $^{37}_{17}$Cl.

Problem 2.3

Name the elements given in Problem 2.2. Write the symbols of their atomic nuclei.

EXAMPLE 2.4

A number of elements have an equal number of protons and neutrons in their nuclei. Among these are oxygen, nitrogen, and neon. What are the atomic numbers of these elements? How many protons and neutrons does an atom of each have? Write the name and the symbol of each of these atomic nuclei.

Solution

Atomic numbers for these elements are found in the list of elements on the inside back cover. This table shows that oxygen (O) has atomic number 8, nitrogen (N) has atomic number 7, and neon (Ne) has atomic number 10. This means that the oxygen has 8 protons and 8 neutrons. Its name is oxygen-16, and its symbol is $^{16}_{8}$O. Nitrogen has 7 protons and 7 neutrons, its name is nitrogen-14, and its symbol is $^{14}_{7}$N. Neon has 10 protons and 10 neutrons, its name is neon-20, and its symbol is $^{20}_{10}$Ne.

Problem 2.4

(a) What are the atomic numbers of mercury (Hg) and lead (Pb)?
(b) How many protons does an atom of each have?
(c) If both Hg and Pb have 120 neutrons in their nuclei, what is the mass number of each?
(d) Write the name and the symbol of each.

D. Isotopes

Although we can say that a carbon atom always has 6 protons and 6 electrons, we cannot say that a carbon atom must have any particular number of neutrons. Some of the carbon atoms found in nature have 6 neutrons; the mass number of these atoms is 12; they are written as carbon-12, and their symbol is $^{12}_{6}$C. Other carbon atoms have 6 protons and 7 neutrons and, therefore, a mass number of 13; they are written as carbon-13, and their symbol is $^{13}_{6}$C. Still other carbon atoms have 6 protons and 8 neutrons; they are written as carbon-14 or $^{14}_{6}$C. Atoms with the same number of protons but different numbers of neutrons are called **isotopes.** All isotopes of carbon contain 6 protons and 6 electrons (or they wouldn't be carbon). Each isotope, however, contains a different number of neutrons and, therefore, has a different mass number.

The fact that isotopes exist means that the second statement of Dalton's atomic theory (Section 2.3) is not correct.

The properties of isotopes of the same element are almost identical, and for most purposes we regard them as identical. They differ, however, in radioactivity properties, which we discuss in Chapter 3.

EXAMPLE 2.5

How many neutrons are in each isotope of oxygen? Write the symbol of each isotope.
(a) Oxygen-16   (b) Oxygen-17   (c) Oxygen-18

Solution

Each oxygen atom has 8 protons. The difference between the mass number and the number of protons gives the number of neutrons.
(a) Oxygen-16 has $16 - 8 = 8$ neutrons. Its symbol is $^{16}_{8}O$.
(b) Oxygen-17 has $17 - 8 = 9$ neutrons. Its symbol is $^{17}_{8}O$.
(c) Oxygen-18 has $18 - 8 = 10$ neutrons. Its symbol is $^{18}_{8}O$.

Problem 2.5

Two iodine isotopes are used in medical treatments: iodine-125 and iodine-131. How many neutrons are in each isotope? Write the symbol for each isotope.

Most elements are found on Earth as mixtures of isotopes, in a more or less constant ratio. For example, all naturally occurring samples of the element chlorine contain 75.77% chlorine-35 (18 neutrons) and 24.23% chlorine-37 (20 neutrons). Silicon exists in nature in a fixed ratio of three isotopes, with 14, 15, and 16 neutrons, respectively. For some elements, the ratio of isotopes may vary slightly from place to place but, for most purposes, we can ignore these slight variations. The atomic masses and isotopic abundances are determined using an instrument called a mass spectrometer.

E. Atomic Weight

The **atomic weight** of an element given in the Periodic Table is a weighted average of the masses (in amu) of its isotopes found on the Earth. As an example of the calculation of atomic weight, let us examine chlorine. As we have just seen, two isotopes of chlorine exist in nature, chlorine-35 and chlorine-37. The mass of a chlorine-35 atom is 34.97 amu, and the mass of a chlorine-37 atom is 36.97 amu. Note that the atomic weight of each chlorine isotope (its mass in amu) is very close to its mass number (the number of protons and neutrons in its nucleus). This statement holds true for the isotopes of chlorine and those of all elements, because protons and neutrons have a mass of approximately (but not exactly) 1 amu.

The atomic weight of chlorine is a weighted average of the masses of the two naturally occurring chlorine isotopes:

Atomic weight The weighted average of the masses of the naturally occurring isotopes of the element. The units of atomic weight are atomic mass units (amu).

GOB
Chemistry Now™
Click *Coached Problems* to see an example of the **Average Atomic Mass from Isotopic Abundances**

Chlorine-35 · Chlorine-37

$$\left(\frac{75.77}{100} \times 34.97\ \text{amu}\right) + \left(\frac{24.23}{100} \times 36.97\ \text{amu}\right) = 35.45\ \text{amu}$$

Atomic weight in the Periodic Table is given to four decimal places

Some elements—for example, gold, fluorine, and aluminum—occur naturally as only one isotope. The atomic weights of these elements are close to whole numbers (gold, 196.97 amu; fluorine, 18.998 amu; aluminum, 26.98 amu). A table of atomic weights is found on the inside back cover of this book.

EXAMPLE 2.6

The natural abundances of the three stable isotopes of magnesium are 78.99% magnesium-24 (23.98 504 amu), 10.00% magnesium-25 (24.9858 amu), and 11.01% magnesium-26 (25.9829 amu). Calculate the atomic weight of magnesium and compare your value with that given in the Periodic Table.

Solution

To calculate the weighted average of the masses of the isotopes, multiply each atomic mass by its abundance and then add.

$$\left(\frac{78.99}{100} \times 23.99 \text{ amu}\right) + \left(\frac{10.00}{100} \times 24.986 \text{ amu}\right) + \left(\frac{11.01}{100} \times 25.983 \text{ amu}\right) =$$

$$18.95 \quad + \quad 2.499 \quad + \quad 2.861 \quad = 24.31 \text{ amu}$$

The atomic weight of magnesium given in the Periodic Table to four decimal places is 24.3050.

Problem 2.6

The atomic weight of lithium is 6.941 amu. Lithium has only two naturally occurring isotopes: lithium-6 and lithium-7. Estimate which isotope is in greater natural abundance.

F. The Mass and Size of an Atom

A typical heavy atom (although not the heaviest) is lead-208, a lead atom with 82 protons, 82 electrons, and 126 neutrons. It has a mass of 3.5×10^{-22} g. You would need 1.3×10^{24} atoms of lead-208 to make 1 lb of lead—a very large number. There are approximately 6 billion people on Earth right now. If you divided 1 lb of these atoms among all the people on Earth, each person would get about 2.2×10^{14} atoms.

An atom of lead-208 has a diameter of about 3.5×10^{-10} m. If you could line them up with the atoms just touching, it would take 73 million lead atoms to make a line 1-inch long. Despite their tiny size, we can actually see atoms, in certain cases, by using a special instrument called a scanning tunneling microscope (Figure 2.7).

Virtually all of the mass of an atom is concentrated in its nucleus (because the nucleus contains the protons and neutrons). The nucleus of a lead-208 atom, for example, has a diameter of about 1.6×10^{-14} m. When you compare this with the diameter of a lead-208 atom, which is about 3.5×10^{-10} m, you see that the nucleus occupies only a tiny fraction of the total volume of the atom. If the nucleus of a lead-208 atom were the size of a baseball, then the entire atom would be much larger than a baseball stadium. In fact, it would be a sphere about one mile in diameter. Because a nucleus has such a relatively large mass concentrated in such a relatively small volume, a nucleus has a very high density. The density of a lead-208 nucleus, for example, is 1.8×10^{14} g/cm^3. Nothing in our daily life has a density anywhere near as high. If a paper clip had this density, it would weigh about 10 million (10^7) tons.

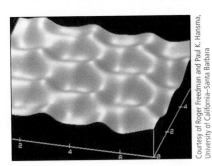

Figure 2.7 The surface of graphite is revealed with a scanning tunneling microscope. The contours represent the arrangement of individual carbon atoms on a crystal surface.

Courtesy of Roger Freedman and Paul K. Hansma, University of California–Santa Barbara

CHEMICAL CONNECTIONS 2C

Isotopic Abundances and Astrochemistry

The term **isotopic abundance** refers to the relative amounts of the isotopes of an element present in a sample of that element. When we speak of isotopic abundances and the atomic weights derived from them, we refer to isotopic abundances on Earth. For example, there are 11 known isotopes of chlorine, Cl, ranging from chlorine-31 to chlorine-41. Only two of these isotopes occur on Earth: chlorine-35 with an abundance of 75.77% and chlorine-37 with an abundance of 24.23%.

With the advance of space exploration, it has become evident that other parts of the solar system—the Sun, the Moon, planets, asteroids, comets, and stars, as well

Mars.

as intergalactic gases—may have different isotopic abundances. For example, the deuterium/hydrogen ratio ($^2H/^1H$) on Mars is five times larger than that on Earth. The $^{17}O/^{18}O$ ratio is also higher on Mars than on Earth. Such differences in isotopic abundances are used to establish theories about the origins and history of the solar system.

Furthermore, a comparison of isotopic abundances in certain meteorites found on Earth enables us to guess their origin. For example, it is now believed that certain meteorites that landed on Earth came from Mars. Scientists speculate that they were ejected from its surface when the Red Planet collided with some other large body—perhaps an asteroid.

2.5 | What Is the Periodic Table?

A. Origin of the Periodic Table

In the 1860s, a number of scientists—most notably the Russian Dmitri Mendeleev (1834–1907), then professor of chemistry at the University of St. Petersburg—produced one of the first periodic tables, the form of which we still use today. Mendeleev started by arranging the known elements in order of increasing atomic weight beginning with hydrogen. He soon discovered that when the elements are arranged according to increasing atomic weight, certain sets of properties recur periodically. Mendeleev then arranged those elements with recurring properties into **periods** (horizontal rows) by starting a new row each time he came to an element with properties similar to hydrogen. In this way, he discovered that lithium, sodium, potassium, and so forth each start new rows. All are metallic solids, all form ions with a charge of +1 (Li^+, Na^+, K^+, and so on), and all react with water to form metal hydroxides (LiOH, NaOH, KOH, and so on). Mendeleev also discovered that elements in other groups (vertical columns) share similar properties. For example, the elements fluorine (atomic number 9), chlorine (17), bromine (35), and iodine (53) all fall in the same column in the table. These elements, which are called **halogens,** are all colored substances, with color deepening as we go down the table (Figure 2.8). All form compounds with sodium that have the formula NaX (for example, NaCl and NaBr), but not NaX_2, Na_2X, Na_3X, or anything else. Only the elements in this column share this property.

At this point, we must say a word about the numbering of the columns of the Periodic Table. Mendeleev gave them numerals and added the letter A for some columns and B for others. This numbering pattern remains in common use in the United States today. An alternative pattern used internationally relies on the numbers 1 to 18, without added letters, going from left to right. Thus, in Mendeleev's numbering system, the halogens are in Group 7A;

Period of the Periodic Table
A horizontal row of the Periodic Table

Dmitri Mendeleev.

"X" is a commonly used symbol for a halogen.

Figure 2.8 Four halogens.
Fluorine and chlorine are gases,
bromine is a liquid, and iodine
is a solid.

Main-group element An element
of any of the A groups of the
Periodic Table

in the new international numbering system, they are in Group 17. Although
this book uses the traditional numbering, both patterns are shown on the
Periodic Table on the inside front cover.

If we look at the Periodic Table on the inside front cover of this book, we
see that columns 1A and 2A extend above the rest, as does column 8A. The
elements in columns 1A to 8A are called **main-group elements.** The elements in the B columns (Groups 3 to 12 in the new numbering system) are
called **transition elements.** Notice that elements 58 to 71 and 90 to 103
are not included in the main body of the table but rather are shown separately at the bottom. These sets of elements, called **inner transition elements,** actually belong in the main body of the Periodic Table, between
columns 3B and 4B (between La and Hf). As is customary, we put them outside the main body solely to make a more compact presentation. If you like,
you may mentally take a pair of scissors, cut through the heavy line
between columns 3B and 4B, move them apart, and insert the inner transition elements. You will now have a table with 32 columns.

CHEMICAL CONNECTIONS 2D

Strontium-90

Elements in the same column of the Periodic Table show similar
properties. One important example is the similarity of strontium (Sr)
and calcium (Ca) (strontium is just below calcium in Group 2A).
Calcium is an important element for humans because our bones
and teeth consist largely of calcium compounds. We need some of
this mineral in our diet every day, and we get it mostly from milk,
cheese, and other dairy products.

One of the products released by test nuclear explosions in the
1950s and 1960s was the isotope strontium-90. This isotope is
radioactive, with a half-life of 28.1 years. (Half-life is discussed in
Section 3.4.) Strontium-90 was present in the fallout from above-
ground nuclear testing. It was carried all over the Earth by winds

and slowly settled to the ground, where it was eaten by cows
and other animals. Strontium-90 got into milk and eventually into
human bodies as well. If it were not so similar to calcium, our
bodies would eliminate it within a few days. Because it is similar,
however, some of the strontium-90 became deposited in bones and
teeth (especially in children), subjecting all of us to a small amount
of radioactivity for long periods of time.

A 1963 treaty between the United States and the former Soviet
Union banned above-ground nuclear testing. Although a few other
countries still conduct occasional above-ground tests, there is
reason to hope that such testing will be completely halted in the
near future.

B. Classification of the Elements

There are three classes of elements: metals, nonmetals, and metalloids. The majority of elements are **metals**—only 24 are not. Metals are solids (except for mercury, which is a liquid), shiny, conductors of electricity, ductile (they can be drawn into wires), and malleable (they can be hammered and rolled into sheets). They also form alloys, which are solutions of one or more metals dissolved in another metal. Brass, for example, is an alloy of copper and zinc. Bronze is an alloy of copper and tin, and pewter is an alloy of tin, antimony, and lead. In their chemical reactions, metals tend to give up electrons (Section 4.2). Figure 2.9 shows a form of the Periodic Table in which the elements are classified by type.

The second class of elements is the **nonmetals.** With the exception of hydrogen, the 18 nonmetals appear to the right side of the Periodic Table. With the exception of graphite, which is one form of carbon, they do not conduct electricity. At room temperature, nonmetals such as phosphorus and iodine are solids. Bromine is a liquid, and the elements of Group 8A (the noble gases)—helium through radon—are gases. In their chemical reactions, nonmetals tend to accept electrons (Section 4.2). Virtually all of the compounds we will encounter in our study of organic and biochemistry are built from just six nonmetals: H, C, N, O, P, and S.

Six elements are classified **metalloids:** boron, silicon, germanium, arsenic, antimony, and tellurium. These elements have some properties of metals and some properties of nonmetals. For example, some metalloids are shiny like metals, but do not conduct electricity. One of these metalloids, silicon, is a semiconductor—that is, it does not conduct electricity under a certain applied voltage, but becomes a conductor at higher applied voltages. This semiconductor property of silicon makes it a vital element for Silicon Valley–based companies and the entire electronics industry (Figure 2.10).

C. Examples of Periodicity in the Periodic Table

Not only do the elements in any particular column of the Periodic Table share similar properties, but the properties also vary in some fairly regular ways as we go up or down a column. For instance, Table 2.2 shows that the melting and boiling points of the **halogens** regularly increase as we go down the column.

Another example involves the Group 1A elements, also called the **alkali metals.** All alkali metals are soft enough to be cut with a knife, and their

Metal An element that is a solid (except for mercury, which is a liquid), shiny, conducts electricity, is ductile and malleable, and forms alloys.

Nonmetal An element that does not have the characteristic properties of a metal and tends to accept electrons

Although hydrogen (H) appears in Group 1A, it is not an alkali metal; it is a nonmetal. Hydrogen is placed in Group 1A because its *s* shell is the one being filled (Section 2.7).

Metalloid An element that displays some of the properties of metals and some of the properties of nonmetals

Halogen An element in Group 7A of the Periodic Table

Alkali metal An element in Group 1A of the Periodic Table

Figure 2.9 Classification of the elements.

≠ Not yet named

Copper, Cu

Magnesium, Mg

Aluminum, Al

(a) Metals

Bromine, Br$_2$ Iodine, I$_2$

(b) Nonmetals

Forms of silicon

(c) Metalloids

Charles D. Winters

Figure 2.10   Representative elements. (a) Magnesium, aluminum, and copper are metals. All can be drawn into wires and conduct electricity. (b) Only 18 or so elements are classified as nonmetals. Shown here are liquid bromine and solid iodine. (c) Only six elements are generally classified as metalloids. This photograph is of solid silicon in various forms, including a wafer on which is printed electronic circuits.

Charles D. Winters

Sodium metal can be cut with a knife.

softness increases in going down the column. They have relatively low melting and boiling points, which decrease in going down the columns (Table 2.3).

All alkali metals react with water to form hydrogen gas, H$_2$, and a metal hydroxide with the formula MOH, where "M" stands for the alkali metal. The violence of their reaction with water increases in going down the column.

$$2Na + 2H_2O \longrightarrow 2NaOH + H_2$$

Sodium   Water   Sodium   Hydrogen
                      hydroxide   gas

They also form compounds with the halogens with the formula MX, where "X" stands for the halogen.

$$2Na + Cl_2 \longrightarrow 2NaCl$$

Sodium   Chlorine   Sodium
                       chloride

Table 2.2   Melting and Boiling Points of the Halogens (Group 7A Elements)

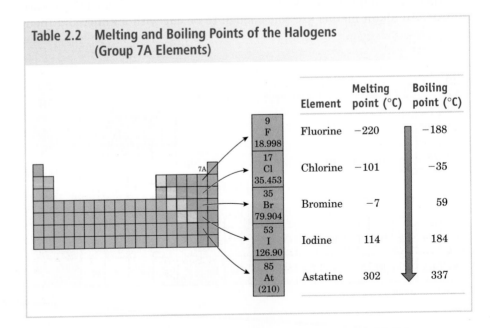

Element	Melting point (°C)	Boiling point (°C)
Fluorine	−220	−188
Chlorine	−101	−35
Bromine	−7	59
Iodine	114	184
Astatine	302	337

9 F 18.998	
17 Cl 35.453	
35 Br 79.904	
53 I 126.90	
85 At (210)	

Table 2.3 **Melting and Boiling Points of the Alkali Metals (Group 1A Elements)**

Element	Melting point (°C)	Boiling point (°C)
Lithium	180	1342
Sodium	98	883
Potassium	63	760
Rubidium	39	686
Cesium	28	669

The elements in Group 8A, often called the **noble gases,** provide yet another example of how properties of elements change gradually within a column. Group 8A elements are gases under normal temperature and pressure, and they form either no compounds or very few compounds. Notice

GOB
Chemistry ⚛ Now™

Click *Chemistry Interactive* to see an example of **Chemical Periodicity**

CHEMICAL CONNECTIONS 2E

The Use of Metals as Historical Landmarks

The malleability of metals played an important role in the development of human society. In the Stone Age, tools were made from stone, which has no malleability. Then, about 11,000 BCE, it was discovered that the pure copper found on the surface of the Earth could be hammered into sheets, which made it suitable for use in vessels, utensils, and religious or artistic objects. This period became known as the Copper Age. Pure copper on the surface of the Earth, however, is scarce. Around 5000 BCE, humans found that copper could be obtained by putting malachite, $Cu_2CO_3(OH)_2$, a green copper-containing stone, into a fire. Malachite yielded pure copper at the relatively low temperature of 200°C.

Copper is a soft metal made of layers of large copper crystals. It can easily be drawn into wires because the layers of crystals can slip past one another. When hammered, the large crystals break into smaller ones with rough edges and the layers can no longer slide past one another. Therefore, hammered copper sheets are harder than drawn copper. Using this knowledge, the ancient profession of coppersmith was born, and beautiful plates, pots, and ornaments were produced.

Celtic shield.

Werner Forman/Art Resource, NY

Around 4000 BCE, it was discovered that an even greater hardness could be achieved by mixing molten copper with tin. The resulting alloy is called bronze. The Bronze Age was born somewhere in the Middle East and quickly spread to China and all over the world. Because hammered bronze takes an edge, knives and swords could be manufactured using it.

An even harder metal was soon to come. The first raw iron was found in meteorites. (The ancient Sumerian name of iron is "metal from heaven.") Around 2500 BCE, it was discovered that iron could be recovered from its ore by smelting. Smelting is the process of recovering a metal from its ore by heating the ore. Thus began the Iron Age. More advanced technology was needed for smelting iron ores because iron melts only at a high temperature (about 1500°C). For this reason, it took a longer time to perfect the smelting process and to learn how to manufacture steel, which is about 90–95% iron and 5–10% carbon. Steel objects appeared first in India around 100 BCE.

Modern historians look back at ancient cultures and use the discovery of a new metal as a landmark for that age.

Table 2.4 Melting and Boiling Points of the Noble Gases (Group 8A Elements)

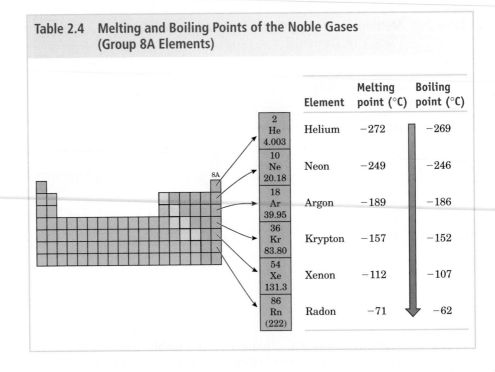

Element	Melting point (°C)	Boiling point (°C)
Helium	−272	−269
Neon	−249	−246
Argon	−189	−186
Krypton	−157	−152
Xenon	−112	−107
Radon	−71	−62

how close the melting and boiling points of the elements in this series are to one another (Table 2.4).

The Periodic Table is so useful that it hangs in nearly every chemistry classroom and chemical laboratory throughout the world. What makes it so useful is that it correlates a vast amount of data about the elements and their compounds and allows us to make many predictions about both chemical and physical properties. For example, if you were told that the boiling point of germane (GeH_4) is −88°C and that of methane (CH_4) is −164°C, could you predict the boiling point of silane (SiH_4)? The position of silicon in the table, between germanium and carbon, might lead you to a prediction of about −125°C. The actual boiling point of silane is −112°C, not far from this prediction.

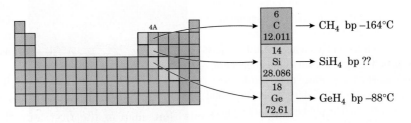

2.6 | How Are the Electrons in an Atom Arranged?

We have seen that the protons and neutrons of an atom are concentrated in the atom's very small nucleus, and that the electrons of an atom are located in the considerably larger space outside the nucleus. We can now ask how the electrons of an atom are arranged in this extranuclear space. Are they arranged randomly like seeds in a watermelon, or are they organized into layers like the layers of an onion?

Let us begin with hydrogen because it has only one electron and is the simplest atom. Before we do so, however, it is necessary to describe a dis-

Figure 2.11 An energy stairway. Unlike our stairways, the spaces between energy levels in an atom are not equal; they become smaller as they go up.

E_3

E_2

E_1 (ground state)

covery made in 1913 by the Danish physicist Niels Bohr. At the time, it was known that an electron is always moving around the nucleus and so possesses kinetic energy. Bohr discovered that only certain values are possible for this energy. This was a very surprising discovery. If you were told that you could drive your car at 23.4 mi/h or 28.9 mi/h or 34.2 mi/h, but never at any speed in between these values, you wouldn't believe it. Yet that is just what Bohr discovered about electrons in atoms. The lowest possible energy level is the **ground state.**

If an electron is to have more energy than it has in the ground state, only certain values are allowed; values in between are not permitted. Bohr was unable to explain why these energy levels of electrons exist in atoms, but the accumulated evidence forced him to the conclusion that they do. We say that the energy of electrons in atoms is quantized. We can liken quantization to walking up a flight of stairs (Figure 2.11). You can put your foot on any step, but you cannot stand any place between two steps.

Ground-state electron configuration The electron configuration of the lowest energy state of an atom

A. Electrons Are Distributed in Shells, Subshells, and Orbitals

One conclusion reached by Bohr is that electrons in atoms do not move freely in the space around the nucleus, but rather are confined to specific regions of space called **principal energy levels,** or more simply, **shells.** These shells are numbered 1, 2, 3, and 4, and so on from the inside out. Table 2.5 gives the number of electrons that each of the first four shells can hold.

Principal energy level An energy level containing orbitals of the same number (1, 2, 3, 4 and so forth)

Shell All orbitals of a principal energy level of an atom

Table 2.5	Distribution of Electrons in Shells	
Shell	Number of Electrons Shell Can Hold	Relative Energies of Electrons in Each Shell
4	32	Higher
3	18	
2	8	↑
1	2	Lower

Table 2.6 Distribution of Orbitals within Shells

Shell	Orbitals Contained in Each Shell	Maximum Number of Electrons Shell Can Hold
4	One 4s, three 4p, five 4d, and seven 4f orbitals	2 + 6 + 10 + 14 = 32
3	One 3s, three 3p, and five 3d orbitals	2 + 6 + 10 = 18
2	One 2s and three 2p orbitals	2 + 6 = 8
1	One 1s orbital	2

Electrons in the first shell are closest to the positively charged nucleus and are held most strongly by it; these electrons are said to be the lowest in energy (hardest to remove). Electrons in higher-numbered shells are farther from the nucleus and are held less strongly to it; these electrons are said to be higher in energy (easier to remove).

Shells are divided into **subshells** designated by the letters s, p, d, and f. Within these subshells, electrons are grouped in **orbitals.** An orbital is a region of space and can hold two electrons (Table 2.6). The second shell contains one s orbital and three p orbitals. All p orbitals come in sets of three and can hold six electrons. The third shell contains one s orbital, three p orbitals, and five d orbitals. All d orbitals come in sets of five and can hold ten electrons. All f orbitals come in sets of seven and can hold 14 electrons.

Subshell All of the orbitals of an atom having the same principal energy level and the same letter designation (either s, p, d, or f)

Orbital A region of space around a nucleus that can hold a maximum of two electrons

The d and f orbitals are less important to us, so we will not discuss their shapes.

B. Orbitals Have Definite Shapes and Orientations in Space

All s orbitals have the shape of a sphere with the nucleus at the center of the sphere. Figure 2.12 shows the shapes of the 1s and 2s orbitals. Of the s orbitals, the 1s is the smallest sphere, the 2s is a larger sphere, and the 3s (not shown) is a still larger sphere. Figure 2.12 also shows the three-dimensional shapes of the three 2p orbitals. Each 2p orbital has the shape of a dumbbell with the nucleus at the midpoint of the dumbbell. The three 2p orbitals are at right angles to each other, with one orbital on the x axis, the second on the y axis, and the third on the z axis. The shapes of 3p orbitals are similar, but larger.

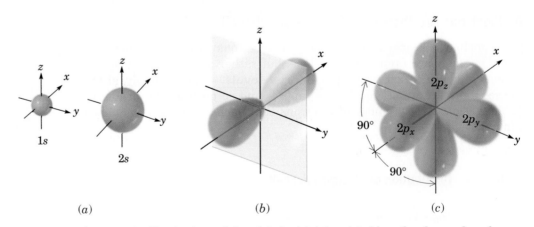

(a) (b) (c)

Figure 2.12 The 1s, 2s, and 2p orbitals. (a) A 1s orbital has the shape of a sphere, with the nucleus at the center of the sphere. A 2s orbital is a larger sphere than a 1s orbital, and a 3s orbital is larger still. (b) A 2p orbital has the shape of a dumbbell, with the nucleus at the midpoint of the dumbbell. (c) Each 2p orbital is perpendicular to the other two. The 3p orbitals are similar in shape but larger. To make it easier for you to see the two lobes of each 2p orbital, one lobe is colored red and the other is colored blue.

Because the vast majority of organic molecules and biomolecules consist of the elements H, C, N, O, P, and S, which use only $1s$, $2s$, $2p$, $3s$, and $3p$ orbitals for bonding, we will concentrate on just these and other elements of the first, second, and third periods of the Periodic Table.

C. Electron Configurations of Atoms are Governed by Three Rules

The **electron configuration** of an atom is a description of the orbitals that its electrons occupy. The orbitals available to all atoms are the same—namely, $1s$, $2s$, $2p$, $3s$, $3p$, and so on. In the ground state of an atom, only the lowest-energy orbitals are occupied; all other orbitals are empty. We determine the ground-state electron configuration of an atom using the following rules:

Rule 1. Orbitals fill in the order of increasing energy from lowest to highest.

> **Example:** In this book we are concerned primarily with elements of the first, second, and third periods of the Periodic Table. Orbitals in these elements fill in the order $1s$, $2s$, $2p$, $3s$, and $3p$. Figure 2.13 shows the order of filling through the third period.

Rule 2. Each orbital can hold up to two electrons with spins paired.

> **Example:** With four electrons, the $1s$ and $2s$ orbitals are filled and we write them as $1s^2 2s^2$. With an additional six electrons, the three $2p$ orbitals are filled and we write them either in the expanded form of $2p_x^2 2p_y^2 2p_z^2$, or in the condensed form $2p^6$. Spin pairing means that the electrons spin in opposite directions (Figure 2.14).

Rule 3. When there is a set of orbitals of equal energy, each orbital becomes half-filled before any of them becomes completely filled.

> **Example:** After the $1s$ and $2s$ orbitals are filled, a fifth electron is put into the $2p_x$ orbital, a sixth into the $2p_y$ orbital, and a seventh into the $2p_z$ orbital. Only after each $2p$ orbital has one electron is a second added to any $2p$ orbital.

D. Showing Electron Configurations: Orbital Box Diagrams

To illustrate how these rules are used, let us write ground-state electron configurations for several of the elements in periods 1, 2, and 3. In the following **orbital box diagrams,** we use a box to represent an orbital, an arrow with its head up to represent a single electron, and a pair of arrows with heads in opposite directions to represent two electrons with paired spins. In addition, we show both expanded and condensed electron configurations. Table 2.7 gives the complete condensed ground-state electron configurations for elements 1 through 18.

Hydrogen (H) The atomic number of hydrogen is 1, which means that it has a single electron. In the ground state, this electron is placed in the $1s$ orbital. Shown first is its orbital box diagram and then its electron configuration. The ground-state electron configuration of a hydrogen atom has one unpaired electron.

Electron configuration A description of the orbitals of an atom or ion occupied by electrons

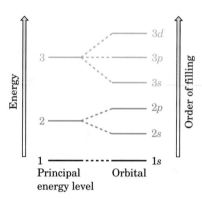

Figure 2.13 Energy levels for orbitals through the third shell.

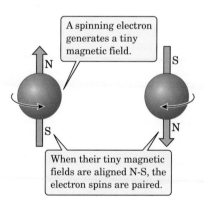

GOB
Chemistry ⚛ Now™
Active Figure 2.14 The pairing of electron spins. **See a simulation based on this figure, and take a short quiz on the concepts at http://www .now.brookscole.com/gob8 or on the CD.**

There is one electron in this orbital

H (1) [↑] Electron configuration: $1s^1$
      $1s$

Table 2.7 Ground-State Electron Configurations of the First 18 Elements

	Orbital Box Diagram									Electron Configuration (Condensed)	Noble Gas Notation
	1s	2s	2p$_x$	2p$_y$	2p$_z$	3s	3p$_x$	3p$_y$	3p$_z$		
H (1)	↑									$1s^1$	
He (2)	↑↓									$1s^2$	
Li (3)	↑↓	↑								$1s^2\,2s^1$	[He] $2s^1$
Be (4)	↑↓	↑↓								$1s^2\,2s^2$	[He] $2s^2$
B (5)	↑↓	↑↓	↑							$1s^2\,2s^2\,2p^1$	[He] $2s^2\,2p^1$
C (6)	↑↓	↑↓	↑	↑						$1s^2\,2s^2\,2p^2$	[He] $2s^2\,2p^2$
N (7)	↑↓	↑↓	↑	↑	↑					$1s^2\,2s^2\,2p^3$	[He] $2s^2\,2p^3$
O (8)	↑↓	↑↓	↑↓	↑	↑					$1s^2\,2s^2\,2p^4$	[He] $2s^2\,2p^4$
F (9)	↑↓	↑↓	↑↓	↑↓	↑					$1s^2\,2s^2\,2p^5$	[He] $2s^2\,2p^5$
Ne (10)	↑↓	↑↓	↑↓	↑↓	↑↓					$1s^2\,2s^2\,2p^6$	[He] $2s^2\,2p^6$
Na (11)	↑↓	↑↓	↑↓	↑↓	↑↓	↑				$1s^2\,2s^2\,2p^6\,3s^1$	[Ne] $3s^1$
Mg (12)	↑↓	↑↓	↑↓	↑↓	↑↓	↑↓				$1s^2\,2s^2\,2p^6\,3s^2$	[Ne] $3s^2$
Al (13)	↑↓	↑↓	↑↓	↑↓	↑↓	↑↓	↑			$1s^2\,2s^2\,2p^6\,3s^2\,3p^1$	[Ne] $3s^2\,3p^1$
Si (14)	↑↓	↑↓	↑↓	↑↓	↑↓	↑↓	↑	↑		$1s^2\,2s^2\,2p^6\,3s^2\,3p^2$	[Ne] $3s^2\,3p^2$
P (15)	↑↓	↑↓	↑↓	↑↓	↑↓	↑↓	↑	↑	↑	$1s^2\,2s^2\,2p^6\,3s^2\,3p^3$	[Ne] $3s^2\,3p^3$
S (16)	↑↓	↑↓	↑↓	↑↓	↑↓	↑↓	↑↓	↑	↑	$1s^2\,2s^2\,2p^6\,3s^2\,3p^4$	[Ne] $3s^2\,3p^4$
Cl (17)	↑↓	↑↓	↑↓	↑↓	↑↓	↑↓	↑↓	↑↓	↑	$1s^2\,2s^2\,2p^6\,3s^2\,3p^5$	[Ne] $3s^2\,3p^5$
Ar (18)	↑↓	↑↓	↑↓	↑↓	↑↓	↑↓	↑↓	↑↓	↑↓	$1s^2\,2s^2\,2p^6\,3s^2\,3p^6$	[Ne] $3s^2\,3p^6$

Helium (He) The atomic number of helium is 2, which means that it has two electrons. In the ground state, both electrons are placed in the 1s orbital with paired spins, which fills the 1s orbital. All electrons in helium are paired.

This orbital is now filled with two electrons

He (2) ↑↓   Electron configuration: $1s^2$
1s

Lithium (Li) Lithium, atomic number 3, has three electrons. In the ground state, two electrons are placed in the 1s orbital with paired spins, and the third electron is placed in the 2s orbital. The ground-state electron configuration of a lithium atom has one unpaired electron.

Li has one unpaired electron

Li (3) $1s$ $2s$ Electron configuration: $1s^22s^1$

Carbon (C)   Carbon, atomic number 6, has six electrons. Two are placed in the $1s$ orbital with paired spins and two are placed in the $2s$ orbital with paired spins. The fifth and sixth electrons are placed one each in the $2p_x$ and $2p_y$ orbitals. The ground-state electron configuration of carbon has two unpaired electrons.

All orbitals of equal energy have at least one electron before any of them is filled

In a condensed electron configuration, orbitals of equal energy are grouped together

C (6)

$1s$ $2s$ $2p_x$ $2p_y$ $2p_z$

Electron configuration
Expanded: $1s^22s^22p_x^{\,1}2p_y^{\,1}$
Condensed: $1s^22s^22p^2$

Oxygen (O)   Oxygen, atomic number 8, has eight electrons. The first four electrons fill the $1s$ and $2s$ orbitals. The next three electrons are placed in the $2p_x$, $2p_y$, and $2p_z$ orbitals so that each $2p$ orbital has one electron. The remaining electron now fills the $2p_x$ orbital. Oxygen has two unpaired electrons.

Oxygen has two unpaired electrons

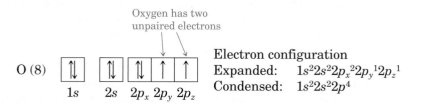

O (8)

$1s$ $2s$ $2p_x$ $2p_y$ $2p_z$

Electron configuration
Expanded: $1s^22s^22p_x^{\,2}2p_y^{\,1}2p_z^{\,1}$
Condensed: $1s^22s^22p^4$

Neon (Ne)   Neon, atomic number 10, has ten electrons, which completely fill all orbitals of the first and second shells. Neon has no unpaired electrons.

Ne (10)

$1s$ $2s$ $2p_x$ $2p_y$ $2p_z$

Electron configuration
Expanded: $1s^22s^22p_x^{\,2}2p_y^{\,2}2p_z^{\,2}$
Condensed: $1s^22s^22p^6$

Sodium (Na)   Sodium, atomic number 11, has 11 electrons. The first 10 fill the $1s$, $2s$, and $2p$ orbitals. The 11th electron is placed in the $3s$ orbital. Sodium has one unpaired electron.

Na (11)

$1s$ $2s$ $2p_x$ $2p_y$ $2p_z$ $3s$

Electron configuration
Expanded: $1s^22s^22p_x^{\,2}2p_y^{\,2}2p_z^{\,2}3s^1$
Condensed: $1s^22s^22p^63s^1$

Phosphorus (P)   Phosphorus, atomic number 15, has 15 valence electrons. The first 12 fill the $1s$, $2s$, $2p$, and $3s$ orbitals. Electrons 13, 14, and 15 are placed one each in the $3p_x$, $3p_y$, and $3p_z$ orbitals. Phosphorus has three unpaired electrons.

P (15) $1s$ $2s$ $2p_x$ $2p_y$ $2p_z$ $3s$ $3p_x$ $3p_y$ $3p_z$

Electron configuration
Expanded: $1s^2 2s^2 2p_x^2 2p_y^2 2p_z^2 3s^2 3p_x^1 3p_y^1 3p_z^1$
Condensed: $1s^2 2s^2 2p^6 3s^2 3p^3$

E. Showing Electron Configurations: Noble Gas Notations

In an alternative way of writing electron configurations, we use the symbol of the noble gas immediately preceding the particular atom to indicate the electron configuration of all filled shells. The first shell of lithium, for example, is abbreviated [He] and the single electron in its $2s$ shell is indicated by $2s^1$. Thus the electron configuration of lithium is [He]$2s^1$.

F. Showing Electron Configurations: Lewis Dot Structures

Valence electron An electron in the outermost occupied (valence) shell of an atom

Valence shell The outermost occupied shell of an atom

Lewis dot structure The symbol of the element surrounded by a number of dots equal to the number of electrons in the valence shell of an atom of that element

When discussing the physical and chemical properties of an element, chemists often focus on the outermost shell of its electrons because electrons in this shell are the ones involved in the formation of chemical bonds (Chapter 4) and in chemical reactions (Chapter 5). Outer-shell electrons are called **valence electrons,** and the energy level in which they are found is called the **valence shell.** Carbon, for example, with a ground-state electron configuration of $1s^2 2s^2 2p^2$, has four valence (outer-shell) electrons.

To show the outermost electrons of an atom, we commonly use a representation called a **Lewis dot structure,** named after the American chemist Gilbert N. Lewis (1875–1946), who devised this notation. A Lewis structure shows the symbol of the element surrounded by a number of dots equal to the number of electrons in the outer shell of an atom of that element. In a Lewis structure, the atomic symbol represents the nucleus and all filled inner shells. Table 2.8 shows Lewis structures for the first 18 elements of the Periodic Table.

The noble gases helium and neon have filled valence shells. The valence shell of helium is filled with two electrons; that of neon is filled with eight electrons (Table 2.7). Neon and argon have in common an electron configuration in which the s and p orbitals of their valence shells are filled with eight electrons. The valence shells of all other elements shown in Table 2.8 contain fewer than eight electrons.

At this point, let us compare the Lewis structures given in Table 2.8 with the ground-state electron configurations given in Table 2.7. The Lewis structure of boron (B), for example, is shown in Table 2.8 with three valence

Table 2.8   Lewis Dot Structures for Elements 1–18 of the Periodic Table

1A	2A	3A	4A	5A	6A	7A	8A
H·							He:
Li·	Be:	B:	·C:	·N:	:O:	:F:	:Ne:
Na·	Mg:	Al:	·Si:	·P:	:S:	:Cl:	:Ar:

Each dot represents one valence electron.

electrons; these are the paired $2s$ electrons and the single $2p_x$ electron shown in Table 2.7. The Lewis structure of carbon (C) is shown in Table 2.8 with four valence electrons; these are the two paired $2s$ electrons and the unpaired $2p_x$ and $2p_y$ electrons shown in Table 2.7.

EXAMPLE 2.7

The Lewis dot structure for nitrogen shows five valence electrons. Write the expanded electron configuration for nitrogen and show to which orbitals its five valence electrons are assigned.

Solution

Nitrogen, atomic number 7, has the following ground-state electron configuration:

$$1s^2 2s^2 2p_x^{\,1} 2p_y^{\,1} 2p_z^{\,1}$$

The five valence electrons of the Lewis dot structure are the two paired electrons in the $2s$ orbital and the three unpaired electrons in the $2p_x$, $2p_y$, and $2p_z$ orbitals.

Problem 2.7

Write the Lewis dot structure for the element that has the following ground-state electron configuration. What is the name of this element?

$$1s^2 2s^2 2p_x^{\,2} 2p_y^{\,2} 2p_z^{\,2} 3s^2 3p_x^{\,1}$$

2.7 | How Are Electron Configuration and Position in the Periodic Table Related?

When Mendeleev published his first Periodic Table in 1869, he could not explain why it worked—that is, why elements with similar properties became aligned in the same column. Indeed, no one had a good explanation for this phenomenon. It was not until the discovery of electron configurations that chemists finally understood why the Periodic Table works. The answer is very simple: Elements in the same column have the same configuration of electrons in their outer shells. Figure 2.15 shows the relationship between shells (principal energy levels) and orbitals being filled.

All main-group elements have in common the fact that either their s or p orbitals are being filled. Notice that the $1s$ shell is filled with two electrons; there are only two elements in the first period. The $2s$ and $2p$ orbitals are filled with eight electrons; there are eight elements in period 2. Similarly, the $3s$ and $3p$ orbitals are filled with eight electrons; there are eight elements in period 3.

To create the elements of period 4, one $4s$, three $4p$, and five $3d$ orbitals are available. These orbitals can hold a total of 18 electrons; there are 18 elements in period 4. Similarly, there are 18 elements in period 5. Inner transition elements are created by filling f orbitals, which come in sets of seven and can hold a total of 14 electrons; there are 14 inner transition elements in the lanthanide series and 14 in the actinide series.

GOB
Chemistry•♦•Now™

Click *Chemistry Interactive* to see an example of **Electron Configurations and Periodic Blocks**

Figure 2.15 Electron configuration and the Periodic Table.

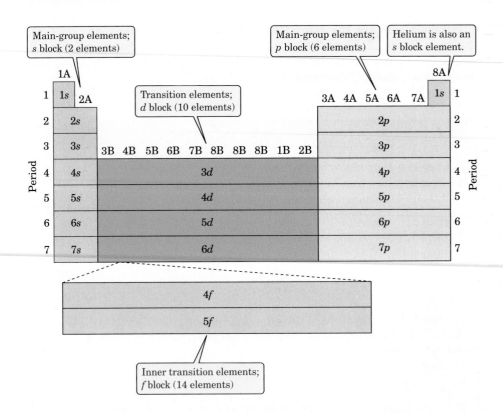

To see the similarities in electron configurations within the Periodic Table, let us look at the elements in column 1A. We already know the configurations for lithium, sodium, and potassium (Table 2.7). To this list we can add rubidium and cesium. All elements in column 1A have one electron in their valence shell (Table 2.9).

All Group 1A elements are metals, with the exception of hydrogen, which is a nonmetal. The properties of elements largely depend on the electron configuration of their outer shell. As a consequence, it is not surprising that Group 1A elements, all of which have the same outer-shell configuration, are metals (except for hydrogen) and have such similar physical and chemical properties.

Table 2.9 Noble Gas Notation and Lewis Dot Structures for the Alkali Metals (Group 1A Elements)

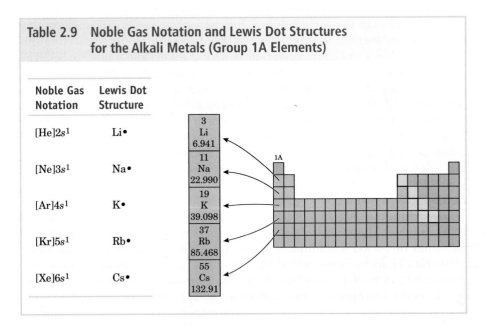

Noble Gas Notation	Lewis Dot Structure
$[He]2s^1$	Li•
$[Ne]3s^1$	Na•
$[Ar]4s^1$	K•
$[Kr]5s^1$	Rb•
$[Xe]6s^1$	Cs•

2.8 | What Are Periodic Properties?

As we have now seen, the Periodic Table was constructed on the basis of trends (periodicity) in chemical properties. With an understanding of electron configurations, chemists realized that the periodicity in chemical properties could be understood in terms of periodicity in electron configuration. As we noted in the opening of Section 2.7 "The Periodic Table works because elements in the same column have the same configuration of electrons in their outer shells." Thus, chemists could now explain why certain chemical and physical properties of elements changed in predictable ways in going down a column or going across a row of the Periodic Table. In this section, we will concentrate on the periodicity of one physical property (atomic size), and one chemical property (ionization energy) to illustrate how periodicity is related to position in the Periodic Table.

A. Atomic Size

The size of an atom is determined by the size of its outermost occupied orbital. The size of a sodium atom, for example, is the size of its singly-occupied $3s$ orbital. The size of a chlorine atom is determined by the size of its three $3p$ orbitals. The simplest way to determine the size of an atom is to determine the distance between atoms in a sample of the element. A chlorine molecule, for example, has a diameter of 198 pm (pm = picometer; 1 pm = 10^{-12} meter). The radius of a chlorine atom is thus 99 pm, which is one-half of the distance between the two chlorine atoms in Cl_2.

Similarly, the distance between carbon atoms in diamond is 154 pm, and so the radius of a carbon atom is 77 pm.

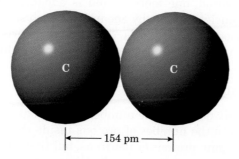

From measurements such as these we can assemble a set of atomic radii (Figure 2.16). From the information in this figure, we can see that for main group elements, (1) atomic radii increase in going down a group and (2) decrease in going across a period. Let us examine the correlation between each of these trends and electron configuration.

Figure 2.16 Atomic radii of main-group elements (in picometers, 1 pm = 10^{-12} m).

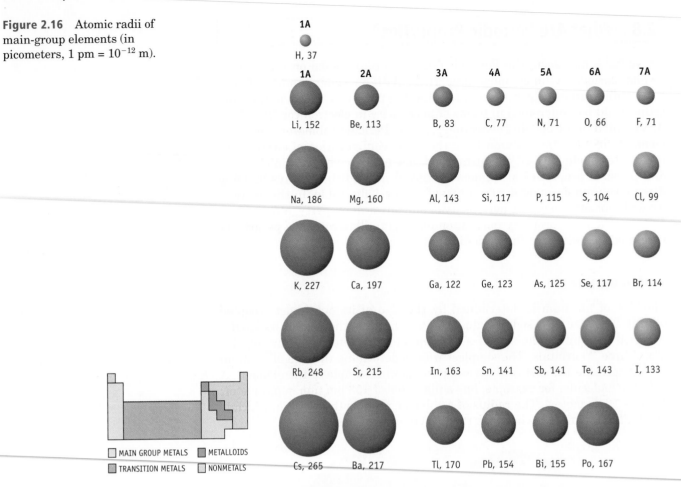

1. The size of an atom is determined by its outermost electrons. In going down a column, the outermost electrons are assigned to higher and higher principal energy levels. The electrons of lower principal energy levels (those lying below the valence shell) must occupy some space, so the outer-shell electrons must be farther and farther from the nucleus, and therefore, the increase in size in going down a column is rationalized.

2. For elements in the same period, the principal energy level remains the same (for example, the valence electrons of all period 2 elements occupy the second principal energy level). But in going from one element to the next across a period, one more proton is added to the nucleus, thus increasing the nuclear charge by one unit for each step from left to right across the period. The result is the nucleus exerts an increasingly stronger pull on the valence electrons and atomic radii decreases.

B. Ionization Energy

Atoms are electrically neutral—the number of electrons outside the nucleus of an atom is equal to the number of protons inside the nucleus. Atoms do not normally lose or gain protons or neutrons, but they can lose or gain electrons. When a lithium atom, for example, loses one electron, it becomes a lithium **ion.** A lithium atom has three protons in its nucleus and three electrons outside the nucleus. When a lithium atom loses one of these electrons, it still has three protons in its nucleus (and, therefore, is still lithium), but now it has only two electrons outside the nucleus. The charge of two of the protons is cancelled by the two remaining electrons, but there is no third

Ion An atom with an unequal number of protons and electrons electron to cancel the charge of the third proton. Therefore, a lithium ion has a charge of +1 and we write it as Li^+. The ionization energy for a lithium atom in the gas phase is 0.52 kJ/mol.

$$Li + energy \longrightarrow Li^+ + e^-$$
Lithium   Ionization   Lithium   Electron
          energy   ion

Ionization energy is a measure of how difficult it is to remove the most loosely held electron from an atom in the gaseous state. The more difficult it is to remove the electron, the higher the ionization energy required to remove an electron from an atom. Ionization energies are always positive because energy must be supplied to overcome the attractive force between the electron and the positively charged nucleus. Figure 2.17 shows the ionization energies for the atoms of main-group elements 1 through 37 (hydrogen through rubidium).

As we see in this figure, ionization energy generally increases as we go up a column of the Periodic Table and, with a few exceptions, generally increases as we go from left to right across a row. For example, within the Group 1A metals, rubidium gives up its 5s electron most easily and lithium gives up its 2s electron least easily.

We explain this trend by saying that the 5s electron of rubidium is farther from the positively charged nucleus than is the 4s electron in potassium, which in turn is farther from the positively charged nucleus than is the 3s electron of sodium, and so forth. Furthermore, the 5s electron of rubidium is more "shielded" by inner-shell electrons from the attractive force of the positive nucleus than is the 4s electron of potassium, and so forth. The greater the shielding, the lower the ionization energy. Thus, going down a column of the Periodic Table, the shielding of an atom's outermost electrons increases, and the elements' ionization energies decrease.

We explain the increase in ionization energy across a row by the fact that the valence electrons across a row are in the same shell (principal energy level) and, therefore, are approximately the same distances from the nucleus. The number of protons in the nucleus, however, increases regularly across a row, which means that valence electrons experience an increasingly stronger pull by the nucleus, which makes them more difficult to remove. Thus ionization energy increases from left to right across a row of the periodic table.

Ionization energy The energy required to remove the most loosely held electron from an atom in the gas phase

Ionization energy

Figure 2.17 Ionization energy versus atomic number for elements 1–37.

SUMMARY OF KEY QUESTIONS

SECTION 2.1 What Is Matter Made Of?

- The Greek philosopher Democritus (circa 460–370 BCE) was the first person to propose an atomic theory of matter. He stated that all matter is made of very tiny particles, which he called atoms.

SECTION 2.2 How Do We Classify Matter?

- We classify matter as **elements, compounds, or mixtures.**

SECTION 2.3 What Are the Postulates of Dalton's Atomic Theory?

- (1) All matter is made up of atoms; (2) all atoms of a given element are identical and the atoms of any one element are different from those of any other element; (3) compounds are formed by the chemical combination of two or more different kinds of atoms; and (4) a molecule is a cluster of two or more atoms that acts as a single unit.
- Dalton's theory is based on the **law of conservation of mass** (matter can be neither created nor destroyed) and the **law of constant composition** (any compound is always made up of elements in the same proportion by mass).

SECTION 2.4 What Are Atoms Made Of?

- Atoms consist of protons and neutrons found inside the nucleus and electrons located outside it. An **electron** has a mass of approximately 0.0005 amu and a charge of −1. A **proton** has a mass of approximately 1 amu and a charge of +1. A **neutron** has a mass of approximately 1 amu and no charge.
- The **mass number** of an atom is the sum of the number of its protons and neutrons.
- The **atomic number** of an element is the number of protons in its nucleus.
- **Isotopes** are atoms with the same atomic number but different mass numbers; that is, they have the same number of protons in their nuclei but different numbers of neutrons.
- The **atomic weight** of an element is a weighted average of the masses (in amu) of the isotopes as they occur in nature.
- Atoms are very tiny, with a very small mass, almost all of which is concentrated in the nucleus. The nucleus is tiny, with an extremely high density.

SECTION 2.5 What Is the Periodic Table?

- The **Periodic Table** is an arrangement of elements into columns with similar properties, which gradually change as we move down a column.
- **Metals** are solids (except for mercury, which is a liquid), shiny, conductors of electricity, ductile, malleable, and form alloys, which are solutions of one or more metals dissolved in another metal. In their chemical reactions, metals tend to give up electrons.
- With the exception of hydrogen, the 18 **nonmetals** appear on the right side of the Periodic Table. With the exception of graphite, they do not conduct electricity. In their chemical reactions, nonmetals tend to accept electrons.
- Six elements are classified as **metalloids:** boron, silicon, germanium, arsenic, antimony, and tellurium. These elements have some properties of metals and some properties of nonmetals.

SECTION 2.6 How Are the Electrons in an Atom Arranged?

- Electrons in atoms exist in **principal energy levels** or **shells.**
- All principal energy levels except the first are divided into **subshells** designated by the letters s, p, d, and f. Within each subshell, electrons are grouped into **orbitals.** An orbital is a region of space that can hold two electrons with paired spins. All s orbitals are spherical and can hold two electrons. All p orbitals come in sets of three, and each is shaped like a dumbbell. A set of three p orbitals can hold six electrons.
- (1) Orbitals fill in order of increasing energy; (2) each orbital can hold a maximum of two electrons with paired spins; and (3) when filling orbitals of equivalent energy, each orbital adds one electron before any orbital adds a second electron.
- The electron configuration of an atom may be shown by an orbital notation, an orbital box diagram, or a noble gas notation.
- Electrons in the outermost or **valence shell** of an atom are called **valence electrons.** In a **Lewis dot structure** of an atom, the symbol of the element is surrounded by a number of dots equal to the number of its valence electrons.

SECTION 2.7 How Are Electron Configuration and Position in the Periodic Table Related?

- The Periodic Table works because elements in the same column have the same outer-shell electron configuration.

SECTION 2.8 What Are Periodic Properties?

- The **size of an atom (atomic radius)** is determined by the size of its outermost occupied orbital. Atomic size is a periodic property. For main group elements, (1) atomic size increases in going down a Group, and (2) decreases in going across a Period. In going down a column, the outermost electrons are assigned to higher and higher principal energy levels. For elements in the same Period, the principal energy level remains the same but from one element to the next, the nuclear charge increases by one unit (by one proton). As a result of this increase in nuclear charge across a Period, the nucleus exerts an increasingly stronger pull on the valence electrons and atomic size decreases.

- **Ionization energy** is the energy necessary to remove the most loosely held electron from an atom in the gas phase to form an **ion.** Ionization energy

increases from bottom to top within a column of the Periodic Table because the valence shell of the atom becomes closer to the positively charged nucleus. It increases from left to right within a row because the positive charge on the nucleus increases in this direction.

PROBLEMS

GOB
Chemistry ⚛ Now™

Assess your understanding of this chapter's topics with additional quizzing and conceptual-based problems at **http://now.brookscole.com/gob8** or on the CD.

A blue problem number indicates an applied problem.

■ denotes problems that are available on the GOB ChemistryNow website or CD and are assignable in OWL.

SECTION 2.1 What Is Matter Made Of?

2.8   In what way(s) was Democritus's atomic theory similar to Dalton's atomic theory?

SECTION 2.2 How Do We Classify Matter?

2.9   Classify each of the following as an element, a compound, or a mixture:

(a) Oxygen   (b) Table salt
(c) Sea water   (d) Wine
(e) Air   (f) Silver
(g) Diamond   (h) A pebble
(i) Gasoline   (j) Milk
(k) Carbon dioxide (l) Bronze

2.10   ■ Name these elements (try not to look at a Periodic Table):

(a) O   (b) Pb   (c) Ca   (d) Na
(e) C   (f) Ti   (g) S   (h) Fe
(i) H   (j) K   (k) Ag   (l) Au

2.11   The elements game, Part I. Name and give the symbol of the element that is named for each person.

(a) Niels Bohr (1885–1962), Nobel Prize for physics in 1922

(b) Pierre and Marie Curie, Nobel Prize for chemistry in 1903

(c) Albert Einstein (1879–1955), Nobel Prize for physics in 1921

(d) Enrico Fermi (1901–1954), Nobel Prize for physics in 1938

(e) Ernest Lawrence (1901–1958), Nobel Prize for physics in 1939

(f) Lisa Meitner (1868–1968), codiscoverer of nuclear fission

(g) Dmitri Mendeleev (1834–1907), first person to formulate a workable Periodic Table

(h) Alfred Nobel (1833–1896), discoverer of dynamite

(i) Ernest Rutherford (1871–1937), Nobel Prize for chemistry in 1908

(j) Glen Seaborg (1912–1999), Nobel Prize for chemistry in 1951

2.12   The elements game, Part II. Name and give the symbol of the element that is named for each geographic location.

(a) The Americas

(b) Berkeley, California

(c) The state and University of California

(d) Dubna, location in Russia of the Joint Institute of Nuclear Research

(e) Europe

(f) France

(g) Gallia, Latin name for ancient France

(h) Germany

(i) Hafnia, Latin name for ancient Copenhagen

(j) Hesse, a German state

(k) Holmia, Latin name for ancient Stockholm

(l) Lutetia, Latin name for ancient Paris

(m) Magnesia, a district in Thessaly

(n) Poland, the native country of Marie Curie

(o) Rhenus, Latin name for the river Rhine

(p) Ruthenia, Latin name for ancient Russia

(q) Scandia, Latin name for ancient Scandinavia

(r) Strontian, a town in Scotland

(s) Ytterby, a village in Sweden (three elements)

(t) Thule, the earliest name for Scandinavia

2.13   The elements game, Part III. Give the names and symbols for the four elements named for planets.

2.14   ■ Write the formulas of compounds in which the combining ratios are as follows:

(a) Potassium : oxygen, 2 : 1

(b) Sodium : phosphorus : oxygen, 3 : 1 : 4

(c) Lithium : nitrogen : oxygen, 1 : 1 : 3

2.15   Write the formulas of compounds in which the combining ratios are as follows:

(a) Sodium : hydrogen : carbon : oxygen, 1 : 1 : 1 : 3

(b) Carbon : hydrogen : oxygen, 2 : 6 : 1

(c) Potassium : manganese : oxygen, 1 : 1 : 4

SECTION 2.3 What Are Postulates of Dalton's Atomic Theory?

2.16 How does Dalton's atomic theory explain the following:

(a) The law of conservation of mass?

(b) The law of constant composition?

2.17 When 2.16 g of mercury oxide is heated, it decomposes to yield 2.00 g of mercury and 0.16 g of oxygen. Which law is supported by this experiment?

2.18 The compound carbon monoxide contains 42.9% carbon and 57.1% oxygen. The compound carbon dioxide contains 27.3% carbon and 72.7% oxygen. Does this disprove Proust's law of constant composition?

2.19 ■ Calculate the percentage of hydrogen and oxygen in water, H_2O, and hydrogen peroxide, H_2O_2.

SECTION 2.4 What Are Atoms Made Of?

2.20 Where in an atom are these subatomic particles located?

(a) Protons   (b) Electrons   (c) Neutrons

2.21 It has been said that "the number of protons determines the identity of the element." Do you agree or disagree with this statement? Explain.

2.22 What is the mass number of an atom with

(a) 22 protons, 22 electrons, and 26 neutrons?

(b) 76 protons, 76 electrons, and 114 neutrons?

(c) 34 protons, 34 electrons, and 45 neutrons?

(d) 94 protons, 94 electrons, and 150 neutrons?

2.23 Name and give the symbol for each element in Problem 2.22.

2.24 Given the mass number and the number of neutrons, what is the name and symbol of each element?

(a) Mass number 45; 24 neutrons

(b) Mass number 48; 26 neutrons

(c) Mass number 107; 60 neutrons

(d) Mass number 246; 156 neutrons

(e) Mass number 36; 18 neutrons

2.25 If each of the atoms in Problem 2.24 acquired two more neutrons, what element would each then be?

2.26 How many neutrons are there in

(a) A carbon atom of mass number 13?

(b) A germanium atom of mass number 73?

(c) An osmium atom of mass number 188?

(d) A platinum atom of mass number 195?

2.27 ■ How many protons and how many neutrons does each of these isotopes of radon contain?

(a) Rn-210   (b) Rn-218   (c) Rn-222

2.28 ■ How many neutrons and protons are there in each isotope?

(a) ^{22}Ne (b) ^{104}Pd

(c) ^{35}Cl (d) Tellurium-128

(e) Lithium-7   (f) Uranium-238

2.29 Tin-118 is one of the isotopes of tin. Write the symbols for the isotopes of tin that contain two, three, and six more neutrons than tin-118.

2.30 What is the difference between atomic number and mass number?

2.31 Define:

(a) Ion   (b) Isotope

2.32 ■ There are only two naturally occurring isotopes of antimony: ^{121}Sb (120.90 amu) and ^{123}Sb (122.90 amu). The atomic weight of antimony given in the Periodic Table is 121.75. Which of its two naturally occurring isotopes has the greater natural abundance?

2.33 The two most abundant naturally occurring isotopes of carbon are carbon-12 (98.90%, 12.000 amu) and carbon-13 (1.10%, 13.003 amu). From these abundances, calculate the atomic weight of carbon and compare your calculated value with that given in the Periodic Table.

2.34 Another isotope of carbon, carbon-14, occurs in nature but in such small amounts relative to carbon-12 and carbon-13 that it does not contribute in a significant way to the atomic weight of carbon as recorded in the Periodic Table. Carbon-14 is invaluable in the science of radiocarbon dating (see Chemical Connections 3A). Give the number of protons, neutrons, and electrons in an atom of carbon-14.

2.35 The isotope carbon-11 does not occur in nature but has been made in the laboratory. This isotope is used in a medical imaging technique called positron emission tomography (PET, see Section 3.7A). Give the number of protons, neutrons, and electrons in an atom of carbon-11.

2.36 Other isotopes used in PET therapy are fluorine-18, nitrogen-13, and oxygen-15. None of these isotopes occurs in nature; all must be produced in the laboratory. Give the number of protons, neutrons, and electrons in an atom of each of these artificial isotopes.

2.37 Americium-241 is used in household smoke detectors. This element has 11 known isotopes, none of which occurs in nature, but must be made in the laboratory. Give the number of protons, neutrons, and electrons in an atom of americium-241.

2.38 In dating geological samples, scientists compare the ratio of rubidium-87 to strontium-87. Give the number of protons, neutrons, and electrons in an atom of each element.

SECTION 2.5 What Is the Periodic Table?

2.39 How many metals, metalloids, and nonmetals are there in the third period of the Periodic Table?

2.40 Which group(s) of the Periodic Table contain

(a) Only metals?   (b) Only metalloids?

(c) Only nonmetals?

2.41 Which period(s) in the Periodic Table contain more nonmetals than metals? Which contain more metals than nonmetals?

2.42 Using only the Periodic Table, group the following elements according to similar properties: As, I, Ne, F, Mg, K, Ca, Ba, Li, He, N, P.

2.43 Which of the following are transition elements?
(a) Pd   (b) K   (c) Co
(d) Ce   (e) Br   (f) Cr

2.44 Which element in each pair is more metallic?
(a) Silicon or aluminum
(b) Arsenic or phosphorus
(c) Gallium or germanium
(d) Gallium or aluminum

2.45 ■ Classify these elements as metals, nonmetals, or metalloids:
(a) Argon   (b) Boron   (c) Lead
(d) Arsenic   (e) Potassium   (f) Silicon
(g) Iodine   (h) Antimony   (i) Vanadium
(j) Sulfur   (k) Nitrogen

SECTION 2.6 How Are the Electrons in an Atom Arranged?

2.46 How many periods of the Periodic Table have two elements? How many have eight elements? How many have 18 elements? How many have 32 elements?

2.47 What is the correlation between the group number of the main-group elements (1A, 2A, . . . , 8A) and the number of valence electrons in an element in the group?

2.48 Given your answer to Problem 2.47, write the Lewis dot structures for each of the following elements using no information other than the number of the group in the Periodic Table to which the element belongs.
(a) Carbon (4A)   (b) Silicon (4A)
(c) Oxygen (6A)   (d) Sulfur (6A)
(e) Aluminum (3A)   (f) Bromine (7A)

2.49 ■ Write the condensed ground-state electron configuration for each of the following elements. The element's atomic number is given in parentheses.
(a) Li (3)   (b) Ne (10)   (c) Be (4)
(d) C (6)   (e) Mg (12)

2.50 Write the Lewis dot structure for each element in Problem 2.49.

2.51 ■ Write the condensed ground-state electron configuration for each of the following elements. The element's atomic number is given in parentheses.
(a) He (2)   (b) Na (11)   (c) Cl (17)
(d) P (15)   (e) H (1)

2.52 Write the Lewis dot structure for each element in Problem 2.51.

2.53 What is the same and what is different in the electron configurations of
(a) Na and Cs?   (b) O and Te?
(c) C and Ge?

2.54 Silicon, atomic number 14, is in Group 4A. How many orbitals are occupied by the valence electrons of Si in its ground state?

2.55 You are presented with a Lewis dot structure of element X as X:. To which two groups in the Periodic Table might this element belong?

2.56 The electron configurations for the elements with atomic numbers higher than 36 follow the same rules as given in the text for the first 36 elements. In fact, you can arrive at the correct order of filling of orbitals from Figure 2.15 by starting with H and reading the orbitals from left to right across the first row, then the second row, and so on. Write the condensed ground-state electron configuration for
(a) Rb   (b) Sr   (c) Br

SECTION 2.7 How Are Electron Configuration and Position in the Periodic Table Related?

2.57 Why do the elements in column 1A of the Periodic Table (the alkali metals) have similar but not identical properties?

SECTION 2.8 What Are Periodic Properties?

2.58 Account for the following observations.
(a) The atomic radius of an anion is always larger than the atoms from which it is derived. Examples: Cl 99 pm and Cl$^-$ 181 pm; O 73 pm and O^{2-} 140 pm.
(b) The atomic radius of a cation is always smaller than that of the atom from which it is derived. Examples: Li 152 pm and Li$^+$ 76 pm; Na 156 pm and Na$^+$ 98 pm.

2.59 ■ Using only the Periodic Table, arrange the elements in each set in order of increasing ionization energy:
(a) Li, Na, K   (b) C, N, Ne
(c) O, C, F   (d) Br, Cl, F

2.60 Account for the fact that the first ionization energy of oxygen is less than that of nitrogen.

2.61 Every atom except hydrogen has a series of ionization energies because they have more than one electron that can be removed. Following are the first three ionization energies for magnesium.

$$Mg(g) \longrightarrow Mg^+(g) + e^-(g) \qquad IE_1 = 738 \text{ kJ/mol}$$

$$Mg^+(g) \longrightarrow Mg^{2+}(g) + e^-(g) \qquad IE_2 = 1450 \text{ kJ/mol}$$

$$Mg^{2+}(g) \longrightarrow Mg^{3+}(g) + e^-(g) \qquad IE_3 = 7734 \text{ kJ/mol}$$

(a) Write the ground-state electron configuration for Mg, Mg$^+$, Mg^{2+}, and Mg^{3+}.
(b) Account for the large increase in ionization energy for the removal of the third electron compared with the ionization energies for removal of the first and second electrons.

Chemical Connections

2.62   (Chemical Connections 2A) Why does the body need sulfur, calcium, and iron?

2.63   (Chemical Connections 2B) Which are the two most abundant elements, by weight, in

(a) The Earth's crust?   (b) The human body?

2.64   (Chemical Connections 2C) Consider the isotopic abundance of hydrogen on Mars. Would the atomic weight of hydrogen on Mars be greater than, the same as, or smaller than that on Earth?

2.65   (Chemical Connections 2D) Why is strontium-90 more dangerous to humans than most other radioactive isotopes that were present in the Chernobyl fallout?

2.66   (Chemical Connections 2E) Bronze is an alloy of which two metals?

2.67   (Chemical Connections 2E) Copper is a soft metal. How can one make it harder?

Additional Problems

2.68   Give the designations of all subshells in the

(a) 1 shell   (b) 2 shell

(c) 3 shell   (d) 4 shell

2.69   Tell whether metals or nonmetals are more likely to have each of the following characteristics:

(a) Conduct electricity and heat

(b) Accept electrons

(c) Be malleable

(d) Be a gas at room temperature

(e) Be a transition element

(f) Lose electrons

2.70   What is the outer-shell electron configuration of the elements in

(a) Group 3A?   (b) Group 7A?

(c) Group 5A?

2.71   ■ Determine the number of protons, electrons, and neutrons present in

(a) ^{32}P (b) ^{98}Mo (c) ^{44}Ca

(d) ^{3}H (e) ^{158}Gd (f) ^{212}Bi

2.72   What percentage of the mass of each element is contributed by neutrons?

(a) Carbon-12   (b) Calcium-40

(c) Iron-55   (d) Bromine-79

(e) Platinum-195   (f) Uranium-238

2.73   Do isotopes of the heavy elements (for example, those from atomic number 37 to 53) contain more, the same, or fewer neutrons than protons?

2.74   ■ What is the symbol for each of the following elements? (Try not to look at a Periodic Table.)

(a) Phosphorus   (b) Potassium

(c) Sodium   (d) Nitrogen

(e) Bromine   (f) Silver

(g) Calcium   (h) Carbon

(i) Tin   (j) Zinc

2.75   The natural abundance of boron isotopes is as follows: 19.9% boron-10 (10.013 amu) and 80.1% boron-11 (11.009 amu). Calculate the atomic weight of boron (watch the significant figures) and compare your calculated value with that given in the Periodic Table.

2.76   How many electrons are in the outer shell of each of the following elements?

(a) Si   (b) Br   (c) P

(d) K   (e) He   (f) Ca

(g) Kr   (h) Pb   (i) Se

(j) O

2.77   The mass of a proton is 1.67×10^{-24} g. The mass of a grain of salt is 1.0×10^{-2} g. How many protons would it take to have the same mass as a grain of salt?

2.78   (a) What are the charges of an electron, a proton, and a neutron?

(b) What are the masses (in amu, to one significant figure) of an electron, a proton, and a neutron?

2.79   ■ Rubidium has two natural isotopes: rubidium-85 and rubidium-87. What is the natural abundance of each isotope if the atomic weight of rubidium is 85.47?

2.80   What is the name of this element and how many protons and neutrons does it have in its nucleus: $^{131}_{54}X$?

2.81   Based on the data presented in Figure 2.16, which atom would have the highest ionization energy: I, Cs, Sn, or Xe?

2.82   Assume that a new element has been discovered with atomic number 117. Its chemical properties should be similar to those of At. Predict whether the new element's ionization energy will be greater than, the same as, or smaller than that of (a) At and (b) Ra.

Looking Ahead

2.83   Suppose that you face a problem similar to Mendeleev: You must predict the properties of an element not yet discovered. What will element 118 be like if and when enough of it is made for chemists to study its physical and chemical properties?

2.84   Compare the neutron to proton ratio for heavier and lighter elements. Does the value of this ratio generally increase, decrease, or remain the same as atomic number increases?

Nuclear Chemistry

The Sun's energy is the result of nuclear fusion.

3.1 | How Was Radioactivity Discovered?

Every so often, a scientist makes the kind of discovery that changes the future of the world in some significant way. In 1896 a French physicist, Henri Becquerel (1852–1908), made one of these discoveries. At the time, Becquerel was engaged in a study of phosphorescent materials. In his experiments, he exposed certain salts—among them uranium salts—to sunlight for several hours, whereupon they phosphoresced. He then placed the glowing salts on a photographic plate that had been wrapped in opaque paper. Becquerel observed that, by placing a coin or a metal cutout between the phosphorescing salts and the covered plate, he could create photo-graphic images of the coin or metal cutout. He concluded that besides emit-ting visible light, the phosphorescent materials must be emitting something akin to Xrays, which William Röntgen had discovered just the previous year. What was even more surprising to Becquerel was that his uranium salts continued to emit this same type of penetrating radiation long after their phosphorescence has ceased. What he had discovered was a type of radiation that Marie Curie was to call radioactivity. For this discovery, Bec-querel shared the 1903 Nobel Prize for physics with Pierre and Marie Curie.

In this chapter, we will study the major types of radioactivity, their ori-gin in the nucleus, the uses of radioactivity in the health and biological sci-ences, and its use as a source of power and energy.

3.2 | What Is Radioactivity?

GOB
Chemistry ❄ Now™
Click *Coached Problems* to see **Evidence of Subatomic Particles**

Early experiments identified three kinds of radiation, which were named alpha (α), beta (β), and gamma (γ) rays after the first three letters of the Greek alphabet. Each type of radiation behaves differently when it passes between electrically charged plates. When a radioactive material is placed in a lead container that has a small opening, the emitted radiation passes through the opening and then between charged plates (Figure 3.1). One ray (β) is deflected toward the positive plate, indicating that it consists of negatively charged particles. A second ray (α) is deflected toward the nega-tive plate, indicating that it consists of positively charged particles. A third

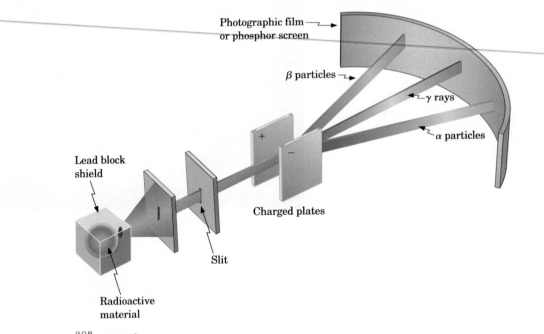

GOB
Chemistry ❄ Now™

Active Figure 3.1 Electricity and radioactivity. Positively charged alpha (α) particles are attracted to the negative plate and negatively charged beta (β) particles are attracted to the positive plate. Gamma (γ) rays have no charge and are not deflected as they pass between the charged plates. Note that beta particles are deflected more than alpha particles. **See a simulation based on this figure, and take a short quiz on the concepts at http://www.now.brookscole.com/ gob8 or on the CD.**

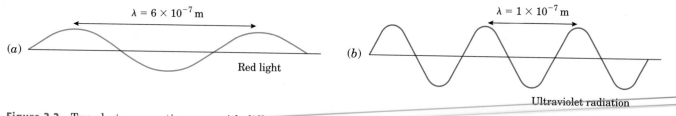

(a) $\lambda = 6 \times 10^{-7}$ m

Red light

(b) $\lambda = 1 \times 10^{-7}$ m

Ultraviolet radiation

Figure 3.2 Two electromagnetic waves with different wavelengths.

ray (γ) passes between the charged plates without deflection, indicating that it has no charge.

Alpha particles are helium nuclei. Each contains two protons and two neutrons; each has an atomic number of 2 and a charge of +2.

Beta particles are electrons. Each has a charge of −1.

Gamma rays are high-energy electromagnetic radiation. They have no mass or charge.

Gamma rays are only one form of electromagnetic radiation. Many others exist, including visible light, radio waves, and cosmic rays. All consist of waves (Figure 3.2).

The only difference between one form of electromagnetic radiation and another is the **wavelength** (λ, Greek letter lambda), which is the distance from one wave crest to the next. The **frequency** (ν, Greek letter nu) of a radiation is the number of crests that pass a given point in one second. Mathematically wavelength and frequency are related by the following equation, where c is the speed of light (3.0×10^8 m/s):

$$\lambda = \frac{c}{\nu}$$

As you can see from this relationship, the lower the frequency (ν), the longer the wavelength (λ); conversely, the shorter the frequency, the longer the wavelength.

A direct relationship also exists between the frequency of electromagnetic and its energy: The higher the frequency, the higher its energy. Electromagnetic radiation comes in packets; the smallest units are called **photons.**

Figure 3.3 shows the wavelengths of various types of radiation of the electromagnetic spectrum. Gamma rays are electromagnetic radiation of very high frequency (and high energy). Humans cannot see them because our eyes are not sensitive to waves of this frequency, but they can be detected

The only radiation known to have an even higher frequency (and energy) than gamma rays is cosmic rays.

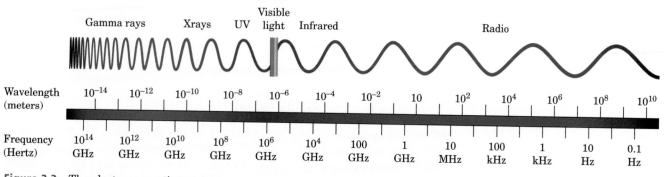

Figure 3.3 The electromagnetic spectrum.

Table 3.1   Particles and Rays Frequently Encountered in Radiation

Particle or Ray	Common Name of Radiation	Symbol	Charge	Atomic Mass Units	Penetrating Power[a]	Energy[b] Range
Proton	Proton beam	^1_1H	+1	1	1–3 cm	60 MeV
Electron	Beta particle	$^0_{+1}e$ or β	−1	$0.00055\left(\dfrac{1}{1835}\right)$	0–4 mm	1–3 MeV
Neutron	Neutron beam	^1_0n	0	1	—	—
Positron	—	$^0_{+1}e$ or β^+	+1	0.00055	—	—
Helium nucleus	Alpha particle	^4_2He or α	+2	4	0.02–0.04 mm	3–9 MeV
Energetic radiation	Gamma ray	γ	0	0	1–20 cm	0.1–10 MeV
	X ray		0	0	0.01–1 cm	0.1–10 keV

[a]Distance at which half the radiation has been stopped in water.
[b]MeV = 1.602×10^{-13} J = 3.829×10^{-14} cal.

UV stands for ultraviolet and IR for infrared.

The electron-volt (eV) is a non-SI energy unit used frequently in nuclear chemistry.
1 eV = 1.60×10^{-19} J

by instruments (Section 3.5). Another kind of radiation, called Xrays, has frequencies (and energies) higher than those of visible light but less than those of gamma rays.

Materials that emit radiation (alpha, beta, or gamma) are called **radioactive.** Radioactivity comes from the atomic nucleus and not from the electron cloud. Table 3.1 summarizes the properties of the particles and rays that come out of radioactive nuclei, along with the properties of some other particles and rays. Note that Xrays are not considered to be a form of radioactivity, because they do not come out of the nucleus but are generated in other ways.

We have said that humans cannot see gamma rays. We cannot see alpha or beta particles either. Likewise, we cannot hear them, smell them, or feel them. They are undetectable by our senses. We can detect radioactivity only by instruments, as discussed in Section 3.5.

3.3 | What Happens When a Nucleus Emits Radioactivity?

As mentioned in Section 2.4D, different nuclei consist of different numbers of protons and neutrons. It is customary to indicate these numbers with subscripts and superscripts placed to the left of the atomic symbol. The atomic number (the number of protons in an atom of the element) is shown as a subscript and the mass number (the number of protons and neutrons) as a superscript. For example, here are the symbols and names for the three isotopes of hydrogen:

^1_1H	hydrogen-1	hydrogen	(not radioactive)
^2_1H	hydrogen-2	deuterium	(not radioactive)
^3_1H	hydrogen-3	tritium	(radioactive)

A. Radioactive and Stable Nuclei

Some isotopes are radioactive, whereas others are stable. Scientists have identified more than 300 naturally occurring isotopes. Of these, 264 are stable, meaning that the nuclei of these isotopes never give off any radioactivity. As far as we can tell, they will last forever. The remainder are

radioactive isotopes—they do give off radioactivity. Furthermore, scientists have made more than 1000 artificial isotopes in laboratories. All artificial isotopes are radioactive.

Isotopes in which the number of protons and neutrons are balanced seem to be stable. In the lighter elements, this balance occurs when the numbers of protons and neutrons are approximately equal. For example, $^{12}_{6}C$ is a stable nucleus (six protons and six neutrons), as are $^{16}_{8}O$ (eight protons and eight neutrons), and $^{20}_{10}Ne$ (ten protons and ten neutrons). Among the heavier elements, stability requires more neutrons than protons. Lead-206, one of the most stable isotopes of lead, contains 82 protons and 124 neutrons.

If there is a serious imbalance in the proton-to-neutron ratio (either too few or too many neutrons), a nucleus will undergo **nuclear reaction** to make the ratio more favorable and the nucleus more stable.

Radioactive isotope A radiation-emitting isotope of an element

The role of neutrons seems to be to provide binding energy to overcome the repulsion between protons.

Nuclear reaction A reaction that changes atomic nuclei of elements (usually to atomic nuclei of other elements)

B. Beta Emission

If a nucleus has more neutrons than it needs for stability, it can stabilize itself by converting a neutron to a proton and an electron.

$$^{1}_{0}n \longrightarrow ^{1}_{1}H + ^{0}_{-1}e$$
Neutron   Proton   Electron

The proton remains in the nucleus but the electron is emitted from it. The emitted electron is called a **beta particle,** and the process is called **beta emission.** Phosphorus-32, for example, is a beta emitter:

$$^{32}_{15}P \longrightarrow ^{32}_{16}S + ^{0}_{-1}e$$

A phosphorus-32 nucleus has 15 protons and 17 neutrons. The nucleus remaining after an electron has been emitted has 16 protons and 16 neutrons; its atomic number is increased by 1 but its mass number is unchanged. The new nucleus is, therefore, sulfur-32. Thus, when the unstable phosphorus-32 (15 protons and 17 neutrons) is converted to sulfur-32 (16 protons and 16 neutrons), nuclear stability is achieved

The changing of one element into another is called **transmutation.** It happens naturally every time an element gives off a beta particle. Every time a nucleus emits a beta particle, it is transformed into another nucleus with the same mass number but an atomic number one unit greater.

HOW TO . . .

Balance a Nuclear Equation

In writing nuclear equations, we consider only the nucleus and disregard the surrounding electrons. There are two simple rules for balancing nuclear equations:

1. The sum of the mass numbers (superscripts) on both sides of the equation must be equal.
2. The sum of the atomic numbers (subscripts) on both sides of the equation must be equal. For the purposes of determining atomic numbers in a nuclear equation, an electron emitted from the nucleus has an atomic number of −1.

To see how to apply these rules, let us look at the decay of phosphorus-32, a beta emitter.

$$^{32}_{15}P \longrightarrow ^{32}_{16}S + ^{0}_{-1}e$$

1. Mass number balance: The total mass number on each side of the equation is 32.
2. Atomic number balance: The atomic number on the left is 15. The sum of the atomic numbers on the right is $16 - 1 = 15$.

Thus, in the phosphorus-32 decay equation, mass numbers are balanced (32 and 32), atomic numbers are balanced (15 and 15), and, therefore, the nuclear equation is balanced.

EXAMPLE 3.1

Carbon-14, $^{14}_{6}C$, is a beta emitter. Write an equation for this nuclear reaction, and identify the product element.

$$^{14}_{6}C \longrightarrow ? + ^{0}_{-1}e$$

Solution
The $^{14}_{6}C$ nucleus has six protons and eight neutrons. One of its neutrons is converted to a proton and an electron. The electron is emitted as a beta particle, but the proton remains in the nucleus. The nucleus now has seven protons and seven neutrons:

$$^{14}_{6}C \longrightarrow ^{14}_{7}? + ^{0}_{-1}e$$

The sum of the mass numbers on each side of the equation is 14, and the sum of the atomic numbers on each side is 6. When we look in the Periodic Table to find out which element has atomic number 7, we see that it is nitrogen. The product of this nuclear reaction is, therefore, nitrogen-14 and we can now write a complete equation.

$$^{14}_{6}C \longrightarrow ^{14}_{7}N + ^{0}_{-1}e$$

Problem 3.1
Iodine-139 is a beta emitter. Write an equation for this nuclear reaction, and identify the product.

C. Alpha Emission

For heavy elements, the loss of alpha (α) particles is an especially important stabilization process. For example,

$$^{238}_{92}U \longrightarrow ^{234}_{90}Th + ^{4}_{2}He$$

$$^{210}_{84}Po \longrightarrow ^{206}_{82}Pb + ^{4}_{2}He + \gamma$$

Note that the radioactive decay of polonium-210 emits both an α particle and gamma rays.

A general rule for alpha emission is this: The new nucleus always has a mass number four units lower and an atomic number two units lower than the original nucleus.

EXAMPLE 3.2

Polonium-218 is an alpha emitter. Write an equation for this nuclear reaction, and identify the product.

Solution

The atomic number of polonium is 84, so the partial equation is

$$^{218}_{84}\text{Po} \longrightarrow ? + {}^{4}_{2}\text{He}$$

The mass number of the new isotope is $218 - 4 = 214$. The atomic number of the new isotope is $84 - 2 = 82$. We can now write

$$^{218}_{84}\text{Po} \longrightarrow {}^{214}_{82}? + {}^{4}_{2}\text{He}$$

In the Periodic Table we find that the element with an atomic number of 82 is lead, Pb. Therefore, the product is $^{214}_{82}\text{Pb}$, and we can now write the complete equation:

$$^{218}_{84}\text{Po} \longrightarrow {}^{214}_{82}\text{Pb} + {}^{4}_{2}\text{He}$$

Problem 3.2

Thorium-223 is an alpha emitter. Write an equation for this nuclear reaction and identify the product.

D. Positron Emission

A positron is a particle that has the same mass as an electron, but a charge of $+1$ rather than -1. Its symbol is β^+ or $^{0}_{+1}\text{e}$. Positron emission is much rarer than alpha or beta emission. Because a positron has no appreciable mass, the nucleus is transmuted into another nucleus with the same mass number but an atomic number that is one unit less. Carbon-11, for example, is a positron emitter:

$$^{11}_{6}\text{C} \longrightarrow {}^{11}_{5}\text{B} + {}^{0}_{+1}\text{e}$$

In this balanced nuclear equation, the mass numbers on both the left and the right are 11. The atomic number on the left is 6; that on the right is also 6 ($5 + 1 = 6$).

EXAMPLE 3.3

Nitrogen-13 is a positron emitter. Write an equation for this nuclear reaction, and identify the product.

Solution

We begin by writing a partial equation:

$$^{13}_{7}\text{N} \longrightarrow ? + {}^{0}_{+1}\text{e}$$

Because a positron has no appreciable mass, the mass number of the new isotope is still 13. The sum of the atomic numbers on each side must be 7, which means that the atomic number of the new isotope must be 6. In the Periodic Table we see that the element with atomic number 6 is carbon. Therefore, the new isotope formed in this nuclear reaction is carbon-13 and the balanced nuclear equation is

$$^{13}_{7}\text{N} \longrightarrow\ ^{13}_{6}\text{C} +\ ^{0}_{+1}\text{e}$$

Problem 3.3

Arsenic-74 is a positron emitter used in locating brain tumors. Write an equation for this nuclear reaction and identify the product.

E. Gamma Emission

Both alpha and beta emissions can be either "pure" or mixed with gamma rays.

Although rare, some nuclei are pure gamma emitters:

$$^{11}_{5}\text{B*} \longrightarrow\ ^{11}_{5}\text{B} + \gamma$$

Gamma emission often accompanies α and β emissions.
In this equation, $^{11}_{5}\text{B*}$ symbolizes a boron nucleus in a high-energy (excited) state. In this case, no transmutation takes place. The boron is still boron, but its nucleus is in a lower-energy (more stable) state after the emission of excess energy in the form of gamma rays. When all excess energy has been emitted, the nucleus returns to its most stable, lowest-energy state.

F. Electron Capture

In electron capture (abbreviated E.C.), an extranuclear electron is captured by the nucleus and there reacts with a proton to form a neutron. Thus, the atomic number of the element is reduced by one, but the mass number is unchanged. Beryllium-7, for example, decays by electron capture to give lithium-7:

$$^{7}_{4}\text{Be} +\ ^{0}_{-1}\text{e} \longrightarrow\ ^{7}_{3}\text{Li}$$

EXAMPLE 3.4

Chromium-51, which is used to create images that show the size and shape of the spleen, decays by electron capture and gamma emission. Write an equation for this nuclear decay and identify the product.

Solution
We begin by writing a partial equation:

$$^{51}_{24}\text{Cr} +\ ^{0}_{-1}\text{e} \longrightarrow\ ? + \gamma$$

There is no change in mass number, so the new nucleus has a mass number of 51. The new nucleus has 23 protons, one less than chromium-51. In the Periodic Table we see that the element with atomic number 23 is vanadium and, therefore, the new element formed is vanadium-51. We can now write the complete equation for this nuclear decay:

$$^{51}_{24}\text{Cr} + {}^{0}_{-1}\text{e} \longrightarrow {}^{51}_{23}\text{V} + \gamma$$

Problem 3.4

Thallium-201, a radioisotope used to evaluate heart function in exercise stress tests, decays by electron capture and gamma emission. Write an equation for this nuclear decay and identify the product.

3.4 | What Is Nuclear Half-life?

Suppose we have 40 g of a radioactive isotope—say, $^{90}_{38}\text{Sr}$. Strontium-90 nuclei are unstable and decay by beta emission to yttrium-90:

$$^{90}_{38}\text{Sr} \longrightarrow {}^{90}_{39}\text{Y} + {}^{0}_{-1}\beta$$

Our 40-g sample of strontium-90 contains about 2.7×10^{23} atoms. We know that these nuclei decay, but at what rate? Do all of the nuclei decay at the same time, or do they decay gradually over time? The answer is that they decay one at a time, at a fixed rate. For strontium-90, the decay rate is such that one half of our sample (about 1.35×10^{23} atoms) will have decayed by the end of 28.1 years. The time it takes for one half of any sample of radioactive material to decay is called the **half-life, $t_{1/2}$**.

It does not matter how big or small a sample is. For example, in the case of our 40 g of strontium-90, 20 g will be left at the end of 28.1 years (the rest will be converted to yttrium-90). It will then take another 28.1 years for half of the remainder to decay, so that at the end of that time we will have 10 g of strontium-90. If we wait for a third span of 28.1 years, then 5 g will be left. If we had begun with 100 g, then 50 g would be left after the first 28.1-year period.

Figure 3.4 shows the radioactive decay curve of iodine-131. Inspection of this graph shows that, at the end of 8 days, half of the original has disappeared. Thus the half-life of iodine-131 is 8 days. It would take a total of 16 days, or two half-lives, for three fourths of the original amount of iodine-131 to decay.

When a nucleus gives off radiation, it is said to *decay*.

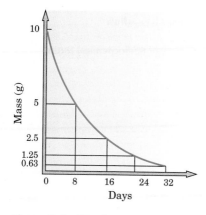

Figure 3.4 The decay curve of iodine-131.

EXAMPLE 3.5

If 10.0 mg of $^{131}_{53}\text{I}$ is administered to a patient, how much is left in the body after 32 days?

Solution

We know from Table 3.2 that $t_{1/2}$ of iodine-131 is eight days. We also know that 32 days corresponds to four half-lives. Thus 5.00 mg is left after one half-life; 2.50 mg after two half-lives; 1.25 mg after three half-lives; and 0.625 mg after four half-lives.

$$\overbrace{10.0 \text{ mg} \times \frac{1}{2} \times \frac{1}{2} \times \frac{1}{2} \times \frac{1}{2}}^{32 \text{ days (4 half-lives)}} = 0.625 \text{ mg}$$

Problem 3.5

Barium-122 has a half-life of 2 minutes. Suppose you obtained a sample weighing 10.0 g. It takes 10 minutes to set up the experiment in which barium-122 will be used. How many grams of barium-122 will be left when you begin the experiment?

In theory, it would take an infinite amount of time for all of a radioactive sample to disappear. In reality, most of the radioactivity disappears after five half-lives; by that time, only 3% of the original remains.

$$\overbrace{\frac{1}{2} \times \frac{1}{2} \times \frac{1}{2} \times \frac{1}{2} \times \frac{1}{2}}^{\text{5 half-lives}} \times 100 = 3.1\%$$

After ten half-lives, less than 0.1% of the activity remains.

The half-life of an isotope is independent of temperature and pressure—indeed, of all other conditions—and is a property of the particular isotope only. It does not depend on what kinds of atoms surround the particular nucleus (that is, what kind of molecule the nucleus is part of). We do not know any way to speed up radioactive decay or to slow it down.

Table 3.2 gives some half-lives. Even this brief sampling indicates the tremendous differences among half-lives. Some isotopes, such as technetium-99m, decay and disappear in a day; others, such as uranium-238, remain radioactive for billions of years. Very short-lived isotopes, especially the artificial heavy elements (Section 3.8) with atomic numbers greater than 100, have half-lives of the order of seconds.

The usefulness or inherent danger in radioactive isotopes is related to their half-lives. In assessing the long-range health effects of atomic-bomb

GOB
Chemistry ⚛ Now™

Click *Coached Problems* to see an example of **Radiochemical Dating**

Table 3.2 Half-lives of Some Radioactive Nuclei

Name	Symbol	Half-life	Radiation
Hydrogen-3 (tritium)	$^{3}_{1}H$	12.26 years	Beta
Carbon-14	$^{14}_{6}C$	5730 years	Beta
Phosphorus-28	$^{28}_{15}P$	0.28 second	Positrons
Phosphorus-32	$^{32}_{15}P$	14.3 days	Beta
Potassium-40	$^{40}_{19}K$	1.28×10^{9} years	Beta + gamma
Scandium-42	$^{42}_{21}Sc$	0.68 second	Positrons
Cobalt-60	$^{60}_{27}Co$	5.2 years	Gamma
Strontium-90	$^{90}_{38}Sr$	28.1 years	Beta
Technetium-99m	$^{99m}_{43}Tc$	6.0 hours	Gamma
Indium-116	$^{116}_{49}In$	14 seconds	Beta
Iodine-131	$^{131}_{53}I$	8 days	Beta + gamma
Mercury-197	$^{197}_{80}Hg$	65 hours	Gamma
Polonium-210	$^{210}_{84}Po$	138 days	Alpha
Radon-205	$^{205}_{86}Rn$	2.8 minutes	Alpha
Radon-222	$^{222}_{86}Rn$	3.8 days	Alpha
Uranium-238	$^{238}_{92}U$	4×10^{9} years	Alpha

CHEMICAL CONNECTIONS 3A

Radioactive Dating

Carbon-14, with a half-life of 5730 years, can be used to date archeological objects as old as 60,000 years. This dating technique relies on the principle that the carbon-12/carbon-14 ratio of an organism—whether plant or animal—remains constant during the lifetime of the organism. When the organism dies, the carbon-12 level remains constant (carbon-12 is not radioactive), but any carbon-14 present decays by beta emission to nitrogen-14.

$$^{14}_{6}C \longrightarrow {}^{14}_{7}N + {}^{0}_{-1}e$$

Using this fact, a scientist can calculate the changed carbon-12/carbon-14 ratio to determine the date of an artifact.

For example, in charcoal made from a tree that has recently died, the carbon-14 gives a radioactive count of 13.70 disintegrations/min per gram of carbon. In a piece of charcoal found in a cave in France near some ancient Cro-Magnon cave paintings, the carbon-14 count was 1.71 disintegrations/min for each gram of carbon. From this information, the cave paintings can be dated. After one half-life, the number of disintegrations/minute per gram is 6.85; after two half-lives, it is 3.42, and after three half-lives, it is 1.71. Therefore, three half-lives have passed since the paintings were created. Given that carbon-14 has a half-life of 5730 years, the paintings are about $3 \times 5730 = 17,190$ years old.

The famous Shroud of Turin, a piece of linen cloth with the image of a man's head on it, was believed by many to be the original cloth that was wrapped around the body of Jesus Christ after his death. However, radioactive dating showed, with 95% certainty, that the plants from which the linen was obtained were alive sometime between AD 1260 and 1380, proving that the cloth could

Cro-Magnon cave painting.

not have been the shroud of Christ. Note that it was not necessary to destroy the shroud to perform the tests. In fact, scientists in different laboratories used just a total of a few square centimeters of cloth from its edge.

Rock samples can be dated on the basis of their lead-206 and uranium-238 content. The assumption is that all of the lead-206 comes from the decay of uranium-238, which has a half-life of 4.5 billion years. One of the oldest rocks found on the earth is a granite outcrop in Greenland, dated at 3.7×10^9 years old.

On the basis of dating of meteorites, the estimated age of the solar system is 4.6×10^9 years.

(a)   (b)

The Shroud of Turin (a) as it appears to the eye and (b) as a photographic image, which brings out the images more clearly.

A Geiger-Müller counter.

GOB
Chemistry ⚛ Now™

Click *Chemistry Interactive* to see the workings of a **Geiger-Müller Counter**

A television picture tube works on a similar principle.

Figure 3.5 A schematic drawing of a Geiger-Müller counter. The counter is made of a glass cylinder, the inner surface of which is coated with a material that conducts electricity. In the center is the collector electrode, made of tungsten wire. The collector electrode is positively charged to about 1200 volts. The space within the tube is filled with helium or argon gas at low pressure. A thin Mylar window at the end keeps the gas in, yet allows the ionizing radiation to pass through. When ionizing radiation enters the tube, it converts argon or helium atoms to positive ions, which go to the negative electrode. The current caused by these positive ions is amplified and used to operate the counter.

damage or of nuclear power plant accidents, like those at Three Mile Island, Pennsylvania, in 1979 and Chernobyl (in the former Soviet Union) in 1986 (Chemical Connections 3E), we can see that radioactive isotopes with long half-lives, such as $^{85}_{36}Kr$ ($t_{1/2} = 10$ years) or $^{60}_{27}Co$ ($t_{1/2} = 5.2$ years) are more important than short-lived ones. Conversely, when a radioactive isotope is used in medical imaging or therapy, short-lived isotopes are more useful because they disappear faster from the body—for example, $^{99m}_{43}Tc$, $^{32}_{15}P$, $^{131}_{53}I$, and $^{197}_{80}Hg$.

3.5 | How Do We Detect and Measure Nuclear Radiation?

As already noted, radioactivity is not detectable by our senses. We cannot see it, hear it, feel it, or smell it. Then how do we know it is there? Alpha, beta, gamma, positron, and Xrays all have a property we can use to detect them: When these rays interact with matter, they usually knock electrons out of the electron cloud surrounding an atomic nucleus, thereby creating positively charged ions from neutral atoms. For this reason, we call all of these rays **ionizing radiation.**

Ionizing radiation is characterized by two physical measurements: (1) its **intensity,** which is the number of particles or photons emerging per unit time, and (2) the **energy** of each particle or photon emitted.

A. Intensity

To measure intensity, we take advantage of the ionizing property of radiation. Instruments such as the **Geiger-Müller counter** (Figure 3.5) and the **proportional counter** contain a gas such as helium or argon. When a radioactive nucleus emits alpha or beta particles or gamma rays, these radiations ionize the gas, and the instrument registers this fact by indicating that an electric current is passing between two electrodes. In this way, the instrument counts particle after particle.

Other measuring devices, such as **scintillation counters,** have a material called a phosphor that emits a unit of light for each alpha or beta particle or gamma ray that strikes it. Once again, the particles are counted one by one. The quantitative measure of radiation intensity can be reported in counts/minute or counts/second.

A commonly used unit of radiation intensity is the **curie** (Ci), named in honor of Marie Curie, whose lifelong work with radioactive materials greatly helped expand our understanding of nuclear phenomena. One curie is defined as 3.7×10^{10} disintegrations per second (dps). This is radiation of very high intensity, the amount a person would get from 1 g of pure $^{226}_{88}Ra$. This intensity is too high for regular medical use, so the units most commonly used in the health sciences are small fractions of it. Another, albeit much smaller, unit of radiation activity (intensity) is the **becquerel** (Bq), which is the SI unit. One becquerel is one disintegration per second (dps).

$$1 \text{ becquerel (Bq)} = 1.0 \text{ dps}$$

$$1 \text{ curie (Ci)} = 3.7 \times 10^{10} \text{ dps}$$

$$1 \text{ millicurie (mCi)} = 3.7 \times 10^{7} \text{ dps}$$

$$1 \text{ microcurie (}\mu\text{Ci)} = 3.7 \times 10^{4} \text{ dps}$$

EXAMPLE 3.6

A radioactive isotope with an intensity (activity) of 100 mCi per vial is delivered to a hospital. The vial contains 10 mL of liquid. The instruction is to administer 2.5 mCi intravenously. How many milliliters of the liquid should be administered?

Solution

The intensity (activity) of a sample is directly proportional to the amount present, so

$$2.5 \text{ mCi} \times \frac{10 \text{ mL}}{100 \text{ mCi}} = 0.25 \text{ mL}$$

Problem 3.6

A radioactive isotope in a 9.0-mL vial has an intensity of 300 mCi. A patient is required to take 50 mCi intravenously. How much liquid should be used for the injection?

The intensity of any radiation decreases with the square of the distance from the source. If, for example, the distance from a radiation source doubles, then the intensity of the received radiation decreases by a factor of four.

$$\frac{I_1}{I_2} = \frac{d_2^2}{d_1^2}$$

EXAMPLE 3.7

If the intensity of radiation is 28 mCi at a distance of 1.0 m, what is the intensity at a distance of 2.0 m?

Solution

From the preceding equation we have

$$\frac{28 \text{ mCi}}{I_2} = \frac{2.0^2}{1.0^2} = 4.0$$

$$I_2 = \frac{28 \text{ mCi}}{4.0} = 7.0 \text{ mCi}$$

Problem 3.7

If the intensity of radiation 1.0 cm from a source is 300 mCi, what is the intensity at 3.0 m?

B. Energy

The energies of different particles or photons vary. As shown in Table 3.1, each particle has a certain range of energy. For example, beta particles have an energy range of 1 to 3 MeV (mega-electron-volts). This range may overlap with the energy range of some other type of radiation—for example, gamma rays. The penetrating power of a radiation depends on its energy as well on the mass of its particles. Alpha particles are the most massive and the most highly

Figure 3.6 Penetration of radioactive emissions. Alpha particles, with a charge of +2 and a mass of 4 amu, interact strongly with matter but penetrate the least; they are stopped by several sheets of paper. Beta particles, with less mass and a lower charge than alpha particles, interact less strongly with matter; they easily penetrate paper but are stopped by a 0.5-cm sheet of lead. Gamma rays, with neither mass nor charge, have the greatest penetrating power. It takes 10 cm of lead to stop them.

Alpha (α)

Beta (β)

Gamma (γ)

Paper

0.5-cm lead

10-cm lead

charged and, therefore, the least penetrating; they can be stopped by several sheets of ordinary paper, by ordinary clothing, and by the skin. Beta particles have less mass and lower charge than alpha particles and, consequently, have greater penetrating power. They can penetrate several millimeters of bone or tissue. Gamma radiation, which has neither mass nor charge, is the most penetrating of the three types of radiation. (Alpha particles, when in the body, are more damaging, however.) They can pass completely through the body. Several centimeters of lead or concrete are required to stop gamma rays (Figure 3.6).

One easy way to protect against ionizing radiation is to wear lead aprons, covering sensitive organs. This practice is followed routinely when diagnostic Xrays are taken. Another way to lessen the damage from ionizing radiation is to move farther away from the source.

3.6 | How Is Radiation Dosimetry Related to Human Health?

The term *activity of a radiation* is the same as *intensity of a radiation*

In studying the effect of radiation on the body, neither the energy of the radiation (in kcal/mol) nor its intensity (in Ci) alone or in combination is of particular importance. Rather, the critical question is what kind of effects such radiation produces in the body. Three different units are used to describe the effects of radiation on the body: roentgens, rads, and rems.

Roentgens (R) Roentgens are a measure of the energy delivered by a radiation source and are, therefore, a measure of exposure to a particular form of radiation. One roentgen is the amount of radiation that produces ions having 2.58×10^{-4} coulomb of ions per kilogram (a coulomb is a unit of electrical charge).

Rads The rad, which stands for *r*adiation *a*bsorbed *d*ose, is a measure of the radiation absorbed from a radiation source. The SI unit is the gray (Gy), where 1 Gy = 100 rad. Roentgens (delivered energy) do not take into account the effect of radiation on tissue and the fact that different tissues absorb different amounts of delivered radiation. Radiation damages body tissue by causing ionization; for such ionization to occur, the tissue must absorb the delivered energy. The relationship between the delivered dose in roentgens and the absorbed dose in rads can be illustrated as follows: Exposure to 1 roentgen yields 0.97 rad of absorbed radiation in water, 0.96 rad in muscle, and 0.93 rad in bone. This relationship holds for high-energy photons. For lower-energy photons, such as "soft" Xrays, each roentgen yields 3 rads of absorbed dose in bone. This principle underlies diagnostic Xrays, wherein soft tissue lets the radiation through to strike a photographic plate but bone absorbs the radiation and casts a shadow on the plate.

Rems   The rem, which stands for *r*oentgen *e*quivalent for *m*an, is a measure of the effect of the radiation when a person absorbs 1 roentgen. Other units are the **millirem** (mrem; 1 mrem = 1×10^{-3} rem) and the **sievert** (Sv; 1 Sv = 100 rem). The sievert is the SI unit. The rem is used because tissue damage from 1 rad of absorbed energy depends on the type of radiation. One rad from alpha rays, for example, causes ten times more damage than 1 rad from Xrays or gamma rays. Table 3.3 summarizes the various radiation units and what each measures.

Although alpha particles cause more damage than Xrays or gamma rays, they have a very low penetrating power (Table 3.1) and cannot pass through the skin. Consequently, they are not harmful to humans or animals as long as they do not enter the body. If they do get in, however, they can prove quite harmful. They can get inside, for example, if a person swallows or inhales a small particle of a substance that emits alpha particles. Beta particles are less damaging to tissue than alpha particles but penetrate farther and so are generally more harmful. Gamma rays, which can completely penetrate the skin, are by far the most dangerous and harmful form of radiation. Remember, of course, that α particles once in the body, as for example, alpha radiation from radon-222, are very damaging. Therefore, for comparison purposes and for determining exposure from all kinds of sources, the equivalent dose is an important measure. If an organ receives radiation from different sources, the total effect can be summed up in rems (or mrem or Sv). For example, 10 mrem of alpha particles and 15 mrem of gamma radiation give a total of 25 mrem absorbed equivalent dose. Table 3.4 shows the amount of radiation exposure that an average person obtains yearly from both natural and artificial sources.

The naturally occurring background radiation varies with the geological location. For example, a tenfold higher than average radiation has been detected at some phosphate mines. People who work in nuclear medicine are, of course, exposed to greater amounts. To ensure that exposures do not get too high, they wear radiation badges. A single whole-body irradiation of 25 rem is noticeable in the blood count, and 100 rem causes the typical symptoms of radiation sickness, which include nausea, vomiting, a decrease in the white blood cell count, and loss of hair. A dose of 400 rem causes death within one month in 50% of exposed persons, and 600 rem is almost invariably

A radiation badge.

Table 3.3   Radiation Dosimetry

Unit	What the Unit Measures	SI Unit	Other Units
Roentgen	The amount of radiation delivered from a radiation source	Roentgen (R)	
Rad	The ratio between radiation absorbed by a tissue and radiation delivered to the tissue	Gray (Gy)	1 rad = 0.01 Gy
Rem	The ratio between the tissue damage caused by a rad of radiation and the type of radiation	Sievert (Sv)	1 rem = 0.01 Sv

CHEMICAL CONNECTIONS 3B

The Indoor Radon Problem

Most of our exposure to ionizing radiation comes from natural sources (Table 3.4), with radon gas being the main source. Radon has more than 20 isotopes, all of which are radioactive. The most important is radon-222, an alpha emitter. Radon-222 is a natural decay product of uranium-238, which is widely distributed in the Earth's crust.

Radon poses a particular health hazard among radioactive elements because it is a gas at normal temperatures and pressures. As a consequence, it can enter our lungs with the air we breathe and become trapped in the mucous lining of the lungs. Radon-222 has a half-life of 3.8 days. It decays naturally and produces, among other isotopes, two harmful alpha emitters: polonium-218 and polonium-214. These polonium isotopes are solids and do not leave the lungs with exhalation. In the long run, they can cause lung cancer.

Testing devices are available to determine whether radon is building up in a home.

The U.S. Environmental Protection Agency has set a standard of 4 pCi/L (one picocurie, pCi, is 10^{-12} Ci) as a safe exposure level. A survey of single-family homes in the United States showed that 7% exceeded this level. Most radon seeps into dwellings through cracks in cement foundations and around pipes, then accumulates in basements. The remedy is to ventilate both basements and houses enough to reduce the radiation levels. In a notorious case, a group of houses in Grand Junction, Colorado was built from bricks made from uranium tailings. Obviously, the radiation levels in these buildings were unacceptably high. Because they could not be controlled, the buildings had to be destroyed. In our modern radiation-conscious age, more and more homebuyers choose to request a certification of radon levels before buying a house.

lethal within a short time. It should be noted that as much as 50,000 rem is needed to kill bacteria and as much as 10^6 rem to inactivate viruses.

Fortunately, most of us never get a single dose of more than a few rem and so never suffer from any form of radiation sickness. This does not mean, however, that small doses are totally harmless. The harm may arise in two ways:

1. Small doses of radioactivity over a period of years can cause cancer, especially blood cancers such as leukemia. Repeated exposure to sunlight also carries a risk of tissue damage. Most of the high-energy UV radiation of the sun is absorbed by the Earth's protective ozone layer in the stratosphere. In tanning, however, the repeated overexposure to UV radiation can cause skin cancer (see Chemical Connections 19D). No one knows how many cancers have resulted from this practice, because the doses are so small and continue for so many years that they cannot be measured accurately. Also, because so many other causes of cancer exist, it is often difficult or impossible to decide whether any particular case is caused by radiation.

2. If any form of radiation strikes an egg or sperm cell, it can cause a change in the genes (see Chemical Connections 25E). Such changes are called mutations. If an affected egg or sperm cell mates, grows, and becomes a new individual, that individual may have mutated characteristics, which are usually harmful and frequently lethal.

Because radiation carries so much potential for harm, it would be nice if we could totally escape it. But can we? Table 3.4 shows that this is impossible. Naturally occurring radiation, called **background radiation,** is present everywhere on Earth. This background radiation vastly outstrips the average radiation level from artificial sources (mostly diagnostic Xrays). If we eliminated all forms of artificial radiation, including medical uses, we would still be exposed to the background radiation.

Table 3.4   Average Exposure to Radiation from Common Sources

Source	Dose (mrem/year)
Naturally Occurring Radiation	
Cosmic rays	27
Terrestrial radiation (rocks, buildings)	28
Inside human body (K-40 and Ra-266 in the bones)	39
Radon in the air	200
Total	294
Artificial Radiation	
Medical Xrays[a]	39
Nuclear medicine	14
Consumer products	10
Nuclear power plants	0.5
All others	1.5
Total	65.0
Grand total	359[b]

[a]Individual medical precedures may expose certain parts of the body to much higher doses. For instance, one chest Xray gives 27 mrem and a diagnostic GI series gives 1970 mrem.

[b]The federal safety standard for allowable occupational exposure is about 5000 mrem/year. It has been suggested that this level be lowered to 4000 mrem/year or even lower to reduce the risk of cancer stemming from low levels of radiation.

SOURCE: *National Council on Radiation Protection and Measurements,* NCRP Report No. 93 (1993).

3.7 | What Is Nuclear Medicine?

When we think of nuclear chemistry, our first thoughts may well be of nuclear power, atomic bombs, and weapons of mass destruction. True as this may be, it is also true that nuclear chemistry and the use of radioactive elements have become invaluable tools in all areas of science. Nowhere is this more important than in nuclear medicine—that is, in the use of radioactive isotopes as tools for both the diagnosis and the treatment of diseases. To describe the full range of medical uses of nuclear chemistry would take far more space than this text allows. Recognizing this fact, we have chosen several examples of each use to illustrate the range of applications of nuclear chemistry to the health sciences.

A. Medical Imaging

Medical imaging is the most widely used aspect of nuclear medicine. The goal of medical imaging is to create a picture of a target tissue. To create a useful image requires three things:

- A radioactive element administered in pure form or in a compound that becomes concentrated in the tissue to be imaged
- A method of detecting radiation from the radioactive source and recording its intensity and location
- A computer to process the intensity–location data and transform those data into a useful image

Chemically and metabolically, a radioactive isotope in the body behaves in exactly the same way as do the nonradioactive isotopes of the same element. In the simplest form of imaging, a radioactive isotope is injected

Cosmic rays High-energy particles, mainly protons, from outer space bombarding the Earth

A scan of radiation released by radioactive iodine concentrated in thyroid tissue gives an image of the thyroid gland.

Table 3.5 Some Radioactive Isotopes Useful in Medical Imaging

	Isotope	Mode of Decay	Half-life	Use in Medical Imaging
$^{11}_{6}C$	Carbon-11	β^+, γ	20.3 m	Brain scan to trace glucose metabolism
$^{18}_{9}F$	Fluorine-18	β^+, γ	109 m	Brain scan to trace glucose metabolism
$^{32}_{15}P$	Phosphorus-32	β	14.3 d	Detect eye tumors
$^{51}_{24}Cr$	Chromium-51	E.C., γ	27.7 d	Diagnose albinism, image the spleen and gastrointestinal tract
$^{59}_{26}Fe$	Iron-59	β, γ	44.5 d	Bone marrow function, diagnose anemias
$^{67}_{31}Ga$	Gallium-67	E.C., γ	78.3 h	Whole-body scan for tumors
$^{75}_{34}Se$	Selenium-75	E.C., γ	118 d	Pancreas scan
$^{81m}_{36}Kr$	Krypton-81m	γ	13.3 s	Lung ventilation scan
$^{81}_{38}Sr$	Strontium-81	β	22.2 m	Scan for bone diseases, including cancer
$^{99m}_{43}Tc$	Technetium-99m	γ	6.01 h	Brain, liver, kidney, bone scans; diagnosis of damaged heart muscle
$^{131}_{53}I$	Iodine-131	β, γ	8.04 d	Diagnosis of thyroid malfunction
$^{197}_{80}Hg$	Mercury-197	E,C, γ	64.1 h	Kidney scan
$^{201}_{81}Tl$	Thallium-201	E,C, γ	3.05 d	Heart scan and exercise stress test

Normal

Meningioma (brain tumor)

"Brain death"

Scalp tumor

CRC Handbook in Clinical Laboratory Science

Figure 3.7 A comparison of dynamic scan patterns for normal and pathological brains. The studies were performed by injecting technetium-99m into blood vessels.

intravenously and a technician uses a detector to monitor how the radiation becomes distributed in the body of the patient. Table 3.5 lists some of the most important radioisotopes used in imaging and diagnosis.

The use of iodine-131, a beta and gamma emitter ($t_{1/2}$ = 8.0 days), to image and diagnose a malfunctioning thyroid gland is a good example. The thyroid gland in the neck produces a hormone, thyroxine, which controls the overall rate of metabolism (use of food) in the body. One molecule of thyroxine contains four iodine atoms. When radioactive iodine-131 is administered into the bloodstream, the thyroid gland takes it up and incorporates it into the thyroxine (Chemical Connections 13C). A normally functioning thyroid absorbs about 12% of the administered iodine within a few hours. An overactive thyroid (hyperthyroidism) absorbs and localizes iodine-131 in the gland faster, and an underactive thyroid (hypothyroidism) does so much more slowly than normal. By counting the gamma radiation emitted from the neck, then, one can determine the rate of uptake of iodine-131 into the thyroid gland and diagnose hyperthyroidism or hypothyroidism.

Most organ scans are similarly based on the preferential uptake of some radioactive isotopes by a particular organ (Figure 3.7).

Another important type of medical imaging is positron emission tomography (PET). This method is based on the property that certain isotopes (such as carbon-11 and fluorine-18) emit positrons (Section 3.3D). Fluorine-18 decays by positron emission to oxygen-18:

$$^{18}_{9}F \longrightarrow {}^{18}_{8}O + {}^{0}_{+1}e$$

Positrons have very short lives. When a positron collides with an electron, the two annihilate each other, resulting in two gamma rays.

$$\underset{\text{Positron}}{^{0}_{+1}e} + \underset{\text{Electron}}{^{0}_{-1}e} \longrightarrow 2\,\gamma$$

Because electrons are present in every atom, there are always lots of them around, so positrons generated in the body cannot live long.

A favorite tagged molecule for following the uptake and metabolism of glucose, $C_6H_{12}O_6$, is 18-fluorodeoxyglucose (FDG), a molecule of glucose in which one of glucose's six oxygen atoms is replaced by fluorine-18. When FDG is administered intravenously, the tagged glucose soon enters blood

CHEMICAL CONNECTIONS 3C

Magnetic Resonance Imaging

Magnetic resonance imaging (MRI) is probably the best known of the various medical imaging techniques. Like those techniques we have already discussed, MRI involves atomic nuclei; unlike the other techniques, it does not involve nuclear decay or the emission of particles or rays from the nucleus.

Nuclear magnetic resonance (NMR) was discovered in the 1950s. By the 1960s, it had become an invaluable analytical tool for chemists. By the early 1970s, it was realized that this technique could be used to image parts of the body and could serve as a valuable addition to diagnostic medicine. Because the term "nuclear magnetic resonance" sounds to many people as if the technique might involve radioactive material, health care personnel call the technique magnetic resonance imaging.

MRI is based on the following principles:

- The nuclei of certain elements behave as if they are tiny bar magnets; that is, they behave as if they have a north pole and a south pole.
- In the presence of a very strong external magnetic field, a majority of these tiny nuclear bar magnets become aligned with the external field; that is, they become aligned so that their north poles align with the south pole of the external field, and their south poles align with the north pole of the external field.
- If energy in the form of radio-frequency waves is focused on the aligned nuclei, their nuclear bar magnets flip so that they become aligned against the external field.
- Resonance occurs when the proper combination of external magnetic field and radio-frequency wave causes a nuclear bar magnet to flip from alignment with the external field to alignment against it.
- A computer records which radio-frequency waves cause the nuclear bar magnets to flip to alignment against the external field, along with their location in the body and, with these data, constructs a useful image.
- The body contains several nuclei that, in principle, could be used for MRI. Of these, hydrogen-1, most of which comes from water, fats, and membrane phospholipids, gives the most useful signals.

In MRI, the patient is placed in a magnetic field gradient that can be varied from one part of the body to another. Nuclei in the

A medical MRI imager.

Beverly March/courtesy of Long Island Jewish Hospital

weaker magnetic field gradient absorb radiation at a lower radio frequency. Nuclei elsewhere, in the stronger magnetic field, absorb radiation at a higher radio frequency. Because a magnetic field gradient along a single axis images a plane, MRI techniques can create views of any part of the body in slice-like sections. Paul Lauterbur and Sir Peter Mansfield were awarded the 2003 Nobel Prize in physiology/medicine for their discoveries that led to the development of this imaging technique.

MRI is especially good for imaging soft tissues that contain many hydrogen atoms. Because bones contain few hydrogen atoms, they are almost transparent in this kind of imaging. Thus, the soft tissues with their different water content stand out in an MRI.

An important application of MRI is in the examination of the brain and spinal cord. White and gray matter, the two different layers of the brain, are easily distinguished by MRI. The resulting information is useful in the study of such diseases as multiple sclerosis.

Magnetic resonance imaging and Xray imaging are in many cases complementary. The hard, outer layer of bone is essentially invisible to MRI but shows up extremely well in Xray images, whereas soft tissue is nearly transparent to Xrays, but shows up well in MRI.

and from there moves to the brain. Detectors for the gamma rays can pick up the signals that come from the areas where the tagged glucose accumulates. In this way, one can see which areas of the brain are involved when we process, for example, visual information (Figure 3.8). PET scans can diagnose early stages of epilepsy and other diseases that involve abnormal glucose metabolism, such as schizophrenia.

Because tumors have high metabolic rates, PET scans using FDG are becoming increasingly preferred as the diagnostic choice for their detection and localization. FDG/PET has been used in the diagnosis of malignant melanoma, malignant lymphoma, and ovarian cancer, among other conditions.

Figure 3.8 Positron emission tomography brain scans. The upper scans show that 18-fluorodeoxyglucose can cross the blood–brain barrier. The lower scans show that visual stimulation increases the blood flow to and the glucose concentration in certain areas of the brain. These areas are shown in red.

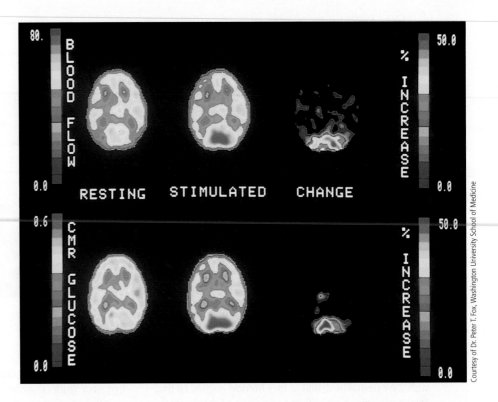

Courtesy of Dr. Peter T. Fox, Washington University School of Medicine

Another important use of radioactive isotopes is to learn what happens to ingested material. The foods and drugs swallowed or otherwise taken in by the body are transformed, decomposed, and excreted. To understand the pharmacology of a drug, it is important to know how these processes occur. For example, a certain drug may be effective in treating certain bacterial infections. Before beginning a clinical trial of the drug, its manufacturer must prove that the drug is not harmful to humans. In a typical case, the drug is first tested in animal studies. It is synthesized, and some radioactive isotope, such as hydrogen-3, carbon-14, or phosphorus-32, is incorporated into its structure. The drug is administered to the test animals, and after a certain period the animals are sacrificed. The fate of the drug is then determined by isolating from the body any radioactive compounds formed.

One typical pharmacological experiment studied the effects of tetracycline. This powerful antibiotic tends to accumulate in bones and is not given to pregnant women because it is transferred to the bones of the fetus. A particular tetracycline was tagged with radioactive hydrogen-3, and its uptake in rat bones was monitored in the presence and the absence of a sulfa drug. With the aid of a scintillation counter, researchers measured the radiation intensity of the maternal and fetal bones. They found that the sulfa drug helped to minimize this undesirable side effect of tetracycline—namely, its accumulation in the fetal bones.

The metabolic fate of essential chemicals in the body can also be followed with radioactive tracers. The use of radioactive isotopes has illuminated a number of normal and pathological body functions as well.

B. Radiation Therapy

The main use of radioactive isotopes in therapy is the selective destruction of pathological cells and tissues. Recall that radiation, whether from gamma rays, Xrays, or other sources, is detrimental to cells. Ionizing radia-

CHEMICAL CONNECTIONS 3D

How Radiation Damages Tissues: Free Radicals

As mentioned earlier, high-energy radiation damages tissue by causing ionization. That is, the radiation knocks electrons out of the molecules that make up the tissue (generally one electron per molecule), thereby forming unstable ions. For example, the interaction of high-energy radiation with water forms an unstable cation.

$$Energy + H_2O \longrightarrow H_2O^+ + e^-$$

$$H_2O^+ \longrightarrow H^+ + \cdot OH$$
$$\text{Hydroxyl}$$
$$\text{radical}$$

The positive charge on the cation means that one of the electrons normally present in the water molecule, either from a covalent bond or from an unshared pair, is missing in this cation; it has been knocked out. Once formed, the H_2O^+ cation is unstable and decomposes or H^+ and a hydroxyl radical:

The unpaired electron is on oxygen

$$[H-\overset{\cdot}{\underset{\cdot\cdot}{O}}-H]^+$$

It is instructive to consider the hydroxide ion and the hydroxyl radical.

An unpaired electron

:ÖH   ·ÖH
Hydroxide   Hydroxyl
ion   radical

Whereas the oxygen atom in the hydroxide ion, OH^-, has a complete octet—it is surrounded by three unshared pairs of electrons and one shared—the oxygen atom in the hydroxyl radical is surrounded by only seven valence electrons,—two unshared pairs, one shared pair and one unpaired electron. Compounds that have unpaired electrons are called **free radicals**, or more simply, just **radicals**.

The fact that the oxygen atom of the $\cdot OH$ radical has an incomplete octet makes the radical extremely reactive. It rapidly interacts with other molecules, causing chemical reactions that damage tissues. These reactions may have especially serious consequences if they occur inside cell nuclei and damage genetic material. In addition, they affect rapidly dividing cells more than they do stationary cells. Thus the damage is greater to embryonic cells, cells of the bone marrow, the intestines, and cells in the lymph. Symptoms of radiation sickness include nausea, vomiting, a decrease in the white blood cell count, and loss of hair.

tion damages cells, especially those that divide rapidly. This damage may be great enough to destroy diseased cells or to sufficiently alter the genes in them so that multiplication of the cells slows down.

In therapy applications, cancerous cells are the main targets for such ionizing radiation. Radiation is typically used when a cancer is highly localized; however, it may also be employed when the cancerous cells spread and are in a metastatic state. In addition, it is used for preventive purposes, to eliminate any possible remaining cancerous cells after surgery has been performed. The idea, of course, is to kill cancerous cells but not normal ones. Therefore, radiation such as high-energy Xrays or gamma rays from a cobalt-60 source is focused at a small part of the body where cancerous cells are suspected to reside. Besides Xrays and gamma rays from cobalt-60, other ionizing radiation is used to treat inoperable tumors. Proton beams from cyclotrons, for instance, have been used to treat ocular melanoma and tumors of the skull base and spine.

Despite this pinpointing technique, the radiation inevitably kills normal cells along with their cancerous counterparts. Because the radiation is most effective against rapidly dividing cancer cells, rather than normal cells, and because the radiation is aimed at a specific location, the damage to healthy tissues is minimized.

Another way to localize radiation damage in therapy is to use specific radioactive isotopes. In the case of thyroid cancer, large doses of iodine-131 are administered, which are taken up by the gland. The isotope, which has high radioactivity, kills all the cells of the gland (cancerous as well as normal ones), but does not appreciably damage other organs.

A metastatic state exists when the cancerous cells break off from their primary site(s) and begin moving to other parts of the body.

Another radioisotope, iodine-125, is used in the treatment of prostate cancer. Seeds of iodine-125, a gamma emitter, are implanted in the cancerous area of the prostate gland while being imaged with ultrasound. The seeds deliver 160 Gy (16,000 rad) over their lifetime.

A newer form of treatment of prostate cancer with great potential relies on actinium-225, an alpha emitter. As discussed in Section 3.6, alpha particles cause more damage to the tissues than any other form of radiation, but they have low penetrating power. Researchers have developed a very clever way to deliver actinium-225 to the cancerous region of the prostate gland without damaging healthy tissues. The cancer has a high concentration of prostate-specific antigen (PSA) on its surface. A monoclonal antibody (Section 30.4) hones in on the PSA and interacts with it. A single actinium-225 atom attached to such a monoclonal antibody can deliver the desired radiation, thereby destroying the cancer. Actinium-225 is especially effective because it has a half-life of ten days and it decays to three nuclides, themselves alpha emitters. In clinical trials, a single injection of antibody with an intensity in the kBq range (nanocuries) led to tumor regression without toxicity.

3.8 | What Is Nuclear Fusion?

An estimated 98% of all matter in the universe is made up of hydrogen and helium. The "big bang" theory of the formation of the universe postulates that our universe started with an explosion (big bang) in which matter was formed out of energy and that, at the beginning, only the lightest element, hydrogen, was in existence. Later, as the universe expanded, stars were born when the hydrogen clouds collapsed under gravitational forces. In the cores of these stars, hydrogen nuclei fused together to form helium.

The fusion of two hydrogen nuclei into a helium nucleus liberates a very large amount of energy in the form of photons, largely by the following reaction:

$$\overset{2}{\underset{1}{}}\text{H} \;+\; \overset{3}{\underset{1}{}}\text{H} \;\longrightarrow\; \overset{4}{\underset{2}{}}\text{He} + \overset{1}{\underset{0}{}}\text{n} + 5.3 \times 10^8 \text{ kcal/mol He}$$

Hydrogen-2   Hydrogen-3
(Deuterium)   (Tritium)

Nuclear fusion Joining together atomic nuclei to form a nucleus heavier than the starting nuclei.

The reactions occurring in the Sun are essentially the same as those in hydrogen bombs.

This process, called **fusion,** is how the Sun makes its energy. Uncontrolled fusion is employed in the "hydrogen bomb." If we can ever achieve a controlled version of this fusion reaction (which is unlikely to happen in the near term), we should be able to solve our energy problems.

As we have just seen, the fusion of deuterium and tritium nuclei to form a helium nucleus gives off a very large amount of energy. What is the source of this energy? When we compare the masses of the reactants and the products we see that there is a loss of $5.0301 - 5.0113 = 0.0188$ g for each mole of helium formed:

$$\overset{2}{\underset{1}{}}\text{H} \;+\; \overset{3}{\underset{1}{}}\text{H} \;\longrightarrow\; \overset{4}{\underset{2}{}}\text{He} \;+\; \overset{1}{\underset{0}{}}\text{n}$$

2.01410 g   3.0161 g   4.0026 g   1.0087 g

5.0301 g   5.0113 g

When the deuterium and tritium nuclei are converted to helium and a neutron, the extra mass has to go somewhere. But where? The answer is that the missing mass is converted to energy. We even know, from the equation developed by Albert Einstein (1879–1955), how much energy we can get from the conversion of any amount of mass:

$$E = mc^2$$

This equation says that the mass (m), in kilograms, that is lost multiplied by the square of the velocity of light (c^2, where $c = 3.0 \times 10^8$ m/s), in meters squared per second squared (m^2/s^2), is equal to the amount of energy created (E), in joules. For example, 1 g of matter completely converted to energy would produce 8.8×10^{13} J, which is enough energy to boil 34,000,000 L of water initially at 20°C. This is equivalent to the amount of water in an Olympic-size swimming pool. As you can see, we get a tremendous amount of energy from a little bit of mass.

All of the **transuranium elements** (elements with atomic numbers greater than 92) are artificial and have been prepared by a fusion process in which heavy nuclei are bombarded with light ones. Many, as their names indicate, were first prepared at the Lawrence Laboratory of the University of California, Berkeley, by Glenn Seaborg (1912–1999; Nobel laureate in chemistry, 1951) and his colleagues:

$$^{244}_{96}\text{Cm} + {}^4_2\text{He} \longrightarrow {}^{245}_{97}\text{Bk} + {}^1_1\text{H} + 2\,{}^1_0\text{n}$$

$$^{238}_{92}\text{U} + {}^{12}_{6}\text{C} \longrightarrow {}^{246}_{98}\text{Cf} + 4\,{}^1_0\text{n}$$

$$^{252}_{98}\text{Cf} + {}^{10}_{5}\text{B} \longrightarrow {}^{257}_{103}\text{Lr} + 5\,{}^1_0\text{n}$$

These transuranium elements are unstable, and most have very short half-lives. For example, the half-life of lawrencium-257 is 0.65 second. Many of the new superheavy elements have been obtained by bombarding lead isotopes with calcium-48 or nickel-64. So far, the creation of elements 110, 111, and 112–116 has been reported, even though their detection was based on the observation of the decay of a single atom.

3.9 | What Is Nuclear Fission and How Is It Related to Atomic Energy?

In the 1930s, Enrico Fermi (1901–1954) and his colleagues in Rome, and Otto Hahn (1879–1968), Lisa Meitner (1878–1968), and Fritz Strassman (1902–1980) in Germany tried to produce new transuranium elements by bombarding uranium-235 with neutrons. To their surprise, they found that, rather than fusion, they obtained **nuclear fission** (fragmentation of large nuclei into smaller pieces):

$$^{235}_{92}\text{U} + {}^1_0\text{n} \longrightarrow {}^{141}_{56}\text{Ba} + {}^{92}_{36}\text{Kr} + 3\,{}^1_0\text{n} + \gamma + \text{energy}$$

In this reaction, a uranium-235 nucleus first absorbs a neutron to become uranium-236 and then breaks into two smaller nuclei. The most important product of this nuclear decay is energy, which results because the products have less mass than the starting materials. This form of energy, called **atomic energy,** has been used for both war (in the atomic bomb) and peace.

Nuclear power plant in Salem, New Jersey.

Courtesy of Public Service Electric & Gas Co.

Figure 3.9 A chain reaction begins when a neutron collides with a nucleus of uranium-235.

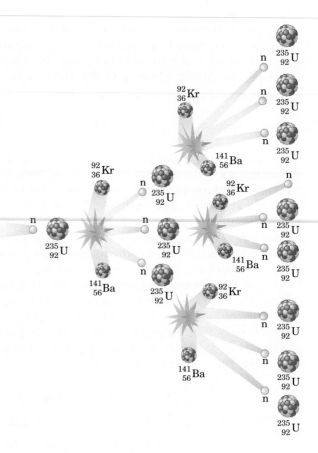

With uranium-235, each fission produces three neutrons, which in turn can generate more fissions by colliding with other uranium-235 nuclei. If even one of these neutrons produces a new fission, the process becomes a self-propagating **chain reaction** (Figure 3.9) that continues at a constant rate. If all three neutrons are allowed to produce new fission, the rate of the reaction increases constantly and eventually culminates in a nuclear explosion. In nuclear plants, the rate of reaction can be controlled by inserting boron control rods to absorb neutrons.

In nuclear power plants, the energy produced by fission is sent to heat exchangers and used to generate steam, which drives a turbine to produce electricity. Today, more than 15% of the electrical energy in the United States is supplied by such plants. The opposition to nuclear plants is based on safety considerations and on the unsolved problems of waste disposal. Although nuclear plants in general have good safety records, accidents such as those at Chernobyl (see Chemical Connections 3E) and Three Mile Island have caused concern.

Waste disposal is a long-range problem. The fission products of nuclear reactors are highly radioactive themselves, with long half-lives. Spent fuel contains these high-level fission products as nuclear wastes, together with uranium and plutonium that can be recovered and reused as mixed oxide (MOX) fuel. Reprocessing is costly: Although done routinely in Europe and Russia, it is not practiced by nuclear plants in the United States for economic reasons. However, this situation may change because cleaner extraction processes have been developed that use supercritical carbon dioxide (Chemical Connections 6F), thereby eliminating the need to dispose of solvent.

The United States has about 50,000 metric tons of spent fuel, stored under water and in dry casks at power plants. The additional nuclear

CHEMICAL CONNECTIONS 3E

Radioactive Fallout from Nuclear Accidents

On April 26, 1986, an accident occurred at the nuclear reactor in the town of Chernobyl in the former Soviet Union. It was a clear reminder of the dangers involved in this industry and of the far-reaching contamination that such accidents can produce. In Sweden, more than 500 miles away from the accident, the radioactive cloud increased the background radiation from 4 to 15 times the normal level. The radioactive cloud reached England, about 1300 miles away, one week later. There, it increased the natural background radiation by 15%. The radioactivity from iodine-131 was measured at 400 Bq/L in milk and 200 Bq/kg in leafy vegetables. Even some 4000 miles away in Spokane, Washington, an iodine-131 activity of 242 Bq/L was found in rainwater, and smaller activities—1.03 Bq/L of ruthenium-103 and 0.66 Bq/L of cesium-137—were recorded as well. These levels are not harmful.

Map showing those areas most affected by the Chernobyl accident.

---

wastes from the nuclear weapons program, research reactors, and other sources are stored by the Department of Energy in three major sites. After 40 years, the level of radioactivity that the wastes had immediately after their removal from the reactor is reduced a thousandfold. Such nuclear waste is a good candidate for underground burial. Recently, the U.S. federal government gave its final approval to a plan to store nuclear waste in Yucca Mountain, Nevada.

Environmental concerns persist, however. The site cannot be guaranteed to stay dry for centuries. Moisture may corrode the steel cylinders and even the inner glass/ceramic cylinders surrounding the nuclear waste. Some fear that leaked materials from such storage tanks may escape if carbon-14 is oxidized to radioactive carbon dioxide or, less likely, that other radioactive nuclides may contaminate the groundwater, which lies far below the desert rock of Yucca Mountain.

To keep these problems in perspective, one must remember that most other ways of generating large amounts of electrical power have their own environmental problems. For example, burning coal or oil contributes to the accumulation of CO_2 in the atmosphere and to acid rain (see Chemical Connections 7A).

Closer to the source of the nuclear accident, in neighboring Poland, potassium iodide pills were given to children. This step was taken to prevent radioactive iodine-131 (which might come from contaminated food) from concentrating in their thyroid glands, which could lead to cancer. In the wake of the September 11, 2001, terrorist attacks, Massachusetts became the first state to authorize the storage of KI pills in case of nuclear-related terrorist activity.

Storage of nuclear wastes in a storage room carved out of an underground salt bed.

U.S. Department of Energy

SUMMARY OF KEY QUESTIONS

SECTION 3.1 How Was Radioactivity Discovered?

- Henri Becquerel discovered radioactivity in 1896.

SECTION 3.2 What Is Radioactivity?

- The four major types of radioactivity are **alpha particles** (helium nuclei), **beta particles** (electrons), **gamma rays** (high-energy photons), and **positrons** (positively charged electrons).

SECTION 3.3 What Happens When a Nucleus Emits Radioactivity?

- When a nucleus emits a **beta particle,** the new element has the same mass number but an atomic number one unit greater.
- When a nucleus emits an **alpha particle,** the new element has an atomic number two units lower and a mass number four units smaller.
- When a nucleus emits a **positron** (positive electron), the new element has the same mass number but an atomic number one unit smaller.
- In **gamma emission,** no transmutation takes place; only the energy of the nucleus is lowered.
- In **electron capture,** the new element has the same mass number but an atomic number one unit smaller.

SECTION 3.4 What Is Nuclear Half-life?

- Each radioactive isotope decays at a fixed rate described by its **half-life,** which is the time required for half of the sample to decay.

SECTION 3.5 How Do We Detect and Measure Nuclear Radiation?

- Radiation is detected and counted by devices such as **Geiger-Müller counters.**

- The main unit of intensity of radiation is the **curie (Ci),** which is equal to 3.7×10^{10} disintegrations per second. Other common units are the millicurie (mCi), the microcurie (μCi), and the becquerel (Bq).

SECTION 3.6 How Is Radiation Dosimetry Related to Human Health?

- For medical purposes and to measure potential radiation damage, the absorbed dose is measured in **rads.** Different particles damage body tissues differently; the **rem** is a measure of relative damage caused by type of radiation.

SECTION 3.7 What Is Nuclear Medicine?

- Nuclear medicine is the use of radionuclei for diagnostic imaging and therapy.

SECTION 3.8 What Is Nuclear Fusion?

- **Nuclear fusion** is the combining (fusing) of two lighter nuclei to form a heavier nucleus. Helium is synthesized in the interiors of stars by fusion of hydrogen nuclei. The energy released in this process is the energy of our Sun.

SECTION 3.9 What Is Nuclear Fission and How Is It Related to Atomic Energy?

- **Nuclear fission** is the splitting of a heavier nucleus into two or more smaller nuclei. Nuclear fission releases large amounts of energy, which can be either controlled (nuclear reactors) or uncontrolled (nuclear weapons).

SUMMARY OF KEY REACTIONS

1. **Beta (β) emission (Section 3.3B)** When a nucleus decays by beta emission, the new element has the same mass number but an atomic number one unit greater.

$$^{32}_{15}\text{P} \longrightarrow {}^{32}_{16}\text{S} + {}^{0}_{-1}\text{e}$$

2. **Alpha (α) emission (Section 3.3C)** When a nucleus decays by alpha emission, the new nucleus has a mass number four units smaller and an atomic number two units smaller.

$$^{238}_{92}\text{U} \longrightarrow {}^{234}_{90}\text{Th} + {}^{4}_{2}\text{He}$$

3. **Positron (β^{+}) emission (Section 3.3D)** When a nucleus decays by positron emission, the new element has the

same mass number but an atomic number one unit smaller.

$$^{11}_{6}\text{C} \longrightarrow {}^{11}_{5}\text{B} + {}^{0}_{+1}\text{e}$$

4. **Gamma (γ) emission (Section 3.3E)** When a nucleus emits gamma radiation, there is no change in either mass number or atomic number of the nucleus.

$$^{11}_{5}\text{B}^{*} \longrightarrow {}^{11}_{5}\text{B} + \gamma$$

5. **Electron capture (Section 3.3F)** When a nucleus decays by electron capture, the product nucleus has the same mass number but an atomic number one unit smaller.

$$^{7}_{4}\text{Be} + {}^{0}_{-1}\text{e} \longrightarrow {}^{7}_{3}\text{Li}$$

6. Nuclear fusion (Section 3.8) In nuclear fusion, two or more nuclei react to form a larger nucleus. In the process, there is a slight decrease in mass; the sum of the masses of the fusion products is less than the sum of the masses of the starting nuclei. The lost mass appears as energy.

$$^2_1H + ^3_1H \longrightarrow ^4_2He + ^1_0n + 5.3 \times 10^8 \text{ kcal/mol He}$$

7. Nuclear fission (Section 3.9) In nuclear fission, a nucleus captures a neutron to form a nucleus with a mass number increased by one unit. The new nucleus then splits into two smaller nuclei.

$$^{235}_{92}U + ^1_0n \longrightarrow ^{141}_{56}Ba + ^{92}_{36}Kr + 3\,^1_0n + \gamma + \text{energy}$$

PROBLEMS

Chemistry Now™

Assess your understanding of this chapter's topics with additional quizzing and conceptual-based problems at **http://now.brookscole.com/gob8** or on the CD.

A blue problem number indicates an applied problem.

■ denotes problems that are available on the GOB ChemistryNow website or CD and are assignable in OWL.

SECTION 3.2 What Is Radioactivity?

3.8 What is the difference between an alpha particle and a proton?

3.9 Microwaves are a form of electromagnetic radiation that is used for the rapid heating of foods. What is the frequency of a microwave with a wavelength of 5.8 cm?

3.10 In each case, given the frequency, give the wavelength in centimeters or nanometers and tell what kind of radiation it is.
(a) 7.5×10^{14}/s
(b) 1.0×10^{10}/s
(c) 1.1×10^{15}/s
(d) 1.5×10^{18}/s

3.11 ■ Red light has a wavelength of 650 nm. What is its frequency?

3.12 ■ Which has the longest wavelength: (a) infrared, (b) ultraviolet, or (c) Xrays? Which has the highest energy?

3.13 ■ Write the symbol for a nucleus with the following components:
(a) 9 protons and 10 neutrons
(b) 15 protons and 17 neutrons
(c) 37 protons and 50 neutrons

3.14 In each pair, tell which isotope is more likely to be radioactive:
(a) Nitrogen-14 or nitrogen-13
(b) Phosphorus-31 or phosphorus-33
(c) Lithium-7 or lithium-9
(d) Calcium-39 or calcium-40

3.15 Which isotope of boron is the most stable: boron-8, boron-10, or boron-12?

3.16 Which isotope of oxygen is the most stable: oxygen-14, oxygen-16, or oxygen-18?

SECTION 3.3 What Happens When a Nucleus Emits Radioactivity?

3.17 Samarium-151 is a beta emitter. Write an equation for this nuclear reaction and identify the product nucleus.

3.18 ■ The following nuclei turn into new nuclei by emitting beta particles. Write an equation for each nuclear reaction and identify the product nucleus.
(a) $^{159}_{63}Eu$ (b) $^{141}_{56}Ba$ (c) $^{242}_{95}Am$

3.19 Chromium-51 is used in diagnosing the pathology of the spleen. The nucleus of this isotope captures an electron according to the following equation. What is the transmutation product?

$$^{51}_{24}Cr + ^0_{-1}e \longrightarrow ?$$

3.20 ■ The following nuclei decay by emitting alpha particles. Write an equation for each nuclear reaction and identify the product nucleus.
(a) $^{210}_{83}Bi$ (b) $^{238}_{94}Pu$ (c) $^{174}_{72}Hf$

3.21 Curium-248 was bombarded, yielding antimony-116 and cesium-160. What was the bombarding nucleus?

3.22 Phosphorus-29 is a positron emitter. Write an equation for this nuclear reaction and identify the product nucleus.

3.23 For each of the following, write a balanced nuclear equation and identify the radiation emitted.
(a) Beryllium-10 changes to boron-10
(b) Europium-151m changes to europium-151
(c) Thallium-195 changes to mercury-195
(d) Plutonium-239 changes to uranium-235

3.24 In the first three steps in the decay of uranium-238, the following isotopic species appear: uranium-238, thorium-234, protactinium-234, which then decays to uranium-234. What kind of emission occurs in each step?

3.25 What kind of emission does *not* result in transmutation?

3.26 ■ Complete the following nuclear reactions.

(a) $^{16}_{8}O + ^{16}_{8}O \longrightarrow ? + ^{4}_{2}He$

(b) $^{235}_{92}U + ^{1}_{0}n \longrightarrow ^{90}_{38}Sr + ? + 3\,^{1}_{0}n$

(c) $^{13}_{6}C + ^{4}_{2}He \longrightarrow ^{16}_{8}O + ?$

(d) $^{210}_{83}Bi \longrightarrow ? + ^{0}_{-1}e$

(e) $^{12}_{6}C + ^{1}_{1}H \longrightarrow ? + \gamma$

3.27 Americium-240 is made by bombarding plutonium-239 with α particles. In addition to americium-240, a proton and two neutrons are formed. Write a balanced equation for this nuclear reaction.

SECTION 3.4 What Is Nuclear Half-life?

3.28 Iodine-125 emits gamma rays and has a half-life of 60 days. If a 20-mg pellet of iodine-125 is implanted onto a prostate gland, how much iodine-125 remains there after one year?

3.29 Polonium-218, a decay product of radon-222 (see Chemical Connections 3B), has a half-life of 3 min. What percentage of the polonium-218 formed will remain in the lung 9 min after inhalation?

3.30 ■ A rock containing 1 mg plutonium-239/kg rock is found in a glacier. The half-life of plutonium-239 is 25,000 years. If this rock was deposited 100,000 years ago during an ice age, how much plutonium-239, per kilogram of rock, was in the rock at that time?

3.31 The element radium is extremely radioactive. If you converted a piece of radium metal to radium chloride (with the weight of the radium remaining the same), would it become less radioactive?

3.32 In what ways can we increase the rate of radioactive decay? Decrease it?

3.33 Suppose 50.0 mg of potassium-45, a beta emitter, was isolated in pure form. After one hour, only 3.1 mg of the radioactive material was left. What is the half-life of potassium-45?

3.34 A patient receives 200 mCi of iodine-131, which has a half-life of eight days.

(a) If 12% of this amount is taken up by the thyroid gland after two hours, what will be the activity of the thyroid after two hours, in millicuries and in counts per minute?

(b) After 24 days, how much activity will remain in the thyroid gland?

SECTION 3.5 How Do We Detect and Measure Nuclear Radiation?

3.35 If you work in a lab containing radioisotopes emitting all kinds of radiation, from which emission should you seek most protection?

3.36 What do Geiger-Müller counters measure: (a) the intensity or (b) the energy of radiation?

3.37 You work in a lab and it is known that radioactivity is being emitted with an intensity of 175 mCi at a distance of 1.0 m from the source. How far, in meters, from the source should you stand if you wish to be subjected to no more than 0.20 mCi?

SECTION 3.6 How Is Radiation Dosimetry Related to Human Health?

3.38 Does a curie (Ci) measure radiation intensity or energy?

3.39 What property is measured each term?

(a) Rad   (b) Rem   (c) Roentgen
(d) Curie   (e) Gray   (f) Becquerel
(g) Sievert

3.40 A radioactive isotope with an activity (intensity) of 80.0 mCi per vial is delivered to a hospital. The vial contains 7.0 cc of liquid. The instruction is to administer 7.2 mCi intravenously. How many cubic centimeters of liquid should be used for one injection?

3.41 Why does exposure of a hand to alpha rays not cause serious damage to the person, whereas entry of an alpha emitter into the lung as an aerosol produces very serious damage to the person's health?

3.42 A certain radioisotope has an intensity of 10^6 Bq at 1-cm distance from the source. What would be the intensity at 20 cm? Give your answer in both Bq and μCi units.

3.43 Assuming the same amount of effective radiation, in rads, from three sources, which would be the most damaging to the tissues: alpha particles, beta particles, or gamma rays?

3.44 In an accident involving radioactive exposure, person A received 3.0 Sv while person B received 0.50 mrem of exposure. Who was hurt more seriously?

SECTION 3.7 What Is Nuclear Medicine?

3.45 In 1986, the nuclear reactor in Chernobyl had an accident and spewed radioactive nuclei that were carried by the winds for hundreds of miles. Today, among the child survivors of the event, the most common damage is thyroid cancer. Which radioactive nucleus do you expect to be responsible for these cancers?

3.46 Cobalt-60, with a half-life of 5.26 years, is used in cancer therapy. The energy of the radiation from cobalt-62 is even higher (half-life = 14 minutes). Why isn't cobalt-60 also used for cancer therapy?

3.47 Match the radioactive isotope with its proper use:

_____ (a) Cobalt-60   1. Heart scan in exercise
_____ (b) Thallium-201   2. Measure water content of body
_____ (c) Tritium   3. Kidney scan
_____ (d) Mercury-197   4. Cancer therapy

SECTION 3.8 What Is Nuclear Fusion?

3.48 What is the product of the fusion of hydrogen-2 and hydrogen-3 nuclei?

3.49 Assuming that one proton and two neutrons will be produced in an alpha-bombardment fusion reaction, which target nucleus would you use to obtain berkelium-249?

3.50 Element 109 was first prepared in 1982. A single atom of this element ($^{266}_{109}$Mt), with a mass number of 266 was made by bombarding a bismuth-209 nucleus with an iron-58 nucleus. What other products, if any, must have been formed besides $^{266}_{109}$Mt?

3.51 A new element was formed when lead-208 was bombarded by krypton-86. One could detect four neutrons as the product of the fusion. Identify the new element.

SECTION 3.9 What Is Nuclear Fission and How Is It Related to Atomic Energy?

3.52 Boron-10 is used as control rods in nuclear reactors. This nucleus absorbs a neutron and then emits an alpha particle. Write an equation for each nuclear reaction and identify each product nucleus.

Chemical Connections

3.53 (Chemical Connections 3A) Why is it accurate to assume that the carbon-14 to carbon-12 ratio in a living plant remains constant over the lifetime of the plant?

3.54 (Chemical Connections 3A) In a recent archeological dig in the Amazon region of Brazil, charcoal paintings were found in a cave. The carbon-14 content of the charcoal was one-fourth of what is found in charcoal prepared from that year's tree harvest. How long ago was the cave settled?

3.55 (Chemical Connections 3A) Carbon-14 dating of the Shroud of Turin indicated that the plant from which the shroud was made was alive around AD 1350. To how many half-lives does this correspond?

3.56 (Chemical Connections 3A) The half-life of carbon-14 is 5730 years. The wrapping of an Egyptian mummy gave off 7.5 counts per minute per gram of carbon. A piece of linen purchased today would give an activity of 15 counts per minute per gram of carbon. How old is the mummy?

3.57 (Chemical Connections 3B) How does radon-222 produce polonium-218?

3.58 (Chemical Connections 3E) In a nuclear accident, one of the radioactive nuclei that concerns people is iodine-131. Iodine is easily vaporized, can be carried by the winds, and can cause radioactive fallout hundreds—even thousands—of miles away. Why is iodine-131 especially harmful?

Additional Problems

3.59 Phosphorus-32 ($t_{1/2}$ = 14.3 d) is used in the medical imaging and diagnosis of eye tumors. Suppose a patient is given 0.010 mg of this isotope. Prepare a graph showing the mass in milligrams remaining in the patient's body after one week. (Assume that none is excreted from the body.)

3.60 During the bombardment of argon-40 with protons, one neutron is emitted for each proton absorbed. What new element is formed?

3.61 Neon-19 and sodium-20 are positron emitters. What products result in each case?

3.62 The half-life of nitrogen-16 is 7 seconds. How long does it take for 100 mg of nitrogen-16 to be reduced to 6.25 mg?

3.63 Do the curie and the becquerel measure the same or different properties of radiation?

3.64 Selenium-75 has a half-life of 120.4 days, so it would take 602 days (five half-lives) to diminish to 3% of the original quantity. Yet this isotope is used for pancreatic scans without any fear that the radioactivity will cause undue harm to the patient. Suggest a possible explanation.

3.65 Use Table 3.4 to determine the percentage of annual radiation we receive from the following sources:

(a) Naturally occurring sources

(b) Diagnostic medical sources

(c) Nuclear power plants

3.66 $^{225}_{89}$Ac is an alpha emitter. In its decay process, it produces three more alpha emitters in succession. Identify each of the decay products.

3.67 Which radiation will cause more ionization, Xrays or radar?

3.68 You have an old wristwatch that still has radium paint on its dial. Measurement of the radioactivity of the watch shows a beta-ray count of 0.50 count/s. If 1.0 microcurie of radiation of this sort produces 1000 mrem/year, how much radiation, in millirems, do you expect from the wristwatch if you wear it for one year?

3.69 Americium-241, which is used in some smoke detectors, has a half-life of 432 years and is an alpha emitter. What is the decay product of americium-241, and approximately what percentage of the original americium-241 will be still around after 1000 years?

3.70 On rare occasions a nucleus, instead of emitting a beta particle, captures one. Berkelium-246 is such a nucleus. What is the product of this nuclear transmutation?

3.71 A patient is reported to have been irradiated by a dose of 1 sievert in a nuclear accident. Is he in mortal danger?

3.72 What is the ground state of a nucleus?

3.73 Explain the following:

(a) It is impossible to have a completely pure sample of any radioactive isotope.

(b) Beta emission of a radioactive isotope creates a new isotope with an atomic number one unit higher than that of the radioactive isotope.

3.74 Yttrium-90, which emits beta particles, is used in radiotherapy. What is the decay product of yttrium-90?

3.75 The half-lives of some oxygen isotopes are as follows:

Oxygen-14: 71 s   Oxygen-15: 124 s

Oxygen-19: 29 s   Oxygen-20: 14 s

Oxygen-16 is the stable, nonradioactive isotope. Do the half-lives indicate anything about the stability of the other oxygen isotopes?

3.76 $^{225}_{89}$Ac is effective in prostate cancer therapy when administered at kBq levels. If an antibody tagged with $^{225}_{89}$Ac has an intensity of 2 million Bq/mg and if a solution contains 5 mg/L tagged antibody, how many milliliters of the solution should you use for an injection to administer 1 kBq intensity?

3.77 When $^{208}_{82}$Pb is bombarded with $^{64}_{28}$Ni, a new element and six neutrons are produced. Identify the new element.

3.78 Americium-241, the isotope used in smoke detectors, has a half-life of 475 years, which is sufficiently long to allow for handling in large quantities. This isotope is prepared in the laboratory by bombarding plutonium-239 with neutrons. In this reaction, plutonium-239 absorbs two neutrons and then decays by emission of a β particle. Write an equation for this nuclear reaction and identify the isotope formed as an intermediate between plutonium-239 and americium-241.

3.79 Boron-10, an effective absorber of neutrons, is used in control rods of uranium-235 fission reactors to absorb neutrons and thereby control the rate of reaction. Boron-10 absorbs a neutron and then emits an α particle. Write a balanced equation for this nuclear reaction and identify the nucleus formed as an intermediate between boron-10 and the final nuclear product.

3.80 Tritium, $^{3}_{1}$H, is a beta emitter widely used as a radioactive tracer in chemical and biochemical research. Tritium is prepared by the bombardment of lithium-6 with neutrons. Complete the following nuclear equation:

$$^{6}_{3}\text{Li} + ^{1}_{0}\text{n} \longrightarrow ^{6}_{3}\text{He} + ?$$

Chemical Bonds

Sodium chloride crystal.

Charles D. Winters

4.1 | What Do We Need to Know Before We Begin?

In Chapter 2, we stated that compounds are tightly bound groups of atoms. In this chapter, we will see that the atoms in compounds are held together by powerful forces of attraction called chemical bonds. There are two main types: ionic bonds and covalent bonds. We begin by examining ionic bonds. To talk about ionic bonds, however, we must first discuss why atoms form the ions they do.

4.2 | What Is the Octet Rule?

In 1916, Gilbert N. Lewis (Section 2.6) devised a beautifully simple model that unified many of the observations about chemical bonding and chemical reactions. He pointed out that the lack of chemical reactivity of the noble gases (Group 8A) indicates a high degree of stability of their electron configurations: helium with a filled valence shell of two electrons ($1s^2$), neon with

Noble Gas	Noble Gas Notation
He	$1s^2$
Ne	$[He]2s^2 2p^6$
Ar	$[Ne]3s^2 3p^6$
Kr	$[Ar]4s^2 4p^6$
Xe	$[Kr]5s^2 5p^6$

Octet rule When undergoing chemical reaction, atoms of Group 1A–7A elements tend to gain, lose, or share sufficient electrons to achieve an election configuration having eight valence electrons

Anion An ion with a negative electric charge

Cation An ion with a positive electric charge

GOB
Chemistry Now™
Click *Chemistry Interactive* to see an example of **Ion Formation**

a filled valence shell of eight electrons ($2s^2 2p^6$), argon with a filled valence shell of eight electrons ($3s^2 3p^6$), and so forth.

The tendency to react in ways that achieve an outer shell of eight valence electrons is particularly common among Group 1A–7A elements and is given the special name of the **octet rule.** An atom with almost eight valence electrons tends to gain the needed electrons to have eight electrons in its valence shell and an electron configuration like that of the noble gas nearest to it in atomic number. In gaining electrons, the atom becomes a negatively charged ion called an **anion.** An atom with only one or two valence electrons tends to lose the number of electrons required to have an electron configuration like the noble gas nearest it in atomic number. In losing electrons, the atom becomes a positively charged ion called a **cation.** When an ion forms, the number of protons and neutrons in the nucleus of the atom remains unchanged; only the number of electrons in the valence shell of the atom changes.

EXAMPLE 4.1

Show how the following chemical changes obey the octet rule:
(a) A sodium atom loses an electron to form a sodium ion, Na^+.

$$Na \longrightarrow Na^+ + \boxed{e^-}$$

A sodium atom   A sodium ion   An electron

(b) A chlorine atom gains an electron to form a chloride ion, Cl^-.

$$Cl + \boxed{e^-} \longrightarrow Cl^-$$

A chlorine atom   An electron   A chloride ion

Solution

To see how each chemical change follows the octet rule, first write the condensed ground-state electron configuration (Section 2.6C) of the atom involved in the chemical change and the ion it forms, and then compare them.

(a) The condensed ground-state electron configurations for Na and Na^+ are

$$Na\ (11\ electrons):\ 1s^2 2s^2 2p^6\ 3s^1$$

$$Na^+\ (10\ electrons):\ 1s^2 2s^2 2p^6$$

A sodium atom has one electron in its valence shell. The loss of this one valence electron changes the sodium atom to a sodium ion, Na^+, which has a complete octet of electrons in its valence shell and the same electron configuration as neon, the noble gas nearest to it in atomic number. We can write this chemical change using Lewis dot structures (Section 2.6F):

$$Na\cdot \longrightarrow Na^+ + \boxed{e^-}$$

(b) The condensed ground-state electron configurations for Cl and Cl^- are as:

$$Cl\ (17\ electrons):\ 1s^2 2s^2 2p^6\ 3s^2 3p^5$$

$$Cl^-\ (18\ electrons):\ 1s^2 2s^2 2p^6\ 3s^2 3p^6$$

A chlorine atom has seven electrons in its valence shell. The gain of one electron changes the chlorine atom to a chloride ion, Cl^-, which has a complete octet of electrons in its valence shell and the same electron configuration as argon, the noble gas nearest to it in atomic number. We can write this chemical change using Lewis dot structures:

$$:\ddot{Cl}\cdot + e^- \longrightarrow :\ddot{Cl}:^-$$

Problem 4.1

Show how the following chemical changes obey the octet rule:
(a) A magnesium atom forms a magnesium ion, Mg^{2+}
(b) A sulfur atom forms a sulfide ion, S^{2-}

The octet rule gives us a good way to understand why Group 1A–7A elements form the ions that they do. It is not perfect, however, for the following two reasons:

1. Ions of period 1 and 2 elements with charges greater than +2 are unstable. Boron, for example, has three valence electrons. If it lost these three electrons, it would become B^{3+} and have a complete outer shell like that of helium. It seems, however, that this is far too large a charge for an ion of this period 2 element; consequently, this ion is not found in stable compounds. By the same reasoning, carbon does not lose its four valence electrons to become C^{4+}, nor does it gain four valence electrons to become C^{4-}. Either of these changes would place too great a charge on this period 2 element.

2. The octet rule does not apply to Group 1B–7B elements (the transition elements), most of which form ions with two or more different positive charges. Copper, for example, can lose one valence electron to form Cu^+; alternatively, it can lose two valence electrons to form Cu^{2+}.

It is important to understand that there are enormous differences between the properties of an atom and those of its ion(s). Atoms and their ions are completely different chemical species and have completely different chemical and physical properties. Consider, for example, sodium and chlorine. Sodium is a soft metal made of sodium atoms that react violently with water. Chlorine atoms are very unstable and even more reactive than sodium atoms. Both sodium and chlorine are poisonous. NaCl, common table salt, is made up of sodium ions and chloride ions. These two ions are quite stable and unreactive. Neither sodium ions nor chloride ions react with water at all.

Because atoms and their ions are different chemical species, we must be careful to distinguish one from the other. Consider the drug commonly known as "lithium," which is used to treat manic depression. The element lithium, like sodium, is a soft metal that reacts with water. The drug used to treat manic depression is not lithium atoms, Li, but rather lithium ions, Li^+, usually given in the form of lithium carbonate, Li_2CO_3. Another example comes from the fluoridation of drinking water, and of toothpastes and dental gels. The element fluorine, F_2, is an extremely poisonous and corrosive gas; it is not what is used for this purpose. Instead, fluoridation uses fluoride ions, F^-, in the form of sodium fluoride, NaF, a compound that is unreactive and nonpoisonous in the concentrations used.

(a) Sodium chloride

(b) Sodium

(c) Chlorine

Charles D. Winters

(*a*) The chemical compound sodium chloride (table salt) is composed of the elements (*b*) sodium and (*c*) chlorine in chemical combination; Salt is very different from the elements that constitute it.

4.3 | How Do We Name Anions and Cations?

Names for anions and cations are formed by using a system developed by the International Union of Pure and Applied Chemistry. We will refer to these names as "systematic" names. Of course, many ions have "common" names that were in use long before chemists undertook an effort to systematize their naming. In this and the following chapters, we will make every effort to use systematic names for ions, but where a long-standing common name remains in use, we will give it as well.

A. Naming Monatomic Cations

Copper(I) oxide and copper(II) oxide. The different copper ion charges result in different colors.

Charles D. Winters

A monatomic (containing only one atom) cation forms when a metal loses one or more valence electrons. Elements of Groups 1A, 2A, and 3A form only one type of cation. For ions from these metals, the name of the cation is the name of the metal followed by the word "ion" (Table 4.1). There is no need to specify the charge on these cations, because only one charge is possible. For example, Na^+ is sodium ion and Ca^{2+} is calcium ion.

Most transition and inner transition elements form more than one type of cation and, therefore, the name of the cation must show its charge. To show the charge we write a Roman numeral immediately following (with no space) the name of the metal (Table 4.2). For example, Cu^+ is copper(I) ion and Cu^{2+} is copper(II) ion. Note that even though silver is a transition metal, it forms only Ag^+; therefore, there is no need to use a Roman numeral to show this ion's charge.

Table 4.1 Names of Cations from Some Metals That Form Only One Positive Ion

Group 1A		Group 2A		Group 3A	
Ion	Name	Ion	Name	Ion	Name
H^+	Hydrogen ion	Mg^{2+}	Magnesium ion	Al^{3+}	Aluminum ion
Li^+	Lithium ion	Ca^{2+}	Calcium ion		
Na^+	Sodium ion	Sr^{2+}	Strontium ion		
K^+	Potassium ion	Ba^{2+}	Barium ion		

Table 4.2 Names of Cations from Four Metals That Form Two Different Positive Ions

Ion	Systematic Name	Common Name	Origin of the Symbol of the Element or the Common Name of the Ion
Cu^+	Copper(I) ion	Cuprous ion	*Cupr-* from *cuprum*, the Latin name for copper
Cu^{2+}	Copper(II) ion	Cupric ion	
Fe^{2+}	Iron(II) ion	Ferrous ion	*Ferr-* from *ferrum*, the Latin name for iron
Fe^{3+}	Iron(III) ion	Ferric ion	
Hg^+	Mercury(I) ion	Mercurous ion	*Hg* from *hydrargyrum*, the Latin name for mercury
Hg^{2+}	Mercury(II) ion	Mercuric ion	
Sn^{2+}	Tin(II) ion	Stannous ion	*Sn* from *stannum*, the Latin name for tin
Sn^{4+}	Tin(IV) ion	Stannic ion	

In the older, common system for naming metal cations with two different charges, the suffix *-ous* is used to show the smaller charge and *-ic* is used to show the larger charge (Table 4.2).

B. Naming Monatomic Anions

A monatomic anion is named by adding *-ide* to the stem part of the name. Table 4.3 gives the names of the monatomic anions we deal with most often.

C. Naming Polyatomic Ions

A **polyatomic ion** contains more than one atom. Examples are the hydroxide ion, OH^-, and the phosphate ion, PO_4^{3-}. We will not be concerned with how these ions are formed, only that they exist and are present in the materials around us. While rules for naming polyatomic ions have been developed, the simplest thing to do is just to memorize them.

The preferred system for naming polyatomic ions that differ in the number of hydrogen atoms is to use the prefixes *di-*, *tri-*, and so forth, to show the presence of more than one hydrogen. For example, HPO_4^{2-} is hydrogen phosphate ion, and $H_2PO_4^-$ is dihydrogen phosphate ion. Table 4.4 lists some of the important polyatomic ions. Because several hydrogen-containing polyatomic anions have common names that are still widely used, you should memorize them as well. In these common names, the prefix *bi* is used to show the presence of one hydrogen.

Table 4.3 Names of the Most Common Monatomic Anions

Anion	Stem Name	Anion Name
F^-	*fluor*	Fluoride
Cl^-	*chlor*	Chloride
Br^-	*brom*	Bromide
I^-	*iod*	Iodide
O^{2-}	*ox*	Oxide
S^{2-}	*sulf*	Sulfide

GOB
Chemistry ❖ Now™
Click *Coached Problems* to see an example of **Polyatomic Ions**

Table 4.4 Names of Common Polyatomic Ions

Polyatomic Ion	Name	Polyatomic Ion	Systematic Name
NH_4^+	Ammonium	HCO_3^-	Hydrogen carbonate (bicarbonate)*
OH^-	Hydroxide	SO_3^{2-}	Sulfite
NO_2^-	Nitrite	HSO_3^-	Hydrogen sulfite (bisulfite)
NO_3^-	Nitrate	SO_4^{2-}	Sulfate
CH_3COO^-	Acetate	HSO_4^-	Hydrogen sulfate (bisulfate)
CN^-	Cyanide	PO_4^{3-}	Phosphate
MnO_4^-	Permanganate	HPO_4^{2-}	Hydrogen phosphate
CrO_4^{2-}	Chromate	$H_2PO_4^-$	Dihydrogen phosphate
CO_3^{2-}	Carbonate		

*Common names, where still widely used, are given in parentheses.

CHEMICAL CONNECTIONS 4A

Coral Chemistry and Broken Bones

Bone is a highly structured matrix consisting of both inorganic and organic materials. The inorganic material is chiefly hydroxyapatite, $Ca_5(PO_4)_3OH$, which makes up about 70% of bone by dry weight. By comparison, the enamel of teeth consists almost entirely of hydroxyapatite. Chief among the organic components of bone are collagen fibers (proteins; see Chemical Connections 21F), which thread their way through the inorganic matrix, providing extra strength and allowing bone to flex under stress. Also weaving through the hydroxyapatite–collagen framework are blood vessels that supply nutrients.

A problem faced by orthopedic surgeons is how to repair bone damage. For a minor fracture, usually a few weeks in a cast suffice for the normal process of bone growth to repair the damaged area. For severe fractures, especially those involving bone loss, a bone graft may be needed. An alternative to a bone graft is an implant of synthetic bone material. One such material, called Pro Osteon, is derived by heating coral with ammonium hydrogen phosphate to form a hydroxyapatite similar to that of bone. Throughout the heating process, the porous structure of the coral, which resembles that of bone, is retained.

A wrist fracture repaired with bone cement (white area).

and/or screws, and let new bone tissue grow into the pores of the implant.

In an alternative process, a dry mixture of calcium dihydrogen phosphate monohydrate, $Ca(H_2PO_4)_2 \cdot H_2O$, calcium phosphate, $Ca_3(PO_4)_2$, and calcium carbonate, $CaCO_3$, is prepared. Just before the surgical implant occurs, these chemicals are mixed with a solution of sodium phosphate to form a paste that is then injected into the bony area to be repaired. In this way, the fractured bony area is held in the desired position by the synthetic material while the natural process of bone rebuilding replaces the implant with living bone tissue.

$$5CaCO_3 + 3(NH_4)_2HPO_4 \xrightarrow[24-60 \text{ hours}]{200°C}$$

Coral

$$Ca_5(PO_4)_3OH + 3(NH_4)_2CO_3 + 2H_2CO_3$$

Hydroxyapatite

The surgeon can shape a piece of this material to match the bone void, implant it, stabilize the area by inserting metal plates

---

4.4 | What Are the Two Major Types of Chemical Bonds?

A. Ionic and Covalent Bonds

According to the Lewis model of chemical bonding, atoms bond together in such a way that each atom participating in a bond acquires a valence-shell electron configuration matching that of the noble gas nearest to it in atomic number. Atoms acquire completed valence shells in two ways:

Ionic bond A chemical bond resulting from the attraction between a positive ion and a negative ion

1. An atom may lose or gain enough electrons to acquire a filled valence shell, becoming an ion as it does so (Section 4.2). An **ionic bond** results from the force of attraction between a cation and an anion.

Covalent bond A bond resulting from the sharing of electrons between two atoms

2. An atom may share electrons with one or more other atoms to acquire a filled valence shell. A **covalent bond** results from the force of attraction between two atoms that share one or more pairs of electrons.

We can now ask how to determine whether two atoms in a compound are bonded by an ionic bond or a covalent bond. One way to do so is to consider the relative positions of the two atoms in the Periodic Table. Ionic bonds usually form between a metal and a nonmetal. An example of an ionic bond is that formed between the metal sodium and the nonmetal chlorine in the compound sodium chloride, Na^+Cl^-. When two nonmetals or a metalloid and a nonmetal combine, the bond between them is usually covalent. Examples of compounds containing covalent bonds between nonmetals include Cl_2, H_2O, CH_4, and NH_3. Examples of compounds containing covalent bonds between a metalloid and a nonmetal include BF_3, $SiCl_4$, and AsH_3.

Another way to determine the bond type is to compare the electronegativities of the atoms involved, which is the subject of the next subsection.

B. Electronegativity and Chemical Bonds

Electronegativity is a measure of an atom's attraction for the electrons it shares in a chemical bond with another atom. The most widely used scale of electronegativities (Table 4.5) was devised by Linus Pauling in the 1930s. On the Pauling scale, fluorine, the most electronegative element, is assigned an electronegativity of 4.0, and all other elements are assigned values relative to fluorine.

As you study the electronegativity values in Table 4.5, note that they generally increase from left to right across a row of the Periodic Table and from bottom to top within a column. Values increase from left to right because of the increasing positive charge on the nucleus, which leads to a stronger attraction for electrons in the valence shell. Values increase going up a column because of the decreasing distance of the valence electrons from the nucleus, which leads to stronger attraction between a nucleus and its valence electrons.

You might compare these trends in electronegativity with the trends in ionization energy (Section 2.8). Each illustrates the periodic nature of elements within the Periodic Table. Ionization energy measures the amount of energy necessary to remove an electron from an atom. Electronegativity measures how tightly an atom holds the electrons that it shares with another atom. Notice that for periods within the Periodic Table, both electronegativity and ionization potential generally increase from left to right across a row of the Periodic Table from Columns 1A to 7A. In addition, both electronegativity and ionization potential increase in going up a column.

Table 4.5   Electronegativity Values of the Elements (Pauling Scale)

1A	2A	3B	4B	5B	6B	7B	8B			1B	2B	3A	4A	5A	6A	7A
H 2.1																
Li 1.0	Be 1.5											B 2.0	C 2.5	N 3.0	O 3.5	F 4.0
Na 0.9	Mg 1.2											Al 1.5	Si 1.8	P 2.1	S 2.5	Cl 3.0
K 0.8	Ca 1.0	Sc 1.3	Ti 1.5	V 1.6	Cr 1.6	Mn 1.5	Fe 1.8	Co 1.8	Ni 1.8	Cu 1.9	Zn 1.6	Ga 1.6	Ge 1.8	As 2.0	Se 2.4	Br 2.8
Rb 0.8	Sr 1.0	Y 1.2	Zr 1.4	Nb 1.6	Mo 1.8	Tc 1.9	Ru 2.2	Rh 2.2	Pd 2.2	Ag 1.9	Cd 1.7	In 1.7	Sn 1.8	Sb 1.9	Te 2.1	I 2.5
Cs 0.7	Ba 0.9	La 1.1	Hf 1.3	Ta 1.5	W 1.7	Re 1.9	Os 2.2	Ir 2.2	Pt 2.2	Au 2.4	Hg 1.9	Tl 1.8	Pb 1.8	Bi 1.9	Po 2.0	At 2.2

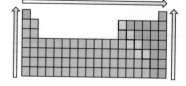

Electronegativity increases

EXAMPLE 4.2

Judging from their relative positions in the Periodic Table, which element in each pair has the larger electronegativity?
(a) Lithium or carbon   (b) Nitrogen or oxygen
(c) Carbon or oxygen

Solution

The elements in each pair are in the second period of the Periodic Table. Within a period, electronegativity increases from left to right.
(a) C > Li   (b) O > N   (c) O > C

Problem 4.2

Judging from their relative positions in the Periodic Table, which element in each pair has the larger electronegativity?
(a) Lithium or potassium   (b) Nitrogen or phosphorus
(c) Carbon or silicon

4.5 | What Is an Ionic Bond and How Does One Form?

A. Forming Ionic Bonds

According to the Lewis model of bonding, an ionic bond forms by the transfer of one or more valence-shell electrons from an atom of lower electronegativity to the valence shell of an atom of higher electronegativity. The more electronegative atom gains one or more valence electrons and becomes an anion; the less electronegative atom loses one or more valence electrons and becomes a cation. The compound formed by the combination of positive and negative ions is called an **ionic compound.**

As a guideline, we say that this type of electron transfer to form an ionic compound is most likely to occur if the difference in electronegativity between two atoms is approximately 1.9 or greater. A bond is more likely to be covalent if this difference is less than 1.9. You should be aware that the value of 1.9 for the formation of an ionic bond is somewhat arbitrary. Some chemists prefer a slightly larger value, others a slightly smaller value. The essential point is that the value of 1.9 gives us a guidepost against which to decide if a bond is more likely to be ionic or more likely to be covalent. Section 4.7 discusses covalent bonding.

An example of an ionic compound is that formed between the metal sodium (electronegativity 0.9) and the nonmetal chlorine (electronegativity 3.0). The difference in electronegativity between these two elements is 2.1. In forming the ionic compound NaCl, the single $3s$ valence electron of a sodium atom is transferred to the partially filled valence shell of a chlorine atom.

$$\text{Na } (1s^2 2s^2 2p^6 3s^1) + \text{Cl } (1s^2 2s^2 2p^6 3s^2 3p^5) \longrightarrow \text{Na}^+ (1s^2 2s^2 2p^6) + \text{Cl}^- (1s^2 2s^2 2p^6 3s^2 3p^6)$$

Sodium atom   Chlorine atom   Sodium ion   Chloride ion

In the following equation, we use a single-headed curved arrow to show this transfer of one electron:

$$\text{Na}\cdot + \cdot\ddot{\underset{\cdot\cdot}{\text{Cl}}}{:} \longrightarrow \text{Na}^+ \; :\ddot{\underset{\cdot\cdot}{\text{Cl}}}{:}^-$$

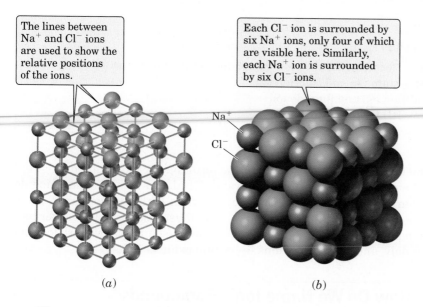

The lines between Na^+ and Cl^- ions are used to show the relative positions of the ions.

Each Cl^- ion is surrounded by six Na^+ ions, only four of which are visible here. Similarly, each Na^+ ion is surrounded by six Cl^- ions.

Na^+

Cl^-

(a)   (b)

Figure 4.1 The structure of a sodium chloride crystal. (a) Ball-and-stick models show the relative positions of the ions. (b) Space-filling models show the relative sizes of the ions.

The ionic bond in solid sodium chloride results from the force of attraction between positive sodium ions and negative chloride ions. In its solid (crystalline) form, sodium chloride consists of a three-dimensional array of Na^+ and Cl^- ions arranged as shown in Figure 4.1.

Although ionic compounds do not consist of molecules, they do have a definite ratio of one kind of ion to another; their formulas give this ratio. For example, NaCl represents the simplest ratio of sodium ions to chloride ions—namely, 1:1.

B. Predicting Formulas of Ionic Compounds

Ions are charged particles, but the matter we see all around us and deal with every day is electrically neutral (uncharged). If ions are present in any sample of matter, the total number of positive charges must equal the total number of negative charges. Therefore, we cannot have a sample containing only Na^+ ions. Any sample that contains Na^+ ions must also contain negative ions, such as Cl^-, Br^-, or S^{2-}, and the sum of the positive charges must equal the sum of the negative charges.

EXAMPLE 4.3

Write the formulas for the ionic compounds formed from the following ions:
(a) Lithium ion and bromide ion
(b) Barium ion and iodide ion
(c) Aluminum ion and sulfide ion

Solution
(a) Table 4.1 shows that the charge on a lithium ion is +1, and Table 4.3 shows that the charge on a bromide ion is −1. The formula for lithium bromide is LiBr.
(b) The charge on a barium ion is +2 and the charge on an iodide ion is −1. Two I^- ions are required to balance the charge of one Ba^{2+} ion. The formula for barium iodide is BaI_2.
(c) The charge on an aluminum ion is +3, and the charge on a sulfide ion is −2. For the compound to have an overall charge of zero, the ions must combine in the ratio of two aluminum ions to three sulfur ions. The formula of aluminum sulfide is Al_2S_3.

Problem 4.3

Write the formulas for the ionic compounds formed from the following ions:
(a) Potassium ion and chloride ion
(b) Calcium ion and fluoride ion
(c) Iron(III) ion and oxide ion

Remember that the subscripts in the formulas for ionic compounds represent the ratio of the ions. Thus a crystal of BaI_2 has twice as many iodide ions as barium ions. For ionic compounds, when both charges are 2, as in the compound formed from Ba^{2+} and O^{2-}, we must "reduce to lowest terms." That is, barium oxide is BaO, not Ba_2O_2. The reason is that we are looking at ratios only, and the ratio of ions in barium oxide is $1:1$.

4.6 | How Do We Name Ionic Compounds?

To name an ionic compound, we give the name of the cation first followed by the name of the anion.

A. Binary Ionic Compounds of Metals That Form Only One Positive Ion

A **binary compound** contains only two elements. In a **binary ionic compound,** both of the elements are present as ions. The name of the compound consists of the name of the metal from which the cation (positive ion) was formed, followed by the name of the anion (negative ion). We ignore subscripts in naming binary ionic compounds. For example, $AlCl_3$ is named aluminum chloride. We know this compound contains three chloride ions because the positive and negative charges must match—that is, one Al^{3+} ion must combine with three Cl^- ions to balance the charges.

EXAMPLE 4.4

Name these binary ionic compounds:
(a) LiBr   (b) Ag_2S (c) NaBr

Solution
(a) Lithium bromide   (b) Silver sulfide   (c) Sodium bromide

Problem 4.4

Name these binary ionic compounds:
(a) MgO   (b) BaI_2 (c) KCl

EXAMPLE 4.5

Write the formulas for these binary ionic compounds:
(a) Barium hydride   (b) Sodium fluoride   (c) Calcium oxide

Solution
(a) BaH_2 (b) NaF   (c) CaO

Problem 4.5

Write the formulas for these binary ionic compounds:
(a) Magnesium chloride   (b) Aluminum oxide   (c) Lithium iodide

B. Binary Ionic Compounds of Metals That Form More Than One Positive Ion

Table 4.2 shows that many transition metals form more than one positive ion. For example, copper forms both Cu^+ and Cu^{2+} ions. For systematic names, we use Roman numerals in the name to show the charge. For common names we use the *-ous/-ic* system.

EXAMPLE 4.6

Give each binary compound a systematic name and a common name.
(a) CuO   (b) Cu_2O

Solution

Remember in answering part (b) that we ignore subscripts in naming binary ionic compounds. Therefore, the 2 in Cu_2O is not indicated. You know that two copper(I) ions are present because two positive charges are needed to balance the two negative charges on an O^{2-} ion.
(a) Systematic name: copper(II) oxide. Common name: cupric oxide.
(b) Systematic name: copper(I) oxide. Common name: cuprous oxide.

Problem 4.6

Give each binary compound a systematic name and a common name.
(a) FeO   (b) Fe_2O_3

C. Ionic Compounds That Contain Polyatomic Ions

To name ionic compounds containing polyatomic ions, name the positive ion first and then the negative ion, each as a separate word.

EXAMPLE 4.7

Name these ionic compounds, each of which contains a polyatomic ion:
(a) $NaNO_3$ (b) $CaCO_3$ (c) $(NH_4)_2SO_3$ (d) NaH_2PO_4

Solution

Recall that the name of the $H_2PO_4^-$ ion is dihydrogen phosphate.
(a) Sodium nitrate   (b) Calcium carbonate
(c) Ammonium sulfite   (d) Sodium dihydrogen phosphate

Problem 4.7

Name these ionic compounds, each of which contains a polyatomic ion:
(a) K_2HPO_4 (b) $Al_2(SO_4)_3$ (c) $FeCO_3$

Ionic Compounds in Medicine

Many ionic compounds have medical uses, some of which are shown in the table.

Formula	Name	Medical Use
$AgNO_3$	Silver nitrate	Astringent (external)
$BaSO_4$	Barium sulfate	Radiopaque medium for Xray work
$CaSO_4$	Calcium sulfate	Plaster casts
$FeSO_4$	Iron(II) sulfate	To treat iron deficiency
$KMnO_4$	Potassium permanganate	Anti-infective (external)
KNO_3	Potassium nitrate (saltpeter)	Diuretic
Li_2CO_3	Lithium carbonate	To treat manic depression
$MgSO_4$	Magnesium sulfate (Epsom salts)	Cathartic
$NaHCO_3$	Sodium bicarbonate (baking soda)	Antacid
NaI	Sodium iodide	Iodine for thyroid hormones
NH_4Cl	Ammonium chloride	To acidify the digestive system
$(NH_4)_2CO_3$	Ammonium carbonate	Expectorant
SnF_2	Tin(II) fluoride	To strengthen teeth (external)
ZnO	Zinc oxide	Astringent (external)

Drinking a "barium cocktail" containing barium sulfate makes the intestinal tract visible on an Xray.

Charles D. Winters

4.7 | What Is a Covalent Bond and How Does One Form?

A. Formation of a Covalent Bond

A covalent bond forms when electron pairs are shared between two atoms whose difference in electronegativity is less than 1.9. As we have already mentioned, the most common covalent bonds occur between two nonmetals or between a nonmetal and a metalloid.

According to the Lewis model, a pair of electrons in a covalent bond functions in two ways simultaneously: It is shared by two atoms and it fills the valence shell of each atom. The simplest example of a covalent bond is that in a hydrogen molecule, H_2. When two hydrogen atoms bond, the single electrons from each atom combine to form an electron pair. A bond formed by sharing a pair of electrons is called a **single bond** and is represented by a single line between the two atoms. The electron pair shared between the two hydrogen atoms in H_2 completes the valence shell of each hydrogen. Thus, in H_2, each hydrogen has two electrons in its valence shell and an electron configuration like that of helium, the noble gas nearest to it in atomic number.

The single line represents a shared pair of electrons

$$H\cdot + \cdot H \longrightarrow H-H$$

Table 4.6 Classification of Chemical Bonds

Electronegativity Difference Between Bonded Atoms	Type of Bond	Most Likely to Form Between
Less than 0.5	Nonpolar covalent	Two nonmetals or a nonmetal
0.5 to 1.9	Polar covalent	and a metalloid
Greater than 1.9	Ionic	A metal and a nonmetal

B. Nonpolar and Polar Covalent Bonds

Although all covalent bonds involve the sharing of electrons, they differ widely in the degree of sharing. We classify covalent bonds into two categories, **nonpolar covalent** and **polar covalent,** depending on the difference in electronegativity between the bonded atoms. In a nonpolar covalent bond, electrons are shared equally. In a polar covalent bond, they are shared unequally. It is important to realize that no sharp line divides these two categories, nor, for that matter, does a sharp line divide polar covalent bonds and ionic bonds. Nonetheless, the rule-of-thumb guidelines given in Table 4.6 will help you decide whether a given bond is more likely to be nonpolar covalent, polar covalent, or ionic.

An example of a polar covalent bond is that in H—Cl, in which the difference in electronegativity between the bonded atoms is $3.0 - 2.1 = 0.9$. A covalent bond between carbon and hydrogen, for example, is classified as nonpolar covalent because the difference in electronegativity between these two atoms is only $2.5 - 2.1 = 0.4$. You should be aware, however, that there is some slight polarity to a C—H bond but, because it is quite small, we arbitrarily say that a C—H bond is nonpolar.

Nonpolar covalent bond A covalent bond between two atoms whose difference in electronegativity is less than 0.5

Polar covalent bond A covalent bond between two atoms whose difference in electronegativity is between 0.5 and 1.9

EXAMPLE 4.8

Classify each bond as nonpolar covalent, polar covalent, or ionic.
(a) O—H   (b) N—H   (c) Na—F
(d) C—Mg   (e) C—S

Solution

Bond	Difference in Electronegativity	Type of Bond
(a) O—H	$3.5 - 2.1 = 1.4$	Polar covalent
(b) N—H	$3.0 - 2.1 = 0.9$	Polar covalent
(c) Na—F	$4.0 - 0.9 = 3.1$	Ionic
(d) C—Mg	$2.5 - 1.2 = 1.3$	Polar covalent
(e) C—S	$2.5 - 2.5 = 0$	Nonpolar covalent

Problem 4.8

Classify each bond as nonpolar covalent, polar covalent, or ionic.
(a) S—H   (b) P—H   (c) C—F   (d) C—Cl

An important consequence of the unequal sharing of electrons in a polar covalent bond is that the more electronegative atom gains a greater fraction of the shared electrons and acquires a partial negative charge, indicated by the symbol $\delta-$ (read "delta minus"). The less electronegative atom has a lesser fraction of the shared electrons and acquires a partial positive charge, indicated by the symbol $\delta+$ (read "delta plus"). This separation of charge produces a **dipole** (two poles). We commonly show the presence of a bond dipole by an arrow, with the head of the arrow near the negative end of the dipole and a cross on the tail of the arrow near the positive end (Figure 4.2).

We can also show the polarity of a covalent bond by an electron density model. In this type of molecular model, a blue color shows the presence of a $\delta+$ charge and a red color shows the presence of a $\delta-$ charge. Figure 4.2 also shows an electron density model of HCl. The ball-and-stick model in the center shows the orientation of atoms in space. The transparent surface surrounding the ball-and-stick model shows the relative sizes of the atoms (equivalent to the size shown by a space-filling model). Colors on the surface show the distribution of electron density. We see by the blue color that hydrogen bears a $\delta+$ charge and by the red color that chlorine bears a $\delta-$ charge.

Dipole A chemical species in which there is a separation of charge; there is a positive pole in one part of the species, and a negative pole in another part

Low electron density (blue)   High electron density (red)

$$\overset{\delta+}{\text{H}}\!-\!\overset{\delta-}{\text{Cl}}$$
$\longmapsto$

Figure 4.2 HCl is a polar covalent molecule. In the electron density model of HCl, red indicates a region of high electron density, and blue indicates a region of low electron density.

EXAMPLE 4.9

Using the symbols $\delta-$ and $\delta+$, indicate the polarity in each polar covalent bond.
(a) C—O
(b) N—H
(c) C—Mg

Solution
For (a), carbon and oxygen are both in period 2 of the Periodic Table. Because oxygen is farther to the right than carbon, it is more electronegative than carbon. For (c), magnesium is a metal located to the far left in the Periodic Table and carbon is a nonmetal located to the right. All nonmetals, including hydrogen, have a greater electronegativity than do the metals in columns 1A and 2A. The electronegativity of each element is given below the symbol of the element.

$$(a)\ \overset{\delta+}{\text{C}}\!-\!\overset{\delta-}{\text{O}} \qquad (b)\ \overset{\delta-}{\text{N}}\!-\!\overset{\delta+}{\text{H}} \qquad (c)\ \overset{\delta-}{\text{C}}\!-\!\overset{\delta+}{\text{Mg}}$$
$$2.5\ \ 3.5 \qquad 3.0\ \ 2.1 \qquad 2.5\ \ 1.2$$

Problem 4.9
Using the symbols δ^- and δ^+, indicate the polarity in each polar covalent bond.
(a) C—N
(b) N—O
(c) C—Cl

C. Drawing Lewis Structures of Covalent Compounds

The ability to write Lewis structures for covalent molecules is a fundamental skill for the study of chemistry. The following How To box will help you with this task.

HOW TO...

Draw Lewis Structures

1. *Determine the number of valence electrons in the molecule.*
Add up the number of valence electrons contributed by each atom. To determine the number of valence electrons, you simply need to know the number of each kind of atom in the molecule. You don't have to know anything about how the atoms are bonded to each other.

Example: The Lewis structure for formaldehyde, CH_2O, must show 12 valence electrons:

$$4 \text{ (from C)} + 2 \text{ (from the two H)} + 6 \text{ (from O)} = 12$$

2. *Determine the connectivity of the atoms (which atoms are bonded to each other) and connect bonded atoms by single bonds.*
Drawing a Lewis structure is often the most challenging task. For some molecules we give you as examples or problems, we ask you to propose a connectivity. For most, however, we give you the experimentally determined connectivity and ask you to complete the Lewis structure.

Example: The atoms in formaldehyde are bonded in the following order. Note that we do not attempt at this point to show bond angles or the three-dimensional shape of the molecule; we just show what is bonded to what.

$$
\begin{array}{c}
O \\
| \\
H - C - H
\end{array}
$$

This partial structure shows six valence electrons in the three single bonds. In it we have accounted for six of the 12 valence electrons.

3. *Arrange the remaining electrons so that each atom has a complete outer shell.*
Each hydrogen atom must be surrounded by two electrons. Each carbon, nitrogen, oxygen, and halogen atom must be surrounded by eight valence electrons (to satisfy the octet rule). The remaining valence electrons may be shared between atoms in bonds or may be unshared pairs on a single atom. A pair of electrons involved in a covalent bond (**bonding electrons**) is shown as a single line; an unshared pair of electrons (**nonbonding electrons**) is shown as a pair of Lewis dots.

By placing two pairs of bonding electrons between C and O, we give carbon a complete octet. By placing the remaining four electrons on oxygen as two Lewis dot pairs, we give oxygen eight valence electrons and a complete octet (octet rule). Note that we placed the two pairs of bonding electrons between C and O before we assigned the unshared pairs of electrons on the oxygen.

$$
\begin{array}{c}
H \\
\diagdown \\
C = \ddot{O} \\
\diagup \\
H
\end{array}
$$

Lewis structure

Unshared pairs of electrons are not shown in ball-and-stick models

Ball-and-stick model of formaldehyde

GOB
Chemistry•♦•Now™
Click Mastering the Essentials *to practice*
How to Draw Lewis Structures

Bonding electrons Valence electrons involved in forming a covalent bond; that is, shared electrons

Nonbonding electrons Valence electrons not involved in forming covalent bonds; that is, unshared electrons

Single bond A bond formed by sharing one pair of electrons and represented by a single line between two atoms

Lewis Structure A formula for a molecule or ion showing all pairs of bonding electrons as single, double, or triple bonds, and all nonbonded electrons as pairs of Lewis dots

Structural formula A formula showing how atoms in a molecule or ion are bonded to each other; similar to a Lewis structure except that a structural formula typically shows only bonding pairs of electrons

Problem 4.10

Draw a Lewis structure for each molecule. Each has only one possible order of attachment of its atoms, which is left for you to determine.
(a) Ethane, C_2H_6
(b) Chloromethane, CH_3Cl
(c) Hydrogen cyanide, HCN

EXAMPLE 4.11

Why does carbon have four bonds and no unshared pairs of electrons in covalent compounds?

Solution
When carbon has four bonds, it has a complete octet. An additional lone pair of electrons would place ten electrons in the valence shell of carbon and violate the octet rule.

Problem 4.11

Draw a Lewis structure of a covalent compound in which carbon has
(a) Four single bonds
(b) Two single bonds and one double bond
(c) Two double bonds
(d) One single bond and one triple bond

D. Exceptions to the Octet Rule

The Lewis model of covalent bonding focuses on valence electrons and the necessity for each atom other than hydrogen to have a completed valence shell containing eight electrons. Although most molecules formed by main-group elements (Groups 1A–7A) have structures that satisfy the octet rule, some important exceptions exist. One exception involves molecules that contain an atom with more than eight electrons in its valence shell. Atoms of period 2 elements use one $2s$ and three $2p$ orbitals for bonding; these four orbitals can contain only eight valence electrons—hence the octet rule. Atoms of period 3 elements, however, have one $3s$ orbital, three $3p$ orbitals, and five $3d$ orbitals; they can accommodate more than eight electrons in their valence shells (Section 2.6A). In phosphine, PH_3, phosphorus has eight electrons in its valence shell and obeys the octet rule. The phosphorus atoms in phosphorus pentachloride, PCl_5, and phosphoric acid, H_3PO_4, have ten electrons in their valence shells and, therefore, are exceptions to the octet rule.

Phosphine

Phosphorus pentachloride

Phosphoric acid

Sulfur, another period 3 element, forms compounds in which it has 8, 10, and even 12 electrons in its valence shell. The sulfur atom in H_2S has 8 electrons in its valence shell and obeys the octet rule. The sulfur atoms in SO_2 and H_2SO_4 have 10 and 12 electrons in their valence shells and are exceptions to the octet rule.

4.8 | How Do We Name Binary Covalent Compounds?

A **binary covalent compound** is a binary (two-element) compound in which all bonds are covalent. In naming a binary covalent compound:

1. First name the less electronegative element (see Table 4.5). Note that the less electronegative element is also generally written first in the formula.

2. Then name the more electronegative element. To name it, add *-ide* to the stem name of the element. Chlorine, for example, becomes chloride, and oxygen becomes oxide.

3. Use the prefixes *di-, tri-, tetra-,* and so on to show the number of atoms of each element. The prefix *mono-* is omitted when it refers to the first atom named, and it is rarely used with the second atom. An exception to this rule is CO, which is named carbon monoxide.

The name is then written as two words.

Name of the first element in the formula; use prefixes *di-* and so forth if necessary	Name of the second element; use prefixes *mono-* and so forth if necessary

EXAMPLE 4.12

Name these binary covalent compounds:
(a) NO   (b) SF_2 (c) N_2O

Solution
(a) Nitrogen oxide (more commonly called nitric oxide)
(b) Sulfur difluoride
(c) Dinitrogen oxide (more commonly called nitrous oxide or laughing gas)

Problem 4.12
Name these binary covalent compounds:
(a) NO_2 (b) PBr_3 (c) SCl_2 (d) BF_3

CHEMICAL CONNECTIONS 4C

Nitric Oxide: Air Pollutant and Biological Messenger

Nitric oxide, NO, is a colorless gas whose importance in the environment has been known for several decades, but whose biological importance is only now being fully recognized. This molecule has 11 valence electrons. Because its number of electrons is odd, it is not possible to draw a structure for NO that obeys the octet rule; there must be one unpaired electron, here shown on the less electronegative nitrogen atom:

An unpaired electron

$:\ddot{N}=\ddot{O}:$

Nitric oxide

Colorless nitric oxide, NO, coming from the tank, bubbles through the water. When it reaches the air, it is oxidized to brown nitrogen dioxide, NO_2.

The importance of NO in the environment arises from the fact that it forms as a by-product during the combustion of fossil fuels. Under the temperature conditions of the internal combustion engine and other combustion sources, nitrogen and oxygen of the air react to form small quantities of NO:

$$N_2 + O_2 \xrightarrow{\text{heat}} 2NO$$
Nitric oxide

When inhaled, NO passes from the lungs into the bloodstream. There it interacts with the iron in hemoglobin, decreasing its ability to carry oxygen.

What makes nitric oxide so hazardous in the environment is that it reacts almost immediately with oxygen to form NO_2. When dissolved in water, NO_2 reacts with water to form nitric acid, which is a major acidifying component of acid rain.

$$2NO + O_2 \longrightarrow 2NO_2$$
Nitric oxide   Nitrogen dioxide

$$NO_2 + H_2O \longrightarrow HNO_3$$
Nitrogen dioxide   Nitric acid

Imagine the surprise when it was discovered within the last two decades that this highly reactive, seemingly hazardous compound is synthesized in humans and plays a vital role as a signaling molecule in the cardiovascular system (Chemical Connections 24F).

4.9 | What Is Resonance?

As chemists developed a deeper understanding of covalent bonding in organic and inorganic compounds, it became obvious that, for a great many molecules and ions, no *single* Lewis structure provides a truly accurate representation. For example, Figure 4.3 shows three Lewis structures for the carbonate ion, CO_3^{2-}. In each structure, carbon is bonded to three oxygen

Figure 4.3 (a–c) Three Lewis structures for the carbonate ion.

(a)   (b)   (c)

atoms by a combination of one double bond and two single bonds. Each Lewis structure implies that one carbon-oxygen bond is different from the other two. However, this is not the case. It has been determined experimentally that all three carbon-oxygen bonds are identical.

The problem for chemists, then, is how to describe the structure of molecules and ions for which no single Lewis structure is adequate and yet still retain Lewis structures. As an answer to this problem, Linus Pauling proposed the theory of resonance.

A. Theory of Resonance

The theory of **resonance,** developed primarily by Pauling in the 1930s, provides an explanation. According to this theory, many molecules and ions are best described by writing two or more Lewis structures and considering the real molecule or ion to be a hybrid of these structures. Individual Lewis structures are called **contributing structures** (or sometimes **resonance structures** or **resonance contributors**). We show that the real molecule or ion is a **resonance hybrid** of the various contributing structures by interconnecting them with **double-headed arrows.** Do not confuse the double-headed arrow with the double arrow used to show chemical equilibrium. As we explain shortly, resonance structures are not in equilibrium with each other.

Figure 4.4 shows three contributing structures for the carbonate ion. These contributing structures are said to be equivalent; all three have identical patterns of covalent bonding.

Resonance A theory that many molecules and ions are best described as a hybrid of two or more Lewis contributing structures

Contributing structure Representations of a molecule or ion that differ only in the distribution of valence electrons

Resonance hybrid A molecule or ion described as a composite or hybrid of a number of contributing structures

Double-headed arrow A symbol used to show that the structures on either side of it are resonance contributing structures

---

HOW TO ...

Draw Curved Arrows and Push Electrons

Notice in Figure 4.4 that the only difference among contributing structures (a), (b), and (c) is the position of the valence electrons. To generate one resonance structure from another, chemists use a **curved arrow.** The arrow indicates where a pair of electrons originates (the tail of the arrow) and where it is repositioned in the next structure (the head of the arrow).

A curved arrow is nothing more than a bookkeeping symbol for keeping track of electron pairs or, as some call it, **electron pushing.** Do not be misled by its simplicity. Electron pushing will help you see the relationship among contributing structures.

Following are contributing structures for the nitrite and acetate ions. Curved arrows show how the contributing structures are interconverted. For each ion, the contributing structures are equivalent: They have the same bonding pattern.

Nitrite ion
(equivalent contributing structures)

Acetate ion
(equivalent contributing structures)

A common mistake is to use curved arrows to indicate the movement of atoms or positive charges. This is never correct. Curved arrows are used only to show the repositioning of electron pairs, when a new resonance hybrid is generated.

Figure 4.4 (*a–c*) The carbonate ion represented as a hybrid of three equivalent contributing structures. Curved arrows (in red) show how electron pairs are redistributed from one contributing structure to the next.

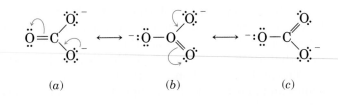

(a)   (b)   (c)

The use of the term "resonance" for this theory of covalent bonding might appear to suggest that bonds and electron pairs are constantly changing back and forth from one position to another over time. This notion is not at all correct. The carbonate ion, for example, has one—and only one—real structure. The problem is ours: How do we draw that one real structure? The resonance method offers a way to describe the real structure while simultaneously retaining Lewis structures with electron-pair bonds and showing all nonbonding pairs of electrons. Thus, although we realize that the carbonate ion is not accurately represented by any one contributing structure shown in Figure 4.4, we continue to represent it by one of these for convenience. We understand, of course, that we are referring to the resonance hybrid.

One consequence of resonance is that it is a stabilizing factor when it exists—that is, a resonance hybrid is more stable than chemists would estimate or calculate compared to any one of its contributing structures. We will see three particularly striking illustrations of the stability of resonance hybrids when we consider the unusual chemical properties of benzene and aromatic hydrocarbons in Chapter 13, the acidity of carboxylic acids in Chapter 18, and the geometry of the amide bonds in proteins in Chapter 19.

EXAMPLE 4.13

Draw the contributing structure indicated by the curved arrows. Be certain to show all valence electrons and all charges.

Solution

Problem 4.13

Draw the contributing structure indicated by the curved arrows. Be certain to show all valence electrons and all charges.

B. Writing Acceptable Contributing Structures

Certain rules must be followed to write acceptable contributing structures:

1. All contributing structures must have the same number of valence electrons.
2. All contributing structures must obey the rules of covalent bonding. In particular, no contributing structure may have more than two electrons in the valence shell of hydrogen or more than eight electrons in the valence shell of a second-period element. Third-period elements, such as phosphorus and sulfur, may have up to 12 electrons in their valence shells.
3. The positions of all nuclei must be the same in all resonance structures; that is, contributing structures must differ only in the distribution of valence electrons.
4. All contributing structures must have the same number of paired and unpaired electrons.

EXAMPLE 4.14

Which sets are valid pairs of contributing structures?

(a) CH₃—C(=O)—CH₃   and   CH₃—C⁺(:Ö:⁻)—CH₃

(b) CH₃—C(=O)—CH₃   and   CH₂=C(:Ö—H)—CH₃

Solution
(a) These are valid contributing structures. They differ only in the distribution of valence electrons.
(b) These are not valid contributing structures. They differ in the arrangement of their atoms.

Problem 4.14

Which sets are valid pairs of contributing structures?

A final note: Do not confuse resonance contributing structures with equilibration among different species. A molecule described as a resonance hybrid is not equilibrating among the individual electron configurations of the contributing structures. Rather, the molecule has only one structure, which is best described as a hybrid of its various contributing structures. The colors on the color wheel provide a good analogy. Purple is not a primary color; the primary colors of blue and red are mixed to make purple.

You can think of molecules represented by resonance hybrids as being purple. Purple is not sometimes blue and sometimes red: Purple is purple. In an analogous way, a molecule described as a resonance hybrid is not sometimes one contributing structure and sometimes another: It is a single structure all the time.

4.10 | How Do We Predict Bond Angles in Covalent Molecules?

In Section 4.7, we used a shared pair of electrons as the fundamental unit of covalent bonds and drew Lewis structures for several small molecules containing various combinations of single, double, and triple bonds (see, for example, Table 4.7). We can predict **bond angles** in these and other molecules by using the **valence-shell electron-pair repulsion (VSEPR) model.**

According to this model, the valence electrons of an atom may be involved in the formation of single, double, or triple bonds, or they may be unshared. Each combination creates a negatively charged region of electron density around the nucleus. Because like charges repel each other, the various regions of electron density around an atom spread out so that each is as far away as possible from the others.

You can demonstrate the bond angles predicted by this model in a very simple way. Imagine that a balloon represents a region of electron density. If two balloons are tied together by their ends, they assume the shapes shown in Figure 4.5. The point where they are tied together represents the atom about which you want to predict a bond angle, and the balloons represent regions of electron density about that atom.

We use the VSEPR model and the balloon model analogy in the following way to predict the shape of a molecule of methane, CH_4. The Lewis structure for CH_4 shows a carbon atom surrounded by four regions of electron density. Each region contains a pair of electrons forming a single covalent bond to a hydrogen atom. According to the VSEPR model, the four regions radiate from carbon so that they are as far away from one another as possible. The maximum separation occurs when the angle between any two regions of electron density is 109.5°. Therefore, we predict all H—C—H bond angles to be 109.5°, and the shape of the molecule to be **tetrahedral**

Bond angle The angle between two atoms bonded to a central atom

GOB Chemistry Now™
Click *Coached Problems* for an example of **Determining Molecular Shapes**

 (a)
 (b)
 (c)

GOB Chemistry Now™
Active Figure 4.5 Balloon models to predict bond angles. (a) Two balloons assume a linear shape with a bond angle of 180° about the tie point. (b) Three balloons assume a trigonal planar shape with bond angles of 120° about the tie point. (c) Four balloons assume a tetrahedral shape with bond angles of 109.5° about the tie point. **See a simulation based on this figure, and take a short quiz on the concepts at http://www.brookscole.com/gob8 or on the CD.**

(a)

$$H - \overset{\underset{\displaystyle |}{\displaystyle |}}{C} - H$$

(b) 109.5°

109.5°

Figure 4.6 The shape of a methane molecule, CH_4. (*a*) Lewis structure and (*b*) ball-and-stick model. The hydrogens occupy the four corners of a regular tetrahedron, and all H—C—H bond angles are 109.5°.

(a) H—N̈—H
       |
       H

(b) —— Unshared electron pair

107.3°

Figure 4.7 The shape of an ammonia molecule, NH_3. (*a*) Lewis structure and (*b*) ball-and-stick model. The H—N—H bond angles are 107.3°, slightly smaller than the H—C—H bond angles of methane.

(Figure 4.6). The H—C—H bond angles in methane have been measured experimentally and found to be 109.5°. Thus the bond angles and shape of methane predicted by the VSEPR model are identical to those observed experimentally.

We can predict the shape of an ammonia molecule, NH_3, in the same way. The Lewis structure of NH_3 shows nitrogen surrounded by four regions of electron density. Three regions contain single pairs of electrons that form covalent bonds with hydrogen atoms. The fourth region contains an unshared pair of electrons [Figure 4.7(a)]. Using the VSEPR model, we predict that the four regions are arranged in a tetrahedral manner and that the three H—N—H bond angles in this molecule are 109.5°. The observed bond angles are 107.3°. We can explain this small difference between the predicted and the observed angles by proposing that the unshared pair of electrons on nitrogen repels adjacent electron pairs more strongly than the bonding pairs repel one another.

The geometry of an ammonia molecule is described as **pyramidal;** that is, the molecule is shaped like a triangular-based pyramid with the three hydrogens located at the base and the single nitrogen located at the apex.

Figure 4.8 shows a Lewis structure and a ball-and-stick model of a water molecule. In H_2O, oxygen is surrounded by four regions of electron density. Two of these regions contain pairs of electrons used to form single covalent bonds to hydrogens; the remaining two regions contain unshared electron pairs. Using the VSEPR model, we predict that the four regions of electron density around oxygen are arranged in a tetrahedral manner and that the H—O—H bond angle is 109.5°. Experimental measurements show that the actual H—O—H bond angle in a water molecule is 104.5°, a value smaller than that predicted. We can explain this difference between the predicted and the observed bond angle by proposing, as we did for NH_3, that unshared pairs of electrons repel adjacent pairs more strongly than bonding pairs do. Note that the distortion from 109.5° is greater in H_2O, which has two unshared pairs of electrons, than it is in NH_3, which has only one unshared pair.

A general prediction emerges from this discussion. If a Lewis structure shows four regions of electron density around an atom, the VSEPR model predicts a tetrahedral distribution of electron density and bond angles of approximately 109.5°.

(a) H—Ö—H

(b)

Unshared electron pairs

104.5°

Figure 4.8 The shape of a water molecule, H_2O. (*a*) Lewis structure and (*b*) ball-and-stick model.

Formaldehyde

Top view   Side view

Ethylene

Top view   Side view

Figure 4.9   The shapes of formaldehyde, CH_2O, and ethylene, C_2H_4.

Ammonia gas is drilled into the soil of a farm field. Most of the ammonia manufactured in the world is used as fertilizer because ammonia supplies the nitrogen needed by green plants.

In many of the molecules we will encounter, an atom is surrounded by three regions of electron density. Figure 4.9 shows Lewis structures and ball-and-stick models for molecules of formaldehyde, CH_2O, and ethylene, C_2H_4.

In the VSEPR model, we treat a double bond as a single region of electron density. In formaldehyde, carbon is surrounded by three regions of electron density. Two regions contain single pairs of electrons, each of which forms a single bond to a hydrogen; the third region contains two pairs of electrons, which form a double bond to oxygen. In ethylene, each carbon atom is also surrounded by three regions of electron density; two contain single pairs of electrons, and the third contains two pairs of electrons.

Three regions of electron density about an atom are farthest apart when they lie in a plane and make angles of 120° with one another. Thus the predicted H—C—H and H—C—O bond angles in formaldehyde and the H—C—H and H—C—C bond angles in ethylene are all 120°. Furthermore, all atoms in each molecule lie in a plane. Thus, both formaldehyde and ethylene are planar molecules. The geometry about an atom surrounded by three regions of electron density, as in formaldehyde and ethylene, is described as **trigonal planar.**

In still other types of molecules, a central atom is surrounded by two regions of electron density. Figure 4.10 shows Lewis structures and ball-and-stick models of molecules of carbon dioxide, CO_2, and acetylene, C_2H_2.

In carbon dioxide, carbon is surrounded by two regions of electron density; each contains two pairs of electrons and forms a double bond to an

Carbon dioxide

Side view   End view

Acetylene

Side view   End view

Figure 4.10   The shapes of carbon dioxide, CO_2, and acetylene, C_2H_2.

Table 4.8   Predicted Molecular Shapes (VSEPR model)

Regions of Electron Density Around Central Atom	Predicted Distribution of Electron Density	Predicted Bond Angles	Examples (Shape of the Molecule)
4	Tetrahedral	109.5°	Methane (tetrahedral) Ammonia (pyramidal) Water (bent)
3	Trigonal planar	120°	Ethylene (planar) Formaldehyde (planar)
2	Linear	180°	Carbon dioxide (linear) Acetylene (linear)

oxygen atom. In acetylene, each carbon is also surrounded by two regions of electron density; one contains a single pair of electrons and forms a single bond to a hydrogen atom, and the other contains three pairs of electrons and forms a triple bond to a carbon atom. In each case, the two regions of electron density are farthest apart if they form a straight line through the central atom and create an angle of 180°. Both carbon dioxide and acetylene are linear molecules.

Table 4.8 summarizes the predictions of the VSEPR model. In this table, three-dimensional shapes are shown using a solid wedge to represent a bond coming toward you, out of the plane of the paper. A broken wedge represents a bond going away from you, behind the plane of the paper. A single line represents a bond in the plane of the paper.

EXAMPLE 4.15

Predict all bond angles and the shape of each molecule:
(a) CH_3Cl (b) $CH_2{=}CHCl$

Solution
(a) The Lewis structure for CH_3Cl shows that carbon is surrounded by four regions of electron density. Therefore, we predict that the distribution of electron pairs about carbon is tetrahedral, all bond angles are 109.5°, and the shape of CH_3Cl is tetrahedral.

(b) In the Lewis structure for $CH_2{=}CHCl$, each carbon is surrounded by three regions of electron density. Therefore, we predict that all bond angles are 120° and that the molecule is planar. The bonding about each carbon is trigonal planar.

(Top view)

(Viewed along the C=C bond)

Problem 4.15

Predict all bond angles for these molecules:

(a) CH_3OH (b) CH_2Cl_2 (c) H_2CO_3 (carbonic acid)

4.11 | How Do We Determine if a Molecule Is Polar?

In Section 4.7B, we used the terms "polar" and "dipole" to describe a covalent bond in which one atom bears a partial positive charge and the other bears a partial negative charge. We also saw that we can use the difference in electronegativity between bonded atoms to determine the polarity of a covalent bond and the direction of its dipole. We can now combine our understanding of bond polarity and molecular geometry (Section 4.10) to predict the polarity of molecules. To discuss the physical and chemical properties of a molecule, it is essential to have an understanding of polarity. Many chemical reactions, for example, are driven by the interaction of the positive part of one molecule with the negative part of another molecule.

A molecule will be polar if (1) it has polar bonds and (2) its centers of partial positive charge and partial negative charge lie at different places within the molecule. Consider first carbon dioxide, CO_2, a molecule with two polar carbon-oxygen double bonds. The oxygen on the left pulls electrons of the O=C bond toward it, giving it a partial negative charge. Similarly, the oxygen on the right pulls electrons of the C=O bond toward it by the same amount, giving it the same partial negative charge as the oxygen on the left. Carbon bears a partial positive charge. We can show the polarity of these bonds by using the symbols $\delta+$ and $\delta-$. Alternatively, we can show that each carbon-oxygen bond has a dipole by using an arrow, where the head of the arrow points to the negative end of the dipole and the crossed tail is positioned at the positive end of the dipole. Because carbon dioxide is a linear molecule, its centers of negative and positive partial charge coincide. Therefore, CO_2 is a nonpolar molecule; that is, it has no dipole.

$$\overset{\longleftarrow \; + \; +\longrightarrow}{\underset{\delta- \quad \delta+ \quad \delta-}{\ddot{O}=C=\ddot{O}}}$$

Carbon dioxide
(a nonpolar molecule)

In a water molecule, each O—H bond is polar. Oxygen, the more electronegative atom, bears a partial negative charge and each hydrogen bears a partial positive charge. The center of partial positive charge in a water molecule is located halfway between the two hydrogen atoms, and the center of partial negative charge is on the oxygen atom. Thus a water molecule has polar bonds and, because of its geometry, is a polar molecule.

Water
(a polar molecule)

The center of partial positive charge is midway between the two hydrogen atoms

Ammonia has three polar N—H bonds. Because of its geometry, the centers of partial positive and partial negative charges are found at different places within the molecule. Thus ammonia has polar bonds and, because of its geometry, is a polar molecule.

Ammonia
(a polar molecule)

The center of partial positive charge is midway between the three hydrogen atoms

EXAMPLE 4.16

Which of these molecules are polar? Show the direction of the molecular dipole by using an arrow with a crossed tail.
(a) CH_2Cl_2 (b) CH_2O (c) C_2H_2

Solution
Both dichloromethane, CH_2Cl_2, and formaldehyde, CH_2O, have polar bonds and, because of their geometry, are polar molecules. Because acetylene, C_2H_2, contains no polar bonds, it is a nonpolar molecule.

This model shows the double bond as only a single cylinder

(a) δ−   δ−
   Cl δ+ Cl
      C
   H H

Dichloromethane

(b) O δ−
    ‖
    C δ+
  H   H

Formaldehyde

(c)
H—C≡C—H

Acetylene

Problem 4.16

Which of these molecules are polar? Show the direction of the molecular dipole by using an arrow with a crossed tail.
(a) CH_3Cl (b) HCN (c) C_2H_6

SUMMARY OF KEY QUESTIONS

SECTION 4.2 What Is the Octet Rule?

- The **octet rule** states that elements of Groups 1A–7A tend to gain or lose electrons so as to achieve an outer shell containing eight valence electrons.
- An atom with almost eight valence electrons tends to gain the needed electrons to have eight electrons in its valence shell—that is, to achieve the same electron configuration as the noble gas nearest to it in atomic number. In gaining electrons, the atom becomes a negatively charged ion called an **anion.**
- An atom with only one or two valence electrons tends to lose the number of electrons required to have eight valence electrons in its next lower shell—that is, to have the same electron configuration as the noble gas nearest it in atomic number. In losing electrons, the atom becomes a positively charged ion called a **cation.**

SECTION 4.3 How Do We Name Anions and Cations?

- For metals that form only one type of cation, the name of the cation is the name of the metal followed by the word "ion."
- For metals that form more than one type of cation, we show the charge on the ion by placing a Roman numeral in parentheses immediately following the name of the metal.
- A **monatomic anion** is named by adding -*ide* to the stem part of the name.
- A **polyatomic ion** contains more than one type of atom.

SECTION 4.4 What Are the Two Major Types of Chemical Bonds?

- The two major types of chemical bonds are ionic bonds and covalent bonds.
- According to the Lewis model of chemical bonding, atoms bond together in such a way that each atom participating in a bond acquires a valence-shell electron configuration matching that of the noble gas nearest to it in atomic number.
- **Electronegativity** is a measure of the force of attraction that an atom exerts on electrons it shares in a chemical bond. It increases from left to right

across a row and from bottom to top in a column of the Periodic Table.
- An **ionic bond** forms between two atoms if the difference in electronegativity between them is greater than 1.9.
- A **covalent bond** forms if the difference in electronegativity between the bonded atoms is 1.9 or less.

SECTION 4.5 What Is an Ionic Bond and How Does One Form?

- An **ionic bond** forms by the transfer of valence-shell electrons from an atom of lower electronegativity to the valence shell of an atom of higher electronegativity.
- In an ionic compound, the total number of positive charges must equal the total number of negative charges.

SECTION 4.6 How Do We Name Ionic Compounds?

- For a **binary ionic compound,** name the cation first, followed by the name of the anion. Where a metal ion may form different cations, use a Roman numeral to show its positive charge. Alternatively, use the suffix -*ous* to show the lower positive charge, and -*ic* to show the higher positive charge. To name an ionic compound that contains polyatomic ions, name the cation first, followed by the name of the anion.

SECTION 4.7 What Is a Covalent Bond and How Does One Form?

- According to the Lewis model, a **covalent bond** forms when pairs of electrons are shared between two atoms whose difference in electronegativity is 1.9 or less.
- A pair of electrons in a covalent bond is shared by two atoms and at the same time fills the valence shell of each atom.
- A **nonpolar covalent bond** is a covalent bond in which the difference in electronegativity between bonded atoms is less than 0.5. A **polar covalent bond** is a covalent bond in which the difference in electronegativity between bonded atoms is between 0.5 and 1.9. In a polar covalent bond, the more elec-

tronegative atom bears a partial negative charge (δ^-) and the less electronegative atom bears a partial positive charge (δ^+). This separation of charge produces a **dipole.**

- A **Lewis structure** for a covalent compound must show (1) the correct arrangement of atoms, (2) the correct number of valence electrons, (3) no more than two electrons in the outer shell of hydrogen, and (4) no more than eight electrons in the outer shell of any period 2 element.
- Exceptions to the octet rule include compounds of period 3 elements, such as phosphorus and sulfur, which may have as many as 10 and 12 electrons, respectively, in their valence shells.

SECTION 4.8 How Do We Name Binary Covalent Compounds?

- To name a **binary covalent compound,** name the less electronegative element first, followed by the name of the more electronegative element. The name of the more electronegative element is derived by adding *-ide* to the stem name of the element. Use the prefixes *di-, tri-, tetra-,* and so on to show the presence of two or more atoms of the same kind.

SECTION 4.9 What Is Resonance?

- According to the theory of resonance, a molecule or ion for which no single Lewis structure is adequate is best described by writing two or more **resonance contributing structures** and considering the real molecule or ion to be a hybrid of these contributing structures. To show how valence electrons are redistributed from one contributing structure to the next, we use curved arrows. A curved arrow extends from where the electrons are initially shown (on an atom or in a covalent bond) to their new location (on an adjacent atom or an adjacent covalent bond).

SECTION 4.10 How Do We Predict Bond Angles in Covalent Molecules?

- The **valence-shell electron-pair repulsion (VSEPR) model** predicts bond angles of 109.5° about atoms surrounded by four regions of electron density; angles of 120° about atoms surrounded by three regions of electron density; and angles of 180° about atoms surrounded by two regions of electron density.

SECTION 4.11 How Do We Determine if a Molecule Is Polar?

- A molecule is polar (has a dipole) if it has polar bonds and the centers of partial positive and partial negative charge do not coincide.
- If a molecule has polar bonds but the centers of partial positive and negative charges coincide, the molecule is nonpolar (it has no dipole).

PROBLEMS

Chemistry⚛Now™

Assess your understanding of this chapter's topics with additional quizzing and conceptual-based problems at **http://now.brookscole.com/gob8** or on the CD.

A blue problem number indicates an applied problem.

■ denotes problems that are available on the GOB ChemistryNow website or CD and are assignable in OWL.

SECTION 4.2 What Is the Octet Rule?

4.17 How many electrons must each atom gain or lose to acquire an electron configuration identical to the noble gas nearest to it in atomic number?
(a) Li (b) Cl (c) P (d) Al
(e) Sr (f) S (g) Si (h) O

4.18 Show how each chemical change obeys the octet rule.
(a) Lithium forms Li^+
(b) Oxygen forms O^{2-}

4.19 Show how each chemical change obeys the octet rule.
(a) Hydrogen forms H^- (hydride ion)
(b) Aluminum forms Al^{3+}

4.20 Write the formula for the most stable ion formed by each element.
(a) Mg (b) F (c) Al
(d) S (e) K (f) Br

4.21 Why is Li^- not a stable ion?

4.22 Predict which ions are stable:
(a) I^- (b) Se^{2+} (c) Na^+
(d) S^{2-} (e) Li^{2+} (f) Ba^{3+}

4.23 Predict which ions are stable:
(a) Br^{2-} (b) C^{4-} (c) Ca^+
(d) Ar^+ (e) Na^- (f) Cs^+

4.24 Why are carbon and silicon reluctant to form ionic bonds?

4.25 Table 4.2 shows the following ions of copper: Cu^+ and Cu^{2+}. Do these violate the octet rule? Explain.

SECTION 4.3 How Do We Name Anions and Cations?

4.26 ■ Name each polyatomic ion.
(a) HCO_3^- (b) NO_2^- (c) SO_4^{2-}
(d) HSO_4^- (e) $H_2PO_4^-$

SECTION 4.4 What Are the Two Major Types of Chemical Bonds?

4.27 Why does electronegativity generally increase in going up a column of the Periodic Table?

4.28 Why does electronegativity generally increase in going from left to right across a row of the Periodic Table?

4.29 Judging from their relative positions in the Periodic Table, which element in each pair has the larger electronegativity?
(a) F or Cl (b) O or S (c) C or N (d) C or F

4.30 Toward which atom are the electrons shifted in a covalent bond between each of the following pairs:
(a) H and Cl (b) N and O (c) C and O
(d) Cl and Br (e) C and S (f) P and S
(g) H and O

4.31 Which of these bonds is the most polar? The least polar?
(a) C—N (b) C—C (c) C—O

4.32 Classify each bond as nonpolar covalent, polar covalent, or ionic.
(a) C—Cl (b) C—Li (c) C—N

4.33 Classify each bond as nonpolar covalent, polar covalent, or ionic.
(a) C—Br (b) S—Cl (c) C—P

SECTION 4.5 What Is an Ionic Bond and How Does One Form?

4.34 Complete the chart by writing formulas for the compounds formed:

	Br^-	ClO_4^-	O^{2-}	NO_3^-	SO_4^{2-}	PO_4^{3-}	OH^-
Li^+							
Ca^{2+}							
Co^{3+}							
K^+							
Cu^{2+}							

4.35 ■ Write a formula for the ionic compound formed from each pair of elements.
(a) Sodium and bromine
(b) Sodium and oxygen
(c) Aluminum and chlorine
(d) Barium and chlorine
(e) Magnesium and oxygen

4.36 Although not a transition metal, lead can form Pb^{2+} and Pb^{4+} ions. Write the formula for the compound formed between each of these lead ions and the following anions:
(a) Chloride ion (b) Hydroxide ion
(c) Oxide ion

4.37 Describe the structure of sodium chloride in the solid state.

4.38 What is the charge on each ion in these compounds?
(a) CaS (b) MgF_2 (c) Cs_2O
(d) $ScCl_3$ (e) Al_2S_3

4.39 ■ Write the formula for the compound formed from the following pairs of ions:
(a) Iron(III) ion and hydroxide ion
(b) Barium ion and chloride ion
(c) Calcium ion and phosphate ion
(d) Sodium ion and permanganate ion

4.40 Write the formula for the ionic compound formed from the following pairs of ions:
(a) Iron(II) ion and chloride ion
(b) Calcium ion and hydroxide ion
(c) Ammonium ion and phosphate ion
(d) Tin(II) ion and fluoride ion

4.41 Which formulas are not correct? For each that is not correct, write the correct formula.
(a) Ammonium phosphate; $(NH_4)_2PO_4$
(b) Barium carbonate; Ba_2CO_3
(c) Aluminum sulfide; Al_2S_3
(d) Magnesium sulfide; MgS

4.42 Which formulas are not correct? For each that is not correct, write the correct formula.
(a) Calcium oxide; CaO_2
(b) Lithium oxide; LiO
(c) Sodium hydrogen phosphate; $NaHPO_4$
(d) Ammonium nitrate; NH_4NO_3

SECTION 4.6 How Do We Name Ionic Compounds?

4.43 Potassium chloride and potassium bicarbonate are used as potassium supplements. Write the formula of each compound.

4.44 Potassium nitrite has been used as a vasodilator and as an antidote for cyanide poisoning. Write the formula of this compound.

4.45 Name the polyatomic ion in each compound.
(a) Na_2SO_3 (b) KNO_3 (c) Cs_2CO_3
(d) NH_4OH (e) K_2HPO_4

4.46 Write the formulas for the ions present in each compound.
(a) NaBr (b) $FeSO_3$
(c) $Mg_3(PO_4)_2$ (d) KH_2PO_4
(e) $NaHCO_3$ (f) $Ba(NO_3)_2$

4.47 Name these ionic compounds:
(a) NaF (b) MgS
(c) Al_2O_3 (d) $BaCl_2$
(e) $Ca(HSO_3)_2$ (f) KI
(g) $Sr_3(PO_4)_2$ (h) $Fe(OH)_2$
(i) NaH_2PO_4 (j) $Pb(CH_3COO)_2$
(k) BaH_2 (l) $(NH_4)_2HPO_4$

4.48 ■ Write formulas for the following ionic compounds:

(a) Potassium bromide (b) Calcium oxide

(c) Mercury(II) oxide (d) Copper(II) phosphate

(e) Lithium sulfate (f) Iron(III) sulfide

4.49 Write formulas for the following ionic compounds:

(a) Ammonium hydrogen sulfite

(b) Magnesium acetate

(c) Strontium dihydrogen phosphate

(d) Silver carbonate

(e) Strontium chloride

(f) Barium permanganate

SECTION 4.7 What Is a Covalent Bond and How Does One Form?

4.50 How many covalent bonds are normally formed by each element?

(a) N (b) F (c) C

(d) Br (e) O

4.51 What is

(a) A single bond?

(b) A double bond?

(c) A triple bond?

4.52 In Section 2.3B, we saw that there are seven diatomic elements.

(a) Draw Lewis structures for each of these diatomic elements.

(b) Which diatomic elements are gases at room temperature? Which are liquids? Which are solids?

4.53 ■ Draw a Lewis structure for each covalent compound.

(a) CH_4 (b) C_2H_2 (c) C_2H_4

(d) BF_3 (e) CH_2O (f) C_2Cl_6

4.54 What is the difference between a molecular formula, a structural formula, and a Lewis structure?

4.55 ■ Write the total number of valence electrons in each molecule.

(a) NH_3 (b) C_3H_6 (c) $C_2H_4O_2$ (d) C_2H_6O

(e) CCl_4 (f) HNO_2 (g) CCl_2F_2 (h) O_2

4.56 ■ Draw a Lewis structure for each of the following molecules. In each case, the atoms can be connected in only one way.

(a) Br_2 (b) H_2S (c) N_2H_4 (d) N_2H_2

(e) CN^- (f) NH_4^+ (g) N_2 (h) O_2

4.57 What is the difference between (a) a bromine atom, (b) a bromine molecule, and (c) a bromide ion? Draw the Lewis structure for each.

4.58 Acetylene (C_2H_2), hydrogen cyanide (HCN), and nitrogen (N_2) each contain a triple bond. Draw a Lewis structure for each molecule.

4.59 Why can't hydrogen have more than two electrons in its valence shell?

4.60 Why can't second-row elements have more than eight electrons in their valence shells? That is, why does the octet rule work for second-row elements?

4.61 Why does nitrogen have three bonds and one unshared pair of electrons in covalent compounds?

4.62 Draw a Lewis structure of a covalent compound in which nitrogen has

(a) Three single bonds and one pair of electrons

(b) One single bond, one double bond, and one unshared pair of electrons

(c) One triple bond and one unshared pair of electrons

4.63 Why does oxygen have two bonds and two unshared pairs of electrons in covalent compounds?

4.64 Draw a Lewis structure of a covalent compound in which oxygen has

(a) Two single bonds and two unshared pairs of electrons

(b) One double bond and two unshared pairs of electrons

4.65 The ion O^{6+} has a complete outer shell. Why is this ion not stable?

4.66 Draw a Lewis structure for a molecule in which a carbon atom is bonded by a double bond to (a) another carbon atom, (b) an oxygen atom, and (c) a nitrogen atom.

4.67 Which of the following molecules have an atom that does not obey the octet rule (not all these are stable molecules)?

(a) BF_3 (b) CF_2 (c) BeF_2 (d) C_2H_4

(e) CH_3 (f) N_2 (g) NO

SECTION 4.8 How Do We Name Binary Covalent Compounds?

4.68 Name these binary covalent compounds.

(a) SO_2 (b) SO_3 (c) PCl_3 (d) CS_2

SECTION 4.9 What Is Resonance?

4.69 Write two acceptable contributing structures for the bicarbonate ion, HCO_3^- and show by the use of curved arrows how the first contributing structure is converted to the second.

4.70 Ozone, O_3, is an unstable blue gas with a characteristic pungent odor. In an ozone molecule, the connectivity of the atoms is O—O—O, and both O—O bonds are equivalent.

(a) How many valence electrons must be present in an acceptable Lewis structure for an ozone molecule?

(b) Write two equivalent resonance contributing structures for ozone. Be certain to show any positive or negative charges that may be present in your contributing structures. By *equivalent contributing structures*, we mean that each has the same pattern of bonding

(c) Show by the use of curved arrows how the first of your contributing structures may be converted to the second.

(d) Based on your contributing structures, predict the O—O—O bond angle in ozone.

(e) Explain why the following is not an acceptable contributing structure for an ozone molecule:

$$\ddot{O}=\ddot{O}=\ddot{O}$$

4.71 ■ Nitrous oxide, N_2O, laughing gas, is a colorless, nontoxic, tasteless, and odorless gas. It is used as an inhalation anesthetic in dental and other surgeries. Because nitrous oxide is soluble in vegetable oils (fats), it is used commercially as a propellant in whipped cream toppings.

Charles D. Winters

Nitrous oxide dissolves in fats. The gas is added under pressure to cans of whipped topping. When the valve is opened, the gas expands, thus expanding (whipping) the topping and forcing it out of the can.

(a) How many valence electrons are present in a molecule of N_2O?

(b) Write two equivalent contributing structures for this molecule. The connectivity in nitrous oxide is N—N—O.

(c) Explain why the following is not an acceptable contributing structure:

$$:N\equiv N=\ddot{O}$$

SECTION 4.10 How Do We Predict Bond Angles in Covalent Molecules?

4.72 State the shape of a molecule whose central atom is surrounded by

(a) Two regions of electron density

(b) Three regions of electron density

(c) Four regions of electron density

4.73 Hydrogen and oxygen combine in different ratios to form H_2O (water) and H_2O_2 (hydrogen peroxide).

(a) How many valence electrons are found in H_2O? In H_2O_2?

(b) Draw Lewis structures for each molecule in part (a). Be certain to show all valence electrons.

(c) Using the VSEPR model, predict the bond angles about the oxygen atom in water and about each oxygen atom in hydrogen peroxide.

4.74 Hydrogen and nitrogen combine in different ratios to form three compounds: NH_3 (ammonia), N_2H_4 (hydrazine), and N_2H_2 (diimide).

(a) How many valence electrons must the Lewis structure of each molecule show?

(b) Draw a Lewis structure for each molecule.

(c) Predict the bond angles about the nitrogen atom(s) in each molecule.

4.75 Predict the shape of each molecule.

(a) CH_4 (b) PH_3 (c) CHF_3 (d) SO_2

(e) SO_3 (f) CCl_2F_2 (g) NH_3 (h) PCl_3

4.76 Predict the shape of each ion.

(a) NO_2^- (b) NH_4^+ (c) CO_3^{2-}

SECTION 4.11 How Do We Determine if a Molecule Is Polar?

4.77 Both CO_2 and SO_2 have polar bonds. Account for the fact that CO_2 is nonpolar and SO_2 is polar.

4.78 Consider the molecule boron trifluoride, BF_3.

(a) Write a Lewis structure for BF_3.

(b) Predict the F—B—F bond angles using the VSEPR model.

(c) Does BF_3 have polar bonds? Is it a polar molecule?

4.79 Is it possible for a molecule to have polar bonds and yet have no dipole? Explain.

4.80 Is it possible for a molecule to have no polar bonds and yet have a dipole? Explain.

4.81 ■ In each case, tell whether the bond is ionic, polar covalent, or nonpolar covalent.

(a) Br_2 (b) BrCl   (c) HCl   (d) SrF_2

(e) SiH_4 (f) CO   (g) N_2 (h) CsCl

4.82 Account for the fact that chloromethane, CH_3Cl, which has only one polar C—Cl bond, is a polar molecule, but carbon tetrachloride, CCl_4, which has four polar C—Cl bonds, is a nonpolar molecule.

Chemical Connections

4.83 (Chemical Connections 4A) What are the three main inorganic components of one dry mixture currently used to create synthetic bone?

4.84 (Chemical Connections 4B) Why is sodium iodide often present in the table salt we buy at the grocery store?

4.85 (Chemical Connections 4B) What is a medical use of barium sulfate?

4.86 (Chemical Connections 4B) What is a medical use of potassium permanganate?

4.87 (Chemical Connections 4A) What is the main metal ion present in bone and tooth enamel?

4.88 (Chemical Connections 4C) In what way does the gas nitric oxide, NO, contribute to the acidity of acid rain?

Additional Problems

4.89 Explain why argon does not form either (a) ionic bonds or (b) covalent bonds.

4.90 Knowing what you do about covalent bonding in compounds of carbon, nitrogen, and oxygen, and given the fact that silicon is just below carbon in the Periodic Table, phosphorus is just below nitrogen, and sulfur is just below oxygen, predict the molecular formula for the compound formed by (a) silicon and chlorine, (b) phosphorus and hydrogen, and (c) sulfur and hydrogen.

4.91 ■ Use the valence-shell electron-pair repulsion model to predict the shape of a molecule in which a central atom is surrounded by five regions of electron density—as, for example, in phosphorus pentafluoride, PF_5. (*Hint:* Use molecular models or, if you do not have a set handy, use a marshmallow or gumdrop and toothpicks.)

4.92 Use the valence-shell electron-pair repulsion model to predict the shape of a molecule in which a central atom is surrounded by six regions of electron density—as, for example, in sulfur hexafluoride, SF_6.

4.93 Chlorine dioxide, ClO_2, is a yellow to reddish yellow gas at room temperature. This strong oxidizing agent is used for bleaching cellulose, paper pulp, and textiles, and for water purification. It was the gas used to kill anthrax spores in the anthrax-contaminated Hart Senate Office Building.

(a) How many valence electrons are present in ClO_2?

(b) Draw a Lewis structure for this molecule. (*Hint:* The order of attachment of atoms in this molecule is O—Cl—O. Chlorine is a period 3 element, and its valence shell may contain more than eight electrons.)

Reading Labels

4.94 Name and write the formula for the fluorine-containing compound present in fluoridated toothpastes and dental gels.

4.95 If you read the labels of sun-blocking lotions, you will find that a common UV-blocking agent is a compound containing zinc. Name and write the formula of this zinc-containing compound.

4.96 On packaged table salt, it is common to see a label stating that the salt "supplies iodide, a necessary nutrient." Name and write the formula of the iodine-containing compound found in iodized salt.

4.97 We are constantly warned about the dangers of "lead-based" paints. Name and write the formula for a lead-containing compound found in lead-based paints.

4.98 If you read the labels of several liquid and tablet antacid preparations, you will find that in many of them, the active ingredients are compounds containing hydroxide ions. Name and write formulas for these hydroxide ion–containing compounds.

4.99 Iron forms Fe^{2+} and Fe^{3+} ions. Which ion is found in the over-the-counter preparations intended to treat "iron-poor blood"?

4.100 Read the labels of several multivitamin/multimineral formulations. Among their components you will find a number of so-called trace minerals—minerals required in the diet of a healthy adult in amounts less than 100 mg per day, or present in the body in amounts less than 0.01 percent of total body weight. Following are 18 trace minerals. Name at least one form of each trace mineral present in multivitamin formulations.

(a) Phosphorus   (b) Magnesium
(c) Potassium   (d) Iron
(e) Calcium   (f) Zinc
(g) Manganese   (h) Titanium
(i) Silicon   (j) Copper
(k) Boron   (l) Molybdenum
(m) Chromium   (n) Iodine
(o) Selenium   (p) Vanadium
(q) Nickel   (r) Tin

4.101 Write formulas for these compounds.

(a) Calcium sulfite, which is used in preserving cider and other fruit juices

(b) Calcium hydrogen sulfite, which is used in dilute aqueous solutions for washing casks in brewing to prevent souring and cloudiness of beer and to prevent secondary fermentation

(c) Calcium hydroxide, which is used in mortar, plaster, cement, and other building and paving materials

(d) Calcium hydrogen phosphate, which is used in animal feeds and as a mineral supplement in cereals and other foods

4.102 Many paint pigments contain transition metal compounds. Name the compounds in these pigments using a Roman numeral to show the charge on the metal ion.

(a) Yellow, CdS   (b) Green, Cr_2O_3
(c) White, TiO_2 (d) Purple, $Mn_3(PO_4)_2$
(e) Blue, Co_2O_3 (f) Ochre, Fe_2O_3

Looking Ahead

4.103 Perchloroethylene, which is a liquid at room temperature, is one of the most widely used solvents for commercial dry cleaning. It is sold for this purpose under several trade names, including Perclene. Does this molecule have polar bonds? Is it a polar molecule? Does it have a dipole?

Perchloroethylene

4.104 Vinyl chloride is the starting material for the production of poly(vinyl chloride), abbreviated PVC. Its recycling code is "V". The major use of PVC is for tubing in residential and commercial construction (Section 12.7).

Vinyl chloride

(a) Complete the Lewis structure for vinyl chloride by showing all unshared pairs of electrons.

(b) Predict the H—C—H, H—C—C, and Cl—C—H, bond angles in this molecule.

(c) Does vinyl chloride have polar bonds? Is it a polar molecule? Does it have a dipole?

4.105 Tetrafluoroethylene is the starting material for the production of poly(tetrafluoroethylene), a polymer that is widely used for the preparation of nonstick coatings on kitchenware (Section 12.7). The most widely known trade name for this product is Teflon.

Tetrafluoroethylene

(a) Complete the Lewis structure for tetrafluoroethylene by showing all unshared pairs of electrons.

(b) Predict the F—C—F and F—C—C bond angles in this molecule.

(c) Does tetrafluoroethylene have polar bonds? Is it a polar molecule? Does it have a dipole?

4.106 Some of the following structural formulas are incorrect because they contain one or more atoms that do not have their normal number of covalent bonds. Which structural formulas are incorrect, and which atom or atoms in each have the incorrect number of bonds?

Chemical Reactions

Joseph Nettis/Photo Researchers, Inc.

Fireworks are spectacular displays of chemical reactions.

5.1 | What Are Chemical Reactions?

In Chapter 1, we learned that chemistry is mainly concerned with two things: the structure of matter and the transformations of one form of matter to another. In Chapters 2, 3, and 4, we discussed the first of these topics, and now we are ready to turn our attention to the second. In a chemical change, also called a **chemical reaction,** one or more **reactants** (starting materials) are converted into one or more **products.**

Chemical reactions occur all around us. They fuel and keep alive the cells of living tissues; they occur when we light a match, cook dinner, start a car, listen to a portable radio, or watch television. Most of the world's manufacturing processes involve chemical reactions; they include petroleum

GOB
Chemistry ⚛ Now™
Look for this logo in the chapter and go to GOB ChemistryNow at **http://now.brookscole.com/gob8** or on the CD for turtorials, simulations, and problems.

A table of atomic weights is given on the inside back cover. Atomic weights can also be found in the Periodic Table on the inside front cover.

Table 5.1 Formula Weights for Two Ionic and Two Covalent Compounds

Ionic Compounds

Sodium chloride (NaCl)	23.0 amu Na + 35.5 amu Cl = 58.5 amu
Nickel(II) chloride hydrate ($NiCl_2 \cdot 6H_2O$)*	58.7 amu Ni + 2(35.5 amu Cl) + 12(1.0 amu H) + 6(16.0 amu O) = 237.7 amu

Covalent Compounds

Water (H_2O)	2(1.0 amu H) + 16.0 amu O = 18.0 amu
Aspirin ($C_9H_8O_4$)	9(12.0 amu C) + 8(1.0 amu H) + 4(16.0 amu O) = 180.0 amu

*Nickel(II) chloride crystallizes from an aqueous solution with six moles of water per mole of $NiCl_2$. The presence of water molecules in the crystal is indicated by the name "hydrate."

refining and food processing as well as the manufacture of drugs, plastics, synthetic fibers, explosives, and many other materials.

In this chapter we discuss four aspects of chemical reactions: (1) how to write and balance chemical equations, (2) mass relationships in chemical reactions, (3) types of chemical reactions, and (4) heat gains and losses.

5.2 | What Are Molecular Weights and Formula Weights, and How Do They Differ?

We begin our study of mass relationships with a discussion of formula weight. The **formula weight (FW)** of any substance is the sum of the atomic weights in atomic mass units (amu) of all the atoms in the compound's formula. The term "formula weight" can be used for both ionic and molecular compounds and tells nothing about whether the compound is ionic or molecular. Table 5.1 gives formula weights for two ionic compounds and two covalent compounds.

Another term, **molecular weight (MW),** is strictly correct only when used for covalent compounds. In this book, we use "formula weight" for both ionic and covalent compounds and "molecular weight" only for covalent compounds.

Molecular weight (MW) The sum of the atomic weights of all atoms in a molecular compound expressed in atomic mass units (amu)

EXAMPLE 5.1

What is the molecular weight of (a) glucose, $C_6H_{12}O_6$, and (b) urea, $(NH_2)_2CO$?

Solution

(a) Glucose, $C_6H_{12}O_8$
- C 6 × 12.0 = 72.0
- H 12 × 1.0 = 12.0
- O 6 × 16.0 = 96.0

  $C_6H_{12}O_6$ = 180.0 amu

(b) Urea, $(NH_2)_2CO$
- N 2 × 14.0 = 28.0
- H 4 × 1.0 = 4.0
- C 1 × 12.0 = 16.0
- O 1 × 16.0 = 16.0

  $(NH_2)_2CO$ = 60.0 amu

Problem 5.1

What is (a) the molecular weight of ibuprofen, $C_{13}H_{18}O_2$, and (b) the formula weight of barium phosphate, $Ba_3(PO_4)_2$?

5.3 | What Is a Mole and How Do We Use It to Calculate Mass Relationships?

Atoms and molecules are so tiny (Section 2.4F) that chemists are seldom able to deal with them one at a time. When we weigh even a very small quantity of a compound, huge numbers of formula units (perhaps 10^{19}) are present. The formula unit may be atoms, molecules, or ions. To overcome this problem, chemists long ago defined a unit called the **mole (mol).** A mole is the amount of substance that contains as many atoms, molecules, or ions as there are atoms in exactly 12 g of carbon-12. The important point here is that whether we are dealing with a mole of iron atoms, a mole of methane molecules, or a mole of sodium ions, a mole always contains the same number of formula units. We are used to scale-up factors in situations where there are large numbers of units involved in counting. We buy eggs by the dozen and pencils by the gross. Just as the dozen (12 units) is a useful scale-up factor for eggs, and the gross (144 units) is a useful scale-up factor for pencils, the mole is a useful scale-up factor for atoms and molecules. We will soon see that the number of units is much larger for a mole than for a dozen or a gross.

The number of formula units in a mole is called **Avogadro's number** after the Italian physicist, Amadeo Avogadro (1776–1856), who first proposed the concept of a mole, but was not able to determine the number experimentally. Note that Avogadro's number is not a defined value, but rather a value that must be determined experimentally. Its value is now known to nine significant figures.

Avogadro's number = $6.02214199 \times 10^{23}$ formula units per mole

For most calculations in this text, we round this number to three significant figures to 6.02×10^{23} formula units per mole.

A mole of hydrogen atoms is 6.02×10^{23} hydrogen atoms, a mole of sucrose (table sugar) molecules is 6.02×10^{23} sugar molecules, a mole of apples is 6.02×10^{23} apples, and a mole of sodium ions is 6.02×10^{23} sodium ions. Just as we call 12 of anything a dozen, 20 a score, and 144 a gross, we call 6.02×10^{23} of anything a mole.

The **molar mass** of any substance (the mass of one mole of any substance) is the formula weight of the substance expressed in grams. For instance, the formula weight of glucose, $C_6H_{12}O_6$ (Example 5.1) is 180 amu; therefore, 180 g of glucose is one mole of glucose. Likewise, the formula weight of urea, $(NH_2)_2CO$, is 60.0 amu and, therefore, 60.0 g of $(NH_2)_2CO$ is one mole of urea. For atoms, one mole is the atomic weight expressed in grams: 12.0 g of carbon is one mole of carbon atoms; 32.1 g of sulfur is one mole of sulfur atoms; and so on. As you see, the important point here is that to talk about the mass of a mole, we need to know the chemical formula of the substance we are considering. Figure 5.1 shows one-mole quantities of several compounds.

GOB
Chemistry Now™
Click *Chemistry Interactive* to explore more about **Moles**

Mole (mol) The formula weight of a substance expressed in grams

Avogadro's number 6.02×10^{23} formula units per mole; the amount of any substance that contains as many formula units as the number of atoms in 12 g of carbon-12.

One mole of pennies placed side by side would stretch for more than 1 million light-years, a distance far outside our solar system and even outside our own galaxy. Six moles of this text would weigh as much as the Earth.

Molar mass The mass of one mole of a substance expressed in grams; the formula weight of a compound expressed in grams

GOB
Chemistry Now™
Click *Coached Problems* for examples of **Determining Molar Mass**

Figure 5.1 One-mole quantities of (*a*) six metals and (*b*) four compounds. (*a*) Top row (left to right): Cu beads (63.5 g), Al foil (27.0 g), and Pb shot (207.2 g). Bottom row (left to right): S powder (32.1 g), Cr chunks (52.0 g), and Mg shavings (24.4 g). (*b*) H_2O (18.0 g); small beaker, NaCl (58.4 g); large beaker, aspirin, $C_9H_8O_4$, (180.2 g); green $NiCl_2 \cdot 6H_2O$ (237.7 g).

(*a*)   (*b*)

Charles D. Winters

Now that we know the relationship between moles and molar mass (g/mol), we can use molar mass as a conversion factor to convert from grams to moles, and from moles to grams. For this calculation, we use molar mass as a conversion factor.

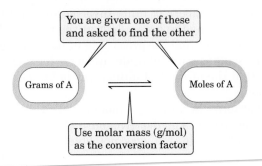

Suppose we want to know the number of moles of water in a graduated cylinder that contains 36.0 g of water. We know that the molar mass of water is 18.0 g/mol. If 18.0 g of water is one mole of water, then 36.0 g must be two moles of water.

$$36.0 \text{ g } H_2O \times \frac{1 \text{ mol } H_2O}{18.0 \text{ g } H_2O} = 2.00 \text{ mol } H_2O$$

Molar mass can also be used to convert from moles to grams. Suppose you have a beaker that contains 0.753 mol of sodium chloride and you are asked to calculate the number of grams of sodium chloride in the beaker. As a conversion factor, use the fact that the molar mass of NaCl is 58.5 g/mol.

$$0.753 \text{ mol NaCl} \times \frac{58.5 \text{ g NaCl}}{1 \text{ mol NaCl}} = 44.1 \text{ g NaCl}$$

This type of calculation can be performed for ionic compounds, such as NaF, as well as for molecular compounds, such as CO_2 and urea.

EXAMPLE 5.2

We have 27.5 g of sodium fluoride, NaF, the form of fluoride ions most commonly used in fluoride toothpastes. How many moles is this?

Solution

The formula weight of NaF = 23.0 + 19.0 = 42.0 amu. Thus each mole of NaF has a mass of 42.0 g, allowing us to use the conversion factor 1 mol NaF = 42.0 g NaF:

$$27.5 \text{ g NaF} \times \frac{1 \text{ mol NaF}}{42.0 \text{ g NaF}} = 0.655 \text{ mol NaF}$$

Problem 5.2

A person drinks 1500 g of water per day. How many moles is this?

EXAMPLE 5.3

We wish to weigh 3.41 mol of ethanol, C_2H_6O. How many grams is this?

Solution

The formula weight of C_2H_6O is 2(12.0) + 6(1.0) + 16.0 = 46.0 amu, so the conversion factor is 1 mol C_2H_6O = 46.0 g C_2H_6O.

$$3.41 \text{ mol } C_2H_6O \times \frac{46.0 \text{ g } C_2H_6O}{1.00 \text{ mol } C_2H_6O} = 157 \text{ g } C_2H_6O$$

Problem 5.3

We wish to weigh 2.84 mol of sodium sulfide, Na_2S. How many grams is this?

EXAMPLE 5.4

How many moles of nitrogen atoms and oxygen atoms are there in 21.4 mol of the explosive trinitrotoluene (TNT), $C_7H_5N_3O_6$?

Solution

The formula $C_7H_5N_3O_6$ tells us that each molecule of TNT contains three nitrogen atoms and six oxygen atoms. It also tells us that each mole of TNT contains three moles of N atoms and six moles of O atoms. Therefore, we have the following conversion factors: 1 mol TNT = 3 mol N atoms, and 1 mol TNT = 6 mol O atoms. The number of moles of N atoms in 21.4 mol of TNT is

$$21.4 \text{ mol TNT} \times \frac{3 \text{ mol N atoms}}{1 \text{ mol TNT}} = 64.2 \text{ mol N atoms}$$

The number of moles of O atoms in 21.4 mol of TNT is

$$21.4 \text{ mol TNT} \times \frac{6 \text{ mol O atoms}}{1 \text{ mol TNT}} = 128 \text{ mol O atoms}$$

Note that we give the answer to three significant figures because we were given the number of moles to three significant figures. The ratio of moles of O atoms to moles of TNT is an exact number.

Problem 5.4

How many moles of C atoms, H atoms, and O atoms are there in 2.5 mol of glucose, $C_6H_{12}O_6$?

EXAMPLE 5.5

How many moles of sodium ions, Na^+, are there in 5.63 g of sodium sulfate, Na_2SO_4?

Solution

First, we need to find out how many moles of Na_2SO_4 we have. The formula weight of Na_2SO_4 is $2(23.0) + 32.1 + 4(16.0) = 142.1$ amu. In the conversion of grams Na_2SO_4 to moles of Na_2SO_4, we can use the conversion factor 1 mol Na_2SO_4 = 142.1 g Na_2SO_4.

$$5.63 \text{ g } Na_2SO_4 \times \frac{1 \text{ mol } Na_2SO_4}{142.1 \text{ g } Na_2SO_4} = 0.0396 \text{ mol } Na_2SO_4$$

The formula Na_2SO_4 shows us that each mole of Na_2SO_4 contains two moles of Na^+ ions. Therefore, we use the conversion factor 1 mol Na_2SO_4 = 2 mol Na^+. The number of moles of Na^+ ions in 5.63 g of Na_2SO_4 is

$$0.0396 \text{ mol } Na_2SO_4 \times \frac{2 \text{ mol } Na^+}{1 \text{ mol } Na_2SO_4} = 0.0792 \text{ mol } Na^+$$

Problem 5.5

How many moles of copper(I) ions, Cu^+, are there in 0.062 g of copper(I) nitrate, $CuNO_3$?

EXAMPLE 5.6

An aspirin tablet, $C_9H_8O_4$, contains 0.360 g of aspirin. (The rest of it is starch or other fillers.) How many molecules of aspirin are present in this tablet?

Solution

The molecular weight of aspirin is $9(12.0) + 8(1.0) + 4(16.0) = 180.0$ amu. First we need to find out how many moles of aspirin are in 0.360 g:

$$0.360 \text{ g aspirin} \times \frac{1 \text{ mol aspirin}}{180.0 \text{ g aspirin}} = 0.00200 \text{ mol aspirin}$$

Each mole of aspirin contains 6.02×10^{23} molecules. Therefore, the number of molecules of aspirin in the tablet is

$$0.00200 \text{ mol} \times 6.02 \times 10^{23} \frac{\text{molecules}}{\text{mol}} = 1.20 \times 10^{21} \text{ molecules}$$

Problem 5.6

How many molecules of water, H_2O, are in a glass of water (235 g)?

5.4 | How Do We Balance Chemical Equations?

When propane, which is the major component in bottled gas or LP (lique-fied petroleum) gas, burns in air, it reacts with the oxygen in the air. These reactants are converted to the products carbon dioxide and water in a chemical reaction called **combustion.** We can write this chemical reaction in the form of a **chemical equation,** using chemical formulas for the reac-tants and products, and an arrow to indicate the direction in which the reaction proceeds. In addition, it is important to show the state of each reactant and product—that is, whether it is a gas, liquid, or solid. We use the symbol (g) for gas, (ℓ) for liquid, (s) for solid, and (aq) for a substance dissolved in water (aqueous). We place the appropriate symbol immedi-ately following each reactant and product. In our combustion equation, propane, oxygen, and carbon dioxide are gases, and the flame produced when propane burns is hot enough that the water forms as a gas (steam).

Combustion Burning in air

Chemical equation A representation using chemical formulas of the process that occurs when reactants are converted to products

$$C_3H_8(g) + O_2(g) \longrightarrow CO_2(g) + H_2O(g)$$
Propane   Oxygen   Carbon   Water
dioxide

The equation we have written is incomplete, however. While it tells us the formulas of the starting materials and products (which every chemical equation must do) and the physical state of each reactant and product, it does not give the amounts correctly. It is not balanced, which means that the number of atoms on the left side of the equation is not the same as the number of atoms on the right side. From the law of conservation of mass (Section 2.3A), we know that, in chemical reactions, atoms are never destroyed or created; they merely shift from one substance to another. Thus all the atoms present at the start of the reaction (on the left side of the equa-tion) must still be present at the end of the reaction (on the right side of the equation). The equation we have just written has three carbon atoms on the left side but only one on the right side.

Propane burning in air.

Charles D. Winters

EXAMPLE 5.7

Balance this equation:

$$Ca(OH)_2(s) + HCl(g) \longrightarrow CaCl_2(s) + H_2O(\ell)$$
Calcium   Hydrogen   Calcium
hydroxide   chloride   chloride

Solution

The calcium is already balanced—there is one Ca on each side. There is one Cl on the left and two on the right. To balance them, we add the coeffi-cient 2 in front of HCl:

$$Ca(OH)_2(s) + 2HCl(g) \longrightarrow CaCl_2(s) + H_2O(\ell)$$

Looking at hydrogens, we see that there are four hydrogens on the left but only two on the right. Placing the coefficient 2 in front of H_2O balances the hydrogens. It also balances the oxygens and completes the balancing of the equation:

$$Ca(OH)_2(s) + 2HCl(g) \longrightarrow CaCl_2(s) + 2H_2O(\ell)$$

Problem 5.7

Following is an unbalanced equation for photosynthesis, the process by which green plants convert carbon dioxide and water to glucose and oxygen. Balance this equation.

$$CO_2(g) + H_2O(\ell) \xrightarrow{\text{Photosynthesis}} C_6H_{12}O_6(aq) + O_2(g)$$
$$\text{Glucose}$$

EXAMPLE 5.8

Balance this equation for the combustion of butane, the fluid most commonly used in pocket lighters:

$$C_4H_{10}(\ell) + O_2(g) \longrightarrow CO_2(g) + H_2O(g)$$
$$\text{Butane}$$

A pocket lighter contains butane in both the liquid and the gaseous state.

Solution

The equation for the combustion of butane is very similar to the one we examined at the beginning of this section for the combustion of propane. Here we put a 4 in front of the CO_2 (because there are four carbons on the left) and a 5 in front of the H_2O (because there are ten hydrogens on the left):

$$C_4H_{10}(g) + O_2(g) \longrightarrow 4CO_2(g) + 5H_2O(g)$$

When we count the oxygens, we find 2 on the left and 13 on the right. We can balance their numbers by putting $\frac{13}{2}$ in front of the O_2.

$$C_4H_{10}(g) + \frac{13}{2}O_2(g) \longrightarrow 4CO_2(g) + 5H_2O(g)$$

Although chemists sometimes have good reason to write equations with fractional coefficients, it is common practice to use only whole-number coefficients. We accomplish this by multiplying everything by 2, which gives the balanced equation:

$$2C_4H_{10}(g) + 13O_2(g) \longrightarrow 8CO_2(g) + 10H_2O(g)$$

Problem 5.8

Balance this equation:

$$C_6H_{14}(g) + O_2(g) \longrightarrow CO_2(g) + H_2O(g)$$

EXAMPLE 5.9

Balance this equation:

$$Na_2SO_3(aq) + H_3PO_4(aq) \longrightarrow H_2SO_3(aq) + Na_3PO_4(aq)$$
$$\text{Sodium} \qquad \text{Phosphoric} \qquad \text{Sulfurous} \qquad \text{Sodium}$$
$$\text{sulfite} \qquad \text{acid} \qquad \text{acid} \qquad \text{phosphate}$$

HOW TO . . .

Balance a Chemical Equation

To balance an equation, we place numbers in front of the formulas until the number of each kind of atom in the products is the same as the number in the starting materials. These numbers are called **coefficients.** As an example, let us balance our propane equation:

$$C_3H_8(g) + O_2(g) \longrightarrow CO_2(g) + H_2O(g)$$

Propane   Oxygen   Carbon   Water
                    dioxide

GOB
Chemistry•⚛•Now™
Click *Chemistry Interactive* for more practice on **How to Balance Chemical Equations**

To balance an equation:

1. Begin with atoms that appear in only one compound on the left and only one compound on the right. In the equation for the reaction of propane and oxygen, begin with either carbon or hydrogen.
2. If an atom occurs as a free element—as, for example, O_2 does in the reaction of propane with oxygen—balance this element last.
3. You can change only coefficients in balancing an equation; you cannot change chemical formulas. For example, if you have H_2O on the left side of an equation but need two oxygens, you can add the coefficient "2" to read $2H_2O$. You cannot, however, get two oxygens by changing the formula to H_2O_2. You can't do this because the reactant is water, H_2O, and not hydrogen peroxide, H_2O_2.

For the equation for the reaction of propane with oxygen, we can begin with carbon. Three carbon atoms appear on the left and one on the right. If we put a 3 in front of the CO_2 (indicating that three CO_2 molecules are formed), three carbons will appear on each side, and the carbons will be balanced:

Three C on each side

$$C_3H_8(g) + O_2(g) \longrightarrow 3CO_2(g) + H_2O(g)$$

Next we look at the hydrogens. There are eight on the left and two on the right. If we put a 4 in front of the H_2O, there will be eight hydrogens on each side and the hydrogens will be balanced:

Eight H on each side

$$C_3H_8(g) + O_2(g) \longrightarrow 3CO_2(g) + 4H_2O(g)$$

The only atom still unbalanced is oxygen. Notice that we saved this reactant for last (rule 2). There are two oxygen atoms on the left and ten on the right. If we put a 5 in front of the O_2 on the left, we both balance the oxygen atoms and arrive at the balanced equation:

Ten O on each side

$$C_3H_8(g) + 5O_2(g) \longrightarrow 3CO_2(g) + 4H_2O(g)$$

At this point, the equation ought to be balanced, but we should always check, just to make sure. In a balanced equation, there must be the same number of atoms of each element on both sides. A check of our work shows three C, ten O, and eight H atoms on each side. The equation is, indeed, balanced.

Solution

The key to balancing equations like this one is to realize that polyatomic ions such as SO_3^{2-} and PO_4^{3-} usually remain intact on both sides of the equation. We can begin by balancing the Na^+ ions. We put a 3 in front of Na_2SO_3 and a 2 in front of Na_3PO_4, giving us six Na^+ ions on each side:

Six Na on each side

$$3Na_2SO_3(aq) + H_3PO_4(aq) \longrightarrow H_2SO_3(aq) + 2Na_3PO_4(aq)$$

There are now three SO_3^{2-} units on the left and only one on the right, so we put a 3 in front of H_2SO_3:

Three SO_3^{2-} units on each side

$$3Na_2SO_3(aq) + H_3PO_4(aq) \longrightarrow 3H_2SO_3(aq) + 2Na_3PO_4(aq)$$

Now let's look at the PO_4^{3-} units. There are two PO_4^{3-} units on the right, but only one on the left. To balance them, we put a 2 in front of H_3PO_4. In doing so, we balance not only the PO_4^{3-} units but also the hydrogens, and arrive at the balanced equation:

Two PO_4^{3-} units on each side

$$3Na_2SO_3(aq) + 2H_3PO_4(aq) \longrightarrow 3H_2SO_3(aq) + 2Na_3PO_4(aq)$$

Problem 5.9

Balance this equation:

$$K_2C_2O_4(aq) + Ca_3(AsO_4)_2(s) \longrightarrow K_3AsO_4(aq) + CaC_2O_4(s)$$

Potassium oxalate   Calcium arsenate   Potassium arsenate   Calcium oxalate

One final point about balancing chemical equations. The following equation for the combustion of propane is correctly balanced:

$$C_3H_8(g) + 5O_2(g) \longrightarrow 3CO_2(g) + 4H_2O(g)$$

Propane

Would it be correct if we doubled all the coefficients?

$$2C_3H_8(g) + 10O_2(g) \longrightarrow 6CO_2(g) + 8H_2O(g)$$

Propane

This equation is mathematically and scientifically correct, but chemists do not normally write equations with coefficients that are all divisible by a common number. A correctly balanced equation is almost always written with the coefficients expressed as the lowest set of whole numbers.

5.5 | How Do We Calculate Mass Relationships in Chemical Reactions?

A. Stoichiometry

As we saw in Section 5.4, a balanced chemical equation tells us not only which substances react and which form, but also the molar ratios in which they react. For example, using the molar ratios in a balanced chemical equation, we can calculate the mass of starting materials we need to produce a particular mass of a product. The study of mass relationships in chemical reactions is called **stoichiometry.**

Let us look once again at the balanced equation for the burning of propane:

$$C_3H_8(g) + 5O_2(g) \longrightarrow 3CO_2(g) + 4H_2O(g)$$
Propane

This equation tells us not only that propane and oxygen are converted to carbon dioxide and water, but also that 1 mol of propane combines with 5 mol of oxygen to produce 3 mol of carbon dioxide and 4 mol of water; that is, we know the mole ratios involved. The same is true for any other balanced equation. This fact allows us to answer questions such as the following:

1. How many moles of any particular product are formed if we start with a given mass of a starting material?

2. How many grams (or moles) of one starting material are necessary to react completely with a given number of grams (or moles) of another starting material?

3. How many grams (or moles) of starting material are needed if we want to form a certain number of grams (or moles) of a certain product?

4. How many grams (or moles) of another product are obtained when a certain amount of main product is produced?

It might seem as if we have four different types of problems here. In fact, we can solve them all by the same very simple procedure summarized in the following diagram:

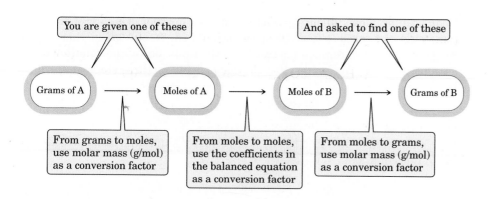

You will always need a conversion factor that relates moles to moles. You will need the conversion factors from grams to moles and from moles to grams according to the way the problem is asked; you may need one or both

---

GOB
Chemistry⚛Now™
Click *Chemistry Interactive* to explore
Weight Relations in Chemical Reactions

Stoichiometry The mass relationships in a chemical reaction

"Stoichiometry" comes from the Greek *stoicheion,* meaning "element," and *metron,* meaning "measure."

In Section 5.4, we saw that the coefficients in an equation represent numbers of molecules. Because moles are proportional to molecules (Section 5.3), the coefficients in an equation also represent numbers of moles.

in some problems and not in others. It is easy to weigh a given number of grams, but the molar ratio determines the amount of substance involved in a particular reaction.

EXAMPLE 5.10

Ammonia is produced on an industrial scale by the reaction of nitrogen gas with hydrogen gas (the Haber process) according to this balanced equation:

$$N_2(g) + 3H_2(g) \longrightarrow 2NH_3(g)$$
$$\text{Ammonia}$$

How many grams of N_2 are necessary to produce 7.50 g of NH_3?

Solution

Step 1 The coefficients in an equation refer to the relative numbers of moles, not grams. Therefore, we must first find out how many moles of NH_3 are in 7.50 g of NH_3. To convert grams of NH_3 to moles of NH_3, we use the conversion factor 1 mol NH_3 = 17.0 g NH_3. At this stage, we don't do the multiplication, but just leave the relationship set up:

$$7.50 \text{ g NH}_3 \times \frac{1 \text{ mol NH}_3}{17.0 \text{ g NH}_3} = \text{mol NH}_3$$

Step 2 Next we turn to the balanced equation, which shows us that 2 mol of NH_3 are produced from 1 mol of N_2. We multiply the number of moles of NH_3 by this conversion factor:

$$7.50 \text{ g NH}_3 \times \frac{1 \text{ mol NH}_3}{17.0 \text{ g NH}_3} \times \frac{1 \text{ mol N}_2}{2 \text{ mol NH}_3} = \text{mol N}_2$$

Step 3 To find the number of grams of N_2, we multiply by the conversion factor 28.0 g N_2 = 1 mol N_2. At this point, the conversion factors are set up properly and we have converted from grams of NH_3 to grams of N_2. Now we do the arithmetic and get an answer of 6.18 g N_2:

$$7.50 \text{ g NH}_3 \times \frac{1 \text{ mol NH}_3}{17.0 \text{ g NH}_3} \times \frac{1 \text{ mol N}_2}{2 \text{ mol NH}_3} \times \frac{28.0 \text{ g N}_2}{1 \text{ mol N}_2} = 6.18 \text{ g N}_2$$

In all such problems, we are given the mass (or number of moles) of one compound and asked to find the mass (or number of moles) of another compound. The two compounds can be on the same side of the equation or on opposite sides. We can do all such problems by the three steps we have just used.

Problem 5.10

Pure aluminum is prepared by the electrolysis of aluminum oxide according to this equation:

$$Al_2O_3(s) \xrightarrow{\text{Electrolysis}} Al(s) + O_2(g)$$
$$\text{Aluminum}$$
$$\text{oxide}$$

(a) Balance this equation.
(b) What mass of aluminum oxide is required to prepare 27 g (1 mol) of aluminum?

EXAMPLE 5.11

Silicon to be used in computer chips is manufactured by a process represented by the following reaction:

$$SiCl_4(s) + 2Mg(s) \longrightarrow Si(s) + 2MgCl_2(s)$$

Silicon
tetrachloride

Magnesium
chloride

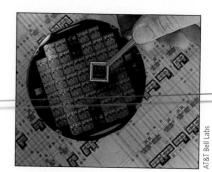

As microprocessor chips became increasingly smaller, the purity of the silicon becomes more important, because impurities can prevent the circuit from working properly.

A sample of 225 g of silicon tetrachloride, $SiCl_4$, is reacted with an excess (more than necessary) amount of Mg. How many moles of Si are produced?

Solution

We are given 225 g of $SiCl_4$ and asked for moles of Si. We must first convert grams of $SiCl_4$ to moles of $SiCl_4$. For this calculation we use the conversion factor 1 mol $SiCl_4$ = 170 g $SiCl_4$:

$$225 \text{ g } SiCl_4 \times \frac{1 \text{ mol } SiCl_4}{170 \text{ g } SiCl_4} = \text{mol } SiCl_4$$

Next we use the conversion factor 1 mol $SiCl_4$ = 1 mol Si, which we obtain from the balanced chemical equation. Now we do the arithmetic and obtain an answer of 1.32 mol Si:

$$225 \text{ g } SiCl_4 \times \frac{1 \text{ mol } SiCl_4}{170 \text{ g } SiCl_4} \times \frac{1 \text{ mole Si}}{1 \text{ mol } SiCl_4} = 1.32 \text{ mol Si}$$

Problem 5.11

In the industrial synthesis of acetic acid, methanol is reacted with carbon monoxide. How many moles of CO are required to produce 16.6 mol of acetic acid?

$$CH_3OH(g) + CO(g) \longrightarrow CH_3COOH(\ell)$$

Methanol   Carbon   Acetic acid
               monoxide

EXAMPLE 5.12

When urea, $(NH_2)_2CO$, is acted on by the enzyme urease in the presence of water, ammonia and carbon dioxide are produced:

$$(NH_2)_2CO(aq) + H_2O(\ell) \xrightarrow{\text{Urease}} 2NH_3(aq) + CO_2(g)$$

Urea   Ammonia

If excess water is present (more than necessary for the reaction), how many grams each of CO_2 and NH_3 are produced from 0.83 mol of urea?

Solution

For grams of CO_2:

We are given moles of urea and asked for grams of CO_2. We first use the conversion factor 1 mol urea = 1 mol CO_2 to get the number of moles of CO_2 that will be formed:

$$0.83 \text{ mol urea} \times \frac{1 \text{ mol } CO_2}{1 \text{ mol urea}} = \text{mol } CO_2$$

Then we use the conversion factor 1 mol CO_2 = 44 g CO_2 to give us the answer:

$$0.83 \text{ mol urea} \times \frac{1 \text{ mol } CO_2}{1 \text{ mol urea}} \times \frac{44 \text{ g } CO_2}{1 \text{ mol } CO_2} = 37 \text{ g } CO_2$$

For grams of NH_3:
We follow the same procedure as for CO_2 but use different conversion factors:

$$0.83 \text{ mol urea} \times \frac{2 \text{ mol } NH_3}{1 \text{ mol urea}} \times \frac{17 \text{ g } NH_3}{1 \text{ mol } NH_3} = 28 \text{ g } NH_3$$

Problem 5.12

Ethanol is produced industrially by the reaction of ethylene with water in the presence of an acid catalyst. How many grams of ethanol are produced from 7.24 mol of ethylene? Assume that excess water is present.

$$C_2H_4(g) + H_2O(\ell) \xrightarrow{\text{Acid catalyst}} CH_3CH_2OH(\ell)$$
Ethylene   Ethanol

B. Limiting Reagents

Frequently reactants are mixed in molar proportions that differ from those that appear in a balanced equation. It often happens that one reactant is completely used up when other reactants have not been used up. At times, we deliberately choose to have an excess of one reagent over another. As an example, consider an experiment in which NO is prepared by mixing five moles of N_2 with one mole of O_2. Only one mole of N_2 will react, consuming the one mole of O_2. The oxygen is used up completely and four moles of nitrogen are left over. These mole relationships are summarized under the balanced equation:

$$N_2(g) + O_2(g) \longrightarrow 2NO(g)$$

Before reaction (moles)	5.0	1.0	0
After reaction (moles)	4.0	0	2.0

Limiting reagent The reactant that is consumed, leaving an excess of another reagent or reagents unreacted

The **limiting reagent** is the reactant that is used up first. In this example, O_2 is the limiting reagent, because it governs how much NO can form. The other reagent, N_2, is in excess.

EXAMPLE 5.13

Suppose 12 g of C is mixed with 64 g of O_2, and the following reaction takes place:

$$C(s) + O_2(g) \longrightarrow CO_2(g)$$

(a) Which reactant is the limiting reagent, and which reactant is in excess?
(b) How many grams of CO_2 will be formed?

Solution

(a) We use the molar mass of each reactant to calculate the number of moles of each present before reaction.

$$12 \text{ g C} \times \frac{1 \text{ mol C}}{12 \text{ g C}} = 1 \text{ mol C}$$

$$64 \text{ g O}_2 \times \frac{1 \text{ mol O}_2}{32 \text{ g O}_2} = 2 \text{ mol O}_2$$

According to the balanced equation, reaction of one mole of C requires one mole of O_2. But two moles of O_2 are present at the start of the reaction. Therefore, C is the limiting reagent and O_2 is in excess.

(b) To calculate the number of grams of CO_2 formed, we use the conversion factor 1 mol CO_2 = 44 g CO_2.

$$12 \text{ g C} \times \frac{1 \text{ mol C}}{12 \text{ g C}} \times \frac{1 \text{ mol CO}_2}{1 \text{ mol C}} \times \frac{44 \text{ g CO}_2}{12 \text{ mol CO}_2} = 44 \text{ g CO}_2$$

We can summarize these numbers in the following table. Note that, as required by the law of conservation of mass, the sum of the masses of the material present after reaction is the same as the amount present before the reaction took place—namely, 76 g of material.

	C	+	O$_2$	⟶	CO$_2$
Before reaction	12 g		64 g		0
Before reaction	1.0 mol		2.0 mol		0
After reaction	0		1.0 mol		1.0 mol
After reaction	0		32 g		44 g

Problem 5.13

Assume that 6.0 g of C and 2.1 g H_2 are mixed and react to form methane according to the following balanced equation:

$$C(s) + 2H_2(g) \longrightarrow CH_4(g)$$
$$\text{Methane}$$

(a) Which reactant is the limiting reagent and which reactant is in excess?
(b) How many grams of CH_4 are produced in the reaction?

C. Percent Yield

When carrying out a chemical reaction, we often get less of a product than we might expect from the type of calculation we discussed in Section 5.5A. For example, suppose we react 32.0 g (1 mol) of CH_3OH with excess CO to form acetic acid:

$$\text{CH}_3\text{OH} + \text{CO} \longrightarrow \text{CH}_3\text{COOH}$$
$$\text{Methanol} \quad \text{Carbon} \quad \text{Acetic acid}$$
$$\text{monoxide}$$

If we calculate the expected yield based on the stoichiometry of the balanced equation, we find that we should get 1 mol (60.0 g) of acetic acid. Suppose, however, we get only 57.8 g of acetic acid. Does this result mean that the law

A reaction that does not give the main product is called a side reaction.

Occasionally, the percent yield is larger than 100%. For example, if a chemist fails to dry a product completely before weighing it, the product weighs more than it should because it also contains water. In such cases, the actual yield may be larger than expected and the percent yield may be greater than 100%.

of conservation of mass is being violated? No, it does not. We get less than 60.0 g of acetic acid either because some of the CH_3OH does not react or because some of it reacts in some other way, or perhaps because our laboratory technique is not perfect and we lose a little in transferring material from one container to another.

At this point, we need to define three terms, all of which relate to yield of product in a chemical reaction:

Actual yield The mass of product formed in a chemical reaction

Theoretical yield The mass of product that should form in a chemical reaction according to the stoichiometry of the balanced equation

Percent yield The actual yield divided by the theoretical yield times 100

$$\text{Percent yield} = \frac{\text{actual yield}}{\text{theoretical yield}} \times 100$$

We summarize the data for the preparation of acetic acid in the following table:

	CH_3OH	+	CO	$\longrightarrow$	CH_3COOH
Before reaction	32.0 g		Excess		0
Before reaction	1.00 mol		Excess		0
Theoretical yield					1.00 mol
Theoretical yield					60.0 g
Actual yield					57.8 g

We calculate the percent yield in this experiment as follows:

$$\text{Percent yield} = \frac{57.8 \text{ g acetic acid}}{60.0 \text{ g acetic acid}} \times 100 = 96.3\%$$

Why is it important to know the percent yield of a chemical reaction or a series of reactions? The most important reason often relates to cost. If the yield of commercial product is, say, only 10%, the chemists will probably be sent back to the lab to tweak the experimental conditions in an attempt to improve the yield. As an example, consider a reaction in which starting material A is converted first to compound B, then to compound C, and finally to compound D.

$$A \longrightarrow B \longrightarrow C \longrightarrow D$$

Suppose the yield is 50% at each step. In this case, the yield of compound D is 13% based on the mass of compound A. If, however, the yield at each step is 90%, the yield of compound D increases to 73%; and if the yield at each step is 99%, the yield of compound D is 97%. These numbers are summarized in the following table:

If the Percent Yield per Step Is	The Percent Yield of Compound D Is
50%	$0.50 \times 0.50 \times 0.50 \times 100 = 13\%$
90%	$0.90 \times 0.90 \times 0.90 \times 100 = 73\%$
99%	$0.99 \times 0.99 \times 0.99 \times 100 = 97\%$

EXAMPLE 5.14

In an experiment forming ethanol, the theoretical yield is 50.5 g. The actual yield is 46.8 g. What is the percent yield?

Solution

$$\text{Percent Yield} = \frac{46.8 \text{ g}}{50.5 \text{ g}} \times 100 = 92.7\%$$

Problem 5.14

In an experiment to prepare aspirin, the theoretical yield is 153.7 g. If the actual yield is 124.5 g, what is the percent yield?

5.6 | How Can We Predict if Ions in Aqueous Solution Will React with Each Other?

Many ionic compounds are soluble in water. As we saw in Section 4.5, ionic compounds always consist of both positive and negative ions. When they dissolve in water, the positive and negative ions become separated from each other by water molecules. We call such a separation a **dissociation.** For example,

$$\text{NaCl(s)} \xrightarrow{\text{H}_2\text{O}} \text{Na}^+\text{(aq)} + \text{Cl}^-\text{(aq)}$$

H$_2$O above the arrow means that the reaction takes place in water.

What happens when we mix **aqueous solutions** of two different ionic compounds? Does a reaction take place between the ions? The answer depends on the ions. If any of the negative and positive ions come together to form a water-insoluble compound, then a reaction takes place and a precipitate forms; otherwise, no reaction occurs.

Aqueous solution A solution in which the solvent is water

As an example, suppose we prepare one solution by dissolving sodium chloride, NaCl, in water and a second solution by dissolving silver nitrate, AgNO$_3$, in water.

$$\text{Solution 1}\quad \text{NaCl(s)} \xrightarrow{\text{H}_2\text{O}} \text{Na}^+\text{(aq)} + \text{Cl}^-\text{(aq)}$$
$$\text{Solution 2}\quad \text{AgNO}_3\text{(s)} \xrightarrow{\text{H}_2\text{O}} \text{Ag}^+\text{(aq)} + \text{NO}_3{}^-\text{(aq)}$$

If we now mix the two solutions, four ions are present in the solution: Ag$^+$, Na$^+$, Cl$^-$, and NO$_3{}^-$. Two of these ions, Ag$^+$ and Cl$^-$, react to form the compound AgCl (silver chloride), which is insoluble in water. A reaction therefore takes place, forming a white precipitate of AgCl that slowly sinks to the bottom of the container (Figure 5.2). We write this reaction as follows:

$$\underset{\substack{\text{Silver}\\\text{ion}}}{\text{Ag}^+\text{(aq)}} + \underset{\substack{\text{Nitrate}\\\text{ion}}}{\text{NO}_3{}^-\text{(aq)}} + \underset{\substack{\text{Sodium}\\\text{ion}}}{\text{Na}^+\text{(aq)}} + \underset{\substack{\text{Chloride}\\\text{ion}}}{\text{Cl}^-\text{(aq)}} \longrightarrow \underset{\substack{\text{Silver}\\\text{chloride}}}{\text{AgCl(s)}} + \text{Na}^+\text{(aq)} + \text{NO}_3{}^-\text{(aq)}$$

Notice that the Na$^+$ and NO$_3{}^-$ ions do not participate in the reaction, but merely remain dissolved in the water. Ions that do not participate in a reaction are called **spectator ions**—certainly an appropriate name.

Spectator ion An ion that appears unchanged on both sides of a chemical equation

Figure 5.2 Adding Cl^- ions to a solution of Ag^+ ions produces a white precipitate of silver chloride, AgCl.

Net ionic equation A chemical equation that does not contain spectator ions

GOB
Chemistry ⚛ Now™
Click *Coached Problems* to explore more about **Net Ionic Equations**

Figure 5.3 When aqueous solutions of $NaHCO_3$ and HCl are mixed, a reaction between HCO_3^- and H_3O^+ ions produces CO_2 gas, which can be seen as bubbles.

We can simplify the equation for the formation of silver chloride by omitting all spectator ions:

Net ionic equation: $\quad Ag^+(aq) + Cl^-(aq) \longrightarrow AgCl(s)$
$\qquad\qquad\qquad$ Silver $\qquad$ Chloride $\qquad$ Silver
$\qquad\qquad\qquad$ ion $\qquad\qquad$ ion $\qquad\qquad$ chloride

This kind of equation that we write for ions in solution is called a **net ionic equation.** Like all other chemical equations, net ionic equations must be balanced. We balance them in the same way as we do other equations, except now we must make sure that charges balance as well as atoms.

Net ionic equations show only the ions that react—no spectator ions are shown. For example, the net ionic equation for the precipitation of arsenic(III) sulfide from aqueous solution is

Net ionic equation: $\quad 2As^{3+}(aq) + 3S^{2-}(aq) \longrightarrow As_2S_3(s)$

Not only are there two arsenic atoms and three sulfur atoms on each side, but the total charge on the left side is the same as the total charge on the right side; they are both zero.

In general, ions in solution react with each other only when one of these four things can happen:

1. Two ions form a solid that is insoluble in water. AgCl is one example, as shown in Figure 5.2.
2. Two ions form a gas that escapes from the reaction mixture as bubbles. An example is the reaction of sodium bicarbonate, $NaHCO_3$, with HCl to form gaseous carbon dioxide, CO_2 (Figure 5.3). The net ionic equation for this reaction is

Net ionic equation: $\quad HCO_3^-(aq) + H_3O^+(aq) \longrightarrow CO_2(g) + 2H_2O(\ell)$
$\qquad\qquad\qquad$ Bicarbonate $\qquad\qquad\qquad\qquad$ Carbon
$\qquad\qquad\qquad$ ion $\qquad\qquad\qquad\qquad\qquad$ dioxide

3. An acid neutralizes a base. Acid–base reactions are so important that we devote Chapter 9 to them.
4. One of the ions can oxidize another. We discuss this type of reaction in Section 5.7.

In many cases, no reaction takes place when we mix solutions of ionic compounds because none of these situations holds. For example, if we mix solutions of copper(II) nitrate, $Cu(NO_3)_2$, and potassium sulfate, K_2SO_4, we merely have a mixture containing Cu^{2+}, K^+, NO_3^-, and SO_4^{2-} ions dissolved in water. None of these ions reacts with the others; therefore, we see nothing happening (Figure 5.4).

Figure 5.4 (*a*) The beaker on the left contains a solution of potassium sulfate (colorless) and the beaker on the right contains a solution of copper(II) nitrate (blue). (*b*) When the two solutions are mixed, the blue color becomes lighter because the copper(II) nitrate is less concentrated, but nothing else happens.

(*a*) $\qquad\qquad\qquad\qquad\qquad\qquad\qquad\qquad$ (*b*)

CHEMICAL CONNECTIONS 5A

Solubility and Tooth Decay

The outermost protective layer of a tooth is the enamel, which is composed of approximately 95% hydroxyapatite, $Ca_{10}(PO_4)_6(OH)_2$, and 5% collagen (Chemical Connections 21G). Like most other phosphates and hydroxides, hydroxyapatite is insoluble in water. In acidic media, however, it dissolves to a slight extent, yielding Ca^{2+}, PO_4^{3-}, and OH^- ions. This loss of enamel creates pits and cavities in the tooth.

Acidity in the mouth is produced by bacterial fermentation of remnants of food, especially carbohydrates. Once pits and cavities

form in the enamel, bacteria can hide there and cause further damage in the underlying softer material called dentin. The fluoridation of water brings F^- ions to the hydroxyapatite. There, F^- ions take the place of OH^- ions, forming the considerably less acid-soluble fluoroapatite, $Ca_{10}(PO_4)_6F_2$. Fluoride-containing toothpastes enhance this exchange process and provide further protection against tooth decay.

EXAMPLE 5.15

When a solution of barium chloride, $BaCl_2$, is added to a solution of sodium sulfate, Na_2SO_4, a white precipitate of barium sulfate, $BaSO_4$, forms. Write the net ionic equation for this reaction.

Solution

Because both barium chloride and sodium sulfate are ionic compounds, each exists in water as its dissociated ions:

$$Ba^{2+}(aq) + 2Cl^-(aq) \quad \text{and} \quad 2Na^+(aq) + SO_4^{2-}(aq)$$

We are told that a precipitate of barium sulfate forms:

$$Ba^{2+}(aq) + 2Cl^-(aq) + 2Na^+(aq) + SO_4^{2-}(aq) \longrightarrow$$
$$BaSO_4(s) + 2Na^+(aq) + 2Cl^-(aq)$$
Barium
sulfate

Because Na^+ and Cl^- ions appear on both sides of the equation (they are spectator ions), we cancel them and are left with the following net ionic equation:

$$\text{Net ionic equation:} \quad Ba^{2+}(aq) + SO_4^{2-}(aq) \longrightarrow BaSO_4(s)$$

Problem 5.15

When a solution of copper(II) chloride, $CuCl_2$, is added to a solution of potassium sulfide, K_2S, a black precipitate of copper(II) sulfide, CuS, forms. Write the net ionic equation for the reaction.

The mixing of solutions of barium chloride, $BaCl_2$, and sodium sulfate, Na_2SO_4, forms a white precipitate of barium sulfate, $BaSO_4$.

Charles D. Winters

Of the four ways for ions to react in water, one of the most common is the formation of an insoluble compound. We can predict when this result will happen if we know the solubilities of the ionic compounds. There is no simple way to remember which ionic solids are soluble in water and which are not, but we can make some useful generalizations.

1. All compounds containing Na^+, K^+, or NH_4^+ ions are soluble in water.
2. All nitrates (NO_3^-) and acetates (CH_3COO^-) are soluble in water.

(*a*)   (*b*)   (*c*)

(*a*) Sea animals of the mollusk family often use insoluble $CaSO_4$ to construct their shells. Both (*b*) iron(III) hydroxide, $Fe(OH)_3$, and (*c*) copper(II) carbonate, $CuCO_3$, are insoluble in water.

3. Most chlorides (Cl^-) and sulfates (SO_4^{2-}) are soluble in water. Important exceptions include silver chloride ($AgCl$), lead(II) chloride ($PbCl_2$), mercury(I) chloride (Hg_2Cl_2), barium sulfate ($BaSO_4$), and lead sulfate ($PbSO_4$), which are insoluble in water.

4. Most carbonates (CO_3^{2-}), phosphates (PO_4^{3-}), sulfides (S^{2-}), and hydroxides (OH^-) are insoluble in water. Important exceptions include LiOH, NaOH, KOH, and aqueous NH_3, which are soluble in water.

5.7 | What Are Oxidation and Reduction, and Why Do They Always Occur Together?

One of the most important and most common types of reaction is oxidation–reduction. **Oxidation** is the loss of electrons. **Reduction** is the gain of electrons. An **oxidation–reduction reaction** (often called a **redox reaction**) involves the transfer of electrons from one species to another. An example is the oxidation of zinc by copper ions, the net ionic equation for which is

$$Zn(s) + Cu^{2+}(aq) \longrightarrow Zn^{2+}(aq) + Cu(s)$$

When we put a piece of zinc metal into a beaker containing copper(II) ions in aqueous solution, three things happen (Figure 5.5):

1. Some of the zinc metal dissolves and goes into solution as Zn^{2+}.
2. Copper metal is deposited on the zinc metal.
3. The blue color of the Cu^{2+} ions gradually disappears.

The zinc metal loses electrons to the copper ions:

$$Zn(s) \longrightarrow Zn^{2+}(aq) + 2e^- \quad \textbf{Zn is oxidized}$$

At the same time, the Cu^{2+} ions gain electrons from the zinc. The copper ions are reduced:

$$Cu^{2+}(aq) + 2e^- \longrightarrow Cu(s) \quad \textbf{Cu}^{2+} \textbf{ is reduced}$$

Oxidation and reduction are not independent reactions. That is, a species cannot gain electrons from nowhere, nor can a species lose electrons

---

Oxidation The loss of electrons; the gain of oxygen atoms or the loss of hydrogen atoms

Reduction The gain of electrons; the loss of oxygen atoms or the gain of hydrogen atoms

Redox reaction An oxidation–reduction reaction

GOB
Chemistry ⚛ Now™
Click *Chemistry Interactive* to explore more about **Oxidation–Reduction Reactions.**

Negative ions such as Cl^- or NO_3^- are present to balance charges, but we do not show them because they are spectator ions.

Figure 5.5 When a piece of zinc is added to a solution containing Cu^{2+} ions, the Zn is oxidized by Cu^{2+} ions, and the Cu^{2+} ions are reduced by the Zn.

Blue solution of Cu^{2+} ions

Coating of copper plating out, and zinc bar partially dissolved

Zinc bar

Charles D. Winters

Colorless solution of Zn^{2+} ions

to nothing. In other words, no oxidation can occur without an accompanying reduction, and vice versa. In the preceding reaction, Cu^{2+} is oxidizing Zn. We call Cu^{2+} an **oxidizing agent.** Similarly, Zn is reducing Cu^{2+}, and we call Zn a **reducing agent.**

We summarize these oxidation–reduction relationships for the reaction of zinc metal with Cu(II) ion in the following way:

Oxidizing agent An entity that accepts electrons in an oxidation–reduction reaction

Reducing agent An entity that donates electrons in an oxidation–reduction reaction

Electrons flow from Zn to Cu^{2+}

$$Zn(s) \quad + \quad Cu^{2+}(aq) \longrightarrow Zn^{2+}(aq) + Cu(s)$$

Loses electrons; is oxidized

Gains electrons; is reduced

Gives electrons to Cu^{2+}; is the reducing agent

Takes electrons from Zn; is the oxidizing agent

Although the definitions we have given for oxidation (loss of electrons) and reduction (gain of electrons) are easy to apply in many redox reactions, they are not so easy to apply in other cases. For example, another redox reaction is the combustion (burning) of methane, CH_4, in which CH_4 is oxidized to CO_2 while O_2 is reduced to CO_2 and H_2O:

$$CH_4(g) + 2O_2(g) \longrightarrow CO_2(g) + 2H_2O(g)$$
Methane

It is not easy to see the electron loss and gain in such a reaction, so chemists developed another definition of oxidation and reduction, one that is easier to apply in many cases, especially where organic compounds are involved:

Oxidation The gain of oxygen and/or the loss of hydrogen

Reduction The loss of oxygen and/or the gain of hydrogen

Applying these alternative definitions to the reaction of methane with oxygen, we find the following:

$$CH_4(g) \quad + \quad 2O_2(g) \longrightarrow CO_2(g) + 2H_2O(g)$$

Gains O and loses H; is oxidized

Gains H; is reduced

Is the reducing agent

Is the oxidizing agent

In fact, this second definition is much older than the one involving electron transfer; it is the definition given by Lavoisier when he first discovered oxidation and reduction more than 200 years ago. Note that we could not apply this definition to our zinc-copper example.

EXAMPLE 5.16

In each equation, identify the substance that is oxidized, the substance that is reduced, the oxidizing agent, and the reducing agent.

(a) $Al(s) + Fe^{3+}(aq) \longrightarrow Al^{3+}(aq) + Fe(s)$

(b) $\underset{\text{Methanol}}{CH_3OH(g)} + O_2(g) \longrightarrow \underset{\text{Formic acid}}{HCOOH(g)} + H_2O(g)$

Solution

(a) $Al(s)$ loses three electrons in going to Al^{3+}; therefore, aluminum is oxidized. In the process of being oxidized, $Al(s)$ gives its electrons to Fe^{3+}, so $Al(s)$ is the reducing agent. Fe^{3+} gains three electrons in going to $Fe(s)$ and is reduced. In the process of being reduced, Fe^{3+} accepts three electrons from $Al(s)$, so Fe^{3+} is the oxidizing agent. To summarize:

$$Al(s) \quad + \quad Fe^{3+}(aq) \longrightarrow Al^{3+}(aq) + Fe(s)$$

Electrons flow from Al to Fe³⁺ · 3e⁻

Loses electrons; aluminum is oxidized — Gains electrons; iron is reduced

Gives electrons to Fe³⁺; Al is the reducing agent — Takes electrons from Al; Fe³⁺ is the oxidizing agent

(b) Because it is not easy to see the loss or gain of electrons in this example, we apply the second set of definitions. In converting CH_3OH to $HCOOH$, CH_3OH both gains oxygens and loses hydrogens; it is oxidized. In being converted to H_2O, O_2 gains hydrogens; it is reduced. The compound that is oxidized is the reducing agent; CH_3OH is the reducing agent. The compound that is reduced is the oxidizing agent; O_2 is the oxidizing agent. To summarize:

$$CH_3OH(g) \quad + \quad O_2(g) \longrightarrow HCOOH(g) + H_2O(g)$$

Is oxidized; methanol is the reducing agent — Is reduced; oxygen is the oxidizing agent

Problem 5.16

In each equation, identify the substance that is oxidized, the substance that is reduced, the oxidizing agent, and the reducing agent:

(a) $Ni^{2+}(aq) + Cr(s) \longrightarrow Ni(s) + Cr^{2+}(aq)$

(b) $\underset{\text{Formaldehyde}}{CH_2O(g)} + H_2(g) \longrightarrow \underset{\text{Methanol}}{CH_3OH(g)}$

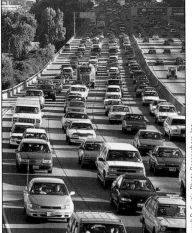

Air pollution is caused by incomplete fuel combustion.

CHEMICAL CONNECTIONS 5B

Voltaic Cells

In Figure 5.5, we see that when a piece of zinc metal is put in a solution containing Cu^{2+} ions, zinc atoms give electrons to Cu^{2+} ions. We can change the experiment by putting the zinc metal in one beaker and the Cu^{2+} ions in another, and then connecting the two beakers by a length of wire and a salt bridge (see the accompanying figure). A reaction still takes place: Zinc atoms still give electrons to Cu^{2+} ions. Now, however, the electrons must flow through the wire to get from the Zn to the Cu^{2+}. This flow of electrons produces an electric current, and the electrons keep flowing until either the Zn or the Cu^{2+} is used up. In this way, the apparatus generates an electric current by using a redox reaction. We call this device a **voltaic cell** or, more commonly, a battery.

The electrons produced at the zinc end carry negative charges. This end of the battery is a negative electrode (the **anode**). The electrons released at the anode as zinc is oxidized go through an outside circuit and, in doing so, produce the battery's electric current. At the other end of the battery, at the positively charged elec-

trode (the **cathode**), electrons are consumed as Cu^{2+} ions are reduced to copper metal.

To see why a salt bridge is necessary, we must look at the Cu^{2+} solution. Because we cannot have positive charges in any place without an equivalent number of negative charges, negative ions must be in the beaker as well—perhaps sulfate, nitrate, or some other anion. When electrons come over the wire, the Cu^{2+} is converted to Cu:

$$Cu^{2+}(aq) + 2e^- \longrightarrow Cu(s)$$

This reaction diminishes the number of Cu^{2+} ions, but the number of negative ions remains unchanged. The salt bridge is necessary to carry some of these negative ions to the other beaker, where they are needed to balance the Zn^{2+} ions being produced by the following reaction:

$$Zn(s) \longrightarrow Zn^{2+}(aq) + 2e^-$$

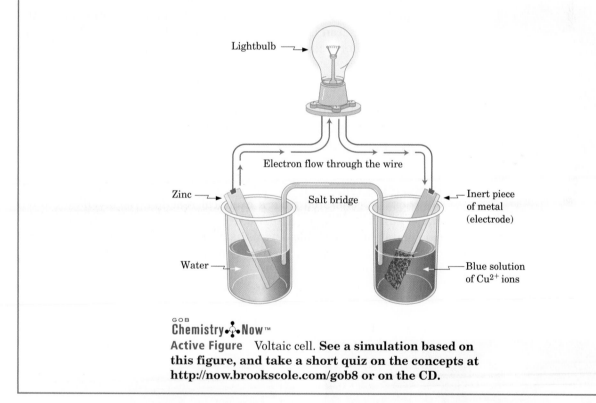

GOB
Chemistry Now™
Active Figure Voltaic cell. **See a simulation based on this figure, and take a short quiz on the concepts at http://now.brookscole.com/gob8 or on the CD.**

We have said that redox reactions are extremely common. Here are some important categories:

1. **Combustion** All combustion (burning) reactions are redox reactions in which the compounds or mixtures that are burned are oxidized by oxygen, O_2. They include the burning of gasoline, diesel oil, fuel oil, natural gas, coal, wood, and paper. All these materials contain carbon, and all except coal also contain hydrogen. If the combustion is complete,

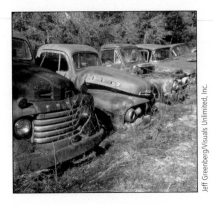

The rusting of iron and steel can be a serious problem in an industrial society.

These household bleaches are oxidizing agents.

carbon is oxidized to CO_2 and hydrogen is oxidized to H_2O. In an incomplete combustion, these elements are oxidized to other compounds, many of which cause air pollution. Unfortunately, much of today's combustion that takes place in gasoline and diesel engines and in furnaces is incomplete, so it contributes to air pollution. In the incomplete combustion of methane, for example, carbon is oxidized to carbon monoxide, CO, due to a lack of sufficient oxygen to complete its oxidation to CO_2:

$$2CH_4(g) + 3O_2(g) \longrightarrow 2CO(g) + 4H_2O(g)$$
Methane

2. **Respiration** Humans and animals get their energy by respiration. The oxygen in the air we breathe oxidizes carbon-containing compounds in our cells to produce CO_2 and H_2O. In fact, this oxidation reaction is really the same as combustion, although it takes place more slowly and at a much lower temperature. We discuss these reactions more fully in Chapter 27. The important product of these reactions is not CO_2 (which the body eliminates) or H_2O, but energy.

3. **Rusting** When iron or steel objects are left out in the open air, they eventually rust (steel is mostly iron but contains certain other elements, too). In rusting, iron is oxidized to a mixture of iron oxides. We can represent the main reaction by the following equation:

$$4Fe(s) + 3O_2(g) \longrightarrow 2Fe_2O_3(s)$$

4. **Bleaching** Most bleaching involves oxidation, and common bleaches are oxidizing agents. The colored compounds being bleached are usually organic compounds; oxidation converts them to colorless compounds.

5. **Batteries** A voltaic cell (Chemical Connections 5B) is a device in which electricity is generated from a chemical reaction. Such cells are often called batteries (Figure 5.6). We are all familiar with batteries in our cars and in such portable devices as radios, flashlights, cell phones, and computers. In all cases, the reaction taking place in the battery is a redox reaction.

(a) (b)

Anode
Cathode
Negative plates: lead grills filled with spongy lead
Positive plates: lead grills filled with PbO_2

Figure 5.6 (a) Dry cell batteries. (b) A lead storage battery.

CHEMICAL CONNECTIONS 5C

Oxidizing Antiseptics

An antiseptic is a compound that kills bacteria. Antiseptics are used to treat wounds—not to heal them any faster, but rather to prevent them from becoming infected by bacteria. Some antiseptics operate by oxidizing (and hence destroying) compounds essential to the normal functioning of the bacteria. One example is iodine, I_2, which was used as a household antiseptic for minor cuts and bruises for many years, not in the pure form but as a dilute solution in ethanol, called a tincture. Pure I_2 is a steel-gray solid that gives off a purple vapor when heated. In contrast, the tincture is a brown liquid. Another example of an oxidizing antiseptic is a dilute solution (commonly 3%) of hydrogen peroxide, H_2O_2, which is used as a mouth rinse to heal gum infections. Oxidizing antiseptics, however, are often regarded as too harsh. They not only kill bacteria but also harm skin and other normal tissues. For this reason, oxidizing antiseptics have largely been replaced by phenolic antiseptics (Section 13.5).

Disinfectants are also employed to kill bacteria, but they are used on inanimate objects rather than on living tissues. Many disinfectants are oxidizing agents. Two important examples are chlorine, Cl_2, a pale green gas, and ozone, O_3, a colorless gas. Both gases are

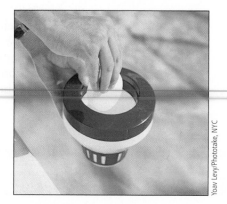

These tablets, which produce chlorine when mixed with water, are added to a swimming pool filter.

added in small quantities to municipal water supplies to kill any harmful bacteria that may be present. Both gases must be handled carefully because they are very poisonous.

5.8 | What Is Heat of Reaction?

In almost all chemical reactions, not only are starting materials converted to products, but heat is also either given off or absorbed. For example, when one mole of carbon is oxidized by oxygen to produce one mole of CO_2, 94.0 kilocalories of heat is given off:

$$C(s) + O_2(g) \longrightarrow CO_2(g) + 94.0 \text{ kcal}$$

The heat given off or gained in a reaction is called the **heat of reaction.** A reaction that gives off heat is **exothermic;** a reaction that absorbs heat is **endothermic.** The amount of heat given off or absorbed is proportional to the amount of material. For example, when 2 mol of carbon is oxidized by oxygen to give carbon dioxide, $2 \times 94.0 = 188$ kcal of heat is given off.

The energy changes accompanying a chemical reaction are not limited to heat. In some reactions, such as in voltaic cells (Chemical Connections 5B), the energy given off takes the form of electricity. In other reactions, such as photosynthesis (the reaction whereby plants convert water and carbon dioxide to sugar and oxygen), the energy absorbed is in the form of light.

An example of an endothermic reaction is the decomposition of mercury(II) oxide:

$$2HgO(s) + 43.4 \text{ kcal} \longrightarrow 2Hg(\ell) + O_2(g)$$
Mercury(II) oxide
(mercuric oxide)

This equation tells us that, if we want to decompose 2 mol of mercury(II) oxide into the elements $Hg(\ell)$ and $O_2(g)$, we must add 43.4 kcal of energy to HgO.

Heat of reaction The heat given off or absorbed in a chemical reaction

Exothermic reaction A chemical reaction that gives off heat

Endothermic reaction A chemical reaction that absorbs heat

Mercury(II) oxide, a red compound, decomposes into two elements when heated: mercury (a metal) and oxygen (a nonmetal). Mercury vapor condenses on the cooler upper portion of the test tube.

Incidentally, the law of conservation of energy tells us that the reverse reaction, the oxidation of mercury, must give off exactly the same amount of heat:

$$2Hg(\ell) + O_2(g) \longrightarrow 2HgO(s) + 43.4 \text{ kcal}$$

Especially important are the heats of reaction for combustion reactions. As we saw in Section 5.7, combustion reactions are the most important heat-producing reactions, because most of the energy required for modern society to function is derived from them. All combustions are exothermic. The heat given off in a combustion reaction is called the **heat of combustion.**

SUMMARY OF KEY QUESTIONS

SECTION 5.2 What Are Molecular Weights and Formula Weights, and How Do They Differ?

- The **formula weight** of a compound is the sum of the atomic weights of all atoms in the compound expressed in atomic mass units (amu). Formula weight applies to both ionic and molecular compounds.
- The term **molecular weight,** also expressed in amu, applies to only molecular compounds.

SECTION 5.3 What Is a Mole and How Do We Use It to Calculate Mass Relationships?

- A **mole (mol)** of any substance is defined as Avogadro's number (6.02×10^{23}) of formula units of the substance.
- The **molar mass** of a substance is its formula weight expressed in grams.

SECTION 5.4 How Do We Balance Chemical Equations?

- A **chemical equation** is an expression showing which reactants are converted to which products. A balanced chemical equation shows how many moles of each starting material are converted to how many moles of each product.

SECTION 5.5 How Do We Calculate Mass Relationships in Chemical Reactions?

- **Stoichiometry** is the study of the mass relationships in chemical reactions.
- The reagent that is used up first in a reaction is called the **limiting reagent.**
- The **percent yield** for a reaction equals the **actual yield** divided by the **theoretical yield** multiplied by 100.

SECTION 5.6 How Can We Predict if Ions in Aqueous Solution Will React with Each Other?

- When ions are mixed in aqueous solution, they are said to react with one another if (1) a precipitate forms, (2) a gas forms, (3) an acid neutralizes a base, or (4) an oxidation–reduction reaction takes place.
- Ions that do not react are called **spectator ions.**
- A **net ionic equation** shows only those ions that react. In a net ionic equation, both the charges and the number of atoms must be balanced.

SECTION 5.7 What Are Oxidation and Reduction, and Why Do They Always Occur Together?

- **Oxidation** is the loss of electrons; **reduction** is the gain of electrons. These two processes must take place together; you cannot have one without the other. The joint process is often called a redox reaction.
- Oxidation can also be defined as the gain of oxygens and/or the loss of hydrogens. Reduction can also be defined as the loss of oxygens and/or the gain of hydrogens.

SECTION 5.8 What Is Heat of Reaction?

- Almost all chemical reactions are accompanied by either a gain or a loss of heat. This heat is called the **heat of reaction.**
- Reactions that give off heat are **exothermic;** those that absorb heat are **endothermic.**
- The heat given off in a combustion reaction is called the **heat of combustion.**

PROBLEMS

GOB
Chemistry·⊹·Now™
Assess your understanding of this chapter's topics with additional quizzing and conceptual-based problems at **http://now.brookscole.com/gob8** or on the CD.

A blue problem number indicates an applied problem.

■ denotes problems that are available on the GOB ChemistryNow website or CD and are assignable in OWL.

SECTION 5.2 What Are Molecular Weights and Formula Weights, and How Do They Differ?

5.17 ■ Calculate the formula weight of
  (a) KCl
  (b) Na_3PO_4
  (c) $Fe(OH)_2$

5.18 Calculate the formula weight of
  (a) $NaAl(SO_3)_2$
  (b) $Al_2(SO_4)_3$
  (c) $(NH_4)_2CO_3$

5.19 Calculate the molecular weight of
  (a) Sucrose, $C_{12}H_{22}O_{11}$
  (b) Glycine, $C_2H_5NO_2$
  (c) DDT, $C_{14}H_9Cl_5$

SECTION 5.3 What Is a Mole and How Do We Use It to Calculate Mass Relationships?

5.20 ■ Calculate the number of moles in
  (a) 32 g of methane, CH_4
  (b) 345.6 g of nitric oxide, NO
  (c) 184.4 g of chlorine dioxide, ClO_2
  (d) 720 g of glycerin, $C_3H_8O_3$

5.21 ■ Calculate the number of grams in
  (a) 1.77 mol of nitrogen dioxide, NO_2
  (b) 0.84 mol of 2-propanol, C_3H_8O (rubbing alcohol)
  (c) 3.69 mol of uranium hexafluoride, UF_6
  (d) 0.348 mol of galactose, $C_6H_{12}O_6$
  (e) 4.9×10^{-2} mol of vitamin C, $C_6H_8O_6$

5.22 Calculate the number of moles of
  (a) O atoms in 18.1 mol of formaldehyde, CH_2O
  (b) Br atoms in 0.41 mol of bromoform, $CHBr_3$
  (c) O atoms in 3.5×10^3 mol of $Al_2(SO_4)_3$
  (d) Hg atoms in 87 g of HgO

5.23 Calculate the number of moles of
  (a) S^{2-} ions in 6.56 mol of Na_2S
  (b) Mg^{2+} ions in 8.320 mol of $Mg_3(PO_4)_2$
  (c) Acetate ion, CH_3COO^-, in 0.43 mol of $Ca(CH_3COO)_2$

5.24 Calculate the number of
  (a) Nitrogen atoms in 25.0 g of TNT, $C_7H_5N_3O_6$
  (b) Carbon atoms in 40.0 g of ethanol, C_2H_6O
  (c) Oxygen atoms in 500 mg of aspirin, $C_9H_8O_4$
  (d) Sodium atoms in 2.40 g of sodium dihydrogen phosphate, NaH_2PO_4

5.25 A single atom of cerium weighs just about twice as much as a single atom of gallium. What is the weight ratio of 25 atoms of cerium to 25 atoms of gallium?

5.26 What is the mass in grams of each number of molecules of formaldehyde, CH_2O?
  (a) 100 molecules   (b) 3000 molecules
  (c) 5.0×10^6 molecules   (d) 2.0×10^{24} molecules

5.27 How many molecules are in
  (a) 2.9 mol of TNT, $C_7H_5N_3O_6$
  (b) one drop (0.0500 g) of water
  (c) 3.1×10^{-1} g of aspirin, $C_9H_8O_4$

5.28 A typical deposit of cholesterol, $C_{27}H_{46}O$, in an artery might have a mass of 3.9 mg. How many molecules of cholesterol are in this mass?

5.29 The molecular weight of hemoglobin is about 68,000 amu. What is the mass in grams of a single molecule of hemoglobin?

SECTION 5.4 How Do We Balance Chemical Equations?

5.30 ■ Balance each equation.
  (a) $HI + NaOH \longrightarrow NaI + H_2O$
  (b) $Ba(NO_3)_2 + H_2S \longrightarrow BaS + HNO_3$
  (c) $CH_4 + O_2 \longrightarrow CO_2 + H_2O$
  (d) $C_4H_{10} + O_2 \longrightarrow CO_2 + H_2O$
  (e) $Fe + CO_2 \longrightarrow Fe_2O_3 + CO$

5.31 ■ Balance each equation.
  (a) $H_2 + I_2 \longrightarrow HI$
  (b) $Al + O_2 \longrightarrow Al_2O_3$
  (c) $Na + Cl_2 \longrightarrow NaCl$
  (d) $Al + HBr \longrightarrow AlBr_3 + H_2$
  (e) $P + O_2 \longrightarrow P_2O_5$

5.32 If you blow carbon dioxide gas into a solution of calcium hydroxide, a milky-white precipitate of calcium carbonate forms. Write a balanced equation for the formation of calcium carbonate in this reaction.

5.33 Calcium oxide is prepared by heating limestone (calcium carbonate, $CaCO_3$) to a high temperature, at which point it decomposes to calcium oxide and carbon dioxide. Write a balanced equation for this preparation of calcium oxide.

5.34 The brilliant white light in some fireworks displays is produced by burning magnesium in air. The magnesium reacts with oxygen in the air to form magnesium oxide. Write a balanced equation for this reaction.

5.35 The rusting of iron is a chemical reaction of iron with oxygen in the air to form iron(III) oxide. Write a balanced equation for this reaction.

5.36 When solid carbon burns in a limited supply of oxygen gas, carbon monoxide gas, CO, forms. This gas is deadly to humans because it combines with hemoglobin in the blood, making it impossible for the blood to transport oxygen. Write a balanced equation for the formation of carbon monoxide.

5.37 Solid ammonium carbonate, $(NH_4)_2CO_3$, decomposes at room temperature to ammonia, carbon dioxide, and water. Because of the ease of decomposition and the penetrating odor of ammonia, ammonium carbonate can be used as smelling salts. Write a balanced equation for this decomposition.

5.38 In the chemical test for arsenic, the gas arsine, AsH_3, is prepared. When arsine is decomposed by heating, arsenic metal deposits as a mirrorlike coating on the surface of a glass container, and hydrogen gas, H_2, is given off. Write a balanced equation for the decomposition of arsine.

5.39 When a piece of aluminum metal is dropped into hydrochloric acid, HCl, hydrogen is released as a gas, and a solution of aluminum chloride forms. Write a balanced equation for this reaction.

5.40 In the industrial chemical preparation of chlorine, Cl_2, an electric current is passed through an aqueous solution of sodium chloride to give $Cl_2(g)$ and $H_2(g)$. The other product of this reaction is sodium hydroxide. Write a balanced equation for this reaction.

SECTION 5.5 How Do We Calculate Mass Relationships in Chemical Reactions?

5.41 ■ For the reaction

$$2N_2(g) + 3O_2(g) \longrightarrow 2N_2O_3(g)$$

(a) How many moles of N_2 are required to react completely with 1 mole of O_2?

(b) How many moles of N_2O_3 are produced from the complete reaction of 1 mole of O_2?

(c) How many moles of O_2 are required to produce 8 moles of N_2O_3?

5.42 Magnesium reacts with sulfuric acid according to the following equation. How many moles of H_2 are produced by the complete reaction of 230 mg of Mg with sulfuric acid?

$$Mg(s) + H_2SO_4(aq) \longrightarrow MgSO_4(aq) + H_2(g)$$

5.43 ■ Chloroform, $CHCl_3$, is prepared industrially by the reaction of methane with chlorine. How many grams of Cl_2 are needed to produce 1.50 moles of chloroform?

$$CH_4(g) + 3Cl_2(g) \longrightarrow CHCl_3(\ell) + 3HCl(g)$$
Methane   Chloroform

5.44 ■ At one time, acetaldehyde was prepared industrially by the reaction of ethylene with air in the presence of a copper catalyst. How many grams of acetaldehyde can be prepared from 81.7 g of ethylene?

$$2C_2H_4(g) + O_2(g) \longrightarrow 2C_2H_4O(g)$$
Ethylene   Acetaldehyde

5.45 Chlorine dioxide, ClO_2, is used for bleaching paper. It is also the gas used to kill the anthrax spores that contaminated the Hart Senate Office Building in 2001. Chlorine dioxide is prepared by treating sodium chlorite with chlorine gas.

$$NaClO_2(aq) + Cl_2(g) \longrightarrow ClO_2(g) + NaCl(aq)$$
Sodium   Chlorine
chlorite   dioxide

(a) Balance the equation for the preparation of chlorine dioxide.

(b) Calculate the weight of chlorine dioxide that can be prepared from 5.50 kg of sodium chlorite.

5.46 Ethanol, C_2H_6O, is added to gasoline to produce "gasohol," a fuel for automobile engines. How many grams of O_2 are required for complete combustion of 421 g of C_2H_6O?

$$C_2H_5OH(\ell) + 3O_2(g) \longrightarrow 2CO_2(g) + 3H_2O$$
Ethanol

5.47 In photosynthesis, green plants convert CO_2 and H_2O to glucose, $C_6H_{12}O_6$. How many grams of CO_2 are required to produce 5.1 g of glucose?

$$6CO_2(g) + 6H_2O(\ell) \xrightarrow{\text{Photosynthesis}} C_6H_{12}O_6(aq) + 6O_2(g)$$
Glucose

5.48 Iron ore is converted to iron by heating it with coal, C, and oxygen according to the following equation:

$$2Fe_2O_3(s) + 6C(s) + 3O_2(g) \longrightarrow 4Fe(s) + 6CO_2(g)$$

If the process is run until 3940 g of Fe is produced, how many grams of CO_2 will also be produced?

5.49 Given the reaction in Problem 5.48, how many grams of C are necessary to react completely with 0.58 g of Fe_2O_3?

5.50 ■ Aspirin is made by the reaction of salicylic acid with acetic anhydride. What is the theoretical yield of aspirin if 85.0 g of salicylic acid is treated with excess acetic anhydride?

$$C_7H_6O_3(s) + C_4H_6O_3(\ell) \longrightarrow C_9H_8O_4(s) + C_2H_4O_2(\ell)$$
Salicylic   Acetic   Aspirin   Acetic
acid   anhydride   acid

5.51 Suppose that the yield of aspirin by the reaction shown in Problem 5.50 is 75%. How many grams of salicylic acid would be required to give an actual yield of 25.0 g of aspirin?

5.52 Benzene reacts with bromine to produce bromobenzene according to the following equation:

$$C_6H_6(\ell) + Br_2(\ell) \longrightarrow C_6H_5Br(\ell) + HBr(g)$$
Benzene   Bromine   Bromobenzene   Hydrogen
                                             bromide

If 60.0 g of benzene is mixed with 135 g of bromine,
(a) Which is the limiting reagent?
(b) How many grams of bromobenzene are formed in the reaction?

5.53 ■ Ethyl chloride is prepared by the reaction of chlorine with ethane according to the following balanced equation. When 5.6 g of ethane is reacted with excess chlorine, 8.2 g of ethyl chloride forms. Calculate the percent yield of ethyl chloride.

$$C_2H_6(g) + Cl_2(g) \longrightarrow C_2H_5Cl(\ell) + HCl(g)$$
Ethane   Ethyl chloride

5.54 Diethyl ether is made from ethanol according to the following reaction:

$$2C_2H_5OH(\ell) \longrightarrow (C_2H_5)_2O(\ell) + H_2O(\ell)$$
Ethanol   Diethyl
                  ether

In an experiment, starting with 517 g of ethanol gave 391 g of diethyl ether. What was the percent yield in this experiment?

SECTION 5.6 How Can We Predict if Ions in Aqueous Solution Will React with Each Other?

5.55 Define (a) spectator ion, (b) net ionic equation, and (c) aqueous solution.

5.56 Balance these net ionic equations.
(a) $Ag^+(aq) + Br^-(aq) \longrightarrow AgBr(s)$
(b) $Cd^{2+}(aq) + S^{2-}(aq) \longrightarrow CdS(s)$
(c) $Sc^{3+}(aq) + SO_4^{2-}(aq) \longrightarrow Sc_2(SO_4)_3(s)$
(d) $Sn^{2+}(aq) + Fe^{2+}(aq) \longrightarrow Sn(s) + Fe^{3+}(aq)$
(e) $K(s) + H_2O(\ell) \longrightarrow K^+(aq) + OH^-(aq) + H_2(g)$

5.57 In the equation

$$2Na^+(aq) + CO_3^{2-}(aq) + Sr^{2+}(aq) + 2Cl^-(aq) \longrightarrow$$
$$SrCO_3(s) + 2Na^+(aq) + 2Cl^-(aq)$$

(a) Identify the spectator ions.
(b) Write the balanced net ionic equation.

5.58 Predict whether a precipitate will form when aqueous solutions of the following compounds are mixed. If a precipitate will form, write its formula, and write a net ionic equation for its formation. To make your predictions, use the solubility generalizations in Section 5.6.
(a) $CaCl_2(aq) + K_3PO_4(aq) \longrightarrow$
(b) $KCl(aq) + Na_2SO_4(aq) \longrightarrow$
(c) $(NH_4)_2CO_3(aq) + Ba(NO_3)_2(aq) \longrightarrow$
(d) $FeCl_2(aq) + KOH(aq) \longrightarrow$
(e) $Ba(NO_3)_2(aq) + NaOH(aq) \longrightarrow$
(f) $Na_2S(aq) + SbCl_3(aq) \longrightarrow$
(g) $Pb(NO_3)_2(aq) + K_2SO_4(aq) \longrightarrow$

5.59 ■ When a solution of ammonium chloride is added to a solution of lead(II) nitrate, $Pb(NO_3)_2$, a white precipitate, lead(II) chloride, forms. Write a balanced net ionic equation for this reaction. Both ammonium chloride and lead nitrate exist as dissociated ions in aqueous solution.

5.60 When a solution of hydrochloric acid, HCl, is added to a solution of sodium sulfite, Na_2SO_3, sulfur dioxide gas is released from the solution. Write a net ionic equation for this reaction. An aqueous solution of HCl contains H^+ and Cl^- ions, and Na_2SO_3 exists as dissociated ions in aqueous solution.

5.61 When a solution of sodium hydroxide is added to a solution of ammonium carbonate, H_2O is formed and ammonia gas, NH_3, is released when the solution is heated. Write a net ionic equation for this reaction. Both NaOH and $(NH_4)_2CO_3$ exist as dissociated ions in aqueous solution.

5.62 ■ Using the solubility generalizations given in Section 5.6, predict which of these ionic compounds are soluble in water.
(a) KCl   (b) NaOH   (c) $BaSO_4$
(d) Na_2SO_4 (e) Na_2CO_3 (f) $Fe(OH)_2$

5.63 ■ Using the solubility generalizations given in Section 5.6, predict which of these ionic compounds are soluble in water.
(a) $MgCl_2$ (b) $CaCO_3$ (c) Na_2SO_3
(d) NH_4NO_3 (e) $Pb(OH)_2$

SECTION 5.7 What Are Oxidation and Reduction, and Why Do They Always Occur Together?

5.64 Give two definitions of oxidation and two definitions of reduction.

5.65 Can reduction take place without oxidation? Explain.

5.66 ■ In the reaction

$$Pb(s) + 2Ag^+(aq) \longrightarrow Pb^2(aq) + 2Ag(s)$$

(a) Which species is oxidized and which species is reduced?

(b) Which species is the oxidizing agent and which species is the reducing agent?

5.67 In the reaction

$$C_7H_{12}(\ell) + 10O_2(g) \longrightarrow 7CO_2(g) + 6H_2O(\ell)$$

(a) Which species is oxidized and which species is reduced?

(b) Which species is the oxidizing agent and which is the reducing agent?

5.68 When a piece of sodium metal is added to water, hydrogen is evolved as a gas and a solution of sodium hydroxide is formed.

(a) Write a balanced equation for this reaction.

(b) What is oxidized in this reaction? What is reduced?

SECTION 5.8 What Is Heat of Reaction?

5.69 What is the difference between *exothermic* and *endothermic*?

5.70 Which of these reactions are exothermic and which are endothermic?

(a) $2NH_3(g) + 22.0 \text{ kcal} \longrightarrow N_2(g) + 3H_2(g)$
(b) $H_2(g) + F_2(g) \longrightarrow 2HF(g) + 124 \text{ kcal}$
(c) $C(s) + O_2(g) \longrightarrow CO_2(g) + 94.0 \text{ kcal}$
(d) $H_2(g) + CO_2(g) + 9.80 \text{ kcal} \longrightarrow H_2O(g) + CO(g)$
(e) $C_3H_3(g) + 5O_2(g) \longrightarrow$
$$3CO_2(g) + 4H_2O(g) + 531 \text{ kcal}$$

5.71 In the following reaction, 9.80 kcal is absorbed per mole of CO_2 undergoing reaction. How much heat is given off if two moles of water is reacted with two moles of carbon monoxide?

$$H_2(g) + CO_2(g) + 9.80 \text{ kcal} \longrightarrow H_2O(g) + CO(g)$$

5.72 Following is the equation for the combustion of acetone:

$$2C_3H_6O(\ell) + 8O_2(g) \longrightarrow 6CO_2(g) + 6H_2O(g) + 853.6 \text{ kcal}$$
Acetone

How much heat will be given off if 0.37 mol of acetone is burned completely?

5.73 The oxidation of glucose, $C_6H_{12}O_6$, to carbon dioxide and water is exothermic. The heat liberated is the same whether glucose is metabolized in the body or burned in air.

$$C_6H_{12}O_6 + 6O_2 \longrightarrow 6CO_2 + 6H_2O + 670 \text{ kcal/mol}$$
Glucose

Calculate the heat liberated when 15.0 g of glucose is metabolized to carbon dioxide and water in the body.

5.74 The heat of combustion of glucose, $C_6H_{12}O_6$, is 670 kcal/mol. The heat of combustion of ethanol, C_2H_6O, is 327 kcal/mol. The heat liberated by oxidation of each compound is the same whether it is burned in air or metabolized in the body. On a kcal/g basis, metabolism of which compound liberates more heat?

5.75 A plant requires approximately 4178 kcal for the production of 1.00 kg of starch (Section 19.6A) from carbon dioxide and water.

(a) Is the production of starch in a plant an exothermic process or an endothermic process?

(b) Calculate the energy in kilocalories required by a plant for the production of 6.32 g of starch.

5.76 To convert 1 mol of iron(III) oxide to its elements requires 196.5 kcal:

$$Fe_2O_3(s) + 196.5 \text{ kcal} \longrightarrow 2Fe(s) + \tfrac{3}{2}O_2(g)$$

How many grams of iron can be produced if 156.0 kcal of heat is absorbed by a large-enough sample of iron(III) oxide?

Chemical Connections

5.77 (Chemical Connections 5A) How does fluoride ion protect the enamel against tooth decay?

5.78 (Chemical Connections 5A) What kind of ions is hydroxyapatite made of?

5.79 (Chemical Connections 5B) A voltaic cell is represented by the following equation:

$$Fe(s) + Zn^{2+}(aq) \longrightarrow Fe^{2+}(aq) + Zn(s)$$

Which electrode is the anode and which electrode is the cathode?

5.80 (Chemical Connections 5C) Hydrogen peroxide is not only an antiseptic but also an oxidizing agent. The following equation shows the reaction of hydrogen peroxide with acetaldehyde to give acetic acid:

$$C_2H_4O(\ell) + H_2O_2(\ell) \longrightarrow C_2H_4O_2(\ell) + H_2O(\ell)$$
Acetaldehyde   Hydrogen   Acetic   Water
                peroxide   acid

In this reaction, which species is oxidized and which species is reduced? Which species is the oxidizing agent and which species is the reducing agent?

Additional Problems

5.81 When gaseous dinitrogen pentoxide, N_2O_5, is bubbled into water, nitric acid, HNO_3, forms. Write a balanced equation for this reaction.

5.82 In a certain reaction, Cu^+ is converted to Cu^{2+}. Is Cu^+ ion oxidized or reduced in this reaction? Is Cu^+ ion an oxidizing agent or a reducing agent in this reaction?

5.83 Using the equation

$$Fe_2O_3(s) + 3CO(g) \longrightarrow 2Fe(s) + 3CO_2(g)$$

(a) Show that this is a redox reaction. Which species is oxidized and which is reduced?

(b) How many moles of Fe_2O_3 are required to produce 38.4 mol of Fe?

(c) How many grams of CO are required to produce 38.4 mol of Fe?

5.84 Explain this statement: Our modern civilization depends on the heat obtained from chemical reactions.

5.85 When an aqueous solution of Na_3PO_4 is added to an aqueous solution of $Cd(NO_3)_2$, a precipitate forms. Write a net ionic equation for this reaction and identify the spectator ions.

5.86 The active ingredient in an analgesic tablet is 488 mg of aspirin, $C_9H_8O_5$. How many moles of aspirin does the tablet contain?

5.87 Chlorophyll, the compound responsible for the green color of leaves and grasses, contains one atom of magnesium in each molecule. If the percentage by weight of magnesium in chlorophyll is 2.72%, what is the molecular weight of chlorophyll?

5.88 If 7.0 kg of N_2 is added to 11.0 kg of H_2 to form NH_3, which reactant is in excess?

$$N_2(g) + 3H_2(g) \longrightarrow 2NH_3(g)$$

5.89 Lead(II) nitrate and aluminum chloride react according to the following equation:

$$3Pb(NO_3)_2 + 2AlCl_3 \longrightarrow 3PbCl_2 + 2Al(NO_3)_3$$

In an experiment, 8.00 g of lead nitrate was reacted with 2.67 g of aluminum chloride, and 5.55 g of lead chloride was formed.

(a) Which reactant was the limiting reagent?

(b) What was the percent yield?

5.90 Assume that the average red blood cell has a mass of 2×10^{-8} g and that 20% of its mass is hemoglobin (a protein whose molar mass is 68,000). How many molecules of hemoglobin are present in one red blood cell?

5.91 Reaction of pentane, C_5H_{12}, with oxygen, O_2, gives carbon dioxide and water.

(a) Write a balanced equation for this reaction.

(b) In this reaction, what is oxidized and what is reduced?

(c) What is the oxidizing agent and what is the reducing agent?

5.92 Ammonia is prepared industrially by the reaction of nitrogen and hydrogen according to the following equation:

$$N_2(g) + 3H_2(g) \longrightarrow 2NH_3(g)$$

Ammonia

If 29.7 kg of N_2 is added to 3.31 kg of H_2,

(a) Which reactant is the limiting reagent?

(b) How many grams of the other reactant are left over?

(c) How many grams of NH_3 are formed if the reaction goes to completion?

Looking Ahead

5.93 The heat of combustion of methane, CH_4, the major component of natural gas, is 213 kcal/mol. The heat of combustion of propane, C_3H_8, the major component of LPG or bottled gas, is 530 kcal/mol.

(a) Write a balanced equation for the complete combustion of each fuel to CO_2 and H_2O.

(b) On a kcal/mol basis, which of these two fuels is the better source of heat energy?

(c) On a kcal/g basis, which of these two fuels is the better source of heat energy?

5.94 The two major sources of energy in our diets are fats and carbohydrates. Palmitic acid, one of the major components of both animal fats and vegetable oils, belongs to a group of compounds called fatty acids. The metabolism of fatty acids is responsible for releasing the energy from fats. The major carbohydrates in our diets are sucrose (table sugar; Section 20.4A) and starch (Section 20.5A). Both starch and sucrose are first converted in the body to glucose, and then glucose is metabolized to produce energy. The heat of combustion of palmitic acid is 2385 kcal/mol, and that of glucose is 670 kcal/mol. Following are equations for the metabolism of each body fuel:

$$C_{16}H_{32}O_2(aq) + O_2(g) \longrightarrow CO_2(g) + H_2O(\ell) + 2385 \text{ kcal/mol}$$

Palmitic acid
(256 g/mol)

$$C_6H_{12}O_6(aq) + O_2(g) \longrightarrow CO_2(g) + H_2O(\ell) + 670 \text{ kcal/mol}$$

Glucose
(180 g/mol)

(a) Balance the equation for the metabolism of each fuel.

(b) Calculate the heat of combustion of each in kcal/g.

(c) In terms of kcal/mol, which of the two is the better source of energy for the body?

(d) In terms of kcal/g, which of the two is the better source of energy for the body?

Gases, Liquids, and Solids

Vince Streano/Corbis

Hot-air balloon, Utah.

GOB
Chemistry ⚛ Now™
Look for this logo in the chapter and go to GOB ChemistryNow at **http://now.brookscole.com/gob8** or on the CD for tutorials, simulations, and problems.

6.1 | What Are the Three States of Matter?

Various forces hold matter together and cause it to take different forms. In an atomic nucleus, very strong forces of attraction keep the protons and neutrons together (Chapter 3). In an atom itself, there are attractions between the positive nucleus and the negative electrons that surround it. Within molecules, atoms are attracted to each other by covalent bonds, the arrangement of which causes the molecules to assume a particular shape. Within an ionic crystal, three-dimensional shapes arise because of electrostatic attractions between ions.

In addition to these forces, there are attractive forces between molecules. These forces, which are the subject of this chapter, are weaker than any of the forces already mentioned; nevertheless, they help to determine whether a particular compound is a solid, a liquid, or a gas at room temperature.

These attractive forces hold matter together; in effect, they counteract another form of energy—kinetic energy—that tends to lead to a number of different ways for molecules to arrange themselves. In the absence of attractive

CHEMICAL CONNECTIONS 6A

Entropy: A Measure of Dispersal of Energy

As noted earlier, the molecules of a gas move randomly. The higher the temperature, the faster they move. In general, random motion means that many more arrangements of molecules are possible. When the temperature decreases, the molecules slow down and fewer different arrangements are possible. When all the molecules of a system become motionless and line up perfectly, we achieve the greatest possible order. At this point, the substance is a solid. The measure of such order is called **entropy**. When the order is perfect, the entropy of the system is zero. When molecules rotate or move from one place to another, the disorder increases, as does the entropy. Thus, when a crystal melts, entropy increases; when a liquid vaporizes, it also increases. When the temperature of a gas, a liquid, a solid, or a mixture of these states increases, the entropy of the system increases because an increase in temperature always increases molecular motions. When we combine two pure substances and they mix, disorder increases and so does the entropy.

We learned about the absolute, or Kelvin, temperature scale in Section 1.4E. Absolute zero (0 K or −273°C) is the lowest possible temperature. Although scientists have not been able to reach absolute zero, they have been able to produce temperatures within

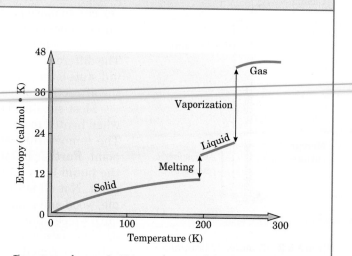

The entropy of ammonia, NH_3, as a function of absolute temperature.

a few billionths of a degree of it. At this low temperature, virtually all molecular motion ceases, almost perfect order reigns, and the entropy of the substance is nearly zero. In fact, a completely ordered pure crystalline solid at 0 K has an entropy of zero.

forces, the kinetic energy that particles possess keeps them constantly moving, mostly in random, disorganized ways. Kinetic energy increases with increasing temperature. Therefore, the higher the temperature, the greater the tendency of particles to have more possible arrangements. The total energy remains the same, but it is more widely dispersed. This dispersal of energy will have some important consequences, as we will see shortly.

The physical state of matter thus depends on a balance between the kinetic energy of particles, which tends to keep them apart, and the attractive forces between them, which tend to bring them together (Figure 6.1).

At high temperatures, molecules possess a high kinetic energy and move so fast that the attractive forces between them are too weak to hold them together. This situation is called the **gaseous state.** At lower temperatures, molecules move more slowly, to the point where the forces of attraction between them become important. When the temperature is low enough, a gas condenses to form a **liquid state.** Molecules in the liquid state still move past

Gas
- Molecules far apart and disordered
- Negligible interactions between molecules

Liquid
- Intermediate situation

Solid
- Molecules close together and ordered
- Strong interactions between molecules

Figure 6.1 The three states of matter. A gas has no definite shape, and its volume is the volume of the container. A liquid has a definite volume but no definite shape. A solid has a definite shape and a definite volume.

Figure 6.2 A mercury barometer.

each other, but they travel much more slowly than they do in the gaseous state. When the temperature is even lower, molecules no longer have enough velocity to move past each other. In the **solid state**, each molecule has a certain number of nearest neighbors, and these neighbors do not change.

The attractive forces between molecules are the same in all three states. The difference is that, in the gaseous state (and to a lesser degree in the liquid state), the kinetic energy of the molecules is great enough to overcome the attractive forces between them.

Most substances can exist in any of the three states. Typically a solid, when heated to a sufficiently high temperature, melts and becomes a liquid. The temperature at which this change takes place is called the melting point. Further heating causes the temperature to rise to the point at which the liquid boils and becomes a gas. This temperature is called the boiling point. Not all substances, however, can exist in all three states. For example, wood and paper cannot be melted. On heating, they either decompose or burn (depending on whether air is present), but they do not melt. Another example is sugar, which does not melt when heated but rather forms a dark substance called caramel.

6.2 | What Is Gas Pressure and How Do We Measure It?

On the earth we live under a blanket of air that presses down on us and on everything else around us. As we all know from weather reports, the **pressure** of the atmosphere varies from day to day.

A gas consists of molecules in rapid, random motion. The pressure a gas exerts on a surface, such as the walls of a container, results from the continual bombardment on the walls of the container by the rapidly moving gas molecules. We use an instrument called a **barometer** (Figure 6.2) to measure atmospheric pressure. A barometer consists of a long glass tube that is completely filled with mercury and then inverted into a pool of mercury in a dish. Because there is no air at the top of the mercury column in the tube (there is no way air could get in), no gas pressure is exerted on the mercury column. The entire atmosphere, however, exerts its pressure on the mercury in the open dish. The difference in the heights of the two mercury levels is the atmospheric pressure.

Pressure is most commonly measured in **millimeters of mercury (mm Hg)**. Pressure is also measured in **torr,** a unit named in honor of the Italian physicist and mathematician Evangelista Torricelli (1608–1647), who invented the barometer. At sea level, the average pressure of the atmosphere is 760 mm Hg. We use this number to define still another unit of pressure, the **atmosphere (atm).**

There are several other units with which to measure pressure. The SI unit is the pascal, and meteorologists report pressure in inches of mercury. In this book we use only mm Hg and atm.

A barometer is adequate for measuring the pressure of the atmosphere, but to measure the pressure of a gas in a container, we use a simpler instrument called a **manometer.** One form of manometer consists of a U-shaped tube containing mercury (Figure 6.3). Arm A has been evacuated, and sealed, and it has zero pressure. Arm B is connected to the container in which the gas sample is enclosed. The pressure of the gas depresses the level of the mercury in arm B. The difference between the two mercury levels gives the pressure directly in mm Hg. If more gas is added to the sample container, the mercury level in B will be pushed down and that in A will rise as the pressure in the bulb increases.

Pressure The force per unit area exerted against a surface

1 atm = 760 mm Hg
= 760 torr
= 101,325 pascals
= 29.92 in. Hg

Figure 6.3 A mercury manometer.

6.3 | What Are the Laws That Govern the Behavior of Gases?

By observing the behavior of gases under different sets of temperatures and pressures, scientists have established a number of relationships. In this section, we study three of the most important.

These gas laws we describe hold not only for pure gases but also for mixtures of gases.

A. Boyle's Law and the Pressure–Volume Relationship

Boyle's law states that for a fixed mass of gas at a constant temperature, the volume of the gas is inversely proportional to the pressure. If the pressure doubles, for example, the volume decreases by one half. This law can be stated mathematically in the following equation, where P_1 and V_1 are the initial pressure and volume, and P_2 and V_2 are the final pressure and volume:

$$PV = \text{constant} \quad \text{or} \quad P_1V_1 = P_2V_2$$

This relationship between pressure and volume accounts for what happens in the cylinder of a car engine (Figure 6.4).

GOB
Chemistry⚛Now™
Click *Chemistry Interactive* for an interactive exploration of a **Boyle's Law** experiment

---

CHEMICAL CONNECTIONS 6B

Breathing and Boyle's Law

Under normal resting conditions, we breathe about 12 times per minute, each time inhaling and exhaling about 500 mL of air. When we inhale, we lower the diaphragm or raise the rib cage, either of which increases the volume of the chest cavity. In accord with Boyle's law, as the volume of the chest cavity increases, the pressure within it decreases and becomes lower than the outside pressure. As a result, air flows from the higher-pressure area outside the body into the lungs. While the difference in these two pressures is only about 3 mm Hg, it is enough to cause air to flow into the lungs. In exhaling, we reverse the process: We raise the diaphragm

or lower the rib cage. The resulting decrease in volume increases the pressure inside the chest cavity, causing air to flow out of the lungs.

In certain diseases the chest becomes paralyzed, and the affected person cannot move either the diaphragm or the rib cage. In such a case, a respirator is used to help the person breathe. The respirator first pushes down on the chest cavity and forces air out of the lungs. The pressure of the respirator is then lowered below atmospheric pressure, causing the rib cage to expand and draw air into the lungs.

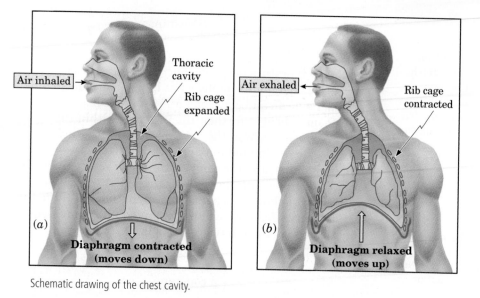

Schematic drawing of the chest cavity.

Figure 6.4 Boyle's law illustrated in an automobile cylinder. When the piston moves up, the volume occupied by the gas (a mixture of air and gasoline vapor) decreases and the pressure increases. The spark plug then ignites the compressed gasoline vapor.

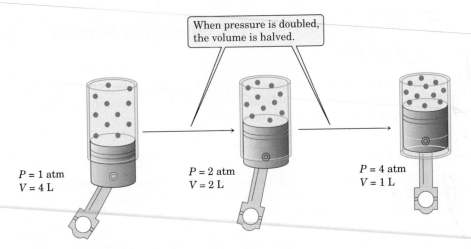

When pressure is doubled, the volume is halved.

$P = 1$ atm
$V = 4$ L

$P = 2$ atm
$V = 2$ L

$P = 4$ atm
$V = 1$ L

GOB
Chemistry·⚛·Now™

Click *Chemistry Interactive* for an interactive exploration of a **Charles's Law** experiment

When using the gas laws, temperature must be expressed in kelvins (K). The zero in this scale is the lowest possible temperature.

Greg Gawlowski/Dembinsky Photo Associates

Figure 6.5 Charles's law illustrated in a hot-air balloon. Because the balloon can stretch, the pressure inside it remains constant. When the air in the balloon is heated, its volume increases, expanding the balloon. As the air in the balloon expands, it becomes less dense than the surrounding air, providing the lift for the balloon. (Charles was one of the first balloonists.)

Combined gas law The pressure, volume, and temperature in kelvins of two samples of the same gas are related by the equation $P_1V_1/T_1 = P_2V_2/T_2$

B. Charles's Law and the Temperature–Volume Relationship

Charles's law states that the volume of a fixed mass of gas at a constant pressure is directly proportional to the temperature in kelvins (K). In other words, as long as the pressure on a gas remains constant, increasing the temperature of the gas causes an increase in the volume occupied by the gas. Charles's law can be stated mathematically this way:

$$\frac{V}{T} = \text{a constant} \quad \text{or} \quad \frac{V_1}{T_1} = \frac{V_2}{T_2}$$

This relationship between volume and temperature is the basis of the hot-air balloon operation (Figure 6.5).

C. Gay-Lussac's Law and the Temperature–Pressure Relationship

Gay-Lussac's law states that, for a fixed mass of a gas at constant volume, the pressure is directly proportional to the temperature in kelvins (K):

$$\frac{P}{T} = \text{a constant} \quad \text{or} \quad \frac{P_1}{T_1} = \frac{P_2}{T_2}$$

As the temperature of the gas increases, the pressure increases proportionately. Consider, for example, what happens inside an autoclave. Steam generated inside an autoclave at 1 atm pressure has a temperature of 100°C. As the steam is heated further, the pressure within the autoclave increases as well. A valve controls the pressure inside the autoclave; if the pressure exceeds the designated maximum, the valve opens, releasing the steam. At maximum pressure, the temperature may reach 120°C to 150°C. All microorganisms in the autoclave are destroyed at such high temperatures.

Table 6.1 shows mathematical expressions of these three laws.

The three gas laws can be combined and expressed by a mathematical equation called the **combined gas law:**

$$\frac{PV}{T} = \text{a constant} \quad \text{or} \quad \frac{P_1V_1}{T_1} = \frac{P_2V_2}{T_2}$$

Table 6.1	Mathematical Expressions of the Three Gas Laws for a Fixed Mass of Gas	
Name	Expression	Constant
Boyle's law	$P_1V_1 = P_2V_2$	T
Charles's law	$\dfrac{V_1}{T_1} = \dfrac{V_2}{T_2}$	P
Gay-Lussac's law	$\dfrac{P_1}{T_1} = \dfrac{P_2}{T_2}$	V

EXAMPLE 6.1

A gas occupies 3.00 L at 2.00 atm pressure. Calculate its volume when we increase the pressure to 10.15 atm at the same temperature.

Solution

First we identify the known quantities. Because T_1 and T_2 are the same in this example and consequently cancel each other, we don't need to know the temperature.

Initial: $P_1 = 2.00$ atm $V_1 = 3.00$ L

Final: $P_2 = 10.15$ atm $V_2 = ?$

We want the final volume, so we solve the combined gas law equation for V_2:

$$V_2 = \frac{P_1 V_1 \cancel{T_2}}{\cancel{T_1} P_2} = \frac{(2.00\ \cancel{atm})(3.00\ L)}{10.15\ \cancel{atm}} = 0.591\ L$$

Problem 6.1

A gas occupies 3.8 L at 0.70 atm pressure. If we expand the volume at constant temperature to 6.5 L, what is the final pressure?

EXAMPLE 6.2

In an autoclave, steam at 100°C is generated at 1.00 atm. After the autoclave is closed, the steam is heated at constant volume until the pressure gauge indicates 1.13 atm. What is the final temperature in the autoclave?

Solution

All temperatures in gas law calculations must be in kelvins; therefore, we must convert the Celsius temperature to kelvins:

$$100°C = 100 + 273 = 373\ K$$

We then identify the known quantities. Because V_1 and V_2 are the same in this example and consequently cancel each other, we don't need to know the volume of the autoclave.

Initial: $P_1 = 1.00$ atm $T_1 = 373$ K

Final: $P_2 = 1.13$ atm $T_2 = ?$

An autoclave used to sterilize hospital equipment.

Because we want a final temperature, we now solve the combined gas law equation for T_2:

$$T_2 = \frac{P_2 V_2 T_1}{P_1 V_1} = \frac{(1.13\ \text{atm})(373\ \text{K})}{1.00\ \text{atm}} = 421\ \text{K}$$

The final temperature is 421 K, or $421 - 273 = 148°C$.

Problem 6.2

A constant volume of oxygen gas, O_2, is heated from 120°C to 212°C. The final pressure is 20.3 atm. What was the initial pressure?

EXAMPLE 6.3

A gas in a flexible container has a volume of 0.50 L and a pressure of 1.0 atm at 393 K. When the gas is heated to 500 K, its volume expands to 3.0 L. What is the new pressure of the gas in the flexible container?

Solution
The known quantities are

Initial: $P_1 = 1.0$ atm $V_1 = 0.50$ L $T_1 = 393$ K

Final: $P_2 = ?$ $V_2 = 3.0$ L $T_2 = 500$ K

Solving the combined gas law for P_2, we find

$$P_2 = \frac{P_1 V_1 T_2}{T_1 V_2} = \frac{(1.0\ \text{atm})(0.50\ \text{L})(500\ \text{K})}{(3.0\ \text{L})(393\ \text{K})} = 0.21\ \text{atm}$$

Problem 6.3

A gas is expanded from an initial volume of 20.5 L at 0.92 atm at room temperature (23.0°C) to a final volume of 340.6 L. During the expansion the gas cools to 12.0°C. What is the new pressure?

6.4 What Are Avogadro's Law and the Ideal Gas Law?

The relationship between the amount of gas present and its volume is described by **Avogadro's law,** which states that equal volumes of gases at the same temperature and pressure contain equal numbers of molecules. Thus, if the temperature, pressure, and volumes of two gases are the same, then the two gases contain the same number of molecules, regardless of their identity (Figure 6.6). Avogadro's law is valid for all gases, no matter what they are.

The actual temperature and pressure at which we compare two or more gases do not matter. It is convenient, however, to select one temperature and one pressure as standard, and chemists have chosen 1 atm as the standard pressure and 0°C (273 K) as the standard temperature. These conditions are called **standard temperature and pressure (STP).**

Avogadro's law Equal volumes of gases at the same temperature and pressure contain the same number of molecules

Standard temperature and pressure (STP) 0°C (273 K) and one atmosphere pressure

All gases at STP or any other combination of temperature and pressure contain the same number of molecules in any given volume. But how many is that? In Chapter 5 we saw that one mole contains 6.02×10^{23} formula units. What volume of a gas at STP contains one mole of molecules? This quantity has been measured experimentally and found to be 22.4 L. Thus one mole of any gas at STP occupies a volume of 22.4 L.

Avogadro's law allows us to write a gas law that is valid not only for any pressure, volume, and temperature, but also for any quantity of gas. This law, called **the ideal gas law,** is

$$PV = nRT$$

where P = pressure of the gas in atmospheres (atm)
V = volume of the gas in liters (L)
n = amount of the gas in moles (mol)
T = temperature of the gas in kelvins (K)
R = a constant for all gases, called the **ideal gas constant**

We can find the value of R by using the fact that one mole of any gas at STP occupies a volume of 22.4 L:

$$R = \frac{PV}{nT} = \frac{(1.00 \text{ atm})(22.4 \text{ L})}{(1.00 \text{ mol})(273 \text{ K})} = 0.0821 \frac{\text{L} \cdot \text{atm}}{\text{mol} \cdot \text{K}}$$

The ideal gas law holds for all ideal gases at any temperature, pressure, and volume. But the only gases we have around us in the real world are real gases. How valid is it to apply the ideal gas law to real gases? The answer is that, under most experimental conditions, real gases behave sufficiently like ideal gases that we can use the ideal gas law for them with little trouble. Thus, using $PV = nRT$, we can calculate any one quantity—P, V, T, or n—if we know the other three quantities.

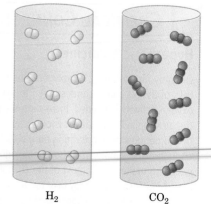

H_2 CO_2

T, P, and V are equal in both containers

Figure 6.6 Avogadro's law. Two tanks of gas of equal volume at the same temperature and pressure contain the same number of molecules.

Ideal gas law $PV = nRT$

Ideal Gas A gas whose physical properties are described accurately by the ideal gas law

Universal gas constant (R) $0.0821 \text{ L} \cdot \text{atm} \cdot \text{mol}^{-1} \cdot \text{K}^{-1}$

Real gases behave most like ideal gases at low pressures (1 atm or less) and high temperatures (300 K or higher).

EXAMPLE 6.4

One mole of CH_4 gas occupies 20.0 L at 1.00 atm pressure. What is the temperature of the gas in kelvins?

Solution
Solve the ideal gas law for T and plug in the given values:

$$T = \frac{PV}{nR} = \frac{PV}{n} \times \frac{1}{R} = \frac{(1.00 \text{ atm})(20.0 \text{ L})}{(1.00 \text{ mol})} \times \frac{\text{mol} \cdot \text{K}}{0.0821 \text{ L} \cdot \text{atm}} = 244 \text{ K}$$

Note that we calculated the temperature for 1.00 mol of CH_4 gas under these conditions. The answer would be the same for 1.00 mol of CO_2, N_2, NH_3, or any other gas under these conditions. Note also that we have shown the gas constant separately to make it clear what is happening with the units attached to all quantities. We are going to do this throughout.

Problem 6.4

If 2.00 mol of NO gas occupies 10.0 L at 295 K, what is the pressure of the gas in atmospheres?

GOB
Chemistry Now™
Click *Coached Problems* to see a simulation of the **Ideal Gas Law**

CHEMICAL CONNECTIONS 6C

Hyperbaric Medicine

Ordinary air contains 21% oxygen. Under certain conditions, the cells of tissues can become starved for oxygen (hypoxia), and quick oxygen delivery is needed. Increasing the percentage of oxygen in the air supplied to a patient is one way to remedy this situation, but sometimes even breathing pure (100%) oxygen may not be enough. For example, in carbon monoxide poisoning, hemoglobin, which normally carries most of the O_2 from the lungs to the tissues, binds CO and cannot take up any O_2 in the lungs. Without any help, tissues would soon become starved for oxygen and the patient would die. When oxygen is administered under a pressure of 2 to 3 atm, it dissolves in the plasma to such a degree that the tissues receive enough of it to recover without the help of the poisoned hemoglobin molecules. Other conditions for which hyperbaric medicine is used are treatment of gas gangrene, smoke inhalation, cyanide poisoning, skin grafts, thermal burns, and diabetic lesions.

Breathing pure oxygen for prolonged periods, however, is toxic. For example, if O_2 is administered at 2 atm for more than 6 hours, it may damage both lung tissue and the central nervous system. In addition, this treatment may cause nuclear cataract formation,

Hyperbaric oxygen chamber at Medical City Dallas Hospital.

necessitating postrecovery eye surgery. Therefore, recommended exposures of O_2 are 2 hours at 2 atm and 90 minutes at 3 atm. The benefits of hyperbaric medicine must be carefully weighed against these and other contraindications.

Gregory G. Dimijian/Photo Researchers, Inc.

Charles D. Winters

Molar volume. The cube has a volume of 22.4 liters, which is the volume of one mole of gas at STP.

EXAMPLE 6.5

If there is 5.0 g of CO_2 gas in a 10-L cylinder at 25°C, what is the gas pressure within the cylinder?

Solution

We are given the quantity of CO_2 in grams but, to use the ideal gas law, we must express the quantity in moles. Therefore, we must first convert grams of CO_2 to moles, and then use this value in the ideal gas law. To convert from grams to moles, we use the conversion factor 1 mol CO_2 = 44 g.

$$5.0 \text{ g } CO_2 \times \frac{1 \text{ mol } CO_2}{44 \text{ g } CO_2} = 0.11 \text{ mol } CO_2$$

We now use this value in the ideal gas equation to solve for the pressure of the gas.

$$P = \frac{nRT}{V}$$

$$= \frac{nT}{V} \times R = \frac{(0.11 \text{ mol } CO_2)(298 \text{ K})}{10 \text{ L}} \times \frac{0.0821 \text{ L} \cdot \text{atm}}{\text{mol} \cdot \text{K}} = 0.27 \text{ atm}$$

Problem 6.5

A certain quantity of neon gas is under 1.05 atm pressure at 303 K in a 10.0-L vessel. How many moles of neon are present?

EXAMPLE 6.6

If 3.3 g of a gas at 40°C and 1.15 atm pressure occupies a volume of 1.0 L, what is the mass of one mole of the gas?

Solution

This problem is more complicated than previous ones. We are given grams of gas and P, T, and V values. We are asked to calculate the mass of one mole of the gas (g/mol). We can solve this problem in two steps.

Step 1 Use the P, V, and T measurements and the ideal gas law to calculate the number of moles of gas present in the sample. To use the ideal gas law, we must first convert 40°C to kelvins: $40 + 273 = 313$ K.

$$n = \frac{PV}{RT} = \frac{PV}{T} \times \frac{1}{R} = \frac{(0.15 \text{ atm})(1.0 \text{ L})}{313 \text{ K}} \times \frac{\text{mol} \cdot \text{K}}{0.0821 \text{ L} \cdot \text{atm}} = 0.0448 \text{ mol}$$

Step 2 Calculate the mass of one mole of the gas by dividing grams by moles.

$$\text{Mass of one mole} = \frac{3.3 \text{ g}}{0.0448 \text{ mol}} = 74 \text{ g} \cdot \text{mol}^{-1}$$

Problem 6.6

An unknown amount of He gas occupies 30.5 L at 2.00 atm pressure and 300 K. What is the weight of the gas in the container?

GOB
Chemistry Now™
Click *Coached Problems* to try a problem calculating pressure **Using the Ideal Gas Law**

6.5 What Is Dalton's Law of Partial Pressures?

In a mixture of gases, each molecule acts independently of all the others, provided that the gases behave as ideal gases and do not react with each other. For this reason, the ideal gas law works for mixtures of gases as well as for pure gases. **Dalton's law of partial pressures** states that the total pressure, P_T, of a mixture of gases is the sum of the partial pressures of each individual gas:

$$P_T = P_1 + P_2 + P_3 + \cdots$$

A corollary to Dalton's law is that the **partial pressure** of a gas in a mixture is the pressure that the gas would exert if it were alone in the container. The equation holds separately for each gas in the mixture as well as for the mixture as a whole.

Partial pressure The pressure that a gas in a mixture of gases would exert if it were alone in the container

EXAMPLE 6.7

To a tank containing N_2 at 2.0 atm and O_2 at 1.0 atm, we add an unknown quantity of CO_2 until the total pressure within the tank is 4.6 atm. What is the partial pressure of the CO_2?

Solution

Dalton's law tells us that the addition of CO_2 does not affect the partial pressures of the N_2 or O_2 already present in the tank. The partial pressures of N_2 and O_2 remain at 2.0 atm and 1.0 atm, respectively, and their sum is 3.0 atm. If the new pressure is 4.6 atm, the partial pressure of the added CO_2 must be 1.6 atm. Thus, when the final pressure is 4.6 atm, the partial pressures are

$$4.6 \text{ atm} = 2.0 \text{ atm} + 1.0 \text{ atm} + 1.6 \text{ atm}$$

| Total pressure | Partial pressure of N_2 | Partial pressure of O_2 | Partial pressure of CO_2 |

Problem 6.7

A vessel under 2.015 atm pressure contains nitrogen, N_2, and water vapor, H_2O. The partial pressure of N_2 is 1.908 atm. What is the partial pressure of the water vapor?

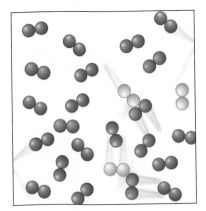

Figure 6.7 The kinetic molecular model of a gas. Molecules of nitrogen (blue) and oxygen (red) are in constant motion and collide with each other and with the walls of the container. Collisions with the walls of the container create the gas pressure. In air at STP, 6.02×10^{23} molecules undergo approximately 10 billion collisions per second.

6.6 | What Is the Kinetic Molecular Theory?

To this point, we have studied the macroscopic properties of gases—namely, the various laws dealing with the relationships among temperature, pressure, volume, and number of molecules in a sample of gas. Now let us examine the behavior of gases at the molecular level and see how we can explain their behavior in terms of molecules and the interactions between them.

The relationship between the observed behavior of gases and the behavior of individual gas molecules within the gas can be explained by the **kinetic molecular theory,** which makes the following assumptions about the molecules of a gas:

1. Gases consist of particles, either atoms or molecules, constantly moving through space in straight lines, in random directions, and with various speeds. Because these particles move in random directions, different gases mix very readily.

2. The average kinetic energy of gas particles is proportional to the temperature in kelvins. The higher the temperature, the faster they move through space, and the greater their kinetic energy.

3. Molecules collide with each other, much as billiard balls do, bouncing off each other and changing direction. Each time they collide, they may exchange kinetic energies (one moves faster than before, the other slower), but the total kinetic energy of the gas sample remains the same.

4. Gas particles have no volume. Most of the volume taken up by a gas is empty space, which explains why gases can be compressed so easily.

5. There are no attractive forces between gas particles. They do not stick together after a collision occurs.

6. Molecules collide with the walls of the container, and these collisions constitute the pressure of the gas. The greater the number of collisions per unit time, the greater the pressure.

These six assumptions of the kinetic molecular theory give us an idealized picture of the molecules of a gas and their interactions with one another (Figure 6.7). In real gases, however, forces of attraction between molecules do exist, and molecules do occupy some volume. Because of these factors, a gas described by these six assumptions of the kinetic molecular theory is called an **ideal gas.** In reality, there is no ideal gas; all gases are real. At STP, however, most real gases behave in much the same way that an ideal gas would, so we can safely use these assumptions.

6.7 | What Types of Attractive Forces Exist Between Molecules?

As noted in Section 6.1, the strength of the intermolecular forces (forces between molecules) in any sample of matter determines whether the sample is a gas, a liquid, or a solid under normal conditions of temperature and pressure. In general, the closer the molecules are to each other, the greater the effect of the intermolecular forces. When the temperature of a gas is high (room temperature or higher) and the pressure is low (1 atm or less), molecules of the gas are so far apart that we can effectively ignore attractions between them and treat the gas as ideal. When the temperature decreases, the pressure increases, or both, the distances between molecules decrease so that we can no longer ignore intermolecular forces. In fact, these forces become so important that they cause **condensation** (change from a gas to a liquid) and **solidification** (change from a liquid to a solid). Therefore, before discussing the structures and properties of liquids and solids, we must look at the nature of these intermolecular forces of attraction.

In this section, we discuss three types of intermolecular forces: London dispersion forces, dipole–dipole interactions, and hydrogen bonding. Table 6.2 shows the strengths of these three forces. Also shown for comparison are the strengths of ionic and covalent bonds, both of which are considerably stronger than the other three types of intermolecular forces. Although intermolecular forces are relatively weak compared to ionic and covalent bonds,

Condensation The change of a substance from the vapor or gaseous state to the liquid state

Table 6.2 Forces of Attraction Between Molecules and Ions

Attractive Force	Example	Typical Energy (kcal/mol)
Ionic bonds	Na^+⫶⫶⫶Cl^-, Mg^{2+}⫶⫶⫶O^{2-}	170–970
Single, double, and triple covalent bonds	C—C C=C C≡C O—H	80–95 175 230 90–120
Hydrogen bonding		2–10
Dipole-dipole interaction		1–6
London dispersion forces	Ne⫶⫶Ne	0.01–2.0

For purposes of comparison, the strengths of single, double, and triple covalent bonds are also given.

it is the former that determines many of the physical properties of molecules, such as melting point, boiling point, and viscosity. As we will see in Chapters 21–31, these forces are also extremely important in determining the three-dimensional shapes of biomolecules such as proteins and nucleic acids, and in deciding how these types of biomolecules recognize and interact with one another.

A. London Dispersion Forces

London dispersion forces Extremely weak attractive forces between atoms or molecules caused by the electrostatic attraction between temporary induced dipoles

Figure 6.8 London dispersion forces. A temporary polarization of electron density in one neon atom creates positive and negative charges, which in turn induce temporary positive and negative charges in an adjacent atom. The intermolecular attractions between the temporary induced positive end of one dipole and the negative end of another temporary dipole are called London dispersion forces.

Attractive forces are present between all molecules, whether they are polar or nonpolar. If the temperature falls far enough, even nonpolar molecules such as He, Ne, H_2, and CH_4 can be liquefied. Neon, for example, is a gas at room temperature and atmospheric pressure. It can be liquefied if cooled to $-246°C$. The fact that these and other nonpolar gases can be liquefied means that some sort of interaction must occur between them to make them stick together in the liquid state. These weak attractive forces are called **London dispersion forces,** after the American chemist Fritz London (1900–1954), who was the first to explain them.

London dispersion forces have their origin in electrostatic interactions. To visualize the origin of these forces, it is necessary to think in terms of instantaneous distributions of electrons within an atom or molecule. Consider, for example, a sample of neon atoms, which can be liquefied if cooled to $-246°C$. Over time, the distribution of electron density in a neon atom is symmetrical, and a neon atom has no dipole; that is, it has no separation of positive and negative charges. However, at any given instant, the electron density in a neon atom may be shifted more toward one part of the atom than another, thus creating a temporary dipole (Figure 6.8). This temporary dipole, which lasts for only tiny fractions of a second, induces temporary dipoles in adjacent neon atoms. The attractions between these temporary induced dipoles are called London dispersion forces, and they make nonpolar molecules stick together to form the liquid state.

London dispersion forces exist between all molecules, but they are the only forces of attraction between nonpolar molecules. They range in strength from 0.01 to 2.0 kcal/mol depending on the mass, size, and shape of the interacting molecules. In general, their strength increases as the mass and number of electrons in a molecule increase. Even though London dispersion forces are very weak, they contribute significantly to the attractive forces between large molecules because they act over large surface areas.

B. Dipole–Dipole Interactions

Dipole–dipole interaction The interaction between the positive end of a dipole of one molecule and the negative end of another dipole in the same or different molecule

As mentioned in Section 4.11, many molecules are polar. The attraction between the positive end of one dipole and the negative end of another dipole is called a **dipole–dipole interaction.** These interactions can exist between two identical polar molecules or between two different polar molecules. To see the importance of dipole–dipole interactions, we can look at the differences in boiling points between nonpolar and polar molecules of comparable molecular weight. Butane, C_4H_{10}, with a molecular weight of 58 amu, is a nonpolar molecule with a boiling point of 0.5°C. Acetone, C_3H_6O, with the same molecular weight, has a boiling point of 58°C. Acetone is a polar molecule, and its molecules are held together in the liquid state by dipole–dipole attractions between the negative end of the C=O dipole of one acetone molecule and the positive end of the C=O dipole of another acetone molecule. Because it requires more energy to overcome the dipole–dipole interactions between acetone molecules than it does to overcome the considerably weaker London dispersion forces between butane molecules, acetone has a higher boiling point than butane.

$$CH_3-CH_2-CH_2-CH_3 \qquad CH_3-\overset{\overset{\displaystyle O}{\|}}{\underset{\delta^+}{C}}-CH_3$$

Butane
(bp 0.5°C)

Acetone
(bp 58°C)

C. Hydrogen Bond

As we have just seen, the attraction between the positive end of one dipole and the negative end of another results in dipole–dipole attraction. When the positive end of one dipole is a hydrogen atom bonded to an O or N (atoms of high electronegativity; see Table 4.5) and the negative end of the other dipole is an O or N atom, the attractive interaction between dipoles is particularly strong and is given a special name: **hydrogen bond.** An example is the hydrogen bonding that occurs between molecules of water in both the liquid and solid states (Figure 6.9).

The strength of hydrogen bonding ranges from 2 to 10 kcal/mol. The strength in liquid water, for example, is approximately 5 kcal/mol. By comparison, the strength of the O—H covalent bond in water is approximately 119 kcal/mol. As can be seen by comparing these numbers, an O---H hydrogen bond is considerably weaker than an O—H covalent bond. Nonetheless, the presence of hydrogen bonds in liquid water has an important effect on the physical properties of water. Because of hydrogen bonding, extra energy is required to separate each water molecule from its neighbors—hence the relatively high boiling point of water. As we will see in later chapters, hydrogen bonds play an especially important role in biological molecules.

Hydrogen bonds are not restricted to water, however. They form between two molecules whenever one molecule has a hydrogen atom covalently bonded to O or N, and the other molecule has an O or N atom.

Hydrogen bond A noncovalent force of attraction between the partial positive charge on a hydrogen atom bonded to an atom of high electronegativity, most commonly oxygen or nitrogen, and the partial negative charge on a nearby oxygen or nitrogen

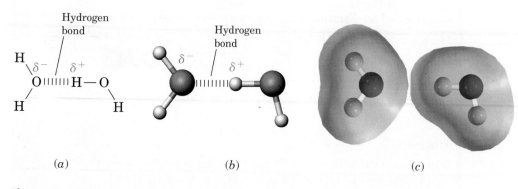

(a)   (b)   (c)

Figure 6.9 Two water molecules joined by a hydrogen bond (*a*) Structural formulas, (*b*) ball-and-stick models, and (*c*) electron density models.

EXAMPLE 6.8

Can a hydrogen bond form between
(a) Two molecules of methanol, CH_3OH?
(b) Two molecules of formaldehyde, CH_2O?
(c) One molecule of methanol, CH_3OH, and one of formaldehyde, CH_2O?

Solution

(a) Yes. Methanol is a polar molecule and has a hydrogen atom covalently bonded to an oxygen atom.

(b) No. Although formaldehyde is a polar molecule, it does not have a hydrogen covalently bonded to an oxygen or nitrogen atom. Its molecules, however, are attracted to each other by dipole–dipole interaction—that is, by the attraction between the negative end of the $C{=}O$ dipole of one molecule and the positive end of the $C{=}O$ dipole of another molecule.

(c) Yes. Methanol has a hydrogen atom bonded to an oxygen atom and formaldehyde has an oxygen atom.

Problem 6.8

Will the molecules in each set form a hydrogen bond between them?
(a) A molecule of water and a molecule of methanol, CH_3OH
(b) Two molecules of methane, CH_4

6.8 | How Do We Describe the Behavior of Liquids at the Molecular Level?

We have seen that we can describe the behavior of gases under most circumstances by the ideal gas law, which assumes that there are no attractive forces between molecules. As pressure increases in a real gas, however, the molecules of a gas become squeezed into a smaller space, with the result that attractions between molecules become increasingly effective in causing molecules to stick together.

If the distances between molecules decrease so that they touch or almost touch each other, the gas condenses to a liquid. Unlike gases, liquids do not fill all the available space, but they do have a definite volume, irrespective of the container. Because gases have a lot of empty space between molecules, it is easy to compress them into a smaller volume. In contrast, there is very little empty space in liquids; consequently, liquids are difficult to compress. A great increase in pressure is needed to cause even a very small decrease in the volume of a liquid. Thus liquids, for all practical purposes, are incompressible. In addition, the density of liquids is much greater than that of gases because the same mass occupies a much smaller volume in liquid form than it does in gaseous form.

The brake system in a car is based on hydraulics. The force you exert with the brake pedal is transmitted to the brake via cylinders filled with liquid. This system works very well until an air leak occurs. Once air gets into the brake line, pushing the brake pedal compresses air instead of moving the brake pads.

The positions of molecules in the liquid state are random, and some irregular empty space is available into which molecules can slide. Molecules in the liquid state are, therefore, constantly changing their positions with respect to neighboring molecules. This property causes liquids to be fluid and explains why liquids have a constant volume but not a constant shape.

A. Surface Tension

Unlike gases, liquids have surface properties, one of which is **surface tension** (Figure 6.10). The surface tension of a liquid is directly related to the strength of the intermolecular attraction between its molecules. Water has a high surface tension because of strong hydrogen bonding among water molecules. As a result, a steel needle can easily be made to float on the surface of water. If, however, the same needle is pushed below the elastic skin into the interior of the liquid, it sinks to the bottom. Similarly, water bugs gliding on the surface of a pond appear to be walking on an elastic skin of water.

B. Vapor Pressure

An important property of liquids is their tendency to evaporate. A few hours after a heavy rain, for example, most of the puddles have dried up; the water has evaporated and gone into the air. The same thing occurs if we leave a container of water or any other liquid out in the open. Let us explore how this change occurs.

In any liquid, there is a distribution of velocities among its molecules. Some of the molecules have high kinetic energy and move rapidly. Others have low kinetic energy and move slowly. Whether fast or slow, a molecule in the interior of the liquid cannot go very far before it hits another molecule and has its speed and direction changed by the collision. A molecule at the

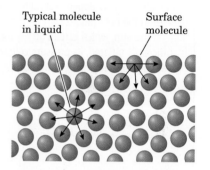

Typical molecule in liquid   Surface molecule

Figure 6.10 Surface tension. Molecules in the interior of a liquid have equal intermolecular attractions in every direction. Molecules at the surface (the liquid–gas interface), however, experience greater attractions toward the interior of the liquid than toward the gaseous state above it. Therefore, molecules on the surface are preferentially pulled toward the center of the liquid. This pull crowds the molecules on the surface, thereby creating a layer, like an elastic skin, that is tough to penetrate.

Hermann Eisenbeiss/Photo Researchers, Inc.

A water-strider standing on water. The surface tension of water supports it.

CHEMICAL CONNECTIONS 6D

Blood Pressure Measurement

Liquids, like gases, exert a pressure on the walls of their containers. Blood pressure, for example, results from pulsating blood pushing against the walls of the blood vessels. When the heart ventricles contract, pushing blood out into the arteries, the blood pressure is high (systolic pressure); when the ventricles relax, the blood pressure is lower (diastolic pressure). Blood pressure is usually expressed as a fraction showing systolic over diastolic pressure—for instance, 120/80. The normal range in young adults is 100 to 120 mm Hg systolic and 60 to 80 mm Hg diastolic. In older adults, the corresponding normal ranges are 115 to 135 and 75 to 85 mm Hg, respectively.

A sphygmomanometer—the instrument used to measure blood pressure—consists of a bulb, a cuff, a manometer, and a stethoscope. The cuff is wrapped around the upper arm and inflated by squeezing the bulb (figure, part a). The inflated cuff exerts a pressure on the arm, which is read on the manometer. When the cuff is sufficiently inflated, its pressure collapses the brachial artery, preventing pulsating blood from flowing to the lower arm. At this

pressure, no sound is heard in the stethoscope because the applied pressure in the cuff is greater than the blood pressure. Next, the cuff is slowly deflated, which decreases the pressure on the arm (figure, part b). The first faint tapping sound is heard when the pressure in the cuff just matches the systolic pressure as the ventricle contracts—that is, when the pressure in the cuff is low enough to allow pulsating blood to begin flowing into the lower arm. As the cuff pressure continues to decrease, the tapping first becomes louder and then begins to fade. At the point when the last faint tapping sound is heard, the cuff pressure matches the diastolic pressure when the ventricle is relaxed, thus allowing continuous blood flow into the lower arm. (figure, part c)

Digital blood pressure monitors are now available for home or office use. In these instruments, the stethoscope and the manometer are combined in a sensory device that records the systolic and diastolic blood pressures together with the pulse rate. The cuff and the inflation bulb are used the same way as in traditional sphygmomanometers.

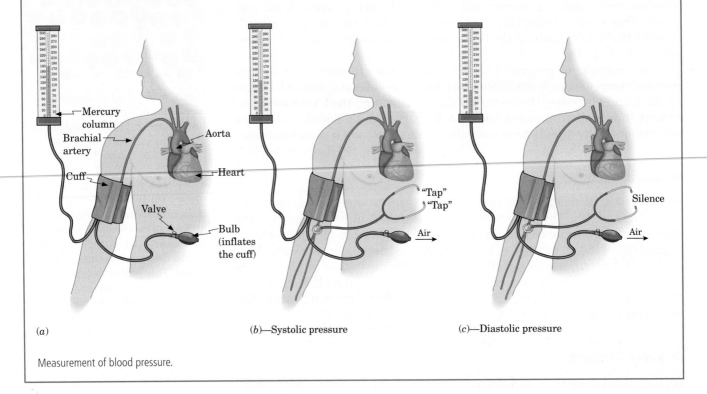

(a) (b)—Systolic pressure (c)—Diastolic pressure

Measurement of blood pressure.

surface, however, is in a different situation (Figure 6.11). If it is moving slowly (has a low kinetic energy), it cannot escape from the liquid because of the attractions of its neighboring molecules. If it is moving rapidly (has a high kinetic energy) and upward, however, it can escape from the liquid and enter the gaseous space above it.

In an open container, this process continues until all molecules have escaped. If the liquid is in a closed container, as in Figure 6.12, the molecules in the gaseous state cannot diffuse away (as they would do if the con-

Molecules from air (O_2, N_2)   Molecules of vapor

Molecules from air   Molecules of vapor

Figure 6.11   Evaporation. Some molecules at the surface of a liquid are moving fast enough to escape into the gaseous space.

Figure 6.12   Evaporation and condensation. In a closed container, molecules of liquid escape into the vapor phase, and the liquid recaptures vapor molecules.

tainer were open). Instead, they remain in the air space above the liquid, where they move rapidly in straight lines until they strike something. Some of these vapor molecules move downward, strike the surface of the liquid, and are recaptured by it.

At this point we have reached an **equilibrium.** As long as the temperature does not change, the number of vapor molecules reentering the liquid equals the number escaping from it. At equilibrium, the space above the liquid shown in Figure 6.12 contains both air and vapor molecules, and we can measure the partial pressure of the vapor, called the **vapor pressure** of the liquid. Note that we measure the partial pressure of a gas but call it the vapor pressure of the liquid.

The vapor pressure of a liquid is a physical property of the liquid and a function of temperature (Figure 6.13). As the temperature of a liquid increases, the average kinetic energy of its molecules increases and the easier it becomes for molecules to escape from the liquid state to the gaseous state. As the temperature of the liquid increases, its vapor pressure continues to increase until it equals the atmospheric pressure. At this point, bubbles of vapor form under the surface of the liquid, and then force their way upward through the surface of the liquid, and the liquid boils.

The molecules that evaporate from a liquid surface are those that have a higher kinetic energy. When they enter the gas phase, the molecules left behind are those with a lower kinetic energy. Because the temperature of a sample is proportional to the average kinetic energy of its molecules, the temperature of the liquid drops as a result of evaporation. This evaporation of the layer of water from your skin produces the cooling effect you feel when you come out of a swimming pool and the layer of water evaporates from your skin.

C. Boiling Point

The **boiling point** of a liquid is the temperature at which its vapor pressure is equal to the pressure of the atmosphere in contact with its surface. The boiling point when the atmospheric pressure is 1 atm is called the

Vapor A gas

Equilibrium A condition in which two opposing physical forces are equal

Vapor pressure The pressure of a gas in equilibrium with its liquid form in a closed container

At equilibrium, the rate of vaporization equals the rate of liquefaction.

Boiling point The temperature at which the vapor pressure of a liquid is equal to the atmospheric pressure

Figure 6.13 The change in vapor pressure with temperature for four liquids. The normal boiling point of a liquid is defined as the temperature at which its vapor pressure equals 760 mm Hg.

Normal boiling point The temperature at which a liquid boils under a pressure of 1 atm

normal boiling point. For example, 100°C is the normal boiling point of water because it is the temperature at which water boils at 1 atm pressure.

The use of a pressure cooker is an example of boiling water at higher temperatures. In this type of pot, food is cooked at, say, 2 atm, where the boiling point of water is 121°C. Because the food has been raised to a higher temperature, it cooks faster than it would in an open pot, in which boiling water cannot get hotter than 100°C. Conversely, at low pressures water boils at lower temperatures. For example, at the top of a mountain where the atmospheric pressure is less than 760 mm Hg, the boiling point of water might be 95°C.

GOB
Chemistry ∙ Now™

Click *Coached Problems* to see a simulation of **How Boiling Point Is Controlled by Molecular Structure**

D. Factors That Affect Boiling Point

As Figure 6.13 shows, different liquids have different normal boiling points. Table 6.3 gives molecular formulas, molecular weights, and normal boiling points for four liquids.

As you study the information in this table, note that chloroform, which has the largest molecular weight of the five compounds, has the lowest boiling point. Water, which has the lowest molecular weight, has the second

Table 6.3	Names, Molecular Formulas, Molecular Weights, and Normal Boiling Points for Hexane and the Four Liquids in Figure 6.13		
Name	**Molecular Formula**	**Molecular Weight (amu)**	**Boiling Point (°C)**
Chloroform	$CHCl_3$	120	62
Hexane	$CH_3CH_2CH_2CH_2CH_2CH_3$	86	69
Ethanol	CH_3CH_2OH	46	78
Water	H_2O	18	100
Acetic acid	CH_3COOH	60	118

highest boiling point. From a study of these and other compounds, chemists have determined that the boiling point of covalent compounds depends primarily on three factors:

1. **Intermolecular forces** Water (H_2O, MW 18) and methane (CH_4, MW 16) have about the same molecular weight. The normal boiling point of water is 100°C, while that of methane is −164°C. The difference in boiling points reflects the fact that CH_4 molecules in the liquid state must overcome only the weak London dispersion forces to escape to the vapor state (low boiling point). In contrast, water molecules, being hydrogen-bonded to each other, need more kinetic energy (and a higher boiling temperature) to escape into the vapor phase. Thus the difference in boiling points between these two compounds is due to the greater strength of hydrogen bonding compared with the much weaker London dispersion forces.

2. **Number of sites for intermolecular interaction (surface area)** Consider the boiling points of methane, CH_4 and hexane, C_6H_{14}. Both are nonpolar compounds with no possibility for hydrogen bonding or dipole–dipole interactions between their molecules. The only force of attraction between molecules of either compound is London dispersion forces. The normal boiling point of hexane is 69°C, and that of methane is −164°C. The difference in their boiling points reflects the fact that hexane has more electrons and a larger surface area than methane. Because of its larger surface area, there are more sites for London dispersion forces to arise between hexane molecules than in methane molecules and, therefore, hexane has the higher boiling point.

3. **Molecular shape** When molecules are similar in every way except shape, the strengths of their London dispersion forces determine their relative boiling points. Consider pentane, bp 36.1°C, and 2,2-dimethyl-propane, bp 9.5°C (Figure 6.14).

Both compounds have the same molecular formula, C_5H_{12}, and the same molecular weight, but the boiling point of pentane is approximately 26° higher than that of 2,2-dimethylpropane. These differences in boiling point are related to molecular shape in the following way. The only forces of attraction between these nonpolar molecules are London dispersion forces. Pentane is a roughly linear molecule, whereas 2,2-dimethylpropane has a spherical shape and a smaller surface area than pentane. As surface area decreases, contact between adjacent molecules, the strength of London dispersion forces, and boiling points all decrease. Consequently, London dispersion forces between molecules of 2,2-dimethylpropane are less than those between molecules of pentane and, therefore, 2,2-dimethylpropane has a lower boiling point.

Pentane
(bp 36.2°C)

2,2-Dimethylpropane
(bp 9.5°C)

Figure 6.14 Pentane and 2,2-dimethylpropane have the same molecular formula, C_5H_{12}, but quite different shapes.

CHEMICAL CONNECTIONS 6E

The Densities of Ice and Water

The hydrogen-bonded superstructure of ice contains empty spaces in the middle of each hexagon because the H_2O molecules in ice are not as closely packed as those in liquid water. For this reason, ice has a lower density (0.917 g/cm³) than does liquid water (1.00 g/cm³). As ice melts, some of the hydrogen bonds are broken and the hexagonal superstructure of ice collapses into the more densely packed organization of water. This change explains why ice floats on top of water instead of sinking to the bottom. Such behavior is highly unusual—most substances are denser in the solid state than they are in the liquid state. The lower density of ice keeps fish and microorganisms alive in many rivers and lakes that would freeze solid each winter if the ice sank to the bottom. The presence of ice on top insulates the remaining water and keeps it from freezing.

The fact that ice has a lower density than liquid water means that a given mass of ice takes up more space than the same mass of liquid water. This factor explains the damage done to biological tissues by freezing. When parts of the body (usually fingers, toes, nose, and ears) are subjected to extreme cold, they develop a condition called frostbite. The water in the cells freezes despite the blood's attempt to keep the temperature at 37°C. As liquid water freezes, it expands and in doing so ruptures the walls of cells containing it, causing damage. In some cases, frostbitten fingers or toes must be amputated.

Cold weather can damage plants in a similar way. Many plants are killed when the air temperature drops below the freezing point of water for several hours. Trees can survive cold winters because they have a low water content inside their trunks and branches.

Slow freezing is often more damaging to plant and animal tissues than quick freezing. In slow freezing, only a few crystals form, and these can grow to large sizes, rupturing cells. In quick freezing, such as can be achieved in liquid nitrogen (at a temperature of −196°C), many tiny crystals form. Because they do not grow much, tissue damage may be minimal.

Snowflakes are six-sided structures reflecting the underlying structure of ice.

Mehau Kulyk/Science Photo Library/Photo Researchers, Inc.

6.9 | What Are the Characteristics of the Various Types of Solids?

Even in the solid state, molecules and ions do not stop moving completely. They vibrate around fixed points.

Crystallization The formation of a solid from a liquid

When liquids are cooled, their molecules come closer together and attractive forces between them become so strong that random motion stops, and a solid forms. Formation of a solid from a liquid is called solidification or, alternatively, **crystallization.**

All solids have a regular shape that, in many cases, is obvious to the eye (Figure 6.15). This regular shape often reflects the arrangement of the particles within the crystal. In table salt, for example, the Na^+ and Cl^- ions are arranged in a cubic system (Figure 4.1). Metals also consist of particles arranged in a regular crystal lattice (generally not cubic), but here the par-

Garnet   Sulfur   Quartz   Pyrite

Figure 6.15   Some crystals.

ticles are atoms rather than ions. Because the particles in a solid are almost always closer together than they are in the corresponding liquid, solids almost always have a higher density than liquids.

As can be seen in Figure 6.15, crystals have characteristic shapes and symmetries. We are familiar with the cubic nature of table salt and the hexagonal ice crystals in snowflakes. A less well-known fact is that some compounds have more than one type of solid state. The best-known example is the element carbon, which has five crystalline forms (Figure 6.16). Diamond occurs when solidification takes place under very high pressure (thousands of atmospheres). Another form of carbon is the graphite in a pencil. Carbon atoms are packed differently in high-density, hard diamonds than they are in low-density, soft graphite.

In a third form of carbon, each molecule contains 60 carbon atoms arranged in a structure having 12 pentagons and 20 hexagons as faces, resembling a soccer ball [Figure 6.16(c)]. Because the famous architect Buckminster Fuller (1895–1983) invented domes of a similar structure (he called them geodesic domes), the C-60 substance was named buckminsterfullerene, or "buckyball" for short. The discovery of buckyballs has generated a whole new area of carbon chemistry. Similar cage-like structures containing 72, 80, and even larger numbers of carbon have been synthesized. As a group, they are called fullerenes.

New variations on the fullerenes are nanotubes [(Figure 6.16(d)]. The *nano*- part of the name comes from the fact that the cross section of each

Richard E. Smalley (1943–), Robert F. Curl, Jr. (1933–), and Harold Kroto (1939–) were awarded the 1996 Nobel Prize in chemistry for the discovery of these compounds.

Figure 6.16   Solid forms of carbon: (*a*) graphite, (*b*) diamond, (*c*) "buckyball," (*d*) nanotube, and (*e*) soot.

(*a*)

(*b*)   (*c*)   (*d*)   (*e*)

Table 6.4 Types of Solids

Type	Made Up Of	Characteristics	Examples
Ionic	Ions in a crystal lattice	High melting point	NaCl, K_2SO_4
Molecular	Molecules in a crystal lattice	Low melting point	Ice, aspirin
Polymeric	Giant molecules; can be crystalline, semicrystalline, or amorphous	Low melting point or cannot be melted; soft or hard	Rubber, plastics, proteins
Network	A very large number of atoms connected by covalent bonds	Very hard; very high melting point or cannot be melted	Diamond, quartz
Amorphous	Randomly arranged atoms or molecules	Mostly soft, can be made to flow, but no melting point	Soot, tar, glass

tube is only nanometers in size (10^{-9} m). Nanotubes come in a variety of forms. Single-walled carbon nanotubes can vary in diameter from 1 to 3 nm and are about 20 mm long. These compounds have generated great industrial interest because of their optical and electronic properties. They may play a role in miniaturization of instruments, giving rise to a new generation of nanoscale devices.

Soot is the fifth form of solid carbon. This substance solidifies directly out of carbon vapor and is an **amorphous solid;** that is, its atoms have no set pattern and are arranged randomly [Figure 6.16(e)]. Another example of an amorphous solid is glass. In essence, glass is an immobilized liquid.

As we have now seen, some crystalline solids consist of orderly arrays of ions (ionic solids; Figure 4.1), and others consist of molecules (molecular solids). Ions are held in the crystal lattice by ionic bonds. Molecules are held only by intermolecular forces, which are much weaker than ionic bonds. Therefore, molecular solids generally have much lower melting points than do ionic solids.

Other types of solids exist as well. Some are extremely large molecules, with each molecule having as many as 10^{23} atoms, all connected by covalent bonds. In such a case, the entire crystal is one big molecule. We call such molecules **network solids** or network crystals. A good example is diamond [Figure 6.16(b)]. When you hold a diamond in your hand, you are holding a gigantic assembly of bonded atoms. Like ionic crystals, network solids have very high melting points—if they can be melted at all. In many cases they cannot be. Table 6.4 summarizes the various types of solids.

6.10 What Is a Phase Change and What Energies Are Involved?

A. The Heating Curve for $H_2O(s)$ to $H_2O(g)$

Imagine the following experiment: We heat a piece of ice that is initially at $-20°C$. At first we don't see any difference in its physical state. The temperature of the ice increases, but its appearance does not change. At $0°C$, the ice begins to melt and liquid water appears. As we continue heating, more and more of the ice melts, but the temperature stays constant at $0°C$ until all the ice has melted and only liquid water remains. After all the ice has

Table 6.5   Energy Required to Heat 1.0 g of Solid Water at −20°C to 120°C

Physical Change	Energy (cal/g)	Basis for Calculation of Energy Required
Warming ice from −20°C to 0°C	9.6	Specific heat of ice = 0.48 cal/g·°C
Melting ice; temperature = 0°C	80	Heat of fusion of ice = 80 cal/g
Warming water from 0°C to 100°C	100	Specific heat of liquid water = 1.0 cal/g·°C
Boiling water; temperature = 100°C	540	Heat of vaporization = 540 cal/g
Warming steam from 100°C to 120°C	9.6	Specific heat of steam = 0.48 cal/g·°C

melted, the temperature of the water again increases as heat is added. At 100°C, the water boils. We continue heating as it continues to evaporate, but the temperature of the remaining liquid water does not change. Only after all the liquid water has changed to gaseous water (steam) does the temperature of the sample rise above 100°C.

These changes in state are called **phase changes.** A **phase** is any part of a system that looks uniform (homogeneous) throughout. Solid water (ice) is one phase, liquid water is another, and gaseous water is still another phase. Table 6.5 summarizes the energies for each step in the conversion of ice to steam.

Let us use the data in Table 6.5 to calculate the heat required to raise the temperature of 1.0 g of ice at −20°C to water vapor at 120°C. We begin with ice, whose specific heat is 0.48 cal/g·°C (Table 1.4). It requires 0.48 × 20 = 9.6 cal to raise the temperature of 1.0 g of ice from −20 to 0°C.

$$0.48 \, \frac{\text{cal}}{\text{g} \cdot {}^{\circ}\text{C}} \times 1.0 \, \text{g} \times 20{}^{\circ}\text{C} = 9.6 \, \text{cal}$$

After the ice reaches 0°C, additional heat causes a phase change: Solid water melts and becomes liquid water. The heat necessary to melt 1.0 g of any solid is called its **heat of fusion.** The heat of fusion of ice is 80 cal/g. Thus it requires 80 cal to melt 1.0 g of ice—that is, to change 1.0 g of ice at 0°C to liquid water at 0°C.

Only after the ice has completely melted does the temperature of the water rise again. The **specific heat** of liquid water is 1 cal/g·°C (Table 1.4). Thus it requires 100 cal to raise the temperature of 1.0 g of liquid water from 0° to 100°C. Contrast this with the 80 cal required to melt 1.0 g of ice.

When liquid water reaches 100°C, the normal boiling point of water, the temperature of the sample remains constant as another phase change takes place: Liquid water vaporizes to gaseous water. The amount of heat necessary to vaporize 1.0 g of a liquid at its normal boiling point is called its **heat of vaporization.** For water, this value is 540 cal/g. Once all of the liquid water has been vaporized, the temperature again rises as the water vapor (steam) is heated. The specific heat of steam is 0.48 cal/g (Table 1.4). Thus it requires 9.6 cal to heat one 1.0 g of steam from 100°C to 120°C. The data for heating 1.0 g of water from −20°C to 120°C can be shown in a graph called a **heating curve** (Figure 6.17).

An important aspect of these phase changes is that each one of them is reversible. If we start with liquid water at room temperature and cool it by immersing the container in a dry ice bath (−78°C), the reverse process is observed. The temperature drops until it reaches 0°C, and then ice begins to crystallize. During this phase change, the temperature of the sample stays constant but heat is given off. The amount of heat given off when 1.0 g of liquid water at 0°C freezes is exactly the same as the amount of heat absorbed when the 1.0 g of ice at 0°C melts.

Phase change A change from one physical state (gas, liquid, or solid) to another

The criterion of uniformity is the way it appears to our eyes and not as it is on the molecular level.

Vapors of gaseous CO_2 are cold enough to cause the moisture in the air above it to condense. The mixture of CO_2 vapors and condensed water vapor is heavier than air and slowly glides along the table or other surface on which the dry ice is placed.

GOB
Chemistry⟐Now™

Active Figure 6.17 The heating curve of ice. The graph shows the effect of adding heat to 1.0 g of ice initially at −20°C and raising its temperature to 120°C. **See a simulation based on this figure, and take a short quiz on the concepts at http://now.brookscole.com/gob8 or on the CD.**

Figure labels: Boiling water: 540 cal; Warming steam: 9.6 cal; Warming water: 100 cal; Melting ice: 80 cal; Warming ice: 9.6 cal. Axes: Temperature (°C) vs Heat added (cal).

A transition from the solid state directly into the vapor state without going through the liquid state is called **sublimation.** Solids usually sublime only at reduced pressures (less than 1 atm). At high altitudes, where the atmospheric pressure is low, snow sublimes. Solid CO_2 (dry ice) sublimes at −78.5°C under 1 atm pressure. At 1 atm pressure, CO_2 can exist only as a solid or as a gas, never as a liquid.

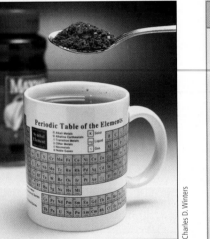

These freeze-dried coffee crystals were prepared by subliming water from frozen coffee.

Charles D. Winters

EXAMPLE 6.9

The heat of fusion of ice is 80 cal/g. How many calories are required to melt 1.0 mol of ice?

Solution
One mole of H_2O has a mass of 18 g. We use the factor-label method to calculate the heat required to melt one mole of ice at 0°.

$$\frac{80 \text{ cal}}{\text{g ice}} \times 18 \text{ g ice} = 1.4 \times 10^3 \text{ cal} = 1.4 \text{ kcal}$$

Problem 6.9
What mass of water at 100°C can be vaporized by the addition of 45.0 kcal of heat?

EXAMPLE 6.10

What will be the final temperature if we add 1000 cal of heat to 10.0 g of ice at 0°C?

Solution
The first thing the added heat will do is to melt the ice. This phase change will use up 10.0 g × 80 cal/g = 800 cal. The remaining 200 cal will be used to heat the liquid water. The specific heat (SH; Section 1.9) of liquid water is 1.00 cal/g · °C (Table 1.4). The amount of heat required to bring about this change is given by the following equation:

$$\text{Amount of heat} = \text{SH} \times m \times (T_2 - T_1)$$

Solving this equation for $T_2 - T_1$ gives

$$T_2 - T_1 = \text{Amount of heat} \times \frac{1}{\text{SH}} \times \frac{1}{m}$$

$$T_2 - T_1 = 200 \text{ cal} \times \frac{g \cdot °C}{1.00 \text{ cal}} \times \frac{1}{10.0 \text{ g}} = 20°C$$

Thus the temperature of the liquid water will rise by 20°C from 0°C, and it is now 20°C.

Problem 6.10

The specific heat of iron is 0.11 cal/g · °C (Table 1.4). The heat of fusion of iron—that is, the heat required to convert iron from a solid to a liquid at its melting point—is 63.7 cal/g. Iron melts at 1530°C. How much heat must be added to 1.0 g of iron at 25°C to completely melt it?

We can show all phase changes for any substance on a **phase diagram.** Figure 6.18 is a phase diagram for water. Temperature is plotted on the *x*-axis and pressure on the *y*-axis. Three areas with different colors are labeled solid, liquid, and vapor. Within these areas water exists either as ice or liquid water or water vapor. The line (A–B) separating the solid phase from the liquid phase contains all the freezing (melting) points of water—for example, 0°C at 1 atm and 0.005°C at 400 mm Hg.

At the melting point, the solid and liquid phases coexist. The line separating the liquid phase from the gas phase (A-C) contains all the boiling points of water—for example, 100°C at 760 mm Hg and 84°C at 400 mm Hg. At the boiling points, the liquid and gas phases coexist.

Finally, the line separating the solid phase from the gas phase (A-D) contains all the sublimation points. At the sublimation points, the solid and gas phases coexist.

At a unique point (A) on the phase diagram, called the **triple point,** all three phases coexist. The triple point for water occurs at 0.01°C and 4.58 mm Hg pressure.

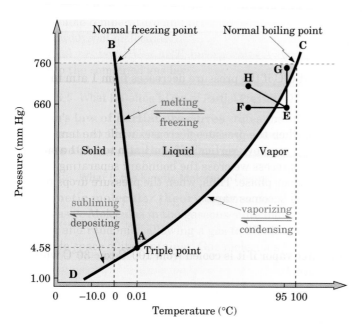

Figure 6.18 Phase diagram of water. Temperature and pressure scales are greatly reduced.

6.51 The partial pressures of a mixture of gases were as follows: oxygen, 210 mm Hg; nitrogen, 560 mm Hg; and carbon dioxide, 15 mm Hg. The total pressure of the gas mixture was 790 mm Hg. Was there another gas present in the mixture?

SECTION 6.6 What Is the Kinetic Molecular Theory?

6.52 Compare and contrast Dalton's atomic theory and the kinetic molecular theory.

SECTION 6.7 What Types of Attractive Forces Exist Between Molecules?

6.53 Which forces are stronger, intramolecular covalent bonds or intermolecular hydrogen bonds?

6.54 Under which condition does water vapor behave most ideally?
(a) 0.5 atm, 400 K   (b) 4 atm, 500 K
(c) 0.01 atm, 500 K

6.55 ■ Can water and dimethyl sulfoxide, $(CH_3)_2S=O$, form hydrogen bonds between them?

6.56 What kind of intermolecular interactions take place in (a) liquid CCl_4 and (b) liquid CO? Which will have the highest surface tension?

6.57 Ethanol, C_2H_5OH, and carbon dioxide, CO_2, have approximately the same molecular weight, yet carbon dioxide is a gas at STP and ethanol is a liquid. How do you account for this difference in physical property?

6.58 Can dipole–dipole interactions ever be weaker than London dispersion forces? Explain.

6.59 ■ Which compound has a higher boiling point: butane, C_4H_{10}, or hexane, C_6H_{14}?

SECTION 6.8 How Do We Describe the Behavior of Liquids at the Molecular Level?

6.60 The melting point of chloroethane, CH_3CH_2Cl, is $-136°C$ and its boiling point is $12°C$. Is chloroethane a gas, a liquid, or a solid at STP?

SECTION 6.9 What Are the Characteristics of the Various Types of Solids?

6.61 Which types of solids have the highest melting points? Which have the lowest melting points?

SECTION 6.10 What Is a Phase Change and What Energies Are Involved?

6.62 Calculate the specific heat (Section 1.9) of gaseous Freon-12, CCl_2F_2, if it requires 170 cal to change the temperature of 36.6 g of Freon-12 from $30°C$ to $50°C$.

6.63 The heat of vaporization of liquid Freon-12, CCl_2F_2, is 4.71 kcal/mol. Calculate the energy required to vaporize 39.2 g of this compound. The molecular weight of Freon-12 is 120.9 amu.

6.64 The specific heat (Section 1.9) of mercury is 0.0332 cal/g·°C. Calculate the energy necessary to raise the temperature of one mole of liquid mercury by $36°C$.

6.65 ■ Using Figure 6.13, estimate the vapor pressure of ethanol at (a) $30°C$, (b) $40°C$, and (c) $60°C$.

6.66 CH_4 and H_2O have about the same molecular weights. Which has the higher vapor pressure at room temperature? Explain.

6.67 The normal boiling point of a substance depends on both the mass of the molecule and the attractive forces between molecules. Arrange the compounds in each set in order of increasing boiling point and explain your answer:
(a) HCl, HBr, HI   (b) O_2, HCl, H_2O_2

6.68 Refer to Figure 6.17. How many calories are required to bring one mole of ice at $0°C$ to a liquid state at room temperature ($23°C$)?

6.69 Compare the number of calories absorbed when 100 g of ice at $0°C$ is changed to liquid water at $37°C$ with the number of calories absorbed when 100 g of liquid water is warmed from $0°C$ to $37°C$.

6.70 (a) How much energy is released when 10 g of steam at $100°C$ is condensed and cooled to body temperature ($37°C$)?
(b) How much energy is released when 100 g of liquid water at $100°C$ is cooled to body temperature ($37°C$)?
(c) Why are steam burns more painful than hot-water burns?

6.71 When iodine vapor hits a cold surface, iodine crystals form. Name the phase change that is the reverse of this condensation.

6.72 If a 156-g block of dry ice, CO_2, is sublimed at $25°C$ and 740 mm Hg, what volume does the gas occupy?

6.73 Trichlorofluoromethane (Freon-11, CCl_3F) as a spray is used to temporarily numb the skin around minor scrapes and bruises. It accomplishes this by reducing the temperature of the treated area, thereby numbing the nerve endings that perceive pain. Calculate the heat in kilocalories that can be removed from the skin by 1.00 mL of Freon-11. The density of Freon-11 is 1.49 g/mL, and its heat of vaporization is 6.42 kcal/mol.

6.74 Using the phase diagram of water (Figure 6.18), describe the process by which you can sublime 1 g of ice at $-10°C$ and at 1 atm pressure to water vapor at the same temperature.

Chemical Connections

6.75 (Chemical Connections 6A) Which has lower entropy, a gas at $100°C$ or one at $200°C$? Explain.

6.76 (Chemical Connections 6A) Which form of carbon presented in Figure 6.16 has the highest entropy?

6.77 (Chemical Connections 6B) What happens when a person lowers the diaphragm in his or her chest cavity?

6.78 (Chemical Connections 6C) In carbon monoxide poisoning, hemoglobin is incapable of transporting oxygen to the body's tissues. How does the oxygen get delivered to the cells when a patient is put into a hyperbaric chamber?

6.79 (Chemical Connections 6D) In a sphygmomanometer one listens to the first tapping sound as the constrictive pressure of the arm cuff is slowly released. What is the significance of this tapping sound?

6.80 (Chemical Connections 6E) Why is the damage by severe frostbite irreversible?

6.81 (Chemical Connections 6E) If you fill a glass bottle with water, cap it, and cool to $-10°C$, the bottle will crack. Explain.

6.82 (Chemical Connections 6F) In what way does supercritical CO_2 have some of the properties of a gas, and some of the properties of a liquid?

Additional Problems

6.83 Why is it difficult to compress a liquid or a solid?

6.84 Explain in terms of the kinetic molecular theory, what causes (a) the pressure of a gas and (b) the temperature of a gas.

6.85 The unit of pressure most commonly used for checking the inflation of automobile and bicycle tires is pounds per square inch (lb/in^2), abbreviated psi. The conversion factor between atm and psi is 1 atm = 14.7 psi. Suppose an automobile tire is filled to a pressure of 34 psi. What is the pressure in atm in the tire?

6.86 The gas in an aerosol can is at a pressure of 3.0 atm at 23°C. What will the pressure of the gas in the can be if the temperature is raised to 400°C?

6.87 Why do aerosol cans carry the warning "Do not incinerate"?

6.88 Under certain weather conditions (just before rain), the air becomes less dense. How does this change affect the barometric pressure reading?

6.89 An ideal gas occupies 387 mL at 275 mm Hg and 75°C. If the pressure changes to 1.36 atm and the temperature increases to 105°C, what is the new volume?

6.90 Which compound has greater intermolecular interactions, CO or CO_2?

6.91 On the basis of what you have learned about intermolecular forces, predict which liquid has the highest boiling point:
(a) Pentane, C_5H_{12}
(b) Chloroform, $CHCl_3$
(c) Water, H_2O

6.92 A 10-L gas cylinder is filled with N_2 to a pressure of 35 in. Hg. How many moles of N_2 do you have to add to your container to raise the pressure to 60 in. Hg? Assume a constant temperature of 27°C.

6.93 When filled, a typical tank for an outdoor grill contains 20 lb of LP (liquefied petroleum) gas, the major component of which is propane, C_3H_8. For this problem, assume that propane is the only substance present.
(a) How do you account for the fact that when propane is put under pressure, it can be liquefied?
(b) How many kilograms of propane does a full tank contain?
(c) How many moles of propane does a full tank contain?
(d) If the propane in a full tank were released into a flexible container, what volume would it occupy at STP?

6.94 Explain why gases are transparent.

6.95 ■ The density of a gas is 0.00300 g/cm^3 at 100°C and 1.00 atm. What is the mass of one mole of the gas?

6.96 The normal boiling point of hexane, C_6H_{14}, is 69°C, and that of pentane, C_5H_{12}, is 36°C. Predict which of these compounds has a higher vapor pressure at 20°C.

6.97 If 60.0 g of NH_3 occupies 35.1 L under a pressure of 77.2 in. Hg, what is the temperature of the gas, in °C?

6.98 Water is a liquid at STP. Hydrogen sulfide, H_2S, a heavier molecule, is a gas under the same conditions. Explain.

6.99 Why does the temperature of a liquid drop as a result of evaporation?

6.100 What volume of air (21% oxygen) measured at 25°C and 0.975 atm is required to completely oxidize 3.42 g of aluminum to aluminum oxide, Al_2O_3?

Tying It Together

6.101 Diving, particularly SCUBA (Self-Contained Underwater Breathing Apparatus) diving, subjects the body to increased pressure. Each 10 m (approximately 33 ft) of water exerts an additional pressure of 1 atm on the body.
(a) What is the pressure on the body at a depth of 100 ft?
(b) The partial pressure of nitrogen gas in air at 1 atm is 593 mm Hg. Assuming a SCUBA diver breathes compressed air, what is the partial pressure of nitrogen entering the lungs from a breathing tank at a depth of 100 ft?
(c) The partial pressure of oxygen gas in the air at 1 atm is 158 mm Hg. What is the partial pressure of oxygen in the air in the lungs at a depth of 100 ft?
(d) Why is it absolutely essential to exhale vigorously in a rapid ascent from a depth of 100 ft?

Solutions and Colloids

GOB
Chemistry ⚛ Now™

Look for this logo in the chapter and go to GOB ChemistryNow at **http://now.brookscole.com/gob8** or on the CD for tutorials, simulations, and problems.

Human blood cells in an isotonic solution.

David Scharf/Peter Arnold, Inc.

Making a homogeneous solution. A green solid, nickel nitrate, is stirred into water, where it dissolves to form a homogeneous solution.

Charles D. Winters

7.1 | What Do We Need to Know as We Begin?

In Chapter 2, we discussed pure substances—compounds made of two or more elements in a fixed ratio. Such systems are the easiest to study, so it was convenient to begin with them. In our daily lives, however, we more frequently encounter mixtures—systems having more than one component. Air, smoke, sea water, milk, blood, and rocks, for example, are mixtures (Section 2.2C).

If a mixture is uniform throughout at the molecular level, we call it a homogeneous mixture or, more commonly, a solution. Air and sea water, for example, are both solutions. They are clear and transparent. In contrast, in most rocks, we can see distinct regions separated from each other by well-defined boundaries. Such rocks are heterogeneous mixtures. Another example is a mixture of sand and sugar. We can easily distinguish between the two components; the mixing does not occur at the molecular level (Figure 2.4). Thus mixtures are classified on the basis of how they look to the unaided eye.

Some systems, however, fall between homogeneous and heterogeneous mixtures. Cigarette smoke, milk, and blood plasma may look homogeneous,

Table 7.1   The Most Common Types of Solutions

Solute		Solvent	Appearance of Solution	Example
Gas	in	Liquid	Liquid	Carbonated water
Liquid	in	Liquid	Liquid	Wine
Solid	in	Liquid	Liquid	Salt water (saline solution)
Gas	in	Gas	Gas	Air
Solid	in	Solid	Solid	14-Carat gold

but they do not have the transparency of air or sea water. These mixtures are classified as colloidal dispersions (suspensions). We will deal with such systems in Section 7.7.

Although mixtures can contain many components, we will generally restrict our discussion here to two-component systems, with the understanding that everything we say can be extended to multicomponent systems.

7.2 | What Are the Most Common Types of Solutions?

When we think of a solution, we normally think of a liquid. Liquid solutions, such as sugar in water, are the most common kind, but there are also solutions that are gases or solids. In fact, all mixtures of gases are solutions. Because gas molecules are far apart from each other and much empty space separates them, two or more gases can mix with each other in any proportions. Because the mixing takes place at the molecular level, a true solution always forms; that is, there are no heterogeneous mixtures of gases.

With solids, we are at the other extreme. Whenever we mix solids, we get a heterogeneous mixture. Because even microscopic pieces of solid still contain many billions of particles (molecules, ions, or atoms), there is no way to achieve mixing at the molecular level. Homogeneous mixtures of solids (or **alloys**), such as brass, do exist, but we make them by melting the solids, mixing the molten components, and allowing the mixture to solidify.

Table 7.1 lists the five most common types of solutions. Examples of other types (gas in solid, liquid in gas, and so on) are also known but are much less important.

When a solution consists of a solid or a gas dissolved in a liquid, the liquid is called the **solvent,** and the solid or gas is called the **solute.** A solvent may have several solutes dissolved in it, even of different types. A common example is spring water, in which gases (carbon dioxide and oxygen) and solids (salts) are dissolved in the solvent, water.

When one liquid is dissolved in another, a question may arise regarding which is the solvent and which is the solute. In most cases, the one present in the greater amount is called the solvent, but no rigid rule applies.

7.3 | What Are the Distinguishing Characteristics of Solutions?

The following are some properties of solutions:

1. **The distribution of particles in a solution is uniform.**
   Every part of the solution has exactly the same composition and properties as every other part. That, in fact, is the definition of "homogeneous."

Beer is a solution in which a liquid (alcohol), a solid (malt), and a gas (CO_2) are dissolved in the solvent, water.

Mixtures can be homogenous, as with brass, which is a solid solution of copper and zinc. Alternatively, they can be heterogeneous, as with granite, which contains discrete regions of different minerals (feldspar, mica, and quartz.)

Alloys A homogeneous mixture of two metals

Many alloys are solid solutions. One example is stainless steel, which is mostly iron but also contains carbon, chromium, and other elements. (See also Chemical Connections 2E.)

We normally do not use the terms "solute" and "solvent" when talking about solutions of gases in gases or solids in solids.

Polar compounds dissolve in polar compounds because the positive end of the dipole of one molecule attracts the negative end of the dipole of the other molecule.

Figure 7.2 The solubilities of some solids in water as a function of temperature. The solubility of glycine increases rapidly, that of NaCl barely increases, and that of Li_2SO_4 decreases with increasing temperature.

Charles D. Winters

Figure 7.3 When a supersaturated aqueous solution of sodium acetate ($CH_3COO^-Na^+$) is disturbed, the excess salt crystallizes rapidly.

Henry's law The solubility of a gas in a liquid is directly proportional to the pressure

When we say "like," we mostly mean similar in terms of polarity. In other words, polar compounds dissolve in polar solvents, and nonpolar compounds dissolve in nonpolar solvents. For example, the liquids benzene (C_6H_6) and carbon tetrachloride (CCl_4) are nonpolar compounds. They dissolve in each other, and other nonpolar materials, such as gasoline, dissolve in them. In contrast, ionic compounds such as sodium chloride (NaCl) and polar compounds such as table sugar ($C_{12}H_{22}O_{11}$) are insoluble in these solvents.

The most important polar solvent is water. We have already seen that most ionic compounds are soluble in water, as are small covalent compounds that can form hydrogen bonds with water. It is worth noting that even polar molecules are usually insoluble in water if they cannot either react with water or form hydrogen bonds with water molecules. Water as a solvent is discussed in Section 7.6.

B. Temperature

For most solids and liquids that dissolve in liquids, the rule is that solubility increases with increasing temperature. Sometimes the increase is great; at other times it is moderate. For a few substances, solubility even decreases with increasing temperature (Figure 7.2).

For example, the solubility of glycine, $H_2N\text{—}CH_2\text{—}COOH$, a white, crystalline solid and a polar building block of proteins, is 52.8 g in 100 g of water at 80°C but only 33.2 g at 30°C. If, for instance, we prepare a saturated solution of glycine in 100 g of water at 80°C, it will hold 52.8 g of glycine. If we then allow it to cool to 30°C where the solubility is 33.2 g, we might expect the excess glycine, 19.6 g, to precipitate from solution as crystals. It often does, but on many occasions it does not. The latter case is an example of a **supersaturated solution.** Even though the solution contains more glycine than the water can normally hold at 30°C, the excess glycine stays in solution because the molecules need a seed—a surface on which to begin crystallizing. If no such surface is available, no precipitate will form.

Supersaturated solutions are not indefinitely stable, however. If we shake or stir the solution, we may find that the excess solid precipitates at once (Figure 7.3). Another way to crystallize the excess solute is to add a crystal of the solute, a process called **seeding.** The seed crystal provides the surface onto which the solute molecules can converge.

For gases, solubility in liquids almost always decreases with increasing temperature. The effect of temperature on the solubility of gases in water can have important consequences for fish, for example. Oxygen is only slightly soluble in water, but fish need that oxygen to live. When the temperature of a body of water increases, perhaps because of the output from a nuclear power plant, the solubility of the oxygen decreases and may become so low that fish die. This situation is called thermal pollution.

C. Pressure

Pressure has little effect on the solubility of liquids or solids. For gases, however, **Henry's law** applies: The higher the pressure, the greater the solubility of a gas in a liquid. This concept is the basis of the hyperbaric medicine discussed in Chemical Connections 6C. When the pressure increases, more O_2 dissolves in the blood plasma and reaches tissues at higher than normal pressures (2 to 3 atm).

Henry's law also explains why a bottle of beer or a carbonated beverage foams when it is opened. The bottle is sealed under greater than 1 atm of pressure. When opened at 1 atm, the solubility of CO_2 in the liquid decreases. The excess CO_2 is released, forming bubbles, and the gas pushes out some of the liquid.

CHEMICAL CONNECTIONS 7B

The Bends

Deep-sea divers encounter high pressures (see Problem 6.101). For them to breathe properly under such conditions, oxygen must be supplied under pressure. At one time, this goal was achieved with compressed air. As pressure increases, the solubility of the gases in the blood increases. This is especially true for nitrogen, which constitutes almost 80% of our air.

When divers come up and the pressure on their bodies decreases, the solubility of nitrogen in their blood decreases as well. As a consequence, the previously dissolved nitrogen in blood and tissues starts to form small bubbles, especially in the veins. The formation of gas bubbles, called the **bends,** can hamper blood circulation. If this condition is allowed to develop uncontrolled, the resulting pulmonary embolism can prove fatal.

If the divers' ascent is gradual, the dissolved gases are removed by regular exhalation and diffusion through the skin. Divers use decompression chambers, where the high pressure is gradually reduced to normal pressure.

If decompression disease develops after a dive, patients are put into a hyperbaric chamber (see Chemical Connections 6C), where they breathe pure oxygen at 2.8 atm pressure. In the standard form of treatment, the pressure is reduced to 1 atm over a period of 6 hours.

Nitrogen also has a narcotic effect on divers when they breathe compressed air below 40 m. This "rapture of the deep" is similar to alcohol-induced intoxication.

Because of the problem caused by nitrogen, divers' tanks often hold a helium–oxygen mixture instead of air. The solubility of helium in blood is affected less by pressure than is the solubility of nitrogen.

Brian Parker/Tom Stack and Associates

Ascending too rapidly will cause dissolved nitrogen bubbles to be released and form bubbles in the blood.

Sudden decompression and ensuing bends are important not only in deep-sea diving but also in high-altitude flight, especially orbital flight.

7.5 | What Are the Most Common Units for Concentration?

GOB
Chemistry⚛Now™
Click *Chemistry Interactive* to see how
Henry's Law explains why soft drinks go flat

We can express the amount of a solute dissolved in a given quantity of solvent—that is, the **concentration** of the solution—in a number of ways. Some concentration units are better suited than others for some purposes. Sometimes qualitative terms are good enough. For example, we may say that a solution is dilute or concentrated. These terms give us little specific information about the concentration, but we know that a concentrated solution contains more solute than a dilute solution does.

For most purposes, however, we need quantitative concentrations. For example, a nurse must know how much glucose to give to a patient. Many methods of expressing concentration exist, but in this chapter we deal with just the three most important: percent concentration, molarity, and parts per million (ppm).

A. Percent Concentration

Chemists represent **percent concentration** in three ways. The most common is weight of solute per volume of solution (w/v):

$$\text{Weight/volume (w/v)}\% = \frac{\text{weight solute}}{\text{volume of solution}} \times 100$$

Percent concentration (% w/v)
The number of grams of solute in 100 mL of solution

If we dissolve 10 g of sucrose (table sugar) in enough water so that the total volume is 100 mL, the concentration is 10% w/v. Note that here we need to know the total volume of the solution, not the volume of the solvent.

EXAMPLE 7.1

The label on a bottle of vinegar says it contains 5.0% acetic acid, CH_3COOH. The bottle contains 240 mL of vinegar. How many grams of acetic acid are in the bottle?

Solution

We are given the volume of the solution and its weight/volume concentration. To find out how many grams of CH_3COOH are in this solution, we use the conversion factor 5.0 g of acetic acid in 100 mL of solution:

$$240 \text{ mL solution} \times \frac{5.0 \text{ g } CH_3COOH}{100 \text{ mL solution}} = 12 \text{ g } CH_3COOH$$

Problem 7.1

How would we make 250 mL of a 4.4% w/v KBr solution in water? Assume that a 250-mL volumetric flask is available.

EXAMPLE 7.2

If 6.0 g of NaCl is dissolved in enough water to make 300 mL of solution, what is the w/v percent of NaCl?

Solution

To calculate the w/v percent, we divide the weight of the solute by the volume of the solution and multiply by 100:

$$\frac{6.0 \text{ g NaCl}}{300 \text{ mL solution}} \times 100 = 2.0\% \text{ w/v}$$

Problem 7.2

If 7.7 g of lithium iodide, LiI, is dissolved in enough water to make 400 mL of solution, what is the w/v percent of LiI?

A second way to represent percent concentration is weight of solute per weight of solution (w/w):

$$\text{Weight/weight (w/w)}\% = \frac{\text{weight solute}}{\text{weight of solution}} \times 100$$

Calculations of w/w are essentially the same as w/v calculations, except that we use the weight of the solution instead of its volume. A volumetric flask is not used for these solutions. (Why not?)

Finally, we can represent percent concentration as volume of solute per volume of solution (v/v):

$$\text{Volume/volume (v/v)\%} = \frac{\text{volume solute}}{\text{volume of solution}} \times 100$$

The unit v/v is used only for solutions of liquids in liquids—most notably, alcoholic beverages. For example, 40% v/v ethanol in water means that 40 mL of ethanol has been added to enough water to make 100 mL of solution. This solution might also be called 80 proof, where proof of an alcoholic beverage is twice the v/v concentration.

A 40% v/v solution of ethanol in water is 80 proof. Proof is twice the percent concentration (v/v) of ethanol in water.

B. Molarity

For many purposes, it is easiest to express concentration by using the weight or volume percentage methods just discussed. When we want to focus on the number of molecules present, however, we need another way. For example, a 5% solution of glucose in water does not contain the same number of solute molecules as a 5% solution of ethanol in water. That is why chemists often use molarity. **Molarity (M),** is defined as the number of moles of solute dissolved in 1 L of solution. The units of molarity are moles per liter.

$$\text{Molarity } (M) = \frac{\text{moles solute } (n)}{\text{volume of solution (L)}}$$

Thus, in the same volume of solution, a 0.2-molar solution of glucose, $C_6H_{12}O_6$, in water contains the same number of molecules of solute as a 0.2-molar solution of ethanol, C_2H_6O, in water. In fact, this relationship holds true for equal volumes of any solution, as long as the molarities are the same.

We can prepare a solution of a given molarity in essentially the same way that we prepare a solution of given w/v concentration, except that we use moles instead of grams in our calculations. We can always find out how many moles of solute are in any volume of a solution of known molarity by using the following formula:

$$\text{Molarity} \times \text{volume in liters} = \text{number of moles}$$

$$\frac{\text{moles}}{\text{liters}} \times \text{liters} = \text{moles}$$

The solution is then prepared as shown in Figure 7.4.

EXAMPLE 7.3

How do we prepare 2.0 L of a 0.15 M aqueous solution of sodium hydroxide, NaOH?

Solution

We are given solid NaOH and want 2.0 L of a 0.15 M solution. First, we find out how many moles of NaOH will be in this solution:

$$\frac{0.15 \text{ mol NaOH}}{1 \text{ L}} \times 2.0 \text{ L} = 0.30 \text{ mol NaOH}$$

GOB
Chemistry Now™

Click *Coached Problems* to try a problem calculating the **Amount of Solute Needed to Prepare a Solution of Known Molarity**

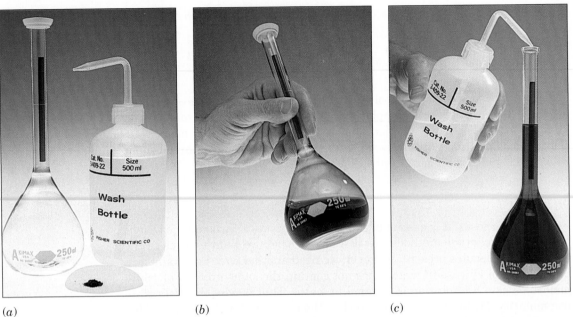

(a)   (b)   (c)

Charles D. Winters

Figure 7.4 Preparation of 250 mL of 0.0100 M potassium permanganate solution. (a) A 250-mL volumetric flask. A 0.0100 M solution of $KMnO_4$ is made by adding enough water to 0.395 g (2.5×10^{-3} mol of $KMnO_4$, FW 158.03 g/mol) to make 0.250 L of solution. (b) 0.395 g of $KMnO_4$ is placed in a volumetric flask and dissolved in a small amount of water. (c) After dissolving is complete, sufficient water is added to fill the flask to the calibration mark.

To convert 0.30 mol of NaOH to grams of NaOH, we multiply by the molar mass of NaOH, which is 40.0 g/mol:

$$0.30 \ \cancel{\text{mol NaOH}} \times \frac{40.0 \text{ g NaOH}}{1 \ \cancel{\text{mol NaOH}}} = 12 \text{ g NaOH}$$

To prepare this solution, we place 12 g of NaOH in a 2-L volumetric flask, add some water, swirl until the solid dissolves, and then fill the flask with water to the 2-L mark.

Problem 7.3

How would we prepare 2.0 L of a 1.06 M aqueous solution of KCl?

Blood (left to right): serum, coagulated, and whole blood.

© Stan Elems/Visuals Unlimited

EXAMPLE 7.4

We dissolve 18.0 g of Li_2O (molar mass = 29.9 g/mol) in sufficient water to make 500 mL of solution. Calculate the molarity of the solution.

Solution

We are given 18.0 g Li_2O in 500 mL of water and want the molarity of this solution. We first calculate the number of moles of Li_2O in this solution. For this calculation, we use two conversion factors: molar mass of Li_2O = 29.9 g and 1000 mL = 1 L.

$$\frac{18.0 \ \text{g} \ \cancel{Li_2O}}{500 \ \cancel{\text{mL}}} \times \frac{1 \text{ mol } Li_2O}{29.9 \ \text{g} \ \cancel{Li_2O}} \times \frac{1000 \ \cancel{\text{mL}}}{1 \text{ L}} = 0.722 \ M$$

Problem 7.4

If we dissolve 0.440 g of KSCN in enough water to make 340 mL of solution, what is the molarity of the solution?

EXAMPLE 7.5

The concentration of sodium chloride in blood serum is approximately 0.14 M. What volume of blood serum contains 2.0 g of NaCl?

Solution

We are given the concentration in moles per liter. We want the volume of blood that contains 2.0 g NaCl. To find the volume of blood, we use two conversion factors: the molar mass of NaCl is 58.44 g and the concentration of NaCl in blood is 0.14 M.

$$2.0 \text{ g NaCl} \times \frac{1 \text{ mol NaCl}}{58.44 \text{ g NaCl}} \times \frac{L}{0.14 \text{ mol NaCl}} = 0.24 \text{ L} = 240 \text{ mL}$$

Blood serum is the liquid part of the blood that remains after removal of the cellular particulates and fibrinogen.

Problem 7.5

If a 0.300 M glucose solution is available for intravenous infusion, how many milliliters are needed to deliver 10.0 g of glucose?

EXAMPLE 7.6

How many grams of HCl are in 225 mL of 6.00 M HCl?

Solution

We are given 225 mL of 6.00 M HCl and asked to find grams of HCl. We use two conversion factors: the molar mass of HCl = 36.5 g and 1000 mL = 1 L.

$$225 \text{ mL} \times \frac{1 \text{ L}}{1000 \text{ mL}} \times \frac{6.00 \text{ mol HCl}}{1 \text{ L}} \times \frac{36.5 \text{ g HCl}}{1 \text{ mol HCl}} = 49.3 \text{ g HCl}$$

Problem 7.6

A certain wine contains 0.010 M NaHSO$_3$ (sodium bisulfite) as a preservative. How many grams of sodium bisulfite must be added to a 100-gallon barrel of wine to reach this concentration? Assume no change in volume of wine upon addition of the sodium bisulfite.

C. Dilution

We frequently prepare solutions by diluting concentrated solutions rather than by weighing out pure solute (Figure 7.5). Because we add only solvent during dilution, the number of moles of solute remains unchanged. Before we dilute, the equation that applies is

$$M_1 V_1 = \text{moles}$$

(a)   *(b)*

Figure 7.9 *(a)* The needle-shaped sodium urate monohydrate crystals that cause gout. *(b)* The pain of gout as depicted by a cartoonist.

anode. The movement of ions constitutes an electric current. The migration of ions completes the circuit initiated by the battery and can cause an electric bulb to light up (see also Chemical Connections 5B).

Substances that conduct an electric current when dissolved in water or when in the molten state are called **electrolytes.** Sodium chloride, for example, is an electrolyte. Hydrated Na^+ ions carry positive charges, and hydrated Cl^- ions carry negative charges; as a result, the bulb in Figure 7.10 lights brightly if these ions are present. Substances that do not conduct electricity are called **nonelectrolytes.** Distilled water, for example, is a nonelectrolyte. The lightbulb shown in Figure 7.10 does not light up if only distilled water is placed in the beaker. However, with tap water in the beaker, the bulb lights dimly. Tap water contains enough ions to carry elec-

Figure 7.10 Conductance by an electrolyte.

Anode
(positive electrode)

Cathode
(negative electrode)

Battery

CHEMICAL CONNECTIONS 7C

Hydrates and Air Pollution: The Decay of Buildings and Monuments

Many buildings and monuments in urban areas throughout the world are decaying, ruined by air pollution. The main culprit in this process is acid rain, an end product of air pollution. The stones most commonly used for buildings and monuments are limestone and marble, both of which are largely calcium carbonate. In the absence of polluted air, these stones can last for thousands of years. Thus many statues and buildings from ancient times (Babylonian, Egyptian, Greek, and others) survived until recently with little change. Indeed, they remain intact in many rural areas.

In urban areas, however, the air is polluted with SO_2 and SO_3, which come mostly from the combustion of coal and petroleum products containing small amounts of sulfur compounds as impurities (see Chemical Connections 7A). They react with the calcium carbonate at the surface of the stones to form calcium sulfate. When calcium sulfate interacts with water in the rain, it forms the dihydrate gypsum. The problem is that gypsum has a larger volume than the original marble or limestone, so it causes the surface of the stone to expand. This activity, in turn, results in flaking. Eventually, statues such as those in the Parthenon (in Athens, Greece) become noseless and later faceless.

Acid rain damage to stonework on the walls of York Minster.

Martin Bond/Science Photo Library/Photo Researchers, Inc.

---

tricity, but their concentration is so low that the solution conducts only a small amount of electricity.

As we see, electric conductance depends on the concentration of ions. The higher the ion concentration, the greater the electric conductance of the solution. Nevertheless, differences in electrolytes exist. If we take a 0.1 M aqueous solution of NaCl and compare it with a 0.1 M acetic acid (CH_3COOH) solution, we find that the NaCl solution lights a bulb brightly, but the acetic acid solution lights it only dimly. We might have expected the two solutions to behave similarly. They have the same concentration, 0.1 M, and each compound provides two ions, a cation and an anion (Na^+ and Cl^-; H^+ and CH_3COO^-). The reason they behave differently is that, whereas NaCl dissociates completely to two ions (each hydrated and each moving independently), in the case of CH_3COOH only a few molecules are dissociated into ions. Most of the acetic acid molecules do not dissociate, and undissociated molecules do not conduct electricity. Compounds that dissociate completely are called **strong electrolytes,** and those that dissociate into ions only partially are called **weak electrolytes.**

Electrolytes are important components of the body because they help to maintain the acid–base balance and the water balance. The most important cations in tissues of the human body are Na^+, K^+, Ca^{2+}, and Mg^{2+}. The most abundant anions in the body are bicarbonate, HCO_3^-; chloride, Cl^-; hydrogen phosphate, HPO_4^{2-}; and dihydrogen phosphate, $H_2PO_4^-$.

Sports drinks help to maintain the body's electrolytic balance.

George Semple

D. How Water Dissolves Covalent Compounds

Water is a good solvent not only for ionic compounds but also for many covalent compounds. In a few cases, the covalent compounds dissolve because they react with water. An important example is the covalent compound HCl.

HCl is a gas (with a penetrating, choking odor) that attacks the mucous membranes of the eyes, nose, and throat. When dissolved in water, HCl molecules react with water to give ions:

$$\text{HCl(g)} + \text{H}_2\text{O}(\ell) \longrightarrow \text{Cl}^-(\text{aq}) + \text{H}_3\text{O}^+(\text{aq})$$

Hydrogen chloride   Hydronium ion

H⁺ does not exist in aqueous solution; it combines with a water molecule and forms a hydronium ion, H_3O^+.

Another example is the gas sulfur trioxide, which reacts as follows:

$$\text{SO}_3(\text{g}) + 2\text{H}_2\text{O}(\ell) \longrightarrow \text{H}_3\text{O}^+(\text{aq}) + \text{HSO}_4{}^-(\text{aq})$$

Sulfur trioxide   Hydronium ion

Because HCl and SO_3 are completely converted to ions in dilute aqueous solution, these solutions are ionic solutions and behave just as other electrolytes do (they conduct a current). Nevertheless, HCl and SO_3 are themselves covalent compounds, unlike salts such as NaCl.

Most covalent compounds that dissolve in water do not, in fact, react with water. They dissolve because the water molecules surround the entire covalent molecule and solvate it. For example, when methanol, CH_3OH, dissolves in water, the methanol molecules are solvated by the water molecules (Figure 7.11).

There is a simple way to predict which covalent compounds will dissolve in water and which will not. Covalent compounds will dissolve in water if they can form hydrogen bonds with water, provided that the solute molecules are fairly small. Hydrogen bonding is possible between two molecules if one of them contains an O or N atom and the other contains O—H or N—H bond. Every water molecule contains an O atom and O—H bonds. Therefore, water can form hydrogen bonds with any molecule that contains an O or N atom or an O—H or N—H bond. If these molecules are small enough, they will be soluble in water. How small? In general, they should have no more than three C atoms for each O or N atom.

For example, acetic acid, CH_3COOH, is soluble in water, but benzoic acid, C_6H_5COOH, is not. Similarly, ethanol, C_2H_5OH, is soluble in water, but dipropyl ether, $C_6H_{14}O$, is not. Table sugar, $C_{12}H_{22}O_{11}$ (Section 20.4A), is very soluble in water. Although it contains a large number of carbon atoms, it has so many oxygen atoms that it forms many hydrogen bonds with water molecules; thus this molecule is very well solvated.

Covalent molecules that do not contain O or N atoms are almost always insoluble in water. For example, methanol, CH_3OH, is infinitely soluble in water, but chloromethane, CH_3Cl, is not. The exception is the rare case where a covalent compound reacts with water—for instance, HCl.

Figure 7.11 Solvation of a polar covalent compound by water. The dotted lines represent hydrogen bonds.

E. Water in the Body

Water is important in the body not only because it dissolves ionic substances as well as some covalent compounds, but also because it hydrates all polar molecules in the body. In this way, it serves as a vehicle to carry most of the organic compounds, nutrients, and fuels used by the body as well as waste material. Blood and urine are two examples of aqueous body fluids.

In addition, the hydration of macromolecules such as proteins, nucleic acids, and polysaccharides allows the proper motions within these molecules, which are necessary for such functions as enzyme activity (described in Chapter 23).

7.7 | What Are Colloids?

Up to now we have discussed only solutions. The maximum diameter of the solute particles in a true solution is about 1 nm. If the diameter of the solute particles exceeds this size, then we no longer have a true solution—we have a **colloid.** In a colloid (also called a colloidal dispersion or colloidal system), the diameter of the solute particles ranges from about 1 to 1000 nm. The term *colloid* has acquired a new name recently. In Chapter 6, we encountered the term *nanotube.* The "nano" part refers to dimensions in the nanometer ($1 \text{ nm} = 10^{-9}$ m) range, which is the size range of colloids. Thus, when we encounter terms such as "nanoparticle" or "nanoscience," they are equivalent to "colloidal particle" or "colloid science," although the former terms refer mostly to particles with well-defined geometrical shape (such as tubes), while the latter terms are more general.

Particles of this size usually have a very large surface area, which accounts for the two basic characteristics of colloidal systems:

1. They scatter light and therefore appear turbid, cloudy, or milky.
2. Although colloidal particles are large, they form stable dispersions— they do not form separate phases that settle out. As with true solutions, colloids can exist in a variety of phases (Table 7.2).

All colloids exhibit the following characteristic effect. When we shine light through a colloid and look at the system from a 90° angle, we see the pathway of the light without seeing the colloidal particles themselves (they are too small to see). Rather, we see flashes of the light scattered by the particles in the colloid (Figure 7.12). The **Tyndall effect** is due to light scattering of colloidal particles. Smoke, serum, and fog, to name a few examples, all exhibit the Tyndall effect. We are all familiar with the sunbeams that can be seen when sunlight passes through dusty air. This, too, is an example of the Tyndall effect. Again, we do not see the particles in dusty air, but only the light scattered by them.

Colloidal systems are stable. Mayonnaise, for example, stays emulsified and does not separate into oil and water. When the size of colloidal particles is larger than about 1000 nm, however, the system is unstable and separates into phases. Such systems are called **suspensions.**

For example, if we take a lump of soil and disperse it in water, we get a muddy suspension. The soil particles are anywhere from 10^3 to 10^9 nm in diameter. The muddy mixture scatters light and, therefore, appears turbid. It is not a stable system, however. If left alone, the soil particles soon settle to the bottom, with clear water found above the sediment. Therefore, soil in water is a suspension, not a colloidal system.

Figure 7.12 The Tyndall effect. A narrow beam of light from a laser is passed through a colloidal mixture (left). Then through a NaCl solution and finally through a colloidal mixture of gelatin and water (right). This illustrates the light-scattering ability of the colloid-sized particles.

Tyndall effect Light passing through and scattered by a colloid viewed at a right angle

Ordinary household dust particles, magnified 2200 times.

Table 7.2	Types of Colloidal Systems
Type	**Example**
Gas in gas	None
Gas in liquid	Whipped cream
Gas in solid	Marshmallows
Liquid in gas	Clouds, fog
Liquid in liquid	Milk, mayonnaise
Liquid in solid	Cheese, butter
Solid in gas	Smoke
Solid in liquid	Jelly
Solid in solid	Dried paint

Table 7.3	Properties of Three Types of Mixtures		
Property	Solutions	Colloids	Suspensions
Particle size (nm)	0.1–1.0	1–1000	>1000
Filterable with ordinary paper	No	No	Yes
Homogeneous	Yes	Borderline	No
Settles on standing	No	No	Yes
Behavior to light	Transparent	Tyndall effect	Translucent or opaque

Table 7.3 summarizes the properties of three types of mixtures: solutions, colloids, and suspensions.

What makes a colloidal dispersion stable? To answer this question, we must first realize that the colloidal particles are in constant motion. Just look at the dust particles dancing in a ray of sunlight that enters your room. Actually, we do not see the dust particles themselves; they are too small. Rather, we see flashes of scattered light. The motion of the dust particles dispersed in air is a random, chaotic motion. This motion of any colloidal particle suspended in a solvent is called **Brownian motion** (Figure 7.13).

The constant buffeting and collisions by solvent molecules cause the colloidal particles to move in random Brownian motion. (In the case of the dust particles, the solvent is air.) This ongoing motion creates favorable conditions for collisions between particles. When such large particles collide, they stick together, combine to give larger particles, and finally settle out of the solution. That is what happens in a suspension.

So why do colloidal particles remain in solution despite all the collisions due to their Brownian motion? Two reasons explain this phenomenon:

1. Most colloidal particles carry a large solvation layer. If the solvent is water, as in the case of protein molecules in the blood, the colloidal particles are surrounded by a large number of water molecules, which move together with the colloidal particles and cushion them. When two colloidal particles collide as a result of Brownian motion, they do not actually touch each other; instead, only the solvent layers collide. As a consequence, the particles do not stick together and thus stay in solution.

2. The large surface area of colloidal particles acquires charges from the solution. All colloids in a particular solution acquire the same kind of charge—for example, a negative charge. This development leaves a net positive charge in the solvent. When a charged colloidal particle encounters another colloidal particle, the two repel each other because of their like charges.

Thus the combined effects of the solvation layer and the surface charge keep colloidal particles in a stable dispersion. By taking advantage of these effects, chemists can either increase or decrease the stability of a colloidal system. If we want to get rid of a colloidal dispersion, we can remove the solvation layer, the surface charge, or both. For example, proteins in the blood form a colloidal dispersion. If we want to isolate a protein from blood, we may want to precipitate it. We can accomplish this task in two ways: by removing the hydration layer or by removing the surface charges. If we add solvents such as ethanol or acetone, each of which has great affinity for water, the water will be removed from the solvation layer of the protein, and then the unprotected protein molecules will stick together when they collide and form sediment. Similarly, by adding electrolytes such as NaCl to the

Figure 7.13 Brownian motion.

Freshly made wines are often cloudy because of colloidal particles (left). Removing the particles clarifies the wine (right).

NALCO Chemical Company

Nanotubes, nanowires, and nanopores in composite coatings have unusual electronic and optical properties because of their enormous surface areas. For example, titanium oxide particles smaller than 20 nm are used to coat surfaces of plastics, glass, and other materials. These thin coatings have self-cleaning, antifogging, antifouling, and sterilizing properties.

CHEMICAL CONNECTIONS 7D

Emulsions and Emulsifying Agents

Oil and water do not mix. Even when we stir them vigorously and the oil droplets become dispersed in the water, the two phases separate as soon as we stop stirring. There are, however, a number of stable colloidal systems made of oil and water, known as **emulsions**. For example, the oil droplets in milk are dispersed in an aqueous solution. This is possible because milk contains a protec-

tive colloid—the milk protein called casein. Casein molecules surround the oil droplets and, because they are polar and carry a charge, they protect and stabilize the oil droplets. Casein is thus an emulsifying agent.

Another emulsifying agent is egg yolk. This ingredient in mayonnaise coats the oil droplets and prevents them from separating.

---

solution, we can remove the charges from the surface of the proteins (by a mechanism too complicated to discuss here). Without their protective charges, two protein molecules will no longer repel each other. Instead, when they collide, they will stick together and precipitate from the solution.

Emulsion A system, such as fat in milk, consisting of a liquid with or without an emulsifying agent in an immiscible liquid, usually as droplets of larger than colloidal size

7.8 | What Is a Colligative Property?

A **colligative property** is any property of a solution that depends only on the number of solute particles dissolved in the solvent and not on the nature of the solute particles. Several colligative properties exist, including freezing-point depression and osmotic pressure. Of these two, osmotic pressure is the most important in biological systems.

Colligative property A property of a solution that depends only on the number of solute particles and not on the chemical identity of the solute

A. Freezing-Point Depression

One mole of any particle dissolved in 1000 g of water lowers the freezing point of the water by 1.86°C. The nature of the solute does not matter, only the number of particles.

Freezing-point depression The decrease in the freezing point of a liquid caused by adding a solute

$$\Delta T = \frac{1.86°C}{mol} \times \text{mol of particles}$$

This principle is used in a number of practical ways. In winter, we use salts (sodium chloride and calcium chloride) to melt snow and ice on our streets. The salts dissolve in the melting snow and ice, which lowers the freezing point of the water. Another application is the use of antifreeze in automobile radiators. Because water expands upon freezing (see Chemical Connections 6F), the ice formed in a car's cooling system when the outside temperature falls below 0°C can crack the engine block. The addition of antifreeze prevents this problem, because it makes the water freeze at a much lower temperature. The most common antifreeze is ethylene glycol, $C_2H_6O_2$.

Note that in preparing a solution for this purpose we do not use molarity. That is, we do not need to measure the total volume of the solution.

EXAMPLE 7.10

If we add 275 g ethylene glycol, $C_2H_6O_2$, a nondissociating molecular compound, per 1000 g of water in a car radiator, what will the freezing point of the solution be?

Solution

We are given 275 g of ethylene glycol (molar mass, 62.0 g) per 1000 g water and want the freezing point of the solution. We first need to calculate the

Some ions are small but still do not go through the membrane because they are solvated by a shell of water molecules (see Figure 7.7).

convenience in calculation, we define a new term, **osmolarity** (in osmol), as molarity (M) multiplied by the number of particles (i) produced by each formula unit of solute.

$$\text{Osmolarity} = M \times i$$

Charles D. Winters

The effect of osmosis on a carrot. The carrot on the left is in a solution that contains a high concentration of NaCl. Water flowing out of the carrot has caused it to shrink. The carrot on the right, in pure water, is slightly swollen.

EXAMPLE 7.12

A 0.89% w/v NaCl solution is referred to as a physiological or isotonic saline solution because it has the same concentration of salts as normal human blood, although blood contains several salts and saline solution has only NaCl. What is the osmolarity of this solution?

Solution
We are given an 0.89% solution—that is, one that contains 0.89 g NaCl per 100 mL of solution. Because osmolarity is based on grams of solute per 1000 grams of solution, we calculate that this solution contains 8.9 g of NaCl per 1000 g of solution. Given this concentration, we can then calculate the molarity of the solution.

$$\frac{0.89 \text{ g NaCl}}{100 \text{ mL}} \times \frac{1000 \text{ mL}}{1 \text{ L}} \times \frac{1 \text{ mol NaCl}}{58.5 \text{ g NaCl}} = \frac{0.15 \text{ mol NaCl}}{1 \text{ L}} = 0.15 \text{ } M$$

Each NaCl dissociates into two particles, Na^+ and Cl^-, in aqueous solutions. Therefore, the osmolarity is two times the molarity.

$$\text{Osmolarity} = 0.15 \times 2 = 0.30 \text{ osmol}$$

Problem 7.12
What is the osmolarity of a 3.3% w/v Na_3PO_4 solution?

As noted earlier, osmotic pressure is a colligative property. The osmotic pressure generated by a solution across a semipermeable membrane—the difference between the heights of the two columns in Figure 7.14(b)—

CHEMICAL CONNECTIONS 7E

Reverse Osmosis and Desalinization

In osmosis, the solvent flows spontaneously from the dilute solution compartment into the concentrated solution compartment. In reverse osmosis, the opposite happens. When we apply pressures greater than the osmotic pressure to the more concentrated solution, solvent flows from the more dilute solution to the more concentrated solution by a process we call **reverse osmosis.**

Reverse osmosis is used to make drinkable water from sea water or brackish water. In large plants in the Persian Gulf countries, for example, more than 100 atm pressure is applied to sea water containing 35,000 ppm salt. The water that passes through the semipermeable membrane under this pressure contains only 400 ppm salt—well within the limits set by the World Health Organization for drinkable water.

Courtesy of Recovery Engineering, Inc.

An emergency hand-operated water desalinator that works by reverse osmosis. It can produce 4.5 L of pure water per hour from sea water, which can save someone adrift at sea.

Figure 7.15 Red blood cells in solutions of different tonicity.

(*a*) Isotonic solution   (*b*) Hypertonic solution   (*c*) Hypotonic solution

depends on the osmolarity of the solution. If the osmolarity increases by a factor of 2, the osmotic pressure will also increase by a factor of 2. Osmotic pressure is very important in biological organisms because cell membranes are semipermeable. Therefore, biological fluids must have the proper osmolarity. For example, red blood cells in the body are suspended in a medium called plasma, which must have the same osmolarity as the red blood cells. Two solutions with the same osmolarity are called **isotonic,** so plasma is said to be isotonic with red blood cells. As a consequence, no osmotic pressure is generated across the cell membrane.

What would happen if we suspended red blood cells in distilled water instead of in plasma? Inside the red blood cells, the osmolarity is approximately the same as in a physiological saline solution—0.30 osmol. Distilled water has zero osmolarity. As a consequence, water flows into the red blood cells. The volume of the cells increases, and the cells swell, as shown in Figure 7.15(c). The membrane cannot resist the osmotic pressure, and the red blood cells eventually burst, spilling their contents into the water. We call this process **hemolysis.**

Solutions in which the osmolarity (and hence osmotic pressure) is lower than that of suspended cells are called **hypotonic solutions.** Obviously, it is very important that we always use isotonic solutions and never hypotonic solutions in intravenous feeding and blood transfusion. Hypotonic solutions would simply kill the red blood cells by hemolysis.

Equally important, we should not use **hypertonic solutions.** A hypertonic solution has a greater osmolarity (and greater osmotic pressure) than the red blood cells. If red blood cells are placed in a hypertonic solution—for example, 0.5 osmol glucose solution—water flows from the cells into the glucose solution through the semipermeable cell membrane. This process, called **crenation,** shrivels the cells, as shown in Figure 7.15(b).

As already mentioned, 0.89% NaCl (physiological saline) is isotonic with red blood cells and is used in intravenous injections.

An isotonic saline solution.

A 5.5% glucose solution is also isotonic and is used in intravenous feeding.

EXAMPLE 7.13

Is a 0.50% w/v solution of KCl (a) hypertonic, (b) hypotonic, or (c) isotonic compared to red blood cells?

Solution

The 0.50% w/v solution of KCl contains 5.0 g KCl in 1.0 L of solution:

$$\frac{5.0 \text{ g KCl}}{1.0 \text{ L}} \times \frac{1.0 \text{ mol KCl}}{74.6 \text{ g KCl}} = \frac{0.067 \text{ mol KCl}}{1.0 \text{ L}} = 0.067 \, M$$

A portable dialysis unit.

Dialysis A process in which a solution containing particles of different sizes is placed in a bag made of a semipermeable membrane. The bag is placed into a solvent or solution containing only small molecules. The solution in the bag reaches equilibrium with the solvent outside, allowing the small molecules to diffuse across the membrane but retaining the large molecules.

Because each KCl yields two particles, the osmolarity is $0.067 \times 2 = 0.13$ osmol; this is smaller than the osmolarity of the red blood cells, which is 0.30 osmol. Therefore, the KCl solution is hypotonic.

Problem 7.13

Which solution is isotonic compared to red blood cells: (a) 0.1 M Na$_2$SO$_4$, (b) 1.0 M Na$_2$SO$_4$, or (c) 0.2 M Na$_2$SO$_4$?

C. Dialysis

An osmotic semipermeable membrane allows only solvent and not solute molecules to pass. If, however, the openings in the membrane are somewhat larger, then small solute molecules can also get through, but large solute molecules, such as macromolecular and colloidal particles, cannot. This process is called **dialysis.**

For example, ribonucleic acids are important biological molecules that we will study in Chapter 25. When biochemists prepare ribonucleic acid solutions, they must remove small molecules, such as NaCl, from the solution to obtain a pure nucleic acid preparation. To do so, they place the nucleic acid solution in a dialysis bag (made of cellophane) of sufficient pore size to allow all the small molecules to diffuse and retain only the large nucleic acid

---

CHEMICAL CONNECTIONS 7F

Hemodialysis

The kidneys' main function is to remove toxic waste products from the blood. When the kidneys are not functioning properly, these waste products may threaten life. **Hemodialysis** is a process that performs the same filtration function.

In hemodialysis, the patient's blood circulates through a long tube of cellophane membrane suspended in an isotonic solution and then returns to the patient's vein. The cellophane membrane retains the large molecules (for example, proteins) but allows the small ones, including the toxic wastes, to pass through. In this way, dialysis removes wastes from the blood.

If the cellophane tube were suspended in distilled water, other small molecules and ions, such as Na$^+$ Cl$^-$ and glucose, would also be removed from the blood. That is something we don't want to happen. The isotonic solution used in hemodialysis includes 0.6% NaCl, 0.04% KCl, 0.2% NaHCO$_3$, and 0.72% glucose (all w/v). It ensures that no glucose or Na$^+$ is lost from the blood.

A patient usually remains on an artificial kidney machine for four to seven hours. During this time, the isotonic bath is changed every two hours. Kidney machines allow people with kidney failure to lead a fairly normal life, although they must take these hemodialysis treatments regularly.

A schematic diagram of the hollow-fiber (or capillary) dialyzer, the most commonly used artificial kidney. The blood flows through small tubes constructed of a semipermeable membrane; these tubes are bathed in the dialyzing solution.

molecules. If the dialysis bag is suspended in flowing distilled water, all of the NaCl and small molecules will leave the bag. After a certain amount of time, the bag will contain only the pure nucleic acids dissolved in water.

Our kidneys work in much the same way. The millions of nephrons, or kidney cells, have very large surface areas in which the capillaries of the blood vessels come in contact with the nephrons. The kidneys serve as a gigantic filtering machine. The waste products of the blood dialyse out through semipermeable membranes in the glomeruli and enter collecting tubes that carry the urine to the ureter. Meanwhile, large protein molecules and cells are retained in the blood.

The glomeruli of the kidneys are fine capillary blood vessels in which the body's waste products are removed from the blood.

SUMMARY OF KEY QUESTIONS

SECTION 7.1 What Do We Need to Know as We Begin?

- Systems containing more than one component are **mixtures.**
- **Homogeneous mixtures** are uniform throughout.
- **Heterogeneous mixtures** exhibit well-defined boundaries between phases.

SECTION 7.2 What Are the Most Common Types of Solutions?

- The most common types of solutions are gas in liquid, liquid in liquid, solid in liquid, gas in gas, and solid in solid.
- When a solution consists of a solid or gas dissolved in a liquid, the liquid acts as the **solvent,** and the solid or gas is the **solute.** When one liquid is dissolved in another, the liquid present in greater amount is considered to be the solvent.

SECTION 7.4 What Factors Affect Solubility?

- The **solubility** of a substance is the maximum amount of the substance that dissolves in a given amount of solvent at a given temperature.
- "Like dissolves like" means that polar molecules are soluble in polar solvents and that nonpolar molecules are soluble in nonpolar solvents. The solubility of solids and liquids in liquids usually increases with increasing temperature; the solubility of gases in liquids usually decreases with increasing temperature.

SECTION 7.5 What Are the Most Common Units for Concentration?

- Percent concentration is given in either weight per unit volume of solution (w/v) or volume per unit volume of solution (v/v).
- Percent weight/volume (w/v) is the weight of solute per unit volume of solvent.
- Percent volume/volume (v/v) is the volume of solute per unit volume of solution (v/v).
- **Molarity** is the number of moles of solute in 1 L of solution.

SECTION 7.6 Why Is Water Such a Good Solvent?

- Water is the most important solvent, because it dissolves polar compounds and ions through hydrogen bonding and dipole–dipole interactions. Hydrated ions are surrounded by water molecules (as a solvation layer) that move together with the ion, cushioning it from collisions with other ions. Aqueous solutions of ions and molten salts are **electrolytes** and conduct electricity.

SECTION 7.7 What Are Colloids?

- Colloids exhibit a chaotic random motion, called **Brownian motion.** Colloids are stable mixtures despite the relatively large size of the colloidal particles (1 to 1000 nm). The stability results from the solvation layer that cushions the colloid's particles from direct collisions and from the electric charge on the surface of colloidal particles.

SECTION 7.8 What Is a Colligative Property?

- A **colligative property** is a property that depends only on the number of solute particles present.
- **Freezing-point depression** and **osmotic pressure** are two examples of colligative properties.
- Osmotic pressure operates across an osmotic semipermeable membrane that allows only solvent molecules to pass but screens out all other molecules. In osmotic pressure calculations, concentration is measured in **osmolarity,** which is the molarity of the solution multiplied by the number of particles produced by dissociation of the solute.
- Red blood cells in **hypotonic solution** swell and burst, a process called **hemolysis.**
- Red blood cells in **hypertonic solution** shrink, a process called **crenation.** Some semipermeable membranes allow small solute molecules to pass through along with solvent molecules.
- In **dialysis,** such membranes are used to separate large molecules from smaller ones.

PROBLEMS

A blue problem number indicates an applied problem.

■ denotes problems that are available on the GOB ChemistryNow website or CD and are assignable in OWL.

SECTION 7.2 What Are the Most Common Types of Solutions?

7.14 Vinegar is a homogeneous aqueous solution containing 6% acetic acid. Which is the solvent?

7.15 A solution is made by dissolving glucose in water. Which is the solvent and which is the solute?

7.16 In each of the following, tell whether the solutes and solvents are gases, liquids, or solids.

(a) Bronze (see Chemical Connections 2E)

(b) Cup of coffee

(c) Car exhaust

(d) Champagne

7.17 Give a familiar example of solutions of each of these types:

(a) Liquid in liquid

(b) Solid in liquid

(c) Gas in liquid

(d) Gas in gas

7.18 Are mixtures of gases true solutions or heterogeneous mixtures? Explain.

SECTION 7.4 What Factors Affect Solubility?

7.19 We dissolved 0.32 g of aspartic acid in 115.0 mL of water and obtained a clear solution. After two days standing at room temperature, we notice a white powder at the bottom of the beaker. What may have happened?

7.20 The solubility of a compound is 2.5 g in 100 mL of aqueous solution at 25°C. If we put 1.12 g of the compound in a 50-mL volumetric flask at 25°C and add sufficient water to fill it to the 50-mL mark, what kind of solution do we get—saturated or unsaturated? Explain.

7.21 To a separatory funnel with two layers—the nonpolar diethyl ether and the polar water—is added a small amount of solid. After shaking the separatory funnel, in which layer will we find each of the following solids?

(a) NaCl

(b) Camphor, $C_{10}H_{16}O$

(c) KOH

7.22 On the basis of polarity and hydrogen bonding, which solute would be the most soluble in benzene, C_6H_6?

(a) CH_3OH (b) H_2O

(c) $CH_3CH_2CH_2CH_3$ (d) H_2SO_4

7.23 Suppose that we have a stain on an oil painting that we want to remove without damaging the painting. The stain is not water-insoluble. Knowing the polarities of the following solvents, which one would we try first and why?

(a) Benzene, C_6H_6

(b) Isopropyl (rubbing) alcohol, C_3H_7OH

(c) Hexane, C_6H_{14}

7.24 Which pairs of liquids are likely to be miscible?

(a) H_2O and CH_3OH

(b) H_2O and C_6H_6

(c) C_6H_{14} and CCl_4

(d) CCl_4 and CH_3OH

7.25 The solubility of aspartic acid in water is 0.500 g in 100 mL at 25°C. If we dissolve 0.251 g of aspartic acid in 50.0 mL of water at 50°C and let the solution cool to 25°C without stirring, shaking, or otherwise disturbing the solution, would the resulting solution be a saturated, unsaturated, or supersaturated solution? Explain.

7.26 Near a power plant, warm water is discharged into a river. Sometimes dead fish are observed in the area. Why do fish die in the warm water?

7.27 If a bottle of beer is allowed to stand for several hours after being opened, it becomes "flat" (it loses CO_2). Explain.

7.28 Would we expect the solubility of ammonia gas in water at 2 atm pressure to be (a) greater, (b) the same, or (c) smaller than at 0.5 atm pressure?

SECTION 7.5 What Are the Most Common Units for Concentration?

7.29 Verify the following statements.

(a) One part per million corresponds to one minute in two years, or a single penny in $10,000.

(b) One part per billion corresponds to one minute in 2000 years, or a single penny in $10 million.

7.30 ■ Describe how we would make the following solutions:

(a) 500.0 g of a 5.32% w/w H_2S solution in water

(b) 342.0 g of a 0.443% w/w benzene solution in toluene

(c) 12.5 g of a 34.2% w/w dimethyl sulfoxide solution in acetone

7.31 ■ Describe how we would make the following solutions:

(a) 280 mL of a 27% v/v solution of ethanol, C_2H_5OH, in water

(b) 435 mL of a 1.8% v/v solution of ethyl acetate, $C_4H_8O_2$, in water

(c) 1.65 L of an 8.00% v/v solution of benzene, C_6H_6, in chloroform, $CHCl_3$

7.32 Describe how we would make the following solutions:

(a) 250 mL of a 3.6% w/v solution of NaCl in water

(b) 625 mL of a 4.9% w/v solution of glycine, $C_2H_5NO_2$, in water

(c) 43.5 mL of a 13.7% w/v solution of Na_2SO_4 in water

(d) 518 mL of a 2.1% w/v solution of acetone, C_3H_6O, in water

7.33 Calculate the w/v percentage of each of these solutes:

(a) 623 mg of casein in 15.0 mL of milk

(b) 74 mg vitamin C in 250 mL of orange juice

(c) 3.25 g of sucrose in 186 mL of coffee

7.34 Describe how we would prepare 250 mL of 0.10 M NaOH from solid NaOH and water.

7.35 Assuming that the appropriate volumetric flasks are available, describe how we would make these solutions:

(a) 175 mL of a 1.14 M solution of NH_4Br in water

(b) 1.35 L of a 0.825 M solution of NaI in water

(c) 330 mL of a 0.16 M solution of ethanol, C_2H_5OH, in water

7.36 ■ What is the molarity of each solution?

(a) 47 g of KCl dissolved in enough water to give 375 mL of solution

(b) 82.6 g of sucrose, $C_{12}H_{22}O_{11}$, dissolved in enough water to give 725 mL of solution

(c) 9.3 g of $(NH_4)_2SO_4$ dissolved in enough water to give 2.35 L of solution

7.37 A teardrop with a volume of 0.5 mL contains 5.0 mg NaCl. What is the molarity of the NaCl in the teardrop?

7.38 ■ The concentration of stomach acid, HCl, is approximately 0.10 M. What volume of stomach acid contains 0.25 mg of HCl?

7.39 The label on a sparkling cider says it contains 22.0 g glucose ($C_6H_{12}O_6$), 190 mg K^+, and 4.00 mg Na^+ per serving of 240 mL cider. Calculate the molarities of these ingredients in the sparkling cider.

7.40 ■ If 3.18 g $BaCl_2$ is dissolved in enough solvent to make 500.0 mL of solution, what is the molarity?

7.41 ■ The label on a jar of jam says it contains 13 g of sucrose, $C_{12}H_{22}O_{11}$ per tablespoon (15 mL). What is the molarity of sucrose in the jam?

7.42 A particular toothpaste contains 0.17 g NaF in 75 mL toothpaste. What are the percent w/v and the molarity of NaF in the toothpaste?

7.43 A student has a bottle labeled 0.750% albumin solution. The bottle contains exactly 5.00 mL. How much water must the student add to make the concentration of albumin become 0.125%?

7.44 ■ How many grams of solute are present in each of the following aqueous solutions?

(a) 575 mL of a 2.00 M solution of HNO_3

(b) 1.65 L of a 0.286 M solution of alanine, $C_3H_7NO_2$

(c) 320 mL of a 0.0081 M solution of $CaSO_4$

7.45 A student has a stock solution of 30.0% w/v H_2O_2 (hydrogen peroxide). Describe how the student should prepare 250 mL of a 0.25% w/v H_2O_2 solution.

7.46 To make 5.0 L of a fruit punch that contains 10% v/v ethanol, how much 95% v/v ethanol must be mixed with how much fruit juice?

7.47 A pill weighing 325 mg contains the following. What is the concentration of each in ppm?

(a) 12.5 mg Captopril, a medication for high blood pressure

(b) 22 mg Mg^{2+}

(c) 0.27 mg Ca^{2+}

7.48 ■ One slice of enriched bread weighing 80 g contains 70 μg of folic acid. What is the concentration of folic acid in ppm and ppb?

7.49 Dioxin is considered to be poisonous in concentrations above 2 ppb. If a lake containing 1×10^7 L has been contaminated by 0.1 g of dioxin, did the concentration reach a dangerous level?

7.50 An industrial wastewater contains 3.60 ppb cadmium, Cd^{2+}. How many mg of Ca^{2+} could be recovered from a ton (1016 kg) of this wastewater?

7.51 According to the label on a piece of cheese, one serving of 28 g provides the following daily values: 2% of Fe, 6% of Ca, and 6% of vitamin A. The recommended daily allowances (RDA) of each of these nutrients are as follows: 15 mg Fe, 1200 mg Ca, and 0.800 mg vitamin A. Calculate the concentrations of each of these nutrients in the cheese in ppm.

SECTION 7.6 Why Is Water Such a Good Solvent?

7.52 Considering polarities, electronegativities, and similar concepts learned in Chapter 4, classify each of the following as a strong electrolyte, a weak electrolyte, or a nonelectrolyte.

(a) KCl   (b) C_2H_5OH (ethanol)   (c) NaOH

(d) HF   (e) $C_6H_{12}O_6$ (glucose)

7.53 Which of the following would produce the brightest light in the conductance apparatus shown in Figure 7.10?

(a) 0.1 M KCl   (b) 0.1 M $(NH_4)_3PO_4$

(c) 0.5 M sucrose

7.54 Ethanol is very soluble in water. Describe how water dissolves ethanol.

7.55 Predict which of these covalent compounds is soluble in water.
(a) C_2H_6
(b) CH_3OH
(c) HF
(d) NH_3
(e) CCl_4

SECTION 7.7 What Are Colloids?

7.56 A type of car tire is made of synthetic rubber in which carbon black particles of the size of 200–500 nm are randomly dispersed. Because carbon black absorbs light, we do not see any turbidity (that is, a Tyndall effect). Do we consider a tire to be a colloidal system and, if so, what kind? Explain.

7.57 On the basis of Tables 7.1 and 7.2, classify the following systems as homogeneous, heterogeneous, or colloidal mixtures.
(a) Physiological saline solution
(b) Orange juice
(c) A cloud
(d) Wet sand
(e) Suds
(f) Milk

7.58 Table 7.2 shows no examples of a gas-in-gas colloidal system. Considering the definition of a colloid, explain why.

7.59 A solution of protein is transparent at room temperature. When it is cooled to 10°C, it becomes turbid. What causes this change in appearance?

7.60 What gives nanotubes their unique optical and electrical properties?

SECTION 7.8 What Is a Colligative Property?

7.61 Calculate the freezing points of solutions made by dissolving one mole of each of the following ionic solutes in 1000 g of H_2O.
(a) NaCl
(b) $MgCl_2$
(c) $(NH_4)_2CO_3$
(d) $Al(HCO_3)_3$

7.62 ■ If we add 175 g of ethylene glycol, $C_2H_6O_2$, per 1000 g of water to a car radiator, what will be the freezing point of the solution?

7.63 Methanol, CH_3OH, is used as an antifreeze. How many grams of methanol do we need per 1000 g of water for an aqueous solution to stay liquid at −20°C?

7.64 ■ In winter, after a snowstorm, salt (NaCl) is spread to melt the ice on roads. How many grams of salt per 1000 g of ice is needed to make it liquid at −5°C?

7.65 A 4 M acetic acid (CH_3COOH) solution lowered the freezing point by 8°C; a 4 M KF solution yields a 15°C freezing-point depression. What can account for this difference?

Osmosis

7.66 In an apparatus using a semipermeable membrane, a 0.005 M glucose (small molecule) solution yielded an osmotic pressure of 10 mm Hg. What kind of osmotic pressure change would we expect if instead of a semipermeable membrane we used a dialysis membrane?

7.67 In each case, tell which side (if either) rises and why. The solvent is water.

Osmotic membrane

	Ⓐ	Ⓑ
(a)	1% glucose	5% glucose
(b)	0.1 M glucose	0.5 M glucose
(c)	1 M NaCl	1 M glucose
(d)	1 M NaCl	1 M K_2SO_4
(e)	3% NaCl	3% NaI
(f)	1 M NaBr	1 M NaCl

7.68 An osmotic semipermeable membrane that allows only water to pass separates two compartments, A and B. Compartment A contains 0.9% NaCl, and compartment B contains 3% glycerol, $C_3H_8O_3$.
(a) In which compartment will the level of solution rise?
(b) Which compartment has the higher osmotic pressure (if either)?

7.69 Calculate the osmolarity of each of the following solutions.
(a) 0.39 M Na_2CO_3
(b) 0.62 M $Al(NO_3)_3$
(c) 4.2 M LiBr
(d) 0.009 M K_3PO_4

7.70 Two compartments are separated by a semipermeable osmotic membrane through which only water molecules can pass. Compartment A contains a 0.3 M KCl solution, and compartment B contains a 0.2 M Na_3PO_4 solution. Predict from which compartment the water will flow to the other compartment.

7.71 A 0.9% NaCl solution is isotonic with blood plasma. Which solution would crenate red blood cells?
(a) 0.3% NaCl
(b) 0.9 M glucose (MW 180)
(c) 0.9% glucose

Chemical Connections

7.72 (Chemical Connections 7A) Oxides of nitrogen (NO, NO$_2$, N$_2$O$_3$) are also responsible for acid rain. Which acids do we suppose can be formed from these oxides?

7.73 (Chemical Connections 7A) What makes normal rainwater slightly acidic?

7.74 (Chemical Connections 7B) Why do deep-sea divers use a helium–oxygen mixture in the tank instead of air?

7.75 (Chemical Connections 7B) What is nitrogen narcosis?

7.76 (Chemical Connections 7C) Write balanced equations (two steps) for the conversion of marble to gypsum dihydrate.

7.77 (Chemical Connections 7C) What is the chemical formula for the main component of limestone and marble?

7.78 (Chemical Connections 7D) What is the protective colloid in milk?

7.79 (Chemical Connections 7E) What is the minimum pressure on sea water that will force water to flow from the concentrated solution into the dilute solution?

7.80 (Chemical Connections 7E) The osmotic pressure generated across a semipermeable membrane by a solution is directly proportional to its osmolarity. Given the data in Chemical Connections 7E on the purification of sea water, estimate what pressure we need to apply to purify brackish water containing 5000 ppm salt by reverse osmosis.

7.81 (Chemical Connections 7F) A manufacturing error occurred in the isotonic solution used in hemodialysis. Instead of 0.2% NaHCO$_3$, 0.2% of KHCO$_3$ was added. Did this error change the labeled tonicity of the solution? If so, is the resulting solution hypotonic or hypertonic? Would such an error create an electrolyte imbalance in the patient's blood? Explain.

7.82 (Chemical Connections 7F) The artificial kidney machine uses a solution containing 0.6% w/v NaCl, 0.04% w/v KCl, 0.2% w/v NaHCO$_3$, and 0.72% w/v glucose. Show that this is an isotonic solution.

Additional Problems

7.83 When a cucumber is put into a saline solution to pickle it, the cucumber shrinks; when a prune is put into the same solution, the prune swells. Explain what happens in each case.

7.84 A solution of As$_2$O$_3$ has a molarity of 2×10^{-5} M. What is this concentration in ppm? (Assume that the density of the solution is 1.00 g/mL.)

7.85 Two bottles of water are carbonated, with CO$_2$ gas being added, under 2 atm pressure and then capped. One bottle is stored at room temperature; the other is stored in the refrigerator. When the bottle stored at room temperature is opened, large bubbles escape, along with a third of the water. The bottle stored in the refrigerator is opened without frothing or bubbles escaping. Explain.

7.86 Both methanol, CH$_3$OH, and ethylene glycol, C$_2$H$_6$O$_2$, are used as antifreeze. Which is more efficient—that is, which produces a lower freezing point if equal weights of each are added to the same weight of water?

7.87 We know that a 0.89% saline (NaCl) solution is isotonic with blood. In a real-life emergency, we run out of physiological saline solution and have only KCl as a salt and distilled water. Would it be acceptable to make a 0.89% aqueous KCl solution and use it for intravenous infusion? Explain.

7.88 Carbon dioxide and sulfur dioxide are soluble in water because they react with water. Write possible equations for these reactions.

7.89 A reagent label shows that it contains 0.05 ppm lead as a contaminant. How many grams of lead are present in 5.0 g of the reagent?

7.90 A concentrated nitric acid solution contains 35% HNO$_3$. How would we prepare 300 mL of 4.5% solution?

7.91 Which will have greater osmotic pressure:
(a) A 0.9% w/v NaCl solution?
(b) A 25% w/v solution of a nondissociating dextran with a molecular weight of 15,000?

7.92 Government regulations permit a 6 ppb concentration of a certain pollutant. How many grams of pollutant are allowed in 1 ton (1016 kg) of water?

7.93 The average osmolarity of sea water is 1.18. How much pure water would have to be added to 1.0 mL of sea water for it to achieve the osmolarity of blood (0.30 osmol)?

7.94 A swimming pool containing 20,000 L of water is chlorinated to have a final Cl$_2$ concentration of 0.00500 M. What is the Cl$_2$ concentration in ppm? How many kilograms of Cl$_2$ were added to the swimming pool to reach this concentration?

Looking Ahead

7.95 Synovial fluid that exists in joints is a colloidal solution of hyaluronic acid (Section 20.6A) in water. When a biochemist wants to isolate hyaluronic acid from synovial fluid, he or she adds ethanol, C$_2$H$_6$O, to bring the solution to 65% ethanol. The hyaluronic acid precipitates upon standing. What makes the hyaluronic acid solution unstable and causes it to precipitate?

Challenge Problem

7.96 To ensure that tap water is safe to drink, the U.S. Environmental Protection Agency (EPA) proscribes regulations that limit the amount of certain contaminants in water provided for public water systems. MCL is the highest level of a contaminant that is allowed in drinking water. MCLs are set at very stringent levels. To understand the possible health effects described for many regulated constituents, a person would have to drink two liters of water every day at the MCL level for a lifetime to have a one-in-a-million chance of having the described health effect. More information about contaminants and potential health effects can be obtained by calling the EPA's Safe Drinking Water hotline at 800-426-4791.

If a person were to consume two liters of MCL water every day for a lifetime, (assume 72 years), how many grams of each contaminant would that person ingest over a lifetime?

Inorganic Contaminants	Maximum Contaminant Levels (MCL)*	Likely Sources of Contaminants
Arsenic	10	Erosion of natural deposits, runoff from orchards, runoff from glass and electronics production wastes
Barium	2	Discharge from drilling water, discharge from metal refineries, erosion of natural deposits
Chromium	100	Discharge from steel and pulp mills, erosion of natural deposits
Copper	1.3	Corrosion of household plumbing systems, erosion of natural deposits, leaching from wood preservatives
Fluoride	4	Erosion of natural deposits, water additive that promotes strong teeth, discharge from fertilizer and aluminum factories
Lead	15	Corrosion of household plumbing systems, erosion of natural deposits
Mercury	2	Erosion of natural deposits, discharge from refineries and factories, runoff from landfills and cropland
Nickel	100	Nickel occurs naturally in soils, ground water, and surface water, and is often used in electroplating, stainless steel, and alloy products
Nitrate	10 ppm[†]	Runoff from fertilizer use, leaching from septic tanks, sewage, erosion from natural deposits
Nitrite	1 ppm[†]	Runoff from fertilizer use, erosion from natural deposits

*Units of MCL are ppb (parts per billion) unless otherwise specified.
[†] Parts per million

CHAPTER 8

Reaction Rates and Chemical Equilibrium

KEY QUESTIONS

Charles D. Winters

When magnesium burns in air, it creates a brilliant white light.

In the course of several years, a few molecules of glucose and O_2 will react, but not enough for us to detect within a laboratory period.

Chemistry·⦿·Now™

Look for this logo in the chapter and go to GOB ChemistryNow at **http://now.brookscole.com/gob8** or on the CD for tutorials, simulations, and problems.

8.1 | How Do We Measure Reaction Rates?

Some chemical reactions take place rapidly; others are very slow. For example, glucose and oxygen gas react with each other to form water and carbon dioxide:

$$C_6H_{12}O_6(s) + 6O_2(g) \longrightarrow 6CO_2(g) + 6H_2O(\ell)$$
Glucose

This reaction is extremely slow, however. A sample of glucose exposed to O_2 in the air shows no measurable change even after many years.

In contrast, consider what happens when you take one or two aspirin tablets for a slight headache. Very often, the pain disappears in half an hour or so. Thus the aspirin must have reacted with compounds in the body within that time.

Many reactions occur even faster. For example, if we add a solution of silver nitrate to a solution of sodium chloride (NaCl), a precipitate of silver chloride (AgCl) forms almost instantaneously.

Net ionic equation: $Ag^+(aq) + Cl^-(aq) \longrightarrow AgCl(s)$

This is a net ionic equation, so it does not show spectator ions.

Chemical kinetics The study of the rates of chemical reactions

The precipitation of AgCl is essentially complete in considerably less than 1 s.

The study of reaction rates is called **chemical kinetics.** The **rate of a reaction** is the change in concentration of a reactant (or product) per unit time. Every reaction has its own rate, which must be measured in the laboratory.

Consider the following reaction carried out in the solvent acetone:

This is a net ionic equation, so it does not show spectator ions.

$$CH_3\text{—}Cl + I^- \xrightarrow{\text{Acetone}} CH_3\text{—}I + Cl^-$$
Chloromethane   Iodomethane

To determine the reaction rate, we can measure the concentration of the product, iodomethane, in the acetone at periodic time intervals—say, every 10 min. For example, the concentration might increase from 0 to 0.12 mol/L over a period of 30 min. The rate of the reaction is the change in the concentration of iodomethane divided by the time interval:

The rate could also be determined by following the decrease in concentration of CH_3Cl or of I^-, if that is more convenient.

$$\frac{(0.12 \text{ mol } CH_3I/L) - (0 \text{ mol } CH_3I/L)}{30 \text{ min}} = \frac{0.0040 \text{ mol } CH_3I/L}{\text{min}}$$

This unit is read "0.0040 mole per liter per minute." During each minute of the reaction, an average of 0.0040 mol of chloromethane is converted to iodomethane for each liter of solution.

The rate of a reaction is not constant over a long period of time. At the beginning, in most reactions, the change in concentration is directly proportional to time. This period is shown as the linear portion of the curve in Figure 8.1. The rate calculated during this period, called the **initial rate,** is constant during this time interval. Later, as the reactant is used up, the rate of reaction decreases.

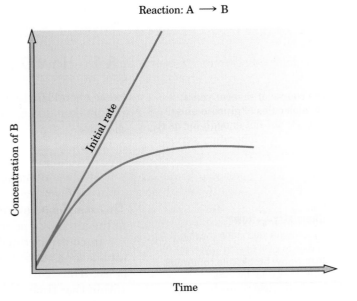

Figure 8.1 Changes in the concentration of B in the A → B system with respect to time. The rate (the change in concentration of B per unit time) is largest at the beginning of the reaction and gradually decreases until it reaches zero at the completion of the reaction.

EXAMPLE 8.1

Another way to determine the rate of the reaction of chloromethane with iodide ion is to measure the disappearance of I^- from the solution. Suppose that the concentration of I^- was 0.24 mol I^-/L at the start of the reaction. At the end of 20 min, the concentration dropped to 0.16 mol I^-/L. This difference is equal to a change in concentration of 0.08 mol I^-/L. What is the rate of reaction?

Solution

The rate of the reaction is

$$\frac{(0.16 \text{ mol } I^-/L) - (0.24 \text{ mol } I^-/L)}{20 \text{ min}} = \frac{-0.0040 \text{ mol } I^-/L}{\text{min}}$$

Because the stoichiometry of the components is 1 : 1 in this reaction, we get the same numerical answer for the rate whether we monitor a reactant or a product. Note, however, that when we measure the concentration of a reactant that disappears with time, the rate of reaction is a negative number.

Problem 8.1

In the reaction

$$2HgO(s) \longrightarrow 2Hg(\ell) + O_2(g)$$

we measure the evolution of gas to determine the rate of reaction. At the beginning of the reaction (at 0 min), 0.020 L of O_2 is present. After 15 min, the volume of O_2 gas is 0.35 L. What is the rate of reaction?

The rates of chemical reactions—both the ones that we carry out in the laboratory and the ones that take place inside our bodies—are very important. A reaction that goes more slowly than we need may be useless, whereas a reaction that goes too fast may be dangerous. Ideally, we would like to know what causes the enormous variety in reaction rates. In the next three sections, we examine this question.

8.2 Why Do Some Molecular Collisions Result in Reaction Whereas Others Do Not?

For two molecules or ions to react with each other, they must first collide. As we saw in Chapter 6, molecules in gases and liquids are in constant motion and frequently collide with each other. If we want a reaction to take place between two compounds A and B, we allow them to mix if they are gases or dissolve them in a solvent if they are liquids. In either case, the constant motion of the molecules will lead to frequent collisions between molecules of A and B. In fact, we can even calculate how many such collisions will take place in a given period of time. Such calculations indicate that so many collisions occur between A and B molecules that most reactions should be over in considerably less than one second. Because the actual reactions generally proceed much more slowly, we must conclude that most collisions do not result in a reaction. Typically, when a molecule of A collides with a molecule of B, the two simply bounce apart without reacting. Every once in a while, though, molecules of A and B collide and react to form a new compound. A collision that results in a reaction is called an **effective collision.**

Effective Collision A collision between two molecules or ions that results in a chemical reaction

Figure 8.2 The energy of molecular collisions varies. (*a*) Two fast-moving molecules colliding head-on have a higher collision energy than (*b*) two slower-moving molecules colliding at an angle.

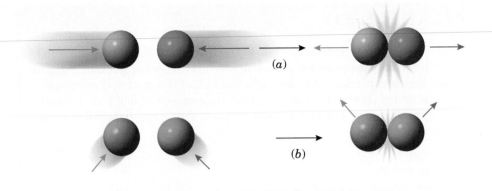

Why are some collisions effective whereas others are not? There are two main reasons:

1. In most cases, for a reaction to take place between A and B, one or more covalent bonds must be broken in A or B or both, and energy is required for this to happen. The energy comes from the collision between A and B. If the energy of the collision is large enough, bonds will break and a reaction will take place. If the collision energy is too low, the molecules will bounce apart without reacting. The minimum energy necessary for a reaction to occur is called the **activation energy.**

   Activation energy The minimum energy necessary to cause a chemical reaction to occur

   The energy of any collision depends on the relative speeds (that is, on the relative kinetic energies) of the colliding objects and on their angle of approach. Much greater damage is done in a head-on collision of two cars both going 40 mi/h than in a collision in which a car going 20 mi/h sideswipes one going 10 mi/h. The same consideration applies with molecules, as Figure 8.2 shows.

2. Even if two molecules collide with an energy greater than the activation energy, a reaction may not take place if the molecules are not oriented properly when they collide. Consider, for example, the reaction between H_2O and HCl:

$$H_2O(\ell) + HCl(g) \longrightarrow H_3O^+(aq) + Cl^-(aq)$$

For this reaction to take place, the molecules must collide in such a way that the H of the HCl hits the O of the water, as shown in Figure 8.3(a). A

(*a*)   (*b*)

GOB
Chemistry · Now™

Active Figure 8.3 Molecules must be properly oriented for a reaction to take place. (*a*) HCl and H_2O molecules are oriented so that the H of HCl collides with the O of H_2O, and a reaction takes place. (*b*) No reaction takes place because Cl, and not H, collides with the O of H_2O. The colored arrows show the path of the molecules. **See a simulation based on this figure, and take a short quiz on the concepts at http://now.brookscole.com/gob8 or on the CD.**

collision in which the Cl hits the O, as shown in Figure 8.3(b), cannot lead to a reaction, even if sufficient energy is available.

Returning to the example given at the beginning of this chapter, we can now see why the reaction between glucose and O_2 is so slow. The O_2 molecules are constantly colliding with glucose molecules, but the percentage of effective collisions is extremely tiny at room temperature.

8.3 | What Is the Relationship Between Activation Energy and Reaction Rate?

Figure 8.4 shows a typical energy diagram for an exothermic reaction. The products have a lower energy than the reactants; we might, therefore, expect the reaction to take place rapidly. As the curve shows, however, the reactants cannot be converted to products without the necessary activation energy. The activation energy is like a hill. If we are in a mountainous region, we may find that the only way to go from one point to another is to climb over a hill. It is the same in a chemical reaction. Even though the products may have a lower energy than the reactants, the products cannot form unless the reactants "go over the hill" or over a high pass—that is, gain the necessary activation energy.

Let us look into this issue more closely. In a typical reaction, existing bonds are broken and new bonds form. For example, when H_2 reacts with N_2 to give NH_3, six covalent bonds (counting a triple bond as three bonds) must break, and six new covalent bonds must form.

$$3H\!-\!H \;+\; N\!\equiv\!N \longrightarrow 2H\!-\!N\!\!\begin{array}{c}H\\ \\H\end{array}$$

Ammonia

Figure 8.4   Energy diagram for the exothermic reaction

$$H_2O(\ell) + HCl(g) \longrightarrow H_3O^+(aq) + Cl^-(aq)$$

The energy of the reactants is greater than the energy of the products. The diagram shows the positions of all atoms before, at, and after the transition state.

Figure 8.5 Energy diagram for an endothermic reaction. The energy of the products is greater than that of the reactants.

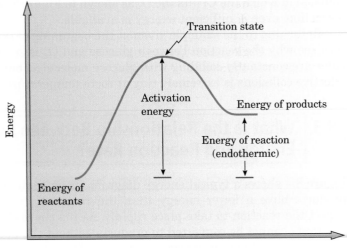

Breaking a bond requires an input of energy, but a bond forming releases energy. In a "downhill" reaction of the type shown in Figure 8.4, the amount of energy released in creating the new bonds is greater than that required to break the original bonds. In other words, the reaction is exothermic. Yet it may well have a substantial activation energy, or energy barrier, because, in most cases, at least one bond must break before any new bonds can form. Thus energy must be put into the system before we get any back. This is analogous to the following situation: Somebody offers to let you buy into a business from which, for an investment of $10,000, you could get an income of $40,000 per year, beginning in one year. In the long run, you would do very well. First, however, you need to put up the initial $10,000 (the activation energy) to start the business.

"Uphill" reactions are endothermic.

Every reaction has a different energy diagram. Sometimes, the energy of the products is higher than that of the reactants (Figure 8.5); that is, the reaction is ("uphill"). For almost all reactions, however, there is an energy "hill"—the activation energy. The activation energy is inversely related to the rate of the reaction. The lower the activation energy, the faster the reaction; the higher the activation energy, the slower the reaction.

The top of the hill on an energy diagram is called the **transition state.** When the reacting molecules reach this point, one or more original bonds are partially broken, and one or more new bonds may be in the process of formation. The transition state for the reaction of iodide ion with chloromethane

Figure 8.6 Transition state for the reaction of CH_3Cl with I^-. In the transition state, iodide ion, I^-, attacks the carbon of chloromethane from the side opposite the C—Cl bond. In this transition state, both chlorine and iodine have partial negative charges.

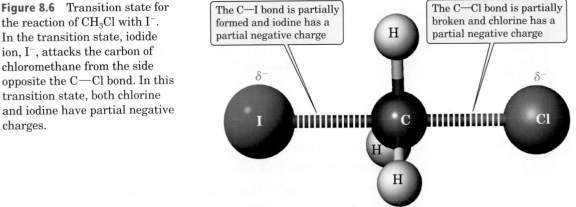

The C—I bond is partially formed and iodine has a partial negative charge

The C—Cl bond is partially broken and chlorine has a partial negative charge occurs as an iodide ion collides with a molecule of chloromethane in such a way that the iodide ion hits the carbon atom (Figure 8.6).

The speed of a reaction is proportional to the probability of effective collisions. In a single-step reaction, the probability that two particles will collide is greater than the probability of a simultaneous collision of five particles. If you consider the net ionic reaction

$$H_2O_2 + 3I^- + 2H^+ \longrightarrow I_3^- + 2H_2O$$

it is highly unlikely that six reactant particles will collide simultaneously; thus this reaction should be slow. In reality, this reaction is very fast. This fact indicates that the reaction does not occur in one step but rather takes place in multiple steps. In each of those steps, the probability is high for collisions between two particles. Even a simple reaction such as

$$H_2(g) + Br_2(g) \longrightarrow 2HBr(g)$$

occurs in three steps:

$$\text{Step 1:} \quad Br_2 \xrightarrow{\text{slow}} 2Br\cdot$$

$$\text{Step 2:} \quad Br\cdot + H_2 \xrightarrow{\text{fast}} HBr + H\cdot$$

$$\text{Step 3:} \quad H\cdot + Br_2 \xrightarrow{\text{fast}} HBr + Br\cdot$$

The dot ($\cdot$) indicates the single unpaired electron in the atom. The overall rate of the reaction will be controlled by the slowest of the three steps, just as the slowest-moving car controls the flow of traffic on a street. In the preceding reaction, step 1 is the slowest, because it has the highest activation energy.

8.4 | How Can We Change the Rate of a Chemical Reaction?

In Section 8.2, we saw that reactions occur as a result of collisions between fast-moving molecules possessing a certain minimum energy (the activation energy). In this section, we examine some of the factors that affect activation energies and reaction rates.

A. Nature of the Reactants

In general, reactions that take place between ions in aqueous solution (Section 5.6) are extremely rapid, occurring almost instantaneously. Activation energies for these reactions are very low because usually no covalent bonds must be broken. As we might expect, reactions between covalent molecules, whether in aqueous solution or not, take place much more slowly. Many of these reactions require 15 min to 24 h or longer for most of the reactants to be converted to the products. Some reactions take a good deal longer, of course, but they are seldom useful.

B. Concentration

Consider the following reaction:

$$A + B \longrightarrow C + D$$

For reactions in the gas phase, an increase in pressure usually increases the rate.

Figure 8.7 The reaction of steel wool with oxygen. (*a*) When heated in air, steel wool glows but does not burn rapidly because the concentration of O_2 in the air is only about 20%. (*b*) When the glowing steel wool is put into 100% O_2, it burns vigorously.

(*a*)   (*b*)

In the case where one of the reactants is a solid, the rate is affected by the surface area of the solid. For this reason, a substance in powder form reacts faster than the same substance in the form of large chunks.

In most cases, the reaction rate increases when we increase the concentration of either or both reactants (Figure 8.7). For many reactions—though by no means all—a direct relationship exists between concentration and reaction rate; that is, when the concentration of a reactant is doubled, the reaction rate also doubles. This outcome is easily understandable on the basis of the collision theory. If we double the concentration of A, there are twice as many molecules of A in the same volume, so the molecules of B in that volume now collide with twice as many A molecules per second than before. Given that the reaction rate depends on the number of effective collisions per second, the rate doubles.

We can express the relationship between rate and concentration mathematically. For example, for the reaction

$$2H_2O_2(\ell) \longrightarrow 2H_2O(\ell) + O_2(g)$$

the rate was determined to be -0.01 mol H_2O_2/L/min at a constant temperature when the initial concentration of H_2O_2 was 1 mol/L. In other words, every minute 0.01 mol/L of hydrogen peroxide was used up. Researchers also found that every time the concentration of H_2O_2 was doubled, the rate also doubled. Thus the rate is directly proportional to the concentration of H_2O_2. We can write this relationship as

The brackets [ ] stand for the concentration of the chemical species whose formula is between the brackets.

$$\text{Rate} = k[H_2O_2]$$

Rate constant A proportionality constant, k, between the molar concentration of reactants and the rate of reaction; rate = k[compound]

where k is a constant, called the **rate constant.** Rate constants are usually calculated from the **initial rates of reaction** (Figure 8.1).

GOB
Chemistry◆Now™
Click *Coached Problems* to see **How Concentration Affects Reaction Rate**

EXAMPLE 8.2

Calculate the rate constant, k, for the reaction

$$2H_2O_2(\ell) \longrightarrow 2H_2O(\ell) + O_2(g)$$

using the rate and the initial concentration mentioned in the preceding discussion:

$$\frac{-0.01 \text{ mol } H_2O_2}{L \cdot \text{min}} \qquad [H_2O_2] = \frac{1 \text{ mol}}{L}$$

Solution

We start with the rate equation, solve it for k, and then insert the appropriate experimental values.

$$\text{Rate} = k[H_2O_2]$$

$$k = \frac{\text{Rate}}{[H_2O_2]}$$

$$= \frac{-0.01 \text{ mol } H_2O_2}{L \cdot \text{min}} \times \frac{L}{1 \text{ mol } H_2O_2}$$

$$= \frac{-0.01}{\text{min}}$$

Problem 8.2

Calculate the rate for the reaction in Example 8.2 when the initial concentration of H_2O_2 is 0.36 mol/L.

C. Temperature

In virtually all cases, reaction rates increase with increasing temperature. A rule of thumb for many reactions is that every time the temperature goes up by 10°C, the rate of reaction doubles. This rule is far from exact, but it is not far from the truth in many cases. As you can see, this effect can be quite large. It says, for example, that if we run a reaction at 90°C instead of at

GOB
Chemistry ⚛ Now ™
Click *Coached Problems* to see **How Temperature and Activation Energy Affect Reaction Rate**

CHEMICAL CONNECTIONS 8A

Why High Fever Is Dangerous

Chemical Connections 1B points out that a sustained body temperature of 41.7°C (107°F) is invariably fatal. We can now see why a high fever is dangerous. Normal body temperature is 37°C (98.6°F), and all the many reactions in the body—including respiration, digestion, and the synthesis of various compounds—take place at that temperature. If an increase of 10°C causes the rates of most reactions to approximately double, then an increase of even 1°C makes them go significantly faster than normal.

Fever is a protective mechanism, and a small increase in temperature allows the body to kill germs faster by mobilizing the immune defense mechanism. This increase must be small, however: A rise of 1°C brings the temperature to 38°C (100.4°F); a rise of 3°C brings it to 40°C (104°F). A temperature higher than 104°F increases reaction rates to the danger point.

One can easily detect the increase in reaction rates when a patient has a high fever. The pulse rate increases and breathing becomes faster as the body attempts to supply increased amounts of oxygen for the accelerated reactions. A marathon runner, for example, may become overheated on a hot and humid day. After

An operating table unit monitors a patient packed in ice.

a time, perspiration can no longer cool his or her body effectively, and the runner may suffer hyperthermia or heat stroke, which, if not treated properly, can cause brain damage.

Figure 8.8 Distribution of kinetic energies (molecular velocities) at two temperatures. The kinetic energy on the *x*-axis designated E_a indicates the energy (molecular velocity) necessary to pass through the activation energy barrier. The shaded areas represent the fraction of molecules that have kinetic energies (molecular velocities) greater than the activation energy.

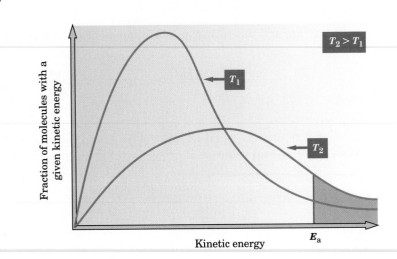

room temperature (20°C), the reaction will go about 128 times faster. There are seven 10° increments between 20°C and 90°C, and $2^7 = 128$. Put another way, if it takes 20 h to convert 100 g of reactant A to product C at 20°C, then it would take only 10 min at 90°C. Temperature, therefore, is a powerful tool that lets us increase the rates of reactions that are inconveniently slow. It also lets us decrease the rates of reactions that are inconveniently fast. For example, we might choose to run reactions at low temperatures because explosions might result or the reactions would otherwise be out of control at room temperature.

What causes reaction rates to increase with increasing temperature? Once again, we turn to collision theory. Here temperature has two effects:

1. In Section 6.6, we learned that temperature is related to the average kinetic energy of molecules. When the temperature increases, molecules move more rapidly, which means that they collide more frequently. More frequent collisions mean higher reaction rates. However, this factor is much less important than the second factor.

2. Recall from Section 8.2 that a reaction between two molecules takes place only if an effective collision occurs—a collision with an energy equal to or greater than the activation energy. When the temperature increases, not only is the average speed (kinetic energy) of the molecules greater, but there is also a different distribution of speeds. The number of very fast molecules increases much more than the number with the average speed (Figure 8.8). As a consequence, the number of effective collisions rises even more than the total number of collisions. Not only do more collisions take place, but the percentage of collisions that have an energy greater than the activation energy also rises. This factor is mainly responsible for the sharp increase in reaction rates with increasing temperature.

In this dish, chloride ion, Cl⁻, acts as a catalyst for the decomposition of NH_4NO_3.

Catalyst A substance that increases the rate of a chemical reaction by providing an alternative pathway with a lower activation energy

D. Presence of a Catalyst

Any substance that increases the rate of a reaction without itself being used up is called a **catalyst.** Many catalysts are known—some that increase the rate of only one reaction and others that can affect several reactions. Although we have seen that we can speed up reactions by increasing the temperature, in some cases they remain too slow even at the highest temperatures we can conveniently reach. In other cases, it is not feasible to increase the temperature—perhaps because other, unwanted reactions would be speeded up, too. In such cases, a catalyst, if we can find the right

CHEMICAL CONNECTIONS 8B

The Effects of Lowering Body Temperature

Like a significant increase in body temperature, a substantial decrease in body temperature below 37°C (98.6°F) can prove harmful because reaction rates are abnormally low. It is sometimes possible to take advantage of this effect. In some heart operations, for example, it is necessary to stop the flow of oxygen to the brain for a considerable time. At 37°C (98.6°F), the brain cannot survive without oxygen for longer than about 5 min without suffering permanent damage. When the patient's body temperature is deliberately lowered to about 28 to 30°C (82.4 to 86°F), however, the oxygen flow can be stopped for a considerable time without causing damage because reaction rates slow down. At 25.6°C (78°F), the body's oxygen consumption is reduced by 50%.

one for a given reaction, can prove very valuable. Many important industrial processes rely on catalysts (see Chemical Connections 8E), and virtually all reactions that take place in living organisms are catalyzed by enzymes (Chapter 22).

Catalysts work by allowing the reaction to take a different pathway, one with a lower activation energy. Without the catalyst, the reactants would have to get over the higher energy hill shown in Figure 8.9. The catalyst provides a lower hill. As we have seen, a lower activation energy means a higher reaction rate.

Each catalyst has its own way of providing an alternative pathway. Many catalysts provide a surface on which the reactants can meet. For example, the reaction between formaldehyde (CH_2O) and hydrogen (H_2) to give methanol (CH_3OH) goes so slowly without a catalyst that it is not practical, even if we increase the temperature to a reasonable level. If the mixture of gases is shaken with finely divided platinum metal, however, the reaction takes place at a convenient rate (Section 12.6D). The formaldehyde and hydrogen molecules meet each other on the surface of the platinum, where the proper bonds can be broken and new bonds form and the reaction can proceed:

Heterogeneous catalyst A catalyst in a separate phase from the reactants—for example, the solid platinum, Pt(s), in the reaction between $CH_2O(g)$ and $H_2(g)$

Homogeneous catalyst A catalyst in the same phase as the reactants—for example, enzymes in body tissues

Formaldehyde   Methanol

We often write the catalyst over or under the arrow.

Figure 8.9 Energy diagram for a catalyzed reaction. The dashed line shows the energy curve for the uncatalyzed process. The catalyst provides an alternative pathway whose activation energy is lower.

Activation energy for catalyzed reaction

Energy

Progress of reaction

CHEMICAL CONNECTIONS 8C

Timed-Release Medication

It is often desirable that a particular medicine act slowly and maintain its action evenly in the body for 24 h. We know that a solid in powder form reacts faster than the same weight in pill form because the powder has a greater surface area at which the reaction can take place. To slow the reaction and to have the drug be delivered evenly to the tissues, pharmaceutical companies coat beads of some of their drugs. The coating prevents the drug from reacting for a time. The thicker the coating, the longer it takes the drug to react. A drug with a smaller bead size has more surface area than a drug with a larger bead size; hence, drugs packaged in a smaller

Two packages of timed-release medications.

bead size will react more rapidly. By combining the proper bead size with the proper amount of coating, the drug can be designed to deliver its effect over a 24-h period. In this way, the patient needs to take only one pill per day.

Coating can also prevent problems related to stomach irritation. For example, aspirin can cause stomach ulceration or bleeding in some people. Enteric (from the Greek *enteron*, which means affecting the intestines) coated aspirin tablets have a polymeric coat that is acid resistant. Such a drug does not dissolve until it reaches the intestines, where it causes no harm.

8.5 | What Does It Mean to Say That a Reaction Has Reached Equilibrium?

Many reactions are irreversible. When a piece of paper is completely burned, the products are CO_2 and H_2O. Anyone who takes pure CO_2 and H_2O and tries to make them react to give paper and oxygen will not succeed.

A tree, of course, turns CO_2 and H_2O into wood and oxygen, and we, in sophisticated factories, make paper from the wood. These activities are not the same as directly combining CO_2, H_2O, and energy in a single process to get paper and oxygen, however. Therefore, we can certainly consider the burning of paper to be an irreversible reaction.

Other reactions are reversible. A **reversible reaction** can be made to go in either direction. For example, if we mix carbon monoxide with water in the gas phase at a high temperature, carbon dioxide and hydrogen are produced:

$$CO(g) + H_2O(g) \longrightarrow CO_2(g) + H_2(g)$$

If we desire, we can also make this reaction take place the other way. That is, we can mix carbon dioxide and hydrogen to get carbon monoxide and water vapor:

$$CO_2(g) + H_2(g) \longrightarrow CO(g) + H_2O(g)$$

Let us see what happens when we run a reversible reaction. We will add some carbon monoxide to water vapor in the gas phase. The two compounds begin to react at a certain rate (the forward reaction):

$$CO(g) + H_2O(g) \longrightarrow CO_2(g) + H_2(g)$$

As the reaction proceeds, the concentrations of CO and H_2O gradually decrease because both reactants are being used up. In turn, the rate of the reaction gradually decreases because it depends on the concentrations of the reactants (Section 8.4B).

But what is happening in the other direction? Before we added the carbon monoxide, no carbon dioxide or hydrogen was present. As soon as the forward reaction began, it produced small amounts of these substances, and we now have some CO_2 and H_2. These two compounds will now, of course, begin reacting with each other (the reverse reaction):

$$CO_2(g) + H_2(g) \longrightarrow CO(g) + H_2O(g)$$

At first, the reverse reaction is very slow. As the concentrations of H_2 and CO_2 (produced by the forward reaction) gradually increase, the rate of the reverse reaction also gradually increases.

We have a situation, then, in which the rate of the forward reaction gradually decreases, while the rate of the reverse reaction (which began at zero) gradually increases. Eventually the two rates become equal. At this point, the process is in **dynamic equilibrium** (or just **equilibrium**).

$$CO_2(g) + H_2(g) \underset{\text{reverse}}{\overset{\text{forward}}{\rightleftharpoons}} CO(g) + H_2O(g)$$

What happens in the reaction container once we reach equilibrium? If we measure the concentrations of the substances in the container, we find that no change in concentration takes place after equilibrium is reached (Figure 8.10). Whatever the concentrations of all the substances are at equilibrium, they remain the same forever unless something happens to disturb the equilibrium (as discussed in Section 8.8). This does not mean that all the concentrations must be the same—all of them can, in fact, be different and usually are—but it does mean that, whatever they are, they no longer change once equilibrium has been reached, no matter how long we wait.

Given that the concentrations of all the reactants and products no longer change, can we say that nothing is happening? No, we know that both reactions are occurring; all the molecules are constantly reacting—the CO and H_2O are being changed to CO_2 and H_2, and the CO_2 and H_2 are

GOB
Chemistry·⚛·Now™

Click Coached Problems to see **How Reactants and Products Come to Equlibrium in a Chemical System**

Dynamic equilibrium A state in which the rate of the forward reaction equals the rate of the reverse reaction

We use a double arrow to indicate that a reaction is reversible.

Another way to look at this situation is to say that the concentration of carbon monoxide (and of the other three compounds) does not change at equilibrium, because it is being used up as fast as it is being formed.

Reaction A + B $\rightleftharpoons$ C + D

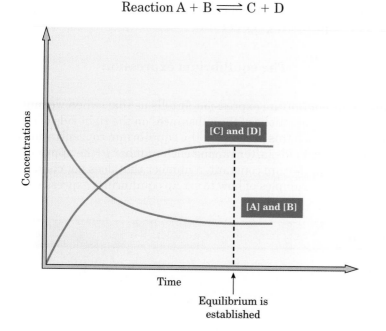

Figure 8.10   Changes in the concentrations of reactants (A and B) and products (C and D) as a system approaches equilibrium. Only A and B are present at the beginning of the reaction.

Figure 8.11 An equilibrium can be approached from either direction.

At the start:
A mixture of
CO and H_2O

At equilibrium:
All four substances
present

At the start:
A mixture of
CO_2 and H_2

being changed to CO and H_2O. Because the rates of the forward and reverse reactions are the same, however, none of the concentrations change.

In the example just discussed, we approached equilibrium by adding carbon monoxide to water vapor. Alternatively, we could have added carbon dioxide to hydrogen. In either case, we eventually get an equilibrium mixture containing the same four compounds (Figure 8.11).

It is not necessary to begin with equal amounts. We could, for example, take 10 moles of carbon monoxide and 0.2 mole of water vapor. We would still arrive at an equilibrium mixture of all four compounds.

8.6 | What Is an Equilibrium Constant and How Do We Use It?

Chemical equilibria can be treated by a simple mathematical expression. First, let us write the following reaction as the general equation for all reversible reactions:

$$aA + bB \rightleftharpoons cC + dD$$

In this equation, the capital letters stand for substances—CO_2, H_2O, CO, and H_2, for instance—and the lowercase letters are the coefficients of the balanced equation. The double arrow shows that the reaction is reversible. In general, any number of substances can be present on either side.

Once equilibrium is reached, the following equation is valid, where K is a constant called the **equilibrium constant:**

Equilibrium constant A value calculated from the equilibrium expression for a given reaction indicating in which direction the reaction goes

$$K = \frac{[C]^c[D]^d}{[A]^a[B]^b}$$ **The equilibrium expression**

Let us examine the equlibrium expression. It tells us that, when we multiply the equilibrium concentrations of the substances on the right side of the chemical equation and divide this product by the equilibrium concentrations of the substances on the left side (after raising each number to the appropriate power), we get the equilibrium constant, a number that does not change.

Let us look at several examples of how to set up equilibrium expressions.

EXAMPLE 8.3

Write the equilibrium expression for the reaction

$$CO(g) + H_2O(g) \rightleftharpoons CO_2(g) + H_2(g)$$

Solution

$$K = \frac{[CO_2][H_2]}{[CO][H_2O]}$$

This expression tells us that, at equilibrium, the concentration of carbon dioxide multiplied by the concentration of hydrogen and divided by the concentrations of water and carbon monoxide is a constant, K. Note that no exponent is written in this equation because all of the coefficients of the chemical equation are 1, and, by convention, an exponent of 1 is not written. Mathematically, it would be just as correct to write the left-side compounds on top, but the universal custom is to write them as shown here, with the products on top and the reactants on the bottom.

Problem 8.3

Write the equilibrium expression for the reaction

$$SO_3(g) + H_2O(\ell) \rightleftharpoons H_2SO_4(aq)$$

It is understood that the concentration of a species within brackets is always expressed in moles per liter.

GOB
Chemistry⚛Now™
Click *Coached Problems* to write some
Equilibrium Expressions

EXAMPLE 8.4

Write the equilibrium expression for the reaction

$$O_2(g) + 4ClO_2(g) \rightleftharpoons 2Cl_2O_5(g)$$

Solution

$$K = \frac{[Cl_2O_5]^2}{[O_2][ClO_2]^4}$$

In this case, the chemical equation has coefficients other than unity, so the equilibrium expression contains exponents.

Problem 8.4

Write the equilibrium expression for the reaction

$$2NH_3(g) \rightleftharpoons N_2(g) + 3H_2(g)$$

Now let us see how K is calculated.

EXAMPLE 8.5

Some H_2 is added to I_2 at 427°C and the following reaction is allowed to come to equilibrium:

$$H_2(g) + I_2(g) \rightleftharpoons 2HI(g)$$

When equilibrium is reached, the concentrations are $[I_2] = 0.42$ mol/L, $[H_2] = 0.025$ mol/L, and $[HI] = 0.76$ mol/L. Calculate K at 427°C.

Solution

The equilibrium expression is

$$K = \frac{[HI]^2}{[I_2][H_2]}$$

Substituting the concentrations, we get

$$K = \frac{[0.76 \, M]^2}{[0.42 \, M][0.025 \, M]} = 55$$

The equilibrium constant of this reaction has no units because the molarities cancel. Depending on the stoichiometry of the reaction, equilibrium constants may have different units, such as M, and so on.

Problem 8.5

What is the equilibrium constant for the following reaction? Equilibrium concentrations are given under the formula of each component.

$$PCl_3 \; + \; Cl_2 \; \rightleftharpoons \; PCl_5$$
$$\text{1.66 } M \qquad \text{1.66 } M \qquad \text{1.66 } M$$

Example 8.5 shows us that the reaction between I_2 and H_2 to give HI has an equilibrium constant of 55. What does this value mean? At constant temperature, equilibrium constants remain the same no matter what concentrations we have. That is, at 427°C, if we begin by adding, say, 5 moles of H_2 to 5 moles of I_2, the forward and backward reactions will take place, and equilibrium will eventually be reached. At that point, the value of K will equal 55. If we begin at 427°C with different numbers of moles of H_2 and I_2, perhaps 7 moles of H_2 and 2 moles of I_2, once equilibrium is reached, the value of $[HI]^2/[I_2][H_2]$ will again be 55. It makes no difference what the initial concentrations of the three substances are. At 427°C, as long as all three substances are present and equilibrium has been reached, the concentrations of the three substances will adjust themselves so that the value of the equilibrium constant equals 55.

The equilibrium constant is different for every reaction. Some reactions have a large K; others have a small K. A reaction with a very large K proceeds almost to completion (to the right). For example, K for the following reaction is about 100,000,000, or 10^8 at 25°C:

$$N_2(g) + 3H_2(g) \rightleftharpoons 2NH_3(g)$$

The symbol $\rightleftharpoons$ means that the equilibrium lies far to the right.

This value of 10^8 for K means that, at equilibrium, $[NH_3]$ must be very large and $[N_2]$ and $[H_2]$ must be very small so that $[NH_3]^2/[N_2][H_2]^3 = 10^8$. Thus, if we add N_2 to H_2, we can be certain that, when equilibrium is reached, an essentially complete reaction has taken place.

On the other hand, a reaction such as the following, which has a very small K, hardly goes forward at all:

$$AgCl(s) \rightleftharpoons Ag^+(aq) + Cl^-(aq)$$

GOB
Chemistry⋅Now™

Click *Coached Problems* to see how the value of the **Equilibrium Constant** relates to the relative amounts of reactants and products

Equilibrium effects are most obvious in reactions with K values between 10^3 and 10^{-3}. In such cases, the reaction goes part of the way, and significant concentrations of all substances are present at equilibrium. An example is the reaction between carbon monoxide and water discussed in Section 8.5, for which K is equal to 10 at 600°C.

In dilute solutions, it is customary to omit the concentration of the solvent from the equilibrium expression. Consider the reaction between ammonia and water:

$$NH_3(aq) + H_2O(\ell) \rightleftharpoons NH_4^+(aq) + OH^-(aq)$$

If the concentration of ammonia is small, and consequently the concentrations of ammonium ion and hydroxide ion are small, the concentration of water molecules remains essentially the same. Because the molar concentration of water is effectively constant, we do not include it in the equilibrium expression:

$$K = \frac{[NH_4^+][OH^-]}{[NH_3]}$$

EXAMPLE 8.6

Calculate the equilibrium constant for the above reaction (a) with and (b) without the inclusion of water in the equilibrium expression. The equilibrium concentrations are as follows: $[NH_3] = 0.0100\ M$; $[NH_4^+] = 0.000400\ M$; $[OH^-] = 0.000400\ M$.

Solution

(a) Because molarity is based on 1 L, first we find the concentration of water in water—that is, how many moles of water are present in 1 L of water.

$$[H_2O] = \frac{1000\ g\ H_2O}{1\ L\ H_2O} \times \frac{1\ mol\ H_2O}{18.00\ g\ H_2O} = \frac{55.56\ mol\ H_2O}{1\ L\ H_2O}$$

$$= 55.56\ M$$

We then use this value and the concentration of the other species to calculate the equilibrium constant for this reaction.

$$K = \frac{[0.000400\ M][0.000400\ M]}{[0.0100\ M][55.56\ M]} = 2.88 \times 10^{-7}$$

(b) $$K = \frac{[0.000400\ M][0.000400\ M]}{[0.0100\ M]} = 1.60 \times 10^{-5}\ M$$

The inclusion of the concentration of the solvent, water, gives an equilibrium constant two magnitudes smaller than without it. The equilibrium constant without the inclusion of the solvent is the accepted one.

Problem 8.6

Ethyl acetate is a common solvent in many industrial products, ranging from nail enamel and remover to liquid cement for plastics. It is prepared by reacting acetic acid with ethanol in the presence of an acid catalyst, HCl. Note that this synthesis starts with acetic acid and pure ethanol. Water is one of the products and, therefore, its concentration must be included in the equilibrium constant. Write the equilibrium expressions for the reaction.

$$\underset{\text{Acetic acid}}{CH_3COOH(\ell)} + \underset{\text{Ethanol}}{C_2H_5OH(\ell)} \overset{HCl}{\rightleftharpoons} \underset{\text{Ethyl acetate}}{CH_3COOC_2H_5(\ell)} + H_2O(\ell)$$

GOB
Chemistry Now™
Click *Coached Problems* to try a problem calculating the **Equilibrium Constant** from concentrations at equilibrium

HCl is a catalyst. It speeds up the attainment of equilibrium, but does not affect the equilibrium position.

The equilibrium constant for a given reaction remains the same no matter what happens to the concentrations, but the same is not true for changes in temperature. The value of K, however, does change when the temperature changes.

8.7 | How Long Does It Take for a Reaction to Reach Equilibrium?

As pointed out in Section 8.6, the equilibrium expression is valid only after equilibrium has been reached. Before that point, there is no equilibrium, and the equilibrium expression is not valid. But how long does it take for a reaction to reach equilibrium? There is no easy answer to this question. Some reactions, if the reactants are well mixed, reach equilibrium in less than one second; others will not get there even after millions of years.

There is no relationship between the rate of a reaction (how long it takes to reach equilibrium) and the value of K. It is possible to have a large K and a slow rate, as in the reaction between glucose and O_2 to give CO_2 and H_2O, which does not reach equilibrium for many years (Section 8.1), or a small K and a fast rate. In other reactions, the rate and K are both large or both small.

8.8 | What Is Le Chatelier's Principle?

When a reaction reaches equilibrium, the forward and reverse reactions take place at the same rate, and the equilibrium concentration of the reaction mixture does not change as long as we don't do anything to the system. But what happens if we do? In 1888, Henri Le Chatelier (1850–1936) put forth the statement known as **Le Chatelier's principle:** If an external stress is applied to a system in equilibrium, the system reacts in such a way as to partially relieve that stress. Let us look at five types of stress that can be put on chemical equilibria: adding a reactant or product, removing a reactant or product, and changing the temperature.

Le Chatelier's principle A principle stating that when a stress is applied to a system in chemical equilibrium, the position of the equilibrium shifts in the direction that will relieve the applied stress

A. Addition of a Reaction Component

Suppose that the reaction between acetic acid and ethanol has reached equilibrium:

$$\underset{\text{Acetic acid}}{CH_3COOH} + \underset{\text{Ethanol}}{C_2H_5OH} \underset{}{\overset{HCl}{\rightleftarrows}} \underset{\text{Ethyl acetate}}{CH_3COOC_2H_5} + H_2O$$

This means that the reaction flask contains all four substances (plus the catalyst) and that their concentrations no longer change.

We now disturb the system by adding some acetic acid.

Adding CH_3COOH

$$\underset{\text{Acetic acid}}{CH_3\overset{\displaystyle O}{\overset{\|}{C}}OH} + \underset{\text{Ethanol}}{HOCH_2CH_3} \overset{HCl}{\rightleftarrows} \underset{\text{Ethyl acetate}}{CH_3\overset{\displaystyle O}{\overset{\|}{C}}OCH_2CH_3} + H_2O$$

Equilibrium shifts to formation of more products

The result is that the concentration of acetic acid suddenly increases, which increases the rate of the forward reaction. As a consequence, the concentrations of the products (ethyl acetate and water) begin to increase. At the same time, the concentrations of reactants decrease. Now, an increase in the concentrations of the products causes the rate of the reverse reaction to increase, but the rate of the forward reaction is decreasing, so eventually the two rates will be equal again and a new equilibrium will be established.

When that happens, the concentrations are once again constant, but they are not the same as they were before the addition of the acetic acid. The concentrations of ethyl acetate and water are higher now, and the concentration of ethanol is lower. The concentration of acetic acid is higher because we added some, but it is less than it was immediately after we made the addition.

When we add more of any component to a system in equilibrium, that addition constitutes a stress. The system relieves this stress by increasing the concentrations of the components on the other side of the equilibrium equation. We say that the equilibrium shifts in the opposite direction. The addition of acetic acid, on the left side of the equation, causes the rate of the forward reaction to increase and the reaction to move toward the right: More ethyl acetate and water form, and some of the acetic acid and ethanol are used up. The same thing happens if we add ethanol.

On the other hand, if we add water or ethyl acetate, the rate of the reverse reaction increases, and the reaction shifts to the left:

$$CH_3COOH + C_2H_5OH \xrightleftharpoons{HCl} CH_3COOC_2H_5 + H_2O \quad \text{Adding ethyl acetate}$$

Acetic acid   Ethanol   Ethyl acetate

←――――――――――― Equilibrium shifts toward formation of reactants

We can summarize by saying that the addition of any component causes the equilibrium to shift to the opposite side.

The tube on the left contains a saturated solution of silver acetate (Ag^+ ions and CH_3COO^- ions) in equilibrium with solid silver acetate. When more silver ions are added in the form of silver nitrate solution, the equilibrium shifts to the right, producing more silver acetate, as can be seen in the tube on the right.

$$Ag^+(aq) + CH_3COO^-(aq) \rightleftharpoons CH_3COOAg(s)$$

Charles D. Winters

EXAMPLE 8.7

When dinitrogen tetroxide, a colorless gas, is enclosed in a vessel, a color indicating the formation of brown nitrogen dioxide, soon appears (see Figure 8.12 later in this chapter). The intensity of the brown color indicates the amount of nitrogen dioxide formed. The equilibrium reaction is

$$N_2O_4(g) \rightleftharpoons 2NO_2(g)$$

Dinitrogen   Nitrogen
tetroxide   dioxide
(colorless)   (brown)

When more N_2O_4 is added to the equilibrium mixture, the brownish color becomes darker. Explain what happened.

Solution

The darker color indicates that more nitrogen dioxide is formed. This happens because the addition of the reactant shifts the equilibrium to the right, forming more product.

Problem 8.7

What happens to the following equilibrium reaction when Br_2 gas is added to the equilibrium mixture?

$$2NOBr(g) \rightleftharpoons 2NO(g) + Br_2(g)$$

B. Removal of a Reaction Component

It is not always as easy to remove a component from a reaction mixture as it is to add one, but there are often ways to do it. The removal of a component, or even a decrease in its concentration, lowers the corresponding reaction rate and changes the position of the equilibrium. If we remove a reactant, the reaction shifts to the left, toward the side from which the reactant was removed. If we remove a product, the reaction shifts to the right, toward the side from which the product was removed.

In the case of the acetic acid–ethanol equilibrium, ethyl acetate has the lowest boiling point of the four components and can be removed by distillation. The equilibrium then shifts to that side so that more ethyl acetate is produced to compensate for the removal. The concentrations of acetic acid and ethanol decrease, and the concentration of water increases. The effect of removing a component is thus the opposite of adding one. The removal of a component causes the equilibrium to shift to the side from which the component was removed.

$$CH_3COOH + C_2H_5OH \underset{}{\overset{HCl}{\rightleftharpoons}} H_2O + CH_3COOC_2H_5 \quad \text{Removing ethyl acetate}$$

Acetic acid   Ethanol   Ethyl acetate

Equilibrium shifts toward ⟶
formation of more products

No matter what happens to the individual concentrations, the value of the equilibrium constant remains unchanged.

This headstone has been damaged by acid rain.

John D. Cunningham/Visuals Unlimited

EXAMPLE 8.8

The beautiful stone we know as marble is mostly calcium carbonate. When acid rain containing sulfuric acid attacks marble, the following equilibrium reaction can be written:

$$CaCO_3(s) + H_2SO_4(aq) \rightleftharpoons CaSO_4(s) + CO_2(g) + H_2O(\ell)$$

Calcium   Sulfuric   Calcium   Carbon
carbonate   acid   sulfate   dioxide

How does the fact that carbon dioxide is a gas influence the equilibrium?

Solution

The gaseous CO_2 diffuses away from the reaction site, meaning that this product is removed from the equilibrium mixture. The equilibrium shifts to the right, so that the statue continues eroding.

Problem 8.8

Consider the following equilibrium reaction for the decomposition of an aqueous solution of hydrogen peroxide:

$$2H_2O_2(aq) \rightleftharpoons 2H_2O(\ell) + O_2(g)$$

Hydrogen
peroxide

Oxygen has limited solubility in water (see the table in Chemical Connections 7A). What happens to the equilibrium after the solution becomes saturated with oxygen?

C. Change in Temperature

The effect of a change in temperature on a reaction that has reached equilibrium depends on whether the reaction is exothermic (gives off heat) or endothermic (requires heat). Let us look first at an exothermic reaction:

$$2H_2(g) + O_2(g) \rightleftharpoons 2H_2O(\ell) + 137,000 \text{ cal per mol } H_2O$$

If we consider heat to be a product of this reaction, then we can use Le Chatelier's principle and the same type of reasoning as we did before. An increase in temperature means that we are adding heat. Because heat is a product, its addition pushes the equilibrium to the opposite side. We can therefore say that, if this exothermic reaction is at equilibrium and we increase the temperature, the reaction goes to the left—the concentrations of H_2 and O_2 increase and that of H_2O decreases. This is true of all exothermic reactions.

- An increase in temperature drives an exothermic reaction toward the reactants (to the left).
- A decrease in temperature drives an exothermic reaction toward the products (to the right).

For an endothermic reaction, of course, the opposite is true.

- An increase in temperature drives an endothermic reaction toward the products (to the right).
- A decrease in temperature drives an endothermic reaction toward the reactants (to the left).

Recall from Section 8.4 that a change in temperature changes not only the position of equilibrium but also the value of K, the equilibrium constant.

CHEMICAL CONNECTIONS 8D

Sunglasses and Le Chatelier's Principle

Heat is not the only form of energy that affects equilibria. The statements made in the text regarding endothermic or exothermic reactions can be generalized to reactions involving other forms of energy. A practical illustration of this generalization is the use of sunglasses with adjustable shading. The compound silver chloride, AgCl, is incorporated in the glasses. This compound, upon exposure to sunlight, produces metallic silver, Ag, and chlorine, Cl_2:

$$\text{Light} + 2Ag^+ + 2Cl^- \rightleftharpoons 2Ag(s) + Cl_2$$

The more silver metal produced, the darker the glasses. At night, or when the wearer goes indoors, the reaction is reversed according to Le Chatelier's principle. In this case, the addition of energy in the form of sunlight drives the equilibrium to the right; its removal drives the equilibrium to the left.

Charles D. Winters

Figure 8.12 Effect of temperature on the N_2O_4—NO_2 system at equilibrium. (top) At 50°C, the deep brown color indicates the predominance of NO_2. (bottom) At 0°C, N_2O_4, which is colorless, predominates.

EXAMPLE 8.9

The conversion of nitrogen dioxide to dinitrogen tetroxide is an exothermic reaction:

$$2NO_2(g) \rightleftharpoons N_2O_4(g) + 13,700 \text{ cal}$$

Nitrogen dioxide   Dinitrogen tetroxide
(brown)   (colorless)

In Figure 8.12 we see that the brown color is darker at 50°C than it is at 0°C. Explain.

Solution
To go from 0°C to 50°C, heat must be added. But heat is a product of this equilibrium reaction, as it is written in the question. The addition of heat, therefore, shifts the equilibrium to the left. This shift produces more $NO_2(g)$, leading to the darker brown color.

Problem 8.9
Consider the following equilibrium reaction:

$$A \rightleftharpoons B$$

Increasing the temperature results in an increase in the equilibrium concentration of B. Is the conversion of A to B an exothermic reaction or an endothermic reaction? Explain.

D. Change in Pressure

A change in pressure influences the equilibrium only if one or more components of the reaction mixture are gases. Consider the following equilibrium reaction:

$$N_2O_4(g) \rightleftharpoons 2NO_2(g)$$

Dinitrogen   Nitrogen
tetroxide   dioxide

In this equilibrium, we have one mole of gas as a reactant and two moles of gas as products. According to Le Chatelier's principle, an increase in pressure shifts the equilibrium in the direction that will decrease the moles in the gas phase and thus decrease the internal pressure. In the preceding reaction, the equilibrium will shift to the left.

- An increase in pressure shifts the reaction toward the side with fewer moles of gas.
- A decrease in pressure shifts the reaction toward the side with more moles of gas.

EXAMPLE 8.10

In the production of ammonia, both reactants and products are gases:

$$N_2(g) + 3H_2(g) \rightleftharpoons 2NH_3(g)$$

What kind of pressure change would increase the yield of ammonia?

Solution

There are four moles of gases on the left side and two moles on the right side. To increase the yield of ammonia, we must shift the equilibrium to the right. An increase in pressure shifts the equilibrium toward the side with fewer moles—that is, to the right. Thus an increase in pressure will increase the yield of ammonia.

Problem 8.10

What happens to the following equilibrium reaction when the pressure is increased?

$$O_2(g) + 4ClO_2(g) \rightleftharpoons 2Cl_2O_5(g)$$

E. The Effects of a Catalyst

As we saw in Section 8.4D, a catalyst increases the rate of a reaction without itself being changed. For a reversible reaction, catalysts always increase the rates of both the forward and reverse reactions to the same extent. Therefore, the addition of a catalyst has no effect on the position of equilibrium. However, adding a catalyst to a system not yet at equilibrium causes it to reach equilibrium faster than it would without the catalyst.

CHEMICAL CONNECTIONS 8E

The Haber Process

Both humans and other animals need proteins and other nitrogen-containing compounds to live. Ultimately, the nitrogen in these compounds comes from the plants that we eat. Although the atmosphere contains plenty of N_2, nature converts it to compounds usable by biological organisms in only one way: Certain bacteria have the ability to "fix" atmospheric nitrogen—that is, convert it to ammonia. Most of these bacteria live in the roots of certain plants such as clover, alfalfa, peas, and beans. However, the amount of nitrogen fixed by such bacteria each year is far less than the amount necessary to feed all the humans and animals in the world.

The world today can support its population only by using fertilizers made by artificial fixing, primarily the **Haber process**, which converts N_2 to NH_3.

$$N_2(g) + 3H_2(g) \rightleftharpoons 2NH_3(g) + 22 \text{ kcal}$$

Early workers who focused on the problem of fixing nitrogen were troubled by a conflict between equilibrium and rate. Because the synthesis of ammonia is an exothermic reaction, an increase in temperature drives the equilibrium to the left, so the best results (largest possible yield) should be obtained at low temperatures. At low temperatures, however, the rate is too slow to produce any meaningful amounts of NH_3. In 1908, Fritz Haber (1868–1934) solved this problem when he discovered a catalyst that permits the reaction to take place at a convenient rate at 500°C.

Ammonia is produced by bacteria in these root nodules.

C. P. Vance/Visuals Unlimited

The NH_3 produced by the Haber process is converted to fertilizers, which are used all over the world. Without these fertilizers, food production would diminish so much that widespread starvation would result.

SUMMARY OF KEY QUESTIONS

SECTION 8.1 How Do We Measure Reaction Rates?

• The **rate of a reaction** is the change in concentration of a reactant or a product per unit time. Some reactions are fast; others are slow.

SECTION 8.2 Why Do Some Molecular Collisions Result in Reaction Whereas Others Do Not?

• The rate of a reaction depends on the number of **effective collisions**—that is, collisions that lead to a reaction.
• The energy necessary for a reaction to take place is the **activation energy.** Effective collisions have (1) more than the activation energy required for the reaction to proceed forward and (2) the proper orientation in space of the colliding particles.

SECTION 8.3 What Is the Relationship Between Activation Energy and Reaction Rate?

• The lower the activation energy, the faster the reaction.
• An energy diagram shows the progress of a reaction.
• The position at the top of the curve in an energy diagram is called the **transition state.**

SECTION 8.4 How Can We Change the Rate of a Chemical Reaction?

• Reaction rates generally increase with increasing concentration and temperature; they also depend on the nature of the reactants.
• The rates of some reactions can be increased by adding a **catalyst,** a substance that provides an alternative pathway with a lower activation energy.
• A rate constant gives the relationship between the rate of the reaction and the concentrations of the reactants at a constant temperature.

SECTION 8.5 What Does It Mean to Say That a Reaction Has Reached Equilibrium?

• Many reactions are reversible and eventually reach equilibrium.
• At **equilibrium,** the forward and reverse reactions take place at equal rates, and concentrations do not change.

SECTION 8.6 What Is an Equilibrium Constant and How Do We Use It?

• Every equilibrium has an **equilibrium expression** and an **equilibrium constant,** K, which does not change when concentrations change but does change when temperature changes.

SECTION 8.7 How Long Does It Take for a Reaction to Reach Equilibrium?

• There is no necessary relationship between the value of the equilibrium constant, K, and the rate at which equilibrium is reached.

SECTION 8.8 What Is Le Chatelier's Principle?

• **Le Chatelier's principle** tells us what happens when we put stress on a system in equilibrium.
• The addition of a component causes the equilibrium to shift to the opposite side.
• The removal of a component causes the equilibrium to shift to the side from which the component is removed.
• Increasing the temperature drives an exothermic equilibrium to the side of the reactants; increasing the temperature drives an endothermic equilibrium to the side of the products.
• Addition of a catalyst has no effect on the position of equilibrium.

PROBLEMS

GOB
Chemistry ⚛ Now™

Assess your understanding of this chapter's topics with additional quizzing and conceptual-based problems at **http://now.brookscole.com/gob8** or on the CD.

A blue problem number indicates an applied problem.

■ denotes problems that are available on the GOB ChemistryNow website or CD and are assignable in OWL.

SECTION 8.1 How Do We Measure Reaction Rates?

8.11 Consider the following reaction:

$$CH_3—Cl \ + I^- \longrightarrow CH_3—I \ + Cl^-$$
Chloromethane   Iodomethane

Suppose we start the reaction with an initial concentration of 0.260 M. This concentration increases to 0.840 M over a period of 1 h 20 min. What is the rate of reaction?

SECTION 8.2 Why Do Some Molecular Collisions Result in Reaction Whereas Others Do Not?

8.12 Two kinds of gas molecules are reacted at a set temperature. The gases are blown into the reaction vessel from two tubes. In setup A, the two tubes are aligned parallel to each other; in setup B, they are 90° to each other; and in setup C, they are aligned against each other. Which setup would yield the most effective collisions?

8.13 Why are reactions between ions in aqueous solution generally much faster than reactions between covalent molecules?

SECTION 8.3 What Is the Relationship Between Activation Energy and Reaction Rate?

8.14 What is the likelihood that the following reaction occurs in a single step? Explain.

$$O_2(g) + 4ClO_2(g) \rightleftharpoons 2Cl_2O_5(g)$$

8.15 A certain reaction is exothermic by 9 kcal/mol and has an activation energy of 14 kcal/mol. Draw an energy diagram for this reaction, and label the transition state.

SECTION 8.4 How Can We Change the Rate of a Chemical Reaction?

8.16 A quart of milk quickly spoils if left at room temperature but keeps for several days in a refrigerator. Explain.

8.17 If a certain reaction takes 16 h to go to completion at 10°C, what temperature should we run it if we want it to go to completion in 1 h?

8.18 In most cases, when we run a reaction by mixing a fixed quantity of substance A with a fixed quantity of substance B, the rate of the reaction begins at a maximum and then decreases as time goes by. Explain.

8.19 If you were running a reaction and wanted it to go faster, what three things might you try to accomplish this goal?

8.20 What factors determine whether a reaction run at a given temperature will be fast or slow?

8.21 Explain how a catalyst increases the rate of a reaction.

8.22 If you add a piece of marble, $CaCO_3$, to a 6 M HCl solution at room temperature, you will see some bubbles form around the marble as gas slowly rises. If you crush another piece of marble and add it to the same solution at the same temperature, you will see vigorous gas formation, so much so that the solution appears to be boiling. Explain.

SECTION 8.5 What Does It Mean to Say That a Reaction Has Reached Equilibrium?

8.23 Burning a piece of paper is an irreversible reaction. Give some other examples of irreversible reactions.

8.24 Suppose the following reaction is at equilibrium:

$$PCl_3 + Cl_2 \rightleftharpoons PCl_5$$

(a) Are the equilibrium concentrations of PCl_3, Cl_2, and PCl_5 necessarily equal? Explain.

(b) Is the equilibrium concentration of PCl_3 necessarily equal to that of Cl_2? Explain.

SECTION 8.6 What is an Equilibrium Constant and How Do We Use It?

8.25 ■ Write equilibrium expressions for these reactions.
(a) $2H_2O_2 \rightleftharpoons 2H_2O + O_2$
(b) $2N_2O_5 \rightleftharpoons 2N_2O_4 + O_2$
(c) $6H_2O + 6CO_2 \rightleftharpoons C_6H_{12}O_6 + 6O_2$

8.26 Write the chemical equations corresponding to the following equilibrium expressions.

(a) $K = \dfrac{[H_2CO_3]}{[CO_2][H_2O]}$

(b) $K = \dfrac{[P_4][O_2]^5}{[P_4O_{10}]}$

(c) $K = \dfrac{[F_2]^3[PH_3]}{[HF]^3[PF_3]}$

8.27 ■ Consider the following equilibrium reaction. Under each species is its equilibrium concentration. Calculate the equilibrium constant for the reaction.

$$CO(g) + H_2O(g) \rightleftharpoons CO_2(g) + H_2(g)$$
$$0.933\,M \quad 0.720\,M \quad\quad 0.133\,M \quad 3.37\,M$$

8.28 When the following reaction reached equilibrium at 325 K, the equilibrium constant was found to be 172. When a sample was taken of the equilibrium mixture, it was found to contain 0.0714 M NO_2. What was the equilibrium concentration of N_2O_4?

$$2NO_2(g) \rightleftharpoons N_2O_4(g)$$

8.29 ■ The following reaction was allowed to reach equilibrium at 25°C. Under each component is its equilibrium concentration. Calculate the equilibrium constant, K, for this reaction.

$$2NOCl(g) \rightleftharpoons 2NO(g) + Cl_2(g)$$
$$2.6\,M \quad\quad 1.4\,M \quad 0.34\,M$$

8.30 Write the equilibrium expression for this reaction:

$$HNO_3(aq) + H_2O(\ell) \rightleftharpoons H_3O^+(aq) + NO_3^-(aq)$$

8.31 Here are equilibrium constants for several reactions. Which of them favor the formation of products and which favor the formation of reactants?
(a) 4.5×10^{-8} (b) 32
(c) 4.5   (d) 3.0×10^{-7}
(e) 0.0032

8.32 A particular reaction has an equilibrium constant of 1.13 under one set of conditions and an equilibrium constant of 1.72 under a different set of conditions. Which conditions would be more advantageous in an industrial process that sought to obtain the maximum amount of products? Explain.

SECTION 8.7 How Long Does It Take for a Reaction to Reach Equilibrium?

8.33 If a reaction is very exothermic—that is, if the products have a much lower energy than the reactants—can we be reasonably certain that it will take place rapidly?

8.34 If a reaction is very endothermic—that is, if the products have a much higher energy than the reactants—can we be reasonably certain that it will take place extremely slowly or not at all?

8.35 A reaction has a high rate constant but a small equilibrium constant. What does this mean in terms of producing an industrial product?

SECTION 8.8 What is Le Chatelier's Principle?

8.36 Complete the following table showing the effects of changing reaction conditions on the equilibrium and value of the equilibrium constant, K.

Change in Condition	How the Reacting System Changes to Achieve a New Equilibrium	Does the Value of K Increase or Decrease?
Addition of a reactant	Shift to product formation	Neither
Removal of a reactant		
Addition of a product		
Removal of a product		
Increasing pressure		

8.37 ■ Assume that the following exothermic reaction is at equilibrium:

$$H_2(g) + I_2(g) \rightleftharpoons 2HI(g)$$

Tell whether the position of equilibrium will shift to the right or the left if we
(a) Remove some HI
(b) Add some I_2
(c) Remove some I_2
(d) Increase the temperature
(e) Add a catalyst

8.38 The following reaction is endothermic:

$$3O_2(g) \rightleftharpoons 2O_3(g)$$

If the reaction is at equilibrium, tell whether the equilibrium will shift to the right or the left if we
(a) Remove some O_3
(b) Remove some O_2
(c) Add some O_3
(d) Decrease the temperature
(e) Add a catalyst
(f) Increase the pressure

8.39 The following reaction is exothermic: After it reaches equilibrium, we add a few drops of Br_2.

$$2NO(g) + Br_2(g) \rightleftharpoons 2NOBr(g)$$

(a) What will happen to the equilibrium?
(b) What will happen to the equilibrium constant?

8.40 Is there any change in conditions that changes the equilibrium constant, K, of a given reaction?

8.41 ■ The equilibrium constant at 1127°C for the following endothermic reaction is 571:

$$2H_2S(g) \rightleftharpoons 2H_2(g) + S_2(g)$$

If the mixture is at equilibrium, what happens to K if we
(a) Add some H_2S?
(b) Add some H_2?
(c) Lower the temperature to 1000°C?

Chemical Connections

8.42 (Chemical Connections 8A) In a bacterial infection, body temperature may rise to 101°F. Does this body defense kill the bacteria directly by heat or by another mechanism? If so, by which mechanism?

8.43 (Chemical Connections 8A and 8B) Why is a high fever dangerous? Why is a low body temperature dangerous?

8.44 (Chemical Connections 8B) Why do surgeons sometimes lower body temperatures during heart operations?

8.45 (Chemical Connections 8C) A painkiller—for example, Tylenol—can be purchased in two forms, each containing the same amount of drug. One form is a solid coated pill, and the other is a capsule that contains tiny beads and has the same coat. Which medication will act faster? Explain.

8.46 (Chemical Connections 8D) What reaction takes place when sunlight hits the compound silver chloride?

8.47 (Chemical Connections 8E) If the equilibrium for the Haber process is unfavorable at high temperatures, why do factories nevertheless use high temperatures?

Additional Problems

8.48 In the reaction between H_2 and Cl_2 to give HCl, a 10°C increase in temperature doubles the rate of reaction. If the rate of reaction at 15°C is 2.8 moles of HCl per liter per second, what are the rates at −5°C and at 45°C?

8.49 Draw an energy diagram for an exothermic reaction that yields 75 kcal/mol. The activation energy is 30 kcal/mol.

8.50 Draw a diagram similar to Figure 8.4. Draw a second line of the energy profile starting and ending at the same level as the first but having a smaller peak than the first line. Label them 1 and 2. What may have occurred to change the energy profile of a reaction from 1 to 2?

8.51 ■ For the reaction

$$2NOBr(g) \rightleftharpoons 2NO(g) + Br_2(g)$$

the rate of the reaction was −2.3 mol NOBr/L/h when the initial NOBr concentration was 6.2 mol NOBr/L. What is the rate constant of the reaction?

8.52 The equilibrium constant for the following reaction is 25:

$$2NOBr(g) \rightleftharpoons 2NO(g) + Br_2(g)$$

A measurement made on the equilibrium mixture found that the concentrations of NO and Br_2 were each 0.80 M. What is the concentration of NOBr at equilibrium?

8.53 In the following reaction, the concentration of N_2O_4 in mol/L was measured at the end of the times shown. What is the rate of the reaction?

$$N_2O_4(g) \longrightarrow 2NO_2(g)$$

Time (s)	[N₂O₄]
0	0.200
10	0.180
20	0.160
30	0.140

8.54 How could you increase the rate of a gaseous reaction without adding more reactants or a catalyst and without changing the temperature?

8.55 In an endothermic reaction, the activation energy is 10.0 kcal/mol. Is the activation energy of the reverse reaction also 10.0 kcal/mol, or would it be more or less? Explain with the aid of a diagram.

8.56 Write the reaction to which the following equilibrium expression applies:

$$K = \frac{[NO_2]^4[H_2O]^6}{[NH_3]^4[O_2]^7}$$

8.57 The rate for the following reaction at 300 K was found to be 0.22 M NO₂/min. What would be the approximate rate at 320 K?

$$N_2O_4(g) \rightleftharpoons 2NO_2(g)$$

8.58 Assume that two different reactions are taking place at the same temperature. In reaction A, two different spherical molecules collide to yield a product. In reaction B, the shape of the colliding molecules is rodlike. Each reaction has the same number of collisions per second and the same activation energy. Which reaction goes faster?

8.59 Is it possible for an endothermic reaction to have zero activation energy?

8.60 In the following reaction, the rate of appearance of I_2 is measured at the times shown. What is the rate of the reaction?

$$2HI(g) \rightleftharpoons H_2(g) + I_2(g)$$

Time (s)	[I₂]
0	0
10	0.30
20	0.60
30	0.90

8.61 (Chemical Connections 8D) You have a recipe to manufacture sunglasses: 3.5 g AgCl/kg glass. A new order comes in to manufacture sunglasses to be used in deserts like the Sahara. How would you change the recipe?

8.62 A reaction occurs in three steps which the following rate constants:

$$A \xrightarrow[\text{Step 1}]{k_1=0.3\ M} B \xrightarrow[\text{Step 2}]{k_2=0.05\ M} C \xrightarrow[\text{Step 3}]{k_3=4.5\ M} D$$

(a) Which step is the rate-determining step?

(b) Which step has the lowest activation energy?

Looking Ahead

8.63 As we shall see in Chapter 18, the reaction of a carboxylic acid with an alcohol in the presence of an acid catalyst to form an ester and water is an equilibrium reaction. An example is the reaction of acetic acid with ethanol in the presence of HCl to form ethyl acetate and water.

$$\text{Adding } CH_3COOH \quad CH_3\overset{\displaystyle O}{\overset{\|}{C}}OH + HOCH_2CH_3 \underset{\text{HCl}}{\rightleftharpoons} CH_3\overset{\displaystyle O}{\overset{\|}{C}}OCH_2CH_3 + H_2O$$

Acetic acid   Ethanol   Ethyl acetate

→ Equilibrium shifts to formation of more products

Initial	1.00 mol	1.00 mol	0 mol	0 mol
At equilibrium	0.33 mol	——	——	——

(a) Given the stoichiometry of this reaction from inspection of the balanced equation, fill in the remaining three equilibrium concentrations.

(b) Calculate the equilibrium constant, K, for this reaction.

8.64 As we shall see in Chapter 25, there are two forms of glucose, designated alpha (α) and beta (β), which are in equilibrium in aqueous solution. The equilibrium constant for the reaction is 1.5 at 30°C.

$$\alpha\text{-D-glucose(aq)} \rightleftharpoons \beta\text{-D-glucose(aq)} \quad K = 1.5$$

(a) If you begin with a fresh 1.0 M solution of α-D-glucose in water, what will be its concentration when equilibrium is reached?

(b) Calculate the percentage of α-glucose and of β-glucose present at equilibrium in aqueous solution at 30°C.

Acids and Bases

Some foods and household products are very acidic while others are basic. From your prior experiences, can you tell which ones belong to which category?

Photo, Charles D. Winters; models, S. H. Young

GOB
Chemistry Now™

Look for this logo in the chapter and go to GOB ChemistryNow at **http://now.brookscole.com/gob8** or on the CD for tutorials, simulations, and problems.

9.1 | What Are Acids and Bases?

We frequently encounter acids and bases in our daily lives. Oranges, lemons, and vinegar are examples of acidic foods, and sulfuric acid is in our automobile batteries. As for bases, we take antacid tablets for heartburn and use household ammonia as a cleaning agent. What do these substances have in common? Why are acids and bases usually discussed together?

In 1884, a young Swedish chemist named Svante Arrhenius (1859–1927) answered the first question by proposing what was then a new definition of acids and bases. According to the Arrhenius definition, an **acid** is a substance that produces H_3O^+ ions in aqueous solution, and a **base** is a substance that produces OH^- ions in aqueous solution.

This definition of acid is a slight modification of the original Arrhenius definition, which stated that an acid produces hydrogen ions, H^+. Today we know that H^+ ions cannot exist in water. An H^+ ion is a bare proton, and a charge of $+1$ is too concentrated to exist on such a tiny particle (Section 4.2).

Therefore, an H^+ ion in water immediately combines with an H_2O molecule to give a **hydronium ion,** H_3O^+.

$$H^+(aq) + H_2O(\ell) \longrightarrow H_3O^+(aq)$$
$$\text{Hydronium ion}$$

Apart from this modification, the Arrhenius definitions of acid and base are still valid and useful today, as long as we are talking about aqueous solutions. Although we know that acidic aqueous solutions do not contain H^+ ions, we frequently use the terms "H^+" and "proton" when we really mean "H_3O^+." The three terms are generally used interchangeably.

When an acid dissolves in water, it reacts with the water to produce H_3O^+. For example, hydrogen chloride, HCl, in its pure state is a poisonous gas. When HCl dissolves in water, it reacts with a water molecule to give hydronium ion and chloride ion:

$$H_2O(\ell) + HCl(aq) \longrightarrow H_3O^+(aq) + Cl^-(aq)$$

Thus a bottle labeled aqueous "HCl" is actually not HCl at all, but rather an aqueous solution of H_3O^+ and Cl^- ions in water.

We can show the transfer of a proton from an acid to a base by using a curved arrow. First we write the Lewis structure of each reactant and product. Then we use curved arrows to show the change in position of electron pairs during the reaction. The tail of the curved arrow is located at the electron pair. The head of the curved arrow shows the new position of the electron pair.

In this equation, the curved arrow on the left shows that an unshared pair of electrons on oxygen forms a new covalent bond with hydrogen. The curved arrow on the right shows that the pair of electrons of the H—Cl bond is given entirely to chlorine to form a chloride ion. Thus, in the reaction of HCl with H_2O, a proton is transferred from HCl to H_2O and, in the process, an O—H bond forms and an H—Cl bond is broken.

With bases, the situation is slightly different. Many bases are metal hydroxides, such as KOH, NaOH, $Mg(OH)_2$, and $Ca(OH)_2$. When these ionic solids dissolve in water, their ions merely separate, and each ion is solvated by water molecules (Section 7.6A). For example,

$$NaOH(s) \xrightarrow{H_2O} Na^+(aq) + OH^-(aq)$$

Other bases are not hydroxides. Instead, they produce OH^- ions in water by reacting with water molecules. The most important example of this kind of base is ammonia, NH_3, a poisonous gas. When ammonia dissolves in water, it reacts with water to produce ammonium ions and hydroxide ions.

$$NH_3(aq) + H_2O(\ell) \rightleftharpoons NH_4^+(aq) + OH^-(aq)$$

As we will see in Section 9.2, ammonia is a weak base, and the position of the equilibrium for its reaction with water lies considerably toward the left. In a 1.0 M solution of NH_3 in water, for example, only about 4 molecules of

NH_3 out of every 1000 react with water to form NH_4^+ and OH^-. Thus, when ammonia is dissolved in water, it exists primarily as NH_3 molecules. Nevertheless, some OH^- ions are produced and, therefore, NH_3 is a base.

Bottles of NH_3 in water are sometimes labeled "ammonium hydroxide" or "NH_4OH," but this gives a false impression of what is really in the bottle. Most of the NH_3 molecules have not reacted with the water, so the bottle contains mostly NH_3 and H_2O and only a little NH_4^+ and OH^-.

We indicate how the reaction of ammonia with water takes place by using curved arrows to show the transfer of a proton from a water molecule to an ammonia molecule. Here, the curved arrow on the left shows that the unshared pair of electrons on nitrogen forms a new covalent bond with a hydrogen of a water molecule. At the same time as the new N—H bond forms, an O—H bond of a water molecule breaks and the pair of electrons forming the H—O bond moves entirely to oxygen, forming OH^-.

Thus ammonia produces an OH^- ion by taking H^+ from a water molecule and leaving OH^- behind.

9.2 How Do We Define the Strength of Acids and Bases?

All acids are not equally strong. According to the Arrhenius definition, a **strong acid** is one that reacts completely or almost completely with water to form H_3O^+ ions. Table 9.1 gives the names and molecular formulas for six of the most common strong acids. They are strong acids because, when they dissolve in water, they dissociate completely to give H_3O^+ ions.

Weak acids produce a much smaller concentration of H_3O^+ ions. Acetic acid, for example, is a weak acid. In water it exists primarily as acetic acid molecules; only a few acetic acid molecules (4 out of every 1000) are converted to acetate ions.

$$CH_3COOH(aq) + H_2O(\ell) \rightleftharpoons CH_3COO^-(aq) + H_3O^+(aq)$$

There are four common **strong bases** (Table 9.1), all of which are metal hydroxides. They are strong bases because, when they dissolve in water,

Strong acid An acid that ionizes completely in aqueous solution

Weak acid An acid that is only partially ionized in aqueous solution

Strong base A base that ionizes completely in aqueous solution

Table 9.1 Strong Acids and Bases

Acid Formula	Name	Base Formula	Name
HCl	Hydrochloric acid	LiOH	Lithium hydroxide
HBr	Hydrobromic acid	NaOH	Sodium hydroxide
HI	Hydroiodic acid	KOH	Potassium hydroxide
HNO_3	Nitric acid	$Ba(OH)_2$	Barium hydroxide
H_2SO_4	Sulfuric acid		
$HClO_4$	Perchloric acid		

CHEMICAL CONNECTIONS 9A

Some Important Acids and Bases

STRONG ACIDS Sulfuric acid, H_2SO_4, is used in many industrial processes. In fact, sulfuric acid is one of the most widely produced single chemicals in the United States.

Hydrochloric acid, HCl, is an important acid in chemistry laboratories. Pure HCl is a gas, and the HCl in laboratories is an aqueous solution. HCl is the acid in the gastric fluid in your stomach, where it is secreted at a strength of about 5% w/v.

Nitric acid, HNO_3, is a strong oxidizing agent. A drop of it causes the skin to turn yellow because the acid reacts with skin proteins. A yellow color upon contact with nitric acid has long been a test for proteins.

WEAK ACIDS Acetic acid, CH_3COOH, is present in vinegar (about 5%). Pure acetic acid is called glacial acetic acid because of its melting point of 17°C, which means that it freezes on a moderately cold day.

Boric acid, H_3BO_3, is a solid. Solutions of boric acid in water were once used as antiseptics, especially for eyes. Boric acid is toxic when swallowed.

Phosphoric acid, H_3PO_4, is one of the strongest of the weak acids. The ions produced from it—$H_2PO_4^-$, HPO_4^{2-}, and PO_4^{3-}—are important in biochemistry (see also Section 27.3).

STRONG BASES Sodium hydroxide, NaOH, also called lye, is the most important of the strong bases. It is a solid whose aqueous

Weak acids are found in many common materials. In the foreground are strips of litmus paper that have been dipped into solutions of these materials. Acids turn litmus paper red.

Weak bases are also common in many household products. These cleaning agents all contain weak bases. Bases turn litmus paper blue.

solutions are used in many industrial processes, including the manufacture of glass and soap. Potassium hydroxide, KOH, also a solid, is used for many of the same purposes as NaOH.

WEAK BASES Ammonia, NH_3, the most important weak base, is a gas with many industrial uses. One of its chief uses is for fertilizers. A 5% solution is sold in supermarkets as a cleaning agent, and weaker solutions are used as "spirits of ammonia" to revive people who have fainted.

Magnesium hydroxide, $Mg(OH)_2$, is a solid that is insoluble in water. A suspension of about 8% $Mg(OH)_2$ in water is called milk of magnesia and is used as a laxative. $Mg(OH)_2$ is also used to treat wastewater in metal-processing plants and as a flame retardant in plastics.

Weak base A base that is only partially ionized in aqueous solution

they ionize completely to give OH^- ions. Another base, $Mg(OH)_2$, dissociates almost completely once dissolved, but it is also very insoluble in water to begin with. As we saw in Section 9.1, ammonia is a **weak base** because the equilibrium for its reaction with water lies far to the left.

It is important to understand that the strength of an acid or a base is not related to its concentration. HCl is a strong acid, whether it is concentrated or dilute, because it dissociates completely in water to chloride ions and hydronium ions. Acetic acid is a weak acid, whether it is concentrated or dilute, because the equilibrium for its reaction with water lies far to the left. When acetic acid dissolves in water, most of it is present as undissociated CH_3COOH molecules.

$$HCl(aq) + H_2O(\ell) \longrightarrow Cl^-(aq) + H_3O^+(aq)$$

$$\underset{\text{Acetic acid}}{CH_3COOH(aq)} + H_2O(\ell) \rightleftharpoons \underset{\text{Acetate ion}}{CH_3COO^-(aq)} + H_3O^+(aq)$$

In Section 7.6C, we saw that electrolytes (substances that produce ions in aqueous solution) can be strong or weak. The strong acids and bases in Table 9.1 are strong electrolytes. Almost all other acids and bases are weak electrolytes.

9.3 | What Are Conjugate Acid–Base Pairs?

The Arrhenius definitions of acid and base are very useful in aqueous solutions. But what if water is not involved? In 1923, the Danish chemist Johannes Brønsted and the English chemist Thomas Lowry independently proposed the following definitions: An **acid** is a proton donor, a **base** is a proton acceptor, and an **acid–base reaction** is a proton transfer reaction. Furthermore, according to the Brønsted-Lowry definitions, any pair of molecules or ions that can be interconverted by transfer of a proton is called a **conjugate acid–base pair.** When an acid transfers a proton to a base, the acid is converted to its **conjugate base.** When a base accepts a proton, it is converted to its **conjugate acid.**

We can illustrate these relationships by examining the reaction between acetic acid and ammonia:

$$CH_3COOH + NH_3 \rightleftharpoons CH_3COO^- + NH_4^+$$

Acetic acid   Ammonia   Acetate   Ammonium
                               ion   ion
(Acid)   (Base)   (Conjugate   (Conjugate
                              base of   acid of
                           acetic acid)   ammonia)

> **Conjugate acid–base pair** A pair of molecules or ions that are related to one another by the gain or loss of a proton
>
> **Conjugate base** In the Brønsted-Lowry theory, a substance formed when an acid donates a proton to another molecule or ion
>
> **Conjugate acid** In the Brønsted-Lowry theory, a substance formed when a base accepts a proton

We can use curved arrows to show how this reaction takes place. The curved arrow on the right shows that the unshared pair of electrons on nitrogen becomes shared to form a new H—N bond. At the same time that the H—N bond forms, the O—H bond breaks and the electron pair of the O—H bond moves entirely to oxygen to form —O⁻ of the acetate ion. The result of these two electron-pair shifts is the transfer of a proton from an acetic acid molecule to an ammonia molecule:

Acetic acid   Ammonia   Acetate   Ammonium
(Proton donor) (Proton acceptor) ion   ion

A box of Arm & Hammer baking soda (sodium bicarbonate). Sodium bicarbonate is composed of Na^+ and HCO_3^-, the amphiprotic bicarbonate ion.

Table 9.2 gives examples of common acids and their conjugate bases. As you study the examples of conjugate acid–base pairs in Table 9.2, note the following points:

1. An acid can be positively charged, neutral, or negatively charged. Examples of these charge types are H_3O^+, H_2CO_3, and $H_2PO_4^-$, respectively.

2. A base can be negatively charged or neutral. Examples of these charge types are PO_4^{3-} and NH_3, respectively.

3. Acids are classified as monoprotic, diprotic, or triprotic depending on the number of protons each may give up. Examples of **monoprotic acids** include HCl, HNO_3, and CH_3COOH. Examples of **diprotic acids**

> **Monoprotic acid** An acid that can give up only one proton
>
> **Diprotic acid** An acid that can give up two protons

Table 9.2 Some Acids and Their Conjugate Bases

	Acid	Name	Conjugate Base	Name	
Strong Acids	HI	Hydroiodic acid	I^-	Iodide ion	Weak Bases
	HCl	Hydrochloric acid	Cl^-	Chloride ion	
	H_2SO_4	Sulfuric acid	HSO_4^-	Hydrogen sulfate ion	
	HNO_3	Nitric acid	NO_3^-	Nitrate ion	
	H_3O^+	Hydronium ion	H_2O	Water	
	HSO_4^-	Hydrogen sulfate ion	SO_4^{2-}	Sulfate ion	
	H_3PO_4	Phosphoric acid	$H_2PO_4^-$	Dihydrogen phosphate ion	
	CH_3COOH	Acetic acid	CH_3COO^-	Acetate ion	
	H_2CO_3	Carbonic acid	HCO_3^-	Bicarbonate ion	
	H_2S	Hydrogen sulfide	HS^-	Hydrogen sulfide ion	
	$H_2PO_4^-$	Dihydrogen phosphate ion	HPO_4^{2-}	Hydrogen phosphate ion	
	NH_4^+	Ammonium ion	NH_3	Ammonia	
	HCN	Hydrocyanic acid	CN^-	Cyanide ion	
	C_6H_5OH	Phenol	$C_6H_5O^-$	Phenoxide ion	
	HCO_3^-	Bicarbonate ion	CO_3^{2-}	Carbonate ion	
	HPO_4^{2-}	Hydrogen phosphate ion	PO_4^{3-}	Phosphate ion	
Weak Acids	H_2O	Water	OH^-	Hydroxide ion	Strong Bases
	C_2H_5OH	Ethanol	$C_2H_5O^-$	Ethoxide ion	

Triprotic acid An acid that can give up three protons

include H_2SO_4 and H_2CO_3. An example of a **triprotic acid** is H_3PO_4. Carbonic acid, for example, loses one proton to become bicarbonate ion, and then a second proton to become carbonate ion.

$$H_2CO_3 + H_2O \rightleftharpoons HCO_3^- + H_3O^+$$
Carbonic acid / Bicarbonate ion

$$HCO_3^- + H_2O \rightleftharpoons CO_3^{2-} + H_3O^+$$
Bicarbonate ion / Carbonate ion

4. Several molecules and ions appear in both the acid and conjugate base columns; that is, each can function as either an acid or a base. The bicarbonate ion, HCO_3^-, for example, can give up a proton to become CO_3^{2-} (in which case it is an acid) or it can accept a proton to become H_2CO_3 (in which case it is a base). A substance that can act as either an acid or a base is called **amphiprotic.** The most important amphiprotic substance in Table 9.2 is water, which can accept a proton to become H_3O^+ or lose a proton to become OH^-.

Amphiprotic A substance that can act as either an acid or a base

5. A substance cannot be a Brønsted-Lowry acid unless it contains a hydrogen atom, but not all hydrogen atoms can be given up. For example, acetic acid, CH_3COOH, has four hydrogens but is monoprotic; it gives up only one of them. Similarly, phenol, C_6H_5OH, gives up only one of its six hydrogens:

$$C_6H_5OH + H_2O \rightleftharpoons C_6H_5O^- + H_3O^+$$
Phenol / Phenoxide ion

This is because a hydrogen must be bonded to a strongly electronegative atom, such as oxygen or a halogen, to be acidic.

6. There is an inverse relationship between the strength of an acid and the strength of its conjugate base: The stronger the acid, the weaker its conjugate base. HI, for example, is the strongest acid listed in Table 9.2 and I^-, its conjugate base, is the weakest base. As another example, CH_3COOH (acetic acid) is a stronger acid than H_2CO_3 (carbonic acid); conversely, CH_3COO^- (acetate ion) is a weaker base than HCO_3^- (bicarbonate ion).

EXAMPLE 9.1

Show how the amphiprotic ion hydrogen sulfate, HSO_4^-, can react as both an acid and a base.

Solution
Hydrogen sulfate reacts like an acid in the equation shown below:

$$HSO_4^- + H_2O \rightleftharpoons H_3O^+ + SO_4^{2-}$$

It can react like a base in the equation shown below:

$$HSO_4^- + H_3O^+ \rightleftharpoons H_2O + H_2SO_4$$

Problem 9.1
Draw the acid and base reactions for the amphiprotic ion, HPO_4^{2-}.

GOB
Chemistry Now™
Click *Chemistry Interactive* to learn more about **Acid–Base Conjugate Pairs**

9.4 | How Can We Tell the Position of Equilibrium in an Acid–Base Reaction?

We know that HCl reacts with H_2O according to the following equilibrium:

$$HCl + H_2O \rightleftharpoons Cl^- + H_3O^+$$

We also know that HCl is a strong acid, which means the position of this equilibrium lies very far to the right. In fact, this equilibrium lies so far to the right that out of every 10,000 HCl molecules dissolved in water, all but one react with water molecules to give Cl^- and H_3O^+.

For this reason, we usually write the acid reaction of HCl with a unidirectional arrow, as follows:

$$HCl + H_2O \longrightarrow Cl^- + H_3O^+$$

As we have also seen, acetic acid reacts with H_2O according to the following equilibrium:

$$\underset{\text{Acetic acid}}{CH_3COOH} + H_2O \rightleftharpoons \underset{\text{Acetate ion}}{CH_3COO^-} + H_3O^+$$

Acetic acid is a weak acid. Only a few acetic acid molecules react with water to give acetate ions and hydronium ions, and the major species present in equilibrium in aqueous solution is CH_3COOH and H_2O. The position of this equilibrium, therefore, lies very far to the left.

In these two acid–base reactions, water is the base. But what if we have a base other than water as the proton acceptor? How can we determine which are the major species present at equilibrium? That is, how can we determine if the position of equilibrium lies toward the left or toward the right?

As an example, let us examine the acid–base reaction between acetic acid and ammonia to form acetate ion and ammonium ion. As indicated by the question mark over the equilibrium arrow, we want to determine whether the position of this equilibrium lies toward the left or toward the right.

$$CH_3COOH + NH_3 \overset{?}{\rightleftharpoons} CH_3COO^- + NH_4^+$$

Acetic acid	Ammonia	Acetate ion	Ammonium ion
(Acid)	(Base)	(Conjugate base of CH_3COOH)	(Conjugate acid of NH_3)

In this equilibrium there are two acids present: acetic acid and ammonium ion. There are also two bases present: ammonia and acetate ion. One way to analyze this equilibrium is to view it as a competition of the two bases, ammonia and acetate ion, for a proton. Which is the stronger base? The information we need to answer this question is found in Table 9.2. We first determine which conjugate acid is the stronger acid and then use this information along with the fact that the stronger the acid, the weaker its conjugate base. From Table 9.2, we see that CH_3COOH is the stronger acid, which means that CH_3COO^- is the weaker base. Conversely, NH_4^+ is the weaker acid, which means that NH_3 is the stronger base. We can now label the relative strengths of each acid and base in this equilibrium:

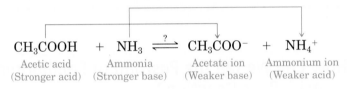

$$CH_3COOH + NH_3 \overset{?}{\rightleftharpoons} CH_3COO^- + NH_4^+$$

Acetic acid	Ammonia	Acetate ion	Ammonium ion
(Stronger acid)	(Stronger base)	(Weaker base)	(Weaker acid)

In an acid–base reaction, the equilibrium position always favors reaction of the stronger acid and stronger base to form the weaker acid and

---

CHEMICAL CONNECTIONS 9B

Acid and Base Burns of the Cornea

The cornea is the outermost part of the eye. This transparent tissue is very sensitive to chemical burns, whether caused by acids or by bases. Acids and bases in small amounts and in low concentrations that would not seriously damage other tissues, such as skin, may cause severe corneal burns. If not treated promptly, these burns can lead to permanent loss of vision.

For acid splashed in the eyes, immediately wash the eyes with a steady stream of cold water. Quickly removing the acid is of utmost importance. Time should not be wasted looking for a solution of some mild base to neutralize it because removing the acid is more helpful than neutralizing it. The extent of the burn can be assessed after all the acid has been washed out.

Concentrated ammonia or sodium hydroxide of the "liquid plumber" type can also cause severe damage when splattered in the eyes. Again, immediately washing the eyes is of utmost importance; however, burns by bases can later develop ulcerations in which the healing wound deposits scar tissue that is not transparent, thereby impairing vision. For this reason, first aid is not sufficient for burns by bases, and a physician should be consulted immediately.

Charles D. Winters

Cleaning products like these contain bases. Care must be taken to make sure that they are not splashed into anyone's eyes.

Fortunately, the cornea is a tissue with no blood vessels and, therefore, can be transplanted without immunological rejection problems. Today, corneal grafts are common.

weaker base. Thus, at equilibrium, the major species present are the weaker acid and the weaker base. In the reaction between acetic acid and ammonia, therefore, the equilibrium lies to the right and the major species present are acetate ion and ammonium ion:

$$CH_3COOH \;+\; NH_3 \;\rightleftharpoons\; CH_3COO^- \;+\; NH_4^+$$

Acetic acid	Ammonia	Acetate ion	Ammonium ion
(Stronger acid)	(Stronger base)	(Weaker base)	(Weaker acid)

To summarize, we use the following four steps to determine the position of an acid–base equilibrium:

1. Identify the two acids in the equilibrium; one is on the left side of the equilibrium, and the other on the right side.
2. Using the information in Table 9.2, determine which acid is the stronger acid and which acid is the weaker acid.
3. Identify the stronger base and the weaker base. Remember that the stronger acid gives the weaker conjugate base and the weaker acid gives the stronger conjugate base.
4. The stronger acid and stronger base react to give the weaker acid and weaker base. The position of equilibrium, therefore, lies on the side of the weaker acid and weaker base.

EXAMPLE 9.2

For each acid–base equilibrium, label the stronger acid, the stronger base, the weaker acid, and the weaker base. Then predict whether the position of equilibrium lies toward the right or toward the left.

(a) $H_2CO_3 \;+\; OH^- \rightleftharpoons HCO_3^- \;+\; H_2O$

(b) $HPO_4^{2-} \;+\; NH_3 \rightleftharpoons PO_4^{3-} \;+\; NH_4^+$

Solution

Arrows connect the conjugate acid–base pairs, with the red arrows showing the stronger acid. The position of equilibrium in (a) lies toward the right. In (b) it lies toward the left.

Problem 9.2

For each acid–base equilibrium, label the stronger acid, the stronger base, the weaker acid, and the weaker base. Then predict whether the position of equilibrium lies toward the right or the left.

(a) $H_3O^+ \;+\; I^- \rightleftharpoons H_2O \;+\; HI$

(b) $CH_3COO^- \;+\; H_2S \rightleftharpoons CH_3COOH \;+\; HS^-$

9.5 | How Do We Use Acid Ionization Constants?

In Section 9.2, we learned that acids vary in the extent to which they produce H_3O^+ when added to water. Because the ionizations of weak acids in water are all equilibria, we can use equilibrium constants (Section 8.6) to tell us quantitatively just how strong any weak acid is. The reaction that takes place when a weak acid, HA, is added to water is

$$HA + H_2O \rightleftharpoons A^- + H_3O^+$$

The equilibrium constant expression for this ionization is

$$K = \frac{[A^-][H_3O^+]}{[HA][H_2O]}$$

Notice that this expression contains the concentration of water. Because water is the solvent and its concentration changes very little when we add HA to it, we can treat the concentration of water, $[H_2O]$, as a constant equal to 1000 g/L or approximately 55.5 mol/L. We can then combine these two constants (K and $[H_2O]$) to define a new constant called an **acid ionization constant, K_a.**

Acid ionization constant (K_a)
An equilibrium constant for the ionization of an acid in aqueous solution to H_3O^+ and its conjugate base; also called an acid dissociation constant

$$K_a = K[H_2O] = \frac{[A^-][H_3O^+]}{[HA]}$$

The value of the acid ionization constant for acetic acid, for example, is 1.8×10^{-5}. Because acid ionization constants for weak acids are numbers with negative exponents, we often use an algebraic trick to turn them into numbers that are easier to use. To do so, we take the negative logarithm of the number. Acid strengths are therefore expressed as $-\log K_a$, which we call the pK_a. The "p" of anything is just the negative logarithm of that thing. The pK_a of acetic acid is 4.75. Table 9.3 gives names, molecular formulas, and values of K_a and pK_a for some weak acids. As you study the entries in this table, note the inverse relationship between the values of K_a and pK_a. The weaker the acid, the smaller its K_a, but the larger its pK_a.

pK_a is $-\log K_a$

One reason for the importance of K_a is that it immediately tells us how strong an acid is. For example, Table 9.3 shows us that although acetic acid, formic acid, and phenol are all weak acids, their strengths as acids are not

Table 9.3 K_a and pK_a Values for Some Weak Acids

Formula	Name	K_a	pK_a
H_3PO_4	Phosphoric acid	7.5×10^{-3}	2.12
HCOOH	Formic acid	1.8×10^{-4}	3.75
$CH_3CH(OH)COOH$	Lactic acid	1.4×10^{-4}	3.86
CH_3COOH	Acetic acid	1.8×10^{-5}	4.75
H_2CO_3	Carbonic acid	4.3×10^{-7}	6.37
$H_2PO_4^-$	Dihydrogen phosphate ion	6.2×10^{-8}	7.21
H_3BO_3	Boric acid	7.3×10^{-10}	9.14
NH_4^+	Ammonium ion	5.6×10^{-10}	9.25
HCN	Hydrocyanic acid	4.9×10^{-10}	9.31
C_6H_5OH	Phenol	1.3×10^{-10}	9.89
HCO_3^-	Bicarbonate ion	5.6×10^{-11}	10.25
HPO_4^{2-}	Hydrogen phosphate ion	2.2×10^{-13}	12.66

Increasing acid strength

the same. Formic acid, with a K_a of 1.8×10^{-4}, is stronger than acetic acid, whereas phenol, with a K_a of 1.3×10^{-10}, is much weaker than acetic acid. Phosphoric acid is the strongest of the weak acids. We can tell that an acid is classified as a weak acid by the fact that we list a pK_a for it, and the pK_a is a positive number. If we tried to take the negative logarithm of the K_a for a strong acid, we would get a negative number.

EXAMPLE 9.3

K_a for benzoic acid is 6.5×10^{-5}. What is the pK_a of this acid?

Solution
Take the logarithm of 6.5×10^{-5} on your scientific calculator. The answer is -4.19. Because pK_a is equal to $-\log K_a$, you must multiply this value by -1 to get pK_a. The pK_a of benzoic acid is 4.19.

Problem 9.3
K_a for hydrocyanic acid, HCN, is 4.9×10^{-10}. What is its pK_a?

EXAMPLE 9.4

Which is the stronger acid:
(a) Benzoic acid with a K_a of 6.5×10^{-5} or hydrocyanic acid with a K_a of 4.9×10^{-10}?
(b) Boric acid with a pK_a of 9.14 or carbonic acid with a pK_a of 6.37?

Solution
(a) Benzoic acid is the stronger acid, it has the larger K_a value.
(b) Carbonic acid is the stronger acid; it has the smaller pK_a value.

Problem 9.4
Which is the stronger acid:
(a) Carbonic acid, $pK_a = 6.37$, or ascorbic acid (vitamin C), $pK_a = 4.1$?
(b) Aspirin, $pK_a = 3.49$, or acetic acid, $pK_a = 4.75$?

All of these fruits and fruit drinks contain organic acids.

9.6 | What Are the Properties of Acids and Bases?

Today's chemists do not taste the substances they work with, but 200 years ago they routinely did so. That is how we know that acids taste sour and bases taste bitter. The sour taste of lemons, vinegar, and many other foods, for example, is due to the acids they contain.

A. Neutralization

The most important reaction of acids and bases is that they react with each other in a process called neutralization. This name is appropriate because, when a strong corrosive acid such as hydrochloric acid reacts with a strong corrosive base such as sodium hydroxide, the product (a solution of ordinary table salt in water) has neither acidic nor basic properties. We call such a solution neutral. Section 9.9 discusses neutralization reactions in detail.

GOB
Chemistry⋅᛭⋅Now™
Active Figure 9.1 A ribbon of magnesium metal reacts with aqueous HCl to give H_2 gas and aqueous $MgCl_2$. **See a simulation based on this figure, and take a short quiz on the concepts at http://www.now.brookscole.com/gob8 or on the CD.**

B. Reaction with Metals

Strong acids react with certain metals (called active metals) to produce hydrogen gas, H_2, and a salt. Hydrochloric acid, for example, reacts with magnesium metal to give the salt magnesium chloride and hydrogen gas (Figure 9.1).

$$Mg(s) + 2HCl(aq) \longrightarrow MgCl_2(aq) + H_2(g)$$

Magnesium   Hydrochloric   Magnesium   Hydrogen
acid   chloride

The reaction of an acid with an active metal to give a salt and hydrogen gas is a redox reaction (Section 5.7). The metal is oxidized to a metal ion and H^+ is reduced to H_2.

C. Reaction with Metal Hydroxides

Acids react with metal hydroxides to give a salt and water.

$$HCl(aq) + KOH(aq) \longrightarrow H_2O(\ell) + KCl(aq)$$

Hydrochloric   Potassium   Water   Potassium
acid   hydroxide   chloride

Both the acid and the metal hydroxide are ionized in aqueous solution. Furthermore, the salt formed is an ionic compound that is present in aqueous solution as anions and cations. Therefore, the actual equation for the reaction of HCl and KOH could be written showing all of the ions present (Section 5.6):

$$H_3O^+(aq) + Cl^-(aq) + K^+(aq) + OH^-(aq) \longrightarrow 2H_2O(\ell) + Cl^-(aq) + K^+(aq)$$

We usually simplify this equation by omitting the spectator ions (Section 5.6), which gives the following equation for the net ionic reaction of any strong acid and strong base to give a salt and water:

$$H_3O^+(aq) + OH^-(aq) \longrightarrow 2H_2O(\ell)$$

D. Reaction with Metal Oxides

Strong acids react with metal oxides to give water and a salt, as shown in the following net ionic equation:

$$2H_3O^+(aq) + CaO(s) \longrightarrow 3H_2O(\ell) + Ca^{2+}(aq)$$

Calcium
oxide

E. Reaction with Carbonates and Bicarbonates

When a strong acid is added to a carbonate such as sodium carbonate, bubbles of carbon dioxide gas are rapidly given off. The overall reaction is a summation of two reactions. In the first reaction, carbonate ion reacts with H_3O^+ to give carbonic acid. Almost immediately, in the second reaction, carbonic acid decomposes to carbon dioxide and water. The following equations show the individual reactions and then the overall reaction:

$$2H_3O^+(aq) + CO_3^{2-}(aq) \longrightarrow H_2CO_3(aq) + 2H_2O(\ell)$$
$$H_2CO_3(aq) \longrightarrow CO_2(g) + H_2O(\ell)$$
$$\overline{2H_3O^+(aq) + CO_3^{2-}(aq) \longrightarrow CO_2(g) + 3H_2O(\ell)}$$

CHEMICAL CONNECTIONS 9C

Drugstore Antacids

Stomach fluid is normally quite acidic because of its HCl content. At some time, you probably have gotten "heartburn" caused by excess stomach acidity. To relieve your discomfort, you may have taken an antacid, which, as the name implies, is a substance that neutralizes acids—in other words, a base.

The word "antacid" is a medical term, not one used by chemists. It is, however, found on the labels of many medications available in drugstores and supermarkets. Almost all of them use bases such as $CaCO_3$, $Mg(OH)_2$, $Al(OH)_3$, and $NaHCO_3$ to decrease the acidity of the stomach.

Also in drugstores and supermarkets are nonprescription drugs labeled "acid reducers." Among these brands are Zantac, Tagamet, Pepcid, and Axid. Instead of neutralizing acidity, these compounds reduce the secretion of acid into the stomach. In larger doses (sold only with a prescription), some of these drugs are used in the treatment of stomach ulcers.

Commercial remedies for excess stomach acid.

Charles D. Winters

Strong acids also react with bicarbonates such as potassium bicarbonate to give carbon dioxide and water:

$$H_3O^+(aq) + HCO_3^-(aq) \longrightarrow H_2CO_3(aq) + H_2O(\ell)$$
$$\underline{H_2CO_3(aq) \longrightarrow CO_2(g) + H_2O(\ell)}$$
$$H_3O^+(aq) + HCO_3^-(aq) \longrightarrow CO_2(g) + 2H_2O(\ell)$$

To generalize, any acid stronger than carbonic acid will react with carbonate or bicarbonate ion to give CO_2 gas.

The production of CO_2 is what makes bread doughs and cake batters rise. The earliest method used to generate CO_2 for this purpose involved the addition of yeast, which catalyzes the fermentation of carbohydrates to produce carbon dioxide and ethanol (Chapter 28):

$$\underset{\text{Glucose}}{C_6H_{12}O_6} \xrightarrow{\text{Yeast}} 2CO_2 + \underset{\text{Ethanol}}{2C_2H_5OH}$$

The production of CO_2 by fermentation, however, is slow. Sometimes it is desirable to have its production take place more rapidly, in which case bakers use the reaction of $NaHCO_3$ (sodium bicarbonate, also called **baking soda**) and a weak acid. But which weak acid? Vinegar (a 5% solution of acetic acid in water) would work, but it has a potential disadvantage—it imparts a particular flavor to foods. For a weak acid that imparts little or no flavor, bakers use either sodium dihydrogen phosphate, NaH_2PO_4, or potassium dihydrogen phosphate, KH_2PO_4. The two salts do not react when they are dry but, when mixed with water in a dough or batter, they react quite rapidly to produce CO_2. The production of CO_2 is even more rapid in an oven!

$$H_2PO_4^-(aq) + H_2O(\ell) \rightleftharpoons HPO_4^{2-}(aq) + H_3O^+(aq)$$
$$\underline{HCO_3^-(aq) + H_3O^+(aq) \longrightarrow CO_2(g) + 2H_2O(\ell)}$$
$$H_2PO_4^-(aq) + HCO_3^-(aq) \longrightarrow HPO_4^{2-}(aq) + CO_2(g) + H_2O(\ell)$$

Baking powder contains a weak acid, either sodium or potassium dihydrogen phosphate, and sodium or potassium bicarbonate. When they are mixed with water, they react to produce the bubbles of CO_2 seen in this picture.

Charles D. Winters

F. Reaction with Ammonia and Amines

Any acid stronger than NH_4^+ (Table 9.2) is strong enough to react with NH_3 to form a salt. In the following reaction, the salt formed is ammonium chloride, NH_4Cl, which is shown as it would be ionized in aqueous solution:

$$HCl(aq) + NH_3(aq) \longrightarrow NH_4^+(aq) + Cl^-(aq)$$

In Chapter 16 we will meet a family of compounds called amines, which are similar to ammonia except that one or more of the three hydrogen atoms of ammonia are replaced by carbon groups. A typical amine is methylamine, CH_3NH_2. The base strength of most amines is similar to that of NH_3, which means that amines also react with acids to form salts. The salt formed in the reaction of methylamine with HCl is methylammonium chloride, shown here as it would be ionized in aqueous solution:

$$\underset{\text{Methylamine}}{HCl(aq) + CH_3NH_2(aq)} \longrightarrow \underset{\substack{\text{Methylammonium}\\\text{ion}}}{CH_3NH_3^+(aq) + Cl^-(aq)}$$

The reaction of ammonia and amines with acids to form salts is very important in the chemistry of the body, as we will see in later chapters.

9.7 | What Are the Acidic and Basic Properties of Pure Water?

We have seen that an acid produces H_3O^+ ions in water and that a base produces OH^- ions. Suppose that we have absolutely pure water, with no added acid or base. Surprisingly enough, even pure water contains a very small number of H_3O^+ and OH^- ions. They are formed by the transfer of a proton from one molecule of water (the proton donor) to another (the proton acceptor).

$$\underset{\text{Acid}}{H_2O} + \underset{\text{Base}}{H_2O} \rightleftharpoons \underset{\substack{\text{Conjugate}\\\text{base of } H_2O}}{OH^-} + \underset{\substack{\text{Conjugate}\\\text{acid of } H_2O}}{H_3O^+}$$

What is the extent of this reaction? We know from the information in Table 9.2 that, in this equilibrium, H_3O^+ is the stronger acid and OH^- is the stronger base. Therefore, as shown by the arrows, the equilibrium for this reaction lies far to the left. We shall soon see exactly how far, but first let us write the equilibrium expression:

$$K = \frac{[H_3O^+][OH^-]}{[H_2O]^2}$$

Because the degree of self-ionization of water is so slight, we can treat the concentration of water, $[H_2O]$, as a constant equal to 1000 g/L or approximately 55.5 mol/L, just as we did in Section 9.5 in developing K_a for a weak acid. We can then combine these two constants (K and $[H_2O]^2$) to define a new constant called the **ion product of water**, K_w. In pure water at room temperature, K_w has a value of 1.0×10^{-14}.

$$K_w = K[H_2O]^2 = [H_3O^+][OH^-]$$

$$K_w = 1.0 \times 10^{-14}$$

K_w is the ion product of water, also called the water constant, and is equal to 1×10^{-14}.

In pure water, H_3O^+ and OH^- form in equal amounts (see the balanced equation for the self-ionization of water), so their concentrations must be equal. That is, in pure water,

$$\left.\begin{array}{l} [H_3O^+] = 1.0 \times 10^{-7} \text{ mol/L} \\ [OH^-] = 1.0 \times 10^{-7} \text{ mol/L} \end{array}\right\} \text{ In pure water}$$

These are very small concentrations, not enough to make pure water a conductor of electricity. Pure water is not an electrolyte (Section 7.6C).

The equation for the ionization of water is important because it applies not only to pure water but also to any water solution. The product of $[H_3O^+]$ and $[OH^-]$ in any aqueous solution is equal to 1.0×10^{-14}. If, for example, we add 0.010 mol of HCl to 1 L of pure water, it reacts completely to give H_3O^+ ions and Cl^- ions. The concentration of H_3O^+ will be 0.010 M, or 1.0×10^{-2} M. This means that $[OH^-]$ must be $1.0 \times 10^{-14}/1.0 \times 10^{-2} = 1.0 \times 10^{-12}$ M.

EXAMPLE 9.5

The $[OH^-]$ of an aqueous solution is 1.0×10^{-4} M. What is its $[H_3O^+]$?

Solution
We substitute into the equation:

$$[H_3O^+][OH^-] = 1.0 \times 10^{-14}$$

$$[H_3O^+] = \frac{1.0 \times 10^{-14}}{1.0 \times 10^{-4}} = 1.0 \times 10^{-10} \text{ } M$$

Problem 9.5
The $[OH^-]$ of an aqueous solution is 1.0×10^{-12} M. What is its $[H_3O^+]$?

Aqueous solutions can have a very high $[H_3O^+]$ but the $[OH^-]$ must then be very low, and vice versa. Any solution with a $[H_3O^+]$ greater than 1.0×10^{-7} M is acidic. In such solutions, of necessity $[OH^-]$ must be less than 1.0×10^{-7} M. The higher the $[H_3O^+]$, the more acidic the solution. Similarly, any solution with an $[OH^-]$ greater than 1.0×10^{-7} M is basic. Pure water, in which $[H_3O^+]$ and $[OH^-]$ are equal (they are both 1.0×10^{-7} M), is neutral—that is, neither acidic nor basic.

GOB
Chemistry ⚛ Now™

Click *Chemistry Interactive* to learn more about the **Self-Ionization of Water**

HOW TO ...

Use Logs and Antilogs

When dealing with acids, bases, and buffers, we often have to use common or base 10 logarithms (logs). To most people, a logarithm is just a button they push on a calculator. Here we describe briefly how to handle logs and antilogs.

1. What is a logarithm and how is it calculated?

A common logarithm is the power to which you raise 10 to get another number. For example, the log of 100 is 2, since you must raise 10 to the second power to get 100.

$$\log 100 = 2 \quad \text{since} \quad 10^2 = 100$$

Other examples are

$$\log 1000 = 3 \quad \text{since} \quad 10^3 = 1000$$
$$\log 10 = 1 \quad \text{since} \quad 10^1 = 10$$
$$\log 1 = 0 \quad \text{since} \quad 10^0 = 0$$
$$\log 0.1 = -1 \quad \text{since} \quad 10^{-1} = 0.1$$

The common logarithm of a number other than a simple power is usually obtained from a calculator by entering the number and then pressing log. For example,

$$\log 52 = 1.72$$
$$\log 4.5 = 0.653$$
$$\log 0.25 = -0.602$$

Try it now. Enter 100 and then press log. Did you get 2? If so you did it right. Try again with 52. Enter 52 and press log. Did you get 1.72 (rounded to two decimal places)?

2. What are antilogarithms (antilogs)?

An antilog is the reverse of a log. It is also called the inverse log. If you take 10 and raise it to a power, you are taking an antilog. For example,

$$\text{antilog } 5 = 100,000$$

because taking the antilog of 5 means raising 10 to the power of 5 or

$$10^5 = 100,000$$

Try it now on your calculator. What is the antilog of 3? Enter 3 on your calculator. Press INV (inverse) or 2ndF (second function), and then press log. The answer should be 1000.

3. What is the Difference between antilog and −log?

There is a huge and very important difference. Antilog 3 means that we take 10 and raise it to the power of 3, so we get 1000. In contrast, −log 3 means that we take the log of 3, which equals 0.477, and take the negative of it. Thus −log 3 equals −0.48. For example,

$$\text{antilog } 2 = 100$$
$$-\log 2 = -0.30$$

In Section 9.8, we will use negative logs to calculate pH. The pH equals $-\log [H^+]$. Thus, if we know that $[H^+]$ is 0.01 M, to find the pH we enter 0.01 into our calculator and press log. That gives an answer of −2. Then we take the negative of that value to give a pH of 2.

In the last example, what answer would we have gotten if we had taken the antilog instead of the negative log? We would have gotten what we started with: 0.01. Why? Because all we calculated was antilog log 0.01. If we take the antilog of the log, we have not done anything at all.

9.8 | What Are pH and pOH?

Because hydronium ion concentrations for most solutions are numbers with negative exponents, these concentrations are more conveniently expressed as pH, where

$$pH = -\log [H_3O^+]$$

similarly to how we expressed pK_a values in Section 9.5.

In Section 9.7, we saw that a solution is acidic if its $[H_3O^+]$ is greater than 1.0×10^{-7}, and that it is basic if its $[H_3O^+]$ is less than 1.0×10^{-7}. We can now state the definitions of acidic and basic solutions in terms of pH.

A solution is acidic if its pH is less than 7.0
A solution is basic if its pH is greater than 7.0
A solution is neutral if its pH is equal to 7.0

The pH of this soft drink is 3.12. Soft drinks are often quite acidic.

EXAMPLE 9.6

(a) The $[H_3O^+]$ of a certain liquid detergent is 1.4×10^{-9} M. What is its pH? Is this solution acidic, basic, or neutral?
(b) The pH of black coffee is 5.3. What is its $[H_3O^+]$? Is it acidic, basic, or neutral?

Solution

(a) On your calculator, take the log of 1.4×10^{-9}. The answer is -8.85. Multiply this value by -1 to give the pH of 8.85. This solution is basic.
(b) Enter 5.3 into your calculator and then press the +/- key to change the sign to minus and give -5.3. Then take the antilog of this number. The $[H_3O^+]$ of black coffee is 5×10^{-6}. This solution is acidic.

Problem 9.6

(a) The $[H_3O^+]$ of an acidic solution is 3.5×10^{-3} M. What is its pH?
(b) The pH of tomato juice is 4.1. What is its $[H_3O^+]$? Is this solution acidic, basic, or neutral?

Just as pH is a convenient way to designate the concentration of H_3O^+, pOH is a convenient way to designate the concentration of OH^-.

$$pOH = -\log [OH^-]$$

As we saw in the previous section, in aqueous solutions, the ion product of water, K_w, is 1×10^{-14}, which is equal to the product of the concentration of H^+ and OH^-:

$$K_w = 1 \times 10^{-14} = [H^+][OH^-]$$

By taking the logarithm of both sides, and the fact that $-\log (1 \times 10^{-14}) =$ 14, we can rewrite this equation as shown below:

$$14 = \text{pH} + \text{pOH}$$

Thus, once we know the pH of a solution, we can easily calculate the pOH.

EXAMPLE 9.7

The [OH⁻] of a strongly basic solution is 1.0×10^{-2}. What are the pOH and pH of this solution?

Solution
The pOH is $-\log 1.0 \times 10^{-2}$ or 2, and the pH is $14 - 2 = 12$.

Problem 9.7
The [OH⁻] of a solution is 1.0×10^{-4} M. What are the pOH and pH of this solution?

The pH of three household substances. The colors of the acid–base indicators in the flasks show that vinegar is more acidic than club soda, and the cleaner is basic.

All fluids in the human body are aqueous; that is, the only solvent present is water. Consequently, all body fluids have a pH value. Some of them have a narrow pH range; others have a wide pH range. The pH of blood, for example, must be between 7.35 and 7.45 (slightly basic). If it goes outside these limits, illness and even death may result (Chemical Connections 9D). In contrast, the pH of urine can vary from 5.5 to 7.5. Table 9.4 gives pH values for some common materials.

One thing you must remember when you see a pH value is that, because pH is a logarithmic scale, an increase (or decrease) of one pH unit means a tenfold decrease (or increase) in the [H₃O⁺]. For example, a pH of 3 does not sound very different from a pH of 4. The first, however, means a [H₃O⁺] of 10^{-3} M, whereas the second means a [H₃O⁺] of 10^{-4} M. The [H₃O⁺] of the pH 3 solution is ten times the [H₃O⁺] of the pH 4 solution.

There are two ways to measure the pH of an aqueous solution. One way is to use pH paper, which is made by soaking plain paper with a mixture of pH indicators. A pH **indicator** is a substance that changes color at a certain pH. When we place a drop of solution on this paper, the paper turns a certain color. To determine the pH, we compare the color of the paper with the colors on a chart supplied with the paper.

Strips of paper impregnated with indicator are used to find an approximate pH.

Table 9.4 pH Values of Some Common Materials

Material	pH	Material	pH
Battery acid	0.5	Saliva	6.5–7.5
Gastric juice	1.0–3.0	Pure water	7.0
Lemon juice	2.2–2.4	Blood	7.35–7.45
Vinegar	2.4–3.4	Bile	6.8–7.0
Tomato juice	4.0–4.4	Pancreatic fluid	7.8–8.0
Carbonated beverages	4.0–5.0	Sea water	8.0–9.0
Black coffee	5.0–5.1	Soap	8.0–10.0
Urine	5.5–7.5	Milk of magnesia	10.5
Rain (unpolluted)	6.2	Household ammonia	11.7
Milk	6.3–6.6	Lye (1.0 M NaOH)	14.0

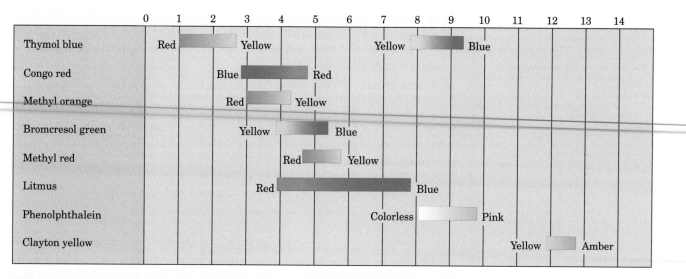

Figure 9.2 Some acid–base indicators. Note that some indicators have two color changes.

One example of an acid–base indicator is the compound methyl orange. When a drop of methyl orange is added to an aqueous solution with a pH of 3.2 or lower, this indicator turns red and the entire solution becomes red. When added to an aqueous solution with a pH of 4.4 or higher, this indicator turns yellow. These particular limits and colors apply only to methyl orange. Other indicators have other limits and colors (Figure 9.2)

The second way of determining pH is more accurate and more precise. In this method, we use a pH meter (Figure 9.3). We dip the electrode of the pH meter into the solution whose pH is to be measured, and then read the pH on a dial. The most commonly used pH meters read pH to the nearest hundredth of a unit. It should be mentioned that the accuracy of a pH meter, like that of any instrument, depends on correct calibration.

9.9 | How Do We Use Titrations to Calculate Concentration?

Laboratories, whether medical, academic, or industrial, are frequently asked to determine the exact concentration of a particular substance in solution, such as the concentration of acetic acid in a given sample of vinegar, or the concentrations of iron, calcium, and magnesium ions in a sample of "hard" water. Determinations of solution concentrations can be made using an analytical technique called a **titration.**

In a titration, we react a known volume of a solution of known concentration with a known volume of a solution of unknown concentration. The solution of unknown concentration may contain an acid (such as stomach acid), a base (such as ammonia), an ion (such as Fe^{2+} ion), or any other substance whose concentration we are asked to determine. If we know the titration volumes and the mole ratio in which the solutes react, we can then calculate the concentration of the second solution.

Titrations must meet several requirements:

1. We must know the equation for the reaction so that we can determine the stoichiometric ratio of reactants to use in our calculations.

2. The reaction must be rapid and complete.

Charles D. Winters

Figure 9.3 A pH meter can rapidly and accurately measure the pH of an aqueous solution.

GOB
Chemistry ⚛ Now™
Click *Coached Problems* to learn more about the **pH Scale**

Titration An analytical procedure whereby we react a known volume of a solution of known concentration with a known volume of a solution of unknown concentration

Equivalence point The point at which there is an equal amount of acid and base in a neutralization reaction

3. When the reactants have combined exactly, there must be a clear-cut change in some measurable property of the reaction mixture. We call the point at which the reactants combine exactly the **equivalence point** of the titration.

4. We must have accurate measurements of the amount of each reactant.

Let us apply these requirements to the titration of a solution of sulfuric acid of known concentration with a solution of sodium hydroxide of unknown concentration. We know the balanced equation for this acid–base reaction, so requirement 1 is met.

$$2NaOH(aq) \ + \ H_2SO_4(aq) \longrightarrow Na_2SO_4(aq) + 2H_2O(\ell)$$
$$\text{(Concentration} \qquad \text{(Concentration}$$
$$\text{not known)} \qquad \text{known)}$$

Sodium hydroxide ionizes in water to form sodium ions and hydroxide ions; sulfuric acid ionizes to form hydronium ions and sulfate ions. The reaction between hydroxide and hydronium ions is rapid and complete, so requirement 2 is met.

To meet requirement 3, we must be able to observe a clear-cut change in some measurable property of the reaction mixture at the equivalence point. For acid–base titrations, we use the sudden pH change that occurs at this point. Suppose we add the sodium hydroxide solution slowly. As it is added, it reacts with hydronium ions to form water. As long as any unreacted hydronium ions are present, the solution is acidic. When the number of hydroxide ions added exactly equals the original number of hydronium ions, the solution becomes neutral. Then, as soon as any extra hydroxide ions are added, the solution becomes basic. We can observe this sudden change in pH by reading a pH meter.

Another way to observe the change in pH at the equivalence point is to use an acid–base indicator (Section 9.8). Such an indicator changes color when the solution changes pH. Phenolphthalein, for example, is colorless in acid solution and pink in basic solution. If this indicator is added to the original sulfuric acid solution, the solution remains colorless as long as excess hydronium ions are present. After enough sodium hydroxide solution has been added to react with all of the hydronium ions, the next drop of base provides excess hydroxide ions, and the solution turns pink (Figure 9.4). Thus we have a clear-cut indication of the equivalence point. The point at which an indicator changes color is called the **end point** of the titration. It is convenient if the end point and the equivalence point are the same, but there are many pH indicators whose end points are not at pH 7.

To meet requirement 4, which is that the volume of each solution used must be known, we use volumetric glassware such as volumetric flasks, burets, and pipets.

Data for a typical acid–base titration are given in Example 9.8. Note that the experiment is run in triplicate, a standard procedure for checking the precision of a titration.

EXAMPLE 9.8

Following are data for the titration of 0.108 M H_2SO_4 with a solution of NaOH of unknown concentration. What is the concentration of the NaOH solution?

	Volume of 0.108 M H_2SO_4	Volume of NaOH
Trial I	25.0 mL	33.48 mL
Trial II	25.0 mL	33.46 mL
Trial III	25.0 mL	33.50 mL

Solution

From the balanced equation for this acid–base reaction, we know the stoichiometry: Two moles of NaOH react with one mole of H_2SO_4. From the three trials, we calculate that the average volume of the NaOH required for complete reaction is 33.48 mL. Because the units of molarity are moles/liter, we must convert volumes of reactants from milliliters to liters. We can then use the factor-label method (Section 1.5) to calculate the molarity of the NaOH solution. What we wish to calculate is the number of moles of NaOH per liter of NaOH.

$$\frac{\text{mol NaOH}}{\text{L NaOH}} = \frac{0.108 \text{ mol } H_2SO_4}{1 \text{ L } H_2SO_4} \times \frac{0.0250 \text{ L } H_2SO_4}{0.03348 \text{ L NaOH}} \times \frac{2 \text{ mol NaOH}}{1 \text{ mol } H_2SO_4}$$

$$= \frac{0.161 \text{ mol NaOH}}{\text{L NaOH}} = 0.161 \ M$$

Problem 9.8

Calculate the concentration of an acetic acid solution using the following data. Three 25.0-mL samples of acetic acid were titrated to a phenolphthalein end point with 0.121 M NaOH. The volumes of NaOH were 19.96 mL, 19.73 mL, and 19.79 mL.

(a)   (b)   (c)

Figure 9.4 An acid–base titration. (*a*) An acid of known concentration is in the Erlenmeyer flask. (*b*) When a base is added from the buret, the acid is neutralized. (*c*) The end point is reached when the color of the indicator changes from colorless to pink.

Charles D. Winters

It is important to understand that a titration is not a method for determining the acidity (or basicity) of a solution. If we want to do that, we must measure the sample's pH, which is the only measurement of solution acidity or basicity. Rather, titration is a method for determining the total acid or base concentration of a solution, which is not the same as the acidity. For example, a 0.1 M solution of HCl in water has a pH of 1, but a 0.1 M solution of acetic acid has a pH of 2.9. These two solutions have the same concentration of acid and each neutralizes the same volume of NaOH solution, but they have very different acidities.

9.10 What Are Buffers?

GOB
Chemistry⚛Now™
Click *Coached Problems* to learn more about **Determining the Concentration of an Unknown Using Titration**

Buffer A solution that resists change in pH when limited amounts of an acid or a base are added to it; an aqueous solution containing a weak acid and its conjugate base

As noted earlier, the body must keep the pH of blood between 7.35 and 7.45. Yet we frequently eat acidic foods such as oranges, lemons, sauerkraut, and tomatoes, and doing so eventually adds considerable quantities of H_3O^+ to the blood. Despite these additions of acidic or basic substances, the body manages to keep the pH of blood remarkably constant. The body manages this feat by using buffers. A **buffer** is a solution whose pH changes very little when small amounts of H_3O^+ or OH^- ions are added to it. In a sense, a pH buffer is an acid or base "shock absorber."

The most common buffers consist of approximately equal molar amounts of a weak acid and a salt of the weak acid. Put another way, they consist of approximately equal amounts of a weak acid and its conjugate base. For example, if we dissolve 1.0 mol of acetic acid (a weak acid) and 1.0 mol of its conjugate base (in the form of CH_3COONa, sodium acetate) in 1.0 L of water, we have a good buffer solution. The equilibrium present in this buffer solution is

A. How Do Buffers Work?

A buffer resists any change in pH upon the addition of small quantities of acid or base. To see how, we will use an acetic acid–sodium acetate buffer as an example. If a strong acid such as HCl is added to this buffer solution, the added H_3O^+ ions react with CH_3COO^- ions and are removed from solution.

$$CH_3COO^- + H_3O^+ \longrightarrow CH_3COOH + H_2O$$

Acetate ion (Conjugate base of a weak acid) / Acetic acid (A weak acid)

There is a slight increase in the concentration of CH_3COOH as well as a slight decrease in the concentration of CH_3COO^-, but there is no appreciable change in pH. We say that this solution is buffered because it resists a change in pH upon the addition of small quantities of a strong acid.

If NaOH or another strong base is added to the buffer solution, the added OH^- ions react with CH_3COOH molecules and are removed from solution:

$$CH_3COOH + OH^- \longrightarrow CH_3COO^- + H_2O$$

Acetic acid   Acetate ion
(A weak acid)   (Conjugate base
                                                            of a weak acid)

Here there is a slight decrease in the concentration of CH_3COOH as well as a slight increase in the concentration of CH_3COO^-, but, again, there is no appreciable change in pH.

The important point about this or any other buffer solution is that when the conjugate base of the weak acid removes H_3O^+, it is converted to the undissociated weak acid. Because a substantial amount of weak acid is already present, there is no appreciable change in its concentration and, because H_3O^+ ions are removed from solution, there is no appreciable change in pH. By the same token, when the weak acid removes OH^- ions from solution, it is converted to its conjugate base. Because OH^- ions are removed from solution, there is no appreciable change in pH.

The effect of a buffer can be quite powerful. Addition of either dilute HCl or NaOH to pure water, for example, causes a dramatic change in pH (Figure 9.5).

When HCl or NaOH is added to a phosphate buffer, the results are quite different. Suppose we have a phosphate buffer solution of pH 7.21 prepared by dissolving 0.10 mol NaH_2PO_4 (a weak acid) and 0.10 mol Na_2HPO_4 (its conjugate base) in enough water to make 1.00 L of solution. If we add 0.010 mol of HCl to 1.0 L of this solution, the pH decreases to only 7.12. If we add 0.01 mol of NaOH, the pH increases to only 7.30.

Phosphate buffer (pH 7.21) + 0.010 mol HCl   pH 7.21 $\longrightarrow$ 7.12

Phosphate buffer (pH 7.21) + 0.010 mol NaOH   pH 7.21 $\longrightarrow$ 7.30

Figure 9.6 shows the effect of adding acid to a buffer solution.

B. Buffer pH

In the previous example, the pH of the buffer containing equal molar amounts of $H_2PO_4^-$ and HPO_4^{2-} is 7.21. From Table 9.3, we see that 7.21 is the pK_a of the acid $H_2PO_4^-$. This is not a coincidence. If we make a buffer

GOB
Chemistry⚛Now™

Click *Chemistry Interactive* to learn more about **Buffers**

(*a*) pH 7.00   (*b*) pH 2.00   (*c*) pH 12.00

Figure 9.5   The addition of HCl and NaOH to pure water. (*a*) The pH of pure water is 7.0. (*b*) The addition of 0.01 mol of HCl to 1 L of pure water causes the pH to decrease to 2. (*c*) The addition of 0.010 mol of NaOH to 1 L of pure water causes the pH to increase to 12.

(a) (b)

Charles D. Winters

Figure 9.6 Buffer solutions. The solution in the Erlenmeyer flask on the right in both (a) and (b) is a buffer of pH 7.40, the same pH as human blood. The buffer solution also contains bromcresol green, an acid–base indicator that is blue at pH 7.40 (see Figure 9.2). (a) The beaker contains some of the pH 7.40 buffer and the bromcresol green indicator to which has been added 5 mL of 0.1 M HCl. After the addition of the HCl, the pH of the buffer solution drops only 0.65 unit to 6.75. (b) The beaker contains pure water and bromcresol green indicator to which has been added 5 mL of 0.10 M HCl. After the addition of the HCl, the pH of the unbuffered solution drops to 3.02.

solution by mixing equimolar concentrations of any weak acid and its conjugate base, the pH of the solution will equal the pK_a of the weak acid.

This fact allows us to prepare buffer solutions to maintain almost any pH. For example, if we want to maintain a pH of 9.14, we could make a buffer solution from boric acid, H_3BO_3, and sodium dihydrogen borate, NaH_2BO_3, the sodium salt of its conjugate base (see Table 9.3).

EXAMPLE 9.9

What is the pH of buffer solution containing equimolar quantities of
(a) H_3PO_4 and NaH_2PO_4? (b) H_2CO_3 and $NaHCO_3$?

Solution
Because we are adding equimolar quantities of a weak acid and its conjugate base, the pH is equal to the pK_a of the weak acid, which we find in Table 9.3:
(a) pH = 2.12 (b) pH = 6.37

Problem 9.9
What is the pH of a buffer solution containing equimolar quantities of
(a) NH_4Cl and NH_3? (b) CH_3COOH and CH_3COONa?

C. Buffer Capacity

Buffer capacity The extent to which a buffer solution can prevent a significant change in pH of a solution upon addition of a strong acid or a strong base

Buffer capacity is the amount of hydronium or hydroxide ions that a buffer can absorb without a significant change in its pH. We have already mentioned that a pH buffer is an acid–base "shock absorber." We now ask what makes one solution a better acid–base shock absorber than another solution. The capacity of a pH buffer depends on both its pH relative to its pK_a and its concentration.

pH:	The closer the pH of the buffer is to the pK_a of the weak acid, the greater the buffer capacity.
Concentration:	The greater the concentration of the weak acid and its conjugate base, the greater the buffer capacity.

An effective buffer has a pH equal to the pK_a of the weak acid ± 1. For acetic acid, for example, the pK_a is 4.75. Therefore, a solution of acetic acid and sodium acetate functions as an effective buffer within the pH range of approximately 3.75–5.75. The most effective buffer is one with equal concentrations of the weak acid and its salt—that is, one in which the pH of the buffer solution equals the pK_a of the weak acid.

Buffer capacity also depends on concentration. The greater the concentration of the weak acid and its conjugate base, the greater the buffer capacity. We could make a buffer solution by dissolving 1.0 mol each of CH_3COONa and CH_3COOH in 1 L of H_2O, or we could use only 0.10 mol of each. Both solutions have the same pH of 4.75. However, the former has a buffer capacity ten times that of the latter. If we add 0.2 mol of HCl to the former solution, it performs the way we expect—the pH drops to 4.57. If we add 0.2 mol of HCl to the latter solution, however, the pH drops to 1.0 because the buffer has been swamped out. That is, the amount of H_3O^+ added has exceeded the buffer capacity. The first 0.10 mol of HCl completely neutralizes essentially all the CH_3COO^- present. After that, the solution contains only CH_3COOH and is no longer a buffer, so the second 0.10 mol of HCl decreases the pH to 1.0.

D. Blood Buffers

The average pH of human blood is 7.4. Any change larger than 0.10 pH unit in either direction may cause illness. If the pH goes below 6.8 or above 7.8, death may result. To hold the pH of the blood close to 7.4, the body uses three buffer systems: carbonate, phosphate, and proteins (proteins are discussed in Chapter 22).

The most important of these systems is the carbonate buffer. The weak acid of this buffer is carbonic acid, H_2CO_3; the conjugate base is the bicarbonate ion, HCO_3^-. The pK_a of H_2CO_3 is 6.37 (from Table 9.3). Because the pH of an equal mixture of a weak acid and its salt is equal to the pK_a of the weak acid, a buffer with equal concentrations of H_2CO_3 and HCO_3^- has a pH of 6.37.

Blood, however, has a pH of 7.4. The carbonate buffer can maintain this pH only if $[H_2CO_3]$ and $[HCO_3^-]$ are not equal. In fact, the necessary $[HCO_3^-]/[H_2CO_3]$ ratio is about 10:1. The normal concentrations of these species in blood are about 0.025 M HCO_3^- and 0.0025 M H_2CO_3. This buffer works because any added H_3O^+ is neutralized by the HCO_3^- and any added OH^- is neutralized by the H_2CO_3.

The fact that the $[HCO_3^-]/[H_2CO_3]$ ratio is 10:1 means that this system is a better buffer for acids, which lower the ratio and thus improve buffer efficiency, than for bases, which raise the ratio and decrease buffer capacity. This is in harmony with the actual functioning of the body because, under normal conditions, larger amounts of acidic than basic substances enter the blood. The 10:1 ratio is easily maintained under normal conditions, because the body can very quickly increase or decrease the amount of CO_2 entering the blood.

The second most important buffering system of the blood is a phosphate buffer made up of hydrogen phosphate ion, HPO_4^{2-}, and dihydrogen phosphate ion, $H_2PO_4^-$. In this case, a 1.6:1 $[HPO_4^{2-}]/[H_2PO_4^-]$ ratio is necessary to maintain a pH of 7.4. This ratio is well within the limits of good buffering action.

9.11 | How Do We Calculate the pH of a Buffer?

Suppose we want to make a phosphate buffer solution of pH 7.00. The weak acid with a pK_a closest to this desired pH is $H_2PO_4^-$; it has a pK_a of 7.21. If we use equal concentrations of NaH_2PO_4 and Na_2HPO_4, however, we will have a buffer of pH 7.21. We want a phosphate buffer that is slightly more acidic than 7.21, so it would seem reasonable to use more of the weak acid, $H_2PO_4^-$, and less of its conjugate base, HPO_4^{2-}. But what proportions of these two salts do we use? Fortunately, we can calculate these proportions using the **Henderson-Hasselbalch** equation.

The Henderson-Hasselbalch equation is a mathematical relationship between pH, the pK_a of a weak acid, and the concentrations of the weak acid and its conjugate base. The equation is derived in the following way. Assume that we are dealing with weak acid, HA, and its conjugate base, A^-.

$$HA + H_2O \rightleftharpoons A^- + H_3O^+$$

$$K_a = \frac{[A^-][H_3O^+]}{[HA]}$$

Taking the logarithm of this equation gives

$$\log K_a = \log [H_3O^+] + \log \frac{[A^-]}{[HA]}$$

Rearranging terms gives us a new expression, in which $-\log K_a$ is, by definition, pK_a, and $-\log [H_3O^+]$ is, by definition, pH. Making these substitutions gives the Henderson-Hasselbalch equation.

$$-\log [H_3O^+] = -\log K_a + \log \frac{[A^-]}{[HA]}$$

$$pH = pK_a + \log \frac{[A^-]}{[HA]} \quad \text{Henderson-Hasselbalch Equation}$$

The Henderson-Hasselbalch equation gives us a convenient way to calculate the pH of a buffer when the concentrations of the weak acid and its conjugate base are not equal.

EXAMPLE 9.10

What is the pH of a phosphate buffer solution containing 1.0 mol/L of sodium dihydrogen phosphate, NaH_2PO_4, and 0.50 mol/L of sodium hydrogen phosphate, Na_2HPO_4?

Solution
The weak acid in this problem is $H_2PO_4^-$; its ionization produces HPO_4^{2-}. The pK_a of this acid is 7.21 (from Table 9.3). Under the weak acid and its conjugate base are shown their concentrations.

$$H_2PO_4^- + H_2O \rightleftharpoons HPO_4^{2-} + H_3O^+ \quad pK_a = 7.21$$
$$\text{1.0 mol/L} \qquad \text{0.50 mol/L}$$

Substituting these values in the Henderson Hasselbalch equation gives a pH of 6.91.

$$pH = 7.21 + \log \frac{0.50}{1.0}$$

$$= 7.21 - 0.30 = 6.91$$

Problem 9.10

What is the pH of a hydrocyanic acid buffer solution containing 0.25 mol/L of hydrocyanic acid, HCN, and 0.50 mol/L of cyanide ion, CN⁻? See Table 9.3 for the pK_a of hydrocyanic acid.

CHEMICAL CONNECTIONS 9D

Respiratory and Metabolic Acidosis

The pH of blood is normally between 7.35 and 7.45. If the pH goes lower than that level, the condition is called **acidosis.** Acidosis leads to depression of the nervous system. Mild acidosis can result in dizziness, disorientation, or fainting; a more severe case can cause coma. If the acidosis persists for a sufficient period of time, or if the pH gets too far away from 7.35 to 7.45, death may result.

Acidosis has several causes. One type, called **respiratory acidosis,** results from difficulty in breathing (hypoventilation). An obstruction in the windpipe or diseases such as pneumonia, emphysema, asthma, or congestive heart failure may diminish the amount of oxygen that reaches the tissues and the amount of CO_2 that leaves the body through the lungs. You can even produce mild acidosis by holding your breath. If you have ever tried to see how long you could swim underwater in a pool without surfacing, you will have noticed a deep burning sensation in all your muscles when you finally came up for air. The pH of the blood decreases because the CO_2, unable to escape fast enough, remains in the blood, where it lowers the [HCO_3^-]/[H_2CO_3] ratio. Rapid breathing as a result of physical exertion is more about getting rid of CO_2 than it is about breathing in O_2.

Acidosis caused by other factors is called **metabolic acidosis.** Two causes of this condition are starvation (or fasting) and heavy exercise. When the body doesn't get enough food, it burns its own fat, and the products of this reaction are acidic compounds that enter the blood. This problem sometimes happens to people on fad diets. Heavy exercise causes the muscles to produce excessive amounts of lactic acid, which makes muscles feel tired and sore. The lowering of the blood pH due to lactic acid is also what leads to the rapid breathing, dizziness, and nausea that athletes feel at the end of a sprint. In addition, metabolic acidosis is caused by a number of metabolic irregularities. For example, the disease diabetes mellitus produces acidic compounds called ketone bodies (Section 28.6).

Both types of acidosis can be related. When cells are deprived of oxygen, respiratory acidosis results. These cells are unable to produce the energy they need through aerobic (*oxygen-requiring*) pathways that we will learn about in Chapters 27 and 28. To survive, the cells must use the anaerobic (*without oxygen*) pathway called

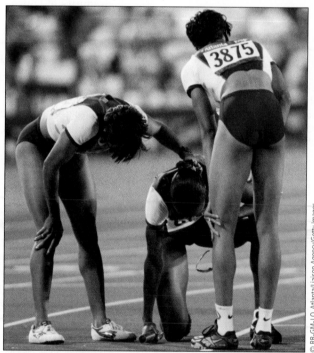

These U.S. runners have just won the gold medal in the 4 × 400 m relay race at the 1996 Olympic Games. The buildup of lactic acid and lowered blood pH has caused severe muscle pain and breathlessness.

glycolysis. This pathway has lactic acid as an end product, leading to metabolic acidosis. The lactic acid is the body's way of buying time and keeping the cells alive and functioning a little longer. Eventually the lack of oxygen, called an oxygen debt, must be repaid, and the lactic acid must be cleared out. In extreme cases, the oxygen debt is too great, and the individual can die. This was the case of a famous cyclist, Tom Simpson, who died on the slopes of Mont Ventoux during the 1967 Tour de France. Under the influence of amphetamines, he rode so hard that he built up a fatal oxygen debt.

Returning to the problem posed at the beginning of this section, how do we calculate the proportions of NaH_2PO_4 and Na_2HPO_4 needed to make up a phosphate buffer of pH 7.00? We know that the pK_a of $H_2PO_4^-$ is 7.21 and that the buffer we wish to prepare has a pH of 7.00. We can substitute these two values in the Henderson-Hasselbalch equation as follows:

$$7.00 = 7.21 + \log \frac{[HPO_4^{2-}]}{[H_2PO_4^-]}$$

Rearranging and solving gives

$$\log \frac{[HPO_4^{2-}]}{[H_2PO_4^-]} = 7.00 - 7.21 = -0.21$$

$$\frac{[HPO_4^{2-}]}{[H_2PO_4^-]} = \frac{0.62}{1}$$

Thus, to prepare a phosphate buffer of pH 7.00, we can use 0.62 mol of Na_2HPO_4 and 1.0 mol of NaH_2PO_4. Alternatively, we can use any other amounts of these two salts, as long as their mole ratio is 0.62 : 1.0.

GOB
Chemistry⋅𝟄⋅Now™
Click *Coached Problems* to learn more about calculating the **pH of Buffer Solutions**

9.12	What Are TRIS, HEPES, and These Buffers with the Strange Names?

The original buffers used in the lab were made from simple weak acids and bases, such as acetic acid, phosphoric acid, and citric acid. It was eventually discovered that many of these buffers had limitations. For example, they often changed their pH too much if the solution was diluted or if the temperature changed. They often permeated cells in solution, thereby changing the chemistry of the interior of the cell. To overcome these shortcomings, a scientist named N. E. Good developed a series of buffers that consist of zwitterions, molecules with both positive and negative charges. Zwitterions do not readily permeate cell membranes. Zwitterionic buffers also are more resistant to concentration and temperature changes.

Most of the common synthetic buffers used today have complicated formulas, such as 3-[*N*-morpholino]propanesulfonic acid, which we abbreviate MOPS. Table 9.5 gives a few examples.

CHEMICAL CONNECTIONS 9E

Alkalosis and the Sprinter's Trick

Reduced pH is not the only irregularity that can occur in the blood. The pH may also be elevated, a condition called **alkalosis** (blood pH higher than 7.45). It leads to overstimulation of the nervous system, muscle cramps, dizziness, and convulsions. It arises from rapid or heavy breathing, called hyperventilation, which may be caused by fever, infection, the action of certain drugs, or even hysteria. In this case, the excessive loss of CO_2 raises both the ratio of $[HCO_3^-]/[H_2CO_3]$ and the pH.

Athletes who compete in short-distance races that take about a minute to finish have learned how to use hyperventilation to their advantage. By hyperventilating right before the start, they force extra CO_2 out of their lungs. This causes more H_2CO_3 to dissociate into CO_2 and H_2O to replace the lost CO_2. In turn, the loss of the HA form of the bicarbonate blood buffer raises the pH of the blood. When an athlete starts an event with a slightly higher blood pH, he or she can absorb more lactic acid before the blood pH drops to the point where performance is impaired. Of course, the timing of this hyperventilation must be perfect. If the athlete artificially raises blood pH and then the race does not start quickly, the same effects of dizziness will occur.

Table 9.5 Acid and Base Forms of Some Useful Biochemical Buffers

Acid Form		Base Form	pK_a
TRIS—H⁺ (protonated form) $(HOCH_2)_3CNH_3^+$	N—*tris*[hydroxymethyl]aminomethane (TRIS) ⇌	TRIS (free amine) $(HOCH_2)_3CNH_2$	8.3
⁻TES—H⁺ (zwitterionic form) $(HOCH_2)_3CNH_2CH_2CH_2SO_3^-$	N—*tris*[hydroxymethyl]methyl-2-aminoethane sulfonate (TES) ⇌	⁻TES (anionic form) $(HOCH_2)_3CNHCH_2CH_2SO_3^-$	7.55
⁻HEPES—H⁺ (zwitterionic form) $HOCH_2CH_2N^+$ $NCH_2CH_2SO_3^-$ H	N—2—hydroxyethylpiperazine-N'-2-ethane sulfonate (HEPES) ⇌	⁻HEPES (anionic form) $HOCH_2CH_2N$ $NCH_2CH_2SO_3^-$	7.55
⁻MOPS—H⁺ (zwitterionic form) O $^+NCH_2CH_2CH_2SO_3^-$ H	3—[N—morpholino]propane-sulfonic acid (MOPS) ⇌	⁻MOPS (anionic form) O $NCH_2CH_2CH_2SO_3^-$	7.2
²⁻PIPES—H⁺ (protonated dianion) $^-O_3SCH_2CH_2N$ $^+NCH_2CH_2SO_3^-$ H	Piperazine—N,N'-*bis*[2-ethanesulfonic acid] (PIPES) ⇌	²⁻PIPES (dianion) $^-O_3SCH_2CH_2N$ $NCH_2CH_2SO_3^-$	6.8

The important thing to remember is that you don't really need to know the structure of these odd-sounding buffers to use them correctly. The important considerations are the pK_a of the buffer and the concentration you want to have. The Henderson-Hasselbalch equation works just fine whether or not you know the structure of the compound in question.

EXAMPLE 9.11

What is the pH of a solution if you mix 100 mL of 0.2 *M* HEPES in the acid form with 200 mL of 0.2 *M* HEPES in the basic form?

Solution

First we must find the pK_a, which we see from Table 9.5 is 7.55. Then we must calculate the ratio of the conjugate base to the acid. The formula calls for the concentration, but in this situation, the ratio of the concentrations will be the same as the ratio of the moles, which will be the same as the ratio of the volumes, because both solutions had the same starting concentration of 0.2. Thus, we can see that the ratio of base to acid is 2 : 1 because we added twice the volume of base.

$$pH = pK_a + \log ([A^-]/[HA]) = 7.55 + \log (2) = 7.85$$

Notice that we did not have to know anything about the structure of HEPES to work out this example.

Problem 9.11

What is the pH of a solution made by mixing 0.2 mol of TRIS acid and 0.05 mol of TRIS base in 500 mL of water?

SUMMARY OF KEY QUESTIONS

SECTION 9.1 What Are Acids and Bases?

- By the **Arrhenius definitions,** acids are substances that produce H_3O^+ ions in aqueous solution.
- Bases are substances that produce OH^- ions in aqueous solution.

SECTION 9.2 How Do We Define the Strength of Acids and Bases?

- A strong acid reacts completely or almost completely with water to form H_3O^+ ions.
- A strong base reacts completely or almost completely with water to form OH^- ions.

SECTION 9.3 What Are Conjugate Acid–Base Pairs?

- The **Brønsted-Lowry definitions** expand the definitions of acid and base to beyond water.
- An acid is a proton donor; a base is a proton acceptor.
- Every acid has a **conjugate base,** and every base has a **conjugate acid.** The stronger the acid, the weaker its conjugate base. Conversely, the stronger the base, the weaker its conjugate acid.
- An **amphiprotic substance,** such as water, can act as either an acid or a base.

SECTION 9.4 How Can We Tell the Position of Equilibrium in an Acid–Base Reaction?

- In an acid–base reaction, the position of equilibrium favors the reaction of the stronger acid and the stronger base to form the weaker acid and the weaker base.

SECTION 9.5 How Do We Use Acid Ionization Constants?

- The strength of a weak acid is expressed by its **ionization constant, K_a.**
- The larger the value of K_a, the stronger the acid. $pK_a = -\log [K_a]$.

SECTION 9.6 What Are the Properties of Acids and Bases?

- Acids react with metals, metal hydroxides, and metal oxides to give **salts,** which are ionic compounds made up of cations from the base and anions from the acid.
- Acids also react with carbonates, bicarbonates, ammonia, and amines to give salts.

SECTION 9.7 What Are the Acidic and Basic Properties of Pure Water?

- In pure water, a small percentage of molecules undergo self-ionization:

$$H_2O + H_2O \rightleftharpoons H_3O^+ + OH^-$$

- As a result, pure water has a concentration of $10^{-7}\ M$ for H_3O^+ and $10^{-7}\ M$ for OH^-.
- The **ion product of water, K_w,** is equal to 1.0×10^{-14}. $pK_W = 14$.

SECTION 9.8 What Are pH and pOH?

- Hydronium ion concentrations are generally expressed in **pH** units, with $pH = -\log [H_3O^+]$.
- $pOH = -\log [OH^-]$.
- Solutions with pH less than 7 are acidic; those with pH greater than 7 are basic. A **neutral solution** has a pH of 7.
- The pH of an aqueous solution is measured with an acid–base indicator or with a pH meter.

SECTION 9.9 How Do We Use Titrations to Calculate Concentration?

- We can measure the concentration of aqueous solutions of acids and bases using **titration.** In an acid–base titration, a base of known concentration is added to an acid of unknown concentration (or vice versa) until an equivalence point is reached, at which point the acid or base being titrated is completely neutralized.

SECTION 9.10 What Are Buffers?

- A **buffer** does not change its pH very much when either hydronium ions or hydroxide ions are added to it.
- Buffer solutions consist of approximately equal concentrations of a weak acid and its conjugate base.
- The **buffer capacity** depends on both its pH relative to its pK_a and its concentration. The most effective buffer solutions have a pH equal to the pK_a of the weak acid. The greater the concentration of the weak acid and its conjugate base, the greater the buffer capacity.
- The most important buffers for blood are bicarbonate and phosphate.

SECTION 9.11 How Do We Calculate the pH of a Buffer?

- The **Henderson-Hasselbalch** equation is a mathematical relationship between pH, the pK_a of a weak acid, and the concentrations of the weak acid and its conjugate base:

$$pH = pK_a + \log \frac{[A^-]}{[HA]}$$

SECTION 9.12 What Are TRIS, HEPES, and These Buffers with the Strange Names?

- Many modern buffers have been designed, and their names are often abbreviated.
- These buffers have qualities useful to scientists, such as not crossing membranes and resisting pH change with dilution or temperature change.
- You do not have to understand the structure of these buffers to use them. The important things to know are the molecular weight and the pK_a of the weak acid form of the buffer.

PROBLEMS

GOB
Chemistry·⚛·Now™

Assess your understanding of this chapter's topics with additional quizzing and conceptual-based problems at **http://now.brookscole.com/gob8** or on the CD.

A blue problem number indicates an applied problem.

■ denotes problems that are available on the GOB ChemistryNow website or CD and are assignable in OWL.

SECTION 9.1 What Are Acids and Bases?

9.12 Define (a) an Arrhenius acid and (b) an Arrhenius base.

9.13 Write an equation for the reaction that takes place when each acid is added to water. For a diprotic or triprotic acid, consider only its first ionization.
(a) HNO_3 (b) HBr   (c) H_2SO_3
(d) H_2SO_4 (e) HCO_3^- (f) H_3BO_3

9.14 Write an equation for the reaction that takes place when each base is added to water.
(a) LiOH   (b) $(CH_3)_2NH$

SECTION 9.2 How Do We Define the Strength of Acids and Bases?

9.15 ■ For each of the following, tell whether the acid is strong or weak.
(a) Acetic acid   (b) HCl
(c) H_3PO_4 (d) H_2SO_4
(e) HCN   (f) H_2CO_3

9.16 ■ For each of the following, tell whether the base is strong or weak.
(a) NaOH   (b) Sodium acetate
(c) KOH   (d) Ammonia
(e) Water

9.17 If an acid has a pK_a of 2.1, is it strong or weak?

SECTION 9.3 What Are Conjugate Acid–Base Pairs?

9.18 Which of these acids are monoprotic, which are diprotic, and which are triprotic? Which are amphiprotic?
(a) $H_2PO_4^-$ (b) HBO_3^{2-} (c) $HClO_4$ (d) C_2H_5OH
(e) HSO_3^- (f) HS^- (g) H_2CO_3

9.19 Define (a) a Brønsted-Lowry acid and (b) a Brønsted-Lowry base.

9.20 Write the formula for the conjugate base of each acid.
(a) H_2SO_4 (b) H_3BO_3 (c) HI   (d) H_3O^+
(e) NH_4^+ (f) HPO_4^{2-}

9.21 Write the formula for the conjugate base of each acid.
(a) $H_2PO_4^-$ (b) H_2S
(c) HCO_3^- (d) CH_3CH_2OH
(e) H_2O

9.22 ■ Write the formula for the conjugate acid of each base.
(a) OH^- (b) HS^- (c) NH_3
(d) $C_6H_5O^-$ (e) CO_3^{2-} (f) HCO_3^-

9.23 Write the formula for the conjugate acid of each base.
(a) H_2O (b) HPO_4^{2-} (c) CH_3NH_2
(d) PO_4^{3-} (e) HBO_3^{2-}

SECTION 9.4 How Can We Tell the Position of Equilibrium in an Acid–Base Reaction?

9.24 For each equilibrium, label the stronger acid, stronger base, weaker acid, and weaker base. For which reaction(s) does the position of equilibrium lie toward the right? For which does it lie toward the left?
(a) $H_3PO_4 + OH^- \rightleftharpoons H_2PO_4^- + H_2O$
(b) $H_2O + Cl^- \rightleftharpoons HCl + OH^-$
(c) $HCO_3^- + OH^- \rightleftharpoons CO_3^{2-} + H_2O$

9.25 ■ For each equilibrium, label the stronger acid, stronger base, weaker acid, and weaker base. For which reaction(s) does the position of equilibrium lie toward the right? For which does it lie toward the left?
(a) $C_6H_5OH + C_2H_5O^- \rightleftharpoons C_6H_5O^- + C_2H_5OH$
(b) $HCO_3^- + H_2O \rightleftharpoons H_2CO_3 + OH^-$
(c) $CH_3COOH + H_2PO_4^- \rightleftharpoons CH_3COO^- + H_3PO_4$

9.26 Will carbon dioxide be evolved as a gas when sodium bicarbonate is added to an aqueous solution of each compound? Explain.
(a) Sulfuric acid
(b) Ethanol, C_2H_5OH
(c) Ammonium chloride, NH_4Cl

SECTION 9.5 How Do We Use Acid Ionization Constants?

9.27 Which has the larger numerical value?
(a) The pK_a of a strong acid or the pK_a of a weak acid
(b) The K_a of a strong acid or the K_a of a weak acid

9.28 ■ In each pair, select the stronger acid.
(a) Pyruvic acid ($pK_a = 2.49$) or lactic acid ($pK_a = 3.08$)
(b) Citric acid ($pK_a = 3.08$) or phosphoric acid ($pK_a = 2.10$)
(c) Benzoic acid ($K_a = 6.5 \times 10^{-5}$) or lactic acid ($K_a = 8.4 \times 10^{-4}$)
(d) Carbonic acid ($K_a = 4.3 \times 10^{-7}$) or boric acid ($K_a = 7.3 \times 10^{-10}$)

9.29 Which solution will be more acidic; that is, which will have a lower pH?
(a) 0.10 M CH_3COOH or 0.10 M HCl
(b) 0.10 M CH_3COOH or 0.10 M H_3PO_4
(c) 0.010 M H_2CO_3 or 0.010 M $NaHCO_3$
(d) 0.10 M NaH_2PO_4 or 0.10 M Na_2HPO_4
(e) 0.10 M aspirin ($pK_a = 3.47$) or 0.10 M acetic acid

9.30 Which solution will be more acidic; that is, which will have a lower pH?

(a) 0.10 M C_6H_5OH (phenol) or 0.10 M C_2H_5OH (ethanol)

(b) 0.10 M NH_3 or 0.10 M NH_4Cl

(c) 0.10 M NaCl or 0.10 M NH_4Cl

(d) 0.10 M $CH_3CH(OH)COOH$ (lactic acid) or 0.10 M CH_3COOH

(e) 0.10 M ascorbic acid (vitamin C, $pK_a = 4.1$) or 0.10 M acetic acid

SECTION 9.6 What Are the Properties of Acids and Bases?

9.31 Write an equation for the reaction of HCl with each compound. Which are acid–base reactions? Which are redox reactions?

(a) Na_2CO_3 (b) Mg   (c) NaOH   (d) Fe_2O_3

(e) NH_3 (f) CH_3NH_2 (g) $NaHCO_3$

9.32 When a solution of sodium hydroxide is added to a solution of ammonium carbonate and is heated, ammonia gas, NH_3, is released. Write a net ionic equation for this reaction. Both NaOH and $(NH_4)_2CO_3$ exist as dissociated ions in aqueous solution.

SECTION 9.7 What Are the Acidic and Basic Properties of Pure Water?

9.33 Given the following values of $[H_3O^+]$, calculate the corresponding value of $[OH^-]$ for each solution.

(a) $10^{-11}\,M$ (b) $10^{-4}\,M$ (c) $10^{-7}\,M$ (d) $10\,M$

9.34 Given the following values of $[OH^-]$, calculate the corresponding value of $[H_3O^+]$ for each solution.

(a) $10^{-10}\,M$ (b) $10^{-2}\,M$ (c) $10^{-7}\,M$ (d) $10\,M$

SECTION 9.8 What Are pH and pOH?

9.35 What is the pH of each solution, given the following values of $[H_3O^+]$? Which solutions are acidic, which are basic, and which are neutral?

(a) $10^{-8}\,M$ (b) $10^{-10}\,M$ (c) $10^{-2}\,M$

(d) $10^0\,M$ (e) $10^{-7}\,M$

9.36 What is the pH and pOH of each solution given the following values of $[OH^-]$? Which solutions are acidic, which are basic, and which are neutral?

(a) $10^{-3}\,M$ (b) $10^{-1}\,M$ (c) $10^{-5}\,M$ (d) $10^{-7}\,M$

9.37 What is the pH of each solution, given the following values of $[H_3O^+]$? Which solutions are acidic, which are basic, and which are neutral?

(a) $3.0 \times 10^{-9}\,M$ (b) $6.0 \times 10^{-2}\,M$

(c) $8.0 \times 10^{-12}\,M$ (d) $5.0 \times 10^{-7}\,M$

9.38 Which is more acidic, a beer with $[H_3O^+] = 3.16 \times 10^{-5}$ or a wine with $[H_3O^+] = 5.01 \times 10^{-4}$?

9.39 ■ What is the $[OH^-]$ and pOH of each solution?

(a) 0.10 M KOH, pH = 13.0

(b) 0.10 M Na_2CO_3, pH = 11.6

(c) 0.10 M Na_3PO_4, pH = 12.0

(d) 0.10 M $NaHCO_3$, pH = 8.4

SECTION 9.9 How Do We Use Titrations to Calculate Concentration?

9.40 What is the purpose of an acid–base titration?

9.41 What is the molarity of a solution made by dissolving 12.7 g of HCl in enough water to make 1.00 L of solution?

9.42 What is the molarity of a solution made by dissolving 3.4 g of $Ba(OH)_2$ in enough water to make 450 mL of solution? Assume that $Ba(OH)_2$ ionizes completely in water to Ba^{2+} and OH^- ions.

9.43 Describe how you would prepare each of the following solutions (in each case assume that you have the solid bases).

(a) 400.0 mL of 0.75 M NaOH

(b) 1.0 L of 0.071 M $Ba(OH)_2$

9.44 If 25.0 mL of an aqueous solution of H_2SO_4 requires 19.7 mL of 0.72 M NaOH to reach the end point, what is the molarity of the H_2SO_4 solution?

9.45 A sample of 27.0 mL of 0.310 M NaOH is titrated with 0.740 M H_2SO_4. How many milliliters of the H_2SO_4 solution are required to reach the end point?

9.46 A 0.300 M solution of H_2SO_4 was used to titrate 10.00 mL of an aqueous NaOH; 15.00 mL of acid was required to neutralize the basic solution. What was the molarity of the base?

9.47 A solution of an aqueous NaOH was titrated with 0.150 M HCl, and 22.0 mL of acid was needed to reach the end point of the titration. How many moles of the base were in the solution?

9.48 The usual concentration of HCO_3^- ions in blood plasma is approximately 24 millimoles per liter (mmol/L). How would you make up 1.00 L of a solution containing this concentration of HCO_3^- ions?

9.49 What is the end point of a titration?

9.50 Why does a titration not tell us the acidity or basicity of a solution?

SECTION 9.10 What Are Buffers?

9.51 Write equations to show what happens when, to a buffer solution containing equimolar amounts of CH_3COOH and CH_3COO^-, we add

(a) H_3O^+ (b) OH^-

9.52 Write equations to show what happens when, to a buffer solution containing equimolar amounts of HPO_4^{2-} and $H_2PO_4^-$, we add

(a) H_3O^+ (b) OH^-

9.53 We commonly refer to a buffer as consisting of approximately equal molar amounts of a weak acid and its conjugate base—for example, CH_3COOH and CH_3COO^-. Is it also possible to have a buffer consisting of approximately equal molar amounts of a weak base and its conjugate acid? Explain.

9.54 What is meant by buffer capacity?

9.55 How can you change the pH of a buffer? How can you change the capacity of a buffer?

9.56 What is the connection between buffer action and Le Chatelier's principle?

9.57 Give two examples of a situation where you would want a buffer to have unequal amounts of the conjugate acid and the conjugate base.

SECTION 9.11 How Do We Calculate the pH of a Buffer?

9.58 What is the pH of a buffer solution made by dissolving 0.10 mol of formic acid, HCOOH, and 0.10 mol of sodium formate, HCOONa, in 1 L of water?

9.59 The pH of a solution made by dissolving 1.0 mol of propanoic acid and 1.0 mol of sodium propanoate in 1.0 L of water is 4.85.

   (a) What would the pH be if we used 0.10 mol of each (in 1 L of water) instead of 1.0 mol?

   (b) With respect to buffer capacity, how would the two solutions differ?

9.60 Show that when the concentration of the weak acid, [HA], in an acid–base buffer equals that of the conjugate base of the weak acid, [A$^-$], the pH of the buffer solution is equal to the pK_a of the weak acid.

9.61 Calculate the pH of an aqueous solution containing the following:

   (a) 0.80 M lactic acid and 0.40 M lactate ion

   (b) 0.30 M NH$_3$ and 1.50 M NH$_4^+$

9.62 The pH of 0.10 M HCl is 1.0. When 0.10 mol of sodium acetate, CH$_3$COONa, is added to this solution, its pH changes to 2.9. Explain why the pH changes, and why it changes to this particular value.

9.63 If you have 100 mL of a 0.1 M buffer made of NaH$_2$PO$_4$ and Na$_2$HPO$_4$ that is at pH 6.8, and you add 10 mL of 1 M HCl, will you still have a usable buffer?

SECTION 9.12 What Are TRIS, HEPES, and These Buffers with the Strange Names?

9.64 Write an equation showing the reaction of TRIS in the acid form with sodium hydroxide (do not write out the chemical formula for TRIS).

9.65 What is the pH of a solution that is 0.1 M in TRIS in the acid form and 0.05 M in TRIS in the basic form?

9.66 Explain why you do not need to know the chemical formula of a buffer compound to use it.

9.67 If you have a HEPES buffer at pH 4.75, will it be a usable buffer? Why or why not?

9.68 Which of the compounds listed in Table 9.5 would be the most effective for making a buffer at pH 8.15? Why?

Chemical Connections

9.69 (Chemical Connections 9A) Which weak base is used as a flame retardant in plastics?

9.70 (Chemical Connections 9B) What is the most important immediate first aid in any chemical burn of the eyes?

9.71 (Chemical Connections 9B) With respect to corneal burns, are strong acids or strong bases more dangerous?

9.72 (Chemical Connections 9C) Name the most common bases used in over-the-counter antacids.

9.73 (Chemical Connections 9D) What causes (a) respiratory acidosis and (b) metabolic acidosis?

9.74 (Chemical Connections 9E) Explain how the sprinter's trick works. Why would an athlete want to raise the pH of his or her blood?

9.75 (Chemical Connections 9E) Another form of the sprinter's trick is to drink a sodium bicarbonate shake before the event. What would be the purpose of doing so? Give the relevant equations.

Additional Problems

9.76 4-Methylphenol, CH$_3$C$_6$H$_4$OH (pK_a = 10.26), is only slightly soluble in water, but its sodium salt, CH$_3$C$_6$H$_4$O$^-$Na$^+$, is quite soluble in water. In which of the following solutions will 4-methylphenol dissolve more readily than in pure water?

   (a) Aqueous NaOH   (b) Aqueous NaHCO$_3$

   (c) Aqueous NH$_3$

9.77 Benzoic acid, C$_6$H$_5$COOH (pK_a = 4.19), is only slightly soluble in water, but its sodium salt, C$_6$H$_5$COO$^-$Na$^+$, is quite soluble in water. In which of the following solutions will benzoic acid dissolve more readily than in pure water?

   (a) Aqueous NaOH   (b) Aqueous NaHCO$_3$

   (c) Aqueous Na$_2$CO$_3$

9.78 Assume that you have a dilute solution of HCl (0.10 M) and a concentrated solution of acetic acid (5.0 M). Which solution is more acidic? Explain.

9.79 Which of the two solutions from Problem 9.78 would take a greater amount of NaOH to hit a phenolphthalein end point? Explain.

9.80 If the [OH$^-$] of a solution is 1 × 10^{-14},

   (a) What is the pH of the solution?

   (b) What is the [H$_3$O$^+$]?

9.81 What is the molarity of a solution made by dissolving 0.583 g of the diprotic acid oxalic acid, H$_2$C$_2$O$_4$, in enough water to make 1.75 L of solution?

9.82 Following are three organic acids and the pK_a of each: butanoic acid, 4.82; barbituric acid, 5.00; and lactic acid, 3.85.

   (a) What is the K_a of each acid?

   (b) Which of the three is the strongest acid and which is the weakest?

   (c) What information would you need to predict which one of the three acids would require the most NaOH to reach a phenolphthalein end point?

9.83 The pK_a value of barbituric acid is 5.0. If the H_3O^+ and barbiturate ion concentrations are each 0.0030 M, what is the concentration of the undissociated barbituric acid?

9.84 If pure water self-ionizes to give H_3O^+ and OH^- ions, why doesn't pure water conduct an electric current?

9.85 Can an aqueous solution have a pH of zero? Explain your answer using aqueous HCl as your example.

9.86 If an acid, HA, dissolves in water such that the K_a is 1000, what is the pK_a of that acid? Is this scenario possible?

9.87 A scale of K_b values for bases could be set up in a manner similar to that for the K_a scale for acids. However, this setup is generally considered unnecessary. Explain.

9.88 Do a 1.0 M CH_3COOH solution and a 1.0 M HCl solution have the same pH? Explain.

9.89 Do a 1.0 M CH_3COOH solution and a 1.0 M HCl solution require the same amount of 1.0 M NaOH to hit a titration end point? Explain.

9.90 Suppose you wish to make a buffer whose pH is 8.21. You have available 1 L of 0.100 M NaH_2PO_4 and solid Na_2HPO_4. How many grams of the solid Na_2HPO_4 must be added to the stock solution to accomplish this task? (Assume that the volume remains 1 L.)

9.91 In the past, boric acid was used to rinse an inflamed eye. How would you make up 1 L of a $H_3BO_3/H_2BO_3^-$ buffer solution that has a pH of 8.40? Start with solid NaH_2BO_3 and 1.00 M H_3BO_3.

9.92 Suppose you want to make a CH_3COOH/CH_3COO^- buffer solution with a pH of 5.60. The acetic acid concentration is to be 0.10 M. What should the acetate ion concentration be?

9.93 For an acid–base reaction, one way to determine the position of equilibrium is to say that the larger of the equilibrium arrow pair points to the acid with the higher value of pK_a. For example,

$$CH_3COOH + HCO_3^- \rightleftharpoons CH_3COO^- + H_2CO_3$$
$$pK_a = 4.75 \qquad\qquad\qquad pK_a = 6.37$$

Explain why this rule works.

9.94 When a solution prepared by dissolving 4.00 g of an unknown acid in 1.00 L of water is titrated with 0.600 M NaOH, 38.7 mL of the NaOH solution is needed to neutralize the acid. What was the molarity of the acid solution?

9.95 Write equations to show what happens when, to a buffer solution containing equal amounts of HCOOH and $HCOO^-$, we add

(a) H_3O^+ (b) OH^-

9.96 If we add 0.10 mol of NH_3 to 0.50 mol of HCl dissolved in enough water to make 1.0 L of solution, what happens to the NH_3? Will any NH_3 remain? Explain.

9.97 Suppose you have an aqueous solution prepared by dissolving 0.050 mol of NaH_2PO_4 in 1 L of water. This solution is not a buffer, but suppose you want to make it into one. How many moles of solid Na_2HPO_4 must you add to this aqueous solution to make it into

(a) A buffer of pH 7.21

(b) A buffer of pH 6.21

(c) A buffer of pH 8.21

9.98 The pH of a 0.10 M solution of acetic acid is 2.93. When 0.10 mol of sodium acetate, CH_3COONa, is added to this solution, its pH changes to 4.74. Explain why the pH changes, and why it changes to this particular value.

9.99 Suppose you have a $H_2PO_4^-/HPO_4^{2-}$ buffer of pH 7.21. If you add more solid NaH_2PO_4 to this buffer, would you expect the pH of the buffer to increase, decrease, or remain unchanged? Explain.

9.100 Suppose you have a bicarbonate buffer containing carbonic acid, H_2CO_3, and sodium bicarbonate, $NaHCO_3$, and that the pH of the buffer is 6.37. If you add more solid $NaHCO_3$ to this buffer solution, would you expect its pH to increase, decrease, or remain unchanged? Explain.

9.101 A student pulls a bottle of TRIS off of a shelf and notes that the bottle says, "TRIS (basic form), $pK_a = 8.3$." The student tells you that if you add 0.1 mol of this compound to 100 mL of water, the pH will be 8.3. Is the student correct? Explain.

Looking Ahead

9.102 Unless under pressure, carbonic acid in aqueous solution breaks down into carbon dioxide and water, and carbon dioxide is evolved as bubbles of gas. Write an equation for the conversion of carbonic acid to carbon dioxide and water.

9.103 Following are pH ranges for several human biological materials. From the pH at the midpoint of each range, calculate the corresponding $[H_3O^+]$. Which materials are acidic, which are basic, and which are neutral?

(a) Milk, pH 6.6–7.6

(b) Gastric contents, pH 1.0–3.0

(c) Spinal fluid, pH 7.3–7.5

(d) Saliva, pH 6.5–7.5

(e) Urine, pH 4.8–8.4

(f) Blood plasma, pH 7.35–7.45

(g) Feces, pH 4.6–8.4

(h) Bile, pH 6.8–7.0

9.104 What is the ratio of $HPO_4^{2-}/H_2PO_4^-$ in a phosphate buffer of pH 7.40 (the average pH of human blood plasma)?

9.105 What is the ratio of $HPO_4^{2-}/H_2PO_4^-$ in a phosphate buffer of pH 7.9 (the pH of human pancreatic fluid)?

CHAPTER 10

Organic Chemistry

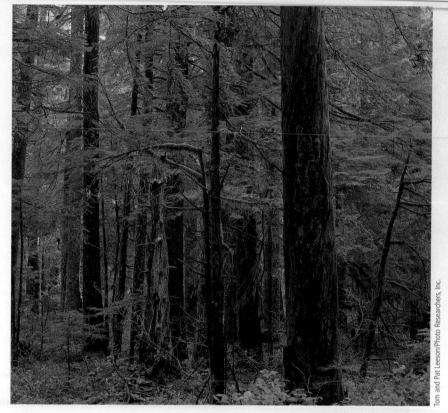

The bark of the Pacific yew contains paclitaxel, a substance that has proven effective in treating certain types of ovarian and breast cancer (see Chemical Connections 10A).

GOB
Chemistry✦Now™

Look for this logo in the chapter and go to GOB ChemistryNow at **http://now.brookscole.com/gob8** or on the CD for tutorials, simulations, and problems.

10.1 | What Is Organic Chemistry?

Organic chemistry is the chemistry of the compounds of carbon. As you study Chapters 10–19 (organic chemistry) and 20–31 (biochemistry), you will see that organic compounds are everywhere around us. They are in our foods, flavors, and fragrances; in our medicines, toiletries, and cosmetics; in our plastics, films, fibers, and resins; in our paints, varnishes, and glues; and, of course, in our bodies and the bodies of all other living organisms.

Perhaps the most remarkable feature of organic compounds is that they involve the chemistry of carbon and only a few other elements—chiefly, hydrogen, oxygen, and nitrogen. While the majority of organic compounds contain carbon and just these three elements, many also contain sulfur, a halogen (fluorine, chlorine, bromine, or iodine), and phosphorus.

As of the writing of this text, there are 116 known elements. Organic chemistry concentrates on carbon, just one of the 116. The chemistry of the other 115 elements comes under the field of inorganic chemistry. As we see

Figure 10.1 Abundance of the elements in the Earth's crust.

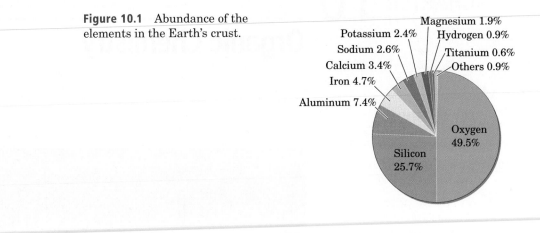

in Figure 10.1, carbon is far from being among the most abundant elements in the Earth's crust. In terms of elemental abundance, approximately 75% of the Earth's crust is composed of just two elements: oxygen and silicon. These two elements are the components of silicate minerals, clays, and sand. In fact, carbon is not even among the ten most abundant elements. Instead, it is merely one of the elements making up the remaining 0.9% of the Earth's crust. Why, then, do we pay such special attention to just one element from among 116?

The first reason is largely historical. In the early days of chemistry, scientists thought organic compounds were those produced by living organisms, and that inorganic compounds were those found in rocks and other nonliving matter. At that time, they believed that a "vital force," possessed only by living organisms, was necessary to produce organic compounds. In other words, chemists believed that they could not synthesize any organic compound by starting with only inorganic compounds. This theory was very easy to disprove if, indeed, it was wrong. It required only one experiment in which an organic compound was made from inorganic compounds. In 1828, Friedrich Wöhler (1800–1882) carried out just such an experiment. He heated an aqueous solution of ammonium chloride and silver cyanate, both inorganic compounds and—to his surprise—obtained urea, an "organic" compound found in urine.

Although this single experiment of Wöhler's was sufficient to disprove the "doctrine of vital force," it took several years and a number of additional experiments for the entire scientific community to accept the fact that organic compounds could be synthesized in the laboratory. This discovery meant that the terms "organic" and "inorganic" no longer had their original meanings because, as Wöhler demonstrated, organic compounds could be obtained from inorganic materials. A few years later, August Kekulé (1829–1896) put forth a new definition—organic compounds are those containing carbon—and his definition has been accepted ever since.

Table 10.1 A Comparison of Properties of Organic and Inorganic Compounds

Organic Compounds	Inorganic Compounds
Bonding is almost entirely covalent.	Most have ionic bonds.
May be gases, liquids, or solids with low melting points (less than 360°C).	Most are solids with high melting points.
Most are insoluble in water.	Many are soluble in water.
Most are soluble in organic solvents such as diethyl ether, toluene, and dichloromethane.	Almost all are insoluble in organic solvents.
Aqueous solutions do not conduct electricity.	Aqueous solutions conduct electricity.
Almost all burn.	Very few burn.
Reactions are usually slow.	Reactions are often very fast.

Organic and inorganic compounds differ in their properties because they differ in their structure and composition—not because they obey different natural laws. One set of natural laws applies to all compounds.

A second reason for the study of carbon compounds as a separate discipline is the sheer number of organic compounds. Chemists have discovered or synthesized more than 10 million of them, and an estimated 10,000 new ones are reported each year. By comparison, chemists have discovered or synthesized an estimated 1.7 million inorganic compounds. Thus approximately 85% of all known compounds are organic compounds.

A third reason—and one particularly important for those of you going on to study biochemistry—is that biochemicals, including carbohydrates, lipids, proteins, enzymes, nucleic acids (DNA and RNA), hormones, vitamins, and almost all other important chemicals in living systems are organic compounds. Furthermore, their reactions are often strikingly similar to those occurring in test tubes. For this reason, knowledge of organic chemistry is essential for an understanding of biochemistry.

One final point about organic compounds. They generally differ from inorganic compounds in many of their properties, some of which are shown in Table 10.1. Most of these differences stem from the fact that the bonding in organic compounds is almost entirely covalent, while most inorganic compounds have ionic bonds.

Of course, the entries in Table 10.1 are generalizations, but they are largely true for the vast majority of compounds of both types.

10.2 | Where Do We Obtain Organic Compounds?

Chemists obtain organic compounds in two principal ways: isolation from nature and synthesis in the laboratory.

A. Isolation from Nature

Living organisms are "chemical factories." Each terrestrial, marine, and freshwater plant (flora) and animal (fauna)—even microorganisms such as bacteria—makes thousands of organic compounds by a process called biosynthesis. One way, then, to get organic compounds is to extract, isolate, and purify them from biological sources. In this book, we will encounter many compounds that are or have been isolated in this way. Some important examples include

CHEMICAL CONNECTIONS 10A

Taxol: A Story of Search and Discovery

In the early 1960s, the National Cancer Institute undertook a program to analyze samples of native plant materials in the hope of discovering substances that would prove effective in the fight against cancer. Among the materials tested was an extract of the bark of the Pacific yew, *Taxus brevifolia,* a slow-growing tree found in the old-growth forests of the Pacific Northwest. This biologically active extract proved to be remarkably effective in treating certain types of ovarian and breast cancer, even in cases where other forms of chemotherapy failed. The structure of the cancer-fighting component of yew bark was determined in 1962, and the compound was named paclitaxel (Taxol).

Paclitaxel
(Taxol)

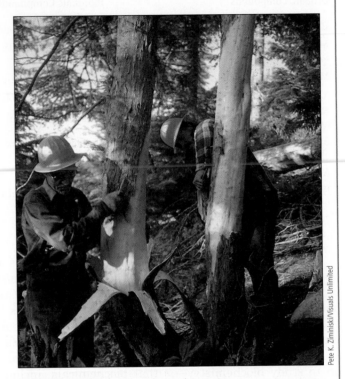

Pacific yew bark being stripped for Taxol extraction.

Unfortunately, the bark of a single 100-year-old yew tree yields only about 1 g of Taxol, not enough for effective treatment of even one cancer patient. Furthermore, isolating Taxol means stripping the bark from trees, which kills them. In 1994, chemists succeeded in synthesizing Taxol in the laboratory, but the cost of the synthetic drug was far too high to be economical. Fortunately, an alternative natural source of the drug was found. Researchers in France discovered that the needles of a related plant, *Taxus baccata,* contain a compound that can be converted in the laboratory to Taxol. Because the needles can be gathered without harming the plant, it is not necessary to kill trees to obtain the drug.

Taxol inhibits cell division by acting on microtubules, a key component of the scaffolding of cells. Before cell division can take place, the cell must disassemble these microtubule units, and Taxol prevents this disassembly. Because cancer cells divide faster than normal cells, the drug effectively controls their spread.

The remarkable success of Taxol in the treatment of breast and ovarian cancer has stimulated research efforts to isolate and/or synthesize other substances that treat human diseases the same way in the body and that may be even more effective anticancer agents than Taxol.

vitamin E, the penicillins, table sugar, insulin, quinine, and the anticancer drug paclitaxel (Taxol; see Chemical Connections 10A). Nature also supplies us with three other important sources of organic compounds: natural gas, petroleum, and coal. We will discuss them in Section 11.10.

B. Synthesis in the Laboratory

Ever since Wöhler synthesized urea, organic chemists have sought to develop more ways to synthesize the same compounds or design derivatives of those found in nature. In recent years, the methods for doing so have become so sophisticated that there are few natural organic compounds, no matter how complicated, that chemists cannot synthesize in the laboratory.

CHEMICAL CONNECTIONS 10B

Combinatorial Chemistry

Since the beginning of synthetic organic chemistry about 200 years ago, the goal has been to produce pure organic compounds with high percentage yields. More often than not, a particular use for a compound was discovered accidentally. Such was the case, for example, with aspirin and the barbiturates. Later, when researchers discovered the mode of action of certain drugs, it became possible to design new compounds that would have a reasonable probability of being more effective or having fewer side effects than some existing drug. Such was the case with ibuprofen and albuterol (see Chemical Connections 16E). Even in such a "rational" approach (as compared to the "shotgun" approach, in which compounds were synthesized at random), the goal was still to obtain a single pure compound, no matter how many steps were involved in the synthesis or how difficult the steps were. In the last few years this strategy has changed, mostly through new trends in pharmaceutical research.

Combinatorial chemistry is a process in which chemists try to produce as many compounds from as few building blocks as possible. For example, if we allow 20 compounds (building blocks) to react at random, and if we could find a method to restrict the combination in such a way that only molecules containing three of these building blocks are produced, then a mixture of 8000 different compounds (20^3) would result. Perhaps this mixture might include two or three candidates with the potential to be useful drugs. The task, then, is to select from this synthetic soup those compounds that promise to be biologically active. Researchers accomplish this goal by screening the mixture for a desired biological activity and then separating out promising candidates for further study. In essence, it is cheaper and faster to produce a large "library" of combinatorial synthetic compounds and then "fish out" the biologically active ones than to pursue the old way of randomly synthesizing new compounds and screening them one at a time.

Compounds made in the laboratory are identical in both chemical and physical properties to those found in nature—assuming, of course, that each is 100% pure. There is no way that anyone can tell whether a sample of any particular compound was made by chemists or obtained from nature. As a consequence, pure ethanol made by chemists has exactly the same physical and chemical properties as pure ethanol prepared by distilling wine. The same is true for ascorbic acid (vitamin C). There is no advantage, therefore, in paying more money for vitamin C obtained from a natural source than for synthetic vitamin C, because the two are identical in every way.

Organic chemists, however, have not been satisfied with merely duplicating nature's compounds. They have also designed and synthesized compounds not found in nature. In fact, the majority of the more than 10 million known organic compounds are purely synthetic and do not exist in living organisms. For example, many modern drugs—Valium, albuterol, Prozac, Zantac, Zoloft, Lasix, Viagra, and Enovid—are synthetic organic compounds not found in nature. Even the over-the-counter drugs aspirin and ibuprofen are synthetic organic compounds not found in nature.

The vitamin C in an orange is identical to its synthetic tablet form.

10.3 How Do We Write Structural Formulas of Organic Compounds?

GOB
Chemistry·ᐩ·Now™
Click *Coached Problems* for an interactive exploration of **Bonding in Organic Compounds**

A structural formula shows all the atoms present in a molecule as well as the bonds that connect the atoms to each other. The structural formula for ethanol, C_2H_6O, for example, shows all nine atoms and the eight bonds that connect them:

Table 10.2 Single, Double, and Triple Bonds in Compounds of Carbon. Bond angles and geometries are predicted using the VSEPR model

Ethane
(bond angles
109.5°)

Ethylene
(bond angles
120°)

Acetylene
(bond angles
180°)

Chloroethane
(bond angles
109.5°)

Methanol
(bond angles
109.5°)

Formaldehyde
(bond angles
120°)

Methylamine
(bond angles
109.5°)

Methyleneimine
(bond angles 120°)

Hydrogen cyanide
(bond angle 180°)

The Lewis model of bonding (Section 4.7C) enables us to see how carbon forms four covalent bonds that may be various combinations of single, double, and triple bonds. Furthermore, the valence-shell electron-pair repulsion (VSEPR) model (Section 4.10) tells us that the most common bond angles about carbon atoms in covalent compounds are approximately 109.5°, 120°, and 180° for tetrahedral, trigonal planar, and linear geometries, respectively.

Table 10.2 shows several covalent compounds containing carbon bonded to hydrogen, oxygen, nitrogen, and chlorine. From these examples, we see the following:

- Carbon normally forms four covalent bonds and has no unshared pairs of electrons.
- Nitrogen normally forms three covalent bonds and has one unshared pair of electrons.
- Oxygen normally forms two covalent bonds and has two unshared pairs of electrons.
- Hydrogen forms one covalent bond and has no unshared pairs of electrons.
- A halogen (fluorine, chlorine, bromine, and iodine) normally forms one covalent bond and has three unshared pairs of electrons.

EXAMPLE 10.1

The structural formulas for acetic acid, CH_3COOH, and ethylamine, $CH_3CH_2NH_2$, are

Acetic acid

Ethylamine

(a) Complete the Lewis structure for each molecule by adding unshared pairs of electrons so that each atom of carbon, oxygen, and nitrogen has a complete octet.

(b) Using the VSEPR model (Section 4.10), predict all bond angles in each molecule.

Solution

(a) Each carbon atom is surrounded by eight valence electrons and therefore has a complete octet. To complete the octet of each oxygen, add two unshared pairs of electrons. To complete the octet of nitrogen, add one unshared pair of electrons.

(b) To predict bond angles about a carbon, nitrogen, or oxygen atom, count the number of regions of electron density (lone pairs and bonding pairs of electrons about it). If the atom is surrounded by four regions of electron density, predict bond angles of 109.5°. If it is surrounded by three regions, predict bond angles of 120°. If it is surrounded by two regions, predict a bond angle of 180°.

Acetic acid

Ethylamine

Problem 10.1

The structural formulas for ethanol, CH_3CH_2OH, and propene, $CH_3CH=CH_2$, are

Ethanol   Propene

(a) Complete the Lewis structure for each molecule showing all valence electrons.

(b) Using the VSEPR model, predict all bond angles in each molecule.

10.4 | What Are Functional Groups?

As noted earlier in this chapter, more than 10 million organic compounds have been isolated and synthesized by organic chemists. It might seem an almost impossible task to learn the physical and chemical properties of so many compounds. Fortunately, the study of organic compounds is not as formidable a task as you might think. While organic compounds can undergo a wide variety of chemical reactions, only certain portions of their structures undergo chemical transformations. We call atoms or groups of atoms of an organic molecule that undergo predictable chemical reactions **functional groups.** As we will see, the same functional group, in whatever organic molecule it occurs, undergoes the same types of chemical reactions. Therefore, we do not have to study the chemical reactions of even a fraction of the 10 million known organic compounds. Instead, we need to identify only a few characteristic functional groups and then study the chemical reactions that each undergoes.

Functional groups are also important because they are the units by which we divide organic compounds into families of compounds. For example, we group those compounds that contain an —OH (hydroxyl) group bonded to a tetrahedral carbon into a family called alcohols; compounds containing a —COOH (carboxyl group) belong to a family called carboxylic acids. Table 10.3 introduces six of the most common functional groups. A complete list of all functional groups that we will study appears on the inside back cover of the text.

At this point, our concern is simply pattern recognition—that is, how to recognize and identify one of these six common functional groups when you see it, and how to draw structural formulas of molecules containing them. We will have more to say about the physical and chemical properties of these and several other functional groups in Chapters 11–19.

Functional groups also serve as the basis for naming organic compounds. Ideally, each of the 10 million or more organic compounds must have a unique name different from the name of every other compound. We will show how these names are derived in Chapters 11–19 as we study individual functional groups in detail.

To summarize, functional groups

- Are sites of predictable chemical behavior—a particular functional group, in whatever compound it is found, undergoes the same types of chemical reactions.
- Determine in large measure the physical properties of a compound.

Functional group An atom or group of atoms within a molecule that shows a characteristic set of predictable physical and chemical behaviors

GOB
Chemistry⋅⚛⋅Now™

Click *Coached Problems* to try a set of problems in Identifying **Functional Groups**

Table 10.3 Six Common Functional Groups

Family	Functional Group	Example	Name
Alcohol	—OH	CH_3CH_2OH	Ethanol
Amine	—NH$_2$	$CH_3CH_2NH_2$	Ethanamine
Aldehyde	$-\overset{\overset{\displaystyle O}{\|\|}}{C}-H$	$CH_3\overset{\overset{\displaystyle O}{\|\|}}{C}H$	Ethanal
Ketone	$-\overset{\overset{\displaystyle O}{\|\|}}{C}-$	$CH_3\overset{\overset{\displaystyle O}{\|\|}}{C}CH_3$	Acetone
Carboxylic acid	$-\overset{\overset{\displaystyle O}{\|\|}}{C}-OH$	$CH_3\overset{\overset{\displaystyle O}{\|\|}}{C}OH$	Acetic acid
Ester	$-\overset{\overset{\displaystyle O}{\|\|}}{C}-OR$	$CH_3\overset{\overset{\displaystyle O}{\|\|}}{C}OCH_2CH_3$	Ethyl acetate

- Serve as the units by which we classify organic compounds into families; and
- Serve as a basis for naming organic compounds.

A. Alcohols

As previously mentioned, the functional group of an **alcohol** is an **—OH (hydroxyl) group** bonded to a tetrahedral carbon atom (a carbon having bonds to four atoms). In the general formula of an alcohol (shown below on the left), the symbol R— indicates either a hydrogen or another carbon group. The important point of the general structure is the —OH group bonded to a tetrahedral carbon atom.

Alcohol A compound containing an —OH (hydroxyl) group bonded to a tetrahedral carbon atom

Hydroxyl group An —OH group bonded to a tetrahedral carbon atom

| Functional group | Structural formula | Condensed structural formula |
| R = H or carbon group | An alcohol (Ethanol) | |

Here we also represent the alcohol by a **condensed structural formula,** CH_3CH_2OH. In a condensed structural formula, CH_3 indicates a carbon bonded to three hydrogens, CH_2 indicates a carbon bonded to two hydrogens, and CH indicates a carbon bonded to one hydrogen. Unshared pairs of electrons are generally not shown in condensed structural formulas.

Alcohols are classified as **primary (1°), secondary (2°),** or **tertiary (3°),** depending on the number of carbon atoms bonded to the carbon bearing the —OH group.

A 1° alcohol   A 2° alcohol   A 3° alcohol

EXAMPLE 10.2

Draw Lewis structures and condensed structural formulas for the two alcohols with molecular formula C_3H_8O. Classify each as primary, secondary, or tertiary.

Solution
Begin by drawing the three carbon atoms in a chain. The oxygen atom of the hydroxyl group may be bonded to the carbon chain at two different positions on the chain: either to an end carbon or to the middle carbon.

$$C-C-C \qquad C-C-C-OH \qquad C-\overset{\overset{\displaystyle OH}{|}}{C}-C$$

Carbon chain   The two locations for the —OH group

Finally, add seven more hydrogens, giving a total of eight as shown in the molecular formula. Show unshared electron pairs on the Lewis structures but not on the condensed structural formulas.

Lewis structures	Condensed structural formulas	Ball-and-stick models

$CH_3CH_2CH_2OH$

A primary alcohol
(1-Propanol)

CH_3CHCH_3 with OH

A secondary alcohol
(2-Propanol)

The secondary alcohol 2-propanol, whose common name is isopropyl alcohol, is the cooling, soothing component in rubbing alcohol.

Problem 10.2

Draw Lewis structures and condensed structural formulas for the four alcohols with the molecular formula $C_4H_{10}O$. Classify each alcohol as primary, secondary, or tertiary. (*Hint:* First consider the connectivity of the four carbon atoms; they can be bonded either four in a chain or three in a chain with the fourth carbon as a branch on the middle carbon. Then consider the points at which the —OH group can be bonded to each carbon chain.)

2-Propanol (isopropyl alcohol) is used to disinfect cuts and scrapes.

Charles D. Winters

B. Amines

Amine An organic compound in which one, two, or three hydrogens of ammonia are replaced by carbon groups; RNH_2, R_2NH, or R_3N

Amino group An —NH_2, RNH_2, R_2NH, or R_3N group

The functional group of an amine is an **amino group**—a nitrogen atom bonded to one, two, or three carbon atoms. In a **primary (1°) amine,** nitrogen is bonded to one carbon group. In a **secondary (2°) amine,** it is bonded to two carbon groups. In a **tertiary (3°) amine,** it is bonded to three carbon groups. The second and third structural formulas can be written in a more abbreviated form by collecting the CH_3 groups and writing them as $(CH_3)_2NH$ and $(CH_3)_3N$. The latter are known as condensed structural formulas.

CH_3NH_2

Methylamine
(a 1° amine)

CH_3NH or $(CH_3)_2NH$ with CH_3

Dimethylamine
(a 2° amine)

CH_3NCH_3 or $(CH_3)_3N$ with CH_3

Trimethylamine
(a 3° amine)

EXAMPLE 10.3

Draw condensed structural formulas for the two primary amines with the molecular formula C_3H_9N.

Solution

For a primary amine, draw a nitrogen atom bonded to two hydrogens and one carbon.

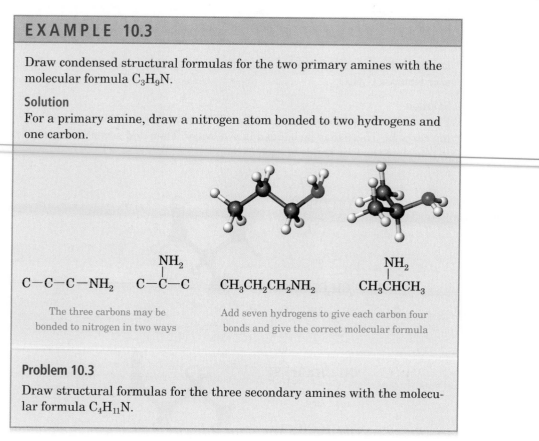

$$C-C-C-NH_2 \qquad C-\underset{\underset{NH_2}{|}}{C}-C$$

The three carbons may be bonded to nitrogen in two ways

$$CH_3CH_2CH_2NH_2 \qquad CH_3\underset{\underset{NH_2}{|}}{C}HCH_3$$

Add seven hydrogens to give each carbon four bonds and give the correct molecular formula

Problem 10.3

Draw structural formulas for the three secondary amines with the molecular formula $C_4H_{11}N$.

C. Aldehydes and Ketones

Both aldehydes and ketones contain a **C=O (carbonyl) group.** The **aldehyde** functional group contains a carbonyl group bonded to a hydrogen. Formaldehyde, CH_2O, the simplest aldehyde, has two hydrogens bonded to its carbonyl carbon. In a condensed structural formula, the aldehyde group may be written showing the carbon–oxygen double bond as CH=O or, alternatively, it may be written —CHO. The functional group of a **ketone** is a carbonyl group bonded to two carbon atoms. In the general structural formula of each functional group, we use the symbol R to represent other groups bonded to carbon to complete the tetravalence of carbon.

Carbonyl group A C=O group

Aldehyde A compound containing a carbonyl group bonded to a hydrogen; a —CHO group

Ketone A compound containing a carbonyl group bonded to two carbon groups

Functional group

Acetaldehyde (an aldehyde)

Functional group

Acetone (a ketone)

EXAMPLE 10.4

Draw condensed structural formulas for the two aldehydes with the molecular formula C_4H_8O.

Solution

First draw the functional group of an aldehyde, and then add the remaining carbons. These may be bonded in two ways. Then add seven hydrogens to complete the four bonds of each carbon.

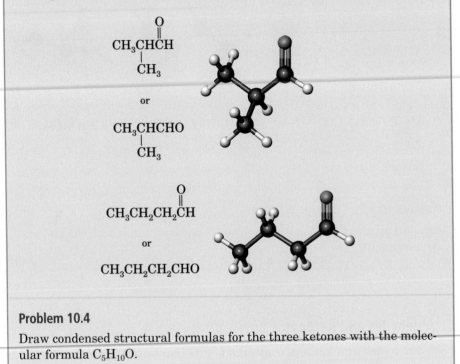

$$\overset{\quad\;\; O}{\underset{\displaystyle \underset{CH_3}{|}}{CH_3CH\overset{||}{C}H}}$$

or

$$\underset{\displaystyle \underset{CH_3}{|}}{CH_3CHCHO}$$

$$CH_3CH_2CH_2\overset{\displaystyle \overset{O}{||}}{C}H$$

or

$$CH_3CH_2CH_2CHO$$

Problem 10.4

Draw condensed structural formulas for the three ketones with the molecular formula $C_5H_{10}O$.

D. Carboxylic Acids

Carboxyl group A —COOH group

Carboxylic acid A compound containing a —COOH group

The functional group of a carboxylic acid is a —**COOH** (**carboxyl:** carbonyl + hydroxyl) **group.** In a condensed structural formula, a carboxyl group may also be written —CO_2H.

$$\overset{\displaystyle \overset{O}{||}}{R}COH$$

Functional group

$$CH_3\overset{\displaystyle \overset{O}{||}}{C}OH$$

Acetic acid
(a carboxylic acid)

EXAMPLE 10.5

Draw a condensed structural formula for the single carboxylic acid with the molecular formula $C_3H_6O_2$.

Solution

The only way the carbon atoms can be written is three in a chain, and the —COOH group must be on an end carbon of the chain.

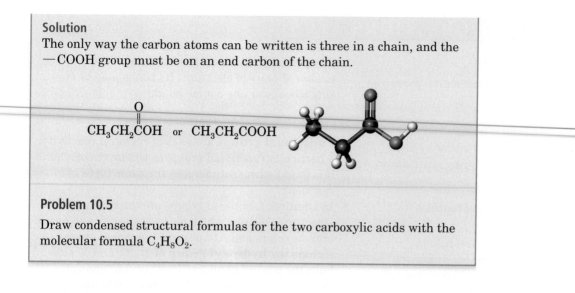

$$CH_3CH_2\overset{\overset{\displaystyle O}{\|}}{C}OH \quad \text{or} \quad CH_3CH_2COOH$$

Problem 10.5

Draw condensed structural formulas for the two carboxylic acids with the molecular formula $C_4H_8O_2$.

E. Carboxylic Esters

A **carboxylic ester,** commonly referred to as simply an **ester,** is a derivative of a carboxylic acid in which the hydrogen of the carboxyl group is replaced by a carbon group. The ester group is written —COOR in this text.

Carboxylic ester A derivative of a carboxylic acid in which the H of the carboxyl group is replaced by a carbon group

Functional group

$$CH_3-\overset{\overset{\displaystyle O}{\|}}{C}-O-CH_3 \quad \text{or} \quad H_3CCOOCH_3$$

Methyl acetate
(an ester)

EXAMPLE 10.6

The molecular formula of methyl acetate is $C_3H_6O_2$. Draw the structural formula of another ester with the same molecular formula.

Solution

There is only one other ester with this molecular formula. Its structural formula is

$$H-\overset{\overset{\displaystyle O}{\|}}{C}-O-CH_2-CH_3$$

Ethyl formate

We will usually write this formula as $HCOOCH_2CH_3$.

Problem 10.6

Draw structural formulas for the four esters with the molecular formula $C_4H_8O_2$.

SUMMARY OF KEY QUESTIONS

SECTION 10.1 What is Organic Chemistry?

- Organic chemistry is the study of compounds containing carbon.

SECTION 10.2 Where Do We Obtain Organic Compounds?

- Chemists obtain organic compounds either by isolation from plant and animal sources or by synthesis in the laboratory.

SECTION 10.3 How Do We Write Structural Formulas of Organic Compounds?

- Carbon normally forms four bonds and has no unshared pairs of electrons. Its four bonds may be four single bonds, two single bonds and one double bond, or one triple bond and one single bond.
- Nitrogen normally forms three bonds and has one unshared pair of electrons. Its bonds may be three single bonds, one single bond and one double bond, or one triple bond.

- Oxygen normally forms two bonds and has two unshared pairs of electrons. Its bonds may be two single bonds or one double bond.

SECTION 10.4 What Are Functional Groups?

- **Functional groups** are sites of chemical reactivity; a particular functional group, in whatever compound it is found, always undergoes the same types of reactions.
- In addition, functional groups are the characteristic structural units by which we both classify and name organic compounds. Important functional groups include the **hydroxyl group** of 1°, 2°, and 3° alcohols; the **amino group** of 1°, 2°, and 3° amines; the **carbonyl group** of aldehydes and ketones; the **carboxyl group** of carboxylic acids, and the **ester group.**

PROBLEMS

Chemistry Now™

Assess your understanding of this chapter's topics with additional quizzing and conceptual-based problems at **http://now.brookscole.com/gob8** or on the CD.

A blue problem number indicates an applied problem.

■ denotes problems that are available on the GOB ChemistryNow website or CD and are assignable in OWL.

SECTION 10.2 Where Do We Obtain Organic Compounds?

10.7 Is there any difference between vanillin made synthetically and vanillin extracted from vanilla beans?

10.8 Suppose that you are told that only organic substances are produced by living organisms. How would you rebut this assertion?

10.9 What important experiment was carried out by Wöhler in 1828?

SECTION 10.3 How Do We Write Structural Formulas of Organic Compounds?

10.10 ■ List the four principal elements that make up organic compounds and give the number of bonds each typically forms.

10.11 Think about the types of substances in your immediate environment, and make a list of those that are organic—for example, textile fibers. We will ask you

to return to this list later in the course and to refine, correct, and possibly expand it.

10.12 ■ How many electrons are in the valence shell of each of the following atoms? Write a Lewis dot structure for an atom of each element. (*Hint:* Use the Periodic Table.)
(a) Carbon   (b) Oxygen
(c) Nitrogen   (d) Fluorine

10.13 What is the relationship between the number of electrons in the valence shell of each of the following atoms and the number of covalent bonds it forms?
(a) Carbon   (b) Oxygen
(c) Nitrogen   (d) Hydrogen

10.14 Write Lewis structures for these compounds. Show all valence electrons. None of them contains a ring of atoms. (*Hint:* Remember that carbon has four bonds, nitrogen has three bonds and one unshared pair of electrons, oxygen has two bonds and two unshared pairs of electrons, and each halogen has one bond and three unshared pairs of electrons.)
(a) H_2O_2 (b) N_2H_4
    Hydrogen peroxide   Hydrazine
(c) CH_3OH (d) CH_3SH
    Methanol   Methylethiol
(e) CH_3NH_2 (f) CH_3Cl
    Methanamine   Chloromethane

10.15 ■ Write Lewis structures for these compounds. Show all valence electrons. None of them contains a ring of atoms.

(a) CH_3OCH_3
Dimethyl ether

(b) C_2H_6
Ethane

(c) C_2H_4
Ethylene

(d) C_2H_2
Acetylene

(e) CO_2
Carbon dioxide

(f) CH_2O
Formaldehyde

(g) H_2CO_3
Carbonic acid

(h) CH_3COOH
Acetic acid

10.16 ■ Write Lewis structures for these ions.

(a) HCO_3^-
Bicarbonate ion

(b) CO_3^{2-}
Carbonate ion

(c) CH_3COO^-
Acetate ion

(d) Cl^-
Chloride ion

10.17 Why are the following molecular formulas impossible?

(a) CH_5

(b) C_2H_7

Review of the VSEPR Model

10.18 Explain how to use the valence-shell electron-pair repulsion (VSEPR) model to predict bond angles and geometry about atoms of carbon, oxygen, and nitrogen.

10.19 Suppose you forget to take into account the presence of the unshared pair of electrons on nitrogen in the molecule NH_3. What would you then predict for the H—N—H bond angle and geometries of ammonia?

10.20 Suppose you forget to take into account the presence of the two unshared pairs of electrons on the oxygen atom of ethanol, CH_3CH_2OH. What would you then predict for the C—O—H bond angle and geometry?

10.21 ■ Use the VSEPR model to predict the bond angles and geometry about each highlighted atom. (*Hint:* Remember to take into account the presence of unshared pairs of electrons.)

10.22 Use the VSEPR model to predict the bond angles about each highlighted atom.

SECTION 10.4 What Are Functional Groups?

10.23 What is meant by the term *functional group?*

10.24 List three reasons why functional groups are important in organic chemistry.

10.25 ■ Draw Lewis structures for each of the following functional groups. Show all valence electrons.

(a) Carbonyl group

(b) Carboxyl group

(c) Hydroxyl group

(d) Primary amino group

(e) Ester group

10.26 ■ Complete the following structural formulas by adding enough hydrogens to complete the tetravalence of each carbon. Then write the molecular formula of each compound.

10.27 ■ Some of the following structural formulas are incorrect (that is, they do not represent real compounds) because they have atoms with an incorrect number of bonds. Which structural formulas are incorrect, and which atoms in them have an incorrect number of bonds?

10.28 What is the meaning of the term *tertiary* (3°) when it is used to classify alcohols?

10.29 Draw a structural formula for the one tertiary (3°) alcohol with the molecular formula $C_4H_{10}O$.

10.30 What is the meaning of the term *tertiary* (3°) when it is used to classify amines?

10.31 Draw a structural formula for the one tertiary (3°) amine with the molecular formula $C_4H_{11}N$.

10.32 Draw condensed structural formulas for all compounds with the molecular formula C_4H_8O that contain a carbonyl group (there are two aldehydes and one ketone).

10.33 Draw structural formulas for each of the following:

(a) The four primary (1°) alcohols with the molecular formula $C_5H_{12}O$

(b) The three secondary (2°) alcohols with the molecular formula $C_5H_{12}O$

(c) The one tertiary (3°) alcohol with the molecular formula $C_5H_{12}O$

10.34 Draw structural formulas for the six ketones with the molecular formula $C_6H_{12}O$.

10.35 Draw structural formulas for the eight carboxylic acids with the molecular formula $C_6H_{12}O_2$.

10.36 Draw structural formulas for each of the following:

(a) The four primary (1°) amines with the molecular formula $C_4H_{11}N$

(b) The three secondary (2°) amines with the molecular formula $C_4H_{11}N$

(c) The one tertiary (3°) amine with the molecular formula $C_4H_{11}N$

Chemical Connections

10.37 (Chemical Connections 10A) How was Taxol discovered?

10.38 (Chemical Connections 10A) In what way does Taxol interfere with cell division?

10.39 (Chemical Connections 10B) What is the goal of combinatorial chemistry?

Additional Problems

10.40 Use the VSEPR model to predict the bond angles about each atom of carbon, nitrogen, and oxygen in these molecules. (*Hint:* First add unshared pairs of electrons as necessary to complete the valence shell of each atom and then predict the bond angles.)

(a) $CH_3CH_2CH_2OH$

(b) $CH_3CH_2\overset{\displaystyle O}{\overset{\|}{C}}H$

(c) $CH_3CH{=}CH_2$

(d) $CH_3C{\equiv}CCH_3$

(e) $CH_3\overset{\displaystyle O}{\overset{\|}{C}}OCH_3$

(f) $CH_3\overset{\displaystyle CH_3}{\overset{|}{N}}CH_3$

10.41 ■ Silicon is immediately below carbon in Group 4A of the Periodic Table. Predict the C—Si—C bond angles in tetramethylsilane, $(CH_3)_4Si$.

10.42 Phosphorus is immediately below nitrogen in Group 5A of the Periodic Table. Predict the C—P—C bond angles in trimethylphosphine, $(CH_3)_3P$.

10.43 Draw the structure for a compound with the molecular formula

(a) C_2H_6O that is an alcohol

(b) C_3H_6O that is an aldehyde

(c) C_3H_6O that is a ketone

(d) $C_3H_6O_2$ that is a carboxylic acid

10.44 Draw structural formulas for the eight aldehydes with the molecular formula $C_6H_{12}O$.

10.45 Draw structural formulas for the three tertiary (3°) amines with the molecular formula $C_5H_{13}N$.

10.46 Which of these covalent bonds are polar and which are nonpolar? (*Hint:* Review Section 4.7B.)

(a) C—C   (b) C=C

(c) C—H   (d) C—O

(e) O—H   (f) C—N

(g) N—H   (h) N—O

10.47 Of the bonds in Problem 10.46, which is the most polar? Which is the least polar?

10.48 Using the symbol δ+ to indicate a partial positive charge and δ− to indicate a partial negative charge, indicate the polarity of the most polar bond (or bonds, if two or more have the same polarity) in each of the following molecules.

(a) CH_3OH (b) CH_3NH_2

(c) $HSCH_2CH_2NH_2$ (d) $CH_3\overset{\overset{\displaystyle O}{\|}}{C}CH_3$

(e) $H\overset{\overset{\displaystyle O}{\|}}{C}H$ (f) $CH_3\overset{\overset{\displaystyle O}{\|}}{C}OH$

Looking Ahead

10.49 Identify the functional group(s) in each compound.

(a)

Lactic acid
(produced in actively respiring muscle; gives sourness to sour cream)

(b) $HO—CH_2—CH_2—OH$

Ethylene glycol
(a component of antifreeze)

(c)

Alanine
(one of the building blocks of proteins)

10.50 Identify the functional group(s) in each compound.

(a) $CH_3CH_2\overset{\overset{\displaystyle O}{\|}}{C}CH_3$

2-Butanone
(a solvent for paints and lacquers)

(b) $HO\overset{\overset{\displaystyle O}{\|}}{C}CH_2CH_2CH_2CH_2\overset{\overset{\displaystyle O}{\|}}{C}OH$

Hexanedioic acid
(the second component of nylon-66)

(c) $H_2NCH_2CH_2CH_2CH_2\overset{\displaystyle CH}{\underset{\displaystyle NH_2}{|}}\overset{\overset{\displaystyle O}{\|}}{C}OH$

Lysine
(one of the 20 amino acid building blocks of proteins)

(d) $HOCH_2\overset{\overset{\displaystyle O}{\|}}{C}CH_2OH$

Dihydroxyacetone
(a component of several artificial tanning lotions)

10.51 Consider molecules with the molecular formula $C_4H_8O_2$. Write the structural formula for a molecule with this molecular formula that contains

(a) A carboxyl group

(b) An ester group

(c) A ketone group and a 2° alcohol group

(d) An aldehyde and a 3° alcohol group

(e) A carbon–carbon double bond and a 1° alcohol group

10.52 Following is a structural formula and a ball-and-stick model of benzene, C_6H_6.

(a) Predict each H—C—C and C—C—C bond angle in benzene.

(b) Predict the shape of a benzene molecule.

CHAPTER 11

Alkanes

GOB
Chemistry⋯Now™

Look for this logo in the chapter and go to GOB ChemistryNow at **http://now.brookscole.com/gob8** or on the CD for tutorials, simulations, and problems.

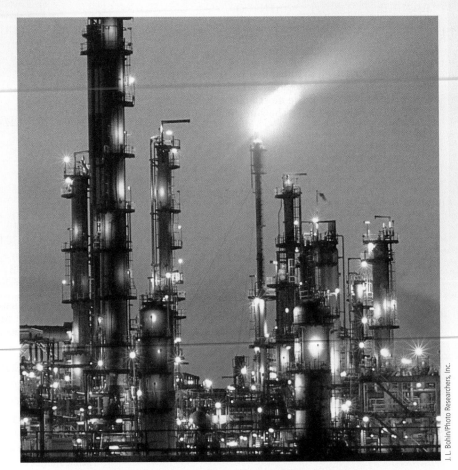

J. L. Bohin/Photo Researchers, Inc.

A petroleum refinery. Petroleum, along with natural gas, provides nearly 90% of the organic raw materials for the synthesis and manufacture of synthetic fibers, plastics, detergents, drugs, dyes, adhesives, and a multitude of other products.

Alkane A saturated hydrocarbon whose carbon atoms are arranged in a chain

Hydrocarbon A compound that contains only carbon and hydrogen atoms

Saturated hydrocarbon A hydrocarbon that contains only carbon–carbon single bonds

11.1 | How Do We Write Structural Formulas of Alkanes?

In this chapter, we examine the physical and chemical properties of **alkanes,** the simplest type of organic compounds. Actually, alkanes are members of a larger class of organic compounds called hydrocarbons. A **hydrocarbon** is a compound composed of only carbon and hydrogen. Figure 11.1 shows the four classes of hydrocarbons, along with the characteristic type of bonding between carbon atoms in each class. Alkanes are **saturated hydrocarbons;** that is, they contain only carbon–carbon single bonds. Saturated in this context means that each carbon in the hydrocarbon has the maximum number of hydrogens bonded to it. A hydrocarbon that contains one or more carbon–carbon double bonds, triple bonds, or benzene rings is classified as an

Figure 11.1   The four classes of hydrocarbons.

unsaturated hydrocarbon. We study alkanes (saturated hydrocarbons) in this chapter and alkenes, alkynes, and arenes (unsaturated hydrocarbons) in Chapters 12 and 13.

We commonly refer to alkanes as **aliphatic hydrocarbons** because the physical properties of the higher members of this class resemble those of the long carbon-chain molecules we find in animal fats and plant oils (Greek: *aleiphar,* fat or oil).

Methane, CH_4, and ethane, C_2H_6, are the first two members of the alkane family. Figure 11.2 shows Lewis structures and ball-and-stick models for these molecules. The shape of methane is tetrahedral, and all H—C—H bond angles are 109.5°. Each carbon atom in ethane is also tetrahedral, and the bond angles in it are all approximately 109.5°. Although the three-dimensional shapes of larger alkanes are more complex than those of methane and ethane, the four bonds about each carbon atom are still arranged in a tetrahedral manner, and all bond angles are still approximately 109.5°.

The next members of the alkane family are propane, butane, and pentane. In the following representations, these hydrocarbons are drawn first as condensed structural formulas, which show all carbons and hydrogens. They can also be drawn in a more abbreviated form called a **line-angle formula.** In this type of representation, a line represents a carbon–carbon bond and a vertex represents a carbon atom. A line ending in space represents a —CH_3 group. To count hydrogens from a line-angle formula, you simply add enough hydrogens in your mind to give each carbon its required four bonds. Chemists use line-angle formulas because they are easier and faster to draw than condensed structural formulas.

Aliphatic hydrocarbon An alkane

GOB
Chemistry ⚛ Now™
Click *Coached Problems* for a set of problems linking **Structural, Molecular, and Line-angle Formulas**

Line-angle formula An abbreviated way to draw structural formulas in which each vertex and line terminus represents a carbon atom and each line represents a bond

Figure 11.2   Methane and ethane.

	Ball-and-stick model		
Line-angle formula			
Condensed structural formula	$CH_3CH_2CH_3$	$CH_3CH_2CH_2CH_3$	$CH_3CH_2CH_2CH_2CH_3$
	Propane	Butane	Pentane

Structural formulas for alkanes can be written in yet another condensed form. For example, the structural formula of pentane contains three CH_2 (**methylene**) groups in the middle of the chain. We can group them together and write the structural formula $CH_3(CH_2)_3CH_3$. Table 11.1 gives the first ten alkanes with unbranched chains. Note that the names of all these alkanes end in "-ane." We will have more to say about naming alkanes in Section 11.3.

Butane, $CH_3CH_2CH_2CH_3$, is the fuel in this lighter. Butane molecules are present in the liquid and gaseous states in the lighter.

Charles D. Winters

EXAMPLE 11.1

Table 11.1 gives the condensed structural formula for hexane. Draw a line-angle formula for this alkane, and number the carbons on the chain beginning at one end and proceeding to the other end.

Solution
Hexane contains six carbons in a chain. Its line-angle formula is

Problem 11.1
Following is a line-angle formula for an alkane. What are the name and molecular formula of this alkane?

Table 11.1   The First Ten Alkanes with Unbranched Chains

Name	Molecular Formula	Condensed Structural Formula	Name	Molecular Formula	Condensed Structural Formula
methane	CH_4	CH_4	hexane	C_6H_{14}	$CH_3(CH_2)_4CH_3$
ethane	C_2H_6	CH_3CH_3	heptane	C_7H_{16}	$CH_3(CH_2)_5CH_3$
propane	C_3H_8	$CH_3CH_2CH_3$	octane	C_8H_{18}	$CH_3(CH_2)_6CH_3$
butane	C_4H_{10}	$CH_3(CH_2)_2CH_3$	nonane	C_9H_{20}	$CH_3(CH_2)_7CH_3$
pentane	C_5H_{12}	$CH_3(CH_2)_3CH_3$	decane	$C_{10}H_{22}$	$CH_3(CH_2)_8CH_3$

11.2 | What Are Constitutional Isomers?

Constitutional isomers are compounds that have the same molecular formula but different structural formulas. By "different structural formulas," we mean that they differ in the kinds of bonds (single, double, or triple) and/or in the connectivity of their atoms. For the molecular formulas CH_4, C_2H_6, and C_3H_8, only one connectivity of their atoms is possible, so there are no constitutional isomers for these molecular formulas. For the molecular formula C_4H_{10}, however, two structural formulas are possible: In butane, the four carbons are bonded in a chain; in 2-methylpropane, three carbons are bonded in a chain with the fourth carbon as a branch on the chain. The two constitutional isomers with molecular formula C_4H_{10} are drawn here both as condensed structural formulas and as line-angle formulas. Also shown are ball-and-stick models of each.

Constitutional isomers
Compounds with the same molecular formula but a different connectivity of their atoms

Constitutional isomers have also been called structural isomers, an older term that is still in use.

GOB
Chemistry •ᯤ• Now™
Click *Coached Problems* for a set of problems in **Identifying Constitutional Isomers**

$CH_3CH_2CH_2CH_3$

Butane
(bp −0.5°C)

$\underset{\displaystyle |}{\overset{\displaystyle CH_3}{CH_3CHCH_3}}$

2-Methylpropane
(bp −11.6°C)

We refer to butane and 2-methylpropane as constitutional isomers; they are different compounds and have different physical and chemical properties. Their boiling points, for example, differ by approximately 11°C.

In Section 10.4, we encountered several examples of constitutional isomers, although we did not call them that at the time. We saw, for example, that there are two alcohols with the molecular formula C_3H_8O, two primary amines with the molecular formula C_3H_9N, two aldehydes with the molecular formula C_4H_8O, and two carboxylic acids with the molecular formula $C_4H_8O_2$.

To determine whether two or more structural formulas represent constitutional isomers, we write the molecular formula of each and then compare them. All compounds that have the same molecular formula but different structural formulas are constitutional isomers.

EXAMPLE 11.2

Do the structural formulas in each of the following sets represent the same compound or constitutional isomers? You will find it helpful to redraw each molecule as a line-angle formula, which will make it easier for you to see similarities and differences in molecular structure.

(a) $CH_3CH_2CH_2CH_2CH_2CH_3$ and $\underset{\displaystyle |}{CH_3CH_2CH_2}$ (Each is C_6H_{14})
$CH_2CH_2CH_3$

(b) $\underset{\displaystyle |}{\overset{\displaystyle CH_3}{CH_3CHCH_2CH}}$ and $\underset{\displaystyle |}{\overset{\displaystyle CH_3}{CH_3CH_2CHCHCH_3}}$ (Each is C_7H_{16})
$\underset{CH_3\ CH_3}{}$ $\underset{CH_3}{}$

Solution
First, find the longest chain of carbon atoms. Note that it makes no differ-ence whether the chain is drawn as straight or bent. As structural formulas are drawn in this problem, there is no attempt to show three-dimensional shapes. Second, number the longest chain from the end nearest the first branch. Third, compare the lengths of the two chains and the sizes and locations of any branches. Structural formulas that have the same molecu-lar formula and the same connectivity of their atoms represent the same compound; those that have the same molecular formula but different con-nectivities of their atoms represent constitutional isomers.

(a) Each structural formula has an unbranched chain of six carbons; they are identical and represent the same compound.

(b) Each structural formula has the same molecular formula, C_7H_{16}. In addi-tion, each has a chain of five carbons with two CH_3 branches. Although the branches are identical, they are at different locations on the chains. Therefore, these structural formulas represent constitutional isomers.

Problem 11.2
Do the line-angle formulas in each of the following sets represent the same compound or constitutional isomers?

EXAMPLE 11.3

Draw line-angle formulas for the five constitutional isomers with the molecular formula C_6H_{14}.

Solution

In solving problems of this type, you should devise a strategy and then follow it. Here is one possible strategy. First, draw a line-angle formula for the constitutional isomer with all six carbons in an unbranched chain. Then, draw line-angle formulas for all constitutional isomers with five carbons in a chain and one carbon as a branch on the chain. Finally, draw line-angle formulas for all constitutional isomers with four carbons in a chain and two carbons as branches. No constitutional isomers for C_6H_{14} having only three carbons in the longest chain are possible.

Six carbons in an unbranched chain

Five carbons in a chain; one carbon as a branch

Four carbons in a chain; two carbons as branches

Problem 11.3

Draw structural formulas for the three constitutional isomers with the molecular formula C_5H_{12}.

The ability of carbon atoms to form strong, stable bonds with other carbon atoms results in a staggering number of constitutional isomers, as the following table shows:

Molecular Formula	Number of Constitutional Isomers
CH_4	1
C_5H_{12}	3
$C_{10}H_{22}$	75
$C_{15}H_{32}$	4347
$C_{25}H_{52}$	36,797,588
$C_{30}H_{62}$	4,111,846,763

Thus, for even a small number of carbon and hydrogen atoms, a very large number of constitutional isomers is possible. In fact, the potential for structural and functional group individuality among organic molecules made from just the basic building blocks of carbon, hydrogen, nitrogen, and oxygen is practically limitless.

11.3 | How Do We Name Alkanes?

A. The IUPAC System

Ideally, every organic compound should have a name from which its structural formula can be drawn. For this purpose, chemists have adopted a set of rules established by the **International Union of Pure and Applied Chemistry (UPAC).**

The IUPAC name for an alkane with an unbranched chain of carbon atoms consists of two parts: (1) a prefix that shows the number of carbon atoms in the chain and (2) the suffix **-ane,** which shows that the compound

GOB
Chemistry Now™
Click *Coached Problems* for a set of problems in **Naming Alkanes**

Table 11.2   Prefixes Used in the IUPAC System to Show the Presence of One to Twenty Carbons in an Unbranched Chain

Prefix	Number of Carbon Atoms	Prefix	Number of Carbon Atoms	Prefix	Number of Carbon Atoms	Prefix	Number of Carbon Atoms
meth-	1	hex-	6	undec-	11	hexadec-	16
eth-	2	hept-	7	dodec-	12	heptadec-	17
prop-	3	oct-	8	tridec-	13	octadec-	18
but-	4	non-	9	tetradec-	14	nonadec-	19
pent-	5	dec-	10	pentadec-	15	eicos-	20

is a saturated hydrocarbon. Table 11.2 gives the prefixes used to show the presence of 1 to 20 carbon atoms.

The IUPAC chose the first four prefixes listed in Table 11.2 because they were well established long before the nomenclature was systematized. For example, the prefix *but-* appears in the name butyric acid, a compound of four carbon atoms formed by the air oxidation of butter fat (Latin: *butyrum,* butter). The prefixes to show five or more carbons are derived from Latin numbers. Table 11.1 gives the names, molecular formulas, and condensed structural formulas for the first ten alkanes with unbranched chains.

IUPAC names of alkanes with branched chains consist of a parent name that shows the longest chain of carbon atoms and substituent names that indicate the groups bonded to the parent chain. For example,

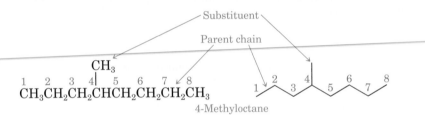

4-Methyloctane

Alkyl group A group derived by removing a hydrogen from an alkane; given the symbol R—

R— A symbol used to represent an alkyl group

A substituent group derived from an alkane by removal of a hydrogen atom is called an **alkyl group** and is commonly represented by the symbol **R—**. Alkyl groups are named by dropping the *-ane* from the name of the parent alkane and adding the suffix *-yl*. Table 11.3 gives the names and condensed structural formulas for eight of the most common alkyl groups. The prefix *sec-* is an abbreviation for "secondary," meaning a carbon bonded to two other carbons. The prefix *tert-* is an abbreviation for "tertiary," meaning a carbon bonded to three other carbons.

The rules of the IUPAC system for naming alkanes are as follows:

1. The name for an alkane with an unbranched chain of carbon atoms consists of a prefix showing the number of carbon atoms in the chain and the suffix *-ane.*

2. For branched-chain alkanes, take the longest chain of carbon atoms as the parent chain and its name becomes the root name.

3. Give each substituent on the parent chain a name and a number. The number shows the carbon atom of the parent chain to which the substituent is bonded. Use a hyphen to connect the number to the name.

$$CH_3CHCH_3$$

with CH₃ substituent shown:

CH₃
|
CH₃CHCH₃

2-Methylpropane

Table 11.3   Names of the Eight Most Common Alkyl Groups

Name	Condensed Structural Formula	Name	Condensed Structural Formula
methyl	$-CH_3$	butyl	$-CH_2CH_2CH_2CH_3$
ethyl	$-CH_2CH_3$	isobutyl	$-CH_2CHCH_3$ $\quad\quad\ \ \|$ $\quad\quad\ CH_3$
propyl	$-CH_2CH_2CH_3$	sec-butyl	$-CHCH_2CH_3$ $\quad\ \|$ $\quad CH_3$
isopropyl	$-CHCH_3$ $\quad\ \|$ $\quad CH_3$	tert-butyl	$\quad\quad CH_3$ $\quad\quad\ \|$ $-CCH_3$ $\quad\quad\ \|$ $\quad\quad CH_3$

4. If there is one substituent, number the parent chain from the end that gives the substituent the lower number.

$$CH_3CH_2CH_2CHCH_3$$

2-Methylpentane
(not 4-methylpentane)

5. If the same substituent occurs more than once, number the parent chain from the end that gives the lower number to the substituent encountered first. Indicate the number of times the substituent occurs by a prefix *di-, tri-, tetra-, penta-, hexa-,* and so on. Use a comma to separate position numbers.

$$CH_3CH_2CHCH_2CHCH_3$$

2,4-Dimethylhexane
(not 3,5-dimethylhexane)

6. If there are two or more different substituents, list them in alphabetical order and number the chain from the end that gives the lower number to the substituent encountered first. If there are different substituents in equivalent positions on opposite ends of the parent chain, give the substituent of lower alphabetical order the lower number.

$$CH_3CH_2CHCH_2CHCH_2CH_3$$

3-Ethyl-5-methylheptane
(not 3-methyl-5-ethylheptane)

7. Do not include the prefixes *di-, tri-, tetra-,* and so on or the hyphenated prefixes *sec-* and *tert-* in alphabetizing. Alphabetize the names of substituents first, and then insert these prefixes. In the following example, the alphabetizing parts are **ethyl** and **methyl,** not *ethyl* and *dimethyl.*

$$CH_3 \quad CH_2CH_3$$
$$CH_3CCH_2CHCH_2CH_3$$
$$CH_3$$

4-Ethyl-2,2-dimethylhexane
(not 2,2-dimethyl-4-ethylhexane)

EXAMPLE 11.4

Write the molecular formula and IUPAC name for each alkane.

(a)   (b)

Solution
The molecular formula of (a) is C_5H_{12}, and that of (b) is $C_{11}H_{24}$. In (a), number the longest chain from the end that gives the methyl substituent the lower number (rule 4). In (b), list isopropyl and methyl substituents in alphabetical order (rule 6).

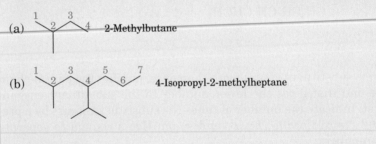

(a) 2-Methylbutane

(b) 4-Isopropyl-2-methylheptane

Problem 11.4
Write the molecular formula and IUPAC name for each alkane.

(a)   (b)

B. Common Names

In the older system of **common nomenclature,** the total number of carbon atoms in an alkane, regardless of their arrangement, determines the name. The first three alkanes are methane, ethane, and propane. All alkanes of formula C_4H_{10} are called butanes, all those of formula C_5H_{12} are called pentanes, and all those of formula C_6H_{14} are called hexanes. For alkanes beyond propane, **iso** indicates that one end of an otherwise unbranched chain terminates in a $(CH_3)_2CH-$ group. Following are examples of common names:

$$CH_3CH_2CH_2CH_3 \qquad CH_3\overset{\overset{\displaystyle CH_3}{|}}{C}HCH_3 \qquad CH_3CH_2CH_2CH_2CH_3 \qquad CH_3CH_2\overset{\overset{\displaystyle CH_3}{|}}{C}HCH_3$$
Butane   Isobutane   Pentane   Isopentane

This system of common names has no way of handling other branching patterns; for more complex alkanes, we must use the more flexible IUPAC system.

In this book, we concentrate on IUPAC names. From time to time, we also use common names, especially when chemists and biochemists use them almost exclusively in everyday discussions. When the text uses both IUPAC and common names, we always give the IUPAC name first, followed by the common name in parentheses. In this way, you should have no doubt about which name is which.

11.4 | What Are Cycloalkanes?

A hydrocarbon that contains carbon atoms joined to form a ring is called a **cyclic hydrocarbon.** When all carbons of the ring are saturated, the hydrocarbon is called a **cycloalkane.** Cycloalkanes of ring sizes ranging from 3 to more than 30 carbon atoms are found in nature, and in principle there is no limit to ring size. Five-membered (cyclopentane) and six-membered (cyclohexane) rings are especially abundant in nature; for this reason, we concentrate on them in this book.

Organic chemists rarely show all carbons and hydrogens when writing structural formulas for cycloalkanes. Rather, we use line-angle formulas to represent cycloalkane rings. We represent each ring by a regular polygon having the same number of sides as there are carbon atoms in the ring. For example, we represent cyclobutane by a square, cyclopentane by a pentagon, and cyclohexane by a hexagon (Figure 11.3).

Cycloalkane A saturated hydrocarbon that contains carbon atoms bonded to form a ring

Cyclobutane   Cyclopentane   Cyclohexane

Figure 11.3 Examples of cycloalkanes.

To name a cycloalkane, prefix the name of the corresponding open-chain alkane with *cyclo-,* and name each substituent on the ring. If there is only one substituent on the ring, there is no need to give it a location number. If there are two substituents, number the ring beginning with the substituent of lower alphabetical order.

EXAMPLE 11.5

Write the molecular formula and IUPAC name for each cycloalkane.

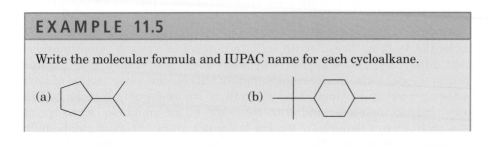

Solution

(a) The molecular formula of this compound is C_8H_{16}. Because there is only one substituent, there is no need to number the atoms of the ring. The IUPAC name of this cycloalkane is isopropylcyclopentane.

(b) The molecular formula is $C_{11}H_{22}$. To name this compound, first number the atoms of the cyclohexane ring beginning with *tert*-butyl, the substituent of lower alphabetical order (remember, alphabetical order here is determined by the *b* of *butyl*, and not by the *t* of *tert*-). The name of this cycloalkane is 1-*tert*-butyl-4-methylcyclohexane.

Problem 11.5

Write the molecular formula and IUPAC name for each cycloalkane.

(a)   (b)   (c)

11.5 | What Are the Shapes of Alkanes and Cycloalkanes?

In this section, we concentrate on ways to visualize molecules as three-dimensional objects and to visualize bond angles and relative distances between various atoms and functional groups within a molecule. We urge you to build molecular models of these compounds and to study and manipulate those models. Organic molecules are three-dimensional objects, and it is essential that you become comfortable in dealing with them as such.

A. Alkanes

Although the VSEPR model (Section 4.10) gives us a way to predict the geometry about each carbon atom in an alkane, it provides us with no information about the three-dimensional shape of an entire molecule. There is, in fact, free rotation about each carbon–carbon bond in an alkane. As a result, even a molecule as simple as ethane has an infinite number of possible three-dimensional shapes, or **conformations.**

Figure 11.4 shows three conformations for a butane molecule. Conformation (a) is the most stable because the methyl groups at the ends of the four-carbon chain are farthest apart. Conformation (b) is formed by a rotation of 120° about the single bond joining carbons 2 and 3. In this conformation, some crowding of groups occurs because the two methyl groups are closer together. Rotation about the C_2—C_3 single bond by another 60° gives conformation (c), which is the most crowded because the two methyl groups face each other.

Figure 11.4 shows only three of the possible conformations for a butane molecule. In fact, there are an infinite number of possible conformations that differ only in the angles of rotation about the various C—C bonds within the molecule. In an actual sample of butane, the conformation of each molecule constantly changes as a result of the molecule's collisions with other butane molecules and with the walls of the container. Even so, at any given time, a majority of butane molecules are in the most stable, fully extended conformation. There are the fewest butane molecules in the most crowded conformation.

At this point, you should review Section 4.10 and use of the valence-shell electron-pair (VSEPR) model to predict bond angles and shapes of molecules.

Conformation Any three-dimensional arrangement of atoms in a molecule that results from rotation about a single bond

GOB
Chemistry‑‑Now™
Click *Coached Problems* for an interactive exploration of the **Conformations of Butane**

(a) Least crowded conformation; methyl groups are farthest apart.

(b) Intermediate crowding; methyl groups are closer to each other.

(c) Most crowded conformation; methyl groups are closest to each other.

GOB
Chemistry⚛Now™
Active Figure 11.4 Three conformations of a butane molecule. **See a simulation based on this figure, and take a short quiz on the concepts at http://www.now.brookscole.com/gob8 or on the CD.**

To summarize, for any alkane (except, of course, for methane), there are an infinite number of conformations. The majority of molecules in any sample will be in the least crowded conformation; the fewest will be in the most crowded conformation.

B. Cycloalkanes

We limit our discussion to the conformations of cyclopentanes and cyclohexanes because they are the carbon rings most commonly found in nature. Nonplanar or puckered conformations are favored in all cycloalkanes larger than cyclopropane.

Cyclopentane

The most stable conformation of cyclopentane is the **envelope conformation** shown in Figure 11.5. In it, four carbon atoms are in a plane, and the fifth carbon atom is bent out of the plane, like an envelope with its flap bent upward. All bond angles in cyclopentane are approximately 109.5°.

Cyclohexane

The most stable conformation of cyclohexane is the **chair conformation** (Figure 11.6), in which all bond angles are approximately 109.5°.

Figure 11.5 The most stable conformation of cyclopentane.

GOB
Chemistry⚛Now™
Click *Coached Problems* to try a set of problems in **Identifying Sites in Cyclic Alkane Structures**

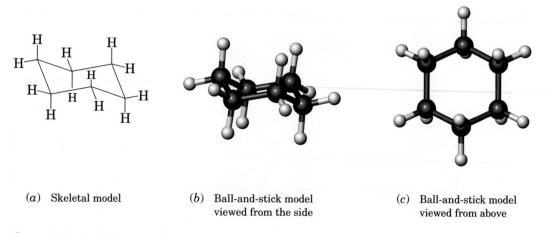

(a) Skeletal model

(b) Ball-and-stick model viewed from the side

(c) Ball-and-stick model viewed from above

Figure 11.6 Cyclohexane. The most stable conformation is a chair conformation.

(a) Ball-and-stick model showing all 12 hydrogens

(b) The six equatorial C—H bonds shown in red

(c) The six axial C—H bonds shown in blue

Figure 11.7 Chair conformation of cyclohexane showing equatorial and axial C—H bonds.

Equatorial position A position on a chair conformation of a cyclohexane ring that extends from the ring roughly perpendicular to the imaginary axis of the ring

Axial position A position on a chair conformation of a cyclohexane ring that extends from the ring parallel to the imaginary axis of the ring

In a chair conformation, the 12 C—H bonds are arranged in two different orientations. Six of them are **axial bonds,** and the other six are **equatorial bonds.** One way to visualize the difference between these two types of bonds is to imagine an axis running through the center of the chair (Figure 11.7). Axial bonds are oriented parallel to this axis. Three of the axial bonds point up; the other three point down. Notice also that axial bonds alternate, first up and then down, as you move from one carbon to the next.

Equatorial bonds are oriented approximately perpendicular to the imaginary axis of the ring and also alternate first slightly up and then slightly down as you move from one carbon to the next. Notice also that if the axial bond on a carbon points upward, the equatorial bond on that carbon points slightly downward. Conversely, if the axial bond on a particular carbon points downward, the equatorial bond on that carbon points slightly upward.

Finally, notice that each equatorial bond is oriented parallel to two ring bonds on opposite sides of the ring. A different pair of parallel C—H bonds is shown in each of the following structural formulas, along with the two ring bonds to which each pair is parallel.

EXAMPLE 11.6

Following is a chair conformation of methylcyclohexane showing a methyl group and one hydrogen. Indicate by a label whether each is equatorial or axial.

Solution

The methyl group is axial, and the hydrogen is equatorial.

Problem 11.6

Following is a chair conformation of cyclohexane with carbon atoms numbered 1 through 6. Draw methyl groups that are equatorial on carbons 1, 2, and 4.

Suppose that —CH₃ or another group on a cyclohexane ring may occupy either an equatorial or an axial position. Chemists have discovered that a six-membered ring is more stable when the maximum number of substituent groups are equatorial. Perhaps the simplest way to confirm this relationship is to examine molecular models. Figure 11.8(a) shows a space-filling model of methylcyclohexane with the methyl group in an equatorial position. In this position, the methyl group is as far away as possible from

CHEMICAL CONNECTIONS 11A

The Poisonous Puffer Fish

Nature is by no means limited to carbon in six-membered rings. Tetrodotoxin, one of the most potent toxins known, is composed of a set of interconnected six-membered rings, each in a chair conformation. All but one of these rings contain atoms other than carbon.

Tetrodotoxin is produced in the liver and ovaries of many species of *Tetraodontidae*, but especially the puffer fish, so called because it inflates itself to an almost spherical spiny ball when alarmed. It is evidently a species highly preoccupied with defense, but the Japanese are not put off by its prickly appearance. They regard the puffer, called *fugu* in Japanese, as a delicacy. To serve it in a public restaurant, a chef must be registered as sufficiently skilled in removing the toxic organs so as to make the flesh safe to eat.

Tetrodotoxin blocks the sodium ion channels, which are essential for neurotransmission (Section 24.3). This blockage prevents communication between neurons and muscle cells and results in weakness, paralysis, and eventual death.

A puffer fish with its body inflated.

Tetrodotoxin

(a) Equatorial methylcyclohexane

(b) Axial methylcyclohexane

Figure 11.8 Methylcyclohexane. The three hydrogens of the methyl group are shown in green to make them stand out more clearly.

other atoms of the ring. When methyl is axial [Figure 11.8(b)], it quite literally bangs into two hydrogen atoms on the top of the ring. Thus the more stable conformation of a substituted cyclohexane ring has the substituent group(s) as equatorial.

11.6 | What Is *Cis-Trans* Isomerism in Cycloalkanes?

Cis-trans isomers Isomers that have the same connectivity of their atoms but a different arrangement of their atoms in space due to the presence of either a ring or a carbon–carbon double bond

Cis-trans isomers have also been called geometric isomers.

Planar representations of five- and six-membered rings are not spatially accurate because these rings normally exist as envelope and chair conformations. Planar representations are, however, adequate for showing *cis-trans* isomerism.

GOB
Chemistry‧‧‧Now™

Click *Coached Problems* to try a set of problems in **Identifying Functional Groups**

Cycloalkanes with substituents on two or more carbons of the ring show a type of isomerism called **cis-trans isomerism.** Cycloalkane *cis-trans* isomers have (1) the same molecular formula and (2) the same connectivity of their atoms, but (3) a different arrangement of their atoms in space because of restricted rotation around the carbon–carbon single bonds of the ring. We study *cis-trans* isomerism in cycloalkanes in this chapter and that of alkenes in Chapter 12.

We can illustrate *cis-trans* isomerism in cycloalkanes by using 1,2-dimethylcyclopentane as an example. In the following structural formulas, the cyclopentane ring is drawn as a planar pentagon viewed through the plane of the ring. (In determining the number of *cis-trans* isomers in a substituted cycloalkane, it is adequate to draw the cycloalkane ring as a planar polygon.) Carbon–carbon bonds of the ring projecting toward you are shown as heavy lines. When viewed from this perspective, substituents are bonded to the cyclopentane ring project above and below the plane of the ring. In one isomer of 1,2-dimethylcyclopentane, the methyl groups are on the same side of the ring (either both above or both below the plane of the ring); in the other isomer, they are on opposite sides of the ring (one above and one below the plane of the ring).

The prefix **cis** (Latin: on the same side) indicates that the substituents are on the same side of the ring; the prefix **trans** (Latin: across) indicates that they are on opposite sides of the ring.

cis-1,2-Dimethyl-cyclopentane

trans-1,2-Dimethyl-cyclopentane

Alternatively, we can view the cyclopentane ring as a regular pentagon seen from above, with the ring in the plane of the page. Substituents on the ring then either project toward you (that is, they project up above the page) and are shown by solid wedges, or they project away from you (project down below the page) and are shown by broken wedges. In the following structural formulas, we show only the two methyl groups; we do not show hydrogen atoms of the ring.

Occasionally hydrogen atoms are written before the carbon, H_3C—, to avoid crowding or to emphasize the C—C bond, as in H_3C—CH_3.

cis-1,2-Dimethyl-cyclopentane

trans-1,2-Dimethyl-cyclopentane

We say that 1,2-dimethylcyclopentane has two stereocenters. A **stereocenter** is a tetrahedral atom, most commonly carbon, about which exchange of two groups produces a stereoisomer. Both carbons 1 and 2 of 1,2-dimethylcyclopentane, for example, are stereocenters; in this molecule, exchange of H and CH_3 groups at either stereocenter converts a *trans* isomer to a *cis* isomer, or vice versa.

Alternatively, we can refer to the stereoisomers of 1,2-dimethylcyclopentane as having either a *cis* or a *trans* configuration. **Configuration** refers to the arrangement of atoms about a stereocenter. We say, for example, that exchange of groups at either stereocenter in the *cis* configuration gives the isomer with the *trans* configuration.

Cis and *trans* isomers are also possible for 1,2-dimethylcyclohexane. We can draw a cyclohexane ring as a planar hexagon and view it through the plane of the ring. Alternatively, we can view it as a regular hexagon viewed from above with substituent groups pointing toward us shown by solid wedges, or pointing away from us shown by broken wedges.

Stereocenter A tetrahedral atom, most commonly carbon, at which exchange of two groups produces a stereoisomer

Configuration Refers to the arrangement of atoms about a stereocenter; that is, to the relative arrangement of parts of a molecule in space

trans-1,4-Dimethylcyclohexane   *cis*-1,4-Dimethylcyclohexane

Because *cis-trans* isomers differ in the orientation of their atoms in space, they are **stereoisomers.** *Cis-trans* isomerism is one type of stereoisomerism. We will study another type of stereoisomerism, called enantiomerism, in Chapter 15.

Stereoisomers Isomers that have the same connectivity of their atoms but a different orientation of their atoms in space

EXAMPLE 11.7

Which of the following cycloalkanes show *cis-trans* isomerism? For each that does, draw both isomers.
(a) Methylcyclopentane   (b) 1,1-Dimethylcyclopentane
(c) 1,3-Dimethylcyclobutane

Solution
(a) Methylcyclopentane does not show *cis-trans* isomerism; it has only one substituent on the ring.
(b) 1,1-Dimethylcyclobutane does not show *cis-trans* isomerism because only one arrangement is possible for the two methyl groups. Because both methyl groups are bonded to the same carbon, they must be *trans* to each other—one above the ring, the other below it.
(c) 1,3-Dimethylcyclobutane shows *cis-trans* isomerism. The two methyl groups may be *cis* or they may be *trans*.

cis-1,3-Dimethylcyclobutane   *trans*-1,3-Dimethylcyclobutane

Problem 11.7

Which of the following cycloalkanes show *cis-trans* isomerism? For each that does, draw both isomers.
(a) 1,3-Dimethylcyclopentane
(b) Ethylcyclopentane
(c) 1,3-Dimethylcyclohexane

11.7 | What Are the Physical Properties of Alkanes?

The most important property of alkanes and cycloalkanes is their almost complete lack of polarity. We saw in Section 4.4B that the electronegativity difference between carbon and hydrogen is 2.5 − 2.1 = 0.4 on the Pauling scale. Given this small difference, we classify a C—H bond as nonpolar covalent. Therefore, alkanes are nonpolar compounds and the only interactions between their molecules are the very weak London dispersion forces (Section 6.7A).

A. Melting and Boiling Points

The boiling points of alkanes are lower than those of almost any other type of compound of the same molecular weight. In general, both boiling and melting points of alkanes increase with increasing molecular weight (Table 11.4).

Alkanes containing 1 to 4 carbons are gases at room temperature. Alkanes containing 5 to 17 carbons are colorless liquids. High-molecular-weight alkanes (those containing 18 or more carbons) are white, waxy solids. Several plant waxes are high-molecular-weight alkanes. The wax found in apple skins, for example, is an unbranched alkane with the molecular formula $C_{27}H_{56}$. Paraffin wax, a mixture of high-molecular-weight alkanes, is used for wax candles, in lubricants, and to seal home-canned jams, jellies, and other preserves. Petrolatum, so named because it is derived from petroleum refining, is a liquid mixture of high-molecular-weight alkanes. It is sold as mineral oil and Vaseline, and is used as an ointment base in pharmaceuticals and cosmetics, and as a lubricant and rust preventive.

Table 11.4 Physical Properties of Some Unbranched Alkanes

Name	Condensed Structural Formula	Molecular Weight (amu)	Melting Point (°C)	Boiling Point (°C)	Density of Liquid (g/mL at 0°C)*
methane	CH_4	16.0	−182	−164	(a gas)
ethane	CH_3CH_3	30.1	−183	−88	(a gas)
propane	$CH_3CH_2CH_3$	44.1	−190	−42	(a gas)
butane	$CH_3(CH_2)_2CH_3$	58.1	−138	0	(a gas)
pentane	$CH_3(CH_2)_3CH_3$	72.2	−130	36	0.626
hexane	$CH_3(CH_2)_4CH_3$	86.2	−95	69	0.659
heptane	$CH_3(CH_2)_5CH_3$	100.2	−90	98	0.684
octane	$CH_3(CH_2)_6CH_3$	114.2	−57	126	0.703
nonane	$CH_3(CH_2)_7CH_3$	128.3	−51	151	0.718
decane	$CH_3(CH_2)_8CH_3$	142.3	−30	174	0.730

*For comparison, the density of H_2O is 1 g/mL at 4°C.

Alkanes that are constitutional isomers are different compounds and have different physical and chemical properties. Table 11.5 lists the boiling points of the five constitutional isomers with a molecular formula of C_6H_{14}. The boiling point of each branched-chain isomer is lower than that of hexane itself; the more branching there is, the lower the boiling point. These differences in boiling points are related to molecular shape in the following way. As branching increases, the alkane molecule becomes more compact, and its surface area decreases. As we learned in Section 6.7A, as surface area decreases, London dispersion forces act over a smaller surface area. Hence the attraction between molecules decreases, and boiling point decreases. Thus, for any group of alkane constitutional isomers, the least-branched isomer generally has the highest boiling point, and the most-branched isomer generally has the lowest boiling point.

Paraffin wax and mineral oil are mixtures of alkanes.

Charles D. Winters

Table 11.5 Boiling Points of the Five Isomeric Alkanes with Molecular Formula C_6H_{14}

Name	bp (°C)
hexane	68.7
3-methylpentane	63.3
2-methylpentane	60.3
2,3-dimethylbutane	58.0
2,2-dimethylbutane	49.7

Larger surface area, an increase in London dispersion forces, and a higher boiling point

Hexane (bp 68.7°)

Smaller surface area, a decrease in London dispersion forces, and a lower boiling point

2,2-Dimethylbutane (bp 49.7°)

B. Solubility: A Case of "Like Dissolves Like"

Because alkanes are nonpolar compounds, they are not soluble in water, which dissolves only ionic and polar compounds. Recall from Section 7.6C that water is a polar substance, and that its molecules associate with one another through hydrogen bonding. Alkanes do not dissolve in water because they cannot form hydrogen bonds with water. Alkanes, however, are soluble in each other, an example of "like dissolves like" (Section 7.4A). Alkanes are also soluble in other nonpolar organic compounds, such as toluene and diethyl ether.

C. Density

The average density of the liquid alkanes listed in Table 11.4 is about 0.7 g/mL; that of higher-molecular-weight alkanes is about 0.8 g/mL. All liquid and solid alkanes are less dense than water (1.0 g/mL) and, because they are insoluble in water, they float on water.

EXAMPLE 11.8

Arrange the alkanes in each set in order of increasing boiling point.
(a) Butane, decane, and hexane
(b) 2-Methylheptane, octane, and 2,2,4-trimethylpentane

Solution
(a) All three compounds are unbranched alkanes. As the number of carbon atoms in the chain increases, London dispersion forces between molecules increase, and boiling points increase (Section 6.7A). Decane has the highest boiling point, and butane has the lowest boiling point.

| Butane | Hexane | Decane |
| bp −0.5°C | bp 69°C | bp 174°C |

(b) These three alkanes are constitutional isomers with the molecular formula C_8H_{18}. Their relative boiling points depend on the degree of branching. 2,2,4-Trimethylpentane is the most highly branched isomer and, therefore, has the smallest surface area and the lowest boiling point. Octane, the unbranched isomer, has the largest surface area and the highest boiling point.

| 2,2,4-Trimethylpentane | 2-Methylheptane | Octane |
| (bp 99°C) | (bp 118°C) | (bp 126°C) |

Problem 11.8

Arrange the alkanes in each set in order of increasing boiling point.
(a) 2-Methylbutane, pentane, and 2,2-dimethylpropane
(b) 3,3-Dimethylheptane, nonane, and 2,2,4-trimethylhexane

11.8 | What Are the Characteristic Reactions of Alkanes?

The most important chemical property of alkanes and cycloalkanes is their inertness. They are quite unreactive toward any of the normal ionic reaction conditions we studied in Chapters 5, 8, and 9. Under certain conditions, however, alkanes do react with oxygen, O_2. By far their most important reaction with oxygen is oxidation (combustion) to form carbon dioxide and water. They also react with bromine and chlorine to form halogenated hydrocarbons.

A. Reaction with Oxygen: Combustion

Oxidation of hydrocarbons, including alkanes and cycloalkanes, is the basis for their use as energy sources for heat [natural gas, liquefied petroleum gas (LPG), and fuel oil] and power (gasoline, diesel fuel, and aviation fuel). Following are balanced equations for the complete combustion of methane, the major component of natural gas, and propane, the major component of LPG or bottled gas. The heat liberated when an alkane is oxidized to carbon dioxide and water is called its heat of combustion (Section 5.8).

$$CH_4 \ + \ 2O_2 \longrightarrow CO_2 + 2H_2O \ + \ 212 \text{ kcal/mol}$$
Methane

$$CH_3CH_2CH_3 + 5O_2 \longrightarrow 3CO_2 + 4H_2O \ + \ 530 \text{ kcal/mol}$$
Propane

B. Reaction with Halogens: Halogenation

If we mix methane with chlorine or bromine in the dark at room temperature, nothing happens. If, however, we heat the mixture to 100°C or higher or expose it to light, a reaction begins at once. The products of the reaction between methane and chlorine are chloromethane and hydrogen chloride. What occurs is a substitution reaction—in this case, the substitution of a chlorine for a hydrogen in methane.

$$CH_4 + Cl_2 \xrightarrow{\text{heat or light}} CH_3Cl + HCl$$
Methane   Chloromethane
                                         (Methyl chloride)

If chloromethane is allowed to react with more chlorine, further chlorination produces a mixture of dichloromethane, trichloromethane, and tetrachloromethane.

$$CH_3Cl + Cl_2 \xrightarrow{\text{heat}} CH_2Cl_2 + HCl$$
Chloromethane   Dichloromethane
(Methyl chloride)   (Methylene chloride)

In the last equation, the reagent Cl_2 is placed over the reaction arrow and the equivalent amount of HCl formed is not shown. Placing reagents over reaction arrows and omitting by-products is commonly done to save space.

We derive IUPAC names of haloalkanes by naming the halogen atom as a substituent (*fluoro-, chloro-, bromo-, iodo-*) and alphabetizing it along with other substituents. Common names consist of the common name of the alkyl group followed by the name of the halogen (chloride, bromide, and so forth) as a separate word. Dichloromethane (methylene chloride) is a widely used solvent for organic compounds.

EXAMPLE 11.9

Write a balanced equation for the reaction of ethane with chlorine to form chloroethane, C_2H_5Cl.

Solution
The reaction of ethane with chlorine results in the substitution of one of the hydrogen atoms of ethane by a chlorine atom.

$$CH_3CH_3 + Cl_2 \xrightarrow{\text{heat or light}} CH_3CH_2Cl + HCl$$

Ethane   Chloroethane
                                            (Ethyl chloride)

Problem 11.9

Reaction of propane with chlorine gives two products with the molecular formula C_3H_7Cl. Draw structural formulas for these two compounds, and give each an IUPAC name and a common name.

11.9 | What Are Some Important Haloalkanes?

One of the major uses of haloalkanes is as intermediates in the synthesis of other organic compounds. Just as we can replace a hydrogen atom of an alkane, we can, in turn, replace the halogen atom by a number of other functional groups. In this way, we can construct more complex molecules. In contrast, alkanes that contain several halogens are often quite unreactive, a fact that has proved especially useful in the design of several classes of consumer products.

A. Chlorofluorocarbons

Of all the haloalkanes, the **chlorofluorocarbons (CFCs)** manufactured under the trade name Freons are the most widely known. CFCs are non-toxic, nonflammable, odorless, and noncorrosive. Originally, they seemed to be ideal replacements for the hazardous compounds such as ammonia and sulfur dioxide formerly used as heat-transfer agents in refrigeration systems. Among the CFCs most widely used for this purpose were trichlorofluoromethane (CCl_3F, Freon-11) and dichlorodifluoromethane (CCl_2F_2, Freon-12). The CFCs also found wide use as industrial cleaning solvents to prepare surfaces for coatings, to remove cutting oils and waxes from millings, and to remove protective coatings. In addition, they were employed as propellants for aerosol sprays.

B. Solvents

Several low-molecular-weight haloalkanes are excellent solvents in which to carry out organic reactions and to use as cleaners and degreasers. Carbon tetrachloride ("carbon tet") was the first of these compounds to find wide

CHEMICAL CONNECTIONS 11B

The Environmental Impact of Freons

Concern about the environmental impact of CFCs arose in the 1970s, when researchers found that more than 4.5×10^5 kg/yr of these compounds were being emitted into the atmosphere. In 1974, Sherwood Rowland and Mario Molina, both of the United States, announced their theory, which has since been amply confirmed, that these compounds destroy the stratospheric ozone layer. When released into the air, CFCs escape to the lower atmosphere. Because of their inertness, however, they do not decompose there. Slowly, they find their way to the stratosphere, where they absorb ultraviolet radiation from the Sun and then decompose. As they do so, they set up chemical reactions that lead to the destruction of the stratospheric ozone layer, which shields the Earth against short-wavelength ultraviolet radiation from the Sun. An increase in short-wavelength ultraviolet radiation reaching the Earth is believed to promote the destruction of certain crops and agricultural species, and even to increase the incidence of skin cancer in light-skinned individuals.

This concern prompted two conventions, one in Vienna in 1985 and one in Montreal in 1987, held by the United Nations Environmental Program. The 1987 meeting produced the Montreal Protocol, which set limits on the production and use of ozone-depleting CFCs and urged the complete phase-out of their production by 1996. This phase-out resulted in enormous costs for manufacturers and is not yet complete in developing countries.

Rowland, Molina, and Paul Crutzen (a Dutch chemist at the Max Planck Institute for Chemistry in Germany) were awarded the 1995 Nobel Prize for chemistry. As noted in the award citation by the Royal Swedish Academy of Sciences, "By explaining the chemical mechanisms that affect the thickness of the ozone layer, these three researchers have contributed to our salvation from a global environmental problem that could have catastrophic consequences."

The chemical industry has responded to this crisis by developing replacement refrigerants that have much lower ozone-depleting potential. The most prominent of these replacements are the hydrofluorocarbons (HFCs) and hydrochlorofluorocarbons (HCFCs). These compounds are much more chemically reactive in the atmosphere than the Freons and are destroyed before they reach the stratosphere. However, they cannot be used in air conditioners in 1994- and earlier-model cars.

HFC-134a   HCFC-141b

---

application, but its use for this purpose has since been discontinued because it is now known that carbon tet is both toxic and a carcinogen. Today, the most widely used haloalkane solvent is dichloromethane, CH_2Cl_2.

11.10 | Where Do We Obtain Alkanes?

The two major sources of alkanes are natural gas and petroleum. **Natural gas** consists of approximately 90 to 95% methane, 5 to 10% ethane, and a mixture of other relatively low-boiling alkanes—chiefly propane, butane, and 2-methylpropane.

Petroleum is a thick, viscous, liquid mixture of thousands of compounds, most of them hydrocarbons, formed from the decomposition of marine plants and animals. Petroleum and petroleum-derived products fuel automobiles, aircraft, and trains. They provide most of the greases and lubricants required for the machinery utilized by our highly industrialized society. Furthermore, petroleum, along with natural gas, provides nearly 90% of the organic raw materials for the synthesis and manufacture of synthetic fibers, plastics, detergents, drugs, dyes, and a multitude of other products.

The fundamental separation process in refining petroleum is fractional distillation (Figure 11.9). Practically all crude petroleum that enters a refinery goes to distillation units, where it is heated to temperatures as high as 370 to 425°C and separated into fractions. Each fraction contains a mixture of hydrocarbons that boils within a particular range.

A petroleum fractional distillation tower.

Ashland Oil Company

Gases
Boiling point range below 20°C
(C_1–C_4 hydrocarbons; used as
fuels and reactants to make plastics)

Gasoline (naphthas) 20–200°C
(C_5–C_{12} hydrocarbons; used as
motor fuels and industrial solvents)

Kerosene 175–275°C (C_{12}–C_{16} hydrocarbons;
used for lamp oil, diesel fuel, starting
material for catalytic cracking)

Fuel oil 250–400°C (C_{15}–C_{18} hydrocarbons;
used for catalytic cracking, heating oil,
diesel fuel)

Lubricating oil above 350°C
(C_{16}–C_{20} hydrocarbons; used as lubricants)

Pipe still

Crude oil
and vapor
are preheated

Residue (asphalt)
($>C_{20}$ hydrocarbons)

Figure 11.9 Practically all crude petroleum that enters a refinery goes to distillation.

Chemical Connections 11C

Octane Rating: What Those Numbers at the Pump Mean

Gasoline is a complex mixture of C_6 to C_{12} hydrocarbons. Its quality as a fuel for internal combustion engines depends on how much it makes an engine "knock." Engine knocking occurs when some of the air–fuel mixture explodes prematurely and independently of ignition by the spark plug.

Two compounds were selected for rating gasoline quality. 2,2,4-Trimethylpentane (isooctane) has very good antiknock properties (the fuel–air mixture burns smoothly in the combustion chamber) and was assigned an octane rating of 100. Heptane, in contrast, has poor antiknock properties and was assigned an octane rating of 0.

Typical octane ratings of commonly available gasolines.

2,2,4-Trimethylpentane
(octane rating 100)

Heptane
(octane rating 0)

The **octane rating** of a particular gasoline is a number equal to the percentage of 2,2,4-trimethylpentane in a mixture of 2,2,4-trimethylpentane and heptane that has the same antiknock proper- ties as the gasoline being rated. For example, the antiknock proper- ties of 2-methylhexane are the same as those of a mixture of 42% 2,2,4-trimethylpentane and 58 percent heptane; therefore, the octane rating of 2-methylhexane is 42. Ethanol, which is added to gasoline to produce gasohol, has an octane rating of 105. Octane itself has an octane rating of −20.

SUMMARY OF KEY QUESTIONS

SECTION 11.1 How Do We Write Structural Formulas of Alkanes?

- A **hydrocarbon** is a compound that contains only carbon and hydrogen.
- A **saturated hydrocarbon** contains only single bonds. An **alkane** is a saturated hydrocarbon whose carbon atoms are arranged in an open chain.

SECTION 11.2 What Are Constitutional Isomers?

- **Constitutional isomers** have the same molecular formula but a different connectivity of their atoms.

SECTION 11.3 How Do We Name Alkanes?

- Alkanes are named according to a set of rules developed by the **International Union of Pure and Applied Chemistry.**
- The IUPAC name of an alkane consists of two parts: a prefix that tells the number of carbon atoms in the parent chain, and the ending **-ane.** Substituents derived from alkanes by removal of a hydrogen atom are called **alkyl groups** and are denoted by the symbol **R—.**

SECTION 11.4 What Are Cycloalkanes?

- An alkane that contains carbon atoms bonded to form a ring is called a **cycloalkane.**
- To name a cycloalkane, prefix the name of the open-chain alkane with **cyclo-.**

SECTION 11.5 What Are the Shapes of Alkanes and Cycloalkanes?

- A **conformation** is any three-dimensional arrangement of the atoms of a molecule that results from rotation about a single bond.
- The lowest-energy conformation of cyclopentane is an **envelope conformation.**
- The lowest-energy conformation of cyclohexane is a **chair conformation.** In a chair conformation, six

C—H bonds are **axial** and six C—H bonds are **equatorial.** A substituent on a six-membered ring is more stable when it is equatorial than when it is axial.

SECTION 11.6 What Is *Cis-Trans* Isomerism in Cycloalkanes?

- ***Cis-trans* isomers** of cycloalkanes have (1) the same molecular formula and (2) the same connectivity of their atoms, (3) but a different orientation of their atoms in space because of the restricted rotation around the C—C bonds of the ring.
- For *cis-trans* isomers of cycloalkanes, *cis* means that substituents are on the same side of the ring; *trans* means that they are on opposite sides of the ring.

SECTION 11.7 What Are the Physical Properties of Alkanes?

- Alkanes are nonpolar compounds, and the only forces of attraction between their molecules are London dispersion forces.
- At room temperature, low-molecular-weight alkanes are gases, higher-molecular-weight alkanes are liquids, and very-high-molecular-weight alkanes are waxy solids.
- For any group of alkane constitutional isomers, the least-branched isomer generally has the highest boiling point, and the most-branched isomer generally has the lowest boiling point.
- Alkanes are insoluble in water but soluble in each other and in other nonpolar organic solvents such as toluene. All liquid and solid alkanes are less dense than water.

SECTION 11.10 Where Do We Obtain Alkanes?

- **Natural gas** consists of 90 to 95 percent methane with lesser amounts of ethane and other lower-molecular-weight hydrocarbons.
- **Petroleum** is a liquid mixture of thousands of different hydrocarbons.

SUMMARY OF KEY REACTIONS

1. **Oxidation of Alkanes (Section 11.8A)** Oxidation of alkanes to carbon dioxide and water, an exothermic reaction, is the basis for our use of them as sources of heat and power.

$$CH_3CH_2CH_3 + 5O_2 \longrightarrow 3CO_2 + 4H_2O + 530 \text{ kcal/mol}$$
Propane

2. **Halogenation of Alkanes (Section 11.8B)** Reaction of an alkane with chlorine or bromine results in the substitution of a halogen atom for a hydrogen.

$$CH_3CH_3 + Cl_2 \xrightarrow{\text{heat or light}} CH_3CH_2Cl + HCl$$
Ethane   Chloroethane
                          (Ethyl chloride)

PROBLEMS

SECTION 11.1 How Do We Write Structural Formulas of Alkanes?

11.10 Define:

   (a) Hydrocarbon

   (b) Alkane

   (c) Saturated hydrocarbon

11.11 Why is it not accurate to describe an unbranched alkane as a "straight-chain" hydrocarbon?

11.12 What is meant by the term *line-angle formula* as applied to alkanes and cycloalkanes?

11.13 For each condensed structural formula, write a line-angle formula.

(a)

$$\underset{\underset{\displaystyle CH(CH_3)_2}{|}}{CH_3CH_2\underset{\overset{\displaystyle |}{\overset{\displaystyle CH_2CH_3}{}}}{CH}CHCH_2\underset{}{CH}CH_3 \quad \overset{CH_3}{|}}$$

(b)

$$CH_3\underset{\underset{\displaystyle CH_3}{|}}{\overset{\overset{\displaystyle CH_3}{|}}{C}}CH_3$$

(c) $(CH_3)_2CHCH(CH_3)_2$

(d)

$$CH_3CH_2\underset{\underset{\displaystyle CH_2CH_3}{|}}{\overset{\overset{\displaystyle CH_2CH_3}{|}}{C}}CH_2CH_3$$

(e) $(CH_3)_3CH$

(f) $CH_3(CH_2)_3CH(CH_3)_2$

11.14 ■ Write the molecular formula for each alkane.

SECTION 11.2 What Are Constitutional Isomers?

11.15 Which statements are true about constitutional isomers?

   (a) They have the same molecular formula.

   (b) They have the same molecular weight.

   (c) They have the same connectivity of their atoms.

   (d) They have the same physical properties.

11.16 ■ Each member of the following set of compounds is an alcohol; that is, each contains an —OH (hydroxyl group; see Section 10.4A). Which structural formulas represent the same compound, and which represent constitutional isomers?

11.17 Each member of the following set of compounds is an amine; that is, each contains a nitrogen bonded to one, two, or three carbon groups (Section 10.4B). Which structural formulas represent the same compound, and which represent constitutional isomers?

11.18 Each member of the following set of compounds is either an aldehyde or ketone (Section 10.4C). Which structural formulas represent the same compound, and which represent constitutional isomers?

(e) [structure] O / H   (f) [structure] O

(g) [structure] O   (h) [structure] O / H

(c) $CH_3(CH_2)_4CHCH_2CH_3$
              |
             CH_2CH_3

(d) [structure]

11.19 For which parts do the pairs of structural formulas represent constitutional isomers?

(a) [square] and [structure]

(b) [structure] and [structure]

(c) [structure] and [structure]

(d) [structure] and [structure]

(e) [structure] and [structure]—NH_2

(f) [structure] and [structure]—OH

11.20 Draw line-angle formulas for the nine constitutional isomers with the molecular formula C_7H_{16}.

SECTION 11.3 How Do We Name Alkanes?

11.21 Name these alkyl groups:

(a) CH_3CH_2—

(b) CH_3CH—
          |
          CH_3

(c) CH_3CHCH_2—
        |
        CH_3

(d) CH_3C—
      |
      CH_3
      |
      CH_3

11.22 Write the IUPAC names for isobutane and isopentane.

SECTION 11.4 What Are Cycloalkanes?

11.23 Write the IUPAC names for these alkanes and cycloalkanes.

(a) $CH_3CHCH_2CH_2CH_3$
      |
      CH_3

(b) $CH_3CHCH_2CH_2CHCH_3$
      | |
      CH_3 CH_3

(e) [structure]

(f) [structure]

11.24 Write line-angle formulas for these alkanes and cycloalkanes.

(a) 2,2,4-Trimethylhexane
(b) 2,2-Dimethylpropane
(c) 3-Ethyl-2,4,5-trimethyloctane
(d) 5-Butyl-2,2-dimethylnonane
(e) 4-Isopropyloctane
(f) 3,3-Dimethylpentane
(g) *trans*-1,3-Dimethylcyclopentane
(h) *cis*-1,2-Diethylcyclobutane

SECTION 11.5 What Are the Shapes of Alkanes and Cycloalkanes?

11.25 Define *conformation*.

11.26 ▪ The condensed structural formula of butane is $CH_3CH_2CH_2CH_3$. Explain why this formula does not show the geometry of the real molecule.

11.27 ▪ Draw a conformation of ethane in which hydrogen atoms on adjacent carbons are as far apart as possible. Also draw a conformation in which they are as close together as possible. In a sample of ethane molecules at room temperature, which conformation is the more likely?

SECTION 11.6 What Is *Cis-Trans* Isomerism in Cycloalkanes?

11.28 What structural feature of cycloalkanes makes *cis-trans* isomerism in them possible?

11.29 Is *cis-trans* isomerism possible in alkanes?

11.30 ▪ Name and draw structural formulas for the *cis* and *trans* isomers of 1,2-dimethylcyclopropane.

11.31 Name and draw structural formulas for the six cycloalkanes with the molecular formula C_5H_{10}. Include *cis-trans* isomers as well as constitutional isomers.

11.32 Why is equatorial methylcyclohexane more stable than axial methylcyclohexane?

SECTION 11.7 What Are the Physical Properties of Alkanes?

11.33 In Problem 11.20, you drew structural formulas for the nine constitutional isomers with molecular formula C_7H_{16}. Predict which isomer has the lowest boiling point and which has the highest boiling point.

11.34 Which unbranched alkane (Table 11.4) has about the same boiling point as water? Calculate the molecular weight of this alkane, and compare it with the molecular weight of water.

11.35 What generalizations can you make about the densities of alkanes relative to the density of water?

11.36 What generalization can you make about the solubility of alkanes in water?

11.37 Suppose that you have samples of hexane and octane. Could you tell the difference by looking at them? What color would they be? How could you tell which is which?

11.38 As you can see from Table 11.4, each CH_2 group added to the carbon chain of an alkane increases its boiling point. This increase is greater going from CH_4 to C_2H_6 and from C_2H_6 to C_3H_8 than it is going from C_8H_{18} to C_9H_{20} or from C_9H_{20} to $C_{10}H_{22}$. What do you think is the reason for this difference?

11.39 How are the boiling points of hydrocarbons during petroleum refining related to their molecular weight?

SECTION 11.8 What Are the Characteristic Reactions of Alkanes?

11.40 ■ Write balanced equations for the combustion of each of the following hydrocarbons. Assume that each is converted completely to carbon dioxide and water.

(a) Hexane   (b) Cyclohexane
(c) 2-Methylpentane

11.41 The heat of combustion of methane, a component of natural gas, is 212 kcal/mol. That of propane, a component of LP gas, is 530 kcal/mol. On a gram-for-gram basis, which hydrocarbon is the better source of heat energy?

11.42 Draw structural formulas for these haloalkanes.

(a) Bromomethane
(b) Chlorocyclohexane
(c) 1,2-Dibromoethane
(d) 2-Chloro-2-methylpropane
(e) Dichlorodifluoromethane (Freon-12)

11.43 The reaction of chlorine with pentane gives a mixture of three chloroalkanes with the molecular formula $C_5H_{11}Cl$. Write a line-angle formula and an IUPAC name for each chloroalkane.

Chemical Connections

11.44 (Chemical Connections 11A) How many rings in tetrodotoxin contain only carbon atoms? How many contain nitrogen atoms? How many contain two oxygen atoms?

11.45 (Chemical Connections 11B) What are Freons? Why were they considered ideal compounds to use as heat-transfer agents in refrigeration systems? Give structural formulas of two Freons used for this purpose.

11.46 (Chemical Connections 11B) In what way do Freons negatively affect the environment?

11.47 (Chemical Connections 11B) What are HFCs and HCFCs? How does their use in refrigeration systems avoid the environmental problems associated with the use of Freons?

11.48 (Chemical Connections 11C) What is an "octane rating"? What two reference hydrocarbons are used for setting the scale of octane ratings?

11.49 (Chemical Connections 11C) Octane has an octane rating of −20. Will it produce more or less engine knocking than heptane does?

11.50 (Chemical Connections 11C) When ethanol is added to gasoline to produce "gasohol," it promotes more complete combustion of the gasoline and is an octane booster. Compare the heats of combustion of 2,2,4-trimethylpentane (1304 kcal/mol) and ethanol (327 kcal/mol). Which has the higher heat of combustion in kcal/mol? In kcal/g?

Additional Problems

11.51 Tell whether the compounds in each set are constitutional isomers.

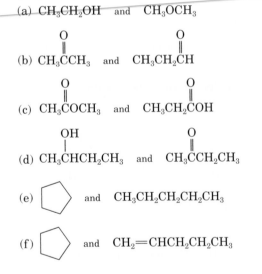

11.52 Explain why each of the following is an incorrect IUPAC name. Write the correct IUPAC name for the compound.

(a) 1,3-Dimethylbutane
(b) 4-Methylpentane
(c) 2,2-Diethylbutane
(d) 2-Ethyl-3-methylpentane
(e) 2-Propylpentane
(f) 2,2-Diethylheptane
(g) 2,2-Dimethylcyclopropane
(h) 1-Ethyl-5-methylcyclohexane

11.53 Which compounds can exist as *cis-trans* isomers? For each that does, draw both isomers using solid and dashed wedges to show the orientation in space of the —OH and —CH₃ groups.

11.54 Tetradecane, C₁₄H₃₀, is an unbranched alkane with a melting point of 5.9°C and a boiling point of 254°C. Is tetradecane a solid, liquid, or gas at room temperature?

11.55 Dodecane, C₁₂H₂₆, is an unbranched alkane. Predict the following:
(a) Will it dissolve in water?
(b) Will it dissolve in hexane?
(c) Will it burn when ignited?
(d) Is it a liquid, solid, or gas at room temperature and atmospheric pressure?
(e) Is it more or less dense than water?

Looking Ahead

11.56 Following is a structural formula for 2-isopropyl-5-methylcyclohexanol:

2-Isopropyl-5-methylcyclohexanol

Using a planar hexagon representation for the cyclohexane ring, draw a structural formula for the *cis-trans* isomer with isopropyl *trans* to —OH and methyl *cis* to —OH. If you answered this part correctly, you have drawn the isomer found in nature and given the name menthol.

11.57 On the left is a representation of the glucose molecule. Convert this representation to the alternative representations using the rings on the right. (We discuss the structure and chemistry of glucose in Chapter 20.)

Planar hexagon representation   Chair conformation

11.58 On the left is a representation for 2-deoxy-D-ribose. This molecule is the "D" of DNA. Convert this representation to the alternative representation using the ring on the right. (We discuss the structure and chemistry of this compound in more detail in Chapter 20.)

2-Deoxy-D-ribose

11.59 As stated in Section 11.7, the wax found in apple skins is an unbranched alkane with the molecular formula C₂₇H₅₆. Explain how the presence of this alkane in apple skins prevents the loss of moisture from within the apple

Alkenes and Alkynes

β-Carotene is a naturally occurring polyene in carrots and tomatoes (Problems 12.57 and 12.58).

Charles D. Winters

Alkene An unsaturated hydrocarbon that contains a carbon–carbon double bond

Alkyne An unsaturated hydrocarbon that contains a carbon–carbon triple bond

Chemistry⚛Now™
Look for this logo in the chapter and go to GOB ChemistryNow at **http://now.brookscole.com/gob8** or on the CD for tutorials, simulations, and problems.

12.1 What Are Alkenes and Alkynes?

In this chapter, we begin our study of unsaturated hydrocarbons. Recall from Section 11.1 that these unsaturated compounds contain one or more carbon–carbon double bonds, triple bonds, or benzene-like rings. In this chapter we study **alkenes** and alkynes. The simplest alkene is ethylene.

Alkynes are unsaturated hydrocarbons that contain one or more carbon–carbon triple bonds. The simplest alkyne is acetylene. Because alkynes are not widespread in nature and have little importance in biochemistry, we will not study their chemistry in depth.

Ethylene
(an alkene)

Acetylene
(an alkyne)

CHEMICAL CONNECTIONS 12A

Ethylene: A Plant Growth Regulator

As we noted earlier, ethylene occurs only in trace amounts in nature. True enough, but scientists have discovered that this small molecule is a natural ripening agent for fruits. Thanks to this knowledge, fruit growers can now pick fruit while it is still green and less susceptible to bruising. Then, when they are ready to pack the fruit for shipment, the grower can treat it with ethylene gas. Alternatively, the fruit can be treated with ethephon (Ethrel), which slowly releases ethylene and initiates fruit ripening.

Ethephon

The next time you see ripe bananas in the market, you might wonder when they were picked and whether their ripening was artificially induced.

Compounds containing carbon–carbon double bonds are especially widespread in nature. Furthermore, several low-molecular-weight alkenes, including ethylene and propene, have enormous commercial importance in our modern, industrialized society. The organic chemical industry world-wide produces more pounds of ethylene than any other chemical. Annual production in the United States alone exceeds 55 billion pounds.

What is unusual about ethylene is that it occurs only in trace amounts in nature. The enormous amounts of it required to meet the needs of the chemical industry are derived the world over by thermal cracking of hydrocarbons. In the United States and other areas of the world with vast reserves of natural gas, the major process for the production of ethylene is thermal cracking of the small quantities of ethane extracted from natural gas. In **thermal cracking,** a saturated hydrocarbon is converted to an unsaturated hydrocarbon plus H_2. Ethane is thermally cracked by heating it in a furnace to 800–900°C for a fraction of a second.

$$CH_3CH_3 \xrightarrow[\text{(thermal cracking)}]{800-900°C} CH_2{=}CH_2 + H_2$$

Ethane   Ethylene

Europe, Japan, and other parts of the world with limited supplies of natural gas depend almost entirely on thermal cracking of petroleum for their ethylene.

From the perspective of the chemical industry, the single most important reaction of ethylene and other low-molecular-weight alkenes is polymerization, which we discuss in Section 12.7. The crucial point to recognize is that ethylene and all of the commercial and industrial products synthesized from it are derived from either natural gas or petroleum—both nonrenewable natural resources!

12.2 | What Are the Structures of Alkenes and Alkynes?

A. Alkenes

GOB
Chemistry ⚛ Now™
Click *Coached Problems* to explore the
Bonding in Alkenes

Using the VSEPR model (Section 4.10), we predict bond angles of 120° about each carbon in a double bond. The observed H—C—C bond angle in ethylene, for example, is 121.7°, close to the predicted value. In other alkenes, deviations from the predicted angle of 120° may be somewhat larger because of interactions between alkyl groups bonded to the doubly bonded carbons. The C—C—C bond angle in propene, for example, is 124.7°.

Ethylene   Propene

If we look at a molecular model of ethylene, we see that the two carbons of the double bond and the four hydrogens bonded to them all lie in the same plane—that is, ethylene is a flat or planar molecule. Furthermore, chemists have discovered that, under normal conditions, no such rotation is possible about the carbon–carbon double bond of ethylene or, for that matter, of any other alkene. Whereas free rotation occurs about each carbon–carbon single bond in an alkane (Section 11.5A), rotation about the carbon–carbon double bond in an alkene does not normally take place. For an important exception to this generalization about carbon–carbon double bonds, see Chemical Connections 12C on *cis-trans* isomerism in vision.

Chemistry•⚛•Now™
Click *Coached Problems* to try a set of problems in **Identifying *Cis-* and *Trans* Isomers**

***Cis-trans* isomers** Isomers that have the same connectivity of their atoms but a different arrangement of their atoms in space. Specifically, *cis* and *trans* stereoisomers result from the presence of either a ring or a carbon–carbon double bond

B. *Cis-Trans* Stereoisomerism in Alkenes

Because of the restricted rotation about a carbon–carbon double bond, an alkene in which each carbon of the double bond has two different groups bonded to it shows *cis-trans* isomerism (one type of stereoisomerism). For example, 2-butene has two *cis-trans* isomers. In *cis*-2-butene, the two methyl groups are located on the same side of the double bond, and the two hydrogens are on the other side. In *trans*-2-butene, the two methyl groups are located on opposite sides of the double bond. *Cis*-2-butene and *trans*-2-butene are different compounds and have different physical and chemical properties.

cis-2-Butene
mp −139°C, bp 4°C

trans-2-Butene
mp −106°C, bp 1°C

12.3 | How Do We Name Alkenes and Alkynes?

Alkenes and alkynes are named using the IUPAC system of nomenclature. As we will see, some are still referred to by their common names.

Chemistry•⚛•Now™
Click *Coached Problems* to explore the relationship between the **Structures and Names of Alkenes**

A. IUPAC Names

The key to the IUPAC system of naming alkenes is the ending **-ene.** Just as the ending *-ane* tells us that a hydrocarbon chain contains only carbon–carbon single bonds, so the ending *-ene* tells us that it contains a carbon–carbon double bond. To name an alkene:

1. Find the longest carbon chain that includes the double bond. Indicate the length of the parent chain by using a prefix that tells the number of carbon atoms in it (see Table 11.2).

2. Number the chain from the end that gives the lower set of numbers to the carbon atoms of the double bond. Designate the position of the double bond by the number of its first carbon.

3. Branched alkenes are named in a manner similar to alkanes; substituent groups are located and named.

6 5 4 3 2 1	6 5 4 3 2 1	5 4 3 2 1
$CH_3CH_2CH_2CH_2CH{=}CH_2$	$CH_3CH_2CHCH_2CH{=}CH_2$	$CH_3CH_2CHC{=}CH_2$
	CH_3	CH_2CH_3
		CH_2CH_3
1-Hexene	4-Methyl-1-hexene	2,3-Diethyl-1-pentene

Note that, although 2,3-diethyl-1-pentene has a six-carbon chain, the longest chain that contains the double bond has only five carbons. The parent alkene is, therefore, a pentene rather than a hexene, and the molecule is named as a disubstituted 1-pentene.

The key to the IUPAC name of an alkyne is the ending **-yne,** which shows the presence of a carbon–carbon triple bond. Thus, $HC{\equiv}CH$ is ethyne (or acetylene) and $CH_3C{\equiv}CH$ is propyne. In higher alkynes, number the longest carbon chain that contains the triple bond from the end that gives the lower set of numbers to the triply bonded carbons. Indicate the location of the triple bond by the number of its first carbon atom.

4 3 2 1	1	1
$CH_3CHC{\equiv}CH$		
CH_3	4 3 2	2 3 4 5
		6 7
		CH_3 1
	1 2 3 4 5 6 7	
	$CH_3CH_2C{\equiv}CCH_2CCH_3$	
	CH_3	
3-Methyl-1-butyne	6,6-Dimethyl-3-heptyne	

EXAMPLE 12.1

Write the IUPAC name of each unsaturated hydrocarbon.

(a) $CH_2{=}CH(CH_2)_5CH_3$

(b)
$$\begin{array}{ccc} H_3C & & CH_3 \\ & C{=}C & \\ H_3C & & H \end{array}$$

(c) $CH_3(CH_2)_2C{\equiv}CCH_3$

Solution

(a) The parent chain contains eight carbons; thus the parent alkene is octene. To show the presence of the carbon–carbon double bond, use the suffix *-ene.* Number the chain beginning with the first carbon of the double bond. This alkene is 1-octene.

(b) Because there are four carbon atoms in the chain containing the carbon–carbon double bond, the parent alkene is butene. The double bond is between carbons 2 and 3 of the chain, and there is a methyl group on carbon 2. This alkene is 2-methyl-2-butene.

(c) There are six carbons in the parent chain, with the triple bond between carbons 2 and 3. This alkyne is 2-hexyne.

Problem 12.1

Write the IUPAC name of each unsaturated hydrocarbon.

B. Common Names

Despite the precision and universal acceptance of IUPAC nomenclature, some alkenes and alkynes—particularly those of low molecular weight—are known almost exclusively by their common names. Three examples follow:

	$CH_2{=}CH_2$	$CH_3CH{=}CH_2$	$CH_3\overset{\displaystyle CH_3}{\underset{\displaystyle |}{C}}{=}CH_2$
IUPAC name:	Ethene	Propene	2-Methylpropene
Common name:	Ethylene	Propylene	Isobutylene

We derive common names for alkynes by prefixing the names of the substituents on the carbon–carbon triple bond to the name acetylene:

	$HC{\equiv}CH$	$CH_3C{\equiv}CH$	$CH_3C{\equiv}CCH_3$
IUPAC name:	Ethyne	Propyne	2-Butyne
Common name:	Acetylene	Methylacetylene	Dimethylacetylene

C. *Cis* and *Trans* Configurations of Alkenes

The orientation of the carbon atoms of the parent chain determines whether an alkene is *cis* or *trans*. If the carbons of the parent chain are on the same side of the double bond, the alkene is *cis*; if they are on opposite sides, it is a *trans* alkene. In the first example below, they are on opposite sides and the compound is a *trans* alkene. In the second example, they are on the same side and the compound is a *cis* alkene.

trans-3-Hexene

cis-3,4-Dimethyl-2-pentene

EXAMPLE 12.2

Name each alkene and specify its configuration by indicating *cis* or *trans* where appropriate.

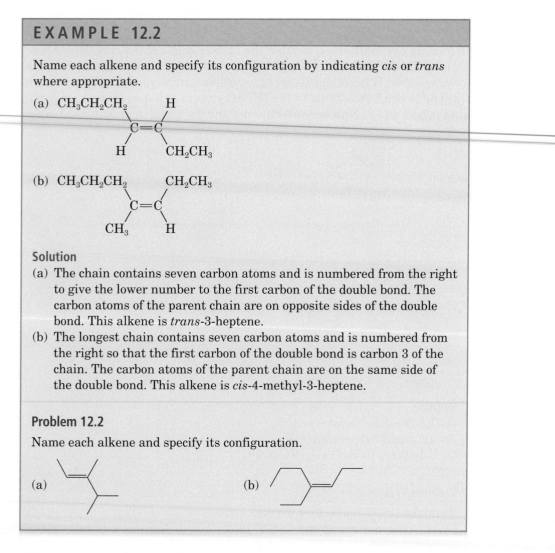

Solution

(a) The chain contains seven carbon atoms and is numbered from the right to give the lower number to the first carbon of the double bond. The carbon atoms of the parent chain are on opposite sides of the double bond. This alkene is *trans*-3-heptene.

(b) The longest chain contains seven carbon atoms and is numbered from the right so that the first carbon of the double bond is carbon 3 of the chain. The carbon atoms of the parent chain are on the same side of the double bond. This alkene is *cis*-4-methyl-3-heptene.

Problem 12.2

Name each alkene and specify its configuration.

(a)   (b)

CHEMICAL CONNECTIONS 12B

The Case of the Iowa and New York Strains of the European Corn Borer

Although humans communicate largely by sight and sound, the vast majority of other species in the animal world communicate by chemical signals. Often, communication within a species is specific for one of two or more stereoisomers. For example, a member of a given species might respond to the *cis* isomer of a chemical but not to the *trans* isomer. Alternatively, it might respond to a precise ratio of *cis* and *trans* isomers but not to other ratios of these same stereoisomers.

Several groups of scientists have studied the components of the sex **pheromones** of both the Iowa and New York strains of the European corn borer. Females of these closely related species secrete the sex attractant 11-tetradecenyl acetate. Males of the Iowa strain show their maximum response to a mixture containing 96% of the *cis* isomer and 4% of the *trans* isomer. When the pure *cis* isomer is used alone, males are only weakly attracted. Males of the New York strain show an entirely different response pattern: They respond maximally to

a mixture containing 3% of the *cis* isomer and 97% of the *trans* isomer.

trans-11-Tetradecenyl acetate

cis-11-Tetradecenyl acetate

Evidence suggests that an optimal response to a narrow range of stereoisomers, as we see here, is widespread in nature and that most insects maintain species isolation for mating and reproduction by the stereochemistry of their pheromones.

D. Cycloalkenes

In naming **cycloalkenes,** number the carbon atoms of the ring double bond 1 and 2 in the direction that gives the substituent encountered first the lower number. It is not necessary to use a location number for the carbons of the double bond. According to the IUPAC system of nomenclature, they will always be 1 and 2. Number substituents and list them in alphabetical order.

3-Methylcyclopentene
(not 5-methylcyclopentene)

4-Ethyl-1-methylcyclohexene
(not 5-ethyl-2-methylcyclohexene)

EXAMPLE 12.3

Write the IUPAC name for each cycloalkene.

(a) (b) (c)

Solution
(a) 3,3-Dimethylcyclohexene
(b) 1,2-Dimethylcyclopentene
(c) 4-Isopropyl-1-methylcyclohexene

Problem 12.3

Write the IUPAC name for each cycloalkene.

(a) (b) (c)

E. Dienes, Trienes, and Polyenes

We name alkenes that contain more than one double bond as alkadienes, alkatrienes, and so on. We often refer to those that contain several double bonds more generally as polyenes (Greek: *poly*, many). Following are three dienes:

$CH_2{=}CHCH_2CH{=}CH_2$
1,4-Pentadiene

$CH_2{=}\overset{\overset{\displaystyle CH_3}{|}}{C}CH{=}CH_2$
2-Methyl-1,3-butadiene
(Isoprene)

1,3-Cyclopentadiene

We saw earlier that for an alkene with one carbon–carbon double bond that can show *cis-trans* isomerism, two stereoisomers are possible. For an alkene with n carbon–carbon double bonds, each of which can show *cis-trans* isomerism, 2^n stereoisomers are possible.

EXAMPLE 12.4

How many stereoisomers are possible for 2,4-heptadiene?

$$CH_3-CH=CH-CH=CH-CH_2-CH_3$$
2,4-Heptadiene

Solution

This molecule has two carbon–carbon double bonds, each of which shows *cis-trans* isomerism. As shown in the following table, $2^2 = 4$ stereoisomers are possible. Line-angle formulas for two of these dienes are drawn here.

Double-Bond	
C_2–C_3	C_4–C_5
trans	*trans*
trans	*cis*
cis	*trans*
cis	*cis*

trans,trans-2,4-Heptadiene   *trans,cis*-2,4-Heptadiene

Problem 12.4

Draw structural formulas for the other two stereoisomers of 2,4-heptadiene.

EXAMPLE 12.5

Draw all stereoisomers that are possible for the following unsaturated alcohol.

$$CH_3C=CHCH_2CH_2C=CHCH_2OH$$
with CH_3 groups above carbons as shown

Solution

Cis-trans isomerism is possible only about the double bond between carbons 2 and 3 of the chain. It is not possible for the other double bond because carbon 7 has two identical groups on it (review Section 12.2B). Thus, $2^1 = 2$ stereoisomers (one *cis-trans* pair) are possible. The *trans* isomer of this alcohol, named geraniol, is a major component of the oils of rose, citronella, and lemon grass.

Lemon grass from Shelton Herb Farm, North Carolina.

David Sieren/Visuals Unlimited

The *trans* isomer

The *cis* isomer

Problem 12.5

How many stereoisomers are possible for the following unsaturated alcohol?

$$CH_3C=CHCH_2CH_2C=CHCH_2CH_2C=CHCH_2OH$$
with three CH_3 groups as shown

CHEMICAL CONNECTIONS 12C

Cis-Trans Isomerism in Vision

The retina, the light-detecting layer in the back of our eyes, contains reddish compounds called visual pigments. Their name, rhodopsin, is derived from the Greek word meaning "rose-colored." Each rhodopsin molecule is a combination of one molecule of a protein called opsin and one molecule of 11-*cis*-retinal, a derivative of vitamin A in which the CH₂OH group of carbon 15 is converted to an aldehyde group, —CH=O.

When rhodopsin absorbs light energy, the less stable 11-*cis* double bond is converted to the more stable 11-*trans* double bond. This isomerization changes the shape of the rhodopsin molecule, which in turn causes the neurons of the optic nerve to fire and produce a visual image.

The retinas of vertebrates contain two kinds of rhodopsin-containing cells: rods and cones. Cones function in bright light and are used for color vision; they are concentrated in the central portion of the retina, called the macula, and are responsible for the greatest visual acuity. The remaining area of the retina consists mostly of rods, which are used for peripheral and night vision. 11-*cis*-Retinal is present in both cones and rods. Rods have one kind of opsin, whereas cones have three kinds: one for blue, one for green, and one for red color vision.

An example of a biologically important polyunsaturated alcohol for which a number of *cis-trans* stereoisomers are possible is vitamin A. Each of the four carbon–carbon double bonds in the chain of carbon atoms bonded to the substituted cyclohexene ring has the potential for *cis-trans* isomerism. There are, therefore, $2^4 = 16$ stereoisomers possible for this structural formula. Vitamin A, the stereoisomer shown here, is the all-*trans* isomer.

Vitamin A (retinol)

12.4 | What Are the Physical Properties of Alkenes and Alkynes?

Alkenes and alkynes are nonpolar compounds, and the only attractive forces between their molecules are very weak London dispersion forces (Section 6.7A). Their physical properties, therefore, are similar to those of alkanes having the same carbon skeletons. Alkenes and alkynes that are liquid at room temperature have densities less than 1.0 g/mL (they float on water). They are insoluble in water but soluble in one another and in other nonpolar organic liquids.

12.5 | What Are Terpenes?

Among the compounds found in the essential oils of plants are a group of substances called **terpenes,** all of which have in common the fact that their carbon skeletons can be divided into two or more carbon units that are identical with the five-carbon skeleton of isoprene. Carbon 1 of an **isoprene unit** is called the head, and carbon 4 is called the tail. A terpene is a compound in which the tail of one isoprene unit becomes bonded to the head of another isoprene unit.

Terpene A compound whose carbon skeleton can be divided into two or more units identical to the five-carbon skeleton of isoprene

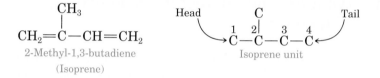

Terpenes are among the most widely distributed compounds in the biological world, and a study of their structure provides a glimpse into the wondrous diversity that nature can generate from a simple carbon skeleton. Terpenes also illustrate an important principle of the molecular logic of living systems: In building large molecules, small subunits are bonded together by a series of enzyme-catalyzed reactions and then chemically modified by additional enzyme-catalyzed reactions. Chemists use the same principles in the laboratory, but their methods cannot match the precision and selectivity of the enzyme-catalyzed reactions of cellular systems.

Probably the terpenes most familiar to you—at least by odor—are components of the so-called essential oils obtained by steam distillation or ether extraction of various parts of plants. Essential oils contain the relatively low-molecular-weight substances that are largely responsible for characteristic plant fragrances. Many essential oils, but particularly those from flowers, are used in perfumes.

One example of a terpene obtained from an essential oil is myrcene (Figure 12.1), a component of bayberry wax and oils of bay and verbena.

Figure 12.1 Four terpenes, each derived from two isoprene units bonded from the tail of the first unit to the head of the second unit. In limonene and menthol, formation of an additional carbon–carbon bond creates a six-membered ring.

California laurel, *Umbelluria californica*, one source of myrcene.

Myrcene is a triene with a parent chain of eight carbon atoms and two one-carbon branches. The two isoprene units in myrcene are joined by bonding the tail of one unit to the head of the other. Figure 12.1 also shows three more terpenes, each consisting of ten carbon atoms. In limonene and menthol, nature has formed an additional bond between two carbons to create a six-membered ring.

Farnesol, a terpene with a molecular formula of $C_{15}H_{26}O$, includes three isoprene units. Derivatives of both farnesol and geraniol are intermediates in the biosynthesis of cholesterol (Section 29.3).

Farnesol
(Lily-of-the-valley)

Vitamin A (Section 12.3E), a terpene with a molecular formula of $C_{20}H_{30}O$, consists of four isoprene units bonded head-to-tail and cross-linked at one point to form a six-membered ring.

12.6 | What Are the Characteristic Reactions of Alkenes?

GOB
Chemistry·꙰·Now™
Click *Coached Problems* to explore the nature of the products of the **Reactions of Alkenes** with different reagents

The most characteristic reaction of alkenes is addition to the carbon–carbon double bond: The double bond is broken and in its place single bonds form to two new atoms or groups of atoms. Table 12.1 shows several examples of alkene addition reactions along with the descriptive name(s) associated with each reaction.

Table 12.1   Characteristic Addition Reactions of Alkenes

Reaction	Descriptive Name(s)
$\diagdown C=C \diagup$ + HCl → H Cl on carbons	hydrochlorination
$\diagdown C=C \diagup$ + H₂O → H OH on carbons	hydration
$\diagdown C=C \diagup$ + Br₂ → Br Br on carbons	bromination
$\diagdown C=C \diagup$ + H₂ → H H on carbons	hydrogenation (reduction)

At this point, you might ask why a carbon–carbon double bond is a site of chemical reactivity, whereas carbon–carbon single bonds are quite unreactive under most experimental conditions. One way to answer this question is to focus on the changes in bonding that occur as a result of an alkene addition reaction. Consider, for example, the addition of hydrogen (H_2) to ethylene. As a result of this addition, one double bond and one single bond (the H—H bond of H_2) are replaced by three single bonds, giving a net conversion of one bond of the double bond to two single bonds.

one double bond
and one single bond

three single bonds

This and almost all other addition reactions of alkenes are exothermic, which means that the products are more stable (have lower energy) than the reactants. Just because an alkene addition reaction is exothermic, however, it doesn't mean that it occurs rapidly. The rate of a chemical reaction depends on its activation energy, not on how exothermic or endothermic it is (Section 8.3). In fact, the addition of H_2 to an alkene is immeasurably slow at room temperature but, as we will see soon, proceeds quite rapidly in the presence of a suitable catalyst.

A. Addition of Hydrogen Halides (Hydrohalogenation)

The hydrogen halides HCl, HBr, and HI add to alkenes to give haloalkanes (alkyl halides). Addition of HCl to ethylene, for example, gives chloroethane (ethyl chloride):

Ethylene

Chloroethane
(Ethyl chloride)

Addition of HCl to propene gives 2-chloropropane (isopropyl chloride); hydrogen adds to carbon 1 of propene and chlorine adds to carbon 2. If the orientation of addition were reversed, 1-chloropropane (propyl chloride) would form. The observed result is that almost no 1-chloropropane forms. Because 2-chloropropane is the observed product, we say that addition of HCl to propene is **regioselective.**

Propene

2-Chloropropane

1-Chloropropane
(not formed)

This regioselectivity was noted by Vladimir Markovnikov (1838–1904), who made the following generalization known as **Markovnikov's rule:** In the addition of HX (where X = halogen) to an alkene, hydrogen adds to the doubly bonded carbon that has the greater number of hydrogens already bonded to it; halogen adds to the other carbon.

Chemistry Now™

Click *Coached Problems* to explore the **Regioselective Nature of Alkene Addition Reactions**

Regioselective reaction A reaction in which one direction of bond forming or bond breaking occurs in preference to all other directions

Markovnikov's rule In the addition of HX or H_2O to an alkene, hydrogen adds to the carbon of the double bond having the greater number of hydrogens

Markovnikov's rule is often paraphrased as "the rich get richer."

EXAMPLE 12.6

Draw a structural formula for the product of each alkene addition reaction.

(a) $CH_3C{=}CH_2$ + HI $\longrightarrow$ (with CH_3 on the middle carbon)

(b) (cyclopentene with CH_3) + HCl $\longrightarrow$

Solution

(a) Markovnikov's rule predicts that the hydrogen of HI adds to carbon 1 and iodine adds to carbon 2 to give 2-iodo-2-methylpropane.

(b) H adds to carbon 2 of the ring and Cl adds to carbon 1 to give 1-chloro-1-methylcyclopentane.

(a) CH_3CCH_3 (with CH_3 on top and I on bottom)

2-Iodo-2-methylpropane

(b)   (cyclopentane ring with Cl and CH_3 on carbon 1, carbon 2 labeled)

1-Chloro-1-methylcyclopentane

Problem 12.6

Draw a structural formula for the product of each alkene addition reaction.

(a) $CH_3CH{=}CH_2$ + HBr $\longrightarrow$

(b) (cyclohexane ring)${=}CH_2$ + HBr $\longrightarrow$

Markovnikov's rule tells us what happens when we add HCl, HBr, or HI to a carbon–carbon double bond. We know that in the addition of HCl or other halogen acid, one bond of the double bond and the H—Cl bond are broken, and that new C—H and C—Cl bonds form. But chemists also want to know how this conversion happens. Are the C=C and H—X bonds broken and both new covalent bonds formed all at the same time? Or does this reaction take place in a series of steps? If the latter, what are these steps and in what order do they take place?

Chemists account for the addition of HX to an alkene by defining a two-step **reaction mechanism,** which we illustrate for the reaction of 2-butene with hydrogen chloride to give 2-chlorobutane. Step 1 is the addition of H^+ to 2-butene. To show this addition, we use a **curved arrow** that shows the repositioning of an electron pair from its origin (the tail of the arrow) to its new location (the head of the arrow). Recall that we used curved arrows in Section 9.1 to show bond breaking and bond formation in proton-transfer reactions. We now use curved arrows in the same way to show bond breaking and bond formation in a reaction mechanism.

Step 1 results in the formation of an organic cation. One carbon atom in this cation has only six electrons in its valence shell, so it carries a charge of $+1$. A species containing a positively charged carbon atom is called a **carbocation** (<u>carb</u>on + <u>cation</u>). Carbocations are classified as primary (1°), secondary (2°), or tertiary (3°), depending on the number of carbon groups bonded to the carbon bearing the positive charge.

Reaction mechanism A step-by-step description of how a chemical reaction occurs

Carbocation A species containing a carbon atom with only three bonds to it and bearing a positive charge

Mechanism: Addition of HCl to 2-Butene

Step 1: Reaction of the carbon–carbon double bond of the alkene with H^+ forms a 2° carbocation intermediate. In forming this intermediate, one bond of the double bond breaks and its pair of electrons forms a new covalent bond with H^+. One carbon of the double bond is left with only six electrons in its valence shell and, therefore, has a positive charge.

Step 2: Reaction of the 2° carbocation intermediate with chloride ion completes the valence shell of carbon and gives 2-chlorobutane.

EXAMPLE 12.7

Propose a two-step mechanism for the addition of HI to methylenecyclohexane to give 1-iodo-1-methylcyclohexane.

Methylenecyclohexane   1-Iodo-1-methylcyclohexane

Solution
The mechanism is similar to that proposed for the addition of HCl to 2-butene.

Step 1: Reaction of H^+ with the carbon–carbon double bond forms a new C—H bond to the carbon bearing the greater number of hydrogens and gives a 3° carbocation intermediate.

A 3° carbocation
intermediate

Step 2: Reaction of the 3° carbocation intermediate with iodide ion completes the valence shell of carbon and gives the product.

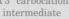

Problem 12.7
Propose a two-step mechanism for the addition of HBr to 1-methylcyclohexene to give 1-bromo-1-methylcyclohexane.

B. Addition of Water: Acid-Catalyzed Hydration

Hydration Addition of water

Most industrial ethanol is made by the acid-catalyzed hydration of ethylene.

In the presence of an acid catalyst, most commonly concentrated sulfuric acid, water adds to the carbon–carbon double bond of an alkene to give an alcohol. Addition of water is called **hydration.** In the case of simple alkenes, hydration follows Markovnikov's rule: H adds to the carbon of the double bond with the greater number of hydrogens and OH adds to the carbon with the smaller number of hydrogens.

$$CH_2{=}CH_2 + H_2O \xrightarrow{H_2SO_4} \overset{\overset{\displaystyle H}{|}}{CH_2}{-}\overset{\overset{\displaystyle OH}{|}}{CH_2}$$
Ethylene   Ethanol

$$CH_3CH{=}CH_2 + H_2O \xrightarrow{H_2SO_4} CH_3\overset{\overset{\displaystyle OH}{|}}{CH}{-}\overset{\overset{\displaystyle H}{|}}{CH_2}$$
Propene   2-Propanol

2-Methylpropene $+ H_2O \xrightarrow{H_2SO_4}$ 2-Methyl-2-propanol

---

EXAMPLE 12.8

Draw a structural formula for the alcohol formed by the acid-catalyzed hydration of 1-methylcyclohexene.

Solution

Markovnikov's rule predicts that H adds to the carbon with the greater number of hydrogens—in this case, carbon 2 of the cyclohexene ring. OH then adds to carbon 1.

1-Methylcyclohexene   1-Methylcyclohexanol

Problem 12.8

Draw a structural formula for the alcohol formed by acid-catalyzed hydration of each alkene:
(a) 2-Methyl-2-butene   (b) 2-Methyl-1-butene

The mechanism for the acid-catalyzed hydration of an alkene is similar to what we proposed for the addition of HCl, HBr, and HI to an alkene and is illustrated by the hydration of propene. This mechanism is consistent with the fact that acid is a catalyst. One H^+ is consumed in Step 1, but another is generated in Step 3.

Mechanism: Acid-Catalyzed Hydration of Propene

Step 1: Addition of H^+ to the carbon of the double bond with the greater number of hydrogens gives a 2° carbocation intermediate.

GOB
Chemistry·⚛·Now™
Click *Coached Problems* to explore an interactive animation of an **Alkene Hydration Reaction Mechanism**

$$CH_3CH{=}CH_2 + H^+ \longrightarrow CH_3\overset{+}{C}H\overset{\overset{\displaystyle H}{|}}{C}H_2$$

A 2° carbocation
intermediate

Step 2: The carbocation intermediate completes its valence shell by forming a new covalent bond with an unshared pair of electrons of the oxygen atom of H_2O to give an **oxonium ion.**

Oxonium ion An ion in which oxygen is bonded to three other atoms and bears a positive charge

$$CH_3\overset{+}{C}HCH_3 + {:}\overset{\overset{\displaystyle }{|}}{\underset{\underset{\displaystyle H}{|}}{\ddot{O}}}{-}H \longrightarrow CH_3CHCH_3$$

An oxonium ion

Step 3: Loss of H^+ from the oxonium ion gives the alcohol and generates a new H^+ catalyst.

$$CH_3CHCH_3 \longrightarrow CH_3CHCH_3 + H^+$$

EXAMPLE 12.9

Propose a three-step reaction mechanism for the acid-catalyzed hydration of methylenecyclohexane to give 1-methylcyclohexanol.

Solution

The reaction mechanism is similar to that for the acid-catalyzed hydration of propene.

Step 1: Reaction of the carbon–carbon double bond with H^+ gives a 3° carbocation intermediate.

A 3° carbocation
intermediate

Step 2: Reaction of the carbocation intermediate with water completes the valence shell of carbon and gives an oxonium ion.

An oxonium ion

Step 3: Loss of H^+ from the oxonium ion completes the reaction and generates a new H^+ catalyst.

Problem 12.9

Propose a three-step reaction mechanism for the acid-catalyzed hydration of 1-methylcyclohexene to give 1-methylcyclohexanol.

C. Addition of Bromine and Chlorine (Halogenation)

Chlorine, Cl_2, and bromine, Br_2, react with alkenes at room temperature by addition of halogen atoms to the carbon atoms of the double bond. This reaction is generally carried out either by using the pure reagents or by mixing them in an inert solvent, such as dichloromethane, CH_2Cl_2.

Addition of bromine is a useful qualitative test for the presence of an alkene. If we dissolve bromine in carbon tetrachloride, the solution is red. In contrast, alkenes and dibromoalkanes are colorless. If we mix a few drops of the red bromine solution with an unknown sample suspected of being an alkene, disappearance of the red color as bromine adds to the double bond tells us that an alkene is, indeed, present.

EXAMPLE 12.10

Complete these reactions.

Solution
In addition of Br_2 or Cl_2 to a cycloalkene, one halogen adds to each carbon of the double bond.

Problem 12.10

Complete these reactions.

(a) CH₃CCH=CH₂ + Br₂ $\xrightarrow{\text{CH}_2\text{Cl}_2}$ (with CH₃ groups on the central carbon)

(b) (cyclohexane ring with =CH₂) + Cl₂ $\xrightarrow{\text{CH}_2\text{Cl}_2}$

D. Addition of Hydrogen: Reduction (Hydrogenation)

Virtually all alkenes react quantitatively with molecular hydrogen, H_2, in the presence of a transition metal catalyst to give alkanes. Commonly used transition metal catalysts include platinum, palladium, ruthenium, and nickel. Because the conversion of an alkene to an alkane involves reduction by hydrogen in the presence of a catalyst, the process is called **catalytic reduction** or, alternatively, **catalytic hydrogenation.**

In Section 21.3 we will see how catalytic hydrogenation is used to solidify liquid vegetable oils to margarines and semisolid cooking fats.

trans-2-Butene + H_2 $\xrightarrow[\text{25°C, 3 atm}]{\text{Pd}}$ $CH_3CH_2CH_2CH_3$ $\xleftarrow[\text{25°C, 3 atm}]{\text{Pd}}$ H_2 + *cis*-2-Butene

Butane

Cyclohexene + H_2 $\xrightarrow[\text{25°C, 3 atm}]{\text{Pd}}$ Cyclohexane

The metal catalyst takes the form of a finely powdered solid. The reaction is carried out by dissolving the alkene in ethanol or another nonreacting organic solvent, adding the solid catalyst, and exposing the mixture to hydrogen gas at pressures ranging from 1 to 150 atm.

Mechanism: Catalytic Reduction

The transition metal catalysts used in catalytic hydrogenation are able to absorb larger quantities of hydrogen onto their surfaces, probably by forming metal–hydrogen bonds. Similarly, alkenes are adsorbed on metal surfaces with the formation of carbon-metal bonds. Addition of hydrogen atoms to an alkene occurs in two steps (Figure 12.2).

(a)   (b)   (c)

Metal surface

Chemistry・Now™

Active Figure 12.2 The addition of hydrogen to an alkene involving a transition metal catalyst. (*a*) Hydrogen and the alkene are adsorbed on the metal surface and (*b*) one hydrogen atom is transferred to the alkene, forming a new C—H bond. The other carbon remains adsorbed on the metal surface. (*c*) A second C—H bond is formed and the alkene is desorbed. **See a simulation based on this figure, and take a short quiz on the concepts at http://www.now.brookscole.com/gob8 or on the CD.**

12.7 | What Are the Important Polymerization Reactions of Ethylene and Substituted Ethylenes?

A. Structure of Polyethylenes

Polymer From the Greek *poly*, "many," and *meros*, "parts"; any long-chain molecule synthesized by bonding together many single parts called monomers

Monomer From the Greek *mono*, "single," and *meros*, "part"; the simplest nonredundant unit from which a polymer is synthesized

From the perspective of the chemical industry, the single most important reaction of alkenes is the formation of **chain-growth polymers** (Greek: *poly*, many, and *meros*, part). In the presence of certain compounds called initiators, many alkenes form polymers made by the stepwise addition of **monomers** (Greek: *mono*, one, and *meros*, part) to a growing polymer chain, as illustrated by the formation of polyethylene from ethylene. In alkene polymers of industrial and commercial importance, n is a large number, typically several thousand.

$$n\,CH_2\!=\!CH_2 \xrightarrow[\text{(polymerization)}]{\text{initiator}} \left(\!CH_2CH_2\!\right)_{\!n}$$

Ethylene   Polyethylene

To show the structure of a polymer, we place parentheses around the repeating monomer unit. The structure of an entire polymer chain can be reproduced by repeating this enclosed structure in both directions. A subscript n is placed outside the parentheses to indicate that this unit is repeated n times, as illustrated for the conversion of propylene to polypropylene.

Monomer units shown in red

polymerization

Propene

$$\underset{\text{Part of an extended polymer chain}}{-CH_2\overset{\overset{\textstyle CH_3}{|}}{CH}-CH_2\overset{\overset{\textstyle CH_3}{|}}{CH}-CH_2\overset{\overset{\textstyle CH_3}{|}}{CH}-CH_2\overset{\overset{\textstyle CH_3}{|}}{CH}-}$$
of polypropylene

$$\left(\!CH_2\overset{\overset{\textstyle CH_3}{|}}{CH}\!\right)_{\!n}$$
The repeating unit

The most common method of naming a polymer is to attach the prefix **poly-** to the name of the monomer from which the polymer is synthesized—for example, polyethylene and polystyrene. When the name of the monomer consists of two words (for example, the monomer vinyl chloride), its name is enclosed in parentheses.

$\overset{\displaystyle}{\underset{Cl}{}}$ $\xrightarrow{\text{polymerization}}$ $\left(\!\!\right)_{\!n}$

Vinyl chloride   Poly(vinyl chloride)
                 (PVC)

Table 12.2 lists several important polymers derived from ethylene and substituted ethylenes, along with their common names and most important uses.

Polyethylene films are produced by extruding molten plastic through a ringlike gap and inflating the film into a balloon.

The Stock Market

Table 12.2   Polymers Derived from Ethylene and Substituted Ethylenes

Monomer Formula	Common Name	Polymer Name(s) and Common Uses
$CH_2{=}CH_2$	ethylene	polyethylene, Polythene; break-resistant containers and packaging materials
$CH_2{=}CHCH_3$	propylene	polypropylene, Herculon; textile and carpet fibers
$CH_2{=}CHCl$	vinyl chloride	poly(vinyl chloride), PVC; construction tubing
$CH_2{=}CCl_2$	1,1-dichloroethylene	poly(1,1-dichloroethylene); Saran Wrap is a copolymer with vinyl chloride
$CH_2{=}CHCN$	acrylonitrile	polyacrylonitrile, Orlon; acrylics and acrylates
$CF_2{=}CF_2$	tetrafluoroethylene	polytetrafluoroethylene, PTFE; Teflon, nonstick coatings
$CH_2{=}CHC_6H_5$	styrene	polystyrene, Styrofoam; insulating materials
$CH_2{=}CHCOOCH_2CH_3$	ethyl acrylate	poly(ethyl acrylate), latex paint
$CH_2{=}CCOOCH_3$ $\quad\mid$ $\quad CH_3$	methyl methacrylate	poly(methyl methacrylate), Lucite; Plexiglas; glass substitutes

Some articles made from chain-growth polymers. (*a*) Saran Wrap, a copolymer of vinyl chloride and 1,1-dichloroethylene. (*b*) Plastic containers for various supermarket products, made mostly from polyethylene and polypropylene. (*c*) Teflon-coated kitchenware. (*d*) Articles made from polystyrene.

(*a*)

(*b*)

a, c, Beverly March; *b, d,* Charles D. Winters

(*c*)

(*d*)

B. Low-Density Polyethylene (LDPE)

Peroxide Any compound that contains an —O—O— bond as, for example, hydrogen peroxide, H—O—O—H

The first commercial process for ethylene polymerization used **peroxide** initiators at 500°C and 1000 atm and yielded a tough, transparent polymer known as **low-density polyethylene (LDPE).** At the molecular level, LDPE chains are highly branched, with the result that they do not pack well together and the London dispersion forces (Section 6.7A) between them are weak. LDPE softens and melts at about 115°C, which means that it cannot be used in products that are exposed to boiling water.

Today approximately 65% of all LDPE is used for the manufacture of films by the blow-molding technique illustrated in Figure 12.3. LDPE film is inexpensive, which makes it ideal for packaging such consumer items as baked goods and vegetables and for the manufacture of trash bags.

C. High-Density Polyethylene (HDPE)

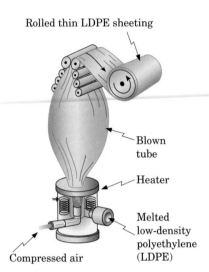

Rolled thin LDPE sheeting

Blown tube

Heater

Melted low-density polyethylene (LDPE)

Compressed air

Figure 12.3 Fabrication of LDPE film. A tube of melted LDPE along with a jet of compressed air is forced through an opening and blown into a gigantic, thin-walled bubble. The film is then cooled and taken up onto a roller. This double-walled film can be slit down the side to give LDPE film or sealed at points along its length to make LDPE bags.

An alternative method for polymerization of alkenes, and one that does not rely on peroxide initiators, was developed by Karl Ziegler of Germany and Giulio Natta of Italy in the 1950s. Polyethylene from Ziegler-Natta systems, termed **high-density polyethylene (HDPE),** has little chain branching. Consequently, its chains pack together more closely than those of LDPE, with the result that the London dispersion forces between chains of HDPE are stronger than those in LDPE. HDPE has a higher melting point than LDPE and is three to ten times stronger.

Approximately 45% of all HDPE products are made by the blow-molding process shown in Figure 12.4. HDPE is used for consumer items such as milk and water jugs, grocery bags, and squeezable bottles.

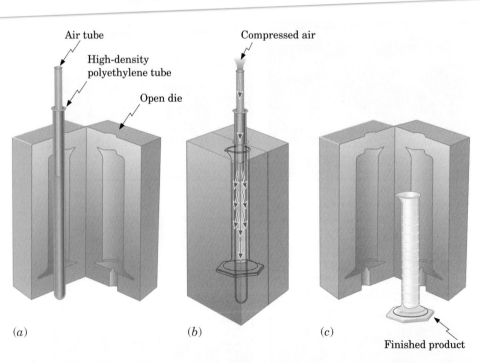

Air tube

Compressed air

High-density polyethylene tube

Open die

Finished product

(a)　　　(b)　　　(c)

Figure 12.4 Blow molding an HDPE container. (a) A short length of HDPE tubing is placed in an open die and the die is closed, sealing the bottom of the tube. (b) Compressed air is forced into the hot polyethylene/die assembly, and the tubing is literally blown up to take the shape of the mold. (c) After the assembly cools, the die is opened, and there is the container!

CHEMICAL CONNECTIONS 12D

Recycling Plastics

Plastics are polymers that can be molded when hot and that retain their shape when cooled. Because they are durable and lightweight, plastics are probably the most versatile synthetic materials in existence. In fact, the current production of plastics in the United States exceeds the U.S. production of steel. Plastics have come under criticism, however, for their role in the solid waste crisis. They account for approximately 21% of the volume and 8% of the weight of solid wastes, with most plastic waste consisting of disposable packaging and wrapping.

Six types of plastics are commonly used for packaging applications. In 1988, manufacturers adopted recycling code letters developed by the Society of the Plastics Industry as a means of identifying them.

Code	Polymer	Common Uses
1 PET	poly(ethylene terephthalate)	soft drink bottles, household chemical bottles, films, textile fibers
2 HDPE	high-density polyethylene	milk and water jugs, grocery bags, squeezable bottles
3 V	poly(vinyl chloride), PVC	shampoo bottles, pipes, shower curtains, vinyl siding, wire insulation, floor tiles
4 LDPE	low-density polyethylene	shrink wrap, trash and grocery bags, sandwich bags, squeeze bottles
5 PP	polypropylene	plastic lids, clothing fibers, bottle caps, toys, diaper linings
6 PS	polystyrene	Styrofoam cups, egg cartons, disposable utensils, packaging materials, appliances
7	all other plastics	various

Currently, only poly(ethylene terephthalate) (PET) and high-density polyethylene are recycled in large quantities. The synthesis and structure of PET, a polyester, is described in Section 19.6B.

The process for recycling most plastics is simple, with separation of the plastic from other contaminants being the most labor-intensive step. For example, PET soft drink bottles usually have a paper label and adhesive that must be removed before the PET can be reused. Recycling begins with hand or machine sorting, after which the bottles are shredded into small chips. An air cyclone then removes paper and other lightweight materials. After any remaining labels and adhesives are eliminated with a detergent wash, the PET chips are dried. The PET produced by this method is 99.9% free of contaminants and sells for about half the price of the virgin material.

Charles D. Winters

These students are wearing jackets made from recycled PET soda bottles.

SUMMARY OF KEY QUESTIONS

SECTION 12.1 What Are Alkenes and Alkynes?

- An **alkene** is an unsaturated hydrocarbon that contains a carbon–carbon double bond.
- An **alkyne** is an unsaturated hydrocarbon that contains a carbon–carbon triple bond.

SECTION 12.2 What Are the Structures of Alkenes and Alkynes?

- The structural feature that makes ***cis-trans* stereoisomerism** possible in alkenes is restricted rotation about the two carbons of the double bond. The *cis* or *trans* configuration of an alkene is determined by the orientation of the atoms of the parent chain about the double bond. If atoms of the parent chain are located on the same side of the double bond, the configuration of the alkene is *cis;* if they are located on opposite sides, the configuration is *trans*.

SECTION 12.3 How Do We Name Alkenes and Alkynes?

- In IUPAC names, the presence of a carbon–carbon double bond is indicated by a prefix showing the number of carbons in the parent chain and the ending **-ene.** Substituents are numbered and named in alphabetical order.
- The presence of a carbon–carbon triple bond is indicated by a prefix that shows the number of carbons in the parent chain and the ending **-yne.**
- The carbon atoms of a cycloalkene are numbered 1 and 2 in the direction that gives the smaller number to the first substituent.
- Compounds containing two double bonds are called dienes, those with three double bonds are called trienes, and those containing four or more double bonds are called polyenes.

SECTION 12.4 What Are the Physical Properties of Alkenes and Alkynes?

- Because alkenes and alkynes are nonpolar compounds and the only interactions between their molecules are London dispersion forces, their physical properties are similar to those of alkanes with similar carbon skeletons.

SECTION 12.5 What Are Terpenes?

- The characteristic structural feature of a **terpene** is a carbon skeleton that can be divided into two or more **isoprene units.**

SECTION 12.6 What Are the Characteristic Reactions of Alkenes?

- A characteristic reaction of alkenes is addition to the double bond.
- In addition, the double bond breaks and bonds to two new atoms or groups of atoms form in its place.
- A **reaction mechanism** is a step-by-step description of how a chemical reaction occurs, including the role of the catalyst (if one is present).
- A **carbocation** contains a carbon with only six electrons in its valence shell and bears a positive charge.

SECTION 12.7 What Are the Important Polymerization Reactions of Ethylene and Substituted Ethylenes?

- **Polymerization** is the process of bonding together many small **monomers** into large, high-molecular-weight **polymers.**

SUMMARY OF KEY REACTIONS

1. **Addition of HX (Section 12.6A)** Addition of HX to the carbon–carbon double bond of an alkene follows Markovnikov's rule. The reaction occurs in two steps and involves formation of a carbocation intermediate.

2. **Acid-Catalyzed Hydration (Section 12.6B)** Addition of H_2O to the carbon–carbon double bond of an alkene follows Markovnikov's rule. Reaction occurs in three steps and involves formation of carbocation and oxonium ion intermediates.

$$CH_3\underset{\underset{CH_3}{|}}{C}=CH_2 + H_2O \xrightarrow{H_2SO_4} CH_3\underset{\underset{OH}{|}}{\overset{\overset{CH_3}{|}}{C}}CH_3$$

3. **Addition of Bromine and Chlorine (Halogenation) (Section 12.6C)** Addition to a cycloalkene gives a 1,2-dihalocycloalkane.

Cyclohexene   1,2-Dibromocyclohexane

4. Reduction: Formation of Alkanes (Hydrogenation) (Section 12.6D) Catalytic reduction involves addition of hydrogen to form two new C—H bonds.

5. Polymerization of Ethylene and Substituted Ethylenes (Section 12.7A) In polymerization of alkenes, monomer units bond together without the loss of any atoms.

$$nCH_2{=}CH_2 \xrightarrow{\text{initiator}} {-}\left(CH_2CH_2\right)_n{-}$$

PROBLEMS

GOB
Chemistry Now™

Assess your understanding of this chapter's topics with additional quizzing and conceptual-based problems at **http://now.brookscole.com/gob8** or on the CD.

A blue problem number indicates an applied problem.

■ denotes problems that are available on the GOB ChemistryNow website or CD and are assignable in OWL.

SECTION 12.2 What Are the Structures of Alkenes and Alkynes?

12.11 What is the difference in structure between a saturated hydrocarbon and an unsaturated hydrocarbon?

12.12 Each carbon atom in ethane and in ethylene is surrounded by eight valence electrons and has four bonds to it. Explain how the VSEPR model (Section 4.10) predicts a bond angle of 109.5° about each carbon in ethane but an angle of 120° about each carbon in ethylene.

12.13 Predict all bond angles about each highlighted carbon atom.

(c) HC≡C—CH=CH₂ (d)

12.14 Predict all bond angles about each highlighted carbon atom.

SECTION 12.3 How Do We Name Alkenes and Alkynes?

12.15 Draw a structural formula for each compound.
(a) *trans*-2-Methyl-3-hexene
(b) 2-Methyl-3-hexyne
(c) 2-Methyl-1-butene
(d) 3-Ethyl-3-methyl-1-pentyne
(e) 2,3-Dimethyl-2-pentene
(f) *cis*-2-Hexene

12.16 Draw a structural formula for each compound.
(a) 3-Chloropropene
(b) 3-Methylcyclohexene
(c) 1,2-Dimethylcyclohexene
(d) *trans*-3,4-Dimethyl-3-heptene
(e) Cyclopropene
(f) 3-Hexyne

12.17 ■ Write the IUPAC name for each unsaturated hydrocarbon.
(a) CH₂=CH(CH₂)₄CH₃

(b)

(c)

(d) (CH₃)₂CHCH=C(CH₃)₂
(e) CH₃(CH₂)₅C≡CH
(f) CH₃CH₂C≡CC(CH₃)₃

12.18 Write the IUPAC name for each unsaturated hydrocarbon.

(c) CH₃CH₂CCH₃ (with CH₂ double-bonded above the third carbon)

(d)
$$\begin{array}{c} CH_3CH_2CH_2 \\ \diagdown \\ C{=}CH_2 \\ \diagup \\ CH_3CH_2CH_2 \end{array}$$

12.19 Explain why each name is incorrect, and then write a correct name.

(a) 1-Methylpropene   (b) 3-Pentene

(c) 2-Methylcyclohexene   (d) 3,3-Dimethylpentene

(e) 4-Hexyne   (f) 2-Isopropyl-2-butene

12.20 Explain why each name is incorrect, and then write a correct name.

(a) 2-Ethyl-1-propene

(b) 5-Isopropylcyclohexene

(c) 4-Methyl-4-hexene

(d) 2-sec-Butyl-1-butene

(e) 6,6-Dimethylcyclohexene

(f) 2-Ethyl-2-hexene

SECTION 12.2 *Cis-Trans* Stereoisomerism in Alkenes

12.21 What structural feature in alkenes makes *cis-trans* isomerism in them possible? What structural feature in cycloalkanes makes *cis-trans* isomerism in them possible? What do these two structural features have in common?

12.22 ■ Which of these alkenes show *cis-trans* isomerism? For each that does, draw structural formulas for both isomers.

(a) 1-Hexene   (b) 2-Hexene

(c) 3-Hexene   (d) 2-Methyl-2-hexene

(e) 3-Methyl-2-hexene   (f) 2,3-Dimethyl-2-hexene

12.23 Which of these alkenes shows *cis-trans* isomerism? For each that does, draw structural formulas for both isomers.

(a) 1-Pentene

(b) 2-Pentene

(c) 3-Ethyl-2-pentene

(d) 2,3-Dimethyl-2-pentene

(e) 2-Methyl-2-pentene

(f) 2,4-Dimethyl-2-pentene

12.24 Name and draw structural formulas for all alkenes with the molecular formula C_5H_{10}. As you draw these alkenes, remember that *cis* and *trans* isomers are different compounds and must be counted separately.

12.25 Arachidonic acid is a naturally occurring polyunsaturated fatty acid. Draw a line-angle formula for arachidonic acid showing the *cis* configuration about each double bond.

$$CH_3(CH_2)_4(CH=CHCH_2)_4CH_2CH_2COOH$$

Arachidonic acid

12.26 Following is the structural formula of a naturally occurring unsaturated fatty acid.

$$CH_3(CH_2)_7CH=CH(CH_2)_7COOH$$

The *cis* stereoisomer is named oleic acid and the *trans* isomer is named elaidic acid. Draw a line-angle formula of each acid, showing clearly the configuration of the carbon–carbon double bond in each.

12.27 ■ For each molecule that shows *cis-trans* isomerism, draw the *cis* isomer.

12.28 Draw structural formulas for all compounds with the molecular formula C_5H_{10} that are

(a) Alkenes that do not show *cis-trans* isomerism

(b) Alkenes that show *cis-trans* isomerism

(c) Cycloalkanes that do not show *cis-trans* isomerism

(d) Cycloalkanes that show *cis-trans* isomerism

12.29 β-Ocimene, a triene found in the fragrance of cotton blossoms and several essential oils, has the IUPAC name *cis*-3,7-dimethyl-1,3,6-octatriene. (*Cis* refers to the configuration of the double bond between carbons 3 and 4, the only double bond in this molecule about which *cis-trans* isomerism is possible). Draw a structural formula for β-ocimene.

SECTION 12.5 What Are Terpenes?

12.30 Which of these terpenes show *cis-trans* isomerism?

(a) Myrcene   (b) Geraniol

(c) Limonene   (d) Farnesol

12.31 Show that the structural formula of vitamin A (Section 12.3E) can be divided into four isoprene units joined by head-to-tail linkages and cross-linked at one point to form the six-membered ring.

SECTION 12.6 What Are the Characteristic Reactions of Alkenes?

12.32 Define *alkene addition reaction*. Write an equation for an addition reaction of propene.

12.33 ■ What reagent and/or catalysts are necessary to bring about each conversion?

12.34 ■ Complete these equations.

(a) —CH$_2$CH$_3$ + HCl $\longrightarrow$

(b) —CH$_2$CH$_3$ + H$_2$O $\xrightarrow{\text{H}_2\text{SO}_4}$

(c) CH$_3$(CH$_2$)$_5$CH=CH$_2$ + HI $\longrightarrow$

(d) ⬡—C(=CH$_2$)(CH$_3$) + HCl $\longrightarrow$

(e) CH$_3$CH=CHCH$_2$CH$_3$ + H$_2$O $\xrightarrow{\text{H}_2\text{SO}_4}$

(f) CH$_2$=CHCH$_2$CH$_2$CH$_3$ + H$_2$O $\xrightarrow{\text{H}_2\text{SO}_4}$

12.35 ■ Draw structural formulas for all possible carbo-cations formed by the reaction of each alkene with HCl. Label each carbocation as primary, secondary, or tertiary.

(a) CH$_3$CH$_2$C(CH$_3$)=CHCH$_3$ (b) CH$_3$CH$_2$CH=CHCH$_3$

(c) cyclopentene with —CH$_3$ (d) cyclohexane ring =CH$_2$

12.36 Draw a structural formula for the product formed by treatment of 2-methyl-2-pentene with each reagent.
(a) HCl   (b) H$_2$O in the presence of H$_2$SO$_4$

12.37 ■ Draw a structural formula for the product of each reaction.
(a) 1-Methylcyclohexene + Br$_2$
(b) 1,2-Dimethylcyclopentene + Cl$_2$

12.38 Draw a structural formula for an alkene with the indicated molecular formula that gives the compound shown as the major product. Note that more than one alkene may give the same compound as the major product.

(a) C$_5$H$_{10}$ + H$_2$O $\xrightarrow{\text{H}_2\text{SO}_4}$ CH$_3$C(CH$_3$)(OH)CH$_2$CH$_3$

(b) C$_5$H$_{10}$ + Br$_2$ $\longrightarrow$ CH$_3$CH(CH$_3$)CHCH$_2$ (Br Br)

(c) C$_7$H$_{12}$ + HCl $\longrightarrow$ cyclohexane with CH$_3$ and Cl

12.39 Draw a structural formula for an alkene with the molecular formula C$_5$H$_{10}$ that reacts with Br$_2$ to give each product.

(a) CH$_3$C(CH$_3$)—CHCH$_3$ (Br Br)   (b) CH$_2$C(CH$_3$)CH$_2$CH$_3$ (Br Br)

(c) CH$_2$CHCH$_2$CH$_2$CH$_3$ (Br Br)

12.40 Draw a structural formula for an alkene with the molecular formula C$_5$H$_{10}$ that reacts with HCl to give the indicated chloroalkane as the major product. More than one alkene may give the same compound as the major product.

(a) CH$_3$C(CH$_3$)(Cl)CH$_2$CH$_3$ (b) CH$_3$CH(CH$_3$)CHCH$_3$ (Cl)

(c) CH$_3$CHCH$_2$CH$_2$CH$_3$ (Cl)

12.41 Draw the structural formula of an alkene that undergoes acid-catalyzed hydration to give the indicated alcohol as the major product. More than one alkene may give each alcohol as the major product.
(a) 3-Hexanol
(b) 1-Methylcyclobutanol
(c) 2-Methyl-2-butanol
(d) 2-Propanol

12.42 Terpin, C$_{10}$H$_{20}$O$_2$, is prepared commercially by the acid-catalyzed hydration of limonene (Figure 12.1).
(a) Propose a structural formula for terpin.
(b) How many *cis-trans* isomers are possible for the structural formula you propose?
(c) Terpin hydrate, the isomer of terpin in which the methyl and isopropyl groups are *trans* to each other, is used as an expectorant in cough medicines. Draw a structural formula for terpin hydrate showing the *trans* orientation of these groups.

12.43 Draw the product formed by treatment of each alkene with H$_2$/Ni.

12.44 Hydrocarbon A, C_5H_8, reacts with 2 moles of Br_2 to give 1,2,3,4-tetrabromo-2-methylbutane. What is the structure of hydrocarbon A?

12.45 Show how to convert ethylene to these compounds.
  (a) Ethane   (b) Ethanol
  (c) Bromoethane   (d) 1,2-Dibromoethane
  (e) Chloroethane

12.46 ■ Show how to convert 1-butene to these compounds.
  (a) Butane   (b) 2-Butanol
  (c) 2-Bromobutane   (d) 1,2-Dibromobutane

Chemical Connections

12.47 (Chemical Connections 12A) What is one function of ethylene as a plant growth regulator?

12.48 (Chemical Connections 12B) What is the meaning of the term *pheromone?*

12.49 (Chemical Connections 12B) What is the molecular formula of 11-tetradecenyl acetate? What is its molecular weight?

12.50 (Chemical Connections 12B) Assume that 1×10^{-12} g of 11-tetradecenyl acetate is secreted by a single corn borer. How many molecules is this?

12.51 (Chemical Connections 12C) What different functions are performed by the rods and cones in the eye?

12.52 (Chemical Connections 12C) In which isomer of retinal is the end-to-end distance longer, the all-*trans* isomer or the 11-*cis* isomer?

12.53 (Chemical Connections 12D) What types of consumer products are made of high-density polyethylene? What types of products are made of low-density polyethylene? One type of polyethylene is currently recyclable and the other is not. Which is which?

12.54 (Chemical Connections 12D) In recycling codes, what do these abbreviations stand for?
  (a) V   (b) PP   (c) PS

Additional Problems

12.55 Write line-angle formulas for all compounds with the molecular formula C_4H_8. Which are sets of constitutional isomers? Which are sets of *cis-trans* isomers?

12.56 Name and draw structural formulas for all alkenes with the molecular formula C_6H_{12} that have these carbon skeletons. Remember to consider *cis* and *trans* isomers.

(a)
$$\begin{array}{c} \quad\quad\quad C \\ \quad\quad\quad | \\ C-C-C-C-C \end{array}$$

(b)
$$\begin{array}{c} \quad\quad C \quad C \\ \quad\quad | \quad | \\ C-C-C-C \end{array}$$

(c)
$$\begin{array}{c} \quad\quad C \\ \quad\quad | \\ C-C-C-C \\ \quad\quad | \\ \quad\quad C \end{array}$$

12.57 Following is the structural formula of lycopene, $C_{40}H_{56}$, a deep-red compound that is partially responsible for the red color of ripe fruits, especially tomatoes. Approximately 20 mg of lycopene can be isolated from 1 kg of fresh, ripe tomatoes.
  (a) Show that lycopene is a terpene; that is, its carbon skeleton can be divided into two sets of four isoprene units with the units in each set joined head-to-tail.
  (b) How many of the carbon–carbon double bonds in lycopene have the possibility for *cis-trans* isomerism? Lycopene is the all-*trans* isomer.

12.58 As you might suspect, β-carotene, $C_{40}H_{56}$, a precursor to vitamin A, was first isolated from carrots. Dilute solutions of β-carotene are yellow—hence its use as a food coloring. In plants, this compound is almost always present in combination with chlorophyll to assist in the harvesting of the energy of sunlight. As tree leaves die in the fall, the green of their chlorophyll molecules is replaced by the yellows and reds of carotene and carotene-related molecules. Compare the carbon skeletons of β-carotene and lycopene. What are the similarities? What are the differences?

12.59 Draw the structural formula for a cycloalkene with the molecular formula C_6H_{10} that reacts with Cl_2 to give each compound.

Lycopene

β-Carotene

12.60 Propose a structural formula for the product(s) when each of the following alkenes is treated with H_2O/H_2SO_4. Why are two products formed in part (b), but only one in parts (a) and (c)?

(a) 1-Hexene gives one alcohol with a molecular formula of $C_6H_{14}O$.

(b) 2-Hexene gives two alcohols, each with a molecular formula of $C_6H_{14}O$.

(c) 3-Hexene gives one alcohol with a molecular formula of $C_6H_{14}O$.

12.61 *cis*-3-Hexene and *trans*-3-hexene are different compounds and have different physical and chemical properties. Yet, when treated with H_2O/H_2SO_4, each gives the same alcohol. What is this alcohol, and how do you account for the fact that each alkene gives the same one?

12.62 Draw the structural formula of an alkene that undergoes acid-catalyzed hydration to give each of the following alcohols as the major product. More than one alkene may give each compound as the major product.

12.63 Show how to convert cyclopentene into these compounds.

(a) 1,2-Dibromocyclopentane

(b) Cyclopentanol

(c) Iodocyclopentane

(d) Cyclopentane

Looking Ahead

12.64 Knowing what you do about the meaning of the terms "saturated" and "unsaturated" as applied to alkanes and alkenes, what do you think these same terms mean when they are used to describe animal fats such as those found in butter and animal meats? What might the term "polyunsaturated" mean in this same context?

12.65 In Chapter 21 on the biochemistry of lipids, we will study the three long-chain unsaturated carboxylic acids shown below. Each has 18 carbons and is a component of animal fats, vegetable oils, and biological membranes. Because of their presence in animal fats, they are called fatty acids. How many stereoisomers are possible for each fatty acid?

Oleic acid $CH_3(CH_2)_7CH=CH(CH_2)_7COOH$

Linoleic acid $CH_3(CH_2)_4(CH=CHCH_2)_2(CH_2)_6COOH$

Linolenic acid $CH_3CH_2(CH=CHCH_2)_3(CH_2)_6COOH$

12.66 The fatty acids in Problem 12.65 occur in animal fats, vegetable oils, and biological membranes almost exclusively as the all-*cis* isomers. Draw line-angle formulas for each fatty acid showing the *cis* configuration about each carbon–carbon double bond.

Benzene and Its Derivatives

Peppers of the *Capsicum* family (see Chemical Connections 13F).

GOB Chemistry ♦ Now™

Look for this logo in the chapter and go to GOB ChemistryNow at **http://now.brookscole.com/gob8** or on the CD for tutorials, simulations, and problems.

Aromatic compound Benzene or one of its derivatives

Arene A compound containing one or more benzene rings

Aryl group A group derived from an arene by removal of a H atom and given the symbol **Ar**—

Ar— The symbol used for an aryl group

13.1 | What Is the Structure of Benzene?

So far we have described three classes of hydrocarbons—alkanes, alkenes, and alkynes—collectively called aliphatic hydrocarbons. More than 150 years ago, organic chemists realized that yet another class of hydrocarbons exists, one whose properties are quite different from those of aliphatic hydrocarbons. Because some of these new hydrocarbons have pleasant odors, they were called **aromatic compounds.** Today we know that not all aromatic compounds share this characteristic. Some do have pleasant odors, but some have no odor at all, and others have downright unpleasant odors. A more appropriate definition of an aromatic compound is any compound that has one or more benzene-like rings.

We use the term **arene** to describe aromatic hydrocarbons. Just as a group derived by removal of an H from an alkane is called an alkyl group and given the symbol R—, so a group derived by removal of an H from an arene is called an **aryl group** and given the symbol **Ar**—.

Benzene, the simplest aromatic hydrocarbon, was discovered by Michael Faraday (1791–1867) in 1825. Its structure presented an immediate problem to chemists of the day. Benzene has the molecular formula C_6H_6, and a compound with so few hydrogens for its six carbons (compare hexane, C_6H_{14}), chemists argued, should be unsaturated. But benzene does not behave like an alkene (the only class of unsaturated hydrocarbons known at that time.) Whereas 1-hexene, for example, reacts instantly with Br_2 (Section 12.6C), benzene does not react at all with this reagent. Nor does benzene react with HBr, H_2O/H_2SO_4, or H_2/Pd—all reagents that normally add to carbon–carbon double bonds.

Benzene is an important compound in both the chemical industry and the laboratory, but it must be handled carefully. Not only is it poisonous if ingested in liquid form, but its vapor is also toxic and can be absorbed either by breathing or through the skin. Long-term inhalation can cause liver damage and cancer.

A. Kekulé's Structure of Benzene

The first structure for benzene was proposed by Friedrich August Kekulé in 1872 and consisted of a six-membered ring with alternating single and double bonds, with one hydrogen bonded to each carbon.

A Kekulé structure showing all atoms

A Kekulé structure as a line-angle drawing

Chemistry⚛Now™
Click *Coached Problems* to explore the **Bonding in Benzene**

Although Kekulé's proposal was consistent with many of the chemical properties of benzene, it was contested for years. The major objection was its failure to account for the unusual chemical behavior of benzene. If benzene contains three double bonds, Kekulé's critics asked, why doesn't it undergo reactions typical of alkenes?

B. Resonance Structure of Benzene

The concept of resonance, developed by Linus Pauling in the 1930s, provided the first adequate description of the structure of benzene. According to the theory of resonance, certain molecules and ions are best described by writing two or more Lewis structures and considering the real molecule or ion to be a **resonance hybrid** of these structures. Each individual Lewis structure is called a **contributing structure.** To show that the real molecule is a resonance hybrid of the two Lewis structures, we position a double-headed arrow between them.

Resonance hybrid A molecule best described as a composite of two or more Lewis structures

The two contributing structures for benzene are often called Kekulé structures.

Alternative Lewis contributing structures for benzene

The resonance hybrid has some of the characteristics of each Lewis contributing structure. For example, the carbon–carbon bonds are neither single nor double but rather something intermediate between the two extremes. It has been determined experimentally that the length of the carbon–carbon bond in benzene is not as long as a carbon–carbon single bond nor as short as a carbon–carbon double bond, but rather midway between the two. The closed loop of six electrons (two from the second bond of each double bond) characteristic of a benzene ring is sometimes called an **aromatic sextet.**

Wherever we find resonance, we find stability. The real structure is generally more stable than any of the hypothetical Lewis contributing structures. The benzene ring is greatly stabilized by resonance, which explains why it does not undergo the addition reactions typical of alkenes.

13.2 | How Do We Name Aromatic Compounds?

A. One Substituent

Monosubstituted alkylbenzenes are named as derivatives of benzene—for example, ethylbenzene. The IUPAC system retains certain common names for several of the simpler monosubstituted alkylbenzenes, including **toluene** and **styrene.**

The IUPAC system also retains common names for the following compounds:

The substituent group derived by loss of an H from benzene is called a **phenyl group,** C_6H_5—, the common symbol for which is **Ph**—. In molecules containing other functional groups, phenyl groups are often named as substituents.

Phenyl group C_6H_5—, the aryl group derived by removing a hydrogen atom from benzene

B. Two Substituents

When two substituents occur on a benzene ring, three isomers are possible. We locate the substituents either by numbering the atoms of the ring or by using the locators **ortho (o)**, **meta (m)**, and **para (p)**. The numbers 1,2- are equivalent to *ortho* (Greek: straight); 1,3- to *meta* (Greek: after); and 1,4- to *para* (Greek: beyond).

1,2- or *ortho-*   1,3- or *meta*   1,4- or *para*

When one of the two substituents on the ring imparts a special name to the compound (for example, —CH_3, —OH, —NH_2, or —COOH), then we name the compound as a derivative of that parent molecule, and assume that the substituent occupies ring position number 1. The IUPAC system retains the common name **xylene** for the three isomeric dimethylbenzenes. Where neither substituent imparts a special name, we locate the two substituents and list them in alphabetical order before the ending "benzene." The carbon of the benzene ring with the substituent of lower alphabetical ranking is numbered C—1.

p-Xylene is a starting material for the synthesis of poly(ethylene terephthalate). Consumer products derived from this polymer include Dacron polyester fibers and Mylar films (Section 19.6B).

4-Bromobenzoic acid
(*p*-Bromobenzoic acid)

3-Chloroaniline
(*m*-Chloroaniline)

1,3-Dimethylbenzene
(*m*-Xylene)

1-Chloro-4-ethylbenzene
(*p*-Chloroethylbenzene)

C. Three or More Substituents

When three or more substituents are present on a benzene ring, specify their locations by numbers. If one of the substituents imparts a special name, then name the molecule as a derivative of that parent molecule. If none of the substituents imparts a special name, then locate the substituents, number them to give the smallest set of numbers, and list them in alphabetical order before the ending "benzene." In the following examples, the first compound is a derivative of toluene, and the second is a derivative of phenol. Because no substituent in the third compound imparts a special name, list its three substituents in alphabetical order followed by the word "benzene."

4-Chloro-2-nitrotoluene   2,4,6-Tribromophenol   2-Bromo-1-ethyl-4-nitrobenzene

CHEMICAL CONNECTIONS 13A

DDT: A Boon and a Curse

Probably the best-known insecticide worldwide is *di*chloro*di*phenyl*tri*chloroethane (not an IUPAC name), commonly abbreviated DDT.

Dichlorodiphenyltrichloroethane
(DDT)

This compound was first prepared in 1874, but it was not until the late 1930s that its potential as an insecticide was recognized. First used for this purpose in 1939, it proved extremely effective in ridding large areas of the world of the insect hosts that transmit malaria and typhus. In addition, it was so effective in killing crop-destroying insect pests that crop yields in many areas of the world increased dramatically.

Widespread use of DDT, however, has been a double-edged sword. Despite DDT's well-known benefits, it has an enormous disadvantage. Because it resists biodegradation, it remains in the soil for years—and this persistence in the environment creates the problem. Scientists estimate that the tissues of adult humans contain, on average, five to ten parts per million of DDT.

The dangers associated with the persistence of DDT in the environment were dramatically portrayed by Rachel Carson in her 1962 book *The Silent Spring*, which documented serious declines in the populations of eagles and other raptors as well as many other kinds of birds. Scientists discovered soon thereafter that DDT

DDT was sprayed on crops in the United States until its use was banned in the 1970s.

inhibits the mechanism by which these birds incorporate calcium into their eggshells; as a result, the shells become so thin and weak that they break during incubation.

Because of these problems, almost all nations have now banned the use of DDT for agricultural purposes. It is still used, however, in some areas to control the population of disease-spreading insects.

Courtesy of the U.S. Department of Agriculture

EXAMPLE 13.1

Write names for these compounds.

(a)   (b)   (c)

Solution
(a) The parent is toluene, and the compound is 3-iodotoluene or *m*-iodotoluene.
(b) The parent is benzoic acid, and the compound is 3,5-dibromobenzoic acid.
(c) The parent is aniline, and the compound is 4-chloroaniline or *p*-chloroaniline.

Problem 13.1

Write names for these compounds.

(a)   (b)   (c)

D. Polynuclear Aromatic Hydrocarbons

Polynuclear aromatic hydrocarbons (PAHs) contain two or more benzene rings, with each pair of rings sharing two adjacent carbon atoms. Naphthalene, anthracene, and phenanthrene, the most common PAHs, and substances derived from them are found in coal tar and high-boiling petroleum residues. At one time, naphthalene was used as mothballs and an insecticide in preserving woolens and furs, but its use decreased after the introduction of chlorinated hydrocarbons such as *p*-dichlorobenzene.

Polynuclear aromatic hydrocarbon A hydrocarbon containing two or more benzene rings, each of which shares two carbon atoms with another benzene ring

Naphthalene   Anthracene   Phenanthrene

CHEMICAL CONNECTIONS 13B

Carcinogenic Polynuclear Aromatics and Smoking

A **carcinogen** is a compound that causes cancer. The first carcinogens to be identified were a group of polynuclear aromatic hydrocarbons, all of which have at least four aromatic rings. Among them is benzo[a]pyrene, one of the most carcinogenic of the aromatic hydrocarbons. It forms whenever incomplete combustion of organic compounds occurs. Benzo[a]pyrene is found, for example, in cigarette smoke, automobile exhaust, and charcoal-broiled meats.

Benzo[a]pyrene causes cancer in the following way. Once it is absorbed or ingested, the body attempts to convert it into a more water-soluble compound that can be excreted easily. By a series of enzyme-catalyzed reactions, benzo[a]pyrene is transformed into a **diol** (two —OH groups) **epoxide** (a three-membered ring, one atom of which is oxygen). This compound can bind to DNA by reacting with one of its amino groups, thereby altering the structure of DNA and producing a cancer-causing mutation.

Benzo[a]pyrene   A diol epoxide

13.3 | What Are the Characteristic Reactions of Benzene and Its Derivatives?

By far the most characteristic reaction of aromatic compounds is substitution at a ring carbon, which we give the name **aromatic substitution.** Groups we can introduce directly on the ring include the halogens, the nitro ($-NO_2$) group, and the sulfonic acid ($-SO_3H$) group.

A. Halogenation

As noted in Section 13.1, chlorine and bromine do not react with benzene, in contrast to their instantaneous reaction with cyclohexene and other alkenes (Section 12.6C). In the presence of an iron catalyst, however, chlorine reacts rapidly with benzene to give chlorobenzene and HCl:

Benzene   Chlorobenzene

Treatment of benzene with bromine in the presence of $FeCl_3$ results in formation of bromobenzene and HBr.

B. Nitration

When we heat benzene or one of its derivatives with a mixture of concentrated nitric and sulfuric acids, one of the hydrogen atoms bonded to the ring is replaced by a nitro ($-NO_2$) group.

Nitrobenzene

CHEMICAL CONNECTIONS 13C

Iodide Ion and Goiter

One hundred years ago, goiter, an enlargement of the thyroid gland caused by iodine deficiency, was common in the central United States and central Canada. This disease results from underproduction of thyroxine, a hormone synthesized in the thyroid gland. Young mammals require this hormone for normal growth and development. A deficiency of thyroxine during fetal development results in mental retardation. Low levels of thyroxine in adults result in hypothyroidism, commonly called goiter, the symptoms of which are lethargy, obesity, and dry skin.

Iodine is an element that comes primarily from the sea. Rich sources of it, therefore, are fish and other seafoods. The iodine in our diets that doesn't come from the sea most commonly is derived from food additives. Most of the iodide ion in the North American diet comes from table salt fortified with sodium iodide, commonly referred to as iodized salt. Another source is dairy products, which accumulate iodide because of the iodine-containing additives used in cattle feeds and the iodine-containing disinfectants used on milking machines and milk storage tanks.

Thyroxine

CHEMICAL CONNECTIONS 13D

The Nitro Group in Explosives

Treatment of toluene with three moles of nitric acid in the presence of sulfuric acid as a catalyst results in nitration of toluene three times to form the explosive 2,4,6-trinitrotoluene, TNT. The presence of these three nitro groups confers the explosive properties to TNT. Similarly, the presence of three nitro groups leads to the explosive properties of nitroglycerin.

In recent years several new explosives have been discovered, all of which contain multiple nitro groups. Among them are RDX and PETN. The plastic explosive Semtex, for example, is a mixture of RDX and PETN. It was used in the destruction of Pan Am flight 103 over Lockerbie, Scotland, in December 1988.

2,4,6-Trinitrotoluene
(TNT)

Trinitroglycerin
(Nitroglycerin)

Cyclonite
(RDX)

Pentaerythritol tetranitrate
(PETN)

A particular value of nitration is that we can reduce the resulting —NO$_2$ group to a primary amino group, —NH$_2$, by catalytic reduction using hydrogen in the presence of a transition-metal catalyst. In the following example, neither the benzene ring nor the carboxyl group is affected by these experimental conditions:

4-Nitrobenzoic acid
(*p*-Nitrobenzoic acid)

4-Aminobenzoic acid
(*p*-Aminobenzoic acid, PABA)

Bacteria require *p*-aminobenzoic acid for the synthesis of folic acid (Section 30.4), which is in turn required for the synthesis of the heterocyclic aromatic amine bases of nucleic acids (Section 25.2). Whereas bacteria can synthesize folic acid from *p*-aminobenzoic acid, folic acid is a vitamin for humans and must be obtained through the diet. Chemical Connections 22D describes how sulfa drugs inhibit the synthesis of this vitamin.

C. Sulfonation

Heating an aromatic compound with concentrated sulfuric acid results in formation of an arenesulfonic acid, all of which are strong acids, comparable in strength to sulfuric acid.

Benzenesulfonic acid

A major use of sulfonation is in the preparation of synthetic detergents, an important example of which is sodium 4-dodecybenzenesulfonate. To prepare this type of detergent, a linear alkylbenzene such as dodecylbenzene is treated with concentrated sulfuric acid to give an alkylbenzenesulfonic acid. The sulfonic acid is then neutralized with sodium hydroxide.

Dodecylbenzene   Sodium 4-dodecylbenzenesulfonate, SDS
(an anionic detergent)

Alkylbenzenesulfonate detergents were introduced in the late 1950s, and today they claim nearly 90% of the market once held by natural soaps. Section 18.4 discusses the chemistry and cleansing action of soaps and detergents.

13.4 | What Are Phenols?

A. Structure and Nomenclature

Phenol A compound that contains an —OH group bonded to a benzene ring

The functional group of a **phenol** is a hydroxyl group bonded to a benzene ring. Substituted phenols are named either as derivatives of phenol or by common names.

Phenol in crystalline form.

Phenol   3-Methylphenol   1,2-Benzenediol   1,3-Benzenediol   1,4-Benzenediol
     (m-Cresol)   (Catechol)   (Resorcinol)   (Hydroquinone)

Phenols are widely distributed in nature. Phenol itself and the isomeric cresols (o-, m-, and p-cresol) are found in coal tar. Thymol and vanillin are important constituents of thyme and vanilla beans, respectively. Urushiol is the main component of the irritating oil of poison ivy. It can cause severe contact dermatitis in sensitive individuals.

Poison ivy.

2-Isopropyl-5-
methylphenol
(Thymol)

4-Hydroxy-3-methoxy-
benzaldehyde
(Vanillin)

Urushiol

B. Acidity of Phenols

Phenols are weak acids, with pK_a values of approximately 10 (Table 9.3). Most phenols are insoluble in water, but they react with strong bases, such as NaOH and KOH, to form water-soluble salts.

Phenol
$pK_a = 9.95$
(stronger acid)

Sodium
hydroxide
(stonger base)

Sodium phenoxide
(weaker base)

Water
$pK_a = 15.7$
(weaker acid)

CHEMICAL CONNECTIONS 13E

FD & C No. 6 (a.k.a. Sunset Yellow)

Did you ever wonder what gives gelatin desserts their red, green, orange, or yellow color? Or what gives margarines their yellow color? Or what gives maraschino cherries their red color? If you read the content labels, you will see code names such as FD & C Yellow No. 6 and FD & C Red No. 40.

At one time, the only colorants for foods were compounds obtained from plant or animal materials. Beginning as early as the 1890s, however, chemists discovered a series of synthetic food dyes that offered several advantages over natural dyes, such as greater brightness, better stability, and lower cost. Opinion remains divided on the safety of their use. No synthetic food colorings are allowed, for example, in Norway and Sweden. In the United States, the Food and Drug Administration (FDA) has certified seven synthetic dyes for use in foods, drugs, and cosmetics (FD & C)—two yellows, two

reds, two blues, and one green. When these dyes are used alone or in combinations, they can approximate the color of almost any natural food.

Following are structural formulas for Allura Red (Red No. 40) and Sunset Yellow (Yellow No. 6). These and the other five food dyes certified in the United States have in common three or more benzene rings and two or more ionic groups, either the sodium salt of a carboxylic acid group, $—COO^-Na^+$, or the sodium salt of a sulfonic acid group, $—SO_3^-Na^+$. These ionic groups make the dyes soluble in water.

To return to our original questions, maraschino cherries are colored with FD & C Red No. 40, and margarines are colored with FD & C Yellow No 6. Gelatin desserts use either one or a combination of the seven certified food dyes to create their color.

Allura Red
(FD & C Red No. 40)

Sunset Yellow
(FD & C Yellow No. 6)

Most phenols are such weak acids that they do not react with weak bases such as sodium bicarbonate; that is, they do not dissolve in aqueous sodium bicarbonate.

C. Phenols as Antioxidants

An important reaction for living systems, foods, and other materials that contain carbon–carbon double bonds is **autoxidation**—that is, oxidation requiring oxygen and no other reactant. If you open a bottle of cooking oil that has stood for a long time, you will notice a hiss of air entering the bottle. This sound occurs because the consumption of oxygen by autoxidation of the oil creates a negative pressure inside the bottle.

Cooking oils contain esters of polyunsaturated fatty acids. We will discuss the structure and chemistry of esters in Chapter 19. The important point here is that all vegetable oils contain fatty acids with long hydrocarbon chains, many of which have one or more carbon–carbon double bonds (see Problems 12.25 and 12.65 for the structures of four of these fatty acids). Autoxidation takes place adjacent to these double bonds.

$$CH_2CH{=}CH{-}\overset{\overset{\displaystyle H}{|}}{CH}{-} \ + \ O_2 \ \xrightarrow[\text{or heat}]{\text{Light}} \ {-}CH_2CH{=}CH{-}\overset{\overset{\displaystyle O{-}O{-}H}{|}}{CH}{-}$$

Section of a fatty
acid hydrocarbon chain

A hydroperoxide

CHEMICAL CONNECTIONS 13F

Capsaicin, for Those Who Like It Hot

Capsaicin, the pungent principle from the fruit of various species of peppers (*Capsicum* and *Solanaceae*), was isolated in 1876, and its structure was determined in 1919. Capsaicin contains both a phenol and a phenol ether.

Capsaicin
(from various types of peppers)

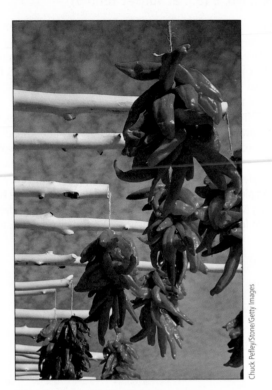

Chuck Pefley/Stone/Getty Images

Red chilies being dried in the sun.

The inflammatory properties of capsaicin are well known; the human tongue can detect as little as one drop in 5 L of water. We all know of the burning sensation in the mouth and sudden tearing in the eyes caused by a good dose of hot chili peppers. For this reason, capsaicin-containing extracts from these flaming foods are used in sprays to ward off dogs or other animals that might nip at your heels while you are running or cycling.

Paradoxically, capsaicin is able both to cause and relieve pain. Currently, two capsaicin-containing creams, Mioton and Zostrix, are prescribed to treat the burning pain associated with postherpetic neuralgia, a complication of the disease known as shingles. They are also prescribed for diabetics to relieve persistent foot and leg pain.

Active Figure Autoxidation is a radical-chain reaction converting a hydrocarbon to a hydroperoxide. **See a simulation based on this figure, and take a short quiz on the concepts at http://www.now.brookscole .com/gob8 or on the CD.**

Autoxidation is a radical-chain process that converts an R—H group to an R—O—O—H group, called a hydroperoxide. This process begins when a hydrogen atom with one of its electrons (H·) is removed from a carbon adjacent to one of the double bonds in a hydrocarbon chain. The carbon losing the H· has only seven electrons in its valence shell, one of which is unpaired. An atom or molecule with an unpaired electron is called a **radical.**

Step 1: Chain Initiation—Formation of a Radical from a Nonradical Compound

Removal of a hydrogen atom (H·) may be initiated by light or heat. The product formed is a carbon radical; that is, it contains a carbon atom with one unpaired electron.

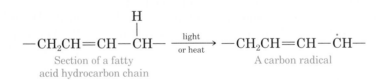

Step 2a: Chain Propagation—Reaction of a Radical to Form a New Radical

The carbon radical reacts with oxygen, itself a diradical, to form a hydroperoxy radical. The new covalent bond of the hydroperoxy radical forms by the combination of one electron from the carbon radical and one electron from the oxygen diradical.

Step 2b: Chain Propagation—Reaction of a Radical to Form a New Radical

The hydroperoxy radical removes a hydrogen atom (H·) from a new fatty acid hydrocarbon chain to complete the formation of a hydroperoxide and at the same time produce a new carbon radical.

The most important point about the pair of chain propagation steps (Steps 2a and 2b) is that they form a continuous cycle of reactions. The new radical formed in Step 2b next reacts with another molecule of O_2 by Step 2a to give a new hydroperoxy radical. This new hydroperoxy radical then reacts with a new hydrocarbon chain to repeat Step 2b, and so forth. This cycle of propagation steps repeats over and over in a chain reaction. Thus, once a radical is generated in Step 1, the cycle of propagation steps may repeat many thousands of times, generating thousands of hydroperoxide molecules. The number of times the cycle of chain propagation steps repeats is called the **chain length.**

Hydroperoxides themselves are unstable and, under biological conditions, degrade to short-chain aldehydes and carboxylic acids with unpleasant "rancid" smells. These odors may be familiar to you if you have ever smelled old cooking oil or aged foods that contain polyunsaturated fats or oils. Similar formation of hydroperoxides in the low-density lipoproteins (Section 21.9) deposited on the walls of arteries leads to cardiovascular disease in humans. In addition, many effects of aging are thought to result from the formation and subsequent degradation of hydroperoxides.

Fortunately, nature has developed a series of defenses against the formation of these and other destructive hydroperoxides, including the phenol vitamin E (Section 30.6). This compound is one of "nature's scavengers." It inserts itself into either Step 2a or 2b, donates an H· from its —OH group to the carbon radical, and converts the carbon radical back to its original hydrocarbon chain. Because the vitamin E radical is stable, it breaks the cycle of chain propagation steps, thereby preventing further formation of destructive hydroperoxides. While some hydroperoxides may form, their numbers are very small and they are easily decomposed to harmless materials by one of several possible enzyme-catalyzed reactions.

Unfortunately, vitamin E is removed in the processing of many foods and food products. To make up for this loss, phenols such as BHT and BHA are added to foods to "retard spoilage" (as they say on the packages) by autoxidation. Likewise, similar compounds are added to other materials, such as plastics and rubber, to protect them against autoxidation.

Vitamin E

Butylated *hydroxy-toluene*
(BHT)

Butylated *hydroxy-anisole*
(BHA)

SUMMARY OF KEY QUESTIONS

SECTION 13.1 What Is the Structure of Benzene?

- **Benzene** and its alkyl derivatives are classified as **aromatic hydrocarbons** or **arenes.**
- The first structure for benzene was proposed by Friederich August Kekulé in 1872.
- The theory of **resonance,** developed by Linus Pauling in the 1930s, provided the first adequate structure for benzene.

SECTION 13.2 How Do We Name Aromatic Compounds?

- Aromatic compounds are named according to the IUPAC system.
- The C_6H_5— group is named **phenyl.**
- Two substituents on a benzene ring may be located by either numbering the atoms of the ring or by using the locators *ortho* (*o*), *meta* (*m*), and *para* (*p*).
- **Polynuclear aromatic hydrocarbons** contain two or more benzene rings, each sharing two adjacent carbon atoms with another ring.

SECTION 13.3 What Are the Characteristic Reactions of Benzene and Its Derivatives?

- A characteristic reaction of aromatic compounds is **aromatic substitution,** in which another atom or group of atoms is substituted for a hydrogen atom of the aromatic ring.
- Typical aromatic substitution reactions are halogenation, nitration, and sulfonation.

SECTION 13.4 What Are Phenols?

- The functional group of a **phenol** is an —OH group bonded to a benzene ring.
- Phenol and its derivatives are weak acids, with pK_a equal to approximately 10.0.
- Vitamin E, a phenolic compound, is a natural antioxidant.
- Phenolic compounds such as BHT and BHA are synthetic antioxidants.

SUMMARY OF KEY REACTIONS

1. **Halogenation (Section 13.3A)** Treatment of an aromatic compound with Cl_2 or Br_2 in the presence of an $FeCl_3$ catalyst substitutes a halogen for an H.

$$\text{C}_6\text{H}_6 + Cl_2 \xrightarrow{\text{FeCl}_3} \text{C}_6\text{H}_5-Cl + HCl$$

2. **Nitration (Section 13.3B)** Heating an aromatic compound with a mixture of concentrated nitric and sulfuric acids substitutes a nitro group for an H.

$$\text{C}_6\text{H}_6 + HNO_3 \xrightarrow[\text{heat}]{\text{H}_2\text{SO}_4} \text{C}_6\text{H}_5-NO_2 + H_2O$$

Nitrobenzene

3. **Sulfonation (Section 13.3C)** Heating an aromatic compound with concentrated sulfuric acid substitutes a sulfonic acid group for an H.

$$\text{⬡} + H_2SO_4 \xrightarrow{\text{heat}} \text{⬡}-SO_3H + H_2O$$

Benzenesulfonic
acid

4. **Reaction of Phenols with Strong Bases (Section 13.4B)** Phenols are weak acids and react with strong bases to form water-soluble salts.

$$\text{⬡}-OH + NaOH \longrightarrow \text{⬡}-O^-Na^+ + H_2O$$

PROBLEMS

Chemistry ⬥ Now™

Assess your understanding of this chapter's topics with additional quizzing and conceptual-based problems at **http://now.brookscole.com/gob8** or on the CD.

A blue problem number indicates an applied problem.

▪ denotes problems that are available on the GOB ChemistryNow website or CD and are assignable in OWL.

SECTION 13.1 What Is the Structure of Benzene?

13.2 What is the difference in structure between a saturated and an unsaturated compound?

13.3 Define *aromatic compound*.

13.4 Why are alkenes, alkynes, and aromatic compounds said to be unsaturated?

13.5 Do aromatic rings have double bonds? Are they unsaturated? Explain.

13.6 Can an aromatic compound be a saturated compound?

13.7 Draw at least two structural formulas for the following. (Several constitutional isomers are possible for each part.)
(a) An alkene of six carbons
(b) A cycloalkene of six carbons
(c) An alkyne of six carbons
(d) An aromatic hydrocarbon of eight carbons

13.8 Write a structural formula and the name for the simplest (a) alkane, (b) alkene, (c) alkyne, and (d) aromatic hydrocarbon.

13.9 Account for the fact that the six-membered ring in benzene is planar but the six-membered ring in cyclohexane is not.

13.10 ▪ The compound 1,4-dichlorobenzene (*p*-dichlorobenzene) has a rigid geometry that does not allow free rotation. Yet no *cis-trans* isomers exist for this structure. Explain why it does not show *cis-trans* isomerism.

13.11 One analogy often used to explain the concept of a resonance hybrid is to relate a rhinoceros to a unicorn and a dragon. Explain the reasoning in this analogy and how it might relate to a resonance hybrid.

SECTION 13.2 How Do We Name Aromatic Compounds?

13.12 ▪ Name these compounds.

(c) $C_6H_5CH_2CH_2CH_2Cl$ (d) $C_6H_5\overset{\overset{\displaystyle Br}{|}}{\underset{\underset{\displaystyle CH_3}{|}}{C}}CH_2CH_3$

13.13 Draw structural formulas for these compounds.
(a) 1-Bromo-2-chloro-4-ethylbenzene
(b) 4-Bromo-1,2-dimethylbenzene
(c) 2,4,6-Trinitrotoluene
(d) 4-Phenyl-1-pentene
(e) *p*-Cresol
(f) 2,4-Dichlorophenol

13.14 We say that naphthalene, anthracene, phenanthrene, and benzo[a]pyrene are polynuclear aromatic hydrocarbons. In this context, what does "polynuclear" mean? What does "aromatic" mean? What does "hydrocarbon" mean?

SECTION 13.3 What Are the Characteristic Reactions of Benzene and Its Derivatives?

13.15 Suppose you have unlabeled bottles of benzene and cyclohexene. What chemical reaction could you use to tell which bottle contains which chemical? Explain what you would do, what you would expect to see, and how you would explain your observations.

13.16 ■ Three possible products with the molecular formula C_6H_4BrCl can form when bromobenzene is treated with chlorine, Cl_2, in the presence of $FeCl_3$ as a catalyst. Name and draw a structural formula for each product.

13.17 The reaction of bromine with toluene in the presence of $FeCl_3$ gives a mixture of three products, all with the molecular formula C_7H_7Br. Name and draw a structural formula for each product.

13.18 ■ What reagents and/or catalysts are necessary to carry out each conversion?

(a) Benzene to nitrobenzene (one step)

(b) 1,4-Dichlorobenzene to 2-bromo-1,4-Dichlorobenzene (one step)

(c) Benzene to aniline (requires two steps)

13.19 What reagents and/or catalysts are necessary to carry out each conversion? Each conversion requires two steps.

(a) Benzene to 3-nitrobenzenesulfonic acid

(b) Benzene to 1-bromo-4-chlorobenzene

13.20 ■ Aromatic substitution can be done on naphthalene. Treatment of naphthalene with concentrated H_2SO_4 gives two (and only two) different sulfonic acids. Draw a structural formula for each.

SECTION 13.4 What Are Phenols?

13.21 Both phenol and cyclohexanol are only slightly soluble in water. Account for the fact that phenol dissolves in aqueous sodium hydroxide but cyclohexanol does not.

13.22 Define *autoxidation*.

13.23 Autoxidation is described as a radical-chain reaction. What is meant by the term "radical" in this context? By the term "chain"? By the term "chain length"?

13.24 Show that, if you add Steps 2a and 2b of the radical-chain mechanism for the autoxidation of a fatty acid hydrocarbon chain, you arrive at the following net equation:

$$\begin{array}{c} H \\ | \\ \text{—CH}_2\text{CH}\text{=}\text{CH—CH—} + \cdot\text{O—O}\cdot \end{array}$$

Section of a fatty acid   Oxygen
hydrocarbon chain

$$\begin{array}{c} \text{O—O—H} \\ | \\ \longrightarrow \text{—CH}_2\text{CH}\text{=}\text{CH—CH—} \end{array}$$

A hydroperoxide

13.25 How does vitamin E function as an antioxidant?

13.26 ■ What structural features are common to vitamin E, BHT, and BHA (the three antioxidants presented in Section 13.4C)?

Chemical Connections

13.27 (Chemical Connections 13A) From what parts of its common name are the letters DDT derived?

13.28 (Chemical Connections 13A) What are the advantages and disadvantages of using DDT as an insecticide?

13.29 (Chemical Connections 13A) Would you expect DDT to be soluble or insoluble in water? Explain.

13.30 (Chemical Connections 13A) One of the degradation products of DDT is dichlorodiphenyldichloroethylene (DDE), which is formed by the loss of HCl from DDT. DDE inhibits the enzyme responsible for the incorporation of calcium ion into bird eggshells. Draw a structural formula for DDE.

13.31 (Chemical Connections 13A) What is meant by the term *biodegradable?*

13.32 (Chemical Connections 13B) What is a carcinogen? What kind of carcinogens are found in cigarette smoke?

13.33 (Chemical Connections 13C) In the absence of iodine in the diet, goiter develops. Explain why goiter is a regional disease.

13.34 (Chemical Connections 13D) Calculate the molecular weight of each explosive in this Chemical Connection. In which explosive do the nitro groups contribute the largest percentage of molecular weight?

13.35 (Chemical Connections 13E) What are the differences in structure between Allura Red and Sunset Yellow?

13.36 (Chemical Connections 13E) Which feature of Allura Red and Sunset Yellow make them water-soluble?

13.37 (Chemical Connections 13E) What color would you get if you mixed Allura Red and Sunset Yellow? (*Hint:* Remember the color wheel.)

13.38 (Chemical Connections 13E) Refer to Problems 12.57 and 12.58.

(a) Write structural formulas for lycopene and β-carotene, two natural food-coloring agents. Compare the structural formulas for these two compounds with those of Allura Red and Sunset Yellow.

(b) What structural feature or features do these four compounds have in common that might account for the fact that each is colored?

(c) Why are these two food and cosmetic dyes soluble in water while lycopene and β-carotene are not?

13.39 (Chemical Connections 13F) From what types of plants is capsaicin isolated?

13.40 (Chemical Connections 13F) Find the phenol group in capsaicin.

13.41 (Chemical Connections 13F) How many *cis-trans* isomers are possible for capsaicin? Is the structural formula shown in this Chemical Connection the *cis* isomer or the *trans* isomer?

13.42 (Chemical Connections 13F) In what ways is cap-saicin used in medicine?

Additional Problems

13.43 ■ The structure for naphthalene given in Section 13.2D is only one of three possible resonance structures. Draw the other two.

13.44 Draw structural formulas for these compounds.
  (a) 1-Phenylcyclopropanol (b) Styrene
  (c) *m*-Bromophenol (d) 4-Nitrobenzoic acid
  (e) Isobutylbenzene (f) *m*-Xylene

13.45 2,6-Di-*tert*-butyl-4-methylphenol (BHT, Section 13.4C) is an antioxidant added to processed foods to "retard spoilage." How does BHT accomplish this goal?

13.46 ■ Write the structural formula for the product of each reaction.

(a) + HNO₃ $\xrightarrow{\text{H}_2\text{SO}_4}$

(b) + Br₂ $\xrightarrow{\text{FeCl}_3}$

(c) + H₂SO₄ ⟶

13.47 Styrene reacts with bromine to give a compound with the molecular formula $C_8H_8Br_2$. Draw a structural formula for this compound.

Looking Ahead

13.48 Benzene, as we have seen in this chapter, is the simplest aromatic compound. Pyridine is an anolog of benzene in which a CH group is replaced by a nitrogen atom. Pyrimidine is an analog of benzene in which two CH groups are replaced by nitrogen atoms. Each nitrogen-containing compound shows the characteristic reactions of benzene and its derivatives—it is highly unsaturated but does not undergo the characteristic addition reactions of alkenes.

Benzene   Pyridine   Pyrimidine

(a) Show by the use of curved arrows, that benzene, pyridine, and pyrimidine can be represented as hybrids of two contributing structures.

(b) Show that each aromatic compound has an aromatic sextet—that is, a loop of six electrons within a cyclic system.

(c) Predict the bond angles in pyridine and pyrimidine and the shape of each molecules.

Alcohols, Ethers, and Thiols

Fermentation vats of wine grapes at the Beaulieu Vineyards, California.

Solutions in which ethanol is the solvent are called tinctures.

14.1 | What Are the Structures, Names, and Properties of Alcohols?

In this chapter, we study the physical and chemical properties of alcohols and ethers, two classes of oxygen-containing organic compounds. We also study thiols, a class of sulfur-containing organic compounds. Thiols are like alcohols in structure, except that they contain an —SH group rather than an —OH group.

$$CH_3CH_2OH \qquad CH_3CH_2OCH_2CH_3 \qquad CH_3CH_2SH$$

Ethanol (An alcohol)   Diethyl ether (An ether)   Ethanethiol (A thiol)

These three compounds are certainly familiar to you. Ethanol is the fuel additive in gasohol, the alcohol in alcoholic beverages, and an important laboratory and industrial solvent. Diethyl ether was the first inhalation anesthetic used in general surgery. It is also an important laboratory and industrial solvent. Ethanethiol, like other low-molecular-weight thiols, has a stench. Traces of ethanethiol are added to natural gas so that gas leaks can be detected by the smell of the thiol.

A. Structure of Alcohols

The functional group of an **alcohol** is an **—OH (hydroxyl) group** bonded to a tetrahedral carbon atom (Section 10.4A). Figure 14.1 shows a Lewis structure and a ball-and-stick model of methanol, CH_3OH, the simplest alcohol.

(a)   (b)

Figure 14.1 Methanol, CH_3OH. (a) Lewis structure and (b) ball-and-stick model. The H—C—O bond angle is 108.6°, very close to the tetrahedral angle of 109.5°.

B. Nomenclature

IUPAC names of alcohols are derived in the same manner as those for alkenes and alkynes, with the exception that the ending of the parent alkane is changed from -e to -ol. The ending -ol tells us that the compound is an alcohol.

1. Select the longest carbon chain that contains the —OH group as the parent alkane and number it from the end that gives —OH the lower number. In numbering the parent chain, the location of the —OH group takes precedence over alkyl groups, aryl groups, and halogens.

2. Change the ending of the parent alkane from -e to **-ol,** and use a number to show the location of the —OH group. For cyclic alcohols, numbering begins at the carbon bearing the —OH group; this carbon is automatically carbon-1.

3. Name and number substituents, and list them in alphabetical order.

To derive common names for alcohols, we name the alkyl group bonded to —OH and then add the word "alcohol." Following are IUPAC names and, in parentheses, common names for eight low-molecular-weight alcohols:

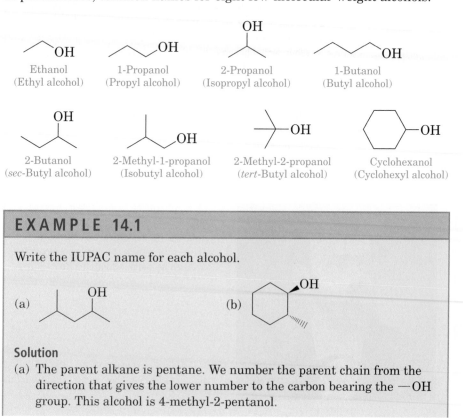

Ethanol
(Ethyl alcohol)

1-Propanol
(Propyl alcohol)

2-Propanol
(Isopropyl alcohol)

1-Butanol
(Butyl alcohol)

2-Butanol
(*sec*-Butyl alcohol)

2-Methyl-1-propanol
(Isobutyl alcohol)

2-Methyl-2-propanol
(*tert*-Butyl alcohol)

Cyclohexanol
(Cyclohexyl alcohol)

EXAMPLE 14.1

Write the IUPAC name for each alcohol.

(a)

(b)

Solution

(a) The parent alkane is pentane. We number the parent chain from the direction that gives the lower number to the carbon bearing the —OH group. This alcohol is 4-methyl-2-pentanol.

(b) The parent cycloalkane is cyclohexane. We number the atoms of the ring beginning with the carbon bearing the —OH group as carbon 1. This alcohol is *trans*-2-methylcyclohexanol.

Problem 14.1

Write the IUPAC name for each alcohol.

(a)   (b)   (c)

CHEMICAL CONNECTIONS 14A

Nitroglycerin: An Explosive and a Drug

In 1847, Ascanio Sobrero (1812–1888) discovered that 1,2,3-propanetriol, more commonly called glycerin, reacts with nitric acid in the presence of sulfuric acid to give a pale yellow, oily liquid called nitroglycerin. Sobrero also discovered the explosive properties of this compound; when he heated a small quantity of it, it exploded!

$$CH_2-OH \quad\quad\quad\quad CH_2-ONO_2$$
$$CH-OH + 3HNO_3 \xrightarrow{H_2SO_4} CH-ONO_2 + 3H_2O$$
$$CH_2-OH \quad\quad\quad\quad CH_2-ONO_2$$

1,2,3-Propanetriol   1,2,3-Propanetriol trinitrate
(Glycerol, Glycerin)   (Nitroglycerin)

Nitroglycerin very soon became widely used for blasting in the construction of canals, tunnels, roads, and mines and, of course, for warfare.

One problem with the use of nitroglycerin was soon recognized: It was difficult to handle safely, and accidental explosions occurred frequently. The Swedish chemist Alfred Nobel (1833–1896), who discovered that a clay-like substance called diatomaceous earth absorbs nitroglycerin so that it will not explode without a fuse, solved this problem. He gave the name *dynamite* to this mixture of nitroglycerin, diatomaceous earth, and sodium carbonate.

Surprising as it may seem, nitroglycerin is used in medicine to treat angina pectoris, the symptoms of which are sharp chest pains caused by reduced flow of blood in the coronary artery. Nitroglycerin, which is available in liquid form (diluted with alcohol to render it nonexplosive), tablet form, or paste form, relaxes the smooth muscles of blood vessels, causing dilation of the coronary artery. This dilation, in turn, allows more blood to reach the heart.

When Nobel became ill with heart disease, his physicians advised him to take nitroglycerin to relieve his chest pains. He refused, saying he could not understand how the explosive could relieve chest pains. It took science more than 100 years to find the answer. We now know that nitric oxide, NO, derived from the nitro groups of nitroglycerin, relieves the pain (see Chemical Connections 24F).

(a)

(b)

(*a*) Nitroglycerin is more stable if absorbed onto an inert solid, a combination called dynamite. (*b*) The fortune of Alfred Nobel, 1833–1896, was built on the manufacture of dynamite and now funds the Nobel Prizes.

We classify alcohols as **primary (1°), secondary (2°),** or **tertiary (3°),** depending on the number of carbon groups bonded to the carbon bearing the —OH group (Section 10.4A).

EXAMPLE 14.2

Classify each alcohol as primary, secondary, or tertiary.

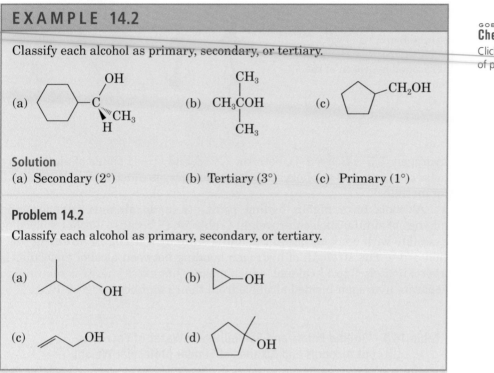

Solution
(a) Secondary (2°)   (b) Tertiary (3°)   (c) Primary (1°)

Problem 14.2

Classify each alcohol as primary, secondary, or tertiary.

In the IUPAC system, a compound containing two hydroxyl groups is named as a **diol,** one containing three hydroxyl groups as a **triol,** and so on. In IUPAC names for diols, triols, and so on, the final *-e* in the name of the parent alkane is retained—for example, 1,2-ethanediol.

As with many other organic compounds, common names for certain diols and triols have persisted. Compounds containing two hydroxyl groups on adjacent carbons are often referred to as **glycols.** Ethylene glycol and propylene glycol are synthesized from ethylene and propylene, respectively—hence their common names.

Diol A compound containing two —OH (hydroxyl) groups

Glycol A compound with hydroxyl (—OH) groups on adjacent carbons

Ethylene glycol is colorless; the color of most antifreezes comes from additives. Review Section 7.8 for a discussion of freezing-point depression.

CH_2—CH_2
| |
OH   OH
1,2-Ethanediol
(Ethylene glycol)

CH_3—CH—CH_2
         | |
        OH   OH
1,2-Propanediol
(Propylene glycol)

CH_2—CH—CH_2
  | | |
 OH   OH   OH
1,2,3-Propanetriol
(Glycerol, Glycerin)

Ethylene glycol is a polar molecule and dissolves readily in the polar solvent water.

Charles D. Winters

C. Physical Properties of Alcohols

The most important physical property of alcohols is the polarity of their —OH groups. Because of the large difference in electronegativity (Table 4.5) between oxygen and carbon $(3.5 - 2.5 = 1.0)$ and between oxygen and

Figure 14.2 Polarity of the
C—O—H bonds in methanol.
(a) There are partial positive
charges on carbon and hydrogen
and a partial negative charge on
oxygen. (b) An electron density
map showing the partial negative
charge (in red) around oxygen
and the partial positive charge
(in blue) around hydrogen of the
OH group.

(a)   (b)

GOB
Chemistry•⚛•Now™

Click *Coached Problems* to explore
the relationship between **Structure
and Boiling Points of Alcohols**

hydrogen (3.5 − 2.1 = 1.4), both the C—O and O—H bonds of alcohols are
polar covalent, and alcohols are polar molecules, as illustrated in Figure 14.2
for methanol.

Alcohols have higher boiling points than do alkanes, alkenes, and
alkynes of similar molecular weight (Table 14.1), because alcohol molecules
associate with one another in the liquid state by **hydrogen bonding** (Section 6.7C). The strength of hydrogen bonding between alcohol molecules is
approximately 2 to 5 kcal/mol, which means that extra energy is required to
separate hydrogen-bonded alcohols from their neighbors (Figure 14.3).

**Table 14.1 Boiling Points and Solubilities in Water of Four Sets
of Alcohols and Alkanes of Similar Molecular Weight**

Structural Formula	Name	Molecular Weight (amu)	Boiling Point (°C)	Solubility in Water
CH₃OH	methanol	32	65	infinite
CH₃CH₃	ethane	30	−89	insoluble
CH₃CH₂OH	ethanol	46	78	infinite
CH₃CH₂CH₃	propane	44	−42	insoluble
CH₃CH₂CH₂OH	1-propanol	60	97	infinite
CH₃CH₂CH₂CH₃	butane	58	0	insoluble
CH₃CH₂CH₂CH₂OH	1-butanol	74	117	8 g/100 g
CH₃CH₂CH₂CH₂CH₃	pentane	72	36	insoluble

Figure 14.3 The association of
ethanol molecules in the liquid
state. Each O—H can participate
in up to three hydrogen bonds
(one through hydrogen and two
through oxygen). Only two of
these three possible hydrogen
bonds per molecule are shown
here.

Hydrogen bonding

Because of increased London dispersion forces (Section 6.7A) between larger molecules, the boiling points of all types of compounds, including alcohols, increase with increasing molecular weight.

Alcohols are much more soluble in water than are hydrocarbons of similar molecular weight (Table 14.1), because alcohol molecules interact by hydrogen bonding with water molecules. Methanol, ethanol, and 1-propanol are soluble in water in all proportions. As molecular weight increases, the water solubility of alcohols becomes more like that of hydrocarbons of similar molecular weight. Higher-molecular-weight alcohols are much less soluble in water because the size of the hydrocarbon portion of their molecules (which decreases water solubility) becomes so large relative to the size of the —OH group (which increases water solubility).

14.2 | What Are the Characteristic Reactions of Alcohols?

In this section, we study the acidity of alcohols, their dehydration to alkenes, and their oxidation to aldehydes, ketones, or carboxylic acids.

A. Acidity of Alcohols

Alcohols have about the same pK_a values as water (Table 9.2), which means that aqueous solutions of alcohols have the same pH as that of pure water. In Section 13.4B, we studied the acidity of phenols, another class of compounds that contains an —OH group. Phenols are weak acids and react with aqueous sodium hydroxide to form water-soluble salts.

Phenol   Sodium phenoxide
                          (a water-soluble salt)

Alcohols are considerably weaker acids than phenols and do not react in this manner.

B. Acid-Catalyzed Dehydration Alkenes

We can convert an alcohol to an alkene by eliminating a molecule of water from adjacent carbon atoms in a reaction called **dehydration.** In the laboratory, dehydration of an alcohol is most often brought about by heating it with either 85% phosphoric acid or concentrated sulfuric acid. Primary alcohols—the most difficult to dehydrate—generally require heating in concentrated sulfuric acid at temperatures as high as 180°C. Secondary alcohols undergo acid-catalyzed dehydration at somewhat lower temperatures. Tertiary alcohols generally undergo acid-catalyzed dehydration at temperatures only slightly above room temperature.

Dehydration Elimination of a molecule of water from an alcohol. In the dehydration of an alcohol, OH is removed from one carbon, and an H is removed from an adjacent carbon.

Thus the ease of acid-catalyzed dehydration of alcohols follows this order:

1° alcohols   2° alcohols   3° alcohols

Ease of dehydration of alcohols

When the acid-catalyzed dehydration of an alcohol yields isomeric alkenes, the alkene having the greater number of alkyl groups on the double bond generally predominates. In the acid-catalyzed dehydration of 2-butanol, for example, the major product is 2-butene, which has two alkyl groups (two methyl groups) on its double bond. The minor product is 1-butene, which has only one alkyl group (an ethyl group) on its double bond.

OH
|
CH₃CH₂CHCH₃ $\xrightarrow[\text{heat}]{\text{H}_3\text{PO}_4}$ CH₃CH=CHCH₃ + CH₃CH₂CH=CH₂ + H₂O
2-Butanol   2-Butene   1-Butene
                                         (80%)   (20%)

EXAMPLE 14.3

Draw structural formulas for the alkenes formed by the acid-catalyzed dehydration of each alcohol. For each part, predict which alkene will be the major product.

(a) 3-Methyl-2-butanol   (b) 2-Methylcyclopentanol

Solution

(a) Elimination of H₂O from carbons 2–3 gives 2-methyl-2-butene; elimination of H₂O from carbons 1–2 gives 3-methyl-1-butene. 2-Methyl-2-butene has three alkyl groups (three methyl groups) on its double bond and is the major product. 3-Methyl-1-butene has only one alkyl group (an isopropyl group) on its double bond and is the minor product.

CH₃
4 3| 2 1
CH₃CHCHCH₃ $\xrightarrow[\text{dehydration}]{\substack{\text{H}_2\text{SO}_4 \\ \text{acid-catalyzed}}}$ CH₃C=CHCH₃ + CH₃CHCH=CH₂ + H₂O
|
OH

3-Methyl-2-butanol   2-Methyl-2-butene   3-Methyl-1-butene
                            (major product)

(b) The major product, 1-methylcyclopentene, has three alkyl groups on its double bond. The minor product, 3-methylcyclopentene, has only two alkyl groups on its double bond.

2-Methylcyclopentanol   1-Methylcyclopentene   3-Methylcyclopentene
                               (major product)

Problem 14.3

Draw structural formulas for the alkenes formed by the acid-catalyzed dehydration of each alcohol. For each part, predict which alkene will be the major product.

(a) 2-Methyl-2-butanol   (b) 1-Methylcyclopentanol

In Section 12.6B, we studied the acid-catalyzed hydration of alkenes to give alcohols. In this section, we study the acid-catalyzed dehydration of alcohols to give alkenes. In fact, hydration–dehydration reactions are reversible. Alkene hydration and alcohol dehydration are competing reactions, and the following equilibrium exists:

An alkene   An alcohol

In accordance with Le Chatelier's principle (Section 8.8), large amounts of water (in other words, using dilute aqueous acid) favor alcohol formation, whereas scarcity of water (using concentrated acid) or experimental conditions where water is removed (heating the reaction mixture above 100°C) favor alkene formation. Thus, depending on the experimental conditions, we can use the hydration–dehydration equilibrium to prepare both alcohols and alkenes, each in high yields.

EXAMPLE 14.4

In part (a), acid-catalyzed dehydration of 2-methyl-3-pentanol gives Compound A. Treatment of Compound A with water in the presence of sulfuric acid in part (b) gives Compound B. Propose structural formulas for Compounds A and B.

(a) $\underset{\underset{\displaystyle OH}{|}}{CH_3\underset{\underset{|}{}}{C}HCHCH_2CH_3} \quad \overset{\displaystyle H_2SO_4}{\underset{\substack{\text{acid-catalyzed} \\ \text{dehydration}}}{\longrightarrow}}$ Compound A (C_6H_{12}) + H_2O

with CH_3 group on the second carbon.

(b) Compound A (C_6H_{12}) + H_2O $\overset{H_2SO_4}{\longrightarrow}$ Compound B $(C_6H_{14}O)$

Solution

(a) Acid-catalyzed dehydration of 2-methyl-3-pentanol gives 2-methyl-2-pentene, an alkene with three substituents on its double bond: two methyl groups and one ethyl group.
(b) Acid-catalyzed addition of water to this alkene gives 2-methyl-2-pentanol in accord with Markovnikov's rule (Section 12.6B).

$\underset{\text{Compound A }(C_6H_{12})}{\overset{\overset{\displaystyle CH_3}{|}}{CH_3C}{=}CHCH_2CH_3} + H_2O \xrightarrow[\substack{\text{acid-catalyzed} \\ \text{hydration}}]{H_2SO_4} \underset{\text{Compound B }(C_6H_{14}O)}{\underset{\underset{\displaystyle OH}{|}}{\overset{\overset{\displaystyle CH_3}{|}}{CH_3C}CH_2CH_2CH_3}}$

Problem 14.4

Acid-catalyzed dehydration of 2-methylcyclohexanol gives Compound C (C_7H_{12}). Treatment of Compound C with water in the presence of sulfuric acid gives Compound D $(C_7H_{14}O)$. Propose structural formulas for Compounds C and D.

C. Oxidation of Primary and Secondary Alcohols

A primary alcohol can be oxidized to an aldehyde or to a carboxylic acid, depending on the experimental conditions. Following is a series of transformations in which a primary alcohol is oxidized first to an aldehyde and then to a carboxylic acid. The letter O in brackets over the reaction arrow indicates that each transformation involves oxidation.

A primary alcohol   An aldehyde   A carboxylic acid

Recall from Section 5.7 that, according to one definition, oxidation is either the loss of hydrogens or the gain of oxygens. Using this definition, conversion of a primary alcohol to an aldehyde is an oxidation reaction because the alcohol loses hydrogens. Conversion of an aldehyde to a carboxylic acid is also an oxidation reaction because the aldehyde gains an oxygen.

 ## CHEMICAL CONNECTIONS 14B

Breath-Alcohol Screening

Potassium dichromate oxidation of ethanol to acetic acid is the basis for the original breath-alcohol screening test used by law enforcement agencies to determine a person's blood alcohol content (BAC). The test is based on the difference in color between the dichromate ion (reddish orange) in the reagent and the chromium(III) ion (green) in the product.

$$CH_3CH_2OH + Cr_2O_7^{2-}$$
Ethanol   Dichromate ion (reddish orange)

$$\xrightarrow[H_2O]{H_2SO_4} CH_3COH + Cr^{3+}$$
Acetic acid   Chromium(III) ion (green)

In its simplest form, breath-alcohol screening uses a sealed glass tube that contains a potassium dichromate–sulfuric acid reagent impregnated on silica gel. To administer the test, the ends of the tube are broken off, a mouthpiece is fitted to one end, and the other

end is inserted into the neck of a plastic bag. The person being tested then blows into the mouthpiece to inflate the plastic bag.

As breath containing ethanol vapor passes through the tube, reddish orange dichromate ion is reduced to green chromium(III) ion. To estimate the concentration of ethanol in the breath, one measures how far the green color extends along the length of the tube. When it reaches beyond the halfway point, the person is judged as having a sufficiently high blood alcohol content to warrant further, more precise testing.

The test described here measures the alcohol content of the breath. The legal definition of being under the influence of alcohol, however, is based on alcohol content in the blood, not in the breath. The correlation between these two measurements is based on the fact that air deep within the lungs is in equilibrium with blood passing through the pulmonary arteries, and thus an equilibrium is established between blood alcohol and breath alcohol. Based on tests in persons drinking alcohol, researchers have determined that 2100 mL of breath contains the same amount of ethanol as 1.00 mL of blood.

Person forces breath through mouthpiece into the tube.

Glass tube containing potassium dichromate–sulfuric acid coated on silica gel particles

As the person blows into the tube, the plastic bag becomes inflated.

The reagent most commonly used in the laboratory for the oxidation of a primary alcohol to a carboxylic acid is potassium dichromate, $K_2Cr_2O_7$, dissolved in aqueous sulfuric acid. Using this reagent, oxidation of 1-octanol, for example, gives octanoic acid. This experimental condition is more than sufficient to oxidize the intermediate aldehyde to a carboxylic acid.

$$CH_3(CH_2)_6CH_2OH \xrightarrow[H_2SO_4]{K_2Cr_2O_7} CH_3(CH_2)_6\overset{\displaystyle O}{\overset{\|}{C}}H \xrightarrow[H_2SO_4]{K_2Cr_2O_7} CH_3(CH_2)_6\overset{\displaystyle O}{\overset{\|}{C}}OH$$

1-Octanol   Octanal   Octanoic acid

Although the usual product of oxidation of a primary alcohol is a carboxylic acid, it is often possible to stop the oxidation at the aldehyde stage by distilling the mixture. That is, the aldehyde, which usually has a lower boiling point than either the primary alcohol or the carboxylic acid, is removed from the reaction mixture before it can be oxidized further.

Secondary alcohols may be oxidized to ketones by using potassium dichromate as the oxidizing agent. Menthol, a secondary alcohol that is present in peppermint and other mint oils, is used in liqueurs, cigarettes, cough drops, perfumery, and nasal inhalers. Its oxidation product, menthone, is also used in perfumes and artificial flavors.

2-Isopropyl-5-methyl-   2-Isopropyl-5-methyl-
cyclohexanol   cyclohexanone
(Menthol)   (Menthone)

Tertiary alcohols resist oxidation because the carbon bearing the —OH is bonded to three carbon atoms and, therefore, cannot form a carbon–oxygen double bond.

1-Methylcyclopentanol

EXAMPLE 14.5

Draw a structural formula for the product formed by oxidation of each alcohol with potassium dichromate.
(a) 1-Hexanol   (b) 2-Hexanol

Solution
1-Hexanol, a primary alcohol, is oxidized to either hexanal or hexanoic acid, depending on the experimental conditions. 2-Hexanol, a secondary alcohol, is oxidized to 2-hexanone, a ketone.

(a)   Hexanal   or   Hexanoic acid   (b)   2-Hexanone

Problem 14.5

Draw the product formed by oxidation of each alcohol with potassium dichromate.

(a) Cyclohexanol   (b) 2-Pentanol

14.3 | What Are the Structures, Names, and Properties of Ethers?

A. Structure

Ether A compound containing an oxygen atom bonded to two carbon atoms

The functional group of an **ether** is an atom of oxygen bonded to two carbon atoms. Figure 14.4 shows a Lewis structure and a ball-and-stick model of dimethyl ether, CH_3OCH_3, the simplest ether.

B. Nomenclature

Although the IUPAC system can be used to name ethers, chemists almost invariably use common names for low-molecular-weight ethers. Common names are derived by listing the alkyl groups bonded to oxygen in alphabetical order and adding the word *ether*. Alternatively, one of the groups on oxygen is named as an alkoxy group. The $-OCH_3$ group, for example, is named "methoxy" to indicate a <u>meth</u>yl group bonded to <u>oxy</u>gen.

(a)

(b)

Figure 14.4 Dimethyl ether, CH_3OCH_3. (a) Lewis structure and (b) ball-and-stick model. The C—O—C bond angle is 110.3°, close to the tetrahedral angle of 109.5°.

GOB
Chemistry Now™
Click *Coached Problems* to try a set of problems in **Naming Ethers**

$$CH_3CH_2OCH_2CH_3$$

Diethyl ether

Cyclohexyl methyl ether
(Methoxycyclohexane)

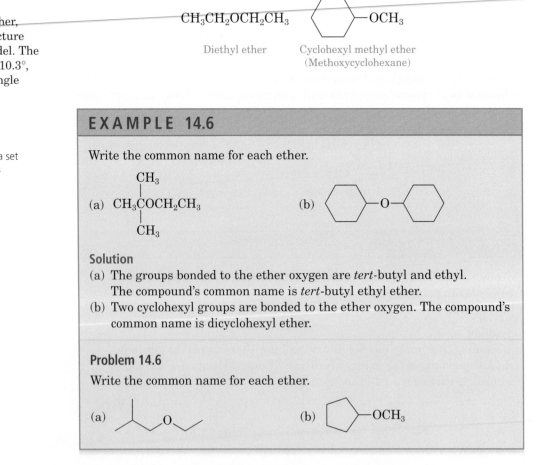

EXAMPLE 14.6

Write the common name for each ether.

(a)
$$\overset{\displaystyle CH_3}{\underset{\displaystyle CH_3}{CH_3\overset{|}{\underset{|}{C}}OCH_2CH_3}}$$

(b)

Solution

(a) The groups bonded to the ether oxygen are *tert*-butyl and ethyl. The compound's common name is *tert*-butyl ethyl ether.

(b) Two cyclohexyl groups are bonded to the ether oxygen. The compound's common name is dicyclohexyl ether.

Problem 14.6

Write the common name for each ether.

(a)   (b)

CHEMICAL CONNECTIONS 14C

Ethylene Oxide: A Chemical Sterilant

Ethylene oxide is a colorless, flammable gas, with a boiling point of 11°C. Because it is such a highly strained molecule (the normal tetrahedral bond angles of both C and O are compressed to approximately 60° in this molecule), ethylene oxide reacts with the amino (—NH$_2$) and sulfhydryl (—SH) groups present in biological materials.

At sufficiently high concentrations, it reacts with enough molecules in cells to cause the death of microorganisms. This toxic property is the basis for ethylene oxide's use as a fumigant in foodstuffs and textiles and its use in hospitals to sterilize surgical instruments.

$$RNH_2 + \triangledown_O \longrightarrow RNH{-}CH_2CH_2O{-}H$$

$$RSH + \triangledown_O \longrightarrow RS{-}CH_2CH_2O{-}H$$

In **cyclic ethers,** one of the atoms in a ring is oxygen. These ethers are also known by their common names. Ethylene oxide is an important building block for the organic chemical industry (Section 14.5). Tetrahydrofuran is a useful laboratory and industrial solvent.

Cyclic ether An ether in which the ether oxygen is one of the atoms of a ring

$$\triangledown_O \qquad \bigcirc_O$$

Ethylene oxide   Tetrahydrofuran (THF)

C. Physical Properties

Ethers are polar compounds in which oxygen bears a partial negative charge and each attached carbon bears a partial positive charge (Figure 14.5). However, only weak forces of attraction exist between ether molecules in the pure liquid. Consequently, the boiling points of ethers are close to those of hydrocarbons of similar molecular weight.

The effect of hydrogen bonding on physical properties is illustrated dramatically by comparing the boiling points of ethanol (78°C) and its constitutional isomer dimethyl ether (−24°C). The difference in boiling point between these two compounds is due to the presence in ethanol of a polar O—H group, which is capable of forming hydrogen bonds. This hydrogen

Figure 14.5 Ethers are polar molecules, but only weak attractive interactions exist between ether molecules in the liquid state. Shown on the right is an electron density map of diethyl ether.

bonding increases intermolecular associations, thereby giving ethanol a higher boiling point than dimethyl ether.

CH₃CH₂OH
Ethanol
bp 78°C

CH₃OCH₃
Dimethyl ether
bp −24°C

Ethers are more soluble in water than hydrocarbons of similar molecular weight and shape. Their greater solubility reflects the fact that the oxygen atom of an ether carries a partial negative charge and forms hydrogen bonds with water.

D. Reactions of Ethers

Ethers resemble hydrocarbons in their resistance to chemical reaction. For example, they do not react with oxidizing agents, such as potassium dichromate. Likewise, they do not react with reducing agents such as H₂ in the presence of a metal catalyst (Section 12.6D). Furthermore, most acids and bases at moderate temperatures do not affect them. Because of their general inertness to chemical reaction and their good solvent properties, ethers are excellent solvents in which to carry out many organic reactions. The most important ether solvents are diethyl ether and tetrahydrofuran.

CHEMICAL CONNECTIONS 14D

Ethers and Anesthesia

Before the mid-1800s, surgery was performed only when absolutely necessary, because no truly effective general anesthetic was available. More often than not, patients were drugged, hypnotized, or simply tied down.

In 1772, Joseph Priestley isolated nitrous oxide, N₂O, a colorless gas. In 1799, Sir Humphry Davy demonstrated this compound's

This painting by Robert Hinckley shows the first use of ether as an anesthetic in 1846. Dr. Robert John Collins was removing a tumor from the patient's neck, and dentist W. T. G. Morton—who discovered its anesthetic properties—administered the ether.

Boston Medical Library in the Francis A. Countway Library of Medicine

anesthetic effect, naming it "laughing gas." In 1844, an American dentist, Horace Wells, introduced nitrous oxide into general dental practice. One patient awakened prematurely, however, screaming with pain; another died during the procedure. Wells was forced to withdraw from practice, became embittered and depressed, and committed suicide at age 33. In the same period, a Boston chemist, Charles Jackson, anesthetized himself with diethyl ether; he also persuaded a dentist, William Morton, to use it. Subsequently, they persuaded a surgeon, John Warren, to give a public demonstration of surgery under anesthesia. The operation was completely successful, and soon general anesthesia by diethyl ether became routine practice for general surgery.

Diethyl ether was easy to use and caused excellent muscle relaxation. Blood pressure, pulse rate, and respiration were usually only slightly affected. Diethyl ether's chief drawbacks are its irritating effect on the respiratory passages and its after-effect of nausea.

Among the inhalation anesthetics used today are several halogenated ethers, the most important of which are enflurane and isoflurane.

Enflurane
(Ethrane)

Isoflurane
(Forane)

<table>
<tr><td>

14.4 | What Are the Structures, Names, and Properties of Thiols?

</td></tr>
</table>

The most notable property of low-molecular-weight **thiols** is their stench. They are responsible for unpleasant odors such as those emanating from skunks, rotten eggs, and sewage. The scent of skunks is due primarily to two thiols:

$$CH_3CH=CHCH_2SH$$
2-Butene-1-thiol

$$CH_3\overset{\overset{\textstyle CH_3}{|}}{C}HCH_2CH_2SH$$
3-Methyl-1-butanethiol

Thiol A compound containing an —SH (sulfhydryl) group bonded to a tetrahedral carbon atom

A. Structure

The functional group of a **thiol** is an **—SH (sulfhydryl) group** bonded to a tetrahedral carbon atom. Figure 14.6 shows a Lewis structure and a ball-and-stick model of methanethiol, CH_3SH, the simplest thiol.

(a)

(b) 100.3°

Figure 14.6 Methanethiol, CH_3SH. (a) Lewis structure and (b) ball-and-stick model. The H—S—C bond angle is 100.3°, somewhat smaller than the tetrahedral angle of 109.5°.

B. Nomenclature

The sulfur analog of an alcohol is called a thiol (*thi-* from the Greek: *theion*, sulfur) or, in the older literature, a **mercaptan,** which literally means "mercury capturing." Thiols react with Hg^{2+} in aqueous solution to give sulfide salts as insoluble precipitates. Thiophenol, C_6H_5SH, for example, gives $(C_6H_5S)_2Hg$.

In the IUPAC system, thiols are named by selecting the longest carbon chain that contains the —SH group as the parent alkane. To show that the compound is a thiol, we add the suffix *-thiol* to the name of the parent alkane. The parent chain is numbered in the direction that gives the —SH group the lower number.

Common names for simple thiols are derived by naming the alkyl group attached to —SH and adding the word *mercaptan*.

$$CH_3CH_2SH$$
Ethanethiol
(Ethyl mercaptan)

$$CH_3\overset{\overset{\textstyle CH_3}{|}}{C}HCH_2SH$$
2-Methyl-1-propanethiol
(Isobutyl mercaptan)

Mercaptan A common name for any molecule containing an —SH group

GOB
Chemistry⚛Now™
Click *Coached Problems* for an exploration **Comparing the Structure and Properties of Thiols to the Properties of Alcohols**

EXAMPLE 14.7

Write the IUPAC name for each thiol.

(a) ~~~~SH

(b) ⟍⟋SH (with SH on second carbon)

Solution
(a) The parent alkane is pentane. We show the presence of the —SH group by adding "thiol" to the name of the parent alkane. The IUPAC name of this thiol is 1-pentanethiol. Its common name is pentyl mercaptan.
(b) The parent alkane is butane. The IUPAC name of this thiol is 2-butanethiol. Its common name is *sec*-butyl mercaptan.

Steven J. Krasemann/Photo Researchers, Inc.

The scent of the spotted skunk, native to the Sonoran Desert, is a mixture of two thiols, 2-butene-1-thiol and 3-methyl-1-butanethiol.

Problem 14.7

Write the IUPAC name for each thiol.

(a)   (b)

C. Physical Properties

Because of the small difference in electronegativity between sulfur and hydrogen ($2.5 - 2.1 = 0.4$), we classify the S—H bond as nonpolar covalent. Because of this lack of polarity, thiols show little association by hydrogen bonding. Consequently, they have lower boiling points and are less soluble in water and other polar solvents than are alcohols of similar molecular weight. Table 14.2 gives boiling points for three low-molecular-weight thiols. Shown for comparison are boiling points of alcohols with the same number of carbon atoms.

    Earlier we illustrated the importance of hydrogen bonding in alcohols by comparing the boiling points of ethanol (78°C) and its constitutional isomer dimethyl ether (−24°C). By contrast, the boiling point of ethanethiol is 35°C and that of its constitutional isomer dimethyl sulfide is 37°C. Because the boiling points of these constitutional isomers are almost identical, we know that little or no association by hydrogen bonding occurs between thiol molecules.

$$CH_3CH_2SH \qquad CH_3SCH_3$$

Ethanethiol   Dimethyl sulfide
bp 35°C   bp 37°C

D. Reactions of Thiols

Thiols are weak acids ($pK_a = 10$) that are comparable in strength to phenols (Section 13.4B). Thiols react with strong bases such as NaOH to form thiolate salts.

$$CH_3CH_2SH + NaOH \xrightarrow{H_2O} CH_3CH_2S^-Na^+ + H_2O$$

Ethanethiol   Sodium
(pK_a 10)   ethanethiolate

Disulfide A compound containing an —S—S— group

    The most common reaction of thiols in biological systems is their oxidation to disulfides, the functional group of which is a **disulfide** (—S—S—) bond. Thiols are readily oxidized to disulfides by molecular oxygen. In fact, they are so susceptible to oxidation that they must be protected from con-

Table 14.2	Boiling Points of Three Thiols and Alcohols with the Same Number of Carbon Atoms			
Thiol	Boiling Point (°C)	Alcohol	Boiling Point (°C)	
methanethiol	6	methanol	65	
ethanethiol	35	ethanol	78	
1-butanethiol	98	1-butanol	117	

tact with air during their storage. Disulfides, in turn, are easily reduced to thiols by several reducing agents. This easy interconversion between thiols and disulfides is very important in protein chemistry, as we will see in Chapters 22 and 23.

$$2HOCH_2CH_2SH \underset{\text{reduction}}{\overset{\text{oxidation}}{\rightleftharpoons}} HOCH_2CH_2S-SCH_2CH_2OH$$
A thiol   A disulfide

We derive the common names of disulfides by listing the names of the groups attached to sulfur and adding the word *disulfide*.

14.5 | What Are the Most Commercially Important Alcohols?

As you study the alcohols described in this section, you should pay particular attention to two key points. First, they are derived almost entirely from petroleum, natural gas, or coal—all nonrenewable resources. Second, many are themselves starting materials for the synthesis of valuable commercial products, without which our modern industrial society could not exist.

At one time **methanol** was derived by heating hard woods in a limited supply of air—hence the name "wood alcohol." Today methanol is obtained entirely from the catalytic reduction of carbon monoxide. Methanol, in turn, is the starting material for the preparation of several important industrial and commercial chemicals, including acetic acid and formaldehyde. Treatment of methanol with carbon monoxide in the presence of a rhodium catalyst gives acetic acid. Partial oxidation of methanol gives formaldehyde. An important use of this one-carbon aldehyde is in the preparation of phenol-formaldehyde and urea-formaldehyde glues and resins, which are used as molding materials and as adhesives for plywood and particle board for the construction industry.

Methanol, CH_3OH, is the fuel used in cars of the type that race in Indianapolis.

The bulk of the **ethanol** produced worldwide is prepared by acid-catalyzed hydration of ethylene, itself derived from the cracking of the ethane separated from natural gas (Section 11.10). Ethanol is also produced by fermentation of the carbohydrates in plant materials, particularly corn and molasses. The majority of the fermentation-derived ethanol is used as an "oxygenate" additive to produce gasohol, which is a blend of up to 10% ethanol in gasoline. Combustion of gasohol produces less air pollution than combustion of gasoline itself.

Acid-catalyzed dehydration of ethanol gives **diethyl ether,** an important laboratory and industrial solvent. Ethylene is also the starting material for the preparation of **ethylene oxide.** This compound itself has few direct uses. Rather, ethylene oxide's importance derives from its role as an intermediate in the production of **ethylene glycol,** a major component of automobile antifreeze. Ethylene glycol freezes at $-12°C$ and boils at $199°C$, which makes it ideal for this purpose. In addition, reaction of ethylene glycol with the methyl ester of terephthalic acid gives the polymer poly(ethylene terephthalate), abbreviated PET or PETE (Section 19.6B). Ethylene glycol is also used as a solvent in the paint and plastics industry, and in the formulation of printer's inks, ink pads, and inks for ballpoint pens.

Isopropyl alcohol, the alcohol in rubbing alcohol, is made by acid-catalyzed hydration of propene. It is also used in hand lotions, after-shave lotions, and similar cosmetics. A multistep process also converts propene to epichlorohydrin, one of the key components in the production of epoxy glues and resins.

Glycerin is a by-product of the manufacture of soaps by saponification of animal fats and tropical oils (Section 21.3). The bulk of the glycerin used for industrial and commercial purposes, however, is prepared from propene. Perhaps the best-known use of glycerin is for the manufacture of nitroglycerin. Glycerin is also used as an emollient in skin care products and cosmetics, in liquid soaps, and in printing inks.

SUMMARY OF KEY QUESTIONS

SECTION 14.1 What Are the Structures, Names, and Properties of Alcohols?

- The functional group of an **alcohol** is an **—OH (hydroxyl)** group bonded to a tetrahedral carbon atom.
- IUPAC names of alcohols are derived by changing the **-e** of the parent alkane to **-ol.** The parent chain is numbered from the end that gives the carbon bearing the —OH group the lower number.
- Common names for alcohols are derived by naming the alkyl group bonded to the —OH group and adding the word "alcohol."

- Alcohols are classified as **1°, 2°,** or **3°,** depending on the number of carbon groups bonded to the carbon bearing the —OH group.
- Compounds containing hydroxyl groups on adjacent carbons are called **glycols.**
- Alcohols are polar compounds in which oxygen bears a partial negative charge and both the carbon and hydrogen bonded to it bear partial positive charges. Alcohols associate in the liquid state by **hydrogen bonding.** As a consequence, their boiling points are higher than those of hydrocarbons of similar molecular weight.

- Because of increased London dispersion forces, the boiling points of alcohols increase with their increasing molecular weight.
- Alcohols interact with water by hydrogen bonding and are more soluble in water than are hydrocarbons of similar molecular weight.
- Alcohols have about the same pK_a values as pure water. For this reason, aqueous solutions of alcohols have the same pH as that of pure water.

SECTION 14.3 What Are the Structures, Names, and Properties of Ethers?

- The functional group of an **ether** is an atom of oxygen bonded to two carbon atoms.
- Common names for ethers are derived by naming the two groups attached to oxygen followed by the word "ether."
- In a **cyclic ether,** oxygen is one of the atoms in a ring. Ethers are weakly polar compounds. Their boiling points are close to those of hydrocarbons of similar molecular weight.

- Because ethers form hydrogen bonds with water, they are more soluble in water than are hydrocarbons of similar molecular weight.

SECTION 14.4 What Are the Structures, Names, and Properties of Thiols?

- A **thiol** contains —**SH** (**sulfhydryl**) **group.**
- Thiols are named in the same manner as alcohols, but the suffix **-e** of the parent alkane is retained, and -**thiol** is added.
- Common names for thiols are derived by naming the alkyl group bonded to —SH and adding the word **"mercaptan."**
- The S—H bond is nonpolar, and the physical properties of thiols resemble those of hydrocarbons of similar molecular weight.

SUMMARY OF KEY REACTIONS

1. **Acid-Catalyzed Dehydration of an Alcohol (Section 14.2B)** When isomeric alkenes are possible, the major product is generally the more substituted alkene.

$$\underset{\text{CH}_3\text{CH}_2\overset{\displaystyle \text{OH}}{\underset{|}{\text{CH}}}\text{CH}_3}{}$$

$$\xrightarrow[\text{heat}]{\text{H}_3\text{PO}_4} \underset{\text{Major product}}{\text{CH}_3\text{CH}=\text{CHCH}_3} + \text{CH}_3\text{CH}_2\text{CH}=\text{CH}_2 + \text{H}_2\text{O}$$

2. **Oxidation of a Primary Alcohol (Section 14.2C)** Oxidation of a primary alcohol by potassium dichromate gives either an aldehyde or a carboxylic acid, depending on the experimental conditions.

$$\text{CH}_3(\text{CH}_2)_6\text{CH}_2\text{OH} \xrightarrow[\text{H}_2\text{SO}_4]{\text{K}_2\text{Cr}_2\text{O}_7} \text{CH}_3(\text{CH}_2)_6\overset{\displaystyle \text{O}}{\overset{\|}{\text{CH}}}$$

$$\xrightarrow[\text{H}_2\text{SO}_4]{\text{K}_2\text{Cr}_2\text{O}_7} \text{CH}_3(\text{CH}_2)_6\overset{\displaystyle \text{O}}{\overset{\|}{\text{C}}}\text{OH}$$

3. **Oxidation of a Secondary Alcohol (Section 14.2C)** Oxidization of a secondary alcohol by potassium dichromate gives a ketone.

$$\text{CH}_3(\text{CH}_2)_4\overset{\displaystyle \text{OH}}{\underset{|}{\text{CH}}}\text{CH}_3 \xrightarrow[\text{H}_2\text{SO}_4]{\text{K}_2\text{Cr}_2\text{O}_7} \text{CH}_3(\text{CH}_2)_4\overset{\displaystyle \text{O}}{\overset{\|}{\text{C}}}\text{CH}_3$$

4. **Acidity of Thiols (Section 14.4D)** Thiols are weak acids, with pK_a values of approximately 10. They react with strong bases to form water-soluble thiolate salts.

$$\underset{\substack{\text{Ethanethiol} \\ (pK_a\ 10)}}{\text{CH}_3\text{CH}_2\text{SH}} + \text{NaOH} \xrightarrow{\text{H}_2\text{O}} \underset{\substack{\text{Sodium} \\ \text{ethanethiolate}}}{\text{CH}_3\text{CH}_2\text{S}^-\text{Na}^+} + \text{H}_2\text{O}$$

5. **Oxidation of a Thiol to a Disulfide (Section 14.4D)** Oxidation of a thiol gives a disulfide. Reduction of a disulfide gives two thiols.

$$2\underset{\text{A thiol}}{\text{HOCH}_2\text{CH}_2\text{SH}} \underset{\text{reduction}}{\overset{\text{oxidation}}{\rightleftharpoons}} \underset{\text{A disulfide}}{\text{HOCH}_2\text{CH}_2\text{S}-\text{SCH}_2\text{CH}_2\text{OH}}$$

PROBLEMS

A blue problem number indicates an applied problem.

■ denotes problems that are available on the GOB ChemistryNow website or CD and are assignable in OWL.

SECTION 14.1 What Are the Structures, Names, and Properties of Alcohols?

14.8 What is the difference in structure between a primary, a secondary, and a tertiary alcohol?

14.9 Which of the following are secondary alcohols?

(a)   (b) $(CH_3)_3COH$

(c)   (d)

14.10 ■ Which of the alcohols in Problem 14.9 are primary? Which are tertiary?

14.11 ■ Write the IUPAC name of each compound.

(a)   (b) HO⏜⏜OH

(c)   (d) HO⏜

(e)   (f)

14.12 Draw a structural formula for each alcohol.
(a) Isopropyl alcohol
(b) Propylene glycol
(c) 5-Methyl-2-hexanol
(d) 2-Methyl-2-propyl-1,3-propanediol
(e) 1-Octanol
(f) 3,3-Dimethylcyclohexanol

14.13 Draw a structural formula for each alcohol.
(a) Isobutyl alcohol
(b) 1,4-Butanediol
(c) 5-Methyl-1-hexanol
(d) 1,3-Pentanediol
(e) *trans*-1,4-Cyclohexanediol
(f) 1-Chloro-2-propanol

14.14 Both alcohols and phenols contain an —OH group. What structural feature distinguishes these two classes of compounds? Illustrate your answer by drawing the structural formulas of a phenol with six carbon atoms and an alcohol with six carbon atoms.

14.15 ■ Name the functional groups in each compound.

(a)

Prednisone
(a synthetic anti-inflammatory steroid)

(b)

Estradiol
(a female sex hormone;
Section 21.10)

14.16 Explain in terms of noncovalent interactions why the low-molecular-weight alcohols are soluble in water but the low-molecular-weight alkanes and alkynes are not.

14.17 Explain in terms of noncovalent interactions why the low-molecular-weight alcohols are more soluble in water than the low-molecular-weight ethers.

14.18 Why does the water solubility of alcohols decrease as molecular weight increases?

14.19 Show hydrogen bonding between methanol and water in the following ways.
(a) Between the oxygen of methanol and a hydrogen of water
(b) Between the hydrogen of methanol's OH group and the oxygen of water

14.20 Show hydrogen bonding between the oxygen of diethyl ether and a hydrogen of water.

14.21 ■ Arrange these compounds in order of increasing boiling point. Values in °C are −42, 78, 117, and 198.
(a) $CH_3CH_2CH_2CH_2OH$
(b) CH_3CH_2OH
(c) $HOCH_2CH_2OH$
(d) $CH_3CH_2CH_3$

14.22 Arrange these compounds in order of increasing boiling point. Values in °C are 0, 35, and 97.

(a) $CH_3CH_2CH_2OH$ (b) $CH_3CH_2OCH_2CH_3$

(c) $CH_3CH_2CH_2CH_3$

14.23 2-Propanol (isopropyl alcohol) is commonly used as rubbing alcohol to cool the skin. 2-Hexanol, also a liquid, is not suitable for this purpose. Why not?

14.24 Explain why glycerol is much thicker (more viscous) than ethylene glycol, which in turn is much thicker than ethanol.

14.25 From each pair, select the compound that is more soluble in water.

(a) CH_3OH or CH_3OCH_3

(b) $CH_3\overset{\underset{|}{OH}}{C}HCH_3$ or $CH_3\overset{\underset{||}{CH_2}}{C}CH_3$

(c) $CH_3CH_2CH_2SH$ or $CH_3CH_2CH_2OH$

14.26 Arrange the compounds in each set in order of decreasing solubility in water.

(a) Ethanol, butane, and diethyl ether

(b) 1-Hexanol, 1,2-hexanediol, and hexane

Synthesis of Alcohols (Review Chapter 12)

14.27 ■ Give the structural formula of an alkene or alkenes from which each alcohol can be prepared.

(a) 2-Butanol

(b) 1-Methylcyclohexanol

(c) 3-Hexanol

(d) 2-Methyl-2-pentanol

(e) Cyclopentanol

SECTION 14.2 What Are the Characteristic Reactions of Alcohols?

14.28 Show how to distinguish between cyclohexanol and cyclohexene by a simple chemical test. Tell what you would do, what you would expect to see, and how you would interpret your observation.

14.29 Compare the acidity of alcohols and phenols, which are both classes of organic compounds that contain an —OH group.

14.30 Both 2,6-diisopropylcyclohexanol and the intravenous anesthetic Propofol are insoluble in water. Show how these two compounds can be distinguished by their reaction with aqueous sodium hydroxide.

2,6-Diisopropylcyclohexanol   2,6-Diisopropylphenol (Propofol)

14.31 ■ Write equations for the reaction of 1-butanol, a primary alcohol, with these reagents.

(a) H_2SO_4, heat

(b) $K_2Cr_2O_7$, H_2SO_4

14.32 Write equations for the reaction of 2-butanol with these reagents.

(a) H_2SO_4, heat

(b) $K_2Cr_2O_7$, H_2SO_4

14.33 Write equations for reaction of each of the following compounds with $K_2Cr_2O_7/H_2SO_4$.

(a) 1-Octanol

(b) 1,4-Butanediol

14.34 ■ Show how to convert cyclohexanol to these compounds.

(a) Cyclohexene

(b) Cyclohexane

(c) Cyclohexanone

(d) Bromocyclohexane

14.35 Show reagents and experimental conditions to synthesize each compound from 1-propanol.

14.36 Show how to convert the alcohol below to compounds (a) and (b).

14.37 Name two important alcohols derived from ethylene and give two important uses of each.

14.38 Name two important alcohols derived from propene and give two important uses of each.

SECTION 14.3 What Are the Structures, Names, and Properties of Ethers?

14.39 Write the common name for each ether.

(a)

(b) $[CH_3(CH_2)_4]_2O$

(c) $\underset{\underset{CH_3}{|}}{CH_3}CH\underset{\underset{CH_3}{|}}{O}CHCH_3$

14.40 Write the common name for each ether.

(a)

(b)

(c)

SECTION 14.4 What Are the Structures, Names, and Properties of Thiols?

14.41 Write the IUPAC name for each thiol.

(a) $CH_3CH_2\underset{\underset{SH}{|}}{CH}CH_3$

(b) $CH_3CH_2CH_2CH_2SH$

(c)

14.42 Write the common name for each thiol in Problem 14.41.

14.43 Following are structural formulas for 1-butanol and 1-butanethiol. One of these compounds has a boiling point of 98°C, and the other has a boiling point of 117°C. Which compound has which boiling point?

$CH_3CH_2CH_2CH_2OH$ $CH_3CH_2CH_2CH_2SH$

1-Butanol   1-Butanethiol

14.44 Explain why methanethiol, CH_3SH, has a lower boiling point (6°C) than methanol, CH_3OH (65°C), even though methanethiol has a higher molecular weight.

Chemical Connections

14.45 (Chemical Connections 14A) When was nitroglycerin discovered? Is this substance a solid, a liquid, or a gas?

14.46 (Chemical Connections 14A) What was Alfred Nobel's discovery that made nitroglycerin safer to handle?

14.47 (Chemical Connections 14A) What is the relationship between the medical use of nitroglycerin to relieve the sharp chest pains (angina) associated with heart disease and the gas nitric oxide, NO?

14.48 (Chemical Connections 14B) What is the color of dichromate ion, $Cr_2O_7^{2-}$? What is the color of chromium(III) ion, Cr^{3+}? Explain how the conversion of one to the other is used in breath-alcohol screening.

14.49 (Chemical Connections 14B) The legal definition of being under the influence of alcohol is based on blood alcohol content. What is the relationship between breath alcohol content and blood alcohol content?

14.50 (Chemical Connections 14C) What does it mean to say that ethylene oxide is a highly strained molecule?

14.51 (Chemical Connections 14D) What are the advantages and disadvantages of using diethyl ether as an anesthetic?

14.52 (Chemical Connections 14D) Show that enflurane and isoflurane are constitutional isomers.

14.53 (Chemical Connections 14D) Would you expect enflurane and isoflurane to be soluble in water? Would you expect them to be soluble in organic solvents such as hexane?

Additional Problems

14.54 Write a balanced equation for the complete combustion of ethanol, the alcohol added to gasoline to produce gasohol.

14.55 Write a balanced equation for the complete combustion of methanol, the alcohol used in the engines of Formula 1 racing cars.

14.56 Knowing what you do about electronegativity, the polarity of covalent bonds, and hydrogen bonding, would you expect an N—H···N hydrogen bond to be stronger than, the same strength as, or weaker than an O—H···O hydrogen bond?

14.57 Draw structural formulas and write IUPAC names for the eight isomeric alcohols with the molecular formula $C_5H_{12}O$.

14.58 Draw structural formulas and write common names for the six isomeric ethers with the molecular formula $C_5H_{12}O$.

14.59 Explain why the boiling point of ethylene glycol (198°C) is so much higher than that of 1-propanol (97°C), even though their molecular weights are about the same.

14.60 1,4-Butanediol, hexane, and 1-pentanol have similar molecular weights. Their boiling points, arranged from lowest to highest, are 69°C, 138°C, and 230°C. Which compound has which boiling point?

14.61 Of the three compounds given in Problem 14.60, one is insoluble in water, another has a solubility of 2.3 g/100 g water, and one is infinitely soluble in water. Which compound has which solubility?

14.62 Each of the following compounds is a common organic solvent. From each pair of compounds, select the solvent with the greater solubility in water.

(a) CH_2Cl_2 or CH_3CH_2OH

(b) $CH_3CH_2OCH_2CH_3$ or CH_3CH_2OH

14.63 Show how to prepare each compound from 2-methyl-1-propanol.

(a) 2-Methylpropene

(b) 2-Methyl-2-propanol

(c) 2-Methylpropanoic acid, $(CH_3)_2CHCOOH$

14.64 ■ Show how to prepare each compound from 2-methylcyclohexanol.

Looking Ahead

14.65 Lipoic acid is a growth factor for many bacteria and protozoa and an essential component of several enzymes involved in human metabolism.

Lipoic acid

(a) Name the two functional groups in lipoic acid.

(b) At one stage in its function in human metabolism, the disulfide bond of lipoic acid is reduced to two thiol groups. Draw a structural formula for this reduced form of lipoic acid.

14.66 Following is a structural formula for the amino acid cysteine:

$$HS-CH_2-CH-\overset{\displaystyle O}{\overset{\displaystyle \|}{C}}-OH$$
$$\underset{NH_2}{|}$$

(a) Name the three functional groups in cysteine.

(b) In the human body, cysteine is oxidized to cystine, a disulfide. Draw a structural formula for cystine.

Chirality: The Handedness of Molecules

Lester Lefkowitz/Stone/Getty Images

Median cross section through the shell of a chambered nautilus found in the deep waters of the Pacific Ocean. The shell shows handedness; this cross section is a left-handed spiral.

15.1 | What Is Enantiomerism?

In Chapters 11 through 14 we studied two types of stereoisomers. Examples of stereoisomers include the *cis-trans* isomers of certain disubstituted cycloalkanes and appropriately substituted alkenes. Recall that stereoisomers have the same connectivity of their atoms, but a different orientation of their atoms in space.

cis-1,4-Dimethyl-cyclohexane   *trans*-1,4-Dimethyl-cyclohexane   *cis*-2-Butene   *trans*-2-Butene

In this chapter, we study the relationship between objects and their **mirror images;** that is, we study stereoisomers called enantiomers and

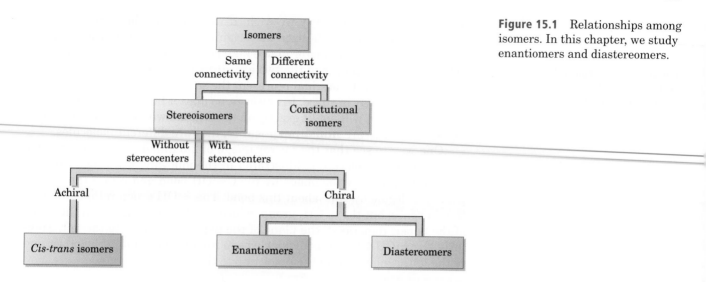

Figure 15.1 Relationships among isomers. In this chapter, we study enantiomers and diastereomers.

diastereomers. Figure 15.1 summarizes the relationship among these isomers and those we studied in Chapters 11 through 14.

The significance of enantiomers is that, except for inorganic compounds and a few simple organic compounds, the vast majority of molecules in the biological world show this type of isomerism. Furthermore, approximately one half of all medications used to treat humans show it as well.

As an example of enantiomerism, let us consider 2-butanol. As we discuss this molecule, we will focus on carbon 2, the carbon bearing the —OH group. What makes this carbon of interest is the fact that it has four different groups bonded to it: CH_3, H, OH, and CH_2CH_3.

2-Butanol

This structural formula does not show the three-dimensional shape of 2-butanol or the orientation of its atoms in space. To do so, we must consider the molecule as a three-dimensional object. On the left is what we will call the "original molecule" and a ball-and-stick model of it. In this drawing, the —OH and —CH_3 groups are in the plane of the paper, —H is behind the plane, and —CH_2CH_3 is in front of it. In the middle is a mirror. On the right is a **mirror image** of the original molecule along with a ball-and-stick model of the mirror image. Every molecule—and, in fact, every object in the world around us—has a mirror image.

The question we now need to ask is, "What is the relationship between the original molecule of 2-butanol and its mirror image?" To answer this question, we need to imagine that we can pick up the mirror image and

GOB
Chemistry⋅◇⋅Now™

Click *Coached Problems* to try a set of problems in **Identifying the Relationship Between Pairs of Structures**

The terms "superposable" and "superimposable" mean the same thing and are both used currently.

move it in space in any way we wish. If we can move the mirror image in space and find that it fits over the original molecule so that every bond, atom, and detail of the mirror image matches exactly the bonds, atoms, and details of the original, then the two are **superposable.** In other words, the mirror image and the original represent the same molecule; they are merely oriented differently in space. If, however, no matter how you turn the mirror image in space, it will not fit exactly on the original with every detail matching, then the two are **nonsuperposable;** they are different molecules.

One way to see that the mirror image of 2-butanol is not superposable on the original molecule is illustrated in the following drawings. Imagine that we hold the mirror image by the C—OH bond and rotate the bottom part of molecule by 180° about this bond. The —OH group retains its position in space, but the —CH₃ group, which was to the right and in the plane of the paper, remains in the plane of the paper but is now to the left. Similarly, the —CH₂CH₃, group, which was in front of the plane of the paper and to the left, is now behind the plane and to the right.

Original molecule   Mirror image of original molecule   Mirror image rotated by 180°

GOB
Chemistry ⚛ Now™

Click *Coached Problems* to examine pairs of **Structures and Their Mirror Images**

Now move the mirror image in space and try to fit it on the original molecule so that all bonds and atoms match.

Mirror image turned by 180° →

Original molecule →

The threads of a drill or screw twist along the axis of the helix. Some plants climb by sending out tendrils that twist helically.

By turning the mirror image as we did, its —OH and —CH₃ groups now fit exactly on top of the —OH and —CH₃ groups of the original molecule. However, the —H and —CH₂CH₃ groups of the two do not match. The —H is away from us in the original but toward us in the mirror image; the —CH₂CH₃ group is toward us in the original but away from us in the mirror image. We conclude that the original molecule of 2-butanol and its mirror image are nonsuperposable and, therefore, different compounds.

To summarize, we can turn and rotate the mirror image of 2-butanol in any direction in space, but as long as no bonds are broken and rearranged, we can make only two of the four groups bonded to carbon-2 of the mirror image coincide with those on its original molecule. Because 2-butanol and its

(a)   (b)

(*a*) Mirror images of two wood carvings. The mirror image cannot be superposed on the actual model. The man's right arm rests on the camera in the mirror image; in the actual statue, the man's left arm rests on the camera. (*b*) Left- and right-handed sea shells. If you cup a right-handed shell in your right hand with your thumb pointing from the narrow end to the wide end, the opening will be on your right.

Charles D. Winters

mirror image are not superposable, they are isomers. Isomers such as these are called **enantiomers.** Enantiomers, like gloves, always occur in pairs.

Objects that are not superposable on their mirror images are said to be **chiral** (pronounced "ki-ral," rhymes with spiral; from the Greek: *cheir*, "hand"); that is, they show handedness. We encounter chirality in three-dimensional objects of all sorts. Our left hand is chiral, and so is our right hand. A spiral binding on a notebook is chiral. A machine screw with a right-handed twist is chiral. A ship's propeller is chiral. As you examine the objects in the world around you, you will undoubtedly conclude that the vast majority of them are chiral.

The most common cause of enantiomerism in organic molecules is the presence of a carbon with four different groups bonded to it. Let us examine this statement further by considering 2-propanol, which has no such carbon atom. In this molecule, carbon-2 is bonded to three different groups, but no carbon is bonded to four different groups.

On the left is a three-dimensional representation of 2-propanol; on the right is its mirror image. Also shown are ball-and-stick models of each molecule.

> **Enantiomers** Stereoisomers that are nonsuperposable mirror images; refers to a relationship between pairs of objects

> **Chiral** From the Greek *cheir*, "hand"; an object that is not superposable on its mirror image

$$\text{OH}$$
$$\text{H}_3\text{C}\overset{|}{\underset{\text{CH}_3}{\text{C}}}\text{H}$$

Original molecule

$$\text{OH}$$
$$\text{H}\overset{|}{\underset{\text{H}_3\text{C}}{\text{C}}}\text{CH}_3$$

Mirror image

The question we now ask is, "What is the relationship of the mirror image to the original?" This time, let us rotate the mirror image by 120° about the C—OH bond, and then compare it to the original. After performing this rotation, we see that all atoms and bonds of the mirror image fit exactly on the original. Thus the structures we first drew for the original molecule and its mirror image are, in fact, the same molecule—just viewed from different perspectives (Figure 15.2).

If an object and its mirror image are superposable, then they are identical and enantiomerism is not possible. We say that such an object is **achiral** (without chirality); that is, it has no handedness. Examples of achiral objects include an undecorated cup, an unmarked baseball bat, a regular tetrahedron, a cube, and a sphere.

To repeat, the most common cause of chirality in organic molecules is a tetrahedral carbon atom with four different groups bonded to it. We call such

GOB
Chemistry ☆ Now™

Click *Coached Problems* to try a set of problems in **Identifying Chiral Carbon Centers**

> **Achiral** An object that lacks chirality; an object that is superposable on its mirror image

GOB
Chemistry⚬Now™

Active Figure 15.2 Rotation of the mirror image about the C—OH bond by 120° does not change the configuration of the stereocenter, but makes it easier to see that the mirror image is superposable on the original molecule. **See a simulation based on this figure, and take a short quiz on the concepts at http://now.brookscole.com/gob8 or on the CD.**

Figure 15.3 Three-dimensional representations of lactic acid and its mirror image.

Stereocenter A tetrahedral carbon atom that has four different groups bonded to it

a carbon atom a **stereocenter.** 2-Butanol has one stereocenter; 2-propanol has none. As another example of a molecule with a stereocenter, let us consider 2-hydroxypropanoic acid, more commonly named lactic acid. Lactic acid is a product of anaerobic glycolysis. (See Section 28.2 and Chemical Connections 28A.) It is also what gives sour cream its sour taste.

Figure 15.3 shows three-dimensional representations of lactic acid and its mirror image. In these representations, all bond angles about the central carbon atom are approximately 109.5°, and the four bonds from it are directed toward the corners of a regular tetrahedron. Lactic acid shows enantiomerism; that is, the original molecule and its mirror image are not superposable but rather are different molecules.

Racemic mixture A mixture of equal amounts of two enantiomers

An equimolar mixture of two enantiomers is called a **racemic mixture,** a term derived from the name racemic acid (Latin: *racemus*, "a cluster of grapes"). Racemic acid is the name originally given to the equimolar mixture of the enantiomers of tartaric acid that forms as a by-product during the fermentation of grape juice.

EXAMPLE 15.1

Each of the following molecules has one stereocenter marked by an asterisk. Draw three-dimensional representations for the enantiomers of each molecule.

(a) CH₃*CHCH₂CH₃ (with Cl on the starred carbon)

(b) (phenyl group)—*CHCH₃ (with NH₂ on the starred carbon)

HOW TO...

Draw Enantiomers

Now that we know what enantiomers are, we can think about how to represent their three-dimensional structures on a two-dimensional surface. Let us take one of the enantiomers of 2-butanol as an example. Following are a molecular model of one enantiomer and four different three-dimensional representations of it:

(1)   (2)   (3)   (4)

In our initial discussions of 2-butanol, we used representation (1) to show the tetrahedral geometry of the stereocenter. In this representation, two groups (OH and CH_3) are in the plane of the paper, one (CH_2CH_3) is coming out of the plane toward us, and one (H) is behind the plane and going away from us. We can turn representation (1) slightly in space and tip it a bit to place the carbon framework in the plane of the paper. Doing so gives us representation (2), in which there are still two groups in the plane of the paper, one coming toward us and one going away from us.

For an even more abbreviated representation of this enantiomer of 2-butanol, we can change representation (2) into the line-angle formula (3). Although we do not normally show hydrogens in a line-angle formula, we do so here just to remind ourselves that the fourth group on the stereocenter is really there and that it is H. Finally, we can carry the abbreviation a step further and write 2-butanol as the line-angle formula (4). Here, we omit the H on the stereocenter, but we know that it must be there (carbon needs four bonds), and we know that it must be behind the plane of the paper. Clearly, the abbreviated formulas (3) and (4) are the easiest to write, and we will rely on this type of representation throughout the remainder of the text.

When you have to write three-dimensional representations of stereocenters, try to keep the carbon framework in the plane of the paper and the other two atoms or groups of atoms on the stereocenter toward and away from you, respectively. Using representation (4) as a model, we get the following alternative representations of its mirror image:

One enantiomer   Alternative representations of
of 2-butanol   its mirror image

Solution

First, draw the carbon stereocenter showing the tetrahedral orientation of its four bonds. One way to do so is to draw two bonds in the plane of the paper, a third bond toward you in front of the plane, and the fourth bond away from you behind the plane. Next, place the four groups bonded to the stereocenter on these positions. To draw an original of (a), for example, place the CH_3 and Cl groups in the plane of the paper. Place H away from you and the CH_2CH_3 toward you; this orientation gives the enantiomer of (a) on the left. Its mirror image is on the right.

(a)

(b)

Problem 15.1

Each of the following molecules has one stereocenter marked by an asterisk. Draw three-dimensional representations for the enantiomers of each molecule.

(a)   (b)

15.2 | How Do We Specify the Configuration of a Stereocenter?

Because enantiomers are different compounds, each must have a different name. The over-the-counter drug ibuprofen, for example, shows enantiomerism and can exist as the pair of enantiomers shown here:

The inactive enantiomer of ibuprofen

The active enantiomer of ibuprofen

Only one enantiomer of ibuprofen is biologically active. It reaches therapeutic concentrations in the human body in approximately 12 minutes, whereas the racemic mixture takes approximately 30 minutes to achieve this feat. However, in this case, the inactive enantiomer is not wasted. The body converts it to the active enantiomer, but this process takes time.

What we need is a way to designate which enantiomer of ibuprofen (or one of any other pair of enantiomers, for that matter) is which without having to draw and point to one or the other of the enantiomers. To do so, chemists have developed the **R,S system.** The first step in assigning an *R* or *S* configuration to a stereocenter is to arrange the groups bonded to it in order of priority. Priority is based on atomic number: The higher the atomic number, the higher the priority. If a priority cannot be assigned on the basis of the atoms bonded directly to the stereocenter, look at the next atom or set of atoms and continue to the first point of difference; that is, continue until you can assign a priority.

Table 15.1 shows the priorities of the most common groups we encounter in organic and biochemistry. In the *R,S* system, a $C=O$ is treated as if carbon were bonded to two oxygens by single bonds; thus $CH=O$ has a higher priority than $-CH_2OH$, in which carbon is bonded to only one oxygen.

GOB
Chemistry Now™
Click *Coached Problems* to try a set of problems in **Identifying the Configuration of a Series of Stereocenters**

R,S system A set of rules for specifying the configuration about a stereocenter

---

EXAMPLE 15.2

Assign priorities to the groups in each set.
(a) $-CH_2OH$ and $-CH_2CH_2OH$
(b) $-CH_2CH_2OH$ and $-CH_2NH_2$

Solution
(a) The first point of difference is O of the $-OH$ group compared to C of the $-CH_2OH$ group.

First point of
difference

$-CH_2OH$ $-CH_2CH_2OH$
Higher priority   Lower priority

(b) The first point of difference is C of the CH_2OH group compared to N of the NH_2 group.

First point of
difference

$-CH_2CH_2OH$ $-CH_2NH_2$
Lower priority   **Higher priority**

Problem 15.2

Assign priorities to the groups in each set.

$$\qquad\qquad\qquad\qquad \overset{\displaystyle O}{\overset{\displaystyle \|}{}}$$

(a) $-CH_2OH$ and $-CH_2CH_2\overset{O}{\overset{\|}{C}}OH$

$$\qquad\qquad\qquad \overset{\displaystyle O}{\overset{\displaystyle \|}{}}$$

(b) $-CH_2NH_2$ and $-CH_2\overset{O}{\overset{\|}{C}}OH$

Table 15.1 R,S Priorities of Some Common Groups

Atom or Group	Reason for Priority: First Point of Difference (Atomic Number)
—I	iodine (53)
—Br	bromine (35)
—Cl	chlorine (17)
—SH	sulfur (16)
—OH	oxygen (8)
—NH$_2$	nitrogen (7)
$\overset{\text{O}}{\overset{\|}{-\text{COH}}}$	carbon to oxygen, oxygen, then oxygen (6 ⟶ 8, 8, 8)
$\overset{\text{O}}{\overset{\|}{-\text{CNH}_2}}$	carbon to oxygen, oxygen, then nitrogen (6 ⟶ 8, 8, 7)
$\overset{\text{O}}{\overset{\|}{-\text{CH}}}$	carbon to oxygen, oxygen, then hydrogen (6 ⟶ 8, 8, 1)
—CH$_2$OH	carbon to oxygen (6 ⟶ 8)
—CH$_2$NH$_2$	carbon to nitrogen (6 ⟶ 7)
—CH$_2$CH$_3$	carbon to carbon (6 ⟶ 6)
—CH$_2$H	carbon to hydrogen (6 ⟶ 1)
—H	hydrogen (1)

Increasing priority ↑

To assign an *R* or *S* configuration to a stereocenter:

1. Assign a priority from 1 (highest) to 4 (lowest) to each group bonded to the stereocenter.

2. Orient the molecule in space so that the lowest-priority group (4) is directed away from you, as would be, for instance, the steering column of a car. The three higher-priority groups (1–3) then project toward you, as would the spokes of a steering wheel.

3. Read the three groups projecting toward you in order from highest (1) to lowest (4) priority.

Group of lowest priority points away from you

R Used in the *R,S* system to show that, when the lowest-priority group is away from you, the order of priority of groups on a stereocenter is clockwise

S Used in the *R,S* system to show that, when the lowest-priority group is away from you, the order of priority of groups on a stereocenter is counterclockwise

4. If reading the groups 1-2-3 proceeds in a clockwise direction, the configuration is designated as **R** (Latin: *rectus*, "straight"); if reading the groups 1-2-3 proceeds in a counterclockwise direction, the configuration is **S** (Latin: *sinister*, "left"). You can also visualize this system as follows: Turning the steering wheel to the right equals *R* and turning it to the left equals *S*.

EXAMPLE 15.3

Assign an R or S configuration to each stereocenter.

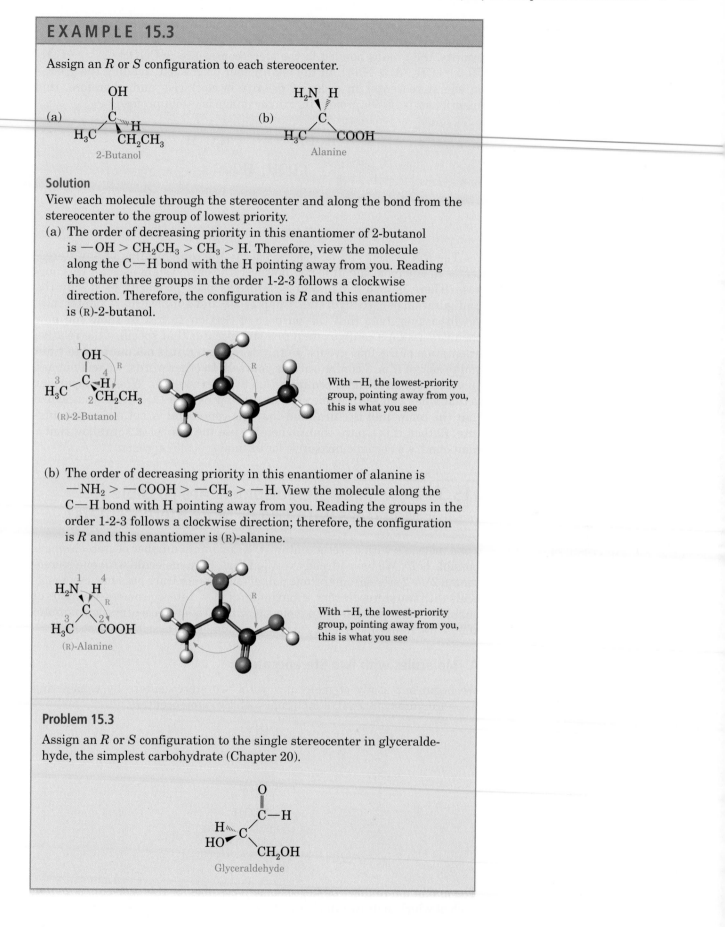

(a) 2-Butanol

(b) Alanine

Solution

View each molecule through the stereocenter and along the bond from the stereocenter to the group of lowest priority.

(a) The order of decreasing priority in this enantiomer of 2-butanol is $—OH > CH_2CH_3 > CH_3 > H$. Therefore, view the molecule along the C—H bond with the H pointing away from you. Reading the other three groups in the order 1-2-3 follows a clockwise direction. Therefore, the configuration is R and this enantiomer is (R)-2-butanol.

(R)-2-Butanol

With —H, the lowest-priority group, pointing away from you, this is what you see

(b) The order of decreasing priority in this enantiomer of alanine is $—NH_2 > —COOH > —CH_3 > —H$. View the molecule along the C—H bond with H pointing away from you. Reading the groups in the order 1-2-3 follows a clockwise direction; therefore, the configuration is R and this enantiomer is (R)-alanine.

(R)-Alanine

With —H, the lowest-priority group, pointing away from you, this is what you see

Problem 15.3

Assign an R or S configuration to the single stereocenter in glyceraldehyde, the simplest carbohydrate (Chapter 20).

Glyceraldehyde

Now let us return to our three-dimensional drawing of the enantiomers of ibuprofen and assign each an R or S configuration. In order of decreasing priority, the groups bonded to the stereocenter are —COOH (1) > —C_6H_5 (2) > —CH_3 (3) > H (4). In the enantiomer on the left, reading the groups on the stereocenter in order of priority is clockwise and, therefore, this enantiomer is (R)-ibuprofen. Its mirror image is (S)-ibuprofen.

(R)-Ibuprofen
(the inactive enantiomer)

(S)-Ibuprofen
(the active enantiomer)

The R,S system can be used to specify the configuration of any stereocenter in any molecule. It is not, however, the only system used for this purpose. There is also a D,L system, which is used primarily to specify the configuration of carbohydrates (Chapter 20) and amino acids (Chapter 22).

In closing, note that the purpose of Section 15.2 is to show you how chemists assign a configuration to a stereocenter that specifies the relative orientation of the four groups on the stereocenter. It is not our goal to train you to assign R or S configurations yourself. In other words, we will not ask you to assign an R and S configuration to a stereocenter. What is important is that when you see a name such as (S)-Naproxen or (R)-Plavix, you realize that the compound is chiral and that the compound is not a racemic mixture. Rather, it is a pure enantiomer. We use the symbol (R,S) to show that a compound is a racemic mixture—for example, (R,S)-Naproxen.

15.3 | How Many Stereoisomers Are Possible for Molecules with Two or More Stereocenters?

For a molecule with n stereocenters, the maximum number of stereoisomers possible is 2^n. We have already verified that, for a molecule with one stereocenter, $2^1 = 2$ stereoisomers (one pair of enantiomers) are possible. For a molecule with two stereocenters, a maximum of $2^2 = 4$ stereoisomers (two pairs of enantiomers) is possible; for a molecule with three stereocenters, a maximum of $2^3 = 8$ stereoisomers (four pairs of enantiomers) is possible; and so forth.

A. Molecules with Two Stereocenters

We begin our study of molecules with two stereocenters by considering 2,3,4-trihydroxybutanal, a molecule with two stereocenters.

2,3,4-Trihydroxybutanal
(2 stereocenters;
4 stereoisomers
are possible)

The maximum number of stereoisomers possible for this molecule is $2^2 = 4$, each of which is drawn in Figure 15.4.

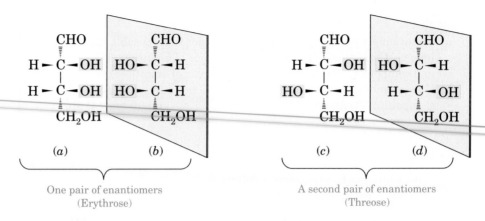

Figure 15.4 The four stereoisomers of 2,3,4-trihydroxybutanal.

One pair of enantiomers
(Erythrose)

A second pair of enantiomers
(Threose)

Stereoisomers (a) and (b) are nonsuperposable mirror images and are, therefore, a pair of enantiomers. Stereoisomers (c) and (d) are also nonsuperposable mirror images and are a second pair of enantiomers. We describe the four stereoisomers of 2,3,4-trihydroxybutanal by saying that they consist of two pairs of enantiomers. Enantiomers (a) and (b) are named **erythrose.** Erythrose is synthesized in erythrocytes (red blood cells); hence the derivation of its name. Enantiomers (c) and (d) are named **threose.** Erythrose and threose belong to the class of compounds called carbohydrates, which we discuss in Chapter 20.

We have specified the relationship between (a) and (b) and that between (c) and (d). What is the relationship between (a) and (c), between (a) and (d), between (b) and (c), and between (b) and (d)? The answer is that they are **diastereomers**—stereoisomers that are not mirror images.

Diastereomers Stereoisomers that are not mirror images

EXAMPLE 15.4

1,2,3-Butanetriol has two stereocenters (carbons 2 and 3); thus $2^2 = 4$ stereoisomers are possible for it. Following are three-dimensional representations for each.

(a) Which stereoisomers are pairs of enantiomers?
(b) Which stereoisomers are diastereomers?

The diagram shows the relationship among these four stereoisomers.

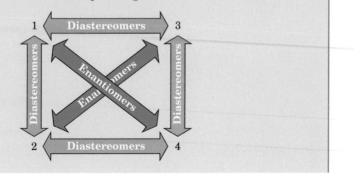

Solution

(a) Enantiomers are stereoisomers that are nonsuperposable mirror images. Compounds (1) and (4) are one pair of enantiomers, and compounds (2) and (3) are a second pair of enantiomers.

(b) Diastereomers are stereoisomers that are not mirror images. Compounds (1) and (2), (1) and (3), (2) and (4), and (3) and (4) are diastereomers.

Problem 15.4

3-Amino-2-butanol has two stereocenters (carbons 2 and 3); thus $2^2 = 4$ stereoisomers are possible for it.

(a) Which stereoisomers are pairs of enantiomers?

(b) Which sets of stereoisomers are diastereomers?

We can analyze chirality in cyclic molecules with two stereocenters in the same way we analyzed it in acyclic compounds.

EXAMPLE 15.5

How many stereoisomers are possible for 3-methylcyclopentanol?

Solution

Carbons 1 and 3 of this compound are stereocenters. There are, therefore, $2^2 = 4$ stereoisomers possible for this molecule. The *cis* isomer exists as one pair of enantiomers; the *trans* isomer exists as a second pair of enantiomers.

cis-3-Methylcyclopentanol
(a pair of enantiomers)

trans-3-Methylcyclopentanol
(a second pair of enantiomers)

Problem 15.5

How many stereoisomers are possible for 3-methylcyclohexanol?

EXAMPLE 15.6

Mark the stereocenters in each compound with an asterisk. How many stereoisomers are possible for each?

Solution

Each stereocenter is marked with an asterisk, and the number of stereoisomers possible appears under each compound. In (a), the carbon bearing the two methyl groups is not a stereocenter; this carbon has only three different groups bonded to it.

Problem 15.6

Mark all stereocenters in each compound with an asterisk. How many stereoisomers are possible for each?

B. Molecules with Three or More Stereocenters

The 2^n rule applies equally well to molecules with three or more stereocenters. The following disubstituted cyclohexanol has three stereocenters, each marked with an asterisk. A maximum of $2^3 = 8$ stereoisomers is possible for this molecule. Menthol, one of the eight, has the configuration shown on the right. Menthol is present in peppermint and other mint oils.

2-Isopropyl-5-methyl-
cyclohexanol
(3 stereocenters;
8 stereoisomers
are possible)

Menthol
(one of the 8
possible stereoisomers)

Menthol drawn as a
chair conformation
(note that the 3 groups
on the cyclohexane ring
are all equatorial)

CHEMICAL CONNECTIONS 15A

Chiral Drugs

Some common drugs used in human medicine—for example, aspirin—are achiral. Others, such as the penicillin and erythromycin classes of antibiotics and the drug captopril, are chiral and are sold as single enantiomers. Captopril is very effective for the treatment of high blood pressure and congestive heart failure. It is manufactured and sold as the (s,s)-stereoisomer.

Captopril

A large number of chiral drugs, however, are sold as racemic mixtures. The popular analgesic ibuprofen (the active ingredient in Motrin, Advil, and many other non-aspirin analgesics) is an example.

Recently, the U.S. Food and Drug Administration established new guidelines for the testing and marketing of chiral drugs. After reviewing these guidelines, many drug companies have decided to develop only single enantiomers of new chiral drugs.

In addition to regulatory pressure, pharmaceutical developers must deal with patent considerations. If a company has a patent on a racemic mixture of a drug, a new patent can often be taken out on one of its enantiomers.

Cholesterol, a more complicated molecule, has eight stereocenters. To identify them, remember to add an appropriate number of hydrogens to complete the tetravalence of each carbon you think might be a stereocenter.

Cholesterol has 8 stereocenters;
256 stereoisomers are possible

This is the stereoisomer
found in human metabolism

15.4 | What Is Optical Activity, and How Is Chirality Detected in the Laboratory?

A. Plane-Polarized Light

As we have already established, the two members of a pair of enantiomers are different compounds, and we must expect, therefore, that some of their properties differ. One such property relates to their effect on the plane of polarized light. Each member of a pair of enantiomers rotates the plane of polarized light; for this reason, each enantiomer is said to be **optically active.** To understand how optical activity is detected in the laboratory, we must first understand plane-polarized light and how a polarimeter, the instrument used to detect optical activity, works.

Ordinary light consists of waves vibrating in all planes perpendicular to its direction of propagation. Certain materials, such as a Polaroid sheet (a plastic film like that used in polarized sunglasses), selectively transmit light waves vibrating only in parallel planes. Electromagnetic radiation vibrating in only parallel planes is said to be **plane polarized.**

Optically active Showing that a compound rotates the plane of polarized light

Plane-polarized light Light with waves vibrating in only parallel planes

Figure 15.5 Schematic diagram of a polarimeter with its sample tube containing a solution of an optically active compound. The analyzer has been turned clockwise by α degrees to restore the light field.

B. A Polarimeter

A **polarimeter** consists of a light source emitting unpolarized light, a polarizer, an analyzer, and a sample tube (Figure 15.5). If the sample tube is empty, the intensity of light reaching the detector (in this case, your eye) is at its maximum when the axes of the polarizer and analyzer are parallel to each other. If the analyzer is turned either clockwise or counterclockwise, less light is transmitted. When the axis of the analyzer is at right angles to the axis of the polarizer, the field of view is dark.

When a solution of an optically active compound is placed in the sample tube, it rotates the plane of the polarized light. If it rotates the plane clockwise, we say it is **dextrorotatory;** if it rotates the plane counterclockwise, we say it is **levorotatory.** Each member of a pair of enantiomers rotates the plane of polarized light by the same number of degrees, but in opposite directions. If one enantiomer is dextrorotatory, the other is levorotatory.

The number of degrees by which an optically active compound rotates the plane of polarized light is called its **specific rotation** and is given the symbol $[\alpha]$. Specific rotation is defined as the observed rotation of an optically active substance at a concentration of 1 g/mL in a sample tube that is 10 cm long. A dextrorotatory compound is indicated by a plus sign in parentheses, (+), and a levorotatory compound is indicated by a minus sign in parentheses, (−). It is common practice to report the temperature (in °C) at which the measurement is made and the wavelength of light used. The most common wavelength of light used in polarimetry is the sodium D line, the same wavelength responsible for the yellow color of sodium-vapor lamps.

Following are specific rotations for the enantiomers of lactic acid measured at 21°C and using the D line of a sodium-vapor lamp as the light source:

Dextrorotatory The clockwise (to the right) rotation of the plane of polarized light in a polarimeter

Levorotatory The counterclockwise (to the left) rotation of the plane of polarized light in a polarimeter

The (+) enantiomer of lactic acid is produced by muscle tissue in humans. The (−) enantiomer is found in sour cream and sour milk.

COOH
|
H_3C—C⸺H
ˋOH

(S)-(+)-Lactic acid
$[\alpha]_D^{21} = +2.6°$

COOH
|
H⸺C
HO   ˋCH_3

(R)-(−)-Lactic acid
$[\alpha]_D^{21} = -2.6°$

15.5 | What Is the Significance of Chirality in the Biological World?

Except for inorganic salts and a few low-molecular-weight organic substances, the majority of molecules in living systems—both plant and animal—are chiral. Although these molecules can exist as a number of stereoisomers, almost invariably only one stereoisomer is found in nature. Of course, instances do occur in which more than one stereoisomer is found, but these isomers rarely exist together in the same biological system.

A. Chirality in Biomolecules

Perhaps the most conspicuous examples of chirality among biological molecules are the enzymes, all of which have many stereocenters. Consider chymotrypsin, an enzyme in the intestines of animals that catalyzes the digestion of proteins (Chapter 23). Chymotrypsin has 251 stereocenters. The maximum number of stereoisomers possible is 2^{251}—a staggeringly large number, almost beyond comprehension. Fortunately, nature does not squander its precious energy and resources unnecessarily; any given organism produces and uses only one of these stereoisomers.

B. How Does an Enzyme Distinguishes Between a Molecule and Its Enantiomer?

An enzyme catalyzes a biological reaction of a molecule by first positioning the molecule at a **binding site** on the enzyme surface. An enzyme with specific binding sites for three of the four groups on a stereocenter can distinguish between a molecule and its enantiomer or one of its diastereomers. Assume, for example, that an enzyme involved in catalyzing a reaction of glyceraldehyde has three binding sites: one specific for —H, a second specific for —OH, and a third specific for —CHO. Assume further that the three sites are arranged on the enzyme surface as shown in Figure 15.6. The enzyme can distinguish (R)-glyceraldehyde (the natural or biologically active form) from its enantiomer because the natural enantiomer can be adsorbed with three groups interacting with their appropriate binding sites;

The horns of this African gazelle show chirality; one horn is the mirror image of the other.

Figure 15.6 A schematic diagram of an enzyme surface that can interact with (R)-glyceraldehyde at three binding sites, but with (S)-glyceraldehyde at only two of these sites.

(R)-Glyceraldehyde fits the three binding sites on surface

(S)-Glyceraldehyde fits only two of the three binding sites

for the *S* enantiomer, at best only two groups can interact with these binding sites.

Because interactions between molecules in living systems take place in a chiral environment, it should come as no surprise that a molecule and its enantiomer or one of its diastereomers elicit different physiological responses. As we have already seen, (s)-ibuprofen is active as a pain and fever reliever, while its *R* enantiomer is inactive. The *S* enantiomer of the closely related analgesic naproxen is also the active pain reliever of this compound, but its *R* enantiomer is a liver toxin!

(s)-Ibuprofen   (s)-Naproxen

SUMMARY OF KEY QUESTIONS

SECTION 15.1 What Is Enantiomerism?

- A **mirror image** is the reflection of an object in a mirror.
- **Enantiomers** are a pair of stereoisomers that are nonsuperposable mirror images.
- A **racemic mixture** contains equal amounts of two enantiomers.
- **Diastereomers** are stereoisomers that are not mirror images.
- An object that is not superposable on its mirror image is said to be **chiral.** An **achiral** object lacks chirality; that is, it has a superposable mirror image.
- The most common cause of chirality in organic molecules is the presence of a tetrahedral carbon atom with four different groups bonded to it. Such a carbon is called a **stereocenter.**

SECTION 15.2 How Do We Specify the Configuration of a Stereocenter?

- We use the ***R,S* system** to specify the configuration at a stereocenter.

SECTION 15.3 How Many Stereoisomers Are Possible for Molecules with Two or More Stereocenters?

- For a molecule with n stereocenters, the maximum number of stereoisomers possible is 2^n.

SECTION 15.4 What Is Optical Activity, and How Is Chirality Detected in the Laboratory?

- Light with waves that vibrate in only parallel planes is said to be **plane polarized.**
- We use a **polarimeter** to measure optical activity. A compound is said to be **optically active** if it rotates the plane of polarized light.
- If a compound rotates the plane clockwise, it is **dextrorotatory.** If it rotates the plane counterclockwise, it is **levorotatory.**
- Each member of a pair of enantiomers rotates the plane of polarized light by an equal number of degrees, but in opposite directions.

SECTION 15.5 What Is the Significance of Chirality in the Biological World?

- An enzyme catalyzes biological reactions of molecules by first positioning them at binding sites on its surface. An enzyme with binding sites specific for three of the four groups on a stereocenter can distinguish between a molecule and its enantiomer or one of its diastereomers.

PROBLEMS

SECTION 15.1 What Is Enantiomerism?

15.7 What does the term "chiral" mean? Give an example of a chiral molecule.

15.8 What does the term "achiral" mean? Give an example of an achiral molecule.

15.9 Define the term "stereoisomer." Name three types of stereoisomers.

15.10 In what way are constitutional isomers different from stereoisomers? In what way are they the same?

15.11 ■ Which of the following objects are chiral (assume that there is no label or other identifying mark)?
(a) Pair of scissors   (b) Tennis ball
(c) Paper clip   (d) Beaker
(e) The swirl created in water as it drains out of a sink or bathtub

15.12 2-Pentanol is chiral but 3-pentanol is not. Explain.

15.13 2-Butene exists as a pair of *cis-trans* isomers. Is *cis*-2-butene chiral? Is *trans*-2-butene chiral? Explain.

15.14 Explain why the carbon of a carbonyl group cannot be a stereocenter.

15.15 ■ Which of the following compounds contain stereocenters?
(a) 2-Chloropentane   (b) 3-Chloropentane
(c) 3-Chloro-1-butene (d) 1,2-Dichloropropane

15.16 Which of the following compounds contain stereocenters?
(a) Cyclopentanol
(b) 1-Chloro-2-propanol
(c) 2-Methylcyclopentanol
(d) 1-Phenyl-1-propanol

15.17 ■ Using only C, H, and O, write structural formulas for the lowest-molecular-weight chiral molecule of each class.
(a) Alkane   (b) Alkene
(c) Alcohol   (d) Aldehyde
(e) Ketone   (f) Carboxylic acid

15.18 ■ Draw the mirror image for each molecule.

15.19 Draw the mirror image for each molecule.

SECTION 15.3 How Many Stereoisomers Are Possible for Molecules with Two or More Stereocenters?

15.20 ■ Mark each stereocenter in these molecules with an asterisk. Note that not all contain stereocenters.

15.21 Mark each stereocenter in these molecules with an asterisk. Note that not all contain stereocenters.

15.22 ■ Label all stereocenters in each molecule with an asterisk. How many stereoisomers are possible for each molecule?

(a) $CH_3CHCHCHCOOH$ with OH and OH groups

(b) CH_2COOH / $CHCOOH$ / $HO-CHCOOH$

(c)   (d)

15.23 Label all stereocenters in each molecule with an asterisk. How many stereoisomers are possible for each molecule?

(a)   (b)

(a)   (b)

15.24 For centuries, Chinese herbal medicine has used extracts of *Ephedra sinica* to treat asthma. The asthma-relieving component of this plant is ephedrine, a very potent dilator of the air passages of the lungs. The naturally occurring stereoisomer is levorotatory and has the following structure:

Ephedrine $[\alpha]_D^{21} = -41°$

(a) Mark each stereocenter in epinephrine with an asterisk.

(b) How many stereoisomers are possible for this compound?

15.25 The specific rotation of naturally occurring ephedrine, shown in Problem 15.24, is −41°. What is the specific rotation of its enantiomer?

15.26 What is a racemic mixture? Is a racemic mixture optically active? That is, will it rotate the plane of polarized light?

Chemical Connections

15.27 (Chemical Connections 15A) What does it mean to say that a drug is *chiral*? If a drug is chiral, will it be optically active? That is, will it rotate the plane of polarized light?

Additional Problems

15.28 Which of the eight alcohols with a molecular formula of $C_5H_{12}O$ are chiral?

15.29 Write the structural formula of an alcohol with the molecular formula $C_6H_{14}O$ that contains two stereocenters.

15.30 Which carboxylic acids with a molecular formula of $C_6H_{12}O_2$ are chiral?

15.31 ■ Following are structural formulas for three of the drugs most widely prescribed to treat depression. Label all stereocenters in each, and state the number of stereoisomers possible for each.

(a)

Fluoxetine
(Prozac)

(b)

Sertraline
(Zoloft)

(c)

Paroxetine
(Paxil)

15.32 ■ Label the four stereocenters in amoxicillin, which belong to the family of semisynthetic penicillins.

Amoxicillin

15.33 Consider a cyclohexane ring substituted with one hydroxyl group and one methyl group. Draw a structural formula for a compound of this composition that

(a) Does not show *cis-trans* isomerism and has no stereocenters.

(b) Shows *cis-trans* isomerism but has no stereocenters.

(c) Shows *cis-trans* isomerism and has two stereocenters.

15.34 The next time you have the opportunity to view a collection of seashells that have a helical twist, study the chirality (handedness) of their twists. For each kind of shell, do you find an equal number of left-handed and right-handed twists or, for example, do they all have the same handedness? What about the handedness among various kinds of shells with helical twists?

15.35 The next time you have an opportunity to examine any of the seemingly endless varieties of blond spiral pasta (rotini, fusilli, radiatori, tortiglione, and so forth), examine their twists. Do the twists of any one kind all have a right-handed twist or a left-handed twist, or are they a racemic mixture?

15.36 Think about the helical coil of a telephone cord or the spiral binding on a notebook. Suppose that you view the spiral from one end and find that it has a left-handed twist. If you view the same spiral from the other end, does it have a left-handed twist from that end as well or does it have a right-handed twist?

Looking Ahead

15.37 ■ Following is a chair conformation of glucose, the most prevalent carbohydrate in the biological world (Chapter 20).

(a) Identify the five stereocenters in this molecule.

(b) How many stereoisomers are possible?

(c) How many pairs of enantiomers are possible?

15.38 Triamcinolone acetonide, the active ingredient in Azmacort Inhalation Aerosol, is a steroid used to treat bronchial asthma.

Triamcinolone acetonide

(a) Label the eight stereocenters in this molecule.

(b) How many stereoisomers are possible for it? (Of these, the stereoisomer with the configuration shown here is the active ingredient in Azmacort.)

Amines

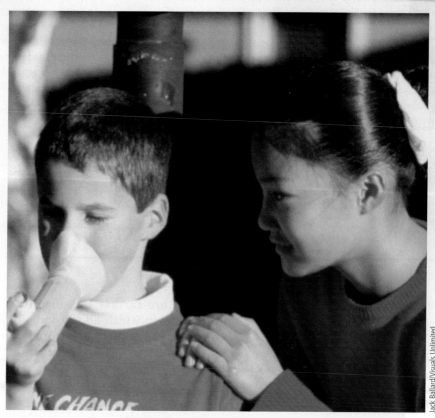

This inhaler delivers puffs of albuterol (Proventil), a potent synthetic bronchodilator whose structure is patterned after that of epinephrine (adrenaline). See Chemical Connections 16E, "Epinephrine: A Prototype for the Development of New Bronchodilators."

16.1 What Are Amines?

Carbon, hydrogen, and oxygen are the three most common elements in organic compounds. Because of the wide distribution of amines in the biological world, nitrogen is the fourth most common element of organic compounds. The most important chemical property of amines is their basicity.

Amines (Section 10.4B) are classified as **primary (1°), secondary (2°),** or **tertiary (3°),** depending on the number of carbon groups bonded to nitrogen.

GOB
Chemistry ❖ Now™
Look for this logo in the chapter and go to GOB ChemistryNow at
http://now.brookscole.com/gob8
or on the CD for tutorials, simulations, and problems.

$$CH_3-NH_2 \qquad CH_3-\overset{\overset{\displaystyle H}{|}}{N}-CH_3 \qquad CH_3-\overset{\overset{\displaystyle CH_3}{|}}{N}-CH_3$$

Methylamine   Dimethylamine   Trimethylamine
(a 1° amine)   (a 2° amine)   (a 3° amine)

CHEMICAL CONNECTIONS 16A

Amphetamines (Pep Pills)

Amphetamine, methamphetamine, and phentermine—all synthetic amines—are powerful stimulants of the central nervous system. Like most other amines, they are stored and administered as their salts. The sulfate salt of amphetamine is prescribed as Benzedrine, the hydrochloride salt of the *S* enantiomer of methamphetamine as Methedrine, and the hydrochloride salt of phentermine as Fastin.

These three amines have similar physiological effects and are referred to by the general name *amphetamines*. Structurally, they have in common a benzene ring with a three-carbon side chain and an amine nitrogen on the second carbon of the side chain. Physiologically, they share an ability to reduce fatigue and diminish hunger by raising the glucose level of the blood. Because of these

properties, amphetamines are widely prescribed to counter mild depression, reduce hyperactivity in children, and suppress appetite in people who are trying to lose weight. They are also used illegally to reduce fatigue and elevate mood.

Abuse of amphetamines can have severe effects on both body and mind. These drugs are addictive, concentrate in the brain and nervous system, and can lead to long periods of sleeplessness, loss of weight, and paranoia.

The action of amphetamines is similar to that of epinephrine (Chemical Connections 16E), the hydrochloride salt of which is named Adrenalin.

Amphetamine
(Benzedrine)

(*s*)-Methamphetamine
(Methedrine)

Phentermine
(Fastin)

Aliphatic amine An amine in which nitrogen is bonded only to alkyl groups

Aromatic amine An amine in which nitrogen is bonded to one or more aromatic rings

GOB
Chemistry⚛Now™
Click *Coached Problems* to classify a **Series of Amines**

Amines are further classified as aliphatic or aromatic. An **aliphatic amine** is one in which all the carbons bonded to nitrogen are derived from alkyl groups. An **aromatic amine** is one in which one or more of the groups bonded to nitrogen are aryl groups.

Aniline
(a 1° aromatic amine)

N-Methylaniline
(a 2° aromatic amine)

Benzyldimethylamine
(a 3° aliphatic amine)

Heterocyclic amine An amine in which nitrogen is one of the atoms of a ring

Heterocyclic aromatic amine An amine in which nitrogen is one of the atoms of an aromatic ring

An amine in which the nitrogen atom is part of a ring is classified as a **heterocyclic amine.** When the ring is saturated, the amine is classified as a **heterocyclic aliphatic amine.** When the nitrogen is part of an aromatic ring (Section 13.1), the amine is classified as a **heterocyclic aromatic amine.** The most important heterocyclic aromatic amines are pyridine and pyrimidine, in which nitrogen atoms replace first one and then two CH groups of a benzene ring. Pyrimidine and purine serve as the building blocks for the amine bases of DNA and RNA (Chapter 25).

Pyrrolidine Piperidine
(heterocyclic aliphatic amines)

Pyridine   Pyrimidine   Imidazole   Purine
(heterocyclic aromatic amines)

EXAMPLE 16.1

How many hydrogen atoms does piperidine have? How many does pyridine have? Write the molecular formula of each amine.

Solution

Recall that hydrogen atoms are not shown in line-angle formulas. Piperidine has 11 hydrogen atoms, and its molecular formula is $C_5H_{11}N$. Pyridine has 5 hydrogen atoms, and its molecular formula is C_5H_5N.

Problem 16.1

How many hydrogen atoms does pyrrolidine have? How many does purine have? Write the molecular formula of each amine.

CHEMICAL CONNECTIONS 16B

Alkaloids

Alkaloids are basic nitrogen-containing compounds found in the roots, bark, leaves, berries, or fruits of plants. In almost all alkaloids, the nitrogen atom is part of a ring. The name "alkaloid" was chosen because these compounds are alkali-like (*alkali* is an older term for a basic substance) and react with strong acids to give water-soluble salts. Thousands of different alkaloids have been extracted from plant sources, many of which are used in modern medicine.

When administered to animals, including humans, alkaloids have pronounced physiological effects. Whatever their individual effects, most alkaloids are toxic in large enough doses. For some, the toxic dose is very small!

ration, paralysis, and eventually death. It was the toxic substance in the "poison hemlock" used in the death of Socrates. Water hemlock is easily confused with Queen Anne's lace, a type of wild carrot—a mistake that has killed numerous people.

(s)-Nicotine occurs in the tobacco plant. In small doses, it is an addictive stimulant. In larger doses, this substance causes depression, nausea, and vomiting. In still larger doses, it is a deadly poison. Solutions of nicotine in water are used as insecticides.

(s)-Coniine   (s)-Nicotine

Cocaine

(s)-Coniine is the toxic principle of water hemlock (a member of the carrot family). Its ingestion can cause weakness, labored respi-

Tobacco plants.

Cocaine is a central nervous system stimulant obtained from the leaves of the coca plant. In small doses, it decreases fatigue and gives a sense of well-being. Prolonged use of cocaine leads to physical addiction and depression.

Inga Spence/Visuals Unlimited

16.2 | How Do We Name Amines?

A. IUPAC Names

IUPAC names for aliphatic amines are derived just as they are for alcohols. The final -**e** of the parent alkane is dropped and replaced by -**amine.** The location of the amino group on the parent chain is indicated by a number.

CH$_3$CHCH$_3$ 2-Propanamine   Cyclohexanamine   1,6-Hexanediamine

IUPAC nomenclature retains the common name **aniline** for C$_6$H$_5$NH$_2$, the simplest aromatic amine. Its simple derivatives are named using numbers to locate substituents or, alternatively, using the locators *ortho* (*o*), *meta* (*m*), and *para* (*p*). Several derivatives of aniline have common names that remain in use. Among them is **toluidine** for a methyl-substituted aniline.

Aniline   4-Nitroaniline   3-Methylaniline
                 (*p*-Nitroaniline)   (*m*-Toluidine)

Unsymmetrical secondary and tertiary amines are commonly named as *N*-substituted primary amines. The largest group bonded to nitrogen is taken as the parent amine; the smaller groups bonded to nitrogen are named, and their locations are indicated by the prefix *N* (indicating that they are bonded to nitrogen).

N-Methylaniline   *N,N*-Dimethyl-
                           cyclopentanamine

EXAMPLE 16.2

Write the IUPAC name for each amine. Try to specify the configuration of the stereocenter in (c).

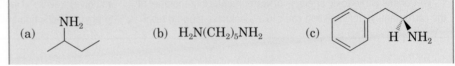

(a)   (b) H$_2$N(CH$_2$)$_5$NH$_2$ (c)

Solution
(a) The parent alkane has four carbon atoms and is butane. The amino group is on carbon 2, giving the IUPAC name 2-butanamine.
(b) The parent chain has five carbon atoms and is pentane. There are amino groups on carbons 1 and 5, giving the IUPAC name 1,5-pentane-diamine. The common name of this diamine is cadaverine, which should give you a hint of where it occurs in nature and of its odor. Cadaverine, one of the end products of decaying flesh, is quite poisonous.
(c) The parent chain has three carbon atoms and is propane. To have the lowest numbers possible, we number from the end that places the phenyl group on carbon 1 and the amino group on carbon 2. The priorities for determining R or S configuration are $NH_2 > C_6H_5CH_2 > CH_3 > H$. This amine's systematic name is (R)-1-phenyl-2-propanamine. Its common name is amphetamine.

Problem 16.2

Write a structural formula for each amine.
(a) 2-Methyl-1-propanamine   (b) Cyclopentanamine
(c) 1,4-Butanediamine

B. Common Names

Common names for most aliphatic amines list the groups bonded to nitrogen in alphabetical order in one word ending in the suffix **-amine.**

Propylamine   *sec*-Butylamine   Diethylmethylamine   Cyclohexylamine

EXAMPLE 16.3

Write a structural formula for each amine.
(a) Isopropylamine   (b) Cyclohexylmethylamine
(c) Triethylamine

Solution

(a) $(CH_3)_2CHNH_2$ (b) ⬡—$NHCH_3$ (c) $(CH_3CH_2)_3N$

        or   or   or

Problem 16.3

Write a structural formula for each amine.
(a) 2-Aminoethanol   (b) Diphenylamine
(c) Diisopropylamine

Several over-the-counter mouthwashes contain an *N*-alkylpyridinium chloride as an antibacterial.

GOB
Chemistry Now™

Click *Coached Problems* to examine the **Relationship Between Structure and Boiling Point for a Series of Amines**

Hydrogen bonding

Figure 16.1 Hydrogen bonding between two molecules of a secondary amine.

When four atoms or groups of atoms are bonded to a nitrogen atom—as, for example, in NH_4^+ and $CH_3NH_3^+$—nitrogen bears a positive charge and is associated with an anion as a salt. The compound is named as a salt of the corresponding amine. The ending **-amine** (or *aniline, pyridine,* or the like) is replaced by **-ammonium** (or *anilinium, pyridinium,* or the like) and the name of the anion (chloride, acetate, and so on) is added.

$$(CH_3CH_2)_3NH^+Cl^-$$

Triethylammonium chloride

16.3 | What Are the Physical Properties of Amines?

Like ammonia, low-molecular-weight amines have very sharp, penetrating odors. Trimethylamine, for example, is the pungent principle in the smell of rotting fish. Two other particularly pungent amines are 1,4-butanediamine (putrescine) and 1,5-pentanediamine (cadaverine).

Amines are polar compounds because of the difference in electronegativity between nitrogen and hydrogen (3.0 − 2.1 = 0.9). Both primary and secondary amines have N—H bonds, and they can form hydrogen bonds with one another (Figure 16.1). Tertiary amines do not have a hydrogen bonded to nitrogen and, therefore, do not form hydrogen bonds with one another.

An N—H----N hydrogen bond is weaker than an O—H----O hydrogen bond, because the difference in electronegativity between nitrogen and hydrogen (3.0 − 2.1 = 0.9) is less than that between oxygen and hydrogen (3.5 − 2.1 = 1.4). To see the effect of hydrogen bonding between molecules of comparable molecular weight, we can compare the boiling points of

CHEMICAL CONNECTIONS 16C

Tranquilizers

Most people face anxiety and stress at some time in their lives, and each person develops various ways to cope with these factors. Perhaps that strategy involves meditation, or exercise, or psychotherapy, or drugs. One modern coping technique is to use tranquilizers, drugs that provide relief from the symptoms of anxiety or tension.

The first modern tranquilizers were derivatives of a compound called benzodiazepine. The first of these compounds, chlorodi-

azepoxide, better known as Librium, was introduced in 1960 and was soon followed by more than two dozen related compounds. Diazepam, better known as Valium, became one of the most widely used of these drugs.

Librium, Valium, and other benzodiazepines are central nervous system sedatives/hypnotics. As sedatives, they diminish activity and excitement, thereby exerting a calming effect. As hypnotics, they produce drowsiness and sleep.

Benzodiazepine   Chlorodiazepoxide   Diazepam
(Librium)   (Valium)

ethane, methanamine, and methanol. Ethane is a nonpolar hydrocarbon, and the only attractive forces between its molecules are weak London dispersion forces (Section 6.7A). Both methanamine and methanol have polar molecules that interact in the pure liquid by hydrogen bonding. Methanol has the highest boiling point of the three compounds, because the hydrogen bonding between its molecules is stronger than that between methanamine molecules.

	CH_3CH_3	CH_3NH_2	CH_3OH
Molecular weight (amu)	30.1	31.1	32.0
Boiling point (°C)	−88.6	−6.3	65.0

All classes of amines form hydrogen bonds with water and are more soluble in water than are hydrocarbons of comparable molecular weight. Most low-molecular-weight amines are completely soluble in water, but higher-molecular-weight amines are only moderately soluble in water or are insoluble.

16.4 | How Do We Describe the Basicity of Amines?

Like ammonia, amines are weak bases, and aqueous solutions of amines are basic. The following acid–base reaction between an amine and water is written using curved arrows to emphasize that, in this proton-transfer reaction (Section 9.1), the unshared pair of electrons on nitrogen forms a new covalent bond with hydrogen and displaces a hydroxide ion.

Methylamine
(a base)

Methylammonium
hydroxide

The base dissociation constant, K_b, for the reaction of an amine with water has the following form, illustrated here for the reaction of methylamine with water to give methylammonium hydroxide. pK_b is defined as the negative logarithm of K_b.

$$K_b = \frac{[CH_3NH_3^+][OH^-]}{[CH_3NH_2]} = 4.37 \times 10^{-4}$$

$$pK_b = -\log 4.37 \times 10^{-4} = 3.360$$

All aliphatic amines have about the same base strength, pK_b 3.0–4.0, and are slightly stronger bases than ammonia (Table 16.1). Aromatic amines and heterocyclic aromatic amines (pK_b 8.5–9.5) are considerably weaker bases than aliphatic amines. One additional point about the basicities of amines: While aliphatic amines are weak bases by comparison with inorganic bases such as NaOH, they are strong bases among organic compounds.

Table 16.1 Approximate Base Strengths of Amines

Class	pK_a	Example	Name	
Aliphatic	3.0–4.0	$CH_3CH_2NH_2$	Ethanamine	Stronger base
Ammonia	4.74			↑
Aromatic	8.5–9.5	⬡—NH_2	Aniline	Weaker base

Given the basicities of amines, we can determine which form of an amine exists in body fluids—say, blood. In a normal, healthy person, the pH of blood is approximately 7.40, which is slightly basic. If an aliphatic amine is dissolved in blood, it is present predominantly as its protonated or conjugate acid form.

Dopamine   Conjugate acid of dopamine
                            (the major form present
                            in blood plasma)

We can show that an aliphatic amine such as dopamine dissolved in blood is present largely as its protonated or conjugate acid form in the following way. Assume that the amine, RNH_2, has a pK_b of 3.50 and that it is dissolved in blood, pH 7.40. We first write the base dissociation constant for the amine and then solve for the ratio of RNH_3^+ to RNH_2.

$$RNH_2 + H_2O \rightleftharpoons RNH_3^+ + OH^-$$

$$K_b = \frac{[RNH_3^+][OH^-]}{[RNH_2]}$$

$$\frac{K_b}{[OH^-]} = \frac{[RNH_3^+]}{[RNH_2]}$$

We now substitute the appropriate values for K_b and $[OH^-]$ in this equation. Taking the antilog of 3.50 gives a K_b of 3.2×10^{-4}. Calculating the concentration of hydroxide requires two steps. First recall from Section 9.8 that pH + pOH = 14. If the pH of blood is 7.40, then its pOH is 6.60 and its $[OH^-]$ is 2.5×10^{-7}. Substituting these values in the appropriate equation gives a ratio of 1300 parts RNH_3^+ to 1 part RNH_2.

$$\frac{3.2 \times 10^{-4}}{2.5 \times 10^{-7}} = \frac{[RNH_3^+]}{[RNH_2]} = 1300$$

As this calculation demonstrates, an aliphatic amine present in blood is more than 99.9% in the protonated form. Thus, even though we may write the structural formula for dopamine as the free amine, it is present as the protonated form. It is important to realize, however, that the amine and ammonium ion forms are always in equilibrium, so some of the unprotonated form is nevertheless present in solution.

Aromatic amines, by contrast, are considerably weaker bases than aliphatic amines and are present in blood largely in the unprotonated form. Performing the same type of calculation for an aromatic amine, $ArNH_2$, with pK_b of approximately 10, we find that the aromatic amine is more than 99.0% in its unprotonated ($ArNH_2$) form.

EXAMPLE 16.4

Select the stronger base in each pair of amines.

Solution

(a) Morpholine (B), a 2° aliphatic amine, is the stronger base. Pyridine (A), a heterocyclic aromatic amine, is the weaker base.

(b) Benzylamine (D), a 1° aliphatic amine, is the stronger base. Even though it contains an aromatic ring, it is not an aromatic amine because the amine nitrogen is not bonded to the aromatic ring. *o*-Toluidine (C), a 1° aromatic amine, is the weaker base.

Problem 16.4

Select the stronger base from each pair of amines.

16.5 | What Are the Characteristic Reactions of Amines?

The most important chemical property of amines is their basicity. Amines, whether soluble or insoluble in water, react quantitatively with strong acids to form water-soluble salts, as illustrated by the reaction of (*R*)-norepinephrine (noradrenaline) with aqueous HCl to form a hydrochloride salt.

CHEMICAL CONNECTIONS 16D

The Solubility of Drugs in Body Fluids

Many drugs have "• HCl" or some other acid as part of their chemical formulas and occasionally as part of their generic names. Invariably these drugs are amines that are insoluble in aqueous body fluids such as blood plasma and cerebrospinal fluid. For the administered drug to be absorbed and carried by body fluids, it must be treated with an acid to form a water-soluble ammonium salt. Methadone, a narcotic analgesic, is marketed as its water-soluble hydrochloride salt. Novocain, one of the first local anesthetics, is the hydrochloride salt of procaine.

Methadone ·HCl

These two drugs are amine salts and are labeled as hydrochlorides.

Procaine ·HCl
(Novocain, a local anesthetic)

There is another reason besides increased water solubility for preparing these and other amine drugs as salts. Amines are very susceptible to oxidation and decomposition by atmospheric oxygen, with a corresponding loss of biological activity. By comparison, their amine salts are far less susceptible to oxidation; they retain their effectiveness for a much longer time.

EXAMPLE 16.5

Complete the equation for each acid–base reaction, and name the salt formed.

(a) $(CH_3CH_2)_2NH + HCl \longrightarrow$

(b) [pyridine] $+ CH_3COOH \longrightarrow$

Solution

(a) $(CH_3CH_2)_2NH_2^+Cl^-$
Diethylammonium chloride

(b) [pyridinium] CH_3COO^-

Pyridinium acetate

Problem 16.5

Complete the equation for each acid–base reaction and name the salt formed.

(a) $(CH_3CH_2)_3N + HCl \longrightarrow$

(b) [cyclohexyl]$NH + CH_3COOH \longrightarrow$

CHEMISTRY IN ACTION 16E

Epinephrine: A Prototype for the Development of New Bronchodilators

Epinephrine was first isolated in pure form in 1897 and its structure determined in 1901. It occurs in the adrenal gland (hence the common name adrenalin) as a single enantiomer with the *R* configuration at its stereocenter. Epinephrine is commonly referred to as a catecholamine: the common name of 1,2-dihydroxybenzene is catechol (Section 13.4A), and amines containing a benzene ring with *ortho*-hydroxyl groups are called catecholamines.

Early on, it was recognized that epinephrine is a vasoconstrictor, a bronchodilator, and a cardiac stimulant. The fact that it has these three major effects has stimulated much research, one line of which sought to develop compounds that are even more effective bronchodilators than epinephrine but that are free from epinephrine's cardiac-stimulating and vasoconstricting effects.

Soon after epinephrine became commercially available, it emerged as an important treatment of asthma and hay fever. It has been marketed for the relief of bronchospasms under several trade names, including Bronkaid Mist and Primatine Mist.

enzyme-catalyzed reaction that converts one of the —OH groups on the catechol unit to an OCH_3 group. A strategy to circumvent this enzyme-catalyzed inactivation was to replace the catechol unit with one that would allow the drug to bind to the catecholamine receptors in the bronchi but would not be inactivated by this enzyme.

In terbutaline (Brethaire), inactivation is prevented by placing the —OH groups *meta* to each other on the aromatic ring. In addition, the isopropyl group of isoproterenol is replaced by a *tert*-butyl group. In albuterol (Proventil), the commercially most successful of the antiasthma medications, one —OH group of the catechol unit is replaced by a —CH_2OH group and the isopropyl group is replaced by a *tert*-butyl group. When terbutaline and albuterol were introduced into clinical medicine in the 1960s, they almost immediately replaced isoproterenol as the drugs of choice for the treatment of asthmatic attacks. The *R* enantiomer of albuterol is 68 times more effective in the treatment of asthma than the *S* enantiomer.

Epinephrine

(*R*)-Isoproterenol

Terbutaline

(*R*)-Albuterol

One of the most important of the first synthetic catecholamines was isoproterenol, the levorotatory enantiomer of which retains the bronchodilating effects of epinephrine but is free from its cardiac-stimulating effects. (*R*)-Isoproterenol was introduced into clinical medicine in 1951; for the next two decades, it was the drug of choice for the treatment of asthmatic attacks. Interestingly, the hydrochloride salt of (*S*)-isoproterenol is a nasal decongestant and was marketed under several trade names, including Sudafed.

A problem with the first synthetic catecholamines (and with epinephrine itself) is the fact that they are inactivated by an

In their search for a longer-acting bronchodilator, scientists reasoned that extending the side chain on nitrogen might strengthen the binding of the drug to the adrenoreceptors in the lungs, thereby increasing the duration of the drug's action. This line of reasoning led to the synthesis and introduction of salmeterol (Serevent), a bronchodilator that is approximately ten times more potent than albuterol and much longer acting.

Salmeterol

Figure 16.2   Separation and purification of an amine and a neutral compound.

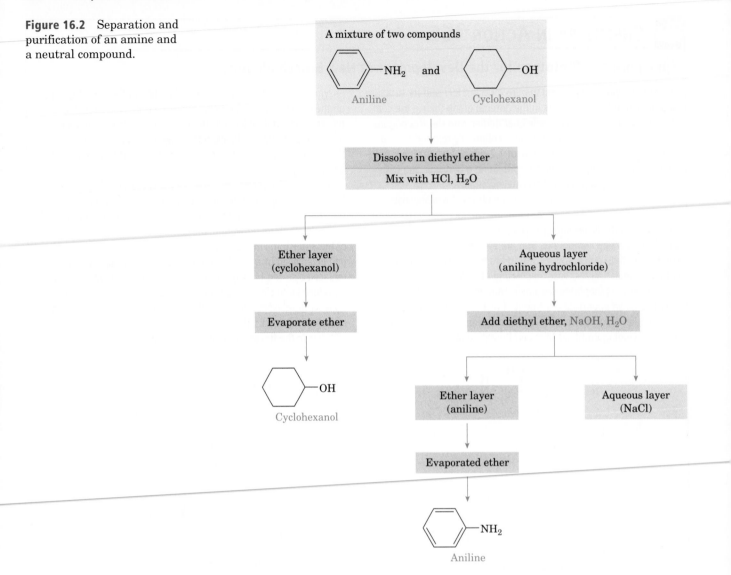

The basicity of amines and the solubility in water of amine salts gives us a way to separate water-insoluble amines from water-insoluble non-basic compounds. Figure 16.2 is a flowchart for the separation of aniline from cyclohexanol, a neutral compound.

SUMMARY OF KEY QUESTIONS

SECTION 16.1 What Are Amines?

- Amines are classified as **primary, secondary,** or **tertiary,** depending on the number of carbon atoms bonded to nitrogen.
- In an **aliphatic amine,** all carbon atoms bonded to nitrogen are derived from alkyl groups.
- In an **aromatic amine,** one or more of the groups bonded to nitrogen are aryl groups.
- In a **heterocyclic amine,** the nitrogen atom is part of a ring.

SECTION 16.2 How Do We Name Amines?

- In IUPAC nomenclature, aliphatic amines are named by changing the final **-e** of the parent alkane to **-amine** and using a number to locate the amino group on the parent chain.
- In the common system of nomenclature, aliphatic amines are named by listing the carbon groups bonded to nitrogen in alphabetical order in one word ending in the suffix **-amine.**

SECTION 16.3 What Are the Physical Properties of Amines?

- Amines are polar compounds, and primary and secondary amines associate by intermolecular hydrogen bonding.
- All classes of amines form hydrogen bonds with water and are more soluble in water than are hydrocarbons of comparable molecular weight.

SECTION 16.4 How Do We Describe the Basicity of Amines?

- Amines are weak bases, and aqueous solutions of amines are basic.

- The base ionization constant for an amine in water is denoted by the symbol K_b. Aliphatic amines are stronger bases than aromatic amines.

SECTION 16.5 What Are the Characteristic Reactions of Amines?

- All amines, whether soluble or insoluble in water, react with strong acids to form water-soluble salts.
- We can use this property to separate water-insoluble amines from water-insoluble nonbasic compounds.

SUMMARY OF KEY REACTIONS

1. **Basicity of Aliphatic Amines (Section 16.4)** Most aliphatic amines have about the same basicity (pK_b 3.0–4.0) and are slightly stronger bases than ammonia (pK_b 4.74).

$$CH_3NH_2 + H_2O \rightleftharpoons CH_3NH_3^+ + OH^- \qquad pK_b = 3.36$$

2. **Basicity of Aromatic Amines (Section 16.4)** Most aromatic amines (pK_b 9.0–10.0) are considerably weaker bases than ammonia and aliphatic amines.

$$\text{(benzene ring)}-NH_2 + H_2O \rightleftharpoons \text{(benzene ring)}-NH_3^+ + OH^- \qquad pK_b = 9.36$$

3. **Reaction with Acids (Section 16.5)** All amines, whether water-soluble or water-insoluble, react quantitatively with strong acids to form water-soluble salts.

Insoluble in water   A water-soluble salt

PROBLEMS

A blue problem number indicates an applied problem.

■ denotes problems that are available on the GOB ChemistryNow website or CD and are assignable in OWL.

SECTION 16.1 What Are Amines?

16.6 We use the terms *primary, secondary,* and *tertiary* to classify both alcohols and amines. Cyclohexanol, for example, is classified as a secondary alcohol, and cyclohexanamine is classified as a primary amine. In each compound, the functional group is bonded to a carbon of the cyclohexane ring. Explain why one compound is classified as secondary whereas the other is primary.

16.7 What is the difference in structure between an aliphatic amine and an aromatic amine?

16.8 In what way are pyridine and pyrimidine related to benzene?

SECTION 16.2 How Do We Name Amines?

16.9 ■ Draw a structural formula for each amine.
  (a) 2-Butanamine
  (b) 1-Octanamine
  (c) 2,2-Dimethyl-1-propanamine
  (d) 1,5-Pentanediamine
  (e) 2-Bromoaniline
  (f) Tributylamine

16.10 Draw a structural formula for each amine.
  (a) 4-Methyl-2-pentanamine
  (b) *trans*-2-Aminocyclohexanol
  (c) *N,N*-Dimethylaniline
  (d) Dicyclohexylamine
  (e) *sec*-Butylamine
  (f) 2,4-Dimethylaniline

16.11 ■ Classify each amino group as primary, secondary, or tertiary, and as aliphatic or aromatic.

(a)

Serotonin
(a neurotransmitter)

(b)

Benzocaine
(a topical anesthetic)

(c)

Diphenhydramine
(the hydrochloride salt is
the antihistamine Benadryl)

16.12 Classify each amino group as primary, secondary, or tertiary, and as aliphatic or aromatic.

(a)

Lysine
(an amino acid)

(b)

Chloroquine
(an antimalaria drug)

(c)

4-Aminobutanoic acid
(a neurotransmitter)

16.13 ■ There are eight constitutional isomers with the molecular formula $C_4H_{11}N$.
   (a) Name and draw a structural formula for each amine.
   (b) Classify each amine as primary, secondary, or tertiary.
   (c) Which are chiral?

16.14 There are eight primary amines with the molecular formula $C_5H_{13}N$.
   (a) Name and draw a structural formula for each amine.
   (b) Which are chiral?

SECTION 16.3 What Are the Physical Properties of Amines?

16.15 ■ Propylamine (bp 48°C), ethylmethylamine (bp 37°C), and trimethylamine (bp 3°C) are constitutional isomers with molecular formula C_3H_9N. Account for the fact that trimethylamine has the lowest boiling point of the three.

16.16 Account for the fact that 1-butanamine (bp 78°C) has a lower boiling point than 1-butanol (bp 117°C).

16.17 2-Methylpropane (bp −12°C), 2-propanol (bp 82°C), and 2-propanamine (bp 32°C) all have approximately the same molecular weight, yet their boiling points are quite different. Explain the reason for these differences.

16.18 Account for the fact that most low-molecular-weight amines are very soluble in water whereas low-molecular-weight hydrocarbons are not.

SECTION 16.4 How Do We Describe the Basicity of Amines?

16.19 Compare the basicities of amines with those of alcohols.

16.20 Write a structural formula for each amine salt.
   (a) Ethyltrimethylammonium hydroxide
   (b) Dimethylammonium iodide
   (c) Tetramethylammonium chloride
   (d) Anilinium bromide

16.21 Name these amine salts.

   (a) $CH_3CH_2NH_3{}^+Cl^-$

   (b) $(CH_3CH_2)_2NH_2{}^+Cl^-$

   (c) ⟨benzene ring⟩$-NH_3{}^+HSO_4{}^-$

16.22 ■ From each pair of compounds, select the stronger base.

(a)

(b) or (c)

16.23 ■ The pK_b of amphetamine is approximately 3.2.

Amphetamine

(a) Which form of amphetamine (the base or its conjugate acid) would you expect to be present at pH 1.0, the pH of stomach acid?

(b) Which form of amphetamine would you expect to be present at pH 7.40, the pH of blood plasma?

SECTION 16.5 What Are the Characteristic Reactions of Amines?

16.24 Suppose you have two test tubes, one containing 2-methylcyclohexanol and the other containing 2-methylcyclohexanamine (both of which are insoluble in water). You do not know which test tube contains which compound. Describe a simple chemical test by which you could tell which compound is the alcohol and which is the amine.

16.25 ■ Complete the equations for the following acid–base reactions.

(a) CH_3COH +

Acetic acid   Pyridine

(b)

+ HCl ⟶

1-Phenyl-2-propanamine
(Amphetamine)

(c)

+ H_2SO_4 ⟶

Methamphetamine

16.26 Pyridoxamine is one form of vitamin B_6.

Pyridoxamine
(Vitamin B_6)

(a) Which nitrogen atom of pyridoxamine is the stronger base?

(b) Draw a structural formula for the salt formed when pyridoxamine is treated with one mole of HCl.

16.27 Many tumors of the breast are correlated with estrogen levels in the body. Drugs that interfere with estrogen binding have antitumor activity and may even help prevent tumor occurrence. A widely used antiestrogen drug is tamoxifen.

Tamoxifen

(a) Name the functional groups in tamoxifen.

(b) Classify the amino group in tamoxifen as primary, secondary, or tertiary.

(c) How many stereoisomers are possible for tamoxifen?

16.28 Classify each amino group in epinephrine and albuterol (Chemical Connections 16E) as primary, secondary, or tertiary. In addition, list the similarities and differences between the structural formulas of these two compounds.

Chemical Connections

16.29 (Chemical Connections 16A) What are the differences in structure between the natural hormone epinephrine (Chemical Connections 16E) and the synthetic pep pill amphetamine? Between amphetamine and methamphetamine?

16.30 (Chemical Connections 16A) What are the possible negative effects of illegal use of amphetamines such as methamphetamine?

16.31 (Chemical Connections 16B) What is an alkaloid? Are all alkaloids basic to litmus?

16.32 (Chemical Connections 16B) Identify all stereocenters in coniine and nicotine. How many stereoisomers are possible for each?

16.33 (Chemical Connections 16B) Of the two nitrogen atoms in nicotine, which is converted to its salt by reaction with one mole of HCl? Draw a structural formula for this salt.

16.34 (Chemical Connections 16B) Cocaine has four stereo-centers. Identify each. Draw a structural formula for the salt formed by treatment of cocaine with one mole of HCl.

16.35 (Chemical Connections 16C) What structural feature is common to all benzodiazepines?

16.36 (Chemical Connections 16C) Is Librium chiral? Is Valium chiral?

16.37 (Chemical Connections 16C) Benzodiazepines affect neural pathways in the central nervous system that are mediated by GABA, whose IUPAC name is 4-aminobutanoic acid. Draw a structural formula for GABA.

16.38 (Chemical Connections 16D) Suppose you saw this label on a decongestant: phenylephrine · HCl. Should you worry about being exposed to a strong acid such as HCl? Explain.

16.39 (Chemical Connections 16D) Give two reasons why amine-containing drugs are most commonly administered as their salts.

Additional Problems

16.40 ■ Draw a structural formula for a compound with each molecular formula.

(a) A 2° aromatic amine, C_7H_9N

(b) A 3° aromatic amine, $C_8H_{11}N$

(c) A 1° aliphatic amine, C_7H_9N

(d) A chiral 1° amine, $C_4H_{11}N$

(e) A 3° heterocyclic amine, $C_5H_{11}N$

(f) A trisubstituted 1° aromatic amine, $C_9H_{13}N$

(g) A chiral quaternary ammonium salt, $C_9H_{22}NCl$

16.41 ■ Arrange these three compounds in order of decreasing ability to form intermolecular hydrogen bonds: CH_3OH, CH_3SH, $(CH_3)_2NH$.

16.42 Consider these three compounds: CH_3OH, CH_3SH, and $(CH_3)_2NH$.

(a) Which is the strongest acid?

(b) Which is the strongest base?

(c) Which has the highest boiling point?

(d) Which forms the strongest hydrogen bonds in the pure state?

16.43 Arrange these compounds in order of increasing boiling point: $CH_3CH_2CH_2CH_3$, $CH_3CH_2CH_2OH$, $CH_3CH_2CH_2NH_2$. Boiling-point values from lowest to highest are $-0.5°C$, $7.2°C$, and $77.8°C$.

16.44 Account for the fact that amines have about the same solubility in water as alcohols of similar molecular weight.

16.45 If you dissolve $CH_3CH_2CH_2OH$ and $CH_3CH_2CH_2NH_2$ in the same container of water and lower the pH of the solution to 2.0 by adding HCl, would anything happen to the structures of these compounds? Write the formula of the species present in solution at pH 2.0.

16.46 The compound phenylpropanolamine hydrochloride is used as both a decongestant and an anorexic. The chemical name of this compound is 1-phenyl-2-amino-1-propanol.

(a) Draw a structural formula for 1-phenyl-2-amino-1-propanol.

(b) How many stereocenters are present in this molecule? How many stereoisomers are possible for it?

16.47 Procaine was one of the first local anesthetics. Its hydrochloride salt is marketed as Novocain.

Procaine

(a) Is procaine chiral? Does it contain a stereocenter?

(b) Which nitrogen atom of procaine is the stronger base?

(c) Draw a structural formula for the salt formed by treating procaine with one mole of HCl, showing which nitrogen is protonated and bears the positive charge.

16.48 Following are two structural formulas for gabapentin, a drug used to treat epilepsy. Is gabapentin better represented by structure (A) or (B)? Explain.

(A)   or   (B)

16.49 Several poisonous plants, including *Atropa belladonna,* contain the alkaloid atropine. The name "belladonna" (which means "beautiful lady") probably comes from the fact that Roman women used extracts from this plant to make themselves more attractive. Atropine is widely used by ophthalmologists and optometrists to dilate the pupils for eye examination.

Atropine

(a) Classify the amino group in atropine as primary, secondary, or tertiary.

(b) Locate all stereocenters in atropine.

(c) Account for the fact that atropine is almost insoluble in water (1 g in 455 mL of cold water), but atropine hydrogen sulfate is very soluble (1 g in 5 mL of cold water).

(d) Account for the fact that a dilute aqueous solution of atropine is basic (pH approximately 10.0).

16.50 Epibatadine, a colorless oil isolated from the skin of the Equadorian poison frog *Epipedobates tricolor*, has several times the analgesic potency of morphine. It is the first chlorine-containing, non-opioid (non-morphine-like in structure) analgesic ever isolated from a natural source.

(a) Which of the two nitrogen atoms in epibatadine is the stronger base?

(b) Mark the three stereocenters in this molecule.

Epibatadine

Poison arrow frog.

Looking Ahead

16.51 ■ Following are two structural formulas for 4-aminobutanoic acid, a neurotransmitter. Is this compound better represented by structural formula (A) or (B)? Explain.

(A)   or   (B)

16.52 Alanine, $C_3H_7O_2N$, is one of the twenty amino acid building blocks of proteins (Chapter 22). Alanine contains a primary amino group ($-NH_2$) and a carboxyl group ($-COOH$), and has one stereocenter. Given this information, draw a structural formula for alanine.

Aldehydes and Ketones

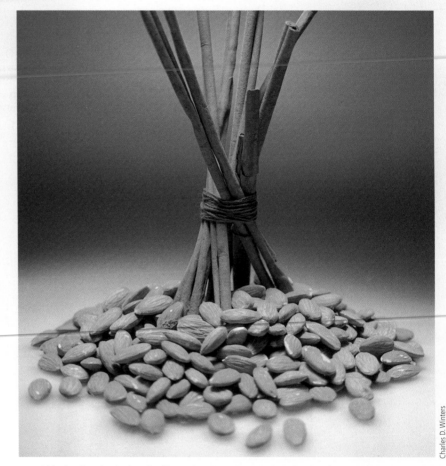

Charles D. Winters

Benzaldehyde is found in the kernels of bitter almonds, and cinnamaldehyde is found in Ceylonese and Chinese cinnamon oils.

GOB
Chemistry☆Now™

Look for this logo in the chapter and go to GOB ChemistryNow at **http://now.brookscole.com/gob8** or on the CD for tutorials, simulations, and problems.

17.1 | What Are Aldehydes and Ketones?

In this and the three following chapters, we study the physical and chemical properties of compounds containing the **carbonyl group, C=O.** Because the carbonyl group is present in aldehydes, ketones, and carboxylic acids and their derivatives, as well as carbohydrates, it is one of the most important functional groups in organic chemistry. Its chemical properties are straightforward, and an understanding of its characteristic reaction patterns leads very quickly to an understanding of a wide variety of organic and biochemical reactions.

The functional group of an **aldehyde** is a carbonyl group bonded to a hydrogen atom (Section 10.4C). In methanal, the simplest aldehyde, the carbonyl group is bonded to two hydrogen atoms. In other aldehydes, it is bonded to one hydrogen atom and one carbon atom. The functional group of a **ketone** is a carbonyl group bonded to two carbon atoms (Section 10.4C).

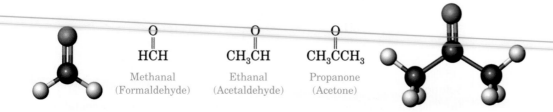

HCH	CH₃CH	CH₃CCH₃
Methanal	Ethanal	Propanone
(Formaldehyde)	(Acetaldehyde)	(Acetone)

Because aldehydes always contain at least one hydrogen bonded to the C=O group, they are often written RCH=O or RCHO. Similarly, ketones are often written RCOR′.

17.2 | How Do We Name Aldehydes and Ketones?

A. IUPAC Names

The IUPAC names for aldehydes and ketones follow the familiar pattern of selecting as the parent alkane the longest chain of carbon atoms that contains the functional group (Section 11.3A). To name an aldehyde, we change the suffix -e of the parent alkane to -al. Because the carbonyl group of an aldehyde can appear only at the end of a parent chain and numbering must start with it as carbon 1, there is no need to use a number to locate the aldehyde group.

For **unsaturated aldehydes,** we show the presence of a carbon–carbon double bond and an aldehyde by changing the ending of the parent alkane from -ane to -enal: "-en-" to show the carbon–carbon double bond, and "-al" to show the aldehyde. We show the location of the carbon–carbon double bond by the number of its first carbon.

GOB
Chemistry Now™

Click *Coached Problems* to try a set of problems in **Naming Aldehydes and Ketones**

Hexanal	3-Methylbutanal	2-Propenal (Acrolein)

In the IUPAC system, we name ketones by selecting as the parent alkane the longest chain that contains the carbonyl group and then indicating the presence of this group by changing the -e of the parent alkane to -one. The parent chain is numbered from the direction that gives the smaller number to the carbonyl carbon. While the systematic name of the simplest ketone is 2-propanone, the IUPAC retains its common name, acetone.

Acetone   5-Methyl-3-hexanone   2-Methylcyclohexanone

EXAMPLE 17.1

Write the IUPAC name for each compound.

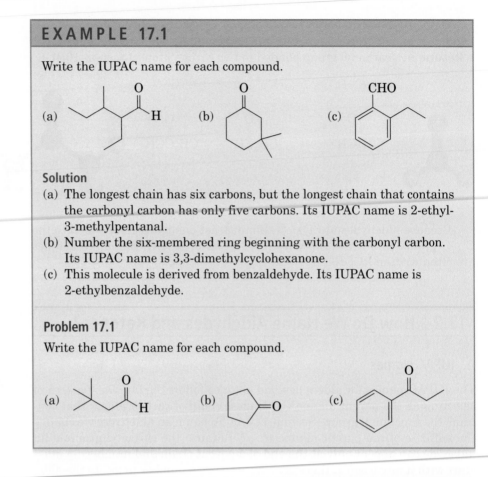

Solution

(a) The longest chain has six carbons, but the longest chain that contains the carbonyl carbon has only five carbons. Its IUPAC name is 2-ethyl-3-methylpentanal.
(b) Number the six-membered ring beginning with the carbonyl carbon. Its IUPAC name is 3,3-dimethylcyclohexanone.
(c) This molecule is derived from benzaldehyde. Its IUPAC name is 2-ethylbenzaldehyde.

Problem 17.1

Write the IUPAC name for each compound.

EXAMPLE 17.2

Write structural formulas for all ketones with the molecular formula $C_6H_{12}O$ and give the IUPAC name of each. Which of these ketones are chiral?

Solution

There are six ketones with this molecular formula: two with a six-carbon chain, three with a five-carbon chain and a methyl branch, and one with a four-carbon chain and two methyl branches. Only 3-methyl-2-pentanone has a stereocenter and is chiral.

2-Hexanone   3-Hexanone   4-Methyl-2-pentanone

3-Methyl-2-pentanone   2-Methyl-3-pentanone   3,3-Dimethyl-2-butanone

Problem 17.2

Write structural formulas for all aldehydes with the molecular formula $C_6H_{12}O$ and give the IUPAC name of each. Which of these aldehydes are chiral?

In naming aldehydes or ketones that also contain an —OH or —NH_2 group elsewhere in the molecule, the parent chain is numbered to give the carbonyl group the lower number. An —OH substituent is indicated by *hydroxy*, and an —NH_2 substituent is indicated by *amino-*. Hydroxy and amino substituents are numbered and alphabetized along with any other substituents that might be present.

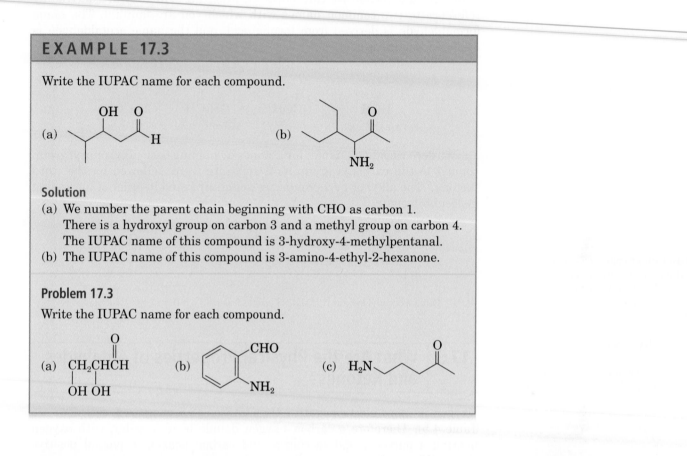

EXAMPLE 17.3

Write the IUPAC name for each compound.

(a)

(b)

Solution

(a) We number the parent chain beginning with CHO as carbon 1. There is a hydroxyl group on carbon 3 and a methyl group on carbon 4. The IUPAC name of this compound is 3-hydroxy-4-methylpentanal.

(b) The IUPAC name of this compound is 3-amino-4-ethyl-2-hexanone.

Problem 17.3

Write the IUPAC name for each compound.

(a) CH₂CHCH
    | |
    OH OH

(b)

(c) H₂N

CHEMICAL CONNECTIONS 17A

Some Naturally Occurring Aldehydes and Ketones

Benzaldehyde
(oil of almonds)

Cinnamaldehyde
(oil of cinnamon)

Citronellal
(citronella oils; also in
lemon and lemon grass oils)

Muscone
(from the musk deer;
used in perfumes)

Vanillin
(vanilla bean)

β-Ionone
(from violets)

B. Common Names

We derive the common name for an aldehyde from the common name of the corresponding carboxylic acid. The word "acid" is dropped and the suffix -ic or -oic is changed to -aldehyde. Because we have not yet studied common names for carboxylic acids, we are not in a position to discuss common names for aldehydes. We can, however, illustrate how they are derived by reference to two common names with which you are familiar. The name formaldehyde is derived from formic acid, and the name acetaldehyde is derived from acetic acid.

| HCH | HCOH | CH₃CH | CH₃COH |
| Formaldehyde | Formic acid | Acetaldehyde | Acetic acid |

We derive common names for ketones by naming each alkyl or aryl group bonded to the carbonyl group as a separate word, followed by the word "ketone." The alkyl or aryl groups are generally listed in order of increasing molecular weight.

Ethyl isopropyl ketone   Methyl ethyl ketone   Dicyclohexyl ketone

2-Butanone, more commonly called methyl ethyl ketone (MEK), is used as a solvent for paints and varnishes.

Polarity of a carbonyl group

Figure 17.1 The polarity of a carbonyl group. The carbonyl oxygen bears a partial negative charge and the carbonyl carbon bears a partial positive charge.

17.3 | What Are the Physical Properties of Aldehydes and Ketones?

Oxygen is more electronegative than carbon (3.5 compared with 2.5; see Table 4.5). Therefore a carbon–oxygen double bond is polar, with oxygen bearing a partial negative charge and carbon bearing a partial positive charge (Figure 17.1).

In liquid aldehydes and ketones, intermolecular attractions occur between the partial positive charge on the carbonyl carbon of one molecule and the partial negative charge on the carbonyl oxygen of another molecule. There is no possibility for hydrogen bonding between aldehyde or ketone molecules, which explains why these compounds have lower boiling points than alcohols (Section 14.1C) and carboxylic acids (Section 18.3), compounds in which hydrogen bonding does occur between molecules.

Table 17.1 lists structural formulas and boiling points of six compounds of similar molecular weight. Of the six, pentane and diethyl ether have the

Table 17.1 Boiling Points of Six Compounds of Comparable Molecular Weight

Name	Structural Formula	Molecular Weight	Boiling Point (°C)
diethyl ether	CH₃CH₂OCH₂CH₃	74	34
pentane	CH₃CH₂CH₂CH₂CH₃	72	36
butanal	CH₃CH₂CH₂CHO	72	76
2-butanone	CH₃CH₂COCH₃	72	80
1-butanol	CH₃CH₂CH₂CH₂OH	74	117
propanoic acid	CH₃CH₂COOH	74	141

lowest boiling points. The boiling point of 1-butanol, which can associate by intermolecular hydrogen bonding, is higher than that of either butanal or 2-butanone. Propanoic acid, in which intermolecular association by hydrogen bonding is the strongest, has the highest boiling point.

Because the oxygen atom of their carbonyl groups are hydrogen bond acceptors, the low-molecular-weight aldehydes and ketones are more soluble in water than are nonpolar compounds of comparable molecular weight. Formaldehyde, acetaldehyde, and acetone are infinitely soluble in water. As the hydrocarbon portion of the molecule increases in size, aldehydes and ketones become less soluble in water.

Most aldehydes and ketones have strong odors. The odors of ketones are generally pleasant, and many are used in perfumes and as flavoring agents. The odors of aldehydes vary. You may be familiar with the smell of formaldehyde; if so, you know that it is not pleasant. Many higher aldehydes, however, have pleasant odors and are used in perfumes.

17.4 | What Are the Characteristic Reactions of Aldehydes and Ketones?

A. Oxidation

Aldehydes are oxidized to carboxylic acids by a variety of oxidizing agents, including potassium dichromate (Section 14.2C).

Hexanal   Hexanoic acid

The body uses nicotinamide adenine dinucleotide, NAD^+, for this type of oxidation (Section 27.3).

Aldehydes are also oxidized to carboxylic acids by the oxygen in the air. In fact, aldehydes that are liquid at room temperature are so sensitive to oxidation that they must be protected from contact with air during storage. Often this is done by sealing the aldehyde in a container under an atmosphere of nitrogen.

Benzaldehyde   Benzoic acid

Ketones, in contrast, resist oxidation by most oxidizing agents, including potassium dichromate and molecular oxygen.

A silver mirror has been deposited on the inside of this flask by the reaction between an aldehyde and Tollens' reagent.

Charles D. Winters

The fact that aldehydes are so easy to oxidize and ketones are not allows us to use simple chemical tests to distinguish between these types of compounds. Suppose that we have a compound we know is either an aldehyde or a ketone. To determine which it is, we can treat the compound with a mild oxidizing agent. If it can be oxidized, it is an aldehyde; otherwise, it is a ketone. One reagent that has been used for this purpose is Tollens' reagent.

Tollens' reagent contains silver nitrate and ammonia in water. When these two compounds are mixed, silver ion combines with NH_3 to form the complex ion $Ag(NH_3)_2^+$. When this solution is added to an aldehyde, the aldehyde acts as a reducing agent and reduces the complexed silver ion to silver metal. If this reaction is carried out properly, the silver metal precipitates as a smooth, mirror-like deposit on the inner surface of the reaction vessel, leading to the name **silver-mirror test.** If the remaining solution is then acidified with HCl, the carboxylic anion, $RCOO^-$, formed during the aldehyde's oxidation is converted to the carboxylic acid, RCOOH.

$$
\underset{\text{Aldehyde}}{R-\overset{\displaystyle O}{\overset{\|}{C}}-H} + \underset{\substack{\text{Tollens'}\\\text{reagent}}}{2Ag(NH_3)_2^+} + 3OH^- \longrightarrow \underset{\substack{\text{Carboxylic}\\\text{anion}}}{R-\overset{\displaystyle O}{\overset{\|}{C}}-O^-} + \underset{\substack{\text{Silver}\\\text{mirror}}}{2Ag} + 4NH_3 + 2H_2O
$$

Today, silver(I) is rarely used for the oxidation of aldehydes because of its high cost and because of the availability of other, more convenient methods for this oxidation. This reaction, however, is still used for making (silvering) mirrors.

EXAMPLE 17.4

Draw a structural formula for the product formed by treating each compound with Tollens' reagent followed by acidification with aqueous HCl.
(a) Pentanal   (b) 4-Hydroxybenzaldehyde

Solution

The aldehyde group in each compound is oxidized to a carboxylic anion, $-COO^-$. Acidification with HCl converts the anion to a carboxylic acid, $-COOH$.

(a) Pentanoic acid   (b) 4-Hydroxybenzoic acid

Problem 17.4

Complete equations for these oxidations.
(a) Hexanedial + $O_2 \longrightarrow$
(b) 3-Phenylpropanal + $Ag(NH_3)_2^+ \longrightarrow$

B. Reduction

In Section 12.6D, we saw that the C=C double bond of an alkene can be reduced by hydrogen in the presence of a transition metal catalyst to a C—C single bond. The same is true of the C=O double bond of an aldehyde or ketone. Aldehydes are reduced to primary alcohols and ketones are reduced to secondary alcohols.

Pentanal   1-Pentanol

Cyclopentanone   Cyclopentanol

The reduction of a C=O double bond under these conditions is slower than the reduction of a C=C double bond. Thus, if the same molecule contains both C=O and C=C double bonds, the C=C double bond is reduced first.

The reagent most commonly used in the laboratory for the reduction of an aldehyde or ketone is sodium borohydride, $NaBH_4$. This reagent contains hydrogens in the form of a hydride ion, $H:^-$. In the hydride ion, hydrogen has two valence electrons and bears a negative charge. In a reduction by sodium borohydride, hydride ion is attracted to and then adds to the partially positive carbonyl carbon, which leaves a negative charge on the carbonyl oxygen. Reaction of this intermediate with aqueous acid gives the alcohol.

Hydride   Alkoxide
ion   ion

Of the two hydrogens added to the carbonyl group in this reduction, one comes from the reducing agent and the other comes from aqueous acid. Reduction of cyclohexanone, for example, with this reagent gives cyclohexanol:

An advantage of using $NaBH_4$ over the H_2/metal reduction is that $NaBH_4$ does not reduce carbon–carbon double bonds. The reason for this selectivity is quite straightforward. There is no polarity (no partial positive or negative charges) on a carbon–carbon double bond. Therefore, a C=C double bond has no site to attract the hydride ion. In the following example, $NaBH_4$ selectively reduces the aldehyde to a primary alcohol:

Cinnamaldehyde   Cinnamyl alcohol

In biological systems, the agent for the reduction of aldehydes and ketones is the reduced form of coenzyme nicotinamide adenine dinucleotide, abbreviated NADH (Section 27.3). This reducing agent, like NaBH₄, delivers a hydride ion to the carbonyl carbon of the aldehyde or ketone. Reduction of pyruvate, for example, by NADH gives lactate:

Pyruvate is the end product of glycolysis, a series of enzyme-catalyzed reactions that converts glucose to two molecules of this ketoacid (Section 28.1). Under anaerobic conditions, NADH reduces pyruvate to lactate. The build-up of lactate in the bloodstream leads to acidosis and in muscle tissue is associated with muscle fatigue. When blood lactate reaches a concentration of about 0.4 mg/100 mL, muscle tissue becomes almost completely exhausted.

EXAMPLE 17.5

Complete the equations for these reductions.

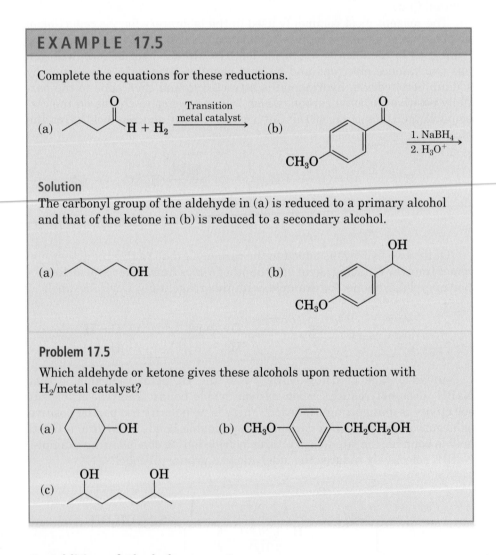

Solution

The carbonyl group of the aldehyde in (a) is reduced to a primary alcohol and that of the ketone in (b) is reduced to a secondary alcohol.

Problem 17.5

Which aldehyde or ketone gives these alcohols upon reduction with H₂/metal catalyst?

Hemiacetal A molecule containing a carbon bonded to one —OH group and one —OR group; the product of adding one molecule of alcohol to the carbonyl group of an aldehyde or ketone

C. Addition of Alcohols

Addition of a molecule of alcohol to the carbonyl group of an aldehyde or ketone forms a **hemiacetal** (a half-acetal). The functional group of a hemiacetal is a carbon bonded to one —OH group and one —OR group. In forming a hemiacetal, the H of the alcohol adds to the carbonyl oxygen and the OR adds to the carbonyl carbon. Shown here are the hemiacetals formed by addition of one molecule of ethanol to benzaldehyde and to cyclohexanone:

Hemiacetals are generally unstable and are only minor components of an equilibrium mixture, except in one very important type of molecule. When a hydroxyl group is part of the same molecule that contains the carbonyl group and a five- or six-membered ring can form, the compound exists almost entirely in a cyclic hemiacetal form. In this case, the —OH group adds to the C=O group of the same molecule. We will have much more to say about cyclic hemiacetals when we consider the chemistry of carbohydrates in Chapter 20.

Hemiacetals can react further with alcohols to form **acetals** plus water. This reaction is acid-catalyzed. The functional group of an acetal is a carbon bonded to two —OR groups.

Acetal A molecule containing two —OR groups bonded to the same carbon

All steps in hemiacetal and acetal formation are reversible. As with any other equilibrium, we can make this one go in either direction by using Le Chatelier's principle (Section 8.8). If we want to drive it to the right (formation of the acetal), we either use a large excess of alcohol or remove water from the equilibrium mixture. If we want to drive it to the left (hydrolysis of the acetal to the original aldehyde or ketone and alcohol), we use a large excess of water.

EXAMPLE 17.6

Show the reaction of 2-butanone with one molecule of ethanol to form a hemiacetal and then with a second molecule of ethanol to form an acetal.

Solution
Given are structural formulas for the hemiacetal and then the acetal.

| 2-Butanone | A hemiacetal | An acetal |

Problem 17.6
Show the reaction of benzaldehyde with one molecule of methanol to form a hemiacetal and then with a second molecule of methanol to form an acetal.

EXAMPLE 17.7

Identify all hemiacetals and acetals in the following structures, and tell whether they are formed from aldehydes or ketones.

$$
\text{(a)} \quad \underset{\overset{|}{\underset{\overset{|}{OCH_3}}{OCH_3}}}{CH_3CH_2CCH_2CH_3} \qquad \text{(b)} \quad CH_3CH_2OCH_2CH_2OH \qquad \text{(c)}
$$

Solution
Compound (a) is an acetal derived from a ketone. Compound (b) is neither a hemiacetal nor an acetal because it does not have a carbon bonded to two oxygens; its functional groups are an ether and a primary alcohol. Compound (c) is a hemiacetal derived from an aldehyde.

$$
\text{(a)} \quad \underset{\overset{|}{\underset{\overset{|}{OCH_3}}{OCH_3}}}{CH_3CH_2CCH_2CH_3} + H_2O \xrightarrow{H^+} CH_3CH_2\overset{\overset{O}{\|}}{C}CH_2CH_3 + 2CH_3OH
$$

| An acetal | 2-Pentanone |

| A hemiacetal | | 5-Hydroxypentanal |

Problem 17.7

Identify all hemiacetals and acetals in the following structures, and tell whether they are formed from aldehydes or ketones.

(a)   CH₃CH₂CCH₂CH₃   (b)   CH₃OCH₂CH₂OCH₃   (c)

17.5 | What Is Keto-Enol Tautomerism?

A carbon atom adjacent to a carbonyl group is called an α-carbon, and a hydrogen atom bonded to it is called an α-hydrogen.

A carbonyl compound that has a hydrogen on an α-carbon is in equilibrium with a constitutional isomer called an **enol.** The name "enol" is derived from the IUPAC designation of it as both an alkene (-*en*-) and an alcohol (-*ol*).

Enol A molecule containing an —OH group bonded to a carbon of a carbon–carbon double bond

Keto and enol forms are examples of **tautomers,** constitutional isomers in equilibrium with each other that differ in the location of a hydrogen atom relative to an O or N atom. This type of isomerism is called **keto-enol tautomerism.** For any pair of keto-enol tautomers, the keto form generally predominates at equilibrium.

Tautomers Constitutional isomers that differ in the location of a hydrogen atom and a double bond relative to an O or N atom

GOB
Chemistry✦Now™

Click *Coached Problems* to explore
Conversions Between Keto and Enol Structures

EXAMPLE 17.8

Draw structural formulas for the two enol forms for each ketone.

(a)   (b)

Solution

(a)

(b)

Problem 17.8

Draw a structural formula for the keto form of each enol.

(a)

(b)

(c)

SUMMARY OF KEY QUESTIONS

SECTION 17.1 What Are Aldehydes and Ketones?

- An **aldehyde** contains a carbonyl group bonded to at least one hydrogen atom.
- A **ketone** contains a carbonyl group bonded to two carbon atoms.

SECTION 17.2 How Do We Name Aldehydes and Ketones?

- We derive the IUPAC name of an aldehyde by changing the -*e* of the parent alkane to -*al*.

- We derive the IUPAC name of a ketone by changing the -*e* of the parent alkane to -*one* and using a number to locate the carbonyl carbon.

SECTION 17.3 What Are the Physical Properties of Aldehydes and Ketones?

- Aldehydes and ketones are polar compounds. They have higher boiling points and are more soluble in water than nonpolar compounds of comparable molecular weight.

SUMMARY OF KEY REACTIONS

1. **Oxidation of an Aldehyde to a Carboxylic Acid (Section 17.4A)** The aldehyde group is among the most easily oxidized functional groups. Oxidizing agents include $K_2Cr_2O_7$, Tollens' reagent, and O_2.

```
        O                              O
        ‖                              ‖
⬡—CH + Ag(NH₃)₂⁺ ⟶ ⬡—CO⁻ + Ag + 2NH₃
   Tollens' reagent
```

2. **Reduction (Section 17.4B)** Aldehydes are reduced to primary alcohols and ketones to secondary alcohols by H_2 in the presence of a transition metal catalyst such as

Pt or Ni. They are also reduced to alcohols by sodium borohydride, $NaBH_4$, followed by protonation.

```
                   Transition
                 metal catalyst
⬡=O + H₂ ⟶ ⬡—OH
```

```
              O
              ‖
⬡—CH=CH—C—H  1. NaBH₄  ⟶ ⬡—CH=CH—CH₂OH
              2. H₂O
```

3. **Addition of Alcohols to Form Hemiacetals (Section 17.4C)**
Hemiacetals are only minor components of an equilibrium mixture of aldehyde or ketone and alcohol, except where the —OH and C=O groups are parts of the same molecule and a five- or six-membered ring can form.

4. **Addition of Alcohols to Form Acetals (Section 17.4C)** Formation of acetals is catalyzed by acid. Acetals are

hydrolyzed in aqueous acid to an aldehyde or ketone and two molecules of an alcohol.

Cyclohexanone   Methanol   An acetal

5. **Keto-Enol Tautomerism (Section 17.5)** The keto form generally predominates at equilibrium.

$$CH_3CCH_3 \rightleftharpoons CH_3C=CH_2$$

Keto form   Enol form
(approximately 99.9%)

PROBLEMS

GOB
Chemistry•⚛•Now™

Assess your understanding of this chapter's topics with additional quizzing and conceptual-based problems at **http://now.brookscole.com/gob8** or on the CD.

A blue problem number indicates an applied problem.

■ denotes problems that are available on the GOB ChemistryNow website or CD and are assignable in OWL.

SECTION 17.1 What Are Aldehydes and Ketones?

17.9   What is the difference in structure between an aldehyde and a ketone?

17.10 What is the difference in structure between an aromatic aldehyde and an aliphatic aldehyde?

17.11 Is it possible for the carbon atom of a carbonyl group to be a stereocenter? Explain.

17.12 ■ Which compounds contain carbonyl groups?

(a) CH_3CHCH_3 with OH

(b) CH_3CH_2CH with O

(c) benzene ring—$COCH_2CH_3$

(d) cyclopentanone =O

(e) tetrahydrofuran ring—OH

(f) $CH_3CH_2CH_2COH$ with O

17.13 ■ Following are structural formulas for two steroid hormones.

Cortisone

Aldosterone

(a) Name the functional groups in each.
(b) Mark all stereocenters in each hormone and state how many stereoisomers are possible for each.

17.14 Draw structural formulas for the four aldehydes with the molecular formula $C_5H_{10}O$. Which of these aldehydes are chiral?

SECTION 17.2 How Do We Name Aldehydes and Ketones?

17.15 ■ Draw structural formulas for these aldehydes.
(a) Formaldehyde   (b) Propanal
(c) 3,7-Dimethyloctanal   (d) Decanal
(e) 4-Hydroxybenzaldehyde
(f) 2,3-Dihydroxypropanal

17.16 ■ Draw structural formulas for these ketones.

(a) Ethyl isopropyl ketone

(b) 2-chlorocyclohexanone

(c) 2,4-Dimethyl-3-pentanone

(d) Diisopropyl ketone

(e) Acetone

(f) 2,5-Dimethylcyclohexanone

17.17 ■ Write IUPAC names for these compounds.

(a) $(CH_3CH_2CH_2)_2C=O$ (b)

(c)   (d) $CH_3-\overset{OH}{\underset{|}{CH}}-\overset{O}{\overset{\|}{C}}-H$

(e)   (f) $HC(CH_2)_4CH$

17.18 Write IUPAC names for these compounds.

(a)   (b)

(c)   (d)

17.19 Explain why each name is incorrect. Write the correct IUPAC name for the intended compound.

(a) 3-Butanone   (b) 1-Butanone

(c) 4-Methylbutanal   (d) 2,2-Dimethyl-3-butanone

17.20 Explain why each name is incorrect. Write the correct IUPAC name for the intended compound.

(a) 2-Pentanal   (b) Cyclopentanal

(c) 3-Ethyl-2-butanone   (d) 5-Aminobenzaldehyde

SECTION 17.3 What Are the Physical Properties of Aldehydes and Ketones?

17.21 ■ In each pair of compounds, select the one with the higher boiling point.

(a) Acetaldehyde or ethanol

(b) Acetone or 3-pentanone

(c) Butanal or butane

(d) Butanone or 2-butanol

17.22 Acetone is completely soluble in water, but 4-heptanone is completely insoluble in water. Explain.

17.23 Account for the fact that acetone has a higher boiling point (56°C) than ethyl methyl ether (11°C), even though their molecular weights are almost the same.

17.24 Pentane, 1-butanol, and butanal all have approximately the same molecular weights but different boiling points. Arrange them in order of increasing boiling point. Explain the basis for your ranking.

17.25 Show how acetaldehyde can form hydrogen bonds with water.

17.26 Why can't two molecules of acetone form a hydrogen bond with each other?

SECTION 17.4 What Are the Characteristic Reactions of Aldehydes and Ketones?

17.27 Draw a structural formula for the principal organic product formed when each compound is treated with $K_2Cr_2O_7/H_2SO_4$. If there is no reaction, say so.

(a) Butanal   (b) Benzaldehyde

(c) Cyclohexanone   (d) Cyclohexanol

17.28 Draw a structural formula for the principal organic product formed when each compound in Problem 17.27 is treated with Tollens' reagent. If there is no reaction, say so.

17.29 What simple chemical test could you use to distinguish between the members of each pair of compounds? Tell what you would do, what you would expect to observe, and how you would interpret your experimental observation.

(a) Pentanal and 2-pentanone

(b) 2-Pentanone and 2-pentanol

17.30 Explain why liquid aldehydes are often stored under an atmosphere of nitrogen rather than in air.

17.31 Suppose that you take a bottle of benzaldehyde (a liquid, bp 179°C) from a shelf and find a white solid in the bottom of the bottle. The solid turns litmus red; that is, it is acidic. Yet aldehydes are neutral compounds. How can you explain these observations?

17.32 Explain why the reduction of an aldehyde always gives a primary alcohol and the reduction of a ketone always gives a secondary alcohol.

17.33 ■ Write a structural formula for the principal organic product formed by treating each compound with H_2/transition metal catalyst. Which products are chiral?

(a) $CH_3\overset{O}{\overset{\|}{C}}CH_2CH_3$ (b) $CH_3(CH_2)_4\overset{O}{\overset{\|}{CH}}$

(c)   (d)

17.34 ■ Write a structural formula for the principal organic product formed by treating each compound in Problem 17.33 with $NaBH_4$ followed by H_2O.

17.35 1,3-Dihydroxy-2-propanone, more commonly known as dihydroxyacetone, is the active ingredient in artificial tanning agents such as Man-Tan and Magic Tan.
(a) Write a structural formula for this compound.
(b) Would you expect it to be soluble or insoluble in water?
(c) Write a structural formula for the product formed by its reduction with $NaBH_4$.

17.36 ■ Draw a structural formula for the product formed by treatment of butanal with each set of reagents.
(a) H_2/metal catalyst
(b) $NaBH_4$, then H_2O
(c) $Ag(NH_3)_2^+$ (Tollens' reagent)
(d) $K_2Cr_2O_7/H_2SO_4$

17.37 Draw a structural formula for the product formed by treatment of acetophenone, $C_6H_5COCH_3$, with each set of reagents given in Problem 17.36.

SECTION 17.5 What Is Keto-Enol Tautomerism?

17.38 ■ Which of these compounds undergo keto-enol tautomerism?

(a) CH_3CH with O double bond

(b) CH_3CCH_3 with O double bond

(c) benzaldehyde

(d) acetophenone

(e) benzophenone

(f) cyclohexanone

17.39 ■ Draw all enol forms of each aldehyde and ketone.

(a) CH_3CH_2CH with O double bond

(b) $CH_3CCH_2CH_3$ with O double bond

(c) 2-methylcyclopentanone

17.40 Draw a structural formula for the keto form of each enol.

(a) cyclopentene with OH

(b) $CH_3C=CHCH_2CH_2CH_3$ with OH

(c) benzene ring $-CH=CCH_3$ with OH

Addition of Alcohols

17.41 What is the characteristic structural feature of a hemiacetal? Of an acetal?

17.42 Which compounds are hemiacetals, which are acetals, and which are neither?

(a) benzene ring $-CHOCH_3$ with OCH_3

(b) $CH_3CH_2CHOCH_3$ with OH

(c) $CH_3OCH_2OCH_3$

(d) dioxolane ring with CH_2CH_3

(e) tetrahydrofuran ring with OCH_2CH_3

(f) tetrahydrofuran ring with OH

17.43 Which compounds are hemiacetals, which are acetals, and which are neither?

(a) HO OCH_3 structure

(b) benzene with CH_3O OCH_3

(c) lactone (five-membered ring with O and =O)

(d) cyclopentane with OCH_3 and OH

(e) tetrahydrofuran with OCH_3

(f) cyclopentane with OCH_3 and OCH_3

17.44 Draw the hemiacetal and then the acetal formed in each reaction. In each case, assume an excess of the alcohol.
(a) Propanal + methanol $\longrightarrow$
(b) Cyclopentanone + methanol $\longrightarrow$

17.45 Draw the structures of the aldehydes or ketones and alcohols formed when these acetals are treated with aqueous acid and hydrolyzed.

(a) cyclic acetal structure

(b) benzene with OCH_3, OCH_3, OCH_3

(c) cyclohexane with OCH_3 and OCH_3

(d) tetrahydropyran with OCH_3

17.46 The following compound is a component of the fragrance of jasmine:

From which carbonyl-containing compound and alcohol is this compound derived?

17.47 What is the difference in meaning between the terms "hydration" and "hydrolysis?" Give an example of each.

17.48 What is the difference in meaning between the terms "hydration" and "dehydration?" Give an example of each.

17.49 Show reagents and experimental conditions to convert cyclohexanone to each of the following compounds.

17.50 Draw a structural formula for an aldehyde or ketone that can be reduced to produce each alcohol. If none exists, say so.

(a) CH₃CHCH₃ (OH)

(b) benzene-CH₂OH

(c) CH₃OH

(d) cyclohexane-OH, CH₃

17.51 Draw a structural formula for an aldehyde or ketone that can be reduced to produce each alcohol. If none exists, say so.

(a) cyclopentane-OH

(b) cyclohexane-CH₂OH

(c) CH₃COH with CH₃, CH₃

(d) HO~~~~~OH

17.52 1-Propanol can be prepared by the reduction of an aldehyde, but it cannot be prepared by the acid-catalyzed hydration of an alkene. Explain why it cannot be prepared from an alkene.

17.53 Show how to bring about these conversions. In addition to the given starting material, use any other organic or inorganic reagents as necessary.

17.54 Show how to bring about these conversions. In addition to the given starting material, use any other organic or inorganic reagents as necessary.

(a) 1-Pentene to 2-pentanone

(b) Cyclohexene to cyclohexanone

17.55 Describe a simple chemical test by which you could distinguish between the members of each pair of compounds.

(a) Cyclohexanone and aniline

(b) Cyclohexene and cyclohexanol

(c) Benzaldehyde and cinnamaldehyde

Additional Problems

17.56 Indicate the aldehyde or ketone group in these compounds.

17.57 Draw a structural formula for the product formed by treating each compound in Problem 17.56 with sodium borohydride, NaBH₄.

17.58 Draw structural formulas for the (a) one ketone and (b) two aldehydes with the molecular formula C₄H₈O.

17.59 Draw structural formulas for these compounds.

(a) 1-Chloro-2-propanone

(b) 3-Hydroxybutanal

(c) 4-Hydroxy-4-methyl-2-pentanone

(d) 3-Methyl-3-phenylbutanal

(e) 1,3-Cyclohexanedione

(f) 5-Hydroxyhexanal

17.60 Why does acetone have a lower boiling point (56°C) than 2-propanol (82°C), even though their molecular weights are almost the same?

17.61 Propanal (bp 49°C) and 1-propanol (bp 97°C) have about the same molecular weight, yet their boiling points differ by almost 50°C. Explain this fact.

17.62 What simple chemical test could you use to distinguish between the members of each pair of compounds? Tell what you would do, what you would expect to observe, and how you would interpret your experimental observation.

(a) Benzaldehyde and cyclohexanone

(b) Acetaldehyde and acetone

17.63 5-Hydroxyhexanal forms a six-membered cyclic hemiacetal, which predominates at equilibrium in aqueous solution.

(a) Draw a structural formula for this cyclic hemiacetal.

(b) How many stereoisomers are possible for 5-hydroxyhexanal?

(c) How many stereoisomers are possible for this cyclic hemiacetal?

17.64 The following molecule is an enediol; each carbon of the double bond carries an —OH group. Draw structural formulas for the α-hydroxyketone and the α-hydroxyaldehyde with which this enediol is in equilibrium.

$$\alpha\text{-hydroxyaldehyde} \rightleftharpoons \begin{array}{c} HC-OH \\ \| \\ C-OH \\ | \\ CH_3 \end{array} \rightleftharpoons \alpha\text{-hydroxyketone}$$

An enediol

17.65 Alcohols can be prepared by the acid-catalyzed hydration of alkenes (Section 12.6B) and by the reduction of aldehydes and ketones (Section 17.4B). Show how you might prepare each of the following alcohols by (1) acid-catalyzed hydration of an alkene and (2) reduction of an aldehyde or ketone.

(a) Ethanol (b) Cyclohexanol

(c) 2-Propanol (d) 1-Phenylethanol

Looking Ahead

17.66 Glucose, $C_6H_{12}O_6$, contains an aldehyde group but exists predominantly in the form of the cyclic hemiacetal shown here. We discuss this cyclic form of glucose in Chapter 20.

β-D-Glucose

(a) A cyclic hemiacetal is formed when the —OH group of one carbon bonds to the carbonyl group of another carbon. Which carbon in glucose provides the —OH group and which provides the CHO group?

(b Write a structural formula for the free aldehyde form of glucose.

17.67 Ribose, $C_5H_{10}O_5$, contains an aldehyde group but exists predominantly in the form of the cyclic hemiacetal shown here. We discuss this cyclic form of ribose in Chapter 20.

β-D-Ribose

(a) Which carbon of ribose provides the —OH group and which provides the CHO group for formation of this cyclic hemiacetal?

(b) Write a structural formula for the free aldehyde form of ribose.

17.68 Name the functional groups in each steroid hormone. Methandrostenolone is a synthetic anabolic steroid; that is, it promotes tissue and muscle growth and development.

(i) Testosterone   (ii) Methandrostenolone

(a) What are the differences in structural formula between testosterone and methandrostenolone?

(b) Mark all stereocenters in each molecule and state the number of stereoisomers possible for each. The stereochemistry shown on the structural formulas is that of the biologically active stereoisomer.

17.69 Sodium borohydride is a laboratory reducing agent. NADH is a biological reducing agent. In what way is the chemistry by which they reduce aldehydes and ketones similar?

Tying It Together

17.70 Complete equations for these reactions.

(a) $\xrightarrow[H_2SO_4]{K_2Cr_2O_7}$

(b) $\xrightarrow[H_2SO_4]{K_2Cr_2O_7}$

17.71 ■ Write an equation for each conversion.

(a) 1-Pentanol to pentanal

(b) 1-Pentanol to pentanoic acid

(c) 2-Pentanol to 2-pentanone

(d) 2-Propanol to acetone

(e) Cyclohexanol to cyclohexanone

CHAPTER 18

Carboxylic Acids

Citrus fruits are sources of citric acid, a tricarboxylic acid.

Charles D. Winters

GOB
Chemistry Now™
Look for this logo in the chapter and go to GOB ChemistryNow at
http://now.brookscole.com/gob8
or on the CD for tutorials, simulations, and problems.

18.1 | What Are Carboxylic Acids?

In this chapter, we study carboxylic acids, another class of organic compounds containing the carbonyl group. The functional group of a **carboxylic acid** is a **carboxyl group** (Section 10.4D), which can be represented in any one of three ways:

$$\overset{\displaystyle O}{\underset{\displaystyle \|}{-C}}-OH \quad -COOH \quad -CO_2H$$

18.2 | How Do We Name Carboxylic Acids?

A. IUPAC Names

We derive the IUPAC name of an acyclic carboxylic acid from the name of the longest carbon chain that contains the carboxyl group. Drop the final **-e** from the name of the parent alkane and replace it by **-oic acid.** Number the chain beginning with the carbon of the carboxyl group. Because the carboxyl carbon is understood to be carbon 1, there is no need to give it a number. In the following examples, the common name is given in parentheses.

GOB
Chemistry·ᛣ·Now™
Click *Coached Problems* to try a set of problems in **Naming Carboxylic Acids**

Hexanoic acid
(Caproic acid)

3-Methylbutanoic acid
(Isovaleric acid)

When a carboxylic acid also contains an alcohol, we indicate the presence of the —OH group by adding the prefix *hydroxy-*. When it contains a primary (1°) amine, we indicate the presence of the —NH$_2$ group by *amino-*.

5-Hydroxyhexanoic acid

4-Aminobenzoic acid
(*p*-Aminobenzoic acid)

To name dicarboxylic acids, we add the suffix **-dioic acid** to the name of the parent alkane that contains both carboxyl groups. The numbers of the carboxyl carbons are not indicated because they can be only at the ends of the parent chain.

Formic acid was first obtained in 1670 from the destructive distillation of ants, whose Latin genus is *Formica*. It is one of the components of the venom injected by stinging ants.

Ethanedioic acid
(Oxalic acid)

Propanedioic acid
(Malonic acid)

Butanedioic acid
(Succinic acid)

Pentanedioic acid
(Glutaric acid)

Hexanedioic acid
(Adipic acid)

The name *oxalic acid* is derived from one of its sources in the biological world—plants of the genus *Oxalis*, one of which is rhubarb. Oxalic acid also occurs in human and animal urine, and calcium oxalate is a major component of kidney stones. Succinic acid is an intermediate in the citric acid cycle (Section 27.4). Adipic acid is one of the two monomers required for the synthesis of the polymer nylon-66 (Section 19.6B).

B. Common Names

Common names for aliphatic carboxylic acids, many of which were known long before the development of IUPAC nomenclature, are often derived from the name of a natural substance from which the acid can be isolated. Table 18.1 lists several of the unbranched aliphatic carboxylic acids found in the biological world along with the common name of each. Those with 16, 18, and 20 carbon atoms are particularly abundant in animal fats and vegetable oils (Section 21.2), and the phospholipid components of biological membranes (Section 21.5).

GOB
Chemistry·ᛣ·Now™
Click *Coached Problems* to explore the
Names and Structures of Commonly Encountered Carboxylic Acids

The unbranched carboxylic acids having between 12 and 20 carbon atoms are known as fatty acids. We study them further in Chapter 21.

Table 18.1 Several Aliphatic Carboxylic Acids and Their Common Names

Structure	IUPAC Name	Common Name	Derivation
HCOOH	methanoic acid	formic acid	Latin: *formica*, ant
CH$_3$COOH	ethanoic acid	acetic acid	Latin: *acetum*, vinegar
CH$_3$CH$_2$COOH	propanoic acid	propionic acid	Greek: *propion*, first fat
CH$_3$(CH$_2$)$_2$COOH	butanoic acid	butyric acid	Latin: *butyrum*, butter
CH$_3$(CH$_2$)$_3$COOH	pentanoic acid	valeric acid	Latin: *valere*, to be strong
CH$_3$(CH$_2$)$_4$COOH	hexanoic acid	caproic acid	Latin: *caper*, goat
CH$_3$(CH$_2$)$_6$COOH	octanoic acid	caprylic acid	Latin: *caper*, goat
CH$_3$(CH$_2$)$_8$COOH	decanoic acid	capric acid	Latin: *caper*, goat
CH$_3$(CH$_2$)$_{10}$COOH	dodecanoic acid	lauric acid	Latin: *laurus*, laurel
CH$_3$(CH$_2$)$_{12}$COOH	tetradecanoic acid	myristic acid	Greek: *myristikos*, fragrant
CH$_3$(CH$_2$)$_{14}$COOH	hexadecanoic acid	palmitic acid	Latin: *palma*, palm tree
CH$_3$(CH$_2$)$_{16}$COOH	octadecanoic acid	stearic acid	Greek: *stear*, solid fat
CH$_3$(CH$_2$)$_{18}$COOH	eicosanoic acid	arachidic acid	Greek: *arachis*, peanut

When common names are used, the Greek letters alpha (α), beta (β), gamma (γ), and so forth, are often added as a prefix to locate substituents.

GABA is a neurotransmitter in the central nervous system.

4-Aminobutanoic acid
(γ-Aminobutyric acid; GABA)

2-Hydroxypropanoic acid
(Lactic acid)

EXAMPLE 18.1

Write the IUPAC name for each carboxylic acid:

Solution

(a) The longest carbon chain that contains the carboxyl group has five carbons and, therefore, the parent alkane is pentane. The IUPAC name is 2-ethylpentanoic acid.

(b) 4-Hydroxybenzoic acid

(c) *trans*-3-Phenyl-2-propenoic acid (cinnamic acid)

Problem 18.1

Each of the following compounds has a well-recognized common name. A derivative of glyceric acid is an intermediate in glycolysis (Section 28.2). β-Alanine is a building block of pantothenic acid (Section 27.5). Mevalonic

acid is an intermediate in the biosynthesis of steroids (Section 27.4). Write
the IUPAC name for each compound. Which of them compounds are chiral?

(a) Glyceric acid

(b) H₂NCH₂CH₂COOH
     β-Alanine

(c) Mevalonic acid

18.3 | What Are the Physical Properties of Carboxylic Acids?

A major feature of carboxylic acids is the polarity of the carboxyl group (Figure 18.1). This group contains three polar covalent bonds: C=O, C—O, and
O—H. It is the polarity of these bonds that determines the major physical
properties of carboxylic acids.

Carboxylic acids have significantly higher boiling points than other
types of organic compounds of comparable molecular weight (Table 18.2).
Their higher boiling points result from their polarity and the fact that
hydrogen bonding between two carboxyl groups creates a dimer that
behaves as a higher-molecular-weight compound.

A hydrogen-bonded
dimer of acetic acid

Carboxylic acids are more soluble in water than are alcohols, ethers,
aldehydes, and ketones of comparable molecular weight. This increased solubility is due to their strong association with water molecules by hydrogen
bonding through both their carbonyl and hydroxyl groups. The first four
aliphatic carboxylic acids (formic, acetic, propanoic, and butanoic) are infinitely soluble in water. As the size of the hydrocarbon chain increases relative to that of the carboxyl group, however, water solubility decreases. The
solubility of hexanoic acid (six carbons) in water is 1.0 g/100 mL water.

We must mention two other properties of carboxylic acids. First, the liquid carboxylic acids from propanoic acid to decanoic acid have sharp, often
disagreeable odors. Butanoic acid is found in stale perspiration and is a

Figure 18.1   Polarity of
a carboxyl group.

Acetic acid

Table 18.2 Boiling Points and Solubilities in Water of Two Groups of Compounds of Comparable Molecular Weight

Structure	Name	Molecular Weight	Boiling Point (°C)	Solubility (g/100 mL H$_2$O)
CH$_3$COOH	acetic acid	60.5	118	infinite
CH$_3$CH$_2$CH$_2$OH	1-propanol	60.1	97	infinite
CH$_3$CH$_2$CHO	propanal	58.1	48	16
CH$_3$(CH$_2$)$_2$COOH	butanoic acid	88.1	163	infinite
CH$_3$(CH$_2$)$_3$CH$_2$OH	1-pentanol	88.1	137	2.3
CH$_3$(CH$_2$)$_3$CHO	pentanal	86.1	103	slight

major component of "locker room odor." Pentanoic acid smells even worse, and goats, which secrete C$_6$, C$_8$, and C$_{10}$ acids (Table 18.1), are not famous for their pleasant odors. Second, carboxylic acids have a characteristic sour taste. The sour taste of pickles and sauerkraut, for example, is due to the presence of lactic acid. The sour tastes of limes (pH 1.9), lemons (pH 2.3), and grapefruit (pH 3.2) are due to the presence of citric and other acids.

18.4 | What Are Soaps and Detergents?

A. Fatty Acids

Fatty acids are long, unbranched carboxylic acids, most commonly consisting of 12 to 20 carbons. They are derived from the hydrolysis of animal fats, vegetable oils, or the phospholipids of biological membranes. More than 500 different fatty acids have been isolated from various cells and tissues. Given in Table 18.3 are common names and structural formulas for the most abundant of them. The number of carbons in a fatty acid and the number of carbon–carbon double bonds in its hydrocarbon chain are shown by two numbers separated by a colon. In this notation, linoleic acid, for example,

Chemistry·ᐧ·Now™

Click *Coached Problems* to examine the **Relationship Between Structure and Function in Soaps and Detergents**

Table 18.3 The Most Abundant Fatty Acids in Animal Fats, Vegetable Oils, and Biological Membranes

Carbon Atoms: Double Bonds*	Structure	Common Name	Melting Point (°C)
Saturated Fatty Acids			
12 : 0	CH$_3$(CH$_2$)$_{10}$COOH	lauric acid	44
14 : 0	CH$_3$(CH$_2$)$_{12}$COOH	myristic acid	58
16 : 0	CH$_3$(CH$_2$)$_{14}$COOH	palmitic acid	63
18 : 0	CH$_3$(CH$_2$)$_{16}$COOH	stearic acid	70
20 : 0	CH$_3$(CH$_2$)$_{18}$COOH	arachidic acid	77
Unsaturated Fatty Acids			
16 : 1	CH$_3$(CH$_2$)$_5$CH=CH(CH$_2$)$_7$COOH	palmitoleic acid	1
18 : 1	CH$_3$(CH$_2$)$_7$CH=CH(CH$_2$)$_7$COOH	oleic acid	16
18 : 2	CH$_3$(CH$_2$)$_4$(CH=CHCH$_2$)$_2$(CH$_2$)$_6$COOH	linoleic acid	−5
18 : 3	CH$_3$CH$_2$(CH=CHCH$_2$)$_3$(CH$_2$)$_6$COOH	linolenic acid	−11
20 : 4	CH$_3$(CH$_2$)$_4$(CH=CHCH$_2$)$_4$(CH$_2$)$_2$COOH	arachidonic acid	−49

*The first number is the number of carbons in the fatty acid; the second number is the number of carbon–carbon double bonds in its hydrocarbon chain.

is designated as an 18:2 fatty acid; its 18-carbon chain contains two carbon–carbon double bonds.

Following are several characteristics of the most abundant fatty acids in higher plants and animals:

1. Nearly all fatty acids have an even number of carbon atoms, most between 12 and 20, in an unbranched chain.

2. The three most abundant fatty acids in nature are palmitic acid (16:0), stearic acid (18:0), and oleic acid (18:1).

3. In most unsaturated fatty acids, the *cis* isomer predominates; the *trans* isomer is rare.

4. Unsaturated fatty acids have lower melting points than their saturated counterparts. The greater the degree of unsaturation, the lower the melting point. Compare, for example, the melting points of these 18-carbon fatty acids: Linolenic acid, with three carbon–carbon double bonds, has the lowest melting point of the three fatty acids.

Stearic acid (18:0)
(mp 70°C)

Oleic acid (18:1)
(mp 16°C)

Linoleic acid (18:2)
(mp –5°C)

Linolenic acid (18:3)
(mp –11°C)

Fatty acids can be divided into two groups: saturated and unsaturated. Saturated fatty acids have only single carbon–carbon bonds in their hydrocarbon chains. Unsaturated fatty acids have at least one C=C double bond in the chain. All unsaturated fatty acids listed in Table 18.3 are the *cis* isomer. This fact explains their physical properties, as reflected in their melting points.

Saturated fatty acids are solids at room temperature, because the regular nature of their hydrocarbon chains allows their molecules to pack together in close parallel alignment. When packed in this manner, the interactions between adjacent hydrocarbon chains (London dispersion forces; Section 6.7A) are maximized. Although London dispersion forces are weak interactions, the regular packing of hydrocarbon chains allows these forces to operate over a large portion of the chain, ensuring that a considerable amount of energy is needed to separate and melt them.

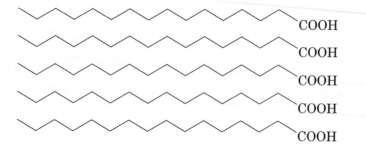

CHEMICAL CONNECTIONS 18A

Trans Fatty Acids: What Are They and How Do You Avoid Them?

Animal fats are rich in saturated fatty acids, whereas plant oils (for example, corn, soybean, canola, olive, and palm oil) are rich in unsaturated fatty acids. Fats are added to processed foods to provide a desirable firmness along with a moist texture and pleasant taste. To supply the demand for dietary fats of the appropriate consistency, the *cis* double bonds of vegetable oils are partially hydrogenated. The greater the extent of hydrogenation, the higher the melting point of the triglyceride. The extent of hydrogenation is carefully controlled, usually by employing a Ni catalyst and a calculated amount of H_2 as a limiting reagent. Under these conditions, the H_2 is used up before all double bonds are reduced, so that only partial hydrogenation and the desired overall consistency is achieved. For example, by controlling the degree of hydrogenation, an oil with a melting point below room temperature can be converted to a semisolid or even a solid product.

The mechanism of catalytic hydrogenation of alkenes was discussed in Section 12.6D. Recall that a key step in this mechanism involves interaction of the carbon–carbon double bond of the alkene with the metal catalyst to form a carbon–metal bond. Because the interaction of a carbon–carbon double bond with the Ni catalyst is reversible, many of the double bonds remaining in the oil may be isomerized from the less stable *cis* configuration to the more stable *trans* configuration. Thus equilibration between the

cis and *trans* configurations may occur when H_2 is the limiting reagent. For example, elaidic acid is the *trans* C_{18} fatty acid analog of oleic acid, a common C_{18} *cis* fatty acid.

The oils used for frying in fast-food restaurants are usually partially hydrogenated plant oils and, therefore, contain substantial amounts of *trans* fatty acids that are transferred to the foods cooked in them. Other major sources of *trans* fatty acids in the diet include stick margarine, certain commercial bakery products, creme-filled cookies, potato and corn chips, frozen breakfast foods, and cake mixes.

Recent studies have shown that consuming a significant amount of *trans* fatty acids can lead to serious health problems related to serum cholesterol levels. Low overall serum cholesterol and a decreased ratio of low-density lipoprotein (LDL) cholesterol to high-density lipoprotein (HDL) cholesterol are associated with good cardiovascular health. High serum cholesterol levels and an elevated ratio (LDL)-cholesterol to (HDL)-cholesterol are linked to a high incidence of cardiovascular disease, especially atherosclerosis. Research has indicated that consuming a diet high in either saturated fatty acids or *trans* fatty acids substantially increases the risk of cardiovascular disease.

The FDA recently announced that processed foods must list the amount of *trans* fatty acids they contain, so that consumers can

Elaidic acid
mp 46°C
(a *trans* C_{18} fatty acid)

In contrast, all unsaturated fatty acids are liquids at room temperature because the *cis* double bonds interrupt the regular packing of the chains and the London dispersion forces act over shorter segments of the chains, so that less energy is required to melt them. The greater the degree of unsaturation, the lower the melting point, because each double bond introduces more disorder into the packing of the fatty acid molecules.

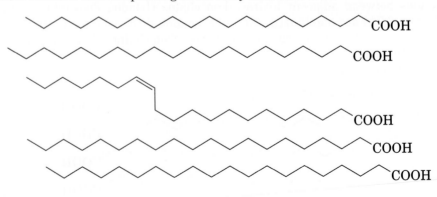

Trans Fatty Acids: What Are They and How Do You Avoid Them? *(continued)*

make better choices about the food they eat. A diet low in saturated and *trans* fatty acids is recommended, along with consumption of more fish, whole grains, fruits, and vegetables, and especially daily exercise, which is tremendously beneficial regardless of diet.

Monounsaturated and polyunsaturated fatty acids have not produced similar health risks in most studies, although too much fat of any kind in the diet can lead to obesity, a major health problem that is associated with several diseases, one of which is diabetes. Some polyunsaturated (*cis*) fatty acids, such as those found in certain types of fish, have even been shown to have beneficial effects in some studies. These are the so-called omega-3 fatty acids.

In omega-3 fatty acids, the last carbon of the last double bond of the hydrocarbon chain ends three carbons in from the methyl terminal end of the chain. The last carbon of the hydrocarbon chain

is called the omega (the last letter of the Greek alphabet) carbon—hence the designation of omega-3. The two most commonly found in health food supplements are eicosapentaenoic acid and docosahexaenoic acid.

Eicosapentaenoic acid, $C_{20}H_{30}O_2$ is an important fatty acid in the marine food chain and serves as a precursor in humans of several members of the prostacyclin and thromboxane families. Note how the name of this fatty acid is derived. *Eicosa-* is the prefix, indicating 20 carbons in the chain; *-pentaene-* indicates five carbon–carbon double bonds; and *-oic acid* shows the carboxyl functional group.

Docosahexaenoic acid, $C_{22}H_{32}O_2$, is found in fish oils and many phospholipids. It is a major structural component of excitable membranes in the retina and brain, and is synthesized in the liver from linoleic acid.

Eicosapentaenoic acid

Eicosapentaenoic acid, drawn in a more compact form

Docosahexaenoic acid

B. Structure and Preparation of Soaps

Natural **soaps** are prepared most commonly from a blend of tallow and coconut oils. In the preparation of tallow, the solid fats of cattle are melted with steam, and the tallow layer that forms on the top is removed. The preparation of soaps begins by boiling these triglycerides with sodium hydroxide. The reaction that takes place is called **saponification** (Latin: *saponem*, "soap"):

Saponification The hydrolysis of an ester in aqueous NaOH or KOH to an alcohol and the sodium or potassium salt of a carboxylic acid

A triglyceride   1,2,3-Propanetriol (Glycerol; Glycerin)   Sodium soaps

Figure 18.2 Soap micelles. (*a*) Soaps are sodium salts of fatty acids. (*b*) Nonpolar (hydrophobic) hydrocarbon chains are clustered in the interior of the micelle and polar (hydrophilic) carboxylate groups are on the surface of the micelle. Soap micelles repel each other because of their negative surface charges.

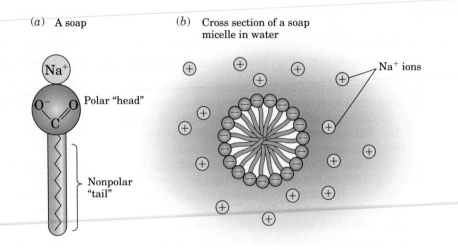

(*a*) A soap

Polar "head"

Nonpolar "tail"

(*b*) Cross section of a soap micelle in water

Na⁺ ions

At the molecular level, saponification corresponds to base-promoted hydrolysis (Section 19.4A) of the ester groups in triglycerides. A **triglyceride** is a triester of glycerol. The resulting soaps contain mainly the sodium salts of palmitic, stearic, and oleic acids from tallow and the sodium salts of lauric and myristic acids from coconut oil.

After hydrolysis is complete, sodium chloride is added to precipitate the sodium salts as thick curds of soap. The water layer is then drawn off, and glycerol is recovered by vacuum distillation. The crude soap contains sodium chloride, sodium hydroxide, and other impurities that are removed by boiling the curd in water and reprecipitating with more sodium chloride. After several purifications, the soap can be used as an inexpensive industrial soap without further processing. Other treatments transform the crude soap into pH-controlled cosmetic soaps, medicated soaps, and the like.

Micelle A spherical arrangement of molecules in aqueous solution such that their hydrophobic parts are shielded from the aqueous environment and their hydrophilic parts are on the surface of the sphere and in contact with the aqueous environment

C. How Soap Cleans

Soap owes its remarkable cleansing properties to its ability to act as an emulsifying agent. Because the long hydrocarbon chains of natural soaps are insoluble in water, they tend to cluster in such a way as to minimize their contact with surrounding water molecules. The polar carboxylate groups, by contrast, tend to remain in contact with the surrounding water molecules. Thus, in water, soap molecules spontaneously cluster into **micelles** (Figure 18.2).

Most of the things we commonly think of as dirt (such as grease, oil, and fat stains) are nonpolar and insoluble in water. When soap and this type of dirt are mixed together, as in a washing machine, the nonpolar hydrocarbon inner parts of the soap micelles "dissolve" the nonpolar dirt molecules. In effect, new soap micelles are formed, this time with nonpolar dirt molecules in the center (Figure 18.3). In this way, nonpolar organic grease, oil, and fat are "dissolved" and washed away in the polar wash water.

Soaps, however, have their disadvantages, foremost among which is the fact that they form water-insoluble salts when used in water containing Ca(II), Mg(II), or Fe(III) ions (hard water):

Soap micelle with "dissolved" grease

Grease

Soap

Figure 18.3 A soap micelle with a "dissolved" oil or grease droplet.

$$2CH_3(CH_2)_{14}COO^-Na^+ + Ca^{2+} \longrightarrow [CH_3(CH_2)_{14}COO^-]_2Ca^{2+} + 2Na^+$$

A sodium soap
(soluble in water as micelles)

Calcium salt of a fatty acid
(insoluble in water)

These water-insoluble calcium, magnesium, and iron salts of fatty acids create problems, including rings around the bathtub, films that spoil the luster of hair, and grayness and roughness that build up on textiles after repeated washings.

D. Synthetic Detergents

After the cleansing action of soaps was understood, chemists were in a position to design a synthetic detergent. Molecules of a good detergent, they reasoned, must have a long hydrocarbon chain—preferably 12 to 20 carbon atoms long—and a polar group at one end of the molecule that does not form insoluble salts with the Ca(II), Mg(II), or Fe(III) ions that are present in hard water. These essential characteristics of a soap, they recognized, could be produced in a molecule containing a sulfonate ($-SO_3^-$) group instead of a carboxylate ($-COO^-$) group. Calcium, magnesium, and iron salts of monoalkylsulfuric and sulfonic acids ($R-SO_3H$) are much more soluble in water than comparable salts of fatty acids.

The most widely used synthetic detergents today are the linear alkylbenzenesulfonates (LAS). One of the most common of these is sodium 4-dodecylbenzenesulfonate. To prepare this type of detergent, a linear alkylbenzene is treated with sulfuric acid to form an alkylbenzenesulfonic acid (Section 13.3C), followed by neutralization of the sulfonic acid with sodium hydroxide:

$$CH_3(CH_2)_{10}CH_2- \bigcirc \xrightarrow[\text{2. NaOH}]{\text{1. } H_2SO_4} CH_3(CH_2)_{10}CH_2-\bigcirc-SO_3^- \ Na^+$$

Dodecylbenzene

Sodium 4-dodecylbenzenesulfonate
(an anionic detergent)

The product is mixed with builders and then spray-dried to give a smooth-flowing powder. The most common builder is sodium silicate. Alkylbenzenesulfonate detergents were introduced in the late 1950s, and today they account for close to 90% of the market that was once held by natural soaps.

Among the most common additives to detergent preparations are foam stabilizers, bleaches, and optical brighteners. A common foam stabilizer added to liquid soaps, but not to laundry detergents (for obvious reasons: think of a top-loading washing machine with foam spewing out of the lid!), is the amide prepared from dodecanoic acid (lauric acid) and 2-aminoethanol (ethanolamine). The most common bleach is sodium perborate tetrahydrate, which decomposes at temperatures higher than 50°C to give hydrogen peroxide, the actual bleaching agent.

$$CH_3(CH_2)_{10}\overset{\overset{\displaystyle O}{\|}}{C}NHCH_2CH_2OH \qquad O=B-O-O^- \ Na^+ \cdot \ 4H_2O$$

N-(2-Hydroxyethyl)dodecanamide
(a foam stabilizer)

Sodium perborate tetrahydrate
(a bleach)

Also added to laundry detergents are optical brighteners (optical bleaches). These substances are absorbed into fabrics and, after absorbing ambient light, fluoresce with a blue color, offsetting the yellow color that develops in fabric as it ages. Optical brighteners produce a "whiter-than-white" appearance. You most certainly have observed their effects if you have seen the glow of white T-shirts or blouses when they are exposed to black light (UV radiation).

18.5 | What Are the Characteristic Reactions of Carboxylic Acids?

A. Acidity

Carboxylic acids are weak acids. Values of K_a for most unsubstituted aliphatic and aromatic carboxylic acids fall within the range of 10^{-4} to 10^{-5} ($pK_a = 4.0–5.0$). The value of K_a for acetic acid, for example, is 1.74×10^{-5}; its pK_a is 4.76.

$$CH_3\overset{\displaystyle O}{\overset{\|}{C}}OH + H_2O \rightleftharpoons CH_3\overset{\displaystyle O}{\overset{\|}{C}}O^- + H_3O^+ \quad K_a = \frac{[CH_3COO^-][H_3O^+]}{[CH_3COOH]} = 1.74 \times 10^{-5}$$

$$pK_a = 4.76$$

Dichloroacetic acid is used as a topical astringent and as a treatment for genital warts in males.

Dentists use a 50% aqueous solution of trichloroacetic acid to cauterize gums. This strong acid stops the bleeding, kills diseased tissue, and allows the growth of healthy gum tissue.

Substituents of high electronegativity (especially —OH, —Cl, and —NH$_3^+$) near the carboxyl group increase the acidity of carboxylic acids, often by several orders of magnitude. Compare, for example, the acidities of acetic acid and the chlorine-substituted acetic acids. Both dichloroacetic acid and trichloroacetic acid are stronger acids than H_3PO_4 (pK_a 2.1).

Formula:	CH_3COOH	$ClCH_2COOH$	$Cl_2CHCOOH$	Cl_3CCOOH
Name:	Acetic acid	Chloroacetic acid	Dichloroacetic acid	Trichloroacetic acid
pK_a:	4.76	2.86	1.48	0.70

Increasing acid strength →

Electronegative atoms on the carbon adjacent to a carboxyl group increase acidity because they pull electron density away from the O—H bond, thereby facilitating ionization of the carboxyl group and making it a stronger acid.

One final point about carboxylic acids: When a carboxylic acid is dissolved in an aqueous solution, the form of the carboxylic acid present depends on the pH of the solution in which it is dissolved. Consider typical carboxylic acids, which have pK_a values in the range of 4.0 to 5.0. When the pH of the solution is equal to the pK_a of the carboxylic acid (that is, when the pH of the solution is in the range 4.0–5.0), the acid, RCOOH, and its conjugate base, RCOO$^-$, are present in equal concentrations. If the pH of the solution is adjusted to 2.0 or lower by the addition of a strong acid, the carboxylic acid then is present in solution almost entirely as RCOOH. If the pH of the solution is adjusted to 7.0 or higher, the carboxylic acid is present almost entirely as its anion. Thus, even in a neutral solution (pH 7.0), a carboxylic acid is present predominantly as its anion.

B. Reaction with Bases

All carboxylic acids, whether soluble or insoluble in water, react with NaOH, KOH, and other strong bases to form water-soluble salts.

Charles D. Winters

Benzoic acid
(slightly soluble in water)

Sodium benzoate
(60 g/100 mL water)

Sodium benzoate, a fungal growth inhibitor, is often added to baked goods "to retard spoilage." Calcium propanoate is also used for the same purpose. Carboxylic acids also form water-soluble salts with ammonia and amines.

Sodium benzoate and calcium propanoate are fungal growth inhibitors, and are added to baked goods "to retard spoilage."

Benzoic acid
(slightly soluble in water)

Ammonium benzoate
(20 g/100 mL water)

Carboxylic acids react with sodium bicarbonate and sodium carbonate to form water-soluble sodium salts and carbonic acid (a weaker acid). Carbonic acid, in turn, decomposes to give water and carbon dioxide, which evolves as a gas.

$$CH_3COOH + NaHCO_3 \longrightarrow CH_3COO^- Na^+ + H_2CO_3$$

$$H_2CO_3 \longrightarrow CO_2 + H_2O$$

$$CH_3COOH + NaHCO_3 \longrightarrow CH_3COO^- Na^+ + CO_2 + H_2O$$

Salts of carboxylic acids are named in the same manner as the salts of inorganic acids: The cation is named first and then the anion. The name of the anion is derived from the name of the carboxylic acid by dropping the suffix *-ic acid* and adding the suffix *-ate*.

GOB
Chemistry·⚛·Now™
Click *Coached Problems* to explore
Common Reactions of Carboxylic Acids

EXAMPLE 18.2

Complete each acid–base reaction and name the carboxylic salt formed.

(a) [structure] COOH + NaOH ⟶

(b) [structure with OH] COOH + NaHCO₃ ⟶
(s)-Lactic acid

Solution

Each carboxylic acid is converted to its sodium salt. In (b), carbonic acid is formed; it decomposes to carbon dioxide and water.

(a) [structure] COOH + NaOH ⟶ [structure] COO⁻Na⁺ + H₂O

Butanoic acid   Sodium butanoate

(b) (s)-Lactic acid + NaHCO$_3$ ⟶ Sodium (s)-lactate + H$_2$O + CO$_2$

Problem 18.2

Write equations for the reaction of each acid in Example 18.2 with ammonia, and name the carboxylic salt formed.

A consequence of the water solubility of carboxylic acid salts is that water-insoluble carboxylic acids can be converted to water-soluble ammonium or alkali metal salts and extracted into aqueous solution. The salt, in turn, can be transformed back to the free carboxylic acid by treatment with HCl, H$_2$SO$_4$, or another strong acid. These reactions allow an easy separation of water-insoluble carboxylic acids from water-insoluble nonacidic compounds.

Shown in Figure 18.4 is a flowchart for the separation of benzoic acid, a water-insoluble carboxylic acid, from benzyl alcohol, a nonacidic compound. First, the mixture of benzoic acid and benzyl alcohol is dissolved in diethyl

Figure 18.4 Flowchart for separation of benzoic acid from benzyl alcohol.

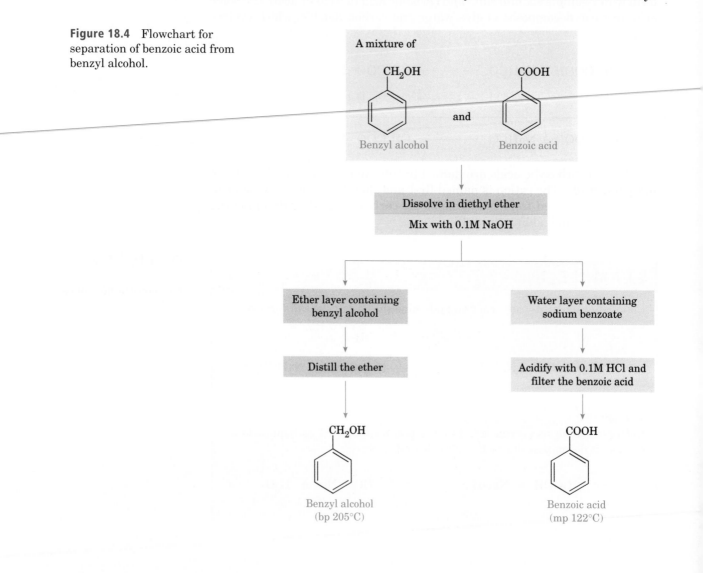

ether. When the ether solution is shaken with aqueous NaOH or another strong base, benzoic acid is converted to its water-soluble sodium salt. Then the ether and aqueous phases are separated. The ether solution is distilled, yielding first diethyl ether (bp 35°C) and then benzyl alcohol (bp 205°C). The aqueous solution is acidified with HCl, and benzoic acid precipitates as a white, crystalline solid (mp 122°C), which is recovered by filtration.

C. Reduction

The carboxyl group is one of the organic functional groups that is most resistant to reduction. It is not affected by catalytic reduction under conditions that readily reduce alkenes to alkanes (Section 12.6D) or by sodium borohydride, (NaBH$_4$), which readily reduces aldehydes to 1° alcohols and ketones to 2° alcohols (Section 17.4B).

The most common reagent for the reduction of a carboxylic acid to a 1° alcohol is the very powerful reducing agent lithium aluminum hydride, LiAlH$_4$. Reduction of a carboxyl group with this reagent is commonly carried out in diethyl ether. The initial product is an aluminum alkoxide, which is then treated with water to give the primary alcohol and lithium and aluminum hydroxides. These two hydroxides are insoluble in diethyl ether and are removed by filtration. Evaporation of the ether solvent yields the primary alcohol.

3-Cyclopentene-
carboxylic acid

4-Hydroxymethyl-
cyclopentene

D. Fischer Esterification

Treatment of a carboxylic acid with an alcohol in the presence of an acid catalyst—most commonly, concentrated sulfuric acid—gives an **ester.** This method of forming an ester is given the special name **Fischer esterification,** after the German chemist Emil Fischer (1852–1919). As an example of Fischer esterification, treating acetic acid with ethanol in the presence of concentrated sulfuric acid gives ethyl acetate and water:

Ester A compound in which the —OH of a carboxyl group, RCOOH, is replaced by an alkyl or an aryl group

Fischer esterification The process of forming an ester by refluxing a carboxylic acid and an alcohol in the presence of an acid catalyst, commonly sulfuric acid

Removal of OH and
H gives the ester

Ethanoic acid
(Acetic acid)

Ethanol
(Ethyl alcohol)

Ethyl ethanoate
(Ethyl acetate)

We study the structure, nomenclature, and reactions of esters in detail in Chapter 19. In this chapter, we discuss only their preparation from carboxylic acids.

In the process of Fischer esterification, the alcohol adds to the carbonyl group of the carboxylic acid to form a tetrahedral carbonyl addition intermediate. Note how closely this step resembles the addition of an alcohol to the carbonyl group of an aldehyde or ketone to form a hemiacetal (Section 17.4C).

In the case of Fischer esterification, the intermediate loses a molecule of water to give the ester.

A tetrahedral carbonyl addition intermediate

Acid-catalyzed esterification is reversible, and, at equilibrium, the quantities of remaining carboxylic acid and alcohol are generally appreciable. By controlling the experimental conditions, however, we can use Fischer esterification to prepare esters in high yields. If the alcohol is inexpensive compared with the carboxylic acid, we can use a large excess of the alcohol (one of the starting reagents) to drive the equilibrium to the right and achieve a high conversion of carboxylic acid to its ester. Alternatively, we can remove water (one of the products of the reaction) as it is formed and drive the equilibrium to the right.

CHEMICAL CONNECTIONS 18B

Esters as Flavoring Agents

Flavoring agents are the largest class of food additives. At present, more than 1000 synthetic and natural flavors are available. The majority of these are concentrates or extracts from the material whose flavor is desired; they are often complex mixtures of tens to hundreds of compounds. A number of ester flavoring agents are synthesized industrially. Many have flavors very close to the target flavor, so adding only one or a few of them is sufficient to make ice cream, soft drinks, or candies taste natural. The table shows the structures of a few of the esters used as flavoring agents.

Ester Flavoring Agents

Structure	Name	Flavor
	Ethyl formate	Rum
	Isopentyl acetate	Banana
	Octyl acetate	Orange
	Methyl butanoate	Apple
	Ethyl butanoate	Pineapple
	Methyl 2-aminobenzoate (Methyl anthranilate)	Grape

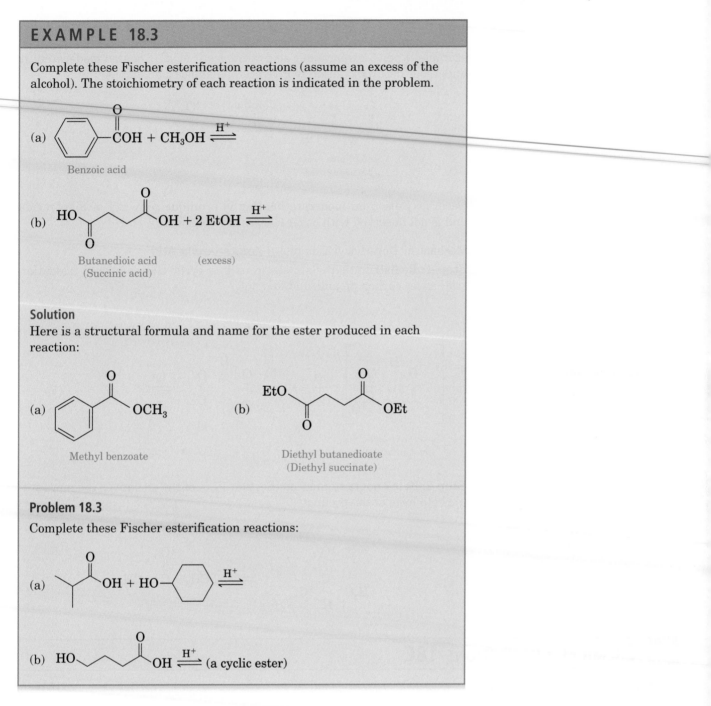

E. Decarboxylation

Decarboxylation is the loss of CO_2 from a carboxyl group. Almost any carboxylic acid, when heated to a very high temperature, undergoes thermal decarboxylation:

$$\text{RCOH} \xrightarrow{\text{decarboxylation}} \text{RH} + CO_2$$

Most carboxylic acids, however, are quite resistant to moderate heat and melt or even boil without decarboxylation. Exceptions are carboxylic acids that have a carbonyl group β to the carboxyl group. This type of carboxylic acid undergoes decarboxylation quite readily on mild heating. For example, when 3-oxobutanoic acid (acetoacetic acid) is heated moderately, it undergoes decarboxylation to give acetone and carbon dioxide:

3-Oxobutanoic acid
(Acetoacetic acid)   Acetone

Decarboxylation on moderate heating is a unique property of β-ketoacids and is not observed with other classes of ketoacids.

Mechanism: Decarboxylation of a β-Ketocarboxylic Acid

Step 1: Redistribution of six electrons in a cyclic six-membered transition state gives carbon dioxide and an enol:

Cyclic six-membered
transition state

Step 2: Keto-enol tautomerism (Section 17.5) of the enol gives the more stable keto form of the product:

CHEMICAL CONNECTIONS 18C

Ketone Bodies and Diabetes

3-Oxobutanoic acid (acetoacetic acid) and its reduction product, 3-hydroxybutanoic acid, are synthesized in the liver from acetyl-CoA (Section 28.5), a product of the metabolism of fatty acids and certain amino acids.

3-Oxobutanoic acid   3-Hydroxybutanoic acid
(Acetoacetic acid)   (β-Hydroxybutyric acid)

3-Hydroxybutanoic acid and 3-oxobutanoic acid are known collectively as ketone bodies.

The concentration of ketone bodies in the blood of healthy, well-fed humans is approximately 0.01 mM/L. However, in persons suffering from starvation or diabetes mellitus, the concentration of ketone bodies may increase to as much as 500 times the normal level. Under these conditions, the concentration of acetoacetic acid increases to the point where it undergoes spontaneous decarboxylation to form acetone and carbon dioxide. Acetone is not metabolized by humans and is excreted through the kidneys and the lungs. The odor of acetone is responsible for the characteristic "sweet smell" on the breath of severely diabetic patients.

An important example of decarboxylation of a β-ketoacid in the biological world occurs during the oxidation of foodstuffs in the tricarboxylic acid (TCA) cycle. Oxalosuccinic acid, one of the intermediates in this cycle, undergoes spontaneous decarboxylation to produce α-ketoglutaric acid. Only one of the three carboxyl groups of oxalosuccinic acid has a carbonyl group in the position β to it, and it is this carboxyl group that is lost as CO₂:

Oxalosuccinic acid   α-Ketoglutaric acid

Note that thermal decarboxylation is a reaction unique to β-keto acids—it does not occur with α-ketoacids. In the biochemistry chapters that follow, however, we will see examples of decarboxylation of α-ketoacids—for example, the decarboxylation of α-ketoglutarate. Because the decarboxylation of α-ketoacids requires an oxidizing agent (NAD⁺), this reaction is called oxidative decarboxylation.

α-Ketoglutarate   Nicotinamide   Succinate
                      adenine dinucleotide
                      (an oxidizing agent)

SUMMARY OF KEY QUESTIONS

SECTION 18.1 What Are Carboxylic Acids?
- The functional group of a **carboxylic acid** is the **carboxyl group, —COOH.**

SECTION 18.2 How Do We Name Carboxylic Acids?
- IUPAC names of carboxylic acids are derived from the name of the parent alkane by dropping the suffix -e and adding -oic acid.

- Dicarboxylic acids are named as -dioic acids. Common names for many carboxylic and dicarboxylic acids are still widely used.

SECTION 18.3 What Are the Physical Properties of Carboxylic Acids?
- Carboxylic acids are polar compounds. Consequently, they have higher boiling points and are more soluble in water than alcohols, aldehydes, ketones, and ethers of comparable molecular weight.

SUMMARY OF KEY REACTIONS

1. **Acidity of Carboxylic Acids (Section 18.5A)** Values of pK_a for most unsubstituted carboxylic acids are within the range of 4 to 5.

$$K_a = \frac{[CH_3COO^-][H_3O^+]}{[CH_3COOH]} = 1.74 \times 10^{-5}$$

$$pK_a = 4.76$$

$$CH_3COOH + H_2O \rightleftharpoons CH_3CO^- + H_3O^+$$

2. **Reaction of Carboxylic Acids with Bases (Section 18.5B)**
Carboxylic acids, whether water soluble or insoluble, react with alkali metal hydroxides, carbonates, and bicarbonates to form water-soluble salts.

Benzoic acid
(slightly soluble in water)

Sodium benzoate
(60 g/100 mL water)

$$CH_3COOH + NaHCO_3 \longrightarrow$$

$$CH_3COO^- Na^+ + CO_2 + H_2O$$

3. **Reduction by Lithium Aluminum Hydride (Section 18.5C)**
Lithium aluminum hydride reduces a carboxyl group to a primary alcohol. This reagent does not normally reduce carbon–carbon double bonds, but it does reduce aldehydes to 1° alcohols and ketones to 2° alcohols.

3-Cyclopentene-
carboxylic acid

4-Hydroxymethyl-
cyclopentene

4. **Fischer Esterification (Section 18.5D)** Fischer esterification is reversible.

$$CH_3COH + CH_3CH_2OH \xrightarrow{H_2SO_4}$$

Ethanoic acid   Ethanol
(Acetic acid)   (Ethyl alcohol)

$$CH_3COCH_2CH_3 + H_2O$$

Ethyl ethanoate
(Ethyl acetate)

One way to force the equilibrium to the right is to use an excess of the alcohol. Alternatively, water can be removed from the reaction mixture as it is formed.

5. **Decarboxylation (Section 18.5E)** Thermal decarboxylation is a unique property of β-ketoacids. The immediate products of thermal decarboxylation of β-ketoacids are carbon dioxide and an enol. Loss of CO_2 is followed immediately by keto-enol tautomerism.

3-Oxobutanoic acid   Acetone
(Acetoacetic acid)

PROBLEMS

GOB
Chemistry⊹Now™

Assess your understanding of this chapter's topics with additional quizzing and conceptual-based problems at **http://now.brookscole.com/gob8** or on the CD.

A blue problem number indicates an applied problem.

■ denotes problems that are available on the GOB ChemistryNow website or CD and are assignable in OWL.

SECTION 18.1 What Are Carboxylic Acids?

18.4 Name and draw structural formulas for the four carboxylic acids with molecular formula $C_5H_{10}O_2$. Which of these carboxylic acids are chiral?

18.5 ■ Write the IUPAC name for each carboxylic acid.

18.6 Write the IUPAC name for each carboxylic acid.

(c) CCl_3COOH

18.7 Draw a structural formula for each carboxylic acid.
(a) 4-Nitrophenylacetic acid
(b) 4-Aminobutanoic acid
(c) 4-Phenylbutanoic acid
(d) cis-3-Hexenedioic acid

18.8 Draw a structural formula for each carboxylic acid.
(a) 2-Aminopropanoic acid
(b) 3,5-Dinitrobenzoic acid
(c) Dichloroacetic acid
(d) o-Aminobenzoic acid

18.9 Draw a structural formula for each salt.
(a) Sodium benzoate   (b) Lithium acetate
(c) Ammonium acetate   (d) Disodium adipate
(e) Sodium salicylate   (f) Calcium butanoate

18.10 Calcium oxalate is a major component of kidney stones. Draw a structural formula for this compound.

18.11 The monopotassium salt of oxalic acid is present in certain leafy vegetables, including rhubarb. Both oxalic acid and its salts are poisonous in high concentrations. Draw a structural formula for monopotassium oxalate.

SECTION 18.3 What Are the Physical Properties of Carboxylic Acids?

18.12 Draw a structural formula for the dimer formed when two molecules of formic acid interact by hydrogen bonding.

18.13 Propanedioic (malonic) acid forms an internal hydrogen bond in which the H of one COOH group forms a hydrogen bond with an O of the other COOH group. Draw a structural formula to show this internal hydrogen bonding. (There are two possible answers.)

18.14 Hexanoic (caproic) acid has a solubility in water of about 1 g/100 mL water. Which part of the molecule contributes to water solubility, and which part prevents solubility?

18.15 Propanoic acid and methyl acetate are constitutional isomers, and both are liquids at room temperature. One of these compounds has a boiling point of 141°C; the other has a boiling point of 57°C. Which compound has which boiling point? Explain.

18.16 ▪ The following compounds have approximately the same molecular weight: hexanoic acid, heptanal, and 1-heptanol. Arrange them in order of increasing boiling point.

18.17 The following compounds have approximately the same molecular weight: propanoic acid, 1-butanol, and diethyl ether. Arrange them in order of increasing boiling point.

18.18 Arrange these compounds in order of increasing solubility in water: acetic acid, pentanoic acid, decanoic acid.

SECTION 18.5A Acidity of Carboxylic Acids

18.19 Alcohols, phenols, and carboxylic acids all contain an —OH group. Which are the strongest acids? Which are the weakest acids?

18.20 ▪ Arrange these compounds in order of increasing acidity: benzoic acid, benzyl alcohol, phenol.

18.21 ▪ Complete the equations for these acid–base reactions.

18.22 Complete the equations for these acid–base reactions.

(a) [structure: 2-methylphenol] + NaOH ⟶

(b) [structure: sodium 2-hydroxybenzoate] + HCl ⟶

(c) [structure: 2-methoxybenzoic acid] + $H_2NCH_2CH_2OH$ ⟶

(d) [cyclohexane]—COOH + $NaHCO_3$ ⟶

18.23 Formic acid is one of the components responsible for the sting of biting ants and is injected under the skin by bee and wasp stings. The pain can be relieved by rubbing the area of the sting with a paste of baking soda ($NaHCO_3$) and water that neutralizes the acid. Write an equation for this reaction.

18.24 Starting with the definition of K_a of a weak acid, HA, as

$$HA + H_2O \rightleftharpoons A^- + H_3O^+ \quad K_a = \frac{[A^-][H_3O^+]}{[HA]}$$

show that

$$\frac{[A^-]}{[HA]} = \frac{K_a}{[H_3O^+]}$$

18.25 ■ Using the equation from Problem 18.24 that shows the relationship between K_a, $[H^+]$, $[A^-]$, and $[HA]$, calculate the ratio of $[A^-]$ to $[HA]$ in a solution whose pH is

(a) 2.0   (b) 5.0   (c) 7.0

(d) 9.0   (e) 11.0

Assume that the pK_a of the weak acid is 5.0.

18.26 The pK_a of acetic acid is 4.76. What form(s) of acetic acid are present at pH 2.0 or lower? At pH 4.76? At pH 8.0 or higher?

18.27 ■ The normal pH range for blood plasma is 7.35 to 7.45. Under these conditions, would you expect the carboxyl group of lactic acid (pK_a 4.07) to exist primarily as a carboxyl group or as a carboxylic anion? Explain.

18.28 The pK_a of ascorbic acid (Chemical Connections 20B) is 4.10. Would you expect ascorbic acid dissolved in blood plasma, pH 7.35–7.45, to exist primarily as ascorbic acid or as ascorbate anion? Explain.

18.29 Complete the equations for the following acid–base reactions. Assume one mole of NaOH per mole of amino acid.

(a) CH₃CHCOOH + NaOH $\xrightarrow{H_2O}$
      |
      NH₃⁺

(b) CH₃CHCOO⁻Na⁺ + NaOH $\xrightarrow{H_2O}$
      |
      NH₃⁺

18.30 Which is the stronger base: $CH_3CH_2NH_2$ or $CH_3CH_2COO^-$? Explain.

18.31 Complete the equations for the following acid–base reactions. Assume one mole of HCl per mole of amino acid.

(a) CH₃CHCOO⁻Na⁺ + HCl $\xrightarrow{H_2O}$
      |
      NH₂

(b) CH₃CHCOO⁻Na⁺ + HCl $\xrightarrow{H_2O}$
      |
      NH₃⁺

SECTION 18.5D Fischer Esterification

18.32 Define and give an example of Fischer esterification.

18.33 Complete these examples of Fischer esterification. In each case, assume an excess of the alcohol.

(a) CH_3COOH + HO⟨...⟩ $\overset{H^+}{\rightleftharpoons}$

(b) CH_3COOH + HO—⟨...⟩ $\overset{H^+}{\rightleftharpoons}$

(c) ⟨...⟩ + CH_3CH_2OH $\overset{H^+}{\rightleftharpoons}$

18.34 From what carboxylic acid and alcohol is each ester derived?

(a) $CH_3\overset{O}{\overset{\|}{C}}O$—⟨...⟩—$O\overset{O}{\overset{\|}{C}}CH_3$

(b) ⟨cyclohexyl⟩—$\overset{O}{\overset{\|}{C}}OCH_3$

(c) $CH_3O\overset{O}{\overset{\|}{C}}CH_2CH_2\overset{O}{\overset{\|}{C}}OCH_3$

(d) ⟨...⟩

18.35 Methyl 2-hydroxybenzoate (methyl salicylate) has the odor of oil of wintergreen. This compound is prepared by Fischer esterification of 2-hydroxybenzoic acid (salicylic acid) with methanol. Draw a structural formula for methyl 2-hydroxybenzoate.

18.36 ■ Show how you could convert cinnamic acid to each compound.

trans-3-Phenyl-2-propenoic acid
(Cinnamic acid)

Additional Problems

18.37 Give the expected organic product formed when phenylacetic acid, $C_6H_5CH_2COOH$, is treated with each of the following reagents.

(a) $NaHCO_3$, H_2O
(b) $NaOH$, H_2O
(c) NH_3, H_2O
(d) $LiAlH_4$, H_2O
(e) $NaBH_4$, H_2O
(f) CH_3OH + H_2SO_4 (catalyst)
(g) H_2/Ni

18.38 ■ Methylparaben and propylparaben are used as preservatives in foods, beverages, and cosmetics.

4-Aminobenzoic acid
(p-Aminobenzoic acid)

Methyl 4-aminobenzoate
(Methylparaben)

Propyl 4-aminobenzoate
(Propylparaben)

Show how each of these preservatives can be prepared from 4-aminobenzoic acid.

18.39 4-Aminobenzoic acid is prepared from benzoic acid by the following two steps:

Benzoic acid

4-Nitrobenzoic acid

4-Aminobenzoic acid

Show reagents and experimental conditions to bring about each step.

Looking Ahead

18.40 When 5-hydroxypentanoic acid is treated with an acid catalyst, it forms a lactone (a cyclic ester). Draw a structural formula for this lactone.

18.41 We have seen that esters can be prepared by treatment of a carboxylic acid and an alcohol in the presence of an acid catalyst. Suppose you start instead with a dicarboxylic acid, a diol such as 1,6-hexanedioic acid (adipic acid), and a diol such as 1,2-ethanediol (ethylene glycol). Show how Fischer esterification in this case can produce a polymer (a macromolecule with molecular weight several thousand times that of the starting materials).

1,6-Hexanedioic acid
(Adipic acid)

$HOCH_2$—CH_2—OH ⟶ A polyester

1,2-Ethanediol
(Ethylene glycol)

Tying It Together

18.42 Draw the structural formula of a compound of the given molecular formula that, on oxidation by potassium dichromate in aqueous sulfuric acid, gives the carboxylic acid or dicarboxylic acid shown.

(a) $C_6H_{14}O$ $\xrightarrow{\text{oxidation}}$

(b) $C_6H_{12}O$ $\xrightarrow{\text{oxidation}}$

(c) $C_6H_{14}O_2$ $\xrightarrow{\text{oxidation}}$

18.43 Complete the equations for these oxidations:

(a) $CH_3(CH_2)_4CH_2OH$ $\xrightarrow[H_2SO_4]{K_2Cr_2O_7}$

(b) + $Ag(NH_3)_2^+$ ⟶

(c) HO—⬡—CH_2OH $\xrightarrow[H_2SO_4]{K_2Cr_2O_7}$

Carboxylic Anhydrides, Esters, and Amides

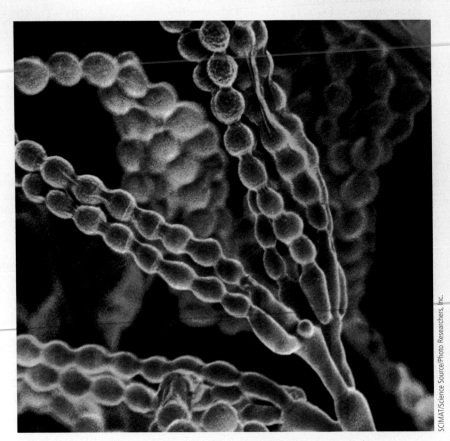

Colored scanning electron micrograph of *Penicillium* fungus. The stalk-like objects are condiophores to which are attached numerous round condia. The condia are the fruiting bodies of the fungus. See Chemical Connections 19B, "The Penicillins and Cephalosporins: β-Lactam Antibiotics."

SCIMAT/Science Source/Photo Researchers, Inc.

Chemistry⋅☆⋅Now™

Look for this logo in the chapter and go to GOB ChemistryNow at **http://now.brookscole.com/gob8** or on the CD for tutorials, simulations, and problems.

19.1 | What Are Carboxylic Anhydrides, Esters, and Amides?

In Chapter 18, we studied the structure and preparation of esters, a class of organic compounds derived from carboxylic acids. In this chapter, we study anhydrides and amides, two more classes of carboxylic derivatives. Below, under the general formula of each carboxylic acid derivative, is a drawing to help you see how the functional group of the derivative is formally related to a carboxyl group. Loss of —OH from a carboxyl group and H— from an alcohol, for example, gives an ester. Loss of —OH from a carboxyl group and —H from ammonia or an amine gives an amide.

Of these three carboxylic derivatives, anhydrides are so reactive that they are rarely found in nature. Esters and amides, however, are widespread in the biological world.

A. Anhydrides

The functional group of an **anhydride** consists of two carbonyl groups bonded to the same oxygen atom. The anhydride may be symmetrical (from two identical **acyl groups**) or mixed (from two different acyl groups). To name anhydrides, we drop the word "acid" from the name of the carboxylic acid from which the anhydride is derived and add the word "anhydride."

Acetic anhydride

B. Esters

The functional group of an **ester** is a carbonyl group bonded to an —OR group. Both IUPAC and common names of esters are derived from the names of the parent carboxylic acids (Chapter 18). The alkyl group bonded to oxygen is named first, followed by the name of the acid in which the suffix **-ic acid** is replaced by the suffix **-ate.**

CH₃COCH₂CH₃
Ethyl ethanoate
(Ethyl acetate)

Diethyl pentanedioate
(Diethyl glutarate)

Recall that cyclic esters are called **lactones.**

C. Amides

The functional group of an **amide** is a carbonyl group bonded to a nitrogen atom. We name amides by dropping the suffix **-oic acid** from the IUPAC name of the parent acid, or **-ic acid** from its common name, and adding **-amide.** If the nitrogen atom of the amide is bonded to an alkyl or aryl group, the group is named and its location on nitrogen is indicated by *N*-. Two alkyl groups are indicated by *N,N*-di-.

CH₃CNH₂
Acetamide
(a 1° amide)

CH₃CNHCH₃
N-Methylacetamide
(a 2° amide)

HCN(CH₃)₂
N,N-Dimethylformamide
(a 3° amide)

Chemistry Now™
Click *Coached Problems* to examine the **Structural Aspects of Carboxylic Acid Derivatives**

Chemistry Now™
Click *Coached Problems* to try a set of problems in **Naming Carboxylic Acid Derivatives**

CHEMICAL CONNECTIONS 19A

The Pyrethrins: Natural Insecticides of Plant Origin

Pyrethrym is a natural insecticide obtained from the powdered flower heads of several species of *Chrysanthemum,* particularly *C. cinerariaefolium.* The active substances in pyrethrum, principally pyrethrins I and II, are contact poisons for insects and cold-blooded vertebrates. Because their concentrations in the pyrethrum powder used in *Chrysanthemum*-based insecticides are nontoxic to plants and higher animals, pyrethrum powder has found wide application in household and livestock sprays as well as in dusts for edible plants. Natural pyrethrins are esters of chrysanthemic acid.

While pyrethrum powders are effective insecticides, the active substances in them are destroyed rapidly in the environment. In an effort to develop synthetic compounds that are as effective as these natural insecticides but that offer greater biostability, chemists have prepared a series of esters related in structure to chrysanthemic acid. Permethrin is one of the most commonly used synthetic pyrethrin-like compounds in household and agricultural products.

Pyrethrin I   Permethrin

Cyclic amides are called **lactams.** Following are structural formulas for a four-membered and a seven-membered lactam. A four-membered lactam is essential to the function of the penicillin and cephalosporin antibiotics (Chemical Connections 19B).

A four-membered   A seven-membered
lactam   lactam

EXAMPLE 19.1

Write the IUPAC name for each amide.

(a) $CH_3CH_2CH_2\overset{\displaystyle O}{\overset{\displaystyle \|}{C}}NH_2$

(b) H_2N ... NH_2

Solution

Given is the IUPAC name and then, in parentheses, the common name.
(a) Butanamide (butyramide, from butyric acid)
(b) Hexanediamide (adipamide, from adipic acid)

Problem 19.1

Draw a structural formula for each amide.
(a) *N*-Cyclohexylacetamide   (b) Benzamide

CHEMICAL CONNECTIONS 19B

The Penicillins and Cephalosporins: β-Lactam Antibiotics

The **penicillins** were discovered in 1928 by the Scottish bacteriologist Sir Alexander Fleming. Thanks to the brilliant experimental work of Sir Howard Florey, an Australian pathologist, and Ernst Chain, a German chemist who fled Nazi Germany, penicillin G was introduced into the practice of medicine in 1943. For their pioneering work in developing one of the most effective antibiotics of all time, Fleming, Florey, and Chain were awarded the Nobel Prize for medicine or physiology in 1945.

The mold from which Fleming discovered penicillin was *Penicillium notatum,* a strain that gives a relatively low yield of penicillin. It was replaced in commercial production of the antibiotic by *P. chrysogenum,* a strain cultured from a mold found growing on a grapefruit in a market in Peoria, Illinois. The structural feature common to all penicillins is the four-membered lactam ring, called a **β-lactam,** bonded to a five-membered, sulfur-containing ring. The penicillins owe their antibacterial activity to a common mechanism that inhibits the biosynthesis of a vital part of bacterial cell walls.

Penicillin G

Amoxicillin

Soon after the penicillins were introduced into medical practice, however, penicillin-resistant strains of bacteria began to appear. They have since proliferated dramatically. One approach to combating resistant strains is to synthesize newer, more effective penicillins, such as ampicillin, methicillin, and amoxicillin.

Another approach is to search for newer, more effective β-lactam antibiotics. The most effective of these agents discovered so far are the **cephalosporins,** the first of which was isolated from the fungus *Cephalosporium acremonium.*

Cephalexin
(a β-lactam antibiotic)

This class of β-lactam antibiotics has an even broader spectrum of antibacterial activity than the penicillins and is effective against many penicillin-resistant bacterial strains. Cephalexin (Keflex) is presently one of the most widely prescribed of the cephalosporin antibiotics.

The commonly prescribed formulation Augmentin is a combination of amoxicillin trihydrate, a penicillin, and clavulanic acid, a β-lactamase inhibitor that is isolated from *Streptomyces clavuligerus.*

Clavulanic acid

Clavulanic acid, which also contains a β-lactam ring, reacts with and inhibits the β-lactamase enzyme before the enzyme can catalyze the inactivation of the penicillin. Augmentin is used as a second line of defense against childhood ear infections when penicillin resistance is suspected. Most children know it as a white liquid with a banana taste.

19.2 | How Do We Prepare Esters?

The most common method for the preparation of esters is Fischer esterification (Section 18.5D). As an example of Fischer esterification, treating acetic acid with ethanol in the presence of concentrated sulfuric acid gives ethyl acetate and water:

$$CH_3\overset{O}{\overset{\|}{C}}OH + CH_3CH_2OH \underset{}{\overset{H_2SO_4}{\rightleftharpoons}} CH_3\overset{O}{\overset{\|}{C}}OCH_2CH_3 + H_2O$$

Ethanoic acid   Ethanol   Ethyl ethanoate
(Acetic acid)   (Ethyl alcohol)   (Ethyl acetate)

CHEMICAL CONNECTIONS 19C

From Willow Bark to Aspirin and Beyond

The story of the development of this modern pain reliever goes back more than 2000 years. In 400 BCE, the Greek physician Hippocrates recommended chewing the bark of the willow tree to alleviate the pain of childbirth and to treat eye infections.

The active component of willow bark was found to be salicin, a compound composed of salicyl alcohol bonded to a unit of β-D-glucose (Section 20.4A). Hydrolysis of salicin in aqueous acid followed by oxidation gave salicylic acid. Salicylic acid proved to be an even more effective reliever of pain, fever, and inflammation than salicin, and without the latter's extremely bitter taste. Unfortunately, patients quickly recognized salicylic acid's major side effect: It causes severe irritation of the mucous membrane lining of the stomach.

Salicin

Salicylic acid

In the search for less irritating but still effective derivatives of salicylic acid, chemists at the Bayer division of I. G. Farben in Germany in 1897 treated salicylic acid with acetic anhydride and prepared acetylsalicylic acid. They gave this new compound the name aspirin.

Acetylsalicylic acid (Aspirin)

Aspirin proved to be less irritating to the stomach than salicylic acid as well as more effective in relieving the pain and inflamma-

tion of rheumatoid arthritis. Aspirin, however, remains irritating to the stomach and frequent use of it can cause duodenal ulcers in susceptible persons.

In the 1960s, in a search for even more effective and less irritating analgesics and nonsteroidal anti-inflammatory drugs (NSAIDs), chemists at the Boots Pure Drug Company in England, who were studying compounds structurally related to salicylic acid, discovered an even more potent compound, which they named ibuprofen. Soon thereafter, Syntex Corporation in the United States developed naproxen, the active ingredient in Aleve. Both ibuprofen and naproxen have one stereocenter and can exist as a pair of enantiomers. For each drug, the active form is the *S* enantiomer. Naproxen is administered as its water-soluble sodium salt.

(*S*)-Ibuprofen

(*S*)-Naproxen

In the 1960s, researchers discovered that aspirin acts by inhibiting cyclooxygenase (COX), a key enzyme in the conversion of arachidonic acid to prostaglandins (Chemical Connections 21H). With this discovery, it became clear why only one enantiomer of ibuprofen and naproxen is active: Only the *S* enantiomer of each has the correct handedness to bind to COX and inhibit its activity.

GOB
Chemistry ·❄·Now™

Click *Coached Problems* to explore the
Common Reactions of Carboxylic Acid Derivatives

19.3 | How Do We Prepare Amides?

In principle, we can form an amide by treating a carboxylic acid with an amine and removing —OH from the acid and an —H from the amine. In practice, mixing these two leads to an acid–base reaction that forms an ammonium salt. If this salt is heated to a high enough temperature, water splits out and an amide forms.

It is much more common, however, to prepare amides by treating an anhydride with an amine (Section 19.4C).

$$\underset{\substack{\text{Acetic} \\ \text{anhydride}}}{CH_3\overset{\displaystyle O}{\overset{\|}{C}}-O-\overset{\displaystyle O}{\overset{\|}{C}}CH_3} + H_2NCH_2CH_3 \longrightarrow \underset{\text{An amide}}{CH_3\overset{\displaystyle O}{\overset{\|}{C}}-NHCH_2CH_3} + CH_3\overset{\displaystyle O}{\overset{\|}{C}}OH$$

19.4 | What Are the Characteristic Reactions of Anhydrides, Esters, and Amides?

The most common reaction of each of these three functional groups is with compounds that contain either an —OH group, as in water (H—OH) or an alcohol (H—OR), or an H—N group, as in ammonia (H—NH$_2$) or in a 1° or 2° amine (H—NR$_2$ or H—NHR). These reactions have in common the addition of the oxygen or nitrogen atom to the carboxyl carbon and the hydrogen atom to the carbonyl oxygen to give a tetrahedral carbonyl addition intermediate. This intermediate then collapses to regenerate the carbonyl group and give either a new carboxyl derivative or a carboxylic acid itself. We illustrate here with the reaction of an ester with water:

$$\underset{OCH_3}{R-\overset{\displaystyle O}{\overset{\|}{C}}} + \overset{\displaystyle H}{\overset{|}{OH}} \rightleftharpoons \left[\underset{OCH_3}{R-\overset{\displaystyle O-H}{\overset{|}{\underset{|}{C}}}-OH}\right] \rightleftharpoons R-\overset{\displaystyle O}{\overset{\|}{C}}-OH + H-OCH_3$$

Compare the formation of this tetrahedral carbonyl addition intermediate with that formed by the addition of an alcohol to the carbonyl group of an aldehyde or ketone and formation of a hemiacetal (Section 17.4C), and that formed by the addition of an alcohol to the carbonyl group of a carboxylic acid during Fischer esterification (Section 18.5D).

A. Reaction with Water: Hydrolysis

Hydrolysis is a chemical decomposition involving breaking a bond and the addition of the elements of water.

Anhydrides

Carboxylic anhydrides, particularly the low-molecular-weight ones, react readily with water to give two carboxylic acids. In the hydrolysis of an anhydride, one of the C—O bonds breaks, and OH is added to carbon and H is added to oxygen of what was the C—O bond. Hydrolysis of acetic anhydride gives two molecules of acetic acid.

$$\underset{\text{Acetic anhydride}}{CH_3\overset{\displaystyle O}{\overset{\|}{C}}O\overset{\displaystyle O}{\overset{\|}{C}}CH_3} + H_2O \longrightarrow \underset{\text{Acetic acid}}{CH_3\overset{\displaystyle O}{\overset{\|}{C}}OH} + \underset{\text{Acetic acid}}{HO\overset{\displaystyle O}{\overset{\|}{C}}CH_3}$$

Esters

Esters are hydrolyzed only very slowly, even in boiling water. Hydrolysis becomes considerably more rapid, however, when the ester is heated in aqueous acid or base. When we discussed acid-catalyzed Fischer esterification in

GOB
Chemistry ⋅ Now™
Click *Coached Problems* to try a set of problems in **Predicting the Products of Reaction of Carboxylic Acid Derivatives**

Section 18.5D, we pointed out that it is an equilibrium reaction. Hydrolysis of esters in aqueous acid, also an equilibrium reaction, is the reverse of Fischer esterification. A large excess of water drives the equilibrium to the right to form the carboxylic acid and alcohol (Le Chatelier's principle; Section 8.8).

$$\underset{\text{Ethyl acetate}}{CH_3\overset{O}{\overset{\|}{C}}OCH_2CH_3} + H_2O \xrightleftharpoons{H^+} \underset{\text{Acetic acid}}{CH_3\overset{O}{\overset{\|}{C}}OH} + \underset{\text{Ethanol}}{CH_3CH_2OH}$$

Hydrolysis of an ester can also be carried out using a hot aqueous base, such as aqueous NaOH. This reaction is often called **saponification,** a reference to its use in the manufacture of soaps (Section 18.4B). The carboxylic acid formed in the hydrolysis reacts with hydroxide ion to form a carboxylic acid anion. Thus each mole of ester hydrolyzed requires one mole of base, as shown in the following balanced equation:

$$\underset{\text{Ethyl acetate}}{CH_3\overset{O}{\overset{\|}{C}}OCH_2CH_3} + \underset{\substack{\text{Sodium}\\\text{hydroxide}}}{NaOH} \xrightarrow{H_2O} \underset{\substack{\text{Sodium}\\\text{acetate}}}{CH_3\overset{O}{\overset{\|}{C}}O^-Na^+} + \underset{\text{Ethanol}}{CH_3CH_2OH}$$

There are two major differences between hydrolysis of esters in aqueous acid and aqueous base.

1. For hydrolysis of an ester in aqueous acid, acid is required in only catalytic amounts. For hydrolysis in aqueous base, base is required in stoichiometric amounts (one mole of base per mole of ester), because base is a reactant, not merely a catalyst.

2. Hydrolysis of an ester in aqueous acid is reversible. Hydrolysis in aqueous base is irreversible because the carboxylate anion does not react with water or hydroxide ion.

EXAMPLE 19.2

Complete the equation for each hydrolysis reaction. Show the products as they are ionized under these experimental conditions.

(a)

$$\text{(b) } CH_3\overset{O}{\overset{\|}{C}}OCH_2CH_2O\overset{O}{\overset{\|}{C}}CH_3 + 2NaOH \xrightarrow{H_2O}$$

Solution

The products of hydrolysis of compound (a) are benzoic acid and 2-propanol. In aqueous NaOH, benzoic acid is converted to its sodium salt. In this reaction, one mole of NaOH is required for hydrolysis of each mole of this ester. Compound (b) is a diester of ethylene glycol and requires two moles of NaOH for its complete hydrolysis.

(a)

Sodium
benzoate

2-Propanol
(Isopropyl alcohol)

(b) $2CH_3CO^-Na^+ + HOCH_2CH_2OH$

Sodium acetate 1,2-Ethanediol
(Ethylene glycol)

Problem 19.2

Complete the equation for each hydrolysis reaction. Show all products as they are ionized under these experimental conditions.

(a) ... $+ 2NaOH \xrightarrow{H_2O}$

(b) ... $+ H_2O \xrightarrow{HCl}$

Amides

Amides require more vigorous conditions for hydrolysis in both acid and base than do esters. Hydrolysis in hot aqueous acid gives a carboxylic acid and an ammonium ion. This reaction is driven to completion by the acid–base reaction between ammonia or the amine and the acid to form an ammonium ion. Complete hydrolysis requires one mole of acid per mole of amide.

$$CH_3CH_2CH_2CNH_2 + H_2O + HCl \xrightarrow[\text{heat}]{H_2O} CH_3CH_2CH_2COH + NH_4^+ Cl^-$$

Butanamide Butanoic acid

The products of amide hydrolysis in aqueous base are a carboxylic acid salt and ammonia or an amine. The acid–base reaction between the carboxylic acid and base to form a carboxylic salt drives this hydrolysis to completion. Thus complete hydrolysis of an amide requires one mole of base per mole of amide.

Acetanilide Sodium acetate Aniline

CHEMICAL CONNECTIONS 19D

Ultraviolet Sunscreens and Sunblocks

Ultraviolet (UV) radiation (Section 3.2) penetrating the Earth's ozone layer is arbitrarily divided by wavelength into two regions: UVB (290–320 nm) and UVA (320–400 nm). UVB, a more energetic form of radiation than UVA, creates more radicals and hence does more oxidative damage to tissue (Section 25.7). UVB radiation interacts directly with biomolecules of the skin and eyes, causing skin cancer, skin aging, eye damage leading to cataracts, and delayed sunburn that appears 12–24 hours after exposure. UVA radiation, by contrast, causes tanning. It also damages skin, albeit much less efficiently than UVB. Its role in promoting skin cancer is less well understood.

Commercial sunscreen products are rated according to their sun protection factor (SPF), which is defined as the minimum effective dose of UV radiation that produces a delayed sunburn on protected skin compared to unprotected skin. Two types of active ingredients are found in commercial sunblocks and sunscreens. The most common sunblock agent is zinc oxide, ZnO, which reflects and scatters UV radiation. Sunscreens, the second type of active ingredient, absorb UV radiation and then reradiate it as heat. These compounds are most effective in screening UVB radiation, but they do not screen UVA radiation. Thus they allow tanning but prevent the UVB-associated damage. Given here are structural formulas for three common esters used as UVB-screening agents, along with the name by which each is most commonly listed in the Active Ingredients labels on commercial products:

Octyl *p*-methoxycinnamate   Homosalate   Padimate A

EXAMPLE 19.3

Write a balanced equation for the hydrolysis of each amide in concentrated aqueous HCl. Show all products as they exist in aqueous HCl.

$$
\text{(a)} \quad CH_3\overset{\displaystyle O}{\overset{\displaystyle \|}{C}}N(CH_3)_2
$$

(b)

Solution

(a) Hydrolysis of *N,N*-dimethylacetamide gives acetic acid and dimethylamine. Dimethylamine, a base, reacts with HCl to form dimethylammonium ion, shown here as dimethylammonium chloride.

$$
CH_3\overset{\displaystyle O}{\overset{\displaystyle \|}{C}}N(CH_3)_2 + H_2O + HCl \xrightarrow{\text{Heat}} CH_3\overset{\displaystyle O}{\overset{\displaystyle \|}{C}}OH + (CH_3)_2NH_2{}^+Cl^-
$$

(b) Hydrolysis of this lactam gives the protonated form of 5-aminopentanoic acid.

Problem 19.3

Write a balanced equation for the hydrolysis of each amide in Example 19.3 in concentrated aqueous NaOH. Show all products as they exist in aqueous NaOH.

B. Reaction with Alcohols

Anhydrides

Anhydrides react with alcohols to give one mole of ester and one mole of a carboxylic acid.

CH₃COCCH₃   +   HOCH₂CH₃   ⟶   CH₃COCH₂CH₃   +   HOCCH₃

Acetic anhydride   Ethanol   Ethyl acetate   Acetic acid

Thus, the reaction of an alcohol with an anhydride is a useful method for the synthesis of esters. Aspirin (Chemical Connections 19C) is synthesized on an industrial scale by the reaction of acetic anhydride with salicylic acid.

Salicylic acid   Acetic anhydride   Acetylsalicylic acid   Acetic acid
                                                        (Aspirin)

C. Reaction with Ammonia and Amines

Anhydrides

Anhydrides react with ammonia and with 1° and 2° amines to form amides. Two moles of amine are required: one to form the amide and one to neutralize the carboxylic acid by-product. We show this reaction here in two steps: (1) formation of the amide and the carboxylic acid by-product, and (2) an acid–base reaction of the carboxylic acid by-product with the second mole of ammonia to give an ammonium salt.

Acetic anhydride   Acetamide   Ammonium acetate

CHEMICAL CONNECTIONS 19E

Barbiturates

In 1864, Adolph von Baeyer (1835–1917) discovered that heating the diethyl ester of malonic acid with urea in the presence of sodium ethoxide (like sodium hydroxide, a strong base) gives a cyclic compound that he named barbituric acid. Some say that Baeyer named it after a friend of his named Barbara. Others claim that he named it after St. Barbara, the patron saint of artillerymen.

Pentobarbital

Diethyl propanedioate   Urea
(Diethyl malonate)

Barbituric acid

Sodium pentobarbital
(Nembutal)

Phenobarbital

A number of derivatives of barbituric acid have powerful sedative and hypnotic effects. One such derivative is pentobarbital. Like other derivatives of barbituric acid, pentobarbital is quite insoluble in water and body fluids. To increase its solubility in these fluids, pentobarbital is converted to its sodium salt, which is given the name Nembutal. Phenobarbital, also administered as its sodium salt, is an anticonvulsant, sedative, and hypnotic.

Technically speaking, only the sodium salts of these compounds should be called barbiturates. In practice, however, all derivatives of barbituric acid are called barbiturates, whether they are the un-ionized form or the ionized, water-soluble salt form.

Barbiturates have two principal effects. In small doses, they are sedatives (tranquilizers); in larger doses, they induce sleep. Barbituric acid, in contrast, has neither of these effects. Barbiturates are dangerous because they are addictive, which means that a regular user will suffer withdrawal symptoms when their use is stopped. They are especially dangerous when taken with alcohol because the combined effect (called a synergistic effect) is usually greater than the sum of the effects of either drug taken separately.

Esters

Esters react with ammonia and with 1° and 2° amines to form amides.

Ethyl 2-phenyl acetate   2-Phenylacetamide

Thus, as seen in this example, amides can be prepared readily from esters. Because carboxylic acids are easily converted to esters by Fischer esterification, we have a good way to convert a carboxylic acid to an amide. This method of amide formation is, in fact, much more useful and applicable than converting a carboxylic acid to an ammonium salt and then heating the salt to form an amide.

Amides

Amides do not react with ammonia or primary or secondary amines.

19.5 | What Are Phosphoric Anhydrides and Phosphoric Esters?

GOB
Chemistry♣Now™
Click *Coached Problems* to examine the **Structural Aspects of Phosphoric Anhydrides and Esters**

A. Phosphoric Anhydrides

Because of the special importance of phosphoric anhydrides in biochemical systems, we discuss them here to show the similarity between them and the anhydrides of carboxylic acids. The functional group of a **phosphoric anhydride** consists of two phosphoryl (P=O) groups bonded to the same oxygen atom. Shown here are structural formulas for two anhydrides of phosphoric acid and the ions derived by ionization of the acidic hydrogens of each:

Diphosphoric acid   Diphosphate ion   Triphosphoric acid   Triphosphate ion
(Pyrophosphoric acid)   (Pyrophosphate ion)

B. Phosphoric Esters

Phosphoric acid has three —OH groups and forms mono-, di-, and triphosphoric esters, which we name by giving the name(s) of the alkyl group(s) bonded to oxygen followed by the word "phosphate" (for example, dimethyl phosphate). In more complex **phosphoric esters,** it is common practice to name the organic molecule and then indicate the presence of the phosphoric ester by including either the word "phosphate" or the prefix phospho-. Dihydroxyacetone phosphate, for example, is an intermediate in glycolysis (Section 28.2). Pyridoxal phosphate is one of the metabolically active forms of vitamin B_6. The last two phosphoric esters are shown here as they are ionized at pH 7.4, the pH of blood plasma.

Dimethyl phosphate   Dihydroxyacetone phosphate   Pyridoxal phosphate

19.6 | What Is Step-Growth Polymerization?

Step-growth polymers form from the reaction of molecules containing two functional groups, with each new bond being created in a separate step. In this section, we discuss three types of step-growth polymers: polyamides, polyesters, and polycarbonates.

A. Polyamides

In the early 1930s, chemists at E. I. DuPont de Nemours & Company began fundamental research into the reactions between dicarboxylic acids and diamines to form **polyamides.** In 1934, they synthesized nylon-66, the first

GOB
Chemistry♣Now™
Click *Coached Problems* to explore the process of **Polymerization**

CHEMICAL CONNECTIONS 19F

Stitches That Dissolve

As the technological capabilities of medicine have expanded, the demand for synthetic materials that can be used inside the body has increased as well. Polymers already have many of the characteristics of an ideal biomaterial: They are lightweight and strong, are inert or biodegradable depending on their chemical structure, and have physical properties (softness, rigidity, elasticity) that are easily tailored to match those of natural tissues.

Even though most medical uses of polymeric materials require biostability, some applications require them to be biodegradable. An example is the polyester of glycolic acid and lactic acid used in

absorbable sutures, which are marketed under the trade name of Lactomer.

A health care specialist must remove traditional suture materials such as catgut after they have served their purpose. Stitches of Lactomer, however, are hydrolyzed slowly over a period of approximately 2 weeks. By the time the torn tissues have healed, the stitches have hydrolyzed, and no suture removal is necessary. The body metabolizes and excretes the glycolic and lactic acids formed during this hydrolysis.

Glycolic acid   Lactic acid   A polymer of
                                         glycolic acid and lactic acid

purely synthetic fiber. Nylon-66 is so named because it is synthesized from two different monomers, each containing six carbon atoms.

In the synthesis of nylon-66, hexanedioic acid and 1,6-hexanediamine are dissolved in aqueous ethanol and then heated in an autoclave to 250°C and an internal pressure of 15 atm. Under these conditions, —COOH and —NH$_2$ groups react by loss of H$_2$O to form a polyamide, similar to the formation of amides we described in Section 19.3.

Hexanedioic acid   1,6-Hexanediamine
(Adipic acid)   (Hexamethylenediamine)

Nylon-66
(a polyamide)

Based on extensive research into the relationships between molecular structure and bulk physical properties, scientists at DuPont reasoned that a polyamide containing benzene rings would be even stronger than nylon-66. This line of reasoning eventually produced a polyamide that DuPont named Kevlar.

In forming a polycarbonate, each mole of phosgene reacts with two moles of the sodium salt of a phenol called bisphenol A.

Lexan is a tough, transparent polymer that has high impact and tensile strength and retains its properties over a wide temperature range. It is used in sporting equipment (helmets and face masks); to make light, impact-resistant housings for household appliances; and in the manufacture of safety glass and unbreakable windows.

SUMMARY OF KEY QUESTIONS

SECTION 19.1 What Are Carboxylic Anhydrides, Esters, and Amides?
- A **carboxylic anhydride** contains two carbonyl groups bonded to the same oxygen.
- A carboxylic **ester** contains a carbonyl group bonded to an —OR group derived from an alcohol or a phenol.
- A carboxylic **amide** contains a carbonyl group bonded to a nitrogen atom derived from an amine.

SECTION 19.2 How Do We Prepare Esters?
- The most common laboratory method for the preparation of esters is Fischer esterification (Section 18.5D).

SECTION 19.3 How Do We Prepare Amides?
- Amides can be prepared by the reaction of an amine with a carboxylic anhydride.

SECTION 19.4 What Are the Characteristic Reactions of Anhydrides, Esters, and Amides?
- Hydrolysis is a chemical process in which a bond is split and the elements of H_2O are added.

- Hydrolysis of a carboxylic anhydride gives two molecules of carboxylic acid.
- Hydrolysis of a carboxylic ester requires the presence of either aqueous acid or base. Acid is a catalyst and the reaction is the reverse of Fischer esterification. Base is a reactant and is required in stoichiometric amounts.
- Hydrolysis of a carboxylic amide requires the presence of either aqueous acid or base. Both acid and base are reactants and are required in stoichiometric amounts.

SECTION 19.6 What Is Step-Growth Polymerization?
- Step-growth polymerization involves the stepwise reaction of difunctional monomers. Important commercial polymers synthesized through step-growth processes include polyamides, polyesters, and polycarbonates.

SUMMARY OF KEY REACTIONS

1. **Fischer Esterification (Section 19.2)** Fischer esterification is reversible. To achieve high yields of ester, it is necessary to force the equilibrium to the right. One way to maximize the yield of ester is to use an excess of the alcohol. Another way is to remove the water as it is formed.

$$CH_3COOH + CH_3CH_2OH \underset{}{\overset{H_2SO_4}{\rightleftharpoons}} CH_3\overset{O}{\overset{\|}{C}}OCH_2CH_3 + H_2O$$

2. **Preparation of an Amide (Section 19.3)** Reaction of an anhydride with an amine gives an amide.

$$CH_3\overset{O}{\overset{\|}{C}}-O-\overset{O}{\overset{\|}{C}}CH_3 + H_2NCH_2CH_3 \longrightarrow CH_3\overset{O}{\overset{\|}{C}}-NHCH_2CH_3 + CH_3\overset{O}{\overset{\|}{C}}OH$$

One remarkable feature of Kevlar is that it weighs less than other materials of similar strength. For example, a cable woven of Kevlar has a strength equal to that of a similarly woven steel cable. Yet the Kevlar cable has only 20% of the weight of the steel cable! Kevlar now finds use in such articles as anchor cables for offshore drilling rigs and reinforcement fibers for automobile tires. It is also woven into a fabric that is so tough that it can be used for bulletproof vests, jackets, and raincoats.

1,4-Benzenedicarboxylic acid (Terephthalic acid) + 1,4-Benzenediamine (p-Phenylenediamine) → Kevlar (a polyaromatic amide)
Remove H_2O / Heat $-H_2O$

B. Polyesters

The first **polyester**, developed in the 1940s, involved polymerization of benzene 1,4-dicarboxylic acid with 1,2-ethanediol to give poly(ethylene terephthalate), abbreviated PET. Virtually all PET is now made from the dimethyl ester of terephthalic acid by the following reaction:

Dimethyl terephthalate + 1,2-Ethanediol (Ethylene glycol) → Poly(ethylene terephthalate) (Dacron, Mylar)
Remove CH_3OH / Heat $-CH_3OH$

The crude polyester can be melted, extruded, and then drawn to form the textile fiber Dacron polyester. Dacron's outstanding features include its stiffness (about four times that of nylon-66), very high tensile strength, and remarkable resistance to creasing and wrinkling. Because the early Dacron polyester fibers were harsh to the touch due to their stiffness, they were usually blended with cotton or wool to make acceptable textile fibers. Newly developed fabrication techniques now produce less harsh Dacron polyester textile fibers. PET is also fabricated into Mylar films and recyclable plastic beverage containers.

Mylar can be made into extremely strong films. Because the film has a very small pore size, it is used for balloons that can be inflated with helium; the helium atoms diffuse only slowly through the pores of the film.

Chemistry Now™ GOB

Click *Coached Problems* to try a set of problems in predicting the **Products of Step-Growth Polymerization**.

C. Polycarbonates

A **polycarbonate**, the most familiar of which is Lexan, forms from the reaction between the disodium salt of bisphenol A and phosgene. Phosgene is a derivative of carbonic acid, H_2CO_3, in which both —OH groups have been replaced by chlorine atoms. An ester of carbonic acid is called a carbonate.

Carbonic acid (H_2CO_3)   Phosgene   Diethyl carbonate (a carbonate ester)

A polycarbonate hockey mask.

Charles D. Winters

3. **Hydrolysis of an anhydride (Section 19.4A)** Anhydrides, particularly low-molecular-weight ones, react readily with water to give two carboxylic acids.

$$\underset{\substack{\text{O} \quad \text{O} \\ \| \quad \|}}{CH_3COCCH_3} + H_2O \longrightarrow \underset{\substack{\text{O} \\ \|}}{CH_3COH} + \underset{\substack{\text{O} \\ \|}}{HOCCH_3}$$

4. **Hydrolysis of an Ester (Section 19.4A)** Esters are hydrolyzed rapidly only in the presence of acid or base. Acid-catalyzed hydrolysis is the reverse of Fischer esterification. Acid is a catalyst. Base is required in an equimolar amount.

$$\underset{\substack{\text{O} \\ \|}}{CH_3COCH_2CH_3} + H_2O \underset{}{\overset{H^+}{\rightleftharpoons}} \underset{\substack{\text{O} \\ \|}}{CH_3COH} + HOCH_2CH_3$$

$$\underset{\substack{\text{O} \\ \|}}{CH_3COCH_2CH_3} + NaOH \xrightarrow{H_2O} \underset{\substack{\text{O} \\ \|}}{CH_3CO^-Na^+} + CH_3CH_2OH$$

5. **Hydrolysis of an Amide (Section 19.4A)** Amides require more vigorous conditions for hydrolysis than do esters. Either acid or base is required in an amount equivalent to that of the amide; acid to convert the resulting amine to an ammonium salt, and base to convert the resulting carboxylic acid to a carboxylate salt.

$$\underset{\substack{\text{O} \\ \|}}{CH_3CH_2CH_2CNH_2} + H_2O + HCl$$

$$\xrightarrow[\text{heat}]{H_2O} \underset{\substack{\text{O} \\ \|}}{CH_3CH_2CH_2COH} + NH_4^+Cl^-$$

$$\underset{\substack{\text{O} \\ \|}}{CH_3CH_2CH_2CNH_2} + NaOH$$

$$\xrightarrow[\text{heat}]{H_2O} \underset{\substack{\text{O} \\ \|}}{CH_3CH_2CH_2CO^-Na^+} + NH_3$$

6. **Reaction of Anhydride with Alcohol (Section 19.4B)** Anhydrides reacts with alcohol to give one mole of ester and one mole of a carboxylic acid.

$$\underset{\substack{\text{O} \quad \text{O} \\ \| \quad \|}}{CH_3COCCH_3} + HOCH_2CH_3$$

$$\longrightarrow \underset{\substack{\text{O} \\ \|}}{CH_3COCH_2CH_3} + \underset{\substack{\text{O} \\ \|}}{HOCCH_3}$$

7. **Reaction of an Anhydride with Ammonia or an Amine (Section 19.4C)** Anhydrides react with ammonia and with 1° and 2° amines to give amides. Two moles of amine are required: one mole to give the amide and one mole to neutralize the carboxylic acid by-product.

$$\underset{\substack{\text{O} \quad \text{O} \\ \| \quad \|}}{CH_3C-O-CCH_3} + 2NH_3$$

$$\longrightarrow \underset{\substack{\text{O} \\ \|}}{CH_3CNH_2} + \underset{\substack{\text{O} \\ \|}}{CH_3CO^-NH_4^+}$$

8. **Reaction of an Ester with Ammonia and with 1° and 2° amines (Section 19.4C)** Esters react with ammonia and with 1° and 2° amines to give an amide and an alcohol.

$$\longrightarrow$$

PROBLEMS

SECTION 19.1 What Are Carboxylic Anhydrides, Esters, and Amides?

19.4 ■ Draw a structural formula for each compound.
  (a) Dimethyl carbonate
  (b) *p*-Nitrobenzamide
  (c) Ethyl 3-hydroxybutanoate
  (d) Diethyl oxalate
  (e) Ethyl *trans*-2-pentenoate
  (f) Butanoic anhydride

19.5 ■ Write the IUPAC name for each compound.

(a)

(b) $CH_3(CH_2)_8COCH_3$

(c) $CH_3(CH_2)_4CNHCH_3$

(d) H_2N—⬡—CNH_2

(e) CH_3CO—⬠

(f) $CH_3CHCH_2COCH_2CH_3$ (OH, O)

SECTION 19.4 What Are the Characteristic Reactions of Anhydrides, Esters, and Amides?

19.6 ■ What product forms when ethyl benzoate is treated with each reagent?

(a) H_2O, NaOH, heat (b) H_2O, HCl, heat

19.7 ■ What product forms when benzamide, $C_6H_5CONH_2$, is treated with each reagent?

(a) H_2O, NaOH, heat (b) H_2O, HCl, heat

19.8 Which of these types of compounds will produce bubbles of CO_2 when added to an aqueous solution of sodium bicarbonate?

(a) A carboxylic acid (b) A carboxylic ester
(c) The sodium salt of a carboxylic acid

19.9 ■ Complete the equations for these reactions.

(a) CH_3O—⬡—$NH_2 + CH_3COCCH_3$ ⟶

(b) ⬡$NH + CH_3COCCH_3$ ⟶

19.10 The analgesic phenacetin is synthesized by treating 4-ethoxyaniline with acetic anhydride. Draw a structural formula for phenacetin.

CH_3CH_2O—⬡—NH_2

4-Ethoxyaniline

19.11 Phenobarbital is a long-acting sedative, hypnotic, and anticonvulsant.

(a) Name all functional groups in this compound.
(b) Draw structural formulas for the products from complete hydrolysis of all amide groups in aqueous NaOH.

Phenobarbital

19.12 ■ Following is a structural formula for aspartame, an artificial sweetener about 180 times as sweet as sucrose (table sugar).

Aspartame

(a) Is aspartame chiral? If so, how many stereoisomers are possible for it?
(b) Name each functional group in aspartame.
(c) Estimate the new charge on an aspartame molecule in aqueous solution at pH 7.0.
(d) Would you expect aspartame to be soluble in water? Explain.
(e) Draw structural formulas for the products of complete hydrolysis of aspartame in aqueous HCl. Show each product as it would be ionized in this solution.
(f) Draw structural formulas for the products of complete hydrolysis of aspartame in aqueous NaOH. Show each product as it would be ionized in this solution.

19.13 Why are nylon-66 and Kevlar referred to as polyamides?

19.14 Draw short sections of two parallel chains of nylon-66 (each chain running in the same direction) and show how it is possible to align them such that there is hydrogen bonding between the N—H groups of one chain and the C=O groups of the parallel chain.

19.15 Why are Dacron and Mylar referred to as polyesters?

SECTION 19.5 What are Phosphoric Anhydrides and Phosphoric Esters?

19.16 ■ What type of structural feature do the anhydrides of phosphoric acid and carboxylic acids have in common?

19.17 Draw structural formulas for the mono-, di-, and tri-ethyl esters of phosphoric acid.

19.18 ■ 1,3-Dihydroxy-2-propanone (dihydroxyacetone) and phosphoric acid form a monoester called dihydroxyacetone phosphate, which is an intermediate in glycolysis (Section 28.2). Draw a structural formula for this monophosphate ester.

19.19 Show how triphosphoric acid can form from three molecules of phosphoric acid. How many H_2O molecules are split out?

19.20 Write an equation for the hydrolysis of trimethyl phosphate to dimethyl phosphate and methanol in aqueous base. Show each product as it would be ionized in this solution.

Chemical Connections

19.21 (Chemical Connections 19A) Locate the ester group in pyrethrin I and draw a structural formula for chrysanthemic acid, the carboxylic acid from which this ester is derived.

19.22 (Chemical Connections 19A) What structural features do pyrethrin I (a natural insecticide) and permethrin (a synthetic pyrethrenoid) have in common?

19.23 (Chemical Connections 19A) A commercial Clothing & Gear Insect Repellant gives the following information about permethrin, its active ingredient:

Cis/*trans* ratio: Minimum 35% (+/−) *cis* and maximum 65% (+/−) *trans*

(a) To what does the *cis*/*trans* ratio refer?

(b) To what does the designation "(+/−)" refer?

19.24 (Chemical Connections 19B) Identify the β-lactam portion of amoxicillin and cephalexin.

19.25 (Chemical Connections 19C) What is the compound in willow bark that is responsible for its ability to relieve pain? How is this compound related to salicylic acid?

19.26 ■ (Chemical Connections 19C) Name the two functional groups in aspirin.

19.27 (Chemical Connections 19C) Once it has been opened, and particularly if it has been left open to the air, a bottle of aspirin may develop a vinegar-like odor. Explain how this might happen.

19.28 (Chemical Connections 19C) What is the structural relationship between aspirin and ibuprofen? Between aspirin and naproxen?

19.29 (Chemical Connections 19D) What is the difference in meaning between *sunblock* and *sunscreen?*

19.30 (Chemical Connections 19D) How do sunscreens prevent UV radiation from reaching the skin?

19.31 (Chemical Connections 19D) What structural features do the three sunscreens given in this Chemical Connection have in common?

19.32 (Chemical Connections 19E) Barbiturates are derived from urea. Identify the portion of the structure of pentobarbital and phenobarbital that is derived from urea.

19.33 (Chemical Connections 19F) Why do Lactomer stitches dissolve within 2 to 3 weeks following surgery?

Additional Problems

19.34 Benzocaine, a topical anesthetic, is prepared by treating 4-aminobenzoic acid with ethanol in the presence of an acid catalyst followed by neutralization. Draw a structural formula for benzocaine.

19.35 The analgesic acetaminophen is synthesized by treating 4-aminophenol with one equivalent of acetic anhydride. Write an equation for the formation of acetaminophen. (*Hint:* The —NH_2 group is more reactive with acetic anhydride than the —OH group.)

19.36 Procaine (its hydrochloride is marketed as Novocain) was one of the first local anesthetics developed for infiltration and regional anesthesia. It is synthesized by the following Fischer esterification. Draw a structural formula for procaine.

p-Aminobenzoic acid   2-Diethylaminoethanol

$$\xrightarrow[\text{esterification}]{\text{Fischer}} \text{Procaine}$$

19.37 1,3-Diphosphoglycerate, an intermediate in glycolysis (Section 28.2), contains a mixed anhydride (an anhydride of a carboxylic acid and phosphoric acid) and a phosphoric ester. Draw structural formulas for the products formed by hydrolysis of the anhydride and ester bonds in this molecule. Show each product as it would exist in solution at pH 7.4.

1,3-Diphosphoglycerate

Looking Ahead

19.38 In Chapter 22, we discuss a class of compounds called amino acids, so named because they contain both an amino group and a carboxyl group. Following is a structural formula for the amino acid alanine.

Alanine

What would you expect to be the major form of alanine present in aqueous solution (a) at pH 2.0, (b) at pH 5–6, and (c) at pH 11.0? Explain.

19.39 We have seen that an amide can be formed from a carboxylic acid and an amine. Suppose that you start instead with an amino acid such as alanine. Show how amide formation in this case can lead to a macromolecule of molecular weight several thousands of times that of the starting materials. We will study these polyamides in Chapter 22 (proteins).

19.40 We will encounter the following molecule in our discussion of glycolysis, the biochemical pathway that converts glucose to pyruvic acid (Section 28.1).

Phosphoenolpyruvate

(a) Draw structural formulas for the products of hydrolysis of the ester bond in phosphoenolpyruvate.

(b) Why are the letters *enol* a part of the name of this compound?

Tying it Together

19.41 From what carboxylic acid and amine or ammonia can each amide be synthesized?

(a)

$$O$$
$$\parallel$$
$$-NHC(CH_2)_4CH_3$$

(b) $(CH_3)_2CHCN(CH_3)_2$

(c) $H_2NC(CH_2)_4CNH_2$

19.42 *N,N*-Diethyl *m*-toluamide (DEET) is the active ingredient in several common insect repellents. From what acid and amine can DEET by synthesized?

N,N-Diethyl *m*-toluamide
(DEET)

19.43 Following are structural formulas for two local anesthetics. Lidocaine was introduced in 1948 and is now the most widely used local anesthetic for infiltration and regional anesthesia. Its hydrochloride salt is marketed under the name Xylocaine. Mepivacaine is faster-acting and somewhat longer in duration than lidocaine. Its hydrochloride salt is marketed under the name Carbocaine.

Lidocaine (Xylocaine)   Mepivacaine (Carbocaine)

(a) Name the functional groups in each anesthetic.

(b) What similarities in structure do you find between these compounds?

Carbohydrates

Breads, grains, and pasta are sources of carbohydrates.

Charles D. Winters

GOB
Chemistry Now™

Look for this logo in the chapter and go to GOB ChemistryNow at **http://now.brookscole.com/gob8** or on the CD for tutorials, simulations, and problems.

20.1 | Carbohydrates: What Are Monosaccharides?

Carbohydrates are the most abundant organic compounds in the plant world. They act as storehouses of chemical energy (glucose, starch, glycogen); are components of supportive structures in plants (cellulose), crustacean shells (chitin), and connective tissues in animals (acidic polysaccharides); and are essential components of nucleic acids (D-ribose and 2-deoxy-D-ribose). Carbohydrates account for approximately three fourths of the dry weight of plants. Animals (including humans) get their carbohydrates by eating plants, but they do not store much of what they consume. In fact, less than 1% of the body weight of animals is made up of carbohydrates.

The word *carbohydrate* means "hydrate of carbon" and derives from the formula $C_n(H_2O)_m$. Two examples of carbohydrates with this general molecular formula that can be written alternatively as hydrates of carbon are

- Glucose (blood sugar): $C_6H_{12}O_6$, which can be written as $C_6(H_2O)_6$
- Sucrose (table sugar): $C_{12}H_{22}O_{11}$, which can be written as $C_{12}(H_2O)_{11}$

Not all carbohydrates, however, have this general formula. Some contain too few oxygen atoms to fit it; some contain too many oxygens. Some also contain nitrogen. The term *carbohydrate* has become so firmly rooted in the chemical nomenclature that, although not completely accurate, it persists as the name for this class of compounds.

At the molecular level, most **carbohydrates** are polyhydroxyaldehydes, polyhydroxyketones, or compounds that yield them after hydrolysis. The simpler members of the carbohydrate family are often referred to as **saccharides** because of their sweet taste (Latin: *saccharum,* "sugar"). Carbohydrates are classified as monosaccharides, oligosaccharides, or polysaccharides depending on the number of simple sugars they contain.

Carbohydrate A polyhydroxyaldehyde or polyhydroxyketone, or a substance that gives these compounds on hydrolysis

GOB
Chemistry-♣-Now™
Click *Coached Problems* to examine the **Structural Aspects of Carbohydrates**

Monosaccharide A carbohydrate that cannot be hydrolyzed to a simpler compound

Aldose A monosaccharide containing an aldehyde group

Ketose A monosaccharide containing a ketone group

A. Structure and Nomenclature

Monosaccharides have the general formula $C_nH_{2n}O_n$, with one of the carbons being the carbonyl group of either an aldehyde or a ketone. The most common monosaccharides have three to nine carbon atoms. The suffix **-ose** indicates that a molecule is a carbohydrate, and the prefixes **tri-, tetr-, pent-,** and so forth indicate the number of carbon atoms in the chain. Monosaccharides containing an aldehyde group are classified as **aldoses;** those containing a ketone group are classified as **ketoses.**

There are only two trioses: the aldotriose glyceraldehyde and the ketotriose dihydroxyacetone.

Glyceraldehyde   Dihydroxyacetone
(an aldotriose)   (a ketotriose)

Often the designations *aldo-* and *keto-* are omitted, and these molecules are referred to simply as trioses, tetroses, and the like.

GOB
Chemistry-♣-Now™
Click *Coached Problems* to try a set of problems in **Classifying Carbohydrates**

Fischer projection A two-dimensional representation showing the configuration of a stereocenter; horizontal lines represent bonds projecting forward from the stereocenter, and vertical lines represent bonds projecting toward the rear

Emil Fischer, who in 1902 became the second Nobel Prize winner in chemistry, made many fundamental discoveries in the chemistry of carbohydrates, proteins, and other areas of organic and biochemistry.

B. Fischer Projection Formulas

Glyceraldehyde contains a stereocenter and therefore exists as a pair of enantiomers (Figure 20.1).

Chemists commonly use two-dimensional representations called **Fischer projections** to show the configuration of carbohydrates. To draw a Fischer projection, you draw a three-dimensional representation of the molecule oriented so that the vertical bonds from the stereocenter are directed away from you and the horizontal bonds from it are directed toward you (none of the bonds to stereocenter are in the plane of the paper). Then write the molecule as a cross, with the stereocenter indicated by the point at which the bonds cross.

(R)-Glyceraldehyde   (R)-Glyceraldehyde
(three-dimensional   (Fischer projection)
representation)

Figure 20.1 The enantiomers of glyceraldehyde.

The horizontal segments of this Fischer projection represent bonds directed toward you, and the vertical segments represent bonds directed away from you. The only atom in the plane of the paper is the stereocenter.

C. D- and L-Monosaccharides

Even though the R,S system is widely accepted today as a standard for designating configuration, the configuration of carbohydrates is commonly designated using the D,L system proposed by Emil Fischer in 1891. At that time, it was known that one enantiomer of glyceraldehyde has a specific rotation (Section 15.4B) of $+13.5°$; the other has a specific rotation of $-13.5°$. Fischer proposed that these enantiomers be designated D and L, but he had no experimental way to determine which enantiomer has which specific rotation. Fischer, therefore, did the only possible thing—he made an arbitrary assignment. He assigned the dextrorotatory enantiomer the following configuration and named it D-glyceraldehyde. He named its enantiomer L-glyceraldehyde. Fischer could have been wrong, but by a stroke of good fortune, he wasn't. In 1952, scientists proved that his assignment of the D,L-configuration to the enantiomers of glyceraldehyde is correct.

D-Glyceraldehyde   L-Glyceraldehyde
$[\alpha]_D^{25} = +13.5°$ $[\alpha]_D^{25} = -13.5°$

D-glyceraldehyde and L-glyceraldehyde serve as reference points for the assignment of relative configurations to all other aldoses and ketoses. The reference point is the penultimate carbon—that is, the next-to-the-last carbon on the chain. A **D-monosaccharide** has the same configuration at its penultimate carbon as D-glyceraldehyde (its —OH group is on the right) in a Fischer projection; an **L-monosaccharide** has the same configuration at its penultimate carbon as L-glyceraldehyde (its —OH group is on the left).

Tables 20.1 and 20.2 show names and Fischer projections for all D-aldo- and D-2-ketotetroses, pentoses, and hexoses. Each name consists of three parts. The D specifies the configuration at the stereocenter farthest from the carbonyl group. Prefixes such as *rib-*, *arabin-*, and *gluc-* specify the configuration of all other stereocenters in the monosaccharide relative to one another. The suffix *-ose* indicates that the compound is a carbohydrate.

The three most abundant hexoses in the biological world are D-glucose, D-galactose, and D-fructose. The first two are D-aldohexoses; the third is a D-2-ketohexose. Glucose, by far the most abundant of the three, is also known

D-Monosaccharide A monosaccharide that, when written as a Fischer projection, has the —OH group on its penultimate carbon to the right

L-Monosaccharide A monosaccharide that, when written as a Fischer projection, has the —OH group on its penultimate carbon to the left

Table 20.1 Configurational Relationships among the Isomeric D-Aldotetroses, D-Aldopentoses, and D-Aldohexoses

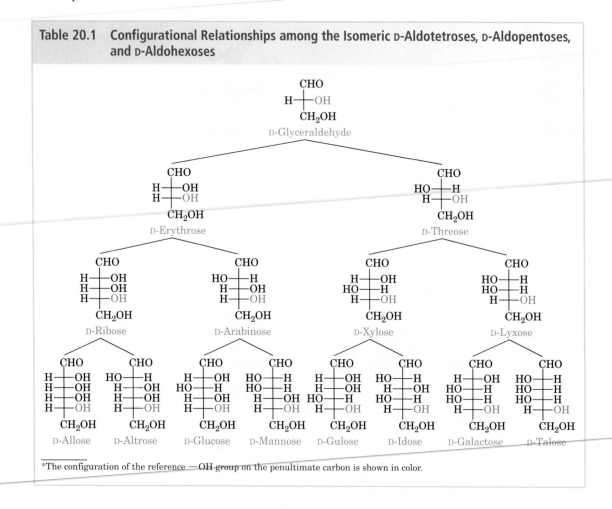

*The configuration of the reference —OH group on the penultimate carbon is shown in color.

as dextrose because it is dextrorotatory. Other names for this monosaccharide include grape sugar and blood sugar. Human blood normally contains 65–110 mg of glucose/100 mL of blood.

EXAMPLE 20.1

Draw Fischer projections for the four aldotetroses. Which are D-monosaccharides, which are L-monosaccharides, and which are enantiomers? Refer to Table 20.1 and write the name of each aldotetrose.

Solution
Following are Fischer projections for the four aldotetroses. The D- and L-refer to the configuration of the penultimate carbon, which, in the case of aldotetroses, is carbon 3. In the Fischer projection of a D-aldotetrose, the —OH group on carbon 3 is on the right; in an L-aldotetrose, it is on the left.

Table 20.2 Configurational Relationships among the D-2-Ketopentoses and D-2-Ketohexoses

Gloved hand holding an intravenous (IV) drip bag containing 0.15% potassium chloride (saline) and 5% glucose.

Claire Paxton & Jacqui Farrow/Science Photo Library/Photo Researchers Inc.

GOB
Chemistry❖Now™

Click *Coached Problems* to try a set of problems in **Classifying D- and L-Monosaccharides**

Problem 20.1

Draw Fischer projections for all 2-ketopentoses. Which are D-2-ketopentoses, which are L-2-ketopentoses, and which are enantiomers? Refer to Table 20.2 and write the name of each 2-ketopentose.

D. Amino Sugars

Amino sugars contain an —NH₂ group in place of an —OH group. Only three amino sugars are common in nature: D-glucosamine, D-mannosamine, and D-galactosamine.

Amino sugar A monosaccharide in which an —OH group is replaced by an —NH₂ group

CHEMICAL CONNECTIONS 20A

Galactosemia

One out of every 18,000 infants is born with a genetic defect that renders the child unable to utilize the monosaccharide galactose. Galactose is part of lactose (milk sugar, Section 20.4B). When the body cannot absorb galactose, it accumulates in the blood and in the urine. This buildup in the blood is harmful because it can lead to mental retardation, failure to grow, cataract formation in the eye, and, in severe cases, death due to liver damage. When galactose accumulation results from a deficiency of the enzyme galactokinase, the disorder, known as galactosuria, has only mild symptoms. When the enzyme galactose-1-phosphate uridinyltrans-

ferase is defective, however, the disorder is called galactosemia, and its symptoms are severe.

The deleterious effects of galactosemia can be avoided by giving the infant a milk formula in which sucrose is substituted for lactose. Because sucrose contains no galactose, the infant consumes a galactose-free diet. A galactose-free diet is critical only in infancy. With maturation, most children develop another enzyme capable of metabolizing galactose. As a consequence, they are able to tolerate galactose as they mature.

N-Acetyl-D-glucosamine, a derivative of D-glucosamine, is a component of many polysaccharides, including connective tissue such as cartilage. It is also a component of chitin, the hard, shell-like exoskeleton of lobsters, crabs, shrimp, and other shellfish. Several other amino sugars are components of naturally occurring antibiotics.

E. Physical Properties of Monosaccharides

Monosaccharides are colorless, crystalline solids. Because hydrogen bonding is possible between their polar —OH groups and water, all monosaccharides are very soluble in water. They are only slightly soluble in ethanol and are insoluble in nonpolar solvents such as diethyl ether, dichloromethane, and benzene.

20.2	What Are the Cyclic Structures of Monosaccharides?

In Section 17.4C, we saw that aldehydes and ketones react with alcohols to form **hemiacetals.** We also saw that cyclic hemiacetals form very readily when hydroxyl and carbonyl groups are part of the same molecule and that their interaction produces a ring. For example, 4-hydroxypentanal forms a five-membered cyclic hemiacetal. Note that 4-hydroxypentanal contains one stereocenter and that hemiacetal formation generates a second stereocenter at carbon 1.

GOB
Chemistry⚛Now™

Click *Coached Problems* to try a set of problems in examining the **Cyclic Structures of Monosaccharides**

Monosaccharides have hydroxyl and carbonyl groups in the same molecule. As a result, they exist almost exclusively as five- and six-membered cyclic hemiacetals.

A. Haworth Projections

A common way of representing the cyclic structure of monosaccharides is the **Haworth projection,** named after the English chemist Sir Walter N. Haworth (Nobel Prize for chemistry, 1937). In a Haworth projection, a five- or six-membered cyclic hemiacetal is represented as a planar pentagon or hexagon, respectively, lying roughly perpendicular to the plane of the paper. Groups bonded to the carbons of the ring then lie either above or below the plane of the ring. The new carbon stereocenter created in forming the cyclic structure is called an **anomeric carbon.** Stereoisomers that differ in configuration only at the anomeric carbon are called **anomers.** The anomeric carbon of an aldose is carbon 1; that of the most common ketoses is carbon 2.

Typically, Haworth projections are most commonly drawn with the anomeric carbon to the right and the hemiacetal oxygen to the back (Figure 20.2).

In the terminology of carbohydrate chemistry, the designation β means that the —OH on the anomeric carbon of the cyclic hemiacetal lies on the same side of the ring as the terminal —CH₂OH. Conversely, the designation α means that the —OH on the anomeric carbon of the cyclic hemiacetal lies on the side of the ring opposite from the terminal —CH₂OH.

A six-membered hemiacetal ring is indicated by **-pyran-,** and a five-membered hemiacetal ring is indicated by **-furan-.** The terms **furanose** and **pyranose** are used because monosaccharide five- and six-membered rings correspond to the heterocyclic compounds furan and pyran.

Furan   Pyran

Because the α and β forms of glucose are six-membered cyclic hemiacetals, they are named α-D-glucopyranose and β-D-glucopyranose, respectively. The designations *-furan-* and *-pyran-* are not always used in monosaccharide names, however. Thus, the glucopyranoses, for example, are often named simply α-D-glucose and β-D-glucose.

Haworth projection A way to view furanose and pyranose forms of monosaccharides; the ring is drawn flat and viewed through its edge, with the anomeric carbon on the right and the oxygen atom to the rear

Anomeric carbon The hemiacetal carbon of the cyclic form of a monosaccharide

Anomers Monosaccharides that differ in configuration only at their anomeric carbons

Furanose A five-membered cyclic hemiacetal form of a monosaccharide

Pyranose A six-membered cyclic hemiacetal form of a monosaccharide

Figure 20.2 Haworth projections for α-D-glucopyranose and β-D-glucopyranose.

You would do well to remember the configurations of the groups on the Haworth projections of α-D-glucopyranose and β-D-glucopyranose as reference structures. Knowing how the open-chain configuration of any other aldohexose differs from that of D-glucose, you can construct Haworth projections for the aldohexose by referring to the Haworth projection of D-glucose.

EXAMPLE 20.2

Draw Haworth projections for the α and β anomers of D-galactopyranose.

Solution

One way to arrive at these projections is to use the α and β forms of D-glucopyranose as references and to remember (or discover by looking at Table 20.1) that D-galactose differs from D-glucose only in the configuration at carbon 4. Thus you can begin with the Haworth projection shown in Figure 20.2 and then invert the configuration at carbon 4.

α-D-Galactopyranose
(α-D-Galactose)

β-D-Galactopyranose
(β-D-Galactose)

Problem 20.2

D-Mannose exists in aqueous solution as a mixture of α-D-mannopyranose and β-D-mannopyranose. Draw Haworth projections for these molecules.

Aldopentoses also form cyclic hemiacetals. The most prevalent forms of D-ribose and other pentoses in the biological world are furanoses. Following are Haworth projections for α-D-ribofuranose (α-D-ribose) and β-2-deoxy-D-ribofuranose (β-2-Deoxy-D-ribose). The prefix *2-deoxy* indicates the absence of oxygen at carbon 2. Units of D-ribose and 2-deoxy-D-ribose in nucleic acids and most other biological molecules are found almost exclusively in the β configuration.

α-D-Ribofuranose
(α-D-Ribose)

β-2-Deoxy-D-ribofuranose
(β-2-Deoxy-D-ribose)

Fructose also forms five-membered cyclic hemiacetals. β-D-Fructofuranose, for example, is found in the disaccharide sucrose (Section 20.4A).

α-D-Fructofuranose
(α-D-Fructose)

D-Fructose

β-D-Fructofuranose
(β-D-Fructose)

Anomeric carbon

B. Conformation Representations

A five-membered furanose ring is so close to being planar that Haworth projections provide adequate representations of furanoses. For pyranoses, however, the six-membered ring is more accurately represented as a **chair conformation** (Section 11.5B). Following are structural formulas for α-D-glucopyranose and β-D-glucopyranose, both drawn as chair conformations. Also shown is the open-chain or free aldehyde form with which the cyclic hemiacetal forms are in equilibrium in aqueous solution. Notice that each group, including the anomeric —OH group, on the chair conformation of β-D-glucopyranose is equatorial. Notice also that the —OH group on the anomeric carbon is axial in α-D-glucopyranose. Because the —OH on the anomeric carbon of β-D-glucopyranose is in the more stable equatorial position (Section 11.5B), the β anomer predominates in aqueous solution.

We do not show hydrogen atoms bonded to the ring in chair conformations. We often show them, however, in Haworth projections.

β-D-Glucopyranose
$[\alpha]_D^{25} = +18.7°$

D-Glucose

Anomeric carbon

α-D-Glucopyranose
$[\alpha]_D^{25} = 112°$

At this point, let us compare the relative orientations of groups on the D-glucopyranose ring in the Haworth projection and the chair conformation. The orientations of groups on carbons 1 through 5 of β-D-glucopyranose, for example, are up, down, up, down, and up in both representations.

β-D-Glucopyranose
(Haworth projection)

β-D-Glucopyranose
(chair conformation)

CHEMICAL CONNECTIONS 20B

L-Ascorbic Acid (Vitamin C)

The structure of L-ascorbic acid (vitamin C) resembles that of a monosaccharide. In fact, this vitamin is synthesized both biochemically by plants and some animals and commercially from D-glucose. Humans do not have the enzymes required to carry out this synthe-

sis. For this reason, we must obtain vitamin C in the food we eat or as a vitamin supplement. Approximately 66 million kg of vitamin C is synthesized annually in the United States.

L-Ascorbic acid
(Vitamin C) ⇌ (oxidation / reduction) ⇌ L-Dehydroascorbic acid

Oranges are a major source of vitamin C.

Charles D. Winters

EXAMPLE 20.3

Draw chair conformations for α-D-galactopyranose and β-D-galactopyranose. Label the anomeric carbon in each.

Solution

The configuration of D-galactose differs from that of D-glucose only at carbon 4. Therefore, draw the α and β forms of D-glucopyranose and then interchange the positions of the —OH and —H groups on carbon 4.

β-D-Galactopyranose (β-D-Galactose) ⇌ D-Galactose ⇌ α-D-Galactopyranose (α-D-Galactose)

Problem 20.3

Draw chair conformations for α-D-mannopyranose and β-D-mannopyranose. Label the anomeric carbon in each.

C. Mutarotation

Mutarotation is the change in specific rotation that accompanies the equilibration of α and β anomers in aqueous solution. For example, a solution prepared by dissolving crystalline α-D-glucopyranose in water has a specific rotation of $+112°$, which gradually decreases to an equilibrium value of $+52.7°$ as α-D-glucopyranose reaches equilibrium with β-D-glucopyranose. A solution of β-D-glucopyranose also undergoes mutarotation, during which the specific rotation changes from $+18.7°$ to the same equilibrium value of $+52.7°$. The equilibrium mixture consists of 64% β-D-glucopyranose and 36% α-D-glucopyranose, with only a trace (0.003%) of the open-chain form. Mutarotation is common to all carbohydrates that exist in hemiacetal forms.

Mutarotation The change in specific rotation that occurs when an α or β form of a carbohydrate is converted to an equilibrium mixture of the two forms

β-D-Glucopyranose
$[\alpha]_D^{25} = +18.7°$

Open-chain form

α-D-Glucopyranose
$[\alpha]_D^{25} = +112°$

20.3 | What Are the Characteristic Reactions of Monosaccharides?

Chemistry·Now™

Click *Coached Problems* to explore the **Common Structures of Monosaccharides**

A. Formation of Glycosides (Acetals)

As we saw in Section 17.4C, treatment of an aldehyde or ketone with one molecule of alcohol yields a hemiacetal, and treatment of the hemiacetal with a molecule of alcohol yields an acetal. Treatment of a monosaccharide—all forms of which exist almost exclusively as cyclic hemiacetals—with an alcohol also yields an acetal, as illustrated by the reaction of β-D-glucopyranose with methanol.

Glycoside A carbohydrate in which the —OH group on its anomeric carbon is replaced by an —OR group

β-D-Glucopyranose
(β-D-Glucose)

Methyl β-D-glucopyranoside
(Methyl β-D-glucoside)

Methyl α-D-glucopyranoside
(Methyl α-D-glucoside)

A cyclic acetal derived from a monosaccharide is called a **glycoside,** and the bond from the anomeric carbon to the —OR group is called a **glycosidic bond.** Mutarotation is not possible in a glycoside because an

Glycosidic bond The bond from the anomeric carbon of a glycoside to an —OR group acetal—unlike a hemiacetal—is no longer in equilibrium with the open-chain carbonyl-containing compound. Glycosides are stable in water and aqueous base; like other acetals (Section 17.4C), however, they are hydrolyzed in aqueous acid to an alcohol and a monosaccharide.

We name glycosides by listing the alkyl or aryl group bonded to oxygen, followed by the name of the carbohydrate in which the ending **-e** is replaced by **-ide**. For example, the methyl glycoside derived from β-D-glucopyranose is named methyl β-D-glucopyranoside; that derived from β-D-ribofuranose is named methyl β-D-ribofuranoside.

EXAMPLE 20.4

Draw a structural formula for methyl β-D-ribofuranoside (methyl β-D-riboside). Label the anomeric carbon and the glycosidic bond.

Solution

Methyl β-D-ribofuranoside
(Methyl β-D-riboside)

Problem 20.4

Draw a Haworth projection and a chair conformation for methyl α-D-mannopyranoside (methyl α-D-mannoside). Label the anomeric carbon and the glycosidic bond.

B. Reduction to Alditols

Oxidations and reductions of monosaccharides in nature are catalyzed by specific enzymes classified as oxidases—for example, glucose oxidase.

Alditol The product formed when the CHO group of a monosaccharide is reduced to a CH_2OH group

The carbonyl group of a monosaccharide can be reduced to a hydroxyl group by a variety of reducing agents, including hydrogen in the presence of a transition metal catalyst and sodium borohydride (Section 17.4C). The reduction products are known as **alditols.** Reduction of D-glucose gives D-glucitol, more commonly known as D-sorbitol. Here, D-glucose is shown in the open-chain form. Only a small amount of this form is present in solution but, as it is reduced, the equilibrium between cyclic hemiacetal forms (only the β form is shown here) and the open-chain form shifts to replace it.

β-D-Glucopyranose   D-Glucose   D-Glucitol (D-Sorbitol)

We name alditols by dropping the **-ose** from the name of the monosaccharide and adding **-itol.** Sorbitol is found in the plant world in many berries and in cherries, plums, pears, apples, seaweed, and algae. It is about 60% as sweet as sucrose (table sugar) and is used in the manufacture of candies and as a sugar substitute for diabetics. Other alditols commonly found in the biological world include erythritol, D-mannitol, and xylitol. Xylitol is used as a sweetening agent in "sugarless" gum, candy, and sweet cereals.

Many "sugar-free" products contain sugar alcohols, such as sorbitol and xylitol.

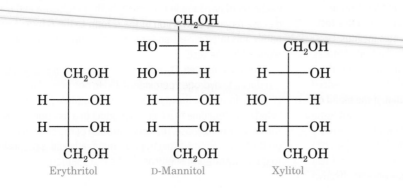

Erythritol   D-Mannitol   Xylitol

C. Oxidation to Aldonic Acids (Reducing Sugars)

As we saw in Section 17.4A, aldehydes (RCHO) are oxidized to carboxylic acids (RCOOH) by several agents, including oxygen, O_2. Similarly, the aldehyde group of an aldose can be oxidized, under basic conditions, to a carboxylate group. Under these conditions, the cyclic form of the aldose is in equilibrium with the open-chain form, which is then oxidized by the mild oxidizing agent. D-Glucose, for example, is oxidized to D-gluconate (the anion of D-gluconic acid).

β-D-Glucopyranose   D-Glucose   D-Gluconate
(β-D-Glucose)   (an aldonic acid)

Any carbohydrate that reacts with an oxidizing agent to form an aldonic acid is classified as a **reducing sugar** (it reduces the oxidizing agent).

Surprisingly, 2-ketoses are also reducing sugars. Carbon 1 (a CH_2OH group) of a ketose is not oxidized directly. Instead, under the basic conditions of this oxidation, a 2-ketose exists in equilibrium with an aldose by way of an enediol intermediate. The aldose is then oxidized by the mild oxidizing agent.

Reducing sugar A carbohydrate that reacts with a mild oxidizing agent under basic conditions to give an aldonic acid; the carbohydrate reduces the oxidizing agent

A 2-ketose   An enediol   An aldose   An aldonate

CHEMICAL CONNECTIONS 20C

Testing for Glucose

The analytical procedure most often performed in a clinical chemistry laboratory is the determination of glucose in blood, urine, or other biological fluids. The high frequency with which this test is performed reflects the high incidence of diabetes mellitus. Approximately 2 million known diabetics live in the United States, and it is estimated that another 1 million remain undiagnosed.

Diabetes mellitus (Chemical Connections 24G) is characterized by insufficient blood levels of the hormone insulin. If the blood concentration of insulin is too low, muscle and liver cells do not absorb glucose from the blood; this problem, in turn, leads to increased levels of blood glucose (hyperglycemia), impaired metabolism of fats and proteins, ketosis, and possible diabetic coma. A rapid test for blood glucose levels is critical for early diagnosis and effective management of this disease. In addition to giving results rapidly, a test must be specific for D-glucose; that is, it must give a positive test for glucose but not react with any other substances normally present in biological fluids.

Today, blood glucose levels are measured by an enzyme-based procedure using the enzyme glucose oxidase. This enzyme catalyzes the oxidation of β-D-glucose to D-gluconic acid.

Glucose oxidase is specific for β-D-glucose. Therefore, complete oxidation of any sample containing both β-D-glucose and α-D-glucose requires conversion of the α form to the β form. Fortunately, this interconversion is rapid and complete in the short time required for the test.

Molecular oxygen, O_2, is the oxidizing agent in this reaction and is reduced to hydrogen peroxide, H_2O_2. In one procedure, hydrogen peroxide formed in the glucose oxidase–catalyzed reaction oxidizes colorless o-toluidine to a colored product in a reaction catalyzed by the enzyme peroxidase. The concentration of the colored oxidation product is determined spectrophotometrically and is proportional to the concentration of glucose in the test solution.

2-Methylaniline
(o-Toluidine)

Several commercially available test kits use the glucose oxidase reaction for qualitative determination of glucose in urine.

Chemstrip kit for blood glucose.

D. Oxidation to Uronic Acids

Enzyme-catalyzed oxidation of the primary alcohol at carbon 6 of a hexose yields a uronic acid. Enzyme-catalyzed oxidation of D-glucose, for example, yields D-glucuronic acid, shown here in both its open-chain and cyclic hemiacetal forms:

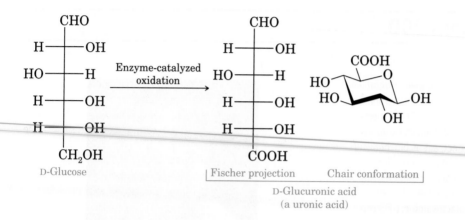

Fischer projection   Chair conformation

D-Glucuronic acid
(a uronic acid)

D-Glucuronic acid is widely distributed in both the plant and animal worlds. In humans, it serves as an important component of the acidic poly-saccharides of connective tissues (Section 20.6A). The body also uses D-glu-curonic acid to detoxify foreign phenols and alcohols. In the liver, these compounds are converted to glycosides of glucuronic acid (glucuronides) and excreted in the urine. The intravenous anesthetic propofol, for example, is converted to the following glucuronide and then excreted in urine:

Propofol   A urine-soluble D-glucuronide

E. The Formation of Phosphoric Esters

Mono- and diphosphoric esters are important intermediates in the metabo-lism of monosaccharides. For example, the first step in glycolysis (Section 28.1) involves conversion of glucose to glucose 6-phosphate. Note that phos-phoric acid is a strong enough acid so that at the pH of cellular and inter-cellular fluids, both acidic protons of a phosphoric ester are ionized, giving the ester a charge of −2.

D-Glucose   D-Glucose 6-phosphate   α-D-Glucose 6-phosphate

CHEMICAL CONNECTIONS 20D

A, B, AB, and O Blood Types

Membranes of animal plasma cells have large numbers of relatively small carbohydrates bound to them. In fact, the outsides of most plasma cell membranes are literally "sugar-coated." These membrane-bound carbohydrates are part of the mechanism by which cell types recognize one another and, in effect, act as biochemical markers. Typically, they contain from 4 to 17 monosaccharide units consisting primarily of relatively few monosaccharides, the most common of which are D-galactose, D-mannose, L-fucose, N-acetyl-D-glucosamine, and N-acetyl-D-galactosamine. L-Fucose is a 6-deoxyaldohexose.

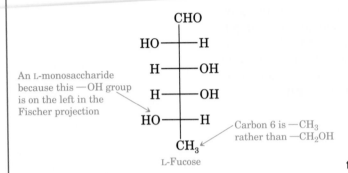

An L-monosaccharide because this —OH group is on the left in the Fischer projection

Carbon 6 is —CH₃ rather than —CH₂OH

L-Fucose

Bag of blood showing blood type.

To see the importance of these membrane-bound carbohydrates, consider the ABO blood group system, discovered in 1900 by Karl Landsteiner (1868–1943). Whether an individual belongs to type A, B, AB, or O is genetically determined and depends on the type of trisaccharide or tetrasaccharide bound to the surface of the red blood cells. These surface-bound carbohydrates, designated as A, B, and O, act as antigens. The type of glycosidic bond joining each monosaccharide is shown in the figure.

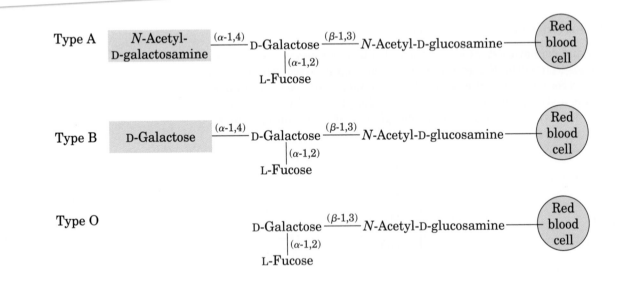

The blood carries antibodies against foreign substances. When a person receives a blood transfusion, the antibodies clump together (aggregate) the foreign blood cells. Type A blood, for example, has A antigens (N-acetyl-D-galactosamine) on the surfaces of its red blood cells and carries anti-B antibodies (against B antigen). B-type blood carries B antigen (D-galactose) and has anti-A antibodies (against A antigens). Transfusion of type A blood into a person with type B blood can be fatal, and vice versa. The relationships between blood type and donor–receiver relationships are summarized in the figure.

A, B, AB, and O Blood Types *(continued)*

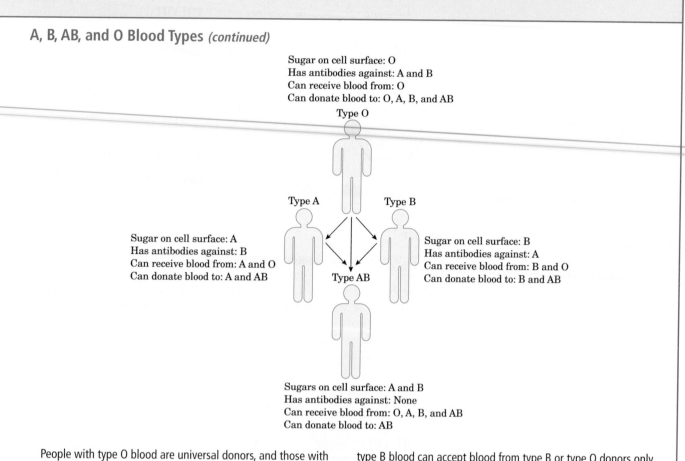

Sugar on cell surface: O
Has antibodies against: A and B
Can receive blood from: O
Can donate blood to: O, A, B, and AB

Type O

Type A

Sugar on cell surface: A
Has antibodies against: B
Can receive blood from: A and O
Can donate blood to: A and AB

Type B

Sugar on cell surface: B
Has antibodies against: A
Can receive blood from: B and O
Can donate blood to: B and AB

Type AB

Sugars on cell surface: A and B
Has antibodies against: None
Can receive blood from: O, A, B, and AB
Can donate blood to: AB

People with type O blood are universal donors, and those with type AB blood are universal acceptors. People with type A blood can accept blood from type A or type O donors only. Those with type B blood can accept blood from type B or type O donors only. Type O persons can accept blood only from type O donors.

20.4 | What Are Disaccharides and Oligosaccharides?

Disaccharide A carbohydrate containing two monosaccharide units joined by a glycosidic bond

Oligosaccharide A carbohydrate containing from six to ten monosaccharide units, each joined to the next by a glycosidic bond

Polysaccharide A carbohydrate containing a large number of monosaccharide units, each joined to the next by one or more glycosidic bonds

Most carbohydrates in nature contain more than one monosaccharide unit. Those that contain two units are called **disaccharides,** those that contain three units are called **trisaccharides,** and so forth. We use the general term **oligosaccharide** to describe carbohydrates that contain from six to ten monosaccharide units. Carbohydrates containing larger numbers of monosaccharide units are called **polysaccharides.**

In a disaccharide, two monosaccharide units are joined by a glycosidic bond between the anomeric carbon of one unit and an —OH group of the other unit. Three important disaccharides are sucrose, lactose, and maltose.

A. Sucrose

Sucrose (table sugar) is the most abundant disaccharide in the biological world. It is obtained principally from the juice of sugar cane and sugar beets. In sucrose, carbon 1 of α-D-glucopyranose bonds to carbon 2 of D-fructofuranose by an α-1,2-glycosidic bond. Because the anomeric carbons of both the glucopyranose and fructofuranose units are involved in formation of the glycosidic bond, neither monosaccharide unit is in equilibrium with its open-chain form. Thus sucrose is a nonreducing sugar.

In the production of sucrose, sugar cane or sugar beet is boiled with water, and the resulting solution is cooled. Sucrose crystals separate and are collected. Subsequent boiling to concentrate the solution followed by cooling yields a dark, thick syrup known as molasses.

Sucrose

B. Lactose

Lactose is the principal sugar present in milk. It accounts for 5 to 8% of
human milk and 4 to 6% of cow's milk. This disaccharide consists of D-galactopyranose bonded by a β-1,4-glycosidic bond to carbon 4 of D-glucopyranose. Lactose is a reducing sugar, because the cyclic hemiacetal of the
D-glucopyranose unit is in equilibrium with its open-chain form and can be
oxidized to a carboxyl group.

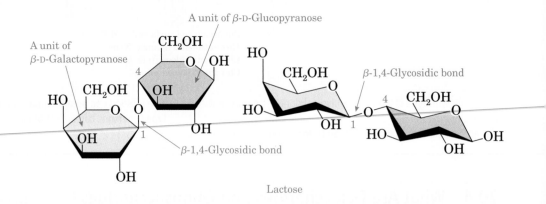

Lactose

C. Maltose

Maltose is an ingredient in most
syrups.

Maltose derives its name from its presence in malt, the juice from sprouted
barley and other cereal grains. It consists of two units of D-glucopyranose
joined by a glycosidic bond between carbon 1 (the anomeric carbon) of one
unit and carbon 4 of the other unit. Because the oxygen atom on the
anomeric carbon of the first glucopyranose unit is alpha, the bond joining
the two units is an α-1,4-glycosidic bond. Following are a Haworth projection and a chair conformation for β-maltose, so named because the —OH
groups on the anomeric carbon of the glucose unit on the right are beta.

Maltose

Table 20.3 Relative Sweetness of Some Carbohydrate and Artificial Sweetening Agents

Carbohydrate	Sweetness Relative to Sucrose	Artificial Sweetener	Sweetness Relative to Sucrose
fructose	1.74	saccharine	450
sucrose (table sugar)	1.00	acesulfame-K	200
honey	0.97	aspartame	180
glucose	0.74		
maltose	0.33		
galactose	0.32		
lactose (milk sugar)	0.16		

Maltose is a reducing sugar; the hemiacetal group on the right unit of D-glucopyranose is in equilibrium with the free aldehyde and can be oxidized to a carboxylic acid.

D. Relative Sweetness

Among the disaccharide sweetening agents, D-fructose tastes the sweetest—even sweeter than sucrose (Table 20.3). The sweet taste of honey is due largely to D-fructose and D-glucose. Lactose has almost no sweetness and is sometimes added to foods as a filler. Some people cannot tolerate lactose well, however, and should avoid these foods.

We have no mechanical way to measure sweetness. Such testing is done by having a group of people taste and rate the sweetness of solutions of varying sweetening agents.

EXAMPLE 20.5

Draw a chair conformation for the β anomer of a disaccharide in which two units of D-glucopyranose are joined by an α-1,6-glycosidic bond.

Solution

First draw a chair conformation of α-D-glucopyranose. Then connect the anomeric carbon of this monosaccharide to carbon 6 of a second D-glucopyranose unit by an α-1,6-glycosidic bond. The resulting molecule is either α or β depending on the orientation of the —OH group on the reducing end of the disaccharide. The disaccharide shown here is the β form.

Problem 20.5

Draw a chair conformation for the α form of a disaccharide in which two units of D-glucopyranose are joined by a β-1,3-glycosidic bond.

20.5 | What Are Polysaccharides?

Polysaccharides consist of large numbers of monosaccharide units bonded together by glycosidic bonds. Three important polysaccharides, all made up of glucose units, are starch, glycogen, and cellulose.

A. Starch: Amylose and Amylopectin

Starch is used for energy storage in plants. It is found in all plant seeds and tubers and is the form in which glucose is stored for later use. Starch can be separated into two principal polysaccharides: amylose and amylopectin. Although the starch from each plant is unique, most starches contain 20 to 25% amylose and 75 to 80% amylopectin.

Complete hydrolysis of both amylose and amylopectin yields only D-glucose. Amylose is composed of continuous, unbranched chains of as many as 4000 D-glucose units joined by α-1,4-glycosidic bonds. Amylopectin contains chains of as many as 10,000 D-glucose units also joined by α-1,4-glycosidic bonds. In addition, considerable branching from this linear network occurs. New chains of 24 to 30 units are started at branch points by α-1,6-glycosidic bonds (Figure 20.3).

B. Glycogen

Glycogen acts as the energy-reserve carbohydrate for animals. Like amylopectin, it is a branched polysaccharide containing approximately 10^6 glucose units joined by α-1,4- and α-1,6-glycosidic bonds. The total amount of glycogen in the body of a well-nourished adult human is about 350 g, divided almost equally between liver and muscle.

C. Cellulose

Cellulose, the most widely distributed plant skeletal polysaccharide, constitutes almost half of the cell-wall material of wood. Cotton is almost pure cellulose.

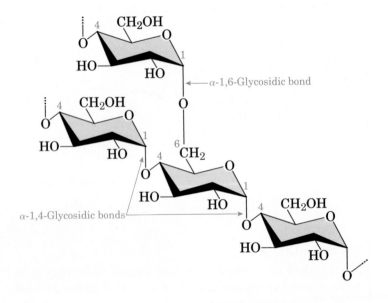

Figure 20.3 Amylopectin is a branched polymer of approximately 10,000 D-glucose units joined by α-1,4-glycosidic bonds. Branches consist of 24–30 D-glucose units started by α-1,6-glycosidic bonds.

β-1,4-Glycosidic bonds

Figure 20.4 Cellulose is a linear polysaccharide containing as many as 3000 units of D-glucose joined by β-1,4-glycosidic bonds.

Cellulose is a linear polysaccharide of D-glucose units joined by β-1,4-glycosidic bonds (Figure 20.4). It has an average molecular weight of 400,000 g/mol, corresponding to approximately 2200 glucose units per molecule.

Cellulose molecules act much like stiff rods, a characteristic that enables them to align themselves side by side into well-organized, water-insoluble fibers in which the OH groups form numerous intermolecular hydrogen bonds. This arrangement of parallel chains in bundles gives cellulose fibers their high mechanical strength. It also explains why cellulose is insoluble in water. When a piece of cellulose-containing material is placed in water, there are not enough —OH groups on the surface of the fiber to pull individual cellulose molecules away from the strongly hydrogen-bonded fiber.

Humans and other animals cannot use cellulose as food because our digestive systems do not contain β-glucosidases, enzymes that catalyze the hydrolysis of β-glucosidic bonds. Instead, we have only α-glucosidases; hence, we use the polysaccharides starch and glycogen as sources of glucose. In contrast, many bacteria and microorganisms do contain β-glucosidases and so can digest cellulose. Termites (much to our regret) have such bacteria in their intestines and can use wood as their principal food. Ruminants (cud-chewing animals) and horses can also digest grasses and hay because β-glucosidase–containing microorganisms are present in their alimentary systems.

CHEMICAL CONNECTIONS 20E

High-Fructose Corn Syrup

If you read the labels of soft drinks and other artificially sweetened food products, you will find that many of them contain high-fructose corn syrup. Looking at Table 20.3, you will see one reason for its use: Fructose is more than 70% sweeter than sucrose (table sugar).

The production of high-fructose corn syrup begins with the partial hydrolysis of corn starch catalyzed by the enzyme α-amylase.

This enzyme catalyzes the hydrolysis of α-glucosidic bonds, which breaks corn starch into small polysaccharides called dextrins. The enzyme glucoamylase then catalyzes the hydrolysis of the dextrins to D-glucose. Finally, enzyme-catalyzed isomerization of D-glucose gives D-fructose. Several billion pounds of high-fructose corn syrup are produced each year in this way for use by the food processing industry.

$$\text{Corn starch} \xrightarrow{\alpha\text{-Amylase}} \text{Dextrins} \xrightarrow{\text{Glucoamylase}} \text{D-Glucose} \xrightarrow[\text{isomerase}]{\text{Glucose}} \text{D-Fructose}$$

(A polysaccharide of α-D-glucose units)   Oligosaccharides of 6–10 α-D-glucose units

20.6 | What Are Acidic Polysaccharides?

Acidic polysaccharides are a group of polysaccharides that contain carboxyl groups and/or sulfuric ester groups. Acidic polysaccharides play important roles in the structure and function of connective tissues. Because they contain amino sugars, a more current name for these substances is glycosaminoglycans. There is no single general type of connective tissue. Rather, a large number of highly specialized forms exist, such as cartilage, bone, synovial fluid, skin, tendons, blood vessels, intervertebral disks, and cornea. Most connective tissues consist of collagen, a structural protein, combined with a variety of acidic polysaccharides (glycosaminoglycans) that interact with collagen to form tight or loose networks.

In rheumatoid arthritis, inflammation of the synovial tissue results in swelling of the joints.

A. Hyaluronic Acid

Hyaluronic acid is the simplest acidic polysaccharide present in connective tissue. It has a molecular weight of between 10^5 and 10^7 g/mol and contains from 300 to 100,000 repeating units, depending on the organ in which it occurs. It is most abundant in embryonic tissues and in specialized connective tissues such as synovial fluid, the lubricant of joints in the body, and the vitreous of the eye, where it provides a clear, elastic gel that holds the retina in its proper position. Hyaluronic acid is also a common ingredient in lotions, moisturizers, and cosmetics.

Hyaluronic acid is composed of D-glucuronic acid joined by a β-1,3-glycosidic bond to N-acetyl-D-glucosamine, which is in turn linked to D-glucuronic acid by a β-1,4-glycosidic bond.

The repeating unit of hyaluronic acid

B. Heparin

Heparin is a heterogeneous mixture of variably sulfonated polysaccharide chains, ranging in molecular weight from 6000 to 30,000 g/mol. This acidic polysaccharide is synthesized and stored in mast cells of various tissues—particularly the liver, lungs, and gut. Heparin has many biological functions, the best known and fully understood of which is its anticoagulant activity. It binds strongly to antithrombin III, a plasma protein involved in terminating the clotting process. A heparin preparation with good anticoagulant activity contains a minimum of eight repeating units (Figure 20.5). The larger the molecule, the greater its anticoagulant activity. Because of this anticoagulant activity, it is widely used in medicine.

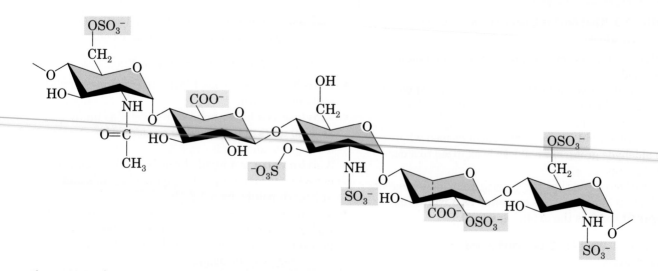

Figure 20.5 A repeating pentasaccharide unit of heparin.

SUMMARY OF KEY QUESTIONS

SECTION 20.1 Carbohydrates: What Are Monosaccharides?

- **Monosaccharides** are polyhydroxyaldehydes or polyhydroxyketones.
- The most common have the general formula $C_nH_{2n}O_n$, where n varies from 3 to 8.
- Their names contain the suffix *-ose,* and the prefixes *tri-, tetr-,* and so on indicate the number of carbon atoms in the chain. The prefix *aldo-* indicates an aldehyde, and the prefix *keto-* indicates a ketone.

SECTION 20.2 What Are the Cyclic Structures of Monosaccharides?

- In a **Fischer projection** of a carbohydrate, we write the carbon chain vertically with the most highly oxidized carbon toward the top. Horizontal lines represent groups projecting above the plane of the page; vertical lines represent groups projecting behind the plane of the page.
- The penultimate carbon of a monosaccharide is the next-to-last carbon of a Fischer projection.
- A monosaccharide that has the same configuration at the penultimate carbon as D-glyceraldehyde is called a **D-monosaccharide;** one that has the same configuration at the penultimate carbon as L-glyceraldehyde is called an **L-monosaccharide.**

- Monosaccharides exist primarily as cyclic hemiacetals.
- A six-membered cyclic hemiacetal is a **pyranose;** a five-membered cyclic hemiacetal is a **furanose.**
- The new stereocenter resulting from hemiacetal formation is called an **anomeric carbon,** and the stereoisomers formed in this way are called **anomers.**
- The symbol **β-** indicates that the —OH group on the anomeric carbon lies on the same side of the ring as the terminal —CH₂OH.
- The symbol **α-** indicates that the —OH group on the anomeric carbon lies on the opposite side from the terminal —CH₂OH.
- Furanoses and pyranoses can be drawn as **Haworth projections.**
- Pyranoses can also be drawn as **chair conformations.**
- **Mutarotation** is the change in specific rotation that accompanies formation of an equilibrium mixture of α and β anomers in aqueous solution.

SECTION 20.3 What Are the Characteristic Reactions of Monosaccharides?

- A **glycoside** is a cyclic acetal derived from a monosaccharide.
- An **alditol** is a polyhydroxy compound formed when the carbonyl group of a monosaccharide is reduced to a hydroxyl group.
- An **aldonic acid** is a carboxylic acid formed when the aldehyde group of an aldose is oxidized to a carboxyl group.
- Any carbohydrate that reacts with an oxidizing agent to form an aldonic acid is classified as a **reducing sugar** (it reduces the oxidizing agent).

SECTION 20.4 What Are Disaccharides and Oligosaccharides?

- A **disaccharide** contains two monosaccharide units joined by a glycosidic bond.
- Terms applied to carbohydrates containing larger numbers of monosaccharides are **trisaccharide, tetrasaccharide, oligosaccharide,** and **polysaccharide.**
- **Sucrose** is a disaccharide consisting of D-glucose joined to D-fructose by an α-1,2-glycosidic bond.
- **Lactose** is a disaccharide consisting of D-galactose joined to D-glucose by a β-1,4-glycosidic bond.

- **Maltose** is a disaccharide of two molecules of D-glucose joined by an α-1,4-glycosidic bond.

SECTION 20.5 What Are Polysaccharides?

- **Starch** can be separated into two fractions: amylose and amylopectin.
- **Amylose** is a linear polysaccharide of as many as 4000 units of D-glucopyranose joined by α-1,4-glycosidic bonds.
- **Amylopectin** is a highly branched polysaccharide of D-glucose joined by α-1,4-glycosidic bonds and, at branch points, by α-1,6-glycosidic bonds.
- **Glycogen,** the reserve carbohydrate of animals, is a highly branched polysaccharide of D-glucopyranose joined by α-1,4-glycosidic bonds and, at branch points, by α-1,6-glycosidic bonds.
- **Cellulose,** the skeletal polysaccharide of plants, is a linear polysaccharide of D-glucopyranose joined by β-1,4-glycosidic bonds.

SECTION 20.6 What Are Acidic Polysaccharides?

- The carboxyl and sulfate groups of **acidic polysaccharides** are ionized to —COO^- and —SO_3^- at the pH of body fluids, which gives these polysaccharides net negative charges.

SUMMARY OF KEY REACTIONS

1. **Formation of Cyclic Hemiacetals (Section 20.2)** A monosaccharide existing as a five-membered ring is a furanose; one existing as a six-membered ring is a pyranose. A pyranose is most commonly drawn as either a Haworth projection or a chair conformation.

β-D-Glucopyranose
$[\alpha]_D^{25} = +18.7°$

Open-chain form

D-Glucose

β-D-Glucopyranose
(β-D-Glucose)

α-D-Glucopyranose
$[\alpha]_D^{25} = 112°$

2. **Mutarotation (Section 20.2C)** Anomeric forms of a monosaccharide are in equilibrium in aqueous solution. Mutarotation is the change in specific rotation that accompanies this equilibration.

3. **Formation of Glycosides (Section 20.3A)** Treatment of a monosaccharide with an alcohol in the presence of an acid catalyst forms a cyclic acetal called a glycoside. The bond to the new —OR group is called a glycosidic bond.

+ CH₃OH

5. Oxidation to an Aldonic Acid (Section 20.3C) Oxidation of the aldehyde group of an aldose to a carboxyl group by a mild oxidizing agent gives a polyhydroxycarboxylic acid called an aldonic acid.

4. Reduction to Alditols (Section 20.3B) Reduction of the carbonyl group of an aldose or ketose to a hydroxyl group yields a polyhydroxy compound called an alditol.

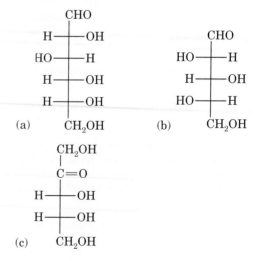

PROBLEMS

A blue problem number indicates an applied problem.

■ denotes problems that are available on the GOB ChemistryNow website or CD and are assignable in OWL.

SECTION 20.1 Carbohydrates: What Are Monosaccharides?

20.6 Define *carbohydrate*.

20.7 ■ What is the difference in structure between an aldose and a ketose? Between an aldopentose and a ketopentose?

20.8 Of the eight D-aldohexoses, which is the most abundant in the biological world?

20.9 Name the three most abundant hexoses in the biological world. Which are aldohexoses, and which are ketohexoses?

20.10 ■ Which hexose is also known as "dextrose"?

20.11 What does it mean to say that D- and L-glyceraldehyde are enantiomers?

20.12 Explain the meaning of the designations D and L as used to specify the configuration of a monosaccharide.

20.13 ■ Which carbon of an aldopentose determines whether the pentose has a D or L configuration?

20.14 How many stereocenters are present in D-glucose? In D-ribose? How many stereoisomers are possible for each monosaccharide?

20.15 ■ Which of the following compounds are D-monosaccharides, and which are L-monosaccharides?

(a)
```
      CHO
   H──┼──OH
  HO──┼──H
   H──┼──OH
   H──┼──OH
      CH₂OH
```

(b)
```
      CHO
  HO──┼──H
   H──┼──OH
  HO──┼──H
      CH₂OH
```

(c)
```
      CH₂OH
      │
      C═O
   H──┼──OH
   H──┼──OH
      CH₂OH
```

20.16 ■ Draw Fischer projections for L-ribose and L-arabinose.

20.17 Draw a Fischer projection for a D-2-ketoheptose.

20.18 Explain why all mono- and disaccharides are soluble in water.

20.19 What is an amino sugar? Name the three amino sugars most commonly found in nature.

SECTION 20.2 What Are the Cyclic Structures of Monosaccharides?

20.20 Define the term *anomeric carbon*. Which carbon is the anomeric carbon in glucose? In fructose?

20.21 Define (a) pyranose and (b) furanose.

20.22 Explain the conventions for using α and β to designate the configurations of the cyclic forms of monosaccharides.

20.23 Are α-D-glucose and β-D-glucose anomers? Explain. Are they enantiomers? Explain.

20.24 Are the hydroxyl groups on carbons 1, 2, 3, and 4 of α-D-glucose all in equatorial positions?

20.25 In what way are chair conformations a more accurate representation of the molecular shapes of hexopyranoses than Haworth projections?

20.26 ■ Convert each of the following Haworth projections to an open-chain form and to a Fischer projection. Name the monosaccharides you have drawn.

20.27 ■ Convert each of the following chair conformations to an open-chain form and to a Fischer projection. Name the monosaccharides you have drawn.

20.28 Explain the phenomenon of mutarotation. How is it detected?

20.29 The specific rotation of α-D-glucose is $+112.2°$. What is the specific rotation of α-L-glucose?

20.30 When α-D-glucose is dissolved in water, the specific rotation of the solution changes from $+112.2°$ to $+52.7°$. Does the specific rotation of α-L-glucose also change when it is dissolved in water? If so, to what value does it change?

20.31 Define the terms *glycoside* and *glycosidic bond*.

20.32 What is the difference in meaning between the terms *glycosidic bond* and *glucosidic bond*?

20.33 Do glycosides undergo mutarotation?

SECTION 20.3 What Are the Characteristic Reactions of Monosaccharides?

20.34 Draw Fischer projections for the product formed by treating each of the following monosaccharides with sodium borohydride, $NaBH_4$, in water.
(a) D-Galactose   (b) D-Ribose

20.35 ■ Reduction of D-glucose by $NaBH_4$ gives D-sorbitol, a compound used in the manufacture of sugar-free gums and candies. Draw a structural formula for D-sorbitol.

20.36 Reduction of D-fructose by $NaBH_4$ gives two alditols, one of which is D-sorbitol. Name and draw a structural formula for the other alditol.

20.37 Ribitol and β-D-ribose 1-phosphate are derivatives of D-ribose. Draw a structural formula for each compound.

SECTION 20.4 What Are Disaccharides and Oligosaccharides?

20.38 Name three important disaccharides. From which monosaccharides is each derived?

20.39 What does it mean to describe a glycosidic bond as β-1,4-? To describe it as α-1,6-?

20.40 Both maltose and lactose are reducing sugars, but sucrose is a nonreducing sugar. Explain why.

20.41 ■ Following is a structural formula for a disaccharide.

(a) Name the two monosaccharide units in the disaccharide.
(b) Describe the glycosidic bond.
(c) Is this disaccharide a reducing sugar or a nonreducing sugar?
(d) Will this disaccharide undergo mutarotation?

20.42 The disaccharide trehalose is found in young mushrooms and is the chief carbohydrate in the blood of certain insects.

Trehalose

(a) Name the two monosaccharide units in trehalose.

(b) Describe the glycosidic bond in trehalose.

(c) Is trehalose a reducing sugar or a nonreducing sugar?

(d) Will trehalose undergo mutarotation?

SECTION 20.5 What Are Polysaccharides?

20.43 What is the difference in structure between oligosaccharides and polysaccharides?

20.44 Name three polysaccharides that are composed of units of D-glucose. In which polysaccharide are the glucose units joined by α-glycosidic bonds? In which are they joined by β-glycosidic bonds?

20.45 Starch can be separated into two principal polysaccharides, amylose and amylopectin. What is the major difference in structure between the two?

20.46 Where is glycogen stored in the human body?

20.47 Why is cellulose insoluble in water?

20.48 How is it possible that cows can digest grass but humans cannot?

20.49 A Fischer projection of N-acetyl-D-glucosamine is given in Section 20.1D.

(a) Draw a Haworth projection and a chair conformation for the β-pyranose form of this monosaccharide.

(b) Draw a Haworth projection and a chair conformation for the disaccharide formed by joining two units of the pyranose form of N-acetyl-D-glucosamine by a β-1,4-glycosidic bond. If you draw them correctly, you will have a structural formula for the repeating dimer of chitin, the structural polysaccharide component of the shell of lobsters and other crustaceans.

20.50 Propose structural formulas for the repeating disaccharide unit in these polysaccharides.

(a) Alginic acid, isolated from seaweed, is used as a thickening agent in ice cream and other foods. Alginic acid is a polymer of D-mannuronic acid in the pyranose form joined by β-1,4-glycosidic bonds.

(b) Pectic acid is the main component of pectin, which is responsible for the formation of jellies from fruits and berries. Pectic acid is a polymer of D-galacturonic acid in the pyranose form joined by α-1,4-glycosidic bonds.

D-Mannuronic acid   D-Galacturonic acid

SECTION 20.6 What Are Acidic Polysaccharides?

20.51 Hyaluronic acid acts as a lubricant in the synovial fluid of joints. In rheumatoid arthritis, inflammation breaks hyaluronic acid down to smaller molecules. Under these conditions, what happens to the lubricating power of the synovial fluid?

20.52 ■ The anticlotting property of heparin is partly due to the negative charges it carries.

(a) Identify the functional groups that provide the negative charges.

(b) Which type of heparin is a better anticoagulant, one with a high degree or a low degree of polymerization?

Chemical Connections

20.53 (Chemical Connections 20A) Why does congenital galactosemia appear only in infants? Why can galactosemia be relieved by feeding an affected infant a formula containing sucrose as the only carbohydrate?

20.54 (Chemical Connections 20B) What is the difference in structure between L-ascorbic acid and L-dehydroascorbic acid? What does the designation L indicate in these names?

20.55 (Chemical Connections 20B) When L-ascorbic acid participates in a redox reaction, it is converted to L-dehydroascorbic acid. In this reaction, is L-ascorbic acid oxidized or reduced? Is L-ascorbic acid a biological oxidizing agent or a biological reducing agent?

20.56 (Chemical Connections 20C) Why is the glucose assay one of the most common analytical tests performed in clinical chemistry laboratories?

20.57 (Chemical Connections 20D) What monosaccharides do types A, B, and O blood have in common? In which monosaccharides do they differ?

20.58 (Chemical Connections 20D) L-Fucose is a monosaccharide unit common to A, B, AB, and O blood types.

(a) Is this monosaccharide an aldose or a ketose? A hexose or a pentose?

(b) What is unusual about this monosaccharide?

(c) If L-fucose were to undergo a reaction in which its terminal $-CH_3$ group was converted to a $-CH_2OH$ group, which monosaccharide would result?

20.59 (Chemical Connections 20D) Why can't a person with type A blood donate to a person with type B blood?

20.60 (Chemical Connections 20E) What is high-fructose corn syrup, and how is it produced?

Additional Problems

20.61 2,6-Dideoxy-D-altrose, also known as D-digitoxose, is a monosaccharide obtained from the hydrolysis of digitoxin, a natural product extracted from purple foxglove (*Digitalis purpurea*). Digitoxin has found wide use in cardiology because it reduces pulse rate, regularizes heart rhythm, and strengthens heart beat. Draw a structural formula for the open-chain form of 2,6-dideoxy-D-altrose.

20.62 In making candy or sugar syrups, sucrose is boiled in water with a little acid, such as lemon juice. Why does the product mixture taste sweeter than the initial sucrose solution?

20.63 Hot-water extracts of ground willow bark are an effective pain reliever (Chemical Connections 19C). Unfortunately, the liquid is so bitter that most persons refuse it. The pain reliever in these infusions is salicin. Name the monosaccharide unit in salicin.

Salicin

Looking Ahead

20.64 One pathway for the metabolism of D-glucose 6-phosphate is its enzyme-catalyzed conversion to D-fructose 6-phosphate. Show that this transformation can be regarded as two enzyme-catalyzed keto-enol tautomerizations (Section 17.5).

D-Glucose 6-phosphate   D-Fructose 6-phosphate

20.65 One step in glycolysis, the pathway that converts glucose to pyruvate (Section 28.2), involves an enzyme-catalyzed conversion of dihydroxyacetone phosphate to D-glyceraldehyde 3-phosphate. Show that this transformation can be regarded as two enzyme-catalyzed keto-enol tautomerizations (Section 17.5).

Dihydroxyacetone   D-Glyceraldehyde
phosphate   3-phosphate

20.66 Following is a Haworth projection and a chair conformation for the repeating disaccharide unit in chondroitin 6-sulfate. This biopolymer acts as the flexible connecting matrix between the tough protein filaments in cartilage. It is available as a dietary supplement, often combined with D-glucosamine sulfate. Some believe this combination can strengthen and improve joint flexibility.

(a) From what two monosaccharide units is the repeating disaccharide unit of chondroitin 6-sulfate derived?

(b) Describe the glycosidic bond between the two units.

20.67 Following is the structural formula of coenzyme A, an important biomolecule.

(a) Is coenzyme A chiral?

(b) Name each functional group in coenzyme A

(c) Would you expect coenzyme A to be soluble in water? Explain.

(d) Draw structural formulas for the products of complete hydrolysis of coenzyme A in aqueous HCl. Show each product as it would be ionized in this solution.

(e) Draw structural formulas for the products of complete hydrolysis of coenzyme A in aqueous NaOH. Show each product as it would be ionized in this solution.

Coenzyme A

CHAPTER 21

Lipids

KEY QUESTIONS

Sea lions are marine mammals that require a heavy layer of fat to survive in cold waters.

Chemistry Now™

Look for this logo in the chapter and go to GOB ChemistryNow at **http://now.brookscole.com/gob8** or on the CD for tutorials, simulations, and problems.

21.1 | What Are Lipids?

Found in living organisms, **lipids** are a family of substances that are insoluble in water but soluble in nonpolar solvents and solvents of low polarity, such as diethyl ether. Unlike the case with carbohydrates, we define lipids in terms of a property and not in terms of their structure.

A. Classification by Function

Lipids play three major roles in human biochemistry: (1) They store energy within fat cells, (2) they are parts of membranes that separate compartments of aqueous solutions from each other, and (3) they serve as chemical messengers.

Storage

An important use for lipids, especially in animals, is the storage of energy. As we saw in Section 20.5, plants store energy in the form of starch. Animals (including humans) find it more economical to use fats instead. Although our bodies do store some carbohydrates in the form of glycogen for quick energy when we need it, energy stored in the form of fats has much greater importance for us. The reason is simple: The burning of fats produces more than twice as much energy (about 9 kcal/g) as the burning of an equal weight of carbohydrates (about 4 kcal/g).

Membrane Components

The lack of water solubility of lipids is an important property because our body chemistry is so heavily based on water. Most body constituents, including carbohydrates and proteins, are soluble in water. However, the body also needs insoluble compounds for the membranes that separate compartments containing aqueous solutions, whether they are cells or organelles within the cells. Lipids provide these membranes. Their water insolubility derives from the fact that the polar groups they contain are much smaller than their alkane-like (nonpolar) portions. These nonpolar portions provide the water-repellent, or *hydrophobic,* property.

Messengers

Lipids also serve as chemical messengers. Primary messengers, such as steroid hormones, deliver signals from one part of the body to another part. Secondary messengers, such as prostaglandins and thromboxanes, mediate the hormonal response.

B. Classification by Structure

For purposes of study, we can classify lipids into four groups: (1) simple lipids, such as fats and waxes; (2) complex lipids; (3) steroids; and (4) prostaglandins, thromboxanes, and leukotrienes.

21.2 | What Are the Structures of Triglycerides?

Animal fats and plant oils are triglycerides. **Triglycerides** are triesters of glycerol and long-chain carboxylic acids called fatty acids. In Section 19.1, we saw that esters are made up of an alcohol part and an acid part. As the name indicates, the alcohol of triglycerides is always glycerol.

$$CH_2-OH$$
$$CH-OH$$
$$CH_2-OH$$

Glycerol

In contrast to the alcohol part, the acid component of triglycerides may be any number of fatty acids (Section 18.4A). These fatty acids do, however, have certain things in common:

1. Fatty acids are practically all unbranched carboxylic acids.
2. They range in size from about 10 to 20 carbons.
3. They contain an even number of carbon atoms.
4. Apart from the —COOH group, they have no functional groups, except that some do have double bonds.
5. In most fatty acids that have double bonds, the *cis* isomers predominate.

Only even-numbered acids are found in triglycerides because the body builds these acids entirely from acetate units and therefore puts the carbons in two at a time (Section 29.2).

In **triglycerides** (also called **triacylglycerols**), all three groups of glycerol are esterified. Thus a typical triglyceride molecule might be

A triglyceride

Triglycerides are the most common lipid materials, although **mono-** and **diglycerides** are not infrequent. In the latter two types, only one or two —OH groups of the glycerol are esterified by fatty acids.

Triglycerides are complex mixtures. Although some of the molecules contain three identical fatty acids, in most cases two or three different acids are present. The hydrophobic character of triglycerides is caused by the long

$$\overset{\text{O}}{\overset{\|}{}}$$

hydrocarbon chains. The ester groups (—C—O—), although polar themselves, are buried in a nonpolar environment, which makes the triglycerides insoluble in water.

21.3 | What Are Some Properties of Triglycerides?

A. Physical State

With some exceptions, **fats** that come from animals are generally solids at room temperature, and those from plants or fish are usually liquids. Liquid fats are often called **oils,** even though they are esters of glycerol just like solid fats and should not be confused with petroleum, which is mostly alkanes.

What is the structural difference between solid fats and liquid oils? In most cases, it is the degree of unsaturation. The physical properties of the fatty acids are carried over to the physical properties of the triglycerides. Solid animal fats contain mainly saturated fatty acids, whereas vegetable oils contain high amounts of unsaturated fatty acids. Table 21.1 shows the average fatty acid content of some common fats and oils. Note that even solid fats contain some unsaturated acids and that liquid fats contain some saturated acids. Some unsaturated fatty acids (linoleic and linolenic acids) are called *essential fatty acids* because the body cannot synthesize them from precursors; they must, therefore, be consumed as part of the diet.

Although most vegetable oils contain high amounts of unsaturated fatty acids, there are exceptions. Coconut oil, for example, has only a small amount of unsaturated acids. This oil is a liquid not because it contains many double bonds, but rather because it is rich in low-molecular-weight fatty acids (chiefly lauric acid).

Oils with an average of more than one double bond per fatty acid chain are called *polyunsaturated.* Their role in the human diet is discussed in Section 30.4.

Pure fats and oils are colorless, odorless, and tasteless. This statement may seem surprising because we all know the tastes and colors of such fats

Fat A mixture of triglycerides containing a high proportion of long-chain, saturated fatty acids

Oil A mixture of triglycerides containing a high proportion of long-chain, unsaturated fatty acids or short-chain, saturated fatty acids

Table 21.1 Average Percentage of Fatty Acids of Some Common Fats and Oils

	Saturated				Unsaturated			
	Lauric	Myristic	Palmitic	Stearic	Oleic	Linoleic	Linolenic	Other
Animal Fats								
Beef tallow	—	6.3	27.4	14.1	49.6	2.5	—	0.1
Butter	2.5	11.1	29.0	9.2	26.7	3.6	—	17.9
Human	—	2.7	24.0	8.4	46.9	10.2	—	7.8
Lard	—	1.3	28.3	11.9	47.5	6.0	—	5.0
Vegetable Oils								
Coconut	45.4	18.0	10.5	2.3	7.5	—	—	16.3
Corn	—	1.4	10.2	3.0	49.6	34.3	—	1.5
Cottonseed	—	1.4	23.4	1.1	22.9	47.8	—	3.4
Linseed	—	—	6.3	2.5	19.0	24.1	47.4	0.7
Olive	—	—	6.9	2.3	84.4	4.6	—	1.8
Peanut	—	—	8.3	3.1	56.0	26.0	—	6.6
Safflower	—	—	6.8	—	18.6	70.1	3.4	1.1
Soybean	0.2	0.1	9.8	2.4	28.9	52.3	3.6	2.7
Sunflower	—	—	5.6	2.2	25.1	66.2	—	—

and oils as butter and olive oil. The tastes, odors, and colors are caused by small amounts of other substances dissolved in the fat or oil.

B. Hydrogenation

In Section 12.6D, we learned that we can reduce carbon–carbon double bonds to single bonds by treating them with hydrogen and a catalyst. It is, therefore, not difficult to convert unsaturated liquid oils to solids. For example:

CHEMICAL CONNECTIONS 21B

Waxes

Plant and animal waxes are simple esters. They are solids because of their high molecular weights. As in fats, the acid portions of the esters consist of a mixture of fatty acids; the alcohol portions are not glycerol, however, but rather simple long-chain alcohols. For example, a major component of beeswax is 1-triacontyl palmitate:

Palmitic acid portion   O   1-Triacontanol portion

$$CH_3(CH_2)_{13}CH_2C{-}OCH_2(CH_2)_{28}CH_3$$
1-Triacontyl palmitate

Waxes generally have higher melting points than fats (60 to 100°C) and are harder. Animals and plants often use them for pro-

tective coatings. For example, the leaves of most plants are coated with wax, which helps to prevent microorganisms from attacking them and allows the plants to conserve water. The feathers of birds and the fur of animals are also coated with wax.

Important waxes include carnauba wax (from a Brazilian palm tree), lanolin (from lamb's wool), beeswax, and spermaceti (from whales). These substances are used to make cosmetics, polishes, candles, and ointments. Paraffin waxes are not esters, but rather mixtures of high-molecular-weight alkanes. Neither is ear wax a simple ester. This gland secretion contains a mixture of fats (triglycerides), phospholipids, and esters of cholesterol.

This hydrogenation is carried out on a large scale to produce the solid shortening sold in stores under such brand names as Crisco, Spry, and Dexo. In making such products, manufacturers must be careful not to hydrogenate all of the double bonds, because a fat with no double bonds would be *too* solid. Partial, but not complete, hydrogenation results in a product with just the right consistency for cooking. Margarine is also made by partial hydrogenation of vegetable oils. Because less hydrogen is used, margarine contains more unsaturation than fully hydrogenated shortenings. The hydrogenation process is the source of *trans* fatty acids, as we have already seen (Chemical Connections 18A). The food-processing industry is taking steps to address this issue.

Many common products contain hydrogenated vegetable oils.

C. Saponification

Glycerides, being esters, are subject to hydrolysis, which can be carried out with either acids or bases. As we saw in Section 19.4, the use of bases is more practical. An example of the saponification of a typical fat is

A triglyceride   1,2,3-Propanetriol (Glycerol; Glycerin)   Sodium soaps

Thus saponification is the base-promoted hydrolysis of fats and oils producing glycerol and a mixture of fatty acid salts called soaps. **Soap** has been used for thousands of years, and saponification is one of the oldest known chemical reactions.

21.4 What Are the Structures of Complex Lipids?

The triglycerides discussed in the previous sections are significant components of fat storage cells. Other kinds of lipids, called complex lipids, are important in a different way. They constitute the main components of

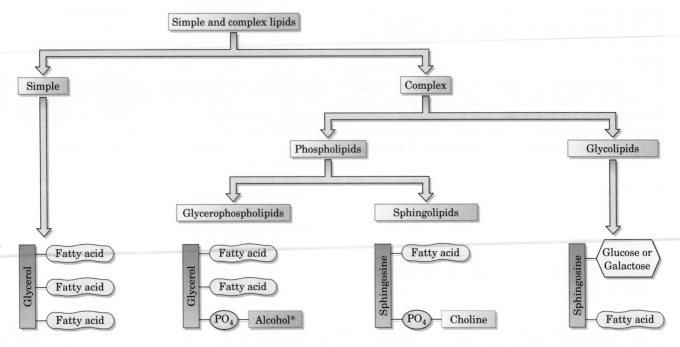

Figure 21.1 Schematic diagram of simple and complex lipids.*

*The alcohol can be choline, serine, ethanolamine, inositol, or certain others.

membranes (Section 21.5). Complex lipids can be classified into two groups: phospholipids and glycolipids.

Phospholipids contain an alcohol, two fatty acids, and a phosphate group. There are two types: **glycerophospholipids** and **sphingolipids.** In glycerophospholipids, the alcohol is glycerol (Section 21.6). In sphingolipids, the alcohol is sphingosine (Section 21.7).

Glycolipids are complex lipids that contain carbohydrates (Section 21.8). Figure 21.1 shows schematic structures for all of these lipids.

21.5 | What Role Do Lipids Play in the Structure of Membranes?

The complex lipids mentioned in Section 21.4 form the **membranes** around body cells and around small structures inside the cells. (These small structures inside the cell are called *organelles.*) Unsaturated fatty acids are important components of these lipids. Most lipid molecules in the bilayer contain at least one unsaturated fatty acid. The cell membranes separate cells from the external environment and provide selective transport for nutrients and waste products into and out of cells.

These membranes are made of **lipid bilayers** (Figure 21.2). In a lipid bilayer, two rows (layers) of complex lipid molecules are arranged tail to tail. The hydrophobic tails point toward each other, which enables them to get as far away from the water as possible. This arrangement leaves the hydrophilic heads projecting to the inner and outer surfaces of the membrane. Cholesterol (Section 21.9), another membrane component, also positions the hydrophilic portion of the molecule on the surface of the membranes and the hydrophobic portion inside the bilayer.

The unsaturated fatty acids prevent the tight packing of the hydrophobic chains in the lipid bilayer, thereby providing a liquid-like character to the membranes. This effect is similar to the one that causes unsaturated

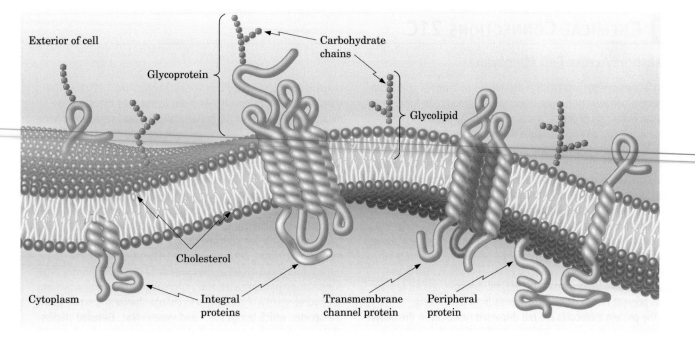

Figure 21.2 The fluid mosaic model of membranes. Note that proteins are embedded in the lipid matrix.

fatty acids to have lower melting points than saturated fatty acids. This property of membrane fluidity is of extreme importance because many products of the body's biochemical processes must cross the membrane, and the liquid nature of the lipid bilayer allows such transport.

The lipid part of the membrane serves as a barrier against any movement of ions or polar compounds into and out of the cells. In the lipid bilayer, protein molecules are either suspended on the surface (peripheral proteins) or partly or fully embedded in the bilayer (integral proteins). These proteins stick out either on the inside or on the outside of the membrane. Others are thoroughly embedded, going through the bilayer and projecting from both sides. The model shown in Figure 21.2, called the **fluid mosaic model** of membranes, allows the passage of nonpolar compounds by diffusion, as these compounds are soluble in the lipid membranes. The term *mosaic* refers to the topography of the bilayers: protein molecules dispersed in the lipid. The term *fluid* is used because the free lateral motion in the bilayers makes it liquid-like. In contrast, polar compounds are transported either via specific channels through the protein regions or by a mechanism called active transport (see Chemical Connections 21C). For any transport process, the membrane must behave like a nonrigid liquid so that the proteins can move sideways within the membrane.

21.6 | What Are Glycerophospholipids?

The structure of glycerophospholipids (also called phosphoglycerides) is very similar to that of fats. Glycerophospholipids are membrane components of cells throughout the body. The alcohol is glycerol. Two of the three groups are esterified by fatty acids. As with the simple fats, these fatty acids may be any long-chain carboxylic acids, with or without double bonds. In all glycerophospholipids, lecithins, cephalins, and phosphatidylinositols, the fatty acid on carbon 2 of glycerol is always unsaturated. The third group is

CHEMICAL CONNECTIONS 21C

Transport Across Cell Membranes

Membranes are not just random assemblies of complex lipids that provide a nondescript barrier. In human red blood cells, for example, the outer part of the bilayer is made largely of phosphatidylcholine and sphingomyelin, while the inner part consists mostly of phosphatidylethanolamine and phosphatidylserine (Sections 21.6 and 21.7). In another example, in the membrane called sarcoplasmic reticulum in the heart muscles, phosphatidylethanolamine is found in the outer part of the membrane, phosphatidylserine is found in the inner part, and phosphatidylcholine is equally distributed in the two layers of the membrane.

Membranes are not static structures, either. In many processes, they fuse with one another; in other processes, they disintegrate and their building blocks are used elsewhere. When membranes fuse in vacuole fusions inside of cells, for example, certain restrictions prevent incompatible membranes from intermixing.

The protein molecules are not dispersed randomly in the bilayer. Sometimes they cluster in patches; at other times they appear in regular geometric patterns. An example of the latter are **gap junctions,** channels made of six proteins that create a central pore. These channels allow neighboring cells to communicate. Gap junctions are an example of **passive transport.** Small polar molecules—which include such essential nutrients as inorganic ions, sugars, amino acids, and nucleotides—can readily pass through gap junctions. Large molecules, such as proteins, polysaccharides, and nucleic acids, cannot.

In **facilitated transport,** a specific interaction takes place between the transporter and the transported molecule. Consider the **anion transporter** of the red blood cells, through which chloride and bicarbonate ions are exchanged in a 1 : 1 ratio. The transporter is a protein with 14 helical structures that span the membrane. One side of the helices contains the hydrophobic parts of the protein, which can interact with the lipid membrane. The other side of the helices forms a channel. This channel contains the hydrophilic portions of the protein, which can interact with the hydrated ions. In this manner, the anion passes through the erythrocyte membrane.

Active transport involves the passage of ions through a concentration gradient. For example, a higher concentration of K^+ is found inside cells than outside the cells in the surrounding environment. Nevertheless, potassium ions can be transported from the outside into a cell, albeit at the expense of energy. The transporter, a membrane protein called Na^+, K^+, ATPase, uses the energy from the hydrolysis of ATP molecules to change the conformation of the transporter, which brings in K^+ and exports Na^+. Detailed studies of K^+ ion channels have revealed that K^+ ions enter the channel in pairs. Each of the hydrated ions carries eight water molecules in its solvation layers, with the negative pole of the water molecules (the oxygen atoms) surrounding the positive ion. Deeper in the channel, K^+ ions encounter a constriction, called a selectivity filter. To pass through, the K^+ ions must shed their water molecules. The pairing of the naked K^+ ions generates enough electrostatic repulsion to keep the ions moving through the channel. The channel itself is lined with oxygen atoms to provide an attractive environment similar to that offered by the stable hydrated form before entry.

Polar compounds, in general, are transported through specific **transmembrane channels.**

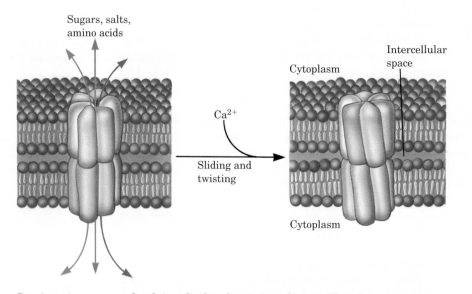

Gap junctions are made of six cylindrical protein subunits. They line up in two plasma membranes parallel to each other, forming a pore. The pores of gap junctions are closed by a sliding and twisting motion of the cylindrical subunits.

esterified not by a fatty acid, but rather by a phosphate group, which is also esterified to another alcohol. If the other alcohol is choline, a quaternary ammonium compound, the glycerophospholipids are called **phosphatidyl-cholines** (common name **lecithin**):

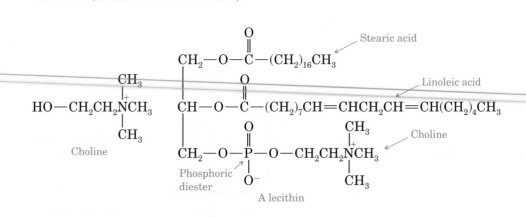

This typical lecithin molecule has stearic acid on one end and linoleic acid in the middle. Other lecithin molecules contain other fatty acids, but the one on the end is always saturated and the one in the middle is always unsaturated. Lecithin is a major component of egg yolk. Because it includes both polar and nonpolar portions within one molecule, it is an excellent emulsifier (see Chemical Connections 7D) and is used in mayonnaise.

Note that lecithin has a negatively charged phosphate group and a positively charged quaternary nitrogen from the choline. These charged parts of the molecule provide a strongly hydrophilic head, whereas the rest of the molecule is hydrophobic. Thus, when a phospholipid such as lecithin is part of a lipid bilayer, the hydrophobic tail points toward the middle of the bilayer, and the hydrophilic heads line both the inner and outer surfaces of the membranes (Figures 21.2 and 21.3).

Lecithins are just one example of glycerophospholipids. Another is the **cephalins,** which are similar to the lecithins in every way except that, instead of choline, they contain other alcohols, such as ethanolamine or serine:

Figure 21.3 Space-filling molecular models of complex lipids in a bilayer.

Another important group of glycerophospholipids is the **phosphatidyli-nositols (PI).** In PI, the alcohol inositol is bonded to the rest of the molecule by a phosphate ester bond. Such compounds not only are integral structural parts of the biological membranes, but also, in their higher phosphorylated form, such as **phosphatidylinositol 4,5-bisphosphates (PIP2),** serve as signaling molecules in chemical communication (see Chapter 24).

Phosphatidylinositols, PI

21.7 | What Are Sphingolipids?

Myelin, the coating of nerve axons, contains a different kind of complex lipid: **sphingolipids.** In sphingolipids, the alcohol portion is sphingosine:

Sphingosine

A long-chain fatty acid is connected to the —NH₂ group by an amide bond, and the —OH group at the end of the chain is esterified by phosphorylcholine:

A sphingomyelin
(a sphingolipid)

Sphingomyelin
(schematic diagram)

The combination of a fatty acid and sphingosine (shown in color) is called the **ceramide** portion of the molecule, because many of these compounds are also found in cerebrosides (Section 21.8). The ceramide part of complex lipids may contain different fatty acids. Stearic acid, for example, occurs mainly in sphingomyelin.

Sphingomyelins are the most important lipids in the myelin sheaths of nerve cells and are associated with diseases such as multiple sclerosis (Chemical Connections 21D). Sphingolipids are not randomly distributed in membranes. In viral membranes, for example, most of the sphingomyelin appears on the inside of the membrane. Johann Thudichum, who discovered sphingolipids in 1874, named these brain lipids after a monster of Greek mythology, the sphinx. Part woman and part winged lion, the sphinx devoured all who could not provide the correct answer to her riddles. Sphingolipids appeared to Thudichum as part of a dangerous riddle of the brain.

21.8 | What Are Glycolipids?

Glycolipids are complex lipids that contain carbohydrates and ceramides. One group, the **cerebrosides,** consists of ceramide mono- or oligosaccharides. Other groups, such as the **gangliosides,** contain a more complex carbohydrate structure (see Chemical Connections 21E). In cerebrosides, the fatty acid of the ceramide part may contain either 18-carbon or 24-carbon chains; the latter form is found only in these complex lipids. A glucose or galactose carbohydrate unit forms a beta-glycosidic bond with the ceramide portion of the molecule. The cerebrosides occur primarily in the brain (accounting for 7% of the brain's dry weight) and at nerve synapses.

Glucocerebroside

EXAMPLE 21.1

A lipid isolated from the membrane of red blood cells had the following structure:

$$CH_2-O-\overset{\overset{\displaystyle O}{\|}}{C}-(CH_2)_{14}CH_3$$
$$CH-O-\overset{\overset{\displaystyle O}{\|}}{C}-(CH_2)_7CH=CH(CH_2)_7CH_3$$
$$CH_2-O-\overset{\overset{\displaystyle O}{\|}}{P}-O-CH_2CH_2NH_3{}^+$$
$$\underset{\displaystyle O^-}{|}$$

(a) To what group of complex lipids does this compound belong?
(b) What are the components?

CHEMICAL CONNECTIONS 21D

The Myelin Sheath and Multiple Sclerosis

The human brain and spinal cord can be divided into gray and white regions. Forty percent of the human brain is white matter. Microscopic examination reveals that the white matter consists of nerve axons wrapped in a white lipid coating, called the **myelin sheath,** which provides insulation and allows the rapid conduction of electrical signals. The myelin sheath consists of 70% lipids and 30% proteins in the usual lipid bilayer structure.

Specialized cells, called **Schwann cells,** wrap themselves around the peripheral nerve axons to form numerous concentric layers. In the brain, other cells perform the wrapping in a similar manner.

Multiple sclerosis affects 250,000 people in the United States. In this disease, the myelin sheath gradually deteriorates. Symptoms, which include muscle weariness, lack of coordination, and loss of vision, may vanish for a time but later return with greater severity. Autopsy of multiple sclerotic brains shows scar-like plaques of white matter, with bare axons not covered by myelin sheaths. The symptoms occur because the demyelinated axons cannot conduct nerve impulses. A secondary effect of the demyelination is damage to the axon itself.

Similar demyelination occurs in the Guillain-Barré syndrome that follows certain viral infections. In 1976, fears of a "swine flu" epidemic prompted a vaccination program that precipitated a number of cases of Guillain-Barré syndrome. This disease can lead to paralysis, which can cause death unless artificial breathing is sup-

Myelination of a nerve axon outside the brain by a Schwann cell. The myelin sheath is produced by the Schwann cell and is rolled around the nerve axon for insulation.

plied. In the 1976 cases, the U.S. government assumed legal responsibility for the few bad vaccines and paid compensation to the victims and their families.

Solution

(a) The molecule is a triester of glycerol and contains a phosphate group; therefore, it is a glycerophospholipid.

(b) Besides glycerol and phosphate, it has palmitic acid and oleic acid components. The other alcohol is ethanolamine. Therefore, it belongs to the subgroup of cephalins.

Problem 21.1

A complex lipid had the following structure:

(a) To what group of complex lipids does this compound belong?

(b) What are the components?

21.9 | What Are Steroids?

The third major class of lipids is the **steroids,** which are compounds containing the following ring system:

In this structure, three cyclohexane rings (A, B, and C) are connected in the same way as in phenanthrene (Section 13.2D); a fused cyclopentane ring (D) is also present. Steroids are thus completely different in structure from the lipids already discussed. Note that they are not necessarily esters, although some of them are.

A. Cholesterol

The most abundant steroid in the human body, and the most important, is **cholesterol:**

Cholesterol

Cholesterol serves as a plasma membrane component in all animal cells—for example, in red blood cells. Its second important function is to serve as a raw material for the synthesis of other steroids, such as the sex and adrenocorticoid hormones (Section 21.10) and bile salts (Section 21.11).

Cholesterol exists both in the free form and esterified with fatty acids. Gallstones contain free cholesterol (Figure 21.4).

Because the correlation between high serum cholesterol levels and such diseases as atherosclerosis has received so much publicity, many people are afraid of cholesterol and regard it as some kind of poison. Far from being poisonous, cholesterol is, in fact, necessary for human life. In essence, our livers manufacture cholesterol that satisfies our needs even without dietary intake. When the cholesterol level exceeds 150 mg/100 mL, cholesterol synthesis in the liver is reduced to half the normal rate of production. The amount of cholesterol is regulated, and it is an excess—rather than the presence—of cholesterol that is associated with disease.

Cholesterol in the body is in a dynamic state. It constantly circulates in the blood. Cholesterol and esters of cholesterol, being hydrophobic, need a water-soluble carrier to circulate in the aqueous medium of blood.

B. Lipoproteins: Carriers of Cholesterol

Cholesterol, along with fat, is transported by **lipoproteins.** Most lipoproteins contain a core of hydrophobic lipid molecules surrounded by a shell of

Figure 21.4 A human gallstone is almost pure cholesterol. This gallstone measures 5 mm in diameter.

© Carolina Biological Supply Company/Phototake, NYC

Lipoproteins Spherically shaped clusters containing both lipid molecules and protein molecules

CHEMICAL CONNECTIONS 21E

Lipid Storage Diseases

Complex lipids are constantly being synthesized and decomposed in the body. In several genetic diseases classified as lipid storage diseases, some of the enzymes needed to decompose the complex lipids are defective or missing. As a consequence, the complex lipids accumulate and cause an enlarged liver and spleen, mental retardation, blindness, and, in certain cases, early death. The table summarizes some of these diseases and indicates the missing enzyme and the accumulating complex lipid in each.

At present, no treatment is available for these diseases. The best way to prevent them is by genetic counseling. Some of the diseases can be diagnosed during fetal development. For example, Tay-Sachs disease, which affects about 1 in every 30 Jewish Americans (versus 1 in 300 in the non-Jewish population), can be diagnosed from amniotic fluid obtained by amniocentesis.

The accumulation of glucocerebrosides in the cell of a patient with Gaucher's disease. These Gaucher cells infiltrate the bone marrow.

Table 21E   Lipid Storage Diseases

Name	Accumulating Lipid	Missing or Defective Enzyme
Gaucher's disease	Glucocerebroside	β-Glucosidase
Krabbe's leukodystrophy	Galactocerebroside	β-Galactosidase
Fabry's disease	Ceramide trihexoside	α-Galactosidase

Lipid Storage Diseases *(continued)*

Table 21E Lipid Storage Diseases

Name	Accumulating Lipid	Missing or Defective Enzyme

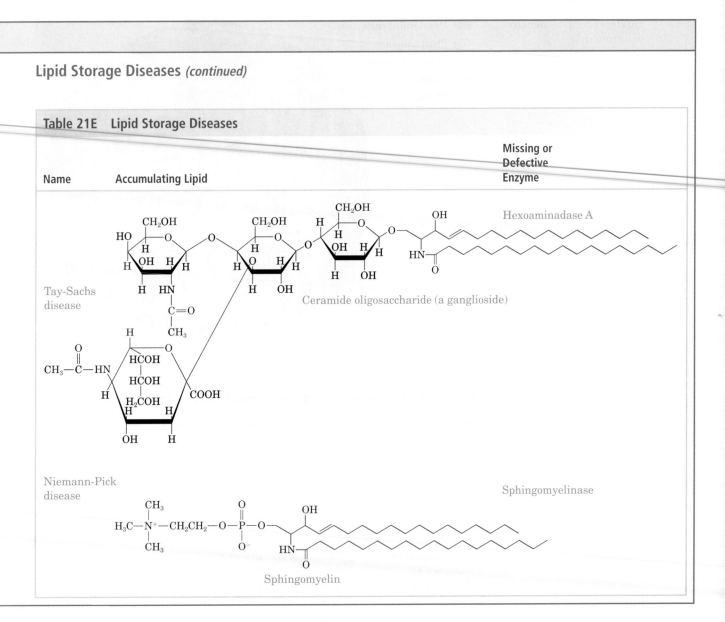

Tay-Sachs disease

Ceramide oligosaccharide (a ganglioside)

Hexoaminadase A

Niemann-Pick disease

Sphingomyelin

Sphingomyelinase

hydrophilic molecules such as proteins and phospholipids (Figure 21.5). As summarized in Table 21.2, there are four kinds of lipoproteins:

- **High-density lipoprotein (HDL) ("good cholesterol"),** which consists of about 33% protein and about 30% cholesterol
- **Low-density lipoprotein (LDL) ("bad cholesterol"),** which contains only 25% protein but 50% cholesterol
- **Very-low-density lipoprotein (VLDL),** which mostly carries triglycerides (fats) synthesized by the liver
- **Chylomicrons,** which carry dietary lipids synthesized in the intestines

C. Transport of Cholesterol in LDL

The transport of cholesterol from the liver starts out as a large VLDL particle (55 nanometers in diameter). The core of VLDL contains triglycerides and cholesteryl esters, mainly cholesteryl linoleate. It is surrounded by a

Figure 21.5 Low-density lipoprotein.

Unesterified cholesterol
Phospholipid
Cholesteryl ester
Protein
Hydrophobic tails

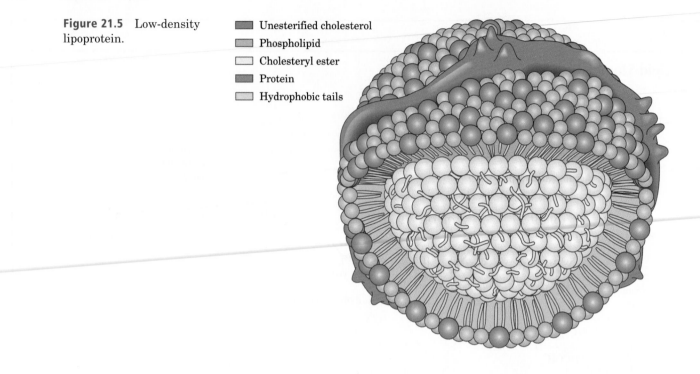

Table 21.2 Compositions and Properties of Human Lipoproteins

Property	HDL	LDL	VLDL	Chylomicrons
Core				
Cholesterol and cholesteryl esters (%)	30	50	22	8
Triglycerides (%)	8	4	50	84
Surface				
Phospholipids (%)	29	21	18	7
Proteins (%)	33	25	10	1–2
Density (g/mL)	1.05–1.21	1.02–1.06	0.95–1.00	<0.95
Diameter (nm)	5–15	18–28	30–80	100–500

Percentages are given as % dry weight.

polar coat of phospholipids and proteins (Figure 21.5). The VLDL is carried in the serum. When the capillaries reach muscle or fat tissues, the triglycerides and all proteins except apoB-100 are removed from the VLDL. At this point, the diameter of the lipoprotein shrinks to 22 nanometers and its core contains only cholesteryl esters. Because of the removal of fat, its density increases and it becomes LDL. Low-density lipoprotein stays in the plasma for about 2.5 days.

The LDL carries cholesterol to the cells, where specific LDL-receptor molecules line the cell surface in certain concentrated areas called **coated pits.** The apoB-100 protein on the surface of the LDL binds specifically to the LDL-receptor molecules in the coated pits. After such binding, the LDL is taken inside the cell (endocytosis), where enzymes break down the lipoprotein. In the process, they liberate free cholesterol from cholesteryl esters. In this manner, the cell can, for example, use cholesterol as a component of a membrane. This is the normal fate of LDL and the normal course of cholesterol transport. Michael Brown and Joseph Goldstein of the University of Texas shared the Nobel Prize in medicine in 1986 for the discovery of the LDL-receptor–mediated pathway. If the LDL receptors are not sufficient in number, cholesterol accumulates in the blood; this accumulation can happen even with low intake of dietary cholesterol. Both genetics and diet play a role in determining cholesterol levels in the blood.

D. Transport of Cholesterol in HDL

High-density lipoprotein transports cholesterol from peripheral tissues to the liver and transfers cholesterol to LDL. While in the serum, the free cholesterols in HDL are converted to cholesteryl esters. These esterified cholesterols are delivered to the liver for synthesis of bile acids and steroid hormones. The cholesterol uptake from HDL differs from that noted with LDL. This process does not involve endocytosis and degradation of the lipoprotein particle. Instead, in a selective lipid uptake, the HDL binds to the liver cell surface and transfers its cholesteryl ester to the cell. The HDL, depleted from its lipid content, then reenters the circulation. It is desirable to have a high level of HDL in the blood because of the way it removes cholesterol from the bloodstream.

E. Levels of LDL and HDL

Like all lipids, cholesterol is insoluble in water. If its level is elevated in the blood serum, plaque-like deposits may form on the inner surfaces of the arteries. The resulting decrease in the diameter of the blood vessels may, in turn, decrease the flow of blood. This **atherosclerosis,** along with accompanying high blood pressure, may lead to heart attack, stroke, or kidney dysfunction.

Atherosclerosis may exacerbate the blockage of some arteries by a clot at the point where the arteries are constricted by plaque. Furthermore, blockage may deprive cells of oxygen, causing them to cease to function. The death of heart muscles due to lack of oxygen is called *myocardial infarction.*

Most cholesterol is transported by low-density lipoproteins. If a sufficient number of LDL receptors are found on the surface of the cells, LDL is effectively removed from the circulation and its concentration in the blood plasma drops. The number of LDL receptors on the surface of cells is controlled by a feedback mechanism (see Section 23.6). That is, when the concentration of cholesterol molecules inside the cells is high, the synthesis of the LDL receptor is suppressed. As a consequence, less LDL is taken into the cells from the plasma and the LDL concentration in the plasma rises. Conversely, when the cholesterol level inside the cells is low, the synthesis of the LDL receptor increases. As a consequence, the LDL is taken up more rapidly and its level in the plasma falls.

In certain cases, however, there are not enough LDL receptors. In the disease called *familial hypercholesterolemia,* the cholesterol level in the plasma may be as high as 680 mg/100 mL, compared to 175 mg/100 mL in normal subjects. These high levels of cholesterol can lead to premature atherosclerosis and heart attack. The high plasma cholesterol levels in affected patients occur because the body lacks enough functional LDL receptors or, if enough are produced, they are not concentrated in the coated pits.

In general, high LDL content means high cholesterol content in the plasma because LDL cannot enter the cells and be metabolized. Therefore, a high LDL level combined with a low HDL level is a symptom of faulty cholesterol transport and a warning for possible atherosclerosis.

The serum cholesterol level controls the amount of cholesterol synthesized by the liver. When serum cholesterol is high, synthesis is at a low level. Conversely, when the serum cholesterol level is low, synthesis of cholesterol increases.

Diets low in cholesterol and saturated fatty acids usually reduce the serum cholesterol level, and a number of drugs can inhibit the synthesis of cholesterol in the liver. Commonly used statin drugs, such as atorvastatin (Lipitor) and simvastatin (Zocor), inhibit one of the key enzymes (biocatalysts) in cholesterol synthesis, HMG-CoA reductase (Section 29.4). In this way, they block the synthesis of cholesterol inside the cells and stimulate the synthesis of LDL-receptor proteins. More LDL then enters the cells, diminishing the amount of cholesterol that will be deposited on the inner walls of arteries.

It is generally considered desirable to have high levels of HDL and low levels of LDL in the bloodstream. High-density lipoproteins carry cholesterol from plaques deposited in the arteries to the liver, thereby reducing the risk of atherosclerosis. Premenopausal women have more HDL than men, which is why women have a lower risk of coronary heart disease. HDL levels can be increased by exercise and weight loss.

21.10 | What Are Some of the Physiological Roles of Steroid Hormones?

Cholesterol is the starting material for the synthesis of steroid hormones. In this process, the aliphatic side chain on the D ring is shortened by the removal of a six-carbon unit, and the secondary alcohol group on carbon 3 is oxidized to a ketone. The resulting molecule, *progesterone,* serves as the starting compound for both the sex hormones and the adrenocorticoid hormones (Figure 21.6).

A. Adrenocorticoid Hormones

The adrenocorticoid hormones (Figure 21.6) are products of the adrenal glands. The term *adrenal* means "adjacent to the renal" (which refers to the kidney). We classify these hormones into two groups according to function: *Mineralocorticoids* regulate the concentrations of ions (mainly Na^+ and K^+), and *glucocorticoids* control carbohydrate metabolism. The term *corticoid* indicates that the site of the secretion is the cortex (outer part) of the gland.

Aldosterone is one of the most important mineralocorticoids. Increased secretion of aldosterone enhances the reabsorption of Na^+ and Cl^- ions in the kidney tubules and increases the loss of K^+. Because Na^+ concentration controls water retention in the tissues, aldosterone controls tissue swelling.

Cortisol is the major glucocorticoid. Its function is to increase the glucose and glycogen concentrations in the body. This accumulation occurs at the expense of other nutrients. Fatty acids from fat storage cells and amino acids from body proteins are transported to the liver, which, under the influence of cortisol, manufactures glucose and glycogen from these sources.

Cortisol and its ketone derivative, *cortisone,* have remarkable anti-inflammatory effects. These or similar synthetic derivatives, such as prednisolone, are used to treat inflammatory diseases of many organs, rheumatoid arthritis, and bronchial asthma.

B. Sex Hormones

The most important male sex hormone is testosterone (Figure 21.6). This hormone, which promotes the normal growth of the male genital organs, is synthesized in the testes from cholesterol. During puberty, increased testos-

Figure 21.6   The biosynthesis of hormones from progesterone.

terone production leads to such secondary male sexual characteristics as deep voice and facial and body hair.

Female sex hormones, the most important of which is estradiol (Figure 21.6), are synthesized from the corresponding male hormone (testosterone) by aromatization of the A ring:

Estradiol, together with its precursor progesterone, regulates the cyclic changes occurring in the uterus and ovaries known as the *menstrual cycle.* As the cycle begins, the level of estradiol in the body rises, which in turn causes the lining of the uterus to thicken. Another hormone, called luteinizing

CHEMICAL CONNECTIONS 21F

Anabolic Steroids

Testosterone, the principal male hormone, is responsible for the buildup of muscles in men. Recognizing this fact, many athletes have taken this drug in an effort to increase their muscular development. The practice is especially common among athletes in sports in which strength and muscle mass are important, including weight lifting, shot put, and hammer throw. Participants in other sports, such as running, swimming, and cycling, would also like larger and stronger muscles.

Although used by many athletes, testosterone has two disadvantages:

1. Besides its effect on muscles, it affects secondary sexual characteristics, and too much of it can result in undesired side effects.
2. It is not very effective when taken orally and must be injected to achieve the best results.

For these reasons, a large number of other anabolic steroids, all of them synthetic, have been developed. Examples include the following compounds:

Methandienone

Methenolone

Nandrolone decanoate

Former home run champion Mark McGwire used androstenedione, a muscle-building dietary supplement that is allowed in baseball, but is banned in professional football, college athletics, and the Olympic Games.

4-Androstene-3,17-dione

Some women athletes use anabolic steroids, just as their male counterparts do. Because their bodies produce only small amounts of testosterone, women actually have much more to gain from anabolic steroids than men.

Another way to increase testosterone concentration is to use prohormones, which the body converts to testosterone. One such prohormone is 4-androstenedione, or "andro." Some athletes have used it to enhance performance.

The use of anabolic steroids is forbidden in many sporting events, especially in international competition, largely for two reasons: (1) It gives some competitors an unfair advantage, and (2) these drugs can have many unwanted and even dangerous side effects, ranging from acne to liver tumors. Side effects can be especially disadvantageous for women; they can include growth of facial hair, baldness, deepening of the voice, and menstrual irregularities.

All athletes participating in the Olympic Games are required to pass a urine test for anabolic steroids. A number of medal-winning athletes have had their victories taken away because they tested positive for steroid use. For example, the Canadian Ben Johnson, a world-class sprinter, was stripped of both his world record and his gold medal in the 1988 Olympiad. A positive test for andro resulted in the U.S. shot put champion Randy Barnes being banned from competition in 1998. Prohormones such as andro are not listed under the Anabolic Steroid Act of 1990; hence, their nonmedical use is not a federal offense, as is the case with anabolic steroids. Mark McGwire hit his record-breaking home runs in 1998 while taking andro, because baseball rules did not prohibit its usage. Even so, the International Olympic Committee has banned the use of both prohormones and anabolic steroids.

hormone (LH), then triggers ovulation. If the ovum is fertilized, increased progesterone levels will inhibit any further ovulation. Both estradiol and progesterone promote further preparation of the uterine lining to receive the fertilized ovum. If no fertilization takes place, progesterone production stops altogether, and estradiol production decreases. This halt decreases the thickening of the uterine lining, which is sloughed off with accompanying bleeding during menstruation (Figure 21.7).

Because progesterone is essential for the implantation of the fertilized ovum, blocking its action leads to termination of pregnancy (see Chemical

Figure 21.7 Events of the menstrual cycle. (*a*) Levels of sex hormones in the bloodstream during the phases of one menstrual cycle in which pregnancy does not occur. (*b*) Development of an ovarian follicle during the cycle. (*c*) Phases of development of the endometrium, the lining of the uterus. The endometrium thickens during the proliferative phase. In the secretory phase, which follows ovulation, the endometrium continues to thicken and the glands secrete a glycogen-rich nutritive material in preparation to receive an embryo. If no embryo is implanted, the new outer layers of the endometrium disintegrate and the blood vessels rupture, producing the menstrual flow.

CHEMICAL CONNECTIONS 21G

Oral Contraception

Because progesterone prevents ovulation during pregnancy, it occurred to investigators that progesterone-like compounds might be used for birth control. Synthetic analogs of progesterone proved to be more effective than the natural compound. In "the Pill," a synthetic progesterone-like compound is supplied together with an estradiol-like compound (the latter prevents irregular menstrual flow). Triple-bond derivatives of testosterone, such as norethindrone, norethynodrel, and ethynodiol diacetate, are used most often in birth-control pills:

Norethindone   Norethynodrel   Ethynodiol diacetate

Connections 21G). Progesterone interacts with a receptor (a protein molecule) in the nucleus of cells. The receptor changes its shape when progesterone binds to it (see Section 24.7).

A drug, now widely used in France and China, called mifepristone or RU486 acts as a competitor to progesterone:

Mifepristone
(RU486)

Mifepristone blocks the action of progesterone by binding to the same receptor sites. Because the progesterone molecule cannot reach the receptor molecule, the uterus is not prepared for the implantation of the fertilized ovum, and the ovum is aborted. Once pregnancy has been established, RU486 can be taken up through 49 days of gestation. This chemical form of abortion has been approved by the U.S. Food and Drug Administration (FDA) and in recent years has found clinical application as a supplement to surgical abortion. RU486 binds to the receptors of glucocorticoid hormones as well. Its use as an antiglucocorticoid is also recommended to alleviate a disease known as Cushing syndrome, involving the overproduction of cortisone.

A completely different approach is the "morning after pill," which can be taken orally up to 72 hours after unprotected intercourse. The "morning after pill" is not an abortion pill, because it acts before pregnancy takes place. Actually, the components of the pill are regular contraceptives. Two kinds are on the market as prescription drugs: a progesterone-like compound, called levonogestrel; and a combination of levonogestrel and ethynil estradiol marketed as Preven.

Estradiol and progesterone also regulate secondary female sex characteristics, such as the growth of breasts. Thanks to this property, RU486, as an antiprogesterone, has been reported to be effective against certain types of breast cancer.

Testosterone and estradiol are not exclusive to either males or females. A small amount of estradiol production occurs in males, and a small amount of testosterone production is normal in females. Only when the proportion of these two hormones (hormonal balance) becomes upset can one observe symptoms of abnormal sexual differentiation.

21.11 | What Are Bile Salts?

Bile salts are oxidation products of cholesterol. First the cholesterol is oxidized to the trihydroxy derivative, and the end of the aliphatic chain is oxidized to the carboxylic acid. The latter, in turn, forms an amide bond with an amino acid, either glycine or taurine:

Glycocholate   Taurocholate

Taurine has developed a certain amount of commercial importance in recent years as an ingredient in sports drinks. The drink marketed under the trade name Red Bull (*taurus* is the Latin word for "bull") contains various sugars (Chapter 20), caffeine, and B vitamins (Section 30.6) in addition to taurine.

Bile salts are powerful detergents. One end of the molecule is strongly hydrophilic because of the negative charge, and the rest of the molecule is largely hydrophobic. As a consequence, bile salts can disperse dietary lipids in the small intestine into fine emulsions, thereby facilitating digestion. The dispersion of dietary lipids by bile salts is similar to the action of soap on dirt.

Because they are eliminated in the feces, bile salts remove excess cholesterol in two ways: (1) They are themselves breakdown products of cholesterol (so cholesterol is eliminated via bile salts), and (2) they solubilize deposited cholesterol in the form of bile salt–cholesterol particles.

21.12 | What Are Prostaglandins, Thromboxanes, and Leukotrienes?

Prostaglandins, a group of fatty-acid-like substances, were discovered by Kurzrok and Leib in the 1930s, when they demonstrated that seminal fluid caused a hysterectomized uterus to contract. Ulf von Euler of Sweden (winner of the Nobel Prize in physiology and medicine in 1970) isolated these compounds from human semen and, thinking that they had come from the prostate gland, named them **prostaglandin.** Even though the seminal gland secretes 0.1 mg of prostaglandin per day in mature males, small amounts of prostaglandins are present throughout the body in both sexes.

Prostaglandins are synthesized in the body from arachidonic acid by a ring closure at carbons 8 and 12. The enzyme catalyzing this reaction is called **cyclooxygenase** (COX, for short). The product, known as PGG_2, is the common precursor of other prostaglandins, including PGE and PGF. The prostaglandin E group (PGE) has a carbonyl group at carbon 9; the subscript indicates the number of double bonds in the hydrocarbon chain. The prostaglandin F group (PGF) has two hydroxyl groups on the ring at carbons 9 and 11. Other prostaglandins (PGAs and PGBs) are derived from PGE.

The COX enzyme comes in two forms in the body: COX-1 and COX-2. COX-1 catalyzes the normal physiological production of prostaglandins, which are always present in the body. For example, PGE_2 and $PGF_{2\alpha}$ stimulate uterine contractions and induce labor. PGE_2 lowers blood pressure by relaxing the muscles around blood vessels. In aerosol form, this prostaglandin is used to treat asthma; it opens up the bronchial tubes by relaxing the surrounding muscles. PGE_1 is used as a decongestant; it opens up nasal passages by constricting blood vessels.

COX-2, by contrast, is responsible for the production of prostaglandins in inflammation. When a tissue is injured or damaged, special inflammatory cells invade the injured tissue and interact with resident cells—for example, smooth muscle cells. This interaction activates the COX-2 enzyme, and prostaglandins are synthesized. Such tissue injury may occur in a heart attack (myocardial infarction), rheumatic arthritis, and ulcerative colitis. Nonsteroidal anti-inflammatory drugs (NSAIDs), such as aspirin, inhibit both COX enzymes (see Chemical Connections 21H).

CHEMICAL CONNECTIONS 21H

Action of Anti-inflammatory Drugs

Anti-inflammatory steroids (such as cortisone; Section 21.10) exert their function by inhibiting phospholipase A_2, the enzyme that releases unsaturated fatty acids from complex lipids in the membranes. For example, arachidonic acid, one of the components of membranes, is made available to the cell through this process. Because arachidonic acid is the precursor of prostaglandins, thromboxanes, and leukotrienes, inhibiting its release stops the synthesis of these compounds and prevents inflammation.

Steroids such as cortisone are associated with many undesirable side effects (duodenal ulcer and cataract formation, among others). Therefore, their use must be controlled. A variety of nonsteroidal anti-inflammatory agents, including aspirin, ibuprofen, ketoprofen, and indomethacin, are available to serve this function.

Aspirin and other NSAIDs (see Chemical Connections 19C) inhibit the cyclooxygenase enzymes, which synthesize prostaglandins and thromboxanes. Aspirin (acetyl salicylic acid), for example, acetylates the enzymes, thereby blocking the entrance of arachidonic acid to the active site. This inhibition of both COX-1 and COX-2 explains why aspirin and the other anti-inflammatory agents have undesirable side effects. NSAIDs also interfere with the COX-1 isoform of the enzyme, which is needed for normal physiological function. Their side effects include stomach and duodenal ulceration and renal (kidney) toxicity.

Obviously, it would be desirable to have an anti-inflammatory agent without such side effects, and one that inhibits only the COX-2 isoform. To date, the FDA has approved two COX-2 inhibitor drugs: Celebrex, which quickly became the most frequently prescribed drug, and Vioxx, a more recent entrant. Despite their selective inhibition of COX-2, however, these drugs also have ulcer-causing side effects. Many other COX-2 inhibitors remain in the clinical trial stage of development.

The use of COX-2 inhibitors is not limited to rheumatoid arthritis and osteoarthritis. Celebrex has been approved by FDA to treat a type of colon cancer called familial adenomateous polyposis, in an approach called chemoprevention. All anti-inflammatory agents reduce pain and relieve fever and swelling by reducing the prostaglandin production, but they do not affect the leukotriene production. As a consequence, asthmatic patients must beware of using these anti-inflammatory agents. Even though they inhibit the prostaglandin synthesis, these drugs may shift the available arachidonic acid to leukotriene production, which could precipitate a severe asthma reaction.

During the fall of 2004, studies indicated that high doses of Vioxx correlate with higher incidence of heart attacks and strokes; concerns were also raised about other COX-2 inhibitors, particularly Celebrex. The inhibition of prostaglandin synthesis allows formation of other lipids, including those that build up in atherosclerotic plaque. Vioxx was taken off the U.S. market, and some physicians became hesitant to prescribe Celebrex. These events caused consternation among patients who had come to depend on these drugs as well as among the pharmaceutical companies that produced them. In February 2005, a panel looked into the controversy for the FDA. This group concluded that COX-2 inhibitors should stay on the market, but their use should be strictly monitored. Warning labels must now appear on the packaging for these drugs.

Another class of arachidonic acid derivatives is the **thromboxanes.** Their synthesis also includes a ring closure. These substances are derived from PGH_2, but their ring is a cyclic acetal. Thromboxane is known to induce platelet aggregation. When a blood vessel is ruptured, the first line of defense is the platelets circulating in the blood, which form an incipient clot. Thromboxane A_2 causes other platelets to clump, thereby increasing the size of the blood clot. Aspirin and similar anti-inflammatory agents inhibit the COX enzyme. Consequently, PGH_2 and thromboxane synthesis is inhibited, and blood clotting is impaired. This effect has prompted many physicians to recommend a daily dose of 81 mg aspirin for people at risk for heart attack or stroke. It also explains why physicians forbid patients to use aspirin and other anti-inflammatory agents for a week before a planned surgery—aspirin and other NSAIDs may cause excessive bleeding.

PGH$_2$ Thromboxane A$_2$

A variety of NSAIDs inhibit COX enzymes. Ibuprofen and indomethacin, both powerful painkillers, can block the inhibitory effect of aspirin and thus eliminate its anticlotting benefits. Therefore, the use of these NSAIDs together with aspirin is not recommended. Other painkillers, such as acetaminophen and diclofenac, do not interfere with aspirin's anticlotting ability and therefore can be taken together.

The **leukotrienes** are another group of substances that act to mediate hormonal responses. Like prostaglandins, they are derived from arachidonic acid by an oxidative mechanism. However, in this case, there is no ring closure.

Arachidonic acid   Leukotriene B4

Leukotrienes occur mainly in white blood cells (leukocytes) but are also found in other tissues of the body. They produce long-lasting muscle contractions, especially in the lungs, and can cause asthma-like attacks. In fact, they are 100 times more potent than histamines. Both prostaglandins and leukotrienes cause inflammation and fever, so the inhibition of their production in the body is a major pharmacological concern. One way to counteract the effects of leukotrienes is to inhibit their uptake by leukotriene receptors (LTRs) in the body. A new antagonist of LTRs, zafirlukast (brand name Accolate), is used to treat and control chronic asthma. Another antiasthmatic drug, zileuton, inhibits 5-lipoxygenase, which is the initial enzyme in leukotriene biosynthesis from arachidonic acid.

SUMMARY OF KEY QUESTIONS

SECTION 21.1 What Are Lipids?

- **Lipids** are water-insoluble substances.
- Lipids are classified into four groups: fats (triglycerides); complex lipids; steroids; and prostaglandins, thromboxanes, and leukotrienes.

SECTION 21.2 What Are the Structures of Triglycerides?

- **Fats** consist of fatty acids and glycerol. In saturated fatty acids, the hydrocarbon chains have only single bonds; unsaturated fatty acids have hydrocarbon chains with one or more double bonds, all in the *cis* configuration.

SECTION 21.3 What Are Some Properties of Triglycerides?

- Solid fats contain mostly saturated fatty acids, whereas **oils** contain substantial amounts of unsaturated fatty acids.
- The alkali salts of fatty acids are called **soaps.**

SECTION 21.4 What Are the Structures of Complex Lipids?

- **Complex lipids** can be classified into two groups: phospholipids and glycolipids.
- **Phospholipids** are made of a central alcohol (glycerol or sphingosine), fatty acids, and a nitrogen-containing phosphate ester, such as phosphorylcholine or inositol phosphate.
- **Glycolipids** contain sphingosine and a fatty acid, collectively known as the ceramide portion of the molecule, and a carbohydrate portion.

SECTION 21.5 What Role Do Lipids Play in the Structure of Membranes?

- Many phospholipids and glycolipids are important components of cell **membranes.**
- Membranes are made of a **lipid bilayer,** in which the hydrophobic parts of phospholipids (fatty acid residues) point toward the middle of the bilayer, and the hydrophilic parts point toward the inner and outer surfaces of the membrane.

SECTION 21.6 What Are Glycerophospholipids?

- **Glycerophospholipids** are complex lipids that consist of a central glycerol moiety to which two fatty acids are esterified. The third alcohol group of the glycerol is esterified to a nitrogen-containing phosphate ester.

SECTION 21.7 What Are Sphingolipids?

- **Sphingolipids** are complex lipids that consist of the long-chain alcohol sphingosine esterified to a fatty acid (the ceramide moiety). Nitrogen-containing phosphate esters are also bonded to the sphingosine moiety.

SECTION 21.8 What Are Glycolipids?

- **Glycolipids** are complex lipids that consist of two parts: a ceramide portion and a carbohydrate portion.

SECTION 21.9 What Are Steroids?

- The third major group of lipids comprises the **steroids.** The characteristic feature of the steroid structure is a fused four-ring nucleus.
- The most common steroid, **cholesterol,** serves as a starting material for the synthesis of other steroids, such as bile salts and sex and other hormones. Cholesterol is also an integral part of membranes, occupying the hydrophobic region of the lipid bilayer. Because of its low solubility in water, cholesterol deposits are implicated in the formation of gallstones and the plaque-like deposits of atherosclerosis.
- Cholesterol is transported in the blood plasma mainly by two kinds of lipoproteins: **HDL** and **LDL.** LDL delivers cholesterol to the cells to be used mostly as a membrane component. HDL delivers cholesteryl esters mainly to the liver to be used in the synthesis of bile acids and steroid hormones.
- High levels of LDL and low levels of HDL are symptoms of faulty cholesterol transport, indicating greater risk of atherosclerosis.

SECTION 21.10 What Are Some of the Physiological Roles of Steroid Hormones?

- An oxidation product of cholesterol is progesterone, a **sex hormone.** It also gives rise to the synthesis of other sex hormones, such as testosterone and estradiol.
- Progesterone is also a precursor of the **adrenocorticoid hormones.** Within this group, cortisol and cortisone are best known for their anti-inflammatory action.

SECTION 21.11 What Are Bile Salts?

- **Bile salts** are oxidation products of cholesterol. They emulsify all kinds of lipids, including cholesterol, and are essential in the digestion of fats.

SECTION 21.12 What Are Prostaglandins, Thromboxanes, and Leukotrienes?

- **Prostaglandins, thromboxanes,** and **leukotrienes** are derived from arachidonic acid. They have a wide variety of effects on body chemistry. Among other things, they can lower or raise blood pressure, cause inflammation and blood clotting, and induce labor. In general, they mediate hormone action.

PROBLEMS

GOB Chemistry Now™

Assess your understanding of this chapter's topics with additional quizzing and conceptual-based problems at **http://now.brookscole.com/gob8** or on the CD.

A blue problem number indicates an applied problem.

■ denotes problems that are available on the GOB ChemistryNow website or CD and are assignable in OWL.

SECTION 21.1 What Are Lipids?

21.2 Why are fats a good source of energy for storage in the body?

21.3 What is the meaning of the term *hydrophobic?* Why is the hydrophobic nature of lipids important?

SECTION 21.2 What Are the Structures of Triglycerides?

21.4 ■ Draw the structural formula of a fat molecule (triglyceride) made of myristic acid, oleic acid, palmitic acid, and glycerol.

21.5 Oleic acid has a melting point of 16°C. If you converted the *cis* double bond into a *trans* double bond, what would happen to the melting point? Explain.

21.6 Draw schematic formulas for all possible 1,3-diglycerides made up of glycerol, oleic acid, or stearic acid. How many are there? Draw the structure of one of the diglycerides.

SECTION 21.3 What Are Some Properties of Triglycerides?

21.7 For the diglycerides in Problem 21.6, predict which two will have the highest melting points and which two will have the lowest melting points.

21.8 ■ Predict which acid in each pair has the higher melting point and explain why.
 (a) Palmitic acid or stearic acid
 (b) Arachidonic acid or arachidic acid

21.9 Which has the higher melting point: (a) a triglyceride containing only lauric acid and glycerol or (b) a triglyceride containing only stearic acid and glycerol?

21.10 Explain why the melting points of the saturated fatty acids increase as we move from lauric acid to stearic acid.

21.11 Predict the order of the melting points of triglycerides containing fatty acids, as follows:

(a) Palmitic, palmitic, stearic

(b) Oleic, stearic, palmitic

(c) Oleic, linoleic, oleic

21.12 Look at Table 21.1. Which animal fat has the highest percentage of unsaturated fatty acids?

21.13 Rank the following in order of increasing solubility in water (assuming that all are made with the same fatty acids): (a) triglycerides, (b) diglycerides, and (c) monoglycerides. Explain your answer.

21.14 How many moles of H_2 are used up in the catalytic hydrogenation of one mole of a triglyceride containing glycerol, palmitic acid, oleic acid, and linoleic acid?

21.15 ■ Name the products of the saponification of this triglyceride:

$$
\begin{array}{l}
\quad\quad\quad\quad O \\
\quad\quad\quad\quad \| \\
CH_2-O-C-(CH_2)_{14}CH_3 \\
| \quad\quad\quad O \\
| \quad\quad\quad \| \\
CH-O-C-(CH_2)_{16}CH_3 \\
| \quad\quad\quad O \\
| \quad\quad\quad \| \\
CH_2-O-C-(CH_2)_7(CH{=}CHCH_2)_3CH_3
\end{array}
$$

21.16 Using the equation in Section 21.3C as a guideline for stoichiometry, calculate the number of moles of NaOH needed to saponify 5 mol of (a) triglycerides, (b) diglycerides, and (c) monoglycerides.

SECTION 21.4 What Are the Structures of Complex Lipids?

21.17 What are the main types of complex lipids, and what are the main characteristics of their structures?

SECTION 21.5 What Role Do Lipids Play in the Structure of Membranes?

21.18 Which portion of the phosphatidylinositol molecule contributes to (a) the fluidity of the bilayer and (b) the surface polarity of the bilayer?

21.19 How do the unsaturated fatty acids of the complex lipids contribute to the fluidity of a membrane?

21.20 Which type of lipid molecules is most likely to be present in membranes?

21.21 What is the difference between an integral and a peripheral membrane protein?

SECTION 21.6 What Are Glycerophospholipids?

21.22 Which glycerophospholipid has the most polar groups capable of forming hydrogen bonds with water?

21.23 Draw the structure of a phosphatidylinositol that contains oleic acid and arachidonic acid.

21.24 ■ Among the glycerophospholipids containing palmitic acid and linolenic acid, which will have the greatest solubility in water: (a) phosphatidylcholine, (b) phosphatidylethanolamine, or (c) phosphatidylserine? Explain.

SECTION 21.7 What Are Sphingolipids?

21.25 Name all the groups of complex lipids that contain ceramides.

21.26 Are the various phospholipids randomly distributed in membranes? Give an example.

SECTION 21.8 What Are Glycolipids?

21.27 ■ Enumerate the functional groups that contribute to the hydrophilic character of (a) glucocerebroside and (b) sphingomyelin.

SECTION 21.9 What Are Steroids?

21.28 Cholesterol has a fused four-ring steroid nucleus and is a part of body membranes. The —OH group on carbon 3 is the polar head, and the rest of the molecule provides the hydrophobic tail that does not fit into the zig-zag packing of the hydrocarbon portion of the saturated fatty acids. Considering this structure, tell whether small amounts of cholesterol that are well dispersed in the membrane contribute to the stiffening (rigidity) or the fluidity of a membrane. Explain.

21.29 Where can pure cholesterol crystals be found in the body?

21.30 ■ (a) Find all of the carbon stereocenters in a cholesterol molecule.

(b) How many total stereoisomers are possible?

(c) How many of these stereoisomers do you think are found in nature?

21.31 Look at the structures of cholesterol and the hormones shown in Figure 21.6. Which ring of the steroid structure undergoes the most substitution?

21.32 What makes LDL soluble in blood plasma?

21.33 How does LDL deliver its cholesterol to the cells?

21.34 How does lovastatin reduce the severity of atherosclerosis?

21.35 How does VLDL become LDL?

21.36 How does HDL deliver its cholesteryl esters to liver cells?

21.37 How does the serum cholesterol level control both cholesterol synthesis in the liver and LDL uptake?

SECTION 21.10 What Are Some of the Physiological Roles of Steroid Hormones?

21.38 What physiological functions are associated with cortisol?

21.39 Estradiol in the body is synthesized starting from progesterone. What chemical modifications occur when estradiol is synthesized?

21.40 Describe the difference in structure between the male hormone testosterone and the female hormone estradiol.

21.41 Considering that RU486 can bind to the receptors of progesterone as well as to the receptors of cortisone and cortisol, what can you say regarding the importance of the functional group on carbon 11 of the steroid ring in drug and receptor binding?

21.42 (a) How does the structure of RU486 resemble that of progesterone?

(b) How do the two structures differ?

21.43 What are the structural features common to oral contraceptive pills, including mifepristone?

SECTION 21.11 What Are Bile Salts?

21.44 List all of the functional groups that make taurocholate water-soluble.

21.45 Explain how the constant elimination of bile salts through the feces can reduce the danger of plaque formation in atherosclerosis.

SECTION 21.12 What Are Prostaglandins, Thromboxanes, and Leukotrienes?

21.46 What is the basic structural difference between:
(a) Arachidonic acid and prostaglandin PGE_2?
(b) PGE_2 and $PGF_{2\alpha}$?

21.47 ■ Find and name all of the functional groups in (a) glycocholate, (b) cortisone, (c) prostaglandin PGE_2 and (d) leukotriene B4.

21.48 What are the chemical and physiological functions of the COX-2 enzyme?

21.49 How does aspirin, an anti-inflammatory drug, prevent strokes caused by blood clots in the brain?

Chemical Connections

21.50 (Chemical Connections 21A) What causes rancidity? How can it be prevented?

21.51 (Chemical Connections 21B) What makes waxes harder and more difficult to melt than fats?

21.52 (Chemical Connections 21C) How do the gap junctions prevent the passage of proteins from cell to cell?

21.53 (Chemical Connections 21C) How does the anion transporter provide a suitable environment for the passage of hydrated chloride ions?

21.54 (Chemical Connections 21C) In what sense is the active transport of K^+ selective? How does K^+ pass through the transporter?

21.55 (Chemical Connections 21D)
(a) What role does sphingomyelin play in the conductance of nerve signals?
(b) What happens to this process in multiple sclerosis?

21.56 (Chemical Connections 21E) Compare the complex lipid structures listed for the lipid storage diseases with the missing or defective enzymes. Explain why the missing enzyme in Fabry's disease is α-galactosidase and not β-galactosidase.

21.57 (Chemical Connections 21E) Identify the monosaccharides in the accumulating glycolipid of Fabry's disease.

21.58 (Chemical Connections 21F) How does the oral anabolic steroid methenolone differ structurally from testosterone?

21.59 (Chemical Connections 21G) What is the role of progesterone and similar compounds in contraceptive pills?

21.60 (Chemical Connections 21H) How does cortisone prevent inflammation?

21.61 (Chemical Connections 21H) How does indomethacin act in the body to reduce inflammation?

21.62 (Chemical Connections 21H) What kind of prostaglandins are synthesized by COX-1 and COX-2 enzymes?

21.63 (Chemical Connections 21H) Steroids prevent asthma-causing leukotriene synthesis as well as inflammation-causing prostaglandin synthesis. Nonsteroidal anti-inflammatory agents (NSAIDs) such as aspirin reduce only prostaglandin production. Why do NSAIDs not affect leukotriene production?

Additional Problems

21.64 What is the role of taurine in lipid digestion?

21.65 Draw a schematic diagram of a lipid bilayer. Show how the bilayer prevents the passage by diffusion of a polar molecule such as glucose. Show why nonpolar molecules, such as CH_3CH_2—O—CH_2CH_3, can diffuse through the membrane.

21.66 How many different triglycerides can you create using three different fatty acids (A, B, and C) in each case?

21.67 Prostaglandins have a five-membered ring closure; thromboxanes have a six-membered ring closure. The synthesis of both groups of compounds is prevented by COX inhibitors; the COX enzymes catalyze ring closure. Explain this fact.

21.68 Which lipoprotein is instrumental in removing the cholesterol deposited in the plaques on arteries?

21.69 What are coated pits? What is their function?

21.70 What are the constituents of sphingomyelin?

21.71 (Chemical Connections 21C) What is the difference between a facilitated transporter and an active transporter?

21.72 Which part of LDL interacts with the LDL receptor?

21.73 What is the major difference between aldosterone and the other hormones listed in Figure 21.6?

21.74 (Chemical Connections 21H) The anti-inflammatory drug Celebrex does not have the usual side effect of stomach upset or ulceration commonly observed with the other NSAIDs. Why?

21.75 How many grams of H_2 are needed to saturate 100.0 g of a triglyceride made of glycerol and one unit each of lauric, oleic, and linoleic acids?

21.76 Prednisolone is the synthetic glucocorticoid medicine most frequently prescribed to combat autoimmune diseases. Compare its structure to the natural glucocorticoid hormone, cortisone. What are the similarities and differences in structure?

Tying It Together

21.77 Lipids and carbohydrates are both vehicles for energy storage. How are they similar in terms of molecular structure, and how do they differ? What does the molecular structure of each class of substance imply about the polarity of typical molecules?

21.78 To what extent do lipids and carbohydrates play structural roles in living organisms? Do these roles differ in plants and in animals?

21.79 Which substances would you expect to consist primarily of carbohydrates and which primarily of lipids: olive oil, butter, cotton, cotton candy?

21.80 To what extent would you expect to find the following functional groups in lipids and in carbohydrates: aldehyde groups, carboxylic acid groups, ester bonds, hydroxyl groups?

Looking Ahead

21.81 Sports drinks tend to contain large amounts of sugars, and some contain taurine in small amounts. Would you expect more of the effect of these performance aids to come from dietary carbohydrates or from the role of taurine in breaking down fats?

21.82 ■ Which of the following foods consist primarily of carbohydrates, and which of fats: soft drinks (not diet drinks), salad dressing, canned fruit, cream cheese?

21.83 The ester bonds in lipids do not give rise to macromolecules, but the amide bonds in proteins do. Comment on the underlying reason for this difference.

21.84 Given the structural differences between steroids and other kinds of lipids, would you expect the synthesis of steroids in living organisms to differ from the synthesis of other lipids?

Challenge Problems

21.85 Some of the lipid molecules that occur in membranes are bulkier than others. Are the bulkier molecules more or less likely to be found on the cytoplasmic side of the cell membrane or on the side facing the exterior of the cell?

21.86 What are the functions of a cell membrane? To what extent is a bilayer that consists entirely of lipids able to carry out these functions?

21.87 Glycerophospholipids tend to have both a positive charge and a negative charge in their hydrophilic portions. Does this fact help or hinder lipid packing in membranes, and why?

21.88 Leukotrienes differ from prostaglandins and thromboxanes in that they lack a ring closure. They also differ from prostaglandins and thromboxanes (and from all other lipids) in another feature of their structure. What is that structural feature? (*Hint:* It has to do with the position of their double bonds.)

CHAPTER 22

Proteins

Hans Strand/Stone/Getty Images

Spider silk is a fibrous protein that exhibits unmatched strength and toughness.

Protein A large biological molecule made of numerous amino acids linked by amide bonds

GOB
Chemistry ⚛ Now™
Look for this logo in the chapter and go to GOB ChemistryNow at **http://now.brookscole.com/gob8** or on the CD for tutorials, simulations, and problems.

22.1 | What Are the Many Functions of Proteins?

Proteins are by far the most important of all biological compounds. The very word "protein" is derived from the Greek *proteios,* meaning "of first importance," and the scientists who named these compounds more than 100 years ago chose an appropriate term. Many types of proteins exist, and they perform a variety of functions, including the following roles:

1. **Structure** In Section 20.5, we saw that the main structural material for plants is cellulose. For animals, it is structural proteins, which are the chief constituents of skin, bones, hair, and nails. Two important structural proteins are collagen and keratin.

2. **Catalysis** Virtually all the reactions that take place in living organisms are catalyzed by proteins called enzymes. Without enzymes, the reactions would take place so slowly as to be useless. We will discuss enzymes in depth in Chapter 23.

3. **Movement** Every time we crook a finger, climb stairs, or blink an eye, we use our muscles. Muscle expansion and contraction are involved in every movement we make. Muscles are made up of protein molecules called myosin and actin.

4. **Transport** A large number of proteins perform transportation duties. For example, hemoglobin, a protein in the blood, carries oxygen from the lungs to the cells in which it is used and carbon dioxide from the cells to the lungs. Other proteins transport molecules across cell membranes.

5. **Hormones** Many hormones are proteins, including insulin, erythropoietin, and human growth hormone.

6. **Protection** When a protein from an outside source or some other foreign substance (called an antigen) enters the body, the body makes its own proteins (called antibodies) to counteract the foreign protein. This antibody production is one of the major mechanisms that the body uses to fight disease. Blood clotting is another protective function carried out by a protein, this one called fibrinogen. Without blood clotting, we would bleed to death from any small wound.

7. **Storage** Some proteins store materials in the way that starch and glycogen store energy. For example, casein in milk and ovalbumin in eggs store nutrients for newborn mammals and birds. Ferritin, a protein in the liver, stores iron.

8. **Regulation** Some proteins not only control the expression of genes, thereby regulating the kind of proteins synthesized in a particular cell, but also dictate when such manufacture takes place.

These are not the only functions of proteins, but they are among the most important. Clearly, any individual needs a great many proteins to carry out these varied functions. A typical cell contains about 9000 different proteins; the entire human body has about 100,000 different proteins.

We can classify proteins into two major types: **fibrous proteins,** which are insoluble in water and are used mainly for structural purposes, and **globular proteins,** which are more or less soluble in water and are used mainly for nonstructural purposes.

22.2 | What Are Amino Acids?

Alpha (α) amino acid An amino acid in which the amino group is linked to the carbon atom next to the —COOH carbon

Although a wide variety of proteins exist, they all have basically the same structure: They are chains of amino acids. As its name implies, an **amino acid** is an organic compound containing an amino group and a carboxyl group. Organic chemists can synthesize many thousands of amino acids, but nature is much more restrictive and uses 20 common amino acids to make up proteins. Furthermore, all but one of the 20 fit the formula:

Even the one amino acid that doesn't fit this formula (proline) comes fairly close: It differs only in that it has a bond between the R and the N. The

Table 22.1   The 20 Amino Acids Commonly Found in Proteins

Name	3-Letter Abbreviation	1-Letter Abbreviation	Isoelectric Point
Alanine	Ala	A	6.01
Arginine	Arg	R	10.76
Asparagine	Asn	N	5.41
Aspartic acid	Asp	D	2.77
Cysteine	Cys	C	5.07
Glutamic acid	Glu	E	3.22
Glutamine	Gln	Q	5.65
Glycine	Gly	G	5.97
Histidine	His	H	7.59
Isoleucine	Ile	I	6.02
Leucine	Leu	L	5.98
Lysine	Lys	K	9.74
Methionine	Met	M	5.74
Phenylalanine	Phe	F	5.48
Proline	Pro	P	6.48
Serine	Ser	S	5.68
Threonine	Thr	T	5.87
Tryptophan	Trp	W	5.88
Tyrosine	Tyr	Y	5.66
Valine	Val	V	5.97

20 amino acids commonly found in proteins are called **alpha amino acids.** They are listed in Table 22.1, which also shows the one- and three-letter abbreviations that chemists and biochemists use for them.

The most important aspect of the R groups is their polarity. On that basis we can classify amino acids into four groups, as shown in Figure 22.1: nonpolar, polar but neutral, acidic, and basic. Note that the nonpolar side chains are *hydrophobic* (they repel water), whereas polar but neutral, acidic, and basic side chains are *hydrophilic* (attracted to water). This aspect of the R groups is very important in determining both the structure and the function of each protein molecule.

When we look at the general formula for the 20 amino acids, we see at once that all of them (except glycine, in which R = H) are chiral with (carbon) stereocenters, since R, H, COOH, and NH₂ are four different groups. Thus each of the amino acids with one stereocenter exists as two enantiomers. As is the case for most examples of this kind, nature makes only one of the two possible enantiomers for each amino acid, and it is virtually always the L-isomer. Except for glycine, which is achiral, all the amino acids in all the proteins in your body are the L-isomer. D amino acids are extremely rare in nature; some are found, for example, in the cell walls of a few types of bacteria.

In Section 20.1C, we learned about the systematic use of the D,L system. There we used glyceraldehyde as a reference point for the assignment of relative configuration. Here again, we can use glyceraldehyde as a reference point with amino acids, as shown in Figure 22.2. The spatial relationship of the functional groups around the carbon stereocenter in L-amino acids, as in L-alanine, can be compared to that of L-glyceraldehyde. When we put the carbonyl groups of both compounds in the same position (top), the —OH of L-glyceraldehyde and the NH₃⁺ of L-alanine lie to the left of the carbon stereocenter.

GOB
Chemistry · Now™

Click *Chemistry Interactive* to learn more about **Amino Acid Structure**

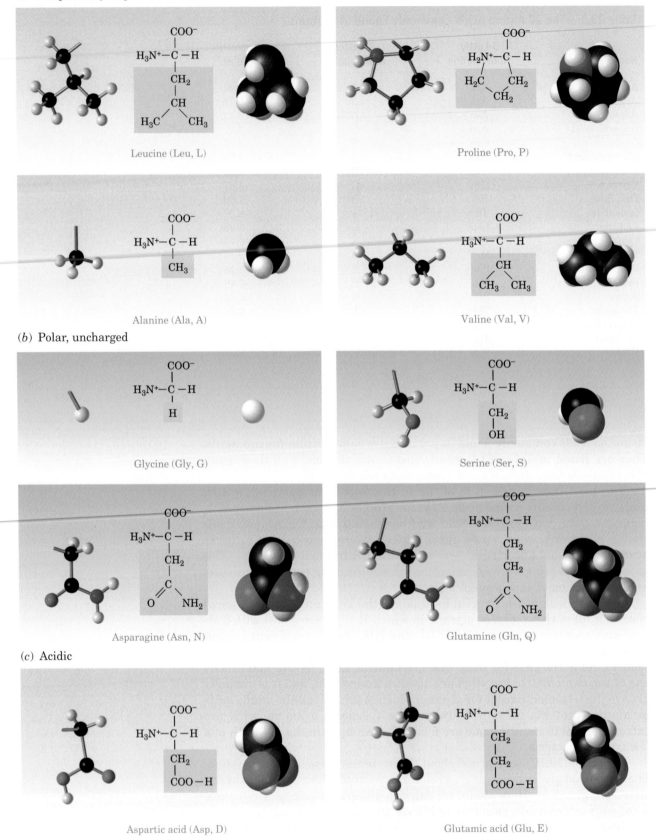

(a) Nonpolar (hydrophobic)

Leucine (Leu, L)

Proline (Pro, P)

Alanine (Ala, A)

Valine (Val, V)

(b) Polar, uncharged

Glycine (Gly, G)

Serine (Ser, S)

Asparagine (Asn, N)

Glutamine (Gln, Q)

(c) Acidic

Aspartic acid (Asp, D)

Glutamic acid (Glu, E)

Figure 22.1 The 20 amino acids that are the building blocks of proteins can be classified as (a) nonpolar (hydrophobic), (b) polar but neutral, (c) acidic, or (d) basic. Also shown here are the one-letter and three-letter codes used to denote amino acids. For each amino acid, the ball-and-stick (left) and space-filling (right) models show only the side chain. (Irving Geis)

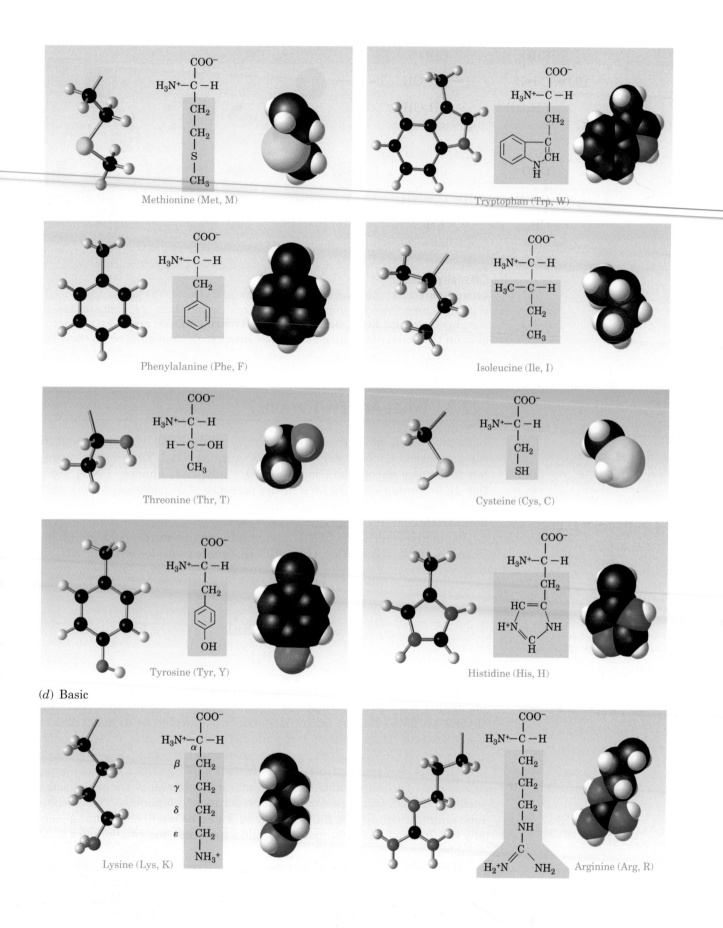

Methionine (Met, M)

Tryptophan (Trp, W)

Phenylalanine (Phe, F)

Isoleucine (Ile, I)

Threonine (Thr, T)

Cysteine (Cys, C)

Tyrosine (Tyr, Y)

Histidine (His, H)

(d) Basic

Lysine (Lys, K)

Arginine (Arg, R)

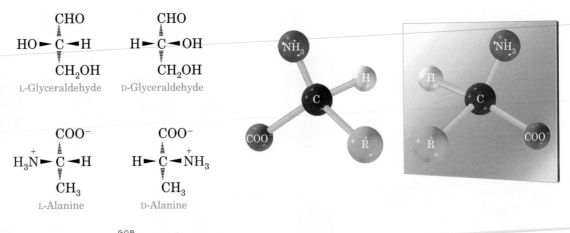

CHO

HO—C◄H

CH₂OH

L-Glyceraldehyde

CHO

H◄C—OH

CH₂OH

D-Glyceraldehyde

COO⁻

H₃N⁺—C◄H

CH₃

L-Alanine

COO⁻

H►C◄NH₃⁺

CH₃

D-Alanine

Chemistry Now™

Active Figure 22.2 Stereochemistry of alanine and glycine. The amino acids found in proteins have the same chirality as L-glyceraldehyde, which is opposite to that of D-glyceraldehyde. **See a simulation based on this figure, and take a short quiz on the concepts at http://www.now.brookscole.com/gob8 or on the CD.**

22.3 | What Are Zwitterions?

In Section 18.5B we learned that carboxylic acids, RCOOH, cannot exist in the presence of a moderately weak base (such as NH_3). They donate a proton to become carboxylate ions, $RCOO^-$. Likewise, amines, RNH_2 (Section 16.5), cannot exist as such in the presence of a moderately weak acid (such as acetic acid). They gain a proton to become substituted ammonium ions, RNH_3^+.

An amino acid has —COOH and —NH_2 groups in the same molecule. Therefore, in water solution, the —COOH donates a proton to the —NH_2 so that an amino acid actually has the structure

$$R-\overset{\overset{\displaystyle H}{|}}{\underset{\underset{\displaystyle NH_3^+}{|}}{C}}-COO^-$$

Compounds that have a positive charge on one atom and a negative charge on another are called **zwitterions,** from the German word *zwitter,* meaning "hybrid." Amino acids are zwitterions, not only in water solution but also in the solid state. They are therefore ionic compounds—that is, internal salts. *Un-ionized $RCH(NH_2)COOH$ molecules do not actually exist, in any form.*

The fact that amino acids are zwitterions explains their physical properties. All of them are solids with high melting points (for example, glycine melts at 262°C), just as we would expect for ionic compounds. The 20 common amino acids are also fairly soluble in water, as ionic compounds generally are. If they had no charges, we would expect only the smaller ones to be soluble.

If we add an amino acid to water, it dissolves and then has the same zwitterionic structure that it has in the solid state. Let us see what happens if we change the pH of the solution, as we can easily do by adding a source of H_3O^+, such as HCl solution (to lower the pH), or a strong base, such as NaOH (to raise the pH). Because H_3O^+ is a stronger acid than a typical carboxylic acid (Section 18.1), it donates a proton to the —COO⁻ group, turning the zwitterion into a positive ion. This happens to all amino acids if the pH is sufficiently lowered—say, to pH 2.

$$\underset{\underset{NH_3^+}{|}}{R-\overset{\overset{H}{|}}{C}-COO^-} + H_3O^+ \longrightarrow \underset{\underset{NH_3^+}{|}}{R-\overset{\overset{H}{|}}{C}-COOH} + H_2O$$

Addition of OH⁻ to the zwitterion causes the —NH₃⁺ to donate its proton to OH⁻, turning the zwitterion into a negative ion. This happens to all amino acids if the pH is sufficiently raised—say, to pH 10.

$$\underset{\underset{NH_3^+}{|}}{R-\overset{\overset{H}{|}}{C}-COO^-} + OH^- \longrightarrow \underset{\underset{NH_2}{|}}{R-\overset{\overset{H}{|}}{C}-COO^-} + H_2O$$

In both cases, the amino acid is still an ion so it is still soluble in water. There is no pH at which an amino acid has no ionic character at all. If the amino acid is a positive ion at low pH and a negative ion at high pH, there must be some pH at which all the molecules have equal positive and negative charges. This pH is called the **isoelectric point (pI).**

Every amino acid has a different isoelectric point, although most of them are not very far apart (see the values in Table 22.1). Fifteen of the 20 amino acids have isoelectric points near 6. However, the three basic amino acids have higher isoelectric points, and the two acidic amino acids have lower values.

At or near the isoelectric point, amino acids exist in aqueous solution largely or entirely as zwitterions. As we have seen, they react with either a strong acid, by taking a proton (the —COO⁻ becomes —COOH), or a strong base, by giving a proton (the —NH₃⁺ becomes —NH₂). To summarize:

Isoelectric point (pI) A pH at which a sample of amino acids or protein has an equal number of positive and negative charges

$$\underset{\underset{NH_3^+}{|}}{R-\overset{\overset{H}{|}}{C}-COOH} \underset{H_3O^+}{\overset{OH^-}{\rightleftharpoons}} \underset{\underset{NH_3^+}{|}}{R-\overset{\overset{H}{|}}{C}-COO^-} \underset{H_3O^+}{\overset{OH^-}{\rightleftharpoons}} \underset{\underset{NH_2}{|}}{R-\overset{\overset{H}{|}}{C}-COO^-}$$

In Section 9.3, we learned that a compound that is both an acid and a base is called amphiprotic. We also learned, in Section 9.10, that a solution that neutralizes both acid and base is a buffer solution. Amino acids are therefore *amphiprotic* compounds, and aqueous solutions of them are *buffers*.

GOB
Chemistry•♣•Now™
Click *Chemistry Interactive* to learn more about **Zwitterions**

22.4 | What Determines the Characteristics of Amino Acids?

The side chain of an amino acid is responsible for the unique characteristics of these molecules. Ultimately the functions of amino acids and their polymers, proteins, are determined by the side chains. For example, one of the 20 amino acids in Table 22.1 has a chemical property not shared by any of the others. This amino acid, cysteine, can easily be dimerized by many mild oxidizing agents:

A disulfide bond

Cysteine   Cystine

The dimer of cysteine, which is called **cystine,** can in turn be fairly easily reduced to give two molecules of cysteine. As we shall see, the presence of cystine has important consequences for the chemical structure and shape of the protein molecules of which it is part. The bond (shown in color) is also called a **disulfide bond** (Section 14.5D).

Several of the amino acids have acidic or basic properties. Two amino acids, glutamic acid and aspartic acid, have carboxyl groups in their side chains in addition to the one present in all amino acids. A carboxyl group can lose a proton, forming the corresponding carboxylate anion—glutamate and aspartate, respectively, in the case of these two amino acids. Because of the presence of the carboxylate, the side chains of these two amino acids are negatively charged at neutral pH. Three amino acids—histidine, lysine, and arginine—have basic side chains. The side chains of lysine and arginine are positively charged at or near neutral pH. In lysine, the side-chain amino group is attached to an aliphatic hydrocarbon tail. In arginine, the side-chain basic group, the guanidino group, is more complex in structure than the amino group, but it is also bonded to an aliphatic hydrocarbon tail. In free histidine, the pK_a of the side-chain imidazole group is 6.0, which is not far from physiological pH. The pK_a values for amino acids depend on the environment and can change significantly within the confines of a protein. Histidine can be found in the protonated or unprotonated forms in proteins, and the properties of many proteins depend on whether individual histidine residues are charged or not. The charged amino acids are often found in the active sites of enzymes, which we will study in Chapter 23.

The amino acids phenylalanine, tryptophan, and tyrosine have aromatic rings in their side chains. They are important for a number of reasons. As a matter of practicality, these amino acids allow us to locate and measure proteins because the aromatic rings absorb strongly at 280 nm and can be detected using a spectrophotometer. These amino acids are also very important physiologically because they are both key precursors to neurotransmitters (substances involved in the transmission of nerve impulses). Tryptophan is converted to serotonin, more properly called 5-hydroxytryptamine, which has a calming effect. Very low levels of serotonin are associated with depression, whereas extremely high levels produce a manic state. Manic-depressive schizophrenia (also called bipolar disorder) can be managed by controlling the levels of serotonin and its further metabolites.

Tyrosine, itself normally derived from phenylalanine, is converted to the neurotransmitter class called catecholamines, which includes epinephrine, commonly known by its proprietary name, adrenalin.

L-Dihydroxyphenylalanine (L-dopa) is an intermediate in the conversion of tyrosine. Lower than normal levels of L-dopa are involved in Parkinson's disease. Tyrosine or phenylalanine supplements might increase the levels of dopamine, although L-dopa, the immediate precursor, is usually prescribed because it passes into the brain quickly through the blood–brain barrier.

Tyrosine and phenylalanine are precursors to norepinephrine and epinephrine, both of which are stimulatory. Epinephrine is commonly known as the "flight or fight" hormone. It causes the release of glucose and other nutrients into the blood and stimulates brain function.

It has been suggested that tyrosine and phenylalanine may have unexpected effects in some people. For example, a growing body of evidence indicates that some people get headaches from the phenylalanine in aspartame, an artificial sweetner commonly found in diet soft drinks. Some people insist that supplements of tyrosine give them a morning lift and that tryptophan helps them sleep at night. Milk proteins have high levels of tryptophan; a glass of warm milk before bed is widely believed to be an aid in inducing sleep.

22.5 | What Are Uncommon Amino Acids?

Many other amino acids in addition to the ones listed in Table 22.1 are known to exist. They occur in some, but by no means all, proteins. Figure 22.3 shows some examples of the many possibilities. These uncommon amino acids are derived from the common amino acids and are produced by modification of the parent amino acid after the protein is synthesized by the organism in a process called post-translational modification (Chapter 26). Hydroxyproline and hydroxylysine differ from their parent amino acids in that they have hydroxyl groups on their side chains; they are found only in a few connective tissue proteins, such as collagen. Thyroxine differs from tyrosine in that it has an extra iodine-containing aromatic group on the side chain; it is found only in the thyroid gland, where it is formed by post-translational modification of tyrosine residues in the protein thyroglobulin. Thyroxine is then released as a hormone by proteolysis of thyroglobulin.

Figure 22.3 Structures of hydroxyproline, hydroxylysine, and thyroxine. The structures of the parent amino acids—proline for hydroxyproline, lysine for hydroxylysine, and tyrosine for thyroxine—are shown for comparison. All amino acids are shown in their predominant ionic forms at pH 7.

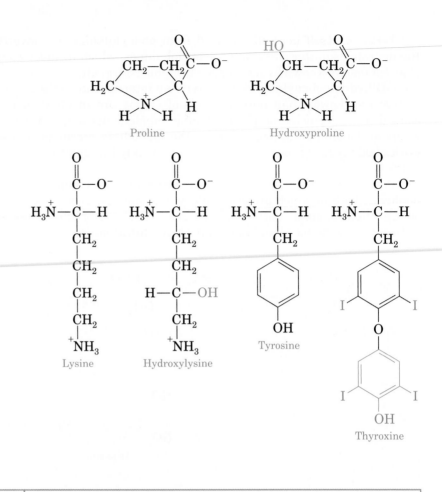

Proline

Hydroxyproline

Lysine

Hydroxylysine

Tyrosine

Thyroxine

22.6 | How Do Amino Acids Combine to Form Proteins?

Each amino acid has a carboxyl group and an amino group. In Chapter 19, we saw that a carboxylic acid and an amine could be combined to form an amide:

$$R-\overset{\overset{\displaystyle O}{\|}}{C}-O^- + R'-NH_3^+ \longrightarrow R-\overset{\overset{\displaystyle O}{\|}}{C}-NH-R' + H_2O$$

In the same way, the —COO⁻ group of one amino acid molecule—say, glycine—can combine with the —NH₃⁺ group of a second molecule—say, alanine:

Glycine

Alanine

Glycylalanine
(Gly—Ala)

This reaction takes place in the cells by a mechanism that we will examine in Section 26.5. The product is an amide. The two amino acids are joined together by a **peptide bond** (also called a **peptide linkage**). The product is a **dipeptide.**

Peptide bond An amide bond that links two amino acids

It is important to realize that glycine and alanine could also be linked the other way:

Alanine   Glycine   Alanylglycine (Ala—Gly)

In this case, we get a *different* dipeptide. The two dipeptides are constitutional isomers, of course; they are different compounds in all respects, with different properties. The phrase "do much, talk little" has the same words as "do little, talk much," but the meaning of the two are quite different. In the same way, the order of amino acids in a peptide or protein is critical to both the structure and function.

EXAMPLE 22.1

Show how to form the dipeptide aspartylserine (Asp—Ser).

Solution
The name implies that this dipeptide is made of two amino acids, aspartic acid (Asp) and serine (Ser), with the amide being formed between the α-carboxyl group of aspartic acid and the α-amino group of serine. Therefore, we write the formula of aspartic acid with its amino group on the left side. Next, we place the formula of serine to the right, with its amino group facing the α-carboxyl group of aspartic acid. Finally, we eliminate a water molecule between the $-COO^-$ and $-NH_3^+$ groups that are next to each other, forming the peptide bond:

Problem 22.1
Show how to form the dipeptide valylphenylalanine (Val—Phe).

Any two amino acids, whether the same or different, can be linked together to form dipeptides in a similar manner. But the possibilities do not end there. Each dipeptide still contains a $-COO^-$ and an $-NH_3^+$ group.

CHEMICAL CONNECTIONS 22A

Glutathione

A very important tripeptide present in high concentrations in all tissues is glutathione. Glutathione contains L-glutamic acid, L-cysteine, and glycine. Its structure is unusual because the glutamic acid is linked to the cysteine by its γ-carboxyl group (the side-chain carboxyl) rather than by the α-carboxyl group, as is usual in most peptides and proteins:

Glutathione functions in the cells as a general protective agent. Oxidizing agents that would damage the cells, such as peroxides, will oxidize glutathione instead (the cysteine portion; see also Section 22.4), thereby protecting the proteins and nucleic acids. Many foreign chemicals also become bonded to glutathione, so in a sense it acts as a detoxifying agent.

$$\overset{+}{H_3}N-CH-CH_2-CH_2-\underset{\underset{O}{\|}}{C}-NH-CH-\underset{\underset{O}{\|}}{C}-NH-CH_2-COO^-$$

Glutathione
(Glu—Cys—Gly)

We can, therefore, add a third amino acid to alanylglycine—say, lysine: The product is a **tripeptide.** Because it also contains a —COO⁻ and an —NH₃⁺ group, we can continue the process to get a tetrapeptide, a pentapeptide, and so on, until we have a chain containing hundreds or even thousands of amino acids. These chains of amino acids are the proteins that serve so many important functions in living organisms.

Ala—Gly

Lys

Ala—Gly—Lys
A tripeptide

C-terminus The amino acid at the end of a peptide that has a free α-carboxyl group

N-terminus The amino acid at the end of a peptide that has a free α-amino group

A word must be said about the terms used to describe these compounds. The shortest chains are often simply called **peptides,** longer ones are **polypeptides,** and still longer ones are **proteins,** but chemists differ about where to draw the line. Many chemists use the terms "polypeptide" and "protein" almost interchangeably. In this book, we will consider a protein to be a polypeptide chain that contains a minimum of 30 to 50 amino acids. The amino acids in a chain are often called **residues.** It is customary to use either the one-letter or the three-letter abbreviations shown in Table 22.1 to represent peptides and proteins. For example, the tripeptide shown on the previous page, alanylglycyllysine, is AGK or Ala—Gly—Lys. The **C-terminal amino acid** or **C-terminus** is the residue with the free α-COO⁻ group (lysine in Ala—Gly—Lys), and the **N-terminal amino acid** or **N-terminus** is the residue with the free α-NH$_3^+$ group (alanine in Ala—Gly—Lys). It is the universal custom to write peptide and protein chains with the N-terminal residue on the left. This decision is not as arbitrary as it might seem. We read left to right, and proteins are synthesized from N-terminus to C-terminus, as we will see in Chapter 26.

22.7 | What Are the Properties of Proteins?

The continuing pattern of peptide bonds forms the backbone of the peptide or protein. The R groups are called the **side chains.** The six atoms of the peptide backbone lie in the same plane (Figure 22.4), and two adjacent peptide bonds can rotate relative to one another about the C—N and C—C bonds.

The 20 different amino acid side chains supply variety and determine the physical and chemical properties of proteins. Among these properties, acid–base behavior is one of the most important. Like amino acids (Section 22.3), proteins behave as zwitterions. The side chains of glutamic and aspartic acids provide acidic groups, whereas lysine and arginine provide

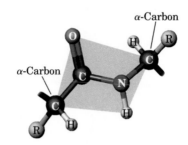

Figure 22.4   Six atoms of the peptide backbone lie in an imaginary (shaded) plane.

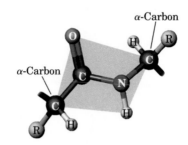

CHEMICAL CONNECTIONS 22B

AGE and Aging

A reaction can take place between a primary amine and an aldehyde or a ketone, linking the two molecules (shown here for an aldehyde):

$$R—\overset{\overset{\displaystyle O}{\|}}{C}—H + H_2N—R' \longrightarrow R—CH{=}N—R' + H_2O$$

An imine

Because proteins have NH$_2$ groups and carbohydrates have aldehyde or ketone groups, they can undergo this reaction, establishing a link between a sugar and a protein molecule. When this reaction is not catalyzed by enzymes, it is called *glycation* of proteins. The process, however, does not stop there. When these linked products are heated in a test tube, high-molecular-weight, water-insoluble, brownish complexes form. These complexes are called **advanced glycation end-products (AGE).** In the body, they cannot be heated, but the same result happens over long periods of time.

The longer we live, and the higher the blood sugar concentration becomes, the more AGE products accumulate in the body. These AGEs can alter the function of proteins. Such AGE-dependent changes are thought to contribute to circulation, joint, and vision problems in people with diabetes. People with diabetes have high blood sugar due to a lack of transport of glucose out of the blood and into the cells. AGE products show up in all of the afflicted

organs of diabetic patients: in the lens of the eye (cataracts), in the capillary blood vessels of the retina (diabetic retinopathy), and in the glomeruli of the kidneys (kidney failure). AGEs have been linked to atherosclerosis, as AGE-modified cells can bind to endothelial cells in blood vessels. AGE-modified collagen causes stiffening of arteries.

For people who do not have diabetes, these harmful protein modifications become disturbing only in an individual's advanced years. In a young person, metabolism functions properly and the AGE products decompose and are eliminated from the body. In an older person, metabolism slows and the AGE products accumulate. The AGE products themselves are thought to enhance oxidative damages.

Scientists are searching for ways to combat the harmful effects of AGEs. One approach is to use antioxidants, including the B vitamin, thiamine. A few other anti-AGE drugs have been developed, including aminoguanidine and metformin. Both of these drugs have been studied in animal models, but have not yet reached large-scale use in humans. Another approach that is being studied for several metabolic problems, including normal aging, is caloric restriction. A vast amount of evidence from animal models and human models alike indicates that lifespan can be extended by living a lean existence. This lifestyle has also been shown to reduce the level of AGEs.

basic groups (histidine does as well, but this side chain is less basic than the other two). (See the structures of these amino acids in Figure 22.1.)

The isoelectric point of a protein occurs at the pH at which there are an equal number of positive and negative charges (the protein has no *net* charge). At any pH above the isoelectric point, the protein molecules have a net negative charge; at any pH below the isoelectric point, they have a net positive charge. Some proteins, such as hemoglobin, have an almost equal number of acidic and basic groups; the isoelectric point of hemoglobin is at pH 6.8. Others, such as serum albumin, have more acidic groups than basic groups; the isoelectric point of this protein is at pH 4.9. In each case, however, because proteins behave like zwitterions, they act as buffers—for example, in the blood (Figure 22.5).

The water solubility of large molecules such as proteins often depends on the repulsive forces between like charges on their surfaces. When protein molecules are at a pH at which they have a net positive or negative charge, the presence of these like charges causes the protein molecules to repel one another. These repulsive forces are smallest at the isoelectric point, when the net charges are zero. When there are no repulsive forces, the protein molecules tend to clump together to form aggregates of two or more molecules, reducing their solubility. As a consequence, *proteins are least soluble in water at their isoelectric points and can be precipitated from their solutions.*

As we pointed out in Section 22.1, proteins have many functions. To understand these functions, we must look at four levels of organization in their structures. The *primary structure* describes the linear sequence of amino acids in the polypeptide chain. The *secondary structure* refers to certain repeating patterns, such as the α-helix conformation or the pleated sheet (Section 22.9), or the absence of a repeating pattern, as with the random coil (Section 22.9). The *tertiary structure* describes the overall confor-

Figure 22.5 Schematic diagram of a protein (*a*) at its isoelectric point and its buffering action when (*b*) H⁺ or (*c*) OH⁻ ions are added.

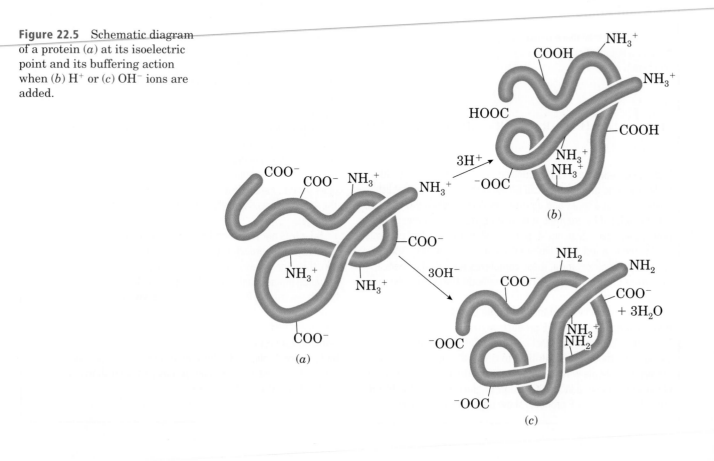

mation of the polypeptide chain (Section 22.10). The *quaternary structure* (Section 22.11) applies mainly to proteins containing more than one polypeptide chain (subunit) and deals with how the different chains are spatially related to one another.

Primary Structure The sequence of amino acids in a protein

22.8 | What Is the Primary Structure of Proteins?

Very simply, the **primary structure** of a protein consists of the sequence of amino acids that makes up the chain. Each of the very large number of peptide and protein molecules in biological organisms has a different sequence of amino acids—and that sequence allows the protein to carry out its function, whatever it may be.

How can so many different proteins arise from different sequences of 20 amino acids? Let us look at a little arithmetic, starting with a dipeptide. How many different dipeptides can be made from 20 amino acids? There are 20 possibilities for the N-terminal amino acid, and for each of these 20 there are 20 possibilities for the C-terminal amino acid. This means that there are $20 \times 20 = 400$ different dipeptides possible from the 20 amino acids. What about tripeptides? We can form a tripeptide by taking any of the 400 dipeptides and adding any of the 20 amino acids. Thus, there are $20 \times 20 \times 20 = 8000$ tripeptides, all different. It is easy to see that we can calculate the total number of possible peptides or proteins for a chain of n amino acids simply by raising 20 to the nth power (20^n).

Taking a typical small protein to be one with 60 amino acid residues, the number of proteins that can be made from the 20 amino acids is $20^{60} = 10^{78}$. This is an enormous number, possibly greater than the total number of atoms in the universe. Clearly, only a tiny fraction of all possible protein molecules have ever been made by biological organisms.

Each peptide or protein in the body has its own unique sequence of amino acids. As with naming of peptides, the *assignment of positions of the amino acids in the sequence starts at the N-terminal end.* Thus, in Figure 22.6, glycine is in the number 1 position on the A chain and phenylalanine is number 1 on the B chain. We mentioned that proteins also have secondary, tertiary, and, in some cases, quaternary structures. We will deal with these in Sections 22.9, 22.10, and 22.11, but here we can say that *the primary structure of a protein determines to a large extent the native* (most frequently occurring) *secondary and tertiary structures.* That is, the particular sequence of amino acids on the chain enables the whole chain to fold and curl in such a way as to assume its final shape. As we will see in Section 22.12, without its particular three-dimensional shape, a protein cannot function.

Just how important is the exact amino acid sequence to the function of a protein? Can a protein perform the same function if its sequence is a little different? The answer to this question is that a change in amino acid sequence may or may not matter, depending on what kind of a change it is. Consider, as an example, cytochrome c, which is a protein of terrestrial vertebrates. Its chain consists of 104 amino acid residues. It performs the same function (electron transport) in humans, chimpanzees, sheep, and other animals. While humans and chimpanzees have exactly the same amino acid sequence of this protein, sheep cytochrome c differs in 10 positions out of the 104. (You can find more about biochemical evolution in Chemical Connections 26F.)

Another example is the hormone insulin. Human insulin consists of two chains having a total of 51 amino acids. The two chains are connected by disulfide bonds. Figure 22.6 shows the sequence of amino acids. Insulin is necessary for proper utilization of carbohydrates (Section 28.1), and people with severe diabetes (Chemical Connections 24G) must take insulin injections.

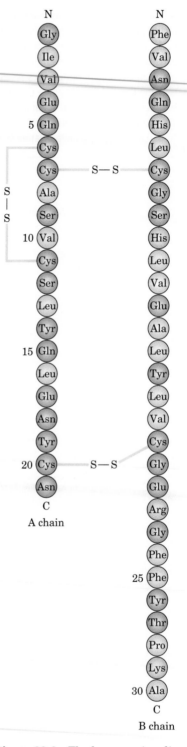

Figure 22.6 The hormone insulin consists of two polypeptide chains, A and B, held together by two disulfide cross-bridges (S—S). The sequence shown is for bovine insulin.

CHEMICAL CONNECTIONS 22C

The Use of Human Insulin

Although human insulin, as manufactured by recombinant DNA techniques (see Section 26.8), is on the market, many diabetic patients continue to use hog or sheep insulin because it is less expensive. Changing from animal to human insulin creates an occasional problem for diabetics. All diabetics experience an insulin reaction (hypoglycemia) when the insulin level in the blood is too high relative to the blood sugar level. Hypoglycemia is preceded by symptoms of hunger, sweating, and poor coordination. These symptoms, called hypoglycemic awareness, signal the patient that hypo-

glycemia is coming and that it must be reversed, which the patient can do by eating sugar.

Some diabetics who changed from animal to human insulin reported that the hypoglycemic awareness from recombinant DNA human insulin is not as strong as that from animal insulin. This lack of recognition can create some hazards, and the effect is probably due to different rates of absorption in the body. The literature supplied with human insulin now incorporates a warning that hypoglycemic awareness may be altered.

The amount of human insulin available is far too small to meet the need for it, so bovine insulin (from cattle) or insulin from hogs or sheep is used instead. Insulin from these sources is similar, but not identical, to human insulin. The differences are entirely in the 8, 9, and 10 positions of the A chain and the C-terminal position (30) of the B chain, as shown in Table 22.2. The remainder of the molecule is the same in all four varieties of insulin. Despite the slight differences in structure, all of these insulins perform the same function and even can be used by humans. However, none of the other three is quite as effective in humans as human insulin. This is one of the reasons that recombinant DNA techniques are now used to produce human insulin from bacteria (Section 26.8 and Chemical Connections 22C).

Another factor showing the effect of substituting one amino acid for another is that sometimes patients become allergic to, say, bovine insulin but can switch to hog or sheep insulin without experiencing an allergic reaction.

In contrast to the previous examples, some small changes in amino acid sequence make a great deal of difference. Consider two peptide hormones, oxytocin and vasopressin (Figure 22.7). These nonapeptides have identical structures, including a disulfide bond, except for different amino acids in positions 2 and 7. Yet their biological functions are quite different. Vasopressin is an antidiuretic hormone. It increases the amount of water reabsorbed by the kidneys and raises blood pressure. Oxytocin has no effect on water in the kidneys and slightly lowers blood pressure. It affects contractions of the uterus in childbirth and the muscles in the breast that aid in the secretion of milk. Vasopressin also stimulates uterine contractions, albeit to a much lesser extent than oxytocin.

Another instance where a minor change makes a major difference is in the blood protein hemoglobin. A change in only one amino acid in a chain

Table 22.2 Amino Acid Sequence Differences for Human, Bovine, Hog, and Sheep Insulin

	A Chain			B Chain
	8	9	10	30
Human	—Thr	—Ser	—Ile—	—Thr
Bovine	—Ala	—Ser	—Val—	—Ala
Hog	—Thr	—Ser	—Ile—	—Ala
Sheep	—Ala	—Gly	—Val—	—Ala

CHEMICAL CONNECTIONS 22D

Sickle Cell Anemia

Normal adult human hemoglobin has two alpha chains and two beta chains (see Figure 22.15). Some people, however, have a slightly different kind of hemoglobin in their blood. This hemoglobin (called HbS) differs from the normal type only in the beta chains and only in one position on these two chains: The glutamic acid in the sixth position of normal Hb is replaced by a valine residue in HbS.

	4	5	6	7	8	9
Normal Hb	—Thr—Pro—		Glu—	Glu—	Lys—	Ala—
Sickle cell Hb	—Thr—Pro—		Val—	Glu—	Lys—	Ala—

This change affects only two positions in a molecule containing 574 amino acid residues, yet it is enough to produce a very serious disease, **sickle cell anemia.**

Red blood cells carrying HbS behave normally when there is an ample oxygen supply. When the oxygen pressure decreases, however, the red blood cells become sickle-shaped, as shown in the figure. This malformation occurs in the capillaries. As a result of this change in shape, the cells may clog the capillaries. The body's defenses destroy the clogging cells, and the loss of the blood cells causes anemia.

This change at only a single position of a chain consisting of 146 amino acids is severe enough to cause a high death rate. A child who inherits two genes programmed to produce sickle cell hemoglobin (a homozygote) has an 80% smaller chance of surviving to adulthood than a child with only one such gene (a heterozygote) or a child with two normal genes. Despite the high mortality of homozygotes, the genetic trait survives. In central Africa, 40% of the population in malaria-ridden areas carry the sickle cell gene, and 4% are homozygotes. It seems that the sickle cell genes help to acquire immunity against malaria in early childhood so that in malaria-ridden areas the transmission of these genes is advantageous.

There is no known cure for sickle cell anemia. Recently the U.S. Food and Drug Administration approved hydroxyurea (sold

Blood cells from a patient with sickle cell anemia. Both normal cells (round) and sickle cells (shriveled) are visible.

under the name Droxia) to treat and control the symptoms of the disease.

Hydroxyurea

Hydroxyurea prompts the bone marrow to manufacture fetal hemoglobin (HbF), which does not have beta chains where the mutation occurs. Thus red blood cells containing HbF do not sickle and do not clog the capillaries. With hydroxyurea therapy, the bone marrow still manufactures mutated HbS, but the presence of cells with fetal hemoglobin dilutes the concentration of the sickling cells, thereby relieving the symptoms of the disease.

$$9 \quad\quad\quad 4 \quad 3 \quad 2 \quad 1$$
Cys—S—S—Cys—Pro—Arg—Gly—NH$_2$
8 Tyr   Asn 5
   Phe—Gln
    7   6

Vasopressin

$$9 \quad\quad\quad 4 \quad 3 \quad 2 \quad 1$$
Cys—S—S—Cys—Pro—Leu—Gly—NH$_2$
8 Tyr   Asn 5
   Ile—Gln
    7   6

Oxytocin

Figure 22.7 The structures of vasopressin and oxytocin. Differences are shown in color.

of 146 is enough to cause a fatal disease—sickle cell anemia (Chemical Connections 22D).

In some cases, slight changes in amino acid sequence make little or no difference to the functioning of peptides and proteins, but most times, the sequence is highly important. The sequences of 10,000 protein and peptide molecules have now been determined. The methods for doing so are complicated and will not be discussed in this book.

22.9 | What Is the Secondary Structure of a Protein?

Secondary Structure A repetitive structure of the protein backbone

Alpha (α)-helix A secondary structure where the protein folds into a coil held together by hydrogen bonds parallel to the axis of the coil

Beta (β)-pleated sheet A secondary protein structure in which the backbone of two protein chains in the same or different molecules is held together by hydrogen bonds

Proteins can fold or align themselves in such a manner that certain patterns repeat themselves. These repeating patterns are referred to as **secondary structures.** The two most common secondary structures encountered in proteins are the α-helix and the β-pleated sheet (Figure 22.8), which were proposed by Linus Pauling and Robert Corey in the 1940s. In contrast, those protein conformations that do not exhibit a repeated pattern are called random coils (Figure 22.9).

In the **α-helix** form, a single protein chain twists in such a manner that its shape resembles a right-handed coiled spring—that is, a helix. The shape of the helix is maintained by numerous **intramolecular hydrogen bonds** that exist between the backbone —C=O and H—N— groups. As shown in Figure 22.8, there is a hydrogen bond between the —C=O oxygen atom of each peptide linkage and the —N—H hydrogen atom of another peptide linkage four amino acid residues farther along the chain. These hydrogen bonds are in just the right position to cause the molecule (or a portion of it) to maintain a helical shape. Each —N—H points upward and each C=O points downward, roughly parallel to the axis of the helix. All the amino acid side chains point outward from the helix.

The other important orderly structure in proteins is the **β-pleated sheet.** In this case, the orderly alignment of protein chains is maintained by **intermolecular** or **intramolecular hydrogen bonds.** The β-sheet structure can occur between molecules when polypeptide chains run parallel (all N-terminal ends on one side) or antiparallel (neighboring N-terminal ends

Figure 22.8 (a) The α-helix. (b) The β-pleated sheet structure.

Figure 22.9   A random coil.

Figure 22.10   Schematic structure of the enzyme carboxypeptidase. The β-pleated sheet portions are shown in blue, the green structures are the α-helix portions, and the orange strings are the random coil areas.

on opposite sides). β-Pleated sheets can also occur intramolecularly, when the polypeptide chain makes a U-turn, forming a hairpin structure, and the pleated sheet is antiparallel (Figure 22.8).

In all secondary structures, the hydrogen bonding is between backbone —C=O and H—N— groups, a characteristic that distinguishes between secondary and tertiary structures. In the latter, as we shall see, the hydrogen bonding can take place between R groups on the side chains.

Few proteins have predominantly α-helix or β-sheet structures. Most proteins, especially globular ones, have only certain portions of their molecules in these conformations. The rest of the molecule consists of **random coil.** Many globular proteins contain all three kinds of secondary structures in different parts of their molecules: α-helix, β-sheet, and random coil. Figure 22.10 shows a schematic representation of such a structure.

Keratin, a fibrous protein of hair, fingernails, horns, and wool, is one protein that does have a predominantly α-helix structure. Silk is made of fibroin, another fibrous protein, which exists mainly in the β-pleated sheet form. Silkworm silk and especially spider silk exhibit a combination of strength and toughness that is unmatched by high-performance synthetic fibers. In its primary structure, silk contains sections that consist of only alanine (25%) and glycine (42%). The formation of β-pleated sheets, largely by the alanine sections, allows microcrystals to orient themselves along the fiber axis, which accounts for the material's superior tensile strength.

Another repeating pattern classified as a secondary structure is the **extended helix** of collagen (Figure 22.11). It is quite different from the α-helix. Collagen is the structural protein of connective tissues (bone, cartilage, tendon, blood vessels, skin), where it provides strength and elasticity. The most abundant protein in humans, it makes up about 30% by weight of all the body's protein. The extended helix structure is made possible by the primary structure of collagen. Each strand of collagen consists of repetitive units that can be symbolized as Gly—X—Y; that is, every third amino acid in the chain is glycine. Glycine, of course, has the shortest side chain (—H) of all amino acids. About one third of the X amino acid is proline, and the Y is often hydroxyproline.

Figure 22.11   The triple helix of collagen.

22.10 | What Is the Tertiary Structure of a Protein?

Tertiary Structure The complete three-dimensional arrangement of the atoms in a protein

The **tertiary structure** of a protein is the three-dimensional arrangement of every atom in the molecule. Unlike the secondary structure, it includes interactions of the side chains, and not just the peptide backbone. In general, tertiary structures are stabilized five ways:

1. **Covalent Bonds** The covalent bond most often involved in stabilization of the tertiary structure of proteins is the disulfide bond. In Section 22.4, we noted that the amino acid cysteine is easily converted to the dimer cystine. When a cysteine residue is in one chain and another cysteine residue is in another chain (or in another part of the same chain), formation of a disulfide bond provides a covalent linkage that binds together the two chains or the two parts of the same chain:

$$\text{—SH HS—} \xrightarrow{[O]} \text{—S—S—}$$

Examples of both types are found in the structure of insulin (Figure 22.6).

2. **Hydrogen Bonding** In Section 22.9, we saw that secondary structures are stabilized by hydrogen bonding between backbone —C=O and —N—H groups. Tertiary structures are stabilized by hydrogen bonding between polar groups on side chains or between side chains and the peptide backbone [Figure 22.12(a)].

3. **Salt Bridges** Salt bridges, also called electrostatic attractions, occur between two amino acids with ionized side chains—that is, between an acidic amino acid ($-COO^-$) and a basic amino acid ($-NH_3^+$ or $=NH_2^+$) side chain. The two are held together by simple ion–ion attraction [Figure 22.12(b)].

4. **Hydrophobic Interactions** In aqueous solution, globular proteins usually turn their polar groups outward, toward the aqueous solvent, and their nonpolar groups inward, away from the water molecules. The nonpolar groups prefer to interact with each other, excluding water from these regions. The result is a series of hydrophobic interactions (see Section 21.1) [Figure 22.12(c)]. Although this type of interaction is

(a) (b) (c) (d)

Figure 22.12 Noncovalent interactions that stabilize the tertiary and quaternary structures of proteins: (*a*) hydrogen bonding, (*b*) salt bridge (electrostatic interaction), (*c*) hydrophobic interaction, and (*d*) metal ion coordination.

CHEMICAL CONNECTIONS 22E

Protein/Peptide Conformation–Dependent Diseases

In a number of diseases, a normal protein or peptide becomes pathological when its conformation changes. A common feature of these proteins is the property to self-assemble into β-sheet–forming amyloid (starch-like) plaques. These amyloid structures appear in several diseases.

One example of this process involves the prion protein, the discovery of which brought Stanley Prusiner of the University of California, San Francisco, the Nobel Prize in 1997. Prions are small proteins found in nerve tissue, although their exact function remains a mystery. When prions undergo conformational change, they can cause diseases such as mad cow disease and scrapie in sheep. During the conformational change, the α-helical content of the normal prion protein unfolds and reassembles in the β-sheet form. This new form has the potential to cause more normal prion proteins to undergo conformational change. In humans, it causes spongiform encephalitis; Creutzfeld-Jakob disease is one variant that mainly afflicts elderly people. Although the transmission of this infection from diseased cows to humans is rare, fear of it caused the wholesale slaughter of British cattle in 1998 and, for a while,

an embargo was placed on the importation of such meat in most of Europe and America. β-Amyloid plaques also appear in the brains of patients with Alzheimer's disease (see Chemical Connections 24D).

The modus operandi of prion diseases stumped scientists for many years. On the one hand, human spongiform encephalopathies behave like inheritable diseases, in that they can be traced through families. On the other hand, they behave like infectious diseases which can be acquired from someone else. It is now believed that the mechanism of spread is a combination of the two. There is a genetic component in that a person could have a 100% wild-type prion protein that would not adopt the alternate form. Several mutations that lead to the abnormal prion form have been identified. However, there appears to be a need for a triggering event as well. This characteristic was seen in studies of sheep in New Zealand, where isolated groups were found to have the right mutations to get a prion disease, but none of the sheep did, generation after generation, because they were never infected with a mutant prion.

Schematic representation of a possible mechanism of amyloid fibril formation. After synthesis, the protein is assumed to fold in native (N) secondary structure aided by chaperones. Under certain conditions, the native structure can partially unfold (I) and form sheets of amyloid fibrils or even completely unfold (U) as a random coil.

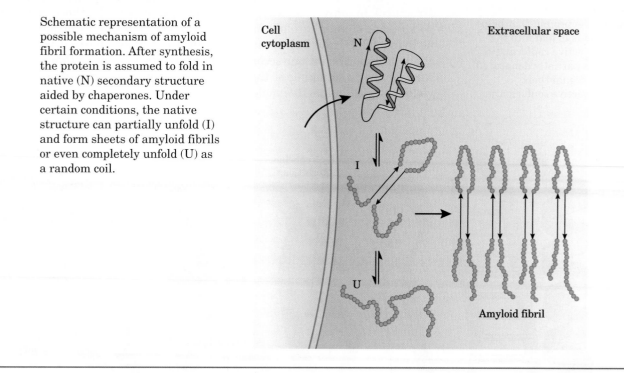

weaker than hydrogen bonding or salt bridges, it usually acts over large surface areas, so that the interactions are collectively strong enough to stabilize a loop or some other tertiary structure formation.

5. **Metal Ion Coordination** Two side chains with the same charge would normally repel each other, but they can also be linked via a metal ion. For example, two glutamic acid side chains ($-COO^-$) would both be attracted to a magnesium ion (Mg^{2+}), forming a bridge. This is one reason that the human body requires certain trace minerals—they are necessary components of proteins [Figure 22.12(d)].

EXAMPLE 22.2

What kind of noncovalent interaction occurs between the side chains of serine and glutamine?

Solution

The side chain of serine ends in an —OH group; that of glutamine ends in an amide, the CO—NH$_2$ group. The two groups can form hydrogen bonds.

Problem 22.2

What kind of noncovalent interaction occurs between the side chains of arginine and glutamic acid?

GOB
Chemistry Now™

Click *Chemistry Interactive* to learn more about **Protein Tertiary Structure**

Chaperone A protein that helps other proteins to fold into the biologically active conformation and enables partially denatured proteins to regain their biologically active conformation

In Section 22.8, we pointed out that the primary structure of a protein largely determines its secondary and tertiary structures. We can now see the reason for this relationship. When the particular R groups are in the proper positions, all of the hydrogen bonds, salt bridges, disulfide linkages, and hydrophobic interactions that stabilize the three-dimensional structure of that molecule can form. Figure 22.13 illustrates the possible combinations of forces that lead to tertiary structure.

The side chains of some proteins allow them to fold (form a tertiary structure) in only one way; other proteins, especially those with long polypeptide chains, can fold in a number of possible ways. Certain proteins in living cells, called **chaperones,** help a newly synthesized polypeptide chain assume the proper secondary and tertiary structures that are necessary for the functioning of that molecule and prevent foldings that would yield biologically inactive molecules.

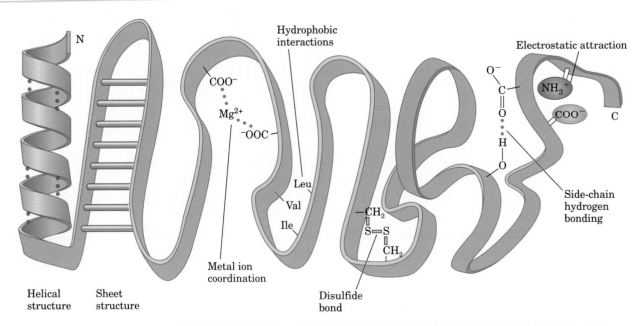

Figure 22.13 Forces that stabilize the tertiary structures of proteins. Note that the helical structure and the sheet structure are two kinds of backbone hydrogen bonding. Although the backbone hydrogen bonding is part of the secondary structure, the conformation of the backbone puts constraints on the possible arrangement of the side chains.

CHEMICAL CONNECTIONS 22F

Proteomics, Ahoy!

Proteins in the body are in a state of dynamic flux. Their multiple functions necessitate that they change constantly: Some are rapidly synthesized, others have their synthesis inhibited; some are degraded, others are modified. The complement of proteins expressed by a genome is called its **proteome.** Today, a concerted effort is being made to catalogue all the proteins in their various forms in a particular cell or a tissue. The name of this venture is *proteomics,* a term coined as an analogy to *genomics* (see Chemical Connections 25E), in which all the genes of an organism and their locations in the chromosomes are determined. The approximately 30,000 genes defined by the Human Genome Project translate into 300,000 to 1 million proteins when alternate splicing and post-translational modifications are considered (Chapter 26).

While a genome remains unchanged to a large extent, the proteins in any particular cell change dramatically as genes are turned on and off in response to its environment. In proteomics, all the proteins and peptides of a cell or tissue are separated and then studied by a variety of procedures, including some very new technologies. The first procedure is necessary to separate proteins from one another. The main way this separation has been achieved is two-dimensional polyacrylamide gel electrophoresis (2-D PAGE; see Chapter 25 for more on electrophoresis). 2-D PAGE can achieve the separation of several thousand different proteins in one gel. High-resolution 2-D PAGE can resolve as many as 10,000 protein spots per gel. In one dimension, the proteins are separated by charge (isoelectric point; Section 22.3); in the second dimension, they are separated by mass. In isoelectric focusing, the proteins migrate in a pH gradient to the pH at which they have no net charge (the isoelectric point). Most commonly, proteins are separated by size in the vertical direction and by isoelectic point in the horizontal direction.

Mass spectrometry is used for mass determination and can be adapted for protein identification. A mass spectrometer separates proteins according to their mass-to-charge (m/Z) ratio. The molecule is ionized by one of several techniques, and the ion is propelled into a mass analyzer by an electric field that resolves each ion according to its m/Z ratio. The detector passes the information to the computer for analysis.

A new technology that holds much promise for the future of protein analysis is **protein microarrays.** They can be used for protein purification, expression profiling, or protein interaction profiling. Many types of substances may be bound to protein arrays, including antibodies, receptors, ligands, nucleic acids, carbohydrates, or chromatographic (cationic, anionic, hydrophobic, hydrophilic) surfaces. Some surfaces have broad specificity and

bind entire classes of proteins; others are highly specific and bind only a few proteins from a complex sample. Some protein arrays contain antibodies (Chapter 31) that are covalently immobilized onto the array surface and that capture corresponding antigens from a complex mixture. Many analyses can follow from this binding. Other proteins of interest can also be immobilized on the array. Bound receptors can reveal ligands, and binding domains for protein–protein interactions can be detected.

The goal of such techniques is to obtain information about the dynamic states of a large number of proteins and the status of a cell or tissue, thereby ascertaining whether it is healthy or pathological.

22.11 What Is the Quaternary Structure of a Protein?

The highest level of protein organization is the **quaternary structure,** which applies to proteins with more than one polypeptide chain. Figure 22.14 summarizes schematically the four levels of protein structure. Quaternary structure determines how the different subunits of the protein fit into an

Quaternary Structure The spatial relationship and interactions between subunits in a protein that has more than one polypeptide chain

Figure 22.14 Primary, secondary, tertiary, and quaternary structures of a protein.

Heme molecules with iron ion

Hemoglobin

Figure 22.15 The quaternary structure of hemoglobin.

Figure 22.16 The structure of heme.

organized whole. The subunits are packed and held together by hydrogen bonds, salt bridges, and hydrophobic interactions—the same forces that operate within tertiary structures.

1. **Hemoglobin** Hemoglobin in adult humans is made of four chains (called globins): two identical α chains of 141 amino acid residues each and two identical β chains of 146 residues each. Figure 22.15 shows how the four chains fit together.

   In hemoglobin, each globin chain surrounds an iron-containing heme unit, the structure of which is shown in Figure 22.16. Proteins that contain non-amino acid portions are called **conjugated proteins.** The non-amino acid portion of a conjugated protein is called a **prosthetic group.** In hemoglobin, the globins are the amino acid portions and the heme units are the prosthetic groups.

   Hemoglobin containing two alpha and two beta chains is not the only kind existing in the human body. In the early development stage of the fetus, the hemoglobin contains two alpha and two gamma chains. The fetal hemoglobin has greater affinity for oxygen than does the adult hemoglobin. In this way, the mother's red blood cells carrying oxygen can pass it to the fetus for its own use. Fetal hemoglobin also alleviates some of the symptoms of sickle cell anemia (see Chemical Connections 22D).

2. **Collagen** Another example of quaternary structure and higher organizations of subunits can be seen in collagen. The triple helix units, called *tropocollagen,* constitute the soluble form of collagen; they are stabilized by hydrogen bonding between the backbones of the three chains. Collagen consists of many tropocollagen units. Tropocollagen is found only in fetal or young connective tissues. With aging, the triple helixes (Figure 22.11) that organize themselves into fibrils cross-link and form insoluble collagen. In collagen, the **cross-linking** consists of covalent bonds that link together two lysine residues on adjacent chains of the helix. This cross-linking of collagen is an example of the tertiary structures that stabilize the three-dimensional conformations of protein molecules.

3. **Integral Membrane Proteins** Integral membrane proteins traverse partly or completely a membrane bilayer. (See Figure 21.2.) An estimated one third of all proteins are integral membrane proteins. To keep the protein stable in the nonpolar environment of a lipid bilayer, it must form quaternary structures in which the outer surface is largely

The designations α and β with respect to hemoglobin have nothing to do with the same designations for the α-helix and β-pleated sheet.

GOB
Chemistry ⚛ Now™
Click *Chemistry Interactive* to learn more about **Protein Quaternary Structures**

Figure 22.17 Integral membrane protein of rhodopsin made of α-helices.

Figure 22.18 A β-barrel of integral membrane protein of the outer membrane of mitochondrion made of eight β-pleated sheets.

nonpolar and interacts with the lipid bilayer. Thus most of the polar groups of the protein must turn inward. Two such quaternary structures exist in integral membrane proteins: (1) often 6 to 10 α-helices crossing the membrane and (2) β-barrels made of 8, 12, 16, or 18 antiparallel β-sheets (Figures 22.17 and 22.18).

22.12 | How Are Proteins Denatured?

Protein conformations are stabilized in their native states by secondary and tertiary structures and through the aggregation of subunits in quaternary structure. Any physical or chemical agent that destroys these stabilizing structures changes the conformation of the protein (Table 22.3). We call this process **denaturation.**

For example, heat cleaves hydrogen bonds, so boiling a protein solution destroys the α-helical and β-pleated sheet structure. In collagen, the triple helices disappear upon boiling, and the molecules have a largely random-coil conformation in the denatured state, which is gelatin. In other proteins, especially globular proteins, heat causes the unfolding of the polypeptide chains; because of subsequent intermolecular protein–protein interactions, precipitation or coagulation then takes place. That is what happens when we boil an egg.

Denaturation The loss of the secondary, tertiary, and quaternary structures of a protein by a chemical or physical agent that leaves the primary structure intact

Table 22.3 Modes of Protein Denaturation (Destruction of Secondary and Higher Structures)

Denaturing Agent	Affected Regions
Heat	H bonds
6 M urea	H bonds
Detergents	Hydrophobic regions
Acids, bases	Salt bridges, H bonds
Salts	Salt bridges
Reducing agents	Disulfide bonds
Heavy metals	Disulfide bonds
Alcohol	Hydration layers

CHEMICAL CONNECTIONS 22G

Quaternary Structure and Allosteric Proteins

Quaternary structure is a property of proteins that consist of more than one polypeptide chain. Each chain is called a subunit. The number of chains can range from two to more than a dozen, and the chains may be identical or different. The chains interact with one another noncovalently via electrostatic attractions, hydrogen bonds, and hydrophobic interactions. As a result of these noncovalent interactions, subtle changes in structure at one site on a protein molecule may cause drastic changes in properties at a distant site. Proteins that exhibit this property are called **allosteric proteins.** Not all multisubunit proteins exhibit allosteric effects, but many do.

A classic illustration of the quaternary structure of proteins and its effect on properties is a comparison of hemoglobin, an allosteric protein, with myoglobin, which consists of a single polypeptide chain. Both hemoglobin and myoglobin bind to oxygen via a heme group (Figure 22.16). As we have seen, hemoglobin is a **tetramer,** a molecule consisting of four polypeptide chains: two α chains and two β chains. The two α chains of hemoglobin are identical, as are the two β chains. The overall structure of hemoglobin is $\alpha_2\beta_2$ in Greek-letter notation. Both the α and β chains of hemoglobin are very similar to the myoglobin chain.

The α chain is 141 residues long, and the β chain is 146 residues long. For comparison, the myoglobin chain is 153 residues long. Many of the amino acids of the α chain, the β chain, and myoglobin are *homologous;* that is, the same amino acid residues occupy the same positions.

The heme group is the same in both myoglobin and hemoglobin. One molecule of myoglobin binds to one oxygen molecule. Four molecules of oxygen can bind to one hemoglobin molecule. Both hemoglobin and myoglobin bind oxygen reversibly, but the binding of oxygen to hemoglobin exhibits **positive cooperativity,** whereas oxygen binding to myoglobin does not. Positive cooperativity means that when one oxygen molecule is bound, it becomes easier for the next molecule to bind. A graph of the oxygen-binding properties of hemoglobin and myoglobin is one of the best ways to illustrate this point.

When the degree of saturation of myoglobin with oxygen is plotted against oxygen pressure, a steady rise is observed until complete saturation is approached and the curve levels off. The oxygen-binding curve of myoglobin is thus said to be **hyperbolic.** In contrast, the oxygen-binding curve for hemoglobin is **sigmoidal.** This shape indicates that the binding of the first oxygen molecule

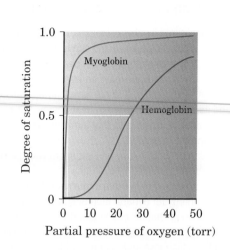

A comparison of the oxygen-binding behavior of myoglobin and hemoglobin. The oxygen-binding curve of myoglobin is hyperbolic, whereas that of hemoglobin is sigmoidal. Myoglobin is 50% saturated with oxygen at 1 torr partial pressure; hemoglobin does not reach 50% saturation until the partial pressure of oxygen reaches 26 torr.

facilitates the binding of the second oxygen, which facilitates the binding of the third oxygen, which in turn facilitates the binding of the fourth oxygen. This is precisely what is meant by the term "cooperative binding."

The two types of behavior are related to the functions of these proteins. Myoglobin has the function of oxygen *storage* in muscle. It must bind strongly to oxygen at very low pressures, and it is 50% saturated at a partial pressure of oxygen of 1 torr (Section 6.2). The function of hemoglobin is oxygen *transport,* and it must be able both to bind strongly to oxygen and to release oxygen easily, depending upon conditions. In the alveoli of lungs (where hemoglobin must bind oxygen for transport to the tissues), the oxygen pressure is 100 torr. At this pressure, hemoglobin is 100% saturated with oxygen. In the capillaries running through active muscles, the pressure of oxygen is 20 torr, corresponding to less than 50% saturation of hemoglobin, which occurs at 26 torr. In other words, hemoglobin gives up oxygen easily in capillaries, where the need for oxygen is great.

The structure of myoglobin.

The structure of hemoglobin.

CHEMICAL CONNECTIONS 22H

Laser Surgery and Protein Denaturation

Proteins can be denatured by physical means, most notably by heat. For instance, bacteria are killed and surgical instruments are sterilized by heat. A special method of heat denaturation that is seeing increasing use in medicine relies on lasers. A laser beam (a highly coherent light beam of a single wavelength) is absorbed by tissues, and its energy is converted to heat energy. This process can be used to cauterize incisions so that a minimal amount of blood is lost during the operation.

Laser beams can be delivered by an instrument called a **fiberscope.** The laser beam is guided through tiny fibers, thousands of which are fitted into a tube only 1 mm in diameter. In this way, the laser delivers the energy for denaturation only where it is needed. It can, for example, seal wounds or join blood vessels without the necessity of cutting through healthy tissues. Fiberscopes have been used successfully to diagnose and treat many bleeding ulcers in the stomach, intestines, and colon.

A novel use of the laser fiberscope is in treating tumors that cannot be reached for surgical removal. A drug called Photofrin, which is activated by light, is given to patients intravenously. The drug in this form is inactive and harmless. The patient then waits 24 to 48 hours, during which time the drug accumulates in the tumor but is removed and excreted from healthy tissues. A laser fiberscope with 630 nm of red light is then directed toward the tumor. An exposure between 10 and 30 minutes is applied. The energy of the laser beam activates the Photofrin, which destroys the tumor.

This technique does not offer a complete cure, because the tumor may grow back or it may have spread before the treatment. The treatment has only one side effect: The patient remains sensitive to exposure to strong light for approximately 30 days (so sunlight must be avoided). Of course, this inconvenience is minor compared to the pain, nausea, hair loss, and other side effects that accompany radiation or chemotherapy of tumors.

In the United States, Photofrin is approved only to treat esophageal cancer. In Europe, Japan, and Canada, it is also used to treat lung, bladder, gastric, and cervical cancers. The light that

Argon krypton laser surgery.

activates Photofrin penetrates only a few millimeters, but the new drugs under development may use radiation in the near-infrared spectrum that can penetrate tumors up to a few centimeters.

The most common use of laser technology in surgery is its application to correct near-sightedness and astigmatism. In a computer-assisted laser surgery process, the curvature of the cornea is changed. Using the energy of the laser beam, physicians remove part of the cornea. In the procedure called photorefractive keratectomy (PRK), the outer layers of the cornea are denatured—that is, burned off. In the LASIK (*laser in situ keratomileusis*) procedure, the surgeon creates a flap or a hinge of the outer layers of the cornea and then with the laser beam burns off a computer-programmed amount under the flap to change the shape of the cornea. After the 5- to 10-minute procedure is complete, the flap is put back, and it heals without stitches. In successful surgeries, patients regain good vision one day after the surgery and no longer need prescription lenses.

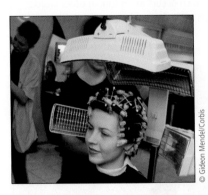

Figure 22.19 Permanent wave alters the shape of hair through reduction and oxidation of disulfides and thiols.

Similar conformational changes can be brought about by the addition of denaturing chemicals. Solutions such as 6 *M* aqueous urea, $H_2N—CO—NH_2$, break hydrogen bonds and cause the unfolding of globular proteins. Surface-active agents (detergents) change protein conformation by opening up the hydrophobic regions, whereas acids, bases, and salts affect both salt bridges and hydrogen bonds.

Reducing agents, such as 2-mercaptoethanol ($HOCH_2CH_2SH$), can break the —S—S— disulfide bonds, reducing them to —SH groups. The processes of permanent waving and straightening of curly hair are examples of the latter effect (Figure 22.19). The protein keratin, which makes up human hair, contains a high percentage of disulfide bonds. These bonds are primarily responsible for the shape of the hair, whether straight or curly. In either permanent waving or straightening, the hair is first treated with a reducing agent that cleaves some of the —S—S— bonds. This treatment allows the molecules to lose their rigid orientations and become more flexible. The hair is then set into the desired shape, using curlers or rollers, and an oxidizing agent is applied. The oxidizing agent reverses the preceding reaction, forming new disulfide bonds, which now hold the molecules together in the desired positions.

Heavy metal ions (for example, Pb^{2+}, Hg^{2+}, and Cd^{2+}) also denature protein by attacking the —SH groups. They form salt bridges, as in —$S^-Hg^{2+-}S$—. This very feature is taken advantage of in the antidote for heavy metal poisoning: raw egg whites and milk. The egg and milk proteins are denatured by the metal ions, forming insoluble precipitates in the stomach. These must be pumped out or removed by inducing vomiting. In this way, the poisonous metal ions are removed from the body. If the antidote is not pumped out of the stomach, the digestive enzymes would degrade the proteins and release the poisonous heavy metal ions, which would then be absorbed into the bloodstream.

Raw egg whites are an antidote to heavy metal poisoning.

Other chemical agents such as alcohol also denature proteins, coagulating them. This process is used in sterilizing the skin before injections. At a concentration of 70%, ethanol penetrates bacteria and kills them by coagulating their proteins, whereas 95% alcohol denatures only surface proteins.

Denaturation changes secondary, tertiary, and quaternary structures. It does not affect primary structures (that is, the sequence of amino acids that make up the chain). If these changes occur to a small extent, denaturation can be reversed. For example, when we remove a denatured protein from a urea solution and put it back into water, it often reassumes its secondary and tertiary structures. This process is called reversible denaturation. In living cells, some denaturation caused by heat can be reversed by chaperones. These proteins help a partially heat-denatured protein to regain its native secondary, tertiary, and quaternary structures. Some denaturation, however, is irreversible. We cannot unboil a hard-boiled egg, for example.

SUMMARY OF KEY QUESTIONS

SECTION 22.1 What Are the Many Functions of Proteins?
- **Proteins** are giant molecules made of amino acids linked together by **peptide bonds.**
- Proteins have many functions: structural (collagen), enzymatic, carrier (hemoglobin), storage (casein), protective (immunoglobulin), and hormonal (insulin).

SECTION 22.2 What Are Amino Acids?
- **Amino acids** are organic compounds containing an amino and a carboxylic acid group.
- The 20 common amino acids found in proteins are classified by their side chains: nonpolar, polar but neutral, acidic, and basic.
- All amino acids in human tissues are L-amino acids.

SECTION 22.3 What Are Zwitterions?
- Amino acids in the solid state, as well as in water, carry both positive and negative charges; they are called **zwitterions.**

- The pH at which the number of positive charges equals the number of negative charges is the **isoelectric point** of an amino acid or protein.

SECTION 22.4 What Determines the Characteristics of Amino Acids?
- Amino acids are nearly identical in most ways except for their side chain (R—) groups.
- It is the unique nature of the side chain that gives an amino acid its particular properties.
- Some amino acids have charged side chains (Glu, Asp, Lys, Arg, His).
- Cysteine is a special amino acid because its side chain (—SH) can form disulfide bridges with another cysteine.
- The aromatic amino acids (Phe, Tyr, Trp) are important physiologically, because they are precursors of neurotransmitters. They also absorb ultraviolet light and allow us to easily measure and locate them.

SECTION 22.5 What Are Uncommon Amino Acids?

- Besides the 20 common amino acids found in proteins, other amino acids are known.
- These amino acids are normally produced after one of the standard amino acids has been incorporated into a protein.
- Examples include hydroxyproline (collagen), hydroxylysine, and thyroxine.

SECTION 22.6 How Do Amino Acids Combine to Form Proteins?

- When the amino group of one amino acid condenses with the carboxyl group of another amino acid, an amide (peptide) bond is formed, with the elimination of water.
- Two amino acids form a dipeptide. Three amino acids form a tripeptide.
- Many amino acids form a **polypeptide chain.** Proteins are made of one or more polypeptide chains.

SECTION 22.8 What Is the Primary Structure of a Protein?

- The linear sequence of amino acids is the **primary structure** of the protein.
- The primary structure is largely responsible for the eventual higher-order structures of proteins.

SECTION 22.9 What Is the Secondary Structure of a Protein?

- The repeating short-range conformations (**α-helix, β-pleated sheet, extended helix of collagen,** and **random coil**) are the **secondary structures** of proteins.

- Secondary structure refers to those repetitive structures that are held together via hydrogen bonds between groups on the peptide backbone only.

SECTION 22.10 What Is the Tertiary Structure of a Protein?

- The **tertiary structure** is the three-dimensional conformation of the protein molecule.
- Tertiary structures are maintained by covalent cross-links such as **disulfide bonds** and by **salt bridges, hydrogen bonds, metal ion coordination,** and **hydrophobic interactions** between the side chains.

SECTION 22.11 What Is the Quaternary Structure of a Protein?

- The precise fit of polypeptide subunits into an aggregated whole is called the **quaternary structure.**
- Not all proteins have a quaternary structure—only those proteins that have subunits.
- Hemoglobin is an example of a protein that exhibits a quaternary structure.

SECTION 22.12 How Are Proteins Denatured?

- Secondary and tertiary structures stabilize the native conformations of proteins.
- Physical and chemical agents, such as heat or urea, destroy these structures and **denature** proteins.
- Protein functions depend on native conformation; when a protein is denatured, it can no longer carry out its function.
- Some (but not all) denaturation is reversible; in some cases, **chaperone** molecules may reverse denaturation.

PROBLEMS

GOB Chemistry ⟡ Now™

Assess your understanding of this chapter's topics with additional quizzing and conceptual-based problems at **http://now.brookscole.com/gob8** or on the CD.

A blue problem number indicates an applied problem.

■ denotes problems that are available on the GOB ChemistryNow website or CD and are assignable in OWL.

SECTION 22.1 What Are the Many Functions of Proteins?

22.3 What are the functions of (a) ovalbumin and (b) myosin?

22.4 The members of which class of proteins are insoluble in water and can serve as structural materials?

22.5 What is the function of an immunoglobulin?

22.6 What are the two basic types of proteins?

SECTION 22.2 What Are Amino Acids?

22.7 What is the difference in structure between tyrosine and phenylalanine?

22.8 ■ Classify the following amino acids as nonpolar, polar but neutral, acidic, or basic.
- (a) Arginine
- (b) Leucine
- (c) Glutamic acid
- (d) Asparagine
- (e) Tyrosine
- (f) Phenylalanine
- (g) Glycine

22.9 Which amino acid has the highest percentage nitrogen (g N/100 g amino acid)?

22.10 ■ Why does glycine have no D or L form?

22.11 Draw the structure of proline. To which class of heterocyclic compounds does this molecule belong? (See Section 16.1.)

22.12 Which amino acid is also a thiol?

22.13 Why is it necessary to have proteins in our diets?

22.14 Which amino acids in Table 22.1 have more than one stereocenter?

22.15 What are the similarities and differences in the structures of alanine and phenylalanine?

22.16 ■ Draw the structures of L- and D-valine.

SECTION 22.3 What are Zwitterions?

22.17 Why are all amino acids solids at room temperature?

22.18 Show how alanine, in solution at its isoelectric point, acts as a buffer (write equations to show why the pH does not change much if we add an acid or a base).

22.19 ■ Explain why an amino acid cannot exist in an un-ionized form at any pH.

22.20 Draw the structure of valine at pH 1 and at pH 12.

SECTION 22.4 What Determines the Characteristics of Amino Acids?

22.21 Which of the three functional groups on histidine is the most unique?

22.22 How are aromatic amino acids related to neurotransmitters?

22.23 Why is histidine considered a basic amino acid when the pK_a of its side chain is 6.0?

22.24 Which are the acidic amino acids?

22.25 Which are the basic amino acids?

22.26 Why does proline not absorb light at 280 nm?

SECTION 22.5 What Are Uncommon Amino Acids?

22.27 Two of the 20 amino acids can be obtained by hydroxylation of other amino acids listed in Table 22.1. What are those two, and what are their precursor amino acids?

22.28 When a protein contains hydroxyproline, at what point in the production of the protein is the proline hydroxylated?

SECTION 22.6 How Do Amino Acids Combine to Form Proteins?

22.29 Show by chemical equations how alanine and glutamine can be combined to give two different dipeptides.

22.30 A tetrapeptide is abbreviated as DPKH. Which amino acid is at the N-terminal and which is at the C-terminal?

22.31 Draw the structure of a tripeptide made of threonine, arginine, and methionine.

22.32 (a) Use the three-letter abbreviations to write the representation of the following tripeptide:

(b) Which amino acid is the C-terminal end, and which is the N-terminal end?

22.33 A polypeptide chain is made of alternating valine and phenylalanine. Which part of the polypeptide is polar (hydrophilic)?

SECTION 22.7 What Are the Properties of Proteins?

22.34 ■ (a) How many atoms of the peptide bond lie in the same plane?

(b) Which atoms are they?

22.35 (a) Draw the structural formula of the tripeptide met—ser—cys.

(b) Draw the different ionic structures of this tripeptide at pH 2.0, 7.0, and 10.0.

22.36 How can a protein act as a buffer?

22.37 Proteins are least soluble at their isoelectric points. What would happen to a protein precipitated at its isoelectric point if a few drops of dilute HCl were added?

SECTION 22.8 What Is the Primary Structure of a Protein?

22.38 ■ How many different tripeptides can be made (a) using one, two, or three residues each of leucine, threonine, and valine and (b) using all 20 amino acids?

22.39 How many different tetrapeptides can be made (a) if the peptides contain the residues of asparagine, proline, serine, and methionine and (b) if all 20 amino acids can be used?

22.40 How many amino acid residues in the A chain of insulin are the same in insulin from humans, cattle (bovine), hogs, and sheep?

22.41 Based on your knowledge of the chemical properties of amino acid side chains, suggest a substitution for leucine in the primary structure of a protein that would probably not change the character of the protein very much.

SECTION 22.9 What Is the Secondary Structure of a Protein?

22.42 Is a random coil a (a) primary, (b) secondary, (c) tertiary, or (d) quaternary structure? Explain.

22.43 Decide whether the following structures that exist in collagen are primary, secondary, tertiary, or quaternary.

(a) Tropocollagen

(b) Collagen fibril

(c) Collagen fiber

(d) The proline—hydroxyproline—glycine repeating sequence

22.44 Proline is often called an α-helix terminator; that is, it is usually in the random-coil secondary structure following an α-helical portion of a protein chain. Why does proline not fit easily into an α-helix structure?

SECTION 22.10 What Is the Tertiary Structure of a Protein?

22.45 Polyglutamic acid (a polypeptide chain made only of glutamic acid residues) has an α-helix conformation below pH 6.0 and a random-coil conformation above pH 6.0. What is the reason for this conformational change?

22.46 Distinguish between intermolecular and intramolecular hydrogen bonding between backbone groups. Where in protein structures do you find one and where do you find the other?

22.47 Identify the primary, secondary, and tertiary structures in the numbered boxes:

SECTION 22.11 What Is the Quaternary Structure of a Protein?

22.48 If both cysteine residues on the B chain of insulin were changed to alanine residues, how would it affect the quaternary structure of insulin?

22.49 (a) What is the difference in the quaternary structure between fetal hemoglobin and adult hemoglobin?

(b) Which can carry more oxygen?

22.50 Where are the nonpolar side chains of proteins located in an integral membrane protein?

22.51 The cytochrome c protein is important in producing energy from food. It contains a heme surrounded by a polypeptide chain. What kind of structure do these two entities form? To which group of proteins does cytochrome c belong?

22.52 Hemoglobin is an important protein for many reasons and has interesting physical characteristics. How would you classify hemoglobin?

SECTION 22.12 How Are Proteins Denatured?

22.53 In a 6 M urea solution, a protein that contained mostly antiparallel β-sheet became a random coil. Which groups and bonds were affected by urea?

22.54 What kind of changes are necessary to transform a protein having a predominantly α-helical structure into one having a β-pleated sheet structure?

22.55 Which amino acid side chain is most frequently involved in denaturation by reduction?

22.56 What does the reducing agent do in straightening curly hair?

22.57 Silver nitrate is sometimes put into the eyes of newborn infants as a preventive measure against gonorrhea. Silver is a heavy metal. Explain how this treatment may work against bacteria.

22.58 Why do nurses and physicians use 70% alcohol to wipe the skin before giving injections?

Chemical Connections

22.59 (Chemical Connections 22A) Write the structure of the oxidation product of glutathione.

22.60 (Chemical Connections 22B) AGE products become disturbing only in elderly people, even though they also form in younger people. Why don't they harm younger people?

22.61 (Chemical Connections 22C) Define *hypoglycemic awareness*.

22.62 (Chemical Connections 22D) How does hydroxyurea therapy alleviate the symptoms of sickle cell anemia?

22.63 (Chemical Connections 22E) What is the difference in the conformation between normal prion protein and the amyloid prion that causes mad cow disease?

22.64 (Chemical Connections 22F) What is the aim of proteomics?

22.65 (Chemical Connections 22G) Explain the difference in oxygen-binding behavior of myoglobin and hemoglobin.

22.66 (Chemical Connections 22H) How does the fiberscope help to heal bleeding ulcers?

Additional Problems

22.67 Which diseases are associated with amyloid plaques?

22.68 How many different dipeptides can be made (a) using only alanine, tryptophan, glutamic acid, and arginine and (b) using all 20 amino acids?

22.69 Denaturation is usually associated with transitions from helical structures to random coils. If an imaginary process were to transform the keratin in your hair from an α-helix to a β-pleated sheet structure, would you call the process denaturation? Explain.

22.70 ■ Draw the structure of lysine (a) above, (b) below, and (c) at its isoelectric point.

22.71 In collagen, some of the chains of the triple helices in tropocollagen are cross-linked by covalent bonds between two lysine residues. What kind of structure is formed by these cross-links? Explain.

22.72 Considering the vast number of animal and plant species on Earth (including those now extinct) and the large variety of protein molecules in each organism, have all possible protein molecules been used already by some species or other? Explain.

22.73 ■ What kind of noncovalent interaction occurs between the following amino acids?
(a) Valine and isoleucine
(b) Glutamic acid and lysine
(c) Tyrosine and threonine
(d) Alanine and alanine

22.74 How many different decapeptides (peptides containing 10 amino acids each) can be made from the 20 amino acids?

22.75 Which amino acid does not rotate the plane of polarized light?

22.76 Write the expected products of the acid hydrolysis of the following tetrapeptide:

22.77 What charges are on aspartic acid at pH 2.0?

22.78 How many ways can you link the two amino acids, lysine and valine, in a dipeptide? Which of these peptide bonds will you find in proteins?

Looking Ahead

22.79 Enzymes are biological catalysts and usually proteins. They catalyze common organic reactions. Why are amino acids, such as histidine, aspartic acid, and serine, found more often near the reaction catalysis site than amino acids such as leucine and valine?

22.80 Hormones are molecules that are released from one tissue but have their effect in another tissue. Give an example of a hormone encountered in this chapter that would be ineffective if taken orally. Give an example of one that could be effective if taken orally.

22.81 Using what you know about protein denaturation, what is one reason that you must maintain a body temperature in a strict range?

22.82 What is the difference between genomics and proteomics?

22.83 Why does knowing the complete genome of an organism not necessarily tell you about the nature of all the proteins in the organism?

22.84 Why is collagen not a very good source of dietary protein?

Enzymes

HIV-1 protease model complexed with VX-478, a commercially available inhibitor of the virus that causes AIDS. By understanding the structure of enzymes, scientists can design novel drugs to fight diseases.

Vertex Pharmaceuticals, Inc.

GOB
Chemistry·Now™
Look for this logo in the chapter and go to GOB ChemistryNow at
http://now.brookscole.com/gob8
or on the CD for tutorials, simulations, and problems.

23.1 | What Are Enzymes?

The cells in your body are chemical factories. Only a few of the thousands of compounds necessary for the operation of the human organism are obtained from the diet. Instead, most of these substances are synthesized within the cells, which means that hundreds of chemical reactions take place in your cells every second of your life.

Nearly all of these reactions are catalyzed by **enzymes,** which are large molecules that increase the rates of chemical reactions without themselves undergoing any change. Without enzymes to act as biological catalysts, life as we know it would not be possible.

The vast majority of all known enzymes are globular proteins, and we will devote most of our study to protein-based enzymes. However, proteins are not the only biological catalysts. **Ribozymes** are enzymes made of ribonucleic acids. They catalyze the self-cleavage of certain portions of their own molecules and have been implicated in the reaction that generates peptide bonds (Chapter 22). Many biochemists believe that during evolution RNA catalysts emerged first, with protein enzymes arriving on the scene later. (We will learn more about RNA catalysts in Section 25.4.)

Like all catalysts, enzymes do not change the position of equilibrium. That is, enzymes cannot make a reaction take place that would not occur without them. Instead, they increase the reaction rate: They cause reactions to take place faster by lowering the activation energy. As catalysts, enzymes are remarkable in two respects:

1. They are extremely effective, increasing reaction rates by anywhere from 10^9 to 10^{20} times.

2. Most of them are extremely **specific.**

> **Substrate specificity** The limitation of an enzyme to catalyze specific reactions with specific substrates

As an example of their effectiveness, consider the oxidation of glucose. A lump of glucose or even a glucose solution exposed to oxygen under sterile conditions would show no appreciable change for months. In the human body, however, the same glucose is oxidized within seconds.

Every organism has many enzymes—many more than 3000 in a single cell. Most enzymes are very specific, each of them speeding up only one particular reaction or class of reactions. For example, the enzyme urease catalyzes only the hydrolysis of urea and not that of other amides, even closely related ones.

$$(NH_2)_2C{=}O + H_2O \xrightarrow{\text{urease}} 2\,NH_3 + CO_2$$
$$\text{Urea}$$

Another type of specificity can be seen with trypsin, an enzyme that cleaves the peptide bonds of protein molecules—but not every peptide bond, only those on the carboxyl side of lysine and arginine residues (Figure 23.1).

The enzyme carboxypeptidase specifically catalyzes the hydrolysis on only the last amino acid on a protein chain—the one at the C-terminal end. Lipases are less specific: They catalyze the hydrolysis of any triglyceride, but they still do not affect carbohydrates or proteins.

The specificity of enzymes also extends to stereospecificity. The enzyme arginase hydrolyzes the amino acid L-arginine (the naturally occurring form) to a compound called L-ornithine and urea (Section 28.8) but has no effect on its mirror image, D-arginine.

Enzymes are distributed according to the body's need to catalyze specific reactions. A large number of protein-splitting enzymes are in the blood, ready to promote clotting. Digestive enzymes, which also catalyze the hydrolysis of proteins, are located in the secretions of the stomach and pancreas. Even within the cells themselves, some enzymes are localized according to the need for specific reactions. The enzymes that catalyze the oxidation of compounds that are part of the citric acid cycle (Section 27.4) are located in the mitochondria, for example, and special organelles such as lysosomes contain an enzyme (lysozyme) that catalyzes the dissolution of bacterial cell walls.

Figure 23.1 A typical amino acid sequence. The enzyme trypsin catalyzes the hydrolysis of this chain only at the points marked with an arrow (the carboxyl side of lysine and arginine).

CHEMICAL CONNECTIONS 23A

Muscle Relaxants and Enzyme Specificity

In the body, nerves transmit signals to the muscles. Acetylcholine is a neurotransmitter (Section 24.1) that operates between the nerve endings and muscles. It attaches itself to a specific receptor in the muscle end plate. This attachment transmits a signal to the muscle to contract; shortly thereafter, the muscles relax. A specific enzyme, acetylcholinesterase, then catalyzes the hydrolysis of the acetylcholine, removing it from the receptor site and preparing it for the next signal transmission—that is, the next contraction.

Succinylcholine is sufficiently similar to acetylcholine so that it, too, binds to the receptor of the muscle end plate. However, acetyl-

cholinesterase can hydrolyze succinylcholine only very slowly. While it remains bound to the receptor, no new signal can reach the muscle to allow it to contract again. Thus the muscle stays relaxed for a long time.

This feature makes succinylcholine a good muscle relaxant during minor surgery, especially when a tube must be inserted into the bronchus (bronchoscopy). For example, after intravenous administration of 50 mg of succinylcholine, paralysis and respiratory arrest are observed within 30 seconds. While respiration is carried on artificially, the bronchoscopy can be performed within minutes.

Acetylcholine   Acetic acid   Choline

Succinylcholine

23.2 | How Are Enzymes Named and Classified?

Enzymes are commonly given names derived from the reaction that they catalyze and/or the compound or type of compound on which they act. For example, lactate dehydrogenase speeds up the removal of hydrogen from lactate (an oxidation reaction). Acid phosphatase catalyzes the hydrolysis of phosphate ester bonds under acidic conditions. As can be seen from these examples, the names of most enzymes end in "-ase." Some enzymes, however, have older names, which were assigned before their actions were clearly understood. Among these are pepsin, trypsin, and chymotrypsin—all enzymes of the digestive tract.

Enzymes can be classified into six major groups according to the type of reaction they catalyze (see also Table 23.1):

1. **Oxidoreductases** catalyze oxidations and reductions.
2. **Transferases** catalyze the transfer of a group of atoms, such as from one molecule to another.
3. **Hydrolases** catalyze hydrolysis reactions.
4. **Lyases** catalyze the addition of two groups to a double bond or the removal of two groups from adjacent atoms to create a double bond.
5. **Isomerases** catalyze isomerization reactions.
6. **Ligases,** or synthetases, catalyze the joining of two molecules.

Table 23.1   Classification of Enzymes

Class	Typical Example	Reaction Catalyzed	Section Number in This Book
1. Oxidoreductases	Lactate dehydrogenase		28.2
2. Transferases	Aspartate amino transferase or aspartate transaminase		28.8
3. Hydrolases	Acetylcholinesterase		24.3
4. Lyases	Aconitase		27.4
5. Isomerases	Phosphohexose isomerase		28.2
6. Ligases	Tyrosine-tRNA synthetase		26.6

pens because, at the saturation point, substrate molecules are bound to all available active sites of the enzymes. Because the reactions take place at the active sites, once they are all occupied, the reaction is proceeding at its maximum rate. Increasing the substrate concentration can no longer increase the rate because the excess substrate cannot find any active sites to which to bind.

B. Temperature

Temperature affects enzyme activity because it changes the conformation of the enzyme. In uncatalyzed reactions, the rate usually increases as the temperature increases (Section 8.4). Changing the temperature has a different effect on enzyme-catalyzed reactions. When we start at a low temperature (Figure 23.5), an increase in temperature first causes an increase in rate. However, protein conformations are very sensitive to temperature changes. Once the optimal temperature is reached, any further increase in temperature alters the enzyme conformation. The substrate may then not fit properly onto the changed enzyme surface, so the rate of reaction actually *decreases*.

After a *small* temperature increase above the optimum, the decreased rate could be increased again by lowering the temperature because, over a narrow temperature range, changes in conformation are reversible. However, at some higher temperature above the optimum, we reach a point where the protein denatures (Section 22.12); the conformation is then altered irreversibly, and the polypeptide chain cannot refold to its native conformation. At this point, the enzyme is completely inactivated. The inactivation of enzymes at low temperatures is used in the preservation of food by refrigeration.

Most enzymes from bacteria and higher organisms have an optimal temperature around 37°C. However, the enzymes of organisms that live at the ocean floor at 2°C have an optimal temperature in that range. Other organisms live in ocean vents under extreme conditions, and their enzymes have optimal conditions at ranges from 90 to 105°C. The enzymes of these hyperthermophile organisms also have other extreme requirements, such as pressures up to 100 atm, and some of them have an optimal pH in the range of 1 to 4. Enzymes from these hyperthermophiles, especially polymerases that catalyze the polymerization of DNA (Section 25.6), have gained commercial importance.

C. pH

As the pH of its environment changes the conformation of a protein (Section 22.12), we would expect pH-related effects to resemble those observed when the temperature changes. Each enzyme operates best at a certain pH (Figure 23.6). Once again, within a narrow pH range, changes in enzyme activity are reversible. However, at extreme pH values (either acidic or basic), enzymes are denatured irreversibly, and enzyme activity cannot be restored by changing back to the optimal pH.

23.5 | What Are the Mechanisms of Enzyme Action?

We have seen that the action of enzymes is highly specific for a substrate. What kind of mechanism can account for such specificity? About 100 years ago, Arrhenius suggested that catalysts speed up reactions by combining with the substrate to form some kind of intermediate compound. In an enzyme-catalyzed reaction, this intermediate is called the **enzyme–substrate complex.**

Figure 23.4 The effect of substrate concentration on the rate of an enzyme-catalyzed reaction. Enzyme concentration, temperature, and pH are constant.

Figure 23.5 The effect of temperature on the rate of an enzyme-catalyzed reaction. Substrate and enzyme concentrations and pH are constant.

Figure 23.6 The effect of pH on the rate of an enzyme catalyzed reaction. Substrate and enzyme concentrations and temperature are constant.

GOB
Chemistry ⚛ Now™

Click *Chemistry Interactive* to learn more about **Factors Affecting Enzyme Activity**

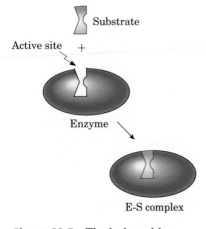

Figure 23.7 The lock-and-key model of the enzyme mechanism.

Lock-and-key model A model explaining the specificity of enzyme action by comparing the active site to a lock and the substrate to a key

Induced-fit model A model explaining the specificity of enzyme action by comparing the active site to a glove and the substrate to a hand

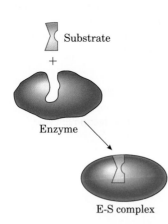

Figure 23.8 The induced-fit model of the enzyme mechanism.

A. Lock-and-Key Model

To account for the high substrate specificity of most enzyme-catalyzed reactions, a number of models have been proposed. The simplest and most frequently referenced is the **lock-and-key model** (Figure 23.7). This model assumes that the enzyme is a rigid, three-dimensional body. The surface that contains the active site has a restricted opening into which only one kind of substrate can fit, just as only the proper key can fit exactly into a lock and turn it open.

According to the lock-and-key model, an enzyme molecule has its particular shape because that shape is necessary to maintain the active site in exactly the conformation required for that particular reaction. An enzyme molecule is very large (typically consisting of 100 to 200 amino acid residues), but the active site is usually composed of only two or a few amino acid residues, which may well be located at different places in the chain. The other amino acids—those that are not part of the active site—are located in the sequence in which we find them because that sequence causes the molecule as a whole to fold up in exactly the required way. This arrangement emphasizes that the shape and the functional groups on the surface of the active site are of utmost importance in recognizing a substrate.

The lock-and-key model was the first to explain the action of enzymes. For most enzymes, however, this model is too restrictive. Enzyme molecules are in a dynamic state, not a static one. There are constant motions within them, so that the active site has some flexibility. Also, while the lock-and-key model does a good job explaining why the enzyme binds to the substrate, if the fit is that perfect, there would be no reason for the reaction to occur, as the enzyme-substrate complex would be too stable.

B. Induced-Fit Model

From x-ray diffraction, we know that the size and shape of the active site cavity change when the substrate enters. To explain this phenomenon, an American biochemist, Daniel Koshland, introduced the **induced-fit model** (Figure 23.8), in which he compared the changes occurring in the shape of the cavity upon substrate binding to the changes in the shape of a glove when a hand is inserted. That is, the enzyme modifies the shape of the active site to accommodate the substrate. Recent experiments during actual catalysis have demonstrated that not only does the shape of the active site change with the binding of substrate, but even in the bound state both the backbone and the side chains of the enzyme are in constant motion.

Both the lock-and-key and the induced-fit models explain the phenomenon of competitive inhibition (Section 23.3). The inhibitor molecule fits into the active site cavity in the same way the substrate does (Figure 23.9), thereby preventing the substrate from entering. The result: Whatever reaction is supposed to take place on the substrate does not occur.

Many cases of noncompetitive inhibition can also be explained by the induced-fit model. In this case, the inhibitor does not bind to the active site but rather binds to a different part of the enzyme. Nevertheless, the binding causes a change in the three-dimensional shape of the enzyme molecule, which so alters the shape of the active site that once the substrate is bound, there can be no catalysis (Figure 23.10).

If we compare enzyme activity in the presence and the absence of an inhibitor, we can tell whether competitive or noncompetitive inhibition is taking place (Figure 23.11). The maximum reaction rate is the same without an inhibitor and in the presence of a competitive inhibitor. The only difference is that this maximum rate is achieved at a low substrate

CHEMICAL CONNECTIONS 23B

Acidic Environment and *Helicobacter*

Certain organisms can live in an extreme pH environment. An example is the bacterium *Helicobacter pylori,* which survives happily in the acidic medium of the stomach wall. These bacteria live under the mucus coating of the stomach and release a toxin that causes ulcers and has been linked to stomach cancer. The role of *H. pylori* in ulceration was discovered only in the last decade, because people did not believe that enzymes of a bacterium could possibly function in such a harsh acidic environment (pH 1.4). In reality, the enzymes in the cytoplasm of *H. pylori* are not exposed to low pH. The bacteria contain a special enzyme, **urease,** which converts urea to the basic ammonia; therefore, despite the acidic environment, *H. pylori* lives in a protective, neutralizing basic cloud. The urease itself, which is found on the surface of the bacteria, is protected against acid denaturation by its quaternary structure. The subunits of urease form a central cavity that contains the active

sites. This cavity is filled with a buffer that protects the active sites before the actual urea hydrolysis starts.

After the destructive role played by *H. pylori* in ulceration was confirmed, new treatments for gastric disorders were designed. Antibiotics such as amoxicillin or tetracyclin can eradicate the bacteria; the ulcers can then heal, and their recurrence is prevented.

H. pylori is also being studied from an evolutionary perspective. It appears that humans co-evolved with the bacteria, and distinct populations of bacteria have been traced to humans' migration across the planet. In addition, doctors are finding that as the incidence of ulcers and stomach cancer has begun decreasing with the gradual eradication of *H. pylori,* the levels of some very aggressive forms of esophageal cancer are increasing, suggesting that there may be some benefit to our coexistence with the bacteria.

concentration with no inhibitor but at a high substrate concentration when an inhibitor is present. This is the true sign of competitive inhibition, because in this scenario the substrate and the inhibitor are competing for the same active site. If the substrate concentration is sufficiently increased, the inhibitor will be displaced from the active site by Le Chatelier's principle.

If the inhibitor is noncompetitive, it cannot be displaced by addition of excess substrate because it is bound to a different site. In this case, the enzyme cannot be restored to its maximum activity, and the maximum rate of the reaction is lower than it would be in the absence of the inhibitor. With a noncompetitive inhibitor, it is always as though less enzyme were

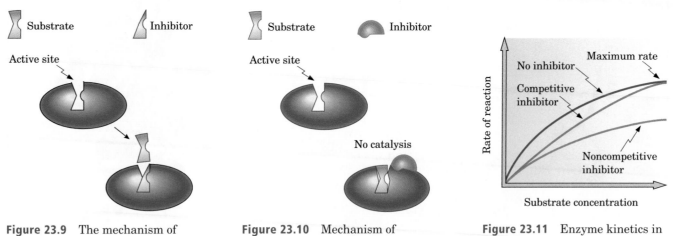

Figure 23.9 The mechanism of competitive inhibition. When a competitive inhibitor enters the active site, the substrate cannot enter.

Figure 23.10 Mechanism of noncompetitive inhibition. The inhibitor binds itself to a site other than the active site (allosterism), thereby changing the conformation of the active site. The substrate still binds but there is no catalysis.

Figure 23.11 Enzyme kinetics in the presence and the absence of inhibitors.

CHEMICAL CONNECTIONS 23C

Active Sites

The perception of the active site as either a rigid cavity (lock-and-key model) or a partly flexible template (induced-fit model) is an oversimplification. Not only is the geometry of the active site important, but so are the specific interactions that take place between the enzyme surface and the substrate. To illustrate, we take a closer look at the active site of pyruvate kinase. This enzyme catalyzes the transfer of the phosphate group from phosphoenol pyruvate (PEP) to ADP, an important step in glycolysis (Section 28.2).

Phosphoenol pyruvate   ADP   Pyruvate   ATP

The active site of the enzyme binds both substrates, PEP and ADP (see figure below left). The rabbit muscle pyruvate kinase has two cofactors, K^+ and either Mn^{2+} or Mg^{2+}. The divalent cation is coordinated to the carbonyl and carboxylate oxygen of the pyruvate substrate and to a glutamate 271 and aspartate 295 residue of the enzyme. (The numbers indicate the position of the amino acid in the sequence.) The nonpolar $=CH_2$ group lies in a hydrophobic pocket formed by an alanine 292, glycine 294, and threonine 327 residue. The K^+ on the other side of the active site is coordinated with the phosphate of the substrate and the serine 76

and asparagine 74 residue of the enzyme. Lysine 269 and arginine 72 are also part of the catalytic apparatus anchoring the ADP. This arrangement of the active site illustrates that specific folding into secondary and tertiary structures is required to bring important functional groups together. The residues of amino acids participating in the active sites are sometimes close in the sequence (asparagine 74 and serine 76), but mostly remain far apart (glutamate 271 and aspartate 295). The figure below right illustrates the secondary and tertiary structures providing such a stable active site.

The active site and substrates of pyruvate kinase.

Ribbon cartoon of pyruvate kinase. Pyruvate, Mg^{2+}, and K^+ are depicted as space-filling models.

available. Competitive and noncompetitive inhibition are the two most common extremes of enzyme inhibition. Many other types of reversible inhibitors exist, but they are beyond the scope of this book.

Enzymes can also be inhibited irreversibly if a compound is bound covalently and permanently to or near the active site. Such inhibition occurs with penicillin, which inhibits the enzyme transpeptidase; this enzyme is necessary for cross-linking bacterial cell walls. Without cross-linking, the bacterial cytoplasm spills out, and the bacteria die (Chemical Connections 18B). Chemical Connections 23D describes two medical applications of inhibitors.

C. Catalytic Power of Enzymes

Both the lock-and-key model and the induced-fit model emphasize the shape of the active site. However, the chemistry at the active site is actually the most important factor. A survey of known active sites of enzymes shows that five amino acids participate in the active sites in more than 65% of all cases. They are, in order of their dominance, His > Cys > Asp > Arg > Glu. A quick glance at Table 22.1 reveals that most of these amino acids have either acidic or basic side chains. Thus acid–base chemistry often underlies the mode of catalysis. The example given in Chemical Connections 23C confirms this relationship. Out of the eleven amino acids in the catalytic site, two are Arg, one is Glu, one is Asp, and two are the related Asn.

We have said that enzymes cannot change the thermodynamic relationships between the substrates and the products of a reaction; rather, they speed up the reaction. But how do they really accomplish this feat? If we look at an energy diagram of a hypothetical reaction, there are reactants on one side and products on the other. The thermodynamic relationship is described by the height difference between the two, as shown in Figure 23.12(a). In any reaction that can be written as follows:

$$A + B \rightleftharpoons C + D$$

before A and B can become C and D, they must pass through a **transition state** where they are something in between. This situation is often thought of as being an "energy hill" that must be scaled. The energy required to climb this hill is the activation energy. Enzymes are powerful catalysts because they lower the energy hill, as shown in Figure 23.12(b). They reduce the activation energy.

How the enzyme reduces the activation energy is very specific to the enzyme and the reaction being catalyzed. As we noted, however, a few amino acids show up in most of the active sites. The specific amino acids in the active site and their exact orientation make it possible for the substrate(s) to bind to the active site and then react to form products. For example, papain is a protease, an enzyme that cleaves peptide bonds as we saw with trypsin. Two critical amino acids are present in the active site of papain (Figure 23.13 on p. 595). The histidine (shown in blue) helps attract the peptide and hold it in the correct orientation via hydrogen bonding (shown as red dashes). The sulfur on the cysteine side chain performs a type of reaction called a **nucleophilic attack** on the carbonyl carbon of the peptide bond, and the C—N bond is broken. Such nucleophilic attacks appear in the vast majority of enzyme mechanisms, and they are possible because of the precise arrangement of the amino acid side chains that can participate in this type of organic reaction.

(a)

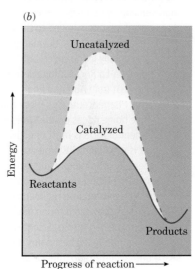

(b)

Figure 23.12 Activation energy profiles. (a) The activation energy profile for a typical reaction. (b) A comparison of the activation energy profiles for catalyzed and uncatalyzed reactions.

Nucleophilic attack A chemical reaction where an electron-rich atom, such as oxygen or sulfur, bonds to an electron-deficient atom, such as carbonyl carbon

Chemistry Now™

Click *Chemistry Interactive* to learn more about **Mechanisms of Enzyme Action**

CHEMICAL CONNECTIONS 23D

Medical Uses of Inhibitors

A key strategy in the treatment of acquired immunodeficiency syndrome (AIDS) has been to develop specific inhibitors that selectively block the actions of enzymes unique to the human immunodeficiency virus (HIV), which causes AIDS. Many laboratories are working on this approach to the development of therapeutic agents.

One of the most important targets is HIV protease, an enzyme essential to the production of new virus particles in infected cells. HIV protease is unique to this virus. It catalyzes the processing of viral proteins in an infected cell. Without these proteins, viable virus particles cannot be released to cause further infection. The structure of HIV protease, including its active site, was elucidated by x-ray crystallography. With this structure in mind, scientists then designed and synthesized competitive inhibitors to bind to the active site. Improvements were made in the drug design by obtaining structures of a series of inhibitors bound to the active site of HIV protease. These structures were also elucidated by x-ray crystallography. This process eventually led to several HIV protease inhibitors: saquinavir from Hoffmann-La Roche, ritonavir from Abbott, indinavir from Merck, viracept from Agouron Pharmaceuticals, and amprenavir from Vertex Pharmaceuticals. (These companies maintain highly informative home pages on the World Wide Web.)

Treatment of AIDS is most effective when a combination of drug therapies is used, and HIV protease inhibitors play an important role. Especially promising results (such as lowering of levels of the virus in the bloodstream) are obtained when HIV protease inhibitors are part of drug regimens for AIDS.

Sometimes the search for an inhibitor as part of drug design leads to unexpected results. Scientists have long sought better drugs to fight *angina* (chest pains due to poor blood flow to the heart) and *hypertension* (otherwise known as high blood pressure), a common ailment these days. Blood flow increases when the smooth muscles in the blood vessels relax. This relaxation is due to a decrease in intracellular Ca^{2+}, which is in turn triggered by an increase in cyclic GMP (cGMP, Chapter 25). Cyclic GMP is degraded by enzymes called phosphodiesterases. Scientists thought that if they could design an inhibitor of these phosphodiesterases, the cGMP would last longer, the blood vessels would stay open longer, and blood pressure would decrease. Scientists developed a drug to mimic cGMP in the hopes of inhibiting phosphodiesterases. The structural name of the drug is sildenafil citrate, but a company called Pfizer marketed it under the name of Viagra.

Active site of HIV-1 protease complexed with VX-478.

Structure of amprenavir (VX-478), an HIV protease inhibitor.

Unfortunately, Viagra showed no significant benefits for reducing the pain of angina or decreasing blood pressure. However, some men in the clinical trials of the drug noted penile erections. Apparently the drug did work to inhibit the phosphodiesterases in the penile vascular tissue, leading to smooth muscle relaxation and increased blood flow. Despite the fact that the drug did not accomplish what it was intended for, this competitive inhibitor became a very big seller for the companies that produce it.

cGMP

Viagra

Note the structural similarity between cGMP (left) and Viagra.

Figure 23.13 Papain is a cysteine protease. A critical cysteine residue is involved in the nucleophilic attack on the peptide bond it hydrolyzes.

23.6 | How Are Enzymes Regulated?

A. Feedback Control

Enzymes are often regulated by environmental conditions. **Feedback control** is an enzyme regulation process in which formation of a product inhibits an earlier reaction in the sequence. The reaction product of one enzyme may control the activity of another, especially in a complex system in which enzymes work cooperatively. For example, in such a system, each step is catalyzed by a different enzyme:

$$A \xrightarrow{E_1} B \xrightarrow{E_2} C \xrightarrow{E_3} D$$

The last product in the chain, D, may inhibit the activity of enzyme E_1 (by competitive, noncompetitive, or some other type of inhibition). When the concentration of D is low, all three reactions proceed rapidly. As the concentration of D increases, however, the action of E_1 becomes inhibited and eventually stops. In this manner, the accumulation of D serves as a message that tells enzyme E_1 to shut down because the cell has enough D for its present needs. Shutting down E_1 stops the entire process.

B. Proenzymes

Some enzymes are manufactured by the body in an inactive form. To make them active, a small part of their polypeptide chain must be removed. These inactive forms of enzymes are called **proenzymes** or **zymogens.** After the excess polypeptide chain is removed, the enzyme becomes active.

For example, trypsin is manufactured as the inactive molecule trypsinogen (a zymogen). When a fragment containing six amino acid residues is removed from the N-terminal end, the molecule becomes a fully active trypsin molecule. Removal of the fragment not only shortens the chain, but also changes the three-dimensional structure (the tertiary structure), thereby allowing the molecule to achieve its active form.

Why does the body go to so much trouble? Why not just make the fully active trypsin to begin with? The reason is very simple. Trypsin is a

Proenzyme (zymogen) A protein that becomes an active enzyme after undergoing a chemical change

GOB
Chemistry ⚛ Now™

Click *Chemistry Interactive* to learn more about **Zymogens**

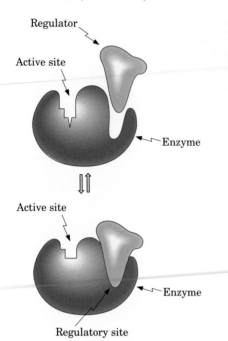

Figure 23.14 The allosteric effect. Binding of a regulator to a site other than the active site changes the shape of the active site.

Allosteric enzyme An enzyme in which the binding of a regulator on one site on the enzyme modifies the enzyme's ability to bind the substrate in the active site

GOB
Chemistry‑⚛‑Now™
Click *Chemistry Interactive* to explore **Allosteric Enzymes**

protease—it catalyzes the hydrolysis of peptide bonds (Figure 23.1)—and is, therefore, an important catalyst for the digestion of the proteins we eat. But it would not be good if it cleaved the proteins of which our own bodies are made! Therefore, the body makes trypsin in an inactive form; only after it enters the digestive tract is it allowed to become active.

C. Allosterism

Sometimes regulation takes place by means of an event that occurs at a site other than the active site but that eventually affects the active site. This type of interaction is called **allosterism,** and any enzyme regulated by this mechanism is called an **allosteric enzyme.** If a substance binds noncovalently and reversibly to a site *other than the active site,* it may affect the enzyme in either of two ways: It may inhibit enzyme action (**negative modulation**) or it may stimulate enzyme action (**positive modulation**).

The substance that binds to the allosteric enzyme is called a **regulator,** and the site to which it attaches is called a **regulatory site.** In most cases, allosteric enzymes contain more than one polypeptide chain (subunits); the regulatory site is on one polypeptide chain and the active site is on another.

Specific regulators can bind reversibly to the regulatory sites. For example, the enzyme depicted in Figure 23.14 is an allosteric enzyme. In this case, the enzyme has only one polypeptide chain, so it carries both the active site and the regulatory site at different points in this chain. The regulator binds reversibly to the regulatory site. As long as the regulator remains bound to the regulatory site, the total enzyme–regulator complex will be inactive. When the regulator is removed from the regulatory site, the enzyme becomes active. In this way, the regulator controls the allosteric enzyme action.

The concepts describing allosteric enzymes include a model of an enzyme that has two forms. One form is more likely to bind the substrate and produce the product than the other form. This more active form is referred to as the **R form,** where "R" stands for *relaxed.* The less active form is referred to as the **T form,** where "T" stands for *taut* (Figure 23.15). There is an equilibrium between the T form and the R form. When the enzyme is

A dimeric protein that can exist in either of two states: R_0 or T_0. This protein can bind three ligands:

1. Substrate (S) ▮: Binds only to R

2. Activator (A) ▲: Binds only to R

3. Inhibitor (I) ▸: Binds only to T

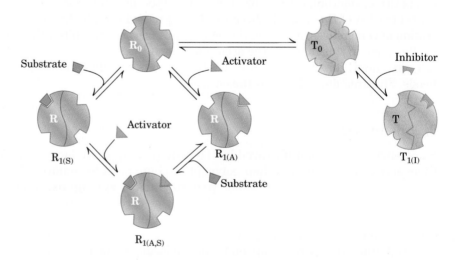

GOB
Chemistry‑⚛‑Now™
Active Figure 23.15 Effects of binding activators and inhibitors to allosteric enzymes. The enzyme has an equilibrium between the T form and the R form. An activator is anything that binds to the regulatory site and favors the R form. An inhibitor binds to the regulatory site and favors the T form. **See a simulation based on this figure, and take a short quiz on the concepts at http://now.brookscole.com/gob8 or on the CD.**

in the R form, it will bind substrate well and catalyze the reaction. Allosteric regulators are seen to function by binding to the enzyme and favoring one form versus the other.

D. Protein Modification

The activity of an enzyme may also be controlled by **protein modification.** The modification is usually a change in the primary structure, typically by addition of a functional group covalently bound to the apoenzyme. The best-known example of protein modification is the activation or inhibition of enzymes by phosphorylation. A phosphate group is often bonded to a serine or tyrosine residue. In some enzymes, such as glycogen phosphorylase (Section 29.1), the phosphorylated form is the active form of the enzyme. Without it, the enzyme would be less active or inactive.

The opposite example is the enzyme pyruvate kinase (PK, discussed in Chemical Connections 23C). Pyruvate kinase from the liver is inactive when it is phosphorylated. Enzymes that catalyze such phosphorylation go by the common name of *kinases*. When the activity of PK is not needed, it is phosphorylated (to PKP) by a protein kinase using ATP as a substrate as well as a source of energy (Section 27.3). When the system wants to turn on PK activity, the phosphate group, Pi, is removed by another enzyme, phosphatase, which renders PK active.

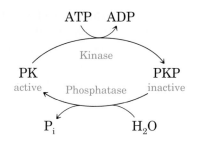

E. Isoenzymes

Another type of regulation of enzyme activity occurs when the same enzyme appears in different forms in different tissues. Lactate dehydrogenase (LDH) catalyzes the oxidation of lactate to pyruvate, and vice versa (Figure 28.3, step 11). The enzyme has four subunits (tetramer). Two kinds of subunits, called H and M, exist. The enzyme that dominates in the heart is an H_4 enzyme, meaning that all four subunits are of the H type, although some M-type subunits are present as well. In the liver and skeletal muscles, the M type dominates. Other types of tetramer combinations exist in different tissues: H_3M, H_2M_2, and HM_3. These different forms of the same enzyme are called **isozymes** or **isoenzymes.**

The different subunits confer subtle, yet important differences to the function of the enzyme in relation to the tissue. The heart is a purely aerobic organ, except perhaps during a heart attack. LDH is used to convert lactate to pyruvate in the heart. The H_4 enzyme is allosterically inhibited by high levels of pyruvate (its product) and has a higher affinity for lactate (its substrate) than does the M_4 enzyme, which is optimized for the opposite reaction. The M_4 isozyme favors the production of lactate.

The distribution of LDH isozymes can be seen using the technique of electrophoresis, where samples are separated in a gel using an electric field. Besides their kinetic differences, the two subunits of LDH carry different charges. Therefore, each combination of subunits travels in the electric field at a different rate (Figure 23.16).

GOB
Chemistry‑❖‑Now™
Click *Chemistry Interactive* to learn more about **Gel Electrophoresis**

Isozymes (Isoenzymes) Enzymes that perform the same function but have different combinations of subunits and thus different quaternary structures

GOB
Chemistry‑❖‑Now™
Click *Chemistry Interactive* to learn more about **Regulation of Enzymes**

Figure 23.16 The isozymes of lactate dehydrogenase (LDH). (*a*) The five combinations possible from mixing two types of subunits, H and M, in all permutations to make a tetramer. (*b*) An electrophoresis gel depiction of the relative isozyme types found in different tissues.

(*a*) The five tetramers of lactate dehydrogenase

M_4

M_3H

M_2H_2

MH_3

H_4

(*b*)

	M_4	M_3H	M_2H_2	MH_3	H_4
Liver					
Muscle					
White cells					
Brain					
Red cells					
Kidney					
Heart					

23.7 | How Are Enzymes Used in Medicine?

Most enzymes are confined within the cells of the body. However, small amounts of them can also be found in body fluids such as blood, urine, and cerebrospinal fluid. The level of enzyme activity in these fluids can easily be monitored. This information can prove extremely useful: Abnormal activity (either high or low) of particular enzymes in various body fluids signals either the onset of certain diseases or their progression. Table 23.2 lists some enzymes used in medical diagnosis and their activities in normal body fluids.

Table 23.2 Enzyme Assays Useful in Medical Diagnosis

Enzyme	Normal Activity	Body Fluid	Disease Diagnosed
Alanine amino-transferase (ALT)	3–17 U/L*	Serum	Hepatitis
Acid phosphatase	2.5–12 U/L	Serum	Prostate cancer
Alkaline phosphatase (ALP)	13–38 U/L	Serum	Liver or bone disease
Amylase	19–80 U/L	Serum	Pancreatic disease or mumps
Aspartate amino-transferase (AST)	7–19 U/L	Serum	Heart attack or hepatitis
	7–49 U/L	Cerebrospinal fluid	
Lactate dehydrog-enase (LDH)	100–350 WU/mL	Serum	Heart attack
Creatine phospho-kinase (CPK)	7–60 U/L	Serum	
Phosphohexose isomerase (PHI)	15–75 U/L	Serum	

*U/L = International units per liter; WU/mL = Wrobleski units per milliliter.

CHEMICAL CONNECTIONS 23E

Glycogen Phosphorylase: A Model of Enzyme Regulation

An excellent example of the subtle elegance of enzyme regulation can be seen in the enzyme glycogen phosphorylase, an enzyme that breaks down glycogen (Chapter 20) into glucose when the human body needs energy. Glycogen phosphorylase is a dimer controlled by modification and by allosterism.

Two forms of phosphorylase exist called phosphorylase *b* and phosphorylase *a* as shown in the figure. Phosphorylase *a* has a phosphate attached to each subunit, which was put there by the enzyme phosphorylase kinase. Phosphorylase *b* does not have the phosphate. The kinase is activated by hormonal signals that indicate a need for quick energy or a need for more blood glucose, depending on the tissue.

Phosphorylase is also controlled allosterically by a variety of regulators. The *b* form is converted to a more active form in the presence of AMP. Glucose-6-phosphate, glucose, and caffeine convert it to the less active form. The *a* form is also converted to the less active form by glucose and caffeine. In general, the equilibrium lies more toward the active form with phosphorylase *a* than with phosphorylase *b*.

This combination of regulation is very beneficial, because it leads to both quick changes and long-lasting ones. When you need quick action for the "fight or flight" response, the first few muscle contractions will cause ATP to be broken down. AMP will increase, which will, in much less than a second, convert some of the phosphorylase to the more active form (R state). At the same time, you will probably experience an adrenaline rush that will cause the activation of phosphorylase kinase and the subsequent phosphorylation of glycogen phosphorylase from the *b* form to the *a* form. This conversion will then cause even more phosphorylase to shift to the active R form (see right side of figure, arrow going down). The hormonal response is a little slower, taking seconds to minutes to take effect, but it is more long-lasting because the equilibrium will stay shifted to the R form until other enzymes (phosphatases) remove the phosphates. Thus the combination of allosteric and covalent control gives us the best of both worlds.

Chemistry ⚛ Now™

Glycogen phosphorylase activity is subject to allosteric control and covalent modification via phosphorylation. The phosphorylated form is more active. The enzyme that puts a phosphate group on phosphorylase is called phosphorylase kinase. **See a simulation based on this figure, and take a short quiz on the concepts at http://now.brookscole.com/gob8 or on the CD.**

For example, a number of enzymes are assayed (measured) during myocardial infarction to diagnose the severity of the heart attack. Dead heart muscle cells spill their enzyme contents into the serum. As a consequence, the level of creatine phosphokinase (CPK) in the serum rises rapidly, reaching a maximum within two days. This increase is followed by a rise in aspartate aminotransferase (AST; formerly called glutamate-oxaloacetate transaminase, or GOT). This second enzyme reaches a maximum two to three days after the heart attack. In addition to CPK and AST, lactate dehydrogenase (LDH) levels are monitored; they peak after five to six days. In infectious hepatitis, the alanine aminotransferase (ALT; formerly called glutamate-pyruvate transaminase, or GPT) level in the serum can rise to 10 times normal. There is also a concurrent increase in AST activity in the serum.

In some cases, administration of an enzyme is part of therapy. After duodenal or stomach ulcer operations, for instance, patients are advised to take tablets containing digestive enzymes that are in short supply in the

CHEMICAL CONNECTIONS 23F

One Enzyme, Two Functions

The enzyme called prostaglandin enderoperoxide synthase (PGHS) catalyzes the conversion of arachidonic acid to PGH$_2$ (Section 21.12) in two steps:

Arachidonic acid

Cyclooxygenase →

PGG$_2$

Peroxidase →

PGH$_2$

In the process, PGHS inserts two molecules of oxygen into arachidonic acid.

The enzyme itself is a single protein molecule. It is associated with the cell membrane and has a heme coenzyme. It has two distinct functions. The enzyme's *cyclooxgenase activity* is to close a substituted cyclopentane ring. The *peroxidase activity* of the same enzyme yields the 15-hydroxy derivative prostaglandin, PGH$_2$.

Commercial painkillers act two ways in inhibiting the formation of prostaglandin, PGH$_2$. First, PGHS's cyclooxygenase activity is inhibited by aspirin and related NSAIDs (Chemical Connections 21H). Aspirin inhibits PGHS by acetylating serine in the active site; as a result, the active site is no longer able to accommodate arachidonic acid. Other NSAIDs also inhibit the cyclooxygenase activity by competitive inhibition, but they do not inhibit the peroxidase activity. Second, another class of inhibitor has antioxidant activity. For example, acetaminophen (Tylenol) acts as a painkiller by inhibiting the peroxidase activity of PGHS.

stomach after surgery. Such enzyme preparations contain lipases, either alone or combined with proteolytic enzymes. They are sold under such names as Pancreatin, Acro-lase, and Ku-zyme.

23.8 | What Are Transition-State Analogs and Designer Enzymes?

As we saw in Section 23.5C, an enzyme lowers the activation energy for a reaction, making the transition state more favorable. It does so by having an active site that actually fits best to the transition state rather than to the substrates or the products. This has been documented by the use of **transition-state analogs,** molecules with a shape that mimics the transition state of the substrate.

Proline racemase, for example, catalyzes a reaction that converts L-proline to D-proline. During this reaction, the α-carbon must change from a tetrahedral arrangement to a planar form, and then back to a tetrahedral form, but with the orientation of two bonds reversed (Figure 23.17). An inhibitor of the reaction is pyrrole-2-carboxylate, a chemical that is structurally similar to what proline would look like at its transition state because it is always planar at the equivalent carbon. This inhibitor binds to proline racemase 160 times more strongly than proline does. Transition-state analogs have been used with many enzymes to help verify a suspected mechanism and structure of the transition state as well as to inhibit an enzyme selectively. They are currently being used for drug design when a specific enzyme is the cause of a disease.

In 1969, William Jencks proposed that an immunogen (a molecule that elicits an antibody response) would elicit antibodies (Chapter 31) with cat-

Transition-state analog A molecule that mimics the transition state of a chemical reaction and that is used as an inhibitor of an enzyme

Figure 23.17 The proline racemase reaction. Pyrrole-2-carboxylate mimics the planar transition state of the reaction.

L-Proline   Planar transition state   D-Proline

Pyrrole-2-carboxylate
(inhibitor and transition-state analog)

alytic activity if the immunogen mimicked the transition state of the reaction. Richard Lerner and Peter Schultz, who created the first catalytic antibodies, verified this hypothesis in 1986. Because an antibody is a protein designed to bind to specific molecules on the immunogen, the antibody will, in essence, serve as a fake active site. For example, the reaction of pyridoxal phosphate and an amino acid to form the corresponding α-ketoacid and pyridoxamine phosphate is a very important reaction in amino acid metabolism. The molecule N^α-(5′-phosphopyridoxyl)-L-lysine serves as a transition-state analog for this reaction. When this antigen molecule was used to elicit antibodies, these antibodies, or **abzymes,** had catalytic activity (Figure 23.18). Thus, in addition to helping us to verify the nature of the transition state or making an inhibitor, transition-state analogs now offer the possibility of creating designer enzymes to catalyze a wide variety of reactions.

Abzyme An antibody that has catalytic ability because it was created by using a transition-state analog as an immunogen

Figure 23.18 (a) N^α-(5′-phosphopyridoxyl)-L-lysine moiety is a transition-state analog for the reaction of an amino acid with pyridoxal 5′-phosphate. When this moiety is attached to a protein and injected into a host, it acts like an antigen and the host produces antibodies that have catalytic activity (abzymes). (b) The abzyme is then used to catalyze the reaction.

CHEMICAL CONNECTIONS 23G

Catalytic Antibodies Against Cocaine

Many addictive drugs, including heroin, operate by binding to a particular receptor in the neurons, mimicking the action of a neuro-transmitter. When a person is addicted to such a drug, a common way to attempt to treat the addiction is to use a compound that blocks the receptor and denies the drug access to it. Cocaine addiction has always been difficult to treat, due primarily to its unique modus operandi. As shown below, cocaine blocks the reuptake of the neurotransmitter dopamine. As a result, dopamine stays in the system longer, overstimulating neurons and leading to the reward signals in the brain that lead to addiction. Using a drug to block a receptor would be of no use with cocaine addiction and would probably just make removal of dopamine even more unlikely.

Cocaine (Section 16.2) can be degraded by a specific esterase, an enzyme that hydrolyzes an ester bond that is part of cocaine's structure. In the process of this hydrolysis, the cocaine must pass through a transition state that changes its shape. Catalytic anti-bodies to the transition state of the hydrolysis of cocaine have now been created. When administered to patients suffering from cocaine addiction, the antibodies successfully hydrolyze cocaine to two harmless degradation products—benzoic acid and ecgonine methyl ester. When degraded, the cocaine cannot block dopamine reuptake. No prolongation of the neuronal stimulus occurs, and the addictive effects of the drug vanish over time.

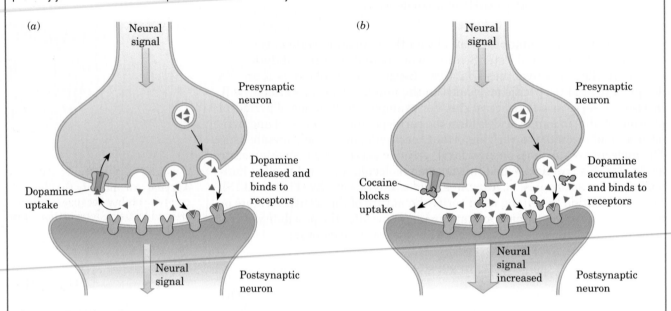

The mechanism of action of cocaine. (*a*) Dopamine acts as a neurotransmitter. It is released from the presynaptic neuron, travels across the synapse, and bonds to dopamine receptors on the postsynaptic neuron. It is later released and taken up into vesicles in the presynaptic neuron. (*b*) Cocaine increases the amount of time that dopamine is available to the dopamine receptors by blocking its uptake. (Adapted from *Immunotherapy for Cocaine Addiction*, by D. W. Landry, *Scientific American*, February 1997, pp 42–45.)

(*a*) Cocaine (*b*) Transition state (*c*) Ecgonine methyl ester Benzoic acid

Site of cleavage. The shaded carbon exists in a brief state where it is bonded to three oxygens as it is hydrolyzed by water.

Degradation of cocaine by esterases or catalytic antibodies. Cocaine (*a*) passes through a transition state (*b*) on its way to being hydrolyzed to benzoic acid and ecgonine methyl ester (*c*). Transition-state analogs are used to generate catalytic antibodies for this reaction. (Adapted from *Immunotherapy for Cocaine Addiction*, by D. W. Landry, *Scientific American*, February 1997, pp 42–45.)

SUMMARY OF KEY QUESTIONS

SECTION 23.1 What Are Enzymes?

- **Enzymes** are macromolecules that catalyze chemical reactions in the body. Most enzymes are very specific—they catalyze only one particular reaction.
- The compound whose reaction is catalyzed by an enzyme is called the **substrate.**
- Most enzymes are proteins, although some are made of RNA.

SECTION 23.2 How Are Enzymes Named and Classified?

- Enzymes are classified into six major groups according to the type of reaction they catalyze.
- Enzymes are typically named after the substrate and the type of reaction they catalyze, by adding the ending "-ase."

SECTION 23.3 What Is the Terminology Used with Enzymes?

- Some enzymes are made of polypeptide chains only. Others have, besides the polypeptide chain (the **apoenzyme**), nonprotein **cofactors,** which are either organic compounds **(coenzymes)** or inorganic ions.
- Only a small part of the enzyme surface, called the **active site,** participates in the actual catalysis of chemical reactions. Cofactors, if any, are part of the active site.
- Compounds that slow enzyme action are called **inhibitors.**
- A **competitive inhibitor** attaches itself to the active site. A **noncompetitive inhibitor** binds to other parts of the enzyme surface.

SECTION 23.4 What Factors Influence Enzyme Activity?

- The higher the enzyme and substrate concentrations, the higher the enzyme activity. At sufficiently high substrate concentrations, however, a saturation point is reached. After this point, increasing substrate concentration no longer increases the reaction rate.

- Each enzyme has an optimal temperature and pH at which it has its greatest activity.

SECTION 23.5 What Are the Mechanisms of Enzyme Action?

- Two closely related mechanisms that seek to explain enzyme activity and specificity are the **lock-and-key model** and the **induced-fit model.**
- Enzymes lower the **activation energy** required for a biochemical reaction to occur.

SECTION 23.6 How Are Enzymes Regulated?

- Enzyme activity is regulated by five mechanisms.
- In **feedback control,** the concentration of products influences the rate of the reaction.
- Some enzymes, called **proenzymes** or **zymogens,** must be activated by removing a small portion of the polypeptide chain.
- In **allosterism,** an interaction takes place at a position other than the active site but affects the active site, either positively or negatively.
- Enzymes can be activated or inhibited by **protein modification.**
- Enzyme activity is also regulated by **isozymes** (isoenzymes), which are different forms of the same enzyme.

SECTION 23.7 How Are Enzymes Used in Medicine?

- Abnormal enzyme activity can be used to diagnose certain diseases.

SECTION 23.8 What Are Transition-State Analogs and Designer Enzymes?

- The active site of an enzyme favors the production of the **transition state.**
- Molecules that mimic the transition state are called **transition-state analogs,** and they make potent enzyme inhibitors.

PROBLEMS

GOB
Chemistry ⚛ Now™

Assess your understanding of this chapter's topics with additional quizzing and conceptual-based problems at **http://now.brookscole.com/gob8** or on the CD.

A blue problem number indicates an applied problem.

■ denotes problems that are available on the GOB ChemistryNow website or CD and are assignable in OWL.

SECTION 23.1 What Are Enzymes?

23.1 What is the difference between a *catalyst* and an *enzyme?*

23.2 What are ribozymes made of?

23.3 Would a lipase hydrolyze two triglycerides, one containing only oleic acid and the other containing only palmitic acid, with equal ease?

23.4 Compare the activation energy in uncatalyzed reactions and in enzyme-catalyzed reactions.

23.5 Why does the body need so many different enzymes?

23.6 Trypsin cleaves polypeptide chains at the carboxyl side of a lysine or arginine residue (Figure 23.1). Chymotrypsin cleaves polypeptide chains on the carboxyl side of an aromatic amino acid residue or any other nonpolar, bulky side chain. Which enzyme is more specific? Explain.

SECTION 23.2 How Are Enzymes Named and Classified?

23.7 Both lyases and hydrolases catalyze reactions involving water molecules. What is the difference in the types of reactions that these two enzymes catalyze?

23.8 ■ Monoamine oxidases are important enzymes in brain chemistry. Judging from the name, which of the following would be a suitable substrate for this class of enzymes:

23.9 ■ On the basis of the classification given in Section 23.2, decide to which group each of the following enzymes belongs:
(a) Phosphoglyceromutase

$^-OOC-CH-CH_2-OPO_3^{2-}$
　　　　|
　　　　OH
3-Phosphoglycerate

$\rightleftharpoons$ $^-OOC-CH-CH_2-OH$
　　　　　　　|
　　　　　　OPO$_3^{2-}$
2-Phosphoglycerate

(b) Urease

$$H_2N-\overset{O}{\overset{||}{C}}-NH_2 + H_2O \rightleftharpoons 2NH_3 + CO_2$$
Urea

(c) Succinate dehydrogenase

$^-OOC-CH_2-CH_2-COO^-$ + FAD
Succinate　　　　　　　　　Coenzyme (oxidized form)

Fumarate　　Coenzyme (reduced form)

(d) Aspartase

Fumarate

$\rightleftharpoons$ $^-OOC-CH_2-CH-COO^-$
　　　　　　　　　　|
　　　　　　　　　NH$_3^+$
L-Asparate

23.10 What kind of reaction does each of the following enzymes catalyze?
(a) Deaminases (b) Hydrolases
(c) Dehydrogenases (d) Isomerases

SECTION 23.3 What Is the Terminology Used with Enzymes?

23.11 What is the difference between a *coenzyme* and a *cofactor*?

23.12 In the citric acid cycle, an enzyme converts succinate to fumarate (see the reaction in Problem 23.9c). The enzyme consists of a protein portion and an organic molecule portion called FAD. What terms do we use to refer to (a) the protein portion and (b) the organic molecule portion?

23.13 What is the difference between reversible and irreversible noncompetitive inhibition?

SECTION 23.4 What Factors Influence Enzyme Activity?

23.14 In most enzyme-catalyzed reactions, the rate of reaction reaches a constant value with increasing substrate concentration. This relationship is described in a saturation curve diagram (Figure 23.4). If the enzyme concentration, on a molar basis, is twice the maximum substrate concentration, would you obtain a saturation curve?

23.15 At a very low concentration of a certain substrate, we find that, when the substrate concentration doubles, the rate of the enzyme-catalyzed reaction also doubles. Would you expect the same finding at a very high substrate concentration? Explain.

23.16 If we wish to double the rate of an enzyme-catalyzed reaction, can we do so by increasing the temperature by 10°C? Explain.

23.17 ■ A bacterial enzyme has the following temperature-dependent activity.

(a) Is this enzyme more or less active at normal body temperature than when a person has a fever?

(b) What happens to the enzyme activity if the patient's temperature is lowered to 35°C?

23.18 The optimal temperature for the action of lactate dehydrogenase is 36°C. It is irreversibly inactivated at 85°C, but a yeast containing this enzyme can survive for months at −10°C. Explain how this can happen.

23.19 ■ The activity of pepsin was measured at various pH values. When the temperature and the concentrations of pepsin and substrate were held constant, the following activities were obtained:

pH	Activity
1.0	0.5
1.5	2.6
2.0	4.8
3.0	2.0
4.0	0.4
5.0	0.0

(a) Plot the pH dependence of pepsin activity.
(b) What is the optimal pH?
(c) Predict the activity of pepsin in the blood at pH 7.4.

23.20 How can the pH profile of an enzyme tell you something about the mechanism if you know the amino acids at the active site?

SECTION 23.5 What Are the Mechanisms of Enzyme Action?

23.21 Urease can catalyze the hydrolysis of urea but not the hydrolysis of diethylurea. Explain why diethylurea is not hydrolyzed.

Urea   Diethylurea

23.22 The following reaction may be represented by the cartoon figures:

Glucose + ATP ⇌ glucose 6-phosphate + ADP

In this enzyme-catalyzed reaction, Mg^{2+} is a cofactor, fluoroglucose is a competitive inhibitor, and Cd^{2+} is a noncompetitive inhibitor. Identify each component of the reaction by a cartoon figure and assemble them to show (a) the normal enzyme reaction, (b) a competitive inhibition, and (c) a noncompetitive inhibition.

23.23 Which amino acids appear most frequently in the active sites of enzymes?

23.24 What kind of chemical reaction occurs most frequently at the active site?

23.25 ■ Which of the following is a correct statement describing the induced-fit model of enzyme action? Substrates fit into the active site
(a) because both are exactly the same size and shape.
(b) by changing their size and shape to match those of the active site.
(c) by changing the size and shape of the active site upon binding.

23.26 What is the maximum rate that can be achieved in competitive inhibition compared with noncompetitive inhibition?

23.27 Enzymes are long protein chains, usually containing more than 100 amino acid residues. Yet the active site contains only a few amino acids. Explain why the other amino acids of the chain are present and what would happen to the enzyme activity if the enzyme's structure were changed significantly.

23.28 ■ Sucrose (common table sugar) is hydrolyzed to glucose and fructose. The reaction is catalyzed by the enzyme invertase. Using the following data, determine whether the inhibition by 2 M urea is competitive or noncompetitive.

Sucrose Concentration (M)	Velocity (arbitrary units)	Velocity + Inhibitor
0.0292	0.182	0.083
0.0584	0.265	0.119
0.0876	0.311	0.154
0.117	0.330	0.167
0.175	0.372	0.192

SECTION 23.6 How Are Enzymes Regulated?

23.29 The hydrolysis of glycogen to yield glucose is catalyzed by the enzyme phosphorylase. Caffeine, which is not a carbohydrate and not a substrate for the enzyme, inhibits phosphorylase. What kind of regulatory mechanism is at work?

23.30 Can the product of a reaction that is part of a sequence act as an inhibitor for another reaction in the sequence? Explain.

23.31 What is the difference between a *zymogen* and a *proenzyme*?

23.32 The enzyme trypsin is synthesized by the body in the form of a long polypeptide chain containing 235 amino acids (trypsinogen), from which a piece must be cut before the trypsin can be active. Why does the body not synthesize trypsin directly?

23.33 Give the structure of a tyrosyl residue of an enzyme modified by a protein kinase.

23.34 What is an *isozyme*?

23.35 The enzyme glycogen phosphorylase initiates the phosphorolysis of glycogen to glucose 1-phosphate. It comes in two forms: Phosphorylase *b* is less active, and phosphorylase *a* is more active. The difference between the *b* and *a* forms is the modification of the apoenzyme. Phosphorylase *a* has two phosphate groups added to the polypeptide chain. In analogy with the pyruvate kinase discussed in the text, give a scheme indicating the transition between the *b* and *a* forms. Which enzymes and which cofactors control this reaction?

23.36 How can you tell if an enzyme is allosteric by plotting velocity versus substrate?

23.37 Explain the nature of the two types of control of glycogen phosphorylase. What is the advantage to having both control types?

23.38 Which type of regulation discussed in Section 23.6 is the least reversible? Explain.

23.39 The enzyme phosphofructokinase (PFK) (Chapter 28) has two types of subunits, M and L, for muscle and liver. These subunits combine to form a tetramer. How many isozymes of PFK exist? What are their designations?

23.40 If you separated PFK using electrophoresis, how would the isozymes migrate if the M subunit has a lower pI than the L subunit?

SECTION 23.7 How Are Enzymes Used in Medicine?

23.41 After a heart attack, the levels of certain enzymes rise in the serum. Which enzyme would you monitor within 24 hours following a suspected heart attack?

23.42 The enzyme formerly known as GPT (glutamate-pyruvate transaminase) has a new name: ALT (alanine aminotransferase). Looking at the equation in Section 28.9, which is catalyzed by this enzyme, what prompted this change of name?

23.43 If an examination of a patient indicated elevated levels of AST but normal levels of ALT, what would be your tentative diagnosis?

23.44 Which LDH isozyme is monitored in the case of a heart attack?

23.45 Chemists who have been exposed for years to organic vapors usually show higher-than-normal activity when given the alkaline phosphatase test. Which organ in the body do organic vapors affect?

23.46 Which enzyme preparation is given to patients after duodenal ulcer surgery?

23.47 Chymotrypsin is secreted by the pancreas and passed into the intestine. The optimal pH for this enzyme is 7.8. If a patient's pancreas cannot manufacture chymotrypsin, would it be possible to supply it orally? What happens to chymotrypsin's activity during its passage through the gastrointestinal tract?

SECTION 23.8 What Are Transition-State Analogs and Designer Enzymes?

23.48 Explain why transition-state analogs are potent inhibitors.

23.49 How do transition-state analogs relate to the idea of the induced-fit model of enzymes?

23.50 Explain the relationship between transition-state analogs and abzymes.

Chemical Connections

23.51 (Chemical Connections 23A) Acetylcholine causes muscles to contract. Succinylcholine, a close relative, is a muscle relaxant. Explain the different effects of these related compounds.

23.52 (Chemical Connections 23A) An operating team usually administers succinylcholine before bronchoscopy. What is achieved by this procedure?

23.53 (Chemical Connections 23B) How does *Helicobacter pylori* protect its enzymes against stomach acid?

23.54 (Chemical Connections 23B) How does the quaternary structure of urease of *H. pylori* protect this enzyme against acid denaturation?

23.55 (Chemical Connections 23B) How do we treat ulcers caused by infection with *H. pylori?*

23.56 (Chemical Connections 23C) What role does Mn^{2+} play in anchoring the substrate in the active site of protein kinase?

23.57 ■ (Chemical Connections 23C) Which amino acids of the active site interact with the $=CH_2$ group of the phosphoenol pyruvate? Do these amino acids provide the same surface environment? What is the nature of the interaction?

23.58 (Chemical Connections 23D) What is the strategy in drug design to fight AIDS?

23.59 (Chemical Connections 23D) Why did scientists want to create a drug to inhibit cGMP diesterases?

23.60 (Chemical Connections 23E) Explain the difference between phosphorylase *a* and phosphorylase *b*. Which is more active and why?

23.61 (Chemical Connections 23E) What is the relationship between protein modification and allosteric control of glycogen phosphorylase?

23.62 (Chemical Connections 23F) Which activity of prostaglandin enderoperoxide synthase (PGHS) is inhibited by Tylenol, and which is inhibited by aspirin?

23.63 (Chemical Connections 23G) Explain how catalytic antibodies are produced to fight cocaine addiction.

23.64 (Chemical Connections 23G) Why can inhibitors not be used to block a cocaine receptor as is often done with other drug addictions?

23.65 (Chemical Connections 23G) How does cocaine work as a drug?

Additional Problems

23.66 Where can one find enzymes that are both stable and active at 90°C?

23.67 ■ Food can be preserved by inactivation of enzymes that would cause spoilage—for example, by refrigeration. Give an example of food preservation in which the enzymes are inactivated (a) by heat and (b) by lowering the pH.

23.68 Why is enzyme activity during myocardial infarction measured in patients' serum rather than in their urine?

23.69 What is the common characteristic of the amino acids of which the carboxyl groups of the peptide bonds can be hydrolyzed by trypsin?

23.70 Many enzymes are active only in the presence of Zn^{2+}. What common term is used for ions such as Zn^{2+} when discussing enzyme activity?

23.71 An enzyme has the following pH dependence:

At what pH do you think this enzyme works best?

23.72 What enzyme is monitored in the diagnosis of infectious hepatitis?

23.73 The enzyme chymotrypsin catalyzes the following type of reaction:

On the basis of the classification given in Section 23.2, to which group of enzymes does chymotrypsin belong?

23.74 Nerve gases operate by forming covalent bonds at the active site of cholinesterase. Is this an example of competitive inhibition? Can the nerve gas molecules be removed by simply adding more substrate (acetylcholine) to the enzyme?

23.75 What would be the appropriate name for an enzyme that catalyzes each of the following reactions:

23.76 In Section 29.5, a reaction between pyruvate and glutamate to form alanine and α-ketoglutarate is given. How would you classify the enzyme that catalyzes this reaction?

23.77 A liver enzyme is made of four subunits: 2A and 2B. The same enzyme, when isolated from the brain, has the following subunits: 3A and 1B. What would you call these two enzymes?

23.78 What is the function of a ribozyme?

Looking Ahead

23.79 Caffeine is a stimulant that is taken by many people in the form of coffee, tea, chocolate, and cola beverages. It is also used by many athletes. Caffeine has many effects, including stimulating lipases. Given its effect on lipases and on glycogen phosphorylase, would you predict caffeine would be more effective as an aid to a runner in a 10 K race or in a 1 mile race?

23.80 Until the discovery of thermophilic bacteria that live in conditions of extreme heat and pressure, it was impossible to have an automated system of DNA synthesis. Explain why this was so, given that it takes temperatures of around 90°C to separate strands of DNA.

23.81 What characteristics of RNA make it likely to have catalytic ability? Why is DNA less likely to have catalytic activity?

CHAPTER 24

Chemical Communications: Neurotransmitters and Hormones

Scanning electron micrograph of neuron touching a muscle cell.

©Don Fawcett/Science Source/Photo Researchers, Inc.

GOB
Chemistry·⚛·Now™

Look for this logo in the chapter and go to GOB ChemistryNow at **http://now.brookscole.com/gob8** or on the CD for tutorials, simulations, and problems.

24.1 | What Molecules Are Involved in Chemical Communications?

The importance of chemical communications in health becomes apparent from a quick glance at Table 24.1. This list contains only a small sample of drugs pertaining to this chapter and it is of utmost interest to health care providers. In fact, a vast number of drugs in the pharmacopoeia act, in one way or another, to influence chemical communications.

Each cell in the body is an isolated entity enclosed in its own membrane. Furthermore, within each cell of higher organisms, organelles, such as the nucleus or the mitochondrion, are enclosed by membranes separating them from the rest of the cell. If cells could not communicate with one another, the thousands of reactions in each cell would be uncoordinated. The same is true for organelles within a cell. Such communication allows the activity of a cell in one part of the body to be coordinated with the activity of cells in a different part of the body. There are three principal types of molecules for communications:

- **Receptors** are protein molecules on the surface of cells embedded in the membrane.
- **Chemical messengers,** also called ligands, interact with the receptors. (Chemical messengers fit into the receptor sites in a manner reminiscent of the lock-and-key model mentioned in Section 23.5.)
- **Secondary messengers** in many cases carry the message from the receptor to the inside of the cell and amplify the message.

If your house is on fire and the fire threatens your life, external signals—light, smoke, and heat—register alarm at specific receptors in your eyes, nose, and skin. From there the signals are transmitted by specific compounds to nerve cells, or **neurons.** Nerve cells are present throughout the body and, together with the brain, constitute the nervous system. In the neurons, the signals travel as electric impulses along the axons (Figure 24.1). When they reach the end of the neuron, the signals are transmitted to adjacent neurons by specific compounds called **neurotransmitters.** Communication between the eyes and the brain, for example, is by neural transmission.

As soon as the danger signals are processed in the brain, other neurons carry messages to the muscles and to the endocrine glands. The message to the muscles is to run away or to take some other action in response to the fire (save the baby or run to the fire extinguisher, for example). To do so, the muscles must be activated. Again, neurotransmitters carry the necessary messages from the neurons to the muscle cells and the endocrine glands. The endocrine glands are stimulated, and a different chemical signal, called a **hormone,** is secreted into your bloodstream. "The adrenaline begins to flow." The danger signal carried by adrenaline makes quick energy available so that the muscles can contract and relax rapidly, allowing your body to take rapid action to avoid the danger.

Without these chemical communicators, the whole organism—you—would not survive because there is a constant need for coordinated efforts to face a complex outside world. The chemical communications between different cells and different organs play a role in the proper functioning of our bodies. Its significance is illustrated by the fact that *a large percentage of the drugs we encounter in medical practice try to influence this communication.* The scope of these drugs covers all fields—from prescriptions against hypertension, heart disease, to antidepressants, to painkillers, just to mention a few. Principally, there are four ways these drugs act in the body. A drug may affect either the **receptor** or the **messenger.**

1. An **antagonist** drug blocks the receptor and prevents its stimulation.
2. An **agonist** drug competes with the natural messenger for the receptor site. Once there, it stimulates the receptor.
3. Other drugs decrease the concentration of the messenger by controlling the release of messengers from their storage.
4. Still other drugs increase the concentration of the messenger by inhibiting its removal from the receptors.

Table 24.1 presents selected drugs and their modes of action that affect neurotransmission.

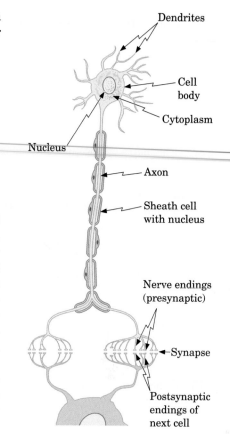

Figure 24.1 Neuron and synapse.

Neurotransmitter A chemical messenger between a neuron and another target cell: neuron, muscle cell, or cell of a gland

Hormone A chemical messenger released by an endocrine gland into the bloodstream and transported there to reach its target cell

Table 24.1   Drugs That Affect Nerve Transmission

Messenger	Drugs That Affect Receptor Sites		Drugs That Affect Available Concentration of the Neurotransmitter or Its Removal from Receptor Sites	
	Agonists (Activate Receptor Sites)	Antagonists (Block Receptor Sites)	Increase Concentration	Decrease Concentration
Acetylcholine (cholinergic)	Nicotine Succinylcholine	Curare Atropine	Malathion Nerve gases Succinylcholine Donepezil (Aricept)	*Clostridium botulinum* toxin
Calcium ion		Nifedipine (Adalat) Diltiazem (Cardiazem)	Digitoxin (Lanoxicaps)	
Epinephrine (α-adrenergic)	Terazosin (Hytrin)			
Norepinephrine (β-adrenergic)	Phenylephrine Epinephrine (Adrenalin)	Propranoloo (Inderal)	Amphetamines	Reserpine Methyldopa (Aldomet) Metyrosine (Demser)
Dopamine (adrenergic)		Clozapine (Clorazil)	Entacapon (Comtan)	
Serotonin (adrenergic)		Ondansetron (Zofran)	Antidepressant Fluoxetine (Prozac)	
Histamine (adrenergic)	2-Methyl-histamine	Fexofenadine (Allegra) Diphenhydramine (Benadryl) Ranitidine (Zantac) Cimetidine (Tagamet)	Histidine	Hydrazino-histidine
Glutamic acid (amino acid)	*N*-Methyl-D-aspartate	Phenylcyclidine		
Enkephalin (peptidergic)	Opiate Morphine Heroin Meperidine (Demerol)		Naloxone (Narcan)	

24.2 | How Are Chemical Messengers Classified as Neurotransmitters and Hormones?

As mentioned earlier, neurotransmitters are compounds that communicate between two nerve cells or between a nerve cell and another cell (such as a muscle cell). A nerve cell (Figure 24.1) consists of a main cell body from which projects a long, fiber-like part called an **axon.** Coming off the other side of the main body are hair-like structures called **dendrites.**

Typically, neurons do not touch each other. Between the axon end of one neuron and the cell body or dendrite end of the next neuron is a space filled with an aqueous fluid, called a **synapse.** If the chemical signal travels, say, from axon to dendrite, we call the nerve ends on the axon the **presynaptic** site. The neurotransmitters are stored at the presynaptic site in **vesicles,** which are small, membrane-enclosed packages. Receptors are located on the **postsynaptic** site of the cell body or the dendrite.

Hormones are diverse compounds secreted by specific tissues (the endocrine glands), released into the bloodstream, and then adsorbed onto specific receptor sites, usually relatively far from their source. (This is the physiological definition of a hormone.) Table 24.2 lists some of the principal

Synapse An aqueous small space between the tip of a neuron and its target cell

Table 24.2   The Principal Hormones and Their Actions

Gland	Hormone	Action	Structures Shown in
Parathyroid	Parathyroid hormone	Increases blood calcium Excretion of phosphate by kidney	
Thyroid	Thyroxine (T$_4$)	Growth, maturation, and metabolic rate	Chemical Connections 14C
	Triiodothyronine (T$_3$)	Metamorphosis	
Pancreatic islets			
Beta cells	Insulin	Hypoglycemic factor	Section 22.8
		Regulation of carbohydrates, fats, and proteins	Chemical Connections 24G
Alpha cells	Glucagon	Liver glycogenolysis	
Adrenal medulla	Epinephrine Norepinephrine	Liver and muscle glycogenolysis	Section 24.5
Adrenal cortex	Cortisol Aldosterone	Carbohydrate metabolism Mineral metabolism	Section 21.10 Section 21.10
	Adrenal androgens	Androgenic activity (especially females)	
Kidney	Renin	Hydrolysis of blood precursor protein to yield angiotensin	
Anterior pituitary	Luteinizing hormone	Causes ovulation	
	Interstitial cell-stimulating hormone	Formation of testosterone and progesterone in interstitial cells	
	Prolactin	Growth of mammary gland	
	Mammotropin	Lactation Corpus luteum function	
Posterior pituitary	Vasopressin	Contraction of blood vessels Kidney reabsorption of water	Section 22.8
	Oxytocin	Stimulates uterine contraction and milk ejection	Section 22.8
Ovaries	Estradiol Progesterone	Estrous cycle Female sex characteristics	Section 21.10 Section 21.10
Testes	Testosterone Androgens	Male sex characteristics Spermatogenesis	Section 21.10

hormones. Figure 24.2 shows the target organs of hormones secreted by the pituitary gland.

The distinction between hormones and neurotransmitters is physiological, not chemical. Whether a certain compound is considered to be a neurotransmitter or a hormone depends on whether it acts over a short distance across a synapse (2×10^{-6} cm), in which case it is a neurotransmitter, or over a long distance (20 cm) from the secretory gland through the bloodstream to the target cell, in which case it is a hormone. For example, epinephrine and norepinephrine are both neurotransmitters and hormones.

There are, broadly speaking, five classes of chemical messengers: *cholinergic, amino acid, adrenergic, peptidergic,* and *steroid* messengers. This classification is based on the chemical nature of the important messenger (ligand) in each group. Neurotransmitters can belong to all five classes, and hormones can belong to the last three classes.

Messengers can also be classified according to how they work. Some of them—epinephrine, for example—*activate enzymes.* Others affect the *synthesis of enzymes and proteins* by working on the transcription of genes

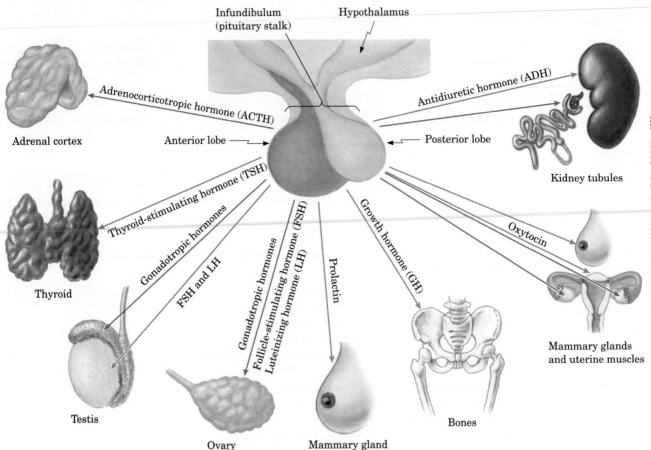

Figure 24.2 The pituitary gland is suspended from the hypothalamus by a stalk of neural tissue. The hormones secreted by the anterior and posterior lobes of the pituitary gland and the target tissues they act on are shown.

Modified from P. W. Davis and E. P. Solomon, The World of Biology. Philadelphia: Saunders College Publishing, 1986

(Section 26.2); steroid hormones (Section 21.10) work in this manner. Finally, some affect the *permeability of membranes;* acetylcholine, insulin, and glucagon belong to this class.

Yet another way of classifying messengers is according to their potential to *act directly* or through a *secondary messenger.* The steroid hormones act directly. They can penetrate the cell membrane and pass through the membrane of the nucleus. For example, estradiol stimulates uterine growth.

Other chemical messengers act through secondary messengers. For example, epinephrine, glucagon, luteinizing hormone, norepinephrine, and vasopressin use cAMP as a secondary messenger (details in Section 24.5C).

In the following sections, we will sample the mode of communication within each of the five chemical categories of messengers.

GOB
Chemistry·❀·Now™

Click *Chemistry Interactive* to **Review the Mode of Action of Chemical Messengers**

24.3 | How Does Acetylcholine Act as a Messenger?

The main **cholinergic neurotransmitter** is acetylcholine:

$$CH_3-\overset{\displaystyle O}{\overset{\|}{C}}-O-CH_2-CH_2-\overset{\displaystyle CH_3}{\overset{|}{N^+}}-CH_3$$
$$\underset{\displaystyle CH_3}{|}$$

Acetylcholine

CHEMICAL CONNECTIONS 24A

Calcium as a Signaling Agent (Secondary Messenger)

The message delivered to the receptors on the cell membranes by neurotransmitters or hormones must be delivered intracellularly to various locations within the cell. The most universal yet most versatile signaling agent is the cation Ca^{2+}.

Calcium ions in the cells come from either extracellular sources or intracellular stores, such as the endoplasmic reticulum. If the ions come from the outside, they enter the cells through specific calcium channels. Calcium ions control our heart beats, our movements through the action of skeletal muscles, and, through the release of neurotransmitters in our neurons, learning and memory. They are also involved in signaling the beginning of life at fertilization and its end at death. Calcium ion signaling controls these functions via two mechanisms: (1) increased concentration (forming sparks and puffs) and (2) duration of the signals.

In the resting state of the neuron, the Ca^{2+} concentration is about 0.1 μM. When neurons are stimulated, this level may increase to 0.5 μM. However, to elicit a fusion between the synaptic vesicles and the plasma membrane of the neuron, much higher concentrations are needed (10–25 μM).

An increase in calcium ion concentration may take the form of sparks or puffs. The source of calcium ions may be external (calcium influx caused by the electric signal of nerve transmission) or internal (calcium released from the stores of the endoplasmic reticulum). Upon receiving the signal of calcium puffs, the vesicles storing acetylcholine travel to the membrane of the presynaptic cleft. There they fuse with the membrane and empty their contents into the synapse.

Calcium ions can also control signaling by controlling the duration of the signal. The signal in arterial smooth muscle lasts for 0.1 to 0.5 s. The wave of Ca^{2+} in the liver lasts 10 to 60 s. The calcium wave in the human egg lasts 1 to 35 min after fertilization. Thus, by combining the concentration, localization, and duration of the signal, calcium ions can deliver messages to perform a variety of functions.

The effects of Ca^{2+} are modulated through specific calcium-binding proteins. In all nonmuscle cells and in smooth muscles, calmodulin serves as the calcium-binding protein. Calmodulin-bound calcium activates an enzyme, protein kinase II, which then phosphorylates an appropriate protein substrate. In this way, the signal is translated into metabolic activity.

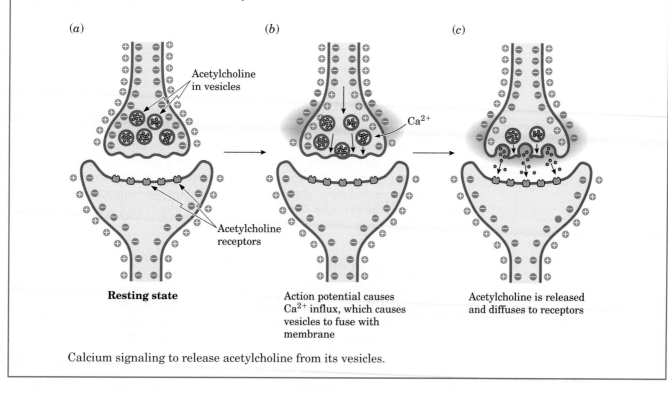

(a) Acetylcholine in vesicles

Acetylcholine receptors

Resting state

(b) Ca^{2+}

Action potential causes Ca^{2+} influx, which causes vesicles to fuse with membrane

(c)

Acetylcholine is released and diffuses to receptors

Calcium signaling to release acetylcholine from its vesicles.

A. Cholinergic Receptors

There are two kinds of receptors for this messenger. We will look at one that exists on the motor end plates of skeletal muscles or in the sympathetic ganglia. The nerve cells that bring messages contain stored acetylcholine in the vesicles in their axons. The receptor on the muscle cells or neurons is also known as nicotinic receptor because nicotine (see Chemical Connections 16B)

Figure 24.3 Acetylcholine in action. The receptor protein has five subunits. When two molecules of acetylcholine bind to the two α subunits, a channel opens to allow the passage of Na$^+$ and K$^+$ ions by facilitated transport (Chemical Connections 21C).

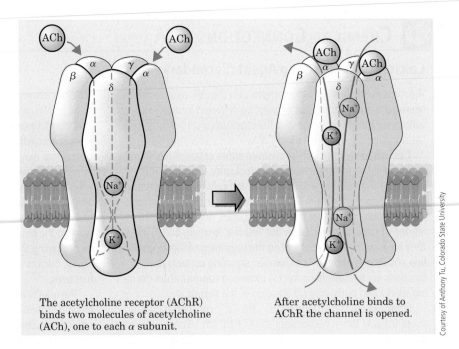

The acetylcholine receptor (AChR) binds two molecules of acetylcholine (ACh), one to each α subunit.

After acetylcholine binds to AChR the channel is opened.

inhibits the neurotransmission of these nerves. The receptor itself is a *transmembrane protein* (Figure 21.2) made of five different subunits. The central core of the receptor is an ion channel through which, when open, Na$^+$ and K$^+$ ions can pass (Figure 24.3). When the ion channels are closed, the K$^+$ ion concentration is higher inside the cell than outside; the reverse is true for the Na$^+$ ion concentration.

B. Storage of Messengers

Events begin when a message is transmitted from one neuron to the next by neurotransmitters. The message is initiated by calcium ions (see Chemical Connections 24A). When the concentration in a neuron reaches a certain level (more than 0.1 μM), the vesicles containing acetylcholine fuse with the presynaptic membrane of the nerve cells. Then they empty the neurotransmitters into the synapse. The messenger molecules travel across the synapse and are adsorbed onto specific receptor sites.

C. The Action of Messengers

The presence of acetylcholine molecules at the postsynaptic receptor site triggers a conformational change (Section 22.10) in the receptor protein. This opens the *ion channel* and allows ions to cross membranes freely. Na$^+$ ions have a higher concentration outside the neuron than do K$^+$ ions; thus more Na$^+$ enters the cell than K$^+$ leaves. Because it involves ions, which carry electric charges, this process is translated into an electric signal. After a few milliseconds, the channel closes again. The acetylcholine still occupies the receptor. For the channel to reopen and transmit a new signal, the acetylcholine must be removed and the neuron must be reactivated.

D. The Removal of Messengers

Acetylcholine is removed rapidly from the receptor site by the enzyme *acetylcholinesterase,* which hydrolyzes it.

This rapid removal enables the nerves to transmit more than 100 signals per second. By this means, the message moves from neuron to neuron until it is finally transmitted, again by acetylcholine molecules, to the muscles or endocrine glands that are the ultimate target of the message.

Obviously, the action of the acetylcholinesterase enzyme is essential to the entire process. When this enzyme is inhibited, the removal of acetylcholine is incomplete, and nerve transmission ceases.

CHEMICAL CONNECTIONS 24B

Nerve Gases and Antidotes

Most nerve gases in the military arsenal exert their lethal effect by binding to acetylcholinesterase. Under normal conditions, this enzyme hydrolyzes the synaptic neurotransmitter acetylcholine within a few milliseconds after it is released at the nerve endings.

Nerve gases such as Sarin (agent GB, also called Tabun), Soman (agent GD), and agent VX are organic phosphonates related to such pesticides as parathion (the latter being much less lethal, of course):

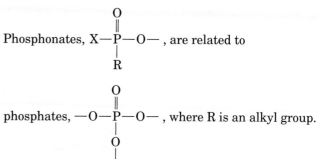

If any one of these phosphonates binds to acetylcholinesterase, the enzyme is irreversibly inactivated, and the transmission of nerve signals stops. The result is a cascade of symptoms: sweating, bronchial constriction due to mucus buildup, dimming of vision, vomiting, choking, convulsions, paralysis, and respiratory failure. Direct inhalation of as little as 0.5 mg can cause death within a few minutes. If the dosage is smaller or if the nerve gas is absorbed through the skin, the lethal effect may take several hours to occur.

In warfare, protective clothing and gas masks are effective countermeasures. An enzyme of the bacteria *Pseudomonas*

diminuta, organophosphorous hydroxylase, is capable of detoxifying nerve gases. Its inclusion in cotton towelettes may help to remove nerve gases from contaminated skin. Also, first-aid kits containing injectable antidotes are available. These antidotes contain the alkaloid atropine and pralidoxime chloride. The two substances must be used in tandem—and very quickly—after exposure to nerve gases.

The nations of the world, with a few exceptions, signed a treaty in 1993 that obliges all signatory governments to destroy their stockpiles of chemical weapons. In addition, the signatories pledged to dismantle the plants that manufacture them within 10 years.

CHEMICAL CONNECTIONS 24C

Botulism and Acetylcholine Release

When meat or fish is improperly cooked or preserved, a deadly type of food poisoning, called botulism, may result. The culprit is the bacterium *Clostridium botulinum,* whose toxin prevents the release of acetylcholine from the presynaptic vesicles. Therefore, no neuro-transmitter reaches the receptors on the surface of muscle cells, and the muscles no longer contract. Left untreated, the affected person may die.

Surprisingly, the botulin toxin has a valuable medical use. It is used in the treatment of involuntary muscle spasms—for example, in facial tics. These tics are caused by the uncontrolled release of acetylcholine. Controlled administration of the toxin, when applied locally to the facial muscles, stops the uncontrolled contractions and relieves the facial distortions.

"Facial distortion" has multiple meanings in the cosmetics industry. "Frown lines" and other wrinkles can also be removed by temporarily paralyzing facial muscles. The U.S. Food and Drug Administration has approved Botox (botulin toxin) for cosmetic use, meaning that its manufacturer can now advertise its rejuve-nating appeal. Its use is spreading fast. Botox has been used on an off-label basis for this application for a number of years, especially

Clostridium botulinum is a food-poisoning bacterium.

in Hollywood. Indeed, several movie directors have complained that some actors have used so much Botox that they can no longer show a variety of facial expressions.

E. Control of Neurotransmission

Acetylcholinesterase is inhibited irreversibly by phosphonates in nerve gases and pesticides (Chemical Connections 24B) or reversibly by succinyl-choline (Chemical Connections 23A) and decamethonium bromide.

$$Br^- \quad \underset{\underset{CH_3}{|}}{\overset{\overset{CH_3}{|}}{CH_3-N^+}}-CH_2(CH_2)_8CH_2-\underset{\underset{CH_3}{|}}{\overset{\overset{Br^- \quad CH_3}{|}}{N^+}}-CH_3$$

Decamethonium bromide

Succinylcholine and decamethonium bromide resemble the choline end of acetylcholine and, therefore, act as competitive inhibitors of acetylcholin-esterase. In small doses, these reversible inhibitors relax the muscles tem-porarily, making them useful as muscle relaxants in surgery. In large doses, they are just as deadly as the irreversible inhibitors.

The inhibition of acetylcholinesterase is but one way in which cholinergic neurotransmission is controlled. Another way is to modulate the action of the receptor. Because acetylcholine enables the ion channels to open and propa-gate signals, this mode of action is called *ligand-gated ion channels.* The attachment of the ligand to the receptor is critical in signaling. Nicotine given in low doses is a stimulant; it is an agonist because it prolongs the receptor's biochemical response. When given in large doses, however, nicotine becomes an antagonist and blocks the action on the receptor. As such, it may cause con-vulsions and respiratory paralysis. Succinylcholine, besides being a reversible inhibitor of acetylcholinesterase, also has this concentration-dependent ago-nist/antagonist effect on the receptor. A strong antagonist, which blocks the receptor completely, can interrupt the communication between neuron and muscle cell. The venom of a number of snakes, especially that of the cobra, cobratoxin, exerts its deadly influence in this manner. The plant extract

CHEMICAL CONNECTIONS 24D

Alzheimer's Disease and Acetylcholine Transferase

Alzheimer's disease is the name given to the symptoms of severe memory loss and other senile behavior that afflict about 1.5 million people in the United States. Postmortem identification of this disease focuses on two pathological hallmarks in the brain: (1) buildup of protein deposits known as β-amyloid plaques outside the nerve cells and (2) neurofibrillar tangles composed of tau proteins. Controversy exists as to which of the two is the primary cause of the neurodegeneration observed in Alzheimer's disease. Each has its advocates. Those who favor the primacy of the plaque are nicknamed β-aptists, and the other camp sports the name of tauists.

Tau protein binds to microtubules, one of the major cytoskeletons. Genetic mutation of tau protein or environmental factors such as hyperphosphorylation may alter the ability of tau to bind to microtubules. These altered tau proteins form tangles in the cytoplasm of neurons. Neurofibrillar tangles have been found in the brains of patients with Alzheimer's disease in the absence of plaque as well, which suggests that tau abnormality may be sufficient to cause neurodegeneration.

In most brains affected by Alzheimer's disease, the dominant feature is plaques. These plaques consist of protein fibers, some 7–10 nm thick, that are mixed with small peptides called β-amyloid peptides. These peptides originate from a water-soluble precursor, amyloid precursor protein (APP); this transmembrane protein has an unknown function. Certain enzymes, called secretases, cut peptides containing 38, 40, and 42 amino acids from the transmembrane region of APP. Mutations of APP proteins in patients with Alzheimer's disease cause preferential accumulation of the 42-amino-acid peptide, forming β-pleated sheets, which precipitates creating plaques.

In Alzheimer's disease, nerve cells in the cerebral cortex die, the brain becomes smaller, and part of the cortex atrophies. The depression of the folds on the brain surface becomes deeper.

People with Alzheimer's disease are forgetful, especially about recent events. As the disease advances, they become confused and, in severe cases, lose their ability to speak; at that point, they need total care. As yet, there is no cure for this disease.

Patients with Alzheimer's disease have significantly diminished acetylcholine transferase activity in their brains. This enzyme synthesizes acetylcholine by transferring the acetyl group from acetyl-CoA to choline:

The diminished concentration of acetylcholine can be partially compensated for by inhibiting the enzyme acetylcholinesterase, which decomposes acetylcholine. Certain drugs, which act as acetylcholinesterase inhibitors, have been shown to improve memory and other cognitive functions in some people with the disease. Drugs such as donepezil (Aricept), rivastigmine (Exelon), and galantamine (Reminyl) belong to this category; they all moderate the symptoms of Alzheimer's disease. The alkaloid huperzine A, an active ingredient of Chinese herb tea that has been used for centuries to improve memory, is also a potent inhibitor of acetylcholinesterase.

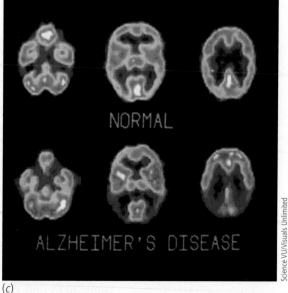

(a) Normal brain. (b) Brain of a person with Alzheimer's disease. (c) PET scan comparing a normal brain with the brain of a person with Alzheimer's disease.

curare, which was used in poisoned arrows by the Amazon Indians, works in the same way. In small doses, curare is used as a muscle relaxant.

Finally, the supply of the acetylcholine messenger can influence the proper nerve transmission. If the acetylcholine messenger is not released from its storage, as in botulism (Chemical Connections 24C), or if its synthesis is impaired, as in Alzheimer's disease (Chemical Connections 24D), the concentration of acetylcholine is reduced, and nerve transmission is impaired.

24.4 | What Amino Acids Act as Neurotransmitters?

A. Messengers

Amino acids are distributed throughout the neurons individually or as parts of peptides and proteins. They can also act as neurotransmitters. Some of them, such as glutamic acid, aspartic acid, and cysteine, act as **excitatory neurotransmitters** similar to acetylcholine and norepinephrine. Others, such as glycine, β-alanine, taurine (Section 21.11), and mainly γ-aminobutyric acid (GABA), are **inhibitory neurotransmitters;** they reduce neurotransmission. Note that some of these neurotransmitter amino acids are not found in proteins.

$$^+H_3NCH_2CH_2SO_3^- \qquad ^+H_3NCH_2CH_2COO^- \qquad ^+H_3NCH_2CH_2CH_2COO^-$$

Taurine   β-Alanine   γ-Aminobutyric acid
                                                    (GABA)
                                             (IUPAC name:
                                          4-aminobutanoic acid)

B. Receptors

Each of these amino acids has its own receptor; in fact, glutamic acid has at least five subclasses of receptors. The best known is the *N*-methyl-D-aspartate (NMDA) receptor. This ligand-gated ion channel is similar to the nicotinic cholinergic receptor discussed in Section 24.3:

$$
\begin{array}{l}
CH_3 \\
| \\
NH_2^+ \\
| \\
CHCH_2{-}COO^- \\
| \\
COO^-
\end{array}
$$

N-Methyl-D-aspartate

When glutamic acid binds to this receptor, the ion channel opens, Na^+ and Ca^{2+} flow into the neuron, and K^+ flows out of the neuron. The same thing happens when NMDA, being an agonist, stimulates the receptor. The gate of this channel is closed by a Mg^{2+} ion.

Phencyclidine (PCP), an antagonist of this receptor, induces hallucination. PCP, known by the street name "angel dust," is a controlled substance; it causes bizarre psychotic behavior and long-term psychological problems.

C. Removal of Messengers

Transporter A protein molecule that carries small molecules, such as glucose or glutamic acid, across a membrane

In contrast to acetylcholine, there is no enzyme that would degrade glutamic acid and thereby remove it from its receptor once the signaling has occurred. Glutamic acid is removed by **transporter** molecules, which bring it back through the presynaptic membrane into the neuron. This process is called **reuptake.**

24.5 | What Are Adrenergic Messengers?

A. Monoamine Messengers

The third class of neurotransmitters/hormones, the adrenergic messengers, includes such monoamines as epinephrine, serotonin, dopamine, and histamine. (Structures of these compounds can be found later in this section and in Chemical Connections 24E.) These monoamines transmit signals by a mechanism whose beginning is similar to the action of acetylcholine. That is, they are adsorbed on a receptor.

B. Action of Messengers

Once the monoamine neurotransmitter/hormone (for example, norepinephrine) is adsorbed onto the receptor site, the signal will be amplified inside the cell. In the example shown in Figure 24.4, the receptor has an associated protein called G-protein. This protein is the key to the cascade that produces many signals inside the cell (amplification). The entire process that occurs after the messenger binds to the receptor is called **signal transduction.** [The name "signal transduction" and much of the pioneering research on this topic come from the work of Martin Rodbell (1925–1998) of the National Institutes of Health, winner of the 1994 Nobel Prize in physiology and medicine.] The active G-protein has an associated nucleotide, guanosine triphosphate (GTP). It is an analog of adenosine triphosphate (ATP), in which the aromatic base adenine is substituted by guanine (Section 25.2). The G-protein becomes inactive when its associated nucleotide is hydrolyzed to guanosine diphosphate (GDP). Signal transduction starts with the active G-protein, which activates the enzyme adenylate cyclase.

> **Signal transduction** A cascade of events through which the signal of a neurotransmitter or hormone delivered to its receptor is carried inside the target cell and amplified into many signals that can cause protein modifications, enzyme activation, and the opening of membrane channels

G-protein also participates in another signal transduction cascade, which involves inositol-based compounds (Section 21.6) as signaling molecules. Phosphatidylinositol diphosphate (PIP2) mediates the action of hormones and neurotransmitters. These messengers can stimulate the phosphorylation of enzymes, in a manner similar to the cAMP cascade. They also play an important role in the release of calcium ions from their storage areas in the endoplasmic reticulum (ER) or sarcoplasmic reticulum (SR).

C. Secondary Messengers

Adenylate cyclase produces a secondary messenger inside the cell, cyclic AMP (cAMP). The manufacture of cAMP activates processes that result in the transmission of an electrical signal. The cAMP is manufactured by adenylate cyclase from ATP:

Adenosine triphosphate (ATP) → (adenylate cyclase) → Cyclic adenosine monophosphate (cAMP) + Pyrophosphate

GOB
Chemistry Now™

Active Figure 24.4   The sequence of events in the postsynaptic membrane when norepinephrine is absorbed onto the receptor site. (*a*) The active G-protein hydrolyzes GTP. The energy of hydrolysis of GTP to GDP activates adenylate cyclase. A molecule of cAMP forms when adenylate cyclase cleaves ATP into cAMP and pyrophosphate. (*b*) Cyclic AMP activates a protein kinase by dissociating the regulatory (R) unit from the catalytic unit (C). A second molecule of ATP, shown in (*b*), has phosphorylated the catalytic unit and been converted to ADP. (*c*) The catalytic unit phosphorylates the ion-translocating protein that blocked the channel for ion flow. The phosphorylated ion-translocating protein changes its shape and position and opens the ion gates.
See a simulation based on this figure, and take a short quiz on the concepts at http://now.brookscole.com/gob8 or on the CD.

The activation of adenylate cyclase accomplishes two important goals:

1. It converts an event occurring at the outer surface of the target cell (adsorption onto receptor site) to a change inside the target cell (formation of cAMP). Thus the primary messenger (neurotransmitter or hormone) does not have to cross the membrane.

2. It amplifies the signal. One molecule adsorbed on the receptor triggers the adenylate cyclase to make many cAMP molecules. In this way, the signal is amplified many thousands of times.

GOB
Chemistry Now™
Click *Chemistry Interactive* to see an example of **How Secondary Messengers Operate**

D. Removal of Signal

How does this signal amplification stop? When the neurotransmitter or hormone dissociates from the receptor, the adenylate cyclase halts the manufacture of cAMP. The cAMP already produced is destroyed by the enzyme phosphodiesterase, which catalyzes the hydrolysis of the phosphoric ester bond, yielding AMP.

The amplification through the secondary messenger (cAMP) is a relatively slow process. It may take from 0.1 s to a few minutes. Therefore, in cases where the transmission of signals must be fast (milliseconds or seconds), a neurotransmitter such as acetylcholine acts on membrane permeability directly, without the mediation of a secondary messenger.

E. Control of Neurotransmission

The G-protein–adenylate cyclase cascade in transduction signaling is not limited to monoamine messengers. A wide variety of peptide hormones and neurotransmitters (Section 24.6) use this signaling pathway. Included among them are glucagon, vasopressin, luteinizing hormone, enkephalins, and P-protein. Neither is the opening of ion channels, depicted in Figure 24.4, the only target of this signaling. A number of enzymes can be phosphorylated by protein kinases, and the phosphorylation controls whether these enzymes will be active or inactive (Section 23.6).

The fine control of the G-protein–adenylate cyclase cascade is essential for health. Consider the toxin of the bacterium *Vibrio cholerae,* which permanently activates G-protein. The result is the symptoms of cholera—namely, severe dehydration as a result of diarrhea. This problem arises because the activated G-proteins overproduce cAMP. This excess, in turn, opens the ion channels, which leads to a large outflow of ions and accompanying water from the epithelial cells to the intestines. Therefore, the first measure taken in treating cholera victims is to replace the lost water and salt.

F. Removal of Neurotransmitters

The inactivation of the adrenergic neurotransmitters differs somewhat from the inactivation of the cholinergic transmitters. While acetylcholine is decomposed by acetylcholinesterase, most of the adrenergic neurotransmitters are inactivated in a different way. *The body inactivates monoamines by oxidizing them to aldehydes.* Enzymes that catalyze these reactions, called monoamine oxidases (MAOs), are very common in the body. For example, one MAO converts both epinephrine and norepinephrine to the corresponding aldehyde:

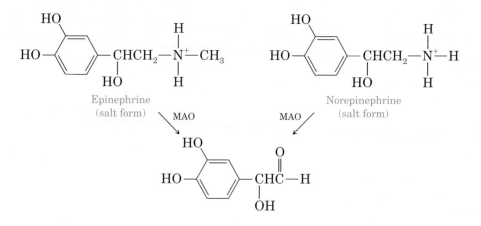

Many drugs that are used as antidepressants or antihypertensive agents are MAO inhibitors—for example, Marplan and Nardil. They prevent MAOs from converting monoamines to aldehydes, thereby increasing the concentration of the active adrenergic neurotransmitters.

CHEMICAL CONNECTIONS 24E

Parkinson's Disease: Depletion of Dopamine

Parkinson's disease is characterized by spastic motion of the eyelids as well as rhythmic tremors of the hands and other parts of the body, often when the patient is at rest. The posture of the patient changes to a forward, bent-over position; walking becomes slow, with shuffling footsteps. The cause of this degenerative nerve disease is unknown, but genetic factors and environmental effects, such as exposure to pesticides or high concentrations of metals such as Mn^{2+} ion, have been implicated.

The neurons affected contain, under normal conditions, mostly dopamine as a neurotransmitter. People with Parkinson's disease have depleted amounts of dopamine in their brains, but the dopamine receptors are not affected. Thus the first line of remedy is to *increase the concentration of dopamine*. Dopamine cannot be administered directly, because it cannot penetrate the blood–brain barrier and therefore does not reach the tissue where its action is needed. L-Dopa, by contrast, is transported through the arterial wall and converted to dopamine in the brain:

(S)-3,4-Dihydroxyphenylalanine
(L-Dopa)

enzyme-catalyzed
decarboxylation

$+ CO_2$

Dopamine

When L-dopa is administered, many patients with Parkinson's disease are able to synthesize dopamine and resume normal nerve transmission. In these individuals, L-dopa reverses the symptoms of their disease, although the respite is only temporary. In other patients, the L-dopa regimen provides little benefit.

Another way to increase dopamine concentration is to *prevent its metabolic elimination*. The drug entacapon (Comtan) inhibits an enzyme that is instrumental in clearing dopamine from the brain. The enzyme (catechol-O-methyl transferase, COMT) converts dopamine to 3-methoxy-4-hydroxy-L-phenylalanine, which is then eliminated. Entacapon is usually administered together with L-dopa. Another drug, (R)-selegiline (L-Deprenyl), is a monoamine oxidase (MAO) inhibitor. L-Deprenyl, which is also given in combination with L-dopa, can reduce the symptoms of Parkinson's disease and even increase the lifespan of patients. It increases the level of dopamine by *preventing its oxidation by MAOs*.

Other drugs may treat only the symptoms of Parkinson's disease: the spastic motions and the tremors. These drugs, such as benzotropin (Cogentin), are similar to atropine (Problem 16.45) and act on cholinergic receptors, thereby preventing muscle spasms.

The real cure for Parkinson's disease may lie in transplanting human embryonic dopamine neurons. In preliminary work, such grafts have been functionally integrated in patients' brains producing dopamine. In the most successful cases, patients have been able to resume normal, independent life following the transplant.

Certain drugs designed to affect one neurotransmitter may also affect another. An example is the drug methylphenidate (Ritalin). In high doses, this drug enhances the dopamine concentration in the brain and acts as a stimulant. In small doses, it is prescribed to calm hyperactive children or to minimize ADD (attention deficit disorder). It seems that in smaller doses Ritalin raises the concentration of serotonin. This neurotransmitter decreases hyperactivity without affecting the dopamine levels of the brain.

Serotonin

The close connection between two monoamine neurotransmitters, dopamine and serotonin, is also evident in their roles in controlling the nausea and vomiting that often follow general anesthesia and chemotherapy. Blockers of dopamine receptors in the brain, such as promethazine (Phenergan), can alleviate the symptoms after anesthesia. A blocker of serotonin receptors in the brain as well as on the terminals of the vagus nerve in the stomach, such as ondansetrone (Zofran), is the drug of choice for preventing chemotherapy-induced vomiting.

Synthesis and degradation of dopamine are not the only way that the brain keeps its concentration at a steady state. The concentration is also controlled by specific proteins, called *transporters*, that ferry the used dopamine from the receptor back across the synapse into the original neuron for reuptake. Cocaine addiction works through such a transporter. Cocaine binds to the dopamine transporter, like a reversible inhibitor, thereby preventing the reuptake of dopamine. As a consequence, dopamine is not transported back to the original neuron and stays in the synapse, increasing the continuous firing of signals, which is the psychostimulatory effect associated with a cocaine "high."

There is also an alternative way to remove adrenergic neurotransmitters. Shortly after adsorption onto the postsynaptic membrane, the neurotransmitter comes off the receptor site and is reabsorbed through the presynaptic membrane and stored again in the vesicles.

CHEMICAL CONNECTIONS 24F

Nitric Oxide as a Secondary Messenger

The toxic effects of the simple gaseous molecule NO have long been recognized (see Chemical Connections 4C). Therefore, it came as a surprise to find that this compound plays a major role in chemical communications. This simple molecule is synthesized in the cells when arginine is converted to citrulline. (These two compounds appear in the urea cycle; see Section 28.8.) Nitric oxide is a relatively nonpolar molecule, and shortly after it has been produced in the nerve cell, it quickly diffuses across the lipid bilayer membrane. During its short half-life (4–6 s), it can reach a neighboring cell. Because NO passes through membranes, it does not need extracellular receptors to deliver its message. NO is very unstable, so there is no need for a special mechanism to carry out its destruction.

NO acts as an intercellular messenger between the endothelial cells surrounding the blood vessels and the smooth muscles encompassing these cells. It relaxes the muscle cells, thereby dilating the blood vessels. The outcome is less-restricted blood flow and a drop in blood pressure. This reason also explains why nitroglycerin (Chemical Connections 13D) is effective against angina: It produces NO in the body.

Another role of NO in dilating blood vessels lies in remedying impotence. The impotence-relieving drug Viagra enhances the activity of NO by inhibiting an enzyme (phosphodiesterase) that otherwise would reduce NO's effect on smooth muscles. When the NO concentration is sufficiently high, the blood vessels dilate, allowing enough blood to flow to provide an erection. In most cases, this happens within an hour after taking the pill.

Sometimes the dilation of blood vessels is not so beneficial. Headaches are caused by dilated arteries in the head. NO-producing compounds in food—nitrites in smoked and cured meats and sodium glutamate in seasoning (Chemical Connections 28D)—can cause such headaches. Nitroglycerin itself often induces headaches.

Nitric oxide is toxic, as discussed in Chemical Connections 4C. This toxicity is used by our immune system (Section 31.2B) to fight infections caused by viruses.

The toxic effect of NO is also evident in strokes. In a stroke, a blocked artery restricts the blood flow to certain parts of the brain; the oxygen-starved neurons then die. Next, neurons in the surrounding area, 10 times larger than the place of the initial attack, release glutamic acid, which stimulates other cells. They, in turn, release NO, which kills all cells in the area. Thus the damage to the brain is spread tenfold. A concentrated effort is under way to find inhibitors of the NO-producing enzyme, nitric oxide synthase, that can be used as antistroke drugs. For the discovery of NO and its role in blood pressure control, three pharmacologists—Robert Furchgott, Louis Ignarro, and Ferid Murad—received the 1998 Nobel Prize in physiology.

G. Histamines

The neurotransmitter histamine is present in mammalian brains. It is synthesized from the amino acid, histidine, by decarboxylation:

The action of histamine as a neurotransmitter is very similar to that of other monoamines. There are two kinds of receptors for histamine. One receptor, H_1, can be blocked by antihistamines such as dimenhydrinate (Dramamine) and diphenhydramine (Benadryl). The other receptor, H_2, can be blocked by ranitidine (Zantac) and cimetidine (Tagamet).

H_1 receptors are found in the respiratory tract. They affect the vascular, muscular, and secretory changes associated with hay fever and asthma. Therefore, antihistamines that block H_1 receptors relieve these symptoms. The H_2 receptors are found mainly in the stomach and affect the secretion of HCl. Cimetidine and ranitidine, both H_2 blockers, reduce acid secretion and, therefore, are effective drugs for ulcer patients. However, the real cure for

Antihistamines block the H_1 receptor for histamine.

ulcers is to kill the bacteria *Helicobacter pylori* (Chemical Connections 23B) by administering either antibiotics or a combination of antibiotics and H_2 blockers. Sir James W. Black of the United Kingdom received the 1988 Nobel Prize in medicine for the invention of cimetidine and such other drugs as propranolol (Table 24.1).

EXAMPLE 24.1

Three enzymes in the adrenergic neurotransmission pathway affect the transduction of the signals. Identify them and describe how they affect the neurotransmission.

Solution
Adenylate cyclase amplifies the signal by producing cAMP secondary messengers. Phosphatase terminates the signal by hydrolyzing cAMP. Monoamine oxidase (MAO) reduces the frequency of signals by oxidizing the monoamine neurotransmitters to the corresponding aldehydes.

Problem 24.1
What is the functional difference between G-protein and GTP?

24.6 | What Is the Role of Peptides in Chemical Communication?

A. Messengers

Many of the most important hormones affecting metabolism belong to the peptidergic messengers group. Among them are insulin (Section 22.8 and Chemical Connections 24G) and glucagons, hormones of the pancreatic islets, and vasopressin and oxytocin (Section 22.8) products of the posterior pituitary gland.

In the last few years, scientists have isolated a number of brain peptides that have affinity for certain receptors and, therefore, act as if they were neurotransmitters. Some 25 or 30 such peptides are now known.

The first brain peptides isolated were the **enkephalins.** These pentapeptides are present in certain nerve cell terminals. They bind to specific pain receptors and seem to control pain perception. Because they bind to the receptor site that also binds the pain-killing alkaloid morphine, it is assumed that the N-terminal end of the pentapeptide fits the receptor (Figure 24.5).

Even though morphine remains the most effective agent in reducing pain, its clinical use is limited because of its side effects. These include respiratory depression, constipation, and, most significantly, addiction. The clinical use of enkephalins has yielded only modest relief. The challenge is to develop analgesic drugs that do not involve the opiate receptors in the brain.

Another brain peptide, **neuropeptide Y,** affects the hypothalamus, a region that integrates the body's hormonal and nervous systems. Neuropeptide Y is a potent orexic (appetite-stimulating) agent. When its receptors are blocked (for example, by leptin, the "thin" protein), appetite is suppressed. Leptin is an anorexic agent.

Yet another peptidergic neurotransmitter is **substance P** (*P* for "pain"). This 11-amino-acid peptide is involved in transmission of pain signals. In injury or inflammation, sensory nerve fibers transmit signals from the peripheral nervous system (where the injury occurred) to the spinal cord, which processes the pain. The peripheral neurons synthesize and release

Morphine   Methionine enkephalin

Figure 24.5 Similarities between the structure of morphine and that of the brain's own pain regulators, the enkephalins.

substance P, which bonds to receptors on the surface of the spinal cord. Substance P, in its turn, removes the magnesium block at the N-methyl-D-aspartate (NMDA) receptor. Glutamic acid, an excitatory amino acid, can then bind to this receptor. In doing so, it amplifies the pain signal going to the brain.

B. Secondary Messengers

All peptidergic messengers, hormones, and neurotransmitters act through secondary messengers. Glucagon, luteinizing hormone, antidiuretic hormone, angiotensin, enkaphalin, and substance P use the G-protein–adenylate cyclase cascade. Others, such as vasopressin, use membrane-derived phosphatidylinositol (PI) derivatives (Section 21.6). As with the cAMP cascade, control of PI secondary messengers occurs through phosphorylation. Activation may be attained via phosphorylation, and inactivation by removing a phosphate. The transition between phosphatidylinositol and its various phosphates is exercised by a kinase/phosphatase mechanism:

Phosphatidylinositol
(PI)

Inositol-1-phosphate   Inositol-1,4-diphosphate
(I-1-P)   (I-1,4-P$_2$)

Still other substances use calcium (Chemical Connections 24A) as secondary messengers.

CHEMICAL CONNECTIONS 24G

Diabetes

The disease diabetes mellitus affects approximately 16 million people in the United States. In a normal person, the pancreas, a large gland behind the stomach, secretes insulin and several other hormones. Diabetes usually results from low insulin secretion. Insulin is necessary for glucose molecules to penetrate such cells as brain, muscle, and fat cells, where they can be used. It accomplishes this task by being adsorbed onto the receptors in the target cells. This adsorption triggers the manufacture of cyclic GMP (not cAMP); this secondary messenger, in turn, increases the transport of glucose molecules into the target cells.

In the resulting cascade of events, the first step is the self-(auto)phosphorylation of the receptor molecule itself, on the cytoplasmic side. The phosphorylated insulin receptor activates enzymes and regulatory proteins by phosphorylating them. As a consequence, glucose transporter molecules (GLUT4) that are stored inside the cells migrate to the plasma membrane. Once there, they facilitate the movement of glucose across the membrane. This transport relieves the accumulation of glucose in blood serum and makes it available for metabolic activity inside the cells. The glucose can then be used as an energy source, stored as glycogen, or even diverted to enter fat and other molecular biosynthetic pathways.

In diabetic patients, the glucose level rises to 600 mg/100 mL of blood or higher (normal is 80 to 100 mg/100 mL). Two kinds of diabetes exist. In insulin-dependent diabetes, patients do not manufacture enough of this hormone in the pancreas. So-called Type I disease develops early, before the age of 20, and must be treated with daily injections of insulin. Even with daily injections of insulin, the blood sugar level fluctuates, which may cause other disorders, such as cataracts, retinal dystrophy leading to blindness, kidney disease, heart attack, and nervous disorders.

One way to counteract these fluctuations is to monitor the blood sugar and, as the glucose level rises, to administer insulin. Such monitoring requires pricking the finger six times per day, an invasive regimen that few diabetics follow faithfully. Recently, noninvasive monitoring techniques have been developed. One of the most promising employs contact lenses. The blood sugar's fall and rise is mimicked by the glucose content of tears. A fluorescent sensor in the contact lens monitors these glucose fluctuations; its data can be read via a hand-held fluorimeter. Thus the patterns of glucose fluctuation can be obtained noninvasively and, if the level rises to the danger zone, it can be counteracted by insulin intake.

The delivery of insulin also has undergone a revolution. The tried-and-true methods of injections and insulin pumps are still widely used, but new delivery is available orally or by nasal sniffing.

In Type II (non–insulin-dependent) diabetes, patients have enough insulin in the blood but cannot utilize it properly because the target cells have an insufficient number of receptors. Such patients typically develop the disease after age 40 and are likely to be obese. Overweight people usually have a lower-than-normal number of insulin receptors in their adipose (fat) cells.

Oral drugs can help patients with Type II diabetes in several ways. For example, sulfonyl urea compounds, such as tolbutamide (Orinase), increase insulin secretion. In addition, insulin concentration in the blood can be increased by enhancing its release from the β-cells of pancreatic islets. The drug repaglinide (Prandin) blocks the K^+-ATP channels of the β-cells, facilitating Ca^{2+} influx, which induces the release of insulin from the cells.

The oral drugs seem to control the symptoms of diabetes, but fluctuations in insulin levels may turn high blood sugar (hyperglycemia), into low blood sugar (hypoglycemia), which is just as dangerous. Other drugs for Type II diabetes that do not elicit hypoglycemia attempt to control the glucose level at its source. Migitol (Glyset), an anti-α-glucosidase drug, inhibits the enzyme that converts glycogen or dietary starch into glucose. The drug metformin (Glucophage) decreases glucose production in the liver, carbohydrate absorption in the intestines, and glucose uptake by fat cells.

Tolbutamide
(Orinase)

Metformin
(Glucophage is the
hydrochloride of metformin)

24.7 | How Do Steroid Hormones Act as Messengers?

In Section 21.10, we saw that a large number of hormones possess steroid ring structures. These hormones, which include the sex hormones, are hydrophobic; therefore, they can cross plasma membranes of the cell by diffusion.

CHEMICAL CONNECTIONS 24H

Tamoxifen and Breast Cancer

Large-scale clinical studies have shown that tamoxifen (Nolvadex) can reduce the recurrence of breast cancer in women who have been successfully treated for the disease. Even more importantly, a study involving 13,000 women showed that tamoxifen reduces the occurrence of breast cancer by 45% in women who are at high risk of developing the disease because of family history, previous breast abnormality, or age. Tamoxifen is a nonsteroidal compound that binds to the estradiol receptors situated on the nuclei of cells. By binding to the receptor, it causes conformational changes.

In this way, it prevents estradiol's binding to its receptor and thus hampers the stimulation of the growth of breast cancer. In addition, tamoxifen has shown promise in treating refractory ovarian cancer.

Although it reduces the risk of breast cancer, tamoxifen increases the risk of both uterine cancer and clot formation. The latter condition can be life-threatening if clots migrate to the lungs, causing pulmonary embolism. However, both diseases are much rarer than breast cancer.

Estradiol

Tamoxifen

Progesterone

There is no need for special receptors embedded in the membrane for these hormones. It has been shown, however, that **steroid hormones** interact inside the cell with protein receptors. Most of these receptors are localized in the nucleus of the cell, but small amounts also exist in the cytoplasm. When they interact with steroids, they facilitate their migration through the aqueous cytoplasm; the proteins themselves are hydrophilic.

Once inside the nucleus, the steroid–receptor complex can either bind directly to the DNA or combine with a transcription factor (Section 26.2), influencing the synthesis of a certain key protein. Thyroid hormones, which also have large hydrophobic domains, have protein receptors as well, which facilitates their transport through aqueous media.

The steroid hormonal response through protein synthesis is not fast. In fact, it takes hours to occur. Steroids can also act at the cell membrane, influencing ligand-gated ion channels. Such a response would take only seconds. An example of such a fast response occurs in fertilization. The sperm head contains proteolytic enzymes, which act on the egg to facilitate its penetration. These enzymes are stored in acrosomes, organelles found on the sperm head. During fertilization, progesterone originating from the follicle cells surrounding the egg acts on the acrosome outer membrane, which disintegrates within seconds and releases the proteolytic enzymes.

The same steroid hormones depicted in Figure 21.6 act as neurotransmitters, too. These neurosteroids are synthesized in the brain cells in both neurons and glia, and they affect receptors—mainly the NMDA and GABA receptors (Section 24.4). Progesterone and progesterone metabolites in brain cells can induce sleep, have analgesic and anticonvulsive effects, and can even serve as natural anesthetics.

SUMMARY OF KEY QUESTIONS

SECTION 24.1 What Molecules Are Involved in Chemical Communications?

- Cell-to-cell communications are carried out by three kinds of molecules.
- **Receptors** are protein molecules embedded in the membranes of cells.
- **Chemical messengers,** or ligands, interact with receptors.
- **Secondary messengers** carry and amplify the signals from the receptor to inside the cell.

SECTION 24.2 How Are Chemical Messengers Classified as Neurotransmitters and Hormones?

- **Neurotransmitters** send chemical messengers across a short distance—the **synapse** between two neurons or between a neuron and a muscle or endocrine gland cell. This communication occurs in milliseconds.
- **Hormones** transmit their signals more slowly and over a longer distance, from the source of their secretion (endocrine gland), through the bloodstream, into target cells.
- **Antagonists** block receptors; **agonists** stimulate receptors.
- Five kinds of chemical messengers exist: **cholinergic, amino acid, adrenergic, peptidergic,** and **steroid.** Neurotransmitters may belong to all five classes, hormones to the last three classes. Acetylcholine is cholinergic, glutamic acid is an amino acid, epinephrine (adrenaline) and norepinephrine are adrenergic, enkephalins are peptidergic, and progesterone is a steroid.

SECTION 24.3 How Does Acetylcholine Act as a Messenger?

- Nerve transmission starts with the neurotransmitters, such as acetylcholine packaged in **vesicles** in the **presynaptic end** of neurons.
- When neurotransmitters are released, they cross the membrane and the synapse and are adsorbed onto receptor sites on the **postsynaptic** membranes. This adsorption triggers an electrical response.
- Some neurotransmitters act directly, whereas others act through a secondary messenger, **cyclic AMP.**
- After the electrical signal is triggered, the neurotransmitter molecules must be removed from the

postsynaptic end. In the case of acetylcholine, this removal is done by an enzyme called acetylcholinesterase.

SECTION 24.4 What Amino Acids Act as Neurotransmitters?

- Amino acids, many of which differ from the amino acids found in proteins, bind to their receptors, which are ligand-gated ion channels.
- Removal of amino acid messengers takes place by **reuptake** through the presynaptic membrane, rather than by hydrolysis.

SECTION 24.5 What Are Adrenergic Messengers?

- The mode of action of monoamines such as epinephrine, serotonin, dopamine, and histamine is similar to that of acetylcholine, in the sense they start with binding to a receptor.
- Cyclic AMP is an important secondary messenger.
- The mode of removal of monoamines differs from the hydrolysis of acetylcholine. In the case of monoamines, enzymes (MAOs) oxidize them to aldehydes.

SECTION 24.6 What Is the Role of Peptides in Chemical Communication?

- Peptides and proteins bind to receptors on the target cell membrane and use secondary messengers to exert their influence.
- **Signal transduction** is the process that occurs after a ligand binds to its receptor. In this process, the signal is carried inside the cell and is amplified.

SECTION 24.7 How Do Steroid Hormones Act as Messengers?

- Steroids penetrate the cell membrane, and their receptors are found in the cytoplasm. Together with their receptors, they penetrate the cell nucleus.
- Steroid hormones can act in three ways: (1) They activate enzymes, (2) they affect the gene transcription of an enzyme or protein, and (3) they change membrane permeability.
- The same steroids can also act as neurotransmitters, when synthesized in neurons.

PROBLEMS

SECTION 24.1 What Molecules Are Involved in Chemical Communications?

24.2 What kind of signal travels along the axon of a neuron?

24.3 What is the difference between a *chemical messenger* and a *secondary messenger?*

SECTION 24.2 How Are Chemical Messengers Classified as Neurotransmitters and Hormones?

24.4 Define the following:
(a) Synapse
(b) Receptor
(c) Presynaptic
(d) Postsynaptic
(e) Vesicle

24.5 What is the role of Ca^{2+} in releasing neurotransmitters into the synapse?

24.6 Which signal takes longer: (a) a neurotransmitter or (b) a hormone? Explain.

24.7 Which gland controls lactation?

24.8 ■ To which of the three groups of chemical messengers do these hormones belong?
(a) Norepinephrine
(b) Thyroxine
(c) Oxytocin
(d) Progesterone

SECTION 24.3 How Does Acetylcholine Act as a Messenger?

24.9 How does acetylcholine transmit an electric signal from neuron to neuron?

24.10 Which end of the acetylcholine molecule fits into the receptor site?

24.11 Cobra venom and botulin are both deadly toxins, but they affect cholinergic neurotransmission differently. How does each cause paralysis?

24.12 Different ion concentrations across a membrane generate a potential (voltage). We call such a membrane *polarized.* What happens when acetylcholine is adsorbed on its receptor?

SECTION 24.4 What Amino Acids Act as Neurotransmitters?

24.13 ■ List two features by which taurine differs from the amino acids found in proteins.

24.14 How is glutamic acid removed from its receptor?

24.15 What is unique in the structure of GABA that distinguishes it from the amino acids that are present in proteins?

24.16 What is the structural difference between NMDA, an agonist of a glutamic acid receptor, and L-aspartic acid?

SECTION 24.5 What Are Adrenergic Messengers?

24.17 (a) Identify two monoamine neurotransmitters in Table 24.1.
(b) Explain how they act.
(c) Which medication controls the diseases caused by the lack of monoamine neurotransmitters?

24.18 ■ What bond is hydrolyzed and what bond is formed in the synthesis of cAMP?

24.19 How is the catalytic unit of protein kinase activated in adrenergic neurotransmission?

24.20 The formation of cyclic AMP is described in Section 24.5. Show by analogy how cyclic GMP is formed from GTP.

24.21 By analogy to the action of MAO on epinephrine, write the structural formula of the product of the corresponding oxidation of dopamine.

24.22 The action of protein kinase is the next-to-last step in the G-protein–adenylate cyclase cascade signal transduction. What kind of effects can elicit the phosphorylation by this enzyme?

24.23 Explain how adrenergic neurotransmission is affected by (a) amphetamines and (b) reserpine. (See Table 24.1.)

24.24 ■ Which step in the events depicted in Figure 24.3 provides an electrical signal?

24.25 What kind of product is the MAO-catalyzed oxidation of epinephrine?

24.26 How is histamine removed from the receptor site?

24.27 ■ Cyclic AMP affects the permeability of membranes for ion flow.
(a) What blocks the ion channel?
(b) How is this blockage removed?
(c) What is the direct role of cAMP in this process?

24.28 Dramamine and cimetidine are both antihistamines. Would you expect Dramamine to cure ulcers and cimetidine to relieve the symptoms of asthma? Explain.

SECTION 24.6 What Is the Role of Peptides in Chemical Communication?

24.29 What is the chemical nature of enkephalins?

24.30 What is the mode of action of Demerol as a painkiller? (See Table 24.1.)

24.31 ■ What enzyme catalyzes the formation of inositol-1,4,5-triphosphate from inositol-1,4-diphosphate? Give the structures of the reactant and the product.

24.32 How is the secondary messenger inositol-1,4,5-triphosphate inactivated?

SECTION 24.7 How Do Steroid Hormones Act as Messengers?

24.33 Where are the receptors for steroid hormones located—on the cell surface or elsewhere?

24.34 Do steroid hormones affect protein synthesis? If so, does this effect have any implications for the time the hormonal response can take?

24.35 Can steroid hormones act as neurotransmitters?

Chemical Connections

24.36 (Chemical Connections 24A) What is the difference between calcium sparks (puffs) and calcium waves?

24.37 (Chemical Connections 24A) What is the role of calmodulin in signaling by Ca^{2+} ions?

24.38 (Chemical Connections 24A) To enable a fusion between the synaptic vesicle and the plasma membrane, the calcium concentration is increased. How many-fold of an increase in Ca^{2+} concentration is needed?

24.39 (Chemical Connections 24B)
   (a) What is the effect of nerve gases?
   (b) What molecular substance do they affect?

24.40 (Chemical Connections 24C) What is the mode of action of botulinum toxin?

24.41 (Chemical Connections 24C) How can a deadly botulinum toxin contribute to the facial beauty of Hollywood actresses?

24.42 (Chemical Connections 24D) What are the neurofibrillar tangles in the brains of patients with Alzheimer's disease made of? How do they affect the cell structure?

24.43 (Chemical Connections 24D) What are the plaques in the brains of patients with Alzheimer's disease made of?

24.44 (Chemical Connections 24D) Alzheimer's disease causes loss of memory. What kind of drugs may provide some relief for—if not cure—this disease? How do they act?

24.45 (Chemical Connections 24E) Why would a dopamine pill be ineffective in treating Parkinson's disease?

24.46 (Chemical Connections 24E) What is the mechanism by which cocaine stimulates the continuous firing of signals between neurons?

24.47 (Chemical Connections 24E) Parkinson's disease is due to a paucity of dopamine neurons, yet its symptoms are relieved by drugs that block cholinergic receptors. Explain.

24.48 (Chemical Connections 24E) In certain cases, embryonic dopamine neurons transplanted into the brains of patients with advanced Parkinson's disease resulted in complete remission. How was this result possible?

24.49 (Chemical Connections 24F) How can NO cause headaches?

24.50 (Chemical Connections 24F) How is NO synthesized in the cells?

24.51 (Chemical Connections 24F) How is the toxicity of NO detrimental in strokes?

24.52 ■ (Chemical Connections 24G) Tolbutamide is called a sulfonyl urea compound. Identify the sulfonyl urea moiety in the structure of the drug.

24.53 (Chemical Connections 24G) What is the difference between insulin-dependent and non–insulin-dependent diabetes?

24.54 (Chemical Connections 24G) How does insulin facilitate the absorption of glucose from blood serum into adipocytes (fat cells)?

24.55 (Chemical Connections 24G) Diabetic patients must frequently monitor the fluctuation of glucose levels in their blood. What is the advantage of the latest technique of monitoring the glucose content in tears over the older technique of obtaining frequent blood samples?

24.56 (Chemical Connections 24H) Tamoxifen is an antagonist. How does it work in reducing the reoccurrence of breast cancer?

Additional Problems

24.57 Considering its chemical nature, how does aldosterone (Section 21.10) affect mineral metabolism (Table 24.2)?

24.58 What is the function of the ion-translocating protein in adrenergic neurotransmission?

24.59 Decamethonium acts as a muscle relaxant. If an overdose of decamethonium occurs, can paralysis be prevented by administering large doses of acetylcholine? Explain.

24.60 Endorphin, a potent painkiller, is a peptide containing 22 amino acids; among them are the same five N-terminal amino acids found in the enkephalins. Does this explain endorphin's pain-killing action?

24.61 How do alanine and beta alanine differ in structure?

24.62 Where is a G-protein located in adrenergic neuro-transmission?

24.63 (Chemical Connections 24F) List a number of effects that are caused when NO, acting as a secondary messenger, relaxes smooth muscles.

24.64 (a) In terms of their action, what do the hormone vasopressin and the neurotransmitter dopamine have in common?

(b) What is the difference in their modes of action?

24.65 What is the difference in the models of action between acetylcholinesterase and acetylcholine transferase?

24.66 How does cholera toxin exert its effect?

24.67 Give the formulas for the following reaction:

$$GTP + H_2O \rightleftharpoons GDP + P_i$$

24.68 Insulin is a hormone that, when it binds to a receptor, enables glucose molecules to enter the cell and be metabolized. If you have a drug that is an agonist, how would the glucose level in the serum change upon administering the drug?

24.69 (Chemical Connections 24E) Ritalin is used to alleviate hyperactivity in attention deficit disorder of children. How does this drug work?

24.70 The pituitary gland releases luteinizing hormone (LH), which enhances the production of progesterone in the uterus. Classify these two messengers and discuss how each delivers its message.

Tying It Together

24.71 Why are receptors proteins rather than any other kind of molecule?

24.72 Why is it useful for organisms to have several different classes of neurotransmitters and hormones?

24.73 What relationship do adrenergic messengers have to amino acid messengers, and what does this relationship say about the biochemical origin of adrenergic messengers?

24.74 ■ What functional groups are found in the structures of chemical messengers? What do these structural features imply about the active sites of the enzymes that process these messengers?

Looking Ahead

24.75 Why is insulin not administered orally in treatment of insulin-dependent diabetes?

24.76 One of the challenges in treating cholera is that of preventing dehydration. What can make this a double challenge? (*Hint:* See Chapter 30; cholera is frequently a water-borne disease.)

24.77 Do any chemical messengers have a *direct* effect on the synthesis of nucleic acids?

24.78 Would you expect the role of chemical messengers to have any bearing on the body's requirements for energy?

Challenge Problems

24.79 Do all chemical messengers require the same time to elicit a response? If there are differences, how do the underlying response mechanisms differ?

24.80 A number of agricultural pesticides are acetyl-cholinesterase inhibitors. Why is their use carefully controlled?

24.81 What benefit is it to an organism to have two different enzymes for the synthesis and breakdown of acetylcholine—acetylcholine transferase and acetyl-cholinesterase, respectively?

24.82 Which would be a better form of therapy for cocaine addiction—an inhibitor for the dopamine transporter or a substance that degrades cocaine?

Nucleotides, Nucleic Acids, and Heredity

While these dogs might appear to be a normal mother and puppy, the latter is really the first cloned dog, Snuppy. The larger dog is a male Afghan whose DNA was used to create the clone.

AP/Wide World Photos

25.1 | What Are the Molecules of Heredity?

Each cell of our bodies contains thousands of different protein molecules. Recall from Chapter 22 that all of these molecules are made up of the same 20 amino acids, just arranged in different sequences. That is, the hormone insulin has a different amino acid sequence than the globin of the red blood cell. Even the same protein—for example, insulin—has a different sequence in different species (Section 22.8). Within the same species, individuals may have some differences in their proteins, although the differences are much less dramatic than those seen between species. This variation is most obvious in cases where people have such conditions as hemophilia, albinism, or color-blindness because they lack certain proteins that normal people have or because the sequence of their amino acids is somewhat different (see Chemical Connections 22D).

After scientists became aware of the differences in amino acid sequences, their next quest was to determine how cells know which proteins to synthesize out of the extremely large number of possible amino acid sequences. The answer is that an individual gets the information from its parents through *heredity.* Heredity is the transfer of characteristics, anatomical as well as biochemical, from generation to generation. We all know that a pig gives birth to a pig and a mouse gives birth to a mouse.

It was easy to determine that the information is obtained from the parent or parents, but what form does this information take? During the last 60 years, revolutionary developments have enabled us to answer this question—the transmission of heredity occurs on the molecular level.

From about the end of the nineteenth century, biologists suspected that the transmission of hereditary information from one generation to another took place in the nucleus of the cell. More precisely, they believed that structures within the nucleus, called **chromosomes,** have something to do with heredity. Different species have different numbers of chromosomes in the nucleus. The information that determines external characteristics (red hair, blue eyes) and internal characteristics (blood group, hereditary diseases) was thought to reside in **genes** located inside the chromosomes.

Chemical analysis of nuclei showed that they are largely made up of special basic proteins called *histones* and a type of compound called *nucleic acids.* By 1940, it became clear through the work of Oswald Avery (1877–1955) that, of all the material in the nucleus, only a nucleic acid called deoxyribonucleic acid (DNA) carries the hereditary information. That is, the genes are located in the DNA. Other work in the 1940s by George Beadle (1903–1989) and Edward Tatum (1909–1975) demonstrated that each gene controls the manufacture of one protein, and that external and internal characteristics are expressed through this gene. Thus the expression of the gene (DNA) in terms of an enzyme (protein) led to the study of protein synthesis and its control. *The information that tells the cell which proteins to manufacture is carried in the molecules of DNA.* We now know that not all genes lead to the production of protein, but all genes do lead to the production of another type of nucleic acid, called ribonucleic acid (RNA).

Human chromosomes magnified about 8000 times.

Gene The unit of heredity; a DNA segment that codes for one protein or one type of RNA

25.2 | What Are Nucleic Acids Made Of?

Two kinds of nucleic acids are found in cells: **ribonucleic acid (RNA)** and **deoxyribonucleic acid (DNA).** Each has its own role in the transmission of hereditary information. As we just saw, DNA is present in the chromosomes of the nuclei of eukaryotic cells. RNA is not found in the chromosomes, but rather is located elsewhere in the nucleus and even outside the nucleus, in the cytoplasm. As we will see in Section 25.4, there are six types of RNA, all with specific structures and functions.

Both DNA and RNA are polymers. Just as proteins consist of chains of amino acids, and polysaccharides consist of chains of monosaccharides, so nucleic acids are also chains. The building blocks (monomers) of nucleic acid chains are *nucleotides.* Nucleotides themselves, however, are composed of three simpler units: a base, a monosaccharide, and a phosphate. We will look at each of these components in turn.

A. Bases

The **bases** found in DNA and RNA are chiefly those shown in Figure 25.1. All of them are basic because they are heterocyclic aromatic amines (Section 16.1). Two of these bases—adenine (A) and guanine (G)—are purines; the other three—cytosine (C), thymine (T), and uracil (U)—are pyrimidines. The two purines (A and G) and one of the pyrimidines (C) are found in both

Bases Purines and pyrimidines, which are components of nucleotides, DNA, and RNA

Purines

Pyrimidines

Figure 25.1 The five principal bases of DNA and RNA. Note how the rings are numbered. The hydrogens shown in blue are lost when the bases bond to monosaccharides.

DNA and RNA, but uracil (U) is found only in RNA, and thymine (T) is found only in DNA. Note that thymine differs from uracil only in the methyl group in the 5 position. Thus both DNA and RNA contain four bases: two pyrimidines and two purines. For DNA, the bases are A, G, C, and T; for RNA, the bases are A, G, C, and U.

B. Sugars

The sugar component of RNA is D-ribose (Section 20.1C). In DNA, it is 2-deoxy-D-ribose (hence the name deoxyribonucleic acid).

The full name of β-D-ribose is β-D-ribofuranose and that of β-2-deoxy-D-ribose is β-2-deoxy-D-ribofuranose (see Section 20.2A).

Nucleoside A compound composed of ribose or deoxyribose and a base

The combination of sugar and base is known as a **nucleoside.** The purine bases are linked to C-1 of the monosaccharide through N-9 (the nitrogen at position 9 of the five-membered ring) by a β-*N*-glycosidic bond:

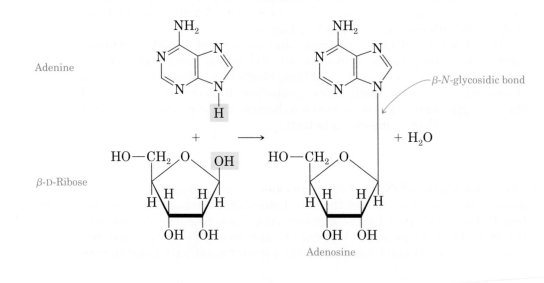

The nucleoside made of guanine and ribose is called **guanosine.** Table 25.1 gives the names of the other nucleosides.

The pyrimidine bases are linked to C-1 of the monosaccharide through their N-1 by a β-N-glycosidic bond.

Uridine

C. Phosphate

The third component of nucleic acids is phosphoric acid. When this group forms a phosphate ester (Section 19.5) bond with a nucleoside, the result is a compound known as a **nucleotide.** For example, adenosine combines with phosphate to form the nucleotide adenosine 5′-monophosphate (AMP):

Nucleotide A nucleoside bonded to one, two, or three phosphate groups

The ′ sign in adenosine 5′-monophosphate is used to distinguish which molecules the phosphate is bound to. Numbers without primes refer to positions on the purine or pyrimidine base. Numbers on the sugar are denoted with primes.

Table 25.1 gives the names of the other nucleotides. Some of these nucleotides play important roles in metabolism. They are part of the structure of key coenzymes, cofactors, and activators (Sections 27.3 and 29.2). Most notably, adenosine 5′-triphosphate (ATP) serves as a common currency into which the energy gained from food is converted and stored. In ATP, two more phosphate groups are joined to AMP in phosphate anhydride bonds (Section 19.5). In adenosine 5′-diphosphate (ADP), only one phosphate group is bonded to the AMP. All other nucleotides have important

Table 25.1 The Eight Nucleosides and Eight Nucleotides in DNA and RNA

Base	Nucleoside	Nucleotide
		DNA
Adenine (A)	Deoxyadenosine	Deoxyadenosine 5'-monophosphate (dAMP)*
Guanine (G)	Deoxyguanosine	Deoxyguanosine 5'-monophosphate (dGMP)*
Thymine (T)	Deoxythymidine	Deoxythymidine 5'-monophosphate (dTMP)*
Cytosine (C)	Deoxycytidine	Deoxycytidine 5'-monophosphate (dCMP)*
		RNA
Adenine (A)	Adenosine	Adenosine 5'-monophosphate (AMP)
Guanine (G)	Guanosine	Guanosine 5'-monophosphate (GMP)
Uracil (U)	Uridine	Uridine 5'-monophosphate (UMP)
Cytosine (C)	Cytidine	Cytidine 5'-monophosphate (CMP)

*The *d* indicates that the sugar is deoxyribose.

multiphosphorylated forms. For example, guanosine exists as GMP, GDP, and GTP.

In Section 25.3, we will see how DNA and RNA are chains of nucleotides. In summary:

A nucleoside = Base + Sugar

A nucleotide = Base + Sugar + Phosphate

A nucleic acid = A chain of nucleotides

EXAMPLE 25.1

GTP is an important store of energy. Draw the structure of guanosine triphosphate.

Solution
The base guanine is linked to a ribose unit by a *β-N*-glycosidic linkage. The triphosphate is linked to C-5' of the ribose by an ester bond.

Problem 25.1
Draw the structure of UMP.

CHEMICAL CONNECTIONS 25A

Anticancer Drugs

A major difference between cancer cells and most normal cells is that cancer cells divide much more rapidly. Rapidly dividing cells require a constant new supply of DNA. One component of DNA is the nucleoside deoxythymidine, which is synthesized in the cell by the methylation of the uracil base.

Fluorouracil

If fluorouracil is administered to a cancer patient as part of chemotherapy, the body converts it to fluorouridine, a compound that irreversibly inhibits the enzyme that manufactures thymidine from uridine, greatly decreasing DNA synthesis. Because this inhibition affects the rapidly dividing cancer cells more than the healthy cells, the growth of the tumor and the spread of the cancer are arrested. Unfortunately, chemotherapy with fluorouracil or other anticancer drugs also weakens the body, because it interferes with DNA synthesis in normal cells.

Chemotherapy is used intermittently to give the body time to recover from the side effects of the drug. During the period after chemotherapy, special precautions must be taken so that bacterial infections do not debilitate the already-weakened body.

25.3 | What Is the Structure of DNA and RNA?

In Chapter 22, we saw that proteins have primary, secondary, and higher-order structures. Nucleic acids, which are chains of monomers, also have primary, secondary, and higher-order structures.

Nucleic acid A polymer composed of nucleotides

A. Primary Structure

Nucleic acids are polymers of nucleotides, as shown schematically in Figure 25.2. Their primary structure is the sequence of nucleotides. Note that it can be divided into two parts: (1) the backbone of the molecule and (2) the bases that are the side-chain groups. The backbone in DNA consists of alternating deoxyribose and phosphate groups. Each phosphate group is linked to the 3′ carbon of one deoxyribose unit and simultaneously to the 5′ carbon of the next deoxyribose unit (Figure 25.3). Similarly, each monosaccharide unit forms a phosphate ester at the 3′ position and another at the 5′ position. The primary structure of RNA is the same except that each sugar is ribose (so an —OH group appears in the 2′ position) rather than deoxyribose and U is present instead of T.

Thus the backbone of the DNA and RNA chains has two ends: a 3′ —OH end and a 5′ —OH end. These two ends have roles similar to those of the C-terminal and N-terminal ends in proteins. The backbone provides structural stability for the DNA and RNA molecules.

As noted earlier, the bases that are linked, one to each sugar unit, are the side chains. They carry all of the information necessary for protein synthesis. Analysis of the base composition of DNA molecules from many different species was done by Erwin Chargaff (1905–2002), who showed that in DNA taken from many different species, the quantity of adenine (in moles) is always approximately equal to the quantity of thymine, and the quantity of guanine is always approximately equal to the quantity of cytosine, although the adenine/guanine ratio varies widely from species to species (see Table 25.2). This important information helped to establish the secondary structure of DNA, as we will soon see.

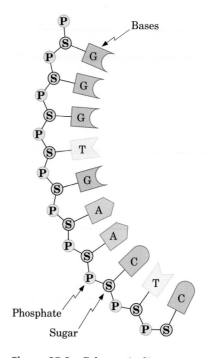

Figure 25.2 Schematic diagram of a nucleic acid molecule. The four bases of each nucleic acid are arranged in various specific sequences.

Figure 25.3 Primary structure of the DNA backbone. The hydrogens shown in blue cause the acidity of nucleic acids. In the body, at neutral pH, the phosphate groups carry a charge of −1 and the hydrogens are replaced by Na⁺ and K⁺.

Table 25.2 Base Composition and Base Ratio in Two Species

Organism	Base Composition (mol %)				Base Ratio	
	A	G	C	T	A/T	G/C
Human	30.9	19.9	19.8	29.4	1.05	1.01
Wheat germ	27.3	22.7	22.8	27.1	1.01	1.00

Just as the order of the amino acid residues of protein side chains determines the primary structure of the protein (for example, —Ala—Gly—Glu—Met—), the order of the bases (for example, —ATTGAC—) provides the primary structure of DNA. As with proteins, we need a convention to tell us which end to start with when we write the sequence of bases. For nucleic acids, the convention is to begin the sequence with the nucleotide that has the free 5′ terminus. Thus the sequence AGT means that adenine is the base at the 5′ terminus and thymine is the base at the 3′ terminus.

Figure 25.4 Three-dimensional structure of the DNA double helix.

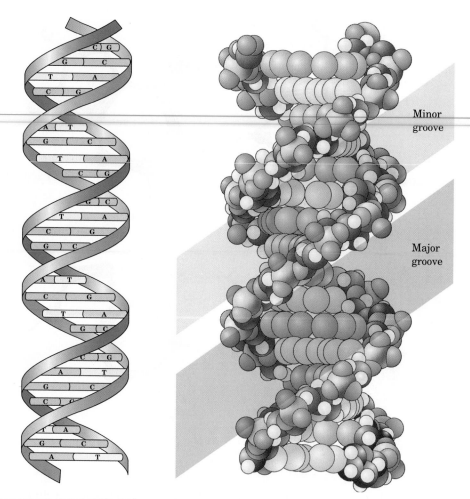

Minor groove

Major groove

B. Secondary Structure of DNA

In 1953, James Watson (1928–) and Francis Crick (1916–2004) established the three-dimensional structure of DNA. Their work is a cornerstone in the history of biochemistry. The model of DNA developed by Watson and Crick was based on two important pieces of information obtained by other workers: (1) the Chargaff rule that (A and T) and (G and C) are present in equimolar quantities and (2) x-ray diffraction photographs obtained by Rosalind Franklin (1920–1958) and Maurice Wilkins (1916–). By the clever use of these facts, Watson and Crick concluded that DNA is composed of two strands entwined around each other in a **double helix,** as shown in Figure 25.4.

In the DNA double helix, the two polynucleotide chains run in opposite directions. Thus, at each end of the double helix, there is one 5′ —OH and one 3′ —OH terminus. The sugar–phosphate backbone is on the outside, exposed to the aqueous environment, and the bases point inward. The bases are hydrophobic, so they try to avoid contact with water. Through their hydrophobic interactions, they stabilize the double helix. The bases are paired according to Chargaff's rule: For each adenine on one chain, a thymine is aligned opposite it on the other chain; each guanine on one chain has a cytosine aligned with it on the other chain. *The bases so paired form hydrogen bonds with each other, two for A—T and three for G—C, thereby stabilizing the double helix* (Figure 25.5). A—T and G—C are **complementary base pairs.**

The important thing, as Watson and Crick realized, is that only adenine could fit with thymine and only guanine could fit with cytosine. Let us consider the other possibilities. Can two purines (AA, GG, or AG) fit opposite each

Double helix The arrangement in which two strands of DNA are coiled around each other in a screw-like fashion

Watson, Crick, and Wilkins were awarded the 1962 Nobel Prize in medicine for their discovery. Franklin died in 1958. The Nobel Committee does not award the Nobel Prize posthumously.

Figure 25.5 A and T pair up by forming two hydrogen bonds; G and C pair up by forming three hydrogen bonds.

Thymine · · · · Adenine

AT pair

Cytosine · · · · Guanine

GC pair

other? Figure 25.6 shows that they would overlap. How about two pyrimidines (TT, CC, or CT)? As shown in Figure 25.6, they would be too far apart. *There must be a pyrimidine opposite a purine.* But could A fit opposite C, or G opposite T? Figure 25.7 shows that the hydrogen bonding would be much weaker.

The entire action of DNA—and of the heredity mechanism—depends on the fact that, *wherever there is an adenine on one strand of the helix, there must be a thymine on the other strand because that is the only base that fits and forms strong hydrogen bonds with adenine, and similarly for G and C.* The entire heredity mechanism rests on these aligned hydrogen bonds (Figure 25.5), as we will see in Section 25.6.

Rosalind Franklin (1920–1958).

Watson and Crick with their model of the DNA molecule.

The form of the DNA double helix shown in Figure 25.4 is called B-DNA. It is the most common and most stable form. Other forms become possible where the helix is wound more tightly or more loosely, or is wound in the opposite direction. With B-DNA, a distinguishing feature is the presence of a **major groove** and a **minor groove,** which arise because the two strands are not equally spaced around the helix. Interactions of proteins and drugs with the major and minor grooves of DNA serve as an active area of research.

C. Higher-Order Structures of DNA

If a human DNA molecule were fully stretched out, its length would be perhaps 1 m. However, the DNA molecules in the nuclei are not stretched out, but rather coiled around basic protein molecules called **histones.** The acidic DNA and the basic histones attract each other by electrostatic (ionic) forces, combining to form units called **nucleosomes.** In a nucleosome, eight histone molecules form a core, around which a 147-base-pair DNA double helix is wound. Nucleosomes are further condensed into **chromatin** when a 30-nm-wide fiber forms in which nucleosomes are wound in a **solenoid** fashion, with six nucleosomes forming a repeating unit (Figure 25.8). Chromatin fibers are organized still further into loops, and loops are arranged into bands to provide the superstructure of chromosomes.

The beauty of establishing the three-dimensional structure of the DNA molecule was that the knowledge of this structure immediately led to the explanation for the transmission of heredity—how the genes transmit traits from one generation to another. Before we look at the mechanism of DNA replication (in Section 25.6), let us summarize the three differences in structure between DNA and RNA:

Two purines overlap

Gap between two pyrimidines

Figure 25.6 The bases of DNA cannot stack properly in the double helix if a purine is opposite a purine or if a pyrimidine is opposite a pyrimidine.

Chromatin The DNA complexed with histone and nonhistone proteins that exists in eukaryotic cells between cell divisions

Solenoid A coil wound in the form of a helix

No hydrogen bonding

Thymine   Guanine

Cytosine   Adenine

No hydrogen bonding

Figure 25.7 Only one hydrogen bond is possible for TG or CA. These combinations are not found in DNA. Compare this figure with Figure 25.5.

Figure 25.8 Superstructure of chromosomes. In nucleosomes, the bandlike DNA double helix winds around cores consisting of eight histones. Solenoids of nucleosomes form 30 nm filament. Loops and minibands are other substructures.

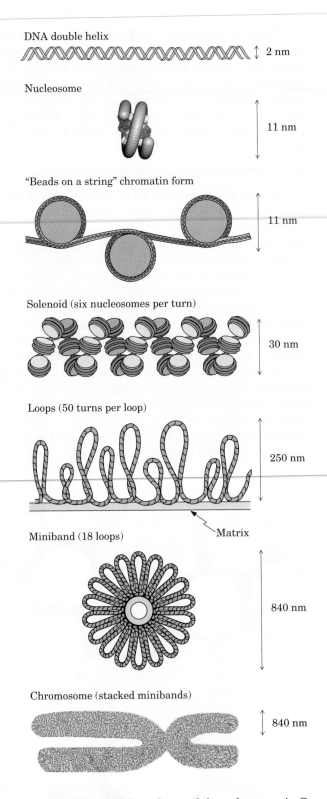

DNA double helix

↕ 2 nm

Nucleosome

11 nm

"Beads on a string" chromatin form

11 nm

Solenoid (six nucleosomes per turn)

30 nm

Loops (50 turns per loop)

250 nm

Matrix

Miniband (18 loops)

840 nm

Chromosome (stacked minibands)

↕ 840 nm

1. DNA has four bases: A, G, C, and T. RNA has three of these bases—A, G, and C—but its fourth base is U, not T.

2. In DNA, the sugar is 2-deoxy-D-ribose. In RNA, it is D-ribose.

3. DNA is almost always double-stranded, with the helical structure shown in Figure 25.4.

There are several kinds of RNA (as we will see in Section 25.4); None of them has a repetitive double-stranded structure like DNA, although base-pairing can occur within a chain (see, for example, Figure 25.10). When it does, adenine pairs with uracil because thymine is not present. Other combinations of hydrogen-bonded bases are also possible outside the confines of a double helix, and Chargaff's rule does not apply.

GOB
Chemistry⚛Now™

Click *Chemistry Interactive* to learn more about **DNA Structure**

25.4 | What Are the Different Classes of RNA?

We previously noted that there are six types of RNA.

1. **Messenger RNA (mRNA)** mRNA molecules are produced in the process called **transcription,** and they carry the genetic information from the DNA in the nucleus directly to the cytoplasm, where most of the protein is synthesized. Messenger RNA consists of a chain of nucleotides whose sequence is exactly complementary to that of one of the strands of the DNA. This type of RNA is not long-lived, however. It is synthesized as needed and then degraded, so its concentration at any given time is rather low. The size of mRNA varies widely, with the average unit containing perhaps 750 nucleotides. Figure 25.9 shows the basic flow of genetic information and the major types of RNA.

Messenger RNA (mRNA)
The RNA that carries genetic information from DNA to the ribosome and acts as a template for protein synthesis

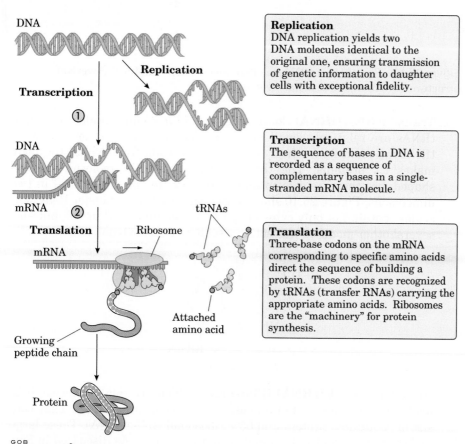

Replication
DNA replication yields two DNA molecules identical to the original one, ensuring transmission of genetic information to daughter cells with exceptional fidelity.

Transcription
The sequence of bases in DNA is recorded as a sequence of complementary bases in a single-stranded mRNA molecule.

Translation
Three-base codons on the mRNA corresponding to specific amino acids direct the sequence of building a protein. These codons are recognized by tRNAs (transfer RNAs) carrying the appropriate amino acids. Ribosomes are the "machinery" for protein synthesis.

GOB
Chemistry⚛Now™

Active Figure 25.9 The fundamental process of information transfer in cells. (1) Information encoded in the nucleotide sequence of DNA is transcribed through synthesis of an RNA molecule whose sequence is dictated by the DNA sequence. (2) As the sequence of this RNA is read (as groups of three consecutive nucleotides) by the protein synthesis machinery, it is translated into the sequence of amino acids in a protein. This information transfer system is encapsulated in the dogma: DNA ⟶ RNA ⟶ protein. **See a simulation based on this figure, and take a short quiz on the concepts at http://www.now.brookscole.com/gob8 or on the CD.**

Figure 25.10 Structure of tRNA. (*a*) Two-dimensional simplified cloverleaf structure. (*b*) Three-dimensional structure.

Transfer RNA (tRNA) The RNA that transports amino acids to the site of protein synthesis in ribosomes

2. **Transfer RNA (tRNA)** Containing from 73 to 93 nucleotides per chain, tRNAs are relatively small molecules. There is at least one different tRNA molecule for each of the 20 amino acids from which the body makes its proteins. The three-dimensional tRNA molecules are L-shaped, but they are conventionally represented as a cloverleaf in two dimensions. Figure 25.10 shows a typical structure. Transfer RNA molecules contain not only cytosine, guanine, adenine, and uracil, but also several other modified nucleotides, such as 1-methylguanosine.

1-Methylguanosine

Ribosomal RNA (rRNA) The RNA complexed with proteins in ribosomes

Ribosome Small spherical bodies in the cell made of protein and RNA; the site of protein synthesis

3. **Ribosomal RNA (rRNA) Ribosomes,** which are small spherical bodies located in the cells but outside the nuclei, contain rRNA. They consist of about 35% protein and 65% ribosomal RNA (rRNA). These large molecules have molecular weights up to 1 million. As discussed in Section 25.5, protein synthesis takes place on the ribosomes.

Dissociation of ribosomes into their components has proved to be a useful way of studying their structure and properties. A particularly important endeavor has been to determine both the number and the kind of RNA and protein molecules that make up ribosomes. This approach has helped elucidate the role of ribosomes in protein synthesis. In both prokaryotes and eukaryotes, a ribosome consists of two subunits, one larger than the other. In turn, the smaller subunit consists of one large RNA molecule and about 20 different proteins; the larger subunit consists of two RNA molecules in prokaryotes (three in eukaryotes) and about 35 different proteins in prokaryotes (about 50 in eukaryotes) (Figure 25.11). The subunits are easily dissociated from one another in the laboratory by lowering the Mg^{2+} concentration of the medium. Raising the Mg^{2+} concentration to its original level reverses the process, and active ribosomes can be reconstituted by this method.

Splicing The removal of an internal RNA segment and the joining of the remaining ends of the RNA molecule

4. **Small Nuclear RNA (snRNA)** A recently discovered RNA molecule is snRNA, which is found, as the name implies, in the nucleus of eukaryotic cells. This type of RNA is small, about 100 to 200 nucleotides long, but it is neither a tRNA molecule nor a small subunit of rRNA. In the cell, it is complexed with proteins to form **small nuclear ribonucleoprotein particles, snRNPs,** pronounced "snurps." Their function is to help with the processing of the initial mRNA transcribed from DNA into a mature form that is ready for export out of the nucleus. This process is often referred to as **splicing,** and it is an active area of research. While studying splicing, scientists realized that part of the splicing reaction involved catalysis by the RNA portion of a snRNP and not the protein portion. This recognition led to the discovery of **ribozymes,** RNA-based enzymes, for which Thomas Cech received the Nobel Prize. Splicing will be discussed further in Chapter 26.

5. **Micro RNA (miRNA)** A very recent discovery is another type of small RNA, miRNA. These RNAs are only 22 nucleotides long but are important in the timing of an organism's development. They bind to sections of mRNA and prevent their translation.

6. **Small Interfering RNA (siRNA)** The process called RNA interference was heralded as the breakthrough of the year in 2002 by *Science* magazine. Short stretches of RNA (20–30 nucleotides long), called small interfering RNA, have been found to have an enormous control over gene expression. This process serves as a protective mechanism in many species, with the siRNAs being used to eliminate expression of an undesirable gene, such as one that causes uncontrolled cell growth or one that came from a virus. siRNAs lead to the degradation of specific mRNA molecules. Scientists who wish to study gene expression are also using these small RNAs. In what has become an explosion of new biotechnology, many companies have been created to produce and market designer siRNAs to knock out hundreds of known genes. This technology also has medical applications, as siRNA has been used to protect mouse liver from hepatitis and to help clear infected liver cells of the disease.

Table 25.3 summarizes the basic types of RNA.

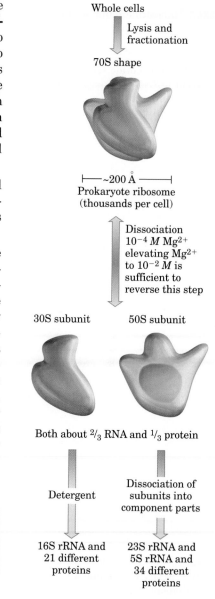

Whole cells
Lysis and fractionation
70S shape

├──~200 Å──┤
Prokaryote ribosome
(thousands per cell)

Dissociation
10^{-4} M Mg^{2+}
elevating Mg^{2+}
to 10^{-2} M is
sufficient to
reverse this step

30S subunit   50S subunit

Both about 2/3 RNA and 1/3 protein

Detergent   Dissociation of subunits into component parts

16S rRNA and   23S rRNA and
21 different   5S rRNA and
proteins   34 different
                proteins

Figure 25.11 The structure of a typical prokaryotic ribosome. The individual components can be mixed, producing functional subunits. Reassociation of subunits gives rise to an intact ribosome. The designation, S, refers to Sredberg, a unit of relative size determined when molecules are separated by centrifugation.

Table 25.3   The Roles of Different Kinds of RNA

RNA Type	Size	Function
Transfer RNA	Small	Transports amino acids to site of protein synthesis
Ribosomal RNA	Several kinds—variable in size	Combines with proteins to form ribosomes, the site of protein synthesis
Messenger RNA	Variable	Directs amino acid sequence of proteins
Small nuclear RNA	Small	Processes initial mRNA to its mature form in eukaryotes
Micro RNA	Small	Affects gene expression; important in growth and development
Small interfering RNA	Small	Affects gene expression; used by scientists to knock out a gene being studied

25.5 | What Are Genes?

A gene is a stretch of DNA, containing a few hundred nucleotides, that carries one particular message—for example, "make a globin molecule" or "make a tRNA molecule." One DNA molecule may have between 1 million and 100 million bases. Therefore, many genes are present in one DNA molecule. In bacteria, this message is continuous; in higher organisms, it is not. That is, stretches of DNA that spell out (encode) the amino acid sequence to be assembled are interrupted by long stretches that seemingly do not code for anything. The coding sequences are called **exons,** short for "expressed sequences," and the noncoding sequences are called **introns,** short for "intervening sequences."

Exon Nucleotide sequence in DNA or mRNA that codes for a protein

Intron A nucleotide sequence in DNA or mRNA that does not code for a protein

For example, the globin gene has three exons broken up by two introns. Because DNA contains both exons and introns, the mRNA transcribed from it also contains both exons and introns. The introns are spliced out by ribozymes, and the exons are spliced together before the mRNA is used to synthesize a protein. In other words, the introns function as spacers and, in rare instances, as enzymes, catalyzing the splicing of exons into mature mRNA. Figure 25.12 shows the difference between prokaryotic and eukaryotic production of proteins.

In prokaryotes, the genes on a stretch of DNA are next to each other. These are turned into a sequence of mRNA, which are then translated by ribosomes to make proteins, all of which happens simultaneously. In eukaryotes, the genes are separated by introns and the processes take place in different compartments. The DNA is turned into RNA in the nucleus, but then the initial mRNA , containing introns, is transported to the cytosol where the exons are spliced out. The final mRNA is then translated to protein. The process of making RNA and protein is the subject of Chapter 26.

In humans, only 3% of the DNA codes for proteins or RNA with clear functions. Introns are not the only noncoding DNA sequences, however. **Satellites** are DNA molecules in which short nucleotide sequences are repeated hundreds or thousands of times. Large satellite stretches appear at the ends and centers of chromosomes and provide stability for the chromosomes.

Smaller repetitive sequences, called **mini-satellites** or **microsatellites,** are associated with cancer when they mutate.

25.6 | How Is DNA Replicated?

The DNA in the chromosomes carries out two functions: (1) It reproduces itself and (2) it supplies the information necessary to make all the RNA and proteins in the body, including enzymes. The second function is covered in Chapter 26. Here we are concerned with the first, **replication.**

Replication The process by which copies of DNA are made during cell division

Each gene is a section of a DNA molecule that contains a specific sequence of the four bases A, G, T, and C, typically comprising about 1000 to 2000 nucleotides. The base sequence of the gene carries the information necessary to produce one protein molecule. If the sequence is changed (for example, if one A is replaced by a G, or if an extra T is inserted), a different protein is produced, which might have an impaired function, as in sickle cell anemia (Chemical Connections 22D).

Consider the monumental task that must be accomplished by the organism. When an individual is conceived, the egg and sperm cells unite to form the zygote. This cell contains only a small amount of DNA, but it nevertheless provides all the genetic information that the individual will ever have.

Prokaryotes:

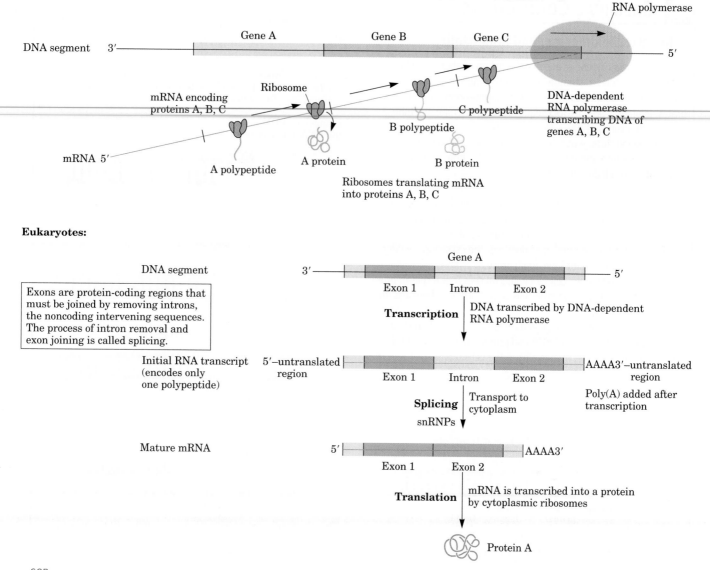

Eukaryotes:

Exons are protein-coding regions that must be joined by removing introns, the noncoding intervening sequences. The process of intron removal and exon joining is called splicing.

Active Figure 25.12   The properties of mRNA molecules in prokaryotic versus eukaryotic cells during transcription and translation. **See a simulation based on this figure, and take a short quiz on the concepts at http://www.now .brookscole.com/gob8 or on the CD.**

In a human cell, some 3 billion base pairs must be duplicated at each cell cycle, and a fully grown human being may contain more than 1 trillion cells. Each cell contains the same amount of DNA as the original single cell. Furthermore, cells are constantly dying and being replaced. Thus there must be a mechanism by which DNA molecules can be copied over and over again without error. In Section 25.7, we will see that such errors sometimes do occur and can have serious consequences. Here, however, we want to examine this remarkable mechanism that takes place every day in billions of organisms, from microbes to whales, and has been taking place for billions of years—with only a tiny percentage of errors.

Replication begins at a point in the DNA called an **origin of replication.** In human cells, the average chromosome has several hundred origins of replication where the copying occurs simultaneously. The DNA double helix

CHEMICAL CONNECTIONS 25B

Telomeres, Telomerase, and Immortality

Every person has a genetic makeup consisting of about 3 billion pairs of nucleotides, distributed over 46 chromosomes. Telomeres are specialized structures at the ends of chromosomes. In vertebrates, telomeres are TTAGGG sequences that are repeated hundreds to thousands of times. In normal **somatic cells** that divide in a cyclic fashion throughout the life of the organism (via mitosis), chromosomes lose about 50 to 200 nucleotides from their telomeres at each cell division.

DNA polymerase, the enzyme that links the fragments, does not work at the end of linear DNA. This fact results in the shortening of the telomeres at each replication. The telomere shortening acts as a clock by which the cells count the number of times they have divided. After a certain number of divisions, the cells stop dividing, having reached the limit of the aging process.

In contrast to somatic cells, all immortal cells (germ cells in proliferative stem cells, normal fetal cells, and cancer cells) possess an enzyme, telomerase, that can extend the shortened telomeres by synthesizing new chromosomal ends. Telomerase is a ribonucleoprotein; that is, it is made of RNA and protein. The activity of this enzyme seems to confer immortality to the cells.

(*b*) **A mechanism by which telomerase may work. (In this case, RNA of the telomerase acts as a template for reverse transcription)**

Extension of DNA on the RNA of telomerase

Telomerase

RNA template in telomerase

3′ end of the leading strand is elongated.

RNA primer　Lagging strand extended by polymerase

Removal of the primer shortens the DNA, but it is now longer by one repeat unit.

The telomerase extension cycle is repeated until there is an adequate number of DNA repeats for the end of the chromosome to survive.

(*a*) **Replication at the end of a linear template**

Template strand

End of linear chromosome

RNA primer　Synthesis (lagging strand)

This portion of the end of the chromosome will be lost when the primer is removed.

---

has two strands running in opposite directions. The point on the DNA where replication is proceeding is called the **replication fork.** (See Figure 25.13.)

If the unwinding of the double helix begins in the middle, the synthesis of new DNA molecules on the old templates continues in both directions until the entire molecule is duplicated. Alternatively, the unwinding can start at one end and proceed in one direction until the entire double helix is unwound.

Replication is bidirectional and takes place at the same speed in both directions. An interesting detail of DNA replication is that the two daughter strands are synthesized in different ways. One of the syntheses is continuous in the 5′ ⟶ 3′ direction (see Section 25.3). It is called the **leading strand.** Along the other strand that runs in the 3′ ⟶ 5′ direction, the synthesis is discontinuous. It is called the **lagging strand.**

The replication process is called **semiconservative** because each daughter molecule has one parental strand (conserved) and one newly synthesized one.

Replication always proceeds from the 5′ to the 3′ direction from the perspective of the chain that is being synthesized. The actual reaction occur-

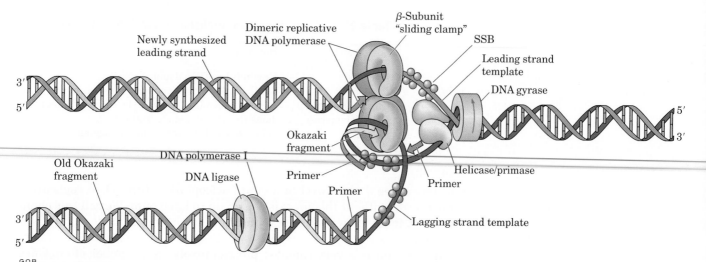

Active Figure 25.13   General features of the replication of DNA. The two strands of the DNA double helix are shown separating at the replication fork. **See a simulation based on this figure, and take a short quiz on the concepts at http://www.now.brookscole.com/gob8 or on the CD.**

Figure 25.14 The addition of a nucleotide to a growing DNA chain. The 3′-hydoxyl group at the end of the growing DNA chain is a nucleophile. It attacks at the phosphorus adjacent to the sugar in the nucleotide, which will be added to the growing chain.

ring is a nucleophilic attack by the 3′ hydroxyl of the deoxyribose of one nucleotide against the first phosphate on the 5′ carbon of the incoming nucleoside triphosphate, as shown in Figure 25.14.

One of the more interesting aspects of DNA replication is that the basic reaction of synthesis always requires an existing chain with a nucleotide

Table 25.4 Components of Replisomes and Their Functions

Component	Function
Helicase	Unwinds the DNA double helix
Primase	Synthesizes short oligonucleotides (primers)
Clamp protein	Allows the leading strand to be threaded through
DNA polymerase	Joins the assembled nucleotides
Ligase	Joins Okazaki fragments in the lagging strand

that has a free 3′- hydroxyl to do the nucleophilic attack. DNA replication cannot begin without this preexisting chain to latch onto. We call this chain a **primer.** In all known forms of replication, the primer is made out of RNA, not DNA.

Replication is a very complex process involving a number of enzymes and binding proteins. A growing body of evidence indicates that these enzymes assemble their products in "factories" through which the DNA moves. Such factories may be bound to membranes in bacteria. In higher organisms, the replication factories are not permanent structures. Instead, they may be disassembled and their parts reassembled in ever-larger factories. These assemblies of enzyme "factories" go by the name of **replisomes,** and they contain key enzymes such as polymerases, helicases, and primases (Table 25.4). The primases can shuttle in and out of the replisomes. Other proteins, such as clamp loaders and clamp proteins, through which the newly synthesized primer is threaded, are also parts of the replisomes.

The replication of DNA occurs in a number of distinct steps. A few of the salient features are enumerated here:

1. **Opening up the superstructure** During replication, the very condensed superstructure of chromosomes must be opened so that it becomes accessible to enzymes and other proteins. A complicated signal transduction mechanism accomplishes this feat. One notable step of the signal transduction is the acetylation and deacetylation of key lysine residues of histones. When histone acetylase, an enzyme, puts acetyl groups on key lysine residues, some positive charges are eliminated and the strength of the DNA–histone interaction is weakened:

$$\text{Histone—}(CH_2)_4\text{—}NH_3^+ + CH_3\text{—}COO^- \underset{\text{deacetylation}}{\overset{\text{acetylation}}{\rightleftharpoons}} \text{Histone—}(CH_2)_4\text{—}NH\text{—}\overset{\overset{\displaystyle O}{\|}}{C}\text{—}CH_3\text{—}H_2O$$

This process allows the opening up of key regions on the DNA molecule. When another enzyme, histone deacetylase, removes these acetyl groups, the positive charges are reestablished. That, in turn, facilitates regaining the highly condensed structure of **chromatin.**

2. **Relaxation of Higher-Order Structures of DNA** Topoisomerases (also called gyrases) are enzymes that facilitate the relaxation of supercoiling in DNA. They do so during replication by temporarily introducing either single- or double-strand breaks in DNA. The transient break forms a phosphodiester linkage between a tyrosyl residue of the enzyme and either the 5′ or 3′ end of a phosphate on the DNA. Once the supercoiling is relaxed, the broken strands are joined together, and the topoisomerase diffuses from the location of the replicating fork. Topoisomerases are also involved in the untangling of the replicated chromosomes, before cell division can occur.

CHEMICAL CONNECTIONS 25C

DNA Fingerprinting

The base sequence in the nucleus of every one of our billions of cells is identical. However, except for people who have an identical twin, the base sequence in the total DNA of one person is different from that of every other person. This uniqueness makes it possible to identify suspects in criminal cases from a bit of skin or a trace of blood left at the scene of the crime and to prove the identity of a child's father in paternity cases.

To do so, the nuclei of the cells of the criminal evidence are extracted. Their DNA is amplified by PCR techniques (Section 25.8). With the aid of restriction enzymes, the DNA molecules are cut at specific points. The resulting DNA fragments are put on a gel and subjected to **electrophoresis**. In this process, the DNA fragments move with different velocities; the smaller fragments move faster and the larger fragments move slower. After a sufficient amount of time, the fragments separate. When they are made visible in the form of an autoradiogram, one can discern bands in a lane. This sequence is called a **DNA fingerprint.**

When the DNA fingerprint made from a sample taken from a suspect matches that from a sample obtained at the scene of the crime, the police have a positive identification. The accompanying figure shows DNA fingerprints derived by using one particular restriction enzyme. Here, a total of nine lanes can be seen. Three (numbers 1, 5, and 9) are control lanes. They contain the DNA fingerprint of a virus, using one particular restriction enzyme.

Three other lanes (2, 3, and 4) were used in a paternity suit: They contain the DNA fingerprints of the mother, the child, and the alleged father. The child's DNA fingerprint (lane 3) contains six bands. The mother's DNA fingerprint (lane 4) has five bands, all of which match those of the child. The alleged father's DNA fingerprint (lane 2) also contains six bands, three of which match those of the child. This is a positive identification. In such cases, one cannot expect a perfect match even if the man is really the father because the child has inherited only half of its genes from the father. Each band in the child's DNA had to come from one of the parents. If the child has a band and the mother does not, then that band must be represented in the DNA of the alleged father; otherwise, he was not the child's father. In the case just described, the paternity suit was won on the basis of the DNA fingerprint matching. However, this gel by itself is not conclusive. Most of the DNA bands are matches between the mother and child, with only one unique match between the alleged father and child. Paternity testing is much more readily used to exclude a potential father than to prove a person is the father. Usually it takes several gels using several different enzymes to get enough data to conclude a positive match.

DNA fingerprint.

In the left area of the radiogram are three more lanes (6, 7, and 8). These DNA fingerprints were used in an attempt to identify a rapist. Lanes 7 and 8 show the DNA fingerprints of semen obtained from the rape victim. Lane 6 is the DNA fingerprint of the suspect. The DNA fingerprints of the semen do not match those of the suspect. This result is a negative identification and excluded the suspect from the case. When a positive identification occurs, the probability that a positive match is due to chance is 1 in 100 billion. Thus, while the identity is not absolutely proven, the law of averages says that there are not enough people on the planet for two of them to have the same DNA pattern.

DNA fingerprints are now routinely accepted in court cases. Many convictions are based on such evidence and, just as importantly, many jailed suspects have been released when DNA fingerprinting proved them innocent. In one bizarre case, a convicted rapist demanded a DNA test. The results showed conclusively that he was not guilty of the rape for which he had been sentenced to prison, and he was subsequently released. However, the police, now having a DNA sample, compared it to evidence in other unsolved crimes. The released prisoner was arrested a week later for three rapes that had previously gone unsolved.

3. **Unwinding the DNA Double Helix** The replication of DNA molecules starts with the unwinding of the double helix, which can occur at either end or in the middle. Special unwinding protein molecules, called **helicases,** attach themselves to one DNA strand (Figure 25.13) and cause the separation of the double helix. Helicases of eukaryotes are made of six different protein subunits. The subunits form a ring with a hollow core, where the single-stranded DNA sits. The helicases hydrolyze ATP

as the DNA strand moves through. The energy of the hydrolysis promotes this movement.

4. **Primers/Primases** Primers are short—4 to 15 nucleotides long—RNA oligonucleotides synthesized from ribonucleoside triphosphates. They are needed to initiate the synthesis of both daughter strands. The enzyme catalyzing this synthesis is called primase. Primases form complexes with DNA polymerase in eukaryotes. Primers are placed about every 50 nucleotides in the lagging-strand synthesis.

5. **DNA Polymerase** The key enzymes in replication are the DNA polymerases. Once the two strands are separated at the replication fork, the DNA nucleotides must be lined up. All four kinds of free DNA nucleotide molecules are present in the vicinity of the replication fork. These nucleotides constantly move into the area and try to fit themselves into new chains. The key to the process is that, as we saw in Section 25.3, *only thymine can fit opposite adenine, and only cytosine can fit opposite guanine.* Wherever a cytosine, for example, is present on one of the strands of an unwound portion of the helix, all four nucleotides may approach, but three of them will be turned away because they do not fit. Only the nucleotide of guanine fits.

In the absence of an enzyme, this alignment is extremely slow. The speed and specificity are provided by DNA polymerase. The active site of this enzyme is quite snug. It surrounds the end of the DNA template–primer complex, creating a specifically shaped pocket for the incoming nucleotide. With such a close contact, the activation energy is lowered and the polymerase enables complementary base pairing with high specificity at a rate of 100 times per second. While the bases of the newly arrived nucleotides are being hydrogen-bonded to their partners, polymerases join the nucleotide backbones.

Along the lagging strand $3' \longrightarrow 5'$, the enzymes can synthesize only short fragments because the only way they can work is from $5'$ to $3'$. These short fragments consist of about 200 nucleotides each, named **Okazaki fragments** after their discoverer.

6. **Ligation** The Okazaki fragments and any nicks remaining are eventually joined together by another enzyme, DNA ligase. At the end of the process, there are two double-stranded DNA molecules, each exactly the same as the original molecule because only thymine fits opposite adenine and only guanine fits against cytosine in the active site of the polymerase.

Okazaki fragment A short DNA segment made of about 200 nucleotides in higher organisms (eukaryotes) and of 2000 nucleotides in prokaryotes

25.7 | How Is DNA Repaired?

The viability of cells depends on DNA repair enzymes that can detect, recognize, and remove mutations from DNA. Such mutations may arise from external or internal sources. Externally, UV radiation or highly reactive oxidizing agents, such as superoxide, may damage a base. Errors in copying or internal chemical reactions—for example, deamination of a base—can create damage internally. Deamination of the base cytosine turns it into uracil (Figure 25.1), which creates a mismatch. The former C—G pair becomes a U—G mispair that must be removed.

The repair can be effected in a number of ways. One of the most common is called BER, *base excision repair* (Figure 25.15). This pathway contains two parts:

1. A specific DNA glycosylase recognizes the damaged base (1). It hydrolyzes the N—C′ β-glycosidic bond between the uracil base and the

Figure 25.15   Base excision repair (BER) pathway. A uracil is replaced by a cytosine.

CHEMICAL CONNECTIONS 25D

Why Does DNA Contain Thymine and Not Uracil?

Given that both uracil and thymine base-pair with adenine, why does RNA contain uracil and DNA contain thymine? Scientists now believe that RNA was the original hereditary molecule, and that DNA developed later. If we compare the structures of uracil and thymine, the only difference is the presence of a methyl group at C-5 of thymine. This group does not appear on the side of the molecule involved in base pairing. Because carbon sources and energy are required to methylate a molecule, there must be a reason that DNA developed with a base that does the same thing as uracil, but that requires more energy to produce.

The answer is that thymine helps guarantee replication fidelity. One of the most common spontaneous mutations of bases is the natural deamination of cytosine.

Cytosine
(2-oxo-4-amino
pyrimidine)

Uracil
(2,4-dioxo-
pyrimidine)

Thymine
(2,4-dioxo-
5-methyl pyrimidine)

Cytosine

Deamination

Uracil

At any moment, a small, but finite number of cytosines lose their amino groups to become uracil. Imagine that during replication, a C—G base pair separates. If at that moment, the cytosine deaminates to uracil, it would have a tendency to base-pair with adenine instead of with guanine. If uracil were a natural base in DNA, the DNA polymerases would just line up an adenine across from the uracil, and there would be no way to know that the uracil was a mistake. This would lead to a much higher rate of mutation during replication. Because uracil is an unnatural base in DNA, DNA polymerases can immediately recognize it as a mistake and replace it. Thus the incorporation of thymine into DNA, while energetically more costly, helps to ensure that the DNA is replicated faithfully.

deoxyribose, then releases the damaged base, completing the excision. The sugar–phosphate backbone is still intact. At the **AP site** (*a*purinic or *a*pyrimidinic site) created in this way, the backbone is cleaved by a second enzyme, endonuclease (2). A third enzyme, exonuclease (3), liberates the sugar–phosphate unit of the damaged site.

2. In the synthesis step, the enzyme DNA polymerase (4) inserts the correct nucleotide, cytidine, and the enzyme DNA ligase seals (5) the backbone to complete the repair.

A second repair mechanism removes not a single mismatched base but rather whole nucleotides—as many as 25 to 32 residue oligonucleotides. Known as NER, for *n*ucleotide *e*xcision *r*epair, it similarly involves a number of repair enzymes.

CHEMICAL CONNECTIONS 25E

Pharmacogenomics: Tailoring Medication to an Individual's Predisposition

The complete DNA sequence of an organism is called its **genome.** The genome of an average human contains approximately 3 billion base pairs, distributed among 22 pairs of chromosomes plus 2 sex chromosomes. Each chromosome consists of a single DNA molecule. Among the 3 billion base pairs are some 90 million that represent 30,000 human genes. It was the task of the Human Genome Project to determine the complete sequence of the genome and, in the process, to identify the sequence and location of the genes. The Human Genome Project was completed in the year 2000. Many other genomes have also been established, ranging from the lowly bacteria *Escherischia coli* (3 million base pairs) to the mouse (3 billion base pairs).

It has always been known that an individual's genetic inheritance plays a role in disease and drug effectiveness. As early as 510 B.C., Pythagoras wrote that some individuals developed hemolytic anemia when eating fava beans, while others thrived on it. Adverse drug reactions are the sixth leading cause of death in the United States (more than 100,000 deaths per year). Problems caused by ineffective drugs are even more numerous. With the new knowledge of an individual's genetic makeup, it is now within our power to prescribe drugs in the dosage that best suits the individual and that minimizes adverse reactions or ineffectiveness. **Pharmacogenomics** is the study of how genetic variation influences individual responses to a given drug or class of drugs.

A case in point is CYP2D6, the gene of a member of the cytochrome P-450 enzyme group. This enzyme detoxifies drugs by adding an — OH group, making them more water-soluble and thus able to be excreted in the urine. The normal or wild type of this enzyme is correlated with *extensive metabolism* (EM) of drugs. A mutation, in which a guanine becomes an adenine in the CYP2D6 gene, exists in approximately 25% of the population. Its presence makes the individual a *poor metabolizer* (PM). Thus a prescribed drug dosage will stay much longer than normal in the body of the individual having this genetic makeup, possibly leading to toxic effects. Another mutation, in which an adenine base is deleted on the CYP2D6 gene, exists in approximately 3% of the population. It is associated with very fast, *ultra-extensive metabolism* (UEM). In these individuals, the drug may be cleared from the body before it can be effective.

These three classes (EM, PM, and UEM) have been known for some time and can be monitored by performing frequent blood sample analyses during a six-week course of therapy with a drug. With the advances of the Human Genome Project, however, an individual can now be tested *before*—rather than during—the administration of a drug therapy. This strategy means that the dosage can be adjusted to the individual's particular needs. Drug companies have developed DNA chips that can read a patient's predisposition to a drug using the DNA of the individual taken in one simple blood test.

Genetic predisposition is merely one factor among many determining the body's total response to a drug, of course. Nevertheless, eliminating errors based on known genetic makeup can only minimize adverse effects or ineffectiveness.

Any defect in the repair mechanisms may lead to harmful or even lethal mutations. For example, individuals with inherited XP (xeroderma pigmentosa), a condition in which one or another enzyme in the NER repair pathway is missing or defective, have a 1000 times greater risk of developing skin cancer than do normal individuals.

25.8 | How Do We Amplify DNA?

To study DNA for basic or applied scientific purposes, we must have enough of it to work with. There are several ways of amplifying DNA. One approach is to allow a rapidly growing organism, like bacteria, to replicate DNA for us. This process, which is usually referred to as **cloning,** will be discussed further in Chapter 26. Millions of copies of selected DNA fragments can also be made within a few hours with high precision by a technique called **polymerase chain reaction (PCR),** which was discovered by Kary B. Mullis (1945–), who shared the 1993 Nobel Prize in chemistry for this achievement.

PCR techniques can be used if the sequence of a gene to be copied is known, or at least a sequence bordering the desired DNA is known. In such a case, one can synthesize two primers that are complementary to the ends of the gene or to the bordering DNA. The primers are polynucleotides

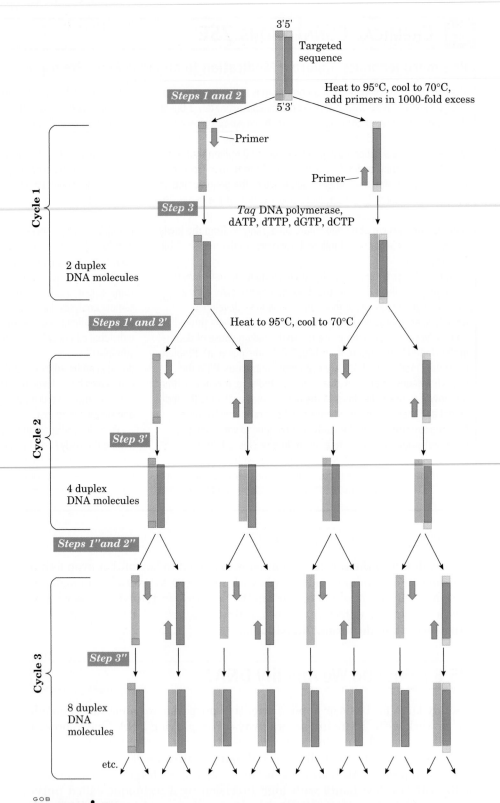

Chemistry ·ᐅ·Now™

Active Figure 25.16 Polymerase chain reaction (PCR). Oligonucleotides complementary to a given DNA sequence prime the synthesis of only that sequence. Heat-stable *Taq* DNA polymerase survives many cycles of heating. Theoretically, the amount of the specific primed sequence is doubled in each cycle. **See a simulation based on this figure, and take a short quiz on the concepts at http://www.now.brookscole.com/gob8 or on the CD.**

consisting of 12 to 16 nucleotides. When added to a target DNA segment, they hybridize with the end of each strand of the gene.

5′CATAGGACAGC—OH   Primer
    | | | | | | | | | | |
3′TACGTATCCTGTCGTAGG—   Gene

In cycle 1 (Figure 25.16), the polymerase extends the primers in each direction as individual nucleotides are assembled and connected on the template DNA. In this way, two new copies are created. The two-step process is repeated (cycle 2) when the primers are **hybridized** with the new strands, and the primers are extended again. At that point, four new copies have been created. The process continues, and in 25 cycles, 2^{25} or some 33 million copies can be made. In practice, only a few million are produced, which is sufficient for the isolation of a gene.

This fast process is practical because of the discovery of heat-resistant polymerases isolated from bacteria that live in hot thermal vents on the sea floor (Section 23.4B). A temperature of 95°C is needed because the double helix must be unwound to hybridize the primer to the target DNA. Once single strands of DNA have been exposed, the mixture is cooled to 70°C. The primers are hybridized and subsequent extensions take place. The 95°C and 70°C cycles are repeated over and over. No new enzyme is required because the polymerase is stable at both temperatures.

PCR techniques are routinely used when a gene or a segment of DNA must be amplified from a few molecules. It is used in studying genomes (Chemical Connections 25E), in obtaining evidence from a crime scene (Chemical Connections 25C), and even in obtaining the genes of long-extinct species found fossilized in amber.

Hybridization The process in which two strands of nucleic acids or segments of nucleic acid strands form a double stranded structure through hydrogen bonding of complementary base pairs

SUMMARY OF KEY QUESTIONS

SECTION 25.1 What Are the Molecules of Heredity?

- Heredity is based on genes located in chromosomes.
- Genes are sections of DNA that encode specific RNA molecules.

SECTION 25.2 What Are Nucleic Acids Made of?

- **Nucleic acids** are composed of sugars, phosphates, and organic bases.
- Two kinds of nucleic acids exist: **ribonucleic acid (RNA)** and **deoxyribonucleic acid (DNA).**
- In DNA, the sugar is the monosaccharide 2-deoxy-D-ribose; in RNA, it is D-ribose.
- In DNA, the heterocyclic amine bases are adenine (A), guanine (G), cytosine (C), and thymine (T).
- In RNA, they are A, G, C, and uracil (U).
- Nucleic acids are giant molecules with backbones made of alternating units of sugar and phosphate. The bases are side chains joined by β-N-glycosidic bonds to the sugar units.

SECTION 25.3 What Is the Structure of DNA and RNA?

- DNA is made of two strands that form a double helix. The sugar–phosphate backbone runs on the outside of the double helix, and the hydrophobic bases point inward.
- **Complementary pairing** of the bases occurs in the double helix, such that each A on one strand is hydrogen-bonded to a T on the other strand, and each G is hydrogen-bonded to a C. No other pairs fit.
- DNA is coiled around basic protein molecules called **histones.** Together they form **nucleosomes,** which are further condensed into chromatin.
- The DNA molecule carries, in the sequence of its bases, all the information necessary to maintain life. When cell division occurs and this information is passed from parent cell to daughter cells, the sequence of the parent DNA is copied.

SECTION 25.4 What Are the Different Classes of RNA?

- There are six kinds of RNA: **messenger RNA (mRNA), transfer RNA (tRNA), ribosomal RNA (rRNA), small nuclear RNA (snRNA), micro RNA (miRNA),** and **small interfering RNA (siRNA).**
- mRNA, tRNA, and rRNA are involved in all protein synthesis.
- Small nuclear RNA is involved in splicing reactions and has been found in some cases to have catalytic activity.
- RNA with catalytic activity is called a **ribozyme.**

SECTION 25.5 What Are Genes?

- A **gene** is a segment of a DNA molecule that carries the sequence of bases that directs the synthesis of one particular protein or RNA molecule.
- DNA in higher organisms contains sequences, called **introns,** that do not code for proteins.
- The sequences that do code for proteins are called **exons.**

SECTION 25.6 How Is DNA Replicated?

- DNA replication occurs in several distinct steps.
- The superstructures of chromosomes are initially loosened by acetylation of histones. **Topoisomerases** relax the higher structures. **Helicases** at the replication fork separate the two strands of DNA.
- RNA primers and primases are needed to start the synthesis of daughter strands. The **leading strand** is synthesized continuously by **DNA polymerase.** The **lagging strand** is synthesized discontinuously as **Okazaki fragments.**
- **DNA ligase** seals the nicks and the Okazaki fragments.

SECTION 25.7 How Is DNA Repaired?

- An important **DNA repair** mechanism is BER, or single base excision repair.

SECTION 25.8 How Do We Amplify DNA?

- The **polymerase chain reaction (PCR)** technique can make millions of copies with high precision in a few hours.

PROBLEMS

A blue problem number indicates an applied problem.

■ denotes problems that are available on the GOB ChemistryNow website or CD and are assignable in OWL.

SECTION 25.1 What Are the Molecules of Heredity?

25.2 What structures of the cell, visible in a microscope, contain hereditary information?

25.3 Name one hereditary disease.

25.4 What is the basic unit of heredity?

SECTION 25.2 What Are Nucleic Acids Made of?

25.5 (a) Where in a cell is the DNA located?
(b) Where in a cell is the RNA located?

25.6 ■ What are the components of (a) a nucleotide and (b) a nucleoside?

25.7 What is the difference between DNA and RNA?

25.8 Draw the structures of ADP and GDP. Are these structures parts of nucleic acids?

25.9 What is the difference in structure between thymine and uracil?

25.10 ■ Which DNA and RNA bases contain a carbonyl group?

25.11 Draw the structures of (a) cytidine and (b) deoxycytidine.

25.12 Which DNA and RNA bases are primary amines?

25.13 What is the difference in structure between D-ribose and 2-deoxy-D-ribose?

25.14 What is the difference between a nucleoside and a nucleotide?

25.15 ■ RNA and DNA refer to nucleic *acids*. Which part of the molecule is acidic?

25.16 What type of bond exists between the ribose and the phosphate in AMP?

25.17 What type of bond exists between the two phosphates in ADP?

25.18 What type of bond connects the base to the ribose in GTP?

SECTION 25.3 What Is the Structure of DNA and RNA?

25.19 In RNA, which carbons of the ribose are linked to the phosphate group and which are linked to the base?

25.20 What constitutes the backbone of DNA?

25.21 Draw the structures of (a) UDP and (b) dAMP.

25.22 In DNA, which carbon atoms of 2-deoxy-D-ribose are bonded to the phosphate groups?

25.23 ■ The sequence of a short DNA segment is ATGGCAATAC.
  (a) What name do we give to the two ends (terminals) of a DNA molecule?
  (b) In this segment, which end is which?

25.24 Chargaff showed that, in samples of DNA taken from many different species, the molar quantity of A was always approximately equal to the molar quantity of T; the same is true for C and G. How did this information help to establish the structure of DNA?

25.25 ■ How many hydrogen bonds can form between uracil and adenine?

25.26 How many histones are present in a nucleosome?

25.27 What is the nature of the interaction between histones and DNA in nucleosomes?

25.28 What are chromatin fibers made of?

25.29 What constitutes the superstructure of chromosomes?

25.30 What is the primary structure of DNA?

25.31 What is the secondary structure of DNA?

25.32 What is the major groove of a DNA helix?

25.33 What are the higher-order structures of DNA that eventually make up a chromosome?

SECTION 25.4 What Are the Different Classes of RNA?

25.34 Which type of RNA has enzyme activity? Where does it function mostly?

25.35 Which has the longest chains: tRNA, mRNA, or rRNA?

25.36 Which type of RNA contains modified nucleotides?

25.37 Which type of RNA has a sequence exactly complementary to that of DNA?

25.38 Where is rRNA located in the cell?

25.39 What kind of functions do ribozymes, in general, perform?

25.40 Which of the RNA types are always involved in protein synthesis?

25.41 What is the purpose of small nuclear RNA?

25.42 What is the purpose of siRNA?

25.43 What is the difference between miRNA and siRNA?

SECTION 25.5 What Are Genes?

25.44 Define:
  (a) Intron   (b) Exon

25.45 Does mRNA also have introns and exons? Explain.

25.46 (a) What percentage of human DNA codes for proteins?
  (b) What is the function of the rest of the DNA?

25.47 Do satellites code for a particular protein?

25.48 Do all genes code for a protein? If not, what do they code for?

SECTION 25.6 How Is DNA Replicated?

25.49 A DNA molecule normally replicates itself millions of times, with almost no errors. What single fact about the structure is most responsible for this fidelity of replication?

25.50 ■ Which functional groups on the bases form hydrogen bonds in the DNA double helix?

25.51 Draw the structures of adenine and thymine, and show with a diagram the two hydrogen bonds that stabilize A—T pairing in DNA.

25.52 Draw the structures of cytosine and guanine, and show with a diagram the three hydrogen bonds that stabilize C—G pairing in nucleic acids.

25.53 How many different bases are present in a DNA double helix?

25.54 What is a replication fork? How many replication forks may exist simultaneously on an average human chromosome?

25.55 Why is replication called semiconservative?

25.56 How does the removal of some positive charges from histones enable the opening of the chromosomal superstructure?

25.57 Write the chemical reaction for the deacetylation of acetyl-histone.

25.58 What is the quaternary structure of helicases in eukaryotes?

25.59 What are helicases? What is their function?

25.60 Can dATP serve as a source for a primer?

25.61 What are the side products of the action of primase in forming primers?

25.62 What do we call the enzymes that join nucleotides into a DNA strand?

25.63 ■ In which direction is the DNA molecule synthesized continuously?

25.64 What kind of bond formation do polymerases catalyze?

25.65 Which enzyme catalyzes the joining of Okazaki fragments?

25.66 What is the nature of the chemical reaction that joins nucleotides together?

25.67 From the perspective of the chain being synthesized, in which direction does DNA synthesis proceed?

SECTION 25.7 How Is DNA Repaired?

25.68 As a result of damage, a few of the guanine residues in a gene are methylated. What kind of mechanism could the cell use to repair the damage?

25.69 What is the function of endonuclease in the BER repair mechanism?

25.70 When cytosine is deaminated, uracil is formed. Uracil is a naturally occurring base. Why would the cell use base excision repair to remove it?

25.71 Which bonds are cleaved by glycosylase?

25.72 What are AP sites? Which enzyme creates them?

25.73 Why are patients with xeroderma pigmentosa 1000 times more likely to develop skin cancer than normal individuals are?

SECTION 25.8 How Do We Amplify DNA?

25.74 What is the advantage of using DNA polymerase from thermophilic bacteria that live in hot thermal vents in PCR?

25.75 What 12-nucleotide primer would you use in the PCR technique when you want to amplify a gene whose end is as follows: 3′ TACCGTCATCCGGTG5—?

Chemical Connections

25.76 (Chemical Connections 25A) Draw the structure of the fluorouridine nucleoside that inhibits DNA synthesis.

25.77 (Chemical Connections 25A) Give an example of how anticancer drugs work in chemotherapy.

25.78 (Chemical Connections 25B) What sequence of nucleotides is repeated many times in telomeres?

25.79 (Chemical Connections 25B) Why are as many as 200 nucleotides lost at each replication?

25.80 (Chemical Connections 25B) How does telomerase make a cancer cell immortal?

25.81 (Chemical Connections 25B) Why is DNA loss with replication not a problem for bacteria? (*Hint:* Bacteria have a circular genome.)

25.82 (Chemical Connections 25C) After having been cut by restriction enzymes, how are DNA fragments separated from each other?

25.83 (Chemical Connections 25C) How is DNA fingerprinting used in paternity suits?

25.84 (Chemical Connections 25C) Why is it easier to exclude someone via DNA fingerprinting than it is to prove that he or she is the person whose sample is being tested?

25.85 (Chemical Connections 25C) What is the principle behind paternity testing via DNA fingerprinting?

25.86 (Chemical Connections 25D) Both thymine and uracil can bond to adenine, yet thymine is more difficult to make than uracil. What is the evolutionary advantage to having DNA with thymine?

25.87 (Chemical Connections 25E) What is the function of cytochrome P-450?

25.88 (Chemical Connections 25E) How does knowledge of the human genome enable patients to be screened for individual drug tolerance?

Additional Problems

25.89 What is the active site of a ribozyme?

25.90 Why is it important that a DNA molecule be able to replicate itself millions of times without error?

25.91 Draw the structures of (a) uracil and (b) uridine.

25.92 How would you classify the functional groups that bond together the three different components of a nucleotide?

25.93 Which nucleic acid molecule is the largest?

25.94 ■ What kind of bonds are broken during replication? Does the primary structure of DNA change during replication?

25.95 In sheep DNA, the mol % of adenine (A) was found to be 29.3. Based on Chargaff's rule, what would be the approximate mol % of G , C, and T?

Looking Ahead

25.96 DNA is the blueprint for the cell, but not all genes in DNA lead to protein. Gene expression is the study of how genes are used to make their particular product. What are some examples of gene products that do not lead to proteins?

25.97 In a process similar to DNA replication, RNA is produced via the process called transcription. The enzyme used is RNA polymerase. When RNA is synthesized, what is the direction of the synthesis reaction?

25.98 The Human Genome Project showed that human DNA is not considerably bigger than much simpler organisms, with about 30,000 total genes. However, humans make over 100, 000 different proteins. How is this possible? (*Hint:* think about splicing.)

25.99 One of the biggest differences between DNA replication and transcription is that RNA polymerase does not require a primer. How does this fact relate to the theory that primordial life was based on RNA and not DNA?

25.100 How could life have evolved if DNA leads to RNA which leads to protein, but it takes many proteins to replicate DNA and to transcribe DNA into RNA?

CHAPTER 26

Gene Expression and Protein Synthesis

KEY QUESTIONS

26.1 How Does DNA Lead to RNA and Protein?

26.2 How Is DNA Transcribed into RNA?

26.3 What Is the Role of RNA in Translation?

26.4 What Is the Genetic Code?

26.5 How Is Protein Synthesized?

26.6 How Are Genes Regulated?

26.7 What Are Mutations?

26.8 How and Why Do We Manipulate DNA?

26.9 What Is Gene Therapy?

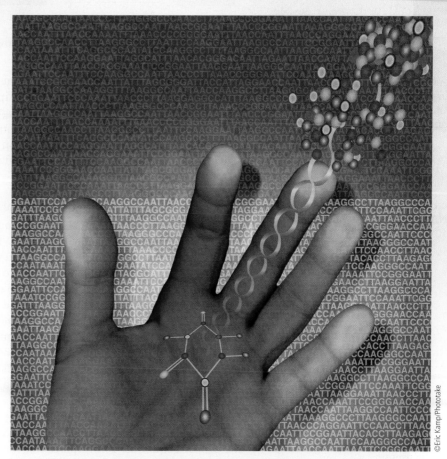

In transcription, the template strand of DNA is used to produce a complementary strand of RNA. Transcription is the most controlled and best understood part of gene regulation.

Chemistry Now™

Look for this logo in the chapter and go to GOB ChemistryNow at **http://now.brookscole.com/gob8** or on the CD for tutorials, simulations, and problems.

26.1 | How Does DNA Lead to RNA and Protein?

We have seen that the DNA molecule is a storehouse of information. We can compare it to a loose-leaf cookbook, each page of which contains one recipe. The pages are the genes. To prepare a meal, we use a number of recipes. Similarly, to provide a certain inheritable trait, a number of genes (Chapter 25)—segments of DNA—are needed.

Of course, the recipe itself is not the meal. The information in the recipe must be expressed in the proper combination of food ingredients. Similarly, the information stored in DNA must be expressed in the proper combination of amino acids representing a particular protein. The way this expression works is now so well established that it is called the **central dogma of molecular biology.** The dogma states that *the information contained in DNA molecules is transferred to RNA molecules, and then from the RNA molecules the information is expressed in the structure of proteins.*

661

Figure 26.1 The central dogma of molecular biology. The yellow arrows represent the general cases, and the blue arrows represent special cases in RNA viruses.

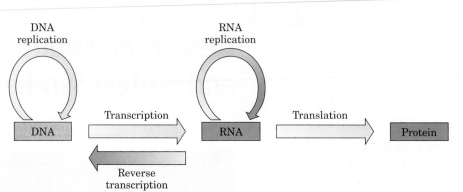

Gene expression The activation of a gene to produce a specific protein; it involves both transcription and translation

Gene expression is the turning on or activation of a gene. Transmission of information occurs in two steps: transcription and translation.

Figure 26.1 shows the central dogma of gene expression. In some viruses (shown in blue), gene expression does not work this way. In some viruses with RNA genomes, replication proceeds from RNA to RNA. In retroviruses, RNA is reverse transcribed to DNA.

Transcription

Transcription The process in which information encoded in a DNA molecule is copied into an mRNA molecule

Because the information (that is, the DNA) is in the nucleus of a eukaryotic cell and the amino acids are assembled outside the nucleus, the information must first be carried out of the nucleus. This step is analogous to copying a recipe from a cookbook. All the necessary information is copied, albeit in a slightly different format, as if we were converting the printed page into handwriting. On the molecular level, this task is accomplished by transcribing the information from the DNA molecule onto a molecule of messenger RNA, so named because it carries the message from the nucleus to the site of protein synthesis. Other RNAs are similarly transcribed. rRNA is needed to form ribosomes, and tRNA is required to carry out the translation into protein language (Chapter 25). The transcribed information on the different RNA molecules is then carried out of the nucleus.

Translation

Translation The process in which information encoded in an mRNA molecule is used to assemble a specific protein

The mRNA serves as a template on which the amino acids are assembled in the proper sequence. To complete the assembly, the information that is written in the language of nucleotides must be translated into the language of amino acids. The translation is done by another type of RNA, transfer RNA (Section 25.4). An exact, word-to-word translation occurs. Each amino acid in the protein language has a corresponding word in the RNA language. Each word in the RNA language is a sequence of three bases. This correspondence between three bases and one amino acid is called the genetic code (we will discuss the code in Section 26.4).

In higher organisms (eukaryotes), transcription and translation occur sequentially. The transcription takes place in the nucleus. After RNA leaves the nucleus and enters the cytoplasm, the translation takes place there. In lower organisms (prokaryotes), there is no nucleus and thus transcription and translation occur simultaneously in the cytoplasm. This extended form of the "central dogma" was challenged in 2001, when it was found that even in eukaryotes about 15% of the proteins are produced in the nucleus itself. Clearly, some simultaneous transcription and translation do occur even in higher organisms.

We know more about bacterial transcription and translation because they are simpler than the processes operating in higher organisms and have been studied for a longer time. Nevertheless, we will concentrate on studying gene expression and protein synthesis in the eukaryotic systems because they are more relevant to human health care.

26.2 | How Is DNA Transcribed into RNA?

Transcription starts when the DNA double helix begins to unwind at a point near the gene that is to be transcribed (Figure 26.2). As we saw in Section 25.3C, nucleosomes form chromatin and higher condensed structures in the chromosomes. To make the DNA available for transcription, these superstructures change constantly. Specific **binding proteins** bind to the nucleosomes, making the DNA become less dense and more accessible. Only then can the enzyme **helicase,** a ring-shaped complex of six protein subunits, unwind the double helix.

Only one strand of the DNA molecule is transcribed. The strand that serves as a template for the formation of RNA has several names, including the **template strand,** the **(−) strand,** and the **antisense strand.** The other strand, while not used as a template, actually has a sequence that matches the RNA that will be produced. This strand is called the **coding strand,** the **(+) strand,** and the **sense strand.** Of these names, coding strand and template strand are the most commonly used.

Ribonucleotides assemble along the unwound DNA strand in the complementary sequence. Opposite each C on the DNA is a G on the growing mRNA; the other complementary bases follow the patterns G ⟶ C, A ⟶ U, and T ⟶ A. The ribonucleotides, when aligned in this way, are linked to form the appropriate RNA.

In eukaryotes, three kinds of **polymerases** catalyze transcription. RNA polymerase I (pol I) catalyzes the formation of most of the rRNA; Pol II catalyzes mRNA formation; and Pol III catalyzes tRNA formation as well as one ribosomal subunit and other small regulatory RNA types, like snRNA. Each enzyme is a complex of 10 or more subunits. Some subunits are unique to each kind of polymerase, whereas other subunits appear in all three polymerases.

The eukaryotic gene has two major parts: the **structural gene** itself, which is transcribed into RNA, and a **regulatory** portion that controls the transcription. The structural gene is made of exons and introns (Figure 26.3). The regulatory portion is not transcribed, but rather has control elements.

One such control element is a **promoter.** On the DNA strand, there is always a sequence of bases that the polymerase recognizes as an **initiation signal,** saying, in essence, "Start here." The promoter is unique to each gene. Besides unique nucleotide sequences, promoters contain **consensus sequences,** such as the TATA box, which gets its name from the sequence beginning, TATAAT. A TATA box lies approximately 26 base pairs upstream—that is, before the beginning of the transcription process (see Figure 26.3).

Template strand The strand of DNA that serves as the template during RNA synthesis

Coding strand The strand of DNA with a sequence that matches the RNA produced during transcription

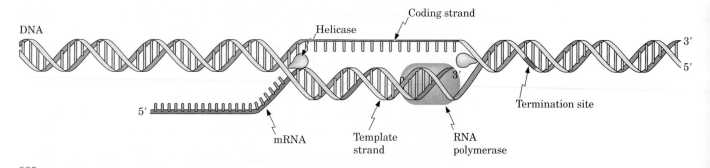

GOB
Chemistry⚛Now™

Figure 26.2   Transcription of a gene. The information in one DNA strand is transcribed to a strand of RNA. The termination site is the locus of termination of transcription. **See a simulation based on this figure, and take a short quiz on the concepts at http://www.now.brookscole.com/gob8 or on the CD.**

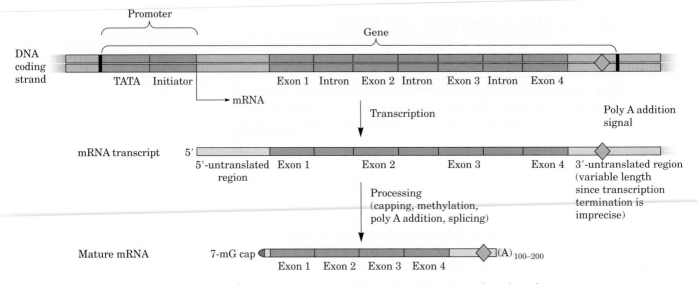

Figure 26.3 Organization and transcription of a split eukaryote gene.

By convention, all sequences of DNA used to describe transcription are given from the perspective of the coding strand. TATA boxes are common to all eukaryotes. All three RNA polymerases interact with their promoter regions via **transcription factors** that are binding proteins.

Another type of control element is an enhancer, a DNA sequence that can be far removed from the promoter region. Enhancers also bind to transcription factors, enhancing transcription above the basal level that would be seen without such binding. Enhancers will be discussed in Section 26.6.

After initiation, the RNA polymerase zips up the complementary bases by forming a phosphate ester bond (Section 19.5) between each ribose and the next phosphate group. This process is called **elongation.**

At the end of the gene is a **termination sequence** that tells the enzyme, "Stop the transcription." The enzyme Pol II has two different forms. At the C-terminal domain, Pol II has serine and threonine repeats that can be phosphorylated. When Pol II starts the initiation, the enzyme is in its unphosphorylated form. Upon phosphorylation, it performs the elongation process. After termination of the transcription, Pol II is dephosphorylated by a phosphatase. In this manner, Pol II is constantly recycled between its initiation and elongation roles.

The enzyme synthesizes the mRNA molecule from the 5′ to the 3′ end (the zipper can move in only one direction). Because complementary nucleotide chains (RNA and DNA) run in opposite directions, however, the enzyme moves along the DNA template strand in the 3′ ⟶ 5′ direction (Figure 26.2). As the RNA is synthesized, it moves away from the DNA template, which then rewinds to the original double-helix form. Transfer RNA and ribosomal RNA are also synthesized on DNA templates in this manner.

The RNA products of transcription are not necessarily the functional RNAs. Previously, we have seen that in higher organisms mRNA contains exons and introns (Section 25.5). To ensure that mRNA is functional, the transcribed product is capped at both ends. The 5′ end acquires a methylated guanine (7-mG cap) and the 3′ end has a poly-A tail that may contain 100 to 200 adenine residues. Once the two ends are capped, the introns are spliced out in a **post-transcription process** (Figure 26.3). Similarly, a transcribed tRNA must be trimmed and capped, and some of its nucleotides methylated, before it becomes functional tRNA. Functional rRNA also undergoes post-transcriptional methylation.

CHEMICAL CONNECTIONS 26A

"Antisense" Makes Sense

Many diseases are caused by faulty or excessive production of certain proteins. As we have seen, the message to make proteins resides in the mRNA. Antisense drugs are chemically modified strands of short stretches of DNA that are either the same as the template DNA strand or exactly opposite—hence "anti"—the coding, or "sense" strand of a mRNA. Thus, if a stretch of mRNA is AACGUU, its antisense will be TTGCAA. When such an antisense drug is delivered to a cell, it strongly binds to the mRNA and prevents the translation of the unwanted protein.

The trick is to prevent the degradation of the antisense drug by the cell's enzymes before or after it attaches to the mRNA. One way to protect the antisense drug is to modify the backbone by substituting a sulfur atom for an oxygen atom in the phosphate group of the backbone, thereby forming phosphorothioate moieties.

Only one antisense drug is currently on the market. Vitravene is used in treatment of cytomegalovirus-induced retinitis in AIDS patients. Vitravene is a stretch of 21 nucleotides. The structure of the first four of these nucleotides is shown in the figure, spelling out GCGT. Many other antisense oligonucleotide drugs are in clinical trials as potential treatments for various cancers, such as lymphoma.

First four of the 21 segments of the oligonucleotide drug Vitravene.

EXAMPLE 26.1

Polymerase II both initiates the transcription and performs the elongation. What are the two forms of the enzyme in these processes? What chemical bond formation occurs in the conversion between these two forms?

Solution
The phosphorylated form of Pol II performs the elongation and the unphosphorylated form initiates the transcription. The chemical bond formed in the phosphorylation is the phosphoric ester between the —OH of serine and threonine residues of the enzyme and phosphoric acid.

Problem 26.1
DNA is highly condensed in the chromosomes. What is the sequence of events that enables the transcription of a gene to begin?

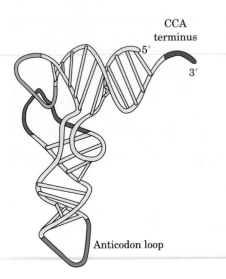

CCA
terminus

5'

3'

Anticodon loop

Figure 26.4 Three-dimensional structure of tRNA.

Codon The sequence of three nucleotides in messenger RNA that codes for a specific amino acid

Anticodon A sequence of three nucleotides on tRNA complementary to the codon in mRNA

Genetic code The sequence of triplets of nucleotides (codons) that determines the sequence of amino acids in a protein

26.3 | What Is the Role of RNA in Translation?

Translation is the process by which the genetic information preserved in the DNA and transcribed into the mRNA is converted to the language of proteins—that is, the amino acid sequence. Three types of RNA (mRNA, rRNA, and tRNA) participate in the process.

The synthesis of proteins takes place on the ribosomes (Section 25.4). These spheres dissociate into two parts—a larger and a smaller body. Each of these bodies contains rRNA and some polypeptide chains that act as enzymes, speeding up the synthesis. In higher organisms, including humans, the larger ribosomal body is called the 60S ribosome, and the smaller one is called the 40S ribosome. The designation "S" refers to *Svedberg*, a measure of density used in centrifugation. In prokaryotes, the corresponding ribosomal subunits are called the 50S and the 30S, respectively. Messenger RNA is attached to the smaller ribosomal body and later joined by the larger body. Together they form a unit on which the mRNA is stretched out. Triplets of bases on the mRNA are called **codons.** After the mRNA is attached to the ribosome in this way, the 20 amino acids are brought to the site, each carried by its own particular tRNA molecule.

The most important segments of the tRNA molecule are (1) the site to which enzymes attach the amino acids and (2) the recognition site. Figure 26.4 shows that the 3' terminus of the tRNA molecule is single-stranded; this end carries the amino acid.

As we have said, each tRNA is specific for one amino acid only. How does the body make sure that alanine, for example, attaches only to the one tRNA molecule that is specific for alanine? The answer is that each cell carries specific enzymes for this purpose. These **aminoacyl-tRNA synthetases** recognize specific tRNA molecules and amino acids. The enzyme then attaches the amino acid to the terminal group of the tRNA, forming an ester bond.

The second important segment of the tRNA molecule carries the **codon recognition site,** which is a sequence of three bases called an **anticodon** located at the opposite end of the molecule in the three-dimensional structure of tRNA (see Figure 26.4). This triplet of bases can align itself in a complementary fashion to the codon triplet on mRNA.

26.4 | What Is the Genetic Code?

By 1961, it was apparent that the order of bases in a DNA molecule corresponds to the order of amino acids in a particular protein, but the code was unknown. Obviously, it could not be a one-to-one correspondence. There are only four bases, so if A coded for glycine, G for alanine, C for valine, and T for serine, there would be 16 amino acids that could not be coded.

In 1961, Marshall Nirenberg (1927–) and his co-workers attempted to break the code in a very ingenious way. They made a synthetic molecule of mRNA consisting of uracil bases only. They put this molecule into a cell-free system that synthesized proteins and then supplied the system with all 20 amino acids. The only polypeptide produced was a chain consisting solely of the amino acid phenylalanine. This experiment showed that the code for phenylalanine must be UUU or some other multiple of U.

A series of similar experiments by Nirenberg and other workers followed, and by 1967 the entire genetic code had been broken. *Each amino acid is coded for by a sequence of three bases,* called a *codon.* Table 26.1 shows the complete code.

The first important aspect of the **genetic code** is that it is almost universal. In virtually every organism, from a bacterium to an elephant to a

Table 26.1 The Genetic Code

First Position (5'-end)	Second Position								Third Position (3'-end)
	U		**C**		**A**		**G**		
U	UUU	Phe	UCU	Ser	UAU	Tyr	UGU	Cys	U
	UUC	Phe	UCC	Ser	UAC	Tyr	UGC	Cys	C
	UUA	Leu	UCA	Ser	UAA	Stop	UGA	Stop	A
	UUG	Leu	UCG	Ser	UAG	Stop	UGG	Trp	G
C	CUU	Leu	CCU	Pro	CAU	His	CGU	Arg	U
	CUC	Leu	CCC	Pro	CAC	His	CGC	Arg	C
	CUA	Leu	CCA	Pro	CAA	Gln	CGA	Arg	A
	CUG	Leu	CCG	Pro	CAG	Gln	CGG	Arg	G
A	AUU	Ile	ACU	Thr	AAU	Asn	AGU	Ser	U
	AUC	Ile	ACC	Thr	AAC	Asn	AGC	Ser	C
	AUA	Ile	ACA	Thr	AAA	Lys	AGA	Arg	A
	AUG*	Met	ACG	Thr	AAG	Lys	AGG	Arg	G
G	GUU	Val	GCU	Ala	GAU	Asp	GGU	Gly	U
	GUC	Val	GCC	Ala	GAC	Asp	GGC	Gly	C
	GUA	Val	GCA	Ala	GAA	Glu	GGA	Gly	A
	GUG	Val	GCG	Ala	GAG	Glu	GGG	Gly	G

*AUG also serves as the principal initiation codon.

human, the same sequence of three bases codes for the same amino acid. The universality of the genetic code implies that all living matter on Earth arose from the same primordial organisms. This finding is perhaps the strongest evidence supporting Darwin's theory of evolution.

Some exceptions to the genetic code in Table 26.1 occur in mitochondrial DNA. Because of that fact and other evidence, it is thought that the mitochondrion may have been an ancient entity. During evolution, it developed a symbiotic relationship with eukaryotic cells. For example, some of the respiratory enzymes located on the cristae of the mitochondrion (see Section 27.2) are encoded in the mitochondrial DNA, and other members of the same respiratory chain are encoded in the nucleus of the eukaryotic cell.

There are 20 amino acids in proteins, but 64 possible combinations of four bases into triplets. All 64 codons (triplets) have been deciphered. Three of them—UAA, UAG, and UGA—are "stop signs." They terminate protein synthesis. The remaining 61 codons code for amino acids. Because there are only 20 amino acids, there must be more than one codon for each amino acid. Indeed, some amino acids have as many as six codons. Leucine, for example, is coded by UUA, UUG, CUU, CUC, CUA, and CUG.

Just as there are three stop signs in the code, there is also an initiation sign. The initiation sign is AUG, which is also the codon for the amino acid methionine. This means that, in all protein synthesis, the first amino acid initially put into the protein is always methionine. Methionine can also be put into the middle of the chain.

Although all protein synthesis starts with methionine, most proteins in the body do not have a methionine residue at the N-terminus of the chain. In most cases, the initial methionine is removed by an enzyme before the polypeptide chain is completed. The code on the mRNA is always read in the 5' ⟶ 3' direction, and the first amino acid to be linked to the initial methionine is the N-terminal end of the translated polypeptide chain.

EXAMPLE 26.2

Which amino acid is represented by the codon CGU? What is its anticodon?

Solution
Looking at Table 26.1, we find that CGU corresponds to arginine; the anticodon is GCA (read 3′ to 5′ to show how the codon and anticodon match up).

Problem 26.2
What are the codons for histidine? What are the anticodons?

26.5 | How Is Protein Synthesized?

So far we have met the molecules that participate in protein synthesis (Section 26.3) and the dictionary of the translation, the genetic code. Now let us look at the actual mechanism by which the polypeptide chain is assembled.

There are four major stages in protein synthesis: activation, initiation, elongation, and termination. At each stage, a number of molecular entities participate in the process (Table 26.2). We will look specifically at prokaryotic translation because it has been studied longer, and we have more complete information on it. The details of eukaryotic translation are very similar, however.

A. Activation

Each amino acid is first activated by reacting with a molecule of ATP:

Table 26.2 Molecular Components of Reactions at Four Stages of Protein Synthesis

Stage	Molecular Components
Activation	Amino acids, ATP, tRNAs, aminoacyl-tRNA synthetases
Initiation	fMet–tRNA^{fMet}, 30S ribosome, initiation factors, mRNA with Shine–Dalgarno sequence, 50S ribosome, GTP
Elongation	30S and 50S ribosomes, aminoacyl-tRNAs, elongation factors, mRNA, GTP
Termination	Release factors, GTP

The activated amino acid is then bound to its own particular tRNA molecule with the aid of an enzyme (a synthetase) that is specific for that particular amino acid and that particular tRNA molecule:

The different synthetases recognize their substrates by stretches of nucleotide sequences on the tRNA. The specific recognition by an enzyme, aminoacyl-tRNA synthetase, of its proper tRNA and amino acid is often referred to as the **second genetic code.** This step is very important because once the amino acid is on the tRNA, there is no other opportunity to check for the correct pairing. In other words, the anticodon of the tRNA will match up with its correct codon on the mRNA regardless of whether it is carrying the correct amino acid, so the aminoacyl-tRNA synthetases have to get it right.

B. Initiation

The initiation stage consists of three steps:

1. **Forming the pre-initiation complex** To initiate the protein synthesis a unique tRNA is used, designated as **tRNA^fMet.** This tRNA carries a formylated methionine (fMet) residue, but it is used solely for the initiation step. It is attached to the 30S ribosomal body and forms the pre-initiation complex, along with GTP, [(Figure 26.5)(1)]. Just as in transcription, each step in translation is aided by a number of factors; these proteins are called **initiation factors.**

2. **Migration to mRNA** Next, the pre-initiation complex binds to the mRNA (2). The ribosome is aligned on the mRNA by recognizing a special RNA sequence called the **Shine-Dalgarno** sequence, which is complementary to a sequence on the 30S ribosomal subunit. The anticodon of the fMet–tRNA^fMet, UAC, lines up against the start codon, AUG.

3. **Forming the full ribosomal complex** The 50S ribosomal body joins the 30S ribosomal complex (3). The complete ribosome carries three sites. The one shown in the middle in Figure 26.5 is called the **P site,** because the growing peptide chain will bind there. The one next to it on the right is called the **A (acceptor) site,** because it accepts the incoming tRNA bringing the next amino acid. As the full initiation complex is completed, the initiation factors dissociate and the GTP is hydrolyzed to GDP.

C. Elongation

1. **Binding to the A site** At this point, the A site is vacant, and each of the aminoacycl-tRNA molecules can try to fit itself in. However, only one of the tRNAs carries the right anticodon that corresponds to the next

Active Figure 26.5 The formation of an initiation complex. The 30S ribosomal subunit binds to mRNA and the fMet-tRNA^fMet in the presence of GTP and the three initiation factors, forming the 30S initiation complex (Steps 1 and 2). The 50S ribosomal subunit is added, forming the full initiation complex (Step 3). **See a simulation based on this figure, and take a short quiz on the concepts at http://www.now.brookscole .com/gob8 or on the CD.**

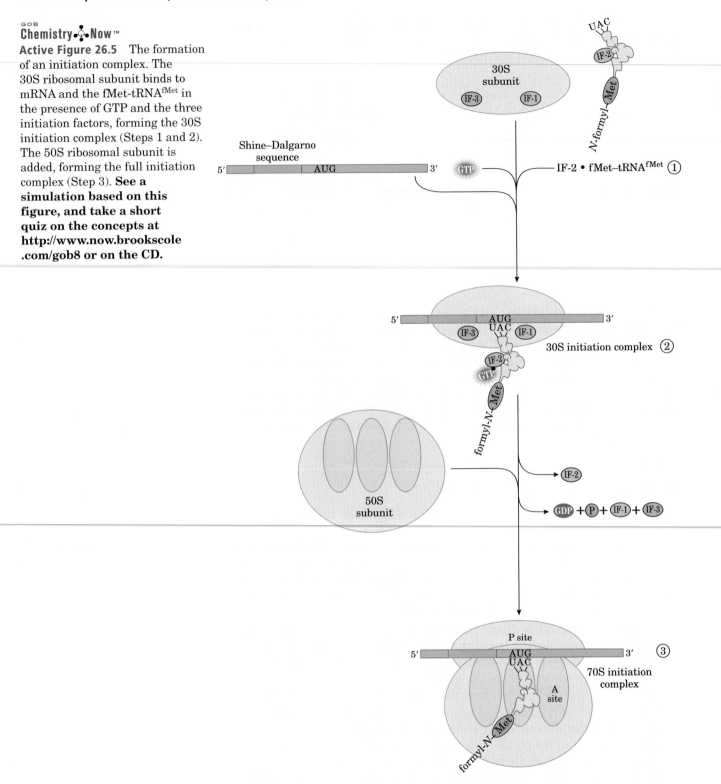

codon on the mRNA. This is an alanine tRNA in Figure 26.6. The binding of this tRNA to the A site takes place with the aid of proteins called **elongation factors** and GTP [Figure 26.6 (2)].

2. **Forming the first peptide bond** At the A site, the new amino acid, alanine (Ala), is linked to the fMet in a peptide bond by the enzyme **peptidyl transferase.** The empty tRNA remains on the P site [Figure 26.6 (3)].

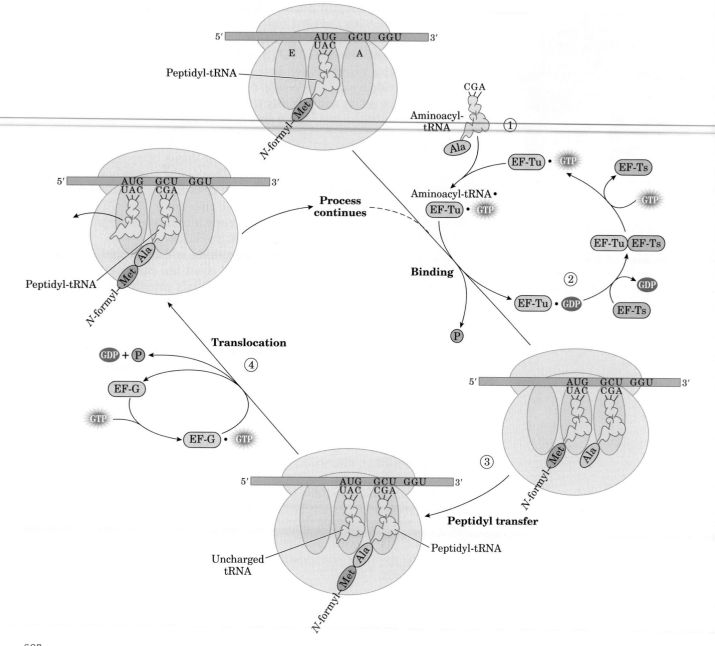

Active Figure 26.6 The steps in chain elongation. (1) An aminoacyl-tRNA is bound to the A site on the ribosome. Elongation factors and GTP are required. The P site on the ribosome is already occupied. (2) Elongation factors are recycled to prepare another incoming tRNA, and GTP is hydrolyzed. (3) The peptide bond is formed, leaving an uncharged tRNA at the P site. (4) In the translocation step, the uncharged tRNA is pushed into the E site and more GTP is hydrolyzed. The A site is now over the next codon on the mRNA. **See a simulation based on this figure, and take a short quiz on the concepts at http://www.now.brookscole.com/gob8 or on the CD.**

3. **Translocation** In the next phase of elongation, the whole ribosome moves one codon along the mRNA. Simultaneously with this move, the dipeptide is **translocated** from the A site to the P site (4). The empty tRNA is moved to the E site. When this cycle occurs one more time, the empty tRNA will be ejected and go back to the pool of tRNA that is available for activation with an amino acid.

Figure 26.7 Ribosome in action. The lower yellow half represents the 30S ribosome; the upper blue half represents the 50S ribosome. The yellow and green twisted cones are tRNAs, and the chain of beads stand for mRNA. The elongation factors are in dark blue.

Courtesy of Dr. J. Frank, Wadsworth Center, Albany, New York

4. **Forming the second peptide bond** After the translocation, the A site is associated with the next codon on the mRNA, which is 5′ GGU 3′ in Figure 26.6. Once again, each tRNA can try to fit itself in, but only the one whose anticodon is 5′ACC 3′ can align itself with GGU. This tRNA, which carries glycine (Gly), now comes in. The transferase establishes a new peptide bond between Gly and Ala, moving the dipeptide from the P site to the A site and forming a tripeptide. These elongation steps are repeated until the last amino acid is attached.

Figure 26.7 shows a three-dimensional model of the translational process, which has been constructed on the basis of recent cryoelectron microscopy and x-ray diffraction studies. This model shows how the elongation factor proteins (in dark blue) fit into a cleft between the 50S (blue) and the 30S (pale yellow) bodies of prokaryotic ribosomes. The tRNAs on the P site (green) and on the A site (yellow) occupy a central cavity in the ribosomal complex. The orange beads represent the mRNA.

The mechanism of peptide bond formation is a nucleophilic attack by the amino group of the A site amino acid upon the carbonyl group of the P site amino acid, as shown in Figure 26.8. While attempting to study this mechanism in detail, researchers discovered a fascinating phenomenon. It turns out that in the vicinity of the nucleophilic attack, there is no protein in the ribosome that could catalyze such a reaction. The only chemical groups nearby that could catalyze a reaction are on a purine of the ribosomal RNA. Thus the ribosome is a ribozyme. Previously, catalytic RNA had been found only in some RNA splicing reactions, but here is a scenario in which RNA catalyzes one of the principal reactions of life.

D. Termination

After the final translocation, the next codon reads "stop" (UAA, UGA, or UAG). At this point, no more amino acids can be added. Releasing factors then cleave the polypeptide chain from the last tRNA via a GTP-requiring mechanism that is not yet fully understood. The tRNA itself is released from the P site. At the end, the whole mRNA is released from the ribosome. This process is shown in Figure 26.9. While the mRNA is attached to the ribosomes, many polypeptide chains are synthesized on it simultaneously.

Figure 26.8 Peptide bond formation in protein synthesis. Nucleophilic attack by the amino group of the A-site aminoacyl-tRNA on the carbonyl carbon of the P-site peptidyl-tRNA is facilitated when a purine moiety of the rRNA abstracts a proton.

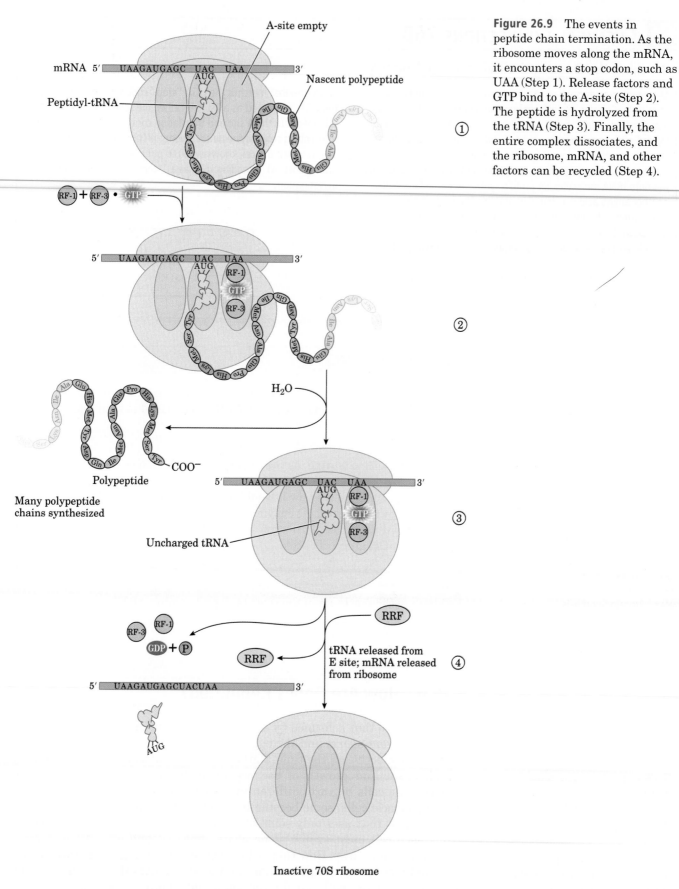

Figure 26.9 The events in peptide chain termination. As the ribosome moves along the mRNA, it encounters a stop codon, such as UAA (Step 1). Release factors and GTP bind to the A-site (Step 2). The peptide is hydrolyzed from the tRNA (Step 3). Finally, the entire complex dissociates, and the ribosome, mRNA, and other factors can be recycled (Step 4).

CHEMICAL CONNECTIONS 26B

Breaking the Dogma: The Twenty-First Amino Acid

Many amino acids, such as citrulline and ornithine found in the urea cycle (Chapter 28), are not building blocks of proteins. Other nonstandard amino acids such as hydroxyproline (Chapter 22) are formed after translation by post-translational modification. When discussing amino acids and translation in the past, the magic number was always 20. That is, only 20 standard amino acids were put onto tRNA molecules for protein synthesis. In the late 1980s, another amino acid was found in proteins from eukaryotes and prokaryotes alike, including humans. It is selenocysteine, a cysteine residue that has the sulfur replaced by selenium.

Selenocysteine is formed by placing a serine onto a special tRNA molecule called tRNA^sec. Once bound, the oxygen in the serine side chain is replaced by selenium. This tRNA molecule has an anticodon that matches the UGA "stop" codon. In special cases, the UGA is not read as a "stop"; rather, the selenocysteine-tRNA^sec is loaded into the A site and translation continues. Some are, therefore, calling selenocysteine the twenty-first amino acid. The methods by which the cell knows when to put selenocysteine into the protein instead of reading UGA as a "stop" codon remain under investigation.

Selenocysteine

EXAMPLE 26.3

A tRNA has an anticodon, 5′ AAG 3′. Which amino acid will this tRNA carry? What are the steps necessary for the amino acid to bind to the tRNA?

Solution
The amino acid is leucine, as the codon is 5′ CUU 3′. Remember that sequences are read from left to right as 5′ ⟶ 3′, so you have to flip the anticodon around to see how it would bind to the codon. Leucine has to be activated by ATP. A specific enzyme, leucine-tRNA synthetase, catalyzes the carboxyl ester bond formation between the carboxyl group of leucine and the —OH group of tRNA.

Problem 26.3
What are the reactants in the reaction forming valine-tRNA?

26.6 | How Are Genes Regulated?

Gene regulation The control process by which the expression of a gene is turned on or off. Because RNA synthesis proceeds in one direction (5′ ⟶ 3′), the gene (DNA) to be transcribed runs from 3′ to 5′. Thus the control sites are in front of, or upstream of, the 3′ end of the structural gene.

Every embryo that is formed by sexual reproduction inherits its genes from both the parent sperm and the egg cells. But the genes in its chromosomal DNA are not active all the time. Rather, they are switched on and off during development and growth of the organism. Soon after formation of the embryo, the cells begin to differentiate. Some cells become neurons, some become muscle cells, some become liver cells, and so on. Each cell is a specialized unit that uses only some of the many genes it carries in its DNA. Thus each cell must switch some of its genes on and off—either permanently or temporarily. How this is done is the subject of **gene regulation.**

We know less about gene regulation in eukaryotes than in the simpler prokaryotes. Even with our limited knowledge, however, we can state that

CHEMICAL CONNECTIONS 26C

Viruses

Nucleic acids are essential for life as we know it. No living thing can exist without them because they carry the information necessary to make protein molecules. The smallest forms of life, the viruses, consist only of a molecule of nucleic acid surrounded by a "coat" of protein molecules. In some viruses, the nucleic acid is DNA; in others, it is RNA. No virus has both. Whether viruses can be considered a true form of life became a topic of debate recently. In the summer of 2002, a group of scientists from the State University of New York, Stony Brook, reported that they had synthesized the polymyelitis virus in the laboratory from fragments of DNA. This new "synthetic" virus caused the same polio symptoms and death as the wild virus.

The shapes and sizes of viruses vary greatly, as shown in the figure. Because their structures are so simple, viruses cannot reproduce themselves in the absence of other organisms. They carry DNA or RNA but do not have the nucleotides, enzymes, amino acids, and other molecules necessary to replicate their nucleic acid (Section 25.6) or to synthesize proteins (Section 26.5). Instead, viruses invade the cells of other organisms and cause those cells (the hosts) to do these tasks for them. Typically, the protein coat of a virus remains outside the host cell, attached to the cell wall, while the DNA or RNA is pushed inside. Once the viral nucleic acid is inside the cell, the cell stops replicating its own DNA and making its own proteins. Instead, it replicates the viral nucleic acid and synthesizes the viral protein, according to the instructions on the viral nucleic acid. One host cell can make many copies of the virus.

In many cases, the cell bursts when a large number of new viruses have been synthesized, sending the new viruses out into the intercellular material, where they can infect other cells. This kind of process causes the host organism to get sick, and perhaps to die. Among the many human diseases caused by viruses are measles, hepatitis, mumps, influenza, the common cold, rabies, and smallpox. There is no cure for most viral diseases. Antibiotics, which can kill bacteria, have no effect on viruses. So far, the best defense against these diseases has been immunization (Chemical Connections 31B), which under the proper circumstances can work spectacularly well. Smallpox, once one of the most dreaded diseases, has been eradicated from this planet by many years of vaccination, and comprehensive programs of vaccination against such diseases as polio and measles have greatly reduced the incidence of these diseases.

Lately, a number of antiviral agents have been developed. They completely stop the reproduction of viral nucleic acids (DNA or RNA) inside infected cells without preventing the DNA of normal cells from replicating. One such drug is called vidarabine, or Ara-A, and is sold under the trade name Vira-A.

Herpes virus

Influenza virus

Poliomyelitis virus

The shape of three viruses. (Illustration by Irving Geis, from *Scientific American*, January 1963. Rights owned by Howard Hughes Medical Institute. Not to be reproduced without permission.)

Vidarabine

Antiviral agents often act like anticancer drugs in that they have structures similar to one of the nucleotides necessary for the synthesis of nucleic acids. Vidarabine is the same as adenosine, except that the sugar is arabinose instead of ribose. Vidarabine is used to fight a life-threatening viral illness, herpes encephalitis. It is also effective in neonatal herpes infection and chickenpox. However, like many other anticancer and antiviral drugs, vidarabine is toxic, causing nausea and diarrhea. In some cases, it has caused chromosomal damage.

organisms do not have a single, unique way of controlling genes. Many gene regulations occur at the **transcriptional level** (DNA ⟶ RNA). Others operate at the **translational level** (mRNA ⟶ protein). A few of these processes are listed here as examples.

A. Control at the Transcriptional Level

In eukaryotes, transcription is regulated by three entities: promoters, enhancers, and response elements.

1. Promoters of a gene are located adjacent to the transcription site. They are defined by an initiator and conserved sequences, such as a TATA box (see also Section 26.2 and Figure 26.3) or one or more copies of other sequences, such as the GGGCGG sequence called a GC box. In eukaryotes, the enzyme RNA-polymerase has little affinity for binding to DNA. Instead, different transcription factors, or binding proteins, bind to the different modules of the promoter.

   There are two basic types of transcription factors. The first is called a **general transcription factor (GTF).** These proteins form a complex with RNA polymerase and the DNA and help to position the RNA polymerase correctly and stimulate the initiation of transcription. For transcription of genes that will lead to mRNA (that is, Pol II transcription), there are six GTFs, all named as TFII and then a letter, for Pol II transcription factor. All of these transcription factors are necessary to establish the initiation of transcription. As can be seen in Figure 26.10, the events in the initiation of Pol II transcription are very complicated. Six transcription factors must bind to the DNA and RNA polymerase to initiate transcription. They first form the **preinitiation complex.** The critical event in starting the transcription is the conversion to the **open complex,** which involves the phosphorylation of the C-terminal end of the RNA polymerase. Only when the open complex has formed can transcription take place. During elongation, three transcription factors (B, E, and H) are released. Transcription factor F remains bound to Pol II with D bound to the TATA box. Only factor F continues with the polymerase.

   With the aid of these transcription factors, the promoter functions to control the transcription on a steady, normal level. Transcription factors may allow the synthesis of mRNA (and from there the target protein) to vary by a factor of 1 million. This wide variation in eukaryotic cells is exemplified by the α-A-crystallin gene, which can be expressed in the lens of the eye, at a rate a millionfold higher than that in the liver cell of the same organism.

2. Another group of transcription factors speeds up the transcription process by binding to DNA sequences that may be located several thousand nucleotides away from the transcription site. These sequences are known as **enhancers.** To stimulate transcription, an enhancer is brought to the vicinity of the promoter by the formation of a loop. Figure 26.11 shows how the transcription factor binds to the enhancer element and forms a bridge to the basal transcription unit. This complex then allows the RNA polymerase II to speed up the transcription when higher-than-normal production of proteins is needed.

   Other DNA sequences bind transcription factors but have the opposite effect—they slow down transcription. These are called **silencers.**

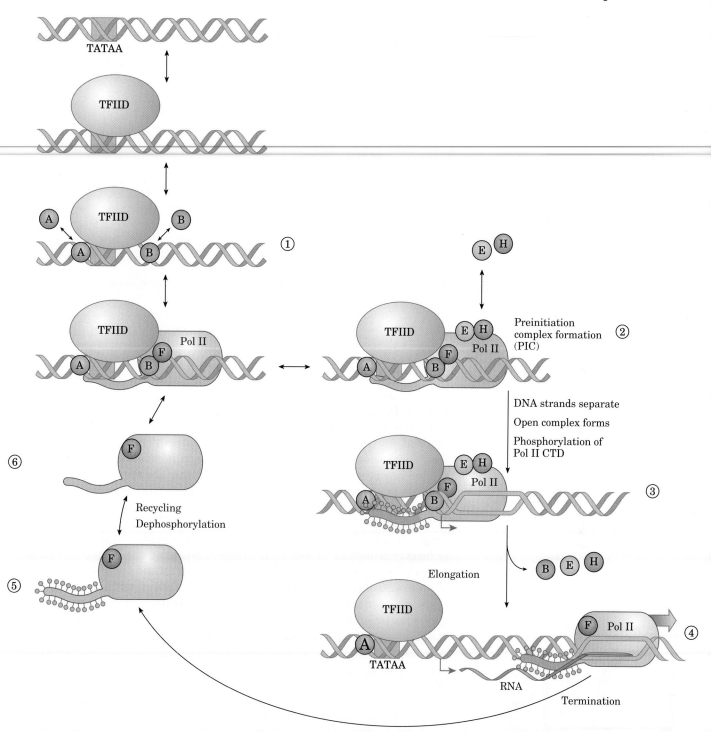

Active Figure 26.10 A schematic representation of the order of events of Pol II transcription. General Transcription Factor TFIID binds to the TATA box on the DNA and recruits TFIIA and TFIIB (Step 1). RNA Polymerase II carrying TFIIF binds to the DNA, followed by TFIIE and TFIIH to form the preinitiation complex (PIC) (Step 2). The C-terminal domain of Pol II is phosphorylated and the DNA strands are separated to form the open complex (Step 3). TFIIB, TFIIE, and TFIIH are released as the polymerase synthesizes RNA in the process of elongation (Step 4). Transcription is terminated when the mRNA is complete and the Pol II is released (Step 5). The Pol II is dephosphorylated and is ready to be recycled for another round of transcription (Step 6). **See a simulation based on this figure, and take a short quiz on the concepts at http://www.now.brookscole.com/gob8 or on the CD.**

Figure 26.11 DNA looping brings enhancers in contact with transcription factors and RNA polymerase.

3. The third type of transcription control involves a type of enhancer called **response elements.** These enhancers are activated by their transcription factors in response to an outside stimulus. The stimulus may be heat shock, heavy metal toxicity, or simply a hormonal signal, such as the binding of a steroid hormone to its receptor. The response element of steroids is in front of, and 260 base pairs upstream from, the starting point of transcription. Only the receptor with the bound steroid hormone can interact with its response element, thereby initiating transcription. The difference between an enhancer and a response element is largely a matter of our own understanding of the system. We call something a response element when we understand the bigger picture of how controlling the gene is related to a pattern of metabolism. Many response elements may be controlling a particular process, and a given gene may be under the control of more than one response element.

4. Transcription does not occur at the same rate throughout the cell's entire life cycle. Instead, it is accelerated or slowed down as the need arises. The signal to speed up transcription may originate from outside of the cell. One such signal, in the GTP–adenylate cyclase–cAMP pathway (Section 24.5B), produces **phosphorylated protein kinase.** This enzyme enters the nucleus, where it phosphorylates transcription factors, which aid in the transcription cascade.

How do these transcription factors find the specific gene control sequences into which they fit, and how do they bind to them? The interaction between the protein and DNA involves nonspecific electrostatic interactions (positive ions attracting negative ions and repelling other positive ions) as well as more specific hydrogen bonding. The transcription factors find their targeted sites by twisting their protein chains so that a certain amino acid sequence is present at the surface. One such conformational twist is provided by **metal-binding fingers** (Figure 26.12). These finger shapes are created by ions, which form covalent bonds with the amino acid side chains of the protein.

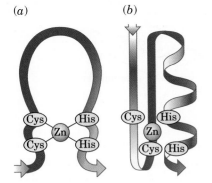

(a)   (b)

Figure 26.12 Cys₂His₂ zinc finger motifs. (*a*) The coordination between zinc and cysteine and histidine residues. (*b*) The secondary structure. (Adapted from Evans R. M. and Hollenberg, S. M., *Cell, 52:1, 1988,* Figure 1)

CHEMICAL CONNECTIONS 26D

CREB: The Most Important Protein You Have Never Heard of?

We will look more closely at the cyclic AMP response element as an example of eukaryotic control of transcription. Hundreds of research papers deal with this topic as more and more genes are found to have this response element as part of their control. Cyclic AMP is produced as a second messenger (Chapter 24) from several hormones such as epinephrine and glucagon. When the levels of cAMP rise, the activity of **cAMP-dependent protein kinase** (protein kinase A) is stimulated. This enzyme phosphorylates many proteins and enzymes inside the cell and is usually associated with switching the cell to a catabolic mode, where macromolecules will be broken down for energy. Protein kinase A phosphorylates a protein called **cyclic AMP response element–binding protein (CREB),** which binds to the cAMP response element and activates the associated genes. CREB does not directly contact the basal transcription machinery (RNA-polymerase and GTFs), however, and the activation requires another protein. **CREB-binding protein (CBP)** binds to CREB after it has been phosphorylated and bridges the response element and the promoter region, as shown in the figure. After this bridge is constructed, transcription is activated above basal levels. CBP is called a *mediator* or *coactivator.* Hundreds of genes are controlled by the cAMP response element. More than 100 known transcription factors also bind to CBP.

Although the research is ongoing, CREB-mediated transcription has been implicated in a tremendous variety of physiological processes, including *cell proliferation, cell differentiation,* and *spermatogenesis.* It controls *release of somatostatin,* a hormone that inhibits growth hormone secretion. It has been shown to be critical for *development of mature T lymphocytes* (immune system cells) and confers *protection to nerve cells* in the brain under hypoxic conditions. It is involved in *metabolism of the pineal gland* and *control of circadian rhythms.* CREB levels have been shown to be elevated during the body's adaptation to strenuous physical exercise. This protein is involved in the *regulation of gluconeogenesis* by the two peptide hormones, glucagon and insulin, and directly affects *transcription of metabolic enzymes,* such as phosphoenolpyruvate carboxykinase (PEPCK) and lactate dehydrogenase. Most interestingly, CREB has been shown to be *critical in learning and storage in long-term memory,* and low levels of CREB have been found in the brains of patients suffering from Alzheimer's disease.

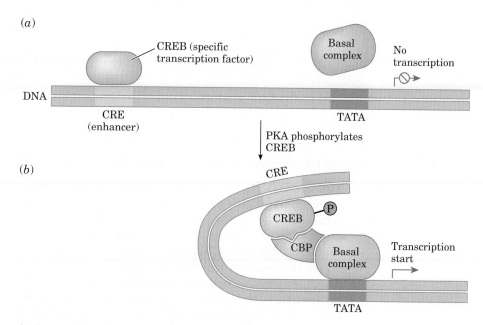

Activation of transcription via CREB and CBP. (*a*) Unphosphorylated CREB does not bind to CREB-binding protein, and no transcription occurs. (*b*) Phosphorylation of CREB causes binding of CREB to CBP, which forms a complex with the basal complex (RNA polymerase and GTFs), thereby activating transcription. (Adapted by permission from *Molecular Biology* by R. F. Weaver, McGraw-Hill, 1999.)

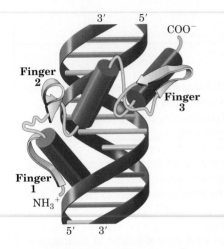

Figure 26.13 Zinc finger proteins follow the major groove of DNA. (Adapted with permission from Pavletich N. and Pabo C. O., *Science* 252:809, 1991 Figure 2. Copyright © 1991 AAAS.)

The zinc fingers interact with specific DNA (or sometimes RNA) sequences. The recognition comes by hydrogen bonding between a nucleotide (for example, guanine) and the side chain of a specific amino acid (for example, arginine). Zinc fingers allow the proteins to bind in the major groove of DNA, as shown in Figure 26.13.

Besides metal-binding fingers, at least two other prominent transcription factors exist: **helix-turn-helix** and **leucine zipper.**

B. Control on the Post-transcriptional Level

In the spring of 2000, scientists were eagerly awaiting the results of the Human Genome Project and an accurate count of the number of genes in the human genome. The odds-on favorite would have been 100,000 to 150,000 genes. After all, it was known that humans produce 90,000 different proteins. The dogma stated that "one gene leads to one mRNA leads to one protein." The only exceptions to this rule were thought to occur in the production of antibodies and other immunoglobulin-based proteins. These proteins were known to undergo alternative splicing of the primary mRNA transcript to give multiple mature mRNAs, and therefore multiple proteins.

It came as a big shock, therefore, when the data revealed that humans have about 30,000 genes, a number close to that of the roundworm or corn. If 30,000 genes can lead to 90,000 proteins, there must be far more alternative splicing to account for it. Scientists now believe that splicing RNA in different ways is a very important process that leads to the differences in species that are otherwise similar. Chimpanzees and humans, for example, share 99% of their DNA. They also produce very similar protein complements. However, significant differences have been found in some tissues, most notably the brain, where certain human genes are more active and others generate different proteins by alternative splicing.

Figure 26.14 summarizes the various ways that alternative splicing can produce many different proteins. Exons may be included in all products, or they may be present on only some of them. Different splicing sites can appear either on the 5′ or 3′ side. In some cases, introns may even be retained in the final product.

C. Control on the Translational Level

During translation, a number of mechanisms ensure quality control.

1. **The specificity of a tRNA for its unique amino acid** First, the attachment of the proper amino acid to the proper tRNA must be achieved. The enzyme that catalyzes this reaction, aminoacyl-tRNA synthetase (AARS), is specific for each amino acid. For those amino acids that have more than one type of tRNA, the same synthetase catalyzes the reaction for all of the tRNA types for that amino acid. The AARS enzymes recognize their tRNAs by specific nucleotide sequences. Furthermore, the active site of the enzyme has two **sieving portions.** For example, in isoleucyl-tRNA synthetases, the first sieve excludes any amino acids that are larger than isoleucine. If a similar amino acid such as valine, which is smaller than isoleucine, arrives at the active site, the second sieve eliminates it. The second sieving site is thus a "proof-reading" site.

2. **Recognition of the stop codon** Another quality-control measure occurs at the termination. The stop codons must be recognized by release factors, leading to the release of the polypeptide chain and allowing the recycling of the ribosomes. Otherwise, a longer polypeptide chain may be toxic. The release factor combines with GTP and binds to

(a)   Skipped exon

Resulting mRNA

5′   3′

(b)   Alternative 5′ splice sites

5′   3′

(c)   Alternative 3′ splice sites

5′   3′

(d)   Retained intron

5′   3′

(e)   Mutually exclusive exon retention

5′   3′

◼ Exon always spliced in   ◼ Exon alternatively spliced   ◻ Intron

Figure 26.14   Alternate Splicing. A gene's primary transcript can be edited in several different ways where splicing activity is indicated by dashed lines. An exon may be left out (a). Splicing machinery may recognize alterative 5′ splice sites for an intron (b) or alternative 3′ splice sites (c). An intron may be retained in the final mRNA transcript (d). Exons may be retained on a mutually exclusive basis (e). (*Scientific American*)

the ribosomal A site when that site is occupied by the termination codon. Both the GTP and the peptidyl-tRNA ester bond are hydrolyzed. This hydrolysis releases the polypeptide chain and the deacylated tRNA. Finally, the ribosome dissociates from the mRNA. As we saw in Chemical Connections 26B, sometimes the stop codon is used to continue translating by inserting a very uncommon amino acid, such as selenocysteine.

3. **Post-translational Controls**

   (a) *Removal of methionine.* In most proteins, the methionine residue at the N-terminus, which was added in the initiation step, is removed. A special enzyme, metaminopeptidase, cleaves the peptide bond. In the case of prokaryotes, when the N-terminus is supposed to be methionine, another enzyme cleaves off the formyl group.

   (b) *Chaperoning.* The tertiary structure of a protein is largely determined by the amino acid sequence (primary structure). Proteins begin to fold even as they are synthesized on ribosomes. Nevertheless, misfolding may occur due to mutation in a gene, lack of fidelity in transcription, or translational errors. All of these errors may lead to aggregation of misfolded proteins that can be detrimental to the cell—for example, causing amyloid diseases, such as Alzheimer's disease or Jakob-Creutzfeld disease. Certain proteins in living cells, called **chaperones,** help the newly synthesized polypeptide chains to fold properly. They recognize hydrophobic regions exposed on unfolded proteins and bind to them. Chaperones then shepherd the proteins to the biologically desirable folding as well as to their local destinations within the cell.

   (c) *Degradation of misfolded proteins.* A third post-translational control exists in the form of **proteasomes.** These cylindrical assemblies include a number of protein subunits with proteolytic activity in the core of the cylinder. If the rescue by chaperones fails, proteases may degrade the misfolded protein first by targeting it with ubiquitination (see Chemical Connections 28E) and finally by proteolysis.

26.7 | What Are Mutations?

All sequences on DNA are given as coding strand sequences. Thus the codon, which is on the mRNA, has the same sequence as the coding strand DNA, except that T is replaced by U.

In Section 25.6, we saw that the base-pairing mechanism provides an almost perfect way to copy a DNA molecule during replication. The key word here is "almost." No machine, not even the copying mechanism of DNA replication, is totally without error. It has been estimated that, on average, one error occurs for every 10^{10} bases (that is, one in 10 billion). An error in the copying of a sequence of bases is called a **mutation.** Mutations can occur during replication. Base errors can also occur during transcription in protein synthesis (a non-inheritable error).

These errors may have widely varying consequences. For example, the codon for valine in mRNA can be GUA, GUG, GUC, or GUU. In DNA, these

---

CHEMICAL CONNECTIONS 26E

Mutations and Biochemical Evolution

We can trace the genetic relationship of different species through the variability of their amino acid sequences in different proteins. For example, the blood of all mammals contains hemoglobin, but the amino acid sequences of the hemoglobins are not identical. In the table below, we see that the first 10 amino acids in the β-globin of humans and gorillas are exactly the same. In fact, there is only one amino acid difference, at position 104, between us and apes. The β-globin of the pig differs from ours at 10 positions, of which 2 are in the N-terminal decapeptide. That of the horse differs from ours in 26 positions, of which 4 are in this decapeptide. β-Globin seems to have gone through many mutations during the evolutionary process, because only 26 of the 146 sites are invariant—that is, exactly the same in all species studied so far.

The relationship between different species can also be established by finding similarities in their mRNA primary structures. Because the mutations actually occurred on the original DNA molecule and then were perpetuated in the progeny by the mutant DNA, it is instructive to learn how a point mutation may occur in different species. Looking at position 4 of the β-globin molecule,

we see a change from serine to threonine. The code for serine is AGU or AGC, whereas that for threonine is ACU or ACC (Table 26.1). Thus a change from G to C in the second position of the codon created the divergence between the β-globins of humans and horses. The genes of closely related species, such as humans and apes, have very similar primary structures, presumably because these two species diverged on the evolutionary tree only recently. In contrast, species far removed from each other diverged long ago and have undergone more mutations, which show up as differences in the primary structures of their DNA, mRNA, and consequently proteins.

The *number* of amino acid substitutions is significant in the evolutionary process caused by mutation, but the *kind* of substitution is even more important. If the substitution involves an amino acid with physicochemical properties similar to those of the amino acid in the ancestor protein, the mutation is most probably viable. For example, in human and gorilla β-globin, position 4 is occupied by threonine, but it is occupied by serine in the pig and horse. Both amino acids provide an —OH carrying side chain.

Amino Acid Sequence of the N-Terminal Decapeptides of β-Globin in Different Species

Species	Position									
	1	2	3	4	5	6	7	8	9	10
Human	Val	His	Leu	Thr	Pro	Glu	Glu	Lys	Ser	Ala
Gorilla	Val	His	Leu	Thr	Pro	Glu	Glu	Lys	Ser	Ala
Pig	Val	His	Leu	Ser	Ala	Glu	Glu	Lys	Ser	Ala
Horse	Val	Glu	Leu	Ser	Gly	Glu	Glu	Lys	Ala	Ala

CHEMICAL CONNECTIONS 26F

Oncogenes

An **oncogene** is a gene that in one way or another participates in the development of cancer. Cancer cells differ from normal cells in a variety of structural and metabolic ways; however, the most important difference is their uncontrolled proliferation. Most ordinary cells are quiescent. When they lose their controls, they give rise to tumors—benign or malignant. The uncontrolled proliferation allows these cells to spread, invade other tissues, and colonize them, in a process called *metastasis*.

Malignant tumors can be caused by a variety of agents: chemical carcinogens such as benzo(a)pyrene (Chemical Connections 13B), ionizing radiation such as Xrays, and viruses.

However, for a normal cell to become a cancer cell, certain transformations must occur. Many of these transformations take place on the level of the genes. Certain genes in normal cells have counterparts in viruses, especially in retroviruses (which carry their genes in the form of RNA rather than DNA). Thus normal cells in a human and in a virus both have genes that code for similar proteins. Usually these genes code for proteins that control cell growth.

One such protein is epidermal growth factor (EGF). When a retrovirus gets into the cell, it takes control of the cell's own EGF gene. At that point, the EGF gene becomes an oncogene, a cancer-causing gene, because it produces large amounts of the EGF protein. This excess production, in turn, allows the epidermal cells to grow in an uncontrolled manner.

Because the normal EGF gene in the normal cell was turned into an oncogene, we call it a **proto-oncogene.** Such conversions of a proto-oncogene to an oncogene can occur not just by viral invasion, but also by a mutation of the gene. Although most cancers are caused by multiple mutations, one form of cancer, called human bladder carcinoma, is caused by a single mutation of a gene—a change from guanine (G) to thymine (T). The resulting protein has a valine in place of a glycine. This change is sufficient to transform the cell into a cancerous state because the protein for which the gene codes amplifies signals (Section 24.5). Thus, because of the mutation that caused the proto-oncogene to become an oncogene, the cell stays "on"—its metabolic process is not shut off—and it is now a cancer cell that proliferates without control.

There are a number of other mechanisms by which a proto-oncogene is transformed into an oncogene, but in each case the gene product has something to do with growth, hormonal action, or some other kind of cell regulation. About 100 viral and cellular oncogenes have been identified to date.

codons correspond to GTA, GTG, GTC, and GTT, respectively. Assume that the original codon in the DNA is GTA. If a mistake is made during replication and GTA is spelled as GTG in the copy, there will be no harmful mutation. Instead, when a protein is synthesized, the GTG will appear on the mRNA as GUG, which also codes for valine. Therefore, although a mutation has occurred, the same protein is manufactured.

Now assume that the original sequence in the gene's DNA is GAA, which will also be GAA in the mRNA and codes for glutamic acid. If a mutation occurs during replication and GAA becomes TAA, a very serious mutation will have occurred. The TAA on the DNA coding strand would be UAA on the mRNA, which does not code for any amino acid but rather is a stop signal. Thus, instead of continuing to build a protein chain with glutamic acid, the synthesis will stop altogether. An important protein will not be manufactured, or at least will be manufactured incorrectly, and the organism may be sick or even die. As we saw in Chemical Connections 22D, sickle cell anemia is caused by a single base mutation that causes a glutamic acid to be replaced with valine.

Ionizing radiation (X rays, ultraviolet light, gamma rays) can cause mutations. Furthermore, a large number of organic compounds can induce mutations by reacting with DNA. Such compounds are called **mutagens.** Many changes caused by radiation and mutagens do not become mutations because the cell has its own repair mechanism, called nucleotide excision repair (NER), which can prevent mutations by cutting out damaged areas and resynthesizing them (Section 25.7). Despite this defense mechanism, certain errors in copying that result in mutations do slip by. Many compounds (both synthetic and natural) are mutagens, and some can cause cancer when introduced into the body. These substances are called **carcinogens** (Chemical Connections 13B). One of the main tasks of the U.S. Food and Drug Administration and the Environmental Protection Agency is to identify

CHEMICAL CONNECTIONS 26G

p53: A Central Tumor Suppressor Protein

Not all cancer-causing gene mutations have their origin in onco-genes. There are some 36 known **tumor suppressor genes,** the products of which are proteins controlling cell growth. None of them is more important than the protein with a molar mass of 53,000, simply named **p53.** This protein responds to a variety of cellular stresses, including DNA damage, lack of oxygen (hypoxia), and aberrantly activated oncogenes (Chemical Connections 26F). In about 40% of all cancer cases, the tumor contains p53 that under-went mutation. Mutated p53 protein can be found in 55% of lung cancers, about half of all colon and rectal cancers, and some 40% of lymphomas, stomach cancers, and pancreatic cancers. In addi-tion, in one third of all soft tissue sarcomas, p53 is inactive, even though it did not undergo mutation.

These statistics indicate that the normal function of the p53 protein is to suppress tumor growth. When it is mutated or other-wise not present in sufficient or active form, it is unable to perform this protective function and cancer spreads. The p53 protein binds to specific sequences of double-stranded DNA. When Xrays or

γ-rays damage DNA, an increase in p53 protein concentration is observed. The increased binding of p53 controls the cell cycle by holding it between cell division and DNA replication. The time gained in this arrested cell cycle allows the DNA to repair its damage. If that fails, the p53 protein triggers apoptosis, the pro-grammed death of the injured cell.

Recently, it was reported that p53 performs a finely tuned func-tion in cells. It suppresses tumor growth, but if p53 is overexpressed (that is, its concentration is high), it contributes to the premature aging of the organism. In those conditions, p53 arrests the cell cycle not only of damaged cells but also of stem cells. These cells nor-mally differentiate into various types (muscle cells, nerve cells, and so forth) and replace those cells that die with aging. Excess p53 slows down this differentiation. In mice, an abundance of p53 pro-tein made the animal cancer-free but at a cost: The mice lost weight and muscle, their bones became brittle, and their wounds took longer to heal. All in all, their average lifespan was 20% shorter than that of normal mice.

carcinogens and eliminate them from our food, drugs, and environment. Even though most carcinogens are mutagens, the reverse is not true.

Not all mutations are harmful. Some are beneficial because they enhance the survival rate of the species. For example, mutation is used to develop new strains of plants that can withstand pests.

If a mutation is harmful, it results in an inborn genetic disease. This condition may be carried as a recessive gene from generation to generation with no individual demonstrating the symptoms of the disease. When both parents carry recessive genes, however, an offspring has a 25% chance of inheriting the disease. If the defective gene is dominant, on the other hand, every carrier will develop symptoms.

26.8 | How and Why Do We Manipulate DNA?

There are no cures for the inborn genetic diseases discussed in Section 26.7. The best we can do is detect the carriers and, through genetic counseling of prospective parents, try not to perpetuate the defective genes. However, **recombinant DNA techniques** give us some hope for the future. At this time, these DNA techniques are being used mostly in bacteria, plants, and test animals (such as mice), but they are slowly being applied to humans as well, as will be described in Section 26.9.

One example of recombinant DNA techniques begins with certain circu-lar DNA molecules found in the cells of the bacterium *Escherichia coli.* These molecules, called **plasmids** (Figure 26.15), consist of double-stranded DNA arranged in a ring. Certain highly specific enzymes called **restriction endonucleases** cleave DNA molecules at specific locations (a different location for each enzyme). For example, one of these enzymes may split a double-stranded DNA as follows:

Figure 26.15 Plasmids from a bacterium used in the recombinant DNA technique.

We use "B" to indicate the remaining DNA in a bacterial plasmid.

$$\text{B} \sim\sim\sim \text{GAATTC} \sim\sim\sim \text{B} \xrightarrow[\text{enzyme}]{\text{restriction}} \text{B}\sim\sim\sim\text{G} \qquad\qquad \text{AATTC}\sim\sim\sim\text{B}$$
$$\text{B} \sim\sim\sim \text{CTTAAG} \sim\sim\sim \text{B} \qquad\qquad \text{B}\sim\sim\sim\text{CTTAA} \quad+\quad \text{G}\sim\sim\sim\text{B}$$

The enzyme is so programmed that whenever it finds this specific sequence of bases in a DNA molecule, it cleaves it as shown. Because a plasmid is circular, cleaving it in this way produces a double-stranded chain with two ends (Figure 26.16). These are called "sticky ends" because on one strand each has several free bases that are ready to pair up with a complementary section if they can find one.

The next step is to give the strands such a section. This is done by adding a gene from some other species. The gene is a strip of double-stranded DNA that contains the necessary base sequence. For example, we can put in the human gene that manufactures insulin, which we can get in two ways:

1. It can be made in a laboratory by chemical synthesis; that is, chemists can combine the nucleotides in the proper sequence to make the gene.

2. We can cut a human chromosome with the same restriction enzyme. Because it is the same enzyme, it cuts the human gene so as to leave the same sticky ends:

$$
\begin{array}{c}
\text{H—GAATTC—H} \\
\text{H—CTTAAG—H}
\end{array}
\xrightarrow[\text{enzyme}]{\text{restriction}}
\begin{array}{c}
\text{H—G} \\
\text{H—CTTAA}
\end{array}
+
\begin{array}{c}
\text{AATTC—H} \\
\text{G—H}
\end{array}
$$

The human gene must be cut at two places so that a piece of DNA that carries two sticky ends is freed. To splice the human gene into the plasmid, the two are mixed in the presence of DNA ligase, and the sticky ends come together:

$$
\begin{array}{c}
\text{H—G} \\
\text{H—CTTAA}
\end{array}
+
\begin{array}{c}
\text{AATTC—B} \\
\text{G—B}
\end{array}
\xrightarrow[\text{ligase}]{\text{DNA}}
\begin{array}{c}
\text{H—GAATTC—B} \\
\text{H—CTTAAG—B}
\end{array}
$$

This reaction takes place at both ends of the human gene, turning the plasmid into a circle once again (Figure 26.16).

The modified plasmid is then put back into a bacterial cell, where it replicates naturally every time the cell divides. Bacteria multiply quickly, so soon we have a large number of bacteria, all containing the modified plasmid. All these cells now manufacture human insulin by transcription and translation. In this way, we can use bacteria as a factory to manufacture specific proteins. This new industry has tremendous potential for lowering the price of drugs that are currently manufactured by isolation from human or animal tissues (for example, human interferon, a molecule that fights infection). Not only bacteria but also viruses can be used to create recombinant DNA (Figure 26.17).

We use "H" to indicate a human gene.

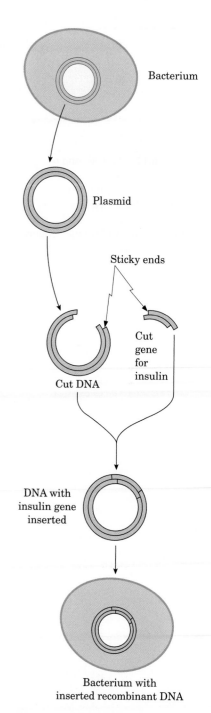

Figure 26.16 The recombinant DNA technique can be used to turn a bacterium into an insulin "factory."

E X A M P L E 26.4

Two different restriction endonucleases act on the following sequence of a double-stranded DNA:

$$\text{〰〰 AATGAATTCGAGGC 〰〰}$$
$$\text{〰〰 TTACTTAAGCTCCG 〰〰}$$

One endonuclease, EcoRI, recognizes the sequence GAATTC and cuts the sequence between G and A. The other endonuclease, TaqI, recognizes the sequence TCGA and cuts the sequence between T and C. What are the sticky ends that each of these endonucleases will create?

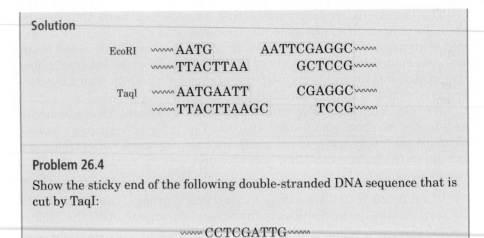

Solution

EcoRI	ᴨᴨAATG	AATTCGAGGCᴨᴨ
	ᴨᴨTTACTTAA	GCTCCGᴨᴨ
Taql	ᴨᴨAATGAATT	CGAGGCᴨᴨ
	ᴨᴨTTACTTAAGC	TCCGᴨᴨ

Problem 26.4

Show the sticky end of the following double-stranded DNA sequence that is cut by TaqI:

ᴨᴨCCTCGATTGᴨᴨ
ᴨᴨGGAGCTAACᴨᴨ

Figure 26.17 The cloning of human DNA fragments with a viral vector. (Adapted with permission from *Dealing with Genes: The Language of Heredity* by Paul Berg and Maxine Singer, 1992 © by University Science Books)

26.9 | What Is Gene Therapy?

While viruses have traditionally been seen as problems for humans, there is one field where they are now being used for good. Viruses can be used to alter somatic cells, where a genetic disease is treated by the introduction of a gene for a missing protein. This process is called **gene therapy.**

The most successful form of gene therapy to date involves the gene for **adenosine deaminase (ADA),** an enzyme involved in purine catabolism (Section 25.8). If this enzyme is missing, dATP builds up in tissues, inhibiting the action of the enzyme ribonucleotide reductase. The result is a deficiency of the other three deoxyribonucleoside triphosphates (dNTPs). The dATP (in excess) and the other three dNTPs (deficient) are precursors for DNA synthesis. This imbalance particularly affects DNA synthesis in lymphocytes, on which much of the immune response depends (Chapter 31). Individuals who are homozygous for adenosine deaminase deficiency develop **severe combined immune deficiency (SCID),** the "bubble boy" syndrome. They are prone to infection because of their highly compromised immune systems. The ultimate goal of the planned gene therapy is to take bone marrow cells from affected individuals, introduce the gene for adenosine deaminase into the cells using a virus as a vector, and then reintroduce the bone marrow cells in the body, where they will produce the desired enzyme. The first clinical trials for a cure to ADA⁻ SCID were simple enzyme replacement therapies begun in 1982. The patients in these trials were given injections of ADA. Later clinical trials sought to correct the gene in mature T cells. In 1990, transformed T cells were given to recipients via transfusions. In trials at the National Institutes of Health (NIH), two girls, aged 4 and 9 years at the start of treatment, showed improvement to the extent that they could attend regular public schools and had no more than the average number of infections. Administration of bone marrow stem cells in addition to T cells was the next step; clinical trials of this procedure were undertaken with two infants, aged 4 months and 8 months in the year 2000. After 10 months, the children were healthy and had restored immune systems.

There are two types of delivery methods in human gene therapy. The first, called **ex vivo,** is the type used to combat SCID. *Ex vivo* means that somatic cells are removed from the patient, altered with the gene therapy, and then returned to the patient. The most common vector for this approach is **Maloney murine leukemia virus (MMLV).** Figure 26.18 shows how the virus is used for gene therapy. Some of the MMLV is altered to remove the *gag, pol,* and *env* genes, rendering the virus unable to replicate. These genes are replaced with an **expression cassette,** which contains the gene being administered, such as the ADA gene, along with a suitable promoter. This mutated virus is used to infect a packaging cell line. Normal MMLV is also used to infect the packaging cell line. The normal MMLV will not replicate in the packaging cell line, but its *gag, pol,* and *env* genes will restore the mutated virus's ability to replicate, albeit only in this cell line. These controls are necessary to keep mutant viruses from escaping to other tissues. The mutated virus particles are collected from the packaging cell line and used to infect the target cells—bone marrow cells, in the case of SCID. MMLV is a retrovirus, so it infects the target cell and produces DNA from its RNA genome; this DNA can then become incorporated into the host genome, along with the promoter and ADA gene. In this way, the target cells that were collected are transformed, and they will produce ADA. These cells are then put back into the patient.

In the second delivery method, called **in vivo,** the virus is used to directly infect the patient's tissues. The most common vector for this delivery

Figure 26.18 Gene therapy via retroviruses. The Maloney murine leukemia virus (MMLV) is used for ex vivo gene therapy. Essential genes (*gag, pol, env*) are removed from the virus and replaced with an expression cassette containing the gene being replaced with gene therapy. Removal of the essential viral genes renders them unable to replicate. The altered virus is grown in a packaging cell line that will allow replication. Viruses are collected and used to infect cultured target cells from the host needing the gene therapy. The altered virus produces RNA, which in turn produces DNA via reverse transcription. The DNA becomes integrated in the host cell's genome, and the host cells produce the desired protein. The cultured cells are returned to the host.

is the DNA virus, **adenovirus.** A particular vector can be chosen based on specific receptors on the target tissue. Adenovirus has receptors in lung and liver cells, and it has been used in clinical trials for gene therapy of cystic fibrosis and ornithine transcarbamoylase deficiency. Clinical trials using gene therapy to combat cystic fibrosis and certain tumors in humans are now under way. In mice, gene therapy has been successful in fighting diabetes.

The field of gene therapy is exciting and full of promise, but many obstacles to its success in humans remain. There are also many risks, such as a dangerous immunological response to the vector carrying the gene, or the danger of a gene becoming incorporated into the host chromosome at a location that activates a cancer-causing gene. Both undesirable outcomes have happened in a limited number of human cases.

Gene therapy has been approved in humans only for the manipulation of somatic cells. It is illegal to tamper with human gametes in an attempt to create a heritable change in the human genome.

SUMMARY OF KEY QUESTIONS

SECTION 26.1 How Does DNA Lead to RNA and Protein?

- A **gene** is a segment of a DNA molecule that carries the sequence of bases that directs the synthesis of one particular protein or RNA.
- The information stored in the DNA is transcribed onto RNA and then expressed in the synthesis of a protein molecule. This process involves two steps: **transcription** and **translation.**

SECTION 26.2 How Is DNA Transcribed into RNA?

- In transcription, the information is copied from DNA onto mRNA by complementary base pairing. There are also start and stop signals.
- The enzyme that synthesizes RNA is called RNA polymerase. In eukaryotes, three kinds of polymerase are used for the different types of RNA.

SECTION 26.3 What Is the Role of RNA in Translation?

- mRNA is strung out along the ribosomes.
- Transfer RNA carries the individual amino acids, with each tRNA going to a specific site on the mRNA.
- A sequence of three bases (a triplet) on mRNA constitutes a **codon.** It spells out the particular amino acid that the tRNA brings to this site.
- Each tRNA has a recognition site, the **anticodon,** that pairs up with the codon.
- When two tRNA molecules are aligned at adjacent sites, the amino acids that they carry are linked by an enzyme, forming a peptide bond.
- The translation process continues until the entire protein is synthesized.

SECTION 26.4 What Is the Genetic Code?

- The **genetic code** provides the correspondence between a codon and an amino acid.
- In most cases, there is more than one codon for each amino acid, but the opposite is not true: A given codon will specify only one amino acid.

SECTION 26.5 How Is Protein Synthesized?

- Protein synthesis takes place in four stages: activation, initiation, elongation, and termination.
- Several of the steps of translation require an input of energy in the form of GTP.
- Ribosomes have three sites: the A site, the P site, and the E site.
- No protein is found in the area where the peptide synthesis is catalyzed. Thus the ribosome is a ribozyme.

SECTION 26.6 How Are Genes Regulated?

- Most of human DNA (96 to 98%) does not code for proteins.

- A number of mechanisms for gene regulation exist on both the transcriptional level and the translational level.
- **Promoters** have initiator and conserved sequences.
- **Transcription factors** bind to the promoter, thereby regulating the rate of transcription.
- **Enhancers** are nucleotide sequences far removed from the transcription sites. They bind to transcription factors and increase the level of transcription.
- Some translational controls, such as **release factors,** act during translation; others, such as **chaperones,** act after translation is completed.

SECTION 26.7 What Are Mutations?

- A change in the sequence of bases is called a **mutation.**
- Mutations can be caused by an internal mistake or induced by chemicals or radiation. In fact, a change in just one base can cause a mutation.
- A mutation may be harmful or beneficial, or it may cause no change in the amino acid sequence. If a mutation is very harmful, the organism may die.
- Chemicals that cause mutations are called **mutagens.** Chemicals that cause cancer are called **carcinogens.** Many carcinogens are mutagens, but the reverse is not true.

SECTION 26.8 How and Why Do We Manipulate DNA?

- With the discovery of restriction enzymes that can cut DNA molecules at specific points, scientists have found ways to splice DNA segments together.
- A human gene (for example, the one that codes for insulin) can be spliced into a bacterial plasmid. The bacteria, when multiplied, can then transmit this new information to the daughter cells so that the ensuing generations of bacteria can manufacture human insulin. This powerful method is called the **recombinant DNA technique.**
- Genetic engineering is the process by which genes are inserted into cells.

SECTION 26.9 What Is Gene Therapy?

- Gene therapy is a technique whereby a missing gene is replaced using a viral vector.
- In **ex vivo** gene therapy, cells are removed from a patient, given the missing gene, and then the cells are given back to the patient.
- In **in vivo** therapy, the patient is given the virus directly.

PROBLEMS

A blue problem number indicates an applied problem.

■ denotes problems that are available on the GOB ChemistryNow website or CD and are assignable in OWL.

SECTION 26.1 How Does DNA Lead to RNA and Protein?

26.5 ■ Does the term *gene expression* refer to

(a) transcription, (b) translation, or (c) transcription plus translation?

26.6 In what part of the eukaryote cell does transcription occur?

26.7 Where does most of the translation occur in a eukaryote cell?

SECTION 26.2 How Is DNA Transcribed into RNA?

26.8 What is the function of RNA polymerase?

26.9 What is the role of helicase in transcription?

26.10 Where is an initiation signal located?

26.11 Which end of the DNA contains the termination signal?

26.12 What would happen to the transcription process if a drug added to a eukaryote cell inhibits the phosphatase?

26.13 Where is the methyl group located in the guanine cap?

26.14 How are the adenine nucleotides linked together in the polyA tails?

SECTION 26.3 What Is the Role of RNA in Translation?

26.15 Where are the codons located?

26.16 ■ What are the two most important sites on tRNA molecules?

26.17 What are the ribosomal subunits for eukaryotic translation?

SECTION 26.4 What Is the Genetic Code?

26.18 ■ (a) If a codon is GCU, what is the anticodon?

(b) For which amino acid does this codon code?

26.19 ■ If a segment of DNA is 981 units long, how many amino acids appear in the protein encoded by this DNA segment? (Assume that the entire segment is used to code for the protein and that there is no methionine at the N-terminal end of the protein.)

26.20 In what sense does the universality of the genetic code support the theory of evolution?

26.21 ■ Which amino acids have the most possible codons? Which have the fewest?

26.22 Using the first column of Table 26.1, explain how changing the second base of the codon is more detrimental to a protein than changing the first or the third base.

SECTION 26.5 How Is Protein Synthesized?

26.23 To which end of the tRNA is the amino acid bonded? Where does the energy come from to form the tRNA–amino acid bond?

26.24 ■ There are three sites on the ribosome, each participating in the translation. Identify them and describe what is happening at each site.

26.25 What is the main role of (a) the 40S ribosome and (b) the 60S ribosome?

26.26 What are the prokaryotic equivalents of the eukaryotic ribosomal subunits?

26.27 What is the function of elongation proteins?

26.28 ■ What are the stages of protein synthesis?

26.29 Explain the nature of the tRNA used to initiate translation.

26.30 Explain what happens to the fMet initially put at the N-terminus.

26.31 Explain why scientists now refer to the ribosome as a ribozyme.

26.32 Why is amino acid activation called the second genetic code?

SECTION 26.6 How Are Genes Regulated?

26.33 Which molecules are involved in gene regulation at the transcriptional level?

26.34 Where are enhancers located? How do they work?

26.35 Where are the sieving portions of AARS enzymes located? What is their function?

26.36 What are the two types of transcription factors, and how do they work?

26.37 What is the difference between an enhancer and a response element?

26.38 How does alternative splicing lead to protein diversity?

26.39 What is the function of proteosomes in quality control?

26.40 What kind of interactions exist between metal-binding fingers and DNA?

SECTION 26.7 What Are Mutations?

26.41 Using Table 26.1, give an example of a mutation that (a) does not change anything in a protein molecule and (b) might cause fatal changes in a protein.

26.42 How do cells repair mutations caused by X rays?

26.43 Can a harmful mutation-causing genetic disease exist from generation to generation without exhibiting the symptoms of the disease? Explain.

26.44 Are all mutagens also carcinogens?

SECTION 26.8 How and Why Do We Manipulate DNA?

26.45 How do restriction endonucleases operate?

26.46 What are sticky ends?

26.47 A new genetically engineered corn has been approved by the Food and Drug Administration. This new corn shows increased resistance to a destructive insect called a corn borer. What is the difference, in principle, between this genetically engineered corn and one that developed insect resistance by mutation (natural selection)?

26.48 EcoRI restriction endonuclease recognizes the sequence GAATTC and cuts it between G and A. What will be the sticky ends of the following double-helical sequence when EcoRI acts on it?

CAAAGAATTCG
GTTTCTTAAGC

Chemical Connections

26.49 (Chemical Connections 26A) What is unique in the backbone of the antisense drug Vitravene?

26.50 (Chemical Connections 26B) Why is selenocysteine called the twenty-first amino acid? Why were amino acids such as hydroxyproline and hydroxylysine not counted as additional amino acids?

26.51 (Chemical Connections 26C) What is a viral "coat"?

26.52 (Chemical Connections 26C) Where do the ingredients—amino acids, enzymes, and so forth—necessary to synthesize the viral coat come from?

26.53 (Chemical Connections 26D) How does CREB work as a transcription factor? What other proteins are necessary?

26.54 (Chemical Connections 26D) With what processes has CREB been associated?

26.55 (Chemical Connections 26E) What is an invariant site?

26.56 (Chemical Connections 26F) What kind of mutation transforms the EGF proto-oncogene into an oncogene in human bladder carcinoma?

26.57 (Chemical Connections 26G) What is p53? Why is its mutated form associated with cancer?

26.58 (Chemical Connections 26G) How does p53 promote DNA repair?

Additional Problems

26.59 In both the transcription and the translation steps of protein synthesis, a number of different molecules come together to act as a factor unit. What are these units of (a) transcription and (b) translation?

26.60 In the tRNA structure, there are stretches where complementary base pairing is necessary and other areas where it is absent. Describe two functionally critical areas (a) where base pairing is mandatory and (b) where it is absent.

26.61 Is there any way to prevent a hereditary disease? Explain.

26.62 How does the cell ensure that a specific amino acid (say, valine) attaches itself only to the one tRNA molecule that is specific for valine?

26.63 (a) What is a plasmid?
(b) How does it differ from a gene?

26.64 Why do we call the genetic code *degenerate?*

26.65 Glycine, alanine, and valine are classified as nonpolar amino acids. Compare their codons. What similarities do you find? What differences do you find?

26.66 Looking at the multiplicity (degeneracy) of the genetic code, you may get the impression that the third base of the codon is irrelevant. Point out where this is not the case. Out of the 16 possible combinations of the first and second bases, in how many cases is the third base irrelevant?

26.67 ■ Which polypeptide is coded for by the mRNA sequence 5′-GCU-GAA-GUC-GAG-GUG-UGG-3′?

26.68 A new endonuclease is found. It cleaves double-stranded DNA at every location where C and G are paired on opposite strands. Could this enzyme be used in producing human insulin by the recombinant DNA technique? Explain.

CHAPTER 27

Bioenergetics: How the Body Converts Food to Energy

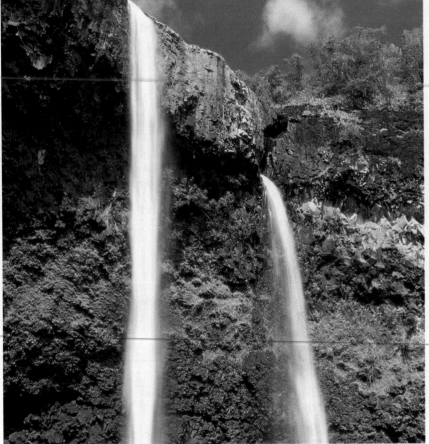

A & L Sinibaldi/Stone/Getty Images

Wailua Falls, Hawaii, is a natural demonstration of two pathways ending in a common pool.

27.1 | What Is Metabolism?

Living cells are in a dynamic state, which means that compounds are constantly being synthesized and then broken down into smaller fragments. Thousands of different reactions take place at the same time. **Metabolism** is the sum total of all the chemical reactions involved in maintaining the dynamic state of the cell.

In general, we can classify metabolic reactions into two broad groups: (1) those in which molecules are broken down to provide the energy needed by cells and (2) those that synthesize the compounds needed by cells—both simple and complex. **Catabolism** is the process of breaking down molecules to supply energy. The process of synthesizing (building up) molecules is **anabolism.** The same compounds may be synthesized in one part of a cell and broken down in a different part of the cell.

In spite of the large number of chemical reactions, a mere handful dominate cell metabolism. In this chapter and Chapter 28, we focus our atten-

Figure 27.1   In this simplified schematic diagram of the common catabolic pathway, an imaginary funnel represents what happens in the cell. The diverse catabolic pathways drop their products into the funnel of the common catabolic pathway, mostly in the form of C_2 fragments (Section 27.4). (The source of the C_4 fragments will be shown in Section 28.9.) The spinning wheel of the citric acid cycle breaks these molecules down further. The carbon atoms are released in the form of CO_2, and the hydrogen atoms and electrons are picked up by special compounds such as NAD^+ and FAD. Then the reduced NADH and $FADH_2$ cascade down into the stem of the funnel, where the electrons are transported inside the walls of the stem and the H^+ ions are expelled to the outside. In their drive to get back, the H^+ ions form the energy carrier ATP. Once back inside, they combine with the oxygen that picked up the electrons and produce water.

tion on the catabolic pathways that yield energy. A **biochemical pathway** is a series of consecutive biochemical reactions. We will see the actual reactions that enable the chemical energy stored in our food to be converted to the energy we use every minute of our lives—to think, to breathe, and to use our muscles to walk, write, eat, and everything else. In Chapter 29, we will look at some synthetic (anabolic) pathways.

The food we eat consists of many types of compounds, largely the ones we discussed in earlier chapters: carbohydrates, lipids, and proteins. All of them can serve as fuel, and we derive our energy from them. To convert those compounds to energy, the body uses a different pathway for each type of compound. *All of these diverse pathways converge to one* **common catabolic pathway,** which is illustrated in Figure 27.1. In the figure, the

Common catabolic pathway
A series of chemical reactions through which foodstuffs are oxidized to yield energy in the form of ATP; the common catabolic pathway consists of (1) the citric acid cycle (Section 27.4) and (2) oxidative phosphorylation (Sections 27.5 and 27.6)

diverse pathways are shown as different food streams. The small molecules produced from the original large molecules in food drop into an imaginary collecting funnel that represents the common catabolic pathway. At the end of the funnel appears the energy carrier molecule, adenosine triphosphate (ATP).

The purpose of catabolic pathways is to convert the chemical energy in foods to molecules of ATP. In the process, foods also yield metabolic intermediates, which the body can use for synthesis. In this chapter, we deal with the common catabolic pathway. In Chapter 28, we discuss the ways in which the different types of food (carbohydrates, lipids, and proteins) feed molecules into the common catabolic pathway.

27.2 | What Are Mitochondria, and What Role Do They Play in Metabolism?

A typical animal cell has many components, as shown in Figure 27.2. Each serves a different function. For example, the replication of DNA (Section 25.6) takes place in the **nucleus; lysosomes** remove damaged cellular components and some unwanted foreign materials; and **Golgi bodies** package and process proteins for secretion and delivery to other cellular compartments. The specialized structures within cells are called **organelles.**

The **mitochondria,** which possess two membranes (Figure 27.3), are the organelles in which the common catabolic pathway takes place in higher

Figure 27.2 Diagram of a rat liver cell, a typical higher animal cell.

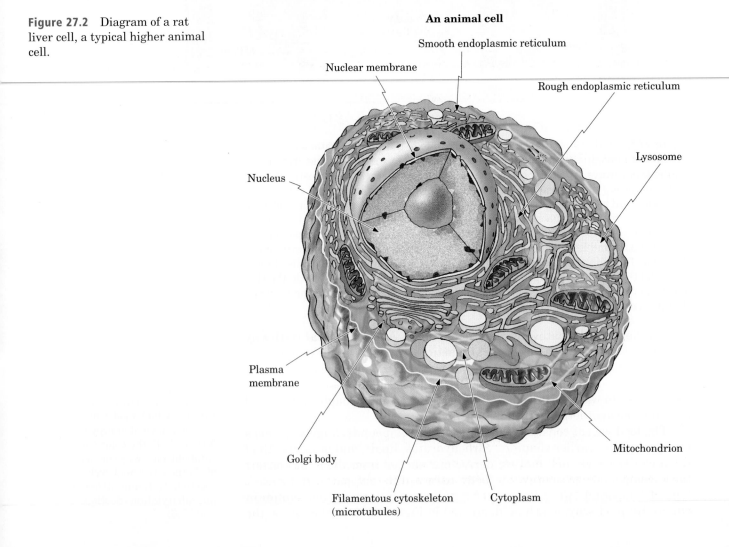

An animal cell

Smooth endoplasmic reticulum

Nuclear membrane

Rough endoplasmic reticulum

Lysosome

Nucleus

Plasma membrane

Golgi body

Filamentous cytoskeleton (microtubules)

Cytoplasm

Mitochondrion

organisms. The enzymes that catalyze the common pathway are all located in these organelles. Because these enzymes are synthesized in the cytosol, they must be imported through the two membranes. They cross the outer membrane through *translocator outer membrane* (TOM) channels and are accepted in the intermembrane space by chaperone-like *translocator inner membrane* (TIM) complexes, which also insert them into the inner membrane.

Because the enzymes are located inside the inner membrane of mitochondria, the starting materials of the reactions in the common pathway must pass through the two membranes to enter the mitochondria. Products must leave the same way.

The inner membrane of a mitochondrion is quite resistant to the penetration of any ions and of most uncharged molecules. (*Mitochondria* is the plural form; *mitochondrion* is the singular form.) However, ions and molecules can still get through the membrane—they are transported across it by the numerous protein molecules embedded in it (Figure 21.2). The outer membrane, by contrast, is quite permeable to small molecules and ions and does not have transporting membrane proteins.

The **matrix** is the inner nonmembranous portion of a mitochondrion (Figure 27.3). The inner membrane is highly corrugated and folded. On the basis of electron microscopic studies, Palade proposed his baffle model of the mitochondrion in 1952. The baffles, which are called **cristae,** project into the matrix like the bellows of an accordion. The enzymes of the oxidative phosphorylation cycle are localized on the cristae. The space between the inner and outer membranes is the **intermembrane space.** This classical baffle model of mitochondria underwent some changes in the late 1990s after three-dimensional pictures were obtained by the new technique called electron-microscope tomography. The 3-D images indicate that the cristae have narrow tubular connections to the inner membrane. These tubular connections may control the diffusion of metabolites from the inside to the intermembrane space. Furthermore, the space between the outer and the inner membrane varies during metabolism, possibly controlling the rate of reactions.

The enzymes of the citric acid cycle are located in the matrix. We will soon see in detail how the specific sequence of these enzymes causes the chain of events in the common catabolic pathway. Furthermore, we will discuss the ways in which nutrients and reaction products move into and out of the mitochondria.

(a)

(b)

Figure 27.3 (*a*) Schematic of a mitochondrion cut to reveal the internal organization. (*b*) Colorized transmission electron micrograph of a mitochondrion in a heart muscle cell.

27.3 | What Are the Principal Compounds of the Common Metabolic Pathway?

The common catabolic pathway has two parts: the **citric acid cycle** (also called the tricarboxylic acid cycle or the Krebs cycle) and the **electron transport chain** and **phosphorylation,** together called the **oxidative phosphorylation pathway.** To understand what actually happens in these reactions, we must first introduce the principal compounds participating in the common catabolic pathway.

A. Agents for Storage of Energy and Transfer of Phosphate Groups

The most important of these agents are three rather complex compounds: **adenosine monophosphate (AMP), adenosine diphosphate (ADP),** and **adenosine triphosphate (ATP)** (Figures 27.4 and 27.5). All three of these molecules contain the heterocyclic amine adenine (Section 25.2) and

Figure 27.4 Adenosine 5'-monophosphate (AMP).

Figure 27.5 Hydrolysis of ATP produces ADP plus dihydrogen phosphate ion plus energy.

the sugar D-ribose (Section 20.2) joined together by a β-N-glycosidic bond, forming adenosine (Section 25.2).

AMP, ADP, and ATP all contain adenosine connected to phosphate groups. The only difference between the three molecules is the number of phosphate groups. As you can see from Figure 27.5, each phosphate is attached to the next by an anhydride bond (Section 19.5A). ATP contains three phosphates—one phosphoric ester and two phosphoric anhydride bonds. In all three molecules, the first phosphate is attached to the ribose by a phosphoric ester bond (Section 19.5B).

A phosphoric anhydride bond contains more chemical energy (7.3 kcal/mol) than a phosphoric ester linkage (3.4 kcal/ mol). Thus, when ATP and ADP are hydrolyzed to yield phosphate ion (Figure 27.5), they release more energy per phosphate group than does AMP. When one phosphate group is hydrolyzed from each, the following energy yields are obtained: AMP = 3.4 kcal/mol; ADP = 7.3 kcal/mol; ATP = 7.3 kcal/mol. (The PO_4^{3-} ion is generally called inorganic phosphate.) Conversely, when inorganic phosphate bonds to AMP or ADP, greater amounts of energy are added to the chemical bond than when it bonds to adenosine. ADP and ATP contain *high-energy* phosphoric anhydride bonds.

ATP releases the most energy and AMP releases the least energy when each gives up one phosphate group. This property makes ATP a very useful compound for energy storage and release. The energy gained in the oxidation of food is stored in the form of ATP, albeit only for a short while. ATP molecules in the cells normally do not last longer than about 1 min. They are hydrolyzed to ADP and inorganic phosphate to yield energy that drives other processes, such as muscle contraction, nerve signal conduction, and biosynthesis. As a consequence, ATP is constantly being formed and decomposed. Its turnover rate is very high. Estimates suggest that the human body manufactures and degrades as much as 40 kg (approximately 88 lb) of ATP every day. Even with these considerations, the body is able to extract only 40 to 60% of the total caloric content of food.

Figure 27.6 The structures of NAD^+ and FAD.

B. Agents for Transfer of Electrons in Biological Oxidation–Reduction Reactions

Two other actors in this drama are the coenzymes (Section 23.3) NAD^+ (nicotinamide adenine dinucleotide) and FAD (flavin adenine dinucleotide), both of which contain an ADP core (Figure 27.6). (The + in NAD^+ refers to the positive charge on the nitrogen.) In NAD^+, the operative part of the coenzyme is the nicotinamide part. In FAD, the operative part is the flavin portion. In both molecules, ADP is the handle by which the apoenzyme holds on to the coenzyme; the other end of the molecule carries out the actual chemical reaction. For example, when NAD^+ is reduced, the nicotinamide part of the molecule gets reduced:

NAD⁺   NADH

The reduced form of NAD$^+$ is called NADH. The same reduction happens on two nitrogens of the flavin portion of FAD:

FAD   FADH$_2$

The reduced form of FAD is called FADH$_2$. We view NAD$^+$ and FAD coenzymes as the **hydrogen ion** and **electron-transporting molecules.**

C. Agent for Transfer of Acetyl Groups

The final principal compound in the common catabolic pathway is **coenzyme A** (CoA; Figure 27.7), which is the **acetyl (CH$_3$CO—)-transporting molecule.** Coenzyme A also contains ADP, but here the next structural unit is pantothenic acid, another B vitamin. Just as ATP can be viewed as an ADP molecule to which a —PO$_3^{2-}$— group is attached by a high-energy bond, so **acetyl coenzyme A** can be considered a CoA molecule linked to an acetyl group by a high-energy thioester bond, for which the energy of hydrolysis is 7.51 kcal/mol. The active part of coenzyme A is the mercaptoethylamine. The acetyl group of acetyl coenzyme A is attached to the SH group:

Acetyl group The group CH$_3$CO—

$$\underset{\text{Acetyl coenzyme A}}{CH_3-\overset{\displaystyle O}{\overset{\|}{C}}-S-CoA}$$

Mercaptoethylamine   Pantothenic acid   Phosphorylated ADP

Figure 27.7 The structure of coenzyme A.

27.4 | What Role Does the Citric Acid Cycle Play in Metabolism?

The common catabolism of carbohydrates and lipids begins when they have been broken down into pieces of two-carbon atoms each. The two-carbon fragments are the acetyl portions of acetyl coenzyme A. The acetyl is now fragmented further in the citric acid cycle.

Figure 27.8 gives the details of the citric acid cycle. A good way to gain an insight into this cycle is to use Figure 27.8 in conjunction with the simplified schematic diagram shown in Figure 27.9, which shows only the carbon balance.

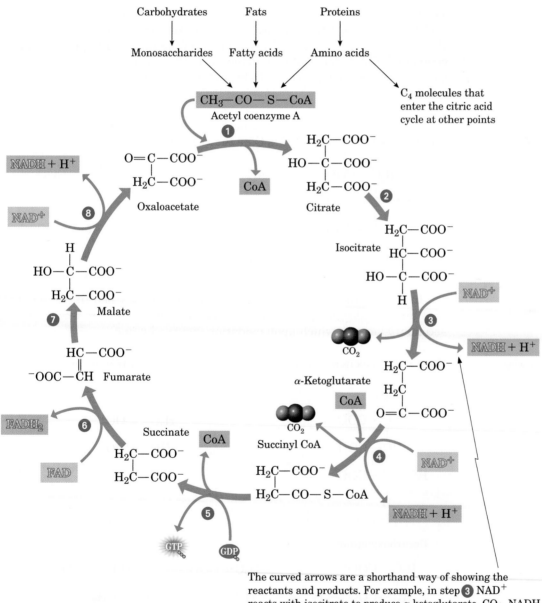

The curved arrows are a shorthand way of showing the reactants and products. For example, in step **3** NAD$^+$ reacts with isocitrate to produce α-ketoglutarate, CO_2, NADH, and H$^+$. The last two then leave the site of the reaction.

Figure 27.8   The citric acid (Krebs) cycle. The numbered steps are explained in detail in the text. [Hans Krebs (1900–1981), Nobel laureate in 1953, established the relationships among the different components of the cycle.]

We will now follow the two carbons of the acetyl group through each step in the citric acid cycle. The circled numbers correspond to those in Figure 27.8.

Step 1 Acetyl coenzyme A enters the cycle by combining with a C_4 compound called oxaloacetate:

The first thing that happens is the addition of the $-CH_3$ group of the acetyl CoA to the $C=O$ of the oxaloacetate, catalyzed by the enzyme citrate synthase. This event is followed by hydrolysis of the thioester to produce the C_6 compound citrate ion and CoA. Therefore, Step ① is a building-up rather than a breaking-down process. In Step ⑧ we will see where the oxaloacetate comes from.

Step 2 The citrate ion is dehydrated to *cis*-aconitate, after which the *cis*-aconitate is hydrated, but this time to isocitrate instead of citrate:

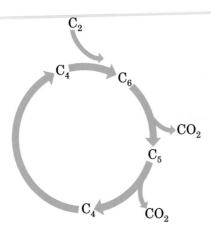

Figure 27.9 A simplified view of the citric acid cycle, showing only the carbon balance.

In citrate, the alcohol is a tertiary alcohol. We learned in Section 14.2 that tertiary alcohols cannot be oxidized. The alcohol in the isocitrate is a secondary alcohol, which upon oxidation yields a ketone.

Decarboxylation The process that leads to the loss of CO_2 from a $-COOH$ group

Step 3 The isocitrate is now oxidized and **decarboxylated** at the same time:

Oxidation:

Decarboxylation:

In oxidizing the secondary alcohol to a ketone, the oxidizing agent NAD^+ removes two hydrogens. One of them is used to reduce NAD^+ to NADH.

(Recall that NAD^+ and NADH are the oxidized and reduced forms, respectively, of nicotinamide adenine dinucleotide [Figure 27.6]). The other hydrogen replaces the COO^- that goes into making CO_2. Note that the CO_2 given off comes from the original oxaloacetate and not from the two carbons of the acetyl CoA. Both of these carbons are still present in the α-ketoglutarate. Also note that we are now down to a C_5 compound, α-ketoglutarate.

Steps 4 and 5 Next, a complex system removes another CO_2 once again from the original oxaloacetate portion rather than from the acetyl CoA portion:

(Recall again that NAD^+ and NADH are the oxidized and reduced forms, respectively, of nicotinamide adenine dinucleotide [Figure 27.6]). The P_i in this equation is the usual notation for inorganic phosphate.

We are now down to a C_4 compound, succinate. This oxidative decarboxylation is more complex than the first. It occurs in many steps and requires a number of cofactors. For our purpose, it is sufficient to know that, during this second oxidative decarboxylation, a high-energy compound called **guanosine triphosphate (GTP)** is also formed.

GTP is similar to ATP, except that guanine replaces adenine. Otherwise, the linkages of the base to ribose and the phosphates are exactly the same as in ATP. The function of GTP is also similar to that of ATP—namely, it stores energy in the form of high-energy phosphoric anhydride bonds (chemical energy). The energy from the hydrolysis of GTP drives many important biochemical reactions—for example, the signal transduction in neurotransmission (Section 24.5).

As a final note on the decarboxylation steps, the CO_2 molecules given off in Steps ③ and ④ are the ones we exhale.

Step 6 In this step, succinate is oxidized by FAD, which removes two hydrogens to give fumarate (the double bond in this molecule is *trans*):

This reaction cannot be carried out in the laboratory, but with the aid of an enzyme catalyst, the body does it easily. (Recall that FAD and $FADH_2$ are the oxidized and reduced forms, respectively, of flavin adenine dinucleotide [Figure 27.6]).

Step 7 The fumarate is now hydrated to give the malate ion:

$$\underset{\text{Fumarate}}{\begin{array}{c} HC-COO^- \\ \parallel \\ {}^-OOC-CH \end{array}} + H_2O \xrightarrow{\text{Fumarase}} \underset{\text{Malate}}{\begin{array}{c} H \\ | \\ HO-C-COO^- \\ | \\ H_2C-COO^- \end{array}}$$

Step 8 In the final step of the cycle, malate is oxidized by to give oxaloacetate:

(Recall that NAD$^+$ and NADH are the oxidized and reduced forms, respectively, of nicotinamide adenine dinucleotide [Figure 27.6]). Thus the final product of the Krebs cycle is oxaloacetate, which is the compound with which we started in Step ①.

In this process, the original two acetyl carbons of acetyl CoA were added to the C$_4$ oxaloacetate to produce a C$_6$ unit, which then lost two carbons in the form of CO$_2$ to produce, at the end of the process, the C$_4$ unit oxaloacetate. The net effect is that one two-carbon acetyl group enters the cycle and two carbon dioxides are given off.

How does the citric acid cycle produce energy? We have already learned that one step in the process produces a high-energy molecule of GTP. However, most of the energy is produced in the other steps that convert NAD$^+$ to NADH and FAD to FADH$_2$. These reduced coenzymes carry the H$^+$ and electrons that eventually will provide the energy for the synthesis of ATP (discussed in detail in Sections 27.5 and 27.6).

This stepwise degradation and oxidation of acetate in the citric acid cycle results in the most efficient extraction of energy. Rather than being generated in one burst, the energy is released in small packets that are carried away step by step in the form of NADH and FADH$_2$.

The cyclic nature of this acetate degradation has other advantages besides maximizing energy yield:

1. The citric acid cycle components provide raw materials for amino acid synthesis as the need arises (Chapter 29). For example, α-ketoglutaric acid is used to synthesize glutamic acid.

2. The many-component cycle provides an excellent method for regulating the speed of catabolic reactions.

The regulation can occur at many different parts of the cycle, so that feedback information can be used at many points to speed up or slow down the process as necessary.

The following equation represents the overall reactions in the citric acid cycle:

$$\text{GDP} + \text{P}_i + \text{CH}_3\text{—CO—S—CoA} + 2\text{H}_2\text{O} + 3\text{NAD}^+ + \text{FAD}$$
$$\longrightarrow \text{CoA} + \text{GTP} + 2\text{CO}_2 + 3\text{NADH} + \text{FADH}_2 + 3\text{H}^+ \qquad \text{(Eq. 27.1)}$$

The citric acid cycle is controlled by feedback mechanisms. When the essential product of this cycle, NADH + H$^+$, and the end product of the common catabolic pathway, ATP, accumulate, they inhibit some of the enzymes in the cycle. Citrate synthase (Step ①), isocitrate dehydrogenase (Step ③), and α-ketoglutarate dehydrogenase (part of the complex enzyme system in Step ④) are inhibited by ATP and/or by NADH + H$^+$. This inhibition slows down or shuts off the cycle. Conversely, when the feed material, acetyl CoA, is in abundance, the cycle accelerates. The enzyme isocitrate dehydrogenase (Step ③) is stimulated by ADP and NAD$^+$, which are the essential reactants from which the end products of the cycle are derived.

27.5 | How Do Electron and H⁺ Transport Take Place?

The reduced coenzymes NADH and $FADH_2$ are end products of the citric acid cycle. They carry hydrogen ions and electrons and, therefore, have the potential to yield energy when these combine with oxygen to form water:

$$4H^+ + 4e^- + O_2 \longrightarrow 2H_2O + energy$$

This simple exothermic reaction is carried out in many steps. The oxygen in this reaction is the oxygen we breathe.

A number of enzymes are involved in this reaction, all of which are embedded in the inner membrane of the mitochondria. These enzymes are situated in a particular *sequence* in the membrane so that the product from one enzyme can be passed on to the next enzyme, in a kind of assembly line. The enzymes are arranged in order of increasing affinity for electrons, so electrons flow through the enzyme system (Figure 27.10).

The sequence of the electron-carrying enzyme systems starts with complex I. The largest complex, it contains some 40 subunits, among them a flavoprotein and several FeS clusters. **Coenzyme Q** (CoQ; also called ubiquinone) is associated with complex I, which oxidizes the NADH produced in the citric acid cycle and reduces the CoQ:

$$NADH + H^+ + CoQ \longrightarrow NAD^+ + CoQH_2$$

Some of the energy released in this reaction is used to move $2H^+$ across the membrane, from the matrix to the intermembrane space. The CoQ is soluble in lipids and can move laterally within the membrane. (The figure of $2H^+$ transported across the membrane is the minimum number that allows

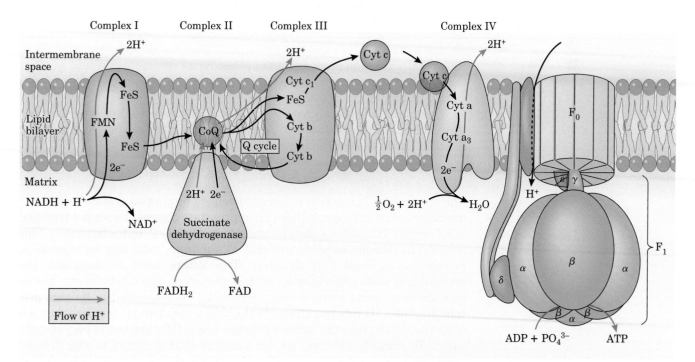

Figure 27.10 Schematic diagram of the electron and H⁺ transport chain and subsequent phosphorylation. The combined processes are also known as oxidative phosphorylation. **See a simulation based on this figure, and take a short quiz on the concepts at http://now.brookscole.com/gob8 or on the CD.**

CHEMICAL CONNECTIONS 27A

Uncoupling and Obesity

The health concerns that surround the growing number of obese people in developed countries have led to research into the causes and alleviation of obesity. A number of weight-reducing drugs exist. Some of them operate as uncouplers of electron transport and oxidative phosphorylation.

The discovery of a role for uncouplers in weight reduction occurred more or less by accident. During World War I, many ammunition workers were exposed to 2,4-dinitrophenol (DNP), a compound used to prepare the explosive picric acid, which is structurally related to the well-known explosive trinitrotoluene (TNT). After it was observed that these workers lost weight, DNP was used as a weight-reducing drug during the 1920s. Unfortunately, DNP eliminated not only the fat but sometimes also the patient, and its use as a diet pill was discontinued after 1929.

Today we know why DNP works as a weight-reducing drug: It is an effective protonophore—a compound that transports ions through cell membranes passively, without the expenditure of energy. As noted earlier, H^+ ions accumulate in the intermembrane space of mitochondria and, under normal conditions, drive the syn-

thesis of ATP while they are going back to the inside. This process is Mitchell's chemiosmotic principle in action. When DNP is ingested, it transfers the H^+ back to the mitochondrion easily, and no ATP is manufactured. The energy of the electron separation is dissipated as heat and is not built in as chemical energy in ATP. The loss of this energy-storing compound makes the utilization of food much less efficient, resulting in weight loss.

A similar mechanism provides heat in hibernating bears. The bears have brown fat; its color is derived from the numerous mitochondria in the tissue. The brown fat also contains an uncoupling protein called thermogenin, a protonophore that allows the ions to stream back into the mitochondrial matrix without manufacturing ATP. The heat generated in this manner keeps the animal alive during cold winter days. In similar fashion, an uncoupling protein is known to be involved in obesity, but it is not known what relationship, if any, exists between this protein and hibernation. The question of obesity and its prevention is important enough, however, to make uncoupling in brown fat a point of departure for obesity research.

the overall oxidation process to take place. Some researchers would say that the number of protons transported by some of these respiratory complexes should be higher.)

Complex II also catalyzes the transfer of electrons to CoQ. The source of the electrons is the oxidation of succinate in the citric acid cycle, producing $FADH_2$. The final reaction is

$$FADH_2 + CoQ \longrightarrow FAD + CoQH_2$$

The energy of this reaction is not sufficient to pump two protons across the membrane, however, nor is there an appropriate channel for such a transfer.

Complex III delivers the electrons from $CoQH_2$ to **cytochrome c.** This integral membrane complex contains 11 subunits, including cytochrome b, cytochrome c_1, and FeS clusters. (The letters used to designate the cytochromes were given in order of their discovery.) Each cytochrome has an iron-ion–containing heme center (Section 22.11) embedded in its own protein. Complex III has two channels through which two H^+ ions are pumped from the $CoQH_2$ into the intermembrane space. The process is very complicated. To simplify matters, we can imagine that it occurs in two distinct steps, as does the electron transfer. Because each cytochrome c can pick up only one electron, two cytochrome c units are needed:

$$CoQH_2 + 2 \text{ cytochrome c (oxydized)}$$
$$\longrightarrow CoQ + 2H^+ + 2 \text{ cytochrome c (reduced)}$$

Cytochrome c is also a mobile carrier of electrons—it can move laterally in the intermembrane space.

Complex IV, known as cytochrome oxidase, contains 13 subunits—most importantly, cytochrome a_3, a heme that has an associated copper center. Complex IV is an integral membrane protein complex. The electron movement flows from cytochrome c to cytochrome a to cytochrome a_3. There, the electrons are transferred to the oxygen molecule, and the O—O bond is cleaved. The oxidized form of the enzyme takes up two H^+ ions from the matrix for each oxygen atom. The water molecule formed in this way is released into the matrix:

$$\tfrac{1}{2}O_2 + 2H^+ + 2e^- \longrightarrow H_2O$$

During this process, two more H^+ ions are pumped out of the matrix and into the intermembrane space. Although the mechanism of pumping out protons from the matrix is not known, the energy driving this process is derived from the energy of water formation. This final pumping into the intermembrane space makes a total of six H^+ ions per NADH + H^+ and four H^+ ions per $FADH_2$ molecule.

27.6 | What Is the Role of the Chemiosmotic Pump in ATP Production?

How do the electron and H^+ transports produce the chemical energy of ATP? In 1961, Peter Mitchell (1920–1992), an English chemist, proposed the **chemiosmotic theory** to answer this question: The energy in the electron transfer chain creates a proton gradient. A **proton gradient** is a continuous variation in the H^+ concentration along a given region. In this case, there is a higher concentration of H^+ in the intermembrane space than inside the mitochondrion. The driving force, which is the result of the spontaneous flow of ions from a region of high concentration to a region of low concentration, propels the protons back to the mitochondrion through a complex known as **proton-translocating ATPase.** This compound is located on the inner membrane of the mitochondrion (Figure 27.10) and is the active enzyme that catalyzes the conversion of ADP and inorganic phosphate to ATP (the reverse of the reaction shown in Figure 27.5):

$$ADP + P_i \overset{ATPase}{\rightleftharpoons} ATP + H_2O$$

Chemiosmotic theory Mitchell's proposal that electron transport is accompanied by an accumulation of protons in the intermembrane space of the mitochondrion, which in turn creates osmotic pressure; the protons driven back to the mitochondrion under this pressure generate ATP

Subsequent studies have confirmed this theory, and Mitchell received the Nobel Prize in chemistry in 1978.

The proton-translocating ATPase is a complex "rotor engine" made of 16 different proteins. The F_0 sector, which is embedded in the membrane, contains the **proton channel** (Figure 27.10). The 12 subunits that form this channel rotate every time a proton passes from the cytoplasmic side (intermembrane) to the matrix side of the mitochondrion. This rotation is transmitted to a "rotor" in the F_1 sector. F_1 contains five kinds of polypeptides. The rotor (γ and ε subunits) is surrounded by the catalytic unit (made of α and β subunits) that synthesizes the ATP. The catalytic unit converts the mechanical energy of the rotor into chemical energy of the ATP molecule. The last unit, the "stator," containing the δ subunit, stabilizes the whole complex. The proton-translocating ATPase can catalyze the reaction in both directions. When protons that have accumulated on the outer surface of the

CHEMICAL CONNECTIONS 27B

Protection Against Oxidative Damage

Most of the oxygen we inhale is used in the final step of the common pathway—namely, in the conversion of our food supply to ATP. This reaction occurs in the mitochondria of cells. However, some 10% of the oxygen we consume is used elsewhere, in specialized oxidation reactions. One such reaction is the hydroxylation of compounds (for example, steroids). It is catalyzed by the enzyme cytochrome P-450, which usually resides outside the mitochondria in the endoplasmic reticulum of cells.

P-450 catalyzes the following reaction, where RH represents a steroid or other molecule:

$$RH + O_2 + \boxed{NADPH} + H^+$$
$$\longrightarrow ROH + H_2O + \boxed{NADP^+}$$

Our bodies use this reaction to detoxify foreign substances (for example, barbiturates). The action of P-450 is not always beneficial, however. Some of the most powerful carcinogens are converted to their active forms by P-450 (see Chemical Connections 13B).

Another use of oxygen for detoxification is in the fight against infection by bacteria. Specialized white blood cells called leukocytes engulf bacteria and kill them by generating superoxide, O_2^-, a highly reactive form of oxygen. But superoxide and other highly reactive forms of oxygen, such as $\cdot OH$ radicals and hydrogen peroxide not only destroy foreign cells but also damage our own cells. They especially attack unsaturated fatty acids, thereby damaging cell membranes (Section 21.5). Our bodies have their own defenses against these very reactive compounds. One example is glutathione (Chemical Connections 22A), which protects us against oxidative damage. Furthermore, our cells contain two powerful enzymes that decompose these highly reactive oxidizing agents:

$$2O_2^- + 2H^+ \xrightarrow{\text{superoxide dismutase}} H_2O_2 + O_2$$

$$2H_2O_2 \xrightarrow{\text{catalase}} 2H_2O + O_2$$

These enzymes also play an important defensive role in heart attacks and strokes. When oxygen deprivation of the heart muscles or the brain occurs because arteries are blocked, a condition known as ischemia reperfusion may arise. This means that when these tissues suddenly become reoxygenated, superoxide (O_2^-) can form. This molecule, and the $\cdot OH$ radicals derived from it, can further damage the tissues. Superoxide dismutase and catalase prevent reperfusion injuries.

Furthermore, the presence of superoxide activates uncoupling proteins (Chemical Connections 27A), allowing the leakage of H^+ ions across the mitochondrial membrane. These H^+ ions then serve as substrates for superoxide dismutase in forming hydrogen peroxide from superoxide.

One theory that has been advanced to explain why humans and animals show symptoms of old age suggests that these oxidizing agents damage healthy cells over the years. All the enzymes involved in reactions with highly reactive forms of oxygen contain a heavy metal: iron in the case of P-450 and catalase, and copper and/or zinc in the case of superoxide dismutase. If the theory is correct, then these protecting agents keep us alive for many years, but the oxidizing processes do so much damage that we eventually exhibit symptoms of aging and die.

---

mitochondrion stream inward, the enzyme manufactures ATP and stores the electrical energy (due to the flow of charges) in the form of chemical energy. In the reverse reaction, the enzyme hydrolyzes ATP and, as a consequence, pumps out H^+ from the mitochondrion. Each pair of protons that is translocated gives rise to the formation of one ATP molecule. Only when the two parts of the proton-translocating ATPase F_1 and F_0 are linked is energy production possible. When the interaction between F_1 and F_0 is disrupted, the energy transduction is lost.

The protons that enter a mitochondrion combine with the electrons transported through the electron transport chain and with oxygen to form water. The net result of the two processes (electron/H^+ transport and ATP formation) is that each oxygen molecule we inhale combines with four H^+ ions and four electrons to give two water molecules. The four H^+ ions and four electrons come from the NADH and $FADH_2$ molecules produced in the citric acid cycle. The oxygen, therefore, has two functions:

- It oxidizes NADH to NAD^+ and $FADH_2$ to FAD so that these molecules can go back and participate in the citric acid cycle.
- It provides energy for the conversion of ADP to ATP.

The latter function is accomplished indirectly, not through the reduction of O_2 to H_2O. The entrance of the H^+ ions into the mitochondrion drives the

ATP formation, but the H^+ ions enter the mitochondrion because the O_2 depleted the H^+ ion concentration when water was formed. This rather complex process involves the transport of electrons along a whole series of enzyme molecules (which catalyze all these reactions); however, the cell cannot utilize the O_2 molecules without it and eventually will die.

The electron and H^+ transport chain and subsequent phosphorylation process are collectively known as oxidative phosphorylation. The following equations represent the overall reactions in oxidative phosphorylation:

$$NADH + 3\ ADP + \tfrac{1}{2}O_2 + 3P_i + H^+ \longrightarrow$$
$$NAD^+ + 3\ ATP + H_2O \qquad \text{(Eq. 27.2)}$$

$$FADH_2 + 2\ ADP + \tfrac{1}{2}O_2 + 2P_i \longrightarrow FAD + 2\ ATP + H_2O \quad \text{(Eq. 27.3)}$$

GOB
Chemistry ⚛ Now ™
Click *Chemistry Interactive* to learn more about the **Mechanism of ATP Synthesis**

27.7 | What Is the Energy Yield Resulting from Electron and H⁺ Transport?

The energy released during electron transport is finally captured in the chemical energy of the ATP molecule. Therefore, it is instructive to look at the energy yield in the universal biochemical currency: the number of ATP molecules.

Each pair of protons entering a mitochondrion results in the production of one ATP molecule. For each NADH molecule, three pairs of protons are pumped into the intermembrane space in the electron transport process. Therefore, for each NADH molecule, we get three ATP molecules, as can be seen in Equation 27.2. For each $FADH_2$ molecule, only four protons are pumped out of the mitochondrion. Therefore, only two ATP molecules are produced for each, as seen in Equation 27.3. Note that ATP production is reported to the nearest whole number. The process is complex, and these numbers represent the least complicated way of looking at it.

Now we can produce the energy balance for the entire common catabolic pathway (citric acid cycle and oxidative phosphorylation combined). For each C_2 fragment entering the citric acid cycle, we obtain three NADH and one $FADH_2$ (Equation 27.1) plus one GTP, which is equivalent in energy to one ATP. Thus the total number of ATP molecules produced per C_2 fragment is

$$
\begin{aligned}
3\ NADH \times 3\ ATP/NADH &= \ 9\ ATP \\
1\ FADH_2 \times 2\ ATP/FADH_2 &= \ 2\ ATP \\
1\ GTP &= \ \underline{1\ ATP} \\
&= 12\ ATP
\end{aligned}
$$

Each C_2 fragment that enters the cycle produces 12 ATP molecules and uses up two O_2 molecules. The total effect of the energy-production chain of reactions discussed in this chapter (the common catabolic pathway) is to oxidize one C_2 fragment with two molecules of O_2 to produce two molecules of CO_2 and 12 molecules of ATP

$$C_2 + 2O_2 + 12\ ADP + 12P_i \longrightarrow 12\ ATP + 2CO_2$$

The important thing is not the waste product, CO_2, but rather the 12 ATP molecules. These molecules will now release their energy when they are converted to ADP.

27.8	How Is Chemical Energy Converted to Other Forms of Energy?

As mentioned in Section 27.3, the storage of chemical energy in the form of ATP lasts only a short time. Usually, within a minute, the ATP is hydrolyzed (an exothermic reaction) and releases its chemical energy. How does the body use this chemical energy? To answer this question, let us look at the different forms in which energy is needed in the body.

A. Conversion to Other Forms of Chemical Energy

The activity of many enzymes is controlled and regulated by phosphorylation. For example, the enzyme phosphorylase, which catalyzes the breakdown of glycogen (Chemical Connections 28B), occurs in an inactive form, phosphorylase *b*. When ATP transfers a phosphate group to a serine residue, the enzyme becomes active. Thus the chemical energy of ATP is used in the form of chemical energy to activate phosphorylase *b* so that glycogen can be utilized. We will see several other examples of this energy conversion in Chapters 28 and 29.

B. Electrical Energy

The body maintains a high concentration of K^+ ions inside the cells despite the fact that the K^+ concentration is low outside the cells. The reverse is true for Na^+. So that K^+ does not diffuse out of the cells and Na^+ does not enter them, special transport proteins in the cell membranes constantly pump K^+ into and Na^+ out of the cells. This pumping requires energy, which is supplied by the hydrolysis of ATP to ADP. Because of this pumping, the charges inside and outside the cell are unequal, which generates an electric potential. Thus the chemical energy of ATP is transformed into electrical energy, which operates in neurotransmission (Section 24.2).

C. Mechanical Energy

ATP is the immediate source of energy in muscle contraction. In essence, muscle contraction takes place when thick and thin filaments slide past each other (Figure 27.11). The thick filament is myosin, an ATPase enzyme (that is, one that hydrolyzes ATP). The thin filament, actin, binds strongly to myosin in the contracted state. However, when ATP binds to myosin, the actin–myosin complex dissociates, and the muscle relaxes. When myosin hydrolyzes ATP, it interacts with actin once more, and a new contraction occurs. In this way, the hydrolysis of ATP drives the alternating association and dissociation of actin and myosin and, consequently, the contraction and relaxation of the muscle.

D. Heat Energy

One molecule of ATP upon hydrolysis to ADP yields 7.3 kcal/mol. Some of this energy is released as heat and used to maintain body temperature. If we estimate that the specific heat of the body is about the same as that of water, a person weighing 60 kg would need to hydrolyze approximately 99 moles (approximately 50 kg) of ATP to raise the temperature of the body from room temperature, 25°C, to 37°C. Not all body heat is derived from ATP hydrolysis; some other exothermic reactions in the body also make heat contributions.

Thick filament
(myosin)

Thin filament
(actin)

Figure 27.11 Schematic diagram
of muscle contraction.

(*a*) Relaxed muscle

(*b*) Contracted muscle

SUMMARY OF KEY QUESTIONS

SECTION 27.1 What Is Metabolism?

- The sum total of all the chemical reactions involved in maintaining the dynamic state of cells is called **metabolism.**
- The breaking down of molecules is **catabolism;** the building up of molecules is **anabolism.**

SECTION 27.2 What Are Mitochondria, and What Role Do They Play in Metabolism?

- Many metabolic activities in cells take place in specialized structures called **organelles.**
- **Mitochondria** are the organelles in which the reactions of the **common catabolic pathway** take place.

SECTION 27.3 What Are the Principal Compounds of the Common Metabolic Pathway?

- The common metabolic pathway oxidizes a two-carbon C_2 fragment (acetyl) from different foods. The products of C_2 oxidation are water and carbon dioxide.
- The energy from oxidation is built into the high-chemical-energy-storing molecule **ATP.** As the C_2 fragments are oxidized, protons (H^+) and electrons are released and passed along to carriers.
- The principal carriers in the common catabolic pathway are as follows: ATP is a phosphate carrier; **CoA** is the C_2 fragment carrier; and **NAD^+** and **FAD** carry the hydrogen ions (protons) and electrons. The unit common to all of these carriers is **ADP.** The nonactive end of the carriers acts as a handle that fits into the active sites of the enzymes.

SECTION 27.4 What Role Does the Citric Acid Cycle Play in Metabolism?

- In the **citric acid cycle,** the C_2 fragment first combines with a C_4 fragment (oxaloacetate) to yield a C_6 fragment (citrate). An oxidative decarboxylation yields a C_5 fragment. One CO_2 is released, and one $NADH + H^+$ is passed to the **electron transport chain** for further oxidation.
- Another oxidative decarboxylation provides a C_4 fragment. Once again, a CO_2 is released, and another $NADH + H^+$ is passed to the electron transport chain.
- The enzymes of the citric acid cycle are located in the mitochondrial matrix. Control of this cycle takes place by a feedback mechanism.

SECTION 27.5 How Do Electron and H^+ Transport Take Place?

- The electrons of NADH enter the electron transport chain at the complex I stage. The coenzyme Q (CoQ) of this complex picks up the electrons and the H^+ and becomes $CoQH_2$. The energy of this reduction reaction is used to expel two H^+ ions from the matrix into the intermembrane space.
- Complex II also has CoQ. Electrons and H^+ are passed to this complex, and complex II catalyzes the transfer of electrons from $FADH_2$. However, no H^+ ions are pumped into the intermembrane space at this point.
- Electrons are passed along by $CoQH_2$ to complex III of the electron transport chain. At complex III, the two H^+ ions from the $CoQH_2$ are expelled into the intermembrane space. Cytochrome c of complex III transfers electrons to complex IV through redox reactions.
- As the electrons are transported from cytochrome c to complex IV, two more H^+ ions are expelled from the matrix of the mitochondrion to the intermembrane space.
- For each NADH, six H^+ ions are expelled. For each $FADH_2$, four H^+ ions are expelled.

- The electrons passed to complex IV return to the matrix, where they combine with oxygen and H^+ to form water.

SECTION 27.6 What Is the Role of the Chemiosmotic Pump in ATP Production?

- Both the citric acid cycle and **oxidative phosphorylation** take place in the mitochondria. The enzymes of the citric acid cycle are found in the mitochondrial matrix, whereas the enzymes of the electron transport chain and oxidative phosphorylation are located on the inner mitochondrial membrane. Some of them project into the intermembrane space.
- When the H^+ ions expelled by electron transport stream back into the mitochondrion, they drive a complex enzyme called **proton-translocating ATPase,** which makes one ATP molecule for each two H^+ ions that enter the mitochondrion.
- The proton-translocating ATPase is a complex "rotor engine." The proton channel part (F_0) is embedded in the membrane, and the catalytic unit (F_1) converts mechanical energy to chemical energy of the ATP molecule.

SECTION 27.7 What Is the Energy Yield Resulting from Electron and H^+ Transport?

- For each NADH + H^+ coming from the citric acid cycle, three ATP molecules are formed. For each $FADH_2$, two ATP molecules are formed. The overall result: For each C_2 fragment that enters the citric acid cycle, 12 ATP molecules are produced.

SECTION 27.8 How Is Chemical Energy Converted to Other Forms of Energy?

- Chemical energy is stored in ATP only for a short time—ATP is quickly hydrolyzed, usually within a minute.
- This chemical energy is used to do chemical, mechanical, and electrical work in the body and to maintain body temperature.

PROBLEMS

GOB
Chemistry Now™

Assess your understanding of this chapter's topics with additional quizzing and conceptual-based problems at **http://now.brookscole.com/gob8** or on the CD.

A blue problem number indicates an applied problem.

■ denotes problems that are available on the GOB ChemistryNow website or CD and are assignable in OWL.

SECTION 27.1 What Is Metabolism?

27.1   To what end product is the energy of foods converted in the catabolic pathways?

27.2   ■ (a) How many sequences are in the common catabolic pathway?
        (b) Name these sequences.

SECTION 27.2 What Are Mitochondria, and What Role Do They Play in Metabolism?

27.3   (a) How many membranes do mitochondria have?
        (b) Which membrane is permeable to ions and small molecules?

27.4   How do the enzymes of the common pathway find their way into the mitochondria?

27.5   What are cristae, and how are they related to the inner membrane of mitochondria?

27.6   (a) Where are the enzymes of the citric acid cycle located?
        (b) Where are the enzymes of oxidative phosphorylation located?

SECTION 27.3 What Are the Principal Compounds of the Common Metabolic Pathway?

27.7   How many high-energy phosphate bonds are in the ATP molecule?

27.8   What are the products of the following reaction? Complete the equation.

$$AMP + H_2O \xrightarrow{H^+}$$

27.9   Which yields more energy, (a) the hydrolysis of ATP to ADP or (b) the hydrolysis of ADP to AMP?

27.10 How much ATP is needed for normal daily activity in humans?

27.11 What kind of chemical bond exists between the ribitol and the phosphate group in FAD?

27.12 ■ When NAD^+ is reduced, two electrons enter the molecule, together with one H^+ ion. Where in the product will the two electrons be located?

27.13 Which atoms in the flavin portion of FAD are reduced to yield $FADH_2$?

27.14 NAD^+ has two ribose units in its structure; FAD has a ribose and a ribitol. What is the relationship between these molecules?

27.15 ■ In the common catabolic pathway, a number of important molecules act as carriers (transfer agents).
        (a) Which is the carrier of phosphate groups?
        (b) Which are the coenzymes transferring hydrogen ions and electrons?
        (c) What kind of groups does coenzyme A carry?

27.16 The ribitol in FAD is bound to phosphate. What is the nature of this bond? On the basis of the energies of the different bonds in ATP, estimate how much energy (in kcal/mol) would be obtained from the hydrolysis of this bond.

27.17 What kind of chemical bond exists between the pantothenic acid and mercaptoethylamine in the structure of CoA?

27.18 ■ Name the vitamin B molecules that are part of the structure of (a) NAD$^+$, (b) FAD, and (c) coenzyme A.

27.19 In both NAD$^+$ and FAD, the vitamin B portion of the molecule is the active part. Is this also true for CoA?

27.20 What type of compound is formed when coenzyme A reacts with acetate?

27.21 The fats and carbohydrates metabolized by our bodies are eventually converted to a single compound. What is it?

SECTION 27.4 What Role Does the Citric Acid Cycle Play in Metabolism?

27.22 The first step in the citric acid cycle is abbreviated as

$$C_2 + C_4 = C_6$$

(a) What do these symbols stand for?
(b) What are the common names of the three compounds involved in this reaction?

27.23 What is the only C_5 compound in the citric acid cycle?

27.24 Identify by number those steps of the citric acid cycle that are not redox reactions.

27.25 Which substrate in the citric acid cycle is oxidized by FAD? What is the oxidation product?

27.26 In Steps ③ and ⑤ of the citric acid cycle, the compounds are shortened by one carbon each time. What is the form of this one-carbon compound? What happens to it in the body?

27.27 According to Table 23.1, to what class of enzymes does fumarase belong?

27.28 ■ List all the enzymes or enzyme systems of the citric acid cycle that could be classified as oxidoreductases.

27.29 Is ATP directly produced during any step of the citric acid cycle? Explain.

27.30 ■ There are four dicarboxylic acid compounds, each containing four carbons, in the citric acid cycle. Which is (a) the least oxidized and (b) the most oxidized?

27.31 Why is a many-step cyclic process more efficient in utilizing energy from food than a single-step combustion?

27.32 Did the two CO_2 molecules given off in one turn of the citric acid cycle originate from the entering acetyl group?

27.33 Which intermediates of the citric acid cycle contain C=C double bonds?

27.34 The citric acid cycle can be regulated by the body; that is, it can be slowed down or speeded up. What mechanism controls this process?

27.35 Oxidation is defined as loss of electrons. When oxidative decarboxylation occurs, as in Step ④, of the citric acid cycle, where do the electrons of the α-ketoglutarate go?

SECTION 27.5 How Do Electron and H$^+$ Transport Take Place?

27.36 What is the main function of oxidative phosphorylation (the electron transport chain)?

27.37 What are the mobile electron carriers of the oxidative phosphorylation?

27.38 In each complex of the electron transport system, the redox reaction occurs around Fe ions.
(a) Identify the compounds that contain such Fe centers.
(b) Identify the compounds that contain ion centers other than iron.

27.39 What kind of motion is set up in the proton-translocating ATPase by the passage of H$^+$ from the intermembrane space into the matrix?

27.40 The following reaction is a reversible reaction:

$$NADH \rightleftharpoons NAD^+ + H^+ + 2e^-$$

(a) Where does the forward reaction occur in the common catabolic pathway?
(b) Where does the reverse reaction occur?

27.41 In oxidative phosphorylation, water is formed from H$^+$, e$^-$, and O_2. Where does this take place?

27.42 At what points in oxidative phosphorylation are the H$^+$ ions and the electrons separated from each other?

27.43 How many ATP molecules are generated (a) for each H$^+$ translocated through the ATPase complex and (b) for each C_2 fragment that goes through the complete common catabolic pathway?

27.44 ■ When H$^+$ is pumped out into the intermembrane space, is the pH there increased, decreased, or unchanged compared with that in the matrix?

SECTION 27.6 What Is the Role of the Chemiosmotic Pump in ATP Production?

27.45 What is the channel through which ions reenter the matrix of mitochondria?

27.46 The proton gradient accumulated in the intermembrane area of a mitochondrion drives the ATP manufacturing enzyme, ATPase. Why do you think Mitchell called this concept the "chemiosmotic theory"?

27.47 Which part of the proton-translocating ATPase machinery is the catalytic unit? What chemical reaction does it catalyze?

27.48 When the interaction between the two parts of proton-translocating ATPase, F_0 and F_1, are disrupted, no energy production is possible. Which subunits maintain connections between F_0 and F_1, and what names are designated for these subunits?

SECTION 27.7 What Is the Energy Yield Resulting from Electron and H^+ Transport?

27.49 If each mole of ATP yields 7.3 kcal of energy upon hydrolysis, how many kilocalories of energy would you get from 1 g of CH_3COO^- entering the citric acid cycle?

27.50 ■ A hexose (C_6) enters the common metabolic pathway in the form of two C_2 fragments.

    (a) How many molecules of ATP are produced from one hexose molecule?

    (b) How many O_2 molecules are used up in the process?

SECTION 27.8 How Is Chemical Energy Converted to Other Forms of Energy?

27.51 (a) How do muscles contract?

    (b) Where does the energy in muscle contraction come from?

27.52 Give an example of the conversion of the chemical energy of ATP to electrical energy.

27.53 How is the enzyme phosphorylase activated?

Chemical Connections

27.54 (Chemical Connections 27A) What is a protonophore?

27.55 (Chemical Connections 27A) Oligomycin is an antibiotic that allows electron transport to continue but stops phosphorylation in both bacteria and humans. Would you use it as an antibacterial drug for people? Explain.

27.56 (Chemical Connections 27B) How does superoxide dismutase prevent damage to the brain during a stroke?

27.57 (Chemical Connections 27B)

    (a) What kind of reaction is catalyzed by cytochrome P-450?

    (b) Where does the oxygen reactant come from?

Additional Problems

27.58 (a) What is the difference in structure between ATP and GTP?

    (b) Compared with ATP, would you expect GTP to carry more, less, or about the same amount of energy?

27.59 How many grams of CH_3COOH (from acetyl CoA) molecules must be metabolized in the common metabolic pathway to yield 87.6 kcal of energy?

27.60 What is the basic difference in the functional groups between citrate and isocitrate?

27.61 The passage of ions from the cytoplasmic side into the matrix generates mechanical energy. Where is this energy of motion exhibited first?

27.62 What kind of reaction occurs in the citric acid cycle when a C_6 compound is converted to a C_5 compound?

27.63 What structural characteristics do citric acid and malic acid have in common?

27.64 Two ketoacids are important in the citric acid cycle. Identify them, and tell how they are manufactured.

27.65 Which filament of muscles is an enzyme, catalyzing the reaction that converts ATP to ADP?

27.66 One of the end products of food metabolism is water. How many molecules of H_2O are formed from the entry of each molecule of (a) NADH + H^+ and (b) $FADH_2$? (*Hint:* Use Figure 27.10.)

27.67 How many stereocenters are in isocitrate?

27.68 Acetyl CoA is labeled with radioactive carbon as shown: CH_3^*CO—S—CoA. This compound enters the citric acid cycle. If the cycle is allowed to progress to only the α-ketoglutarate level, will the CO_2 expelled by the cell be radioactive?

27.69 Where is the H^+ ion channel located in the proton-translocating ATPase complex?

27.70 Is the passage of H^+ ion through the channel converted directly into chemical energy?

27.71 Does all the energy used in ATP synthesis come from the mechanical energy of rotation?

27.72 (a) In the citric acid cycle, how many steps can be classified as decarboxylation reactions?

    (b) In each case, what is the concurrent oxidizing agent? (*Hint:* See Table 23.1.)

27.73 What is the role of succinate dehydrogenase in the citric acid cycle?

27.74 How many stereocenters are in malate?

Tying It Together

27.75 Why does citrate isomerize to isocitrate before any oxidation steps take place in the citric acid cycle?

27.76 Why is the material in this chapter called the common catabolic pathway, rather than giving that designation to any other metabolic reactions?

27.77 What are two ways in which iron is part of the structure of the proteins of the electron transport pathway?

27.78 Why is it necessary for the proteins of the electron transport chain to be integral membrane proteins?

27.79 Why is it necessary to have mobile electron carriers as part of the electron transport chain?

27.80 Why does the loss of CO_2 make the citric acid cycle irreversible?

Looking Ahead

27.81 Why is the citric acid cycle central to biosynthetic pathways as well as to catabolism?

27.82 Is there a significant difference in the energy yield of the central catabolic pathway if FAD is used as an electron carrier rather than NAD^+?

27.83 Are biosynthetic pathways likely to involve oxidation, like the common catabolic pathway, or reduction? Why?

27.84 Are biosynthetic pathways likely to release energy, like the common catabolic pathway, or require energy? Why?

Challenge Problems

27.85 In a typical human, body weight fluctuates very little during the course of a day. How can this statement be consistent with the estimate that the human body manufactures as much as 40 kg of ATP every day?

27.86 When the electron transport pathway was first studied, researchers used inhibitors to block the flow of electrons in their work. Why is it likely that such inhibitors could help establish the order of carriers?

27.87 Oxygen does not appear in any of the reactions of the citric acid cycle, but it is considered part of aerobic metabolism. Why?

27.88 Is it likely that some of the important molecules for transfer of phosphate groups, electrons, and acetyl groups will appear in other metabolic pathways discussed in future chapters?

CHAPTER 28

Specific Catabolic Pathways: Carbohydrate, Lipid, and Protein Metabolism

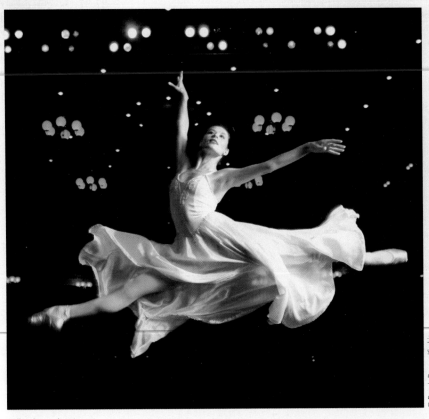

Ballet dancer leaping.

© Dennis Degnan/Corbis

GOB
Chemistry ⚛ Now™
Look for this logo in the chapter and go to GOB ChemistryNow at
http://now.brookscole.com/gob8
or on the CD to view tutorials, simulations, and problems.

28.1 | What Is the General Outline of Catabolic Pathways?

The food we eat serves two main purposes: (1) It fulfills our energy needs, and (2) it provides the raw materials to build the compounds our bodies need. Before either of these processes can take place, food—carbohydrates, fats, and proteins—must be broken down into small molecules that can be absorbed through the intestinal walls.

A. Carbohydrates

Complex carbohydrates (di- and polysaccharides) in the diet are broken down by enzymes and stomach acid to produce monosaccharides, the most important of which is glucose (Section 30.3). Glucose also comes from the enzymatic breakdown of glycogen that is stored in the liver and muscles until needed. Once monosaccharides are produced, they can be used either to build new oligo- and polysaccharides or to provide energy. The specific pathway by which energy is extracted from monosaccharides is called glycolysis (Sections 28.2 and 28.3).

B. Lipids

Ingested fats are hydrolyzed by lipases to glycerol and fatty acids or to monoglycerides, which are absorbed through the intestine (Section 30.4). In a similar fashion, complex lipids are hydrolyzed to smaller units before their absorption. As with carbohydrates, these smaller molecules (fatty acids, glycerol, and so on) can be used to build the complex molecules needed in membranes; they can be oxidized to provide energy; or they can be stored in **fat storage depots** (Figure 28.1). The stored fats can later be hydrolyzed to glycerol and fatty acids whenever they are needed as fuel.

The specific pathway by which energy is extracted from glycerol involves the same glycolysis pathway as that used for carbohydrates (Section 28.4). The specific pathway used by the cells to obtain energy from fatty acids is called β-oxidation (Section 28.5).

C. Proteins

As you might expect from your knowledge of their structures, proteins are hydrolyzed by HCl in the stomach and by digestive enzymes in the stomach (pepsin) and intestines (trypsin, chymotrypsin, and carboxypeptidases) to produce their constituent amino acids. The amino acids absorbed through the intestinal wall enter the **amino acid pool.** They serve as building blocks for proteins as needed and, to a smaller extent (especially during starvation), as a fuel for energy. In the latter case, the nitrogen of the amino acids is catabolized through oxidative deamination and the urea cycle and is expelled from the body as urea in the urine (Section 28.8). The carbon skeletons of the amino acids enter the common catabolic pathway (Chapter 27) as either α-ketoacids (pyruvic, oxaloacetic, and α-ketoglutaric acids) or acetyl coenzyme A (Section 28.9).

In all cases, *the specific pathways of carbohydrate, triglyceride (fat), and protein catabolism converge to the common catabolic pathway* (Figure 28.2). In this way, the body needs fewer enzymes to get energy from diverse food materials. Efficiency is achieved because a minimal number of chemical steps are required and because the energy-producing factories of the body are localized in the mitochondria.

Figure 28.1 Storage of fat in a fat cell. As more and more fat droplets accumulate in the cytoplasm, they coalesce to form a very large globule of fat. Such a fat globule may occupy most of the cell, pushing the cytoplasm and the organelles to the periphery. (Modified from C. A. Villee, E. P. Solomon, and P. W. Davis, *Biology,* Philadelphia, Saunders College Publishing, 1985.)

Amino acid pool The free amino acids found both inside and outside cells throughout the body

28.2 | What Are the Reactions of Glycolysis?

Glycolysis is the specific pathway by which the body gets energy from monosaccharides. The detailed steps in glycolysis are shown in Figure 28.3, and the most important features are shown schematically in Figure 28.4.

A. Glycolysis of Glucose

In the first steps of glucose metabolism, energy is consumed rather than released. At the expense of two molecules of ATP (which are converted to ADP), glucose is phosphorylated. First, glucose 6-phosphate is formed in Step ①, then, after isomerization to fructose 6-phosphate in Step ②, a second phosphate group is bonded to yield fructose 1,6-bisphosphate in Step ③. We can consider these steps to be the activation process.

Glycolysis The biochemical pathway that breaks down glucose to pyruvate, which yields chemical energy in the form of ATP and reduced coenzymes

Figure 28.2 The convergence of the specific pathways of carbohydrate, fat, and protein catabolism into the common catabolic pathway, which is made up of the citric acid cycle and oxidative phosphorylation.

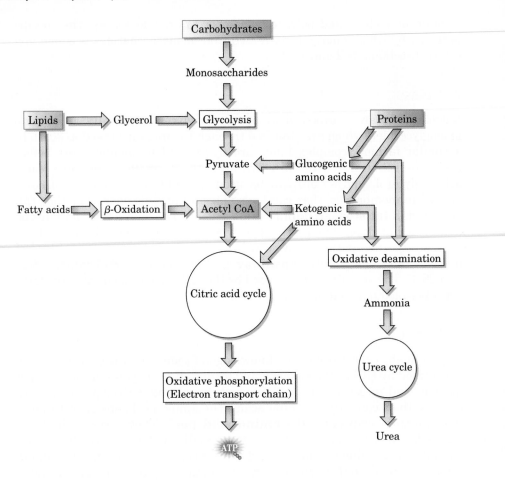

In the second stage, the C_6 compound, fructose 1,6-bisphosphate, is broken into two C_3 fragments in Step ④. The two C_3 fragments, glyceraldehyde 3-phosphate and dihydroxyacetone phosphate, are in equilibrium (they can be converted to each other). Only glyceraldehyde 3-phosphate is oxidized in glycolysis, but as this species is removed from the equilibrium mixture, the equilibrium shifts (see the discussion of Le Chatelier's principle in Section 8.8) and dihydroxyacetone phosphate is converted to glyceraldehyde 3-phosphate.

In the third stage, glyceraldehyde 3-phosphate is oxidized to 1,3-bisphosphoglycerate in Step ⑤. The hydrogen of the aldehyde group is removed by the NAD^+ coenzyme. In Step ⑥, the phosphate from the carboxyl group is transferred to ADP, yielding ATP and 3-phosphoglycerate. The latter compound, after isomerization in Step ⑦ and dehydration in Step ⑧, is converted to phosphoenolpyruvate, which loses its remaining phosphate in Step ⑨ and yields pyruvate and another ATP molecule. [In Step ⑨, after hydrolysis of the phosphate, the resulting enol of pyruvic acid tautomerizes to the more stable keto form (Section 17.5).] Step ⑨ is also the "payoff" step, as the two ATP molecules produced here (one for each C_3 fragment) represent the net yield of ATPs in glycolysis. Step ⑨ is catalyzed by an enzyme, pyruvate kinase, whose active site was depicted in Chemical Connections 23C. This enzyme plays a key role in the regulation of glycolysis. For example, pyruvate kinase is inhibited by ATP and activated by AMP. Thus, when plenty of ATP is available, glycolysis is shut down; when ATP is scarce and AMP levels are high, the glycolytic pathway is speeded up.

All of these glycolysis reactions occur in the cytoplasm outside the mitochondria. Because they occur in the absence of O_2, they are also called

Figure 28.3   Glycolysis, the pathway of glucose metabolism. (Steps ⑩, ⑫, and ⑬ are shown in Figure 28.4.) Some of the steps are reversible, but equilibrium arrows are not shown (they appear in Figure 28.4).

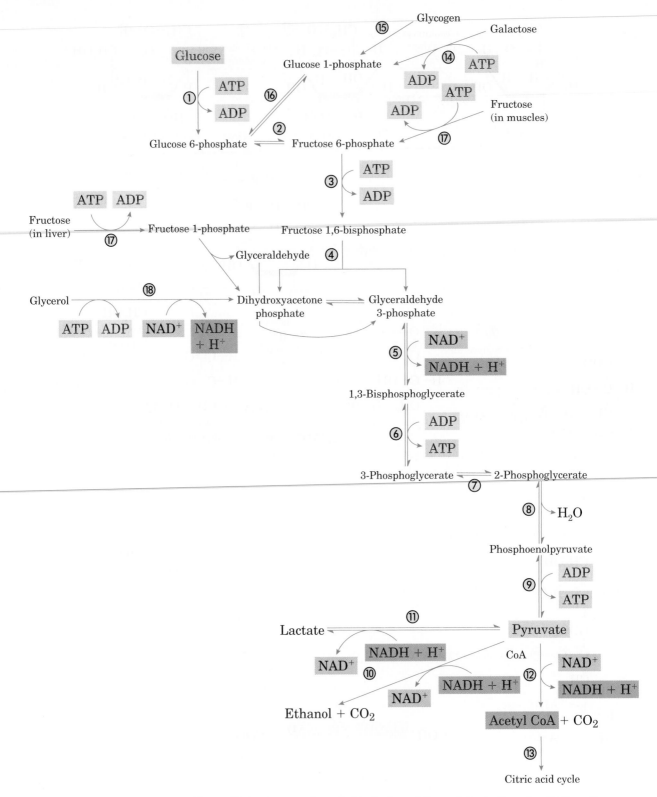

Figure 28.4 An overview of glycolysis and the entries to it and exits from it.

reactions of the **anaerobic pathway.** As indicated in Figure 28.4, the end product of glycolysis, pyruvate, does not accumulate in the body. In certain bacteria and yeast, pyruvate undergoes decarboxylation in Step ⑩ to produce ethanol. In some bacteria, and in mammals in the absence of oxygen, pyruvate is reduced to lactate in Step ⑪. The reactions that produce ethanol

CHEMICAL CONNECTIONS 28A

Lactate Accumulation

Many athletes suffer muscle cramps when they engage in strenuous exercise (see Chapter 9). This problem results from a shift from normal glucose catabolism (glycolysis : citric acid cycle : oxidative phosphorylation) to that of lactate production (see Step ⑪ in Figure 28.4). During exercise, oxygen is used up rapidly, which slows down the rate of the common catabolic pathway. The demand for energy makes anaerobic glycolysis proceed at a high rate, but because the aerobic (oxygen-demanding) pathways are slowed down, not all the pyruvate produced in glycolysis can enter the citric acid cycle. The excess pyruvate ends up as lactate, which causes painful muscle contractions.

The same shift in catabolism occurs in heart muscle when coronary thrombosis leads to cardiac arrest. The blockage of the artery

leading to the heart muscles cuts off the oxygen supply. The common catabolic pathway and its ATP production are consequently shut off. Glycolysis proceeds at an accelerated rate, causing lactate to accumulate. The heart muscle contracts, producing a cramp. Just as in skeletal muscle, massage of heart muscles can relieve the cramp and start the heart beating. Even if heartbeat is restored within 3 minutes (the amount of time the brain can survive without being damaged), acidosis may develop as a result of the cardiac arrest. Therefore, at the same time that efforts are under way to start the heart beating by chemical, physical, or electrical means, an intravenous infusion of 8.4% bicarbonate solution is given to combat acidosis.

in organisms capable of alcoholic fermentation operate in reverse when humans metabolize ethanol. Acetaldehyde (Section 17.2), which is a product of one of these reactions, is a toxic substance that's responsible for much of the damage in fetal alcohol syndrome. Transfer of nutrients and oxygen to the fetus is depressed, with tragic consequences.

B. Entrance to the Citric Acid Cycle

Pyruvate is not the end product of glucose metabolism. Most importantly, pyruvate goes through an oxidative decarboxylation in the presence of coenzyme A in Step ⑫ to produce acetyl CoA:

$$NAD^+ + CH_3 - \overset{\overset{O}{\|}}{C} - COO^- + CoA - SH \longrightarrow CH_3 - \overset{\overset{O}{\|}}{C} - S - CoA + CO_2 + NADH + H^+$$

Pyruvate   Acetyl coenzyme A

This reaction is catalyzed by a complex enzyme system, pyruvate dehydrogenase, that sits on the inner membrane of the mitochondrion. The reaction produces acetyl CoA, CO_2, and NADH + H^+. The acetyl CoA then enters the citric acid cycle in Step ⑬ and goes through the common catabolic pathway.

In summary, after converting complex carbohydrates to glucose, the body gets energy from glucose by converting it to acetyl CoA (by way of pyruvate) and then using the acetyl CoA as a starting material for the common catabolic pathway.

C. Pentose Phosphate Pathway

As we saw in Figure 28.4, glucose 6-phosphate plays a central role in several different entries into the glycolytic pathway. However, glucose 6-phosphate can also be used by the body for other purposes, not just for the production of energy in the form of ATP. Most importantly, glucose 6-phosphate can be shunted to the **pentose phosphate pathway** in Step ⑲ (Figure 28.5). This pathway has the capacity to produce NADPH and ribose in Step ⑳ as well as energy.

NADPH is needed in many biosynthetic processes, including synthesis of unsaturated fatty acids (Section 29.3), cholesterol, and amino acids as well as photosynthesis (Chemical Connections 29A) and the reduction of

Pentose phosphate pathway The biochemical pathway that produces ribose and NADPH from glucose 6-phosphate or, alternatively, releases energy

Figure 28.5 Simplified schematic representation of the pentose phosphate pathway, also called a shunt.

α-D-Glucose 6-phosphate   Ribulose 5-phosphate   Ribose 5-phosphate

Glucose 6-phosphate $+$ 2NADP$^+$ $\xrightarrow{\text{(19)}}$ Ribulose 5-phosphate $+$ 2NADPH $+$ CO$_2$

(20)

Ribose 5-phosphate

(21)

Glyceraldehyde 3-phosphate

Nicotinamide adenine dinucleotide phosphate (NADP$^+$)

ribose to deoxyribose for DNA. Ribose is needed for the synthesis of RNA (Section 25.3). Therefore, when the body needs these synthetic ingredients more than energy, glucose 6-phosphate is used in the pentose phosphate pathway. When energy is needed, glucose 6-phosphate remains in the glycolytic pathway—even ribose 5-phosphate can be channeled back to glycolysis through glyceraldehyde 3-phosphate. Through this reversible reaction, the cells can also obtain ribose directly from the glycolytic intermediates. In addition, NADPH is badly needed in red blood cells as a defense against oxidative damages. Glutathione (Chemical Connections 22A) is the main agent used to keep the hemoglobin in its reduced form. It is regenerated by NADPH, so an insufficient supply of NADPH leads to the destruction of red blood cells, causing severe anemia.

28.3 | What Is the Energy Yield from Glucose Catabolism?

Using Figure 28.4, let us sum up the energy derived from glucose catabolism in terms of ATP production. First, however, we must take into account the fact that glycolysis takes place in the cytoplasm, whereas oxidative phosphorylation occurs in the mitochondria. Therefore, the NADH + H$^+$ produced in glycolysis in the cytoplasm must be converted to NADH in the mitochondria before it can be used in oxidative phosphorylation.

NADH is too large to cross the mitochondrial membrane. Two routes are available to get the electrons into the mitochondria; they have different efficiencies. In glycerol 3-phosphate transport, which operates in muscle and nerve cells, only two ATP molecules are produced for each NADH + H$^+$. In the other transport route, which operates in the heart and the liver, three ATP molecules are produced for each NADH + H$^+$ produced in the cytoplasm, just as is the case in the mitochondria (Section 27.7). Because most energy production takes place in skeletal muscle cells, when we construct the energy balance sheet, we use two ATP molecules for each NADH + H$^+$

Table 28.1 ATP Yield from Complete Glucose Metabolism

Step Numbers in Figure 28.4	Chemical Steps	Number of ATP Molecules Produced
① ② ③	Activation (glucose ⟶ 1,6-fructose bisphosphate)	−2
⑤	Phosphorylation 2(glyceraldehyde 3-phosphate ⟶ 1,3-bisphosphoglycerate), producing 2(NADH + H⁺) in cytosol	4
⑥ ⑨	Dephosphorylation 2(1,3-bisphosphoglycerate ⟶ pyruvate)	4
⑫	Oxidative decarboxylation 2(pyruvate ⟶ acetyl CoA), producing 2(NADH + H⁺) in mitochondrion	6
⑬	Oxidation of two C_2 fragments in citric acid cycle and oxidative phosphorylation common pathways, producing 12 ATP for each C_2 fragment	24
	Total	36

produced in the cytoplasm. (Muscles attached to bones are called skeletal muscles; cardiac muscle is in a different category.)

Armed with this knowledge, we are ready to calculate the energy yield of glucose in terms of ATP molecules produced in skeletal muscles. Table 28.1 shows this calculation. In the first stage of glycolysis (Steps ①, ②, and ③), two ATP molecules are used up, but this loss is more than compensated for by the production of 14 ATP molecules in Steps ⑤, ⑥, ⑦, and ⑫, and in the conversion of pyruvate to acetyl CoA. The net yield of these steps is 12 ATP molecules. As we saw in Section 27.7, the oxidation of one acetyl CoA molecule produces 12 ATP molecules, and one glucose molecule provides two acetyl CoA molecules. Therefore, the total net yield from metabolism of one glucose molecule in skeletal muscle is 36 molecules of ATP, or 6 ATP molecules per carbon atom.

$$C_6H_{12}O_6 + 6O_2 \longrightarrow 6CO_2 + 6H_2O$$

If the same glucose is metabolized in the heart or liver, the electrons of the two NADH molecules produced in glycolysis are transported into the mitochondrion by the malate-aspartate shuttle. Through this shuttle, the two NADH molecules yield a total of 6 ATP molecules, so that, in this case, 38 ATP molecules are produced for each glucose molecule. It is instructive to note that most of the energy (in the form of ATP) from glucose is produced in the common metabolic pathway. Recent investigations suggest that approximately 30 to 32 ATP molecules are actually produced per glucose molecule (2.5 ATP/NADH and 1.5 ATP/$FADH_2$). Further research into the complexity of the oxidative phosphorylation pathway is needed to verify these numbers. We alluded to this point briefly in Chapter 27 when we said that the ATP yields were reported to the nearest whole number. Here we see how the complexity of oxidative phosphorylation affects energy metabolism.

Glucose is not the only monosaccharide that can be used as an energy source. Other hexoses, such as galactose (Step ⑭) and fructose (Step ⑰), enter the glycolysis pathway at the stages indicated in Figure 28.4. They also yield 36 molecules of ATP per hexose molecule. Furthermore, glycogen stored in liver and muscle cells and elsewhere can be converted by enzymatic breakdown and phosphorylation to glucose 1-phosphate (Step ⑮). This compound, in turn, isomerizes to glucose 6-phosphate, providing an entry into the glycolytic pathway (Step ⑯). The pathway in which glycogen breaks down to glucose is called **glycogenolysis.**

Glycogenolysis The biochemical pathway for the breakdown of glycogen to glucose

28.4 | How Does Glycerol Catabolism Take Place?

The glycerol hydrolyzed from fats or complex lipids (Chapter 21) can also be a rich energy source. The first step in glycerol utilization is an activation step. The body uses one ATP molecule to form glycerol 1-phosphate, which is the same as glycerol 3-phosphate:

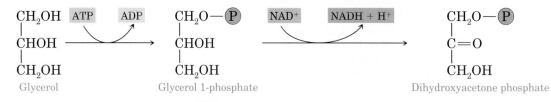

Glycerol   Glycerol 1-phosphate   Dihydroxyacetone phosphate

The glycerol phosphate is oxidized by NAD$^+$ to dihydroxyacetone phosphate, yielding NADH + H$^+$ in the process. Dihydroxyacetone phosphate then enters the glycolysis pathway (Step ⑱ in Figure 28.4) and is isomerized to glyceraldehyde 3-phosphate. A net yield of 20 ATP molecules is produced from each glycerol molecule, or 6.7 ATP molecules per carbon atom.

28.5 | What Are the Reactions of β-Oxidation of Fatty Acids?

As early as 1904, Franz Knoop, working in Germany, proposed that the body utilizes fatty acids as an energy source by breaking them down into fragments. Prior to fragmentation, the β-carbon (the second carbon atom from the COOH group) is oxidized:

$$-C-C-C-\overset{\beta}{C}-\overset{\alpha}{C}-COOH$$

β-oxidation The biochemical pathway that degrades fatty acids to acetyl CoA by removing two carbons at a time and yielding energy

The name **β-oxidation** has its origin in Knoop's prediction. It took about 50 years to establish the mechanism by which fatty acids are utilized as an energy source.

Figure 28.6 depicts the overall process of fatty acid metabolism. As is the case with the other foods we have seen, the first step involves activation. This activation occurs in the cytosol, where the fat was previously hydrolyzed to glycerol and fatty acids. It converts ATP to AMP and inorganic phosphate (Step ①), which is equivalent to the cleavage of two high-energy phosphate bonds. The chemical energy derived from the hydrolysis of ATP is built into the compound acyl CoA, which forms when the fatty acid combines with coenzyme A. The fatty acid oxidation occurs inside the mitochondrion, so the acyl group of acyl CoA must pass through the mitochondrial membrane. Carnitine is the acyl group transporter. The enzyme system that catalyzes the process is carnitine acyltransferase.

Chemistry⚛Now™
Active Figure 28.6 The β-oxidation spiral of fatty acids. Each loop in the spiral contains two dehydrogenations, one hydration, and one fragmentation. At the end of each loop, one acetyl CoA molecule is released. **See a simulation based on this figure, and take a short quiz on the concepts at http://now.brookscole.com/gob8 or on the CD.**

CHEMICAL CONNECTIONS 28B

Effects of Signal Transduction on Metabolism

The ligand-binding/G-protein/adenylate cyclase cascade, which activates proteins by phosphorylation, has a wide range of effects other than opening or closing ion-gated channels (Figure 24.4). An example of such a target is the glycogen phosphorylase enzyme that participates in the breakdown of glycogen stored in muscles. This enzyme cleaves units of glucose 1-phosphate from glycogen, which enters into the glycolytic pathway and yields quick energy (Section 28.2). The active form of the enzyme, phosphorylase *a*, is phosphorylated. When it is dephosphorylated (phosphorylase *b*), it is inactive. When epinephrine signals danger to a muscle cell, phosphorylase is activated through the cascade, and quick energy is produced. In this way, the signal is converted to a metabolic event,

allowing the muscles to contract rapidly and enabling the person in danger to fight or run away.

Not all phosphorylation of enzymes results in activation. Consider glycogen synthase. In this case, the phosphorylated form of the enzyme is inactive, and the dephosphorylated form is active. This enzyme participates in glycogenesis, the conversion of glucose to, and storage in the form of, glycogen. The action of glycogen synthase is the opposite of phosphorylase. Nature provides a beautiful balance, inasmuch as the danger signal of the epinephrine hormone has a dual target: It activates phosphorylase to get quick energy, but simultaneously it inactivates glycogen synthase so that the available glucose will be used solely for energy and will not be stored.

Once the fatty acid in the form of acyl CoA is inside the mitochondrion, the β-oxidation starts. In the first oxidation (dehydrogenation; Step ②), two hydrogens are removed, creating a *trans* double bond between the alpha and beta carbons of the acyl chain. The hydrogens and electrons are picked up by FAD.

In Step ③, the double bond is hydrated. An enzyme specifically places the hydroxyl group on C-3, the beta carbon. The second oxidation (dehydrogenation; Step ④) requires NAD^+ as a coenzyme. The two hydrogens and electrons removed are transferred to the NAD^+ to form $NADH + H^+$. In the process, a secondary alcohol is oxidized to a ketone at the beta carbon. In Step ⑤, the enzyme thiolase cleaves the terminal C_2 fragment (an acetyl CoA) from the chain, and the rest of the molecule is bonded to a new molecule of coenzyme A.

The cycle then starts again with the remaining acyl CoA, which is now two carbon atoms shorter. At each turn of the cycle, one acetyl CoA is produced. Most fatty acids contain an even number of carbon atoms. The cyclic spiral continues until it reachs the last four carbon atoms. When this fragment enters the cycle, two acetyl CoA molecules are produced in the fragmentation step.

The β-oxidation of unsaturated fatty acids proceeds in the same way. An extra step is involved, in which the *cis* double bond is isomerized to a *trans* bond, but otherwise the spiral is the same.

28.6	What Is the Energy Yield from Stearic Acid Catabolism?

To compare the energy yield from fatty acids with that of other foods, let us select a typical and quite abundant fatty acid—stearic acid, the C_{18} saturated fatty acid.

We start with the initial step, in which energy is used up rather than produced. The reaction breaks two high-energy phosphoric anhydride bonds:

$$ATP \longrightarrow AMP + 2\,P_i + energy$$

This reaction is equivalent to hydrolyzing two molecules of ATP to ADP. In each cycle of the spiral, we obtain one $FADH_2$, one $NADH + H^+$, and one acetyl CoA. Stearic acid (C_{18}) goes through seven cycles in the spiral before

Table 28.2   ATP Yield from Complete Stearic Acid Metabolism

Step Number in Figure 28.6	Chemical Steps	Happens	Number of ATP Molecules Produced
①	Activation (stearic acid ⟶ stearyl CoA)	Once	−2
②	Dehydrogenation (acyl CoA ⟶ trans-enoyl CoA), producing $FADH_2$	8 times	16
④	Dehydrogenation (hydroxyacyl CoA ⟶ ketoacyl CoA), producing $NADH + H^+$	8 times	24
	C_2 fragment (acetyl CoA ⟶ common catabolic pathway), producing 12 ATP for each C_2 fragment	9 times	108
		Total	146

it reaches the final stage. In the last (eighth) cycle, one $FADH_2$, one $NADH + H^+$, and two acetyl CoA molecules are produced. Now we can add up the energy. Table 28.2 shows that, for a C_{18} fatty acid, we obtain a total of 146 ATP molecules.

It is instructive to compare the energy yield from fats with that from carbohydrates, as both are important constituents of the diet. In Section 28.2 we saw that glucose produces 36 ATP molecules—that is, 6 ATP molecules for each carbon atom. For stearic acid, there are 146 ATP molecules and 18 carbons, or $146/18 = 8.1$ ATP molecules per carbon atom. Because additional ATP molecules are produced from the glycerol portion of fats (Section 28.4), fats have a higher caloric value than carbohydrates.

EXAMPLE 28.1

Ketone bodies are a source of energy, especially during dieting and starvation. If acetoacetate (see Section 28.7) is metabolized through β-oxidation, how many ATP molecules would be produced?

Solution
In Step ①, activation of acetoacetate to acetoacetyl-CoA requires 2 ATPs. Step ⑤ yields two acetyl CoA molecules, which enter the common catabolic pathway, yielding 12 ATPs for each acetyl CoA, for a total of 24 ATPs. The net yield, therefore, is 22 ATP molecules.

Problem 28.1
Which fatty acid yields more ATP molecules per carbon atom: (a) stearic acid or (b) lauric acid?

28.7 | What Are Ketone Bodies?

In spite of the high caloric value of fats, the body preferentially uses glucose as an energy supply. When an animal is well fed (plenty of sugar intake), fatty acid oxidation is inhibited, and fatty acids are stored in the form of

Test kit for the presence of ketone bodies in the urine.

neutral fat in fat depots. When physical exercise demands energy, when the glucose supply dwindles (as in fasting or starvation), or when glucose cannot be utilized (as in the case of diabetes), the β-oxidation pathway of fatty acid metabolism is mobilized. In some pathological conditions, glucose may not be available at all, giving added importance to this point.

Unfortunately, low glucose supply also slows down the citric acid cycle. This lag happens because some oxaloacetate is essential for the continuous operation of the citric acid cycle (Figure 27.8). Oxaloacetate is produced from malate, but it is also produced by the carboxylation of phosphoenolpyruvate (PEP):

If there is no glucose, there will be no glycolysis, no PEP formation, and, therefore, greatly reduced oxaloacetate production.

Thus, even though the fatty acids are oxidized, not all of the resulting fragments (acetyl CoA) can enter the citric acid cycle because not enough oxaloacetate is present. As a result, acetyl CoA builds up in the body, with the following consequences.

The liver is able to condense two acetyl CoA molecules to produce acetoacetyl CoA:

Ketone bodies A collective name for acetone, acetoacetate and β-hydroxybutyrate; compounds produced from acetyl CoA in the liver that are used as a fuel for energy production by muscle cells and neurons

When the acetoacetyl CoA is hydrolyzed, it yields acetoacetate, which can be reduced to form β-hydroxybutyrate:

These two compounds, together with smaller amounts of acetone, are collectively called **ketone bodies.** Under normal conditions, the liver sends these

CHEMICAL CONNECTIONS 28C

Ketoacidosis in Diabetes

In untreated diabetes, the glucose concentration in the blood is high because a lack of insulin prevents utilization of glucose by the cells. Regular injections of insulin can remedy this situation. However, in some stressful conditions, **ketoacidosis** can still develop.

A typical case was a diabetic patient admitted to the hospital in a semi-comatose state. He showed signs of dehydration, his skin was inelastic and wrinkled, his urine showed high concentrations of glucose and ketone bodies, and his blood contained excess glucose and had a pH of 7.0, a drop of 0.4 pH unit from normal, which is an indication of severe acidosis. The patient's urine also contained the bacterium *Escherichia coli.* This indication of urinary tract infection explained why the normal doses of insulin were insufficient to prevent ketoacidosis.

The stress of infection can upset the normal control of diabetes by changing the balance between administered insulin and other hormones produced in the body. This imbalance happened during the patient's infection, and his body started to produce ketone bodies in large quantities. Both glucose and ketone bodies appear in the blood before they show up in the urine.

The acidic nature of ketone bodies (acetoacetic acid and β-hydroxybutyric acid) lowers the blood pH. A large drop in pH is prevented by the bicarbonate/carbonic acid buffer (Section 9.10), but even a drop of 0.3 to 0.5 pH unit is sufficient to decrease the Na^+ concentration. Such a decrease of Na^+ ions in the interstitial fluids draws out K^+ ions from the cells. This, in turn, impairs brain function and leads to coma. During the secretion of ketone bodies and glucose in the urine, a lot of water is lost, the body becomes dehydrated, and the blood volume shrinks. As a consequence, the blood pressure drops, and the pulse rate increases to compensate for it. Smaller quantities of nutrients reach the brain cells, which can also cause coma.

The patient mentioned here was infused with physiological saline solution to remedy his dehydration. Extra doses of insulin restored his glucose level to normal, and antibiotics cured the urinary infection.

compounds into the bloodstream to be carried to tissues and utilized there as a source of energy via the common catabolic pathway. The brain, for example, normally uses glucose as an energy source. During periods of starvation, however, ketone bodies may serve as the major energy source for the brain. Normally, the concentration of ketone bodies in the blood is low. During starvation and in untreated diabetes mellitus, however, ketone bodies accumulate in the blood and can reach high concentrations. When this buildup occurs, the excess is secreted in the urine. A check of urine for ketone bodies is used in the diagnosis of diabetes.

28.8 | How Is the Nitrogen of Amino Acids Processed in Catabolism?

The proteins of our foods are hydrolyzed to amino acids in digestion. These amino acids are primarily used to synthesize new proteins. Unlike carbohydrates and fats, however, they cannot be stored, so excess amino acids are catabolized for energy production. Section 28.9 explains what happens to the carbon skeletons of the amino acids. Here we discuss the catabolic fate of the nitrogen. Figure 28.7 gives an overview of the entire process of protein catabolism.

In the tissues, amino groups ($-NH_2$) freely move from one amino acid to another. The enzymes that catalyze these reactions are the transaminases. In essence, nitrogen catabolism in the liver occurs in three stages: transamination, oxidative deamination, and the urea cycle.

Figure 28.7 The overview of pathways in protein catabolism.

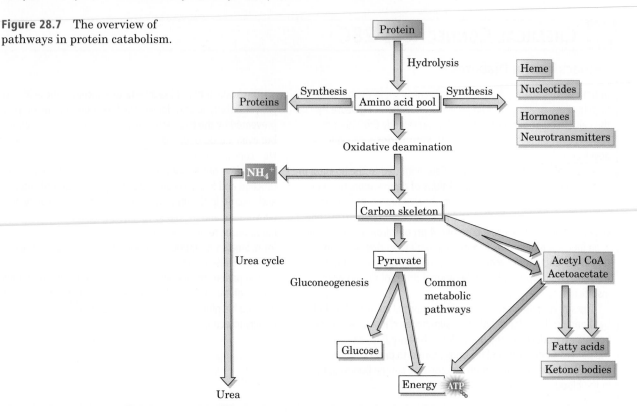

A. Transamination

Transamination The exchange of the amino group of an amino acid and a keto group of an α-ketoacid

In the first stage, **transamination,** amino acids transfer their amino groups to α-ketoglutarate:

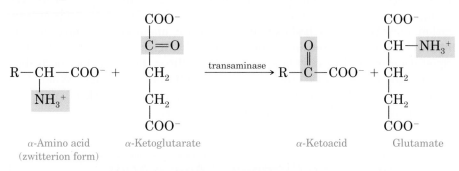

The carbon skeleton of the amino acid remains behind as an α-ketoacid, the catabolism of which is discussed in the next section.

B. Oxidative Deamination

Oxidative deamination The reaction in which the amino group of an amino acid is removed and an α-ketoacid is formed

The second stage of nitrogen catabolism is the **oxidative deamination** of glutamate, which occurs in the mitochondrion:

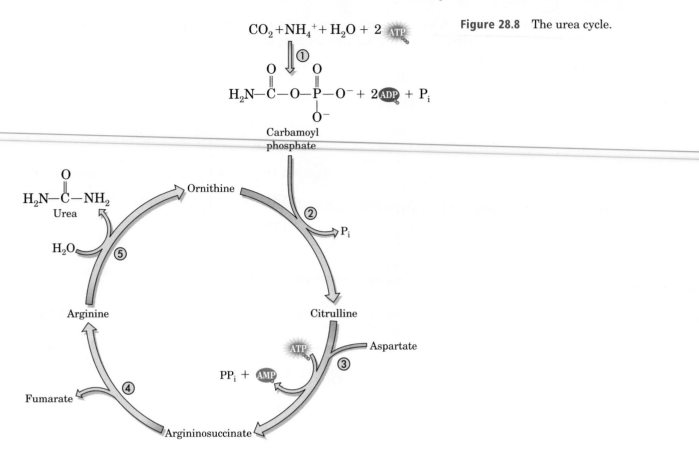

Figure 28.8 The urea cycle.

The oxidative deamination yields NH_4^+ and regenerates α-ketoglutarate, which can again participate in the first stage (transamination). The $NADH + H^+$ produced in the second stage enters the oxidative phosphorylation pathway and eventually produces three ATP molecules. The body must get rid of NH_4^+ because both it and NH_3 are toxic.

C. Urea Cycle

In the third stage, the NH_4^+ is converted to urea through the **urea cycle** (Figure 28.8). In Step ①, NH_4^+ is condensed with CO_2 in the mitochondrion to form an unstable compound, carbamoyl phosphate. This condensation occurs at the expense of two ATP molecules. In Step ②, carbamoyl phosphate is condensed with ornithine, a basic amino acid similar in structure to lysine, but which does not occur in proteins:

Urea cycle A cyclic pathway that produces urea from ammonia and carbon dioxide

The resulting citrulline diffuses out of the mitochondrion into the cytoplasm.

A second condensation reaction in the cytoplasm takes place between citrulline and aspartate, forming argininosuccinate (Step ③):

Citrulline   Aspartate   Argininosuccinate

The energy for this reaction comes from the hydrolysis of ATP to AMP and pyrophosphate (PP_i).

In Step ④, the argininosuccinate is split into arginine and fumarate:

Argininosuccinate   Arginine   Fumarate

In Step ⑤, the final step, arginine is hydrolyzed to urea and ornithine:

Arginine   Ornithine   Urea

The final product of the three stages is urea, which is excreted in the urine of mammals. The ornithine reenters the mitochondrion, completing

---

CHEMICAL CONNECTIONS 28D

Ubiquitin and Protein Targeting

In previous sections, we dealt with the catabolism of dietary proteins, especially with their use as an energy source. The body's own cellular proteins are also broken down and degraded. This occurs sometimes in response to a stress such as starvation, but more often it is done to maintain a steady-state level of regulatory proteins or to eliminate damaged proteins. But how does the cell know which proteins to degrade and which to leave alone?

One pathway that targets proteins for destruction depends on an ancient protein, called **ubiquitin.** (As the name implies, ubiquitin is present in all cells belonging to higher organisms.) Its sequence of 76 amino acids is essentially identical in yeast and in humans. The C-terminal amino acid of ubiquitin is glycine, the carboxyl group of which forms an amide bond with the side chain of a lysine residue in the protein that is targeted for destruction.

This process, known as **ubiquitinylation,** requires three enzymes and consumes energy by using up one ATP molecule per ubiquitin molecule. In many cases, a polymer of ubiquitin molecules (polyubiquitin) is bonded to the target protein molecule. In attaching one ubiquitin molecule to the next, the same bond is used as with the targeted protein: The glycine at the C-terminal of one ubiquitin forms an amide bond with the amino group on the side chain of a lysyl residue of a second ubiquitin. Once a protein molecule becomes "flagged" by several ubiquitin molecules, it is delivered to an organelle containing the machinery for proteolysis. Part of the proteolytic system is proteasome, a large, water-soluble complex assembly of proteases and regulatory proteins. In this assembly, the polyubiquitin is removed, and the protein is fully or partially degraded. Hydrolysis of a protein is called **proteolysis.**

One role of the ubiquitin targeting is that of "garbage collector." Damaged proteins that chaperones (Section 22.10) cannot salvage by proper refolding must be removed. For example, oxidation of hemoglobin causes unfolding, and rapid degradation by ubiquitin targeting follows. A second role is the control of the concentration of some regulatory proteins. The protein products of some oncogenes (Chemical Connections 26F) are also removed by ubiquitin targeting.

Both of these functions result in complete degradation of the targeted proteins. A third role involves only a partial degradation of the target protein. As an example, suppose a virus invades a cell. The infected cell recognizes a foreign protein from the virus and targets it for destruction through the ubiquitin system. The targeted protein is partially degraded. A peptide segment of the viral protein is incorporated into a system that rises to the surface of the infected cell and displays the viral peptide as an **antigen.** An antigen is a foreign body recognized by the immune system (Section 31.3). The T cells (Section 31.3) can now recognize the infected cell as a foreign cell; they therefore attack and kill it.

The excessive zeal of ubiquitinylation as a garbage collector lies at the heart of cystic fibrosis, a genetic disease. Cystic fibrosis causes severe bronchopulmonary disorders and pancreatic insufficiency. The molecular culprit is a protein in the membranes that serves as a chloride channel, providing for the transport of chloride ions into and out of cells. In cystic fibrosis, this protein has a mutation that does not allow it to fold properly. The system perceives these chloride-channel proteins as faulty and targets them for destruction by ubiquitinylation. The result is an insufficient number of functioning chloride channels. This shortage produces obstructions in the respiratory and intestinal tracts by clogging up the secretion pathways with highly viscous mucins.

---

the cycle. It is then ready to pick up another carbamoyl phosphate. An important aspect of carbamoyl phosphate's role as an intermediate is that it can be used for synthesis of nucleotide bases (Chapter 25). Furthermore, the urea cycle is linked to the citric acid cycle in that both involve fumarate. In fact, Hans Krebs, who elucidated the citric acid cycle, was also instrumental in establishing the urea cycle.

Not all organisms dispose of metabolic nitrogen in the form of urea. Bacteria and fish, for example, release ammonia directly into the surrounding water. Birds and reptiles secrete nitrogen in the form of uric acid, the concentrated white solid so familiar in bird droppings.

GOB
Chemistry ⚛ Now™

Click *Chemistry Interactive* to learn more about the **Urea Cycle** in more detail

D. Other Pathways of Nitrogen Catabolism

The urea cycle is not the only way that the body can dispose of the toxic NH_4^+ ions. The oxidative deamination process, which produced the NH_4^+ in the first place, is reversible. Therefore, the buildup of glutamate from α-ketoglutarate and NH_4^+ is always possible. A third possibility for disposing of NH_4^+ is the ATP-dependent amidation of glutamate to yield glutamine:

28.9 How Are the Carbon Skeletons of Amino Acids Processed in Catabolism?

After transamination of amino acids (Section 28.8A) to glutamate, the alpha amino group is removed from glutamate by oxidative deamination (Section 28.8B). The remaining carbon skeletons are used as an energy source (Figure 28.8). We will not study the pathways involved except to point out the eventual fate of the skeletons. Not all of the carbon skeletons of amino acids are used as fuel. Some may be degraded up to a certain point, and the resulting intermediate may then be used as a building block to construct another needed molecule.

For example, if the carbon skeleton of an amino acid is catabolized to pyruvate, the body has two possible choices: (1) use the pyruvate as an energy supply via the common catabolic pathway or (2) use it as a building block to synthesize glucose (Section 29.1). Those amino acids that yield a carbon skeleton that is degraded to pyruvate or another intermediate capable of conversion to glucose (such as oxaloacetate) are called **glucogenic.**

CHEMICAL CONNECTIONS 28E

Hereditary Defects in Amino Acid Catabolism: PKU

Many hereditary diseases involve missing or malfunctioning enzymes that catalyze the breakdown of amino acids. The oldest known of such diseases is cystinuria, which was described as early as 1810. In this disease, cystine shows up as flat hexagonal crystals in the urine. Stones form because of cystine's low solubility in water. This problem leads to blockage in the kidneys or the ureters and requires surgery to resolve it. One way to reduce the amount of cystine secreted is to remove as much methionine as possible from the diet. Beyond that, an increased fluid intake increases the volume of the urine, reducing the solubility problem. In addition, penicillamine can prevent cystinuria.

An even more important genetic defect is the absence of the enzyme phenylalanine hydroxylase, which causes a disease called phenylketonuria (PKU). In normal catabolism, this enzyme helps degrade phenylalanine by converting it to tyrosine. If the enzyme is defective, phenylalanine is converted to phenylpyruvate (see the discussion of the conversion of alanine to pyruvate in Section 28.9). Phenylpyruvate (an α-ketoacid) accumulates in the body and inhibits the conversion of pyruvate to acetyl CoA, thereby depriving the cells of energy via the common catabolic pathway. This effect is most important in the brain, which gets its energy from the utilization of glucose. PKU results in mental retardation.

This genetic defect can be detected early because phenylpyruvic acid appears in the urine. A federal regulation requires that all infants be tested for this disease. When PKU is detected, mental retardation can be prevented by restricting the intake of phenylalanine in the diet. In particular, patients with PKU should avoid the artificial sweetener aspartame because it yields phenylalanine when hydrolyzed in the stomach.

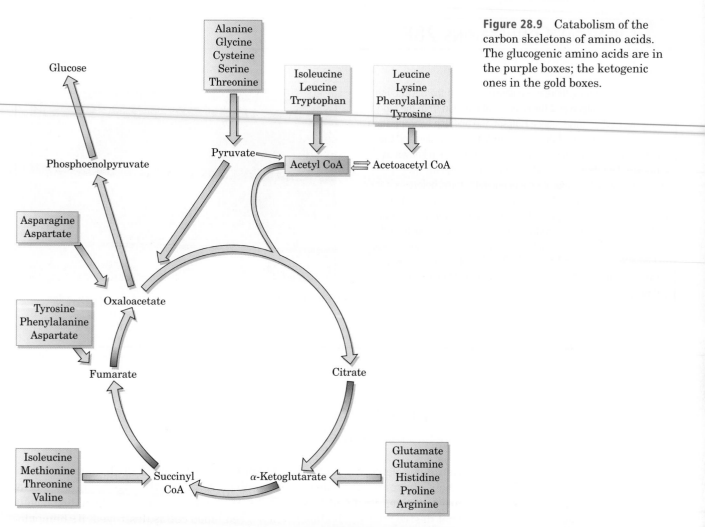

Figure 28.9 Catabolism of the carbon skeletons of amino acids. The glucogenic amino acids are in the purple boxes; the ketogenic ones in the gold boxes.

One example is alanine (Figure 28.9). When alanine reacts with α-ketoglutaric acid, the transamination produces pyruvate directly:

In contrast, many amino acids are degraded to acetyl CoA and acetoacetic acid. These compounds cannot form glucose but are capable of yielding ketone bodies; they are called **ketogenic.** Leucine is an example of a ketogenic amino acid. Some amino acids are both glucogenic and ketogenic—for example, phenylalanine.

Both glucogenic and ketogenic amino acids, when used as an energy supply, enter the citric acid cycle at some point (Figure 28.9) and are eventually oxidized to CO_2 and H_2O. The oxaloacetate (a C_4 compound) produced in this manner enters the citric acid cycle, adding to the oxaloacetate produced from PEP and in the cycle itself.

CHEMICAL CONNECTIONS 28F

Jaundice

In normal individuals, there is a balance between the destruction of heme in the spleen and the removal of bilirubin from the blood by the liver. When this balance is upset, the bilirubin concentration in the blood increases. If this condition persists, the skin and the whites of the eyes both become yellow—a condition known as **jaundice**. Jaundice can signal a malfunctioning of the liver, the spleen, or the gallbladder, where bilirubin is stored before excretion.

Hemolytic jaundice is the acceleration of the destruction of red blood cells in the spleen to such a level that the liver cannot cope with the excess bilirubin production. When gallstones obstruct the bile ducts and bilirubin cannot be excreted in the feces, the backup of bilirubin causes jaundice. In infectious hepatitis, the liver is incapacitated, and bilirubin is not removed from the blood, causing jaundice.

Yellowing of the sclera (white outer coat of eyeball) in jaundice. The yellowing is caused by an excess of bilirubin, a bile pigment, in the blood.

28.10 What Are the Reactions of Catabolism of Heme?

Red blood cells are continuously being manufactured in the bone marrow. Their life span is relatively short—about four months. Aged red blood cells are destroyed in the phagocytic cells. (Phagocytes are specialized blood cells that destroy foreign bodies.) When a red blood cell is destroyed, its hemoglobin is metabolized: The globin (Section 22.11) is hydrolyzed to amino acids, and the heme is first oxidized to biliverdin and finally reduced to bilirubin (Figure 28.10). The color change observed in bruises signal the redox reactions occurring in heme catabolism: Black and blue are due to the congealed blood, green to the biliverdin, and yellow to the bilirubin. The iron is preserved in ferritin, an iron-carrying protein, and reused. The bilirubin enters the liver via the blood and is then transferred to the gallbladder, where it is stored in the bile and finally excreted via the small intestine. The color of feces is provided by urobilin, an oxidation product of bilirubin.

Figure 28.10 Heme degradation from heme to biliverdin to bilirubin.

Figure 28.11 A summary of catabolism showing the role of the common metabolic pathway. Note that the end products of the catabolism of carbohydrates, lipids, and amino acids all appear. (TA is transamination; $\rightarrow \rightarrow \rightarrow$ is a pathway with many steps.)

Postscript

It is useful to summarize the main points of catabolic pathways by showing how they are related. Figure 28.11 shows how all catabolic pathways lead to the citric acid cycle, producing ATP by the reoxidation of NADH and $FADH_2$. We saw the common metabolic pathway in Chapter 27, and here we see how it is related to all of catabolism.

SUMMARY OF KEY QUESTIONS

SECTION 28.1 What Is the General Outline of Catabolic Pathways?

- The foods we eat consist of carbohydrates, lipids, and proteins.
- There are specific breakdown pathways for each kind of nutrient.

SECTION 28.2 What Are the Reactions of Glycolysis?

- The specific pathway of carbohydrate catabolism is **glycolysis.**

- Hexose monosaccharides are activated by ATP and eventually converted to two C_3 fragments, dihydroxy-acetone phosphate and glyceraldehyde phosphate.
- The glyceraldehyde phosphate is further oxidized and eventually ends up as pyruvate. All of these reactions occur in the cytosol.
- Pyruvate is converted to acetyl CoA, which is further catabolized in the common pathway.
- When the body needs intermediates for synthesis rather than energy, the glycolytic pathway can be shunted to the **pentose phosphate pathway.**

NADPH, which is necessary for reduction, is obtained in this way.
- The pentose phosphate pathway also yields ribose, which is necessary for synthesis of RNA.

SECTION 28.3 What Is the Energy Yield from Glucose Catabolism?

- When completely metabolized, a hexose molecule yields the energy of 36 ATP molecules.

SECTION 28.4 How Does Glycerol Catabolism Take Place?

- Fats are broken down to glycerol and fatty acids.
- Glycerol is catabolized in the glycolysis pathway and yields 20 ATP molecules.

SECTION 28.5 What Are the Reactions of β-Oxidation of Fatty Acids?

- Fatty acids are broken down into fragments in the **β-oxidation** spiral.
- At each turn of the spiral, one acetyl CoA is released along with one $FADH_2$ and one $NADH + H^+$. These products go through the common catabolic pathway.

SECTION 28.6 What Is the Energy Yield from Stearic Acid Catabolism?

- Stearic acid, a C_{18} compound, yields 146 molecules of ATP.

SECTION 28.7 What Are Ketone Bodies?

- In starvation and under certain pathological conditions, not all of the acetyl CoA produced in the β-oxidation of fatty acids enters the common catabolic pathway.

- Some acetyl CoA forms acetoacetate, β-hydroxybutyrate, and acetone, commonly called **ketone bodies.**
- Excess ketone bodies in the blood are secreted in urine.

SECTION 28.8 How Is the Nitrogen of Amino Acids Processed in Catabolism?

- Proteins are broken down to amino acids. The nitrogen of the amino acids is first transferred to glutamate.
- Glutamate is **oxidatively deaminated** to yield ammonia.
- Mammals get rid of the toxic ammonia by converting it to urea in the **urea cycle**; urea is secreted in urine.

SECTION 28.9 How Are the Carbon Skeletons of Amino Acids Processed in Catabolism?

- The carbon skeletons of amino acids are catabolized via the citric acid cycle.
- Some amino acids, called **glucogenic amino acids,** enter as pyruvate or other intermediates of the citric acid cycle.
- Other amino acids are incorporated into acetyl CoA or ketone bodies and are called **ketogenic amino acids.**

SECTION 28.10 What Are the Reactions of Catabolism of Heme?

- Heme is catabolized to bilirubin, which is excreted in the feces.

PROBLEMS

GOB
Chemistry ❖ Now™

Assess your understanding of this chapter's topics with additional quizzing and conceptual-based problems at **http://now.brookscole.com/gob8** or on the CD.

A blue problem number indicates an applied problem.

■ denotes problems that are available on the GOB ChemistryNow website or CD and are assignable in OWL.

SECTION 28.1 What Is the General Outline of Catabolic Pathways?

28.2 What are the products of lipase-catalyzed hydrolysis of fats?

28.3 What is the main use of amino acids in the body?

SECTION 28.2 What Are the Reactions of Glycolysis?

28.4 Although catabolism of a glucose molecule eventually produces a lot of energy, the first step uses up energy. Explain why this step is necessary.

28.5 In one step of the glycolysis pathway, a chain is broken into two fragments, only one of which can be further degraded in the glycolysis pathway. What happens to the other fragment?

28.6 Kinases are enzymes that catalyze the addition (or removal) of a phosphate group to (or from) a substance. ATP is also involved. How many kinases are in glycolysis? Name them.

28.7 ■ (a) Which steps in glycolysis of glucose need ATP?
(b) Which steps in glycolysis yield ATP directly?

28.8 At which intermediate of the glycolytic pathway does oxidation—and hence energy production—begin? In what form is the energy produced?

28.9 ■ At what point in glycolysis can ATP act as an inhibitor? What kind of enzyme regulation occurs in this inhibition?

28.10 The end product of glycolysis, pyruvate, cannot enter as such into the citric acid cycle. Which process converts this C_3 compound to a C_2 compound?

28.11 What essential compound is produced in the pentose phosphate pathway that is needed for synthesis as well as for defense against oxidative damages?

28.12 ■ Which of the following steps yields energy and which consumes energy?

(a) Pyruvate $\longrightarrow$ lactate

(b) Pyruvate $\longrightarrow$ acetyl CoA + CO_2

28.13 How many moles of lactate are produced from 3 moles of glucose?

28.14 How many moles of net NADH + H^+ are produced from 1 mole of glucose going to

(a) acetyl CoA? (b) Lactate?

SECTION 28.3 What Is the Energy Yield from Glucose Catabolism?

28.15 Of the 36 molecules of ATP produced by the complete metabolism of glucose, how many are produced directly in glycolysis alone—that is, before the common pathway?

28.16 ■ How many net ATP molecules are produced in the skeletal muscles for each glucose molecule

(a) In glycolysis alone (up to pyruvate)?

(b) In converting pyruvate to acetyl CoA?

(c) In the total oxidation of glucose to CO_2 and H_2O?

28.17 (a) If fructose is metabolized in the liver, how many moles of net ATP are produced from each mole during glycolysis?

(b) How many moles are produced if the same thing occurs in a muscle cell?

28.18 In Figure 28.3, Step ⑤ yields one NADH. Yet in Table 28.1, the same step indicates a yield of 2 NADH + H^+. Is there a discrepancy between these two statements? Explain.

SECTION 28.4 How Does Glycerol Catabolism Take Place?

28.19 Based on the names of the enzymes participating in glycolysis, what would be the name of the enzyme catalyzing the activation of glycerol?

28.20 Which yields more energy upon hydrolysis, ATP or glycerol 1-phosphate? Why?

SECTION 28.5 What Are the Reactions of β-Oxidation of Fatty Acids?

28.21 ■ Two enzymes participating in β-oxidation have the word "thio" in their names.

(a) Name the two enzymes.

(b) To which chemical group does this name refer?

(c) What is the common feature in the action of these two enzymes?

28.22 (a) Which part of the cells contains the enzymes needed for β-oxidation of fatty acids?

(b) How does the activated fatty acid get there?

28.23 Assume that lauric acid (C_{12}) is metabolized through β-oxidation. What are the products of the reaction after three turns of the spiral?

28.24 Is the β-oxidation of fatty acid (without the subsequent metabolism of C_2 fragments via the common metabolic pathway) more efficient with a short-chain fatty acid than with a long-chain fatty acid? Is more ATP produced per carbon atom in a short-chain fatty acid than in a long-chain fatty acid during β-oxidation?

SECTION 28.6 What Is the Energy Yield from Stearic Acid Catabolism?

28.25 ■ Calculate the number of ATP molecules obtained in the β-oxidation of myristic acid, $CH_3(CH_2)_{12}COOH$.

28.26 Assume that the *cis-trans* isomerization in the β-oxidation of unsaturated fatty acids does not require energy. Which fatty acid yields the greater amount of energy, saturated (stearic acid) or monounsaturated (oleic acid)? Explain.

28.27 Assuming that both fats and carbohydrates are available, which does the body preferentially use as an energy source?

28.28 ■ If equal weights of fats and carbohydrates are eaten, which will give more calories? Explain.

SECTION 28.7 What Are Ketone Bodies?

28.29 Acetoacetate is the common source of acetone and β-hydroxybutyrate. Name the type of reactions that yield these ketone bodies from acetoacetate.

28.30 Do ketone bodies have nutritional value?

28.31 What happens to the oxaloacetate produced from carboxylation of phosphoenolpyruvate?

SECTION 28.8 How Is the Nitrogen of Amino Acids Processed in Catabolism?

28.32 What kind of reaction is the following, and what is its function in the body?

28.33 Write an equation for the oxidative deamination of alanine.

28.34 Ammonia, NH_3, and ammonium ion, NH_4^+, are both soluble in water and could easily be excreted in the urine. Why does the body convert them to urea rather than excreting them directly?

28.35 What are the sources of the nitrogen in urea?

28.36 What compound is common to both the urea and citric acid cycles?

28.37 (a) What is the toxic product of the oxidative deamination of glutamate?

(b) How does the body get rid of it?

28.38 If the urea cycle is inhibited, in what other ways can the body get rid of NH_4^+ ions?

SECTION 28.9 How Are the Carbon Skeletons of Amino Acids Processed in Catabolism?

28.39 The metabolism of the carbon skeleton of tyrosine yields pyruvate. Why is tyrosine a glucogenic amino acid?

SECTION 28.10 What Are the Reactions of Catabolism of Heme?

28.40 Why is a high bilirubin content in the blood an indication of liver disease?

28.41 When hemoglobin is fully metabolized, what happens to the iron in it?

28.42 Describe which groups on the biliverdin (Figure 28.10) are the oxidation products and which are the reduction products in the degradation of heme.

Chemical Connections

28.43 (Chemical Connections 28A) What causes cramps of the muscles when a person is fatigued?

28.44 (Chemical Connections 28B) How does the signal of epinephrine result in depletion of glycogen in the muscle?

28.45 (Chemical Connections 28C) What system counteracts the acidic effect of ketone bodies in the blood?

28.46 (Chemical Connections 28C) The patient whose condition is described in Chemical Connections 28C was transferred to a hospital in an ambulance. Could a nurse in the ambulance tentatively diagnose his diabetic condition without running blood and urine tests? Explain.

28.47 (Chemical Connections 28D) Why is it necessary to have a mechanism for tagging proteins that are to be degraded?

28.48 (Chemical Connections 28D) What is meant by the following statement: "The amino acid sequence of ubiquitin is highly conserved"?

28.49 (Chemical Connections 28D) How does ubiquitin attach itself to a target protein?

28.50 (Chemical Connections 28F) How does the body eliminate damaged (oxidized) hemoglobin from the blood?

28.51 (Chemical Connections 28E) Draw structural formulas for each reaction component, and complete the following equation:

$$\text{Phenylalanine} \longrightarrow \text{Phenylpyruvate} + ?$$

28.52 (Chemical Connections 28F) What compound causes the yellow coloration of jaundice?

Additional Problems

28.53 If you receive a laboratory report showing the presence of a high concentration of ketone bodies in the urine of a patient, which disease would you suspect?

28.54 Which compounds give the sequence in coloration of bruises from black and blue to green and yellow?

28.55 ■ (a) At which step of the glycolysis pathway does NAD^+ participate (see Figures 28.3 and 28.4)?

(b) At which step does $NADH + H^+$ participate?

(c) As a result of the overall pathway, is there a net increase of NAD^+, of $NADH + H^+$, or of neither?

28.56 What is the net energy yield in moles of ATP produced when yeast converts one mole of glucose to ethanol?

28.57 Can the intake of alanine, glycine, and serine relieve hypoglycemia caused by starvation? Explain.

28.58 How can glucose be utilized to produce ribose for RNA synthesis?

28.59 Write the products of the transamination reaction between alanine and oxaloacetate:

28.60 Phosphoenolpyruvate (PEP) has a high-energy phosphate bond that has more energy than the anhydride bonds in ATP. Which step in glycolysis suggests that this is so?

28.61 Suppose that a fatty acid labeled with radioactive carbon-14 is fed to an experimental animal. Where would you look for the radioactivity?

28.62 Which functional groups are present in carbamoyl phosphate?

28.63 Is the urea cycle an energy-producing or an energy-consuming pathway?

28.64 Which intermediate of the glycolytic pathway can replenish oxaloacetate in the citric acid cycle?

28.65 How many turns of the spiral are there in the β-oxidation of (a) lauric acid and (b) palmitic acid?

Tying It Together

28.66 The equations of glycolysis indicate that there is a net gain of two ATP molecules for each molecule of glucose processed. Why is it that we see the figure of 36 ATP molecules in Table 28.1?

28.67 What reactions can pyruvate undergo once it is formed? Are these reactions aerobic, anaerobic, or both?

28.68 Is lactate a dead-end product of metabolism, or does it play some role in generating (or regenerating) some useful compound?

28.69 Why do ketone bodies occur in the blood of people who are on severely restricted diets?

28.70 Can amino acids be catabolized to yield energy?

28.71 Suggest a reason why the carbon skeletons and nitrogen-containing portions of amino acids are catabolized separately.

Looking Ahead

28.72 Put the following words into two related groups: energy-yielding, oxidative, anabolism, reductive, energy-requiring, catabolism.

28.73 Would you expect the biosynthesis of a protein from the constituent amino acids to require energy or to release energy? Explain.

28.74 In what ways can the production of glucose from CO_2 and H_2O in photosynthesis be considered the exact reversal of complete aerobic catabolism of glucose? In what ways is it different?

28.75 Why is the citric acid cycle *the* central pathway in metabolism?

Challenge Problems

28.76 With their oxygen-containing functional groups, sugars are more oxidized than the hydrocarbon side chains of fatty acids. Does this fact have any bearing on the energy yield of carbohydrates compared to that of fats?

28.77 Many soft drinks contain citric acid to add flavor. Is it likely to be a good nutrient?

28.78 The intermediates of glycolysis carry phosphate groups, which are charged. The intermediates of the citric acid cycle are not phosphorylated. Suggest a reason for this difference. (*Hint:* In what parts of the cell do these pathways occur?)

CHAPTER 29

Biosynthetic Pathways

Algae on mudflats.

© Bruce M. Herman/Photo Researchers, Inc.

29.1 | What Is the General Outline of Biosynthetic Pathways?

In the human body, and in most other living tissues, the pathways by which a compound is synthesized (anabolism) are usually different from the pathways by which it is degraded (catabolism). (Anabolic pathways are also called biosynthetic pathways, and we will use these terms interchangeably.) There are several reasons why it is biologically advantageous for anabolic and catabolic pathways to be different. We will give two of them here:

1. **Flexibility** If the normal **biosynthetic pathway** is blocked, the body can often use the reverse of the degradation pathway (recall that most steps in degradation are reversible), thereby providing another way to make the necessary compounds.

2. **Overcoming the effect of Le Chatelier's principle** This point can be illustrated by the cleavage of a glucose unit from a glycogen molecule, an equilibrium process:

$$(\text{Glucose})_n + P_i \xrightleftharpoons{\text{phosphorylase}} (\text{Glucose})_{n-1} + \text{Glucose 1-phosphate} \qquad (29.1)$$

Glycogen   Glycogen
(one unit smaller)

Phosphorylase catalyzes not only glycogen degradation (the forward reaction), but also glycogen synthesis (the reverse reaction). However, the body contains a large excess of inorganic phosphate, P_i. This excess would drive the reaction, on the basis of Le Chatelier's principle, to the right, which represents glycogen degradation. To provide a method for the synthesis of glycogen even in the presence of excess inorganic phosphate, a different pathway is needed in which P_i is not a reactant. Thus the body uses the following synthetic pathway:

$$(\text{Glucose})_{n-1} + \text{UDP-glucose} \longrightarrow (\text{Glucose})_n + \text{UDP} \qquad (29.2)$$

Glycogen   Glycogen
(one unit larger)

Not only do the synthetic pathways differ from the catabolic pathways, but the energy requirements are also different, as are the pathways' locations. Most catabolic reactions occur in the mitochondria, whereas anabolic reactions generally take place in the cytoplasm. We will not describe the energy balances of the biosynthetic processes in detail as we did for catabolism. However, keep in mind that, while energy (in the form of ATP) is *obtained* in the degradative processes, biosynthetic processes *consume* energy.

29.2 | How Does the Biosynthesis of Carbohydrates Take Place?

We discuss the biosynthesis of carbohydrates by looking at three examples:

- Conversion of atmospheric CO_2 to glucose in plants
- Synthesis of glucose in animals and humans
- Conversion of glucose to other carbohydrate molecules in animals and humans

A. Conversion of Atmospheric to Glucose in Plants

The most important biosynthesis of carbohydrates takes place in plants, green algae, and cyanobacteria, with the last two representing an important part of the marine food web. In the process of **photosynthesis,** the energy of the sun is built into chemical bonds of carbohydrates. The overall reaction is

Photosynthesis The process in which plants synthesize carbohydrates from CO_2 and H_2O with the help of sunlight and chlorophyll

$$6H_2O + 6CO_2 \xrightarrow[\text{chlorophyll}]{\substack{\text{energy in} \\ \text{the form of} \\ \text{sunlight}}} C_6H_{12}O_6 + 6O_2 \qquad (29.3)$$

Glucose

Although the primary product of photosynthesis is glucose, it is largely converted to other carbohydrates, mainly cellulose and starch. The very complicated process of glucose biosynthesis takes place in large protein–cofactor complexes (Chemical Connections 29A). We will not discuss it further here except to note that the carbohydrates of plants—starch, cellulose, and other mono- and polysaccharides—serve as the basic carbohydrate supply of all animals, including humans.

B. Synthesis of Glucose in Animals

Gluconeogenesis The process by which glucose is synthesized in the body

In Chapter 28, we saw that when the body needs energy, carbohydrates are broken down via glycolysis. When energy is not needed, glucose can be synthesized from the intermediates of the glycolytic and citric acid pathways. This process is called **gluconeogenesis.** As shown in Figure 29.1, a large number of intermediates—pyruvate, lactate, oxaloacetate, malate, and several amino acids (the glucogenic amino acids we met in Section 28.9)—can serve as starting compounds. Gluconeogenesis proceeds in the reverse order from glycolysis, and many of the enzymes of glycolysis also catalyze gluconeogenesis. At four points, however, unique enzymes (marked in Figure 29.1) catalyze only gluconeogenesis and not the breakdown reactions. These four enzymes make *gluconeogenesis a pathway that is distinct from glycolysis.* Note that ATP is used up in gluconeogenesis and produced in glycolysis, another difference between the two pathways.

During periods of strenuous exercise, the body needs to replenish its carbohydrate supply. The Cori cycle makes use of lactate produced in glycolysis (Section 28.2) as the starting point for gluconeogenesis. Lactate produced in working muscle is then transported in the bloodstream to the liver, where

Figure 29.1 Gluconeogenesis. All reactions take place in cytosol, except for those shown in the mitochondria.

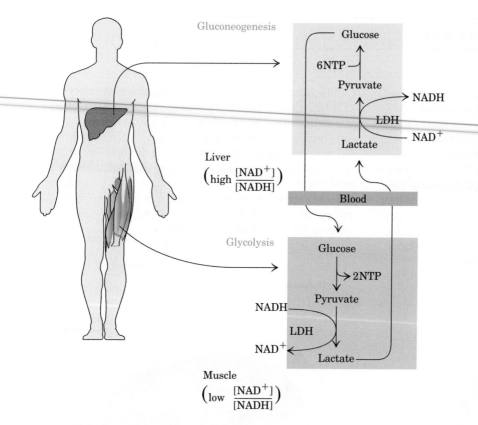

Figure 29.2 The Cori cycle is named for its discovers, Gerty and Carl Cori. Lactate produced in muscles by glycolysis is transported by the blood to the liver. Gluconeogenesis in the liver converts the lactate back to glucose, which can be carried back to the muscles by the blood. (NTP stands for nucleoside triphosphate.)

gluconeogenesis converts it to glucose (Figure 29.2). The newly formed glucose is then transported back to the muscle by the blood, where it can fuel further exercise. Note that the two different pathways, glycolysis and gluconeogenesis, take place in different organs. This division of labor ensures that both pathways are not simultaneously active in the same tissues, which is highly inefficient.

C. Conversion of Glucose to Other Carbohydrates in Animals

The third important biosynthetic pathway for carbohydrates is the conversion of glucose to other hexoses and hexose derivatives and the synthesis of di-, oligo-, and polysaccharides. The common step in all of these processes is the activation of glucose by uridine triphosphate (UTP) to form UDP-glucose:

CHEMICAL CONNECTIONS 29A

Photosynthesis

Photosynthesis requires sunlight, water, CO_2, and pigments found in plants, mainly chlorophyll. The overall reaction shown in Equation 29.3 actually occurs in two distinct steps. First, light interacts with the pigments that are located in highly membranous organelles of plants, called **chloroplasts.** Chloroplasts resemble mitochondria (Section 27.2) in many respects: They contain a whole chain of oxidation–reduction enzymes similar to the cytochrome and iron–sulfur complexes of mitochondrial membranes, and they contain a proton-translocating ATPase. In a manner similar to mitochondria, the proton gradient accumulated in the intermembrane region drives the synthesis of ATP in chloroplasts (see the discussion of the chemiosmotic pump in Section 27.6).

The chlorophyll is the central part of a complex machinery called photosystem I and II. The detailed structure of photosystem I was elucidated in 2001. The photosystem consists of three monomeric entities, which are designated as I, II and III. Each monomer contains 12 different proteins, 96 chlorophyll molecules, and 30 cofactors that include iron clusters, lipids, and Ca^{2+} ions. Its most important feature shows that a central Mg^{2+} is bound to the sulfur of a methionine residue of a surrounding protein. This Mg-S linkage makes this whole assembly a powerful oxidizing agent, meaning that it can readily accept electrons.

Side view of monomer III in photosystem I.

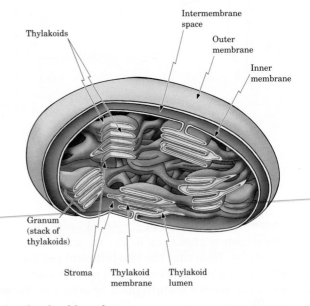

Inside of a chloroplast.

Courtesy of Dr. Petra Fromme

UDP is similar to ADP except that the base is uracil instead of adenine. UTP, an analog of ATP, contains two high-energy phosphate anhydride bonds. For example, when the body has excess glucose and wants to store it as glycogen (a process called **glycogenesis**), the glucose is first converted to glucose 1-phosphate, but then a special enzyme catalyzes the reaction:

Glycogenesis The conversion of glucose to glycogen

$$\text{glucose 1-phosphate} + \boxed{\text{UTP}} \longrightarrow \text{UDP-glucose} + \; {}^{-}O-\overset{\overset{\displaystyle O}{\|}}{\underset{\underset{\displaystyle O^{-}}{|}}{P}}-O-\overset{\overset{\displaystyle O}{\|}}{\underset{\underset{\displaystyle O^{-}}{|}}{P}}-O^{-}$$

$$\text{UDP-glucose} + (\text{glucose})_n \longrightarrow \boxed{\text{UDP}} + (\text{glucose})_{n+1}$$
Glycogen   Glycogen (one unit larger)

The biosynthesis of many other di- and polysaccharides and their derivatives also uses the common activation step: forming the appropriate UDP compound.

Photosynthesis *(continued)*

The chlorophyll itself, buried in a complex protein that traverses the chloroplast membranes, is a molecule similar to the heme we have already encountered in hemoglobin (Figure 22.14). In contrast to heme, however, chlorophyll contains Mg^{2+} instead of Fe^{2+}.

Chlorophyll a

The reactions in photosynthesis, collectively called the light reactions, are the ones in which chlorophyll captures the energy of sunlight and, with its aid, strips the electrons and protons from water to form oxygen, ATP, and NADPH + H$^+$ (see Section 28.2C):

$$H_2O + ADP + P_i + NADP^+ + \text{sunlight}$$
$$\longrightarrow \tfrac{1}{2}O_2 + ATP + NADPH + H^+$$

Another group of reactions, called the dark reactions because they do not need light, in essence converts CO_2 to carbohydrates:

$$CO_2 + ATP + NADPH + H^+$$
$$\longrightarrow (CH_2O)_n + ADP + P_i + NADP^+$$
$$\text{Carbohydrates}$$

Energy, now in the form of ATP, is used to help NADPH + H$^+$ reduce carbon dioxide to carbohydrates. Thus the protons and electrons stripped in the light reactions are added to the carbon dioxide in the dark reactions. This reduction takes place in a multistep cyclic process called the **Calvin cycle,** named after its discoverer, Melvin Calvin (1911–1997), who was awarded the 1961 Nobel Prize in chemistry for his work. In this cycle, CO_2 is first attached to a C_5 fragment and breaks down to two C_3 fragments (triose phosphates). Through a complex series of steps, these fragments are converted to a C_6 compound and, eventually, to glucose.

$$CO_2 + C_5 = 2C_3 = C_6$$

The critical step in the dark reactions (Calvin cycle) is the attachment of CO_2 to ribulose 1,5-bisphosphate, a compound derived from ribulose (Table 20.2). The enzyme that catalyzes this reaction, ribulose-1,5-bisphosphate carboxylase-oxygenase, nicknamed RuBisCO, is one of the slowest in nature. As in traffic, the slowest-moving vehicle determines the overall flow, so RuBisCO is the main factor in the low efficiency of the Calvin cycle. Because of the low efficiency of this enzyme, most plants convert less than 1% of the absorbed radiant energy into carbohydrates. To overcome its inefficiency, plants must synthesize large quantities of this enzyme. More than half of the soluble proteins in plant leaves are RuBisCO enzymes, whose synthesis requires a large energy expenditure.

29.3	How Does the Biosynthesis of Fatty Acids Take Place?

The body can synthesize all the fatty acids it needs except for linoleic and linolenic acids (essential fatty acids; see Section 21.2). The source of carbon in this synthesis is acetyl CoA. Because acetyl CoA is also a degradation product of the β-oxidation spiral of fatty acids (Section 28.5), we might expect the synthesis to be the reverse of the degradation. This is not the case. For one thing, the majority of fatty acid synthesis occurs in the cytoplasm, whereas degradation takes place in the mitochondria. Fatty acid synthesis is catalyzed by a multienzyme system.

However, one aspect of fatty acid synthesis is the same as in fatty acid degradation: Both processes involve acetyl CoA, so both proceed in units of two carbons. Fatty acids are built up two carbons at a time, just as they are broken down two carbons at a time (Section 28.5).

Most of the time, fatty acids are synthesized when excess food is available. That is, when we eat more food than we need for energy, our bodies turn the excess acetyl CoA (produced by catabolism of carbohydrates; see

Figure 29.3 The biosynthesis of fatty acids. The ACP (central blue sphere) has a long side chain (⤙) that carries the growing fatty acid (⌇). The ACP rotates counterclockwise, and its side chain sweeps over a multienzyme system (empty spheres). As each cycle is completed, a C_2 fragment is added to the growing fatty acid chain.

Section 28.2) into fatty acids and then to fats. The fats are stored in the fat depots, which are specialized fat-carrying cells (see Figure 28.1).

The key to fatty acid synthesis is an **acyl carrier protein (ACP)**. It can be looked upon as a merry-go-round—a rotating protein molecule to which the growing chain of fatty acids is bonded. As the growing chain rotates with the ACP, it sweeps over the multienzyme complex; at each enzyme, one reaction of the chain is catalyzed (Figure 29.3).

At the beginning of this cycle, the ACP picks up an acetyl group from acetyl CoA and delivers it to the first enzyme, fatty acid synthase, here called synthase for short:

$$CH_3\overset{\displaystyle O}{\overset{\|}{C}}\!-\!S\!-\!CoA + HS\!-\!ACP \longrightarrow HS\!-\!CoA + CH_3\overset{\displaystyle O}{\overset{\|}{C}}\!-\!S\!-\!ACP$$

Acetyl CoA   Acetyl ACP

$$CH_3\overset{\displaystyle O}{\overset{\|}{C}}\!-\!S\!-\!ACP + \text{synthase}\!-\!SH \longrightarrow CH_3\overset{\displaystyle O}{\overset{\|}{C}}\!-\!S\!-\!\text{synthase} + HS\!-\!ACP$$

The —SH group is the site of acyl binding as a thioester.

The C_2 fragment on the synthase is condensed with a C_3 fragment attached to the ACP in a process in which CO_2 is given off:

$$CH_3\overset{\displaystyle O}{\overset{\|}{C}}\!-\!S\!-\!\text{synthase} + \underset{\underset{\textstyle COO^-}{|}}{CH_2}\!-\!\overset{\displaystyle O}{\overset{\|}{C}}\!-\!S\!-\!ACP$$

Malonyl-ACP

$$\longrightarrow CH_3\overset{\displaystyle O}{\overset{\|}{C}}\!-\!CH_2\!-\!\overset{\displaystyle O}{\overset{\|}{C}}\!-\!S\!-\!ACP + CO_2 + \text{synthase}\!-\!SH$$

Acetoacetyl-ACP

The result is a C_4 fragment, which is reduced twice and dehydrated before it becomes a fully saturated C_4 group. This event marks the end of one cycle of the merry-go-round. These three steps are the reverse of what we saw in the β-oxidation of fatty acids in Section 28.5.

In the next cycle, the fragment is transferred to the synthase, and another malonyl-ACP (C_3 fragment) is added. When CO_2 splits out, a C_6 fragment is obtained. The merry-go-round continues to turn. At each turn, another C_2 fragment is added to the growing chain. Chains up to C_{16} (palmitic acid) can be obtained in this process. If the body needs longer fatty acids—for example, stearic (C_{18})—another fragment is added to palmitic acid by a different enzyme system.

Unsaturated fatty acids are obtained from saturated fatty acids by an oxidation step in which hydrogen is removed and combined with O_2 to form water:

$$R\!-\!CH_2\!-\!CH_2\!-\!(CH_2)_n COOH + O_2 + \boxed{NADPH} + H^+$$

$$\xrightarrow{\text{enzyme}} \overset{\displaystyle H \qquad H}{\underset{\displaystyle R \qquad (CH_2)_n COOH}{C\!=\!C}} + 2H_2O + \boxed{NADP^+}$$

An example of such elongation and unsaturation is docosahexenoic acid, a 22-carbon fatty acid with 6 *cis* double bonds (22:6). Docosahexenoic acid is part of the glycerophospholipids prevalent in the membranes where the visual pigment rhodopsin resides. Its presence is necessary to provide fluidity to the membrane so that it can process the light signals reaching our retina.

29.4 | How Does the Biosynthesis of Membrane Lipids Take Place?

The various membrane lipids (Sections 21.6 to 21.8) are assembled from their constituents. We just saw how fatty acids are synthesized in the body. These fatty acids are activated by CoA, forming acyl CoA. Glycerol 3-phosphate, which is obtained from the reduction of dihydroxyacetone phosphate (a C_3 fragment of glycolysis; see Figure 28.4), is the second building block of glycerophospholipids. This compound combines with two acyl CoA molecules, which can be the same or different:

Glycerol 3-phosphate   Acyl CoA   A phosphatidate

We encountered glycerol 3-phosphate as a vehicle for transporting electrons in and out of mitochondria (Section 28.3). To complete the molecule, an activated serine or choline or ethanolamine is added to the $-OPO_3^{2-}$ group, forming phosphate esters (see the structures in Section 21.6; Figure 29.4 shows a model of phosphatidylcholine). Choline is activated by cytidine triphosphate (CTP), yielding CDP-choline. This process is similar to the activation of glucose by UTP (Section 29.2C) except that the base is cytosine rather than uracil (Section 25.2). Sphingolipids (Section 21.7) are similarly built up from smaller molecules. An activated phosphocholine is added to the sphingosine part of ceramide (Section 21.7) to make sphingomyelin.

The glycolipids are constructed in a similar fashion. Ceramide is assembled as described above, and the carbohydrate is added one unit at a time in the form of activated monosaccharides (UDP-glucose and so on).

Cholesterol, the molecule that controls the fluidity of membranes and is a precursor of all steroid hormones and bile salts, is also synthesized by the human body. It is assembled in the liver from fragments that come from the acetyl group of acetyl CoA. All of the carbon atoms of cholesterol come from the carbon atoms of acetyl CoA molecules (Figure 29.5). Cholesterol in the brain is synthesized in nerve cells themselves; its presence is necessary to form synapses. Cholesterol from our diets and cholesterol that is synthesized in the liver and circulates in the plasma as LDL (see Section 21.9) are not available for synapse formation because LDL cannot cross the blood–brain barrier.

Figure 29.4 A model of phosphatidylcholine, commonly called lecithin.

Figure 29.5 Biosynthesis of cholesterol. The circled carbon atoms come from the —CH₃ group, and the other carbon atoms come from the —CO— group of the acetyl group of acetyl CoA.

Cholesterol synthesis starts with the sequential condensation of three acetyl CoA molecules to form a compound 3-hydroxy-3-methylglutaryl CoA (HMGCoA):

A key enzyme, HMG reductase, controls the rate of cholesterol synthesis. It reduces the thioester of HMGCoA to a primary alcohol, yielding CoA in the process. The resulting compound, mevalonate, undergoes phosphorylation and decarboxylation to yield a C_5 compound, isopentenyl pyrophosphate:

From this basic C_5 unit, the other multiple-C_5 compounds are formed that eventually lead to cholesterol synthesis. These intermediates are geranyl, C_{10}, and farnesyl, C_{15}, pyrophosphates:

Finally, cholesterol is synthesized from the condensation of two farnesyl pyrophosphate molecules.

The statin drugs, such as lovastatin, competitively inhibit the key enzyme HMG reductase and, thereby, the biosynthesis of cholesterol. They are frequently prescribed to control the cholesterol level in the blood so as to prevent atherosclerosis (Section 21.9E).

Note that the intermediates in cholesterol synthesis, the geranyl and farnesyl pyrophosphates, are made of isoprene units; we discussed these C_5 compounds in Section 12.5. The C_{10} and C_{15} compounds are also used to enable protein molecules to be dispersed in the lipid bilayers of membranes. When these multiple-isoprene units are attached to a protein in a process called **prenylation,** the protein becomes more hydrophobic and is able to move laterally within the bilayer with greater ease. (The name *prenylation* originates from isoprene, the five-carbon unit from which the cholesterol intermediates C_{10}, C_{15}, and C_{30} are made.) Prenylation marks proteins to be associated with membranes and to perform other cellular functions, such as signal transduction of G-protein (Chemical Connections 29B).

CHEMICAL CONNECTIONS 29B

Prenylation of Ras Protein and Cancer

Prenylation, or the attachment of multiple isoprene units to a protein, enables the protein to be part of a membrane, an integral protein. The C_{10} and C_{15} units are attached to the protein via a thioether linkage, $—C—S—C—$. For example, farnesyl pyrophosphate interacts with a cysteine residue that is located near the carboxyl terminal of the target protein. As a thioether is formed, the pyrophosphate is liberated simultaneously. The resultant modified protein acquires a hydrophobic tail that makes its association with membranes feasible.

A protein named Ras is a GTP-binding protein that plays a key role in the signal transduction (Section 24.5) that regulates cell growth. It is encoded by an oncogene (Chemical Connections 25E), which indicates that a Ras-related mutation may cause cancer. The signaling activity of Ras depends on its prenylation. Mutations that inhibit prenylation of Ras cause the cell to die. Other mutations of prenylated Ras may cause uncontrolled growth; indeed, one third of all human cancers feature Ras mutations.

Farnesyl
(or geranyl)
　　　　　　　　Cysteine

29.5 How Does the Biosynthesis of Amino Acids Take Place?

The human body needs 20 different amino acids to make its protein chains—all 20 are found in a normal diet. Some of the amino acids can be synthesized from other compounds; these are the nonessential amino acids. Others cannot be synthesized by the human body and must be supplied in the diet; these are the **essential amino acids** (see Section 30.6). Most nonessential amino acids are synthesized from some intermediate of either glycolysis (Section 28.2) or the citric acid cycle (Section 27.4). Glutamate plays a central role in the synthesis of five nonessential amino acids. Glutamate itself is synthesized from α-ketoglutarate, one of the intermediates in the citric acid cycle:

The forward reaction is the synthesis, and the reverse reaction is the oxidative deamination (degradation) reaction we encountered in the catabolism of amino acids (Section 28.8B). In this case, the synthetic and degradative pathways are exactly the reverse of each other.

CHEMICAL CONNECTIONS 29C

Amino Acid Transport and Blue Diaper Syndrome

Amino acids, both essential and nonessential, are usually more concentrated inside cells than in the cells' surroundings. In the body, special transport mechanisms are available to carry the amino acids into cells. Sometimes these transport mechanisms are faulty, however, and the cells lack an amino acid despite the fact that the diet provides the particular amino acid.

An interesting example of this phenomenon is blue diaper syndrome, in which infants excrete blue urine. Indigo blue, a dye, is an oxidation product of the amino acid tryptophan. But how does this oxidation product get into the system to be excreted in the urine?

The patient's tryptophan transport mechanism is faulty. Although enough tryptophan is being supplied by the diet, most of it is not absorbed through the intestine. Instead, it accumulates there and s oxidized by bacteria in the gut. The oxidation product is moved into the cells, but because the cells cannot use it, it is excreted in the urine.

Much of blue diaper syndrome can be eliminated with antibiotics. This treatment kills many of the gut's bacteria, and the nonabsorbed tryptophan is then excreted in the feces because the bacteria can no longer oxidize it to indigo blue.

Structure of tryptophan and indigo.

Glutamate can serve as an intermediate in the synthesis of alanine, serine, aspartate, asparagine, and glutamine. For example, the transamination reaction we saw in Section 28.8A leads to alanine formation:

Besides being the building blocks of proteins, amino acids serve as intermediates for a large number of biological molecules. We have already seen that serine is needed in the synthesis of membrane lipids (Section 29.4). Certain amino acids are also intermediates in the synthesis of heme and of the purines and pyrimidines that are the raw materials for DNA and RNA (Chapter 25).

Postscript

It is useful to summarize the main points of anabolic pathways by considering how they are related. Figure 29.6 shows how all catabolic pathways start at the citric acid cycle, using ATP and the reducing power of NADH and FADH$_2$. Chapter 28 introduced the central metabolic pathway, and here we see how it is related to all of anabolism.

CHEMICAL CONNECTIONS 29D

Essential Amino Acids

The biosynthesis of proteins requires the presence of all of the protein's constituent amino acids. If any of the 20 amino acids is missing or in short supply, protein biosynthesis is inhibited.

Some organisms, including bacteria, can synthesize all the amino acids that they need. Other species, including humans, must obtain some amino acids from dietary sources. The essential amino acids in human nutrition are listed in Table 29.1. The body can synthesize some of these amino acids, but not in sufficient quantities for its needs, especially in the case of growing children (particularly children's requirement for arginine and histidine).

Amino acids are not stored (except in proteins), so dietary sources of essential amino acids are needed at regular intervals. Protein deficiency—especially a prolonged deficiency in sources that contain essential amino acids—leads to the disease

Table 29.1   Amino Acid Requirements in Humans

Essential		Nonessential	
Arginine	Methionine	Alanine	Glutamine
Histidine	Phenylalanine	Asparagine	Glycine
Isoleucine	Threonine	Aspartate	Proline
Leucine	Tryptophan	Cysteine	Serine
Lysine	Valine	Glutamate	Tyrosine

kwashiorkor. The problem in this disease, which is particularly severe in growing children, is not simply starvation, but the breakdown of the body's own proteins.

Figure 29.6   A summary of anabolism showing the role of the central metabolic pathway. Note that carbohydrates, lipids, and amino acids all appear as products. (OAA is oxaloacetate; ALA is a derivative of succinyl CoA; TA is transamination; → → → is a pathway with many steps.)

SUMMARY OF KEY QUESTIONS

SECTION 29.1 What Is the General Outline of Biosynthetic Pathways?

- For most biochemical compounds, the biosynthetic pathways are different from the degradation pathways.

SECTION 29.2 How Does the Biosynthesis of Carbohydrates Take Place?

- In **photosynthesis,** carbohydrates are synthesized in plants from CO_2 and H_2O, using sunlight as an energy source.
- Glucose can be synthesized by animals from the intermediates of glycolysis, from the intermediates of the citric acid cycle, and from glucogenic amino acids. This process is called **gluconeogenesis.**
- When glucose or other monosaccharides are built into di-, oligo-, and polysaccharides, each monosaccharide unit in its activated form is added to a growing chain.

SECTION 29.3 How Does the Biosynthesis of Fatty Acids Take Place?

- Fatty acid biosynthesis is accomplished by a multienzyme system.
- The key to fatty acid biosynthesis is the **acyl carrier protein (ACP),** which acts as a merry-go-round transport system: It carries the growing fatty acid chain over a number of enzymes, each of which catalyzes a specific reaction.

- With each complete turn of the merry-go-round, a C_2 fragment is added to the growing fatty acid chain.
- The source of the C_2 fragment is malonyl-ACP, a C_3 compound bonded to ACP. It becomes C_2 with the loss of CO_2.

SECTION 29.4 How Does the Biosynthesis of Membrane Lipids Take Place?

- Glycerophospholipids are synthesized from glycerol 3-phosphate, fatty acids that are activated by conversion to acyl CoA, and activated alcohols such as choline.
- Cholesterol is synthesized from acetyl CoA. Three C_2 fragments are condensed to form C_6, hydroxymethylglutaryl CoA.
- After reduction and decarboxylation, isoprene C_5 units are formed that condense to the C_{10} and C_{15} intermediates from which cholesterol is built.

SECTION 29.5 How Does the Biosynthesis of Amino Acids Take Place?

- Many nonessential amino acids are synthesized in the body from the intermediates of glycolysis or the intermediates of the citric acid cycle.
- In half of these cases, glutamate is the donor of the amino group in transamination.
- Amino acids serve as building blocks for proteins.

PROBLEMS

GOB
Chemistry◦Now™

Assess your understanding of this chapter's topics with additional quizzing and conceptual-based problems at **http://now.brookscole.com/gob8** or on the CD.

A blue problem number indicates an applied problem.

■ denotes problems that are available on the GOB ChemistryNow website or CD and are assignable in OWL.

SECTION 29.1 What Is the General Outline of Biosynthetic Pathways?

29.1 Why are the pathways that the body uses for anabolism and catabolism mostly different?

29.2 How does the large excess of inorganic phosphate in a cell affect the amount of glycogen present? Explain.

29.3 Glycogen can be synthesized in the body by the same enzymes that degrade it. Why is this process utilized in glycogen synthesis only to a small extent, while most glycogen biosynthesis occurs via a different synthetic pathway?

29.4 ■ Do most anabolic and catabolic reactions take place in the same location?

SECTION 29.2 How Does the Biosynthesis of Carbohydrates Take Place?

29.5 What is the difference in the overall chemical equations for photosynthesis and for respiration?

29.6 ■ In photosynthesis, what are the sources of (a) carbon, (b) hydrogen, and (c) energy?

29.7 Name a compound that can serve as a raw material for gluconeogenesis and is (a) from the glycolytic pathway, (b) from the citric acid cycle, and (c) an amino acid.

29.8 How is glucose activated for glycogen synthesis?

29.9 Glucose is the only carbohydrate compound that the brain can use for energy. Which pathway is mobilized to supply the need of the brain during starvation: (a) glycolysis, (b) gluconeogenesis, or (c) glycogenesis? Explain.

29.10 Are the enzymes that combine two C_3 compounds into a C_6 compound in gluconeogenesis the same as or different from those that cleave the C_6 compound into two C_3 compounds in glycolysis?

29.11 Devise a scheme in which maltose is formed starting with UDP-glucose.

29.12 ■ Glycogen is written as (glucose)$_n$.
   (a) What does n stand for?
   (b) What is the approximate value of n?

29.13 What are the constituents of UTP?

SECTION 29.3 How Does the Biosynthesis of Fatty Acids Take Place?

29.14 What is the source of carbon in fatty acid synthesis?

29.15 (a) Where in the body does fatty acid synthesis occur?
   (b) Does fatty acid degradation occur in the same location?

29.16 Is ACP an enzyme?

29.17 In fatty acid biosynthesis, which compound is added repeatedly to the synthase?

29.18 (a) What is the name of the first enzyme in fatty acid synthesis?
   (b) What does it do?

29.19 From which compound is CO_2 released in fatty acid synthesis?

29.20 ■ What are the common functional groups in CoA, ACP, and synthase?

29.21 In the synthesis of unsaturated fatty acids, $NADPH + H^+$ is converted to $NADP^+$, yet this is an oxidation step and not a reduction step. Explain.

29.22 ■ Which of these fatty acids can be synthesized by the multienzyme fatty acid synthesis complex alone?
   (a) Oleic   (b) Stearic
   (c) Myristic   (d) Arachidonic
   (e) Lauric

29.23 Some enzymes can use NADH as well as NADPH as a coenzyme. Other enzymes use one or the other exclusively. Which features would prevent NADPH from fitting into the active site of an enzyme that otherwise can accommodate NADH?

29.24 Are fatty acids for energy, in the form of fat, synthesized in the same way as fatty acids for the lipid bilayer of membrane?

29.25 Linoleic and linolenic acids cannot be synthesized in the human body. Does this mean that the human body cannot make an unsaturated fatty acid from a saturated one?

SECTION 29.4 How Does the Biosynthesis of Membrane Lipids Take Place?

29.26 When the body synthesizes the following membrane lipid, from which building blocks is it assembled?

29.27 Name the activated constituents necessary to form the glycolipid glucoceramide.

29.28 Why is HMG reductase a key enzyme in cholesterol synthesis?

29.29 Describe by carbon skeleton designation how a C_2 compound ends up as a C_5 compound.

SECTION 29.5 How Does the Biosynthesis of Amino Acids Take Place?

29.30 Which reaction is the reverse of the synthesis of glutamate from α-ketoglutarate, ammonia, and $NADH + H^+$?

29.31 ■ Which amino acid will be synthesized by the following process?

29.32 Draw the structure of the compound needed to synthesize asparagine from glutamate by transamination.

29.33 Name the products of the following transamination reaction:

$$(CH_3)_2CH-\overset{\displaystyle O}{\overset{\displaystyle \|}{C}}-COO^-$$

$$+ \ ^-OOC-CH_2-CH_2-\underset{\underset{\displaystyle NH_3^+}{|}}{CH}-COO^- \longrightarrow$$

Chemical Connections

29.34 (Chemical Connections 29A) Photosystem I and II are complex factories of proteins, chlorophyll, and many cofactors. Where are these photosystems located in plants, and in which reaction of photosynthesis do they participate?

29.35 (Chemical Connections 29A) Which coenzyme reduces CO_2 in the Calvin cycle?

29.36 (Chemical Connections 29B) Why would an unprenylated Ras protein fail to transduct signals across a membrane?

29.37 (Chemical Connections 29C) (a) Which compound produces the colored urine of blue diaper syndrome?

(b) What is the source of this compound?

29.38 (Chemical Connections 29D) What is the result of eating only proteins that do not contain all 20 amino acids?

Additional Problems

29.39 In the structure of $NADP^+$, what are the bonds that connect nicotinamide and adenine to the ribose units?

29.40 Which C_3 fragment carried by ACP is used in fatty acid synthesis?

29.41 When glutamate transaminates phenylpyruvate, which amino acid is produced?

$$C_6H_5-CH_2-\overset{\overset{O}{\|}}{C}-COO^-$$

$$+\ ^-OOC-CH_2-CH_2-\underset{\underset{NH_3^+}{|}}{CH}-COO^- \longrightarrow$$

29.42 Does the prenylation of Ras protein itself cause cancer?

29.43 ■ Each activation step in the synthesis of complex lipids occurs at the expense of one ATP molecule. How many ATP molecules are used in the synthesis of one molecule of lecithin?

29.44 ■ Consider the fact that the deamination of glutamic acid and its synthesis from α-ketoglutaric acid are equilibrium reactions. Which way will the equilibrium shift when the human body is exposed to cold temperature?

29.45 Which compound reacts with glutamate in a transamination process to yield serine?

29.46 What are the names of the C_{10} and C_{15} intermediates in cholesterol biosynthesis?

29.47 Which is carbon 1 in HMGCoA (3-hydroxy-3-methylglutaryl CoA)?

29.48 In most biosynthetic processes, the reactant is reduced to obtain the desired product. Verify whether this statement holds for the overall reaction of photosynthesis.

29.49 What is the major difference in structure between chlorophyll and heme?

29.50 Can the complex enzyme system participating in every fatty acid synthesis manufacture fatty acids of any length?

Tying It Together

29.51 How does fatty acid biosynthesis differ from catabolism of fatty acids?

29.52 How does the energy source differ in carbohydrate biosynthesis in plants and in animals?

29.53 The enzyme that catalyzes carbon dioxide fixation in photosynthesis is one of the least efficient known. What does this fact imply about energy requirements in photosynthesis?

Looking Ahead

29.54 A vegan diet is one that excludes all animal products. Is it possible to get all essential nutrients from such a diet? Will it be more difficult or easier to achieve this goal with a diet that allows animal products?

29.55 Many key proteins in the immune system are glycoproteins (proteins that incorporate sugars in their structure). Would you expect the biosynthesis of such proteins to be affected by a lack of essential amino acids, a low-carbohydrate diet, or both? Explain.

Challenge Problems

29.56 The foods that we eat supply carbohydrates, fats, and proteins. Based on what you have learned in this chapter, which would you predict that we could do without? Explain.

29.57 In general, catabolic and biosynthetic processes do not take place in the same part of the cell. Why is this separation advantageous?

CHAPTER 30

Nutrition

KEY QUESTIONS

30.1 How Do We Measure Nutrition?

30.2 Why Do We Count Calories?

30.3 How Does the Body Process Dietary Carbohydrates?

30.4 How Does the Body Process Dietary Fats?

30.5 How Does the Body Process Dietary Protein?

30.6 What Is the Importance of Vitamins, Minerals, and Water?

Foods high in fiber include whole grains, legumes, fruits, and vegetables.

Charles D. Winters

GOB
Chemistry✦Now™

Look for this logo in the chapter and go to GOB ChemistryNow at **http://now.brookscole.com/gob8** or on the CD for tutorials, simulations, and problems.

30.1 | How Do We Measure Nutrition?

In Chapters 27 and 28, we saw what happens to the food that we eat in its final stages—after the proteins, lipids, and carbohydrates have been broken down into their components. In this chapter, we discuss the earlier stages—nutrition and diet—and then the digestive processes that break down these large molecules into the small ones that undergo metabolism. Food provides energy and new molecules to replace those that the body uses. This synthesis of new molecules is especially important for the period during which a child becomes an adult.

The components of food and drink that provide for growth, replacement, and energy are called **nutrients.** Not all components of food are nutrients. Some components of food and drinks, such as those that provide flavor, color, or aroma, enhance our pleasure in the food but are not themselves nutrients.

Nutritionists classify nutrients into six groups:

1. Carbohydrates
2. Lipids

755

Digestion The process in which the body breaks down large molecules into smaller ones that can be absorbed and metabolized

Discriminatory curtailment diet A diet that avoids certain food ingredients that are considered harmful to the health of an individual—for example, low-sodium diets for people with high blood pressure

3. Proteins
4. Vitamins
5. Minerals
6. Water

For food to be used in our bodies, it must be absorbed through the intestinal walls into the bloodstream or lymph system. Some nutrients, such as vitamins, minerals, glucose, and amino acids, can be absorbed directly. Others, such as starch, fats, and proteins, must first be broken down into smaller components before they can be absorbed. This breakdown process is called **digestion.**

A healthy body needs the proper intake of all nutrients. However, nutrient requirements vary from one person to another. For example, more energy is needed to maintain the body temperature of an adult than that of a child. For this reason, nutritional requirements are usually given per kilogram of body weight. Furthermore, the energy requirements of a physically active individual are greater than those of a person in a sedentary occupation. Therefore, when average values are given, as in **Dietary Reference Intakes (DRI)** and in the former guidelines called **Recommended Daily Allowances (RDA),** one should be aware of the wide range that these average values represent.

The public interest in nutrition and diet changes with time and geography. Seventy or eighty years ago, the main nutritional interest of most Americans was getting enough food to eat and avoiding diseases caused by vitamin deficiencies, such as scurvy or beriberi. This issue is still the main concern of the large majority of the world's population. In affluent societies such as the industrialized nations, however, today's nutritional message is no longer "eat more," but rather "eat less and discriminate more in your selection of food." Dieting to reduce body weight is a constant effort in a sizable percentage of the American population. Many people discriminate in their selection of food to avoid cholesterol (Section 21.9E) and saturated fatty acids to reduce the risk of heart attacks.

Along with such **discriminatory curtailment diets** came many faddish diets. **Diet faddism** is an exaggerated belief in the effects of nutrition upon health and disease. This phenomenon is not new; it has been prevalent for many years. Many times it is driven by visionary beliefs, but backed by little science. In the nineteenth century, Dr. Kellogg (of cornflakes fame) recommended a largely vegetarian diet based on his belief that meat produces sexual excess. Eventually his religious fervor withered and his brother made a commercial success of his inventions of grain-based food. Another fad is raw food, which bans any application of heat higher than 118°F to food. Heat, these faddists maintain, depletes the nutritional value of proteins and vitamins and concentrates pesticides in food. Obviously raw food diet is vegetarian, as it excludes meat and meat products.

A recommended food is rarely as good, and a condemned food is never as bad, as faddists claim. Scientific studies have proven, for instance, that food grown with chemical fertilizers is just as healthy as food grown with organic (natural) fertilizers.

Each food contains a large variety of nutrients. For example, a typical breakfast cereal lists the following items as its ingredients: milled corn, sugar, salt, malt flavoring, and vitamins A, B, C, and D, plus flavorings and preservatives. U.S. consumer laws require that most packaged food be labeled in a uniform manner to show the nutritional values of the food. Figure 30.1 shows a typical label of the type found on almost every can, bottle, or box of food that we buy.

Such labels must list the percentages of **Daily Values** for four key vitamins and minerals: vitamins A and C, calcium, and iron. If other vitamins or minerals have been added, or if the product makes a nutritional claim about other nutrients, their values must be shown as well. The percent

Nutrition Facts

Serving size 1 Bar (28g)
Servings per Container 6

Amount Per Serving

Calories 120   Calories from Fat 35

	% Daily Value*
Total Fat 4g	6%
Saturated Fat 2g	10%
Cholesterol 0mg	0%
Sodium 45mg	2%
Potassium 100mg	3%
Total Carbohydrate 19g	6%
Dietary fiber 2g	8%
Sugars 13g	
Protein 2g	

Vitamin A 15%	•	Vitamin C 15%
Calcium 15%	•	Iron 15%
Vitamin D 15%	•	Vitamin E 15%
Thiamin 15%	•	Riboflavin 15%
Niacin 15%	•	Vitamin B6 15%
Folate 15%	•	Vitamin B12 15%
Biotin 10%	•	Pantothenic Acid 10%
Phosphorus 15%	•	Iodine 2%
Magnesium 4%	•	Zinc 4%

*Percent Daily Values are based on a 2,000 calorie diet. Your daily values may be higher or lower depending on your calorie needs.

	Calories:	2,000	2,500
Total Fat	Less than	65g	80g
Sat Fat	Less than	20g	25g
Cholesterol	Less than	300mg	300mg
Sodium	Less than	2,400mg	2,400mg
Potassium		3,500mg	3,500mg
Total Carbohydrate		300g	375g
Dietary Fiber		25g	30g

Figure 30.1 A food label for a peanut butter crunch bar. The portion at the bottom (following the asterisk) is the same on all labels that carry it.

daily values on the labels are based on a daily intake of 2000 Cal. For anyone who eats more than that amount, the actual percentage figures would
be lower (and higher for those who eat less). Note that each label specifies
the serving size; the percentages are based on that portion, not on the contents of the entire package. The section at the bottom of the label is exactly
the same on all labels, no matter what the food; it shows the daily amounts
of nutrients recommended by the government, based on consumption of
either 2000 or 2500 Cal. Some food packages are allowed to carry shorter
labels, either because they have only a few nutrients or because the package
has limited label space. The uniform labels make it much easier for consumers to know exactly what they are eating.

In 1992, the U.S. Department of Agriculture (USDA) issued a set of
guidelines regarding what constitutes a healthy diet, depicted in the form of
a pyramid (Figure 30.2). These guidelines considered the basis of a healthy

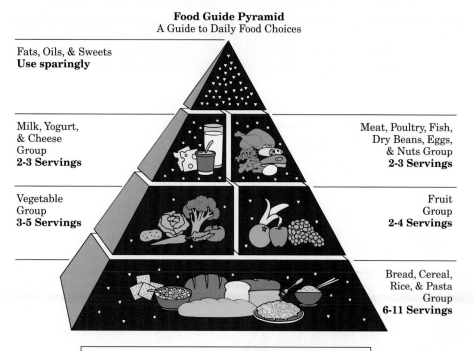

Figure 30.2 The Food Guide
Pyramid developed by the U.S.
Department of Agriculture as a
general guide to a healthful diet.

USDA, 1992

FOOD GROUP SERVING SIZES

Bread, Cereal, Rice, and Pasta
  1/2 cup cooked cereal, rice, pasta
  1 ounce dry cereal
  1 slice bread
  2 cookies
  1/2 medium doughnut

Vegetables
  1/2 cup cooked or raw chopped
    vegetables
  1 cup raw leafy vegetables
  3/4 cup vegetable juice
  10 french fries

Fruit
  1 medium apple, banana, or orange
  1/2 cup chopped, cooked, or canned
    fruit
  3/4 cup fruit juice
  1/4 cup dried fruit

Milk, Yogurt, and Cheese
  1 cup milk or yogurt
  $1\frac{1}{2}$ ounces natural cheese
  2 ounces processed cheese
  2 cups cottage cheese
  $1\frac{1}{2}$ cups ice cream
  1 cup frozen yogurt

Meat, Poultry, Fish, Dry Beans, Eggs, and Nuts
  2–3 ounces cooked lean meat, fish,
    or poultry
  2–3 eggs
  4–6 tablespoons peanut butter
  $1\frac{1}{2}$ cups cooked dry beans
  1 cup nuts

Fats, Oils, and Sweets
  Butter, mayonnaise, salad dressing,
    cream cheese, sour cream, jam, jelly

CHEMICAL CONNECTIONS 30A

The New Food Guide Pyramid

Over time, scientists began questioning some aspects of the original food pyramid shown in Figure 30.2. For example, certain types of fat are known to be essential to health and actually reduce the risk of heart disease. Also, little evidence exists to back up the claim that a high intake of carbohydrates is beneficial, although for certain sports it is essential. The original pyramid glorified carbohydrates while casting all fats as the "bad guys." In fact, plenty of evidence does link the consumption of saturated fat with high cholesterol and risk of heart disease—but mono- and polyunsaturated fats have the opposite effect. While many scientists recognized the distinction between the various types of fats, they felt that the average person would not understand them, so the original pyramid was designed to just send a simple message: "Fat is bad." The implied corollary to "fat is bad" was that "carbohydrates are good." However, after years of study, no evidence proved that a diet in which 30% or fewer of the calories come from fat is healthier than one with a higher level of fat consumption.

In an attempt to reconcile the latest nutritional data while presenting them in a form understandable to the average person, the USDA has created a website at www.mypyramid.gov. This interactive website allows visitors to take a brief tutorial about the new pyramid as well as to calculate their ideal amount of the various food types. In essence, the new pyramid starts with the old pyramid and tips it on its side. This view does not suggest that one food group is better than another; rather, it shows that proper nutrition is a blend of all the food groups. The nutrients are classified into six groups: grains, vegetables, fruits, oils, milk, and meat and beans. For each group, the pyramid describes the amounts and varieties that should be consumed. A big difference between this pyramid and the original version is that this one includes a section dedicated to exercise. There is also an expanded section on foods whose intakes should be limited, such as certain fats, sugars, and salts.

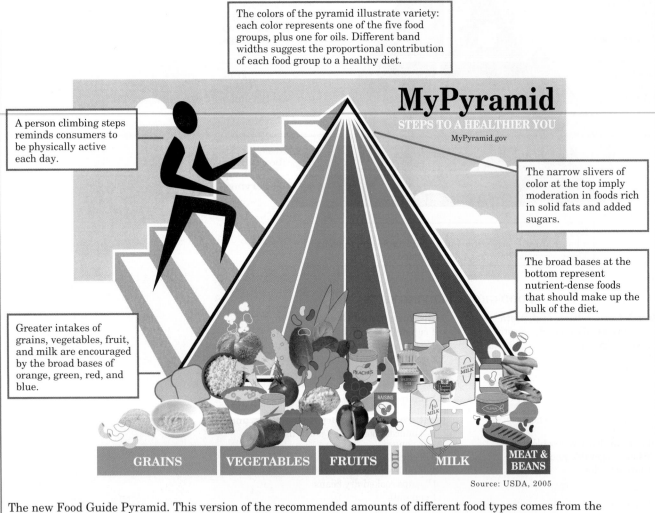

The colors of the pyramid illustrate variety: each color represents one of the five food groups, plus one for oils. Different band widths suggest the proportional contribution of each food group to a healthy diet.

A person climbing steps reminds consumers to be physically active each day.

The narrow slivers of color at the top imply moderation in foods rich in solid fats and added sugars.

The broad bases at the bottom represent nutrient-dense foods that should make up the bulk of the diet.

Greater intakes of grains, vegetables, fruit, and milk are encouraged by the broad bases of orange, green, red, and blue.

MyPyramid
STEPS TO A HEALTHIER YOU
MyPyramid.gov

GRAINS   VEGETABLES   FRUITS   OIL   MILK   MEAT & BEANS

Source: USDA, 2005

The new Food Guide Pyramid. This version of the recommended amounts of different food types comes from the latest research compiled by the USDA. See its website at www.mypyramid.gov for a tutorial. (*USDA, 2005*)

diet to be the foods richest in starch (bread, rice, and so on), plus lots of fruits and vegetables (which are rich in vitamins and minerals). Protein-rich foods (meat, fish, dairy products) were to be consumed more sparingly, and fats, oils, and sweets were not considered necessary at all. The pyramid shape demonstrated the relative importance of each type of food group, with the most important forming the base and the least important or unnecessary appearing at the top. This pictorial description has been used in many textbooks and taught in schools to children of all ages since its initial publication. However, the USDA has recently revised the information and appearance of the food pyramid. Chemical Connections 30A discusses the most recent version of the pyramid.

An important non-nutrient in some foods is **fiber,** which generally consists of the indigestible portion of vegetables and grains. Lettuce, cabbage, celery, whole wheat, brown rice, peas, and beans are all high in fiber. Chemically, fiber is made of cellulose, which, as we saw in Section 20.5C, cannot be digested by humans. Although we cannot digest it, fiber is necessary for proper operation of the digestive system; without it, constipation may result. In more serious cases, a diet lacking sufficient fiber may lead to colon cancer. The DRI recommendation is to ingest 35 g/day for men age 50 years and younger, and 25 g/day for women of the same age.

Fiber The cellulose-based non-nutrient component in our food

Basal caloric requirement The caloric requirement for a resting body

30.2 | Why Do We Count Calories?

The largest part of our food supply goes to provide energy for our bodies. As we saw in Chapters 27 and 28, this energy comes from the oxidation of carbohydrates, fats, and proteins. The energy derived from food is usually measured in calories. One nutritional calorie (Cal) equals 1000 cal or 1 kcal. Thus, when we say that the average daily nutritional requirement for a young adult male is 3000 Cal, we mean the same amount of energy needed to raise the temperature of 3000 kg of water by 1°C or of 30 kg (64 lb) of water by 100°C (Section 1.9B). A young adult female needs 2100 Cal/day. These are peak requirements—children and older people, on average, require less energy. Keep in mind that these energy requirements apply to active people. For bodies completely at rest, the corresponding energy requirement for young adult males is 1800 Cal/day, and that for females is 1300 Cal/day. The requirement for a resting body is called the **basal caloric requirement.**

An imbalance between the caloric requirement of the body and the caloric intake creates health problems. Chronic caloric starvation exists in many parts of the world where people simply do not have enough food to eat because of prolonged drought, the devastation of war, natural disasters, or overpopulation. Famine particularly affects infants and children. Chronic starvation, called **marasmus,** increases infant mortality by as much as 50%. It results in arrested growth, muscle wasting, anemia, and general weakness. Even if starvation is later alleviated, it leaves permanent damage, insufficient body growth, and lowered resistance to disease.

At the other end of the caloric spectrum is excessive caloric intake. It results in *obesity,* or the accumulation of body fat. Obesity is becoming an epidemic in the U.S. population, with important consequences: It increases the risk of hypertension, cardiovascular disease, and diabetes. Obesity is defined by the National Institutes of Health as applying to a person who has a body mass index (BMI) of 30 or greater. The BMI is a measure of body fat based on height and weight that applies to both adult men and women. For example, a person 70 inches tall is *normal* (BMI less than 25) if he or she weighs 174 lb or less. A person of the same height is *overweight* if he or she

Different activity levels are associated with different caloric needs.

weighs more than 174 lb but less than 209 lb; an individual is *obese* if he or she weighs more than 209 lb. More than 97 million Americans are overweight or obese.

Reducing diets aim at decreasing caloric intake without sacrificing any essential nutrients. A combination of exercise and lower caloric intake can eliminate obesity, but usually these diets must achieve their goal over an extended period. Crash diets give the illusion of quick weight loss, but most of this decrease is due to loss of water, which can be regained very quickly. To reduce obesity, we must lose body fat, not water. Achieving this goal takes a lot of effort, because fats contain so much energy. A pound of body fat is equivalent to 3500 Cal. Thus, to lose 10 lb, it is necessary either to consume 35,000 fewer Cal, which can be achieved if one reduces caloric intake by 350 Cal every day for 100 days (or by 700 Cal daily for 50 days) or uses up, through exercise, the same number of food calories.

30.3 | How Does the Body Process Dietary Carbohydrates?

Carbohydrates are the major source of energy in the diet. They also furnish important compounds for the synthesis of cell components (Chapter 29). The main dietary carbohydrates are the polysaccharide starch, the disaccharides lactose and sucrose, and the monosaccharides glucose and fructose. Before the body can absorb carbohydrates, it must break down di-, oligo-, and polysaccharides into monosaccharides, because only monosaccharides can pass into the bloodstream.

The monosaccharide units are connected to each other by glycosidic bonds. Glycosidic bonds are cleaved by hydrolysis. In the body, this hydrolysis is catalyzed by acids and by enzymes. When a metabolic need arises, storage polysaccharides—amylose, amylopectin, and glycogen—are hydrolyzed to yield glucose and maltose.

This hydrolysis is aided by a number of enzymes:

- **α-Amylase** attacks all three storage polysaccharides at random, hydrolyzing the α-1,4-glycosidic bonds,
- **β-Amylase** also hydrolyzes the α-1,4-glycosidic bonds but in an orderly fashion, cutting disaccharidic maltose units one by one from the nonreducing end of a chain.
- The **debranching enzyme** hydrolyzes the α-1,6-glycosidic bonds (Figure 30.3).

In acid-catalyzed hydrolysis, storage polysaccharides are cut at random points. At body temperature, acid catalysis is slower than the enzyme-catalyzed hydrolysis.

The digestion (hydrolysis) of starch and glycogen in our food supply starts in the mouth, where α-amylase is one of the main components of saliva. Hydrochloric acid in the stomach and other hydrolytic enzymes in the intestinal tract hydrolyze starch and glycogen to produce mono- and disaccharides (D-glucose and maltose).

D-glucose enters the bloodstream and is carried to the cells to be utilized (Section 28.2). For this reason, D-glucose is often called blood sugar. In healthy people, little or none of this sugar ends up in the urine except for short periods of time (binge eating). In diabetes mellitus, however, glucose is not completely metabolized and does appear in the urine. As a consequence, it is necessary to test the urine of diabetic patients for the presence of glucose (Chemical Connections 20C).

Figure 30.3 The action of different enzymes on glycogen and starch.

CHEMICAL CONNECTIONS 30B

Why Is It So Hard to Lose Weight?

One of the great tragedies of being human is that it is far too easy to gain weight and far too difficult to lose it. If we had to analyze the specific chemical reactions involved in this reality, we would look very carefully at the citric acid cycle, especially the decarboxylation reactions. Of course, all foods consumed in excess can be stored as fat. This is true for carbohydrates, proteins, and, of course, fats. In addition, these molecules can be interconverted, with the exception that fats cannot give a net yield of carbohydrates. Why can fats not yield carbohydrates? The only way that a fat molecule could make glucose would be to enter the citric acid cycle as acetyl-CoA and then be drawn off as oxaloacetate for gluconeogenesis (Section 29.2). Unfortunately, the two carbons that enter (in the citric acid cycle) are effectively lost by the decarboxylations (Section 27.4). This leads to an imbalance in the catabolic versus anabolic pathways.

All roads lead to fats, but fats cannot lead back to carbohydrates. Humans are very sensitive to glucose levels in the blood because so much of our metabolism is geared toward protecting our brain cells, which prefer glucose as a fuel. If we eat more carbohydrates than we need, the excess carbohydrates will turn to fats. As we know, it is very easy to put on fat especially as we age.

What about the reverse? Why don't we just stop eating? Won't that reverse the process? Yes and no. When we start eating less, fat stores become mobilized for energy production. Fat is an excellent source of energy because it forms acetyl-CoA and gives a steady

influx for the citric acid cycle. Thus we can lose some weight by reducing caloric intake. Unfortunately, our blood sugar will also drop as soon as our glycogen stores run out. We have very little stored glycogen that could be tapped to maintain our blood glucose levels.

When the blood glucose drops, we become depressed, sluggish, and irritable. We start having negative thoughts like, "this dieting thing is really stupid. I should eat a pint of Oreo cookie ice cream." If we continue the diet, and given that we cannot turn fats into carbohydrates, where will the blood glucose come from? Only one source is left—proteins. Proteins will be degraded to amino acids, and eventually be converted to pyruvate for gluconeogenesis. Thus we will begin to lose muscle as well as fat.

There is a bright side to this process, however. Using our knowledge of biochemistry, we can see that there is a better way to lose weight than dieting—exercise! If you exercise correctly, you can train your body to use fats to supply acetyl-CoA for the citric acid cycle. If you consume a normal diet, you will maintain your blood glucose and not degrade proteins for that purpose; your ingested carbohydrates will be sufficient to maintain both blood glucose and carbohydrate stores. With the proper ratio of exercise to food intake, and the proper balance of the right types of nutrients, we can increase the breakdown of fats without sacrificing carbohydrate stores or proteins. In essence, it is easier and healthier to train off the weight than to diet off the weight. This fact has been known for a long time. Now we are in a position to see why it is so, biochemically.

The latest DRI guideline, issued by the National Academy of Sciences in 2002, recommends a minimum carbohydrate intake of 130 g/day. Most people exceed this value. Artificial sweeteners (Chemical Connections 30C) can be used to reduce mono- and disaccharide intake.

Not everyone considers this recommendation to be the final word. Certain diets, such as the Atkins diet, recommend reducing carbohydrate intake to force the body to burn stored fat for energy supply. For example, during the introductory two-week period in the Atkins diet, only 20 g of carbohydrates per day are recommended in the form of salad and vegetables; no fruits or starchy vegetables are allowed. For the longer term, an additional 5 g of carbohydrates per day is added in the form of fruits. This restriction induces ketosis, the production of ketone bodies (Section 28.7) that may create muscle weakness and kidney problems.

Many fad diets exist today. The Atkins diet was preceded by the Zone Diet and the Sugar Buster's Diet, both of which attempted to limit carbohydrate intake. Another diet suggests that you match the foods you eat to your ABO blood type. To date, little scientific evidence supports any of these approaches, although some aspects of many diets have merit.

30.4 | How Does the Body Process Dietary Fats?

Fats are the most concentrated source of energy. About 98% of the lipids in our diet are fats and oils (triglycerides); the remaining 2% consist of complex lipids and cholesterol.

The lipids in the food we eat must be hydrolyzed into smaller components before they can be absorbed into the blood or lymph system through the intestinal walls. The enzymes that promote this hydrolysis are located in the small intestine and are called *lipases*. However, because lipids are insoluble in the aqueous environment of the gastrointestinal tract, they must be dispersed into fine colloidal particles before the enzymes can act on them.

Bile salts (Section 21.11) perform this important function. Bile salts are manufactured in the liver from cholesterol and stored in the gallbladder. From there, they are secreted through the bile ducts into the intestine. Lipases act on the emulsion produced by bile salts and dietary fats, breaking the fats into glycerol and fatty acids and the complex lipids into fatty acids, alcohols (glycerol, choline, ethanolamine, sphingosine), and carbohydrates. These hydrolysis products are then absorbed through the intestinal walls.

Only two fatty acids are essential in higher animals, including humans: linolenic and linoleic acids (Section 21.3). Nutritionists occasionally list arachidonic acid as an **essential fatty acid.** In reality, our bodies can synthesize arachidonic acid from linoleic acid.

30.5 | How Does the Body Process Dietary Protein?

Although the proteins in our diet can be used for energy (Section 28.9), their main use is to furnish amino acids from which the body synthesizes its own proteins (Section 26.5).

The digestion of dietary proteins begins with cooking, which denatures proteins. (Denatured proteins are hydrolyzed more easily by hydrochloric acid in the stomach and by digestive enzymes than are native proteins.) *Stomach acid* contains about 0.5% HCl. This HCl both denatures the proteins and hydrolyzes the peptide bonds randomly. *Pepsin,* the proteolytic enzyme of stomach juice, hydrolyzes peptide bonds on the amino side of the aromatic amino acids: tryptophan, phenylalanine, and tyrosine (see Figure 30.4).

Most protein digestion occurs in the small intestine. There, the enzyme *chymotrypsin* hydrolyzes internal peptide bonds at the same amino acids as does pepsin, except it does so on the other side, leaving these amino acids as the carboxyl termini of their fragments. Another enzyme, *trypsin,* hydrolyzes them only on the carboxyl side of arginine and lysine. Other enzymes, such as *carboxypeptidase,* hydrolyze amino acids one by one from the C-terminal end of the protein. The amino acids and small peptides are then absorbed through the intestinal walls.

Figure 30.4 Different enzymes hydrolyze peptide chains in different but specific ways. Note that both chymotrypsin and pepsin would hydrolyze all of the same amino acids, but they are shown here hydrolyzing separate ones for comparison of the side they hydrolyze.

The human body is incapable of synthesizing ten of the amino acids in sufficient quanities needed to make proteins. These ten **essential amino acids** must be obtained from our food; they are shown in Table 29.1. The body hydrolyzes food proteins into their amino acid constituents and then puts the amino acids together again to make body proteins. For proper nutrition, the human diet should contain about 20% protein.

A dietary protein that contains all of the essential amino acids is called a **complete protein.** Casein, the protein of milk, is a complete protein, as are most other animal proteins—those found in meat, fish, and eggs. People who eat adequate quantities of meat, fish, eggs, and dairy products get all the amino acids they need to keep healthy. About 50 g/day of complete proteins constitutes an adequate quantity.

An important animal protein that is not complete is gelatin, which is made by denaturing collagen (Section 22.12). Gelatin lacks tryptophan and is low in several other amino acids, including isoleucine and methionine. Many people on quick-reducing diets consume "liquid protein." This substance is simply denatured and partially hydrolyzed collagen (gelatin). Therefore, if it is the only protein source in the diet, some essential amino acids will be lacking.

Most plant proteins are incomplete. For example, corn protein lacks lysine and tryptophan; rice protein lacks lysine and threonine; wheat protein lacks lysine; and legumes are low in methionine and cysteine. Even soy protein, one of the best plant proteins, is very low in methionine. Adequate amino acid nutrition is possible with a vegetarian diet, but only if a wide range of vegetables is eaten. **Protein complementation** is one such diet. In protein complementation, two or more foods complement the others' deficiencies. For example, grains and legumes complement each other, with grains being low in lysine but high in methionine. Over time, such protein complementation in vegetarian diets became the staple in many parts of the world—corn tortillas and beans in Central and South America, rice and lentils in India, and rice and tofu in China and Japan.

In many developing countries, protein deficiency diseases are widespread because the people get their protein mostly from plants. Among these diseases is **kwashiorkor,** whose symptoms include a swollen stomach, skin discoloration, and retarded growth.

Proteins are inherently different from carbohydrates and fats when it comes to their relationship to the diet. Unlike the other two fuel sources, proteins have no storage form. If you eat a lot of carbohydrate, you will store glucose in the form of glycogen. If you eat a lot of anything, you will store fat. However, if you eat a lot of protein (more than required for your needs), there is no place to store extra protein. Protein in excess will be metabolized to other substances, such as fat. For this reason, you must consume adequate protein every day. This requirement is especially critical in athletes and growing children. If an athlete works out intensely one day but eats incomplete protein, he or she cannot repair the damaged muscles. The fact that the athlete may have eaten an excess of a complete protein the day before will not help.

Essential amino acid An amino acid that the body cannot synthesize in the required amounts and so must be obtained in the diet

30.6 | What Is the Importance of Vitamins, Minerals, and Water?

Vitamins and **minerals** are essential for good nutrition. Animals maintained on diets that contain sufficient carbohydrates, fats, and proteins and provided with an ample water supply cannot survive on these alone; they

CHEMICAL CONNECTIONS 30C

Dieting and Artificial Sweeteners

Obesity is a serious problem in the United States. Many people would like to lose weight but are unable to control their appetites enough to do so, which explains why artificial sweeteners are so popular. These substances have a sweet taste but do not add calories. Many people restrict their sugar intake. Some are forced to do so by diseases such as diabetes; others, by the desire to lose weight. Because most of us like to eat sweet foods, artificial sweeteners are added to many foods and drinks for those who must (or want to) restrict their sugar intake.

Four noncaloric artificial sweeteners are approved by the U.S. Food and Drug Administration (FDA). The oldest of these substances, saccharin, is 450 times sweeter than sucrose. Saccharin has been in use for about 100 years. Unfortunately, some tests have shown that this sweetener, when fed in massive quantities to rats, caused some of these rats to develop cancer. Other tests have given negative results. Recent research has shown that this cancer in rats is not relevant to human consumption. Saccharin continues to be sold on the market.

A newer artificial sweetener, aspartame, does not have the slight aftertaste that saccharin does. Aspartame is the methyl ester of a dipeptide, Asp-Phe (aspartylphenylalanine). Its sweetness was discovered in 1969. After extensive biological testing, it was approved by the FDA in 1981 for use in cold cereals, drink mixes, gelatins, and as tablets or powder to be used as a sugar substitute. Aspartame is 100 to 150 times sweeter than sucrose. It is made from natural amino acids, so both the aspartic acid and the phenylalanine have the L configuration. Other possibilities have also been synthesized: the L-D, D-L and D-D configurations. However, they are all bitter rather than sweet.

Aspartame is sold under such brand names as Equal and NutraSweet. A newer version of aspartame, neotame, was approved by the FDA in 2002. Twenty-five times sweeter than aspartame, it is essentially the same compound but with one —H substituted by a —CH₂—CH₂—C(CH₃)₃ group at the amino terminus.

A third artificial sweetener, acesulfame-K, is 200 times sweeter than sucrose and is used (under the name Sunette) mostly in dry mixes. The fourth option, Sucralose, is used in sodas, baked goods, and tabletop packets. This trichloro derivative of a disaccharide is 600 times sweeter than sucrose and has no aftertaste. Neither Sucralose nor acesulfame-K is metabolized in the body; that is, they pass through unchanged.

Sucralose

Of course, calories in the diet do not come solely from sugar. Dietary fat is an even more important source (Section 30.4). It has long been hoped that some kind of artificial fat, which would taste the same but have no (or only a few) calories, would help in losing weight. Procter & Gamble Company has developed such a product, called Olestra. Like natural fats, this molecule is a carboxylic ester; instead of glycerol, however, the alcohol component is sucrose (Section 20.4A). All eight of sucrose's OH groups are converted to ester groups; the carboxylic acids are the fatty acids containing eight to ten carbons. The one illustrated here is made with the C₁₀ decanoic (capric) acid.

Olestra

Although Olestra has a chemical structure similar to a fat, the human body cannot digest it because the enzymes that digest ordinary fats are not designed to work with the particular size and shape of this molecule. Therefore, it passes through the digestive system unchanged, and we derive no calories from it. Olestra can be used instead of an ordinary fat in cooking such items as cookies and potato chips. People who eat these foods will then be consuming fewer calories.

Saccharin   Aspartame

Acesulfame-K

also need the essential organic components called vitamins and the inorganic ions called minerals. Many vitamins, especially those in the B group, act as coenzymes and inorganic ions as cofactors in enzyme-catalyzed reactions (Table 30.1). Table 30.2 lists the structures, dietary sources, and functions of the vitamins and minerals. Deficiencies in vitamins and minerals lead to many nutritionally controllable diseases (one example is shown in Figure 30.5); these conditions are also listed in Table 30.2.

The recent trend in vitamin appreciation is connected to their general role rather than to any specific action they have against a particular disease. For example, today the role of vitamin C in prevention of scurvy is barely mentioned but it is hailed as an important antioxidant. Similarly, other antioxidant vitamins or vitamin precursors dominate the medical literature. As an example, it has been shown that consumption of carotenoids (other than β-carotene) and vitamins E and C contributes significantly to respiratory health. The most important of the three is vitamin E. Furthermore, the loss of vitamin C during hemodialysis contributes significantly to oxidative damage in patients, leading to accelerated atherosclerosis.

Figure 30.5   Symptoms of rickets, a vitamin D deficiency in children. The nonmineralization of the bones of the radius and the ulna results in prominence of the wrist.

Table 30.1   Vitamins and Trace Elements as Coenzymes and Cofactors

Vitamin/Trace Element	Form of Coenzyme	Representative Enzyme	Reference
B$_1$, thiamine	Thiamine pyrophosphate, TPP	Pyruvate dehydrogenase	Step 12, Section 28.2
B$_2$, riboflavin	Flavin adenine dinucleotide, FAD	Succinate dehydrogenase	Step 6, Section 27.4
Niacin	Nicotinamide adenine dinucleotide, NAD$^+$	D-Glyceraldehyde-3-phosphate dehydrogenase	Step 5, Section 28.2
Panthotenic acid	Coenzyme A, CoA	Fatty acid synthase	Step 1, Section 29.3
B$_6$, pyridoxal	Pyridoxal phosphate, PLP	Aspartate amino transferase	Class 2, Section 23.2
B$_{12}$, cobalamine		Ribose reductase	Step 1, Section 25.2
Biotin	N-carboxybiotin	Acetyl-CoA carboxylase	Malonyl-CoA, Section 29.3
Folic acid		Purine biosynthesis	Section 25.2
Mg		Pyruvate kinase	Chemical Connections 23C
Fe		Cytochrome oxidase	Section 27.5
Cu		Cytochrome oxidase	Section 27.5
Zn		DNA polymerase	Section 25.6
Mn		Arginase	Step 5, Section 28.8
K		Pyruvate kinase	Chemical Connections 23C
Ni		Urease	Section 23.1
Mo		Nitrate reductase	
Se		Glutathione peroxidase	Chemical Connections 22A

Table 30.2 Vitamins and Minerals: Sources, Functions, Deficiency Diseases, and Daily Requirements

Name	Structure	Best Food Source	Function	Deficiency Symptoms and Diseases	Recommended Dietary Allowance[a]
Fat-Soluble Vitamins					
A	(structure)	Liver, butter, egg yolk, carrots, spinach, sweet potatoes	Vision; to heal eye and skin injuries	Night blindness; blindness; keratinization of epithelium and cornea	800 μg (1500 μg)[b]
D	(structure)	Salmon, sardines, cod liver oil, cheese, eggs, milk	Promotes calcium and phosphate absorption and mobilization	Rickets (in children): pliable bones; osteomalacia (in adults); fragile bones	5–10 μg; exposure to sunlight
E	(structure)	Vegetable oils, nuts, potato chips, spinach	Antioxidant	In cases of malabsorption such as in cystic fibrosis: anemia. In premature infants: anemia	8–10 mg
K	(structure)	Spinach, potatoes, cauliflower, beef liver	Blood clotting	Uncontrolled bleeding (mostly in newborn infants)	65–80 μg
Water-Soluble Vitamins					
B$_1$ (thiamine)	(structure)	Beans, soybeans, cereals, ham, liver	Coenzyme in oxidative decarboxylation and in pentose phosphate shunt	Beriberi. In alcoholics: heart failure; pulmonary congestion	1.1 mg
B$_2$ (riboflavin)	(structure)	Kidney, liver, yeast, almonds, mushrooms, beans	Coenzyme of oxidative processes	Invasion of cornea by capillaries; cheilosis; dermatitis	1.4 mg

Vitamin	Food sources	Function	Deficiency disease	Amount
Nicotinic acid (niacin)	Chickpeas, lentils, prunes, peaches, avocados, figs, fish, meat, mushrooms, peanuts, bread, rice, beans, berries	Coenzyme of oxidative processes	Pellagra	15–18 mg
B_6 (pyridoxal)	Meat, fish, nuts, oats, wheat germ, potato chips	Coenzyme in transamination; heme synthesis	Convulsions; chronic anemia; peripheral neuropathy	1.6–2.2 mg
Folic acid	Liver, kidney, eggs, spinach, beets, orange juice, avocados, cantaloupe	Coenzyme in methylation and in DNA synthesis	Anemia	400 μg
B_{12}	Oysters, salmon, liver, kidney	Part of methyl-removing enzyme in folate metabolism	Patchy demyelination; degradation of nerves, spinal cord, and brain	1–3 μg

Table 30.2 Vitamins and Minerals: Sources, Functions, Deficiency Diseases, and Daily Requirements (continued)

Name	Structure	Function	Deficiency Symptoms and Diseases	Recommended Dietary Allowance[a]
Pantothenic acid	$\text{HOCH}_2\text{C}-\text{CHCNHCH}_2\text{CH}_2\text{COOH}$ with CH_3, CH_3, O, OH groups	Part of CoA; fat and carbohydrate metabolism	Gastrointestinal disturbances; depression	4–7 mg
Biotin	(bicyclic structure with HN, NH, S, $(\text{CH}_2)_4\text{COOH}$)	Synthesis of fatty acids	Dermatitis; nausea; depression	30–100 μg
C (ascorbic acid)	(ring structure with OH, OH, $\text{CH}-\text{OH}$, CH_2OH, O)	Hydroxylation of collagen, wound healing; bond formation; antioxidant	Scurvy; capillary fragility	60 mg
Minerals				
Potassium		Provides membrane potential	Muscle weakness	3500 mg
Sodium		Osmotic pressure	None	2000–2400 mg
Calcium		Bone formation; hormonal function; blood coagulation; muscle contraction	Muscle cramps; osteoporosis; fragile bones	800–1200 mg
Chloride		Osmotic pressure	None	1700–5100 mg
Phosphorus		Balancing calcium in diet	Excess causes structural weakness in bones	800–1200 mg

Best Food Source:
- Pantothenic acid: Peanuts, buckwheat, soybeans, broccoli, lima beans, liver kidney, brain, heart
- Biotin: Yeast, liver, kidney, nuts, egg yolk
- C (ascorbic acid): Citrus fruit, berries, broccoli, cabbage, peppers, tomatoes
- Potassium: Apricots, bananas, dates, figs, nuts, raisins, beans, chickpeas, cress, lentils
- Sodium: Meat, cheese, cold cuts, smoked fish, table salt
- Calcium: Milk, cheese, sardines, caviar
- Chloride: Meat, cheese, cold cuts, smoked fish, table salt
- Phosphorus: Lentils, nuts, oats, grain flours, cocoa, egg yolk, cheese, meat (brain, sweetbreads)

Element	Sources	Function	Deficiency	RDA[a]
Magnesium	Cheeses, cocoa, chocolate, nuts, soybeans, beans	Cofactor in enzymes	Hypocalcemia	280–350 mg
Iron	Raisins, beans, chickpeas, parsley, smoked fish, liver, kidney, spleen, heart, clams, oysters	Oxidative phosphorylation; hemoglobin	Anemia	15 mg
Zinc	Yeast, soybeans, nuts, corn, cheese, meat, poultry	Cofactor in enzymes, insulin	Retarded growth; enlarged liver	12–15 mg
Copper	Oysters, sardines, lamb, liver	Oxidative enzymes cofactor	Loss of hair pigmentation, anemia	1.5–3 mg
Manganese	Nuts, fruits, vegetables, whole-grain cereals	Bone formation	Low serum cholesterol levels; retarded growth of hair and nails	2.0–5.0 mg
Chromium	Meat, beer, whole wheat and rye flours	Glucose metabolism	Glucose not available to cells	0.05–0.2 mg
Molybdenum	Liver, kidney, spinach, beans, peas	Protein synthesis	Retarded growth	0.075–0.250 mg
Cobalt	Meat, dairy products	Component of vitamin B_{12}	Pernicious anemia	0.05 mg (20–30 mg)[b]
Selenium	Meat, seafood	Fat metabolism	Muscular disorders	0.05–0.07 mg (2.4–3.0 mg)[b]
Iodine	Seafood, vegetables, meat	Thyroid glands	Goiter	150–170 μg (1000 μg)[b]
Fluorine	Fluoridated water; fluoridated toothpaste	Enamel formation	Tooth decay	1.5–4.0 mg (8–20 mg)[b]

[a]The U.S. RDAs are set by the Food and Nutrition Board of the National Research Council. The numbers given here are based on the latest recommendations (*National Research Council Recommended Dietary Allowances*, 10th ed., 1989, National Academy Press, Washington, D.C.). The RDA varies with age, sex, and level of activity; the numbers given are average values for both sexes between the ages of 18 and 54.

[b]Toxic if doses above the level shown in parentheses are taken.

CHEMICAL CONNECTIONS 30D

Food for Performance Enhancement

Athletes do whatever they can to enhance their performance. While the press focuses on those illegal methods employed by some athletes, such as using steroids or erythropoietin (EPO), many athletes continue to seek legal ways of enhancing their performance through diet and diet supplements. Any substance that aids performance is called an **ergogenic aid.**

After vigorous exercise lasting 30 minutes or more, performance typically declines because the glycogen stores in the muscle are depleted. After 90 minutes to 2 hours, liver glycogen stores also become depleted. There are two ways to combat this outcome. First, one can start the event with a full load of glycogen in the muscle and liver. This is why many athletes load up on carbohydrates in the form of pasta or other high-carbohydrate meals in the days before the event. Second, one can maintain the blood glucose levels during the event, so that liver glycogen does not have to be used for this purpose and so that ingested sugars can help support the athlete's energy needs, sparing some of the muscle and liver glycogen. This is why athletes often consume energy bars and sport drinks during an event.

The most often used ergogenic aid, although many athletes might not realize it, is caffeine. Caffeine works in two different ways. First, it acts as a general stimulant of the central nervous system, giving the athlete the sensation of having a lot of energy. Second, it stimulates the breakdown of fatty acids for fuel via its role as an activator of the lipases that break triacylglycerols down (Section 30.4). Caffeine is a "double-edged sword," however, because it can also cause dehydration and actually inhibits the breakdown of glycogen.

In recent years, a new performance-enhancer food has appeared on the market and quickly become a best-seller: creatine. It is sold over the counter. Creatine is a naturally occurring amino acid in the muscles, which store energy in the form of high-energy phosphocreatine. During a short and strenuous bout of exercise, such as a 100-meter sprint, the muscles first use up the ATP obtainable from the reaction of phosphocreatine with ADP; only then do they rely on the glycogen stores.

Creatine is sold over the counter.

Both creatine and carbohydrates are natural food and body components and, therefore, cannot be considered equivalent to banned performance enhancers such as anabolic steroids or "andro" (Chemical Connections 21F). They are beneficial in improving performance in sports where short bursts of energy are needed, such as weight lifting, jumping, and sprinting. Lately, creatine has been used experimentally to preserve muscle neurons in degenerative diseases such as Parkinson's disease, Huntington's disease, and muscular dystrophy. Creatine also has few known hazards, even with long-term use. It is a highly nitrogenated compound, however, and overuse of creatine leads to the same problems as eating a diet characterized by excessive protein. The molecule must be hydrated, so water can be tied up with creatine that should be hydrating the body. The kidneys must also deal with excretion of extra nitrogen.

While athletes spend a lot of time and money on commercial ergogenic aids, the most important ergogenic aid remains water, the elixir of life that has been almost forgotten as it is replaced by its more expensive sports-drink cousins. A 1% level of dehydration during an event can adversely affect athletic performance. For a 150-pound athlete, this means losing 1.5 pounds of water as sweat. An athlete could easily lose much more than that in running a 10-kilometer race on all but the coldest days. To make things more difficult, performance is affected before thirst becomes noticeable. Only by drinking often during a long event can dehydration be prevented. Water must be consumed either alone or as part of a sports drink before you notice that you are thirsty.

Phosphocreatine   Creatine

Some vitamins have unusual effects besides their normal participation as coenzymes in many metabolic pathways or action as antioxidants in the body. Among these are the well-known effect of riboflavin, vitamin B_2, as a photosensitizer. Niacinamide, the amide form of vitamin B (niacin), is used in megadoses (2 g/day) to treat an autoimmune disease called bullus pemphigoid, which causes blisters on the skin. Conversely, the same megadoses in healthy patients may cause harm.

Although the concept of RDA has been used since 1940 and periodically updated as new knowledge dictated, a new concept is being developed in the field of nutrition. The DRI, currently in the process of completion, is designed to replace the RDA and is tailor-made to different ages and genders. It gives a set of two to four values for a particular nutrient in RDI:

- The estimated average requirement
- The recommended dietary allowance
- The adequate intake level
- The tolerable upper intake level

For example, the RDA for vitamin D is 5–10 μg, the adequate intake listed in the DRI for vitamin D for a person between age 9 and 50 years is 5 μg, and the tolerable upper intake level for the same age group is 50 μg.

A third set of standards appears on food labels in the United States, the Daily Values discussed in Section 30.2. Each gives a single value for each nutrient and reflects the need of an average healthy person eating 2000 to 2500 calories per day. The Daily Value for vitamin D, as it appears on vitamin bottles, is 400 International Units, which is 10 μg, the same as the RDA.

Water makes up 60% of our body weight. Most of the compounds in our body are dissolved in water, which also serves as a transporting medium to carry nutrients and waste materials. We must maintain a proper balance between water intake and water excretion via urine, feces, sweat, and exhalation of breath. A normal diet requires about 1200 to 1500 mL of water per day, in addition to the water consumed as part of our foods. Public drinking water systems in the United States are regulated by the Environmental Protection Agency (EPA), which sets minimum standards for protection of public health. Public water supplies are treated with disinfectants (typically chlorine) to kill microorganisms. Chlorinated water may have both an aftertaste and an odor. Private wells are not regulated by EPA standards (see end of chapter problem 7.96).

Bottled water may come from springs, streams, or public water sources. Because bottled water is classified as a food, it comes under the FDA supervision, which requires it to meet the same standards for purity and sanitation as tap water. Most bottled water is disinfected with ozone, which leaves no flavor or odor in water.

GOB
Chemistry ⚛ Now™

Click *Chemistry Interactive* to learn more about **Vitamins**

SUMMARY OF KEY QUESTIONS

SECTION 30.1 How Do We Measure Nutrition?

- **Nutrients** are components of foods that provide for growth, replacement, and energy.
- Nutrients are classified into six groups: carbohydrates, lipids, proteins, vitamins, minerals, and water.
- Each food contains a variety of nutrients. The largest part of our food intake is used to provide energy for our bodies.

SECTION 30.2 Why Do We Count Calories?

- A typical young adult needs 3000 Cal (male) or 2100 Cal (female) as an average daily caloric intake.
- **Basal caloric requirements,** the energy needed when the body is completely at rest, are less than the normal requirements.

- An imbalance between energy needs and caloric intake may create health problems. For example, chronic starvation increases infant mortality, whereas obesity leads to hypertension, cardiovascular disease, and diabetes.

SECTION 30.3 How Does the Body Process Dietary Carbohydrates?

- Carbohydrates are the major source of energy in the human diet.
- Monosaccharides are directly absorbed in the intestines, while oligo- and polysaccharides, such as starch, are digested with the aid of stomach acid, α- and β-amylases, and debranching enzymes.

SECTION 30.4 How Does the Body Process Dietary Fats?

- Fats are the most concentrated source of energy.
- Fats are emulsified by bile salts and digested by lipases before being absorbed as fatty acids and glycerol in the intestines.
- Essential fatty and amino acids are needed as building blocks because the human body cannot synthesize them.

SECTION 30.5 How Does the Body Process Dietary Protein?

- Proteins are hydrolyzed by stomach acid and further digested by enzymes such as pepsin and trypsin before being absorbed as amino acids.
- There is no storage form for proteins, so good protein sources must be consumed in the diet every day.

SECTION 30.6 What Is the Importance of Vitamins, Minerals, and Water?

- Vitamins and minerals are essential constituents of the diet that are needed in small quantities.
- The fat-soluble vitamins are A, D, E, and K.
- Vitamins C and the B group are water-soluble vitamins.
- Most of the B vitamins are essential coenzymes.
- The most important dietary minerals are Na^+, Cl^-, K^+, PO_4^{3-}, Ca^{2+}, Fe^{2+}, and Mg^{2+}, but trace minerals are also necessary.
- Water makes up 60% of body weight.

PROBLEMS

GOB Chemistry ⚛ Now™

Assess your understanding of this chapter's topics with additional quizzing and conceptual-based problems at **http://now.brookscole.com/gob8** or on the CD.

A blue problem number indicates an applied problem.

■ denotes problems that are available on the GOB ChemistryNow website or CD and are assignable in OWL.

SECTION 30.1 How Do We Measure Nutrition?

30.1 Are nutrient requirements uniform for everyone?

30.2 Is banana flavoring, isopentyl acetate, a nutrient?

30.3 ■ If sodium benzoate, a food preservative, is excreted as such, and if calcium propionate, another food preservative, is metabolized to CO_2 and H_2O, would you consider either of these preservatives to be a nutrient? If so, why?

30.4 Is corn grown solely with organic fertilizers more nutritious than that grown with chemical fertilizers?

30.5 ■ Which part of the Nutrition Facts label found on food packages is the same for all labels carrying it?

30.6 Of which kinds of food does the U.S. government recommend that we have the most servings each day?

30.7 What is the importance of fiber in the diet?

30.8 Can a chemical that, in essence, goes through the body unchanged be an essential nutrient? Explain.

SECTION 30.2 Why Do We Count Calories?

30.9 A young adult female needs a caloric intake of 2100 Cal/day. Her basal caloric requirement is only 1300 Cal/day. Why is the extra 800 Cal needed?

30.10 What ill effects may obesity bring?

30.11 ■ Assume that you want to lose 20 lb of body fat in 60 days. Your present dietary intake is 3000 Cal/day. What should your caloric intake be, in Cal/day, to achieve this goal, assuming no change in exercise habits?

30.12 What is marasmus?

30.13 Diuretics help to secrete water from the body. Would diuretic pills be a good way to reduce body weight?

SECTION 30.3 How Does the Body Process Dietary Carbohydrates?

30.14 Humans cannot digest wood; termites do so with the aid of bacteria in their digestive tract. Is there a basic difference in the digestive enzymes present in humans and termites?

30.15 What is the product of the reaction when α-amylase acts on amylose?

30.16 Does HCl in the stomach hydrolyze both the 1,4- and 1,6-glycosidic bonds?

30.17 Beer contains maltose. Can beer consumption be detected by analyzing the maltose content of a blood sample?

SECTION 30.4 How Does the Body Process Dietary Fats?

30.18 ■ Which nutrient provides energy in its most concentrated form?

30.19 What is the precursor of arachidonic acid in the body?

30.20 How many (a) essential fatty acids and (b) essential amino acids do humans need in their diets?

30.21 ■ Do lipases degrade (a) cholesterol or (b) fatty acids?

30.22 What is the function of bile salts in digestion of fats?

SECTION 30.5 How Does the Body Process Dietary Protein?

30.23 Is it possible to get a sufficient supply of nutritionally adequate proteins by eating only vegetables?

30.24 Suggest a way to cure kwashiorkor.

30.25 What is the difference between protein digestion by trypsin and by HCl?

30.26 ■ Which one will be digested faster: (a) a raw egg or (b) a hard-boiled egg? Explain.

SECTION 30.6 What Is the Importance of Vitamins, Minerals, and Water?

30.27 In a prison camp during a war, the prisoners are fed plenty of rice and water but nothing else. What would be the result of such a diet in the long run?

30.28 (a) How many milliliters of water per day does a normal diet require?
   (b) How many calories does this amount of water contribute?

30.29 Why did British sailing ships carry a supply of limes?

30.30 What are the symptoms of vitamin A deficiency?

30.31 What is the function of vitamin K?

30.32 (a) Which vitamin contains cobalt?
   (b) What is the function of this vitamin?

30.33 Vitamin C is recommended in megadoses by some people for prevention of all kinds of diseases, ranging from colds to cancer. What disease has been scientifically proven to be prevented when sufficient daily doses of vitamin C are in the diet?

30.34 Why is the Recommended Dietary Allowance (RDA) being phased out in favor of the Daily Reference Intake (DRI)?

30.35 What are the nonspecific effects of vitamin E, C, and carotenoids?

30.36 What are the best dietary sources of calcium, phosphorus, and cobalt?

30.37 Which vitamins contain a sulfur atom?

30.38 What are the symptoms of vitamin B_{12} deficiency?

Chemical Connections

30.39 (Chemical Connections 30A) What is the difference between the original Food Guide Pyramid published in 1992 and the revised version presented here?

30.40 (Chemical Connections 30A) With respect to carbohydrates and fats, why are there multiple locations in the new Food Guide Pyramid for these nutrients?

30.41 (Chemical Connections 30B) Explain what is meant by this statement: "All nutrients in excess can turn into fat, but fat cannot be turned into carbohydrate."

30.42 (Chemical Connections 30B) What does blood glucose have to do with dieting?

30.43 (Chemical Connections 30B) What is the most effective weight loss method?

30.44 (Chemical Connections 30C) Describe the difference between the structure of aspartame and the methyl ester of phenylalanylaspartic acid.

30.45 ■ (Chemical Connections 30C)
   (a) Which artificial sweeteners are not metabolized in the body?
   (b) What could be products of digested aspartame?

30.46 ■ (Chemical Connections 30C) What is common in the structures of Olestra and Sucralose?

30.47 (Chemical Connections 30D) Looking in Table 22.1, which lists the common amino acids found in proteins, which amino acid most resembles creatine?

30.48 (Chemical Connections 30D) Which single compound will have the greatest effect on athletic performance?

30.49 (Chemical Connections 30D) Identify two ways carbohydrates are used for athletic performance.

30.50 (Chemical Connections 30D) Why is creatine an effective ergogenic aid?

30.51 (Chemical Connections 30D) How is caffeine used as an ergogenic aid?

Additional Problems

30.52 Which two chemicals are used most frequently to disinfect public water supplies?

30.53 Which vitamin is part of coenzyme A (CoA)? List a step (or the enzyme) that has CoA as a coenzyme in (a) glycolysis and (b) fatty acid synthesis.

30.54 Which vitamin is prescribed in megadoses to combat autoimmune blisters?

30.55 Why is it necessary to have protein in our diets?

30.56 Which chemical processes take place during digestion?

30.57 According to the U.S. government's Food Guide Pyramid, are there any foods that we can completely omit from our diets and still be healthy?

30.58 Does the debranching enzyme help in digesting amylose?

30.59 As an employee of a company that markets walnuts, you are asked to provide information for an ad that would stress the nutritional value of walnuts. What information would you provide?

30.60 In diabetes, insulin is administered intravenously. Explain why this hormone protein cannot be taken orally.

30.61 Egg yolk contains a lot of lecithin (a phosphoglyceride). After ingesting a hard-boiled egg, would you find an increase in the lecithin level of your blood? Explain.

30.62 What would you call a diet that scrupulously avoids phenylalanine-containing compounds? Could aspartame be used in such a diet?

30.63 What kind of supplemental enzyme would you recommend for a patient after a peptic ulcer operation?

30.64 In a trial, a woman was accused of poisoning her husband by adding arsenic to his meals. Her attorney stated that this supplementation was done to promote her husband's health, as arsenic is an essential nutrient. Would you accept this argument?

Immunochemistry

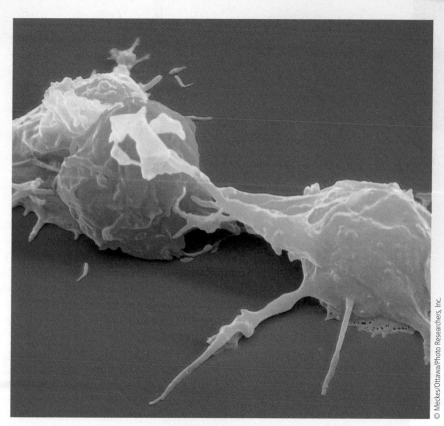

© Meckes/Ottawa/Photo Researchers, Inc.

Two natural killer (NK) cells, shown in yellow-orange, attacking a leukemia cell, shown in red.

31.1 | How Does the Body Defend Itself from Invasion?

When you were in elementary school, you may have had chickenpox along with many other children. The viral disease passed from one person to the next and ran its course, but after the children recovered, they never had chickenpox again. Those who were infected became *immune* to this disease. Humans and other vertebrates possess a highly developed, complex immune system that defends the body against foreign invaders. Figure 31.1, which presents an overview of this complex system, could serve as a road map as you read the different sections in this chapter. Referring back to Figure 31.1, you will find the precise location of the topic under discussion and its relationship to the immune system as a whole. As you can see, the immune system is very involved with multiple layers of protection against invading organisms. In this section, we will briefly introduce the major parts of the immune system. These topics will then be expanded in the sections that follow.

Figure 31.1 Overview of the immune system: its components and their interactions.

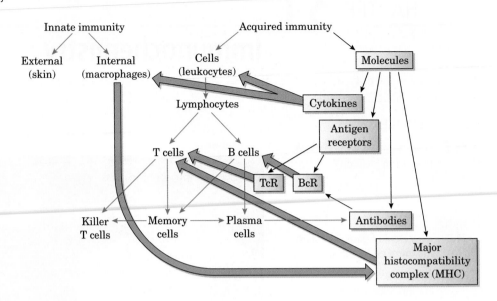

A. Innate Immunity

When one considers the tremendous number of bacteria, viruses, parasites, and toxins that our bodies encounter, it is a wonder that we are not continually sick. Most students learn about antibodies in high school, and these days everyone learns about T cells due to their relationship to AIDS. When discussing immunity, however, many more weapons of defense exist other than T cells and antibodies. In reality, you only discover that you are sick once a pathogen has managed to beat the front-line defense, which is called **innate immunity.**

Innate immunity includes several components. One part, called **external innate immunity,** includes physical barriers, such as skin, mucus, and tears. All of these barriers act to hinder penetration by pathogens and do not require specialized cells to fight a pathogen. If a pathogen—whether a bacterium, virus, or parasite—is able to breach this outer layer of defense, the cellular warriors of the innate system come into play. The cells of the innate immune system that we will discuss are **dendritic cells, macrophages, and natural killer (NK) cells.** Among the first and most important cells to join the fight are the dendritic cells, so called due to their long, tentacle-like projections (Figure 31.2).

Natural killer cells function as the police officers of internal innate immunity. When they encounter cancerous cells, cells infected by a virus, or any other suspicious cells, they attach themselves to those abnormal cells (see the chapter-opening photo). Other nonspecific responses include the proliferation of macrophages, which engulf and digest bacteria and reduce inflammation. In the inflammatory response to injury or infection, the capillaries dilate to allow greater flow of blood to the site of the injury, enabling agents of the internal innate immunity system to congregate there.

B. Adaptive Immunity

Vertebrates have a second line of defense, called **adaptive** or **acquired immunity.** We refer to this type of immunity when we talk colloquially about the **immune system.** The key features of the immune system are *specificity* and *memory.* The key cellular components of acquired immunity are **T cells** and **B cells.** The immune system uses antibodies and cell receptors designed specifically for each type of invader. In a second encounter

Innate immunity The natural nonspecific resistance of the body against foreign invaders, which has no memory

Eye of Science/Photo Researchers, Inc.

Figure 31.2 Dendritic cells get their name from their tentacle-like arms. The one shown here is from a human.

with the same invader, the response is more rapid, more vigorous, and more prolonged than it was in the first case, because the immune system remembers the nature of the invader from the first encounter.

The invaders may be bacteria, viruses, molds, or pollen grains. A body with no defense against such invaders could not survive. In a rare genetic disease, a person is born without a functioning immune system. Attempts have been made to bring up such children in an enclosure totally sealed from the environment. While in this environment, they can survive; when the environment is removed, however, such people always die within a short time. The severity of this disease, called severe combined immunodeficiency, explains why it was the first disease treated with gene therapy (Section 26.9). The disease AIDS (Section 31.8) slowly destroys the immune system, particularly a type of T cell, leaving its victims to die from some invading organism that a person without AIDS would easily be able to fight off.

As we shall see, the beauty of the body's immune system lies in its flexibility. The system is capable of making millions of potential defenders, so it can almost always find just the right one to counter the invader, even when it has never seen that particular organism before.

C. Components of the Immune System

Foreign substances that invade the body are called **antigens.** The immune system is made of both cells and molecules. Two types of **white blood cells,** called lymphocytes, fight against the invaders: (1) T cells kill the invader by contact and (2) B cells manufacture **antibodies,** which are soluble immunoglobulin molecules that immobilize antigens.

The basic molecules of the immune system belong to the **immunoglobulin superfamily.** All molecules of this class have a certain portion of the molecule that can interact with antigens, and all are glycoproteins. The polypeptide chains in this superfamily have two domains: a constant region and a variable region. The constant region has the same amino acid sequence in each of the same class molecules. In contrast, the variable region is antigen-specific, which means that the amino acid sequence in this region is unique for each antigen. The variable regions are designed to recognize only one specific antigen.

There are three representatives of the immunoglobulin superfamily in the immune system:

1. Antibodies are soluble immunoglobulins secreted by the plasma cells.

2. Receptors on the surface of T cells (**TcR**) recognize and bind antigens presented to them.

3. Molecules that present antigens also belong to this superfamily. They reside inside the cells. These protein molecules are known as major histocompatibility complexes (MHC).

When a cell is infected by an antigen, MHC molecules interact with it and bring a characteristic portion of the antigen to the surface of the cell. Such a surface presentation then marks the diseased cell for destruction. This can happen in a cell that was infected by a virus, and it can happen in macrophages that engulf and digest bacteria and virus.

D. The Speed of the Immune Response

The process of finding and making just the right immunoglobulin to fight a particular invader is relatively slow, compared to the actions of the chemical messengers discussed in Chapter 24. While neurotransmitters act within a

Antigen A substance foreign to the body that triggers an immune response

Antibody A glycoprotein molecule that interacts with an antigen

Immunoglobulin superfamily Glycoproteins that are composed of constant and variable protein segments having significant homologies to suggest that they evolved from a common ancestry

Macrophage-ingesting bacteria (the rod-shaped structures). The bacteria will be pulled inside the cell within a membrane-bound vesicle and quickly killed.

© David M. Phillips/Photo Researchers, Inc.

millisecond and hormones within seconds, minutes, or hours, immunoglobulins respond to an antigen over a longer span of time—weeks and months.

Although the immune system can be considered another form of chemical communication (Chapter 24), it is much more complex than neurotransmission because it involves molecular signals and interplay between various cells. In these constant interactions, the major elements are as follows: (1) the cells of the immune system, (2) the antigens and their perception by the immune system, (3) the antibodies (immunoglobulin molecules) that are designed to immobilize antigens, (4) the receptor molecules on the surface of the cells that recognize antigens, and (5) the cytokine molecules that control these interactions. Because the immune system is the cornerstone of the body's defenses, its importance for students in health-related sciences is undeniable.

31.2 | What Organs and Cells Make Up the Immune System?

The blood plasma circulates in the body and comes in contact with the other body fluids through the semipermeable membranes of the blood vessels. Therefore, blood can exchange chemical compounds with other body fluids and, through them, with the cells and organs of the body (Figure 31.3).

A. Lymphoid Organs

The lymphatic capillary vessels drain the fluids that bathe the cells of the body. The fluid within these vessels is called **lymph.** Lymphatic vessels circulate throughout the body and enter certain organs, called **lymphoid organs,** such as the thymus, spleen, tonsils, and lymph nodes (Figure 31.4). The cells primarily responsible for the functioning of the immune system are the specialized white blood cells called **lymphocytes.** As their name implies, these cells are mostly found in the lymphoid organs. Lymphocytes may be either specific for a given antigen or nonspecific.

T cells are lymphocytes that originate in the bone marrow but mature in the thymus gland. B cells are lymphocytes that originate and mature in the bone marrow. Both B and T cells are found mostly in the lymph, where they circulate looking for invaders. Small numbers of lymphocytes are also found in the blood. To get there, they must squeeze through tiny openings between the endothelial cells. This process is aided by signaling molecules called **cytokines** (Section 31.6). The sequence of the response of the body to a foreign invader is depicted schematically in Table 31.1.

Figure 31.3 Exchange of compounds among three body fluids: blood, interstitial fluid, and lymph. (After Holum, J. R., *Fundamentals of General, Organic and Biological Chemistry,* New York: John Wiley & Sons, 1978, p. 569)

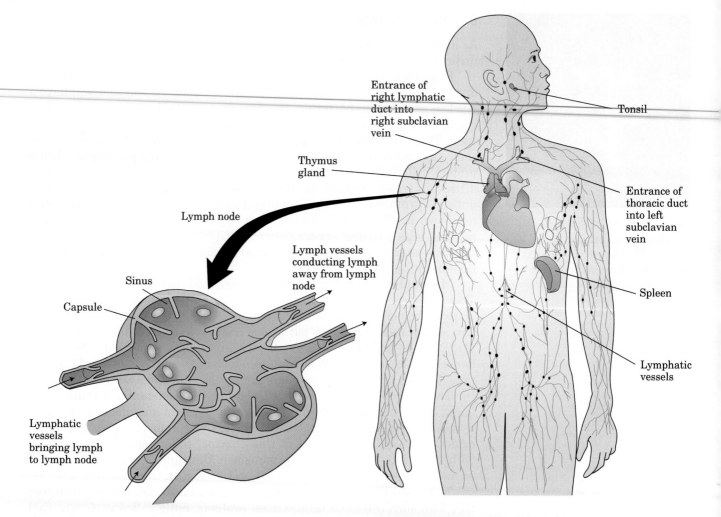

Figure 31.4 The lymphatic system is a web of lymphatic vessels containing a clear fluid, called lymph, and various lymphatic tissues and organs located throughout the body. Lymph nodes are masses of lymphatic tissue covered with a fibrous capsule. Lymph nodes filter the lymph. In addition, they are packed with macrophages and lymphocytes.

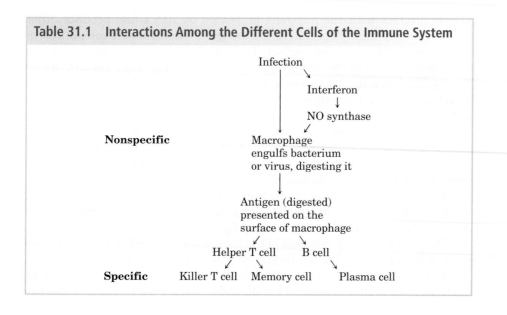

Table 31.1	Interactions Among the Different Cells of the Immune System

Infection

Interferon

NO synthase

Nonspecific   Macrophage engulfs bacterium or virus, digesting it

Antigen (digested) presented on the surface of macrophage

Helper T cell   B cell

Specific   Killer T cell   Memory cell   Plasma cell

Figure 31.5 Dendritic cells and the other cells of the immune system. This figure shows a rat dendritic cell interacting with a T cell. Through these interactions, the dendritic cells teach the acquired immunity system what to attack.

Courtesy of Steve Haley, William Bowers, and Richard Hunt, University of South Carolina Medical School

B. Cells of the Internal Innate Immunity

As mentioned in Section 31.1A, the major cells of innate immunity are dendritic cells, macrophages, and natural killer cells.

Dendritic cells are found in the skin, mucus membranes, the lungs, and the spleen. They are the first cells of the innate system that will have a crack at any virus or bacterium that wanders into their path. Using suction-cup like receptors, they grab onto invaders and then engulf them by endocytosis. These cells then chop up the devoured pathogens and bring parts of their proteins to the surface. Here the protein fragments are displayed on a protein called a **major histocompatibility complex (MHC).** The dendritic cells travel through the lymph to the spleen, where they present these antigens to other cells of the immune system, the **helper T cells (T_H cells)** as shown in Figure 31.5. Dendritic cells are members of a class of cells referred to as **antigen presenting cells (APCs),** and they are the starting point in most of the responses that are traditionally associated with the immune system.

Macrophages are the first cells in the blood that encounter an antigen. They belong to the internal innate immune system; inasmuch as they are nonspecific, macrophages attack virtually anything that is not recognized as part of the body, including pathogens, cancer cells, and damaged tissues. Macrophages engulf an invading bacterium or virus and kill it. The "magic bullet" in this case is the NO molecule, which is both toxic (Chemical Connections 4C) and can act as a secondary messenger (Chemical Connections 24F).

The NO molecule is short-lived and must be constantly manufactured anew. When an infection begins, the immune system manufactures the protein interferon. It, in turn, activates a gene that produces an enzyme, nitric oxide synthase. With the aid of this enzyme, the macrophages, endowed with NO, kill the invading organisms. Macrophages then digest the engulfed antigen and display a small portion of it on their surface.

Natural killer (NK) cells target abnormal cells. Once in physical contact with such cells, NK cells release proteins, aptly called perforins, that perforate the target cell membranes and create pores. The membrane of the target cell becomes leaky, allowing hypotonic (Section 7.8B) liquid from the surroundings to enter the cell, which swells and eventually bursts.

C. Cells of Adaptive Immunity: T and B Cells

T cells interact with the antigens presented by APCs and produce other T cells that are highly specific to the antigen. When these T cells differentiate, some of them become **killer T cells,** also called **cytotoxic T cells (T_c cells),** which kill the invading foreign cells by cell-to-cell contact. Killer T cells, like NK cells, act through perforins, which attach themselves to the target cell, in effect punching holes in its membranes. Through these holes water rushes into the target cell; it swells and eventually bursts.

Other T cells become **memory cells.** They remain in the bloodstream, so that if the same antigen enters the body again, even years after the primary infection, the body will not need to build up its defenses anew but is ready to kill the invader instantly.

A third type of T cell is the **helper T cell (T_H cell).** These cells do not kill other cells directly, but rather are involved in recognizing antigens on APCs and recruiting other cells to help fight the infection.

The production of antibodies is the task of **plasma cells.** These cells are derived from B cells after the B cells have been exposed to an antigen.

The lymphatic vessels, in which most of the antigen attacking takes place, flow through a number of lymph nodes (Figure 31.4). These nodes are

Both natural killer cells and killer T cells act the same way, through perforin. The T_c cells attack specific targets; the NK cells attack all suspicious targets.

essentially filters. Most plasma cells reside in lymph nodes, so most antibodies are produced there. Each lymph node is also packed with millions of other lymphocytes. More than 99% of all invading bacteria and foreign particles are filtered out in the lymph nodes. As a consequence, the outflowing lymph is almost free of invaders and is packed with antibodies produced by the plasma cells. All lymphocytes derive from stem cells in the bone marrow. Stem cells are undifferentiated cells that can become many different cell types. As shown in Figure 31.6, they can differentiate into T cells in the thymus or B cells in the bone marrow.

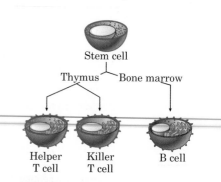

31.3 | How Do Antigens Stimulate the Immune System?

A. Antigens

Antigens are foreign substances that elicit an immune response; for this reason, they are also called immunogens. Three features characterize an antigen. The first condition is foreignness—molecules of your own body should not elicit an immune response. The second condition is that the antigen must be of molecular weight greater than 6000. The third condition is that the molecule must have sufficient complexity. A polypeptide made of lysine only, for example, is not immunogenic.

Antigens can be proteins, polysaccharides, or nucleic acids, as all of these substances are large molecules. Antigens may be soluble in the cytoplasm, or they may be found at the surface of cells, either embedded in the membrane or just absorbed on it. An example of polysaccharidic antigenicity is ABO blood groups (Chemical Connections 20D).

In protein antigens, only part of the primary structure is needed to cause an immune response. Between 5 and 7 amino acids are needed to interact with an antibody, and between 10 and 15 amino acids are necessary to bind to a receptor on a T cell. The smallest unit of an antigen capable of binding with an antibody is called the **epitope.** The amino acids in an epitope do not have to be in sequence in the primary structure, as folding and secondary structures may bring amino acids that are not in sequence into each other's proximity. For example, the amino acids in positions 20 and 28 may form part of an epitope. *Antibodies can recognize all types of antigens, but T-cell receptors recognize only peptide antigens.*

As noted earlier, antigens may be in the interior of an infected cell or on the surface of a virus or bacterium that penetrated the cell. To elicit an immune reaction, the antigen or its epitope must be brought to the surface of the infected cell. Similarly, after a macrophage swallows up and partially digests an antigen, the macrophage must bring the epitope back to its surface to elicit an immune response from T cells (Table 31.1).

B. Major Histocompatibility Complexes

The task of bringing an antigen's epitope to the infected cell's surface is performed by the major histocompatibility complex (MHC). The name derives from the fact that its role in the immune response was first discovered in organ transplants. MHC molecules are transmembrane proteins belonging to the immunoglobulin superfamily. Two classes of MHC molecules exist (Figure 31.7), both of which have peptide-binding variable domains. Class I MHC is made of a single polypeptide chain, whereas class II MHC is a dimer. Class I MHC molecules seek out antigen molecules that have been *synthesized inside a virus-infected cell.* Class II MHC molecules pick up

Figure 31.6 The development of lymphocytes. All lymphocytes are ultimately derived from the stem cells of the bone marrow. In the thymus, T cells develop into helper T cells and killer T cells. B cells develop in the bone marrow.

There are rare exceptions: In an autoimmune disease, the body mistakes its own protein as foreign

Epitope The smallest number of amino acids on an antigen that elicits an immune response

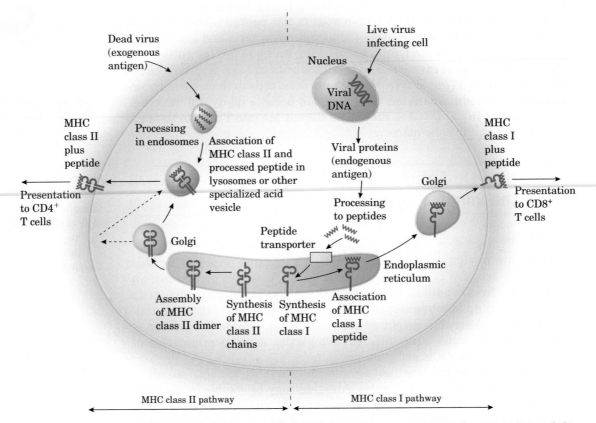

Figure 31.7 Differential processing of antigens in the MHC class II pathway (left) or MHC class I pathway (right). Cluster determinants (CD) are parts of the T-cell receptor complex (see Section 31.5B).

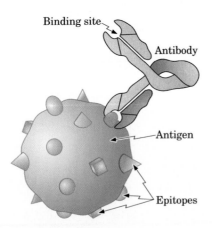

Antibody binding to the epitope of an antigen.

exogenous *"dead" antigens.* In each case, the epitope attached to the MHC is brought to the cell surface to be presented to T cells.

For example, if a macrophage engulfed and digested a virus, the result would be dead antigens. The digestion occurs in several steps. First, the antigen is processed in lysosomes, special organelles of cells that contain proteolytic enzymes. An enzyme called GILT (gamma-interferon inducible lysosomal thiol reductase) breaks the disulfide bridges of the antigen by reduction. The reduced peptide antigen unfolds and is exposed to proteolytic enzymes that hydrolyze it to smaller peptides. These peptides serve as epitopes that are recognized by class II MHC. The difference between MHC I and MHC II becomes significant when we look at the functions of T cells. Antigens bound to MHC I will interact with killer T cells, while those bound to MHC II will interact with helper T cells.

31.4 | What Are Immunoglobulins?

A. Classes of Immunoglobulins

Immunoglobulins are glycoproteins—that is, carbohydrate-carrying protein molecules. Not only do the different classes of immunoglobulins vary in molecular weight and carbohydrate content, but their concentration in the blood differs dramatically as well (Table 31.2). The IgG and IgM antibodies are the most important antibodies in the blood. They interact with antigens and trigger the swallowing up (phagocytosis) of these cells by phagocytes. Inside the phagocytes, the antigens are destroyed by a complicated proce-

CHEMICAL CONNECTIONS 31A

The Mayapple and Chemotherapy Agents

Among the life-threatening diseases in the United States, heart disease remains the number one killer. Collectively cancers represent our second leading cause of death and it is estimated that, within the next several decades, they will head the list. Reducing risk factors is a major focus in the war against heart disease. Because risk factors for cancers are not as clearly recognized, however, treatment by surgery, radiation therapy, and chemotherapy remains the major focus in this war. In chemotherapy, a chemical or combination of chemicals is introduced into the body to destroy rapidly dividing cells. While we are only now beginning to understand how these agents work within the body, many such agents have been known for over a century. Among them is the antitumor effect of an extract of the common mayapple, *Podophyllium peltatum*, described in 1861 by Robert Bentley of Kings College, London. The active principle of mayapple extract was identified twenty years

later and given the name picropodophyllin. It was not until 1946 that the mechanism of action of the drug was discovered. We now know that it suppresses tumor growth by inhibiting the formation of the mitotic spindle, which holds cells undergoing nuclear division. The structure of picropodophyllin was determined in 1954 and in the 1970s, scientists at the Sandoz Company succeeded in synthesizing several analogs of picropodophyllin that are even more effective than their mayapple parent drug. Among them is etoposide, which is effective in treating small-cell lung cancer, testicular cancer, lymphomas, leukemias, and several types of brain tumors. Etoposide also inhibits topoisomerase II, an important DNA-regulating enzyme.

Picropodophyllin

Etopside

dure called the complement pathway. The IgA molecules are found mostly in secretions: tears, milk, and mucus. Therefore, these immunoglobulins attack the invading material before it gets into the bloodstream. The IgE molecules play a part in such allergic reactions as asthma and hay fever, and are involved in the body's defense against parasites.

B. Structure of Immunoglobulins

Each immunoglobulin molecule is made of four polypeptide chains: two identical light chains and two identical heavy chains. The four polypeptide chains are arranged symmetrically, forming a Y shape (Figure 31.8). Four

Table 31.2 Immunoglobulin Classes

Class	Molecular Weight (MW)	Carbohydrate Content (%)	Concentration in Serum (mg/100 mL)
IgA	200,000–700,000	7–12	90–420
IgD	160,000	<1	1–40
IgE	190,000	10–12	0.01–0.1
IgG	150,000	2–3	600–1800
IgM	950,000	10–12	50–190

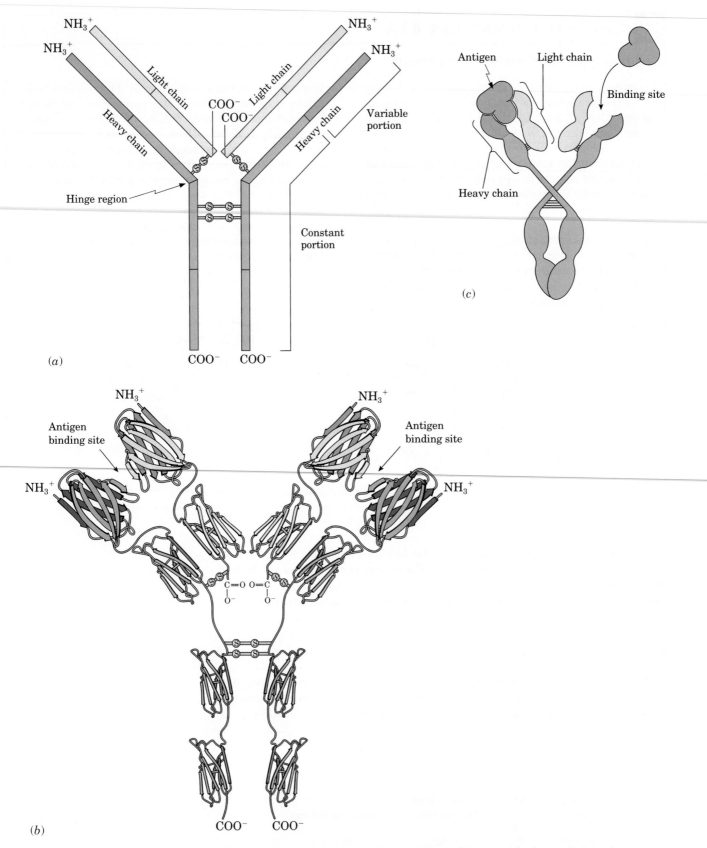

Figure 31.8 (*a*) Schematic diagram of an IgG-type antibody consisting of two heavy chains and two light chains connected by disulfide bonds. The amino terminal end of each chain has the variable portion. (*b*) The same in the ribbon model. (*c*) Model showing how an antibody bonds to an antigen.

disulfide bridges link the four chains into a unit. Both light and heavy chains have constant and variable regions. The constant regions have the same amino acid sequences in different antibodies, and the variable regions have different amino acid sequences in different antibodies.

The variable regions of the antibody recognize the foreign substance (the antigen) and bind to it (Figure 31.9). Because each antibody contains two variable regions, it can bind two antigens, thereby forming a large aggregate, as shown in Figure 31.10.

The binding of the antigen to the variable region of the antibody occurs not by covalent bonds but rather by much weaker intermolecular forces such as London dispersion forces, dipole–dipole interactions, and hydrogen bonds (Section 6.7). This binding is similar to the way in which substrates bind to enzymes or hormones and neurotransmitters bind to a receptor site. That is, the antigen must fit into the antibody surface. Humans have more than 10,000 different antibodies, which enables our bodies to fight a large number of foreign invaders. The potential number of antibodies that can be created by the available genes is in the millions.

C. B Cells and Antibodies

Each B cell synthesizes only one unique immunoglobulin antibody, and that antibody contains a unique antigen-binding site to one epitope. Before encountering an antigen, these antibodies are inserted in the plasma membrane of the B cells, where they serve as receptors. When an antigen interacts with its receptor, it stimulates the B cell to divide and differentiate into plasma cells. These daughter cells secrete soluble antibodies that have the same antigen-binding sites as the original antibody/receptor. The soluble secreted antibodies appear in the serum (the noncellular part of blood) and can react with the antigen. Thus, an immunoglobulin produced in B cells acts both as a receptor to be stimulated by the antigen and as a secreted messenger ready to neutralize and eventually destroy the antigen (Figure 31.11).

D. How Does the Body Acquire the Diversity Needed to React to Different Antigens?

From the moment of conception, an organism contains all of the DNA it will ever have, including that DNA that will lead to antibodies and T-cell receptors. Thus the organism is born with a repertoire of genes necessary to fight infections. During B-cell development, the variable regions of the heavy chains are assembled by a process called V(J)D recombination. A number of exons are present in each of three different areas—V, J, and D—of the immunoglobin gene. Combining one exon from each area yields a *new V(J)D gene*. This process creates a great diversity because of the large number of ways that this combination can be performed (Figure 31.12). For one type of antibody light chain, called kappa, there are roughly 40 V genes and 5 J genes. That alone gives rise to 40 × 5 or 200 combinations of V and J. For another type of light chain, called lambda, about 120 combinations are possible. With the heavy chains, there is even greater diversity: about 50 V genes, 27 D genes, and 6 J genes. By the time one gets done calculating the possible combinations of all of the combinations of V, J, D, and the C regions of both the heavy and light chains, there are more than 2 million possible combinations.

However, that is merely the first step. A second level of diversity is created by mutation of the V(J)D genes in somatic cells. As cells proliferate in response to recognizing an antigen, such mutations can cause a 1000-fold

(a)

(b)

Courtesy of Dr. R. J. Poljak, Pasteur Institute, Paris, France

Figure 31.9 (a) An antigen–antibody complex. The antigen (shown in green) is lysozyme. The heavy chain of the antibody is shown in blue; the light chain in yellow. The most important amino acid residue (glutamine in the 121 position) on the antigen is the one that fits into the antibody groove (shown in red). (b) The antigen–antibody complex has been pulled apart. Note how they fit each other.

Figure 31.10 An antigen–antibody reaction forms a precipitate. An antigen, such as a bacterium or virus, typically has several binding sites for antibodies. Each variable region of an antibody (each prong of the Y) can bind to a different antigen. The aggregate thus formed precipitates and is attacked by phagocytes and the complement system.

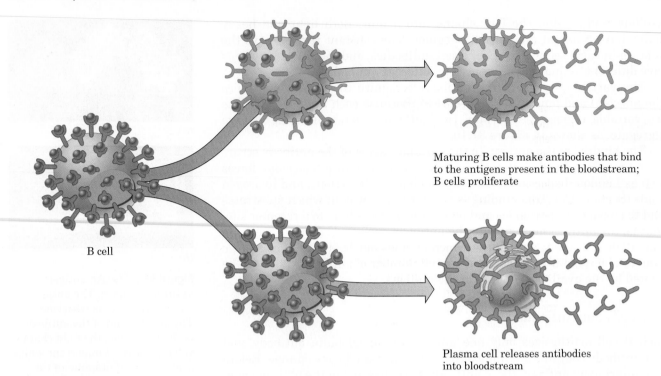

Maturing B cells make antibodies that bind to the antigens present in the bloodstream; B cells proliferate

Plasma cell releases antibodies into bloodstream

GOB
Chemistry·⦿·Now™

Figure 31.11 B cells have antibodies on their surfaces, which allow them to bind to antigens. The B cells with antibodies for the antigens present grow and develop. When B cells develop into plasma cells, they release circulating antibodies into the bloodstream. (Adapted from "How the Immune System Develops," by Irving L. Weissman and Max D. Cooper; illustrated by Jared Schneidman, Scientific American, September 1993.) **See a simulation based on this figure, and take a short quiz on the concepts at http://www.now.brookscole.com/gob8 or on the CD.**

Figure 31.12 Diversification of immunoglobulins by V(J)D recombination. Exons of three genes—the V(ariable) (A, B, C, D); J(oin) (1, 2, 3, 4); and D(iverse) (a, b, c, d) genes—combine to form new V(J)D genes that are transcribed to the corresponding mRNAs. The expression of these new genes results in a wide variety of immunoglobulins having different variable regions on their heavy chains.

CHEMICAL CONNECTIONS 31B

Antibodies and Cancer Therapy

Antibodies come in a variety of forms. As a response to a specific antigen, the body may develop a number of closely related anti-bodies, the so-called **multiclonal antibodies.** Monoclonal anti-bodies, in contrast, are identical copies of a single antibody molecule that bind to a specific antigen.

Monoclonal antibody therapy is the latest advance in fighting cancer. For instance, the drug Rituxan is a monoclonal antibody against the CD20 antigen, which appears in high concentration on the surface of B cells in patients suffering from non-Hodgkin's lymphoma. Rituxan kills tumor cells in two ways. First, by binding to the diseased B cells, it enlists the entire immune system to attack the marked cells and to destroy them. Second, the monoclonal antibody causes the tumor cells to stop growing and die through apoptosis, programmed cell death. Another monoclonal antibody is Herceptin, which is recommended as a treatment for metastatic breast cancers that have a specific antigen, c-erb-B2, on the surface of cancer cells.

One advantage that monoclonal antibody treatment has over chemotherapy is its relatively minimal toxicity. Even when the monoclonal antibody is employed to deliver damaging radiation, as in the case of the antibody against the prostate-specific antigen to which actinium-225 is attached, the antibody delivers the radia-tion to only the specific cancer cells, killing them without signifi-cantly harming healthy tissues.

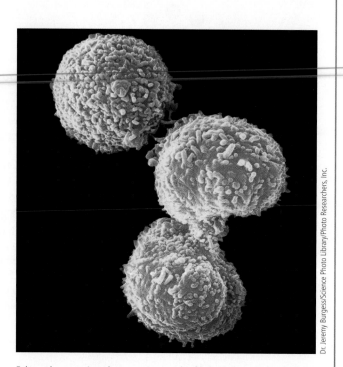

False-color scanning electron micrograph of hybridoma cells producing a monoclonal antibody to cytoskeleton protein.

Dr. Jeremy Burgess/Science Photo Library/Photo Researchers, Inc.

---

increase in the binding affinity of an antigen to an antibody. This process is called **affinity maturation.**

There are three ways to create mutations. Two affect the variable regions, and one alters the constant region

1. Somatic hypermutation (SHM) creates a point mutation (only one nucleotide). The resultant protein of the mutation is able to bind more strongly with the antigen.

2. In a gene conversion (GC) mutation, stretches of nucleotide sequences are copied from a pseudogene V and entered into the V(J)D. This also enables the proteins synthesized from the GC mutation to make stronger contact with the antigen than without the mutation.

3. Mutations on the constant region of the chain can be achieved by class switch recombination (CSR). In this case, exons of the constant regions are swapped between highly repetitive regions.

The diversity of antibodies created by the V(J)D recombination is greatly amplified and finely tuned by the mutations on these genes. Because the response to the antigen occurs on the gene level, it is easily preserved and transmitted from one generation of cells to the next.

While the possible combination of genes leading to antibody diversity seems limitless, it is important to remember that the basis of antibody diversity is the genetic blueprint that the organism was given. Antibodies do not appear because they are needed; rather, they are selected and prolif-erated because they already existed in small quantities before they were stimulated by recognition of an antigen.

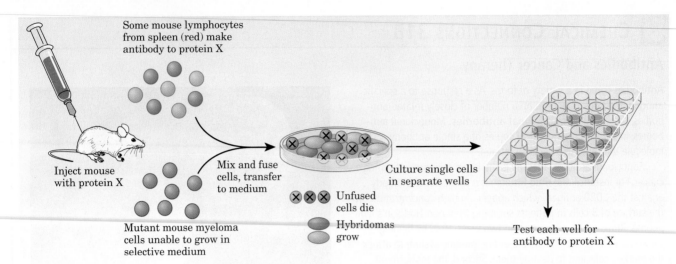

Figure 31.13 A procedure for producing monoclonal antibodies against a protein antigen X. A mouse is immunized against the antigen X, and some of its spleen lymphocytes produce antibody. The lymphocytes are fused with mutant myeloma cells that cannot grow in a given medium because they lack an enzyme found in the lymphocytes. Unfused cells die because lymphocytes cannot grow in culture and the mutant myeloma cells cannot survive in this medium. The individual cells are grown in culture in separate wells and tested for antibody to protein X.

E. Monoclonal Antibodies

When an antigen is injected into an organism (for example, human lysozyme into a rabbit), the initial response is quite slow. It may take from one to two weeks before the anti-lysozyme antibody shows up in the rabbit's serum. Those antibodies, however, are not uniform. The antigen may have many epitopes, and the antisera contains a mixture of immunoglobulins with varying specificity for all the epitopes. Even antibodies to a single epitope usually have a variety of specificities.

Each B cell (and each progeny plasma cell) produces only one kind of antibody. In principle, each such cell should represent a potential source of a supply of homogeneous antibody by cloning. This is not possible in practice because lymphocytes do not grow continuously in culture. In the late 1970s, Georges Köhler and César Milstein developed a method to circumvent this problem, a feat for which they received the Nobel Prize in physiology in 1984. Their technique requires fusing lymphocytes that make the desired antibody with mouse myeloma cells. The resulting **hybridoma** (hybrid myeloma), like all cancer cells, can be cloned in culture (Figure 31.13) and produces the desired antibody. Because the clones are the progeny of a single cell, they produce homogeneous **monoclonal antibodies.** With this technique, it becomes possible to produce antibodies to almost any antigen in quantity. Monoclonal antibodies can, for instance, be used to assay for biological substances that can act as antigens. A striking example of their usefulness is in testing blood for the presence of HIV; this procedure has become routine to protect the public blood supply. Monoclonal antibodies are also commonly used in cancer treatment, as described in Chemical Connections 31B.

31.5 | What Are T Cells and T-Cell Receptors?

A. T-Cell Receptors

Like B cells, T cells carry on their surface unique receptors that interact with antigens. We noted earlier that T cells respond only to protein antigens. An individual has millions of different T cells, each of which carries on its surface a unique T-cell receptor (TcR) that is specific for one antigen only. The TcR is a glycoprotein made of two subunits cross-linked by disulfide bridges. Like immunoglobulins, TcRs have constant (C) and variable (V) regions. The antigen binding occurs on the variable region. The similarity in amino acid sequence between immunoglobulins (Ig) and TcR, as well as the organization of the polypeptide chains, makes TcR molecules members of the immunoglobulin superfamily.

There are, however, some fundamental differences between immuno-globulins and TcRs. For instance, immunoglobulins have four polypeptide chains, whereas TcRs contain only two subunits. Immunoglobulins can interact directly with antigens, but TcRs can interact with them only when the epitope of an antigen is presented by an MHC molecule. Lastly, immuno-globulins can undergo somatic mutation. This kind of mutation can occur in all body cells except the ones involved in sexual reproduction. Thus Ig molecules can increase their diversity by somatic mutation; TcRs cannot.

B. T-Cell Receptor Complex

A TcR is anchored in the membrane by hydrophobic transmembrane segments (Figure 31.14). TcR alone, however, is not sufficient for antigen binding. Also needed are other protein molecules that act as coreceptors and/or signal transducers. These molecules go under the name of CD3, CD4, and CD8, where "CD" stands for **cluster determinant.** TcR and CD together form the **T-cell receptor complex.**

The CD3 molecule adheres to the TcR in the complex not through covalent bonds but rather by intermolecular forces (Section 6.7). It is a signal

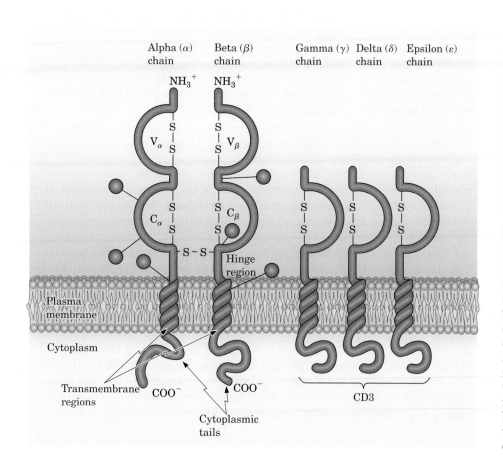

Figure 31.14 Schematic structure of a TcR complex. The TcR consists of two chains: α and β. Each has two extracellular domains: an amino-terminal V domain and a carboxyl-terminal C domain. Domains are stabilized by intrachain disulfide bonds between cysteine residues. The α and β chains are linked by an interchain disulfide bond near the cell membrane (hinge region). Each chain is anchored on the membrane by a hydrophobic transmembrane segment and ends in the cytoplasm with a carboxyl-terminal segment rich in cationic residues. Both chains are glycosylated (red spheres). The cluster determinant (CD) coreceptor consists of three chains: γ, δ, and ε. Each is anchored in the plasma membrane by a hydrophobic transmembrane segment. Each is also cross-linked by a disulfide bridge, and the carboxylic terminal is located in the cytoplasm.

CHEMICAL CONNECTIONS 31C

Immunization

Smallpox was a scourge over many centuries, with each outbreak leaving many people dead and others maimed by deep pits on the face and body. A form of immunization was practiced in ancient China and the Middle East by intentionally exposing people to scabs and fluids of lesions of victims of smallpox. This practice was known as variolation in the Western world, where the disease was called variola. Variolation was introduced to England and the American Colonies in 1721.

Edward Jenner, an English physician, noted that milkmaids who had contracted cowpox from infected cows seemed to be immune to smallpox. Cowpox was a mild disease, whereas smallpox could be lethal. In 1796, Jenner performed a potentially deadly experiment: He dipped a needle into the pus of a cowpox-infected milkmaid and then scratched a boy's hand with the needle. Two months later, Jenner injected the boy with a lethal dose of smallpox-carrying agent. The boy survived and did not develop any symptoms of the disease. The word spread, and Jenner was soon established in the immunization business. When the news reached France, skeptics there coined a derogatory term, *vaccination,* which literally means "encowment." The derision did not last long, and the practice was soon adopted worldwide.

A century later, in 1879, Louis Pasteur found that tissue infected with rabies has much weakened virus in it. When injected into patients, it elicits an immune response that protects against rabies. Pasteur named these attenuated, protective antigens *vaccines* in honor of Jenner's work. Today, immunization and vaccination are synonymous.

Vaccines are available for several diseases, including polio, measles, and smallpox, to name a few. A vaccine may be made of either dead viruses or bacteria or weakened ones. For example, the Salk polio vaccine is a polio virus that has been made harmless by treatment with formaldehyde; it is given by intramuscular injection. In contrast, the Sabin polio vaccine is a mutated form of the wild virus; the mutation makes the virus sensitive to temperature. The mutated live virus is taken orally. The body temperature and the gastric juices render it harmless before it penetrates the bloodstream.

Many cancers have specific carbohydrate cell-surface tumor markers. If such antigens could be introduced by injection without endangering the individual, they could provide an ideal vaccine. Obviously, one cannot use live or even attenuated cancer cells for vaccination. However, the expectation is that a synthetic analog of a natural tumor marker will elicit the same immune reaction that the original cancer-surface markers do. Thus injecting such innocuous synthetic compounds may prompt the body to produce immunoglobulins that can provide a cure for the cancer—or at least prevent the occurrence of it. A compound called 12:13 dEpoB, a derivative of the macrolide epothilon B, is currently being explored for its potential to become an anticancer vaccine.

Vaccines change lymphocytes into plasma cells that produce large quantities of antibodies to fight any invading antigens. This is only the immediate, short-term response, however. Some lymphocytes become memory cells rather than plasma cells. These memory cells do not secrete antibodies, but instead store them to serve as a detecting device for future invasion of the same foreign cells. In this way, long-term immunity is conferred. If a second invasion occurs, these memory cells divide directly into both antibody-secreting plasma cells and more memory cells. This time the response is faster because it does not have to go through the process of activation and differentiation into plasma cells, which usually takes two weeks.

Smallpox, which was once one of humanity's worst scourges, has been totally wiped out, and smallpox vaccination is no longer required. Because smallpox is a potential weapon of bioterrorism, the U.S. government has recently begun gearing up its smallpox vaccine production.

Adhesion molecules Various protein molecules that help to bind an antigen to the T-cell receptor and dock the T cell to another cell via an MHC

transducer because, upon antigen binding, CD3 becomes phosphorylated. This event sets off a signaling cascade inside the cell, which is carried out by different kinases. We saw a similar signaling cascade in neurotransmission (Section 24.5).

The CD4 and CD8 molecules act as **adhesion molecules** as well as signal transducers. A T cell has either a CD4 or a CD8 molecule to help bind the antigen to the receptor and to dock the T cell to an APC or B cell (Figure 31.15). A unique characteristic of the CD4 molecule is that it binds strongly to a special glycoprotein that has a molecular weight of 120,000 (gp120). This glycoprotein exists on the surface of the human immunodeficiency virus (HIV). Through this binding to CD4, HIV can enter and infect helper T cells and cause AIDS. Helper T-cells die as a result of HIV infection, which depletes the T cell population so drastically that the immune system can no longer function. As a consequence, the body succumbs to opportunistic pathogen infections (Section 31.8).

31.6 | How Is the Immune Response Controlled?

A. Nature of Cytokines

Cytokines are glycoprotein molecules that are produced by one cell but alter the function of another cell. They have no antigen specificity. Cytokines transmit intercellular communications between different types of cells at diverse sites in the body. They are short-lived and are not stored in cells.

Cytokines facilitate a coordinated and appropriate inflammatory response by controlling many aspects of the immune reaction. They are released in bursts, in response to all manner of insult or injury (real or perceived). They travel and bind to specific cytokine receptors on the surface of macrophages and B and T cells and induce cell proliferation.

One set of cytokines are called **interleukins** (ILs) because they communicate between and coordinate the actions of leukocytes (all kinds of white blood cells). Macrophages secrete IL-1 upon a bacterial infection. The presence of IL-1 then induces shivering and fever. The elevated body temperature both reduces the bacterial growth and speeds up the mobilization of the immune system (Chemical Connections 8A). One leukocyte may make many different cytokines, and one cell may be the target of many cytokines.

B. Classes of Cytokines

Cytokines can be classified according to their mode of action, origin, or target. The best way to classify them is by their structure—namely, the secondary structure of their polypeptide chains.

1. One class of cytokines is made of four α-helical segments. A typical example is interleukin-2 (IL-2), which is a 15,000-MW polypeptide chain. A prominent source of IL-2 is T cells. IL-2 activates other B and T cells and macrophages. Its action is to enhance proliferation and differentiation of the target cells.

2. Another class of cytokines has only β-pleated sheets in its secondary structure. Tumor necrosis factor (TNF), for example, is produced mostly by T cells and macrophages. Its name comes from its ability to destroy susceptible tumor cells through lysis, after it binds to receptors on the tumor cell.

3. A third class of cytokines has both α-helix and β-sheet secondary structures. A representative of this class is epidermal growth factor (EGF), which is a cysteine-rich protein. As its name indicates, EGF stimulates the growth of epidermal cells; its main function is in wound healing.

4. A subgroup of cytokines, the chemotactic cytokines, are also called **chemokines.** Humans have some 40 chemokines, all low-molecular-weight proteins with distinct structural characteristics. They attract leukocytes to a site of infection or inflammation. All chemokines have four cysteine residues, which form two disulfide bridges: Cys1-Cys3 and Cys2-Cys4. Chemokines have a variety of names, such as interleukin-8 (IL-8) and monocyte chemotactic proteins (MCP-1 to MCP-4). Chemokines interact with specific receptors, which consist of seven helical segments coupled to GTP-binding proteins.

C. Mode of Action of Cytokines

When a tissue is injured, leukocytes are rushed to the inflamed area. Chemokines help leukocytes migrate out of the blood vessels to the site of

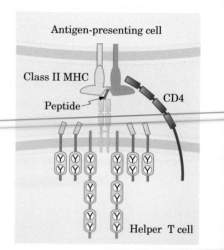

Figure 31.15 Interaction between helper T cells and antigen-presenting cells. Foreign peptides are displayed on the surface by MHC II proteins. These bind to the T-cell receptor of a helper T cell. A docking protein called CD4 helps link the two cells together.

Cytokine A glycoprotein that traffics between cells and alters the function of a target cell

Chemokine A low-molecular-weight polypeptide that interacts with special receptors on target cells and alters their functions injury. There leukocytes, in all their forms— neutrophils, monocytes, lymphocytes—accumulate and attack the invaders by engulfing them (phagocytosis) and later killing them. Other phagocytic cells, the macrophages, which reside in the tissues and thus do not have to migrate, do the same. These activated phagocytic cells destroy their prey by releasing endotoxins that kill bacteria and/or by producing such highly reactive oxygen intermediates as superoxide, singlet oxygen, hydrogen peroxide, and hydroxyl radicals (Chemical Connections 27B).

Chemokines are also major players in chronic inflammation, autoimmune diseases, asthma, and other forms of allergic inflammation, and even in transplant rejection.

31.7 | How Does the Body Distinguish "Self" from "Nonself"?

One major problem facing body defenses is how to recognize the foreign body as "not self" and thereby avoid attacking the "self"—that is, the healthy cells of the organism.

A. Selection of T and B Cells

The members of the adaptive immunity system, B cells and T cells, are all specific and have memory, so they target only truly foreign invaders. The T cells mature in the thymus gland. During the maturation process, those T cells that fail to recognize and interact with MHC and thus cannot respond to foreign antigens are eliminated through a selection process. They essentially die by neglect. T cells that express receptors (TcR) that are prone to interact with normal self antigens are also eliminated through the selection process (Figure 31.16). Thus the activated T cells leaving the thymus gland carry TcRs that can respond to foreign antigens. Even if some T cells prone to react with self antigen escape the selection detection, they can be deactivated through the signal transduction system that, among other functions, performs tyrosine kinase activation and phosphatase deactivation, similar to those processes seen in adrenergic neurotransmitter signaling (Section 24.5).

Similarly, the maturation of B cells in the bone marrow depends on the engagement of their receptors, BcR, with antigen. Those B cells that are prone to interact with self antigen are also eliminated before they leave the bone marrow. As with T cells, many signaling pathways control the proliferation of B cells. Among them, the activation by tyrosine kinase and the deactivation by phosphatase provide a secondary control.

B. Discrimination of the Cells of the Innate Immunity System

The first line of defense is the innate immunity system, in which cells such as natural killer cells or macrophages have no specific targets and no memory of which epitope signals danger. Nevertheless, these cells must somehow discriminate between normal and abnormal cells to identify targets. The mechanism by which this identification is accomplished has only recently been explored and is not yet fully understood. The main point is that the cells of innate immunity have two kinds of receptors on their surface: an **activating receptor** and an **inhibitory receptor.** When a healthy cell of the body encounters a macrophage or a natural killer cell, the inhibitory receptor on the latter's surface recognizes the epitope of the normal cell, binds to it, and prevents the activation of the killer cell or macrophage.

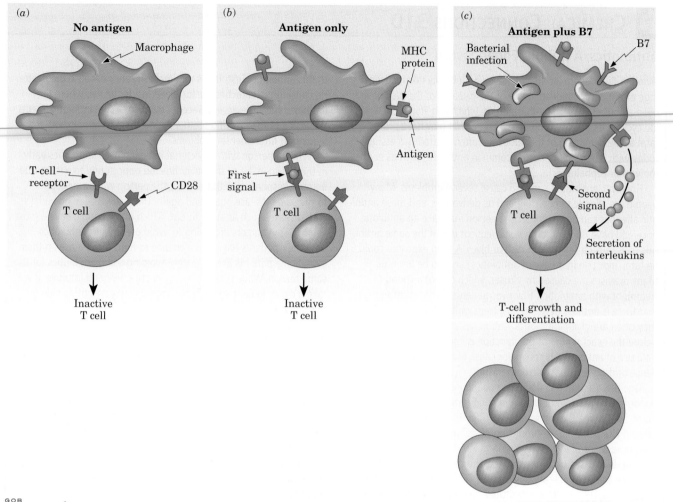

GOB
Chemistry Now™

Figure 31.16 A two-stage process leads to the growth and differentiation of T cells. (*a*) In the absence of antigen, proliferation of T cells does not take place. Those T-cell lines die by neglect. (*b*) In the presence of antigen alone, the T-cell receptor binds to antigen presented on the surface of a macrophage cell by the MHC protein. There is still no proliferation of T cells because the second signal is missing. In this way the body can avoid an inappropriate response to its own antigens. This process occurs early in the development of T cells, effectively eliminating those cells that would otherwise be activated by the self antigens. (*c*) When an infection occurs, a B7 protein is produced in response to the infection. The B7 protein on the surface of the infected cell binds to a CD28 protein on the surface of the immature T cell, giving the second signal that allows it to grow and proliferate. (Adapted from "How the Immune System Recognizes Invaders," by Charles A. Janeway, Jr.; illustration by Ian Warpole, *Scientific American,* September 1993.) **See a simulation based on this figure, and take a short quiz on the concepts at http://www.now.brookscole.com/gob8 or on the CD.**

Conversely, when a macrophage encounters a bacterium with a foreign antigen on its surface, the antigen binds to the activating receptor of the macrophage. This ligand binding prompts the macrophage to engulf the bacterium by phagocytosis. Such foreign antigens may be the lipopolysaccharides of gram-negative bacteria or the peptidoglycans of gram-positive bacteria.

When a cell is infected, damaged, or transformed into a malignant cell, the epitopes that signaled healthy cells diminish greatly, and unusual epitopes are presented on the surface of these cancer cells. The effect is that fewer inhibitory receptors of macrophages or natural killer cells can bind to the surface of the target cell and more activating receptors find inviting ligands. As a consequence, the balance shifts in favor of activation and the macrophages or killer cells will do their job.

CHEMICAL CONNECTIONS 31D

Antibiotics: A Double-Edged Sword

Living in the modern world, you have undoubtedly taken advantage of antibiotics. Indeed, many of the diseases of the past have been all but eradicated by these drugs, which can stop a bacterial life cycle in its tracks. Common infections that may have proved fatal at the beginning of the 1900s are often treated successfully today with penicillin or another common antibiotic, such as erythromycin or cephalosporin.

However, antibiotics can also cause several problems. Many people are allergic to penicillin and its derivatives, and these antibiotic allergies can be very potent. A person may take an antibiotic once and have no symptoms. A subsequent use of the same antibiotic may cause a severe skin rash or hives. A third exposure could be fatal. Indiscriminate use of antibiotics can also be harmful. Many diseases are caused by viruses, which do not respond to treatment with antibiotics. However, patients do not want to hear that there is nothing they can do except wait out the disease, so they often are given antibiotics. Antibiotics are also prescribed before the exact nature of the infection is known. This indiscriminate use of antibiotics is the major cause of the increasing incidence of drug-resistant microorganisms.

One disease that has flourished due to misuse of antibiotics is gonorrhea. One strain of *Neisseria gonorrhoea* produces β-lactamase, an enzyme that degrades penicillin. These strains are referred to as PPNG, penicillinase-producing *N. gonorrhoeae*. Before 1976, almost no cases of PPNG were reported in the United States. Today, thousands of cases occur nationwide. The source of the problem has been traced to military bases in the Philippines, where soldiers acquired the disease from prostitutes. Among prostitutes, there is a common practice of using low doses of antibiotics in an attempt to prevent the spread of sexually transmitted diseases. In reality, constant use of antibiotics has just the opposite effect—it causes development of drug-resistant strains.

An antibiotic commonly given to young children for earaches is amoxicillin, a penicillin derivative. Severe and repeated earaches can cause hearing loss, so parents are often quick to put their children on antibiotics. There are two downsides, however, to overuse

of this antibiotic. First, it has minimal efficiency because the bacteria causing the earaches are localized inside the ear, where the antibiotic has little access. Second, overuse of the antibiotic affects the inside of the growing teeth, causing them to remain soft and leading to future dental problems.

When a person with a bacterial infection uses antibiotics early in the infection, he or she never has the opportunity to elicit a true immune response. For this reason, the person will be susceptible to the same disease again and again. This issue is seen nowadays with diseases such as strep throat, which many people have almost yearly. Some doctors are trying to avoid prescribing antibiotics until their patients have had a chance to fight the disease on their own. Some patients are purposely avoiding using antibiotics for the same reason. While such a strategy is attractive and intuitive, if we take a closer look at strep throat, we can see yet another side to this story.

Rheumatic fever is a complication of untreated strep throat. It is characterized by fever and widespread inflammation of the joints and heart. These effects are produced by the body's immune response to the M protein of the group A streptococci. The M protein resembles a major protein in heart tissue. As a result, the antibodies attack the heart valves as well as the bacterial M protein. About 3% of individuals who do not get treatment with antibiotics when they have strep throat develop rheumatic fever. About 40% of patients with rheumatic fever develop heart valve damage, which may not become evident for 10 years or more. The best way to avoid this complication is early treatment of strep throat with antibiotics.

In summary, antibiotics are a very important weapon in our arsenal against disease, but they should not be used indiscriminately. If they are used, the entire course of treatment should be taken to completion. The last thing you would want to do is kill off most—but not all—of the bacteria you were infected with. This would leave behind a few "superbacteria," which would then be drug-resistant.

C. Autoimmune Diseases

In spite of the safeguards in the body intended to prevent acting against "self," in the form of healthy cells, many diseases exist in which some part of a pathway in the immune system goes awry. The skin disease psoriasis is thought to be a T-cell–mediated disease in which cytokines and chemokines play an essential role. Other autoimmune diseases, such as myasthenia gravis, rheumatoid arthritis, multiple sclerosis (Chemical Connections 21D), and insulin-dependent diabetes (Chemical Connections 24G), also involve cytokines and chemokines. Allergies are another example of malfunctioning of the immune system. Pollens and animal furs are allergens that can provoke asthma attacks. Some people are so sensitive to certain food allergens that even a knife used in smearing peanut butter may prove fatal in a person known to be allergic to peanuts.

The major drug treatment for autoimmune diseases involves the glucocorticoids, the most important of which is cortisol (Section 21.10A). It is a standard therapy for rheumatoid arthritis, asthma, inflammatory bone diseases, psoriasis, and eczema. The beneficial effects of glucocorticoids are overshadowed, however, by their undesirable side effects, which include osteoporosis, skin atrophy, and diabetes. Glucocorticoids regulate the synthesis of cytokines either directly, by interacting with their genes, or indirectly, through transcription factors.

Macrolide drugs are used to suppress the immune system during tissue transplantation or in the case of certain autoimmune diseases. Drugs such as cyclosporin A or rapamycin bind to receptors in the cytosol and, through secondary messengers, inhibit the entrance of nuclear factors into the nucleus. Normally those nuclear factors signal a need for transcription, so their absence prevents the transcription of cytokines—for example, interleukin-2.

31.8 | How Does the Human Immunodeficiency Virus Cause AIDS?

Human immunodeficiency virus (HIV) is the most infamous of the retroviruses, as it is the causative agent of acquired immunodeficiency syndrome (AIDS). This disease affects more than 40 million people worldwide and has continually thwarted attempts to eradicate it. The best medicine today can slow it down, but nothing has been able to stop AIDS.

The HIV genome is a single-stranded RNA that has a number of proteins packed around it, including the virus-specific reverse transcriptase and protease. A protein coat surrounds the RNA–protein assemblage, giving the overall shape of a truncated cone. Finally, a membrane envelope encloses the protein coat. The envelope consists of a phospholipid bilayer formed from the plasma membrane of cells infected earlier in the life cycle of the virus, as well as some specific glycoproteins, such as gp41 and gp120, as shown in Figure 31.17.

(a)

gp120
gp41
Lipid membrane
p17 matrix
p24 capsid
Reverse transcriptase
RNA in p7 nucleocapsid
Protease
Integrase

(b)

Eye of Science/Photo Researchers, Inc.

Figure 31.17 The architecture of HIV. (*a*) The RNA genome is surrounded by P7 nucleocapsid proteins and by several viral enzymes—namely, reverse transcriptase, integrase, and protease. The truncated cone consists of P24 capsid protein subunits. The P17 matrix (another layer of protein) lies inside the envelope, which consists of a lipid bilayer and glycoproteins such as gp41 and gp120. (*b*) An electron micrograph shows both mature virus particles, in which the core (the truncated cone) is visible, and immature virus particles, in which it is not.

Figure 31.18 HIV infection begins when the virus particle binds to CD4 receptors on the surface of the cell (Step 1). The viral core is inserted into the cell and partially disintegrates (Step 2). The reverse transcriptase catalyzes the production of DNA from the viral RNA. The viral DNA is integrated into the DNA of the host cell (Step 3). The DNA, including the integrated viral DNA, is transcribed to RNA (Step 4). Smaller RNAs are produced first, specifying the amino acid sequence of viral regulatory proteins (Step 5). Larger RNAs, which specify the amino acid sequences of viral enzymes and coat proteins, are made next (Step 6). The viral protease assumes particular importance in the budding of new virus particles (Step 7). Both the viral RNA and viral proteins are included in the budding virus, as is some of the membrane of the infected cell (Step 8).

HIV offers a classic example of the mode of operation of retroviruses. The HIV infection begins when the virus particle binds to receptors on the surface of a cell (Figure 31.18). The viral core is inserted into the cell and partially disintegrates. The reverse transcriptase catalyzes the production of DNA from the viral RNA. The viral DNA is integrated into the DNA of the host cell. The DNA, including the integrated viral DNA, is transcribed to RNA. Smaller RNAs are produced first, specifying the amino acid sequences of viral regulatory proteins. Larger RNAs, which specify the amino acid sequences of viral enzymes and coat proteins, are made next. The viral protease assumes particular importance in the budding of new virus particles.

Both the viral RNA and viral proteins are included in the budding virus, as is some of the membrane of the infected cell.

A. HIV's Ability to Confound the Immune System

Why is HIV so deadly and so hard to stop? Many viruses, such as adenovirus, cause nothing more than the common cold; others, such as the virus that causes severe acute respiratory syndrome (SARS), are deadly. At the same time, we have seen the complete eradication of the deadly SARS virus, whereas adenovirus is still with us. HIV has several characteristics that lead to its persistence and eventual lethality. Ultimately, it is deadly because of its targets, the helper T cells. The immune system is under constant attack by the virus, and millions of helper T cells and killer T cells are called up to fight billions of virus particles. Through degradation of the T-cell membrane via budding and the activation of enzymes that lead to cell death, the T-cell count diminishes to the point that the infected person is no longer able to mount a suitable immune response. As a result, the individual eventually succumbs to pneumonia or another opportunistic disease.

There are many reasons that the disease is so persistent. For example, it is slow acting. SARS was eradicated quickly because the virus was quick to act, making it easy to find infected people before they had a chance to spread the disease. In contrast, HIV-infected individuals can go years before they become aware that they have the disease. However, this is only a small part of what makes HIV so difficult to kill.

HIV is difficult to kill because it is difficult to find. For an immune system to fight a virus, it needs to locate specific macromolecules that can be bound to antibodies or T-cell receptors. The reverse transcriptase of HIV is very inaccurate in replication. The result is rapid mutation of HIV, a situation that presents a considerable challenge to those who want to devise treatments for AIDS. The virus mutates so rapidly that multiple strains of HIV may be present in a single individual.

Another trick the virus plays is a conformational change of the gp120 protein when it binds to the CD4 receptor on a T cell. The normal shape of the gp120 monomer may elicit an antibody response, but these antibodies are largely ineffective. The gp120 forms a complex with gp41 and changes shape when it binds to CD4. It also binds to a secondary site on the T cell that normally binds to a cytokine. This change exposes a part of gp120 that was previously hidden and, therefore, cannot elicit antibodies.

HIV is also adept at evading the innate immunity system. Natural killer cells attempt to attack the virus, but HIV binds a particular cell protein, called cyclophilin, to its capsid, which blocks the antiviral agent restriction factor-1. Another of HIV's proteins blocks the viral inhibitor called CEM-15, which normally disrupts the viral life cycle.

Lastly, HIV hides from the immune system by cloaking its outer membrane in sugars that are very similar to the natural sugars found on most of its host's cells, rendering the immune system blind to it.

B. The Search for a Vaccine

The attempt to find a vaccine for HIV is akin to the search for the Holy Grail, and it has met with about as much success. One strategy for using a vaccine to stimulate the body's immunity to HIV is shown in Figure 31.19. DNA for a unique HIV gene, such as the *gag* gene, is injected into muscle. The *gag* gene leads to the Gag protein, which is taken up by antigen-presenting cells and displayed on their cell surfaces. This then elicits the

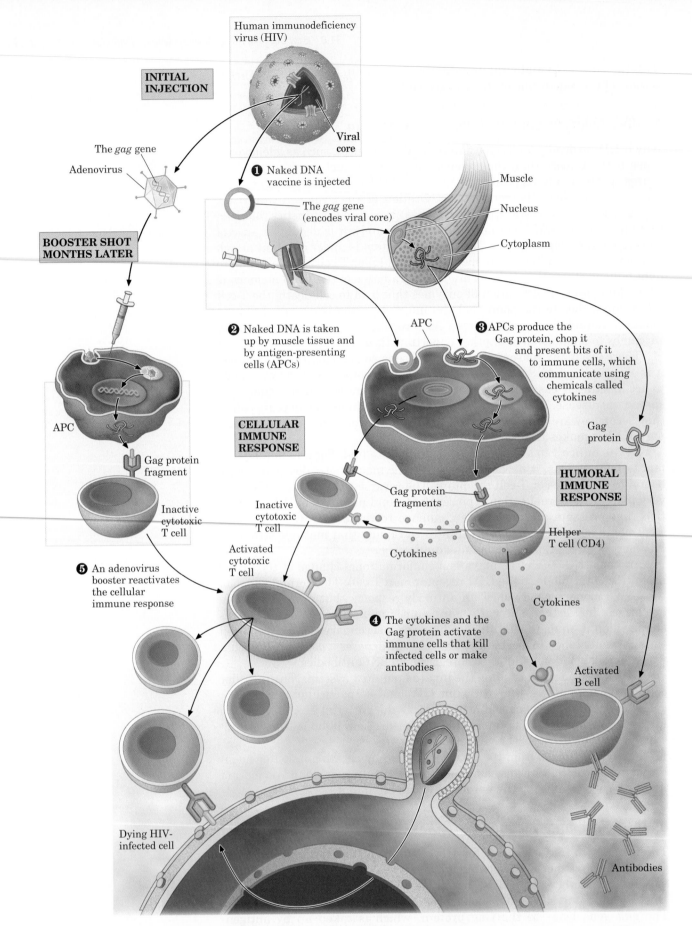

INITIAL INJECTION

Human immunodeficiency virus (HIV)

Viral core

❶ Naked DNA vaccine is injected

The *gag* gene

Adenovirus

BOOSTER SHOT MONTHS LATER

The *gag* gene (encodes viral core)

Muscle

Nucleus

Cytoplasm

❷ Naked DNA is taken up by muscle tissue and by antigen-presenting cells (APCs)

APC

❸ APCs produce the Gag protein, chop it and present bits of it to immune cells, which communicate using chemicals called cytokines

Gag protein

APC

Gag protein fragment

CELLULAR IMMUNE RESPONSE

Inactive cytotoxic T cell

Inactive cytotoxic T cell

Gag protein fragments

HUMORAL IMMUNE RESPONSE

Helper T cell (CD4)

Cytokines

❺ An adenovirus booster reactivates the cellular immune response

Activated cytotoxic T cell

❹ The cytokines and the Gag protein activate immune cells that kill infected cells or make antibodies

Cytokines

Activated B cell

Dying HIV-infected cell

Antibodies

Figure 31.19   One strategy for an AIDS vaccine. (Reprinted by permission from Ezzell C., "Hope in a Vial," *Sci. Am.* 39–45 June 2002.)

cellular immune response, stimulating killer and helper T cells. It also stimulates the humoral immune response, spurring production of antibodies. Figure 31.19 also shows a second phase of the treatment, a booster shot of an altered adenovirus that carries the *gag* gene.

Unfortunately, most attempts at making antibodies have proved unsuccessful to date. The most thorough attempt was made by the VaxGen Company, which carried the research through the third stage of clinical trials, testing the vaccine on more than 1000 high-risk people and comparing them to 1000 individuals who did not receive the vaccine. In the study, 5.7% of the people who received the vaccine eventually became infected, compared to 5.8% of the placebo group. Many people analyzed the data and, despite attempts to show a better response in certain ethnic groups, the trials had to be declared a failure. The vaccine, called AIDSVAX, was based on gp120.

C. Antiviral Therapy

While the search for an effective vaccine continued with little to minor success, pharmaceutical companies flourished by designing drugs that would inhibit retroviruses. By 1996, there were 16 drugs used to inhibit either the HIV reverse transcriptase or the protease. Several others are in clinical trials, including drugs that target gp41 and gp120 in an attempt to prevent entry of the virus. A combination of drugs to inhibit retroviruses has been dubbed **highly active antiretroviral therapy (HAART).** Initial attempts at HAART were very successful, driving the viral load almost to the point of being undetectable, with the concomitant rebounding of the CD4 cell population. However, as always seems to be the case with HIV, it later turned out that while the virus was knocked down, it was not knocked out. HIV remained in hiding in the body and would bounce back as soon as the therapy was stopped. Thus the best-case scenario for an AIDS patient was a lifetime of expensive drug therapies. In addition, long-term exposure to HAART was found to cause constant nausea and anemia, as well as diabetes symptoms, brittle bones, and heart disease.

D. A Second Chance for Antibodies

In the wake of the realization that patients could not stay on the HAART program indefinitely, several researchers attempted to combine HAART with vaccination. Even though most of the vaccines were not found to be effective alone, they proved more effective in combination with HAART. In addition, once on the vaccine, patients were able to take a rest from the other drugs, giving their bodies and minds time to recover from the side effects of the antiviral therapy.

E. The Future of Antibody Research

Early attempts at creating a vaccine appear to have failed because the vaccine elicited too many useless antibodies. What patients need is a **neutralizing antibody,** one capable of completely eliminating its target. Researchers discovered a patient who had HIV for six years, but never developed AIDS. They then studied his blood and found a rare antibody, which they labeled **b12.** In laboratory trials, b12 was found to stop most strains of HIV. What made b12 different from the other antibodies? Structural analysis revealed that this antibody has a different shape from a normal immunoglobulin. It has sections of long tendrils that fit into a fold in gp120. This fold in gp120 cannot mutate very much; otherwise, the protein will not be able to dock properly with the CD4 receptor.

CHEMICAL CONNECTIONS 31E

Why Are Stem Cells Special?

Stem cells are the precursors of all other cell types, including T and B lymphocytes. These undifferentiated cells have the ability to form any cell type as well as to replicate into more stem cells. Stem cells are often called **progenitor cells** due to their ability to differentiate into many cell types. A **pluripotent** stem cell is able to give rise to all cell types in an embryo or an adult. Some cells are called **multipotent** because they can differentiate into more than one cell type, but not into all cell types. The further from a zygote a cell is in the course of its development, the less potent the cell type. The use of stem cells, especially **embryonic stem cells,** has been an exciting field of research for several years.

History of Stem Cell Research

Stem cell research began in the 1970s with studies on teratocarcinoma cells, which are found in testicular cancers. These cells are bizarre blends of differentiated and undifferentiated cells. Such **embryonal carcinoma cells (EC)** were found to be pluripotent, which led to the idea of using them for therapy. However, this line of research was suspended because the cells came from tumors, which made their use dangerous, and they were **aneuploid,** which means they had the wrong number of chromosomes.

Early work with embryonic stem cells (ES) involved cells that were grown in culture after being taken from embryos. It was found that these stem cells could be maintained for long periods. In contrast, most differentiated cells will not grow for extended periods in culture. Stem cells are maintained in culture by addition of certain factors, such as leukemia-inhibiting factor (LIF) or feeder cells (nonmitotic cells such as fibroblasts). Once released from these controls, ES cells will differentiate into all kinds of cells.

Stem Cells Offer Hope

Stem cells placed into a particular tissue, such as blood, will differentiate and grow into blood cells. Others, when placed into brain tissue, will grow brain cells. This discovery is very exciting because it had been believed that there was little hope for patients with spinal cord and other nerve damage, because these cells do not normally regenerate. In theory, neurons could be produced to treat neurodegenerative diseases, such as Alzheimer's or Parkinson's disease. Muscle cells could be produced to treat muscular dystrophies and heart disease. In one study, mouse stem cells were injected into a mouse heart that had undergone a myocardial infarction. The cells spread from an unaffected region into the infarcted zone and began to grow new heart tissue. Human pluripotent stem cells have been used to regenerate nerve tissue in rats with nerve injuries and have been shown to improve motor and cognitive ability in rats that underwent a stroke. Results such as these have led some scientists to claim that stem cell technology will be the most important advancement in science since cloning.

Truly pluripotent stem cells have been harvested primarily from embryonic tissue, and these cells show the greatest ability to differentiate into various tissues and to reproduce in cell culture. Stem cells have also been taken from adult tissues, as some stem cells are always present in an organism, even at the adult stage. These cells are usually multipotent, as they can form several different cell types, but are not as versatile as ES cells. Many scientists believe that the ES cells represent a better source for tissue therapy than adult stem cells for this reason.

The acquisition and use of stem cells can be related to a technique called **cell reprogramming,** which is a necessary compo-

Pluripotent embryonic stem cells can be grown in cell culture. They can be maintained in an undifferentiated state by growing them on certain feeder cells, such as fibroblasts or by using leukemia inhibitory factor (LIF). When removed from the feeder cells or when the LIF is removed, they begin to differentiate in a wide variety of tissue types, which could then be harvested and grown for tissue therapy. (Taken from Donovan P. J., and Gearhart J., *Nature*, 414:92–97, 2001)

Why Are Stem Cells Special? *(continued)*

nent of whole-mammal cloning, such as the process that created the world's most famous sheep, Dolly. Most somatic cells in an organism contain the same genes, but the cells develop as different tissues with wildly different patterns of gene expression. A mechanism that alters expression of genes without changing the actual DNA sequence is called an **epigenetic** mechanism. An epigenetic state of the DNA in a cell is a heritable trait that allows a "molecular memory" to exist in the cells. In essence, a liver cell remembers where it came from and will continue to divide and remain a liver cell. These epigenetic states involve methylation of cytosine–guanine dinucleotides and interactions with proteins of chromatin (Section 25.3C). Mammalian genes possess an additional level of epigenetic information called **imprinting,** which allows the DNA to retain a molecular memory of its germ-line origin. The paternal DNA is imprinted differently than the maternal DNA. In normal development, only DNA that came from both parents would be able to combine and lead to a viable offspring.

The epigenetic states of somatic cells are generally locked in a way that the differentiated tissues remain stable. The key to whole-organism cloning was the ability to erase the epigenetic state and return to the state of a fertilized egg, which has the potential to produce all cell types. If the nucleus of a somatic cell is injected into a recipient oocyte, the epigenetic state of the DNA can be reprogrammed, or at least "partially" reprogrammed. The molecular memory is erased, and the cell begins to behave like a true zygote. This development can be used to derive pluripotent stem cells or to transfer a blastocyst into a mother-carrier for growth and development. In November 2001, the first human cloned blastocyst was created in this way, with the aim of growing enough cells to harvest pluripotent stem cells for research.

Currently, debate continues to rage worldwide over the use of embryonic stem cells. The issue at hand is one of ethics and the definition of life. Embryonic stem cells come from many sources, including aborted fetuses, umbilical cords, and embryos from in vitro fertilization clinics. The report of the cloned human embryonic cells merely added to the controversy. The U.S. government has banned funding for stem cell research, but allows research to continue on all existing embryonic cell lines. Some big questions that people will have to answer are the following: Do a few cells created by therapeutic cloning of your own somatic cells constitute life? If these cells do constitute life, do they have the same rights as a human being conceived naturally? If it were possible, should someone be allowed to grow his or her own therapeutic clone into an adult?

Another antibody was found in a different patient who seemed resistant to HIV. This antibody was actually a dimer and had a shape more like an "I" than the traditional "Y". This antibody, called **2G12,** recognizes some of the sugars on HIV's outer membrane that are unique to HIV.

By identifying a few such antibodies, researchers have been able to search for a vaccine in the opposite direction from the normal way. In **retrovaccination,** researchers have the antibody and need to find a vaccine to elicit it, instead of injecting vaccines and noting which antibodies they produce.

SUMMARY OF KEY QUESTIONS

SECTION 31.1 How Does the Body Defend Itself from Invasion?

- The human immune system protects us against foreign invaders. It consists of two parts: (1) the natural resistance of the body, called innate immunity, and (2) adaptive or acquired immunity.
- **Innate immunity** is nonspecific. **Macrophages** and **natural killer (NK) cells** are cells of innate immunity that function as police officers.
- **Adaptive** or **acquired immunity** is highly specific, being directed against one particular invader.
- Acquired immunity (known as the immune system) also has memory, unlike innate immunity.

SECTION 31.2 What Organs and Cells Make Up the Immune System?

- The principal cellular components of the immune system are white blood cells, or **leukocytes.** The specialized leukocytes in the lymph system are called **lymphocytes.** They circulate mostly in the **lymphoid organs.**
- The **lymph** system is a collection of vessels extending throughout the body and connected to the interstitial fluid on the one hand, and to the blood vessels on the other hand.

- Lymphocytes that mature in the bone marrow and produce soluble immunoglobulins are **B cells.** Lymphocytes that mature in the thymus gland are **T cells.**

SECTION 31.3 How Do Antigens Stimulate the Immune System?

- **Antigens** are large, complex molecules of foreign origin. An antigen may be a bacterium, a virus, or a toxin.
- An antigen may interact with antibodies, T-cell receptors (TcR), or major histocompatibility complex (MHC) molecules. All three types of molecules belong to the **immunoglobulin superfamily.**
- An **epitope** is the smallest part of an antigen that binds to antibodies, TcRs, and MHCs.

SECTION 31.4 What Are Immunoglobulins?

- Antibodies are **immunoglobulins.** These water-soluble glycoproteins are made of two heavy chains and two light chains. All four chains are linked together by disulfide bridges.
- Immunoglobulins contain variable regions in which the amino acid composition of each antibody is different. These regions interact with antigens to form insoluble large aggregates.
- A large variety of antibodies are synthesized by a number of processes in the body.
- During B-cell development, the **variable regions** of the heavy chains are assembled by a process called V(J)D recombination.
- Mutations on these new genes create even greater diversity. Somatic hypermutation (SHM) that creates a point mutation (only one nucleotide) is one way. Patches of mutation introduced into the V(J)D gene constitute gene conversion (GC) mutation.
- Immunoglobulins respond to an antigen over a long span of time, lasting for weeks and even months.
- All antigens—whether proteins, polysaccharides, or nucleic acids—interact with soluble immunoglobulins produced by B cells.

SECTION 31.5 What Are T Cells and T-Cell Receptors?

- Protein antigens interact with T cells. The binding of the epitope to the TcR is facilitated by MHC, which carries the epitope to the T-cell surface, where it is presented to the receptor.
- Upon epitope binding to the receptor, the T cell is stimulated. It proliferates and can differentiate into (1) killer T cells, (2) memory cells, or (3) helper T cells.
- The TcR has a number of helper molecules, such as CD4 or CD8, that enable it to bind the epitope tightly and to bind to other cells via MHC proteins.

- CD (**cluster determinant**) molecules also belong to the immunoglobulin superfamily.
- Antibodies can recognize all types of antigens, but TcRs recognize only peptide antigens.

SECTION 31.6 How Is the Immune Response Controlled?

- The control and coordination of the immune response are handled by **cytokines,** which are small protein molecules.
- Chemotactic cytokines, the **chemokines,** such as interleukin-8, facilitate the migration of leukocytes from blood vessels into the site of injury or inflammation. Other cytokines activate B and T cells and macrophages, enabling them to engulf the foreign body, digest it, or destroy it by releasing special toxins.
- Some cytokines, such as tumor necrosis factor (TNF), can lyse tumor cells.

SECTION 31.7 How Does the Body Distinguish "Self" from "Nonself"?

- A number of mechanisms in the body ensure that the body recognizes "self."
- In adaptive immunity, the selection of T and B cells that are prone to interact with "self" antigens are eliminated.
- In innate immunity, two kinds of receptors exist on the surfaces of T and B cells: an **activating receptor** and an **inhibitory receptor.** The inhibitory receptor recognizes the epitope of a normal cell, binds to it, and prevents the activation of the killer T cell or macrophage.
- Many autoimmune diseases are T-cell–mediated diseases, in which cytokines and chemokines play an essential role.
- The standard drug treatment for autoimmune diseases involves glucocorticoids, which prevent the transcription and hence the synthesis of cytokines.

SECTION 31.8 How Does the Human Immunodeficiency Virus Cause AIDS?

- HIV is a retrovirus that enters helper T cells.
- The virus weakens the immune system by destroying these helper T cells through damage to their cell membranes and activation of enzymes that cause apoptosis.
- HIV has been studied for 25 years in an attempt to find a cure, but none has been discovered as yet. The virus hides from the host's immune system and mutates so frequently that no effective antibody response can be mounted.
- A combination of therapies combining enzyme inhibitors and antibodies has achieved the most success.

PROBLEMS

A blue problem number indicates an applied problem.

■ denotes problems that are available on the GOB ChemistryNow website or CD and are assignable in OWL.

SECTION 31.1 How Does the Body Defend Itself from Invasion?

31.1 Give two examples of external innate immunity in humans.

31.2 ■ Which form of immunity is characteristic of vertebrates only?

31.3 How does the skin fight bacterial invasion?

31.4 T-cell receptors and MHC molecules both interact with antigens. What is the difference in the mode of interaction between the two?

31.5 What differentiates innate immunity from adaptive (acquired) immunity?

SECTION 31.2 What Organs and Cells Make Up the Immune System?

31.6 ■ Where in the body do you find the largest concentration of antibodies as well as T cells?

31.7 Where do T and B cells mature and differentiate?

31.8 What are memory cells? What is their function?

31.9 What are the favorite targets of macrophages? How do they kill the target cells?

SECTION 31.3 How Do Antigens Stimulate the Immune System?

31.10 Would a foreign substance, such as aspirin (MW 180), be considered an antigen by the body?

31.11 What kind of antigen does a T cell recognize?

31.12 What is the smallest unit of an antigen that is capable of binding to an antibody?

31.13 How does the body process antigens to be recognized by class II MHC?

31.14 What role do MHC molecules play in the immune response of the ABO blood groups?

31.15 ■ To which class of compounds do MHCs belong? Where would you find them?

31.16 What is the difference in function between class I and class II MHC molecules?

SECTION 31.4 What are Immunoglobulins?

31.17 When a foreign substance is injected in a rabbit, how long does it take to find antibodies against the foreign substance in the rabbit serum?

31.18 Distinguish among the roles of the IgA, IgE, and IgG immunoglobulins.

31.19 (a) Which immunoglobulin has the highest carbohydrate content and the lowest concentration in the serum?

(b) What is its main function?

31.20 Chemical Connections 20D states that the antigen in the red blood cells of a person with B-type blood is a galactose unit. Show schematically how the antibody of a person with A-type blood would aggregate the red blood cells of a B-type person if such a transfusion were made by mistake.

31.21 In the immunoglobulin structure, the "hinge region" joins the stem of the Y to the arms. The hinge region can be cleaved by a specific enzyme to yield one F_c fragment (the stem of the Y) and two F_{ab} fragments (the two arms). Which of these two kinds of fragments can interact with an antigen? Explain.

31.22 How are the light and heavy chains of an antibody held together?

31.23 What do we mean by the term *immunoglobulin superfamily*?

31.24 If you could isolate two monoclonal antibodies from a certain population of lymphocytes, in what sense would they be similar to each other and in what sense would they differ?

31.25 What kind of interaction takes place between an antigen and an antibody?

31.26 How is a new protein created on the variable portion of a heavy chain by V(J)D recombination?

31.27 What accounts for antibody diversity?

SECTION 31.5 What Are T Cells and T-Cell Receptors?

31.28 T-cell receptor molecules are made of two polypeptide chains. Which part of the chain acts as a binding site, and what binds to it?

31.29 What is the difference between a T-cell receptor (TcR) and a TcR complex?

31.30 ■ What kind of tertiary structure characterizes the TcR?

31.31 What are the components of the TcR complex?

31.32 ■ By what chemical process does CD3 transduct signals inside the cell?

31.33 Which adhesion molecule in the TcR complex helps HIV infect a leukocyte?

31.34 Three kinds of molecules in the T cell belong to the immunoglobulin superfamily. List them and indicate briefly their functions.

31.35 What functions do CD4 and CD8 serve in the immune response?

SECTION 31.6 How Is the Immune Response Controlled?

31.36 What kind of molecules are cytokines?

31.37 With what do cytokines interact? Do they bind to antigens?

31.38 As in most biochemistry literature, one encounters a veritable alphabet soup when reading about cytokines. Identify these cytokines by their full names: (a) TNF, (b) IL, and (c) EGF.

31.39 What are chemokines? How do they deliver their message?

31.40 What is the characteristic chemical signature in the structure of chemokines?

31.41 What are the chemical characteristics of cytokines that allow their classification?

31.42 ■ Which amino acid appears in all chemokines?

SECTION 31.7 How Does the Body Distinguish "Self" from "Nonself"?

31.43 How does the body prevent T cells from being active against a "self" antigen?

31.44 What makes a tumor cell different from a normal cell?

31.45 ■ Name a signaling pathway that controls the maturation of B cells and prevents those with an affinity for self antigen from becoming active.

31.46 How does the inhibitory receptor on a macrophage prevent an attack on normal cells?

31.47 Which components of the immune system are principally involved in autoimmune diseases?

31.48 How do glucocorticoids make individuals with autoimmune disease feel more comfortable?

SECTION 31.8 How Does the Human Immunodeficiency Virus Cause AIDS?

31.49 Which cells are attacked by HIV?

31.50 How does HIV gain entry into the cells it attacks?

31.51 How does HIV confound the human immune system?

31.52 What types of therapy are used to fight AIDS?

31.53 Why have vaccines been relatively unsuccessful in stopping AIDS?

31.54 What are the structural features of the two types of neutralizing antibodies that have been the most successful at combating AIDS? What makes these antibodies more effective?

Chemical Connections

31.55 (Chemical Connections 31A) How are the mayapple and chemotherapy related?

31.56 (Chemical Connections 31B) How does Rituxan, a monoclonal antibody, find its way to the surface of non-Hodgkin's lymphoma cells?

31.57 (Chemical Connections 31B) Why is monoclonal antibody treatment better than chemotherapy?

31.58 (Chemical Connections 31B) Why is a monoclonal antibody treatment preferable to a multiclonal antibody treatment?

31.59 (Chemical Connections 31C) What made Edward Jenner "the father of immunization"? In your opinion, could one do such an experiment today legally?

31.60 (Chemical Connections 31C) What was the observation that led Edward Jenner to attempt his experiment?

31.61 (Chemical Connections 31C) What is the derivation of the word *vaccination?*

Additional Problems

31.62 Which immunoglobulins form the first line of defense against invading bacteria?

31.63 Which cells of the innate immunity system are the first to interact with an invading pathogen?

31.64 Where do all of the lymphatic vessels come together?

31.65 Which compound or complex of compounds of the immune system is mostly responsible for the proliferation of leukocytes?

31.66 Name a process beside V(J)D recombination that can enhance immunoglobulin diversity in the variable region.

31.67 Name a tumor cell marker, a synthetic analog of which may be the first anticancer vaccine.

31.68 Is the light chain of an immunoglobulin the same as the V region?

31.69 Where are TNF receptors located?

31.70 ■ The variable regions of immunoglobulins bind the antigens. How many polypeptide chains carry variable regions in one immunoglobulin molecule?

Exponential Notation

The **exponential notation** system is based on powers of 10 (see table). For example, if we multiply $10 \times 10 \times 10 = 1000$, we express this as 10^3. The 3 in this expression is called the **exponent** or the **power,** and it indicates how many times we multiplied 10 by itself and how many zeros follow the 1.

There are also negative powers of 10. For example, 10^{-3} means 1 divided by 10^3:

$$10^{-3} = \frac{1}{10^3} = \frac{1}{1000} = 0.001$$

Numbers are frequently expressed like this: 6.4×10^3. In a number of this type, 6.4 is the **coefficient** and 3 is the exponent, or power of 10. This number means exactly what it says:

$$6.4 \times 10^3 = 6.4 \times 1000 = 6400$$

Similarly, we can have coefficients with negative exponents:

$$2.7 \times 10^{-5} = 2.7 \times \frac{1}{10^5} = 2.7 \times 0.00001 = 0.000027$$

For numbers greater than 10 in exponential notation, we proceed as follows: *Move the decimal point to the left,* to just after the first digit. The (positive) exponent is equal to the number of places we moved the decimal point.

Exponential notation is also called scientific notation.

For example, 10^6 means a one followed by six zeros, or 1,000,000, and 10^2 means 100.

APP. I.1	Examples of Exponential Notation
	$10,000 = 10^4$
	$1000 = 10^3$
	$100 = 10^2$
	$10 = 10^1$
	$1 = 10^0$
	$0.1 = 10^{-1}$
	$0.01 = 10^{-2}$
	$0.001 = 10^{-3}$

EXAMPLE

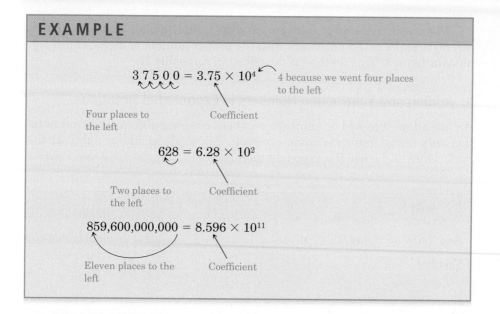

$37500 = 3.75 \times 10^4$ 4 because we went four places to the left

Four places to the left   Coefficient

$628 = 6.28 \times 10^2$

Two places to the left   Coefficient

$859{,}600{,}000{,}000 = 8.596 \times 10^{11}$

Eleven places to the left   Coefficient

We don't really have to place the decimal point after the first digit, but by doing so we get a coefficient between 1 and 10, and that is the custom.

Using exponential notation, we can say that there are 2.95×10^{22} copper atoms in a copper penny. For large numbers, the exponent is always *positive.* Note that we do not usually write out the zeros at the end of the number.

For small numbers (less than 1), we move the decimal point *to the right*, to just after the first nonzero digit, and use a *negative exponent*.

EXAMPLE

$$0.00346 = 3.46 \times 10^{-3}$$

Three places to the right

$$0.000004213 = 4.213 \times 10^{-6}$$

Six places to the right

In exponential notation, a copper atom weighs 2.3×10^{-25} pounds.

To convert exponential notation into fully written-out numbers, we do the same thing backward.

EXAMPLE

Write out in full: (a) 8.16×10^7 (b) 3.44×10^{-4}

Solution

(a) $8.16 \times 10^7 = 81,600,000$ (b) $3.44 \times 10^{-4} = 0.000344$

Seven places to the right (add enough zeros)   Four places to the left

When scientists add, subtract, multiply, and divide, they are always careful to express their answers with the proper number of digits, called significant figures. This method is described in Appendix II.

A. Adding and Subtracting Numbers in Exponential Notation

We are allowed to add or subtract numbers expressed in exponential notation *only if they have the same exponent*. All we do is add or subtract the coefficients and leave the exponent as it is.

EXAMPLE

Add 3.6×10^{-3} and 9.1×10^{-3}.

Solution

$$\begin{array}{r} 3.6 \times 10^{-3} \\ + \ 9.1 \times 10^{-3} \\ \hline 12.7 \times 10^{-3} \end{array}$$

The answer could also be written in other, equally valid ways:

$$12.7 \times 10^{-3} = 0.0127 = 1.27 \times 10^{-2}$$

When it is necessary to add or subtract two numbers that have different exponents, we first must change them so that the exponents are the same.

A calculator with exponential notation changes the exponent automatically.

EXAMPLE

Add 1.95×10^{-2} and 2.8×10^{-3}.

Solution

To add these two numbers, we make both exponents -2. Thus, $2.8 \times 10^{-3} = 0.28 \times 10^{-2}$. Now we can add:

$$\begin{array}{r} 1.95 \times 10^{-2} \\ + \ 0.28 \times 10^{-2} \\ \hline 2.23 \times 10^{-2} \end{array}$$

B. Multiplying and Dividing Numbers in Exponential Notation

To multiply numbers in exponential notation, we first multiply the coefficients in the usual way and then algebraically *add* the exponents.

EXAMPLE

Multiply 7.40×10^5 by 3.12×10^9.

Solution

$$7.40 \times 3.12 = 23.1$$

Add exponents:

$$10^5 \times 10^9 = 10^{5+9} = 10^{14}$$

Answer:

$$23.1 \times 10^{14} = 2.31 \times 10^{15}$$

EXAMPLE

Multiply 4.6×10^{-7} by 9.2×10^4

Solution

$$4.6 \times 9.2 = 42$$

Add exponents:

$$10^{-7} \times 10^4 = 10^{-7+4} = 10^{-3}$$

Answer:

$$42 \times 10^{-3} = 4.2 \times 10^{-2}$$

To divide numbers expressed in exponential notation, the process is reversed. We first divide the coefficients and then algebraically *subtract* the exponents.

EXAMPLE

Divide: $\dfrac{6.4 \times 10^8}{2.57 \times 10^{10}}$

Solution

$$6.4 \div 2.57 = 2.5$$

Subtract exponents:

$$10^8 \div 10^{10} = 10^{8-10} = 10^{-2}$$

Answer:

$$2.5 \times 10^{-2}$$

EXAMPLE

Divide: $\dfrac{1.62 \times 10^{-4}}{7.94 \times 10^7}$

Solution

$$1.62 \div 7.94 = 0.204$$

Subtract exponents:

$$10^{-4} \div 10^7 = 10^{-4-7} = 10^{-11}$$

Answer:

$$0.204 \times 10^{-11} = 2.04 \times 10^{-12}$$

Scientific calculators do these calculations automatically. All that is necessary is to enter the first number, press $+$, $-$, $\times$, or $\div$, enter the second number, and press $=$. (The method for entering numbers of this form varies; consult the instructions that come with the calculator.) Many scientific calculators also have a key that will automatically convert a number such as 0.00047 to its scientific notation form (4.7×10^{-4}), and vice versa. For problems relating to exponential notation, see Chapter 1, Problems 1.17 through 1.24.

APPENDIX II

Significant Figures

If you measure the volume of a liquid in a graduated cylinder, you might find that it is 36 mL, to the nearest milliliter, but you cannot tell if it is 36.2, or 35.6, or 36.0 mL because this measuring instrument does not give the last digit with any certainty. A buret gives more digits, and if you use one you should be able to say, for instance, that the volume is 36.3 mL and not 36.4 mL. But even with a buret, you could not say whether the volume is 36.32 or 36.33 mL. For that, you would need an instrument that gives still more digits. This example should show you that *no measured number can ever be known exactly*. No matter how good the measuring instrument, there is always a limit to the number of digits it can measure with certainty.

We define the number of **significant figures** as the number of digits of a measured number that have uncertainty only in the last digit.

What do we mean by this definition? Assume that you are weighing a small object on a laboratory balance that can weigh to the nearest 0.1 g, and you find that the object weighs 16 g. Because the balance weighs to the nearest 0.1 g, you can be sure that the object does not weigh 16.1 g or 15.9 g. In this case, you would write the weight as 16.0 g. To a scientist, there is a difference between 16 g and 16.0 g. Writing 16 g says that you don't know the digit after the 6. Writing 16.0 g says that you do know it: It is 0. However, you don't know the digit after that. Several rules govern the use of significant figures in reporting measured numbers.

A. Determining the Number of Significant Figures

1. Nonzero digits are always significant.

For example, 233.1 m has four significant figures; 2.3 g has two significant figures.

2. Zeros at the beginning of a number are never significant.

For example, 0.0055 L has two significant figures; 0.3456 g has four significant figures.

3. Zeros between nonzero digits are always significant.

For example, 2.045 kcal has four significant figures; 8.0506 g has five significant figures.

4. Zeros at the end of a number that contains a decimal point are always significant.

For example, 3.00 L has three significant figures; 0.0450 mm has three significant figures.

5. Zeros at the end of a number that contains no decimal point may or may not be significant.

We cannot tell whether they are significant without knowing something about the number. This is the ambiguous case. If you know that a certain small business made a profit of $36,000 last year, you can be sure that the 3 and 6 are significant, but what about the rest? The profit might have been $36,126 or $35,786.53, or maybe even exactly $36,000. We just don't know because it is customary to round off such numbers. On the other hand, if the profit were reported as $36,000.00, then all seven digits would be significant.

In science, to get around the ambiguous case we use exponential notation (Section 1.3). Suppose a measurement comes out to be 2500 g. If we made the measurement, we of course know whether the two zeros are significant, but we need to tell others. If these digits are *not* significant, we write our number as 2.5×10^3. If one zero is significant, we write 2.50×10^3. If both zeros are significant, we write 2.500×10^3. Since we now have a decimal point, all the digits shown are significant.

B. Multiplying and Dividing

The rule in multiplication and division is that the final answer should have the *same* number of significant figures as there are in the number with the *fewest* significant figures.

EXAMPLE

Do the following multiplications and divisions:
(a) 3.6×4.27
(b) 0.004×217.38
(c) $\dfrac{42.1}{3.695}$
(d) $\dfrac{0.30652 \times 138}{2.1}$

Solution
(a) 15 (3.6 has two significant figures)
(b) 0.9 (0.004 has one significant figure)
(c) 11.4 (42.1 has three significant figures)
(d) 2.0×10^1 (2.1 has two significant figures)

C. Adding and Subtracting

In addition and subtraction, the rule is completely different. The number of significant figures in each number doesn't matter. The answer is given to the *same number of decimal places* as the term with the fewest decimal places.

EXAMPLE

Add or subtract:

(a) 320.084	(b) 61.4532	
80.47	13.7	
200.23	22	(c) 14.26
20.0	0.003	−1.05041
620.8	97	13.21

Solution
In each case, we add or subtract in the normal way but then round off so that the only digits that appear in the answer are those in the columns in which every digit is significant.

D. Rounding Off

When we have too many significant figures in our answer, it is necessary to round off. In this book we have used the rule that if *the first digit dropped* is 5, 6, 7, 8, or 9, we raise *the last digit kept* to the next number; otherwise, we do not.

EXAMPLE

In each case, drop the last two digits:
(a) 33.679   (b) 2.4715   (c) 1.1145   (d) 0.001309   (e) 3.52

Solution
(a)   33.679 = 33.7
(b)   2.4715 = 2.47
(c)   1.1145 = 1.11
(d) 0.001309 = 0.0013
(e)   3.52 = 4

E. Counted or Defined Numbers

All of the preceding rules apply to *measured* numbers and **not** to any numbers that are *counted* or *defined*. Counted and defined numbers are known exactly. For example, a triangle is defined as having 3 sides, not 3.1 or 2.9. Here, we treat the number 3 as if it has an infinite number of zeros following the decimal point.

EXAMPLE

Multiply 53.692 (a measured number) × 6 (a counted number).

Solution

$$322.15$$

Because 6 is a counted number, we know it exactly, and 53.692 is the number with the fewest significant figures. All we really are doing is adding 53.692 six times.

For problems relating to significant figures, see Chapter 1, Problems 1.25 to 1.30.

Answers

Chapter 1 Matter, Energy, and Measurement

1.1 multiplication (a) 4.69×10^5 (b) 2.8×10^{-15};
(a) 1.94×10^{18} (b) 1.37×10^5
1.2 (a) 147°F (b) 8.3°C
1.3 109 kg
1.4 13.8 km
1.5 743 mph
1.6 78.5 g
1.7 2.43 g/mL
1.8 1.016 g/mL
1.9 4.8×10^4 cal = 48 kcal
1.10 46°C
1.11 0.0430 cal/g·deg
1.13 (a) Matter is anything that has mass and takes up space. (b) Chemistry is the science that studies matter.
1.15 Dr. X's claim that the extract cured diabetes would be classified as (c) a hypothesis. No evidence had been provided to prove or disprove the claim.
1.17 (a) 3.51×10^{-1} (b) 6.021×10^2 (c) 1.28×10^{-4}
(d) 6.28122×10^5
1.19 (a) 6.65×10^{17} (b) 1.2×10^1 (c) 3.89×10^{-16}
(d) 3.5×10^{-23}
1.21 (a) 1.3×10^5 (b) 9.40×10^4 (c) 5.137×10^{-3}
1.23 4.45×10^6
1.25 (a) 2 (b) 5 (c) 5 (d) 5 (e) 3 (f) 3 (g) 2
1.27 (a) 92 (b) 7.3 (c) 0.68 (d) 0.0032 (e) 5.9
1.29 (a) 1.53 (b) 2.2 (c) 0.00048
1.31 330 min = 5.6 h
1.33 (a) 20 mm (b) 1 inch (c) 1 mile
1.35 Weight would change slightly. Mass is independent of location, but weight is a force exerted by a body influenced by gravity. The influence of the Earth's gravity decreases with increasing distance from sea level.
1.37 (a) 77°F, 298 K (b) 104°F, 313 K (c) 482°F, 523 K,
(d) −459°F, 0 K
1.39 (a) 0.0964 L (b) 27.5 cm (c) 4.57×10^4 g
(d) 4.75 m (e) 21.64 mL (f) 3.29×10^3 cc (g) 44 mL
(h) 0.711 kg (i) 63.7 cc (j) 7.3×10^4 mg
(k) 8.34×10^4 mm (l) 0.361 g
1.41 50 mi/h
1.43 solids and liquids
1.45 No, melting is a physical change.
1.47 bottom: manganese; top: sodium acetate;
middle: calcium chloride
1.49 1.023 g/mL
1.51 water
1.53 One should raise the temperature of water to 4°C. During this temperature change, the density of the crystals decreases, while the density of water increases. This brings the less dense crystals to the surface of the more dense water.
1.55 The motion of the wheels of the car generates kinetic energy, which is stored in your battery as potential energy.
1.57 0.34 cal/g·C°
1.59 334 mg
1.61 The body shivers. Further temperature lowering results in unconsciousness and then death.
1.63 Methanol, because its higher specific heat allows it to retain the heat longer.
1.65 0.732
1.67 kinetic: (b), (d), (e); potential: (a), (c)
1.69 the European car
1.71 kinetic energy
1.73 The largest is 41 g. The smallest is 4.1310×10^{-8} kg.
1.75 10.9 h ≈ 11 h
1.77 The heavy water. When converting the specific heat given in J/g·°C to cal/g·°C, one finds that the specific heat of heavy water is 1.008 cal/g·°C, which is somewhat greater than that of ordinary water.
1.79 (a) 1.57 g/mL (b) 1.25 g/mL
1.81 two
1.83 60 J would raise the temperature by 4.5°C; thus the final temperature will be 24.5°C.
1.85 Weigh the solid urea on a balance, and measure the liquid with a graduated cylinder.
1.87 New medications would consist of the same elements, and many of the fundamental patterns would be similar. The differences would arise in the specific patterns that will be discussed in Chapters 10–20.
1.89 Diethyl ether is less dense than water and insoluble in it. Put the water sample and the diethyl ether in the separatory funnel. The contaminant will dissolve in the diethyl ether, leaving the water sample purified.

Chapter 2 Atoms

2.1 (a) $NaClO_3$ (b) AlF_3
2.2 (a) The mass number is 31.
(b) The mass number is 222.
2.3 (a) The element is phosphorus (P); its symbol is $^{31}_{15}P$.
(b) The element is radon (Rn); its symbol is $^{222}_{86}Rn$.
2.4 (a) The atomic number of mercury (Hg) is 80; that of lead (Pb) is 82.
(b) An atom of Hg has 80 protons; an atom of Pb has 82 protons.
(c) The mass number of this isotope of Hg is 200; the mass number of this isotope of Pb is 202.
(d) The symbols of these isotopes are $^{200}_{80}Hg$ and $^{202}_{82}Pb$.
2.5 The atomic number of iodine (I) is 53. The number of neutrons in each isotope is 72 for iodine-125 and 78 for iodine-131. The symbols for these two isotopes are $^{125}_{53}I$ and $^{131}_{53}I$, respectively.
2.6 Lithium-7 is the more abundant isotope.
2.7 The element is aluminum (Al). Its Lewis dot structure shows three valence electrons.

$$\text{Äl·}$$

2.9 (a) element (b) compound (c) mixture
(d) mixture (e) mixture (f) element
(g) elemen (h) mixture (i) mixture
(j) mixture (k) compound (l) mixture
2.11 (a) bohrium (Bh, 107) (b) curium (Cm, 96)
(c) einsteinium (Es, 99) (d) fermium (Fm, 100)
(e) lawrencium (Lr, 103) (f) meitnerium (Mt, 109)
(g) mendelevium (Md, 101) (h) nobelium (No, 102)
(i) rutherfordium (Rf, 104) (j) seaborgium (Sg, 106)
2.13 The four elements named for planets are mercury (Hg, 80), uranium (U, 92), neptunium (Np, 93), and plutonium (Pu, 94).

2.15 (a) $NaHCO_3$ (b) C_2H_6O (c) $KMnO_4$
2.17 the law of conservation of mass
2.19 H_2O: H = 11.2%, O = 88.8%; H_2O_2: H = 5.9%, O = 94.1%
2.21 The statement is true in the sense that the number of protons (the atomic number) determines the identity of the atom.
2.23 (a) The element is titanium-48. Its symbol is $^{48}_{22}$Ti.
(b) The element is osmium-190. Its symbol is $^{190}_{76}$Os.
(c) The element is selenium-79. Its symbol is $^{79}_{34}$Se.
(d) The element is plutonium-244. Its symbol is $^{244}_{94}$Pu.
2.25 Each would still be the same element because the number of protons has not changed.
2.27 Radon (Rn) has an atomic number of 86, so each isotope has 86 protons.
(a) Radon-210 has 210 − 86 = 124 neutrons.
(b) Radon-218 has 218 − 86 = 132 neutrons.
(c) Radon-222 has 222 − 86 = 136 neutrons.
2.29 Two more neutrons: tin-120; symbol $^{120}_{50}$Sn;
three more neutrons: tin-121; symbol $^{121}_{50}$Sn;
six more neutrons: tin-124; symbol $^{124}_{50}$Sn
2.31 (a) An ion is an atom with an unequal number of protons and electrons.
(b) Isotopes are atoms with the same number of protons, but a different number of neutrons in their nuclei.
2.33 Rounded to four significant figures, the calculated value is 12.01 amu. The value given in the Periodic Table is 12.011 amu.

$$\left(\frac{98.90}{100} \times 12.000\right) + \left(\frac{1.10}{100} \times 13.003\right) = 11.87 + 0.143 = 12.01$$

2.35 Carbon-11 has 6 protons, 6 electrons, and 5 neutrons.
2.37 Americium-241 (Am) has atomic number 95. This isotope has 95 protons, 95 electrons, and 241 − 95 = 146 neutrons.
2.39 In Period 3, there are three metals (Na, Mg, Al), one metalloid (Si), and four nonmetals (P, S, Cl, Ar).
2.41 Periods 1, 2, and 3 contain more nonmetals than metals. Periods 4, 5, 6, and 7 contain more metals than nonmetals.
2.43 (a) Pd, (c) Co, (d) Ce, and (f) Cr are transition metals.
2.45 (a) Argon is a nonmetal.   (b) Boron is a metalloid.
(c) Lead is a metal.   (d) Arsenic is a metalloid.
(e) Potassium is a metal.   (f) Silicon is a metalloid.
(g) Iodine is a nonmetal.   (h) Antimony is a metalloid.
(i) Vanadium is a metal.   (j) Sulfur is a nonmetal.
(k) Nitrogen is a nonmetal.
2.47 The group number tells the number of electrons in the valence shell of the element in the group.
2.49 (a) Li (3): $1s^2 2s^1$ (b) Ne (10): $1s^2 2s^2 2p^6$
(c) Be (4): $1s^2 2s^2$ (d) C (6): $1s^2 2s^2 2p^2$
(e) Mg (12): $1s^2 2s^2 2p^6 3s^2$
2.51 (a) He (2): $1s^2$ (b) Na (11): $1s^2 2s^2 2p^6 3s^1$
(c) Cl (17): $1s^2 2s^2 2p^6 3s^2 3p^5$ (d) P (15): $1s^2 2s^2 2p^6 3s^2 3p^3$
(e) H (1): $1s^1$
2.53 In all three parts of this problem, the outer-shell electron configurations are the same. The only difference is the number of the valence shell being filled.
2.55 The element might be in Group 2A, whose members have two valence electrons. It might also be helium in Group 8A.
2.57 The properties are similar because all of them have similar outer-shell electron configurations. They are not identical because each has a different number of filled inner shells.

2.59 Ionization energy generally increases from left to right within a period in the Periodic Table and from bottom to top within a column:
(a) K, Na, Li (b) C, N, Ne (c) C, O, F (d) Br, Cl, F
2.61 Following are the ground-state electron configurations of the Mg atom, Mg^+, Mg^{2+}, and Mg^{3+}.

	Mg	$\longrightarrow$ Mg^+ + e^- IE = 738 kJ/mol
Electron configuration	$1s^2 2s^2 2p^6 3s^2$	$1s^2 2s^2 2p^6 3s^1$

	Mg^+	$\longrightarrow$ Mg^{2+} + e^- IE = 1450 kJ/mol
Electron configuration	$1s^2 2s^2 2p^6 3s^1$	$1s^2 2s^2 2p^6$

	Mg^{2+}	$\longrightarrow$ Mg^{3+} + e^- IE = 7734 kJ/mol
Electron configuration	$1s^2 2s^2 2p^6$	$1s^2 2s^2 2p^5$

The first electron is removed from the $3s$ orbital. The removal of each subsequent electron requires more energy because, after the first electron is removed, each subsequent electron is removed from a positive ion, which strongly attracts the extranuclear electrons. The third ionization energy is especially large because the electron is removed from the filled second principal energy level, meaning that it is removed from an ion that has the same electron configuration as neon.
2.63 The most abundant elements by weight (a) in the Earth's crust are oxygen and silicon, and (b) in the human body are oxygen and carbon.
2.65 Calcium is an essential element in human bones and teeth. Because strontium behaves chemically much like calcium, strontium-90 gets into our bones and teeth and gives off radioactivity for many years directly into our bodies.
2.67 Copper can be made harder by hammering it.
2.69 (a) metals (b) nonmetals (c) metals
(d) nonmetals (e) metals (f) metals
2.71 (a) Phosphorus-32 has 15 protons, 15 electrons, and 32 − 15 = 17 neutrons.
(b) Molybdenum-98 has 42 protons, 42 electrons, and 98 − 42 = 56 neutrons.
(c) Calcium-44 has 20 protons, 20 electrons, and 44 − 20 = 24 neutrons.
(d) Hydrogen-3 has 1 proton, 1 electron, and 3 − 1 = 2 neutrons.
(e) Gadolinium-158 has 64 protons, 64 electrons, and 158 − 64 = 94 neutrons.
(f) Bismuth-212 has 83 protons, 83 electrons, and 212 − 83 = 129 neutrons.
2.73 For elements with atomic numbers less than that of iron (Fe), the number of neutrons is close to the number of protons. Elements with atomic numbers greater than that of Fe have more neutrons than protons. Therefore, heavy elements have more neutrons than protons.
2.75 Rounded to three significant figures, the atomic weight of naturally occurring boron is 10.8. The value given in the Periodic Table is 10.811.

$$\left(\frac{19.9}{100} \times 10.013\right) + \left(\frac{80.1}{100} \times 11.009\right)$$

$$= 1.993 + 8.818 = 10.811 = 10.81$$

2.77 It would require 6.0×10^{21} protons to equal the mass of 1×10^{-2} g of salt.

$$\frac{1.0 \times 10^{-2}\ \text{g}}{1.67 \times 10^{-24}\ \text{g/proton}} = 6.0 \times 10^{21}\ \text{protons}$$

2.79 The abundances are 23.5% for rubidium-87 and 76.5% for rubidium-85.

2.81 Predict that xenon (Xe) has the highest ionization energy.

2.83 Element 118 will be in Group 8A. Expect it to be a gas that forms either no compounds or very few compounds.

Chapter 3   Nuclear Chemistry

3.1 $^{139}_{53}I \longrightarrow ^{139}_{54}Xe + ^{0}_{-1}e$

3.2 $^{223}_{90}Th \longrightarrow ^{4}_{2}He + ^{219}_{88}Ra$

3.3 $^{74}_{33}As \longrightarrow ^{0}_{+1}e + ^{74}_{32}Ge$

3.4 $^{201}_{81}Tl + ^{0}_{-1}e \longrightarrow ^{201}_{80}Hg + ^{0}_{0}\gamma$

3.5 Barium-122 (10 g) has decayed through 5 half-lives, leaving 0.31 g.
$10.0\,g \longrightarrow 5.0\,g \longrightarrow 2.5\,g \longrightarrow 1.25\,g \longrightarrow 0.625\,g \longrightarrow 0.31\,g$

3.6 1.5 mL

3.7 3.3×10^{-3} mCi

3.9 5.2×10^{9}/s

3.11 Red light has a frequency of 4.6×10^{14}/s.

3.13 (a) $^{19}_{9}F$ (b) $^{32}_{15}P$ (c) $^{87}_{37}Rb$

3.15 Predict that boron-10 is the most stable isotope because it has an equal number of neutrons and protons in its nucleus.

3.17 Samarium-151 decays by beta emission to europium-151.
$$^{151}_{62}Sm \longrightarrow ^{151}_{63}Eu + ^{0}_{-1}e$$

3.19 The new nucleus is vandium-51.
$$^{51}_{24}Cr + ^{0}_{-1}e \longrightarrow ^{51}_{23}V$$

3.21 The bombarding nucleus was neon-28.
$$^{248}_{96}Cm + ^{28}_{10}Ne \longrightarrow ^{116}_{51}Sb + ^{160}_{55}Cs$$

3.23 (a) $^{10}_{4}Be \longrightarrow ^{10}_{5}B + ^{0}_{-1}e$ (beta emission)

(b) $^{151}_{63}Eu^{*} \longrightarrow ^{151}_{63}Eu + \gamma$ (gamma emission)

(c) $^{195}_{81}Tl \longrightarrow ^{195}_{80}Hg + ^{0}_{+1}e$ (positron emission)

(d) $^{239}_{94}Pu \longrightarrow ^{235}_{92}U + ^{4}_{2}He$ (alpha emission)

3.25 Gamma emission does not result in transmutation.

3.27 $^{239}_{94}Pu + ^{4}_{2}He \longrightarrow ^{240}_{95}Am + 2^{1}_{0}n + ^{1}_{1}H$

3.29 After three half-lives, the percentage remaining will be $\frac{1}{2} \times \frac{1}{2} \times \frac{1}{2} = \frac{1}{8}$ or 12.5%.

3.31 No, the conversion of Ra to Ra^{2+} involves loss of valence electrons, which is not a nuclear process and, therefore, does not involve a change in radioactivity.

3.33 $50.0\,mg \longrightarrow 25.0\,mg \longrightarrow 12.5\,mg$
$\longrightarrow 6.25\,mg \longrightarrow 3.12\,mg$;
4 half-lives in 60 min = 60 min/4 = 15 min

3.35 Gamma radiation has the greatest penetrating power; therefore, it requires the largest amount of shielding.

3.37 3.0×10^{1} m = 30 m

3.39 (a) amount of radiation absorbed from the radiation source (b) effective dose absorbed by humans or tissue (c) effective energy delivered (d) intensity of radiation (e) amount of radiation absorbed by tissues from the radiation source (f) intensity of radiation (g) effective dose absorbed by humans or tissues

3.41 Alpha particles have so little penetrating power that they are stopped by the thin skin on the hand. If they get into the lungs, however, the thin membranes offer little resistance to them, and they then damage the cells of the lungs.

3.43 Alpha particles are the most damaging to tissue.

3.45 Iodine-131, which is concentrated in the thyroid, where the radiation can induce thyroid cancer.

3.47 (a) Cobalt-60 is used for (4) cancer therapy.
(b) Thallium-201 is used in (1) heart scans and exercise stress tests.
(c) Tritium is used to (2) measure the water content of the body
(d) Mercury-197 is used for (3) kidney scans.

3.49 Use curium-248 as the target nucleus.
$$^{248}_{96}Cm + ^{4}_{2}He \longrightarrow ^{249}_{97}Bk + 2^{1}_{0}n + ^{1}_{1}H$$

3.51 The new element is the sought-after element 118.
$$^{208}_{82}Pb + ^{86}_{36}Kr \longrightarrow ^{290}_{118}X + 4^{1}_{0}n$$

3.53 The assumption of a constant carbon-14 to carbon-12 ratio rests on two assumptions: (1) that carbon-14 is continually generated in the upper atmosphere by the production and decay of nitrogen-14,
$$^{14}_{7}N + ^{1}_{0}n \longrightarrow ^{14}_{6}C + ^{1}_{1}H$$
and (2) that carbon-14 is incorporated into carbon dioxide, CO_2, and other carbon compounds are then distributed worldwide as part of the carbon cycle. The continual formation of carbon-14; transfer of the isotope within the oceans, atmosphere, and biosphere; and the decay of living matter keep the supply of carbon-14 constant.

3.55 a bit more than 0.1 (or 10%) of the half-life of carbon-14

3.57 Radon-122 decays by alpha emission to polonium-218.
$$^{222}_{86}Rn \longrightarrow ^{218}_{84}Po + ^{4}_{2}He$$

3.59

Decay of P-32

3.61 Neon-19 decays to fluorine-19 and sodium-20 decays to neon-20.
$$^{19}_{10}Ne \longrightarrow ^{19}_{9}F + ^{0}_{+1}e$$
$$^{20}_{11}Na \longrightarrow ^{20}_{10}Ne + ^{0}_{+1}e$$

3.63 Yes. Both the curie and the becquerel have units of disintegrations/second, a measurement of radiation intensity.

3.65 (a) naturally occurring sources = 82%
(b) diagnostic medical sources = 15%
(c) nuclear power plants = 0.1%

3.67 X rays will cause more ionization than radar waves because X rays have higher energy than radar waves.

3.69 The decay product is neptunium-237. 1000/432 = 2.3 half-lives. $\frac{1}{2} \times \frac{1}{2} = \frac{1}{4}$, so somewhat less than 25% of the original americium will remain after 1000 years.

3.71 One sievert is 100 rem. This dose is sufficient to cause radiation sickness but not certain death.

3.73 (a) Radioactive elements are constantly decaying to other isotopes and other elements that are now mixed in with the original isotope.
(b) Beta emission results from the decay of a neutron in the nucleus to a proton (the increase in atomic number) and an electron (the beta particle).

3.75 Oxygen-16 is stable because it has an equal number of protons and neutrons. The other isotopes are unstable because the numbers of protons and neutrons are unequal. In this case, the greater the neutron/proton ratio, the shorter the half-life of the isotope.

3.77 The new element is darmstadtium-266.
$$^{208}_{82}Pb + ^{64}_{28}Ni \longrightarrow ^{266}_{110}Ds + 6^{1}_{0}n$$

3.79 Boron-10 absorbs a neutron to become boron-11, which in turn decays by alpha emission to lithium-7.
$$^{10}_{5}B + ^{1}_{0}n \longrightarrow ^{11}_{5}B$$
$$^{11}_{5}B \longrightarrow ^{7}_{3}Li + ^{4}_{2}He$$

Chapter 4 Chemical Bonds

4.1 By losing two electrons, Mg acquires a complete octet. By gaining two electrons, sulfur acquires a complete octet.
(a) Mg (12 electrons): $1s^2 2s^2 2p^6 3s^2 \longrightarrow Mg^{2+}$
(10 electrons): $1s^2 2s^2 2p^6 + 2e^-$
(b) S (16 electrons): $1s^2 2s^2 2p^6 3s^2 3p^4 + 2e^- \longrightarrow S^{2-}$
(18 electrons): $1s^2 2s^2 2p^6 3s^2 3p^6$

4.2 Each pair of elements is in the same column (group) of the Periodic Table and electronegativity increases from bottom to top within a column.
(a) Li > K (b) N > P   (c) C > Si

4.3 (a) KCl (b) CaF_2 (c) Fe_2O_3

4.4 (a) magnesium oxide   (b) barium iodide
(c) potassium chloride

4.5 (a) $MgCl_2$ (b) Al_2O_3 (c) LiI

4.6 (a) iron(II) oxide, ferrous oxide
(b) iron(III) oxide, ferric oxide

4.7 (a) potassium hydrogen phosphate
(b) aluminum sulfate
(c) iron(II) carbonate, ferrous carbonate

4.8 (a) S—H (2.5 − 2.1 = 0.4); nonpolar covalent
(b) P—H (2.1 − 2.1 = 0.0); nonpolar covalent
(c) C—F (4.0 − 2.5 = 1.5); polar covalent
(d) C—Cl (3.0 − 2.5 = 0.5); polar covalent

4.9 (a) $\overset{\delta+}{C}{-}\overset{\delta-}{N}$ (b) $\overset{\delta+}{N}{-}\overset{\delta-}{O}$ (c) $\overset{\delta+}{C}{-}\overset{\delta-}{Cl}$

4.10 (a) H—C—C—H (with H's above and below each C)   (b) H—C—Cl: (with H above and below)

(c) H—C≡N:

4.12 (a) nitrogen dioxide
(b) phosphorus tribromide
(c) sulfur dichloride
(d) boron trifluoride

4.13 (a) H—C(=O:)—O:⁻ ⟷ H—C(—O:)=O:⁻ (+)

(b) H—C(=O:)—O:⁻ ⟷ H—C(—O:⁻)=O:

(c) CH₃—C(=O:)—O:—CH₃ ⟷ CH₃—C(—O:)=O—CH₃ (+)

4.14 (a) A valid set of contributing structures.
(b) Not a valid set of contributing structures because the central carbon on the right has five bonds to it and, therefore, 10 electrons in its valence shell. The valence shell of carbon can contain only eight valence electrons.

4.15 Given are three-dimensional structures showing all unshared electron pairs and all bond angles.

(a) [C—O: structure, 109.5°]   (b) [C with Cl, 109.5°]

(c) [O=C structure with O's, 109.5°, 120°]

4.16 (a) CH_3Cl contains a polar C—Cl bond and is a polar molecule.
(b) HCN contains a polar C—N bond and is a polar molecule.
(c) C_2H_6 contains no polar bonds and is not a polar molecule.

4.17 (a) A lithium atom has the electron configuration $1s^2 2s^1$. When Li loses its single 2s electron, it forms Li^+, which has the electron configuration $1s^2$. This configuration is the same as that of helium, the noble gas nearest Li in atomic number.
(b) Chlorine must gain one electron to attain an electron configuration that is the same as that of argon, the noble gas nearest it in atomic number.
(c) Phosphorus must gain three electrons to attain an electron configuration that is the same as that of argon, the noble gas nearest it in atomic number.
(d) Aluminum must lose three electrons to attain an electron configuration that is the same as that of neon, the noble gas nearest it in atomic number.
(e) Strontium must lose two electrons to attain an electron configuration that is the same as that of krypton, the noble gas nearest it in atomic number.
(f) Sulfur must gain two electrons to attain the electron configuration that is the same as that of argon, the noble gas nearest it in atomic number.
(g) Silicon must gain four electrons to attain an electron configuration that is the same as that of argon, the noble gas nearest it in atomic number, or it must lose four electrons to attain an electron configuration that is the same as that of neon.
(h) An oxygen atom has the electron configuration $1s^2 2s^2 2p^4$. When O gains two electrons, it forms O^{2-}, which has the electron configuration $1s^2 2s^2 2p^6$. This configuration is the same as that of neon, the noble gas nearest oxygen in atomic number.

4.19 (a) H $1s^1 + 1e^- \longrightarrow H^- 1s^2$
    (an electron configuration the same as that of helium)
(b) Al $1s^2 2s^2 2p^6 3s^2 3p^1 \longrightarrow Al^{3+} 1s^2 2s^2 2p^6 + 3e^-$
    (an electron configuration the same as that of neon)

4.21 Li$^-$ is not a stable ion because it has an unfilled second shell.

4.23 The only stable ion in this set is (f) Cs$^+$.

4.25 No. Copper is a transition metal so the octet rule does not apply. Transition metals can expand their octets using *d* orbitals.

4.27 Electronegativity increases going up a column (group) of the Periodic Table because valence electrons are in shells closer to the electropositive nucleus. The decreasing distance of the valence electrons from the positively charged nucleus causes the valence electrons to experience an increasing force of attraction.

4.29 (a) F (b) O (c) N (d) F

4.31 The C—O bond is the most polar; the C—C bond is the least polar.

4.33 (a) C—Br is nonpolar covalent.
(b) S—Cl is polar covalent.
(c) C—P is nonpolar covalent.

4.35 (a) NaBr (b) Na$_2$O (c) AlCl$_3$ (d) BaCl$_2$ (e) MgO

4.37 Sodium chloride in the solid state has each Na$^+$ ion surrounded by six Cl$^-$ ions and each Cl$^-$ ion surrounded by six Na$^+$ ions.

4.39 (a) Fe(OH)$_3$ (b) BaCl$_2$ (c) Ca$_3$(PO$_4$)$_2$ (d) NaMnO$_4$

4.41 (a) and (b) are incorrect; the correct formulas are (a) (NH$_4$)$_3$PO$_4$ and (b) BaCO$_3$. (c) and (d) are correct.

4.43 KCl and KHCO$_3$

4.45 (a) sulfite ion (b) nitrate ion (c) carbonate ion
(d) ammonium ion and hydroxide ion
(e) hydrogen phosphate ion

4.47 (a) sodium fluoride (b) magnesium sulfide
(c) aluminum oxide (d) barium chloride
(e) calcium hydrogen sulfite (calcium bisulfite)
(f) potassium iodide (g) strontium phosphate
(h) iron(II) hydroxide (ferrous hydroxide)
(i) sodium dihydrogen phosphate
(j) lead(II) acetate (plumbous acetate)
(k) barium hydride (l) ammonium hydrogen phosphate

4.49 (a) NH$_4$HSO$_3$ (b) Mg(C$_2$H$_3$O$_2$)$_2$ (c) Sr(H$_2$PO$_4$)$_2$
(d) Ag$_2$CO$_3$ (e) SrCl$_2$ (f) Ba(MnO$_4$)$_2$

4.51 (a) A single bond is a bond in which two atoms share one pair of electrons.
(b) A double bond is a bond in which two atoms share two pairs of electrons.
(c) A triple bond is a bond in which two atoms share three pairs of electrons.

4.53 Following are Lewis structures for each compound.

(a) H—C—H (with H above and H below central C)

(b) H—C≡C—H

(c) C=C with H, H on left carbon and H, H on right carbon

(d) :F̈—B—F̈: with :F̈: below B

(e) Ö=C with H, H on right

(f) :C̈l—C—C—C̈l: with :C̈l: :C̈l: above and :C̈l: :C̈l: below

4.55 Following the compound is its number of valence electrons.
(a) NH$_3$, 8 (b) C$_3$H$_6$, 18 (c) C$_2$H$_4$O$_2$, 24
(d) C$_2$H$_6$O, 20 (e) CCl$_4$, 32 (f) HNO$_2$, 18
(g) CCl$_2$F$_2$, 32 (h) O$_2$, 12

4.57 A bromine atom contains seven electrons in its valence shell. A bromine molecule contains two bromine atoms bonded by a single covalent bond and with three lone pairs of electrons on each bromine atom. A bromide ion is a bromine atom that has gained one electron in its valence shell and has a complete octet and a charge of −1.

(a) :B̈r· (b) :B̈r—B̈r: (c) :B̈r:$^-$

4.59 Hydrogen has the electron configuration 1s^1. Hydrogen's valence shell has only the 1s orbital, which can hold a maximum of two electrons.

4.61 Nitrogen has five valence electrons. By sharing three more electrons with another atom or atoms, nitrogen can achieve an outer-shell electron configuration that is the same as that of neon, the noble gas nearest it in atomic number. The three shared pairs of electrons may be in the form of three single bonds, one double bond and one single bond, or one triple bond. With each of these bonding combinations, there is one unshared pair of electrons on nitrogen.

4.63 Oxygen has six valence electrons. By sharing two electrons with another atom or atoms, oxygen can achieve an outer-shell electron configuration that is the same as that of neon, the noble gas nearest it in atomic number. The two shared pairs of electrons may be in the form of one double bond or two single bonds. With either of these bonding combinations, there are two unshared pairs of electrons on oxygen.

4.65 O^{6+} has a charge too concentrated for such a small ion. It would take a tremendous amount of energy to form such a small, highly charged species as O^{6+}.

4.67 (a) BF$_3$ has only six electrons in the valence shell of boron. Thus it does not obey the octet rule.
(b) CF$_2$ does not obey the octet rule because carbon has only four electrons in its valence shell.
(c) BeF$_2$ does not obey the octet rule because beryllium has only four electrons in its valence shell.
(d) C$_2$H$_4$ obeys the octet rule.
(e) CH$_3$ does not obey the octet rule because carbon has only seven electrons in its valence shell.
(f) N$_2$ obeys the octet rule.
(g) NO does not obey the octet rule. The Lewis structures that can be drawn for this molecule show either nitrogen or oxygen with only seven electrons in its valence shell.

4.69 (a) The Lewis structure for the bicarbonate ion, HCO$_3$$^-$, must show 24 valence electrons.

H—Ö—C(=O)—Ö:$^-$ ⇌ H—Ö—C=Ö with :Ö:$^-$

4.71 (a) An N$_2$O molecule has 16 valence electrons.

$^-$:N̈=N$^+$=Ö ⇌ :N≡N$^+$—Ö:$^-$

(c) The given structure places 10 electrons in the valence shell of the central nitrogen, which is not possible. The valence shell of nitrogen contains 2s and 2p orbitals that can hold a maximum of 8 valence electrons.

4.73 (a) H_2O has 8 valence electrons, and H_2O_2 has 14 valence electrons.

(b)
$$H-\ddot{O}-H \qquad H-\ddot{O}-\ddot{O}-H$$

Water   Hydrogen peroxide

(c) Predict bond angles of 109.5° about each oxygen atom.
4.75 The shape of each molecule and approximate bond angles about each central atom are as follows:
(a) tetrahedral, 109.5° (b) pyramidal, 109.5°
(c) tetrahedral, 109.5° (d) bent, 120°
(e) trigonal planar, 120° (f) tetrahedral, 109.5°
(g) pyramidal, 109.5° (h) pyramidal, 109.5°
4.77 The difference in polarity arises because of their differences in shapes. CO_2 is a linear molecule, and SO_2 is a bent molecule.

(a) $\ddot{O}=C=\ddot{O}$ (b) $:\!O\!\!=\!\!\overset{\ddot{S}}{\diagup}\!\!=\!\!O:$

4.79 Yes, it is possible for a molecule to have polar bonds and yet no permanent dipole. This result occurs when the individual polar bonds act in equal but opposite directions, as in CO_2.
4.81 Use differences in electronegativity to predict bond polarity.
(a) nonpolar covalent (b) nonpolar covalent
(c) polar covalent (d) ionic (e) nonpolar covalent
(f) polar covalent (g) nonpolar covalent (h) ionic
4.83 calcium dihydrogen phosphate, calcium phosphate, and calcium carbonate
4.85 barium sulfate is used to visualize the gastrointestinal tract by x-ray examination.
4.87 Ca^{2+} is the main metal ion present in bone and tooth enamel.
4.89 Argon already has an octet with eight valence electrons in its outer shell; therefore, (a) it does not donate or accept electrons to form ions, and (b) it does not need to form covalent bonds by sharing electrons.
4.91 The two possibilities for a structure are square pyramidal and trigonal pyramidal.

Square pyramidal   Trigonal bipyramidal

The square pyramidal structure contains eight F—F 90° bonding electron pair repulsions. These repulsions are minimized in a trigonal bipyramidal structure, where there are only six F—F 90° bonding electron-pair repulsions. The VSEPR theory predicts molecules like PF_5 would adopt the least strained configuration, a trigonal bipyramidal structure.
4.93 ClO_2 has 19 valence electrons. In a Lewis structure, either chlorine or one of the oxygens must have only seven valence electrons. In the following structure, the odd electron is placed on chlorine, the less electronegative element.

$$\ddot{O}=\dot{Cl}=\ddot{O}$$

4.95 The compound is zinc oxide, ZnO.
4.97 Lead(IV) oxide, PbO_2, and lead(IV) carbonate, $Pb(CO_3)_2$, are used as white pigments in paints.
4.99 Fe^{2+}
4.101 (a) $CaSO_3$ (b) $Ca(HSO_3)_2$ (c) $Ca(OH)_2$
(d) $CaHPO_4$
4.103 Perchloroethylene has four polar covalent C—Cl bonds, but it is not a polar compound. The molecule lacks a dipole because the polar covalent C—Cl bonds act in equal, but opposite directions.

4.105 (a) Following is a Lewis structure for tetrafluoroethylene.

(b) All bond angles are predicted to be 120°.
(c) Tetrafluoroethylene has four polar covalent C—F bonds but is a nonpolar molecule, because, as in Problem 4.103, the polar covalent bonds act in equal, but opposite directions.

Chapter 5   Chemical Reactions

5.1 (a) $C_{13}H_{18}O_2 = 206$ amu (b) $Ba_3(PO_4)_2 = 602$ amu
5.2 1500 g water is 83 mol water.
5.3 2.84 mol Na_2S is 222 g Na_2S.
5.4 In 2.5 mol of glucose, $C_6H_{12}O_6$, there are 15 mol C atoms, 30 mol H atoms, and 15 mol O atoms.
5.5 0.062 g $CuNO_3$ contains 4.9×10^{-4} mol Cu^+.
5.6 235 g H_2O contains 7.86×10^{24} molecules H_2O.
5.7 The balanced equation is

$$6CO_2(g) + 6H_2O(\ell) \xrightarrow{\text{Photosynthesis}} C_6H_{12}O_6(aq) + 6O_2(g)$$

5.8 The balanced equation is
$$2C_6H_{14}(g) + 19O_2(g) \longrightarrow 12CO_2(g) + 14H_2O(g)$$

5.9 The balanced equation is
$$3K_2C_2O_4(aq) + Ca_3(AsO_4)_2(s)$$
$$\longrightarrow 2K_3AsO_4(aq) + 3CaC_2O_4(s)$$

5.10 (a) The balanced equation is

$$2Al_2O_3(s) \xrightarrow{\text{Electrolysis}} 4Al(s) + 3O_2(g)$$

(b) It requires 51 g of Al_2O_3 to prepare 27 g of Al.
5.11 It requires 16.6 mol of CO to produce 16.6 mol of acetic acid.
5.12 334 g of ethanol is produced.
5.13 (a) C (0.50 mol) is the limiting reagent and H_2 (1.04 mol) is in slight excess.
(b) 8.0 g (0.50 mol) CH_4 is produced.
5.14 The percent yield is 81.00%.
5.15 The net ionic equation is
$Cu^{2+}(aq) + S^{2-}(aq) \longrightarrow CuS(s)$.
5.16 (a) Ni^{2+} gains two electrons, so it is reduced. Cr loses two electrons, so it is oxidized. Ni^{2+} is the oxidizing agent and Cr is the reducing agent.

(b) CH_2O gains hydrogens, so it is reduced. H_2 gains oxygen in being converted to CH_3OH, so it is oxidized. CH_2O is the oxidizing agent and H_2 is the reducing agent.

5.17 (a) $KCl = 74.6$ amu (b) $Na_3PO_4 = 163.9$ amu
(c) $Fe(OH)_2 = 89.9$ amu

5.19 (a) $C_{12}H_{22}O_{11} = 342.3$ amu (b) $C_2H_5NO_2 = 75.1$ amu
(c) $C_{14}H_9Cl_5 = 354.5$ amu

5.21 (a) 81.4 g NO_2 (b) $50.$ g C_3H_8O (c) 1.30×10^3 g UF_6
(d) 62.7 g $C_6H_{12}O_6$ (d) 8.6 g $C_6H_8O_6$

5.23 (a) 6.56 mol S^{2-} ions (b) 24.96 mol Mg^{2+} ions
(c) 0.86 mol acetate ions, CH_3COO^-

5.25 the same; that is, $2:1$

5.27 (a) 2.9 mol TNT $= 1.7 \times 10^{24}$ TNT molecules
(b) 0.0500 g $H_2O = 1.67 \times 10^{21}$ H_2O molecules
(c) 3.1×10^{-3} g aspirin $= 1.0 \times 10^{21}$ aspirin molecules

5.29 1.1×10^{-19} grams per molecule.

5.31 (a) $H_2 + I_2 \longrightarrow 2\,HI$
(b) $4Al + 3O_2 \longrightarrow 2Al_2O_3$
(c) $2Na + Cl_2 \longrightarrow 2NaCl$
(d) $2Al + 6HBr \longrightarrow 2AlBr_3 + 3H_2$
(e) $4P + 5O_2 \longrightarrow 2P_2O_5$

5.33 $CaCO_3(s) \xrightarrow{Heat} CaO(s) + CO_2(g)$

5.35 $4Fe(s) + 3O_2(g) \longrightarrow 2Fe_2O_3(s)$

5.37 $(NH_4)_2CO_3(s) \longrightarrow 2NH_3(g) + CO_2(g) + H_2O(\ell)$

5.39 $2Al(s) + 6HCl(aq) \longrightarrow 2AlCl_3(aq) + 3H_2(g)$

5.41 (a) 0.67 mol N_2 required
(b) 0.67 mol N_2O_3 produced
(c) 12 mol O_2 required

5.43 319 g of Cl_2 required

5.45 (a) The balanced equation is

$$2NaClO_2(aq) + Cl_2(g) \longrightarrow 2ClO_2(g) + 2NaCl(aq)$$

(b) 4.10 kg of ClO_2

5.47 7.5 g CO_2 required

5.49 0.13 g C needed

5.51 26 g of salicylic acid required

5.53 The theoretical yield is 12 g of ethyl chloride. The percentage yield in this problem is 68%.

5.55 (a) spectator ion: an ion that does not take part in a chemical reaction
(b) net ionic equation: a balanced equation with spectator ions removed
(c) aqueous solution: a solution using water as the solvent

5.57 (a) The spectator ions are Na^+ and Cl^-.
(b) $Sr^{2+}(aq) + CO_3^{2-}(aq) \longrightarrow SrCO_3(s)$

5.59 The net ionic equation is

$$Pb^{2+}(aq) + 2Cl^-(aq) \longrightarrow PbCl_2(s)$$

5.61 $NH_4^+(aq) + OH^-(aq) \longrightarrow NH_3(g) + H_2O(\ell)$

5.63 (a) $MgCl_2$ (soluble)
(b) $CaCO_3$ (insoluble)
(c) Na_2SO_3 (soluble)
(d) NH_4NO_3 (soluble)
(e) $Pb(OH)_2$ (insoluble)

5.65 No. In oxidation there is a loss of electrons, and there must be some species to accept those electrons—that is, some species that is reduced.

5.67 (a) C_7H_{14} is oxidized and O_2 is reduced.
(b) O_2 is the oxidizing agent, and C_7H_{14} is the reducing agent.

5.69 An exothermic chemical reaction or process releases heat as a product.
An endothermic chemical reaction or process absorbs heat as a reactant.

5.71 19.6 kcal

5.73 The heat liberated is $670/180 = 3.72$ kcal/g of glucose. Therefore, metabolism of 15.0 grams of glucose will liberate 15.0 g $\times 3.72$ kcal/g $= 55.8$ kcal.

5.75 (a) The synthesis of starch is endothermic.
(b) 26.4 kcal

5.77 Fluoride reacts with $Ca_{10}(PO_4)_6(OH)_2$ in the enamel by exchanging the F^- ions for OH^- ions, forming $Ca_{10}(PO_4)_6F_2$, which is less soluble under the acidic conditions found in the mouth.

5.79 Oxidation occurs at the anode, where $Fe \longrightarrow Fe^{2+} + 2e^-$. Reduction occurs at the cathode, where $Zn^{2+} + 2e^- \longrightarrow Zn$.

5.81 $N_2O_5(g) + H_2O(\ell) \longrightarrow 2HNO_3(aq)$

5.83 (a) Fe_2O_3 loses oxygen; it is reduced. CO gains oxygen; it is oxidized.
(b) 38.4 mol Fe requires 19.2 mol Fe_2O_3.
(c) 1.61×10^3 g CO is required.

5.85 $3Cd^{2+}(aq) + 2PO_4^{3-}(aq) \longrightarrow Cd_3(PO_4)_2(s)$

5.87 894 amu

5.89 (a) $Pb(NO_3)_2$ is the limiting reagent.
(b) 82.6% $PbCl_2$

5.91 (a) $C_5H_{12}(g) + 8O_2(g) \longrightarrow 5CO_2(g) + 6H_2O(g)$
(b) Pentane is oxidized and oxygen is reduced.
(c) Oxygen is the oxidizing agent and pentane is the reducing agent.

5.93 (a) $CH_4(g) + 2O_2(g) \longrightarrow CO_2(g)$
    Methane
$$+ 2H_2O(g) + 213 \text{ kcal/mol}$$

$$C_3H_8(g) + 5O_2(g) \longrightarrow 3CO_2(g)$$
    Propane
$$+ 4H_2O(g) + 530 \text{ kcal/mol}$$

(b) On the basis of kcal/mol, propane (530 kcal/mol) is a better source of heat energy than methane (213 kcal/mol).
(c) On the basis of kcal/gram, methane (13.3 kcal/g) is a better source of heat energy than propane (12.0 kcal/g).

Chapter 6   Gases, Liquids, and Solids

6.1 0.41 atm
6.2 16.4 atm
6.3 0.053 atm
6.4 4.84 atm
6.5 0.422 mol Ne
6.6 9.91 g He
6.7 0.107 atm H_2O vapor
6.8 (a) Yes, there can be hydrogen bonding between molecules of water and methanol.
(b) No, C—H bonds cannot form hydrogen bonds.
6.9 83 g H_2O can be vaporized.
6.10 Heat required to heat 1.0 g of iron to melting $= 230$ cal
6.11 According to the phase diagram of water (Figure 6.18), the vapor will condense to liquid water and then freeze at $0°$. After all is frozen, the temperature will drop to $-30°C$.
6.13 1.5 atm CH_4
6.15 2.51 atm of air in the tire
6.17 1.87 L of ethane gas upon cooling
6.19 748 K ($475°C$)
6.21 1.17 atm

6.23

V_1	T_1	P_1	V_2	T_2	P_2
546 L	43°C	6.5 atm	2.0×10^3 mL	65°C	1.9 atm
43 mL	−56°C	865 torr	48 mL	43°C	1.5 atm
4.2 L	234 K	0.87 atm	3.2 L	29°C	1.5 atm
1.3 L	25°C	740 mm Hg	1.2 L	0°C	1.0 atm

6.25 0.31 L is the balloon's final volume.
6.27 4.35 atm
6.29 61.0 mL
6.31 (a) 2.33 mol
(b) No. The only information that you need to know about the gas to solve this problem is that it is an ideal gas.
6.33 4.00 g/mol
6.35 Density increases as pressure increases and decreases as temperature increases.
6.37 (a) 24.7 mol O_2 (b) 790 g O_2
6.39 One lung volume of air (to two significant figures) contains 1.2 L (0.050 mol) of O_2, which is 3.0×10^{22} molecules of O_2.
6.41 (a) The mass of 1 mol of air (to two significant figures) is 29 g.
(b) The density of air is 1.29 g/L.
6.43 (a) SO_2 = 2.86 g/L
(b) CH_4 = 0.715 g/L
(c) H_2 = 0.0902 g/L
(d) He = 0.178 g/L
(e) CO_2 = 1.96 g/L
Gas comparison: SO_2 and CO_2 are more dense than air, and He, H_2, and CH_4 are less dense than air.
6.45 0.197 L
6.47 The densities would be the same. The density of a substance does not depend on its quantity.
6.49 (a) P_{N_2} = 593.4 mm Hg
P_{O_2} = 159.2 mm Hg
P_{Ar} = 7.1 mm Hg
P_{total} = 759.7 mm Hg or 760 mm Hg
(b) The total pressure exerted by the components is the sum of their partial pressures: 760. mm Hg (1.00 atm).
6.51 Yes. The partial pressures given in the problem total 785 mm Hg and the total pressure given is 790 mm Hg. The difference is 5 mm Hg. Therefore, there must be another gas present.
6.53 Intramolecular covalent bonds are stronger than intermolecular hydrogen bonds.
6.55 yes
6.57 Ethanol is a polar molecule and engages in intermolecular hydrogen bonding through its —OH group. Carbon dioxide is a nonpolar molecule, and the only forces of attraction between its molecules are weak London dispersion forces. The stronger hydrogen bonds require more energy (higher temperature) to break before vaporizing.
6.59 Hexane has the higher boiling point.
6.61 Ionic and network solids have the highest melting points. Molecular solids have the lowest melting points.
6.63 1.53 kcal to vaporize 39.2 g CF_2Cl_2
6.65 (a) approximately 90 mm Hg
(b) approximately 120 mm Hg
(c) approximately 330 mm Hg at 60°C

6.67 Boiling points are HCl = 84.9°C, HBr = −67°C, and HI = −35.4°C.
(a) HCl < HBr < HI; HCl is the most polar of the three acids, and hydrogen bonding would be strongest between molecules of HCl. The increasing size of each molecule increases the intermolecular London dispersion forces. London dispersion forces are the more important factor in determining the boiling points of this group of compounds.
(b) O_2 < HCl < H_2O_2; O_2 has only weak London dispersion forces between its molecules. HCl is a polar molecule with stronger dipole–dipole interactions between its molecules. H_2O_2 has the strongest intermolecular forces (hydrogen bonding) between its molecules.
6.69 The difference between heating water from 0°C to 37°C and heating ice from 0°C to 37°C is the heat of fusion of ice. 11,700 (1.2) kcal is required to heat ice from 0°C to 37°C. 3700 (3.7 kcal) is required to heat water from 0°C to 37°C.
6.71 sublimation
6.73 6.96×10^{-2} kcal
6.75 When the temperature of a substance increases, so does its entropy. Therefore, a gas at 100°C has a lower entropy than the same gas at 200°C.
6.77 When the diaphragm is lowered, the volume of the chest cavity increases, lowering the pressure in the lungs relative to the atmospheric pressure. Air at atmospheric pressure then rushes into the lungs, beginning the breathing process.
6.79 The first tapping sound one hears is the systolic pressure, which occurs when the sphygmomanometer pressure matches the blood pressure when the ventricle contracts, pushing blood into the arm.
6.81 When water freezes, it expands (water is one of the few substances that expands on freezing) and will crack the bottle when the ice expansion exceeds the volume of the bottle.
6.83 It is difficult to compress liquids and solids because their molecules are already very close together and there is very little empty space between them.
6.85 2.3 atm
6.87 Aerosol cans already contain gases under high pressure. Gay-Lussac's law predicts that the pressure inside the can will increase with increasing temperature, with the potential of explosive rupture of the can, causing injury.
6.89 112 mL
6.91 In order of increasing boiling point:
C_5H_{12} < $CHCl_3$ < H_2O
6.93 (a) When a gas is compressed, its molecules are forced closer together, at which point intermolecular forces are able to draw the molecules close enough to stick together and to form a liquid.
(b) 9.1 kg propane
(c) 2.1×10^2 mol propane
(d) 4.6×10^3 L propane
6.95 The density of the gas is 3.00 g/L and its molar mass is 91.9 g/mol.
6.97 313 K (40°C)
6.99 The temperature of a liquid decreases during evaporation because the molecules with higher kinetic energy leave the liquid, which decreases the average kinetic energy of those molecules remaining in the liquid. The temperature of the liquid is directly proportional to the average kinetic energy of the molecules in the liquid state. Therefore the temperature decreases as the average kinetic energy decreases.

6.101 (a) The pressure on the body is 4 atm.
(b) The partial pressure of N_2 entering the lungs is 3.1 atm.
(c) The partial pressure of O_2 entering the lungs is 0.83 atm.
(d) At 100 feet, the partial pressure of N_2 in the blood and fatty tissue is significantly increased. It takes time to release this N_2 by transfer to the blood and then to the lungs for exhalation. If a diver rises too quickly, too much N_2 remains in the blood and fatty tissue, affecting the nervous system. In addition, if bubbles of N_2 form in blood vessels, they can impede the flow of blood, which reduces the flow of O_2 to tissues.

Chapter 7   Solutions and Colloids

7.1   To 11 g of KBr, add a quantity of water sufficient to dissolve the KBr. Following dissolution of the KBr, add water to the 250 mL mark, stopper, and mix.
7.2   1.9% w/v
7.3   Place 158 g (2.12 mol) of KCl in a 2-L volumetric flask, add some water, swirl until the solid has dissolved, and then fill the flask with water to the 2.0 L mark.
7.4   0.0133 M KSCN
7.5   185 mL of glucose solution
7.6   Add 3.9×10^2 g $NaHSO_3$ to a 100-gallon barrel.
7.7   Place 50–100 mL of water in a 300-mL flask, add 15.0 mL of 12. 0 M HCl, swirl until completely mixed, and then fill the flask with water to the 300 mL mark.
7.8   Place 0.13 mL of the 15% KOH solution in a 20-mL volumetric flask, add some water, swirl until completely dissolved, and then fill the flask with water to the 20 mL mark.
7.9   The Na^+ ion concentration is 0.24 ppm Na^+.
7.10 $-12.5°C$
7.11 Solution (c), 4.3 M K_2SO_3, has the highest concentration of solute particles and, therefore, has the lowest freezing point.
7.12 A Na_3PO_4 molecule gives 3 Na^+ ions and 1 PO_4^{3-} ion for a total of 4 solute particles. This solution is 0.81 osmol.
7.13 Solution (a) has the same osmolarity as red blood cells and, therefore, is isotonic compared to red blood cells.
7.15 Glucose is being dissolved; therefore, it is the solute. Water is dissolving the glucose; therefore, water is the solvent.
7.17 (a) wine (ethanol in water)
(b) saline solution (NaCl dissolved in water)
(c) carbonated water (CO_2 dissolved in water)
(d) air (O_2 and N_2)
7.19 One possible explanation is that the prepared aspartic acid solution was unsaturated. Over two days, some of the solvent (water) may have evaporated and the solution became supersaturated, precipitating the excess aspartic acid as a white solid.
7.21 (a) NaCl (an ionic solid) will dissolve in the polar water layer.
(b) Camphor (a nonpolar molecule) will dissolve in the nonpolar diethyl ether layer.
(c) KOH (an ionic solid) will dissolve in the polar water layer.
7.23 Isopropyl alcohol would be a good first choice. The oil base in the paint is nonpolar. Both benzene and hexane are nonpolar solvents and may dissolve the paint, destroying the painting.
7.25 The solubility of aspartic acid at 25°C in 50.0 mL of water is 0.250 g of solute. Assuming that no precipitation

occurs, the cooled solution will contain 0.251 g of aspartic acid in 50.0 mL of water and will be supersaturated by 0.001 g of aspartic acid.
7.27 According to Henry's law, the solubility of a gas in a liquid is directly proportional to the pressure. A closed bottle of a carbonated beverage is under excess CO_2 pressure. After the bottle is opened, the excess CO_2 pressure is released and the CO_2 escapes from solution as bubbles of gas.

7.29 (a) $\dfrac{1 \text{ min}}{1.05 \times 10^6 \text{ min}} \times 10^6 = 1$ ppm

$\dfrac{1 \text{ p}}{1.05 \times 10^6 \text{ p}} \times 10^6 = 1$ ppm

(b) $\dfrac{1 \text{ min}}{1.05 \times 10^9 \text{ min}} \times 10^9 = 1$ ppb

$\dfrac{1 \text{ p}}{1.05 \times 10^9 \text{ p}} \times 10^9 = 1$ ppm

7.31 (a) 76 mL of ethanol is dissolved in 204 mL of water to give 280 mL of solution.
(b) 7.8 mL of ethyl acetate is dissolved in 427 mL of water to give 435 mL of solution.
(c) 0.13 L of benzene is dissolved in 1.52 L of chloroform to give 1.65 L of solution.
7.33 (a) 4.15% w/v casein
(b) 0.030% w/v vitamin C
(c) 1.75% w/v sucrose
7.35 (a) Place 19.5 g of NH_4Br into a 175-mL volumetric flask, add some water, swirl to mix completely, and then fill the flask with water to the 175 mL mark.
(b) Place 167 g of NaI in a 1.35-L volumetric flask, add some water, swirl to mix completely, and then fill the flask with water to the 1.35 L mark.
(c) Place 2.4 g of ethanol in a 330-mL volumetric flask, add some water, swirl to mix completely, and then fill the flask with water to the 330 mL mark.
7.37 0.2 M NaCl
7.39 (a) 0.509 M glucose   (b) 0.0202 M K^+
(c) 7.25×10^{-4} M Na^+
7.41 2.5 M sucrose
7.43 The total volume of the diluted solution will be 30.0 mL. Start with 5.0 mL of solution and add 25.0 mL of water to reach a final volume of 30.0 mL.
7.45 Place 2.1 mL of 30.0% w/v H_2O_2 into a 250-mL volumetric flask, add some water, swirl to mix completely, and fill the flask with water to the 250 mL mark.
7.47 (a) 3.85×10^4 ppm Captopril
(b) 6.8×10^4 ppm Mg^{2+} (c) 8.3×10^2 ppm Ca^{2+}
7.49 Dioxin concentration = 0.01 ppb dioxin.
No, the dioxin level in the lake did not reach a dangerous level.
7.51 (a) 10 ppm Fe or 1×10^1 ppm
(b) 3×10^3 ppm Ca
(c) 2 ppm vitamin A
7.53 (b) 0.1 M $(NH_4)_3PO_4$
7.55 The polar covalent compounds (b) CH_3OH, (c) HF, and (d) NH_3 dissolve in water by forming hydrogen bonds with water molecules.
7.57 (a) homogeneous   (b) heterogeneous   (c) colloidal
(d) heterogeneous   (e) colloidal   (f) colloidal
7.59 As the temperature of the solution decreases, the protein molecules aggregate and form a colloidal mixture. The turbid appearance is the result of the Tyndall effect.
7.61 (a) $-3.72°C$ (b) $-5.58°C$ (c) $-5.58°C$ (d) $-7.44°C$

7.63 3.4×10^2 g CH_3OH

7.65 Acetic acid is a weak acid and, therefore, not completely dissociated into ions. KF is a strong electrolyte and ionizes completely in aqueous solution, thus doubling the effect on the freezing-point depression compared to acetic acid.

7.67 Assume that the first entry is column A, and that the second entry is column B. In each case, the side with the greater osmolarity rises.
(a) B (b) B (c) A (d) B (e) A (f) same

7.69 (a) 1.2 osmol (b) 2.5 osmol (c) 8.4 osmol
(d) 0.04 osmol

7.71 Red blood cells will undergo crenation (shrink) in a hypertonic solution. A 0.9% NaCl solution is 0.3 osmol. Solution (b) is 0.9 osmol (hypertonic) and will crenate red blood cells.

7.73 Carbon dioxide (CO_2) dissolves in rainwater to form a dilute solution of carbonic acid (H_2CO_3), which is a weak acid.

7.75 Nitrogen dissolved in the blood can lead to a narcotic effect referred to as "rapture of the deep," which is similar to alcohol-induced intoxication.

7.77 The main component of limestone and marble is calcium carbonate, $CaCO_3$.

7.79 The minimum pressure required for reverse osmosis in the desalinization of sea water exceeds 100 atm (the osmotic pressure of sea water).

7.81 Yes, the change in composition made a change in the tonicity. The error in replacing $NaHCO_3$ with $KHCO_3$ resulted in a hypotonic solution and an electrolyte imbalance by reducing the number of ions (osmolarity) in solution.

7.83 When a cucumber is placed in a saline solution, the osmolarity of the saline solution is greater than that of the water in the cucumber, so water moves from the cucumber to the saline solution and the cucumber shrinks. When a prune (a partially dehydrated plum) is placed in the same solution, it expands because the osmolarity inside the prune is greater than that of the saline solution, so water moves from the saline solution to inside the prune, causing the prune to expand.

7.85 The solubility of a gas in a solvent is directly proportional to the pressure (Henry's law) of the gas and is inversely proportional to the temperature. The dissolved carbon dioxide formed a saturated solution in water when bottled under 2 atm pressure. When the bottles are opened at atmospheric pressure, the gas becomes less soluble. The carbon dioxide becomes supersaturated in water at room temperature and 1 atm, and thus escapes as bubbles and frothing. In the other bottle, the solution of carbon dioxide in water is unsaturated at a lower temperature and does not lose carbon dioxide as readily.

7.87 No, it would not be acceptable to use an 0.89% KCl solution for intravenous infusions because it will not be isotonic with blood. KCl has a higher molecular weight than NaCl; consequently, its osmolarity will be smaller.

7.89 3×10^{-7} g Pb

7.91 (a) 0.9% w/v NaCl has the greater osmotic pressure by a factor of 18 times.

7.93 2.9 mL of pure water added to 1.0 mL of sea water gives a solution with the same osmolarity as blood.

7.95 The ethanol molecules displace water molecules from the solvation layer of the hyaluronic acid, allowing the hyaluronic acid molecules to stick together upon collision and form aggregates that precipitate.

Chapter 8   Reaction Rates and Chemical Equilibrium

8.1 rate of O_2 formation = 0.022 L O_2/min

8.2 rate $= 4 \times 10^{-3}$ mol H_2O_2/L · min for disappearance of H_2O_2

8.3 $K = \dfrac{[H_2SO_4]}{[SO_3][H_2O]}$

8.4 $K = \dfrac{[N_2][H_2]^3}{[NH_3]^2}$

8.5 $K = 0.602 \ M^{-1}$

8.6 $K = \dfrac{[CH_3COOCH_2CH_3][H_2O]}{[CH_3COOH][HOCH_2CH_3]}$

8.7 Le Chatelier's principle predicts that adding Br_2 (a product) will shift the equilibrium to the left—that is, toward the formation of more NOBr(g).

8.8 Because oxygen's solubility in water is exceeded, oxygen bubbles out of the solution, driving the equilibrium toward the right.

8.9 If the equilibrium shifts to the right with the addition of heat, heat must have been a reactant, and the reaction is endothermic.

8.10 The equilibrium in a reaction where there is an increase in pressure favors the side with fewer moles of gas. Therefore, this equilibrium shifts to the right.

8.11 rate of formation of $CH_3I = 7.3 \times 10^{-3} \ M \ CH_3I$/min

8.13 Reactions involving ions in aqueous solution of ions are faster because they do not require bond breaking and have low activation energies. In addition, the attractive force between positive and negative ions provides energy to drive the reaction. Reactions between covalent compounds require the breaking of covalent bonds and have higher activation energies and, therefore, slower reaction rates.

8.15 The activation energy for the reverse reaction will be 14 + 9 = 23 kcal/mol.

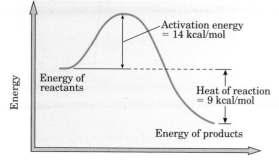

8.17 A general rule for the effect of temperature on the rate of reaction states that for every temperature increase of 10°C, the reaction rate doubles. In this case, a reaction temperature of 50°C would predict completion of the reaction in 1 h.

8.19 You might (a) increase the temperature, (b) increase the concentration of reactants, or (c) add a catalyst.

8.21 A catalyst increases the rate of a reaction by providing an alternative reaction pathway with lower activation energy.

8.23 Other examples of irreversible reactions include digesting a piece of candy, rusting of iron, exploding TNT, and the reaction of Na or K metal with water.

8.25 (a) $K = [H_2O]^2[O_2]/[H_2O_2]^2$
(b) $K = [N_2O_4]^2[O_2]/[N_2O_5]^2$
(c) $K = [C_6H_{12}O_6][O_2]^6/[H_2O]^6[CO_2]^6$

8.27 $K = 0.667$

8.29 $K = 0.099 M$

8.31 Products are favored in (b) and (c). Reactants are favored in (a), (d), and (e).

8.33 No. The rate of reaction is independent of the energy difference between products and reactants—that is, it is independent of the heat of reaction.

8.35 The reaction reaches equilibrium quickly, but the position of equilibrium favors the reactants; therefore, it would not be a very good industrial process.

8.37 (a) right (b) right (c) left (d) left (e) no shift

8.39 (a) Adding Br_2 (a reactant) will shift the equilibrium to the right.
(b) The equilibrium constant will remain the same.

8.41 (a) no change (b) no change (c) smaller

8.43 As temperatures increase, the rates of most chemical processes increase. A high body temperature is dangerous because metabolic processes (including digestion, respiration, and biosynthesis of essential compounds) take place at a faster rate than is safe for the body. As temperatures decrease, so do the rates of most chemical reactions. As body temperature decreases below normal, the vital chemical reactions will slow to rates slower than is safe for the body.

8.45 The capsule with the tiny beads will act faster than the solid coated-pill form. The small bead size increases the drug's surface area, allowing the drug to react faster and deliver its therapeutic effects more quickly.

8.47 The addition of heat is used to increase the rate of reaction. The addition of a catalyst permits the reaction to take place at a more convenient temperature.

8.49 The following energy diagram can be drawn for this exothermic reaction.

8.51 rate = 0.37/h

8.53 rate = 2.0×10^{-3} mol N_2O_4/L·s for disappearance of reactants

8.55 The activation energy of the reverse reaction will be 10.0 kcal/mol − energy of reaction.

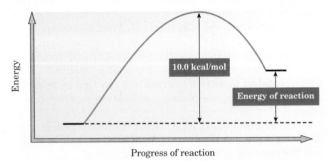

8.57 The temperature increase is 20°C, and the reaction rate would be predicted to double twice. The rate of reaction at 320 K is predicted to be 0.88 M/min.

8.59 No. An endothermic reaction must have an energy of activation greater than zero. By definition, an endothermic reaction is one in which the products have a greater energy than the reactants.

8.61 Assuming that there is an excess of AgCl from the previous recipe, the recipe does not need to be changed. The desert conditions add nothing that would affect the coating process.

8.63

$$\overset{O}{\overset{\|}{}}$$

(a) $CH_3COH + HOCH_2CH_3$
          Initial   1.00 mol   1.00 mol
    At equilibrium   0.33 mol   **0.33 mol**

$$\overset{O}{\overset{\|}{}}$$

$\rightleftharpoons CH_3COCH_2CH_3 + H_2O$
                    0 mol   0 mol
                    0.67 mol   **0.67 mol**

(b) $K = 4.1$

Chapter 9   Acids and Bases

9.1 acid reaction: $HPO_4^{2-} + H_2O \rightleftharpoons PO_4^{3-} + H_3O^+$;
base reaction: $HPO_4^{2-} + H_2O \rightleftharpoons H_2PO_4^- + OH^-$

9.2 (a) toward the left;

$H_3O^+ + I^- \leftrightharpoons H_2O + HI$
Weaker   Weaker   Stronger   Stronger
acid   base   base   acid

(b) toward the right;

$CH_3COO^- + H_2S \leftrightharpoons CH_3COOH + HS^-$
Weaker   Weaker   Stronger   Stronger
base   acid   acid   base

9.3 pK_a is 9.31

9.4 (a) ascorbic acid (b) aspirin

9.5 1.0×10^{-2}

9.6 (a) 2.46 (b) 7.8×10^{-5}, acidic

9.7 pOH = 4, pH = 10

9.8 0.0960 M

9.9 (a) 9.25 (b) 4.74

9.10 9.61

9.11 7.7

9.13 (a) $HNO_3(aq) + H_2O(\ell) \longrightarrow NO_3^-(aq) + H_3O^+(aq)$
(b) $HBr(aq) + H_2O(\ell) \longrightarrow Br^-(aq) + H_3O^+(aq)$
(c) $H_2SO_3(aq) + H_2O(\ell) \longrightarrow HSO_3^-(aq) + H_3O^+(aq)$
(d) $H_2SO_4(aq) + H_2O(\ell) \longrightarrow HSO_4^-(aq) + H_3O^+(aq)$
(e) $HCO_3^-(aq) + H_2O(\ell) \longrightarrow CO_3^{2-}(aq) + H_3O^+(aq)$
(f) $H_3BO_3(aq) + H_2O(\ell) \longrightarrow H_2BO_3^-(aq) + H_3O^+(aq)$

9.15 (a) weak (b) strong (c) weak (d) strong
(e) weak (f) weak

9.17 weak

9.19 (a) A Brønsted-Lowry acid is a proton donor.
(b) A Brønsted-Lowry base is a proton acceptor.

9.21 (a) HPO_4^{2-} (b) HS^- (c) CO_3^{2-}
(d) $CH_3CH_2O^-$ (e) OH^-

9.23 (a) H_3O^+ (b) $H_2PO_4^-$ (c) $CH_3NH_3^+$
(d) HPO_4^{2-} (e) $H_2BO_3^-$

9.25 The equilibrium favors the side with the weaker acid–weaker base combination. Equilibria (b) and (c) lie to the left; equilibrium (a) lies to the right.

(a) $C_6H_5OH + C_2H_5O^- \rightleftharpoons C_6H_5O^- + C_2H_5OH$

  Stronger   Stronger   Weaker   Weaker
   acid   base   base   acid

(b) $HCO_3^- + H_2O \rightleftharpoons H_2CO_3 + OH^-$

  Weaker   Weaker   Stronger   Stronger
   base   acid   acid   base

(c) $CH_3COOH + H_2PO_4^- \rightleftharpoons CH_3COO^- + H_3PO_4$

  Weaker   Weaker   Stronger   Stronger
   acid   base   base   acid

9.27 (a) the pK_a of a weak acid (b) the K_a of a strong acid
9.29 (a) 0.10 M HCl (b) 0.10 M H_3PO_4 (c) 0.010 M H_2CO_3 (d) 0.10 M NaH_2PO_4 (e) 0.10 M aspirin
9.31 Only (b) is a redox reaction. The others are acid–base reactions.
(a) $Na_2CO_3 + 2HCl \longrightarrow 2NaCl + CO_2 + H_2O$
(b) $Mg + 2HCl \longrightarrow MgCl_2 + H_2$
(c) $NaOH + HCl \longrightarrow NaCl + H_2O$
(d) $Fe_2O_3 + 6HCl \longrightarrow 2FeCl_3 + 3H_2O$
(e) $NH_3 + HCl \longrightarrow NH_4Cl$
(f) $CH_3NH_2 + HCl \longrightarrow CH_3NH_3Cl$
(g) $NaHCO_3 + HCl \longrightarrow NaCl + H_2O + CO_2$
9.33 (a) $10^{-3}\ M$ (b) $10^{-10}\ M$ (c) $10^{-7}\ M$ (d) $10^{-15}\ M$
9.35 (a) pH = 8 (basic) (b) pH = 10 (basic) (c) pH = 2 (acidic) (d) pH = 0 (acidic) (e) pH = 7 (neutral)
9.37 (a) pH = 8.5 (basic) (b) pH = 1.2 (acidic) (c) pH = 11.1 (basic) (d) pH = 6.3 (acidic)
9.39 (a) pOH = 1.0, $[OH^-]$ = 0.10 M
(b) pOH = 2.4, $[OH^-] = 4.0 \times 10^{-3}\ M$
(c) pOH = 2.0, $[OH^-] = 1.0 \times 10^{-2}\ M$
(d) pOH = 5.6, $[OH^-] = 2.5 \times 10^{-6}\ M$
9.41 0.348 M
9.43 (a) 12 g of NaOH diluted to 400 mL of solution:

$$400\ \text{mL sol}\left(\frac{1\text{L sol}}{1000\ \text{mL sol}}\right)\left(\frac{0.75\ \text{mol NaOH}}{1\ \text{L sol}}\right)$$

$$\times \left(\frac{40.0\ \text{g NaOH}}{1\ \text{mol NaOH}}\right) = 12\ \text{g NaOH}$$

(b) 12 g of $Ba(OH)_2$ diluted to 1.0 L of solution:

$$\left(\frac{0.071\ \text{mol Ba(OH)}_2}{1\ \text{L sol}}\right)\left(\frac{171.4\ \text{Ba(OH)}_2}{1\ \text{mol Ba(OH)}_2}\right) = 12\ \text{g Ba(OH)}_2$$

9.45 5.66 mL
9.47 3.30×10^{-3} mol
9.49 The point at which the observed change occurs during a titration. It is usually so close to the equivalence point that the difference between the two becomes insignificant.
9.51 (a)
$H_3O^+ + CH_3COO^- \rightleftharpoons CH_3COOH + H_2O$ (removal of H_3O^+)
(b)
$HO^- + CH_3COOH \rightleftharpoons CH_3COO^- + H_2O$ (removal of OH^-)
9.53 Yes, the conjugate acid becomes the weak acid and the weak base becomes the conjugate base.
9.55 The pH of a buffer can be changed by altering the weak acid/conjugate base ratio, according to the Henderson-Hasselbalch equation. The buffer capacity can be changed without a change in pH by increasing or decreasing the amount of weak acid/conjugate base mixture while keeping the ratio of the two constant.
9.57 This would occur in a couple of cases. One is very common: You are using a buffer, such as Tris with a pK_a of 8.3, but you do not want the solution to have a pH of 8.3. If you wanted a pH of 8.0, for example, you would need unequal amounts of the conjugate acid and base, with there being more conjugate acid. Another case might be a situation where you are performing a reaction that you know will generate H^+ but you want the pH to be stable. In that situation, you might start with a buffer that was initially set to have more of the conjugate base so that it could absorb more of the H^+ that you know will be produced.
9.59 (a) According to the Henderson-Hasselbalch equation, no change in pH will be observed as long as the weak acid/conjugate base ratio remains the same.
(b) The buffer capacity increases with increasing amounts of weak acid/conjugate base concentrations; therefore, 1.0-mol amounts of each diluted to 1 L would have a greater buffer capacity than 0.1 mol of each diluted to 1 L.
9.61 (a) 3.55 (b) 8.55
9.63 no
9.65 8.0
9.67 No. Hepes has a pK_a of 7.55, which means it is a usable buffer between pH 6.55 and 8.55.
9.69 $Mg(OH)_2$ is a weak base used in flame-retardant plastics.
9.71 Strong bases are more harmful to the cornea because the healing of the wounds can deposit nontransparent scar tissue that impairs vision.
9.73 (a) Respiratory acidosis is caused by hypoventilation, which occurs due to a variety of breathing difficulties, such as a windpipe obstruction, asthma, or pneumonia.
(b) Metabolic acidosis is caused by starvation or heavy exercise.
9.75 Sodium bicarbonate is the weak base form of one of the blood buffers. It tends to raise the pH of blood, which is the purpose of the sprinter's trick, so that the person can absorb more H^+ during the event. By putting $NaHCO_3$ into the system, the following reaction will occur:
$HCO_3^- + H^+ \rightleftharpoons H_2CO_3$. The loss of H^+ means that the blood pH will rise.
9.77 (a) Benzoic acid is soluble in aqueous NaOH.
$C_6H_5COOH + NaOH \rightleftharpoons C_6H_5COO^- + H_2O$
$pK_a = 4.19$ $pK_a = 15.56$
(b) Benzoic acid is soluble in aqueous $NaHCO_3$.
$C_6H_5COOH + NaHCO_3 \rightleftharpoons CH_3C_6H_4O^- + H_2CO_3$
$pK_a = 4.19$ $pK_a = 6.37$
(c) Benzoic acid is soluble in aqueous Na_2CO_3.
$C_6H_5COOH + CO_3^{2-} \rightleftharpoons CH_3C_6H_4O^- + HCO_3^-$
$pK_a = 4.19$ $pK_a = 10.25$
9.79 The strength of an acid is not important to the amount of NaOH that would be required to hit a phenolphthalein endpoint. Therefore, the more concentrated acid, the acetic acid, would require more NaOH.
9.81 3.70×10^{-3}
9.83 0.9 M
9.85 Yes, a pH of 0 is possible. A 1.0 M solution of HCl has $[H_3O^+]$ = 1.0 M. pH = $-\log[H_3O^+]$ = $-\log[1.0\ M]$ = 0
9.87 The qualitative relationship between acids and their conjugate bases states that the stronger the acid, the weaker its conjugate base. This can be quantified in the equation $K_b \times K_a = K_w$ or $K_b = 1.0 \times 10^{-14}/K_a$, where K_b is the base dissociation equilibrium constant for the conjugate base, K_a is the acid dissociation equilibruim constant for the acid, and K_w is the ionization equilibrium constant for water.
9.89 Yes. The strength of the acid is irrelevant. Both acetic acid and HCl have one H^+ to give up, so equal moles of either will require equal moles of NaOH to titrate to an end point.

9.91 need 0.182 mol of $H_2BO_3^-$ and 1.00 mol of H_3BO_3 in 1.00 L solution

9.93 An equilibrium will favor the side of the weaker acid/weaker base. The larger the pK_a value, the weaker the acid.

9.95 (a) $HCOO^- + H_3O^+ \rightleftharpoons HCOOH + H_2O$
(b) $HCOOH + HO^- \rightleftharpoons HCOO^- + H_2O$

9.97 (a) 0.050 mol (b) 0.0050 mol (c) 0.50 mol

9.99 According to the Henderson-Hasselbalch equation,

$$pH = 7.21 + \log \frac{[HPO_4^{2-}]}{[H_2PO_4^-]}$$

As the concentration of $H_2PO_4^-$ increases, the $\log \frac{[HPO_4^{2-}]}{[H_2PO_4^-]}$ becomes negative, lowering the pH and becoming more acidic.

9.101 No. A buffer will have a pH equal to its pK_a only if equimolar amounts of the conjugate acid and base forms are present. If this is the basic form of Tris, then just putting any amount of it into water will give a pH much higher than the pK_a value.

9.103 (a) pH = 7.1, $[H_3O^+] = 7.9 \times 10^{-8} M$, basic
(b) pH = 2.0, $[H_3O^+] = 7.9 \times 10^{-2} M$, acidic
(c) pH = 7.4, $[H_3O^+] = 4.0 \times 10^{-8} M$, basic
(d) pH = 7.0, $[H_3O^+] = 1.0 \times 10^{-7} M$, neutral
(e) pH = 6.6, $[H_3O^+] = 2.5 \times 10^{-7} M$, acidic
(f) pH = 7.4, $[H_3O^+] = 4.0 \times 10^{-8} M$, basic
(g) pH = 6.5, $[H_3O^+] = 3.2 \times 10^{-7} M$, acidic
(h) pH = 6.9, $[H_3O^+] = 1.3 \times 10^{-7} M$, acidic

9.105 4.9 : 1, or 5 : 1 to one significant figure

Chapter 10   Organic Chemistry

10.1 Following are Lewis structures showing all bond angles.

10.2 Of the four alcohols with the molecular formula $C_4H_{10}O$, two are 1°, one is 2°, and one is 3°. For the Lewis structures of the 3° alcohol and one of the 1° alcohols, some $C-CH_3$ bonds are drawn longer to avoid crowding in the formulas.

10.3 The three secondary (2°) amines with the molecular formula $C_4H_{11}N$ are

$$CH_3CH_2CH_2NHCH_3 \qquad CH_3\overset{\displaystyle CH_3}{\underset{}{CHNHCH_3}} \qquad CH_3CH_2NHCH_2CH_3$$

10.4 The three ketones with the molecular formula $C_5H_{10}O$ are

$$CH_3CH_2CH_2\overset{O}{\overset{\|}{C}}CH_3 \quad CH_3CH_2\overset{O}{\overset{\|}{C}}CH_2CH_3 \quad CH_3\overset{O}{\overset{\|}{C}}\underset{CH_3}{CHCH_3}$$

10.5 The two carboxylic acids with the molecular formula $C_4H_8O_2$ are

$$CH_3CH_2CH_2\overset{O}{\overset{\|}{C}}OH \qquad CH_3\overset{O}{\overset{\|}{\underset{CH_3}{CH}}}COH$$

10.6 The four esters with the molecular formula $C_4H_8O_2$ are

$$\overset{O}{\overset{\|}{HC}}OCH_2CH_2CH_3 \qquad \overset{O}{\overset{\|}{HC}}O\underset{CH_3}{CHCH_3}$$
(1) (2)

$$CH_3\overset{O}{\overset{\|}{C}}OCH_2CH_3 \qquad CH_3CH_2\overset{O}{\overset{\|}{C}}OCH_3$$
(3) (4)

10.7 Assuming each is pure, there is no differences in chemical or physical properties.

10.9 Wöhler heated ammonium chloride and silver cyanate, both inorganic compounds, and obtained urea, an organic compound.

10.11 Among the textile fibers, think of natural fibers such as cotton, wool, and silk. Think also of synthetic textile fibers such as Nylon, Dacron polyester, and polypropylene.

10.13 For hydrogen, the number of valence electrons plus the number of bonds equals 2. For carbon, nitrogen, oxygen, and the halogens, the number of valence elecrons plus the number of bonds equals 8. Carbon, with four valence electrons, forms four bonds. Nitrogen, with five valence electrons, forms three bonds. Oxygen, with six valence electrons, forms two bonds. Following is a Lewis dot structure for each element.

(a) ·Ċ·   (b) :Ö: (c) ·N̈:   (d) H·

10.15 Following is a Lewis structure for each compound.

(a) H—C—Ö—C—H (with H above and below each C)

(b) H—C—C—H (with H H above, H H below)

(c) C=C (with H, H on left; H, H on right)

(d) H—C≡C—H

(e) Ö=C=Ö

(f) C=Ö (with H above and H below on C)

(g) H—Ö—C—Ö—H (with :O: double bonded to C)

(h) H—C—C—Ö—H (with H H on left C, :O: double bonded to right C, H below left C)

10.17 In stable organic compounds, carbon must have four covalent bonds to other atoms. In (a), carbon has bonds to five other atoms. In (b), one of the carbons has four bonds to other atoms, but the second carbon has five bonds to other atoms.

10.19 You would find three regions of electron density around nitrogen and, therefore, predict 120° for the H—N—H bond angles.

10.21 (a) 109.5° about C and O   (b) 120° about C
(c) 180° about C

10.23 A functional group is a part of an organic molecule that undergoes a predictable set of chemical reactions.

10.25

(a) —C— (with O double bonded)
(b) —C—Ö—H (with O double bonded)
(c) —Ö—H
(d) —N̈—H (with H below)
(e) —C—Ö— (with :O: double bonded)

10.27 (a) Incorrect. The carbon on the left has five bonds.
(b) Incorrect. The carbon bearing the Cl has five bonds.
(c) Correct.
(d) Incorrect. The oxygen has three bonds and one carbon has five bonds.
(e) Correct.
(f) Incorrect. The carbon on the right has five bonds.

10.29 The one tertiary alcohol with the molecular formula $C_4H_{10}O$ is

$$CH_3\overset{\overset{\displaystyle CH_3}{|}}{\underset{\underset{\displaystyle CH_3}{|}}{C}}OH$$

10.31 The one tertiary amine with the molecular formula $C_4H_{11}N$ is

$$CH_3CH_2\overset{\overset{\displaystyle CH_3}{|}}{N}CH_3$$

10.33 (a) The four primary alcohols with the molecular formula $C_5H_{12}O$ are

$CH_3CH_2CH_2CH_2CH_2OH$ $CH_3CHCH_2CH_2OH$ (CH₃ branch)

$CH_3CH_2CHCH_2OH$ (CH₃ branch) CH_3CCH_2OH (CH₃ branches)

(b) The three secondary alcohols with the molecular formula $C_5H_{12}O$ are

$CH_3CHCH_2CH_2CH_3$ (OH) $CH_3CH_2CHCH_2CH_3$ (OH) $CH_3CHCHCH_3$ (OH and CH₃)

(c) The one tertiary alcohol with the molecular formula $C_5H_{12}O$ is

$$CH_3CH_2\overset{\overset{\displaystyle CH_3}{|}}{\underset{\underset{\displaystyle CH_3}{|}}{C}}OH$$

10.35 The eight carboxylic acids with the molecular formula $C_6H_{12}O_2$ are

$CH_3CH_2CH_2CH_2CH_2COOH$ $CH_3CHCH_2CH_2COOH$ (CH₃) $CH_3CH_2CHCH_2COOH$ (CH₃)

$CH_3CH_2CH_2CHCOOH$ (CH₃) CH_3CCH_2COOH (CH₃ branches) $CH_3CHCHCOOH$ (CH₃ branches)

CH_3CH_2CCOOH (CH₃ branches) $CH_3CH_2CHCOOH$ (CH₂CH₃)

10.37 Taxol was discovered during a survey of indigenous plants sponsored by the National Cancer Institute with the goal of discovering new chemicals for fighting cancer.
10.39 The goal of combinatorial chemistry is to synthesize large numbers of closely related compounds in one reaction mixture.

10.41 Predict 109.5° for all C—Si—C bond angles.

10.43

(a) CH_3CH_2OH

(b) $CH_3CH_2\overset{\displaystyle O}{\overset{\|}{C}}H$

(c) $CH_3\overset{\displaystyle O}{\overset{\|}{C}}CH_3$

(d) $CH_3CH_2\overset{\displaystyle O}{\overset{\|}{C}}OH$

10.45 The three tertiary amines with the molecular formula $C_5H_{13}N$ are

10.47 (a) O—H is the most polar bond. (b) C—C and C=C are the least polar bonds.

10.49 (a) one secondary hydroxyl group and one carboxyl group
(b) two primary hydroxyl groups
(c) one primary amino group and one carboxyl group

10.51 There are several possible answers for some parts of this problem; only one is given here.

(a) $CH_3CH_2CH_2\overset{\displaystyle O}{\overset{\|}{C}}OH$

(b) $CH_3\overset{\displaystyle O}{\overset{\|}{C}}OCH_2CH_3$

(c) $CH_3\underset{\displaystyle OH}{CH}\overset{\displaystyle O}{\overset{\|}{C}}CH_3$

(d) $CH_3\underset{\displaystyle CH_3}{\overset{\displaystyle HOO}{\overset{|\ \|}{C}}}CH$

(e) $HOCH_2CH=CHCH_2OH$

Chapter 11 Alkanes

11.1 The alkane is octane, and its molecular formula is C_8H_{18}.

11.2 (a) constitutional isomers (b) the same compound

11.3 The three constitutional isomers with the molecular formula C_5H_{12} are

11.4 (a) 5-isopropyl-2-methyloctane, $C_{12}H_{26}$
(b) 4-isopropyl-4-propyloctane, $C_{14}H_{30}$

11.5 (a) isobutylcyclopentane, C_9H_{18}
(b) sec-butylcycloheptane, $C_{11}H_{22}$
(c) 1-ethyl-1-methylcyclopropane, C_6H_{12}

11.6 The structure with the three methyl groups equatorial is

11.7 Cycloalkanes (a) and (c) show *cis-trans* isomerism.

(a)

cis-1,3-Dimethylcyclopentane

trans-1,3-Dimethylcyclopentane

(c)

cis-1,3-Dimethylcyclohexane

trans-1,3-Dimethylcyclohexane

11.8 In order of increasing boiling point, they are
(a) 2,2-dimethylpropane (9.5°C), 2-methylbutane (27.8°C), pentane (36.1°C)
(b) 2,2,4-trimethylhexane, 3,3-dimethylheptane, nonane

11.9 The two chloroalkanes with their systematic and common names are

$CH_3CH_2CH_2Cl$ $CH_3\overset{\displaystyle Cl}{\overset{|}{C}}HCH_3$
1-Chloropropane   2-Chloropropane
(Propyl chloride)   (Isopropyl chloride)

11.11 The carbon chain of an alkane is not straight; it is bent with all C—C—C bond angles of approximately 109.5°.

11.13

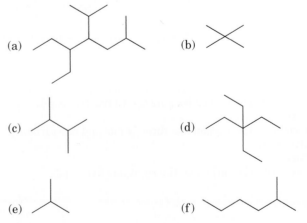

11.15 (a) true (b) true (c) false (d) false
11.17 Structures (a) and (g) represent the same compound. Structures (a), (c), (d), (e), (f), and (g) represent constitutional isomers.
11.19 The structural formulas in parts (b), (c), (e), and (f) represent pairs of constitutional isomers.
11.21 (a) ethyl (b) isopropyl (c) isobutyl (f) *tert*-butyl
11.23 (a) 2-methylpentane (b) 2,5-dimethylhexane (c) 3-ethyloctane (d) 1-isopropyl-2-methylcyclohexane (e) isobutylcyclopentane (f) 1-ethyl-2,4-dimethylcyclohexane
11.25 A conformation is any three-dimensional arrangement of the atoms in a molecule that results from rotation about a single bond.
11.27 In the staggered conformation, the hydrogen atoms on one carbon are as far apart as possible from the hydrogen atoms on the other carbon. In the eclipsed conformation, they are as close as possible. The staggered conformation is more likely at room temperature.

Hydrogens are as far apart as possible

Staggered

Hydrogens are as close as possible

Eclipsed

11.29 no
11.31 Structural formulas for the six cycloalkanes with the molecular formula C_5H_{10} are

Cyclopentane   Methylcyclobutane   1,1-Dimethyl-cyclopropane

trans-1,2-Dimethyl-cyclopropane   *cis*-1,2-Dimethyl-cyclopropane   Ethylcyclopropane

11.33 The isomer with the highest boiling point is heptane (98°C), which is unbranched. The isomer with the lowest boiling point is 2,2-dimethylpentane (79°C), one of the most branched isomers.
11.35 Alkanes are less dense than water. Those that are liquid at room temperature and float on water.
11.37 Both hexane and octane are colorless liquids, and you cannot tell the difference between them by looking at them. One way to tell which compound is which is to determine their boiling points. Hexane has the lower molecular weight, the smaller molecules, and the lower boiling point (69°C). Octane has the higher molecular weight, the larger molecules, and the higher boiling point (126°C).

11.39 The greater the molecular weight, the higher the boiling point.
11.41 On a gram-per-gram basis, methane is the better source of heat energy.

Hydrocarbon	Component of	Heat Combustion (kcal/g/mol)	Mass (g/mol)	Heat of Combustion (kcal/g)
CH_4	natural gas	212	16.0	13.3
$CH_3CH_2CH_3$	LPG	530	44.1	12.0

11.43

1-Chloropentane   2-Chloropentane   3-Chloropentane

11.45 (a) The Freons are a class of chlorofluorocarbons. (b) They were ideal for use as heat transfer agents in refrigeration systems because they are nontoxic, not corrosive, and nonflammable. (c) Two Freons used for this purpose were CCl_3F (Freon-11) and CCl_2F_2 (Freon-12).
11.47 They are hydrofluorocarbons and hydrochlorofluorocarbons, respectively. These compounds are much more chemically reactive in the lower atmosphere than the Freons and are destroyed before they reach the stratosphere.
11.49 Octane will produce more engine knocking than heptane.
11.51 (a) constitutional isomers (b) constitutional isomers (c) constitutional isomers (d) not isomers (e) not isomers (f) constitutional isomers
11.53 Compounds (a) and (c) show *cis-trans* isomerism.

(a) *cis* and *trans* isomers of 3-methylcyclohexanol

(c) *cis* and *trans* isomers of 4-methylcyclohexanol

11.55 Dodecane (a) does not dissolve in water, (b) does dissolve in hexane, (c) burns when ignited, (d) is a liquid at room temperature, and (e) is less dense than water.
11.57 Note that not all hydrogen atoms are shown in the planar hexagon representation.

Planar hexagon
representation

Chair conformation

11.59 Water, a polar molecule, cannot penetrate the surface layer created by this nonpolar hydrocarbon.

Chapter 12   Alkenes and Alkynes

12.1 (a) 3,3-dimethyl-1-pentene   (b) 2,3-dimethyl-2-butene
(c) 3,3-dimethyl-1-butyne
12.2 (a) *trans*-3,4-dimethyl-2-pentene
(b) *cis*-4-ethyl-3-heptene
12.3 (a) 1-isopropyl-4-methylcyclohexene   (b) cyclooctene
(c) 4-*tert*-butylcyclohexene
12.4 Line-angle formulas for the other two heptadienes are

cis,trans-2,4-Heptadiene   *cis,cis*-2,4-Heptadiene

12.5 Four stereoisomers are possible (two *cis-trans* pairs).
12.6

(a) CH₃CHCH₃   (b)

12.7 Propose a two-step mechanism similar to that for the addition of HCl to propene.
Step 1: Reaction of H⁺ with the carbon–carbon double bond gives a 3° carbocation intermediate.

A 3° carbocation
intermediate

Step 2: Reaction of the 3° carbocation intermediate with bromide ion completes the valence shell of carbon and gives the product.

12.8 The product from each acid-catalyzed hydration is the same alcohol.

$$CH_3\overset{\overset{\displaystyle CH_3}{|}}{\underset{\underset{\displaystyle OH}{|}}{C}}CH_2CH_3$$

12.9 Propose a three-step mechanism similar to that for the acid-catalyzed hydration of propene.

Step 1: Reaction of the carbon–carbon double bond with H⁺ gives a 3° carbocation intermediate.

A 3° carbocation
intermediate

Step 2: Reaction of the 3° carbocation intermediate with water completes the valence shell of carbon and gives an oxonium ion.

An oxonium ion

Step 3: Loss of H⁺ from the oxonium ion completes the reaction and generates a new H⁺ catalyst.

12.10 (a)

(b)

12.11 A saturated hydrocarbon contains only carbon–carbon single bonds. An unsaturated hydrocarbon contains one or more carbon–carbon double or triple bonds.
12.13

(a) 109.5° 120°
(b) 120° CH₂OH

(c) HC≡C—CH=CH₂   180° 120° 120°
(d) 120°

12.15

12.17 (a) 1-heptene (b) 1,4,4-trimethylcyclopentene
(c) 1,3-dimethylcyclohexene (d) 2,4-dimethyl-2-pentene
(e) 1-octyne (f) 2,2-dimethyl-3-hexyne

12.19 (a) The longest chain is four carbons. The correct
name is 2-butene.
(b) The chain is not numbered correctly. The correct name
is 2-pentene.
(c) The ring is not numbered correctly. The correct name
is 1-methylcyclohexene.
(d) The double bond must be located. The correct name
is 3,3-dimethyl-1-pentene.
(e) The chain is not numbered correctly. The correct name
is 2-hexyne.
(f) The longest chain is five carbon atoms. The correct
name is 3,4-dimethyl-2-pentene.

12.21 The structural feature that makes *cis-trans* isomerism possible in alkenes is restricted rotation about the
carbon–carbon double bond. In cycloalkanes, it is restricted
rotation about the carbon–carbon bonds of the ring. These
two structural features have in common restricted rotation
about carbon–carbon bonds.

12.23 Only part (b), 2-pentene, shows *cis-trans* isomerism.

cis-2-Pentene   trans-2-Pentene

12.25

Arachidonic acid

12.27 Only compounds (b) and (d) show *cis-trans* isomerism. The *cis* isomer of each is drawn here.

(b)   (d)

12.29 The structural formula of β-ocimene is drawn on
the left showing all atoms and on the right as a line-angle
formula.

β-Ocimene

12.31 The four isoprene units in vitamin A are shown in
bold.

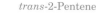

12.33 (a) HBr (b) H₂O/H₂SO₄ (c) HI (d) Br₂

12.35

(a) CH₃CH₂C⁺—CH₂CH₃ and CH₃CH₂CH⁺—CHCH₃
        Tertiary   Secondary

(b) CH₃CH₂CH⁺=CH₂CH₃ and CH₃CH₂CH₂⁺=CHCH₃
        Secondary   Secondary

(c)   and
       Tertiary   Secondary

(d)   —CH₃ and   —CH₂⁺
      Tertiary   Primary

12.37

(a)   (b)

12.39

(a)   and

(b)   (c)

12.41

(a) CH₃CH₂CH=CHCH₂CH₃

(b) [cyclobutene with CH₃] or [methylenecyclobutane CH₂]

(c) CH₂=CCH₂CH₃ or CH₃C=CHCH₃
           | |
          CH₃   CH₃

(d) CH₃CH=CH₂

12.43

(a) CH₃CH₂CH₂CH₂CH₃   (b) CH₃CH₂CH₂CH₂CH₃

(c) [cyclopentane]   (d) [methylcyclopentane CH₃]

12.45 Reagents are shown over each reaction arrow.

CH₃—CH₃ ←H₂/M— CH₂=CH₂ —HBr→ CH₃—CH₂—Br
   (a)   (c)
                  —Br₂/HCl→ Br—CH₂—CH₂—Br
         —H₂O/H₂SO₄↙   (d)
CH₃—CH₂—OH   CH₃—CH₂—Cl
   (b)   (e)

12.47 Ethylene is a natural ripening agent for fruits.

12.49 Its molecular formula is $C_{16}H_{30}O_2$. Its molecular weight is 254.4 amu, and its molar mass is 254.4 g/mol.

12.51 Rods are primarily responsible for peripheral and black-and-white vision. Cones are responsible for color vision.

12.53 The most common consumer items made of high-density polyethylene (HDPE) are milk and water jugs, grocery bags, and squeeze bottles. The most common consumer items made of low-density polyethylene (LDPE) are packaging for baked goods, vegetables, and other produce, as well as trash bags. Currently, only HDPE materials are recyclable.

12.55 There are five compounds with the molecular formula C_4H_8. All are constitutional isomers. The only *cis-trans* isomers are *cis*-2-butene and *trans*-2-butene.

Cyclobutane   Methyl-   1-Butene   cis-2-   trans-2-
               cyclopropane   Butene   Butene

12.57 (a) The carbon skeleton of lycopene can be divided into eight isoprene units, here shown in bold bonds.

(b) Eleven of the 13 double bonds have the possibility for *cis-trans* isomerism. The double bonds at either end of the molecule cannot show *cis-trans* isomerism.

12.59

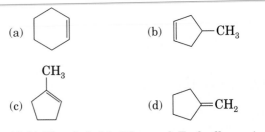

(a)   (b)

(c)   (d)

12.61 The alcohol is 3-hexanol. Each alkene gives the same 2° carbocation intermediate and the same alcohol.

12.63 Reagents are shown over the arrows.

12.65 (a) Oleic acid has one double bond about which *cis-trans* isomerism is possible; For this fatty acid, two stereoisomers are possible: one *cis* isomer and one *trans* isomer.

(b) Linoleic acid has two carbon–carbon double bonds about which *cis-trans* isomerism is possible. This fatty acid has four possible stereoisomers (two *cis-trans* pairs).

(c) Linolenic acid has three carbon–carbon double bonds about which *cis-trans* isomerism is possible. This fatty acid has eight possible stereoisomers.

Chapter 13   Benzene and Its Derivatives

13.1 (a) 2,4,6-tri-*tert*-butylphenol   (b) 2,4-dichloroaniline
(c) 3-nitrobenzoic acid

13.3 An aromatic compound is one that contains one or more benzene rings.

13.5 Yes, they have double bonds, at least in the contributing structures we normally use to represent them. Yes, they are unsaturated because they have fewer hydrogens than a cycloalkane with the same number of carbons.

13.7 (a) An alkene with six carbons has the molecular formula C_6H_{12} and contains one carbon–carbon double bond. Three examples are

1-Hexene   *trans*-3-Hexene   *cis*-3-Hexene

(b) A cycloalkene with six carbons has the molecular formula C_6H_{10} and contains one ring and one carbon–carbon double bond. Three examples are

Cyclohexene   4-Methylcyclopentene   1-Methylcyclopentene

(c) An alkyne with six carbons has the molecular formula C_6H_{10} and contains one carbon–carbon triple bond. Three examples are

1-Hexyne   2-Hexyne   4-Methyl-2-pentyne

(d) An aromatic hydrocarbon with the molecular formula C_8H_{10} contains one benzene ring.

Ethylbenzene   1,2-Dimethylbenzene
                (*o*-xylene)

1,3-Dimethylbenzene   1,4-Dimethylbenzene
(*m*-xylene)   (*p*-xylene)

13.9 Benzene consists of carbons, each surrounded by three regions of electron density, which gives 120° for all bond angles. Bond angles of 120° in benzene can be maintained only if the molecule is planar. Cyclohexane, by contrast, consists of carbons, each surrounded by four regions of electron density, which gives 109.5° for all bond angles. Angles of 109.5° in cyclohexane can be maintained only if the molecule is nonplanar.

13.11 It works this way. Neither a unicorn, which has a horn like a rhinoceros, nor a dragon, which has a tough, leathery hide like a rhinoceros, exists. If they did and you made a hybrid of them, you would have a rhinoceros. Furthermore, a rhinoceros is not a dragon part of the time and a unicorn the rest of the time; a rhinoceros is a rhinoceros all of the time. To carry this analogy to aromatic compounds, resonance-contributing structures for them do not exist; they are imaginary. If they did exist and you could make a hybrid of them, you would have the real aromatic compound.

13.13

13.15 Only cyclohexene will react with a solution of bromine in dichloromethane. A solution of Br_2/CH_2Cl_2 is red, whereas a dibromocycloalkane is colorless. To tell which bottle contains which compound, place a small quantity of each compound in a test tube, and to each add a few drops of Br_2/CH_2Cl_2 solution. If the red color disappears, the compound is cyclohexene. If it remains, the compound is benzene.

13.17

2-Bromotoluene	3-Bromotoluene	4-Bromotoluene
(*o*-Bromotoluene)	(*m*-Bromotoluene)	(*p*-Bromotoluene)

13.19 (a) Nitration using HNO_3/H_2SO_4 followed by sulfonation using H_2SO_4. The order of the steps may also be reversed.
(b) Bromination using $Br_2/FeCl_3$ followed by chlorination using $Cl_2/FeCl_3$. The order of the steps may also be reversed.

13.21 Phenol is a sufficiently strong acid that it reacts with strong bases such as sodium hydroxide to form sodium phenoxide, a water-soluble salt. Cyclohexanol has no comparable acidity and does not react with sodium hydroxide.

13.23 *Radical* indicates a molecule or ion with an unpaired electron. *Chain* means a cycle of two or more steps, called propagation steps that repeat over and over. The net effect of a radical chain reaction is the conversion of starting materials to products. *Chain length* refers to the number of times the cycle of chain propagation steps repeats.

13.25 Vitamin E participates in one or the other of the chain propagation steps and forms a stable radical, which breaks the cycle of propagation steps.

13.27 The abbreviation DDT is derived from *DichloroDiphenylTrichloroethane.*

13.29 Insoluble in water. It is entirely a hydrophobic molecule.

13.31 *Biodegradable* means that a substance can be broken down into one or more harmless compounds by living organisms in the environment.

13.33 Iodine is an element that is found primarily in sea water; thus, seafoods are rich sources of it. Individuals in inland areas where seafood is only a limited part of the diet are the most susceptible to developing goiter.

13.35 One of the aromatic rings in Allura Red has a —CH_3 group and an —OCH_3 group. These groups are not present in Sunset Yellow.

13.37 orange

13.39 Capsaicin is isolated from the fruit of various species of *Capsicum,* otherwise known as chili peppers.

13.41 Two *cis-trans* isomers are possible for capsaicin. The double bond in capsaicin has the *trans* configuration.

13.43 Following are the three contributing structures for naphthalene.

(1)   (2)   (3)

13.45 BHT participates in one of the chain propagation steps of autoxidation, forms a stable radical, and thus terminates autoxidation.

13.47

Chapter 14 Alcohols, Ethers, and Thiols

14.1 (a) 2-heptanol (b) 2,2-dimethyl-1-propanol
(c) *cis*-3-isopropylcyclohexanol
14.2 (a) primary (b) secondary (c) primary
(d) tertiary

14.3 In each case, the major product (circled) contains the more substituted double bond. The structure of the major alkene product from each reaction is enclosed in a box.

(a) $\boxed{CH_3C\!=\!CHCH_3}$ (with CH$_3$ substituent) $+\ CH_2\!=\!CCH_2CH_3$ (with CH$_3$ substituent)

(b) [boxed cyclopentene with CH$_3$] $+$ [cyclopentane with =CH$_2$]

14.4

2-Methyl-cyclohexanol $\xrightarrow{H_2SO_4}$ 1-Methyl-cyclohexene (C) $\xrightarrow[H_2SO_4]{H_2O}$ 1-Methyl-cyclohexanol (D)

14.5 Each secondary alcohol is oxidized to a ketone.

(a) [cyclohexanone structure, =O]

(b) $CH_3\overset{O}{\overset{\|}{C}}CH_2CH_2CH_3$

14.6 (a) ethyl isobutyl ether (b) cyclopentyl methyl ether
14.7 (a) 3-methyl-1-butanethiol (b) 3-methyl-2-butanethiol
14.9 Only (c) and (d) are secondary alcohols.
14.11 (a) 1-pentanol (b) 1,3-propanediol
(c) 1,2-butanediol (d) 3-methyl-1-butanol
(e) *cis*-1,2-cyclohexanediol (f) 2,6-dimethylcyclohexanol
14.13

(a) [branched chain with OH]
(b) HO—[chain]—OH
(c) HO—[branched chain]
(d) HO—[chain with OH]
(e) HO·····[cyclohexane]···OH
(f) [structure with OH and Cl]

14.15 Prednisone contains three ketones, one primary alcohol, one tertiary alcohol, one disubstituted carbon–carbon double bond, and one trisubstituted carbon–carbon double bond. Estradiol contains one secondary alcohol and one disubstituted phenol.
14.17 Low-molecular-weight alcohols form hydrogen bonds with water molecules through both the oxygen and hydrogen atoms of their —OH groups. Low-molecular-weight ethers form hydrogen bonds with water molecules only through the oxygen atom of their —O— groups. The greater extent of hydrogen bonding between alcohol and water molecules makes the low-molecular-weight alcohols more soluble in water than the low-molecular-weight ethers.
14.19 Both types of hydrogen bonding are shown in the following illustration.

[Hydrogen bonding diagram: $CH_3-\overset{\delta-}{O}$ ... H—O—CH$_3$, H$_{\delta+}$, $\overset{\delta+}{H}$—O $\delta-$, H]

14.21 In order of increasing boiling point, they are

$CH_3CH_2CH_3$ (−42°C) CH_3CH_2OH (78°C)

$CH_3CH_2CH_2CH_2OH$ (117°C) $HOCH_2CH_2OH$ (198°C)

14.23 Evaporation of a liquid from the surface of the skin cools because heat is absorbed from the skin in converting molecules from the liquid state to the gaseous state. 2-Propanol (isopropyl alcohol), which has a boiling point of 82°C, absorbs heat from the skin, evaporates rapidly, and has a cooling effect. 2-Hexanol, which has a boiling point of 140°C, also absorbs heat from the surface of the skin but, because of its higher boiling point, evaporates much more slowly and, therefore, does not have the same cooling effect as 2-propanol.
14.25 The more water-soluble compound is circled.

(a) $\boxed{CH_3OH}$ or CH_3OCH_3

(b) $\boxed{CH_3\overset{OH}{\overset{|}{C}HCH_3}}$ or $CH_3\overset{CH_2}{\overset{\|}{C}CH_3}$

(c) $CH_3CH_2CH_2SH$ or $\boxed{CH_3CH_2CH_2OH}$

14.27 For parts (a), (b), and (d), two constitutional isomers will give the desired alcohol. For parts (c) and (e), only one alkene will give the desired alcohol.

(a) $CH_2\!=\!CHCH_2CH_3$ or $CH_3CH\!=\!CHCH_3$

(b) [cyclohexane with CH$_3$ double bond]   or   [cyclohexane with =CH$_2$]

(c) $CH_3CH_2CH\!=\!CHCH_2CH_3$

(d) $CH_2\!=\!\overset{CH_3}{\overset{|}{C}}CH_2CH_2CH_3$ or $CH_3\overset{CH_3}{\overset{|}{C}}\!=\!CHCH_2CH_3$

(e) [cyclopentene structure]

14.29 Phenols are weak acids, with pK_a values approximately equal to 10. Alcohols, which are considerably weaker acids, have about the same acidity as water.
14.31 The first reaction is an acid-catalyzed dehydration; the second is an oxidation.

(a) $CH_3CH_2CH_2CH_2OH \xrightarrow[\text{heat}]{H_2SO_4} CH_3CH_2CH\!=\!CH_2 + H_2O$

(b) $CH_3CH_2CH_2CH_2OH \xrightarrow[H_2SO_4]{K_2Cr_2O_7} CH_3CH_2CH_2\overset{O}{\overset{\|}{C}}OH$

14.33 Oxidation converts each primary alcohol to a carboxyl group.

(a) $CH_3(CH_2)_6CH_2OH \xrightarrow[H_2SO_4]{K_2Cr_2O_7} CH_3(CH_2)_6\overset{O}{\overset{\|}{C}}OH$

1-Octanol → Octanoic acid

(b) HOCH$_2$CH$_2$CH$_2$CH$_2$OH $\xrightarrow[\text{H}_2\text{SO}_4]{\text{K}_2\text{Cr}_2\text{O}_7}$ HOCCH$_2$CH$_2$COH
1.4-Butanediol   Butanedioic acid

14.35 Following are reagents to bring about each step shown in the flowchart.
(a) H$_2$SO$_4$, heat (b) H$_2$O/H$_2$SO$_4$ (c) K$_2$Cr$_2$O$_7$/H$_2$SO$_4$
(d) HBr (e) Br$_2$/CH$_2$Cl$_2$ (f) H$_2$/Ni (g) K$_2$Cr$_2$O$_7$/H$_2$SO$_4$
(h) K$_2$Cr$_2$O$_7$/H$_2$SO$_4$ (i) K$_2$Cr$_2$O$_7$/H$_2$SO$_4$
14.37 Ethanol and ethylene glycol are derived from ethylene. Ethanol is a solvent and is the starting material for the synthesis of diethyl ether, also an important solvent. Ethylene glycol is used in automotive antifreezes and is one of the two starting materials required for the synthesis of poly(ethylene terephthalate), better known as PET (Section 19.6B).
14.39 (a) dicyclopentyl ether (b) dipentyl ether
(c) diisopropyl ether
14.41 (a) 2-butanethiol (b) 1-butanethiol
(c) cyclohexanethiol
14.43 Because 1-butanol molecules associate in the liquid state by hydrogen bonding, it has the higher boiling point (117°C). There is very little polarity to an S—H bond. The only interactions among 1-butanethiol molecules in the liquid state are the considerably weaker London dispersion forces. For this reason, 1-butanethiol has the lower boiling point (98°C).
14.45 Nitroglycerin was discovered in 1847. It is a pale yellow, oily liquid.
14.47 One of the products the body derives from metabolism of nitroglycerin is nitric oxide, NO, which causes the coronary artery to dilate, thus relieving angina.
14.49 The relationship is that 2100 mL of breath contains the same amount of ethanol as 1.00 mL of blood.
14.51 Diethyl ether is easy to use and causes excellent muscle relaxation. Blood pressure, pulse rate, and respiration are usually only slightly affected. Diethyl ether's chief drawbacks are its irritating effect on the respiratory passages and its after effect of nausea.
14.53 Both enflurane and isoflurane have the molecular formula C$_3$H$_2$ClF$_5$O, but they have different structurals formulas and, therefore, they are constitutional isomers. Predict that they are soluble in hexane.
14.55 2CH$_3$OH + 3O$_2$ ⟶ 2CO$_2$ + 4H$_2$O
14.57 The eight isomeric alcohols with the molecular formula C$_5$H$_{12}$O follow.

1-Pentanol   2-Pentanol   3-Pentanol   2-Methyl-1-butanol

2-Methyl-2-butanol   3-Methyl-2-butanol   3-Methyl-1-butanol   2,2-Dimethyl-1-propanol

14.59 Ethylene glycol has two —OH groups by which each molecule participates in intermolecular hydrogen bonding, whereas 1-propanol has only one OH group. The stronger intermolecular forces of attraction between molecules of ethylene glycol give it the higher boiling point.

14.61 Arranged in order of increasing solubility in water, they are

CH$_3$CH$_2$CH$_2$CH$_2$CH$_2$CH$_3$ CH$_3$CH$_2$CH$_2$CH$_2$CH$_2$OH
Hexane   1-Pentanol
(Insoluble)   (2.3 g/mL water)

HOCH$_2$CH$_2$CH$_2$CH$_2$OH
1,4-Butanediol
(Infinitely soluble)

14.63 Each can be prepared from 2-methyl-1-propanol (circled) as shown in this flowchart.

14.65 (a) The two functional groups in lipoic acid are a carboxyl group and a disulfide group.
(b) Its reduction product is dihydrolipoic acid.

Lipoic acid

Dihydrolipoic acid

Chapter 15   Chirality: The Handedness of Molecules

15.1 The enantiomers of each part are drawn with two groups in the plane of the paper, a third group toward you in front of the plane, and the fourth group away from you behind the plane.

(a)

(b)

15.2 The group of higher priority in each set is circled.

(a) —CH$_2$OH and —CH$_2$CH$_2$COOH

(b) —CH$_2$NH$_2$ and —CH$_2$COOH

15.3 The configuration is *R* and the compound is
(R)-glyceraldehyde.
15.4 (a) Structures 1 and 3 are one pair of enantiomers.
Structures 2 and 4 are a second pair of enantiomers.
(b) Compounds 1 and 2, 1 and 4, 2 and 3, and 3 and 4 are
diastereomers.
15.5 Four stereoisomers are possible for 3-methylcyclo-
hexanol. The *cis* isomers are one pair of enantiomers; the
trans isomers are a second pair of enantiomers.
15.6 Stereocenters are marked by an asterisk and the
number of stereoisomers possible is shown under the structural formula.

15.7 Chirality is a property of an object—the object is not
superposable on its mirror image—that is, the object has
handedness. 2-Butanol is a chiral molecule.
15.9 Stereoisomers are isomers that have the same molecular formula and the same connectivity, but a different
orientation of their atoms in space. Three examples are
cis-trans isomers, enantiomers, and diastereomers.
15.11 (a) chiral (b) achiral (c) achiral (d) achiral
(e) chiral, unless you are on the equator, in which case it
goes straight down and no swirl is present.
15.13 Neither *cis*-2-butene nor *trans*-2-butene is chiral.
Each is superposable on its mirror image.
15.15 Compounds (a), (c), and (d) contain stereocenters,
here marked by asterisks.

15.17 The stereocenter in each chiral compound is marked
by an asterisk.

15.19 Following are mirror images of each molecule.

15.21 Only parts (b) and (c) contain stereocenters.

15.23 Each stereocenter is marked by an asterisk. Under
each structural formula is the number of stereoisomers
possible. Compound (b) has no stereocenter.

15.25 The specific rotation of its enantiomer is +41°.
15.27 To say that a drug is chiral means that it exhibits
handedness—that is, it has one or more stereocenters and
the possibility for two or more stereoisomers. Just because
a compound is chiral does not mean that it will be optically
active. It may be chiral and present as a racemic mixture,
in which case it will have no effect on the plane of polarized
light. If it is present as a single enantiomer, it will rotate
the plane of polarized light.
15.29 The two stereocenters in this alcohol are marked by
asterisks.

3-Methyl-2-pentanol

15.31 All three compounds are chiral. The stereocenters in each are marked by asterisks. Under the name of each compound is the number of stereoisomers possible for it.

(a)

Fluoxetine
(Prozac)
($2^1 = 2$)

(b)

Sertraline
(Zoloft)
($2^2 = 4$)

(c)

Paroxetine
(Paxil)
($2^2 = 4$)

15.33 (a) The only possible compound that does not show *cis-trans* isomerism and has no stereocenter is 1-methylcyclohexanol.

1-Methylcyclohexanol

(b) Only 4-methylcyclohexanol shows *cis-trans* isomerism but has no stereocenter. Drawn here are the *cis* and *trans* isomers.

trans-4-Methylcyclohexanol   *cis*-4-Methylcyclohexanol

(c) Both 2-methylcyclohexanol and 3-methylcyclohexanol have two stereocenters and can exist as *cis* and *trans* isomers. The stereocenters in each are marked by asterisks.

2-Methylcyclohexanol   3-Methylcyclohexanol

15.35 The majority have a right-handed twist because the machines that make them all impart the same twist.

15.37 (a) Carbons 1, 2, 3, 4, and 5 are stereocenters.
(b) $2^5 = 32$ stereoisomers are possible.
(c) 16 pairs of enantiomers are possible.

Chapter 16   Amines

16.1 Pyrrolidine has nine hydrogens; its molecular formula is C_4H_9N. Purine has four hydrogens; its molecular formula is $C_5H_4N_4$.

16.2

(a) NH_2 (b) NH_2

(c) H_2N NH_2

16.3

(a) HO NH_2 (b)

(c)

16.4 The stronger base is circled.

(a)

(b) NH_3 or NH_2

16.5 The product of each reaction is an ammonium salt.

(a) $(CH_3CH_2)_3\overset{+}{N}HCl^-$
Triethylammonium chloride

(b)
Piperidinium acetate

16.7 In an aliphatic amine, all carbon groups bonded to nitrogen are alkyl groups. In an aromatic amine, one or more of the carbon groups bonded to nitrogen are aryl (aromatic) groups.

16.9 Following is a structural formula for each amine.

(a)
NH₂

(b) CH₃(CH₂)₆CH₂NH₂

(c) NH₂

(d) H₂N(CH₂)₅NH₂

(e) NH₂
Br

(f) (CH₃CH₂CH₂CH₂)₃N

16.11 Each amine is classified by type.

(a) 1° aliphatic amine
HO
NH₂
N
H heterocyclic aromatic amine

(b) O
O
H₂N
1° aromatic amine

(c) 3° aliphatic amine CH₃
O N
CH₃

16.13 For this molecular formula, there are four primary amines, three secondary amines, and one tertiary amine. Only 2-butanamine is chiral.

1° amines:

NH₂
1-Butanamine (Butylamine)

NH₂
2-Butanamine (sec-Butylamine)

NH₂
2-Methyl-1-propanamine (Isobutylamine)

NH₂
2-Methyl-2-propanamine (tert-Butylamine)

2° amines:

H
N
CH₃
Methylpropyl-amine

H
N
CH₃
Methylisopropyl-amine

H
N
Diethylamine

3° amine:

CH₃
N
CH₃
Dimethylethylamine

16.15 Both propylamine (a 1° amine) and ethylmethyl-amine (a 2° amine) have an N—H group, and hydrogen bonding occurs between their molecules in the liquid state. Because of this intermolecular force of attraction, these two amines have higher boiling points than trimethylamine, which has no N—H bond and, therefore, cannot participate in intermolecular hydrogen bonding.

16.17 2-Methylpropane is a nonpolar hydrocarbon, and the only attractive forces between its molecules in the liquid state are very weak London dispersion forces. Both 2-propanol and 2-propanamine are polar molecules and associate in the liquid state by hydrogen bonding. Hydrogen bonding is stronger between alcohol molecules than between amine molecules because of the greater strength of an O—H---O hydrogen bond compared with an N—H---N hydrogen bond. It takes more energy (a higher temperature) to separate an alcohol molecule in the liquid state from its neighbors than to separate an amine molecule from its neighbors; thus the alcohol has the higher boiling point.

16.19 Nitrogen is less electronegative than oxygen and, therefore, more willing to donate its unshared pair of electrons to H⁺ in an acid–base reaction to form a salt. Therefore, amines are stronger bases than alcohols.

16.21 (a) ethylammonium chloride
(b) diethylammonium chloride
(c) anilinium hydrogen sulfate

16.23 The form of amphetamine present at both pH 1.0 and pH 7.4 is the ammonium ion—that is, the conjugate acid of the amine.

16.25

(c)

16.27 Tamoxifen contains three aromatic (benzene) rings, one carbon–carbon double bond, one ether, and one 3° aliphatic amine. Two stereoisomers are possible, a pair of *cis-trans* isomers.

16.29 (a) Epinephrine has two phenolic —OH groups, whereas amphetamine has none. Epinephrine has a 2° alcohol on its carbon side chain, whereas amphetamine has none. Epinephrine is a 2° amine, whereas amphetamine is a 1° amine. Finally, both epinephrine and amphetamine are chiral, but their stereocenters are in different locations within the molecule.
(b) Methamphetamine is a 2° aliphatic amine, whereas amphetamine is a 1° aliphatic amine. Both compounds are chiral at the same carbon.

16.31 An alkaloid is a basic nitrogen-containing compound found in the roots, bark, leaves, berries, or fruits of plants. In almost all alkaloids, the nitrogen atom is present as a member of a ring; that is, it is present in a heterocyclic ring. By definition, alkaloids are nitrogen-containing bases and, therefore, all turn red litmus blue.

16.33 The tertiary heterocyclic aliphatic amine in the five-membered ring is the stronger base. It reacts with HCl to give this salt.

16.35 The common structural feature is a benzene ring fused to a seven-membered ring containing two nitrogen atoms. This parent structural feature is named benzodiazepine.

16.37 The structural formula of 4-aminobutanoic acid is drawn on the left, showing the 1° amino group (a base) and the carboxyl group (an acid). It is drawn on the right as an internal salt.

A primary amine · A carboxylic acid

$H_2NCH_2CH_2CH_2COH$
Un-ionized amino and carboxyl groups

$H_3\overset{+}{N}CH_2CH_2CH_2CO^-$
Internal salt

16.39 Their salts are more soluble in water and in body fluids, and are more stable (less reactive) toward oxidation by atmospheric oxygen.

16.41 In order of decreasing ability to form intermolecular hydrogen bonds, they are $CH_3OH > (CH_3)_2NH > CH_3SH$. An O—H bond is more polar than an N—H bond, which is in turn, more polar than an S—H bond.

16.43 Butane, the least polar molecule, has the lowest boiling point; 1-propanol, the most polar molecule, has the highest boiling point.

$CH_3CH_2CH_2CH_3$ $CH_3CH_2CH_2NH_2$
−0.5°C   7.6°C

$CH_3CH_2CH_2OH$
77.8°C

16.45 The alcohol will be unchanged. The amine will react with HCl to form a salt.

$CH_3CH_2CH_2NH_3^+ Cl^-$

16.47 (a) Procaine is achiral; it has no stereocenter.
(b) The 3° aliphatic amine is the stronger base.
(c) Following is a structural formula for its hydrochloride salt.

Procaine • HCl

16.49 (a) The amino group is a 3° aliphatic amine.
(b) There are three stereocenters, marked by asterisks.

Atropine hydrogen sulfate

(c) Because it is an ammonium salt, atropine hydrogen sulfate is more soluble in water than atropine.
(d) Dilute aqueous solutions of atropine are basic because its 3° aliphatic amine reacts with water to produce hydroxide ions.

16.51 The compound is better represented by a structural formula (B), which is an internal salt formed by the acid–base reaction between the —NH2 and —COOH groups.

Chapter 17   Aldehydes and Ketones

17.1 (a) 3,3-dimethylbutanal   (b) cyclopentanone
(c) 1-phenyl-1-propanone
17.2 Following are line-angle formulas for the seven aldehydes with the molecular formula $C_6H_{12}O$. In the three that are chiral, the stereocenter is marked by an asterisk.

CHO
Hexanal

CHO
4-Methylpentanal

3-Methylpentanal

2-Methylpentanal

2,3-Dimethylbutanal

3,3-Dimethylbutanal

2,2-Dimethylbutanal

17.3 (a) 2,3-dihydroxypropanal   (b) 2-aminobenzaldehyde
(c) 5-amino-2-pentanone
17.4 Each aldehyde is oxidized to a carboxylic acid.

(a)

Hexanedioic acid
(Adipic acid)

(b)

3-Phenylpropanoic acid

17.5 Each primary alcohol comes from reduction of an
aldehyde. Each secondary alcohol comes from reduction
of a ketone.

(a)

(b) CH_3O— —CH_2CH

(c)

17.6 Shown first is the hemiacetal and then the acetal.

Benzaldehyde   Hemiacetal

Acetal

17.7 (a) A hemiacetal formed from 3-pentanone (a ketone)
and ethanol.
(b) Neither a hemiacetal nor an acetal. This compound is
the dimethyl ether of ethylene glycol.
(c) An acetal derived from 5-hydroxypentanal and methanol.
17.8 Following is the keto form of each enol.

(a)   (b)   (c)

17.9 The carbonyl carbon of an aldehyde is bonded to at
least one hydrogen. The carbonyl carbon of a ketone is
bonded to two carbon groups.
17.11 No. To be a carbon stereocenter, the carbon atom
must have four different groups bonded to it. The carbon
atom of a carbonyl group has only three groups bonded to it.
17.13 (a) Cortisone has three ketones, one primary alcohol,
one tertiary alcohol, and one carbon–carbon double bond.
Aldosterone has two ketones, one aldehyde, one primary
alcohol, one secondary alcohol, and one carbon–carbon
double bond.
(b) Cortisone has 6 stereocenters and $2^6 = 64$ stereoisomers
are possible.
Aldosterone has 7 stereocenters and $2^7 = 128$ stereoisomers
are possible.

Cortisone

Aldosterone

17.15

O
‖
(a) HCH

(b)

(c)

(d) $CH_3(CH_2)_8CH$

(e) HO— —C—H

(f)

17.17 (a) 4-heptanone (b) 2-methylcyclopentanone
(c) *cis*-2-methyl-2-butenal (d) 2-hydroxypropanal
(e) 1-phenyl-2-propanone (f) hexanedial
17.19 (a) The chain is not numbered correctly. Its name is
2-butanone.
(b) The compound is an aldehyde. Its name is butanal.
(c) The longest chain is five carbons. Its name is pentanal.
(d) The location of the ketone takes precedence. Its name
is 3,3-dimethyl-2-butanone.
17.21 (a) ethanol (b) 3-pentanone (c) butanal
(d) 2-butanol
17.23 Acetone has the higher boiling point because of the
intermolecular attraction between the carbonyl groups of
acetone molecules.
17.25 Acetaldehyde forms hydrogen bonds with water
primarily through its carbonyl oxygen.

17.27 Aldehydes are oxidized by this reagent to carboxylic
acids. Ketones are not oxidized under these conditions.
Secondary alcohols are oxidized to ketones.

(a) $CH_3CH_2CH_2COH$ (b) [benzaldehyde–COH structure]

(c) no reaction (d) [cyclohexanone structure]

17.29 (a) Treat each with Tollens' reagent. Only pentanal
will give a silver mirror.
(b) Treat each with $K_2Cr_2O_7/H_2SO_4$. Only 2-pentanol is
oxidized (to 2-pentanone), which causes the red color of
$Cr_2O_7{}^{2-}$ ion to disappear and be replaced by the green color
of Cr^{3+} ion.
17.31 The white solid is benzoic acid, formed by air oxidation of benzaldehyde.
17.33 These experimental conditions reduce an aldehyde to
a primary alcohol and a ketone to a secondary alcohol.

(a) $CH_3\overset{*}{C}HCH_2CH_3$ (b) $CH_3(CH_2)_4CH_2OH$
    (chiral)

(c) [cyclopentanol with CH₃, chiral] (d) [benzene ring with CH₂OH and OH]
    (chiral)

17.35 (a) Following is its structural formula.

1,3-Dihydroxy-2-propanone   1,2,3-Propanetriol
(Dihydroxyacetone)   (Glycerol, glycerin)

(b) Because it has two hydroxyl groups and one carbonyl
group, all of which can interact with water molecules by
hydrogen bonding, predict that it is soluble in water.
(c) Its reduction gives 1,2,3-propanetriol, better known as
glycerol or glycerin.
17.37 The first two reactions involve reduction of the carbonyl group of this ketone to a secondary alcohol. There is
no reaction in (c) or (d); ketones are not oxidized by these
reagents.

(a, b) [benzene ring–CHCH₃ with OH]

17.39 Compound (a) has one enol. Compounds (b) and (c)
have two enols each.

(a) $CH_3CH{=}CH$ [with OH]

(b) $CH_3C{=}CHCH_3$ [with OH] and $CH_2{=}CCH_2CH_3$ [with OH]

(c) [cyclopentene with OH and CH₃] and [cyclopentene with OH and CH₃]

17.41 The characteristic structural feature of a hemiacetal
is a tetrahedral carbon atom bonded to one —OH group and
one —OR group, where R may be either an alkyl group or
aryl group. The characteristic structural feature of an acetal
is a tetrahedral carbonyl group bonded to two OR groups.
17.43 (a) hemiacetal (b) acetal (c) neither
(d) hemiacetal (e) acetal (f) acetal
17.45 Following are structural formulas for the products of
each hydrolysis.

(a) $CH_3CH_2CCH_2CH_3$ + HO———OH

(b) [benzaldehyde with OCH₃] + $2CH_3OH$

(c) [cyclohexanone] $=O$ + $2CH_3OH$

(d) [cyclohexane C=O with H and OH] + CH_3OH

17.47 *Hydration* refers to the addition of one or more molecules of water to a substance. An example of hydration is
the acid-catalyzed addition of water to propene to give
2-propanol. *Hydrolysis* refers to the reaction of a substance
with water, generally with the breaking (lysis) of one or
more bonds in the substance. An example of hydrolysis is
the acid-catalyzed reaction of an acetal with a molecule
of water to give an aldehyde or ketone and two molecules
of alcohol. We will discuss the hydrolysis of esters in detail
in Chapter 18.

17.49 A reagent or reagents to bring about each step are as follows:
(a) NaBH$_4$ or H$_2$/Ni (b) H$_2$SO$_4$/heat (c) HBr
(d) H$_2$/Ni (e) Br$_2$/CH$_2$Cl$_2$

17.51 Compounds (a), (b), and (d) can be formed by reduction of the aldehyde or ketone shown below. Compound (c) is a 3° alcohol and cannot be formed in this manner.

17.53 Reagents for each conversion are shown over the reaction arrow.

17.55 (a) Each compound is insoluble in water. Treat each with dilute aqueous HCl. Aniline, an aromatic amine, reacts with HCl to form a water-soluble salt. Cyclohexanone does not react with this reagent and is insoluble in aqueous HCl.
(b) Treat each with a solution of Br$_2$/CH$_2$Cl$_2$. Cyclohexene reacts to discharge the red color of Br$_2$ and to form 1,2-dibromocyclohexane, a colorless compound. Cyclohexanol does not react with this reagent.
(c) Treat each with a solution of Br$_2$/CH$_2$Cl$_2$. Cinnamaldehyde, which contains a carbon–carbon double bond, reacts to discharge the red color of Br$_2$ and to form 2,3-dibromo-3-phenylpropanal, a colorless compound. Benzaldehyde does not react with this reagent.

17.57 Using this reducing agent, each aldehyde is converted to a 1° alcohol and each ketone is converted to a 2° alcohol.

(a) HOCH$_2$CH$_2$CH$_2$CH$_2$CHCH$_3$ with OH

17.59

(a) CH$_2$CCH$_3$ with O and Cl

(b) CH$_3$CHCH$_2$CH with OH and O

(c) CH$_3$CCH$_2$CCH$_3$ with O, OH, CH$_3$

(d) ring—CCH$_2$CH with CH$_3$ and O

(e)

(f)

17.61 1-Propanol has the higher boiling point because of the greater attraction between its molecules due to hydrogen bonding through its hydroxyl group.

17.63 (a) The hydroxyaldehyde is first redrawn to show the OH group nearer the CHO group. Closing the ring in the hemiacetal formation gives the cyclic hemiacetal.
(b) 5-Hydroxyhexanal has one stereocenter, and two stereoisomers (one pair of enantiomers) are possible.
(c) The cyclic hemiacetal has two stereocenters, and four stereoisomers (two pairs of enantiomers) are possible.

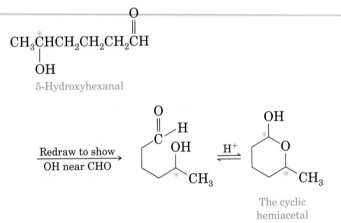

5-Hydroxyhexanal

Redraw to show OH near CHO

The cyclic hemiacetal

17.65 Given first is the alkene that undergoes acid-catalyzed hydration to give the desired alcohol, then the aldehyde or ketone that gives it upon reduction using either NaBH$_4$ or H$_2$ in the presence of a transition metal catalyst.

(a) CH$_2$=CH$_2$ CH$_3$CH with O

(b)

(c) CH$_2$=CHCH$_3$ CH$_3$CCH$_3$ with O

(d)

17.67 (a) Carbon 4 provides the —OH group and carbon 1 provides the —CHO group.
(b) Following is a structural formula for the free aldehyde.

17.69 Each compound reduces a carbonyl group of an aldehyde or ketone to an alcohol by delivering a hydride ion (H:⁻) to the carbonyl carbon. The hydrogen of the resulting —OH group comes from an —OH group of the solvent.
17.71 Each conversion is an oxidation and can be carried out using $K_2Cr_2O_7$ in H_2SO_4.

Chapter 18 Carboxylic Acids

18.1 (a) 2,3-dihydroxypropanoic acid (chiral)
(b) 3-aminopropanoic acid (achiral)
(c) 3,5-dihydroxy-3-methylpentanoic acid (chiral)
18.2 Each acid is converted to its ammonium salt.

18.3

18.5 (a) 3,4-dimethylpentanoic acid
(b) 2-aminobutanoic acid   (c) hexanoic acid
18.7

18.9

18.11 Oxalic acid (IUPAC name: ethanedioic acid) is a dicarboxylic acid. In monopotassium oxalate, each of the carboxyl groups is present as its carboxylic anion, giving a charge of −2. The structural formula is drawn here to show K^+ interacting with each carboxylic anion.

18.13 If you drew this molecule correctly to show the internal hydrogen bonding, you will see that the hydrogen-bonded part of the molecule forms a six-membered ring.

18.15 Propanoic acid has the higher boiling point (141°C). It can participate in intermolecular hydrogen bonding through its carboxyl group. This intermolecular attractive force must be overcome before a propanoic acid molecule can escape from the liquid phase to the vapor phase. There is no comparable intermolecular attractive force between molecules of methyl acetate in the liquid state.

18.17 In order of increasing boiling point, they are

Diethyl ether (bp 35° C)   1-Butanol (bp 117° C)   Propanoic acid (bp 141° C)

18.19 Carboxylic acids are the strongest acids (pK_a approximately 4.0–5.0). Phenols are weaker acids (pK_a approximately 10), and alcohols (pK_a approximately 16–18) are the weakest acids of this group of compounds.

18.21

(a) [benzene]—CH_2COOH + NaOH

$\longrightarrow$ [benzene]—$CH_2COO^-Na^+$ + H_2O

(b) $\diagdown\diagup$COOH + $NaHCO_3$

$\longrightarrow$ $\diagdown\diagup$COO$^-$Na$^+$ + H_2O + CO_2

(c) [benzene with COOH and OCH$_3$] + $NaHCO_3$

$\longrightarrow$ [benzene with COO$^-$Na$^+$ and OCH$_3$] + H_2O + CO_2

(d) $CH_3\overset{\displaystyle OH}{\underset{|}{C}}HCOOH$ + $H_2NCH_2CH_2OH$

$\longrightarrow$ $CH_3\overset{\displaystyle OH}{\underset{|}{C}}HCOO^-H_3\overset{+}{N}CH_2CH_2OH$

(e) $\diagdown\diagup$COO$^-$Na$^+$ + HCl

$\longrightarrow$ $\diagdown\diagup$COOH + NaCl

18.23 The acid–base reaction is the neutralization of formic acid by sodium bicarbonate.

$$HCOOH + NaHCO_3 \longrightarrow HCOO^-Na^+ + H_2O + CO_2$$

Formic acid   Sodium formate

18.25

	pH	$\dfrac{[A^-]}{[HA]}$
(a)	2.0	10^{-3}
(b)	5.0	1.0
(c)	7.0	10^2
(d)	9.0	10^4
(e)	11.0	10^6

18.27 The pK_a of lactic acid is 4.07. At this pH, lactic acid is present as 50% $CH_3CH(OH)COOH$ and 50% $CH_3CH(OH)COO^-$. At pH 7.45, which is more basic than pH 4.07, lactic acid is present primarily as the anion, $CH_3CH(OH)COO^-$.

18.29 In part (a), the —COOH group is a stronger acid than the —NH_3^+ group.

(a) $CH_3\overset{\displaystyle}{\underset{\underset{NH_3^+}{|}}{C}}HCOOH$ + NaOH $\longrightarrow$ $CH_3\overset{\displaystyle}{\underset{\underset{NH_3^+}{|}}{C}}HCOO^-Na^+$ + H_2O

(b) $CH_3\overset{\displaystyle}{\underset{\underset{NH_3^+}{|}}{C}}HCOO^-Na^+$ + NaOH

$\longrightarrow$ $CH_3\overset{\displaystyle}{\underset{\underset{NH_2}{|}}{C}}HCOO^-Na^+$ + H_2O

18.31 In part (a), the —NH_2 group is a stronger base than the —COO^- group.

(a) $CH_3\overset{\displaystyle}{\underset{\underset{NH_2}{|}}{C}}HCOO^-Na^+$ + HCl $\longrightarrow$ $CH_3\overset{\displaystyle}{\underset{\underset{NH_3^+}{|}}{C}}HCOO^-Na^+$ + NaCl

(b) $CH_3\overset{\displaystyle}{\underset{\underset{NH_3^+}{|}}{C}}HCOO^-Na^+$ + HCl $\longrightarrow$ $CH_3\overset{\displaystyle}{\underset{\underset{NH_3^+}{|}}{C}}HCOOH$ + NaCl

18.33 Following is a structural formula for the ester formed in each reaction.

(a) [structure]   (b) $CH_3\overset{\displaystyle O}{\overset{||}{C}}O$—[cyclohexyl]

(c) [benzene with COOCH$_2$CH$_3$ and COOCH$_2$CH$_3$]

18.35 Following is a structural formula for methyl 2-hydroxybenzoate.

[benzene with COOCH$_3$ and OH]

18.37 Following is a structural formula for the product of treating phenylacetic acid with each reagent.
(a) $C_6H_5CH_2COO^-Na^+$ + CO_2 + H_2O
(b) $C_6H_5CH_2COO^-Na^+$ + H_2O
(c) $C_6H_5CH_2COO^-NH_4^+$ (d) $C_6H_5CH_2CH_2OH$ (e) no reaction   (f) $C_6H_5CH_2CO_2CH_3$ + H_2O (g) no reaction

18.39 Step 1: Nitration of the aromatic ring using HNO₃/H₂SO₄ (Section 13.3B). If there is a mixture of nitration products, assume that the desired 4-nitrobenzoic acid can be separated and purified.
Step 2: Catalytic reduction of the nitro group to an amino group (Section 13.3B).

Benzoic acid   4-Nitrobenzoic
                          acid

4-Aminobenzoic
acid

18.41 Each starting material is difunctional and each functional group can participate in formation of an ester, giving rise to a polyester.

From hexanedoic acid   From ethylene
(adipic acid)   glycol

18.43 Following are structural formulas for each starting material.

(a) $CH_3(CH_2)_4COH$ (b)

(c)

Chapter 19 Carboxylic Anhydrides, Esters, and Amides

19.1

(a) CH_3CNH—⬡   (b) ⬡—CNH_2

19.2 Under basic conditions, as in part (a), each carboxyl group is present as a carboxylic anion. Under acidic conditions, as in part (b), each carboxyl group is present in its un-ionized form.

(a)   + 2NaOH

          + 2CH₃OH

(b)   + H₂O

          + HO⌐

19.3 In aqueous NaOH, each carboxyl group is present as a carboxylic anion, and each amine is present in its unprotonated form.

(a) $CH_3CN(CH_3)_2$ + NaOH $\xrightarrow[\text{heat}]{H_2O}$ $CH_3CO^-Na^+$ + $(CH_3)_2NH$

(b)   + NaOH $\xrightarrow[\text{heat}]{H_2O}$ H₂N⌐⌐⌐CO⁻Na⁺

19.5 (a) benzoic anhydride   (b) methyl decanoate
(c) N-methylhexanamide
(d) 4-aminobenzamide or p-aminobenzamide
(e) cyclopentyl ethanoate or cyclopentyl acetate
(f) ethyl 3-hydroxybutanoate
19.7 Each reaction brings about hydrolysis of the amide. Each product is shown as it would exist under the specified reaction conditions.

(a)   —CNH₂ + NaOH

$\xrightarrow{H_2O}$ —CO⁻Na⁺ + NH₃

(b)   —CNH₂ + HCl

$\xrightarrow{H_2O}$ —COH + NH₄⁺ + Cl⁻

19.9 Each reaction gives an amide and a carboxylic acid.

(a)

(b)

19.11 (a) Phenobarbital contains four amide bonds.

(b) The products of hydrolysis are shown as they would be ionized in aqueous NaOH.

Sodium carbonate
(Na_2CO_3)

19.13 Nylon-66 and Kevlar are referred to as polyamides because the repeating functional group in the polymer chain of each is an amide group.

19.15 Dacron and Mylar are referred to as polyesters because the repeating functional group in the polymer chain is an ester group.

19.17

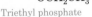

Ethyl phosphate   Diethyl phosphate

Triethyl phosphate

19.19 Two molecules of water are split out.

19.21 The yellow box encloses the COO of the ester group. The structural formula of chrysanthemic acid is shown below.

Pyrethrin I

Chrysanthemic acid

19.23 (a) The *cis/trans* ratio refers to the *cis-trans* relationship between the ester group and the four-carbon chain on the cyclopropane ring. Specifically, the repellant in the commercial preparation consists of a minimum of 35% of the *cis* isomer and a maximum of 75% of the *trans* isomer.
(b) Permethrin has two stereocenters, and four stereoisomers (two pairs of enantiomers) are possible.
The designation "(+/−)" refers to the fact that the *cis* isomer is present as a racemic mixture, as is the *trans* isomer.

19.25 The compound is salicin. Removal of the glucose unit and oxidation of the primary alcohol to a carboxyl group gives salicylic acid.

19.27 The moisture present in humid air may be sufficient to bring about hydrolysis of the ester group in aspirin to give salicylic acid and acetic acid. The vinegar-like odor is due to the presence of acetic acid.

19.29 A *sunblock* prevents all ultraviolet radiation from reaching protected skin by reflecting it away from the skin. A *sunscreen* prevents a portion of the ultraviolet radiation from reaching protected skin. Its effectiveness is related to its skin protection factor (SPF).

19.31 They all contain an ester group bonded to an alkyl group and a benzene ring. The benzene ring has either a nitrogen atom or an oxygen atom on it.

19.33 Lactomer stitches dissolve when the ester groups in the polymer chains are hydrolyzed until only glycolic and lactic acids remain. These small molecules are metabolized and excreted by existing biochemical pathways.

19.35

4-Aminophenol   Acetic anhydride

Acetaminophen   Acetic acid

19.37 Among the hydrolysis products, the carboxyl group is present as its anion and phosphoric acid is present as its dianion.

1,3-Diphosphoglycerate

19.39 Following is a short section of the polyamide formed in this type of polymerization.

$$-NH-CH-\overset{\overset{\displaystyle O}{\|}}{C}-NH-CH-\overset{\overset{\displaystyle O}{\|}}{C}-NH-CH-\overset{\overset{\displaystyle O}{\|}}{C}-$$
$$\qquad \underset{CH_3}{|} \qquad\qquad \underset{CH_3}{|} \qquad\qquad \underset{CH_3}{|}$$

19.41 Following is the combination of carboxylic acid and amine from which each amide can be synthesized.

(a)   ⬡—NH₂ + HOC(CH₂)₄CH₃
(with C=O double bond shown on the HOC group)

(b)   CH₃CHCOH + HN(CH₃)₂
        |
        CH₃
(with C=O double bond shown)

(c)   NH₃ + HOC(CH₂)₄COH + NH₃
(with both C=O double bonds shown)

19.43 (a) Both lidocaine and mepivicaine contain an amide group and a tertiary aliphatic amine.
(b) Both local anesthetics share the structural features shown in the box.

Lidocaine
(Xylocaine)

Mepivacaine
(Carbocaine)

Structural features
common to both
local anesthetics

Chapter 20   Carbohydrates

20.1 Following are Fischer projections for the four 2-ketopentoses. They consist of two pairs of enantiomers.

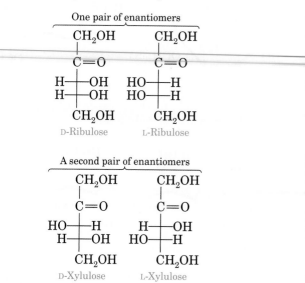

20.2 D-Mannose differs in configuration from D-glucose only at carbon 2. One way to arrive at the structures of the α and β forms of D-mannopyranose is to draw the corresponding α and β forms of D-glucopyranose, and then invert the configuration in each at carbon 2.

β-D-Mannopyranose
(β-D-Mannose)

α-D-Mannopyranose
(α-D-Mannose)

20.3 D-Mannose differs in configuration from D-glucose only at carbon 2.

β-D-Mannopyranose
(β-D-Mannose)

α-D-Mannopyranose
(α-D-Mannose)

20.4 Following is a Haworth projection and a chair conformation for this glycoside.

20.5 The β-glycosidic bond is between carbon 1 of the left unit and carbon 3 of the right unit.

20.7 The carbonyl group in an aldose is an aldehyde. In a ketose, it is a ketone.
20.9 The three most abundant hexoses in the biological world are D-glucose, D-galactose, and D-fructose. The first two are aldohexoses. The third is a 2-ketohexose.
20.11 To say that they are enantiomers means that they are nonsuperposable mirror images.
20.13 The D or L configuration in an aldopentose is determined by its configuration at carbon 4.
20.15 Compounds (a) and (c) are D-monosaccharides. Compound (b) is an L-monosaccharide.
20.17 A 2-ketoheptose has four stereocenters and 16 possible stereoisomers. Eight of these are D-2-ketoheptoses and eight are L-2-ketoheptoses. Following is one of the eight possible D-2-ketoheptoses.

20.19 In an amino sugar, one or more —OH groups are replaced by —NH_2 groups. The three most abundant amino sugars in the biological world are D-glucosamine, D-galactosamine, and N-acetyl-D-glucosamine.
20.21 (a) A pyranose is a six-membered cyclic hemiacetal form of a monosaccharide.
(b) A furanose is a five-membered cyclic hemiacetal form of a monosaccharide.
20.23 Yes, they are anomers. No, they are not enantiomers; that is, they are not mirror images. They differ in configuration only at carbon 1 and, therefore, are diastereomers.
20.25 A Haworth projection shows the six-membered ring as a planar hexagon. In reality, the ring is puckered and its most stable conformation is a chair conformation with all bond angles approximately 109.5°.

20.27 Compound (a) differs from D-glucose only in the configuration at carbon 4. Compound (b) differs only at carbon 3.

20.29 The specific rotation of α-L-glucose is −112.2°.
20.31 A glycoside is a cyclic acetal of a monosaccharide. A glycosidic bond is the bond from the anomeric carbon to the —OR group of the glycoside.
20.33 No, glycosides cannot undergo mutarotation because the anomeric carbon is not free to interconvert between α and β configurations via the open-chain aldehyde or ketone.
20.35 Following are Fischer projections of D-glucose and D-sorbitol. The configurations at the four stereocenters of D-glucose are not affected by this reduction.

20.37 Ribitol is the reduction product of D-ribose. β-D-ribose 1-phosphate is the phosphoric ester of the OH group on the anomeric carbon of β-D-ribofuranose.

20.39 To say that it is a β-1,4-glycosidic bond means that the configuration at the anomeric carbon (carbon 1 in this problem) of the monosaccharide unit forming the glycosidic bond is beta and that it is bonded to carbon 4 of the second monosaccharide unit. To say that it is an α-1,6-glycosidic bond means that the configuration at the anomeric carbon

(carbon 1 in this problem) of the monosaccharide unit forming the glycosidic bond is alpha and that it is bonded to carbon 6 of the second monosaccharide unit.

20.41 (a) Both monosaccharide units are D-glucose.
(b) They are joined by a β-1,4-glycosidic bond.
(c) It is a reducing sugar and
(d) it undergoes mutarotation.

20.43 An oligosaccharide contains approximately six to ten monosaccharide units. A polysaccharide contains more—generally many more—than ten monosaccharide units.
20.45 The difference lies in the degree of chain branching. Amylose is composed of unbranched chains, whereas amylopectin is a branched network with the branches started by β-1,6-glycosidic bonds.
20.47 Cellulose fibers are insoluble in water because the strength of hydrogen bonding of a cellulose molecule in the fiber with surface water molecules is not sufficient to overcome the intermolecular forces that hold it in the fiber.
20.49 (a) In these structural formulas, the CH_3CO (the acetyl group) is abbreviated Ac.

Following are Haworth and chair structures for this repeating disaccharide.

20.51 Its lubricating power decreases.
20.53 With maturation, children develop an enzyme capable of metabolizing galactose. Thus they are able to tolerate galactose as they mature.
20.55 L-Ascorbic acid is oxidized (there is loss of two hydrogen atoms) when it is converted to L-dehydroascorbic acid. L-Ascorbic acid is a biological reducing agent.
20.57 Types A, B, and O have in common D-galactose and L-fucose. Only type A has N-acetyl-D-glucosamine.
20.59 Mixing types A and B blood will result in coagulation.
20.61 Consult Table 20.1 for the structural formula of D-altrose and draw it. Then replace the —OH groups on carbons 2 and 6 with hydrogens.

20.63 The monosaccharide unit of salicin is D-glucose.
20.65 The intermediate in this transformation is an enediol formed by keto-enol tautomerism of dihydroxyacetone phosphate. Keto-enol tautomerism of this intermediate gives D-glyceraldehyde-3-phosphate.

20.67 (a) Coenzyme A is chiral. It has five stereocenters.
(b) The functional groups, starting from the left, are a thiol (—SH), two amides, a secondary alcohol, a phosphate ester, a phosphate anhydride, a phosphate ester, another phosphate ester, a unit of 2-deoxyribose, and a β-glycosidic bond to adenine, a heterocyclic amine.
(c) Yes, it is soluble in water because of the presence of a number of polar C=O groups, one —OH group and three phosphate groups, all of which will interact with water molecules by hydrogen bonding.

(d) Following are the products of hydrolysis of all amide, ester, and glycosidic bonds.

2-Amino-ethanethiol   4-Aminobutanoic acid (β-Alanine)

2,3-Dihydroxy-3-methyl-butanoic acid

Adenine   β-2-Deoxy-D-ribose   Dihydrogen phosphate

Chapter 21   Lipids

21.1 (a) It is an ester of glycerol and contains a phosphate group; therefore it is a glycerophospholipid. Besides glycerol and phosphate, it has a myristic acid and a linoleic acid component. The other alcohol is serine. Therefore, it belongs to the subgroup of cephalins.
(b) The components present are glycerol, myristic acid, linoleic acid, phosphate, and serine.
21.3 *Hydrophobic* means "water hating." If the body did not have such molecules, there could be no structure because the water would dissolve everything.
21.5 The melting point would increase. The *trans* double bonds would fit more in the packing of the long hydrophobic tails, creating more order and therefore more interaction between chains. This would require more energy to disrupt, and hence a higher melting point.
21.7 The diglycerides with the highest melting points will be the ones with two stearic acids (a saturated fatty acid). The lowest melting points will be the ones with two oleic acids (a monounsaturated fatty acid).
21.9 (b), because its molecular weight is higher.
21.11 lowest (c); then (b); highest (a)
21.13 The more long-chain groups, the lower the solubility; lowest (a); then (b); highest (c).
21.15 glycerol, sodium palmitate, sodium stearate, sodium linolenate
21.17 Complex lipids can be classified into two groups: phospholipids and glycolipids. Phospholipids contain an alcohol, two fatty acids, and a phosphate group. There are two types: glycerophospholipids and sphingolipids. In glycerophospholipids, the alcohol is glycerol. In sphingolipids, the alcohol is sphingosine. Glycolipids are complex lipids that contain carbohydrates.

21.19 The presence of *cis* double bonds in fatty acids produces greater fluidity because they cannot pack together as closely as saturated fatty acids.
21.21 Integral membrane proteins are embedded in the membrane. Peripheral membrane proteins are found on membrane surfaces.
21.23 A phosphatidyl inositol containing oleic acid and arachidonic acid:

21.25 Complex lipids that contain ceramides include sphingomyelin, sphingolipids, and the cerebroside glycolipids.
21.27 The hydrophilic functional groups of (a) glucocerebroside: carbohydrate; hydroxyl and amide groups of the cerebroside. (b) Sphingomyelin: phosphate group; choline; hydroxyl and amide of ceramide.
21.29 Cholesterol crystals may be found in (1) gallstones, which are sometimes pure cholesterol, and (2) joints of people suffering from bursitis.
21.31 Carbon 17 of the steroid D ring undergoes the most substitution.
21.33 LDL from the bloodstream enters the cells by binding to LDL receptor proteins on the surface. After binding, the LDL is transported inside the cells, where cholesterol is released by enzymatic degradation of the LDL.
21.35 Removing lipids from the triglyceride cores of VLDL particles increases the density of the particles and converts them from VLDL to LDL particles.
21.37 When serum cholesterol concentration is high, the synthesis of cholesterol in the liver is inhibited and the synthesis of LDL receptors in the cell is increased. Serum cholesterol levels control the formation of cholesterol in the liver by regulating enzymes that synthesize cholesterol.
21.39 Estradiol (E) is synthesized from progesterone (P) through the intermediate testosterone (T). First the D-ring acetyl group of P is converted to a hydroxyl group and T is produced. The methyl group in T, at the junction of the rings A and B, is removed and ring A becomes aromatic. The keto group in P and T is converted to a hydroxyl group in E.
21.41 Steroid structures are shown in Section 21.10. The major structural differences are at carbon 11. Progesterone has no substituents except hydrogen, cortisol has a hydroxyl group, cortisone has a keto group, and RU-486 has a large *p*-aminophenyl group. The functional group at carbon 11 apparently has little importance in receptor binding.
21.43 They have a steroid ring structure, they have a methyl group at carbon 13, they have a triply bonded group at carbon 17, and all have some unsaturation in the A ring, the B ring, or both.

21.45 Bile salts help solubilize fats. They are oxidation products of cholesterol themselves, and they bind to cholesterol, forming complexes that are eliminated in the feces.
21.47 (a) Glycocholate:

(b) Cortisone:

(c) PGE₂:

(d) Leukotriene B4:

21.49 Aspirin slows the synthesis of thromboxanes by inhibiting the COX enzyme. Because thromboxanes enhance the blood clotting process, the result is that strokes caused by blood clots in the brain will occur less often.
21.51 Waxes consist primarily of esters of long-chain saturated acids and alcohols. Because of the saturated components, wax molecules pack more tightly than those of triglycerides, which frequently have unsaturated components.

21.53 The transporter is a helical transmembrane protein. The hydrophobic groups on the helices are turned outward and interact with the membrane. The hydrophilic groups of the helices are on the inside and interact with the hydrated chloride ions.
21.55 (a) Sphingomyelin acts as an insulator.
(b) The insulator is degraded, impairing nerve conduction.
21.57 α-D-galactose, β-D-glucose, β-D-glucose
21.59 They prevent ovulation.
21.61 It inhibits prostaglandin formation by preventing ring closure.
21.63 NSAIDs inhibit cyclooxygenases (COX enzymes) that are needed for ring closure. Leukotrienes have no ring in their structure; therefore they are not affected by COX inhibitors.
21.65 (See Figure 21.2.) Polar molecules cannot penetrate the bilayer. They are insoluble in lipids. Nonpolar molecules can interact with the interior of the bilayer ("like dissolves like").
21.67 Both groups are derived from a common precursor, PGH₂, in a process catalyzed by the COX enzymes.
21.69 Coated pits are concentrations of LDL receptors on the surface of cells. They bind LDL and by endocytosis transfer it inside the cell.
21.71 In facilitated transport, a membrane protein assists in the movement of a molecule through the membrane with no requirement for energy. In active transport, a membrane protein assists in the process, but energy is required. ATP hydrolysis usually supplies the needed energy.
21.73 Aldosterone has an aldehyde group at carbon 13.
21.75 The formula weight of the triglyceride is about 850 g/mol. This is 0.12 mol (100 g ÷ 850 g/mol = 0.12 mol). One mole of hydrogen is required for each mole of double bonds in the triglyceride. There are three double bonds, so the moles of hydrogen required for each 100 g = 0.12 mol × 3 = 0.36 mol of hydrogen gas. Converting to grams of hydrogen, 0.36 × 2 g/mol = 0.72 g hydrogen gas.
21.77 Both lipids and carbohydrates contain carbon, hydrogen, and oxygen. Carbohydrates have aldehyde and ketone groups, as do some steroids. Carbohydrates have a number of hydroxyl groups, which lipids do not have to a great extent. Lipids have major components that are hydrocarbon in nature. These structural features imply that carbohydrates tend to be significantly more polar than lipids.
21.79 primarily lipid: olive oil and butter; primarily carbohydrate: cotton and cotton candy
21.81 The amounts are the key point here. Large amounts of sugar can provide energy. Fat burning due to the presence of taurine plays a relatively minor role because of the small amount.
21.83 The other ends of the molecules involved in the ester linkages in lipids, such as fatty acids, tend not to form long chains of bonds with other molecules.
21.85 The bulkier molecules tend to be found on the exterior of the cell because the curvature of the cell membrane provides more room for them.
21.87 The charges tend to cluster on membrane surfaces. Positive and negative charges attract each other. Two positive or two negative charges repel each other, so unlike charges do not have this repulsion.

Chapter 22 Proteins

22.1

Valylphenylalanine
(Val-Phe)

22.2 a salt bridge
22.3 (a) storage (b) movement
22.5 protection
22.7 Tyrosine has an additional hydroxyl group on the phenyl side chain.
22.9 arginine
22.11

Pyrrolidines
(heterocyclic
aliphatic amines)

22.13 They supply most of the amino acids we need in our bodies.
22.15 These structures are similar except that one of the hydrogens in the side chain of alanine has been replaced with a phenyl group in phenylalanine.
22.17 Amino acids are zwitterions; therefore they all have positive and negative charges. These molecules are very strongly attracted to each other, so they are solids at low temperatures.
22.19 All amino acids have a carboxyl group with a pK_a around 2 and an amino group with a pK_a between 8 and 10. One group is significantly more acidic and one is more basic. To have an un-ionized amino acid, the hydrogen would have to be on the carboxyl group and have vacated the amino group. Given that the carboxyl group is the stronger acid, this would never happen.
22.21 the side-chain imidazole
22.23 The side chain of histidine is an imidazole with a nitrogen that reversibly binds to a hydrogen. When dissociated, it is neutral; when associated, it is positive. Therefore, chemically it is a base, even though it does have a pK_a in the acidic range.
22.25 histidine, arginine, and lysine

22.27 Serine may be obtained by the hydroxylation of alanine. Tyrosine is obtained by the hydroxylation of phenylalanine.
22.29

Alanylglutamine
(Ala-Gln)

Glutaminylalanine
(Gln-Ala)

22.31

22.33 Only the peptide backbone contains polar units.
22.35 (a)

(b) pH 2 is shown above. At pH 7.0 would look like:

(c) At pH 10:

22.37 It would acquire a net positive charge and become more water-soluble.
22.39 (a) 256 (b) 160,000
22.41 valine or isoleucine
22.43 (a) secondary (b) quaternary (c) quaternary (d) primary
22.45 Above pH 6.0, the COOH groups are converted to COO⁻ groups. The negative charges repel each other, disrupting the compact α-helix and converting it to a random coil.
22.47 (1) C-terminal end (2) N-terminal end (3) pleated sheet (4) random coil (5) hydrophobic interaction (6) disulfide bridge (7) α-helix (8) salt bridge (9) hydrogen bonds
22.49 (a) Fetal hemoglobin has fewer salt bridges between the chains. (b) Fetal hemoglobin has a higher affinity for oxygen.
22.51 The heme and the polypeptide chain form the quaternary structure of cytochrome c. This is a conjugated protein.
22.53 the intramolecular hydrogen bonds between the peptide backbone carbonyl group and the N—H group.
22.55 cysteine
22.57 Ions of heavy metals like silver denature bacterial proteins by reacting with cysteine —SH groups. The proteins, denatured by formation of silver salts, form insoluble precipitates.
22.59 Glutathione is a tripeptide of the amino acids Glu, Cys, and Gly. When glutathione is oxidized, two of the molecules react to form a disulfide bond between the two cysteine side chains:

22.61 The symptoms of hunger, sweating, and poor coordination accompany diabetes when hypoglycemia is coming on.
22.63 The abnormal form has a higher percentage of β-pleated sheet compared to the normal form.
22.65 The oxygen-binding behavior of myoglobin is hyperbolic, while that of hemoglobin is sigmoidal.
22.67 The two most common are prion diseases and Alzheimer's disease.
22.69 Even if it is feasible, it is not completely correct to call the imaginary process that converts α-keratin to β-keratin "denaturation." Any process that changes a protein from α to β requires at least two steps: (1) conversion

from the α form to a random coil, and (2) conversion from the random coil to the β form. The term "denaturation" describes only the first half of the process (Step 1). The second step would be called "renaturation." The overall process is called denaturation followed by renaturation. If we assume that the imaginary process actually occurs without passing through a random coil, then the term "denaturation" does not apply.
22.71 a quaternary structure because subunits are cross-linked
22.73 (a) hydrophobic (b) salt bridge (c) hydrogen bond (d) hydrophobic
22.75 glycine
22.77 one positive charge on the amino group
22.79 These amino acids have side chains that can catalyze organic reactions. They are polar or sometimes charged, and the ability to make hydrogen bonds or salt bridges can help catalyze the reaction.
22.81 Proteins can be denatured when the temperature is only slightly higher than a particular optimum. For this reason, the health of a warm-blooded animal is dependent on the body temperature. If the temperature is too high, proteins could denature and lose function.
22.83 Even if you know all of the genes in an organism, not all the genes code for proteins, nor are all genes expressed all the time.

Chapter 23   Enzymes

23.1 A catalyst is any substance that speeds up the rate of a reaction and is not itself changed by the reaction. An enzyme is a biological catalyst, which is either a protein or an RNA molecule.
23.3 Yes. Lipases are not very specific
23.5 because enzymes are very specific and thousands of reactions must be catalyzed in an organism
23.7 Lyases add water across a double bond or are removed from a molecule, thereby generating a double bond. Hydrolases use water to break an ester or amide bond, thereby generating two molecules.
23.9 (a) isomerase (b) hydrolase (c) oxidoreductase (d) lyase
23.11 *Cofactor* is more generic; it means a nonprotein part of an enzyme. A *coenzyme* is an organic cofactor.
23.13 In reversible inhibition, the inhibitor can bind and then be released. With noncompetitive inhibition, once the inhibitor is bound, no catalysis can occur. With irreversible inhibition, once the inhibitor is bound, the enzyme would be effectively dead, as the inhibitor could not be removed and no catalysis could occur.
23.15 No, at high substrate concentration the enzyme surface is saturated, and doubling of the substrate concentration will produce only a slight increase in the rate of the reaction or no increase at all.
23.17 (a) less active at normal body temperature (b) The activity decreases.
23.19 (a)

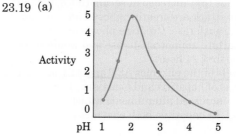

(b) 2

(c) zero activity

23.21 Trypsin is a digestive enzyme that catalyzes the hydrolysis of peptide bonds in dietary proteins in the small intestines. The enzyme is produced in the pancreas in an inactive, zymogen form. If it were produced here in its active form, it would degrade proteins that are essential for normal cell function. It is converted to active trypsin when it is secreted into the small intestines.

23.23 The amino acid residues most often found at enzyme active sites are His, Cys, Asp, Arg, and Glu.

23.25 The correct answer is (c). Initially the enzyme does not have exactly the right shape for strongly binding a substrate, but the shape of the active site changes to better accommodate the substrate molecule.

23.27 Amino acid residues in addition to those at an enzyme active site are present to help form a three-dimensional pocket where the substrate binds. These amino acids act to make the size, shape, and environment (polar or nonpolar) of the active site just right for the substrate.

23.29 Caffeine is an allosteric regulator.

23.31 There is no difference. They are the same.

23.33

23.35

23.37 Glycogen phosphorylase is controlled by allosteric regulation and by phosphorylation. The allosteric controls are very fast, so that when the level of ATP drops, for example, there is an immediate response to the enzyme allowing more energy to be produced. The covalent modification by phosphorylation is triggered by hormone responses. They are a bit slower, but more long lasting and ultimately more effective.

23.39 Just as with lactate dehydrogenase, there are five isozymes of PFK: M_4, M_3L, M_2L_2, ML_3, and L_4.

23.41 Two enzymes that increase in serum concentration following a heart attack are creatine phosphokinase and aspartate aminotransferase. Creatine phosphokinase peaks earlier than aspartate aminotransferase and would be the best choice in the first 24 hours.

23.43 Serum levels of the enzymes AST and ALT are monitored for diagnosis of hepatitis and heart attack. Serum levels of AST are increased after a heart attack, but ALT levels are normal. In hepatitis, both enzymes are elevated.

The diagnosis, until further testing, would indicate the patient may have had a heart attack.

23.45 Chemicals present in organic vapors are detoxified in the liver. The enzyme alkaline phosphatase is monitored to diagnose liver problems.

23.47 It is not possible to administer chymotrypsin orally. The stomach would treat it just as it does all dietary proteins: degrade it by hydrolysis to free amino acids. Even if whole, intact molecules of the enzyme were present in the stomach, the low pH in the region would not allow activity for the enzyme, which prefers an optimal pH of 7.8.

23.49 A transition-state analog is built to mimic the transition state of the reaction. It is not the same shape as the substrate or the product, but rather something in between. The potency with which such analogs can act as inhibitors lends credence to the theory of induced fit.

23.51 Succinylcholine has a chemical structure similar to that of acetylcholine, so both can bind to the acetylcholine receptor of the muscle end plate. The binding of either choline causes a muscle contraction. However, the enzyme acetylcholinesterase hydrolyzes succinylcholine only very slowly compared to acetylcholine. Muscle contraction does not occur as long as succinylcholine is still present and acts as a relaxant.

23.53 The bacterium that causes stomach ulcers, *H. pylori*, protects itself from stomach acid using the enzyme urease. This enzyme catalyzes the hydrolysis of urea, forming ammonia, a weak base. The chemical process continually bathes the bacterial cells with ammonia that neutralizes the stomach acid.

23.55 Antibiotics such as amoxillin and tetracycline are used to treat ulcers caused by *H. pylori*.

23.57 In the enzyme pyruvate kinase, the $=CH_2$ of the substrate phosphoenolpyruvate sits in a hydrophobic pocket formed by the amino acids Ala, Gly, and Thr. The methyl group on the side chain of Thr, rather than the hydroxyl group, is in the pocket. Hydrophobic interactions are at work here to hold the substrate into the active site.

23.59 Researchers were trying to inhibit phosphodiesterases because cGMP acts to relax constricted blood vessels. This approach was hoped to help treat angina and high blood pressure.

23.61 Phosphorylase exists in a phosphorylated form and an unphosphorylated form, with the former being more active. Phosphorylase is also controlled allosterically by several compounds, including AMP and glucose. While the two act semi-independently, they are related to some degree. The phosphorylated form has a higher tendency to assume the R state, which is more active, and the unphosphorylated form has a higher tendency to be in the less active T state.

23.63 In the processing of cocaine by specific esterase enzymes, the cocaine molecule passes through an intermediate state. A molecule was designed that mimics this transition state. This transition-state analog can be given to a host animal, which then produces antibodies to the analog. When these antibodies are given to a person, they act like an enzyme and degrade cocaine.

23.65 Cocaine blocks the reuptake of the neurotransmitter dopamine, leading to overstimulation of the nervous system.

23.67 (a) Vegetables such as green beans, corn, and tomatoes are heated to kill microorganisms before they are preserved by canning. Milk is preserved by the heating process, pasteurization. (b) Pickles and sauerkraut are preserved by storage in vinegar (acetic acid).

23.69 The amino acid residues (Lys and Arg) that are cleaved by trypsin have basic side chains; thus they are positively charged at physiological pH.

23.71 This enzyme works best at a pH of about 7.

23.73 a hydrolase

23.75 (a) The enzyme is called ethanol dehydrogenase or, more generally, alcohol dehydrogenase. It could also be called ethanol oxidoreductase.

(b) ethyl acetate esterase or ethyl acetate hydrolase

23.77 isozymes or isoenzymes

23.79 a 10 km race

23.81 The structure of RNA makes it more likely to be able to adopt a wider range of tertiary structures, so it can fold up to form globular molecules such as protein-based enzymes. It also has an extra oxygen, which gives it an additional reactive group to use in catalysis or an electronegative group, useful in hydrogen bonding.

Chapter 24 Chemical Communications: Neurotransmitters and Hormones

24.1 G-protein is an enzyme; it catalyzes the hydrolysis of GTP to GDP. GTP, therefore, is a substrate.

24.3 A chemical messenger operates between cells; secondary messengers signal inside a cell in the cytoplasm.

24.5 The concentration of Ca^{2+} in neurons controls the process. When it reaches 10^{-4} M, the vesicles release the neurotransmitters into the synapse.

24.7 anterior pituitary gland

24.9 Upon binding of acetylcholine, the conformation of the proteins in the receptor changes and the central core of the ion channel opens.

24.11 The cobra toxin causes paralysis by acting as a nerve system antagonist. It blocks the receptor and interrupts the communication between neuron and muscle cell. The botulin toxin prevents the release of acetylcholine from presynaptic vesicles.

24.13 Taurine is a β-amino acid; its acidic group is —SO_2OH instead of —COOH.

24.15 The amino group in GABA is in the gamma position; proteins contain only alpha amino acids.

24.17 (a) norepinephrine and histamine (b) They activate a secondary messenger, cAMP, inside the cell. (c) amphetamines and histidine

24.19 It is phosphorylated by an ATP molecule.

24.21 Product from the MAO-catalyzed oxidation of dopamine:

24.23 (a) Amphetamines increase and (b) reserpine decreases the concentration of the adrenergic neurotransmitter.

24.25 the corresponding aldehyde

24.27 (a) the ion-translocating protein itself

(b) It gets phosphorylated and changes its shape.

(c) It activates the protein kinase that does the phosphorylation of the ion-translocating protein.

24.29 They are pentapeptides.

24.31 The enzyme is a kinase. The reaction is the phosphorylation of inositol-1,4-diphosphate to inositol-1,4,5-triphosphate:

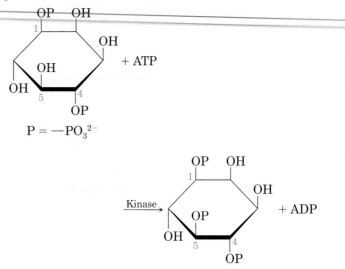

$$P = —PO_3^{2-}$$

24.33 Most receptors for steroid hormones are located in the cell nucleus.

24.35 In the brain, steroid hormones can act as neurotransmitters.

24.37 Calmodulin, a calcium-ion-binding protein, activates protein kinase II, which catalyzes phosphorylation of other proteins. This process transmits the calcium signal to the cell.

24.39 (a) Nerve gases cause a variety of effects: sweating, dizziness, vomiting, convulsions, respiratory failure, and often death.

(b) Many nerve gases bind covalently and irreversibly to acetylcholine esterase, permanently inactivating the enzyme. Acetylcholine is not degraded, and the neurons fire without pause.

24.41 Local injections of the toxin prevent release of acetylcholine in that area. Facial movements are eliminated, lessening the appearance of lines and wrinkles.

24.43 The plaques are caused by the buildup and deposition of β-amyloid proteins in the brain.

24.45 The neurotransmitter dopamine is deficient in Parkinson's disease, but a dopamine pill would not be an effective treatment. Dopamine cannot cross the blood-brain barrier.

24.47 Drugs like Cogentin that block cholinergic receptors are often used to treat the symptoms of Parkinson's disease. These drugs lessen spastic motions and tremors.

24.49 Nitric oxide relaxes the smooth muscle that surrounds blood vessels. This relaxation causes increased blood flow in the brain, which in turn causes headaches.

24.51 Neurons adjacent to those damaged by the stroke begin to release glutamate and NO, which kills other cells in the region.

24.53 Insulin-dependent (type 1) diabetes is caused by insufficient production of insulin by the pancreas. Administration of insulin relieves symptoms of this type of diabetes. Non-insulin-dependent (type 2) diabetes is caused by a deficiency of insulin receptors or by the presence of inactive insulin receptors. Other drugs are used to relieve symptoms.

24.55 Monitoring glucose in the tears relieves the patient of taking many blood samples every day.

24.57 Aldosterone binds to a specific receptor in the nucleus. The aldosterone–receptor complex serves as a transcription factor that regulates gene expression. Proteins for mineral metabolism are produced as a result.

24.59 Large doses of acetylcholine will help. Decamethonium bromide is a competitive inhibitor of acetylcholine esterase. The inhibitor can be removed by increasing substrate concentration.

24.61 Alanine is an α-amino acid in which the amino group is bonded to the same carbon as the carboxyl group. In β-alanine, the amino group is bonded to the carbon adjacent to the one to which the carboxyl group is bonded.

$$CH_3-\overset{\alpha}{CH}-COO^- \qquad \overset{\beta}{CH_2}-CH_2-COO^-$$
$$\overset{|}{NH_3^+} \qquad\qquad \overset{|}{NH_3^+}$$

Alanine (an α-amino acid) β-Alanine (a β-amino acid)

24.63 Effects of NO on smooth muscle are as follows: dilation of blood vessels and increased blood flow; headaches caused by dilation of blood vessels in brain; increased blood flow in the penis, leading to erections.

24.65 Acetylcholine esterase catalyzes the hydrolysis of the neurotransmitter acetylcholine to produce acetate and choline. Acetylcholine transferase catalyzes the synthesis of acetylcholine from acetyl—CoA and choline.

24.67 The reaction shown below is the hydrolysis of GTP:

24.69 Ritalin increases serotonin levels. Serotonin has a calming effect on the brain. One advantage of this drug is that it does not increase levels of the stimulant dopamine.

24.71 Proteins are capable of specific interactions at recognition sites. This ability makes for useful selectivity in receptors.

24.73 Adrenergic messengers, such as dopamine, are derivatives of amino acids. For example, a biochemical pathway exists that produces dopamine from the amino acid tyrosine.

24.75 Insulin is a small protein. It would go through protein digestion if taken orally and would not be taken up as the whole protein.

24.77 Steroid hormones directly affect nucleic acid synthesis.

24.79 Chemical messengers vary in their response times. Those that operate over short distances, such as neurotransmitters, have short response times. Their mode of action frequently consists of opening or closing channels in a membrane or binding to a membrane-bound receptor. Hormones must be transmitted in the bloodstream, which requires a longer time for them to take effect. Some hormones can and do affect protein synthesis, which makes the response time even longer.

24.81 Having two different enzymes for the synthesis and breakdown of acetylcholine means that the rates of formation and breakdown can be controlled independently.

Chapter 25   Nucleotides, Nucleic Acids, and Heredity

25.1

25.3 hemophilia, sickle cell anemia, etc.

25.5 (a) In eukaryotic cells, DNA is located in the cell nucleus and in mitochondria. (b) RNA is synthesized from DNA in the nucleus, but further use of RNA (protein synthesis) occurs on ribosomes in the cytoplasm.

25.7 DNA has the sugar deoxyribose, while RNA has the sugar ribose. Also, RNA has uracil, while DNA has thymine.

25.9 Thymine and uracil are both based on the pyrimidine ring. However, thymine has a methyl substituent at carbon 5, whereas uracil has a hydrogen. All of the other ring substituents are the same.

25.11

Cytidine

"Deoxy" because there is no oxygen at carbon 2'

Deoxycytidine

25.13 D-Ribose and 2-deoxy-D-ribose have the same structure except at carbon 2. D-Ribose has a hydroxyl group and hydrogen on carbon 2, whereas deoxyribose has two hydrogens.

D-Ribose   2-Deoxy-D-ribose

25.15 The name "nucleic acid" derives from the fact that the nucleosides are linked by phosphate groups, which are the dissociated form of phosphoric acid.

25.17 anhydride bonds

25.19 In RNA, carbons 3′ and 5′ of the ribose are linked by ester bonds to phosphates. Carbon 1 is linked to the nitrogen base with an N-glycosidic bond.

25.21 (a)

(b)

25.23 (a) One end will have a free 5′ phosphate or hydroxyl group that is not in phosphodiester linkage. That end is called the 5′ end. The other end, the 3′ end, will have a 3′ free phosphate or hydroxyl group.
(b) By convention, the end drawn to the left is the 5′ end. A is the 5′ end, and C is the 3′ end.

25.25 two

25.27 electrostatic interactions

25.29 The superstructure of chromosomes consists of many elements. DNA and histones combine to form nucleosomes that are wound into chromatin fibers. These fibers are further twisted into loops and minibands to form the chromosome superstructure.

25.31 the double helix

25.33 DNA is wound around histones, collectively forming nucleosomes that are further wound into solenoids, loops, and bands.

25.35 rRNA

25.37 mRNA

25.39 Ribozymes, or catalytic forms of RNA, are involved in post-transcriptional splicing reactions that cleave larger RNA molecules into smaller, more active forms. For example, tRNA molecules are formed in this way.

25.41 Small nuclear RNA is involved in splicing reactions of other RNA molecules.

25.43 Micro RNAs are 22 bases long and prevent transcription of certain genes. Small interfering RNAs vary from 22 to 30 bases and are involved in the degradation of specific mRNA molecules.

25.45 Immediately after transcription, messenger RNA contains both introns and exons. The introns are cleaved out by the action of ribozymes that catalyze splicing reactions on the mRNA.

25.47 no

25.49 the specificity between the base pairs, A-T and G-C

25.51

Thymine   Adenine
AT pair

25.53 four

25.55 In semiconservative DNA replication, the new daughter DNA helix is composed of one strand from the original (or parent) molecule and one new strand.

25.57

$$\text{Histone}-(CH_2)_4-NH_3^+ + CH_3-COO^- \underset{\text{deacetylation}}{\overset{\text{acetylation}}{\rightleftharpoons}}$$

$$\text{Histone}-(CH_2)_4-NH-\overset{O}{\overset{\|}{C}}-CH_3$$

25.59 Helicases are enzymes that break the hydrogen bonds between the base pairs in double-helix DNA and thus help the helix to unwind. This prepares the DNA for the replication process.

25.61 pyrophosphate

25.63 The leading strand or continuous strand is synthesized in the 5′ to 3′ direction.

25.65 DNA ligase

25.67 From the 5′ to the 3′ direction

25.69 One of the enzymes involved in the DNA base excision repair (BER) pathway is an endonuclease that catalyzes the hydrolytic cleavage of the phosphodiester backbone. The enzyme hydrolyzes on the 5′ side of the AP site.

25.71 a β-N-glycosidic bond between the damaged base and the deoxyribose

25.73 Individuals with the inherited disease XP lack an enzyme involved in the NER pathway. They are not able to make repairs in DNA damaged by UV light.

25.75 5′ATGGCAGTAGGC3′

25.77 The anticancer drug fluorouracil causes the inhibition of thymidine synthesis, thereby disrupting replication.
25.79 DNA polymerase, the enzyme that makes the phosphodiester bonds in DNA, is not able to work at the end of linear DNA. This results in the shortening of the telomeres at each replication. The telomere shortening acts as a timer for the cell, allowing it to keep track of the number of divisions.
25.81 Because the genome is circular, even if the 5′ primers are removed, there will always be DNA upstream that can act as a primer for DNA polymerase to use as it synthesizes DNA.
25.83 A DNA fingerprint is made from the DNA of the child, the mother, and any prospective fathers and used to eliminate possible fathers.
25.85 Once a DNA fingerprint is made, each band in the child's DNA must come from one of the parents. Therefore, if the child has a band, and the mother does not, then the father must have that band. In this way, possible fathers are eliminated.
25.87 It detoxifies drugs and other synthetic chemicals by adding a hydroxyl group to them.
25.89 A three-dimensional pocket of ribonucleotides where substrate molecules are bound for catalytic reaction. Functional groups for catalysis include the phosphate backbone, ribose hydroxyl groups, and the nitrogen bases.
25.91 (a) The structure of the nitrogen base uracil is shown in Figure 25.1. It is a component of RNA.
(b) Uracil with a ribose attached by an N-glycosidic bond is called uridine.
25.93 native DNA
25.95 mol % A = 29.3; mol % T = 29.3; mol % G = 20.7; mol % C = 20.7.
25.97 RNA synthesis is 5′ to 3′
25.99 DNA replication requires a primer, which is RNA. Because RNA synthesis does not require a primer, it makes sense that RNA must have preceded DNA as a genetic material. This, added to the fact that RNA has been shown to be able to catalyze reactions, means that RNA can be both an enzyme and a heredity molecule.

Chapter 26   Gene Expression and Protein Synthesis

26.1 First, binding proteins must make the portion of the chromosome where the gene is less condensed and more accessible. Second, the helicase enzyme must unwind the double helix near the gene. Third, the polymerase must recognize the initiation signal on the gene.
26.2 (a) CAU and CAC   (b) GUA and GUG
26.3 valine + ATP + tRNA$_{val}$
26.4 —CCT CGATTG—
—GGAGC TAAC—
26.5 (c); gene expression refers to both processes—transcription and translation.
26.7 Protein translation occurs on the ribosomes.
26.9 Helicases are enzymes that catalyze the unwinding of the DNA double helix prior to transcription. The helicases break the hydrogen bonds between base pairs.
26.11 The termination signal is at the 5′ end of the template strand that is being transcribed. It can also be said to be at the 3′ end of the coding strand.
26.13 The "guanine cap" methyl group is located on nitrogen number 7 of guanine.
26.15 on the messenger RNA

26.17 The main subunits are the 60S and the 40S ribosomal subunits, although these can be dissociated into even smaller subunits.
26.19 327
26.21 Leucine, arginine, and serine have the most, with six codons. Methionine and tryptophan have the fewest, with one apiece.
26.23 The amino acid for protein translation is linked via an ester bond to the 3′ end of the tRNA. The energy for producing the ester bond comes from breaking two energy-rich phosphate anhydride bonds in ATP (producing AMP and two phosphates).
26.25 (a) The 40S subunit in eukaryotes forms the pre-initiation complex with the mRNA and the Met-tRNA that will become the first amino acid in the protein.
(b) The 60S subunit binds to the pre-initiation complex and brings in the next aminoacyl-tRNA. The 60S subunit contains the peptidyl transferase enzyme.
26.27 Elongation factors are proteins that participate in the process of tRNA binding and movement of the ribosome on the mRNA during the elongation process in translation.
26.29 A special tRNA molecule is used for initiating protein synthesis. In prokaryotes, it is tRNA^{fmet}, which will carry a formyl-methionine. In eukaryotes, there is a similar molecule, but it carries methionine. However, this tRNA carrying methionine for the initiation of synthesis is different from the tRNA carrying methionine for internal positions.
26.31 There are no amino acids in the vicinity of the nucleophilic attack that leads to peptide bond formation. Therefore, the ribosome must be using its RNA portion to catalyze the reaction, so it is a type of enzyme called a ribozyme.
26.33 Parts of the DNA involved are promoters, enhancers, silencers, and response elements. Molecules that bind to DNA include RNA polymerase, transcription factors, and other proteins that may bind the RNA polymerase and a transcription factor.
26.35 The active site of aminoacyl-tRNA synthases (AARS) contains the sieving portions that ensure that each amino acid is linked to its correct tRNA. The two sieving steps work on the basis of the size of the amino acid.
26.37 Both are DNA sequences that bind to transcription factors. The difference is largely due to our own understanding of the big picture. A response element controls a set of responses in a particular metabolic context. For example, a response element may activate several genes when the organism is challenged metabolically by heavy metals, by heat, or by a reduction in oxygen pressure.
26.39 Proteosomes play a role in post-translational degradation of damaged proteins. Proteins that are damaged by age or proteins that have misfolded are degraded by the proteosomes.
26.41 (a) Silent mutation: Assume the DNA sequence is TAT on the coding strand, which will lead to UAU on the mRNA. Tyrosine is incorporated into the protein. Now assume a mutation in the DNA to TAC. This will lead to UAC in mRNA. Again, the amino acid will be tyrosine.
(b) Lethal mutation: the original DNA sequence is GAA on the coding strand, which transcribes into GAA on mRNA. This codes for the amino acid glutamic acid. The DNA mutation TAA will lead to UAA, a stop signal that incorporates no amino acid.
26.43 Yes, a harmful mutation may be carried as a recessive gene from generation to generation, with no individual

demonstrating symptoms of the disease. Only when both parents carry recessive genes does an offspring have a 25% chance of inheriting the disease.

26.45 Restriction endonucleases are enzymes that recognize specific sequences on DNA and catalyze the hydrolysis of phosphodiester bonds in that region, thereby cleaving both strands of the DNA.

26.47 Mutation by natural selection is an exceedingly long, slow process that has occurred for centuries. Each natural change in the gene has been ecologically tested and found usually to have a positive effect or the organism is not viable. Genetic engineering, where a DNA mutation is done very fast, does not provide sufficient time to observe all of the possible biological and ecological consequences of the change.

26.49 A sulfur atom is substituted for an oxygen atom in each phosphate group. This functional group change slows the physiological degradation of the drug.

26.51 The viral coat is a protective protein covering around a virus particle. All of the components necessary to make the coat—for example, amino acids and lipids—come from the host.

26.53 CREB binds to the CRE (cyclic AMP response element). It has a phosphorylated form and an unphosphorylated form. When it is phosphorylated, it can bind the CREB-binding protein (CBP), which links the CREB to the RNA polymerase. This form stimulates transcription of those genes controlled by CRE.

26.55 An invariant site is a location in a protein that has the same amino acid in all species that have been studied. Studies of invariant sites help establish genetic links and evolutionary relationships.

26.57 The protein p53 is a tumor suppressor. When its gene is mutated, the protein no longer controls replication and the cell begins to grow at an increased rate.

26.59 (a) Transcription: The units include the DNA being transcribed, the RNA polymerases, and a variety of transcription factors.
(b) Translation: mRNA, ribosomal subunits, aminoacyl-tRNA, initiation factors, elongation factors.

26.61 Hereditary diseases cannot be prevented, but genetic counseling can help people understand the risks involved in passing a mutated gene to their offspring.

26.63 (a) Plasmid: a small, closed circular piece of DNA found in bacteria. It is replicated in a process independent of the bacterial chromosome. (b) Gene: a section of chromosomal DNA that codes for a particular protein molecule or RNA.

26.65 Each of the amino acids has four codons. All of the codons start with G. The second base is different for each amino acid. The third base may be any of the four possible bases. The distinguishing feature for each amino acid is the second base.

26.67 The hexapeptide is Ala-Glu-Val-Glu-Val-Trp.

Chapter 27 Bioenergetics: How the Body Converts Food to Energy

27.1 ATP

27.3 (a) 2 (b) the outer membrane

27.5 Cristae are folded membranes originating from the inner membrane. They are connected to the inner membrane by tubular channels.

27.7 There are two phosphate anhydride bonds:

27.9 Neither; they yield the same energy.

27.11 It is a phosphate ester bond.

27.13 The two nitrogen atoms that are part of C=N bonds are reduced to form $FADH_2$.

27.15 (a) ATP (b) NAD^+ and FAD (c) acetyl groups

27.17 An amide bond is formed between the amine portion of mercaptoethanolamine and the carboxyl group of pantothenic acid (see Figure 27.7).

27.19 No. The pantothenic acid portion is not the active part. The active part is the —SH group at the end of the molecule.

27.21 Both fats and carbohydrates are degraded to acetyl coenzyme A.

27.23 α-ketoglutarate

27.25 Succinate is oxidized by FAD, and the oxidation product is fumarate.

27.27 Fumarase is a lyase (it adds water across a double bond).

27.29 No, but GTP is produced in Step ⑤.

27.31 It allows the energy to be released in small packets.

27.33 carbon–carbon double bonds occur in cis-aconitate and fumarate.

27.35 α-Ketoglutarate transfers its electrons to NAD^+, which becomes NADH + H^+.

27.37 (a) Mobile electron carriers of the electron transport chain: cytochrome c and CoQ
(b) compounds that do not contain iron: CoQ (II) and FMN (I)

27.39 When H^+ passes through the ion channel, the proteins of the channel rotate. The kinetic energy of this rotatory motion is converted to and stored as the chemical energy in ATP.

27.41 This process takes place in the inner membranes of the mitochondria.

27.43 (a) 0.5 (b) 12

27.45 Ions reenter the mitochondrial matrix through the proton-translocating ATPase.

27.47 The F_1 portion of ATPase catalyzes the conversion of ADP to ATP.

27.49 The molecular weight of acetate = 59 g/mol, so 1 g acetate = 1 ÷ 59 = 0.017 mol acetate. Each mole of acetate produces 12 moles of ATP [see Problem 27.43(b)], so 0.017 mol × 12 = 0.204 mol ATP. This gives 0.204 mol ATP × 7.3 kcal/mol = 1.5 kcal.

27.51 (a) Muscles contract by sliding the thick filaments (myosin) and the thin filaments (actin) past each other.
(b) The energy comes from the hydrolysis of ATP.

27.53 ATP transfers a phosphate group to the serine residue at the active site of glycogen phosphorylase, thereby activating the enzyme.

27.55 No. It would harm humans because they would not synthesize enough ATP molecules.

27.57 (a) Cytochrome P-450 catalyzes a hydroxylation reaction. (b) The air we breathe is the source of the oxygen.

27.59 This amount of energy (87.6 kcal) is obtained from 12 mol of ATP (87.6 kcal ÷ 7.3 kcal/mol ATP = 12 mol ATP). Oxidation of 1 mol of acetate yields 12 mol ATP. The molecular weight of CH_3COOH is 60 g/mol, so the answer is 60 g or 1 mol CH_3COOH.

27.61 The energy of motion appears first in the ion channel, where the passage of H^+ causes the proteins lining the channel to rotate.

27.63 They are both hydroxy acids.

$$CH_2-COO^-$$
$$|$$
$$CH-COO^- + NAD^+$$
$$|$$
$$HO-CH-COO^-$$

Isocitrate

$$CH_2-COO^-$$
$$|$$
$$\longrightarrow \quad CH_2 \qquad + CO_2 + NADH + H^+$$
$$|$$
$$O=C-COO^-$$

α-Ketoglutarate

$$HO-CH-COO^-$$
$$| \qquad\qquad + NAD^+$$
$$CH_2-COO^-$$

Malate

$$\longrightarrow \quad O=C-COO^- \qquad + NADH + H^+$$
$$\qquad\qquad |$$
$$\qquad\qquad CH_2-COO^-$$

Oxaloacetate

27.65 Myosin, the thick filament in muscle, is an enzyme that acts as an ATPase.

27.67 Isocitrate has two stereocenters.

27.69 The ion channel is the F_0 portion of the ATPase; it is made of 12 subunits.

27.71 No, it largely comes from the chemical energy as a result of the breaking of bonds in the O_2 molecule.

27.73 It removes two hydrogens from succinate to produce fumarate.

27.75 Citrate isomerizes to isocitrate to convert a tertiary alcohol to a secondary alcohol. Tertiary alcohols cannot be oxidized, but secondary alcohols can be oxidized to produce a keto group.

27.77 Iron is found in iron-sulfur clusters in proteins and is also part of the heme group of cytochromes.

27.79 Mobile electron carriers transfer electrons on their path from one large, less mobile protein complex to another.

27.81 ATP and reducing agents such as NADH and $FADH_2$, which are generated by the citric acid cycle, are needed for biosynthetic pathways.

27.83 Biosynthetic pathways are likely to feature reduction reactions because their net effect is to reverse catabolism, which is oxidative.

27.85 ATP is not stored in the body. It is hydrolyzed to provide energy for many different kinds of processes and thus turns over rapidly.

27.87 The citric acid cycle generates NADH and $FADH_2$, which are linked to oxygen by the electron transport chain.

Chapter 28 Specific Catabolic Pathways: Carbohydrate, Lipid, and Protein Metabolism

28.1 According to Table 28.2, the ATP yield from stearic acid is 146 ATP. This makes 146/18 = 8.1 ATP/carbon atom. For lauric acid (C_{12}):

Step ① Activation	−2 ATP
Step ② Dehydrogenation five times	10 ATP
Step ③ Dehydrogenation five times	15 ATP
Six C_2 fragments in common pathway	72 ATP
Total	95 ATP

95/12 = 7.9 ATP per carbon atom for lauric acid. Thus stearic acid yields more ATP/C atom.

28.3 They serve as building blocks for the synthesis of proteins.

28.5 The two C_3 fragments are in equilibrium. As the glyceraldehyde phosphate is used up, the equilibrium shifts and converts the other C_3 fragment (dihydroxyacetone phosphate) to glyceraldehyde phosphate.

28.7 (a) Steps ① and ③ (b) Steps ⑥ and ⑨

28.9 ATP inhibition takes place at Step ⑨. It inhibits the pyruvate kinase by feedback regulation.

28.11 NADPH is the compound in question.

28.13 Each mole of glucose produces two moles of lactate, so three moles of glucose give rise to six moles of lactate.

28.15 According to Table 28.1, two moles of ATP are produced directly in the cytoplasm.

28.17 Two net ATP molecules are produced in both cases.

28.19 Enzymes that catalyze the phosphorylation of substrates using ATP are called kinases. Therefore, the enzyme that transforms glycerol to glycerol 1-phosphate is called glycerol kinase.

28.21 (a) The two enzymes are thiokinase and thiolase. (b) "Thio" refers to the presence of a sulfur-containing group, such as —SH. (c) Both enzymes insert a CoA—SH into a compound.

28.23 Each turn of fatty acid β-oxidation yields one acetyl CoA, one $FADH_2$, and one NADH. After three turns, $CH_3(CH_2)_4CO$—CoA remains from the original lauric acid; three acetyl CoA, three $FADH_2$, and three NADH + H^+ are produced.

28.25 Using data from Table 28.2, we obtain a figure of 112 moles of ATP for each mole of myristic acid.

28.27 The body preferentially uses carbohydrates as an energy source.

28.29 (a) The transformation of acetoacetate to β-hydroxybutyrate is a reduction reaction. (b) Acetone is produced by decarboxylation of acetoacetate.

28.31 It enters the citric acid cycle.

28.33 Oxidative deamination of alanine to pyruvate:

$$CH_3-CH-COO^- + NAD^+ + H_2O \longrightarrow$$
$$\qquad\quad |$$
$$\qquad\quad NH_3^+$$

$$\qquad\qquad CH_3-C-COO^- + NADH + H^+ + NH_4^+$$
$$\qquad\qquad\qquad ||$$
$$\qquad\qquad\qquad O$$

28.35 One of the nitrogens comes from ammonium ion through the intermediate carbamoyl phosphate. The other nitrogen comes from aspartate.

28.37 (a) The toxic product is ammonium ion. (b) The body gets rid of it by converting it to urea.

28.39 Tyrosine is considered a glucogenic amino acid because pyruvate can be converted to glucose when the body needs it.

28.41 It is stored in ferritin and reused.

28.43 Muscle cramps come from lactic acid accumulation.

28.45 The bicarbonate/carbonic acid buffer counteracts the acidic effects of ketone bodies.

28.47 It is necessary to tag proteins for destruction so that the ones that are no longer needed can be turned over without degrading proteins that are needed.

28.49 Glycine, the C-terminal amino acid of ubiquitin, forms an amide linkage with a lysine residue in the target protein.

28.51 The reaction is a transamination:

Phenylalanine   α-Ketoglutarate

Phenylpyruvate   Glutamate

28.53 Diabetes is a likely possibility.

28.55 (a) NAD⁺: in Steps ⑤ and ⑫

(b) NADH: in Steps ⑩ and ⑪

(c) If enough O_2 is present in muscles, a net increase of NADH + H⁺ occurs. If not enough O_2 is present (as in yeast), there is no net increase of either NAD⁺ or NADH + H⁺.

28.57 Yes, they are glucogenic and can be converted to glucose.

28.59 Products of the transamination reaction of alanine and oxaloacetate:

Alanine   Oxalo-   Pyruvate   Aspartate
             acetate

28.61 The label will ultimately appear in the carbon dioxide exhaled by the animal. In intermediate stages, it will appear in triglycerides and in intermediates of the β-oxidation cycle.

28.63 The urea cycle is energy consuming. Three ATP molecules are hydrolyzed per cycle.

28.65 (a) lauric acid, 5 turns   (b) palmitic acid, 7 turns

28.67 The conversion of pyruvate to acetyl-CoA is part of aerobic metabolism. The other two possibilities, conversion to lactate or (in some organisms) to ethanol, are anaerobic.

28.69 When people follow severely restricted diets, they have to break down fats for energy needs, producing ketone bodies.

28.71 The nitrogen portion of amino acids tends to be excreted, but carbon skeletons are degraded to yield energy or, alternatively, to serve as building blocks for biosynthetic processes.

28.73 Energy is required to form amide bonds. This process is the opposite of the first step in the digestion of proteins.

28.75 All catabolic pathways produce compounds that eventually enter the citric acid cycle. Many intermediates of the citric acid cycle serve as starting points for biosynthetic pathways.

28.77 Citric acid is an excellent nutrient, because it can be completely degraded to carbon dioxide and water. It enters the mitochondria easily.

Chapter 29   Biosynthetic Pathways

29.1 Different pathways allow for flexibility and overcome unfavorable equilibria. Separate control of anabolism and catabolism becomes possible.

29.3 The main biosynthesis of glycogen does not use inorganic phosphate because the presence of a large inorganic phosphate pool would shift the reaction to the degradation process such that no substantial amount of glycogen would be synthesized.

29.5 Photosynthesis is the reverse of respiration:

$$6CO_2 + 6H_2O \longrightarrow C_6H_{12}O_6 + 6O_2 \quad \text{Photosynthesis}$$

$$C_6H_{12}O_6 + 6O_2 \longrightarrow 6CO_2 + 6H_2O \quad \text{Respiration}$$

29.7 A compound that can be used for gluconeogenesis:

(a) from glycolysis: pyruvate

(b) from the citric acid cycle: oxaloacetate

(c) from amino acid oxidation: alanine

29.9 Glucose needs for the brain are met by gluconeogenesis, because the other pathways metabolize glucose, and only gluconeogenesis manufactures it.

29.11 Maltose is a disaccharide that is composed of two glucose units linked by an α-1,4-glycosidic bond.

UDP-glucose + glucose $\longrightarrow$ maltose + UDP

29.13 UTP consists of uracil, ribose, and three phosphates.

29.15 (a) Fatty acid biosynthesis occurs primarily in the cytoplasm.   (b) No, fatty acid degradation occurs in the mitochondrial matrix.

29.17 In fatty acid biosynthesis, a three-carbon compound, malonyl ACP, is repeatedly added to the synthase.

29.19 Carbon dioxide is released from malonyl ACP, leading to the addition of two carbons to the growing fatty acid chain.

29.21 It is an oxidation step because the substrate is oxidized with concomitant removal of hydrogen. The oxidizing agent is O_2. NADPH is also oxidized during this step.

29.23 NADPH is bulkier than NADH because of its extra phosphate group; it also has two more negative charges.

29.25 No, the body makes other unsaturated fatty acids, such as oleic acid and arachidonic acid.

29.27 The activated components needed are sphingosine, acyl CoA, and UDP-glucose.

29.29 All of the carbons in cholesterol originate in acetyl CoA. A C_5 fragment called isopentenyl pyrophosphate is an important intermediate in steroid biosynthesis.

$$3 \text{ Acetyl-CoA} \longrightarrow \text{mevalonate}$$
$$C_2 \qquad\qquad C_6$$
$$\longrightarrow \text{isopentenyl pyrophosphate} + CO_2$$
$$C_5$$

29.31 The amino acid product is aspartic acid.

29.33 The products of the transamination reaction shown are valine and α-ketoglutarate.

29.35 NADPH is the reducing agent in the process of carbon dioxide being incorporated into carbohydrates.

29.37 Indigo blue, an oxidation product of tryptophan. Indigo blue cannot enter the cells and ends up in the urine.

29.39 The bonds that connect the nitrogen bases to the ribose units are β-N-glycosidic bonds just like those found in nucleotides.

29.41 The amino acid produced by this transamination is phenylalanine.

29.43 The structure of a lecithin (phosphatidyl choline) is shown in Section 21.6. Synthesis of a molecule of this sort requires activated glycerol, two activated fatty acids, and activated choline. Each activation requires one ATP molecule, for a total number of four ATP molecules.

29.45 The compound that reacts with glutamate in a transamination reaction to form serine is 3-hydroxypyruvate. The reverse of the reaction is shown below:

29.47 HMG-CoA is hydroxymethylglutaryl CoA. Its structure is shown in Section 29.4. Carbon 1 is the carbonyl group linked to the thio group of CoA.

29.49 Heme is a porphyrin ring with iron at the center. Chlorophyll is a porphyrin ring with magnesium at the center.

29.51 Fatty acid biosynthesis takes place in the cytoplasm, requires NADPH, and uses malonyl CoA. Fatty acid catabolism takes place in the mitochondrial matrix, produces NADH and $FADH_2$, and has no requirement for malonyl CoA.

29.53 Photosynthesis has high requirements for light energy from the Sun.

29.55 Lack of essential amino acids would hinder the synthesis of the protein part. Gluconeogenesis can produce sugars even under starvation conditions.

29.57 Separation of catabolic and anabolic pathways allows for greater efficiency, especially in control of the pathways.

Chapter 30   Nutrition

30.1 No, nutrient requirements vary from person to person.

30.3 Sodium benzoate is not catabolized by the body; therefore, it does not comply with the definition of a nutrient—components of food that provide growth, replacement, and energy. Calcium propionate enters mainstream metabolism by conversion to succinyl-CoA and catabolism by the citric acid cycle and thus is a nutrient.

30.5 The Nutrition Facts label found on all foods must list the percentage of Daily Values for four important nutrients: vitamins A and C, calcium, and iron.

30.7 Chemically, fiber is cellulose, a polysaccharide that cannot be degraded by humans. It is important for proper operation of dietary processes, especially in the colon.

30.9 The basal caloric requirement is calculated assuming the body is completely at rest. Because most of us perform some activity, we need more calories than this basic minimum.

30.11 1833 Cal

30.13 No. Using diuretics would be a temporary fix at best.

30.15 The product would be different-sized oligosaccharide fragments much smaller than the original amylose molecules.

30.17 No. Dietary maltose, the disaccharide composed of glucose units linked by an α-1,4-glycosidic bond, is rapidly hydrolyzed in the stomach and small intestines. By the time it reaches the blood, it is the monosaccharide glucose.

30.19 linoleic acid

30.21 No. Lipases degrade neither; they degrade triacylglycerols.

30.23 Yes, it is possible for a vegetarian to obtain a sufficient supply of adequate proteins; however, the person must be very knowledgeable about the amino acid content of vegetables, so as to allow for protein complementation.

30.25 Dietary proteins begin degradation in the stomach, which contains HCl in a concentration of about 0.5%. Trypsin is a protease present in the small intestines that continues protein digestion after the stomach. Stomach HCl denatures dietary protein and causes somewhat random hydrolysis of the amide bonds in the protein. Fragments of the protein are produced. Trypsin catalyzes hydrolysis of peptide bonds only on the carboxyl side of the amino acids Arg and Lys.

30.27 It is expected that many of the prisoners will develop deficiency diseases in the near future.

30.29 Limes provided sailors with a supply of vitamin C to prevent scurvy.

30.31 Vitamin K is essential for proper blood clotting.

30.33 The only disease that has been proven scientifically to be prevented by vitamin C is scurvy.

30.35 Vitamins E and C and the carotenoids may have significant effects on respiratory health. This may be due to their activity as antioxidants.

30.37 There is a sulfur atom in biotin and in vitamin B_1 (also called thiamine).

30.39 The original Food Guide Pyramid did not consider the differences between types of nutrients. It assumed that all fats were to be limited and that all carbohydrates were healthy. The new guidelines recognizes that polyunsaturated fats are necessary and that carbohydrates from whole grains are better for you than those from refined sources.

30.41 All proteins, carbohydrates, and fats in excess have metabolic pathways that lead to increased levels of fatty acids. However, there is no pathway that allows fats to generate a net surplus of carbohydrates. Thus fat stores cannot be used to make carbohydrates when a person's blood glucose is low.

30.43 All effective weight loss is based on increasing activity while limiting caloric intake. However, it is more effective to concentrate on increasing activity than on limiting intake.

30.45 (a) Most studies show that the artificial sweeteners Sucralose and acesulfame-K are not metabolized in any measurable amounts. (b) Digestion of aspartame can lead to high levels of phenylalanine.

30.47 arginine

30.49 Carbohydrate loading before the event and consuming carbohydrates during the event

30.51 Caffeine acts as a central nervous system stimulant, which provides a feeling of energy that athletes often enjoy. In addition, caffeine reduces insulin levels and stimulates oxidation of fatty acids, which would be beneficial to endurance athletes.

30.53 The vitamin pantothenic acid is part of CoA. (a) Glycolysis: Pyruvate dehydrogenase uses CoA as a coenzyme. (b) Fatty acid synthesis: The first step involves the enzyme fatty acid synthase.

30.55 Proteins that are ingested in the diet are degraded to free amino acids, which are then used to build proteins that carry out specific functions. Two very important functions include structural integrity and biological catalysis. Our proteins are constantly being turned over—that is, continuously being degraded and rebuilt using free amino acids.

30.57 The very tip of the Food Guide Pyramid displays fats, oils, and sweets, with the cautionary statement, "Use sparingly." We can omit sweets completely from the diet; however, complete omission of fats and oils is dangerous. We must have dietary fats and oils that contain the two essential fatty acids. The essential fatty acids may be present as components in other food groups—that is, the meat, poultry, and fish group.

30.59 Walnuts are not just a tasty snack—they are a healthy one. Walnuts have protein. In fact, nuts are included in a group of the U.S. Department of Agriculture's Food Guide Pyramid. Walnuts are also a good source of vitamins and minerals, including vitamins E and B, biotin, potassium, magnesium, phosphorus, zinc, and manganese.

30.61 No, the lecithin is degraded in the stomach and intestines long before it could get into the blood. The phosphoglyceride is degraded to fatty acids, glycerol, and choline, which are absorbed through the intestinal walls.

30.63 Patients who have undergone ulcer surgery are administered digestive enzymes that may have been lost during the procedure. The enzyme supplement should contain proteases to help break down proteins as well as lipases to assist in fat digestion.

Chapter 31 Immunochemistry

31.1 Examples of external innate immunity include action by the skin, tears, and mucus.

31.3 The skin fights infection by providing a barrier against penetration of pathogens. The skin also secretes lactic acid and fatty acids, both of which create a low pH, thereby inhibiting bacterial growth.

31.5 Innate immunity processes have little ability to change in response to immune dangers. The key features of adaptive (acquired) immunity are specificity and memory. The acquired immune system uses antibody molecules designed for each type of invader. In a second encounter with the same danger, the response is more rapid and more prolonged than the first response.

31.7 T cells originate in the bone marrow, but grow and develop in the thymus gland. B cells originate and grow in the bone marrow.

31.9 Macrophages are the first cells in the blood that encounter potential threats to the system. They attack virtually anything that is not recognized as part of the body, including pathogens, cancer cells, and damaged tissue. Macrophages engulf an invading bacterium or virus and kill it with (NO), nitric oxide, and then digest it.

31.11 protein-based antigens

31.13 Class II MHC molecules pick up damaged antigens. A targeted antigen is first processed in lysosomes, where it is degraded by proteolytic enzymes. An enzyme, GILT, reduces the disulfide bridges of the antigen. The reduced peptide antigens unfold and are further degraded by proteases. The peptide fragments remaining serve as epitopes that are recognized by class II MHC molecules.

31.15 MHC molecules are transmembrane proteins that belong to the immunoglobulin superfamily. They are originally present inside cells until they become associated with antigens and move to the surface membrane.

31.17 If we assume that the rabbit has never been exposed to the antigen, the response will occur 1–2 weeks after the injection of antigen.

31.19 (a) IgE molecules have a carbohydrate content of 10–12%, which is equal to that of IgM molecules. IgE molecules have the lowest concentration in the blood. The blood concentration of IgE is about 0.01–0.1 mg/100 mL of blood. (b) IgE molecules are attached to basophils and mast cells. When they encounter their antigens, the cells release histamines, which produce the effects of hay fever and other allergies.

31.21 The two Fab fragments would be able to bind to an antigen. These fragments contain the variable protein sequence regions and hence are able to change during synthesis against a specific antigen.

31.23 *Immunoglobulin superfamily* refers to all of the proteins that have the standard structure of a heavy chain and a light chain.

31.25 Antibodies and antigens are held together by weak, noncovalent interactions: hydrogen bonds, electrostatic interactions (dipole–dipole), and hydrophobic interactions.

31.27 The DNA for the immunoglobulin superfamily has multiple ways of recombining during cell development. The diversity is a reflection of the number of permutations and ways of combining various constant regions, variable regions, joining regions, and diversity regions.

31.29 T cells carry on their surfaces unique receptor proteins that are specific for antigens. These receptors (TcR), which are members of the immunoglobulin superfamily, have constant and variable regions. They are anchored in the T-cell membrane by hydrophobic interactions. They are not able to bind antigens alone, but rather need additional protein molecules called cluster determinants that act as coreceptors. When TcR molecules combine with cluster determinant proteins, they form T-cell-receptor complexes (TcR complexes).

31.31 The components of the TcR complexes are (1) accessory protein molecules called cluster determinants and (2) the T-cell receptor.

31.33 CD4

31.35 They are adhesion molecules that help dock antigen-presenting cells to T cells. They also act as signal transducers.

31.37 Cytokines are glycoproteins that interact with cytokine receptors on macrophages, B cells, and T cells. They do not recognize and bind antigens.

31.39 Chemokines are a class of cytokines that send messages between cells. They attract leukocytes to the site of injury and bind to specific receptors on the leukocytes.

31.41 All chemokines are low-molecular-weight proteins that have four cysteine residues that are linked in very specific disulfide bonds: Cys1—Cys3 and Cys2—Cys4.

31.43 The T cells mature in the thymus gland. During maturation, those cells that fail to interact with MHC and thus cannot respond to foreign antigens are eliminated by a special selection process. T cells that express receptors that may interact with normal self antigens are eliminated by the same selection process.

31.45 A signaling pathway that controls the maturation of B cells is the phosphorylation pathway activated by tyrosine kinase and deactivated by phosphatase.

31.47 the cytokines and chemokines

31.49 helper T cells

31.51 It is hard to find because the virus mutates quickly. Also, one of its docking proteins changes conformation when it docks, so that antibodies elicited against undocked proteins are ineffective. It binds to several proteins that inhibit antiviral factors, and cloaks its outer membrane with sugars that are very similar to the natural sugars found on host cells.

31.53 Vaccines rely on the immune system's ability to recognize a foreign molecule and make specific antibodies to it. HIV hides from the immune system in a variety of ways, and it changes often. The body makes antibodies, but they are not very effective at finding or neutralizing the virus.

31.55 It had been known since the eighteen hundreds that the mayapple had anticancer properties. It was later found that a chemical found in the mayapple, picropodophyllin, inhibits spindle formation during mitosis in dividing cells. As all chemotherapy agents do, they hinder rapidly dividing cells, like cancer cells, more then regular cells.

31.57 MCA treatment for cancer is better than chemotherapy because MCAs are much more specific for the cancer cells. Treatment is directed toward just the cancerous cells—not all cells in general, as in chemotherapy. In addition, MCA therapy shows minimal toxicity.

31.59 Jenner noticed that people infected with cowpox did not also get smallpox. He tried the experiment of infecting a boy with cowpox, a mild disease, and then later injecting the boy with smallpox. The boy developed no symptoms of the dangerous smallpox. Today this dangerous and unethical experiment could not be carried out by a reputable physician.

31.61 It comes from the experiment done by Jenner. The derisive term means "enCOWment," reflecting his studies of cowpox.

31.63 dendritic cells

31.65 Chemokines (or, more generally, cytokines) help leukocytes migrate out of a blood vessel to the site of injury. Cytokines help the proliferation of leukocytes.

31.67 A compound called 12:13 dEpoB, a derivative of epothilon B, is being studied as an anticancer vaccine.

31.69 Tumor necrosis factor receptors are located on the surfaces of several cell types, but especially on tumor cells.

Glossary

Absolute zero (*Section 1.4*) The lowest possible temperature; the zero point of the Kelvin temperature scale.

Acetal (*Section 17.4C*) A molecule containing two —OR groups bonded to the same carbon.

Achiral (*Section 15.1*) An object that lacks chirality; an object that is superposable on its mirror image.

Acid (*Section 9.1*) An Arrhenius acid is a substance that ionizes in aqueous solution to give a hydronium ion, H_3O^+, and an anion.

Acid–base reaction (*Section 9.3*) A proton-transfer reaction.

Acid ionization constant (K_a) (*Section 9.5*) An equilibrium constant for the ionization of an acid in aqueous solution to H_3O^+ and its conjugate base. K_a is also called an **acid dissociation constant**.

Acid rain (*Chemical Connections 7A*) Rain with acids other than carbonic acid dissolved in it.

Acidosis (*Chemical Connections 9D*) A condition in which the pH of blood is lower than 7.35.

Acquired immunity (*Section 31.1*) The second line of defense that vertebrates have against invading organisms.

Activating receptor (*Section 31.7*) A receptor on a cell of the innate immune system that triggers activation of the immune cell in response to a foreign antigen.

Activation energy (*Section 8.2*) The minimum energy necessary to cause a chemical reaction.

Actual yield (*Section 5.5C*) The mass of product formed in a chemical reaction.

Acyl group (*Section 18.3A*) An RCO— group.

Adaptive immunity (*Section 31.1*) Acquired immunity with specificity and memory.

Adhesion molecules (*Section 31.5*) Various protein molecules that help to bind an antigen to the T-cell receptor.

Advanced glycation end products (*Section 22.7*) A chemical product of sugars and proteins linking together to produce an imine.

Affinity maturation (*Section 31.4*) The process of mutation of T cells and B cells in response to an antigen.

AIDS (*Section 31.8*) *A*cquired *i*mmune *d*eficiency *s*yndrome. The disease caused by the human immunodeficiency virus, which attacks and depletes T cells.

Alcohol (*Section 10.3A*) A compound containing an —OH (hydroxyl) group bonded to a tetrahedral carbon atom.

Aldehyde (*Section 10.3C*) A compound containing a carbonyl group bonded to a hydrogen; a —CHO group.

Aliphatic amine (*Section 16.1*) An amine in which nitrogen is bonded only to alkyl groups.

Aliphatic hydrocarbon (*Section 11.1*) An alkane.

Alkaloid (*Chemical Connections 16B*) A basic nitrogen-containing compound of plant origin, many of which have physiological activity when administered to humans.

Alkalosis (*Chemical Connections 9E*) A condition in which the pH of blood is greater than 7.45.

Alkane (*Section 11.1*) A saturated hydrocarbon whose carbon atoms are arranged in an open chain—that is, not arranged in a ring.

Alkene (*Section 12.1*) An unsaturated hydrocarbon that contains a carbon–carbon double bond.

Alkyl group (*Section 11.3A*) A group derived by removing a hydrogen atom from an alkane; given the symbol R—.

Alkyne (*Section 12.1*) An unsaturated hydrocarbon that contains a carbon–carbon triple bond.

Allosteric proteins (*Chemical Connections 22G*) Proteins that exhibit a behavior where binding of one molecule at one site changes the ability of the protein to bind another molecule at a different site.

Alloys (*Section 7.2*) Homogeneous mixtures of metals.

Alpha (α-) amino acid (*Section 22.2*) An amino acid in which the amino group is bonded to the carbon atom next to the —COOH carbon.

Alpha helix (*Section 22.9*) A type of repeating secondary structure of a protein in which the chain adopts a helical conformation and the structure is held together by hydrogen bonds from the peptide backbone N—H to the backbone C=O four amino acids farther up the chain.

Alpha particle (α) (*Section 3.2*) A helium nucleus, He^{2+}, ^4_2He.

Amino acid (*Section 22.1*) An organic compound containing an amino group and a carboxyl group.

Amino group (*Section 10.3B*) An —NH_2 group.

Amorphous solid (*Section 6.9*) A solid whose atoms, molecules, or ions are not in an orderly arrangement.

Amphiprotic (*Section 9.3*) A substance that can act as either an acid or a base.

Amphoteric (*Section 9.3*) An alternative for amphiprotic.

Amylase (*Section 30.3*) An enzyme that catalyzes the hydrolysis of α-1, 4-glycosidic bonds in dietary starches.

Aneuploid cell (*Chemical Connections 31F*) A cell with the wrong number of chromosomes.

Anion (*Section 4.2*) An ion with a negative electric charge.

Anode (*Chemical Connections 5B*) The negatively charged electrode.

Antibody (*Section 31.1*) A defense glycoprotein synthesized by the immune system of vertebrates that interacts with an antigen. It is also called an immunoglobulin.

Antigen (*Sections 31.1, 31.3*) A substance foreign to the body that triggers an immune response.

Antigen-presenting cells (APCs) (*Section 31.2*) Cells that cleave foreign molecules and present them on their surfaces for binding to T cells or B cells.

Aqueous solution (*Section 5.6*) A solution in which the solvent is water.

Ar— (*Section 13.1*) The symbol used for an aryl group.

Arene (*Section 13.1*) A compound containing one or more benzene rings.

Aromatic amine (*Section 16.1*) An amine in which nitrogen is bonded to one or more aromatic rings.

Aromatic compound (*Section 13.1*) A term used to classify benzene and its derivatives.

Aromatic sextet (*Section 13.1B*) The closed loop of six electrons (two from the second bond of each double bond) characteristic of a benzene ring.

Aryl group (*Section 13.1*) A group derived from an arene by removal of a hydrogen atom. Given the symbol Ar—.

Autoxidation (*Section 13.4C*) The reaction of a C—H group with oxygen, O_2, to form a hydroperoxide, R—OOH.

Avogadro's law (*Section 6.4*) Equal volumes of gases at the same temperature and pressure contain the same number of molecules.

Avogadro's number (*Section 5.3*) 6.02×10^{23} formula units per mole; the amount of any substance that contains the same number of formula units as the number of atoms in 12 g of carbon-12.

Axial position (*Section 11.5B*) A position on a chair conformation of a cyclohexane ring that extends from the ring parallel to the imaginary axis of the ring.

B cell (*Section 31.1*) A type of lymphocyte that is produced in and matures in the bone marrow. B cells produce antibody molecules.

Basal caloric requirement (*Section 30.2*) The caloric requirement for an individual at rest, usually given in Cal/day.

Base (*Section 9.1*) An Arrhenius base is a substance that ionizes in aqueous solution to give hydroxide (OH^-) ions and an cation.

Becquerel (Bq) (*Section 3.5A*) A measure of radioactive decay, equal to one disintegration per second.

Beta (β-) pleated sheet (*Section 22.9*) A type of secondary protein structure in which the backbone of two protein chains in the same or different molecules is held together by hydrogen bonds.

Beta particle (β) (*Section 3.2*) An electron, $_{-1}^{0}\beta$.

Binary compound (*Section 4.6A*) A compound containing two elements.

Boiling point (*Section 6.8C*) The temperature at which the vapor pressure of a liquid is equal to the atmospheric pressure.

Bond angle (*Section 4.10*) The angle between two atoms bonded to a central atom.

Bonding electrons (*Section 4.7C*) Valence electrons involved in forming a covalent bond—that is, shared electrons.

Boyle's Law (*Section 6.3A*) The volume of a gas at constant temperature is inversely proportional to the pressure applied to the gas.

Brønsted-Lowry acid (*Section 9.3*) A proton donor.

Brønsted-Lowry base (*Section 9.3*) A proton acceptor.

Brownian motion (*Section 7.8*) The random motion of colloidal-size particles.

Buffer (*Section 9.10*) A solution that resists change in pH when limited amounts of an acid or a base are added to it; an aqueous solution containing a weak acid and its conjugate base.

Buffer capacity (*Section 9.10*) The extent to which a buffer solution can prevent a significant change in the pH of a solution upon addition of an acid or a base.

Calorie (*Section 1.9*) The amount of heat necessary to raise the temperature of 1 g of liquid water by 1°C.

Carbocation (*Section 12.6A*) A species containing a carbon atom with only three bonds to it and bearing a positive charge.

Carbonyl group (*Section 10.3C*) A C=O group.

Carboxyl group (*Section 10.3D*) A —COOH group.

Carboxylic acid (*Section 10.3D*) A compound containing a —COOH group.

Catalyst (*Section 8.4D*) A substance that increases the rate of a chemical reaction by providing an alternative pathway with a lower activation energy.

Cathode (*Chemical Connections 5B*) The positively charged electrode.

Cation (*Section 4.2*) An ion with a positive electric charge.

Cell reprogramming (*Chemical Connections 31F*) A technique used in whole-mammal cloning, in which a somatic cell is reprogrammed to behave like a fertilized egg.

Celsius scale (°C) (*Section 1.4*) A temperature scale based on 0° as the freezing point of water and 100° as the normal boiling point of water.

Chain reaction, nuclear (*Section 3.9*) A nuclear reaction that results from fusion of a nucleus with another particle (most commonly a neutron) followed by decay of the fused nucleus to smaller nuclei and more neutrons. The newly formed neutrons continue the process and result in a chain reaction.

Chair conformation (*Section 11.5B*) The most stable conformation of a cyclohexane ring; all bond angles are approximately 109.5°.

Chaperones (*Section 22.10*) Protein molecules that help other proteins to fold into the biologically active conformation and enable partially denatured proteins to regain their biologically active conformation.

Charles's law (*Section 6.3B*) The volume of a gas at constant pressure is directly proportional to the temperature in kelvins.

Chemical change (*Section 1.1*) Chemical reaction.

Chemical equation (*Section 5.4*) A representation using chemical formulas of the process that occurs when reactants are converted to products.

Chemical equilibrium (*Section 8.5*) A state in which the rate of the forward reaction equals the rate of the reverse reaction.

Chemical kinetics (*Section 8.1*) The study of the rates of chemical reactions.

Chemical property (*Section 1.1*) A chemical reaction that a substance undergoes.

Chemical reaction (*Section 1.1*) As substances are used up, others appear in their place.

Chemistry (*Section 1.1*) The science that deals with matter.

Chemokine (*Section 31.6*) A chemotactic cytokine that facilitates the migration of leukocytes from the blood vessels to the site of injury or inflammation.

Chiral (*Section 15.1*) From the Greek *cheir*, meaning "hand"; an object that is not superposable on its mirror image.

Cis (*Section 11.6*) A prefix meaning "on the same side."

Cis-trans isomers (*Section 11.6*) Isomers that have the same (1) molecular formula (2) order of attachment (connectivity) of their atoms (3) but a different arrangement of their atoms in space due to the presence of either a ring or a carbon–carbon double bond.

Cluster determinant (*Section 31.5*) A set of membrane proteins on T cells that help the binding of antigens to the T-cell receptors.

Coefficient (*Section 5.4*) A number placed in front of a chemical entity in a balanced chemical equation.

Colligative property (*Section 7.8*) A property of a solution that depends only on the number of solute particles and not on the chemical identity of the solute particles.

Colloid (*Section 7.7*) A two-part mixture in which the suspended solute particles range from 1 to 1000 nm in size.

Combined gas law (*Section 6.3C*) The pressure, volume, and temperature in kelvins of two samples of the same gas are related by the equation $P_1V_1/T_1 = P_2V_2/T_2$.

Combustion (*Section 5.4*) Burning in air.

Complete protein (*Section 30.5*) A protein source that contains sufficient quantities of all amino acids required for normal growth and development.

Condensation (*Section 6.7*) The change of a substance from the vapor or gaseous state to the liquid state.

Configuration (*Section 11.6*) The arrangement of atoms about a stereocenter—that is, the relative arrangements of the parts of a molecule in space.

Conformation (*Section 11.5A*) Any three-dimensional arrangement of atoms in a molecule that results from rotation about a single bond.

Conjugate acid (*Section 9.3*) In the Brønsted-Lowry theory, a substance formed when a base accepts a proton.

Conjugate acid–base pair (*Section 9.3*) A pair of molecules or ions that are related to one another by the gain or loss of a proton.

Conjugate base (*Section 9.3*) In the Brønsted-Lowry theory, a substance formed when an acid donates a proton to another molecule or ion.

Conjugated protein (*Section 22.11*) A protein that contains a nonprotein part, such as the heme part of hemoglobin.

Constitutional isomers (*Section 11.2*) Compounds with the same molecular formula but a different order of attachment (connectivity) of their atoms.

Contributing structure (*Section 4.8B*) Representations of a molecule or ion that differ only in the distribution of valence electrons.

Conversion factor (*Section 1.5*) A ratio of two different units.

Cosmic rays (*Section 3.6*) High-energy particles, mainly protons, from outer space bombarding the Earth.

Covalent bond (*Section 4.4A*) A bond resulting from the sharing of electrons between two atoms.

Crystallization (*Section 6.9*) The formation of a solid from a liquid.

C-terminus (*Section 22.6*) The amino acid at the end of a peptide chain that has a free carboxyl group.

Curie (Ci) (*Section 3.5A*) A measure of radioactive decay equal to 3.7 × 10^{10} disintegrations per second.

Cyclic ether (*Section 14.3A*) An ether in which the ether oxygen is one of the atoms of a ring.

Cycloalkane (*Section 11.4*) A saturated hydrocarbon that contains carbon atoms bonded to form a ring.

Cystine (*Section 22.4*) A dimer of cysteine in which the two amino acids are covalently bonded by disulfide bond between their side chain —SH groups.

Cytokine (*Section 31.6*) A glycoprotein that traffics between cells and alters the function of a target cell.

Daily Values (*Section 30.1*) The values recommended by the Food and Drug Administration for key nutrients in the diet.

Dalton's Law (*Section 6.5*) The pressure of a mixture of gases is equal to the sum of the partial pressures of each gas in the mixture.

Debranching enzyme (*Section 30.3*) The enzyme that catalyzes the hydrolysis of the 1,6-glycosidic bonds in starch and glycogen.

Decarboxylation (*Section 18.5E*) The loss of CO_2 from a carboxyl (—COOH) group.

Decay, nuclear (*Section 3.3B*) The change of a radioactive nucleus of one element into the nucleus of another element.

Dehydration (*Section 14.2B*) The elimination of a molecule of water from an alcohol. An OH is removed from one carbon, and an H is removed from an adjacent carbon.

Denaturation (*Section 22.12*) The loss of the secondary, tertiary, and quaternary structure of a protein by a chemical or physical agent that leaves the primary structure intact.

Dendritic cells (*Sections 31.1, 31.2*) Important cells in the innate immune system that are often the first cells to defend against invaders.

Density (*Section 1.7*) The ratio of mass to volume for a substance.

Detergent (*Section 18.4D*) A synthetic soap. The most common are the linear alkylbenzene sulfonic acids (LAS).

Dextrorotatory (*Section 15.4B*) The clockwise (to the right) rotation of the plane of polarized light in a polarimeter.

Dialysis (*Section 7.8C*) A process in which a solution containing particles of different sizes is placed in a bag made of a semipermeable membrane. The bag is placed into a solvent or solution containing only small molecules. The solution in the bag reaches equilibrium with the solvent outside, allowing the small molecules to diffuse across the membrane but retaining the large molecules.

Diastereomers (*Section 15.3A*) Stereoisomers that are not mirror images of each other.

Diet faddism (*Section 30.1*) An exaggerated belief in the effects of nutrition upon health and disease.

Dietary Reference Intake (DRI) (*Section 30.1*) The current numerical system for reporting nutrient requirements; an average daily requirement for nutrients published by the U.S. Food and Drug Administration.

Digestion (*Section 30.1*) The process in which the body breaks down large molecules into smaller ones that can then be absorbed and metabolized.

Diol (*Section 14.2B*) A compound containing two —OH (hydroxyl) groups.

Dipeptide (*Section 22.6*) A peptide with two amino acids.

Dipole (*Section 4.7B*) A chemical species in which there is a separation of charge; there is a positive pole in one part of the species and a negative pole in another part.

Dipole–dipole attraction (*Section 6.7B*) The attraction between the positive end of one dipole and the negative end of another dipole in the same or different molecule.

Diprotic acid (*Section 9.3*) An acid that can give up two protons.

Discriminatory curtailment diet (*Section 30.1*) A diet that avoids certain food ingredients that are considered harmful to the health of an individual—for example, low-sodium diets for people with high blood pressure.

Disulfide (*Section 14.4D*) A compound containing an —S—S— group.

Double bond (*Section 4.7B*) A bond formed by sharing two pairs of electrons; represented by two lines between the two bonded atoms.

Double-headed arrow (*Section 4.8A*) A symbol used to show that the structures on either side of it are resonance contributing structures.

Dynamic equilibrium (*Section 8.5*) A state in which the rate of the forward reaction equals the rate of the reverse reaction.

Effective collision (*Section 8.2*) A collision between two molecules or ions that results in a chemical reaction.

EGF (*Section 31.6*) Epidermal growth factor; a cytokine that stimulates epidermal cells during healing of wounds.

Electrolyte (*Section 7.4C*) A substance that, when dissolved in water, produces a solution that conducts electricity.

Electromagnetic spectrum (*Section 3.2*) The array of electromagnetic phenomena by wavelength.

Electron capture (*Section 3.3F*) A reaction in which a nucleus captures an extranuclear electron and then undergoes a nuclear decay.

Electron volt (eV) (*Section 3.2*) A non-SI energy unit frequently used in nuclear chemistry. $1.0 \text{ eV} = 1.60 \times 10^{-19} \text{ J}$.

Electronegativity (*Section 4.4B*) A measure of an atom's attraction for the electrons it shares in a chemical bond with another atom.

Embryonal carcinoma cell (*Chemical Connections 31F*) A cell that is multipotent and is derived from carcinomas.

Embryonic stem cell (*Chemical Connections 31F*) Stem cells derived from embryonic tissue. Embryonic tissue is the richest source of stem cells.

Emulsion (*Section 7.7*) A system, such as fat in milk, consisting of a liquid with or without an emulsifying agent in an immiscible liquid, usually as droplets larger than colloidal size.

Enantiomers (*Section 15.1*) Stereoisomers that are nonsuperposable mirror images; refers to a relationship between pairs of objects.

End point (*Section 9.9*) The point in a titration where a visible change occurs.

Endothermic reaction (*Section 5.8*) A chemical reaction that absorbs heat.

Energy (*Section 1.8*) The capacity to do work. The SI base unit is the joule (J).

Enol (*Section 17.5*) A molecule containing an —OH group bonded to a carbon of a carbon–carbon double bond.

Epigenetics (*Chemical Connections 31F*) The study of heritable processes that alter gene expression without altering the actual DNA.

Epitope (*Section 31.3*) The smallest number of amino acids on an antigen that elicits an immune response.

Equatorial position (*Section 11.3B*) A position on a chair conformation of a cyclohexane ring that extends from the ring roughly perpendicular to the imaginary axis of the ring.

Equilibrium (*Section 6.8B*) A condition in which two opposing physical forces are equal.

Equilibrium constant (*Section 8.6*) A value calculated from the equilibrium expression for a given reaction indicating in which direction the reaction goes.

Equivalence point (*Section 9.9*) The point in an acid–base titration at which there is an equal amount of acid and base.

Ergogenic aid (*Chemical Connections 30E*) A substance that can be consumed to enhance athletic performance.

Essential amino acid (*Section 30.5*) An amino acid that the body cannot synthesize in the required amounts and so must be obtained in the diet.

Essential fatty acid (*Section 30.4*) A fatty acid required in the diet.

Ester (*Section 18.5D*) A compound in which the OH of a carboxyl group, RCOOH, is replaced by an —OR' group; RCOOR'.

Ether (*Section 14.3A*) A compound containing an oxygen atom bonded to two carbon atoms.

Exothermic reaction (*Section 5.8*) A chemical reaction that gives off heat.

Exponential notation (*Section 1.3*) The expression of a number in base 10.

Extended helix (*Section 22.9*) A type of helix found in collagen, caused by a repeating sequence.

External innate immunity (*Section 31.1*) The innate protection against foreign invaders characteristic of the skin barrier, tears, and mucus.

Fact (*Section 1.2*) A statement based on experience.

Fatty acid (*Section 18.4A*) A long, unbranched chain carboxylic acid, most commonly with 10–20 carbon atoms, derived from animal fats, vegetable oils, or the phospholipids of biological membranes. The hydrocarbon chain may be saturated or unsaturated. In most unsaturated fatty acids, the *cis* isomer predominates. *Trans* isomers are rare.

Fiber (*Section 30.1*) The cellulosic, non-nutrient component in our food.

Fibrous protein (*Section 22.1*) A protein used for structural purposes. Fibrous proteins are insoluble in water and have a high percentage of secondary structures, such as alpha helices and/or beta-pleated sheets.

Fischer esterification (*Section 18.5D*) The process of forming an ester by refluxing a carboxylic acid and an alcohol in the presence of an acid catalyst, commonly sulfuric acid.

Fission, nuclear (*Section 3.9*) The fragmentation of a heavier nucleus into two or more smaller nuclei.

Formula weight (FW) (*Section 5.2*) The sum of the atomic weights of all atoms is a compound's formula expressed in atomic mass units (amu). Formula weight can be used for both ionic and molecular compounds.

Freezing-point depression (*Section 7.8*) The decrease in the freezing point of a liquid caused by adding a solute.

Frequency (ν) (*Section 3.2*) The number of wave crests that pass a given point per unit of time.

Functional group (*Section 10.3*) An atom or group of atoms within a molecule that shows a characteristic set of physical and chemical properties.

Fusion, nuclear (*Section 3.8*) The combining of two or more nuclei to form a heavier nucleus.

Gamma ray (γ) (*Section 3.2*) A form of electromagnetic radiation characterized by very short wavelength and very high energy.

Gas (*Section 1.6*) A physical state of matter in which there is no definite volume or shape.

Gay-Lussac's law (*Section 6.3C*) The pressure of a gas at constant volume is directly proportional to its temperature in kelvins.

Geiger-Müller counter (*Section 3.5*) An instrument for measuring ionizing radiation.

Globular protein (*Section 22.1*) Protein that is used mainly for non-structural purposes and is largely soluble in water.

Glycol (*Section 14.1B*) A compound with hydroxyl (—OH) groups on adjacent carbons.

Gp120 (*Section 31.5*) A 120,000-molecular-weight glycoprotein on the surface of the human immunodeficiency virus that binds strongly to the CD4 molecules on T cells.

Gray (Gy) (*Section 3.6*) The SI unit of the amount of radiation absorbed from a source. 1 Gy = 100 rad.

Haber process (*Chemical Connections 8E*) An industrial process by which H_2 and N_2 are converted to NH_3.

Half-life, of a radioisotope (*Section 3.4*) The time it takes for one half of a sample of radioactive material to decay.

HDPE (*Section 12.7C*) High-density polyethylene.

Heat of combustion (*Section 5.8*) The heat given off in a combustion reaction.

Heat of reaction (*Section 5.8*) The heat given off or absorbed in a chemical reaction.

Helper T cells (*Section 31.2*) A type of T cell that helps in the response of the acquired immune system against invaders but does not kill infected cells directly.

Hemiacetal (*Section 17.4C*) A molecule containing a carbon bonded to one —OH and one —OR group; the product of adding one molecule of alcohol to the carbonyl group of an aldehyde or ketone.

Henderson-Hasselbalch equation (*Section 9.11*) A mathematical relationship between pH, the pK_a of a weak acid, HA, and the concentrations of the weak acid and its conjugate base.

$$pH = pK_a + \frac{\log [A^-]}{[HA]}$$

Henry's law (*Section 7.4C*) The solubility of a gas in a liquid is directly proportional to the pressure of the gas above the liquid.

Heterocyclic aliphatic amine (*Section 16.1*) A heterocyclic amine in which nitrogen is bonded only to alkyl groups.

Heterocyclic amine (*Section 16.1*) An amine in which nitrogen is one of the atoms of a ring.

Heterocyclic aromatic amine (*Section 16.1*) An amine in which nitrogen is one of the atoms of an aromatic ring.

Heterogeneous catalyst (*Section 8.4D*) A catalyst in a separate phase from the reactants—for example, the solid platinum, Pt(s), in the reaction between CO(g) and H_2(g).

Highly active antiretroviral therapy (**HAART**) (*Section 31.8*) An aggressive treatment against AIDS involving the use of several different drugs.

HIV (*Sections 31.4, 31.8*) Human immunodeficiency virus.

Homogeneous catalyst (*Section 8.4D*) A catalyst in the same phase as the reactants—for example, enzymes in body tissues.

Hybridoma (*Section 31.4*) A combination of a myleoma cell with a B cell to produce monoclonal antibodies.

Hydration (*Section 12.7C*) Addition of water.

Hydrobromination (*Section 12.6A*) The addition of HBr to the carbon–carbon double bond of an alkene.

Hydrocarbon (*Section 11.1*) A compound that contains only carbon and hydrogen atoms.

Hydrogen bond (*Section 6.7C*) A noncovalent force of attraction between the partial positive charge on a hydrogen atom bonded to an atom of high electronegativity, most commonly oxygen or nitrogen, and the partial negative charge on a nearby oxygen or nitrogen.

Hydrogenation (*Section 12.6B*) Addition of hydrogen atoms to a double or triple bond using H_2 in the presence of a transition metal catalyst, most commonly Ni, Pd, or Pt. Also called catalytic reduction or catalytic hydrogenation.

Hydronium ion (*Section 9.1*) The H_3O^+ ion.

Hydrophobic interaction (*Section 22.10*) Interaction by London dispersion forces between hydrophobic groups.

Hydroxyl group (*Section 10.3A*) An —OH group bonded to a tetrahedral carbon atom.

Hygroscopic substance (*Section 7.6B*) A compound able to absorb water vapor from the air.

Hypothermia (*Chemical Connections 1B*) Having a body temperature lower than normal.

Hypothesis (*Section 1.2*) A statement that is proposed, without actual proof, to explain certain facts and their relationship.

Ideal gas law (*Section 6.4*) $PV = nRT$.

Immunogen (*Section 31.3*) Another term for antigen.

Immunoglobulin (*Section 31.4*) An antibody protein generated against and capable of binding specifically to an antigen.

Immunoglobulin superfamily (*Section 31.1*) A family of molecules based on a similar structure that includes the immunoglobulins, T-cell receptors, and other membrane proteins that are involved in cell communications. All molecules in this class have a certain portion that can react with antigens.

Indicator, acid–base (*Section 9.8*) A substance that changes color within a given pH range.

Inhibitory receptor (*Section 31.7*) A receptor on the surface of a cell of the innate immune system that recognizes antigens on healthy cells and prevents activation of the immune system.

Innate immunity (*Section 31.1*) The first line of defense against foreign invaders, which includes skin resistance to penetration, tears, mucus, and nonspecific macrophages that engulf bacteria.

Interleukin (*Section 31.6*) A cytokine that controls and coordinates the action of leukocytes.

Internal innate immunity (*Section 31.1*) The type of innate immunity that is used once a pathogen has already penetrated a tissue.

International System of Units (SI) (*Section 1.4*) A system of units of measurement based in part on the metric system.

Intramolecular hydrogen bonds (*Section 22.9*) Hydrogen bonds that exist within a given molecule, such as those holding an α-helix together in a protein.

Ion product of water, K_w (*Section 9.7*) The concentration of H_3O^+ multiplied by the concentration of OH^-; $[H_3O^+][OH^-] = 1 \times 10^{-14}$.

Ionic bond (*Section 4.4A*) A chemical bond resulting from the attraction between a positive ion and a negative ion.

Ionic compound (*Section 4.5A*) A compound formed by the combination of positive and negative ions.

Ionizing radiation (*Section 3.5*) Radiation that causes one or more electrons to be ejected from an atom or molecule, thereby producing positive ions.

Isoelectric point (*Section 22.3*) The pH at which a molecule has no net charge; abbreviated pI.

Isotonic (*Section 7.8B*) Solutions of the same osmolarity.

Joule (J) (*Section 1.9*) The SI base unit for heat; 1 J is 4.184 cal.

Kelvin scale (K) (*Section 1.4*) The SI base unit for temperature.

Ketone (*Section 10.3C*) A compound containing a carbonyl group bonded to two carbons.

Ketone body (*Chemical Connections 18C*) One of several ketone-based molecules—for example, acetone, 3-hydroxybutanoic acid (β-hydroxybutyric acid) and acetoacetic acid (3-oxobutanoic acid) produced in the liver during over-utilization of fatty acids when the supply of carbohydrates is limited.

Killer T cell (*Section 31.2*) A T cell that kills invading foreign cells by cell-to-cell contact. Also called cytotoxic T cell.

Kinetic energy (*Section 1.8*) The energy of motion; energy that is in the process of doing work.

Kwashiorkor (*Section 30.5*) A disease caused by insufficient protein intake and characterized by a swollen stomach, skin discoloration, and retarded growth.

Law of conservation of energy (*Section 1.8*) Energy can be neither created nor destroyed.

LDPE (*Section 12.7B*) Low-density polyethylene.

Le Chatelier's principle (*Section 8.8*) When a stress is applied to a system in chemical equilibrium, the position of the equilibrium shifts in the direction that will relieve the applied stress.

Leukocytes (*Section 31.2*) White blood cells, which are the principal parts of the acquired immunity system and act via phagocytosis or antibody production.

Levorotatory (*Section 15.4B*) The counterclockwise rotation of the plane of polarized light in a polarimeter.

Lewis structure (*Section 4.7B*) A formula for a molecule or ion showing all pairs of bonding electrons as single, double, or triple lines, and all nonbonding (unshared) electrons as pairs of Lewis dots.

Limiting reagent (*Section 5.5B*) The reactant that is consumed, leaving an excess of another reagent or reagents unreacted.

Line-angle formula (*Section 11.1*) An abbreviated way to draw structural formulas in which each vertex and line terminus represents a carbon atom and each line represents a bond.

Lipase (*Section 30.4*) An enzyme that catalyzes the hydrolysis of an ester bond between a fatty acid and glycerol.

Liquid state (*Section 1.6*) A physical state of matter in which there is a definite volume but no specific shape.

London dispersion forces (*Section 6.7A*) Extremely weak attractive forces between atoms or molecules caused by the electrostatic attraction between temporary induced dipoles.

Lymphocyte (*Sections 31.1, 31.2*) A white blood cell that spends most of its time in the lymphatic tissues. Those that mature in the bone marrow are B cells. Those that mature in the thymus are T cells.

Lymphoid organs (*Section 31.2*) The main organs of the immune system, such as the lymph nodes, spleen, and thymus, which are connected together by lymphatic capillary vessels.

Macrophage (*Sections 31.1, 31.2*) An amoeboid white blood cell that moves through tissue fibers, engulfing dead cells and bacteria by phagocytosis, and then displays some of the engulfed antigens on its surface.

Magnetic resonance imaging (MRI) (*Chemical Connections 3C*) The detection of the flip of a nuclear spin from being aligned with an external magnetic field to being against it when the nucleus is irradiated with radio-frequency radiation.

Major histocompatibility complex (MHC) (*Sections 31.2, 31.3*) A transmembrane protein complex that brings the epitope of an antigen to the surface of the infected cell to be presented to the T cells.

Marasmus (*Section 30.2*) Another term for chronic starvation, whereby the individual does not have adequate caloric intake. It is characterized by arrested growth, muscle wasting, anemia, and general weakness.

Markovnikov's rule (*Section 12.6A*) In the addition of HX or H_2O to an alkene, hydrogen adds to the carbon of the double bond having the greater number of hydrogens.

Mass (*Section 1.4*) The quantity of matter in an object; the SI base unit is the kilogram; often referred to as weight.

Matter (*Section 1.1*) Anything that has mass and takes up space.

Memory cell (*Section 31.2*) A type of T cell that stays in the blood after an infection is over and acts as a quick line of defense if the same antigen is encountered again.

Mercaptan (*Section 14.4A*) A common name for any molecule containing an —SH group.

Meta (m) (*Section 13.2B*) Refers to groups occupying the 1 and 3 positions on a benzene ring.

Metabolic acidosis (*Chemical Connections 9D*) The lowering of the blood pH due to metabolic effects such as starvation or intense exercise.

Meter (*Section 1.4*) The SI base unit of length.

Metric system (*Section 1.4*) A system in which measurements of parameter are related by powers of 10.

Mineral in diet (*Section 30.6*) An inorganic ion that is required for structure and metabolism, such as Ca^{2+}, Fe^{2+}, and Mg^{2+}.

Mirror image (*Section 15.1*) The reflection of an object in a mirror.

Molar mass (*Section 5.3*) The mass of one mole of a substance expressed in grams; the formula weight of a compound expressed in grams.

Molarity (*Section 7.5*) The number of moles of solute dissolved in 1 L of solution.

Mole (mol) (*Section 5.3*) The formula weight of a substance expressed in grams.

Molecular weight (MW) (*Section 5.2*) The sum of the atomic weights of all atoms in a molecular compound expressed in atomic mass units (amu).

Monoclonal antibody (*Section 31.4*) An antibody produced by clones of a single B cell specific to a single epitope.

Monomer (*Section 12.7A*) From the Greek *mono*, "single," and *meros*, "part"; the simplest nonredundant unit from which a polymer is synthesized.

Monoprotic acid (*Section 9.3*) An acid that can give up only one proton.

Multiclonal antibodies (*Chemical Connections 31B*) The type of antibodies found in the serum after a vertebrate is exposed to an antigen.

Multipotent stem cell (*Chemical Connections 31F*) A stem cell capable of differentiating into many, but not all, cell types.

Natural killer cell (*Sections 31.1, 31.2*) A cell of the innate immune system that attacks infected or cancerous cells.

Net ionic equation (*Section 5.6*) A chemical equation that does not contain spectator ions.

Neutralizing antibody (*Section 31.8*) A type of antibody that completely destroys its target antigen.

Nonbonding electrons (*Section 4.7C*) Valence electrons not involved in forming covalent bonds—that is, unshared electrons.

Nonpolar covalent bond (*Section 4.7B*) A covalent bond between two atoms whose difference in electronegativity is less than 0.5.

Normal boiling point (*Section 6.8C*) The temperature at which a liquid boils under a pressure of 1 atm.

N-terminus (*Section 22.6*) The amino acid at the end of a peptide chain that has a free amino group.

Nuclear fission (*Section 3.9*) The process of splitting a nucleus into smaller nuclei.

Nuclear fusion (*Section 3.8*) Joining together atomic nuclei to form a heavier nucleus than the starting nuclei.

Nuclear reaction (*Section 3.3A*) A reaction that changes an atomic nucleus (usually to the nucleus of another element.)

Nutrient (*Section 30.1*) Components of food and drink that provide energy, replacement, and growth.

Obesity (*Section 30.2*) Accumulation of body fat beyond the norm.

Octet rule (*Section 4.2*) When undergoing chemical reactions, atoms of Group 1A–7A elements tend to gain, lose, or share electrons to achieve an election configuration having eight valence electrons.

Optically active (*Section 15.4A*) Showing that a compound rotates the plane of polarized light.

Organic chemistry (*Section 10.1*) The study of the compounds of carbon.

Ortho (o) (*Section 13.2B*) Refers to groups occupying the 1 and 2 positions on a benzene ring.

Osmolarity (*Section 7.8B*) Molarity multiplied by the number of particles produced in solution by each formula unit of solute.

Osmotic pressure (*Section 7.8B*) The amount of external pressure applied to the more concentrated solution to stop the passage of solvent molecules into it from across a semipermeable membrane.

Oxidation (*Section 5.7*) The loss of electrons; the gain of oxygen atoms or the loss of hydrogen atoms.

Oxidizing agent (*Section 5.7*) An entity that accepts electrons in an oxidation–reduction reaction.

Oxonium ion (*Section 12.6B*) An ion in which oxygen is bonded to three other atoms and bears a positive charge.

Para (p) (*Section 13.2B*) Refers to groups occupying the 1 and 4 positions on a benzene ring.

Parenteral nutrition (*Chemical Connections 29A*) The technical term for intravenous feeding.

Partial pressure (*Section 6.5*) The pressure that a gas in a mixture of gases would exert if it were alone in the container.

Peptide (*Section 22.6*) A short chain of amino acids linked via peptide bonds.

Peptide backbone (*Section 22.7*) The repeating pattern of peptide bonds in a polypeptide or protein.

Peptide bond (*Section 22.6*) An amide bond that links two amino acids.

Peptide linkage (*Section 22.6*) Another term for peptide bond.

Percent concentration (% w/v) (*Section 7.5*) The number of grams of solute in 100 mL of solution.

Percent yield (*Section 5.5C*) The actual yield divided by the theoretical yield times 100.

Perforin (*Section 31.2*) A protein produced by killer T cells that punches holes in the membrane of target cells.

Peroxide (*Section 12.7A*) A compound that contains an —O—O— bond—for example, hydrogen peroxide, H—O—O—H.

pH (*Section 9.8*) The negative logarithm of the hydronium ion concentration; $pH = -\log[H_3O^+]$

Phagocytosis (*Section 31.4*) The process by which large particulates, including bacteria, are pulled inside a white cell called a phagocyte.

Phase change (*Section 6.10*) A change from one physical state (gas, liquid, or solid) to another.

Phenol (*Section 13.4*) A compound that contains an —OH group bonded to a benzene ring.

Phenyl group (*Section 13.2*) C_6H_5—, the aryl group derived by removing a hydrogen atom from benzene.

Pheromone (*Chemical Connections 12B*) A chemical secreted by an organism to influence the behavior of another member of the same species.

Photon (*Section 3.2*) The smallest unit of electromagnetic radiation.

Physical change (*Section 1.1*) A change in matter in which it does not lose its identity.

Physical property (*Section 1.1*) Characteristics of a substance that are not chemical properties; those properties that are not a result of a chemical change.

Plane-polarized light (*Section 15.4A*) Light vibrating in only parallel planes.

Plasma cell (*Section 31.2*) A cell derived from a B cell that has been exposed to an antigen.

Plastic (*Chemical Connections 12D*) A polymer that can be molded when hot and that retains its shape when cooled.

Pluripotent stem cell (*Chemical Connections 31F*) A stem cell that is capable of developing into every cell type.

pOH (*Section 9.8*) The negative logarithm of the hydroxide ion concentration; $pOH = -\log[OH^-]$

Polar covalent bond (*Section 4.7C*) A covalent bond between two atoms whose difference in electronegativity is between 0.5 and 1.9.

Polarimeter (*Section 15.4B*) An instrument for measuring the ability of a compound to rotate the plane of polarized light.

Polyatomic ion (*Section 4.3C*) An ion that contains more than one atom.

Polymer (*Section 12.7A*) From the Greek *poly*, "many," and *meros*, "parts"; any long-chain molecule synthesized by bonding together many single parts called monomers.

Polynuclear aromatic hydrocarbon (*Section 13.2D*) A hydrocarbon containing two or more benzene rings, each of which shares two carbon atoms with another benzene ring.

Polypeptide (*Section 22.6*) A long chain of amino acids bonded via peptide bonds.

Positive cooperativity (*Chemical Connections 22G*) A type of allosterism where binding of one molecule of a protein makes it easier to bind another of the same molecule.

Positron ($\beta+$) (*Section 3.3D*) A particle with the mass of an electron but a charge of +1, $^0_{+1}\beta$.

Positron emission tomography (PET) (*Section 3.7A*) The detection of positron-emitting isotopes in different tissues and organs; a medical imaging technique.

Potential energy (*Section 1.8*) Energy that is being stored; energy that is available for later use.

Pressure (*Section 6.2*) The force per unit area exerted against a surface.

Primary (1°) alcohol (*Section 10.3A*) An alcohol in which the carbon atom bearing the —OH group is bonded to only one other carbon group, a —CH_2OH group

Primary (1°) amine (*Section 10.3B*) An amine in which nitrogen is bonded to one carbon group and two hydrogens, a —CH_2NH_2 group.

Primary structure, of proteins (*Section 22.8*) The order of amino acids in a peptide, polypeptide, or protein.

Progenitor cells (*Chemical Connections 31F*) Another term for stem cells.

Prosthetic group (*Section 22.11*) The non-amino-acid part of a conjugated protein.

Protein (*Section 22.1*) A long chain of amino acids linked via peptide bonds. There must usually be 30 to 50 amino acids in a chain before it is considered a protein.

Protein complementation (*Section 30.5*) A diet that combines proteins of varied sources to arrive at a complete protein.

Protein microarray (*Chemical Connection 22F*) An automated technique used to study proteomics that is based on having thousands of protein samples imprinted on a chip.

Proteomics (*Chemical Connections 22F*) The collective knowledge of all the proteins and peptides of a cell or a tissue and their functions.

Quaternary structure The organization of a protein that has multiple polypeptide chains, or subunits; refers principally to the way the multiple chains interact.

R— (*Section 11.3A*) A symbol used to represent an alkyl group.

R- (*Section 15.2*) From the Latin *rectus,* meaning "straight, correct"; used in the *R,S* system to show that, when the lowest-priority group is away from you, the order of priority of groups on a stereocenter is clockwise.

R,S system (*Section 15.2*) A set of rules for specifying configuration about a stereocenter.

Racemic mixture (*Section 15.1*) A mixture of equal amounts of two enantiomers.

Rad (*Section 3.5*) *R*adiation *a*bsorbed *d*ose. The SI unit is the gray (Gy).

Radiation, nuclear (*Section 3.3*) Radiation emitted from a nucleus during nuclear decay. Includes alpha particles, beta particles, gamma rays, and positrons.

Radical (*Chemical Connections 3C*) An atom or molecule with one or more unpaired electrons.

Radioactive (*Section 3.2*) Refers to a substance that emits radiation during nuclear decay.

Radioactive dating (*Chemical Connections 3A*) The process of establishing the age of a substance by analyzing radioisotope abundance as compared with a current amount.

Radioactive isotope (*Section 3.3*) A radiation-emitting isotope of an element.

Radioactivity (*Section 3.2*) Another name for nuclear radiation. Includes alpha particles, beta particles, gamma rays, and positrons.

Random coils (*Section 22.9*) Proteins that do not exhibit any repeated pattern.

Rate constant (*Section 8.4B*) A proportionality constant, *k,* between the molar concentrations of reactants and the rate of reaction; rate = *k*[compound].

Reaction mechanism (*Section 12.6A*) A step-by-step description of how a chemical reaction occurs.

Recommended Daily Allowance (RDA) (*Section 30.1*) **also Recommended Dietary Allowance**; an average daily requirement for nutrients published by the U.S. Food and Drug Administration.

Redox reaction (*Section 5.7*) An oxidation–reduction reaction.

Reducing agent (*Section 5.7*) An entity that donates electrons in an oxidation–reduction reaction.

Reduction (*Section 5.7*) The gain of electrons; the loss of oxygen atoms or the gain of hydrogen atoms.

Regioselective reaction (*Section 12.6A*) A reaction in which one direction of bond forming or bond breaking occurs in preference to all other directions.

Rem (*Section 3.6*) *R*oentgen *e*quivalent for *m*an; a biological measure of radiation.

Residue (*Section 22.6*) Another term for an amino acid in a peptide chain.

Resonance (*Section 4.8*) A theory that many molecules and ions are best represented as hybrids of two or more Lewis contributing structures.

Resonance hybrid (*Section 13.1B*) A molecule best described as a composite of two or more Lewis structures.

Respiratory acidosis (*Chemical Connections 9D*) The lowering of the blood pH due to difficulty in breathing.

Retrovaccination (*Section 31.8*) A process whereby scientists have an antibody they want to use and try to develop molecules to elicit it.

Roentgen (R) (*Section 3.6*) The amount of radiation that produces ions having 2.58×10^{-4} coulomb per kilogram.

s- (*Section 15.2*) From the Latin *sinister*, meaning "left"; used in the *R,S* system to show that, when the lowest-priority group is away from you, the order of priority of groups on a stereocenter is counterclockwise.

Saturated hydrocarbon (*Section 11.1*) A hydrocarbon that contains only carbon–carbon single bonds.

Scientific method (*Section 1.2*) A method of acquiring knowledge by testing theories.

Scintillation counter (*Section 3.5A*) An instrument containing a phosphor that emits light on exposure to ionizing radiation.

Secondary (2°) alcohol (*Section 10.3A*) An alcohol in which the carbon atom bearing the —OH group is bonded to two other carbon groups.

Secondary (2°) amine (*Section 10.3B*) An amine in which nitrogen is bonded to two carbons groups and one hydrogen.

Secondary structure, of proteins (*Section 22.9*) Repeating structures within polypeptides that are based solely on interactions of the peptide backbone. Examples are the alpha helix and the beta-pleated sheet.

SI (*Section 1.4*) International System of Units.

Side chains (*Section 22.7*) The part of an amino acid that varies one from the other. The side chain is attached to the alpha carbon, and the nature of the side chain determines the characteristics of the amino acid.

Sievert (Sv) (*Section 3.6*) A biological measure of radiation. One sievert is the value of 100 rem.

Single bond (*Section 4.7C*) A bond formed by sharing one pair of electrons; represented by a single line between two bonded atoms.

Small nuclear RNA (*Section 25.4*) Small RNA molecules (100–200 nucleotides) located in the nucleus that are distinct from tRNA and rRNA.

Soap (*Section 18.4B*) A sodium or potassium salt of a fatty acid.

Solid (*Section 1.6*) A physical state of matter in which there is a definite volume and shape.

Solubility (*Section 7.4*) The maximum amount of solute that can be dissolved in a solute at a specific temperature and pressure.

Solute (*Section 7.2*) The substance or substances that are dissolved in a solvent to produce a solution.

Solvent (*Section 7.2*) The fraction of a solution in which other components are dissolved.

Specific gravity (*Section 1.7*) The density of a substance compared to water as a standard.

Specific heat (*Section 1.9*) The amount of heat (calories) necessary to raise the temperature of 1 g of a substance by 1°C.

Specificity (*Section 31.1*) A characteristic of acquired immunity based on the fact that cells make specific antibodies to a wide range of pathogens.

Spectator ion (*Section 5.6*) An ion that appears unchanged on both sides of a chemical equation.

Standard temperature and pressure (STP) (*Section 6.4*) One atmosphere pressure and 0°C (273 K).

Step-growth polymerization (*Section 18.8*) A polymerization in which chain growth occurs in a stepwise manner between difunctional monomers—as, for example, between adipic acid and hexamethylenediamine to form nylon-66.

Stereocenter (*Section 15.1*) An atom, most commonly a tetrahedral carbon atom, at which exchange of two groups produces a stereoisomer.

Stereoisomers (*Section 11.6*) Isomers that have the same connectivity (the same order of attachment of their atoms) but different orientations of their atoms in space.

Stoichiometry (*Section 5.5A*) The mass relationships in a chemical reaction.

Strong acid (*Section 9.2*) An acid that ionizes completely in aqueous solution.

Strong base (*Section 9.2*) A base that ionizes completely in aqueous solution.

Structural formula (*Section 4.7C*) A formula showing how atoms in a molecule or ion are bonded to each other. Similar to a Lewis structure except that a structural formula shows only bonding pairs of electrons.

Sublimation (*Section 6.10*) A phase change from the solid state directly to the vapor state.

Supersaturated solution (*Section 7.4*) A solution in which the solvent has dissolved an amount of solute beyond the maximum amount at a specific temperature and pressure.

Surface presentation (*Section 31.1*) The process whereby a portion of an antigen from a foreign pathogen that infected a cell is brought to the surface of the cell.

Surface tension (*Section 6.8A*) The layer on the surface of a liquid produced by uneven intermolecular attractions at its surface.

T cell (*Section 31.1*) A type of lymphoid cell that matures in the thymus and that reacts with antigens via bound receptors on its cell surface. T cells can differentiate into memory T cells or killer T cells.

Tautomers (*Section 17.5*) Constitutional isomers that differ in the location of an H atom and a double bond relative to an O or N atom.

T-cell receptor (*Section 31.1*) A glycoprotein of the immunoglobulin superfamily on the surface of T cells that interacts with the epitope presented by MHC.

T-cell receptor complex (*Section 31.5*) The combination of T-cell receptors, antigens, and cluster determinants (CD) that is involved in the T cell's ability to bind antigen.

Temperature (*Section 1.9*) The measure of the amount of heat in a substance. Measurement is made in a temperature scale (°F, °C, or K); K is the SI base unit.

Terpene (*Section 12.5*) A compound whose carbon skeleton can be divided into two or more units identical to the carbon skeleton of isoprene.

Tertiary = (3°) alcohol (*Section 10.3A*) An alcohol in which the carbon atom bearing the —OH group is bonded to three other carbon groups.

Tertiary (3°) amine (*Section 10.3B*) An amine in which nitrogen is bonded to three carbon groups.

Tertiary structure The overall conformation of a polypeptide chain, including the interactions of the side chains and the position of every atom in the polypeptide.

Theoretical yield (*Section 5.5C*) The mass of product that should be formed in a chemical reaction according to the stoichiometry of the balanced equation.

Theory (*Section 1.2*) A hypothesis that is supported by evidence; a hypothesis that has passed tests.

Thiol (*Section 14.4A*) A compound containing an —SH (sulfhydryl) group bonded to a tetrahedral carbon atom.

Tissue necrosis factor (TNF) (*Section 31.6*) A type of cytokine produced by T cells and macrophages that has the ability to lyse susceptible tumor cells.

Titration (*Section 9.9*) An analytical procedure whereby we react a known volume of a solution of known concentration with a known volume of a solution of unknown concentration.

Trans (*Section 11.6*) A prefix meaning "across from."

Transmutation (*Section 3.3B*) Changing one element into another element.

Transuranium element (*Section 3.8*) An element with an atomic number greater than that of uranium; that is, one with an atomic number greater than 92.

Triple bond (*Section 4.7C*) A bond formed by sharing three pairs of electrons; represented by three lines between the two bonded atoms.

Triple helix (*Section 22.11*) The collagen triple helix is composed of three peptide chains. Each chain is itself a left-handed helix. These chains are twisted around each other in a right-handed helix.

Triprotic acid (*Section 9.3*) An acid that can give up three protons.

Tyndall effect (*Section 7.7*) Light passing through and scattered by a colloid viewed at a right angle.

Universal gas constant (R) (*Section 6.4*) $0.0821 \ \mathrm{L \cdot atm \cdot mol^{-1} \cdot K^{-1}}$.

Vapor (*Section 6.8B*) A gas.

Vapor pressure (*Section 6.8B*) The pressure of gas in equilibrium with its liquid form in a closed container.

Vitamin (*Section 30.6*) An organic substance required in small quantities in the diet of most species, which generally functions as a cofactor in important metabolic reactions.

Volume (*Section 1.4*) The space that a substance occupies; the base SI unit is the cubic meter (m^3).

VSEPR model (*Section 4.10*) Valence-shell electron-pair repulsion model.

Wavelength (λ) (*Section 3.2*) The distance from the crest of one wave to the crest of the next.

Weak acid (*Section 9.2*) An acid that is only partially ionized in aqueous solution.

Weak base (*Section 9.2*) A base that is only partially ionized in aqueous solution.

Weight (*Section 1.4*) The result of a mass acted upon by gravity; the base unit of measure is a gram (g).

X ray (*Section 3.2*) A type of electromagnetic radiation with a wavelength shorter than ultraviolet light but longer than gamma rays.

Zwitterion (*Section 22.3*) A molecule that has equal numbers of positive and negative charges, giving it a net charge of zero.

Credits

Text and Illustrations

This page constitutes an extension of the copyright page. We have made every effort to trace the ownership of all copyrighted material and to secure permission from copyright holders. In the event of any question arising as to the use of any material, we will be pleased to make the necessary corrections in future printings. Thanks are due to the following authors, publishers, and agents for permission to use the material indicated.

Chapter 21 p. 529: From Biochemistry by Lubert Stryer © 1975, 1981, 1988, 1995 by Lubert Stryer; © 2002 by W. H. Freeman and Company. Used with permission of W. H. Freeman and Company.

Chapter 24 p. 614: Courtesy of Anthony Tu, Colorado State University.

Chapter 25 p. 644: Courtesy of Dr. Sung-Hou Kim. 644: From Biochemistry, 2nd ed., by Lubert Stryer. Copyright 1981 by W. H. Freeman and Company. All rights reserved. Used with permission of W. H. Freeman and Company.

Chapter 26 p. 675: Illustration by Irving Geis, from Scientific American, Jan. 1963. Rights owned by Howard Hughes Medical Institute. Not to be reproduced without permission. **p. 681:** Scientific American 292 (4) 2005, p. 61.

Photographs

pp. v–xii: Charles D. Winters.
Chapter 1 p. 1: James Balog/Stone/Getty Images. **2:** National Library of Medicine. **p. 3:** Charles D. Winters. **p. 3:** Charles D. Winters. **p. 3:** Charles D. Winters **p. 3:** National Library of Medicine. **p. 7:** © 2003 Richard Megna/Fundamental Photographs, NYC. **p. 7:** © 1992 Kip Peticolas/Fundamental Photographs, NYC. **p. 9:** Photographer's Choice/Getty Images. **p. 10:** Courtesy of Brinkmann Instruments, Co. **p. 10:** Courtesy of Brinkmann Instruments, Co. **p. 10:** Charles D. Winters. **p. 16:** Charles D. Winters. **p. 16:** Charles D. Winters. **p. 16:** Charles D. Winters. **p. 17:** Ken Graham/Stone/Getty Images. **p. 19:** Eric Lars Baleke/Black Star. **p. 20:** courtesy of Shell Solar Industries.
Chapter 2 p. 29: IBMRL/Visuals Unlimited, Inc. **p. 33:** Charles D. Winters. **p. 33:** Ken Eward/Science Source/Photo Researchers, Inc. **p. 33:** Charles D. Winters. **p. 33:** Charles D. Winters. **p. 33:** Charles D. Winters. **p. 40:** Courtesy Roger Freedman & Paul Hansma, University of California. Santa Barbara. **p. 41:** NASA. **p. 41:** E.F. Smith Memorial Collection, Special Collection Department, Van Pelt Library, University of Pennsylvania. **p. 42:** Charles D. Winters. **p. 42:** Charles D. Winters. **p. 42:** Charles D. Winters. **p. 44:** Charles D. Winters. **p. 44:** Charles D. Winters. **p. 44:** Charles D. Winters. **p. 45:** Werner Forman/Art Resource, NY. **p. 44:** Charles D. Winters.
Chapter 3 p. 63: NASA. **p. 73:** Sci-Vu/Visuals Unlimited. **p. 73:** © Corbis Sygma. **p. 73:** © Bettmann/Corbis. **p. 74:** Charles D. Winters. **p. 77:** Klaus Guldbrandsen/Science Photo Library/Photo Researchers, Inc. **p. 78:** Charles D. Winters. **p. 79:** CNRI/SPL/Photo Researchers, Inc. **p. 80:** CRC Handbook in Clinical Lab Science. **p. 80:** CRC Handbook in Clinical Lab Science. **p. 80:** CRC Handbook in Clinical Lab Science. **p. 82:** Dr. Peter T. Fox, Washington University, School of Medicine. **p. 81:** Beverly March/courtesy of Long Island Jewish Hospital. **p. 85:** Courtesy of Public Service Electric & Gas Co. **p. 87:** US Dept Energy.
Chapter 4 p. 93: Charles D. Winters. **p. 96:** Charles D. Winters. **p. 96:** Charles D. Winters. **p. 96:** Charles D. Winters. **p. 96:** Charles D. Winters. **p. 98:** Norian Corp, Cupertino, CA. **p. 104:** Charles D. Winters. **p. 112:** Charles D. Winters. **p. 116:** Charles D. Winters. **p. 116:** Charles D. Winters. **p. 116:** Charles D. Winters. **p. 118:** Arthur C. Smith, III/Grant Heilman Photography. **p. 126:** Charles D. Winters.
Chapter 5 p. 129: © Joseph Nettis/Photo Researches, Inc. **p. 132:** Charles D. Winters. **p. 132:** Charles D. Winters. **p. 135:** Charles D. Winters. **p. 136:** © Jeff J. Daly/Visuals Unlimited, Inc. **p. 141:** AT&T Bell Labs. **p. 146:** Charles Steele. **p. 146:** Beverly March. **p. 147:** Beverly March. **p. 147:** Beverly March. **p. 147:** Charles D. Winters. **p. 148:** Charles D. Winters. **p. 148:** Beverly March. **p. 148:** Beverly March. **p. 149:** Charles D. Winters. **p. 151:** Russell D. Curtis/Photo Researchers. **p. 152:** © Jeff Greenberg/Visuals Unlimited. **p. 152:** Charles D. Winters. **p. 152:** Charles D. Winters. **p. 153:** Yoav Levy, Phototake, NYC. **p. 153:** Charles D. Winters.
Chapter 6 p. 160: © Vince Streano/Corbis. **p. 164:** Greg Gawlowski/Dembinsky Photo Associates. **p. 165:** Custom Medical Stock Photo. **p. 168:** Charles D. Winters. **p. 168:** Gregory G Dimijian/Photo Researchers, Inc. **p. 175:** Hermann Eisenbeiss/Photo Researchers, Inc. **p. 180:** Mehau Kulyk/Science Photo Library/Photo Researchers, Inc. **p. 181:** Beverly March. **p. 181:** Beverly March. **p. 181:** Beverly March. **p. 181:** Beverly March. **p. 183:** Charles D. Winters. **p. 184:** Charles D. Winters.
Chapter 7 p. 192: David Scharf/Peter Arnold, Inc. **p. 192:** Charles D. Winters. **p. 193:** Beverly March. **p. 193:** Charles D. Winters. **p. 194:** Will McIntyre/Photo Researchers, Inc. **p. 196:** Charles D. Winters. **p. 197:** Brian Parker/Tom Stack and Assoc. **p. 200:** Charles D. Winters. **p. 200:** Charles D. Winters. **p. 200:** Charles D. Winters. **p. 200:** Charles D. Winters. **p. 200:** Stan Elems/Visuals Unlimited. **p. 202:** Charles D. Winters. **p. 202:** Charles D. Winters. **p. 202:** Charles D. Winters. **p. 205:** Charles D. Winters. **p. 206:** Courtesy of P.A. Dieppe, P.A. Bacon, A.N. Bamji, and I Watt and the Gower Medical Publishing Co., Ltd., London. **p. 206:** National Library of Medicine. **p. 207:** Martin Bond/Science Photo Library/Photo Researchers, Inc. **p. 207:** George Semple. **p. 209:** Charles D. Winters. **p. 209:** David Scharf/Peter Arnold, Inc. **p. 210:** NALCO Chemical Company. **p. 212:** Charles D. Winters. **p. 214:** Charles D. Winters. **p. 214:** Courtesy of Recovery Engineering, Inc. **p. 215:** Science Source/Photo Reseachers, Inc. **p. 215:** Science Source/Photo Researchers, Inc. **p. 215:** Science Source/Photo Reseachers, Inc. **p. 215:** Charles D. Winters. **p. 216:** SIU/Photo Researchers, Inc.
Chapter 8 p. 223: Charles D. Winters. **p. 230:** Charles D. Winters. **p. 230:** Leon Lewandowski. **p. 231:** Michael Engleish, M.D./Custom Medical Stock Photo. **p. 232:** Charles D. Winters. **p. 234:** Beverly March. **p. 241:** Charles D. Winters. **p. 242:** John D. Cunningham/Visuals Unlimited. **p. 244:** Charles D. Winters. **p. 244:** Charles D. Winters. **p. 245:** C.P. Vance/Visuals Unlimited.
Chapter 9 p. 251: Charles D. Winters. **p. 254:** Charles D. Winters. **p. 254:** Charles D. Winters. **p. 255:** Charles D. Winters. **p. 259:** Charles D. Winters. **p. 261:** Charles D. Winters. **p. 263:** Charles D. Winters. **p. 263:** Charles D. Winters. **p. 267:** Charles D. Winters. **p. 268:** Charles D. Winters. **p. 268:** Charles D. Winters. **p. 277:** © RB-GM-J.O. Atlanta/Liaison Agency/Getty Images. **p. 262:** Charles D. Winters. **p. 269:** Charles D. Winters. **p. 271:** Charles D. Winters. **p. 271:** Charles D. Winters. **p. 273:** Charles D. Winters. **p. 273:** Charles D. Winters. **p. 273:** Charles D. Winters. **p. 274:** Charles D. Winters. **p. 274:** Charles D. Winters.
Chapter 10 p. 285: Tom and Pat Leeson, Photo Researchers, Inc. **p. 288:** Pete K. Ziminiski/Visuals Unlimited. **p. 289:** George Semple. **p. 294:** Charles D. Winters.
Chapter 11 p. 302: J.L. Bohin/Photo Researchers, Inc. **p. 304:** Charles D. Winters. **p. 315:** Tim Rock/Animals/Animals. **p. 319:** Charles D. Winters. **p. 323:** Ashland Oil Co. **p. 324:** Charles D. Winters.
Chapter 12 p. 330: Charles D. Winters. **p. 337:** David Sieren/Visuals Unlimited. **p. 340:** Don Suzio. **p. 349:** Beverly March. **p. 349:** Charles D. Winters. **p. 349:** Beverly March. **p. 349:** Charles D. Winters. **p. 348:** The Stock Market/Corbis. **p. 351:** Charles D. Winters.
Chapter 13 p. 358: Douglas Brown. **p. 362:** US Dept of Agriculture. **p. 366:** Charles D. Winters. **p. 366:** John D. Cunningham/Visuals Unlimited. **p. 367:** Charles D. Winters. **p. 368:** Chuck Pefley/Stone/Getty Image.
Chapter 14 p. 374: Earl Robber/Photo Researchers, Inc. **p. 377:** Charles D. Winters. **p. 376:** Charles D. Winters. **p. 376:** The Bettmann Archive/Corbis. **p. 386:** Boston Medical Library in the Francis A. Countway Lib of Medicine. **p. 387:** Steven J. Krasemann/Photo Researchers, Inc. **p. 389:** D. Young/Tom Stack and Associates.
Chapter 15 p. 396: Lester Lefkowitz/Stone/Getty Images. **p. 399:** Charles D. Winters. **p. 399:** Charles D. Winters. **p. 398:** Charles D. Winters. **p. 398:** Charles D. Winters. **p. 412:** Wiliam Brown.
Chapter 16 p. 417: Jack Ballard/Visuals Unlimited. **p. 419:** Inga Spence/Visuals Unlimited. **p. 422:** Charles D. Winters. **p. 426:** Beverly March. **p. 433:** Tom McHugh/Photo Researchers, Inc.
Chapter 17 p. 434: Charles D. Winters. **p. 440:** Charles D. Winters.
Chapter 18 p. 452: Charles D. Winters. **p. 453:** Ted Nelson/Dembinsky Photo Associates. **p. 463:** Charles D. Winters.
Chapter 19 p. 474: SCIMAT/Science Source/Photo Researchers, Inc. **p. 487:** Charles D. Winters. **p. 487:** Charles D. Winters.
Chapter 20 p. 493: Charles D. Winters. **p. 497:** Claire Paxton and Jacqui Farrow/Science Photo Library/Photo Researchers, Inc. **p. 502:** Charles D. Winters. **p. 505:** Gregory Smolin. **p. 505:** © Martin Dohrn/SPL/Photo Researchers, Inc. **p. 508:** Larry Mulvehill/Photo Researchers, Inc.
Chapter 21 p. 521: © Doug Perrine/dougperrine.com. **p. 525:** Charles D. Winters. **p. 534:** Drs. P. G. Bullogh and V. J. Vigorita and the Gower Med Publ Co. **p. 533:** Carlina Biological Supply Co/Phototake, NYC. **p. 540:** Jed Jacobson/AllSport USA/Getty Images.
Chapter 22 p. 551: Hans Strand/Stone/Getty Images. **p. 567:** G.W. Willis/Visuals Unlimited. **p. 578:** © Gideon Mendel/Corbis. **p. 579:** Charles D. Winters. **p. 578:** Larry Mulvehill/Photo Researchers.
Chapter 23 p. 584: Vertex Pharmaceuticals, Inc. **p. 594:** Vertex Pharmaceuticals, Inc.

Chapter 24 p. 608: © Don Fawcett/Science Source/Photo Researchers, Inc. **p. 616:** Alfred Pasieka/SPL/Photo Researchers, Inc. **p. 617:** Courtesy of Anatomical Collection, National Museum of Health and Medicine, Armed Forces Institute of Pathology, Washington, DC. **p. 617:** Courtesy of Anatomical Collection, National Museum of Health and Medicine, Armed Forces Institute of Pathology, Washington, DC. **p. 617:** Science VU, Visuals Unlimited. **p. 623:** George Semple.

Chapter 25 p. 632: AP/Wide World Photos. **p. 633:** Biophoto Associates/Photo Researchers, Inc. **p. 640:** Vittorio Luzzati/Centre National de Genetique Moleculaire. **p. 640:** Barrington Brown/Science Source/Photo Researchers, Inc. **p. 651:** Courtesy of Dr. Lawrence Koblinsky.

Chapter 26 p. 661: © Eric Kamp/Phototake. **p. 672:** Dr. J. Frank, Wadsworth Center, Albany, NY. **p. 675:** Illustration by Irving Geis, from Scientific American, January 1963. Rights owned by Howard Hughes Medical Institute. Not to be reproduced without permission. **p. 681:** Scientific American 292 (4) 2005, p. 61. **p. 684:** Thomas Broker/Cold Spring Harbor Lab.

Chapter 27 p. 692: A&L Sinibaldi/Stone/Getty Images. **p. 695:** Dennis Kunkel/Phototake.

Chapter 28 p. 714: Dennis Degnan/Corbis. **p. 726:** Charles D. Winters. **p. 734:** Science Photo Library/Photo Researchers, Inc.

Chapter 29 p. 740: Bruce M. Herman/Photo Researchers, Inc.

Chapter 30 p. 755: Charles D. Winters. **p. 759:** Charles D. Winters. **p. 759:** Charles D. Winters. **p. 765:** Courtesy of Drs. P.G. Bullogh and V.J. Vigorita and the Gower Medical Publishing Co. **p. 770:** George Semple.

Chapter 31 p. 775: Meckes/Ottawa/Photo Researchers, Inc. **p. 776:** Eye of Science/Photo Researchers, Inc. **p. 778:** © David M. Phillips/Photo Researchers, Inc. **p. 780:** Courtesy of Steve Haley, William Bowers, and Richard Hunt, University of South Carolina Medical School. **p. 785:** Courtesy of Dr. R.J. Poijak, Pasteur Institute, Paris, France. **p. 785:** Courtesy of Dr. R.J. Poijak, Pasteur Institute, Paris, France. **p. 787:** Dr. Jeremy Burgess/Science Photo Library/Photo Researchers, Inc. **p. 795:** Eye of Science/Photo Researchers, Inc.

Index

Index page numbers in **boldface** refer to boldface terms in the text. Page numbers in *italics* refer to figures. Tables are indicated by a *t* following the page number. Boxed material is indicated by *b* following the page number.